Prions in Humans and Animals

Edited by
Beat Hörnlimann

in collaboration with
Detlev Riesner and Hans Kretzschmar

de Gruyter
Berlin · New York

Dr. Beat Hörnlimann, MPH
BSE 71-92 Ltd. / SVISS Consulting
on Animal & Public Health
P.O. Box 513
CH-6312 Steinhausen
Switzerland
e-mail: consulting@sviss.net
www.PrionOne.ch

Professor
Dr. Detlev Riesner
Heinrich-Heine-University Duesseldorf
Institute of Physical Biology
D-40225 Duesseldorf
Germany
e-mail: riesner@biophys.uni-duesseldorf.de
www.biophys.uni-duesseldorf.de

Professor
Dr. Hans Kretzschmar
Center of Neuropathology and Prion Research
Ludwig-Maximilians-University Munich
Feodor-Lynen-Strasse 23
D-81377 Munich
Germany
e-mail: hans.kretzschmar@med.uni-muenchen.de
www.znp-muenchen.de

This work contains 162 figures and 88 tables.

ISBN-10: 3-11-018275-0
ISBN-13: 978-3-11-018275-0

Library of Congress Cataloging-in-Publication Data

Prions in humans and animals / edited by Beat Hörnlimann, Detlev Riesner, Hans Kretzschmar.
p. cm.
Includes bibliographical references and index.
ISBN-13: 978-3-11-018275-0 (cloth: alk. paper)
ISBN-10: 3-11-018275-0 (cloth: alk. paper)
1. Prion diseases. I. Hörnlimann, Beat, 1958- . II. Riesner, Detlev, 1941- . III. Kretzschmar, H.A. (Hans A.)
[DNLM: 1. Prion Diseases. 2. Prions. WL 300 P95895 2006]
QR201.P737P767 2006
616.8'3–dc22
2006026775

Bibliographic information published by the Deutsche Nationalbibliothek

The Deutsche Nationalbibliothek lists this publication in the Deutsche Nationalbibliografie; detailed bibliographic data are available in the Internet at http://dnb.d-nb.de.

⊚ Printed on acid-free paper which falls within the guidelines of the ANSI to ensure permanence and durability.

Typesetting and printing: Tutte Druckerei GmbH, Salzweg; Binding: Buchbinderei „Thomas Müntzer" GmbH, Bad Langensalza; Cover design: Hansbernd Lindemann.

To Fabian and Nadine for their encouraging support, and a humble apology for the innumerable hours of neglect

Beat Hörnlimann, Chief Editor

To William J. Hadlow for his important contribution to the interdisciplinary cooperation in the research on prion diseases

The Editors

Wisdom

What lies behind us
and what lies beyond us
are tiny matters compared to what lies within us.
And when we bring what is within out into the
world, miracles happen.

Henry David Thoreau

Editor's Note

The idea of writing the first edition of this book came to me on 1 September 1994 in Paris during a TSE conference at the OIE. Two years later, inspired by the following passage from Goethe's Faust the project began in earnest:

Oh, happy, he who still can hope
To rise from this chaotic sea of errors!
What you don't know, is always what you need
And what you know, you have no use for.

After the successful publication of the German-language edition, and as BSE continued to spread around the globe, the editors made the decision to publish an English-language edition in order to reach more scientists and decision-makers.

The scientific fascination of prions and prion diseases was the motivating force in publishing this book. From the very beginning, our goal was to present the historical dimension as well as the scientific, medical, veterinary, and practical viewpoints on prions, their resulting diseases, and their containment. Given that these fields are closely interwoven, the task of bringing the whole project together has been a complex one. I am proud to say that this goal has been achieved and a solid and scientific source of information has now been realized for the second time.

The book is intended to be a unified text and not just a collection of individual contributions. The first section of the book deals with the fundamental historical, medical, and scientific background, while the second section focuses on the practicalities of the subject. The book demonstrates the complex interactions of the many facets of prion diseases. The extensive text is interspersed with numerous detailed figures and tables.

I wish to express my sincere gratitude to all those who took part in the realization of this mammoth project, throughout the writing of which there has been an excellent *esprit de corps*. First of all, I thank Terry Berger for her unfailing support and friendship during my time devoted to the completion of the present edition.

I have had the pleasure of working with two outstanding co-editors, Hans and Detlev. Furthermore, eighty internationally-renowned co-authors contributed to the book and I wish to express my sincere thanks to each and every one of them.

Scientific reviews were carried out by a group of excellent specialists listed below, who checked the quality of content. By contributing their knowledge, the book has become very much a joint oeuvre. My very special thanks go to Stuart C. MacDiarmid and Michael P. Alpers.

In particular, I would like to thank Jutta Bachmann and her team in Norway for their excellent support in translating numerous chapters of the present book. The cooperation with her was characterized by her expert knowledge of the subject matter as well as by an unparalleled cordiality. I could not have envisaged a more suitable institution than Bachmann Consulting to have worked on these scientific texts.

I also wish to thank various other institutions and colleagues who offered their support in many different ways. Further thanks go to the writer of the introductory text, Nobel laureate Werner Arber.

My special thanks also go to the publishing house Walter de Gruyter, in which we have been fortunate to have found a most competent publisher; particular thanks go to Wolfgang Böttner and Stephanie Dawson.

It has been an honor to oversee the project from its conception. The realization of this and the former edition (Prionen und Prionkrankheiten, 2001) made my role as chief editor a rewarding experience. This venture has only been possible with the help of my family. Therefore, I thank Eva Hörnlimann and our children Fabian and Nadine.

December 2006

Beat Hörnlimann, Chief Editor

Beside the authors, the following persons have supported the editors:

General Support

Berger, Terry, Oberrüti, Switzerland
Gruber, Heidi, Düsseldorf, Germany
Hansen Gerheuser, Linda, Windisch, Switzerland
Hartmann, Diethelm, Bern, Switzerland
Häring, Petra, Auw, Switzerland
Holl, Walter, Ebikon, Switzerland
Hörnlimann, Charlotte[†], Engelswilen, Switzerland
Hörnlimann, Eva, Oberrüti, Switzerland
Hörnlimann, Fabian B., Oberrüti, Switzerland
Hörnlimann, Nadine E., Oberrüti, Switzerland
Hörnlimann, Paul[†], Engelswilen, Switzerland
Käppel, Reinhard, Frankfurt, Germany
Kürsteiner, Dorli, Goldach, Switzerland
Kürsteiner, Georg, Goldach, Switzerland
Lederer, Rosi, Munich, Germany
Rogivue, Colette, Bern, Switzerland
Schenk, Jeannette, Auw, Switzerland
Zeltner, Thomas, Bern, Switzerland

Graphics and Setting

Speidel, Christel, Berlin, Germany
Suter Print AG, Ostermundigen, Switzerland
Tutte Druckerei GmbH, Salzweg, Germany
Zech, Martin, Bremen, Germany

Literature and Index

Müller, Isabelle, Bern, Switzerland
Oesch, Bruno, Schlieren, Switzerland
Stahlkopf, Jens, Berlin, Germany

Publishing

Alexander, Patrick, Ossining, New York, NY, USA
Bach, Martina, Berlin, Germany
Böttner, Wolfgang, Berlin, Germany
Dawson, Stephanie, Berlin, Germany
Dobler, Marie-Rose, Berlin, Germany
Grossmann, Mary Louise, Berlin, Germany
Hirshfeld, Amy, Ossining, New York, NY, USA
Hochbach, Julia, Berlin, Germany
Kleine, Josef, Berlin, Germany
Maes, Hildegard, Berlin, Germany
Noyer-Weidner, Mario, Berlin, Germany
Pietrowicz, Jean, Berlin, Germany
Saur, Klaus G., Berlin, Germany

Scientific reviews[1]

Alpers, Michael P., Curtin University of Technology, Perth, Australia & The Kuru Surveillance Team, Papua New Guinea Institute of Medical Research, Goroka, PNG
Bradley, Ray, Private BSE Consultant, Burpham, Guildford, United Kingdom
Gajdusek[1], D. Carleton, Amsterdam Medical Center, Amsterdam, The Netherlands
Ironside, James W., The National CJD Surveillance Unit, Edinburgh, United Kingdom
Konold, Timm, Veterinary Laboratories Agency New Haw, Addlestone, United Kingdom
Masters[1], Colin L., The University of Melbourne, Melbourne, Australia
MacDiarmid, Stuart C., Biosecurity New Zealand, Wellington, New Zealand
Schulz-Schaeffer, Walter J., Prion and Dementia Research Unit, Göttingen, Germany
van Keulen, Lucien J. M., Wageningen University and Research Centre, Lelystad, The Netherlands
Wells, Gerald A. H., Veterinary Laboratories Agency, New Haw, Addlestone, United Kingdom
Will, Robert G., The National CJD Surveillance Unit, Edinburgh, United Kingdom

Translation and Language Polishing

Bachmann, Jutta
Bachmann Consulting, Nøkkefaret 12,
1450 Nesoddtangen, Norway
www.jbachmann-consulting.com

[1] Except for D. Carleton Gajdusek (reviewed Chapter 2 in the first edition) and Colin L. Masters (reviewed Chapters 1; 26 & wrote epilogue)

The book was made possible through the support of:

Baxter BioScience
Global Pathogen Safety
Industriestr. 67
A-1220 Vienna, Austria

Canadian Ministery of Agriculture
Ottawa
Ontario K1A 0A6, Canada

Dr.Weigert GmbH & Co. KG
Muehlenhagen 85
D-20539 Hamburg, Germany

Desopharmex AG
Pharma- und Medizintechnik
Muttenzstr. 107
CH-4133 Pratteln, Switzerland

Friedrich Oberthür
Stiftung
Wahbruch 1
D-49844 Bawinkel, Germany

Johnson & Johnson Medical
Rotzenbuehlstr. 55
CH-8957 Spreitenbach, Switzerland

Novartis Pharma AG
Policy and External Affairs
PO Box
CH-4070 Basel, Switzerland

OIE
World Organisation for Animal Health
12, Rue de Prony
F-75017 Paris, France

US Department of Agriculture
1400 Independence Ave. SW
Washington DC 20250-1300, USA

Prionics AG
Wagistr. 27a
CH-8952 Schlieren, Switzerland

Swiss Federal Public Health Office
Schwarzenburgstr. 165
PO Box
CH-3000 Bern, Switzerland

Swiss Academy of Sciences
Bärenplatz 2
CH-3011 Bern, Switzerland

Swiss Academy of Medical Sciences
Petersplatz 13
CH-4051 Basel, Switzerland

WHO
World Health Organization
Via Appia
CH-1202 Geneva, Switzerland

Foreword

Until the middle of the 1980s, prions were not accepted as a new infectious entity by the scientific community. At that time, research on prions still remained an area of largely esoteric interest. However, as a result of the BSE crisis, prion research has gained more attention in science, medicine and agriculture and has influenced politics and economics – with an impact in more or less every field related to consumer behavior. To the public the problems became obvious through media headlines about "BSE", "Mad Cow Disease" or "variant Creutzfeldt-Jakob-Disease", and in the scientific community a new field of multidisciplinary research has emerged.

The perspectives on prion research are as manifold as the readership which the editors and authors would like to reach. This book addresses readers from human and veterinary medicine, the biological disciplines such as molecular biology, biochemistry, and biophysics, agronomics and epidemiology – regardless of whether they are practitioners, students or work in research and education. In addition, experts from areas such as public health, governmental control agencies and the food and drug industry will also benefit from this systematic treatment.

To this end, the editors would like to express their gratitude to all authors who have contributed to this project. Most of them have been working in this field for years and are internationally renowned. Such a firm foundation helps to guarantee the scientific integrity of the presentation. The first edition of this book appeared in 2001 in German. Thanks to its wide acceptance, the editors felt prompted to work on a second edition, but this time in English – in order to reach a more international audience. Authors of the first edition updated their contributions and further authors joined the group of contributors.

Our goal was to present an extremely complex field in a language comprehensible to all. Repetitions are restricted to those parts in which new relationships are described or details expanded. We have aimed to focus on clear, definitive statements wherever possible. The book also provides information on existing gaps in our knowledge: hypotheses and assumptions are indicated as such.

We hope that this book will lead to innovative research ideas, further interdisciplinary cooperation in the research on prion diseases, and effective containment of disease.

The Editors

Beat Hörnlimann — Detlev Riesner — Hans Kretzschmar

Preface by Werner Arber

Research into prions and the diseases caused by prions has largely been marked in the past few decades by innovative findings in molecular genetics and the study of protein functions. The results gained from prion research represent a very interesting example of an in-depth insight into the complexity of gene functions and the interactions between biologically-active macromolecules.

Until recently, as a result of poor understanding of the subject, this natural complexity did not receive the attention that it was due, especially from the sciences. With regard to genetic information, this was related to a belief in science that genes "programmed" the processes of life in a very strict way, i. e., exactly "determined" the processes of life. In this extreme formulation, this view is not correct and needs to be corrected both within the scientific community as well as in the increasingly important dialogue with the public.

The limits of genetic predetermination depend, in part, on the natural structural flexibility of biologically active macromolecules. These include proteins, which, as the products of genes, influence the processes of life in many ways. Of course, this influence depends on the structural conformation of the proteins. Depending on their primary structure, i. e., their amino acid sequences, proteins can, in general, adopt several different tertiary structures, each being characterized by a certain, sometimes high, stability. The chances of adopting different conformations can depend on external factors, in particular on the effect of other protein molecules, which are then called chaperones or helper proteins.

The example of prions shows that a given protein molecule can efficiently influence the structure of another protein of the same kind. In addition, it shows that life functions and dysfunctions caused by prions are largely influenced by the conformation of the prion protein. It is, therefore, important that the scientific community make it clear to scientists and the general public that many processes of life are influenced by probabilities and uncertainties.

The present book represents an important contribution to the comprehension of the concept outlined here. At the same time it informs the interested reader on the current state of prion research.

Werner Arber

Professor Emeritus of Molecular Microbiology at the Biozentrum, University of Basel, Switzerland

In 1978, Professor Arber was awarded the Nobel Prize for Medicine or Physiology for his findings on restriction enzymes and their use in molecular genetics.

Table of Contents

Topic III: Portraits of Prion Diseases

Topic V: Surveillance, Clinical Aspects and Diagnostics

Topic VI: Epidemiology

Topic VII: Transmissibility

Topic VIII: Agent Inactivation

Topic IX: Prevention

Abbreviations

14–3–3	Protein that is used as diagnostic marker in the CSF (→)
AA	Alanine / Alanine homozygosity
Ab	Antibody
Acc. to	According to
ACDP	Advisory Committee on Dangerous Pathogens
Ala	Alanine (amino acid)
ALS	Amyotrophic lateral sclerosis (synonym: motor neuron disease)
Arg	Arginine (amino acid)
BAB	BSE in animals born after the feed ban (→ MBM; the first feed ban) [1]
BARB	BSE in animals born after the reinforced feed ban (born after 31 July 1996) [2]
Bar	Unit of pressure; replaced by hectopascal (1 bar = 1,000 hPa)
BBB	BSE in cattle born before the feed ban (→ MBM) [1]
BovPrP	Bovine prion protein
BSE	Bovine spongiform encephalopathy
CaPrP	Caprine prion protein
CD	Circular dichroism
CDI	Conformation-dependent immunoassay
CePrP	Cervid prion protein
CHMP	→ CPMP
CI	Confidence interval
CJD	Creutzfeldt–Jakob disease (→ fCJD, iCJD, vCJD, and sCJD)
CNS	Central nervous system
CPMP	Committee for Proprietary Medicinal Products (→ CVMP); later renamed Committee for Medicinal Products for Human use, CHMP
CSF	Cerebrospinal fluid
CT	Computed tomogram
CVL	Central Veterinary Laboratory in Weybridge, GB; later renamed Veterinary Laboratories Agency (→ VLA)
CVMP	Committee for Veterinary Medicinal Products (→ CPMP)
CWD	Chronic wasting disease (in cervids)
DY	'Drowsy' TME strain (in hamster)
EDQM	European Directorate for the Quality of Medicines
EEG	Electroencephalogram
ELISA	Enzyme-linked immunosorbent assay
EMEA	European Medicines Evaluation Agency
ENN	Ears, nose, neck
ES cell	Embryonic stem cell
ET	Transfer of embryos / Embryo transfer (synonym)
Exp. pot.	Exposition potential
fCJD	Familial CJD (synonyms: hereditary CJD, genetic CJD = gCJD)
FCS	Fluorescence correlation spectroscopy
FDC	Follicular dendritic cells
FePrP	Feline prion protein
FFI	Fatal familial insomnia
FSE	Feline spongiform encephalopathy
GBR	Geographical BSE risk
GdnCl	Guanidiniumchloride
GdnSCN	Guanidiniumthiocyanate
GFAP	Glial fibrillary acid protein
GH	Growth hormone
Glu	Glutamine (amino acid)
GPI	Glycosyl-phosphatidyl-inositol
GSS	Gerstmann–Sträussler–Scheinker disease
HaPrP	Hamster prion protein
HBV	Hepatitis B virus
HCV	Hepatitis C virus
H&E	Hematoxylin and eosin stain used in histological preparations
hGH	Human growth hormone
HH	Histidine / Histidine homozygosity
His	Histidine (amino acid)
HIV	Human immunodeficiency virus

hPa	Hectopascal (pressure unit; 1,000 hPa = 1 bar)
HSV	Herpes simplex virus
HSV-TK	Herpes simplex virus thymidine kinase
HuPrP	Human prion protein
HY	'hyper' TME strain (in hamster)
i.c.	Intracerebral
i.m.	Intramuscular
i.p.	Intraperitoneal
i.v.	Intravenous
iCJD	Iatrogenic CJD
ID_{50}	Infectious dose; 1 ID_{50} causes infection in 50% of inoculated animals (→LD_{50})
IHC	Immunohistochemistry
IPSC	Inhibitory postsynaptic current
LD_{50}	Lethal dose; 1 LD_{50} leads to the death of 50% of inoculated animals (→ID_{50})
Leu	Leucine (amino acid)
LRS	Lymphoreticular system
LTP	Long-term potentiation
MBM	Meat–and–bone–meal (protein residue produced by rendering)
$MBM^{\&}$	MBM (→) marked with an '&' symbol as superscript indicates that the reader can obtain comprehensive information about MBM and other important residual feed-related BSE risks or BSE risks in bovine-derived products in the document mentioned in footnote [3]
MEP	Motor evoked potentials
Met	Methionine (amino acid)
MM	Methionine / Methionine homozygosity; e. g. at codon 129 of *PRNP* (→)
MMBM	Mammalian meat–and–bone–meal
MND	Motor neuron disease (→ ALS)
MoPrP	Murine prion protein
$MoPrP^{C}$	Cellular murine prion protein
$MoPrP^{Sc/BSE}$	PrP^{Sc} derived from experimentally BSE infected mice
$MoPrP^{Sc/CJD}$	PrP^{Sc} derived from experimentally CJD infected mice
MRI	Magnetic resonance imaging
MRM	Mechanically recovered meat
mRNA	Messenger RNA
MV	Methionine / Valine heterozygosity
NaDCC	Sodium dichloroisocyanurate
NaOH	Sodium hydroxide
NFT	Neurofibrillary tangles
NMR	Nuclear magnetic resonance
NOE	Nuclear Overhauser effect
NOESY	NOE spectroscopy
NSE	Neuron-specific enolase
nvCJD	New variant Creutzfeldt–Jakob disease; today called vCJD (→)
OIE	World Organisation for Animal Health, Paris
ORF	Open reading frame
OvPrP	Ovine prion protein
p	Statistical significance level
p.i.	Post inoculationem (after inoculation)
p.o.	Per os, peroral, oral (by way of the digestive apparatus / alimentary tract)
pA	Prolonged incubation allele
PAP	Peroxidase-anti-peroxidase
PCR	Polymerase chain reaction
PER	Perchlorethylene
PET	Positron emission tomography
PK	Proteinase K
PML	Progressive multifocal leukoencephalopathy
Prion	*Pro*teinaceous *in*fectious particle
Prnp	Prion protein gene in animals
PRNP	Prion protein gene in humans
$Prnp^{o/o}$	Genetically engineered ablation of *Prnp*
$Prnp^{o/o}$ mouse	Mouse with the PrP gene ablated (synonym: *Prnp* knock-out mouse)
PrP	Prion protein
PrP27–30	N-terminally truncated PrP of 27–30 kDa
PrP^{C}	Cellular prion protein
PrPres	Proteinase K-resistant part of PrP
PrP^{Sc}	Scrapie isoforms of PrP; used for all species/hosts (instead of PrP^{BSE}, PrP^{CJD}, PrP^{CWD} etc.)
PSWC	Periodic sharp wave complexes
QQ	Glutamine / Glutamine homozygosity
r	Correlation coefficient (statistical value)
REM	Rapid eye movements
RFLP	Restriction fragment length polymorphism
RML	Specific scrapie strain named after the Rocky Mountain Laboratory
RR	Arginine / Arginine homozygosity
S100	A calcium-binding protein, unspecific marker protein for CJD in blood and CSF (→)
S100b	Marker protein for the differential diagnosis of dementia
s.c.	Subcutaneous

SAF	Scrapie-associated fibrils
SBM	Specified bovine materials (→ SBO)
SBO	Specified bovine offal; compare SRM (→ SRM)[4]
sCJD	Sporadic CJD
SDS	Sodiumdodecylsulfate
SEAC	Spongiform Encephalopathy Advisory Committee
SEM	Standard error of the mean
SFI	Sporadic fatal insomnia
SHaPrP	Prion protein of the Syrian hamster
Sinc gene	Scrapie incubation period gene in the mouse; corresponds to *Prnp* (→)
Sip gene	Scrapie incubation period gene of the sheep; corresponds to *Prnp* (→)
SRM	Specified risk material [4]
SSBP	Scrapie sheep brain pool
SSC	Scientific Steering Committee of the EU
SSCP	Single strand conformation polymorphism
Tg	Transgenic
TK gene	Thymidine kinase gene (→ HSV-TK)
TME	Transmissible mink encephalopathy
TSE	Transmissible spongiform encephalopathies (synonym: prion diseases)
Val	Valine (amino acid)
vCJD	Variant Creutzfeldt–Jakob disease; formerly called nvCJD (→)
VLA	Veterinary Laboratories Agency based in Weybridge, GB; formerly known as Central Veterinary Laboratory (→ CVL)
VV	Valine / Valine homozygosity; e. g. at codon 129 of *PRNP* (→)
wt mice	Wild-type mice

[1] Measures taken to prevent the spread of BSE to humans and animals are summarized in Tables 39.2 and 39.3.

[2] Allowing for completion of a feed recall scheme that disposed of UK feed produced before 29 March 1996 (date of reinforced feed ban), the ban on feeding of mammalian MBM to farmed livestock in the UK was considered effective after 31 July 1996.

[3] Compare: Quantitative assessment of the residual BSE risk in bovine-derived products. The EFSA J 2005; 307:1–135. See also Webpage 2006: www.efsa.eu.int/science/biohaz/biohaz_documents/1280/efsaqrareport2004_final20dec051.pdf

[4] SRM-relevant data and further study results related to infectivity distribution in the body are summarized in Chapter 53.3.1, Figure 53.2 and the WHO Tables$_{\text{Appendix}}$IA-IC.

Topic I: History

1 Historical Introduction

Beat Hörnlimann, Detlev Riesner, Hans Kretzschmar, Robert G. Will, Stuart C. MacDiarmid, Gerald A.H. Wells, and Michael P. Alpers

1.1 Introduction

Until the middle of the 20th century, the Fore people in the Eastern Highlands of Papua New Guinea (PNG) and the neighboring groups with whom they intermarried lived a completely traditional existence, "untouched by civilization". In contrast to most other aboriginal societies they practiced ritual cannibalism during mourning rites. During these events they came into close contact with and consumed the brains and other body parts of the deceased clan members [1]. This was done out of respect for the deceased and, in a sense, as a way of keeping the spirit of the dead person among the living [2]. Many illnesses were regarded to be the result of sorcery. One of their ailments was a fatal disorder, which they called "kuru", the Fore word for shivering or trembling. We now know kuru to be the first documented epidemic form of a prion disease of humans. The natural history of this infectious neurodegenerative disease, perpetuated by endocannibalism and characterized by long incubation periods, sometimes exceeding 45 years, remains one of the most remarkable examples of a prion disease.

Today, kuru remains of importance [3] because of certain parallels that can be drawn between it and the variant Creutzfeldt–Jakob disease in humans (vCJD), which is caused by the BSE agent. Bovine spongiform encephalopathy (BSE), popularly referred to as "mad cow disease", and vCJD emerged in the 1980s [4] and 1990s [5], respectively. There are other related diseases, listed in the tables of this chapter, some of which have been recognized for much longer than BSE and vCJD but have recently gained increasing attention due to the intense interest created by the newly emerged diseases. It has now become accepted that all these related diseases are caused by a newly recognized type of infectious agent called a prion; this group of diseases are therefore generally referred to as "prion diseases".

The main focus of this book is to describe the interrelation of prion diseases in humans and animals. Chapter 1 gives a historical overview of prion research and prion diseases. Chapters 2 and 3 also serve to introduce the reader to this fascinating topic.

1.2 The cause of prion diseases

Although prion diseases are experimentally transmissible, not all of them are caused by infections or are transmitted naturally. Some of them may occur spontaneously or have a genetic background. Yet all these diseases do, in fact, have several common traits (see Section 1.11), which is why they can be grouped together in a class. Individual diseases nevertheless differ in clinical signs, epidemiology, pathogenesis, and certain aspects of the pathology – lesion distribution for example. The etiological variety of prion diseases described in this book results from the exceptional nature of prions. Prions are agents which are clearly distinguishable from viruses, bacteria, protozoa and fungi because they do not contain nucleic acid. The prion hypothesis was put forward in 1982 by S.B. Prusiner [6;7]. Since then, it has been possible to investigate the nature of the pathogen in greater detail.

The transmissibility of these diseases and their neuropathology, in the form of spongiform changes of the brain (Fig. 1.1), give rise to the descriptive term transmissible spongi-

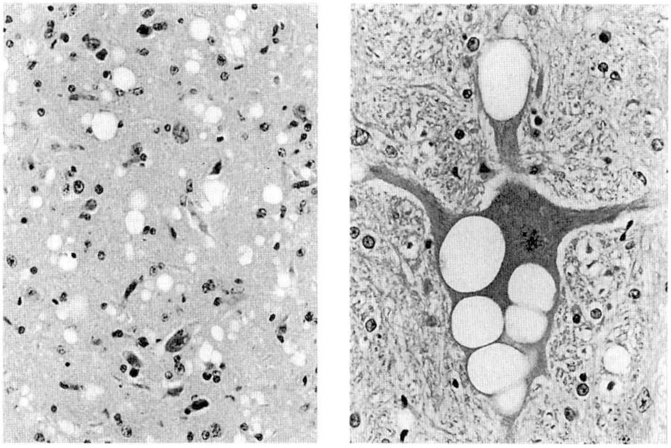

(a) (b)

Fig. 1.1: Spongiform changes in the brain, the characteristic pathological feature of prion disease. Under the light microscope, a sponge-like or spongiform vacuolation can be observed. **(a)** Many vacuoles, located in the neuropil. **(b)** Individual nerve cells can contain multiple intracellular vacuoles. Figure by G.A.Wells and S.Hawkins [8].

form encephalopathies (TSEs). The term prion diseases was later introduced to reflect the causative agent (prion or TSE agent) of these degenerative diseases.

1.3 Scrapie: archetype of all prion diseases

1.3.1 Classical scrapie

Scrapie is the best characterized prion disease and it has strongly influenced related research as well as the nomenclature of all prion diseases; some researchers use the term "scrapie-like agents" as a synonym for "prions" or "TSE agents". According to references in the literature, scrapie – in sheep – has probably been known since the biblical, early Chinese, and the Roman epochs. The first definite year of observation seems to be 1732 [9–13], whereas most other prion diseases were described in the 20th century.

The term scrapie, which has reportedly been in use since 1853 [14], is derived from an important clinical sign of the disease, i.e., to scrape. Other names for the disease are Traberkrankheit (Fig. 1.2) (German: traben, i.e., to trot), and tremblante (French: trembler, i.e., to shake or tremble). There have been many other names by which scrapie has been known in different countries and – historically – because of the regional distribution of sheep breeds in Britain and elsewhere there were also many lo-

Fig. 1.2: Cover and paragraph on scrapie from a book published in 1759. Bottom: Original text in archaic German. English translation below on this page. Source [15].
The translation reflects the content, though it has been modernized linguistically. Translation by J. Bachmann, L. Hansen Gerheuser & L. Atschreiter. Reviewed by M.J. Ulvund & G.A.H. Wells. Figure by B. Hörnlimann ©.

Trotting disease in sheep is contagious

It has also been observed that sheep can get the trotting disease[1] (scrapie[2]), a disease that is identifiable by [certain behavior, for instance,] the affected animal lies down, bites its feet and legs, rubs its back on posts, is generally in poor condition, has a reduced appetite and becomes lame[3]. [The disease] progresses slowly; [the animal] gradually wastes[4] away and eventually dies. Animals that suffer from this form of disease (distemper) never recover. The best solution, once the shepherd has identified a trotting disease-afflicted animal, is to remove it from the healthy ones in the flock and slaughter it for the servants of the estate[5], as the disease is contagious and may wreak havoc on the flock.

[...] Square brackets were used to enclose explanatory/missing words

1 In German: traben, i.e. to trot. In archaic German: "Trab" (of "Traberkrankheit"), i.e., "trotting disease".

2 The term scrapie has reportedly only been in use since 1853 [14]. Therefore in old English, several names that were in use around 1759 would be appropriate but relate to other clinical signs: goggles, rubbers or rickets (explanations see text in Section 1.3.1).

3 It may be assumed that the term "lame" was used to describe "onset of ataxia".

4 In the context of prion diseases, compare (chronic) wasting disease (CWD).

5 *Herrschaftliche(s)* Gesinde = "servants of the estate" ['NB. Readers of today would probably say farmhands/farmworkers']

Nützliche
und auf die Erfahrung gegründete
Einleitung
zu der
Landwirthschaft.
Fünf Theile.

Mit Kupfer und Baurissen.

Durch
Johann George Leopoldt,
Hochreichsgräflich-Promnitzischen Wirthschafts-Amtmann
der Herrschaft Sorau.

Berlin und Glogau,
bey Christian Friedrich Günthern.
1759.

Der Trab ist auch eine Krankheit der Schaafe, und ist ansteckend.

Es bekommen auch manche Schaafe den Trab, welches eine Krankheit ist, die daran zu erkennen, wenn sich das Stück, das solchen bekommt, niederlegt, und beißt mit dem Maule an den Füßen und um die Beine, und reibet sich mit dem Kreuze an denen Stangen, verlieret das Gedeihen, frißt auch nicht recht, und verlahmet endlich; sie schleppen sich lange, verzehren sich nach und nach, und zuletzt müssen sie sterben. Welches Vieh diese Staupe bekommt, wird nicht besser. Daher denn das allerbeste ist, daß ein Schäfer, welcher ein Stück von dem Trabe befallen, gewahr wird, es bald wegschafft, und vors herrschaftliche Gesinde schlachtet. Es muß ein Schäfer ein solches Stück Vieh also gleich von dem gesunden Vieh absondern, denn es steckt an, und kann vielen Schaden unter der Heerde verursachen.

cal names for scrapie, for example: murrain, shewcroft, shakers, scratchie, cuddie trot, goggles (a Wessex term, to goggle; i. e., to stare or squint), rubbers (to rub), rickets [10] or trotting disease (Fig. 1.2), prurigo lumbar (Spanish), and riđa (Icelandic). Scrapie or riđa was apparently brought to Iceland with a ram of English origin in the year 1878.

Because of the economic losses scrapie caused, British farmers tabled a petition in the House of Commons demanding containment of scrapie, on 15 January 1755 [9]. As early as 1759 it was believed by some that scrapie in sheep is "contagious"; for instance, in Germany, where the disease was described as "ansteckend" ("contagious"; Fig. 1.2) [15]. (However, this concept was not generally known or accepted at the time and, even a century later, a number of alternative hypotheses were put forward to explain the cause of scrapie, e. g., sexual hyperactivity of the rams, lightning strikes and so forth [16]). It was only in 1869 that the first evidence pointing to the possibility of infection was presented [17]. Some decades later, in 1898, pathological changes in the brain of scrapie-infected sheep were described for the first time [18;19]; this was a very important breakthrough for the confirmation of diagnosis.

In connection with the complex etiological nature of prion diseases it is interesting that scrapie was already described in 1828 as "epizootic inherited evil" [20]. In 1886 a "hereditary defect" was suggested as being a partial risk factor of scrapie [21;22]. This explanation came extraordinarily close to the current understanding of the disease (see below and Chapter 56), but again, these concepts were not generally known or accepted at the time.

1.3.2 Origin, spread and genetic factors of scrapie

Today it is impossible to determine whether the disease was spread solely by particular sheep breed exports (see below) or whether specific genetic predispositions led to a higher level of susceptibility to the disease in some areas or countries. Indications of a particular susceptibility to scrapie in Merinos, British Cheviots, Suffolks or other breeds can be found in several historical sources [13;14;21;23].

Merinos. Some authors have suggested that scrapie may have been spread from Spain [13;14], and if so, particularly with exports of the transhumant (itinerant) fine-wooled Merino *cabañas* [9;24–26]. However, the disease was not known in the Merino *estantes* that consisted of the non-travelling common settled native sheep breed of Spain [24]. According to the literature [27–31] there is strong evidence that scrapie may have already been present in the indigenous sheep of North-East Europe and the Austro-Hungarian Empire before the importation of Merinos from Spain. H. B. Parry records the crucial importance of the export of Merinos after 1700 to several countries, following which severe and epidemic forms of scrapie occurred in some sheep populations. Examples of importing countries affected by scrapie were Germany and France, but not others, for instance Scandinavia and Britain [9;32]. A disease called scab (which may have been scrapie) occurred in Merinos imported into the USA from Britain between 1812 and 1814 [33].

Cheviots. In some countries, for example Norway [34], British Cheviots probably introduced the disease, the breed being affected by scrapie since 1837 [9] or earlier. Although scrapie was not officially verified in Norway before 1981, there are several reports of clinical cases of "gnave- og travesjuke" (itching and trotting disease) in 1890, 1916 and 1938, all associated with imports of British Cheviot sheep and their descendants (M. J. Ulvund; personal communication).

Suffolks. Between 1920 and 1958 scrapie became a major problem in the Suffolk breed, and there is evidence for the breed spreading the disease with exports from Britain, causing considerable financial loss, for example in Australia [35;36] and New Zealand [37;38]. Both those countries have thoroughly investigated the scrapie risks posed by imports and maintain policies to exclude the disease [39–42].

Until the 1930s nobody knew for certain that scrapie was transmissible and it only became scientifically evident much later that the genetic background of the breed influences the scrapie susceptibility (see Chapter 56). Therefore, historical hypotheses should neither be over-interpreted nor ignored; H. B. Parry regarded scrapie as an inherited disease, at least in part, and he maintained this view until the end of the 1970s [9][1].

Step by step, research has now proven that scrapie is influenced by host genetics in terms of susceptibility (see below; and Chapters 3; 18; 56). So it must be kept in mind that in various countries some local breeds of sheep that are resistant to scrapie disease could prove useful if this disease was to become widespread [43].

In 1936[2] two Frenchmen, J. Cuillé and P. I. Chelle, proved experimentally that scrapie is a transmissible disease [45; 46]; providing for the first time indubitable scientific evidence of the transmissible (French: "inoculable") nature of an animal prion disease.

At the beginning of the 1950s, D. R. Wilson described the characteristics of the "very exceptional scrapie pathogen" [47; 48] and over the course of the second half of the 20th century further strong evidence of transmissibility was provided [49; 50]. Scrapie is spread horizontally and laterally within a flock or indirectly over distances – as has been shown in Iceland [51] – by natural contact with placenta and fetal membranes. Outside the host, scrapie prions may remain infectious and contagious for years [52; 53] and are also extremely resistant to chemical disinfectants (see Chapter 44) and heat (see Chapter 45).

1.4 Transmissible mink encephalopathy

A hitherto unrecognized disease in mink kept for fur production was first observed in the USA in 1947 [54; 55] and later also in the former Soviet Union [56; 57] as well as other countries. The disease was eventually named transmissible mink encephalopathy (TME). Today we believe that in the former Soviet Union [56; 57] and probably also in certain Scandinavian countries, infection was most likely transmitted to mink by feeding scrapie-infected parts of sheep carcasses obtained from abattoirs (see below and Chapters 22; 42.3.3).

TME was the first prion disease diagnosed in carnivores. Other than the primary dietary exposure it is possible that other means of transmission occurred in mink once infection had been established. This could explain the 100% infection rate of animals on several mink farms, a within-farm incidence rate that is unusually high for prion diseases (Table 22.1). The feeding of the cadavers of infected minks to others in the same farm may have amplified transmission within a mink herd. In addition, it can be postulated that this prion disease was sometimes also transmitted parenterally within a cohort since farmed mink tend to be aggressive and bite each other in their communal cages. In particular, mink in the early clinical stages of TME tend to be aggressive. This may have led to an "inoculation" of the agent since their teeth [58] may have been contaminated with infectious tissue. Also, it is likely that female mink, sick with TME and suffering from extreme somnolence in the late clinical stage, were eaten by young mink living in the same cages [55]. In US studies, scrapie has never been successfully transmitted to mink by the oral route and for this reason some researchers have suggested cattle (BSE; see Chapter 22) as a source. For these reasons, some of what is written about the transmissibility of TME is speculation.

1.5 Chronic wasting disease in North American cervids

From the end of the 1960s [59], a new disease was observed in North American cervids (Fig. 1.3) but it was not until the early 1980s that it became clear that such cases belonged to the class of prion diseases (Table 1.1). This disease was called "chronic wasting disease"

---

[1] Until Parry's death in 1980 he seems to have been convinced that scrapie was largely an inherited disease. His book [9], which was edited for posthumous publication by D. R. Oppenheimer, makes this clear.

[2] An excellent bibliography, listing articles from 1934 onwards on scrapie history, genetics and research can be found on the Internet [44].

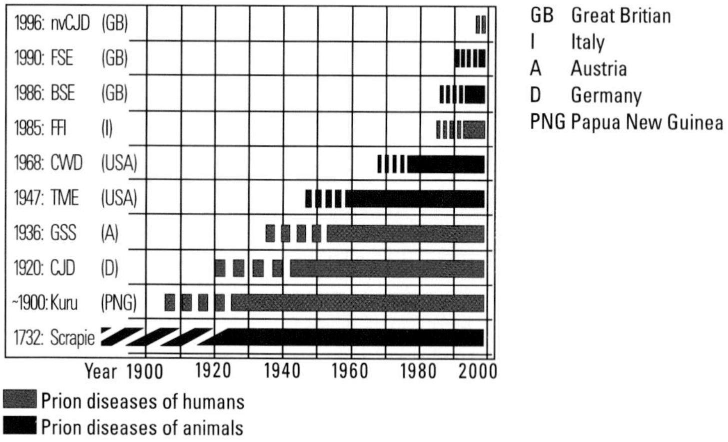

Fig. 1.3: Simplified presentation of the prion diseases as they arose or were discovered, mainly in the 20th century. Scrapie in animals was known long before the 20th century and genetic prion diseases such as GSS (Fig. 16.1) and FFI (Fig. 17.1) in humans also existed before the 20th century but were only recognized in that century. Historical GSS cases can be traced as far back as the late 18th century using preserved tissue samples [60]. In the figure, the country of the first diagnosis is given. Scrapie may have had its origins in Spain. Figure by B. Hörnlimann ©.

(CWD) in reference to one of its most prominent clinical features, emaciation. To date CWD has only been found in certain members of the family Cervidae in North America: wapiti, white-tailed deer, black-tailed deer, mule deer as well as their crossbreeds and moose. CWD occurs among animals living in captivity (commercial game ranches and farms, reserves, zoos) and in the wild, where expanded surveillance since 2001 has revealed endemic foci of the disease in several parts of the USA and Canada. However, it is not clear whether CWD is a disease in its own right or whether it was originally caused by scrapie.

Today we know that CWD is naturally transmissible via horizontal routes and that it is – along with scrapie in sheep and goats – contagious (see Section 1.10.2). Local infection rates can be relatively high: in some areas of Colorado and Wyoming prevalence within wild deer populations can exceed 5 % and even higher rates can occur in captivity (see Chapter 21). Several important aspects of CWD epidemiology, including its natural transmission mechanism(s), are still poorly understood. The CWD agent has been experimentally transmitted on first passage by intracerebral (i.c.) inoculation to cattle.

1.6 Creutzfeldt–Jakob disease and other human prion diseases

In 1920 [61] Hans Gerhard Creutzfeldt and in 1921 [62] Alfons Maria Jakob reported a rare neurodegenerative disease, which Jakob called "spastic pseudosclerosis". This disease is now internationally known as Creutzfeldt–Jakob disease, abbreviated to CJD. Currently, four different forms or variants of CJD are recognized: iatrogenic (iCJD), sporadic (sCJD), familial (fCJD) and the (new) variant CJD (vCJD). These are listed in Table 1.2 and described in more detail in several chapters of this book (see also Fig. 1.4).

1.6.1 Sporadic CJD

Although first reported in 1920 [61] and 1921 [62], the cause of sporadic Creutzfeldt–Jakob disease (sCJD) has remained uncertain. For this reason, the term idiopathic CJD is also used. It has been postulated that sporadic CJD may

Table 1.1: Prion diseases in humans and animals (transmissible spongiform encephalopathies; TSE).

Presence in	Disease / variants	Abbreviation	Year and source of first description[1]
Humans	Creutzfeldt–Jakob disease: four forms or variants exist (Table 1.2)	sCJD iCJD fCJD vCJD	1920/21 [61–63] 1974 [64] 1930/1995 [65;66] 1996 [5]
	Gerstmann–Sträussler–Scheinker syndrome = Gerstmann–Sträussler–Scheinker disease	GSS	1913 [67] 1928 [68] 1936[2] [69]
	Fatal familial insomnia	FFI	1986 [70]
	Sporadic fatal insomnia (thalamic variant of CJD; ?)	SFI	1999 [71]
	Kuru[3]	Kuru	1957 [72]
Animals	Classical scrapie of sheep and goats	Scrapie (Sc)	1732 [11] 1759 [15][4]
	Atypical scrapie[5]	Nor98	1998 [74;75]
	Transmissible mink encephalopathy	TME	1965 [54;55]
	Chronic wasting disease	CWD	1980 [59; 79]
	Bovine spongiform encephalopathy	BSE	1987 [4]
	BSE of nyala and greater kudu[6]	TSE / BSE	1988 [76]; further refs. in Table 1.3
	Feline spongiform encephalopathy	FSE	1990 [77] 1992 [78]

1 The date when the individual forms of the diseases were described for the first time is not always identical with the year in which the disease was diagnosed for the first time.
2 Historical GSS cases can be traced back to the late 18th century.
3 Name given to the disease by the Fore people in Papua New Guinea.
4 First time that an author claims that scrapie is "contagious" (Fig. 1.2).
5 Also referred to as "sporadic scrapie of sheep and goats" or "sporadic TSE of sheep and goats", whereas the term "small ruminant TSE" (SRTSE) summarizes classical and atypical scrapie cases (Chapters 32.2.3; ref. [73]).
6 Possibly also of other Bovinae and Hippotraginae subfamily members (see footnote 1 of Table 1.3).

Table 1.2: The different forms of CJD.

Disease variants	Causes
Sporadic CJD (sCJD)	Idiopathic. The cause of this form is uncertain. Chapter 3 describes how the spontaneous transformation of the cellular prion protein (PrP^{C}) into the disease-causing prion protein (PrP^{Sc}) might occur. Due to its rare but uniform distribution worldwide this variant is called sporadic.
Iatrogenic CJD (iCJD)	Acquired. This form is transmitted through medical procedures (i. e. medical accidents).
Familial CJD (fCJD) synonym*: Genetic CJD (gCJD)	Hereditary. This form is caused by mutations of the prion protein gene (genetically determined, therefore also called genetic CJD).
Variant CJD (vCJD)	Acquired. This variant has been known since 1996 (originally named nvCJD) and is most likely caused by BSE prions from cattle.

* The term "genetic CJD", abbreviated "gCJD", seems more appropriate in the context of the clinical history of a patient or affected family, as there is a family history in only about 50% of all genetic CJD cases.

Note: Sporadic fatal insomnia (SFI) is a newly recognized idiopathic disease possibly identical with the – historically named – "thalamic variant of CJD" (see Chapter 30.9). We did not include this entity in the table.

develop spontaneously (see Chapter 3). The prion hypothesis offers a plausible explanation for a spontaneous origin of sCJD [7]. The term "classical CJD" sometimes found in the literature is not clearly defined; it sometimes summarizes the sporadic, familial and iatrogenic forms of CJD and sometimes refers only to sporadic CJD.

1.6.2 Acquired CJD forms

iCJD. The transmissibility of CJD remained unrecognized until the end of the 1960s, shortly after the transmissibility of kuru was demonstrated [80; 81]. Some cases of CJD were unintentionally transmitted from person to person as a result of medical accidents. These are the iatrogenic forms of CJD, first acknowledged in a publication in 1974 [64]. Iatrogenic CJD (iCJD) cases have been caused by the use of contaminated surgical instruments [82], cornea or dura mater implantations, or the administration of contaminated pituitary gland hormones (Table 14.1) [83].

vCJD. This disease emerged in Great Britain (GB) during the course of the 1990s and to contrast it from sCJD and iCJD it was called (new) variant CJD (vCJD; formerly nvCJD). It is almost certain, but not provable in a direct way, that this new disease was caused by the BSE agent; further details on clinical signs, epidemiology, risk factors and other aspects are described in this book. Although the indications are that a major vCJD epidemic has been averted, it will be another decade before we will have the confidence to assess the final risk, for example of subclinical infection and human-to-human transmission through blood transfusions (see Chapter 51).

1.6.3 Genetic human prion diseases

fCJD/gCJD. "Familial CJD" is hereditary; and may also[3] be referred to as "genetic CJD" (gCJD). Once clinical signs develop the inherited disease is experimentally transmissible: that is, infectivity is generated de novo (see Chapter 3). It should be noted that familial, hereditary and genetic are synonyms in most publications.

GSS. In 1936, Gerstmann, Sträussler and Scheinker described another human prion disease in great detail, namely the Gerstmann–Sträussler–Scheinker disease (GSS), which had already been mentioned in publications in 1913 and 1928 [67; 68] (Fig. 1.3 and see Chapter 16). Indeed, cases from the original Austrian GSS family can be traced back to the late 18th century, although this was only discovered in 1995 when preserved tissue samples were examined [60]. GSS has an exclusively familial background [69]. Nevertheless, it was shown through animal experiments in 1981 that GSS can be transmitted [84].

FFI. Fatal familial insomnia is also a genetic prion disease that was first diagnosed in the mid-1980s in Italy [70; 85], but which actually occurred long before then. As in fCJD and GSS, the family tree [86] may be used to show that there were (unrecognized or unreported) cases of this disease for decades or centuries (Fig. 17.1). The term FFI refers to the observation of a disordered sleep – that is recognized by polysomnography – experienced by patients suffering from the disease. The first experimental transmission of FFI was successfully performed in 1995 [87].

1.6.4 SFI / The thalamic variant of CJD – an idiopathic human prion disease

In 1999, P. Parchi and colleagues described a very rare sporadic form of "fatal insomnia" which is phenotypically indistinguishable from FFI, but without a *PRNP* mutation (see below) [71]. It is possibly caused by a spontaneous conversion of PrP^{C} into PrP^{Sc} (see below) – and was named sporadic fatal insomnia (SFI) – or by an unknown somatic *PRNP* mutation. Until this disease was recognized as a distinct entity it is highly likely that similar cases were previously labeled as "the thalamic variant of CJD"

[3] The term "genetic CJD", abbreviated "gCJD", seems more appropriate in the context of the clinical history of a patient or affected family, since there is a family history only in about 50% of all genetic CJD cases.

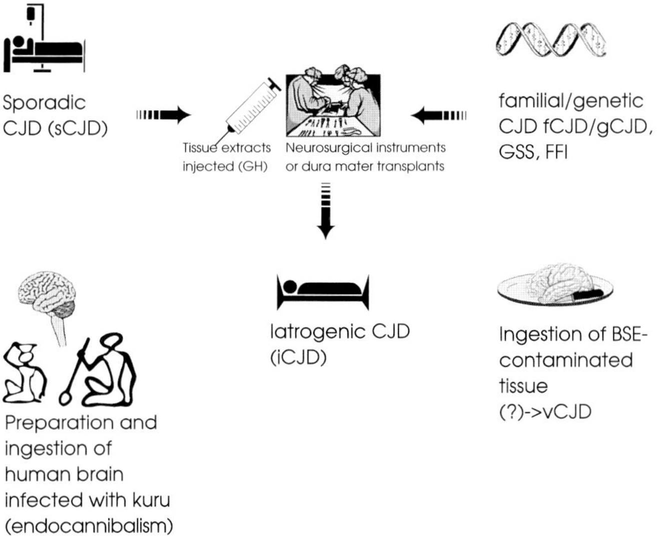

Fig. 1.4: Simplified presentation of the human prion diseases. Sporadic Creutzfeldt–Jakob disease (sCJD) is presumed to develop spontaneously. However, CJD can also be iatrogenically transmitted (iCJD) through the administration of hormones derived from the pituitary gland of sCJD-infected people, the implantation of infected dura mater or through contaminated surgical instruments. Familial prion diseases such as fatal familial insomnia (FFI), Gerstmann–Sträussler–Scheinker disease (GSS) syndrome and familial Creutzfeldt–Jakob disease (fCJD) originate from mutations of the prion protein (PrP) gene. These forms are also transmissible. Kuru was transmitted mainly through the ingestion of infectious human brain. In addition, some individuals were probably infected through skin lesions, mucous membranes, or the eyes. It is conceivable that the first kuru case originated from sCJD. The (new) variant Creutzfeldt–Jakob disease (vCJD) can most probably (? → vCJD) be attributed to the ingestion of BSE-contaminated CNS tissue, e. g. cattle brain or spinal cord. Figure by B. Hörnlimann ©.

or "thalamic dementia" (compare Chapter 30.9; Annex: Historical classification). The clinical features of this disease (see Chapter 17) are similar, if not identical to those of FFI. SFI is of unknown etiology.

1.6.5 Kuru – an acquired human prion disease

In our introduction to the unique world of prion diseases we indicated that kuru was transmitted from person to person through the particular cultural habits of the Fore people and their neighbors. Systematic research on kuru was carried out since the late 1950s by Gajdusek, Alpers (Fig. 38.5), Zigas and others, who were unaware at the time that their studies would open the door to an understanding of a whole series of human prion diseases. A team led by the great pioneer of kuru research, D. Carleton Gajdusek, who was awarded the Nobel Prize in Physiology or Medicine in 1976 [88], discovered the extraordinary mode of transmission of kuru [80; also [1]]. Because of the rather unusual environment in which the kuru epidemic appeared and due to its great importance in initiating research into prion diseases, the topic will be discussed in depth in Chapters 2, 13 and 38.

1.7 The scrapie–kuru connection

Thirty years after the proof of the transmissibility of scrapie [45;46] the first evidence was found that human prion diseases are transmissible [80].

In 1959, William J. Hadlow, a veterinary pathologist, noted the pathological similarities of scrapie and kuru [89; 90]. Up to the 1950s, research on scrapie and human prion diseases was completely separate. Nobody suspected any connection between kuru on the one hand (Fig. 1.4) and scrapie on the other. (There was not even any indication that kuru and CJD were related diseases.)

Hadlow suggested that the similarities between scrapie and kuru should be scientifically investigated and that transmission experiments of kuru to chimpanzees should be performed [90;91]. Subsequently, in 1966, Gajdusek, Gibbs and Alpers described the transmissibility of kuru to chimpanzees [80], proving for the first time the infectious nature of a human prion disease. Today we know that kuru is characterized by a long incubation period [1; 92; 93], which may extend to over 45 years (see Chapter 38.2).

Generally speaking, all acquired human and animal prion diseases are characterized by long incubation periods. Based on the observation of unusually long incubation periods [94] it was suspected that a so-called "slow virus" was the causative agent of the diseases [95].

The concept of "slow virus" diseases, also called slow infections of the central nervous system (CNS), was postulated by Björn Sigurdsson [95] in Iceland during the 1930s and was presented in a series of lectures in London in 1954. The concept was stimulated by the recognition of a number of chronic sheep diseases that threatened the Icelandic sheep economy and trade following the introduction of twenty Karakul sheep from Germany in 1933. Initially the diseases included several diseases, one of which was riđa [96] (Icelandic name for scrapie). The concept later embraced several human diseases for which some of the animal diseases were models. All had long incubation periods (usually years), progressive clinical courses, and fatal outcomes.

1.8 Etiological variety of prion diseases

During the odyssey of research into these diseases, which was largely influenced by kuru, some experts began to doubt whether the term "slow virus" was an accurate description of the causative agent for this group of diseases in humans and animals [97; 98]. Although Sigurdsson's concept of slow virus diseases was initially useful in the context of slow infections of the CNS [99], research showed that they did not have a common cause.

Therefore new ideas and names for the pathogen were proposed: virino [100], unconventional virus [100–104], the agent of infectious amyloidosis [105], postulated exotic agents[4] and finally prion [6; 107] (Fig. 1.5a).

The term "infectious amyloidosis" has been used to convey the light-microscopic observation of plaques, which are found in the brain in some prion diseases and are composed of the

[4] See summary of references in [106].

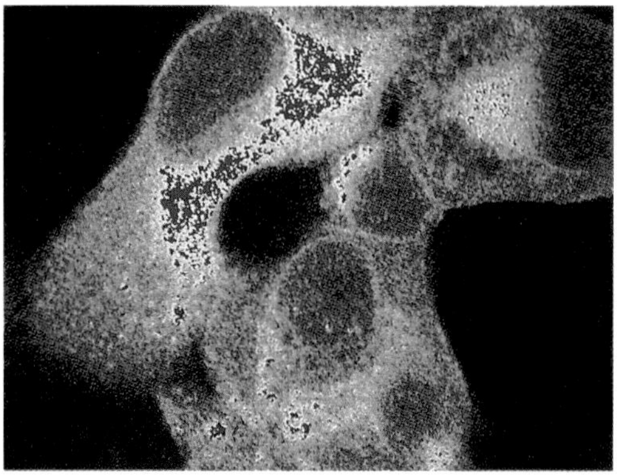

Fig. 1.5: (a) PrP^{Sc} accumulated in a nerve cell. Source [108].

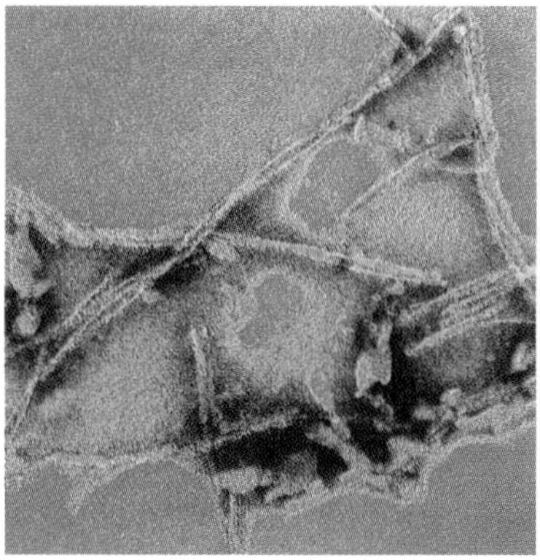

(b) Scrapie-associated fibrils (SAF).

amyloidogenic, misfolded form of the prion protein. Abnormal fibrils, again comprised of the misfolded form of the prion protein and visualized by electron microscopy in brain extracts, were first observed in 1981 by Patricia Merz, who called them scrapie-associated fibrils (SAF; Fig. 1.5b) [109; 110].

Other hypotheses for causative agents have been put forward and have contributed to scientific discussion before being discarded. Some still believe[4] that conventional viruses, which replicate by virtue of their nucleic acid-defined genes, are involved in the causation and progression of the transmissible spongiform encephalopathies (TSEs), but that technical problems have prevented their identification.

In the 1960s, several findings suggested that the infective agent causing TSEs (prion diseases) was not a virus. In 1966, for example, Tikvah Alper found that it was not possible to inactivate the scrapie agent by radiation which usually leads to DNA or RNA damage (Fig. 4.2) [111; 112]. She came to the conclusion that the agent apparently did not contain any genetic information in the form of nucleic acid (see Chapter 4). The notion of a pathogen devoid of nucleic acid has been the cause of much controversy [107].

In 1967, Gibbons and Hunter put forward the idea that a reproductive cell membrane protein was the causative agent [113]. In the same year, Griffith hypothezised that a mere protein might "reproduce" in a kind of autocatalytic process and might therefore be responsible for the transmission of scrapie [114]. These proposals, then purely hypothetical constructs, came quite close to the currently accepted views summarized in this book.

Prusiner took up the question of the nature of the scrapie agent in the 1970s and has developed important insights through years of systematic biochemical research, establishing the model of a protein-containing pathogen without nucleic acid. Prusiner details the basic findings leading to the prion model in Chapter 3. Alper's conclusion that the agent did not contain nucleic acids was further extended by Prusiner's model, which suggests that it is almost impossible for the agent to be a virus.

The term "prion" was published for the first time in 1982 [6], and became increasingly used to denote the pathogen. According to the prion model, the essential component of the scrapie pathogen is a protein. Prusiner called the agent a prion, although the more obvious abbreviation suggested by the term "proteinaceous infectious particle" might have been "proin". However, prion is a better-sounding name – and Prusiner therefore chose 'prion' over 'proin' (M.P. Alpers personal communication recalling conversations he had with Stanley Prusiner in the early 1980s). The hypothesis accommodated the possibilities that the entire pathogen consisted of protein only, or that the protein represented the major component among other components. In 1985, the collaborative research in three laboratories, led by S. Prusiner, C. Weissmann and L. Hood, was reported in a classic article by B. Oesch et al. [115] that describes the sequence of a protein called the prion protein (PrP). It was shown that PrP is the product of a host gene called the PrP gene or *PRNP* in humans and *Prnp* in animals. The gene is present on chromosome No.20 in humans and on chromosome No.13 in cattle. In 1998 it was found that the gene in mice called *Sinc*[5, 6] (defined by disease phenotype – controlling the incubation period) and the PrP gene (*Prnp,* defined by molecular genetics) are one and the same [116]; the *Prnp* varies in individual sheep and in different sheep breeds.

Conceptually it had to be expected, and was indeed shown experimentally, that PrP existed in two different isoforms: in non-infected organisms as a soluble protein called "cellular PrP" or PrP^C, and in infectious tissue as a highly insoluble form of the protein called "scrapie PrP" or PrP^{Sc}, which could be identified due to its unusual resistance to digestion by proteinase K (PK). The physiological function of PrP^C is not yet entirely known (see Chapter 7).

Although the prion model has long been a subject of scientific debate, most researchers

5 *Sinc* gene = Scrapie incubation [period] gene of mice = *Prnp* of mice.

6 The equivalent gene in sheep is called *Sip* gene = Scrapie incubation period gene = *Prnp* of sheep.

have eventually accepted that PrP is consistently found as a component of prion infections, and in the form of PK-resistant PrP^{Sc} can be used as the only known disease-specific marker. PrP^{Sc} is detectable by immunohistochemical means (IHC) in the brain of affected humans and animals [117–119]. All prion diseases in humans and animals are associated with the accumulation of a conformational isomer (PrP^{Sc}) of the host-derived prion protein (PrP^{C}) in the brain. According to the protein-only hypothesis, PrP^{Sc} is the principal or sole component of transmissible prions.

In 1997, Prusiner was awarded the Nobel Prize in Physiology or Medicine [120; also [88]]. In 2004, it became possible to synthetically produce mammalian prions from recombinant PrP; the in vitro synthesis of mammalian prions by G. Legname et al. was an important breakthrough to strengthen Prusiner's hypothesis [121].

The prion model introduced a new era of research into infectious diseases. Biological phenomena such as replication, amplification, species barrier (see below) and strain specificity (see below) have had to be redefined. The traditional understanding of these phenomena was based on the notion of nucleic acids as the carriers of genetic information. However, we now know from the findings in prion research that this can no longer be accepted as dogma [6; 121].

The experimental confirmation that a novel form of human prion disease, variant CJD (see below), is caused by the same prion strain as cattle BSE (see below), has highlighted the pressing need to understand the molecular basis of prion propagation and the transmission barriers that limit their passage between mammalian species. These and other advances in the fundamental biology of prion propagation are described in Topic II of this book (Molecular Biology and Genetics).

1.9 New prion diseases

1.9.1 The species barrier

An important phenomenon in all infectious diseases is the barrier to transmission between different species. Today it is known that in prion diseases the efficiency of the so-called "species barrier", at least in part, depends on the interaction between the intruding PrP^{Sc} and the PrP^{C} of the infected species, i.e., the degree of sequence homology between *Prnp* of the donor and *Prnp* of the recipient. There are several other factors determining the efficiency of the species barrier, such as the route of transmission and the infective dose, but some further factors still remain ill defined. The most probable explanation for BSE is that the species barrier between sheep and cattle had been crossed and that this was likely to have been related to specific feeding practices (see Chapters 19; 39–40; 52). As a further consequence of BSE, barriers between cattle and other species were crossed as outlined in the following sections.

1.9.2 BSE in cattle

Following the reporting of an unusual behavioral and movement disorder in a number of cows in GB in 1985 and 1986, Gerald A.H. Wells and Martin Jeffrey conducted neuropathological examinations. They discovered spongy brain tissue changes (Fig. 1.1) which led to the term 'bovine spongiform encephalopathy' (BSE). The similarity of this new neurological disorder of cattle to scrapie was noted and BSE was identified as a prion disease. The original source of the BSE epidemic may be traced back to sheep scrapie [122; 123] which has been endemic in GB for centuries [9; 124]. The species barrier from sheep to cattle may have been crossed indirectly by way of concentrated animal feed containing inadequately heat-inactivated meat-and-bone-meal (MBM[7]) from sheep infected with scrapie. Important technical alterations in the British animal feed production were introduced in the 1970s (see Chapter 19); subsequently most British rendering plants applied a new fat extraction method (Fig. 19.1), working with temperatures of $< 100\,^{\circ}C$ and with heat exposure times much

[7] MBM (→ the reader can obtain comprehensive information about MBM and other important residual feed-related BSE risks or BSE risks in bovine-derived products in the document [125]).

shorter than before (see Chapters 19.9; 45; 52). Any contaminating scrapie (or BSE; see below) agent would thus not have been fully inactivated (see Chapter 45) leading to a striking increase of infectivity (PrP^{Sc}) in MBM, which served as the vehicle of transmission – now from cattle to cattle (Fig. 19.2).

Alternatively, BSE may have been a rare but hitherto unrecognized indigenous disease of cattle, which became rampant after these technical alterations in British animal feed production methods. Whatever the original source, BSE was amplified in cattle through established feeding practices, which changed in the 1970s (see Chapter 19.9).

In the context of the ongoing debate about the true origin of BSE, the uniform BSE pathology (stereotypic lesion profile; Fig. 27.4) over the course of the epidemic [126; 127] and the fact that, so far, only one (major) BSE-strain has been isolated under UK field conditions [128] are impressive facts which contrast the observation of many different scrapie strains in sheep (Fig. 12.3). However, a few atypical BSE cases have been described in several countries other than the UK (see below).

It is of note here that the epidemiological data are consistent with an origin of BSE from an increased exposure via feed to a cattle-adapted scrapie-like agent from cattle or from a scrapie-like agent from sheep [129] (see Chapter 19.1). The scrapie hypothesis may be correct but the British BSE Inquiry [130] and SSC reports [131] did not come to that conclusion. It is not known for certain which is correct and it is unlikely that this question will ever be answered with certainty (see Chapter 19.9).

As time went by BSE was confirmed in a number of European countries and eventually further afield: e.g., the Falkland Islands, Middle East, Canada, USA, and Japan etc. This was a result of exports of either live incubating, but otherwise healthy, breeding cattle, or more frequently of infectious MBM, primarily from the UK and – indirectly – from other countries [132–134].

BSE has had, in the UK [135] and worldwide, an enormous economic impact [136]. Because of the importance of this cattle disease, we have concentrated here only on a few introductory points, but further details are presented in several chapters of the book.

For a few "atypical" BSE cases identified recently in several countries, it remains unknown whether or not these are caused by different strains of the BSE agent. The possible natural occurrence of phenotypic variation in BSE or novel expression of a TSE of cattle has been raised by reports from Japan [137], Italy [138], France [139] and more recently, in several other countries, of a small number of cattle with molecular and/or pathological features atypical of BSE as defined in the UK. Strikingly – but in the Italian case only – the molecular signature of this previously undescribed atypical bovine PrP^{Sc} was similar to that encountered in a distinct subtype of sporadic Creutzfeldt–Jakob disease, according to C. Casalone et al. [138] (see Chapters 19.2; 27.5).

1.9.3 vCJD: BSE in humans

In the spring of 1996 the details of a newly identified human prion disease with unique and consistent clinical signs, neuropathology and epidemiology in a series of 10 British patients was published by Will et al. [5]. The presence of PrP^{Sc} in the brain was confirmed using IHC (Fig. 15.3), and the disease was classified as a new variant of Creutzfeldt–Jakob disease (nvCJD), today called variant Creutzfeldt–Jakob disease (vCJD).

Up to May 2006 30 vCJD cases have been identified outside the UK and 161 in the UK. Countries other than the UK include France, Ireland, Italy, USA, Canada, Japan, Netherlands, Portugal, Spain and Saudi Arabia. It is of note that all the countries other than Saudi Arabia have reported BSE cases (see Chapter 19.6; Table 19.2). The vCJD cases in the USA and Canada as well as two Irish cases had a history of extended residence in the UK during the relevant period and were likely to have been exposed to BSE in the UK.

The clinical features of this new disease are described in detail in Chapter 30. Based on the temporal and geographical context of BSE, as well as overwhelming experimental evidence (see Chapter 50), it must be assumed that vCJD is caused by BSE prions, most likely from the ingestion of infectious cattle products (see Chapter 15).

In contrast to sCJD, vCJD affects mainly young people with an average age at death of approximately 29 years, which contrasts with the occurrence of sCJD mainly in patients over 60 years of age (Fig. 15.2).

The Valleron group stated in 2004 [140] that the strikingly young age of vCJD cases (Fig. 15.2) remained unexplained. They concluded that, despite the differential dietary exposure to contaminated food products in the UK population according to age and sex during the BSE epidemic, age-dependent susceptibility to infection may also be an important influence on the age distribution of cases. The authors were using published estimates of dietary exposure [References see [140]] in mathematical models of the epidemiology of vCJD.

Boelle et al. also estimated a theoretical mean incubation period that was built on a mathematical model, concluding that this may be around 15 years [140]. To calculate this kind of estimate one inevitably has to make a number of assumptions, some of which are somewhat uncertain. One problem is that the original work [References see [140]] used data on the timing and extent of human exposure to BSE which has some inherent uncertainty. Hence, despite the excellent work behind the conclusion (15 years), it remains a mathematical model, i. e., a theoretical mean incubation period.

In any case, the estimation of 15 years would only be relevant for patients with a homozygous methionine (MM) genotype, whereas individuals with the homozygous valine (VV) genotype or heterozygous (MV) genotype may also be at risk of developing vCJD [141; 142], possibly with longer incubation periods. Alternatively, these people may be asymptomatic carriers who might transmit the condition to other susceptible individuals by blood transfusion or surgery [141; 142].

1.9.4 BSE in exotic animals and in domestic cats

BSE has not only crossed the species barrier to humans but also to some other species. Since the 1980s diseases caused by the BSE prions have been observed in nyala and greater kudu, and a similar or "the same" disease was diagnosed in certain African ungulate species kept in zoos. The complete list of affected species is shown in Tables 1.3 and 19.1 and described in detail in Chapter 20.

In some zoos, cases of feline spongiform encephalopathy (FSE) appeared among exotic big cats such as cheetahs, pumas, ocelots and others. These carnivores most probably acquired infection by ingesting highly infectious tissue (e. g. raw spinal cord) originating from cattle carcasses [158]. Since 1990, FSE has also been diagnosed in domestic cats [159;160] (Fig. 1.3). Moira Bruce was the first to show that the infectious agents of BSE, FSE and prion diseases in nyala and greater kudu are identical in terms of the incubation time and other strain characteristics on transmission to mice (see Chapter 50). Apart from this and further experimental evidence provided by several research groups, it is also the temporal and geographical occurrence of the British BSE cases on the one hand, and prion diseases in zoo animals and domestic cats on the other hand, that supports an epidemiological link between these new emerging prion diseases and the BSE epidemic.

1.9.5 Atypical scrapie – Nor98 (sporadic scrapie)

The origin of atypical scrapie [74] cases (identified in Norway in 1998; designated Nor98) is still speculative; based on the clinical signs, neuropathology, structure of PrP^{Sc}, PrP genotypes [75] and epidemiology, one can not exclude that such cases could represent a spontaneous disorder of sheep, similar to sCJD in humans (see Chapter 1.6).

The number of sporadic or atypical scrapie cases detected increased throughout Europe after the implementation of new diagnostic tests and surveillance strategies. In Norway for example, from 1998 to 2005, a total of 49 atypi-

Table 1.3: 26 species in which the natural occurrence of prion diseases has been reported; experimental animal species are not listed. The different forms of CJD – sporadic (sCJD), familial (fCJD), iatrogenic (iCJD) and variant (vCJD) – are described in Table 1.2. Sources of Table 1.3: see References in last column. Crown copyright ©.

Order	Family	Subfamily	Host	Genus & species	Disease	References[2]
Primates (Omnivora)			Human	*Homo sapiens*	CJD, vCJD, kuru, GSS, FFI and SFI	see Chapters 13–17 & 26 of this book
Artiodactyla (Herbivora)	Bovidae	Bovinae	Domestic cow	*Bos taurus*	BSE	[143;144]
			American bison	*Bison bison*	BSE[1]	[145]
			Nyala	*Tragelaphus angasi*	BSE[1]	[76;146]
			Eland	*Taurotragus oryx*	BSE[1]	[147]
			Greater kudu	*Tragelaphus strepsiceros*	BSE[1]	[147]
		Hippotraginae	Arabian oryx	*Oryx leucoryx*	BSE[1]	[147]
			Scimitar-horned oryx	*Oryx dammah*	BSE[1]	[147]
			Gemsbok	*Oryx gazella*	BSE[1]	[146;147]
		Caprinae	Sheep	*Ovis aries*	Scrapie / Atypical Nor98	[148] / [74;75]
			Moufflon	*Ovis musimon*	Scrapie	[149]
			Goat	*Capra hircus*	Scrapie / Atypical Nor98 / BSE	[150] / [74;75] / [151;152]
	Cervidae	Cervinae	Rocky Mountain elk (wapiti)	*Cervus elaphus nelsoni*	CWD	[153]
		Odocoilinae	Mule deer	*Odocoileus hemionus hemionus* (= subsp. mule deer)	CWD	[153]
			Black-tailed deer	*Odocoileus hemionus columbianus* (= subsp. black-tailed deer)	CWD	[153]
			White-tailed deer	*Odocoileus virginianus*	CWD	[153]
			Moose	*Alces alces*	CWD	[172; 173]
Carnivora	Mustelidae	Mustelinae	Farmed mink	*Mustela vison*	TME	[154]
	Felidae	Felinae	Domestic cat	*Felis catus (domesticus)*	FSE (BSE[1] in cat)	[155]
			Asian golden cat	*Catopuma temmincki*	FSE[1]	[156]
			Puma	*Felis concolor*	FSE[1]	[78]
			Ocelot	*Felis pardalis*	FSE[1]	[145]
			Leopard cat	*Prionailurus*[3] *bengalis*	FSE[1]	[145]
		Acinonichinae	Cheetah	*Acinonyx jubatus*	FSE[1]	[157]
		Pantherinae	Tiger	*Panthera tigris*	FSE[1]	[145]
			Lion	*Panthera leo*	FSE[1]	[145]

[1] Only reported in captive animals (and domestic cats) and probably caused by feed exposure to the BSE agent. Transmission experiments using mice showed that affected nyala, greater kudu and domestic cats have the same prion strain as BSE-affected cattle (see Chap. 50). It is assumed that the disease occurring in all non-domesticated cats and other zoo species of the subfamilies Bovinae and Hippotraginae are attributable to exposure to the BSE agent (see Chap. 20); assumption based on observations related to geographical (mainly in UK zoos) and temporal data (period of BSE epidemic in British cattle; Fig. 20.1).
[2] Only the major reviews or key references on the pathology of the diseases are provided in most cases.
[3] Taxonomists recently changed "*Felis*" to "*Prionailurus*".

cal cases were found in 56 sheep flocks affected with scrapie in this period (see Chapters 54; 57).

Whether these "atypical" cases represent part of the spectrum of disease phenotypes of scrapie, or a new prion disease entity of sheep and goats remains to be established. In the meantime, the differences between such cases and the well known forms of scrapie that have now been termed "classical scrapie" have practical implications for diagnosis of TSE in sheep (see also [161]).

None of the evidence gathered so far suggests that "atypical" cases of scrapie are caused by the BSE agent (see Chapter 27.2).

1.10 Prion diseases and contagion

1.10.1 Human prion diseases

It is reassuring that none of the human prion diseases are contagious by natural routes. That is, normal contact between healthy and infected individuals does not present a risk of infection [162–166]. For example, the care of CJD patients does not present an occupational risk (see Chapter 47). However, unnaturally close contact, e. g., skin contact with brain and spinal cord tissues in association with injury when performing an autopsy, may pose a risk (see Chapters 48; 49).

Because of the rather unusual environment in which the kuru epidemic appeared, the situation regarding contagiousness of this disease is described and discussed in greater detail in Chapters 2, 13 and 38.

1.10.2 Scrapie and CWD

Virtually none of the animal prion diseases are contagious. Nonetheless, there are important exceptions; scrapie and CWD are transmitted naturally from one animal to another [167], which can potentially lead to endemic situations. More details on this particular aspect are described in Chapters 18 and 21.

On the one hand, it is well documented that horizontal and lateral spread are major transmission routes for scrapie agents. On the other hand, despite the speculation in 1913 by Sir Stewart Stockman that an infection with scrapie before birth or between birth and weaning could also occur [14], there is still little convincing scientific evidence, and even some conflicting evidence, for vertical transmission as defined in Chapter 42.1.1. It is not known whether scrapie can be transmitted during the pre-natal phase, i. e., in utero.

Nonetheless, we know that at least one form of maternal transmission in the immediate postnatal period (a form of horizontal transmission) is possible. Further details on the different transmission routes are described in Chapters 18 and 42.2.2.

The i) worldwide occurrence of classical scrapie (with the exception of a few countries), combined with ii) the fact that there is a natural risk of transmission of scrapie agents within sheep or goat flocks, and iii) the hazard that scrapie may be the origin of BSE (see above) has led to one of the biggest challenges for the future control of prion diseases in animals. It is surely only possible to combat scrapie successfully with an enormous effort and excellent cooperation between all countries (see Chapters 32; 54–57). Extensive international surveillance for TSE in small ruminants has been initiated in Europe [73]: over 1,560,000 sheep (of which 8,093[8] tested positive) and 410,000 goats (of which 1,111[9] tested positive) were tested in the EU from April 2002 to 2005 [151]. In January 2005 the first BSE case in a French goat was diagnosed [152] after a new European TSE surveillance program was launched for small ruminants (SRTSE), using discriminatory tests to distinguish between infections with BSE and scrapie strains.

1.11 Summary: traits common to all human and animal prion diseases

1. The diseases are always fatal.
2. The diseases affect both genders, usually with equal frequency. Exceptions in which female

[8] In 3 cases (2 France, 1 Cyprus) discriminatory tests indicated an unusual strain of TSE in sheep (compare Fig. 12.3).

[9] In one case (France) discriminatory tests indicated BSE in a goat [152] (in January 2005).

individuals are more frequently affected include kuru and BSE; this is explained by the specific habits of the Fore women (see Chapter 2) and the gender-specific population structure of cattle, respectively.

3. During the course of the diseases there is a slow degenerative destruction of the brain, accompanied by alterations in behavior and disturbances in movement. In humans there is dementia and movement disorder; in animals severe disturbances in behavior, sensation and movement.

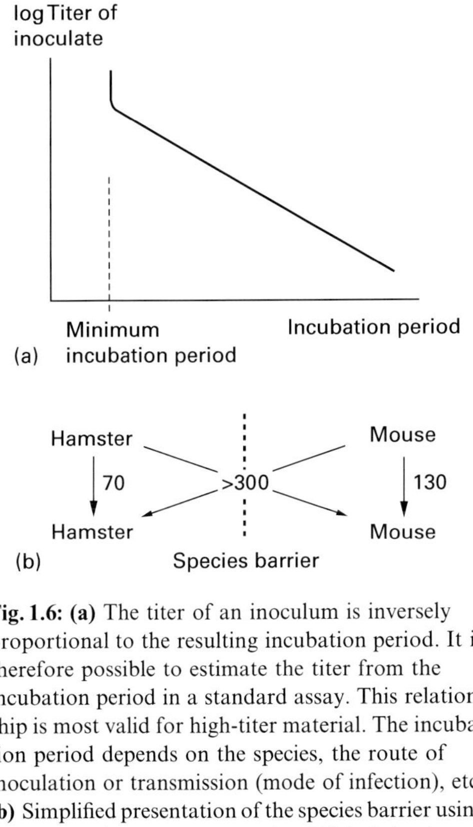

Fig. 1.6: (a) The titer of an inoculum is inversely proportional to the resulting incubation period. It is therefore possible to estimate the titer from the incubation period in a standard assay. This relationship is most valid for high-titer material. The incubation period depends on the species, the route of inoculation or transmission (mode of infection), etc. **(b)** Simplified presentation of the species barrier using hamster and mouse as examples. While infection within one species is quite efficient, infection across species requires longer incubation periods. Therefore, the incubation period is an indirect indicator of the strength of the species barrier. The incubation period in hamsters is 70 days and in mice 130 days. If infection occurs across the species, the incubation period is extended to more than 300 days. Figure by Bruno Oesch [168].

4. The diseases can currently only be diagnosed with certainty after death, unless it is possible to obtain a brain biopsy, which is rarely done. Neuropathological examination shows a degeneration, usually accompanied by vacuolation, termed spongiform encephalopathy. PrP^{Sc} – so far the only disease-specific marker – is detectable by immunohistochemistry (IHC) and other immunochemical methods.
5. Neither during the incubation period nor the phase of clinical manifestation of the disease, is it possible to observe an inflammatory or immunological reaction. In the absence of an immune response, there is no conventional method of diagnosis using antibodies as a disease-specific marker, for instance in blood.
6. Several factors influence the length of the incubation period (Fig. 1.6) and the efficiency of transmission: (i) the route of transmission, (ii) the dose of infectivity, (iii) the genectic factors of the species barrier, or within the same species, intraspecies susceptibility factors (genetic factors, some of which are known for sheep and humans), and (iv) the prion strain.
7. In all humans and animals in late incubation or the clinical stage, the highest infectivity is found in the brain and spinal cord. Therefore, these organs may represent a substantial health risk under certain exposure conditions. A lower infectivity titer can be observed in certain tissues, e. g. after oral infection, in the lymphoid tissue of Peyer's patches. In many tissues and body fluids, e. g. milk [169], infection has not been detected, as summarized in the WHO Tables IA-IC in the appendix of this book.

1.12 Synopsis of events, discoveries and findings since 1732

1732 Scrapie is described in sheep.
1759 A German source mentions that "Trab" (German for "sheep scrapie") is contagious.
1898 First description of pathological changes in the brain of scrapie-infected sheep.
1920 First description of CJD.

1936 First description of GSS as a familial disease in Austria.
1936 Experimental transmission of scrapie to healthy sheep and goats through inoculation.
1947 First appearance of TME on an American mink farm.
1957 First description of kuru cases in medical publications.
1959 Scrapie–kuru connection: first suggestion that scrapie and kuru are related diseases.
1966 Transmission of kuru to chimpanzees: experimental proof that kuru is transmissible.
1966 First publication on the unusual resistance of the scrapie agent to radiation and suggesting the absence of nucleic acid.
1967 First publication of the proposal that a cell membrane protein is responsible for the transmission of scrapie.
1967 First demonstration [170] of the connections between endocannibalistic funerary practices and the transmission of kuru.
1968 First publication on the transmissibility of the sporadic form of CJD (sCJD).
1981 Proof of the transmissibility of GSS.
1982 Publication of the prion hypothesis.
1984 Presentation of a summary showing further evidence [1] of the connection between (endo-) cannibalistic meals and kuru cases.
1985 The PrP gene is detected and identified as a host gene.
1985 First clinical BSE suspects are observed by veterinarians, as revealed retrospectively.
1986 First BSE case is confirmed in the laboratory.
1988 First publication of a hypothesis on the origin of BSE [122].
1988 British ban on the feeding of ruminant protein to ruminants with some exclusions such as milk.
1989 Ban to prevent BSE-specific bovine organs from contaminating human food in the UK.
1990 Prion disease is first detected in a domestic cat (FSE) in GB.
1990 BSE is diagnosed on the European continent.
1993 The British BSE epidemic begins to decline as a consequence of the first feed ban of 1988.
1993 Creation of PrP knock-out mice as a basis for transgenic research on prion diseases.
1994 Experimental evidence that FSE in cats and new prion diseases in zoo animals (wild cats and exotic ruminants) are caused by BSE prions.
1995 Experimental indication that BSE can be orally transmitted to sheep.
1996 Publication of 10 cases of the new variant of CJD (vCJD) in GB.
1997 Various experimental indications for vCJD being caused by BSE prions.
1998 Newly recognized atypical scrapie cases are identified in Norway; designated Nor98.
1999 A very rare sporadic form of "fatal insomnia" (SFI) which is phenotypically indistinguishable from FFI is described for the first time.
1999 First surveillance program using a rapid BSE test to diagnose BSE before the clinical onset of the disease.
2001 BSE surveillance using rapid BSE tests in the EU.
2003 First case of vCJD transmission by blood transfusion (assumption).
2004 First BSE case in the USA.
2004 Synthesis of mammalian prion in vitro [121].
2004 Summarised genetics in relation to kuru are published [171].
2005 First publication of BSE agent signatures in a goat.
2006 First case of Nor98 (atypical scrapie) *in a goat* was diagnosed in Norway on 6th June.
2006 BSE in the UK continues to decline towards zero incidence and the latest forecasts suggest that the epidemic will eventually fade away in the decade to come (see Chapter 39).

References

[1] Klitzman RL, Alpers MP, Gajdusek DC. The natural incubation period of kuru and the episodes of transmission in three clusters of patients. Neuroepidemiology 1984; 3:3–20.

[2] Gajdusek DC. Unconventional viruses and the origin and disappearance of kuru. Science 1977; 197:943–960.

[3] Goldfarb LG. Kuru: the old epidemic in a new mirror. Microbes Infect 2002; 4(8):875–882.

[4] Wells GAH, Scott AC, Johnson CT, et al. A novel progressive spongiform encephalopathy in cattle. Vet Rec 1987; 121:419–420.

[5] Will RG, Ironside JW, Zeidler M, et al. A new variant of Creutzfeldt–Jakob disease in the UK. Lancet 1996; 347:921–925.

[6] Prusiner SB. Novel proteinaceous infectious particles cause scrapie. Science 1982; 216:136–144.

[7] Prusiner SB. The prion diseases. Sci Am 1995; 272:48–57.

[8] SEAC. Transmissible spongiform encephalopathies – a summary of present knowledge and research. HMSO, 1994:21.

[9] Parry HB. Scrapie disease in sheep. Oppenheimer DR, editor. Academic Press, London 1983.

[10] Comber T. Real improvements in agriculture. On the principles of A. Young, Esq. Letters to Reade Peacock, Esq. and to Dr. Hunter, Physician in York, concerning the Rickets in sheep. 1772.

[11] McGowan JP. Scrapie in sheep. Scottish J Agric 1922; 5:365–375.

[12] McGowan JP. Investigation into the disease of sheep called "Scrapie" with special reference to its association with sarcosporidiosis. (Rept. 223, Edinburgh and East of Scotland Coll. of Agric. 1914). Blackwood, Edinburgh, 1914.

[13] Gaiger HS. Scrapie. J Comp Path 1924; 37:259–277.

[14] Stockman S. Scrapie: An obscure disease of sheep. (A lecture delivered before the Yarrow and Ettrick Pastoral Society, 23rd October 1913). J Comp Pathol Ther 1913; 317–327.

[15] Johann Georg Leopoldt. Der Trab ist auch eine Krankheit der Schaafe, und ist ansteckend. [Translation: Trotting disease (scrapie) is also a disease of sheep and is contagious]. In "Nützliche und auf die Erfahrung gegründete Einleitung zu der Landwirtschaft". Fünf Theile. Mit Kupfer und Baurissen. Hofreichsgräflich-Promnitzischen Wirthschafts-Amtmann der Herrschaft Sorau. Christian Friedrich Günthern, Berlin und Glogau, 1759.

[16] Roche-Lubin M. Mémoire pratique sur la maladie des bêtes à laine connue sous les noms de prurigo-lombaire, convulsive, trembleuse, tremblante, etc. Receil de Médecine Vétérinaire pratique 1848; Vol XXV, Tome V, Série 3:698–714.

[17] Dammann C. Landw Zentralblatt v Krocker. August 1869; 87.

[18] Besnoit C and Morel C. Note sur les lésions nerveuses de la tremblante du mouton. Rev Vét (Toulouse) 1898; 23:397–400.

[19] Cassirer R. Ueber die Traberkrankheit der Schafe. Pathologisch-anatomische und bakterielle Untersuchung. Virchows Arch (Pathol Anat Physiol) 1898; 153:89–110.

[20] Korn WG. Entwurf einer kurzen Uebersicht der sogenannten Traberkrankheit der Schaafe als epizootische Erbübel. Wilhelm Gottlieb Korn, Breslau, 1828.

[21] Dammann C. Zweite Vorlesung; Krankheits-Anlagen und Vorbeugung. In: Die Gesundheitspflege der landwirtschaftlichen Haussäugethiere – zwanzig Vorlesungen von Dr. Carl Dammann. Paul Parey, Berlin, 1886:16–25.

[22] Dammann C. Achzehnte Vorlesung; Zucht und Aufzucht. 1886:1146–1170.

[23] Stang V and Wirth D. Tierheilkunde und Tierzucht. Eine Enzyklopädie der praktischen Nutztierkunde. Urban und Schwarzenburg, Berlin, 1931:807–808.

[24] Stumpf G. Versuch einer pragmatischen Geschichte der Schäfereien in Spanien, und der spanischen Schafe in Sachsen. (1785). [An essay on the practical history of sheep farming in Spain and of the Spanish sheep in Saxony]. Transactions of the Royal Dublin Society – Part 1, Vol I, 1800:1–101.

[25] Klein J. The Mesta; a study in Spanish economic history, 1273–1836. [Harvard economic studies]. Harvard University Press Cambridge, Mass., 1920.

[26] Braudel F. The Mediterranean and the Mediterranean world in the age of Philip II. [Transl. from the French by Sian Reynolds]. Harper and Row, New York, 1972.

[27] Oppermann. Lehrbuch der Krankheit des Schafes. Hannover, 1919:234 (also 2nd edition, 1921).

[28] Golf A. Ueber die frühere Einführung der Merinos aus Spanien nach Deutschland und anderen Ländern. In: Doehner H, editor. Handbuch der

Schafzucht und Schafhaltung. Vol I, Ch. 3. Paul Parey, Berlin, 1939:63–72.
[29] Doehner H. Handbuch der Schafzucht und Schafhaltung. Paul Parey, Berlin, 1944.
[30] Belda AS. Letter to H.B. Parry dated 9.7.79. Jefe de la sección de Ganado Bovino. Ministerio de Agricultura, Madrid. According to citation in reference no [9]. 1979.
[31] Behrens H. Discussion on paper by Palmer: on scrapie in Germany. Vet Rec 1957; 69:1328.
[32] Greig JR. Scrapie. Transcript, Highland and Agriculture Society, Scotland 1940; 52:71–90.
[33] Carter HB. His Majesty's Spanish flocks. Sir Joseph Banks and the merinos of George III of England. Angus and Robertson, London, 1964.
[34] Hopp P, Bratberg B, Ulvund MJ. Skrapesjuke hos sau i Norge. Historikk og epidemiologi. (Scrapie in sheep in Norway. History and epidemiology). Norsk veterinærtidsskrift 2000; 112:368–375.
[35] Bull LB and Murnane D. An outbreak of scrapie in British sheep imported into Victoria. Aust Vet J 1958; 34:213–215.
[36] Seddon HR. Scrapie. In: Seddon HR, editor. Diseases of domestic animals in Australia – Part 4. A.J. Arthur, Commonwealth Government Printer, Canberra, 1958:176.
[37] Brash AG. Scrapie in imported sheep in New Zealand. New Zealand Vet J 1952; 1:27–30.
[38] Bruere AN. Scrapie in New Zealand – its history and what it could mean. Proceedings of the New Zealand Veterinary Association's Sheep and Beef Cattle Society's 15th Seminar 1985:9–22.
[39] MacDiarmid SC. Scrapie: the risk of its introduction and effects on trade. Aust Vet J 1996; 73(5):161–164.
[40] AQIS. Requirements for the approval and operation of scrapie freedom assurance programs for imported sheep and goats. Australian Quarantine and Inspection Service, 10 April 1995:1–25.
[41] Roe RT. Scrapie – the Australian experience. Australian Bureau of Animal Health (Report) 1995:1–5.
[42] Anonymous. Import risk analysis report on the revision of import policy related to scrapie. Final report. Australian Quarantine and Inspection Service (AQIS), Canberra, ACT 2601, August 2000. See Webpage 2006: www.affa.gov.au/corporate_docs/publications/pdf/market_access/biosecurity/animal/2000/00-038a.pdf.
[43] Scherf BD. World watch list for domestic animal diversity. FAO (United Nations), Rome, 1995:150–151. See also webpage 2006: www.virtualcentre.org/es/dec/toolbox/Indust/wwl.pdf.
[44] OMIA – online Mendelian inheritance in animals. Spongiform encephalopathy (000944) in *Ovis aries*. Webpage 2006: http://omia.angis.org.au/retrieve.shtml?pid = 1646.
[45] Cuillé J and Chelle PI. La maladie dite tremblante du mouton est-elle inoculable? [Translation: Scrapie in sheep – is it inoculable?] C R Acad Sci Paris 1936; 203:1552–1554.
[46] Cuillé J and Chelle PI. La tremblante du mouton est bien inoculable. [Translation: Scrapie in sheep – it is really inoculable.] C R Acad Sci Paris 1938; 206:78–79.
[47] Pattison IH. A sideways look at the scrapie saga: 1732–1991. In: Prusiner SB, Collinge J, Powell J, et al, editors. Prion diseases of humans and animals. Ellis Horwood, New York, 1992:15–22.
[48] Wilson DR, Anderson RD, Smith W. Studies in scrapie. J Comp Path 1950; 60:267–282.
[49] Pattison IH, Hoare MN, Jebbett JN, et al. Spread of scrapie to sheep and goats by oral dosing with foetal membranes from scrapie-affected sheep. Vet Rec 1972; 90:465–468.
[50] Pattison IH, Hoare MN, Jebbett JN, et al. Further observations on the production of scrapie in sheep by oral dosing with foetal membranes from scrapie-affected sheep. Br Vet J 1974; 130:65–67.
[51] Sigurdarson S. Epidemiology of scrapie in Iceland and experience with control measures. In: Bradley R, Savey M, Marchant B, editors. Subacute spongiform encephalopathies. Commission of European Communities. Kluwer Academic, Netherlands, 1991:233–242.
[52] Brown P and Gajdusek DC. Survival of scrapie virus after 3 years' interment. Lancet 1991; 337:269–270.
[53] Johnson CJ, Phillips KE, Schramm PT, et al. Prions adhere to soil minerals and remain infectious. PLoS Pathog 2006; 2:e32.
[54] Hartsough GR and Burger D. Encephalopathy of mink. I. Epizootiologic and clinical observations. J Infect Dis 1965; 115(4):387–392.
[55] Burger D and Hartsough GR. Encephalopathy of mink. II. Experimental and natural transmission. J Infect Dis 1965; 115(4):393–399.
[56] Gorham JR. Viral and bacterial diseases of mink in Soviet Union. Fur Rancher 1991; 71:10–11.
[57] Gorham JR. Viral and bacterial diseases of mink in Soviet Union. Fur Rancher 1991; 71:3.

[58] Hadlow WJ, Race RE, Kennedy RC. Temporal distribution of transmissible mink encephalopathy virus in mink inoculated subcutaneously. J Virol 1987; 61:3235–3240.
[59] Williams ES and Young S. Chronic wasting disease of captive mule deer: a spongiform encephalopathy. J Wildl Dis 1980; 16:89–98.
[60] Hainfellner JA, Brantner-Inthaler S, Cervenakova L, et al. The original Gerstmann–Sträussler–Scheinker family of Austria: divergent clinicopathological phenotypes but constant PrP genotype. Brain Pathol 1995; 5:201–211.
[61] Creutzfeldt HG. Über eine eigenartige herdförmige Erkrankung des Zentralnervensystems. Z Ges Neurol Psychiat 1920; 57:1–18.
[62] Jakob AM. Über eigenartige Erkrankungen des Zentralnervensystems mit bemerkenswertem anatomischem Befunde (Spastische Pseudosklerose – Encephalomyelopathie mit disseminierten Degenerationsherden). Dtsch Z Nervenheilk 1921; 70:132–146.
[63] Jakob AM. Über eine der multiplen Sklerose klinisch nahestehenden Erkrankung des Zentralnervensystems (spastische Pseudosklerose) mit bemerkenswertem anatomischem Befunde. Med Klin 1921;13:372–376.
[64] Duffy P, Wolf J, Collins G, et al. Possible person-to-person transmission of Creutzfeldt–Jakob disease. N Engl J Med 1974; 290:692–693.
[65] Meggendorfer F. Klinische und genealogische Beobachtungen bei einem Fall von spastischer Pseudosklerose. Z Ges Neurol Psychiat 1930; 128:337.
[66] Kretzschmar HA, Neumann M, Stavrou D. Codon 178 mutation of the human prion protein gene in a German family (Backer family): sequencing data from 72-year-old celloidin-embedded brain tissue. Acta Neuropathol 1995; 89:96–98.
[67] Dimitz L. Bericht des Vereines für Psychiatrie und Neurologie in Wien. (Vereinsjahr 1912/13, Sitzung vom 11. Juni 1912). Jahrb Psychiatr Neurol 1913; 34:384.
[68] Gerstmann J. Über ein noch nicht beschriebenes Reflexphänomen bei einer Erkrankung des zerebellären Systems. Wien Medizin Wochenschr 1928; 78:906–908.
[69] Gerstmann J, Sträussler E, Scheinker I. Über eine eigenartige hereditär-familiäre Erkrankung des Zentralnervensystems. Zugleich ein Beitrag zur Frage des vorzeitigen lokalen Alterns. Z Ges Neurol Psychiat 1936; 154:736–762.
[70] Lugaresi E, Medori R, Montagna P, et al. Fatal familial insomnia and dysautonomia with selective degeneration of thalamic nuclei. N Engl J Med 1986; 315:997–1003.
[71] Parchi P, Capellari S, Chin S, et al. A subtype of sporadic prion disease mimicking fatal familial insomnia. Neurology 1999;52(9):1757–1763.
[72] Zigas V and Gajdusek DC. Kuru: clinical study of a new syndrome resembling paralysis agitans in natives of the Eastern Highlands of Australian New Guinea. Med J Aust 1957; 2:245–254.
[73] EU. Questions and answers on TSEs in sheep [and goat]. Press release of 9 March 2006. Webpage 2006: http://europa.eu.int/rapid/pressReleasesAction.do?reference=MEMO/06/114&format=HTML&aged=0&language=EN.
[74] Benestad SL, Sarradin P, Thu B, et al. Cases of scrapie with unusual features in Norway and designation of a new type, Nor98. Vet Rec 2003; 153(7):202–208.
[75] Moum T, Olsaker I, Hopp P, et al. Polymorphisms at codons 141 and 154 in the ovine prion protein gene are associated with scrapie Nor98 cases. J Gen Virol 2005; 86(1):231–235.
[76] Jeffrey M and Wells GAH. Spongiform encephalopathy in a nyala (*Tragelaphus angasi*). Vet Pathol 1988; 25:398–399.
[77] Wyatt JM, Pearson GR, Smerdon TN, et al. Spongiform encephalopathy in a cat. Vet Rec 1990; 126:513.
[78] Willoughby K, Kelly DF, Lyon DG, et al. Spongiform encephalopathy in a captive puma (*Felis concolor*). Vet Rec 1992; 131:431–434.
[79] Hamir AN, Kunkle RA, Cutlip RC, et al. Experimental transmission of chronic wasting disease agent from mule deer to cattle by the intracerebral route. J Vet Diagn Invest 2005; 17(3):276–281.
[80] Gajdusek DC, Gibbs CJ Jr, Alpers M. Experimental transmission of a kuru-like syndrome to chimpanzees. Nature 1966; 209:794–796.
[81] Gibbs CJ Jr, Gajdusek DC, Asher DM, et al. Creutzfeldt–Jakob disease (spongiform encephalopathy): transmission to the chimpanzee. Science 1968; 161:388–389.
[82] Bernoulli C, Siegfried J, Baumgartner G, et al. Danger of accidental person-to-person transmission of Creutzfeldt–Jakob disease by surgery. Lancet 1977; 1(8009):478–479.
[83] Cooke J. The French catastrophe. In: Cooke J. Cannibals, cows & the CJD catastrophe. Random House, Australia, 1998:236–247.

[84] Masters CL, Gajdusek DC, Gibbs CJ Jr. Creutzfeldt–Jakob disease virus isolations from the Gerstmann–Sträussler syndrome. With an analysis of the various forms of amyloid plaque deposition in the virus-induced spongiform encephalopathies. Brain 1981; 104:559–588.
[85] Lugaresi E, Montagna P, Baruzzi A, et al. Familial insomnia with a malignant course: a new thalamic disease. Rev Neurol 1986; 142(10):791–792.
[86] Almer G, Hainfellner JA, Brücke T, et al. Fatal familial insomnia: a new Austrian family. Brain 1999; 122:5–16.
[87] Tateishi J, Brown P, Kitamoto T, et al. First experimental transmission of fatal familial insomnia. Nature 1995; 376:434–435.
[88] Nobelprize.org. The Nobel prize in physiology or medicine – laureates. Webpage: http://nobelprize.org/medicine/laureates.
[89] Hadlow WJ. The scrapie-kuru connection: recollections of how it came about. In: Prusiner SB, Collinge J, Powell J, et al, editors. Prion diseases of humans and animals. Ellis Horwood, New York, 1992:40–46.
[90] Hadlow WJ. Scrapie and kuru. Lancet 1959; 2:289–290.
[91] Gajdusek DC. Kuru and scrapie. In: Prusiner SB, Collinge J, Powell J, et al, editors. Prion diseases of humans and animals. Ellis Horwood, New York, 1992:47–52.
[92] Alpers MP. Epidemiology and ecology of kuru. In: Prusiner SB, Hadlow WJ, editors. Slow transmissible diseases of the nervous system. Academic Press, New York, Vol 1, 1979:67–90.
[93] Prusiner SB, Gajdusek DC, Alpers MP. Kuru with incubation period exceeding two decades. Ann Neurol 1982; 12:1–9.
[94] Becker LE. Slow infections of the central nervous system. Can J Neurol Sci 1977; 4:81–88.
[95] Palsson PA. Dr. Björn Sigurdsson (1913–1959). A memorial tribute. Ann N Y Acad Sci 1994; 724:1–5.
[96] van Bogaert L, Dewulf A, Palsson PA. Riđa in sheep. Pathological and clinical aspects. Acta Neuropathol (Berl) 1978; 41:201–206.
[97] Houff SA and Sever JL. Slow virus diseases of the central nervous system. Dis Mon 1985; 31:1–71.
[98] Prusiner SB, Cochran SP, Baringer JR, et al. Slow viruses: molecular properties of the agents causing scrapie in mice and hamsters. Prog Clin Biol Res 1980; 39:73–89.
[99] Björnsson J, Carp RI, Löwe A, et al. Slow infections of the central nervous system – the legacy of Dr. Björn Sigurdsson. Ann N Y Acad Sci 1994; 724:1–5.
[100] Dickinson AG and Outram GW. Genetic aspects of unconventional virus infections: the basis of the virino hypothesis. Ciba Found Symp 1988; 135:63–83.
[101] Diringer H and Braig HR. Infectivity of unconventional viruses in dura mater (letter). Lancet 1989; 1:439–440.
[102] Brown F. Unconventional viruses and the central nervous system (editorial). Br Med J (Clin Res) 1987; 295:347–348.
[103] Kingsbury DT, Smeltzer DA, Amyx HL, et al. Evidence for an unconventional virus in mouse-adapted Creutzfeldt–Jakob disease. Infect Immun 1982; 37:1050–1053.
[104] Diringer H, Beekes M, Oberdieck U. The nature of the scrapie agent: the virus theory. In: Björnsson J, Carp RI, Löwe A, et al, editors. Slow infections of the central nervous system. Ann N Y Acad Sci 1994:246–258.
[105] Diringer H, Braig HR, Czub M. Scrapie: a virus-induced amyloidosis of the brain. Ciba Found Symp 1988; 135:135–145.
[106] Darcel C. Reflections on scrapie and related disorders, with consideration of the possibility of a viral aetiology. Vet Res Commun 1995; 19(3):231–252.
[107] Anonym. ASM News, 1995: 327–328.
[108] Prusiner SB, Collinge J, Powell J, et al. Prion diseases of humans and animals. Ellis Horwood, New York, 1992.
[109] Merz PA, Somerville RA, Wisniewski HM, et al. Abnormal fibrils from scrapie-infected brain. Acta Neuropathol 1981; 54:63–74.
[110] Ishikawa K, Doh-ura K, Kudo Y et al. Amyloid imaging probes are useful for detection of prion plaques and treatment of transmissible spongiform encephalopathies. J Gen Virol 2004; 85(6):1785–1790.
[111] Alper T, Haig DA, Clarke MC. The exceptionally small size of the scrapie agent. Biochem Biophys Res Commun 1966; 22:278–284.
[112] Alper T. Photo- and radiobiology of the scrapie agent. In: Prusiner SB, Collinge J, Powell J, et al, editors. Prion diseases of humans and animals. Ellis Horwood, New York, 1992:30–39.
[113] Gibbons RA and Hunter GD. Nature of the scrapie agent. Nature 1967; 215:1041–1043.
[114] Griffith JS. Self-replication and scrapie. Nature 1967; 215:1043–1044.
[115] Oesch B, Westaway D, Walchli M, et al. A cellular gene encodes scrapie PrP27–30 protein. Cell 1985; 40:735–746.

[116] Moore RC, Hope J, McBride PA, et al. Mice with gene targetted prion protein alterations show that *Prnp*, *Sinc* and *Prni* are congruent. Nat Genet 1998; 18:118–125.
[117] Taraboulos A, Jendroska K, Serban D, et al. Regional mapping of prion proteins in brains. Proc Natl Acad Sci U S A 1992; 89:7620–7624.
[118] Meyer RK, McKinley MP, Bowman KA, et al. Separation and properties of cellular and scrapie prion proteins. Proc Natl Acad Sci USA 1986; 83:2310–2314.
[119] Hope J, Reekie LJ, Hunter N, et al. Fibrils from brains of cows with new cattle disease contain scrapie-associated protein. Nature 1988; 336:390–392.
[120] Prusiner SB. Prions (Les Prix Nobel Lecture). In: Frängsmyr T, editor. Les Prix Nobel. Almqvist & Wiksell International, Stockholm, 1998:268–323.
[121] Legname G, Baskakov IV, Nguyen H-OB, et al. Synthetic mammalian prions. Science 2004; 305:673–676.
[122] Wilesmith JW, Wells GAH, Cranwell MP, et al. Bovine spongiform encephalopathy: epidemiological studies. Vet Rec 1988; 123:638–644.
[123] Wilesmith JW. Bovine spongiform encephalopathy (letter). Vet Rec 1988; 122:614.
[124] Morgan KL, Nicholas K, Glover MJ, et al. A questionnaire survey of the prevalence of scrapie in sheep in Britain. Vet Rec 1990; 127:373–376.
[125] EFSA. Quantitative assessment of the residual BSE risk in bovine-derived products. The EFSA Journal 2005; 307:1-135. See also Webpage 2006: www.efsa.eu.int/science/biohaz/biohaz_documents/1280/efsaqrareport2004_final20dec051.pdf
[126] Wells GAH and Wilesmith JW. Bovine spongiform encephalopathy and related diseases. In: Prusiner SB, editor. Prion biology and diseases, 2nd edition. Cold Spring Harbor Laboratory Press, New York, 2004:595–628.
[127] Wells GAH and Wilesmith JW. The neuropathology and epidemiology of bovine spongiform encephalopathy. Brain Pathol 1995; 5:91–103.
[128] Bruce ME, Chree A, McConnell I, et al. 1994. Transmission of bovine spongiform encephalopathy and scrapie to mice: strain variation and the species barrier. Philos Trans R Soc Lond B Bio Sci 1994; 343: 405–411.
[129] Wilesmith JW, Ryan JB, Atkinson MJ. Bovine spongiform encephalopathy: epidemiological studies on the origin. Vet Rec 1991; 128:199–203.
[130] Anonym. The BSE inquiry 1998. Official information from the UK Government. Webpage 2006: www.bseinquiry.gov.uk.
[131] Scientific Steering Committee (SSC). Opinions and Information. Webpage 2006: www.priondata.org/data/A_ECSSC.html.
[132] Hörnlimann B and Guidon D. Import of meat and bone meal as main risk factor for BSE in Switzerland. The Kenya Veterinarian 1994; 18:467–469.
[133] Hörnlimann B, Guidon D, Griot C. Risikoeinschätzung für die Einschleppung von BSE [Risk assessment for the importation of BSE]. Dtsch tierärztl Wschr 1994; 101:295–298.
[134] Hörnlimann B. Geschichte und Epidemiologie der Prionenkrankheiten [History and epidemiology of Prion diseases]. Mitt Gebiete Lebensm Hyg 1996; 87:3–13.
[135] Atkinson N. The impact of BSE on the UK economy. Economics (International) Division, Ministry of Agriculture, Fisheries and Food, London. Webpage 2006: www.iica.org.ar/Bse/14-%20Atkinson.html.
[136] European Commission 2005. The TSE Road Map. COM (2005) 322 final, Brussels 15 July 2005. Webpage 2006: http://europa.eu.int/comm/food/food/biosafety/bse/roadmap_en.pdf.
[137] Yamakawa Y, Hagiwara K, Nohtomi K, et al. Atypical proteinase K-resistant prion protein (PrPres) observed in an apparently healthy 23-month-old Holstein steer. Jpn J Infect Dis 2003; 56(5-6):221–222.
[138] Casalone C, Zanusso G, Acutis P, et al. Identification of a second bovine amyloidotic spongiform encephalopathy: molecular similarities with sporadic Creutzfeldt–Jakob disease. Proc Natl Acad Sci USA 2004; 101(9):3065–3070.
[139] Biacabe AG, Laplanche JL, Ryder S, et al. Distinct molecular phenotypes in bovine prion diseases. EMBO Rep 2004; 5(1):110–115.
[140] Boelle PY, Cesbron JY, Valleron AJ. Epidemiological evidence of higher susceptibility to vCJD in the young. BMC Infect Dis 2004; 4:26.
[141] Ironside JW, Bishop MT, Connolly K, et al. Variant Creutzfeldt–Jakob disease: prion protein genotype analysis of positive appendix samples from a retrospective prevalence study. Br Med J 2006;332:1186–1188.
[142] Peden AH, Head MW, Ritchie DL, et al. Preclinical vCJD after blood transfusion in a *PRNP*

codon 129 heterozygous patient. Lancet 2004;364:527–529.
[143] Wells GAH and Wilesmith JW. The neuropathology and epidemiology of bovine spongiform encephalopathy. Brain Pathol 1995; 5:91–103.
[144] Wells GAH and Wilesmith JW. Bovine spongiform encephalopathy and related diseases. In: Prusiner SB, editor. Prion Biology and Diseases. Cold Spring Harbor Laboratory Press New York, 2004:595–628.
[145] Zoo Prion Disease: Review of scientific literature. Webpage 2006: www.mad-cow.org/zoo_cites_annotated.html.
[146] Kirkwood JK and Cunningham AA. Epidemiological observations on spongiform encephalopathies in captive wild animals in the British Isles. Vet Rec 1994; 135:296–303.
[147] Kirkwood JK and Cunningham AA. Spongiform encephalopathy in captive wild animals in Britain: epidemiological observations. In: Bradley R, Marchant B, editors. Transmissible spongiform encephalopathies. EU: VI/4131/94-EN Document Ref. F. II.3 – JC/0003, Brussels, 1994:29–47.
[148] Jeffrey M and Gonzales L. Pathology and pathogenesis of bovine spongiform encephalopathy and scrapie. In: Harris D, editor. Mad cow disease and related spongiform encephalopathies. Springer-Verlag, Berlin, 2004:65–97.
[149] Wood JL, Lund LJ, Done SH. The natural occurrence of scrapie in moufflon. Vet Rec 1992; 130(2):25–27.
[150] Wood JL and Done SH. Natural scrapie in goats: neuropathology. Vet Rec 1992; 131:93–96.
[151] EC. Food and feed safety. TSE in goats – what is the new information about TSE in goats? Webpage 2006: http://ec.europa.eu/comm/food/food/biosafety/bse/goats_index_en.htm.
[152] Eloit M, Adjou K, Coulpier M et al. BSE agent signatures in a goat. Vet Rec 2005; 156(16):523–524.
[153] Williams ES. Chronic wasting disease. Vet Pathol 2005; 42(5):530–549.
[154] Marsh RF and Hadlow WJ. Transmissible mink encephalopathy. Rev Sci Tech Off Int Epiz 1992; 11:539–550.
[155] Wyatt JM, Pearson GR, Smerdon TN, et al. Naturally occurring scrapie-like spongiform encephalopathy in five domestic cats. Vet Rec 1991; 129(11):233–236.
[156] Young S and Slocombe RF. Prion-associated spongiform encephalopathy in an imported Asiatic golden cat (Catopuma temmincki). Aust Vet J 2003; 81:295–296.
[157] Lezmi S, Bencsik A, Monks E, et al. First case of feline spongiform encephalopathy in a captive cheetah born in France: PrP(Sc) analysis in various tissues revealed unexpected targeting of kidney and adrenal gland. Histochem Cell Biol 2003; 119(5):415–422.
[158] Bruce ME. Strain typing studies of scrapie and BSE. In: Baker HF, Ridley RM, editors. Methods in molecular medicine: Prion diseases. Humana Press, Totowa, NJ, 1996:223–238.
[159] Leggett MM, Dukes J, Pirie HM. A spongiform encephalopathy in a cat. Vet Rec 1990; 127:586–588.
[160] Fraser H, Pearson GR, McConnell I, et al. Transmission of feline spongiform encephalopathy to mice. Vet Rec 1994; 134:449.
[161] Espenes A, Press CM, Landsverk T, et al. Detection of PrP(Sc) in rectal biopsy and necropsy samples from sheep with experimental scrapie. J Comp Pathol 2006; 134(2–3):115–125.
[162] van Duijn CM, Delasnerie-Lauprêtre N, Masullo C, et al. Case-control study of risk factors of Creutzfeldt–Jakob disease in Europe during 1993–1995. Lancet 1998; 351(9109): 1081–1085.
[163] Will RG, Esmonde TFG, Matthews WB. Creutzfeldt–Jakob disease epidemiology. In: Prusiner SB, Collinge J, Powell J, et al, editors. Prion diseases of humans and animals. Ellis Horwood, New York, 1992:188–199.
[164] Cousens SN, Zeidler M, Esmonde TF, et al. Sporadic Creutzfeldt–Jakob disease in the United Kingdom: analysis of epidemiological surveillance data for 1970–1996. Br Med J 1997; 315(7105):389–395.
[165] Will RG, Knight RSG, Zeidler M, et al. Reporting of suspect new variant Creutzfeldt–Jakob disease. Lancet 1997; 349:847.
[166] Wientjens DP, Davanipour Z, Hofman A, et al. Risk factors for Creutzfeldt–Jakob disease: a reanalysis of case-control studies. Neurology 1996; 46:1287–1291.
[167] Sigurdson CJ, Williams ES, Miller MW, et al. Oral transmission and early lymphoid tropism of chronic wasting disease PrPres in mule deer fawns (*Odocoileus hemionus*). J Gen Virol 1999; 80:2757–2764.
[168] Oesch B. Molekulare Grundlagen der Prionerkrankungen. [Molecular base of prion diseases]. Mitt Gebiete Lebensm Hyg 1996; 87:14–26.
[169] EU. Report from the scientific veterinary committee on the risk analysis for colostrum, milk

and milk products. Leg Vet & Zoot 1996; VI/8197/96 Version J. 1–26.
[170] Alpers MP. Kuru: implications of its transmissibility for the interpretation of its changing epidemiologic pattern. In: Bailey OT, Smith DE, editors. The central nervous system, some experimental models of neurological diseases. International Academy of Pathology, Monograph No 9. Proceedings of the fifty-sixth annual meeting of the International Academy of Pathology, Washington, DC, 12–15 Mar 1967. Williams and Wilkins, Baltimore, 1968:234–251.
[171] Goldfarb LG, Cervenakova L, Gajdusek DC. Genetic studies in relation to kuru: an overview. Curr Mol Med 2004; 4(4):375–384.
[172] SEAC. Chronic wasting disease – review of research published since November 2004. CWD Review June 2006. Webpage 2006: www.seac.gov.uk/papers/CWD-review.pdf.
[173] Baeten LA, Powers BE, Jeweill JE, et al. A natural case of chronic wasting disease in a free-ranging moose (*Alces alces shirasi*). J Wildl Dis (in press).

2 History of Kuru Research[1]

Beat Hörnlimann

2.1 Introduction

In 1957, Daniel Carleton Gajdusek initiated the systematic investigation of the disease kuru. Thereafter, many scientists joined the effort and participated in a vast number of epidemiological and medical research projects in the field and in the laboratory.

This chapter is based on extensive anthropological, ethnological, and medical literature of the 1950s and 1960s [1; 2], and focuses on the affected linguistic groups of Papua New Guinea (PNG). The main goal of this chapter is to demonstrate the difficult and exceptional conditions under which kuru researchers were required to work, and how and why various hypotheses about the origins of kuru were developed over the years. While the indigenous people were absolutely convinced that kuru was a disease conjured up by sorcery, Gajdusek and his team were able to prove that kuru was a transmissible infectious disease in 1966. Epidemiological details and their causal interrelation with a special form of cannibalism called endocannibalism (formerly practiced in parts of the Eastern Highlands of PNG), as well as further aspects of kuru, are described in Chapters 13 and 38.

2.2 From the stone age to the present: the Fore people

Today, PNG (Fig. 2.1) is still inhabited by approximately 850 different linguistic groups; these different languages (not dialects) represent 14% of all languages spoken in the world today [3]. In 2000, an estimated five million people lived in PNG and approximately one-third lived in the provinces of the highlands. This is also the home of the Fore people and their neighbors, among whom the kuru epidemic emerged. The kuru epidemic has persisted throughout the entire 20th century (Figs. 1.3, 38.1).

Humans reached the island of New Guinea 50,000–65,000 years ago, in a series of migrations from Southeast Asia [4]. The same migrations populated the continent of Australia. It is presumed that the highlands of New Guinea were first inhabited by people forced out of coastal settlements by subsequent migrants. It may be that the ancestors of the Anga people (Kukukuku) were the first inhabitants of the highlands of New Guinea (Fig. 2.2).

Although Portuguese sailors reached the coast of the island of New Guinea in 1512, the people still lived a traditional lifestyle until well into the 20th century. The highlanders independently developed horticulture 9,000 years ago (when Europeans were still hunters and gatherers) and were fine hunters, as well as accomplished gardeners. Until the twentieth century they used stone axes, stone and bamboo knives as tools, and bows and arrows as well as bone daggers as weapons. Metal was unknown to them, as was the invention of the wheel. There was no written language; the continuity of their culture was based on oral tradition [5].

In 1930, the rest of the world was unaware of their existence. Their first traumatic encounter with modern culture and technology occurred in the 1930s when a propeller airplane flew over the Fore area. This extraordinary apparition and the pervasive noise it made supposedly caused tremendous fear in the Fore people; an event which marked the dawn of a new era in the lives of these people. Michael Alpers and Carleton Gajdusek tabulated all known environmental and cultural changes in the kuru re-

[1] Reviewed by D.C. Gajdusek (German edition, 2001) and M.P. Alpers (English edition, 2006).

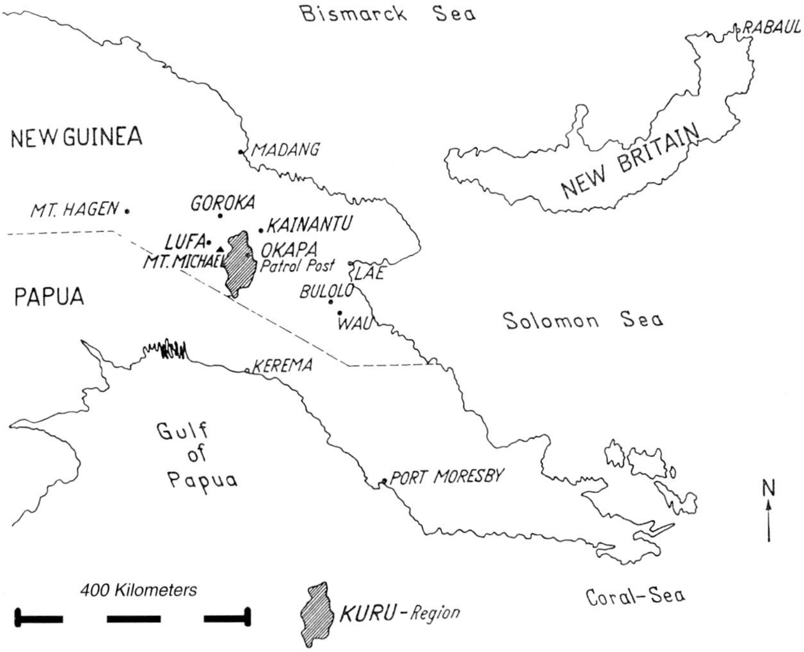

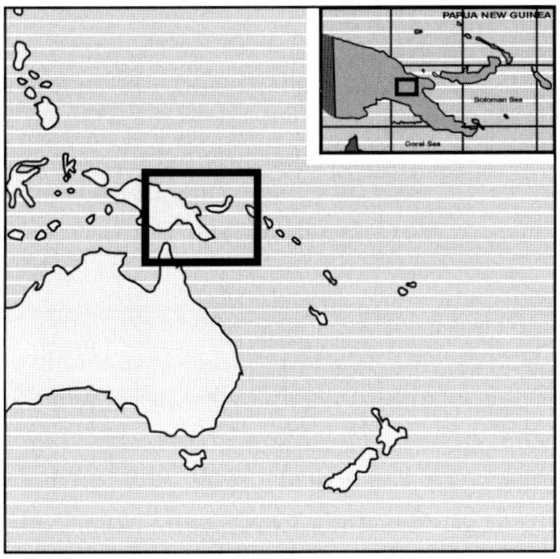

Fig. 2.1: Papua New Guinea (PNG). Kuru exclusively appeared in the Eastern Highlands, in the so-called kuru region (hatched). Reprinted with permission of D.C. Gajdusek [20].

gion chronologically [6], the most important of which are summarized in this chapter.

Ted Ubank, a gold hunter, was one of the first white men to enter the region of the Fore, in the southeast part of the Eastern Highlands of the Territory of New Guinea, as it was then called. Shortly before him, the Australian gold hunters Michael Leahy and Michael Dwyer explored the region north of the main area of the Fore [7].

2.3 Discovery of kuru

Government patrol officers of the Australian Administration reported a strange and fatal disease in the area of the Fore when they first made administrative contact with the people in the 1950s. The Government anthropologist, Charles Julius, was sent in to investigate and wrote a perceptive report [8].

In the early 1950s, Ronald and Catherine Berndt also noticed the occurrence of this

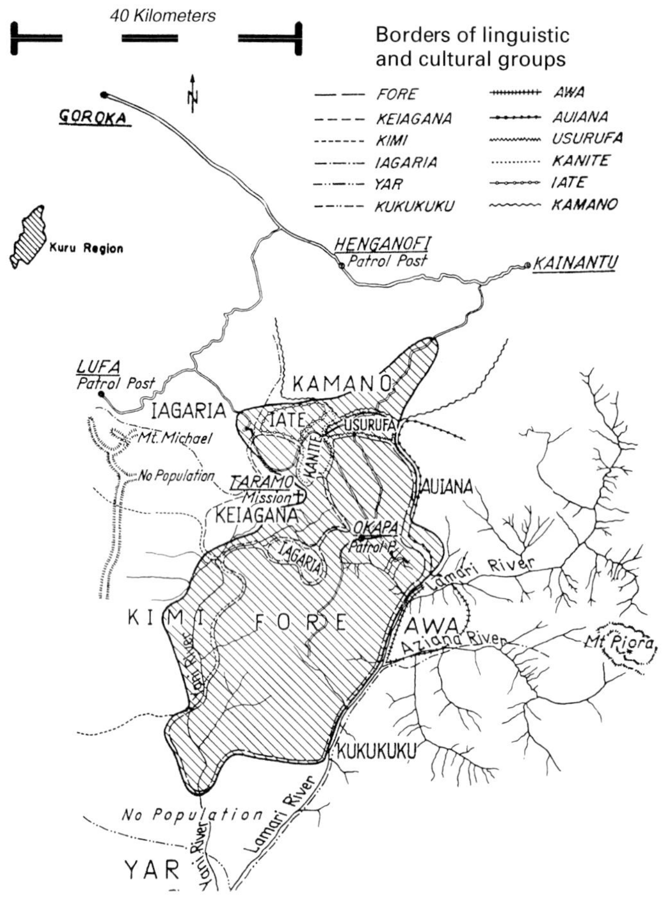

Fig. 2.2: Kuru region (hatched) in the Eastern Highlands of PNG. The borders of linguistic and cultural groups affected by kuru are outlined. Reprinted with permission of D.C. Gajdusek [20].

strange disease while conducting anthropological field studies on the people of the North Fore, Kamano, and surrounding linguistic groups [5]. Ronald Berndt reported that those afflicted suffered from strong tremors and muscle jerks, which he referred to as "Kamano tremor" (D.C. Gajdusek, personal communication). He originally thought that the disease had a psychological origin [9], despite the fact that it was progressively fatal.

According to the beliefs of the local people the disease was caused by sorcery, like many of the locally prevalent diseases. Sorcery, magic, and ritual respect for the dead (Fig. 2.3, bottom right) were important components of the religious customs of the people. Their lives were characterized by a belief in many supernatural powers and by a permanent connection to the spirits of their ancestors [10; 11] (see Section 2.11). In accordance with this view of the world, the affected groups considered the increasingly frequent occurrence of kuru (Fig. 38.1) to be a direct result of the malicious kuru sorcery [9], perpetrated by their own people.

Fig. 2.3: Walking on suspension bridges (above) was particularly difficult for kuru researchers. Papua New Guinean culture, which was passed on orally, was characterized by many ceremonial events (below). Pictured below on the right-hand side, a man can be seen holding a part of a human skull in his hands. The brains of dead relatives were usually dissected from the skull by women and then consumed. Note: The people depicted are not from the Fore. Photograph by K. Joseph ©. ▶

2.4 The "Tukabu" ritual

To avenge an actual or alleged injustice, for example a nonsuccessful courtship, the Fore believed they were able to devise sorcery in the form of kuru. For this it was not necessary for the kuru sorcerer to have been the one who incurred this injustice. It was possible to authorize another member of the group who was an expert sorcerer. To perform this sorcery he needed something from his victim i. e. someone who belonged to the alleged or actual family responsible for the "injustice", for example, hair, a piece of his or her skirt made of tree bark, or excrement, usually feces. The object was wrapped in leaves, tied with vine shoots, and placed in a swamp. The magician shook it daily while focusing his thoughts on the victim, and then cast spells such as: "I will make you tremble until you die...—" [8]. The literal translation of kuru in Fore is "trembling of the body" or "trembling of grass in the wind".

If a member of the clan became a victim of kuru, measures for retaliation were taken against the suspected kuru sorcerer or against any member of the clan to which the supposed sorcerer belonged. According to Charles Julius, such retaliatory measures strictly followed the rules of "tukabu" (usually spelled "tokabu" today), which is a means of ritual killing and an explanation for sudden unexplained death. This was especially common among the South Fore, in whom the disease occurred most frequently [12] (see Chapter 38). In this ritual killing, long thorns or bamboo splinters were drilled into the joints, neck, and other body parts [8]. With few exceptions, this torture led to death, usually by suffocation or bleeding to death. Though such ritual killing was occasionally carried out, tukabu as an explanation based on sorcery was extremely common, e. g., to explain sudden death from pneumonia. Since kuru sorcery was only practiced by the Fore, the neighboring linguistic groups whose members incurred much fewer cases of kuru blamed the Fore for the existence of the disease and for the many tukabu victims [13] (D.C. Gajdusek, personal communication).

2.5 The beginning of kuru research

In 1949, Australia decided that Papua and New Guinea should be combined in one administration resulting in the centralized Territory of Papua and New Guinea (PNG). Until then, both of these regions were independently governed by Australia. However, the legal status of the inhabitants of the Australian Territory of Papua and the United Nations Trust Territory of New Guinea remained different until Papua New Guinean independence on September 16, 1975. The process of pacification and administrative control proceeded gradually over many years.

In 1954 in the region of the North Fore, the first administrative office (patrol post) was set up in Okapa near the village of Moke (Fig. 2.1). At that time 30,000–40,000 indigenous people inhabited the so-called kuru region, a map of which is depicted in Fig. 2.2. Of these inhabitants approximately 11,000 belonged to the Fore [13]. Fore settlements usually consisted of a cluster of hamlets, each containing a few dozen inhabitants, in rare cases a few hundred inhabitants. Smaller Fore families were also found at higher altitudes, hidden in largely inaccessible forests.

According to reports of the census patrols, kuru represented the main medical problem in the entire region. This information resulted in an official report by the administrative official of the Okapa patrol post which mentioned the existence of this disease [13].

Other serious problems [14–16] in the region included numerous fights between settlements and diseases such as chronic obstructive lung disease, right heart failure, pneumonia, tropical ulcers, and yaws. The Fore attributed most deaths to sorcery, of which there were many kinds (e. g., tukabu, see above). The bodies of dead relatives were consumed at mortuary feasts, which was the usual means of disposing of the dead (synonymous with endocannibalism, cannibalism, anthropophagy) [17]. (Editor's note: Michael Alpers uses the term "transumption" for the ritual consumption of the bodies of loved ones who had died; Michael Alpers, personal communication).

In 1956, Vincent Zigas' attention was drawn to kuru. Zigas, who was the physician stationed in Kainantu (Fig. 2.2), was extremely interested in learning more about the disease and its precise cause. On December 26, 1956 he sent a report to John Gunther, the Director of Public Health in Port Moresby, detailing his initial observations. He asked Gunther for permission to spend more of his time scientifically investigating and learning about kuru, a disease which also affected children [18].

2.6 A kuru hospital in Okapa

Soon thereafter, the virologist D.C. Gajdusek, an ambitious 34-year-old scientist and student of three Nobel Laureates, encountered Zigas. On March 10, 1957, they met in Kainantu for the first time and the following day they took a jeep to Okapa. As early as March 13, 1957, Gajdusek wrote his first letter to his former colleague Joe Smadel [14] at the National Institutes of Health (NIH), Bethesda, USA, with detailed descriptions of kuru. Gajdusek was determined to conduct systematic research on kuru [19]. His enthusiasm triggered a cascade of TSE-relevant discoveries within various areas of study, e. g., in medicine, epidemiology, neuropathology, anthropology, ethnology, genetics, and molecular biology. In addition, he made important discoveries in the field of the slow virus diseases, e. g., the understanding of slow measles encephalitis.

The life of the kuru researchers was exceptionally difficult since they were required to live in extensive isolation. Apart from the local people, only Gajdusek and the head of the administrative office in Okapa, Jack Baker, lived in this still very remote and rugged area. Vincent Zigas, Lucy Hamilton, and some other scientists also helped with the kuru project [20] each for relatively short periods of time. A few years later, Michael Alpers also joined them (see Chapter 38). Carleton Gajdusek and the others quickly made friends with the local people and found them cooperative (Michael Alpers, personal communication).

Initially, the numerous patients had to be visited in their hamlets (Fig. 2.4). The primary goal was to gain clinical experience (Fig. 2.5), collect data, and record patients' stories and clinical histories, as well as take blood and other samples. Within the first few months, Gajdusek and his assistants, indigenous scouts, translators and porters, had already walked more than 2,000 km on steep, narrow, and chiefly well-used paths in the humid rain forest. The mountainous and rugged terrain lies approximately 1,000 to > 3,600 m above sea level. It was particularly difficult for them to cross suspension bridges (Fig. 2.3 top), climb steep mountains, deal with injuries, and endure the readily infected cuts caused by the sharp kunai grass.

Another impediment was the fact that there were very few medical laboratories in PNG except in the capital city of Port Moresby. Examination materials, tissue samples, and organs of the deceased were taken to Kainantu, via porters to Okapa and then in jeeps to Kainantu airport. From there they were sent to the laboratories in Port Moresby as well as to laboratories in Australia and the USA. Gajdusek's complex kuru research project was initially financed by money from the research budget of the hospital run by Vincent Zigas in Kainantu and his personal financial resources. When these funds were depleted the project was financially supported by the NIH. With the help of local home builders a simple hospital was built in Okapa (Fig. 13.1a) and operated under the supervision of Gajdusek. Patients suffering from kuru had to be transported to this hospital on the simplest wooden boards, each carried by at least two to four porters over long distances. During the first 10 months, more than 1,000 kuru patients were examined (Figs. 2.5, 13.1). Some of the main symptoms of the disease include: general clumsiness of movement, unsteadiness of gait, trunkal instability and progressive inability to get up or walk without support (Fig. 13.1a), as well as occasional outbursts of laughter characterized by euphoric episodes (Fig. 13.1b). In the early stages of the disease it is possible for relatives, caretakers, doctors and researchers to talk to the kuru patients about their disease as dementia rarely occurs. However, dementia may occur in the terminal phase of the disease. More clinical details are provided in Chapter 13.4. Kuru is characterized by con-

Fig. 2.4: Family in a remote village caring for a relative suffering from kuru. Reprinted with permission of D.C. Gajdusek [20].

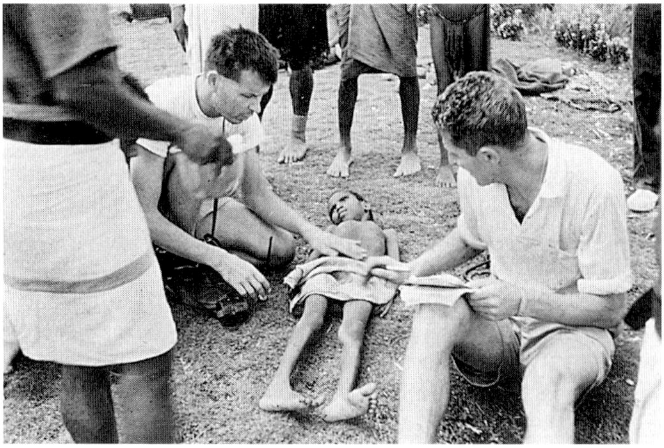

Fig. 2.5: D. Carleton Gajdusek (on the left) and Vincent Zigas (on the right), the first to describe kuru as a disease, examining a South Fore child ill with kuru, in 1957. Reprinted with permission of D.C. Gajdusek; archive no.: 57-369B [20].

stant disease progression and exhibits very distinct signs (Fig. 13.1), signs that relatives of affected patients, except perhaps during the earliest stages, would not mistake as signs of other diseases [13; 21; 22].

2.7 The spread of kuru

By questioning local people, researchers discovered cases of kuru from a very long time ago in most of the afflicted groups [23] (Fig. 2.2).

The origin of the kuru epidemic can probably be dated back to the beginning of the 20th century (Fig. 38.1). According to reliable reports the first case of kuru emerged in Uwami close to Tarabo (D.C. Gajdusek, personal communication). In Awande, the first cases of kuru were discovered around approximately 1910 [22; 24; 25]. The anthropologist Robert Glasse later approximately traced a specific case to the year 1922. Not until the 1940s did kuru take on epidemic proportions, especially among the Fore [14]. Many cases in the first half of the 1950s were found when the first rural census records were obtained.

2.8 The pathological picture

Gajdusek realized early on that kuru was a very unusual brain disease [20] with no detectable signs of extracerebral organ damage. In addition to taking care of the patients, his main concern was to precisely describe as many kuru cases as possible (Fig. 2.5), both clinically and pathologically. His goal was to discover the actual cause in order to be able successfully to combat the disease. It soon became doubtful that kuru could be an infectious disease, since typical clinical (fever) and pathological (inflammation) signs were absent. Bacterial origins were also excluded since therapies with antibiotics did not lead to improvement. Using brain tissue from afflicted patients in animal experiments with rats and mice to test transmissibility yielded negative results.

Therefore, characteristics of the disease pointed towards a non-infectious, degenerative brain disease. In addition to taking other organs, Gajdusek also removed the brain from patients who had died of kuru [14] for laboratory examination. He preferred not to perform autopsies in the houses where the patients lived (whereas later Michael Alpers did all of his autopsies in the houses where the patients lived; M. Alpers, personal communication). Gajdusek performed the autopsies in the hospital, where many patients spent the final phase of the disease before they died. Particularly problematic for the relatives of kuru victims was that patients were not permitted to die at home, although they tolerated autopsies being performed. Initially, 10 formalin-preserved brains of kuru victims were examined in Australian pathological laboratories. On August 5, 1957, Thomas Rivers urged Gajdusek to search for factors that might be specific to the Fore e. g., something special in their daily diet or something ingested during ceremonial rituals [26]. On August 6, 1957, the brains of more women and children who had died of kuru were sent to Bethesda, Maryland, USA [14] for examination. The first examined kuru brain belonged to a woman named Yabaiotu and was examined by the pathologist Igor Klatzo at the NIH on August 15, 1957; he confirmed to his boss, Joe Smadel [27], that in this disease there were no inflammatory signs detectable and that because of this there was no evidence of an infection. According to his observations a toxic metabolite, resulting from nutritional habits, was most likely the cause of kuru. On September 13, 1957, exactly half a year after Gajdusek's first letter about kuru, Igor Klatzo wrote: "It definitely seems to be a new form of disease; something which has not been described in the medical literature so far. The pathological picture, which most closely resembles kuru, was described by Creutzfeldt and Jakob [28] (Creutzfeldt–Jakob disease, CJD). But only about 20 cases have been published in the world" [27]. Among other things, Klatzo detected microscopically small holes (vacuoles) and so-called kuru plaques in brain tissue, which are similar to the amyloidal deposits in the brains of Alzheimer patients [29; 30]. In contrast to the brains of Alzheimer patients however, not all of the "kuru brains" displayed these amyloidal plaques. Therefore, not much importance was attached to this observation by either Klatzo or Gajdusek. Later Gajdusek began to pay more attention to this phenomenon and in connection with it he finally even coined the expression "infectious amyloidosis". It should be mentioned here that essential details of Rivers' as well as Klatzo's observations and interpretations were strikingly similar (see Chapter 13), and important. However, the significance of their observations became obvious only many years later.

2.9 The diet of the Fore

As is usually done when a new disease is discovered, diet is first considered a possible cause for the outbreak. This was investigated for kuru to determine any possible sources of infection, poisoning or any type of deficiency. The local diet consisted mainly of sweet potatoes, green vegetables, bananas, sugar cane, salt, pandanus fruits, leaves, mushrooms, roots, many types of birds, snakes, insects and spiders. In addition, the men hunted several species of animals in the forests and slaughtered the pigs, which were cared for and raised by the women. The women and children always received the innards, and the men had first choice of the desirable fat and muscle (see Chapter 38). Apart from pig, poultry and dogs were the only other domesticated animals. However, nothing pointed to the fact [10; 11; 20; 31] that kuru was an infection that was transmitted from these species or from humans to humans.

The Fore also consumed human flesh regularly [31]. As with pig, the men preferred the muscle and the women and children were left to eat the innards, including the brain of the deceased (see Chapter 38). There is no doubt that such practices took place [17], as Gajdusek actually witnessed these cannibalistic activities several times (D.C. Gajdusek, personal communication). To support his observations it should be noted that before him, R.I. Skinner compiled eyewitness reports on the consumption of humans in this region [17; 31]. The most important evidence surrounding the endocannibalistic practices of the Fore people comes from the many detailed accounts obtained from local people who participated in endocannibalism, as recorded independently by several anthropologists and scientists (M. Alpers, personal communication).

Among the Fore, cannibalism was not based on hostile motives – compared to some regions in PNG, such as the northeastern coastal region [10; 31; 32].

Among the Fore and their neighbors – and in some other regions of the world[2] – a ritualistic cult form of cannibalism was practiced, called endocannibalism, which was the consumption of deceased relatives. In that way the people hoped to integrate the spirit of the dead so that it would, in part, remain in the village community. By doing this, the family also paid a last tribute to their dead relatives. They were proud of this tradition and openly talked about it (D.C. Gajdusek, personal communication).

In the 1950s [16], rumors based on this information spread among medical laymen suggesting that kuru was caused by cannibalism. Gajdusek's research team had repeatedly taken this possibility into consideration since 1957 (D.C. Gajdusek, personal communication). However, the explanatory evidence of the interrelation between cannibalism and the kuru disease was still missing (see Chapter 38). They were not able to detect any infectious or other pathogenic agents in the examined tissues of the deceased. Transmissibility experiments in cell culture yielded negative results. In addition, Gajdusek and his colleague C. Joseph Gibbs, Jr. performed transmissibility experiments on chickens, mice, rats, guinea pigs and rabbits. The experiments were terminated after about 3 months with "no success" [33]. Considering these facts there was no plausible, scientifically sound hypothesis of transmissibility or of cannibalism as the cause. It is of note that "transmissibility of kuru" and "cannibalism being the cause of kuru" were originally quite different hypotheses. They were only put together after kuru had been shown to be transmissible to chimpanzees in 1966 (see below). The first scientific paper to propose this and develop the idea was published by Michael Alpers in 1968) [34].

About 10 years before that, on September 18, 1957, Gajdusek wrote to Smadel: "In the afflicted tribes often the women who are in charge of preparing the food fall ill with kuru..." [14]. At the time the significance of this observation was not recognized [35] (see Section 2.14). An anthropologist specializing in diets, Lucy Hamilton, was then charged with examining the diet of the Fore. However, she was not able to find anything among the numerous examined food samples (including human flesh), which were exclusively consumed by linguistic groups

[2] Aboriginal Australians [54] and people of the Congo delta, Africa [55] also practiced cultic endocannibalism (see Chapter 13).

afflicted with kuru. By October 1957, more than 500 food samples from the groups afflicted with kuru were collected, analyzed and compared to samples taken from kuru-free groups such as the Kukukuku, now called the Anga [36]. Investigators were looking for toxins and infectious agents in plants, bacteria, fungi, roots, cassowaries, cockatoos and other birds, snakes, spiders, insects, possums and other animals. In addition, they also inspected items such as bamboo cane,

Fig. 2.6: Many PNG people used to rub different substances over their bodies. To protect themselves against cold and insects they used, for example, a mixture of ashes, possibly human brain homogenate, and pig grease. During such practices among the Fore the kuru agent may have been parenterally transmitted manually, sometimes through contamination of cuts and scratches or indirect contact with mucous membranes (i. e., of the eye/small inset). The hands were kuru-contaminated while removing the brain from the skull of affected dead relatives, which were prepared to be transumed. The parenteral route of transmission is much more efficient (reflected in a shorter incubation period) than the oral route of infection. In view of the very long incubation periods found in many kuru cases, the oral route was more likely. Note: Here, a woman is shown from a group that is not affected by kuru. Photograph by J. du Boisberranger ©.

which was used by the Fore for cooking dishes and for the preparation of brain (see Section 2.14).

Together with salt, soil, smoke and ash samples, as well as dye materials and fat, which the people used to apply to body and face for many purposes (Fig. 2.6), everything was sent to distant laboratories to be analyzed. The samples were analyzed by A. Price in the capital city Port Moresby or by Gajdusek's colleagues in Bethesda, USA, for toxic substances such as alkaloids, heavy metals, or manganese [37]. However, none of the results of the many analyses were able to provide an explanation for kuru. Diseases resulting from deficiencies, such as vitamin E deficiency, could actually be excluded as being the cause [38; 39]. They also looked at many parasites and analyzed the respective samples [13; 14]. Blood and other samples were sent to Michael Wilson in Melbourne for bacteriological analysis [37]. Nothing pointed to the fact that kuru was transmitted by human bodily fluids or excrements, or that environmental factors were directly or indirectly involved in transmitting or causing the disease. None of these efforts provided satisfactory results and an answer to the question of a possible cause for kuru remained to be found. They had no other choice but to continue with their investigations. Over longer periods of time any food or tissue they could think of was repeatedly analyzed for toxic, microbial, viral and parasitic traces. All possible hypotheses were evaluated and assessed repeatedly [14]. The certainty that kuru was an extraordinary brain disease grew stronger, but the crucial question as to the cause of this affliction remained unanswered.

2.10 The geographical spread of the epidemic and the phylogenetic relations among kuru victims

After analyzing all of the collected information, which will be explained further in this section, Gajdusek suspected that kuru may be a genetically linked disease. Only very few facts did not fit this hypothesis and the majority actually confirmed it; over time it was found that practically all clans and families afflicted with kuru were related to the Fore by marriage. As mentioned earlier, the occurrence of kuru was mostly limited to the region of the Fore [21] (Fig. 2.2). Eventually, within the region of the Fore, almost every village and every settlement reported cases of kuru. However, numerous neighboring linguistic groups were also afflicted – though with notably fewer incidences [40]. Interestingly, in the southeast among the native Kukukuku and in the south among the Yar, kuru never appeared (D.C. Gajdusek, personal communication; Fig. 2.2). Isolated cases of the disease were found among the Iagaria in the northwest and among the Kamano in the north. The original home of the Iagaria was on the western side of Mount Michael and, therefore, separated from the area of the Fore. Several generations ago, however, some of them had left for the eastern side of the mountain and since then strongly intermixed with the Fore by marriage.

Also among the Wana clan (a small group of people), who had been closely connected to the Fore and other tribes by marriage for a long time, cases of kuru were found [13]. Around the end of the 1930s when the Wana were driven out of their homeland, the first cases of kuru were already reported. They sought refuge in the Keiagana area among the Kigupa clan, who had never experienced a case of kuru. The Kigupa, also a small group, did not welcome the Wana and displayed hostility towards them. That is the reason why there was no mixing of the two clans by marriage. For more than 15 years the Wana lived in the area of the Kigupa and reported many cases of kuru while not a single case occurred among the Kigupa. In 1956, the Wana clan returned to their area of origin (Ke‘efu) as the hostilities from former times had ceased. The Kigupa then returned to living in areas that had previously been occupied by the Wana. Even in this area the Kigupa did not suffer from kuru, which, among others, was an important argument against environmental contamination with a kuru-causing agent, at the time unknown but conceivable. Back in the Ke‘efu area, more than approximately 200 Wana people became victims of kuru [13].

These observations led to the preliminary hypothesis by Bennett and colleagues that there was a genetic cause or predisposition [41]; compare [42]. Gajdusek also reasoned that it was probably a genetically caused syndrome based on the geographical limits of the spread of kuru and its co-emergence with intermarriage. Since he was unable to find any additional factor (see above) that had anything to do with the spread of kuru, he submitted the hypothesis for publication after a few months of kuru research [14]. On August 20, 1957, however, Leonard T. Kurland, an epidemiologist from Bethesda, who was reviewing Gajdusek's submitted manuscript, skeptically wrote: "It is very unlikely that a genetically transmitted disease will reach such a high incidence as quickly as is the case for kuru. In addition, it would be extremely unusual to find a genetically transmitted disease which manifests itself among children, adolescents and adults in the same manner" (see Chapter 38). According to his opinion there was just as much evidence against as there was for genetic origins being the cause [42].

2.11 The ancestral cult

Even Gajdusek himself kept discovering contradictory evidence, which did not completely support a genetic hypothesis. In the entire region of the highlands, women always moved to the locality of their husband after marriage. When one looked at clans that were linked genetically by marriage, Fore women who had married into non-Fore communities contracted kuru. However, women who came from kuru-free communities and because of marriage lived with the Fore or other groups suffering from kuru also contracted the disease. The first reported case with such circumstances was found when a young woman fell ill. She originally came from a kuru-free Kimi (Gimi) community (some Gimi communities had kuru, some did not) and as a girl left her home region east of the kuru region (Fig. 2.2) to marry a Fore man. This case caused the plausibility of the genetically caused etiology to crumble for the first time [14]. Later other similar cases were reported (see Chapter 38).

Two anthropologists, Robert Glasse and Shirley Lindenbaum [43], went to live with the Fore people from 1961 to 1963 to consolidate and reevaluate the genetic hypothesis. They painstakingly analyzed all phylogenetic relations of kuru victims. Their records confirmed that kuru could not have a genetic origin. Moreover, in 1964 Gail Williams, Leonard Kurland and their coauthors [44] published a list of valid arguments concluding that the genetic hypothesis was highly unlikely.

Shirley Lindenbaum succeeded in gaining the Fore women's trust. The old women told her about former times, the beginnings of the kuru epidemic, the social impact of this mysterious disease and endocannibalism. If someone died of kuru, his closest relatives – usually maternal kin – claimed the body of the deceased [13; 14; 31]. It is interesting to note that in contrast to kuru victims, contact with certain other dead bodies was avoided, for example contact with those who had died of dysentery[3]. In these cases, the people suspected that the disease was contagious, most likely based on the observation that the time between the outbreak of infection and the beginning of having contracted the disease (incubation time) was very short. However, relatives of alleged victims of the kuru spell showed their usual respect to the dead. They paid their tribute by completely eating the corpse as part of the ritualistic ancestral cult [17]. Specific body parts were saved for the closest relatives [17; 31].

Starting in 1954, the Australian government prohibited and massively suppressed the practice of cannibalism in the Okapa area. They also had help from the Christian Mission at Tarabo (established in 1949), from the Christian Mission at Okasa (established in 1956) and later from the World Mission set up in Purosa in 1958. It should also be mentioned that dubious prison sentences were used to help suppress cannibalism.

Cannibalism was abandoned within a relatively short period of time, not by free will however, and also not because they knew that kuru could be eliminated by ending this practice. It can be assumed that until 1959 isolated endocannibalistic rituals still took place [14].

2.12 Social impact

Kuru led to complex and serious social problems [14]. The tukabu rituals with fatal consequences and ensuing clan feuds (see Section 2.4) have already been mentioned. They arose with the emergence of kuru and were based on superstition. At the height of the epidemic the disease had such a high frequency that more than half of the women died from it. The reason why women, girls and boys contracted kuru more often than adult men will be explained in detail in Chapter 38. In many villages, the ratio of surviving women to men decreased drastically. In the most extreme case it was only 1.0:3.4. Suddenly men had to perform chores such as cooking, childcare, or gardening, traditionally the duties of the women. There were men who practically lost all of their female relatives as a result of kuru. For example Anatu lost his mother, four wives (the Fore lived in polygamy), a sister, a daughter and a son because of kuru. To secure clan survival, girls, who had to marry men with an increasing average age, were of a continually younger age at marriage. They started having children shortly after reaching puberty. Because infants were usually breast-fed for 2 years, toddlers sometimes died of malnutrition after their mothers had died of kuru during the 2-year nursing period [14].

If a family member fell victim to the disease it was a great burden to all members of the extended family as usually the relatives supported their patients in a remarkable way accompanying them in their fate. The relatives lived in direct contact with the sick and without fear of contagion (Fig. 2.4). (It should be mentioned that kuru – although an infectious disease, as was determined later – is *not* contagious through natural human contact). Relatives supported their sick family members until death [13].

During the 1960s, the frequency of kuru cases finally declined perceptively (Figs. 38.1, 38.2). This was later [25; 34] traced back to the pro-

[3] Second World War soldiers introduced dysentery to PNG in 1942.

hibition of cannibalism. (Cannibalism, usually exocannibalism, occurred in many places in PNG, but this is not relevant to the transmission of kuru; M. Alpers, personal communication). As the number of cases decreased, each new case caused great panic in the villages. The people were afraid that the evils of kuru sorcery, in which they still believed, and its consequences, could regress to earlier times. One must keep in mind that an explanation of the actual context and of the origins of kuru (see Chapter 13) was still unavailable.

2.13 The discovery of transmissibility

In 1959, during a kuru exhibition at the Wellcome Medical Museum in London, some light was shed on the disease. The exhibition was organized by Gajdusek as a result of the relevance of the disease to the contemporary situation. Among other items of evidence, Igor Klatzo's microscopic photographs of tissue sections of the brains of kuru patients were displayed. A visitor named William J. Hadlow (author of Chapter 22), who was then working on scrapie projects together with W. Gordon in Compton [28], GB, noticed the neuropathological similarity between scrapie and kuru (scrapie is a prion disease of sheep and goats). In a now famous letter published in the medical journal *The Lancet*, Hadlow pointed out this remarkable similarity. In his letter, he included a recommendation that experiments be performed aiming to transmit kuru to apes, in order to answer central questions concerning the origins of kuru. He was aware of the very long incubation time of scrapie, which was known from laboratory experiments to be transmissible from as early as 1936. (It was not until the 1970s that valid proof for the natural transmission routes of scrapie from animal to animal were shown, and its contagious nature). However, kuru researchers knew nothing about scrapie before Hadlow's letter to *The Lancet* (D.C. Gajdusek, personal communication), with the exception of Joe Smadel, who did not seem to suspect a connection between the two. In 1963, experiments aiming to transmit the disease to chimpanzees were initiated. The first infected chimpanzee to fall sick was a young female, Georgette[4], who started to show signs of kuru in May 1965 after a long incubation period. The chimpanzee Daisy[4] was the first to be inoculated – she fell sick soon after Georgette. The pathologist Elisabeth Beck found the same microscopically small holes (vacuoles, spongiform encephalopathy) and amyloidal deposits or plaques in brain sections that were found in kuru patients. In February 1966, Gajdusek, Gibbs and Alpers published their success in transmitting kuru to chimpanzees. With this, the transmissibility of the spongiform encephalopathies (TSE) in humans was shown for the first time – three decades after the corresponding proof of transmissibility of TSE in animals, with scrapie.

In 1984, Robert Klitzman [17; 45], working with Alpers and Gajdusek, investigated the details of anthropophagy (consumption of human flesh) at the last mortuary feasts and their epidemiological outcomes. Their findings and other detailed interrelations between kuru and the endocannibalistic meals are described in Chapter 38.

Gajdusek was awarded the Nobel Prize for Medicine/Physiology in 1976 for: bringing forth experimental proof that kuru is transmissible; the discovery of the transmissibility of CJD; other important accomplishments pertaining to TSE research and many more relevant discoveries in the field of slow virus diseases. With the Nobel Prize all of his efforts in the fight against kuru were honored. Gajdusek made it his duty to do more for the Fore people than would be expected of an average scientist [20]. He and his colleagues had collected valuable data and countless samples under the most difficult and sometimes extremely adventurous conditions in an environment that was of absolutely no interest to the public. From these analyses they made exceptional scientific findings. On August 6, 1957, Gajdusek already suspected, in a visionary way, that the medical sciences still had much

[4] Note by the editors: Since high ethical standards are demanded from scientists in the field of primates – including respect shown towards each individual animal – the names of two experimental animals are mentioned here.

to learn from kuru [14]. The facts compiled in this book prove he was right. In 1989, the journal *Médecine et Hygiène* emphasized [46] that the success of research in this area was largely due to the persistence and the extraordinary contribution of the remarkable kuru researcher D. Carleton Gajdusek [47]. From a medical, historical, and epidemiological (with respect to the number of cases) point of view, kuru is the most important prion disease of mankind discovered so far.

2.14 The answers to the questions

In the spring of 1960, the anthropologists J. Fischer and A. Fischer from Tulane, USA, drew the following conclusions, which are described in a review article covering the most current knowledge of the kuru disease at the time: "The habit of eating the corpses of the deceased suggests that a viral or toxic agent may be transmitted in the process" [48].

In 1965, the epidemiological evidence pointed in the same direction [6; 49]. The conclusion that ritual mortuary endocannibalism was the most plausible explanation for the transmissibility of kuru [50] was first made in 1967 at a presentation by M. Alpers to the International Academy of Pathology [34]; a similar explanation was proposed by J.D. Mathews [24; 25], R. Glasse [51] and S. Lindenbaum [52].

The practice of endocannibalism as the only risk factor however, does not explain the exclusive presence of kuru in the small region of the Eastern Highlands. Although this risk factor was taken into account by Gajdusek's group repeatedly since 1957, it remained a marginal hypothesis since other groups in PNG, as well as in other parts of the world, practised cannibalism or endocannibalism with no incidences of kuru (see Chapter 13.9).

Moreover, the researchers around Gajdusek paid little attention to this hypothesis (see above) since there was a period of years between the exposure (meal) and the outbreak of the disease. Initially, it was unimaginable that they were dealing with an infectious disease with such a long incubation time. When Gajdusek, Gibbs and Alpers provided proof for the transmissibility of kuru in 1966, the hypothesis that kuru was transmissible on account of endocannibalistic practices became feasible; the epidemiological "circumstantial evidence" for the significance of endocannibalism was strengthened by the work of Klitzman and his colleagues [17; 45] in establishing conclusively interrelated elements (see Chapters 13, 38).

According to reports and oral traditions, confirmed by many independent accounts, it was possible to form a picture of how the Fore took apart the body, removed the brain from the skullcap and prepared human flesh and innards for the meal as part of the ceremony for honoring the dead: among the South Fore, the raw brain was pressed into bamboo cane. This was done manually, and very intense contact – with potentially highly infectious tissue – was plausible. According to oral tradition, the North Fore, however, wrapped the entire corpse in banana leaves, let it cook for hours on hot stones[5] and then opened the body. Interestingly, the incidence of kuru among the North Fore was significantly lower than among the South Fore (see Chapters 13, 38). It was also possible that kuru prions were transmitted by rubbing fingers contaminated with infectious brain tissue on mucous membranes e.g. the eyes [17] (Fig. 2.6, small inset). The relevance of this last mode of transmission was also confirmed in experiments with scrapie in mice. However, the long incubation periods found in kuru were more consistent with an oral than a parenteral route of transmission; this was modeled in hamsters by Prusiner and colleagues [53].

It was only after the revolutionizing laboratory results in 1966 [50], the epidemiological and anthropological findings in 1968 [25; 34] and the epidemiologically informative explanations provided by Klitzman and colleagues in 1984 [17], that the hypothesis involving endocannibalism became the solution to the problem. More details are provided in Chapters 13 and 38.

[5] Treatment using boiling temperature is not sufficient to inactivate prions (the kuru agent). Temperatures of 134–138 °C and steam pressure are required.

References

[1] Alpers MP, Gajdusek DC, Ono SG. Bibliography of kuru. 3rd ed. Nat. Institute of Neurological and Communicative Disorders and Stroke, Nat. Institutes of Health, Bethesda MD, 1975.

[2] Glasse RM. A kuru bibliography. Oceania 1961; 31:294–295.

[3] Foley WA. Language and identity in Papua New Guinea. In: Attenborough RD, Alpers MP, eds. Human biology in Papua New Guinea: the small cosmos. Clarendon, Oxford, 1992:136–149.

[4] Lilley I. Papua New Guinea's human past: the evidence of archaeology. In: Attenborough RD, Alpers MP, editors. Human biology in Papua New Guinea: the small cosmos. Clarendon Press, Oxford, 1992:150–171.

[5] Berndt CH. Social and cultural change in New Guinea: communication and views about "other people". Sociologus 1957; 7:38–57.

[6] Alpers MP. Epidemiol. changes in kuru, 1957 to 1963. In: Gajdusek DC, Gibbs CJ Jr, Alpers M, eds. Slow, latent, and temperate virus infections. NINDB Monograph No 2. US Government Print. Office, Washington DC, 1965:65–82.

[7] Gajdusek DC. A chronology of the kuru area. In: Farquhar J, Gajdusek DC, editors. Kuru – early letters and field notes from the collection of D. Carleton Gajdusek. Raven Press, New York, 1981:293–295.

[8] Julius C. Sorcery among the South Fore with special reference to kuru. Report submitted to the Public Health Department of Papua and New Guinea (28th February 1957). In: Farquhar J, Gajdusek DC, editors. Kuru – early letters and field notes from the collection of D. Carleton Gajdusek. Raven Press, N.Y., 1981:281–288.

[9] Berndt RM. A "devastating disease syndrome" – kuru sorcery in the Eastern Central Highlands of New Guinea. Sociologus 1958; 8:4–28.

[10] Berndt RM. Reaction to contact in the Eastern Highlands of New Guinea. Oceania 1954; 24:256–274.

[11] Berndt RM. Reaction to contact in the Eastern Highlands of New Guinea. Oceania 1954; 24:190–228.

[12] Alpers MP. Epidemiology and ecology of kuru. In: Prusiner SB, Hadlow WJ, editors. Slow transmissible diseases of the nervous system. Academic Press, New York, Vol 1, 1979:67–90.

[13] Gajdusek DC and Zigas V. Untersuchungen über die Pathogenese von Kuru. Klin Wochenschr 1958; 36:445–459.

[14] Gajdusek DC. Letters to Smadel. In: Farquhar J, Gajdusek DC, editors. Kuru – Early letters and field notes from the collection of D. Carleton Gajdusek. Raven Press, New York, 1981: 8–10, 29–31, 66–68, 78–80, 100–104, 120–121, 156–160, 169–171, 277–279.

[15] Gajdusek DC and Zigas V. Degenerative disease of the central nervous system in New Guinea – the epidemic occurrence of kuru in the native population. N Engl J Med 1957; 257:974–978.

[16] Rhodes R. Deadly feasts – tracking the secrets of a terrifying new plague. Simon and Schuster, New York, 1997.

[17] Klitzman RL, Alpers MP, Gajdusek DC. The natural incubation period of kuru and the episodes of transmission in three clusters of patients. Neuroepidemiology 1984; 3:3–20.

[18] Zigas V. Letter to Gunther. In: Farquhar J, Gajdusek DC, editors. Kuru – early letters and field notes from the collection of D. Carleton Gajdusek. Raven Press, New York, 1981:1–2.

[19] Zigas V and Gajdusek DC. Kuru: clinical study of a new syndrome resembling paralysis agitans in natives of the Eastern Highlands of Australian New Guinea. Med J Aust 1957; 2:245–254.

[20] Farquhar J and Gajdusek DC. Kuru – early letters and field notes from the collection of D. Carleton Gajdusek. Raven Press, New York, 1981.

[21] Gajdusek DC. Letter to Burnet. In: Farquhar J, Gajdusek DC, editors. Kuru – early letters and field notes from the collection of D. Carleton Gajdusek. Raven Press, New York, 1981:43–46.

[22] Glasse RM. The spread of kuru among the Fore. A preliminary report. Department of Public Health, Territory of Papua and New Guinea. Typed script, 1962; 1–9.

[23] Gajdusek DC and Zigas V. Letter to Abbott. In: Farquhar J, Gajdusek DC, editors. Kuru – early letters and field notes from the collection of D. Carleton Gajdusek. Raven Press, New York, 1981:84–87.

[24] Mathews JD. The changing face of kuru – an analysis of pedigrees collected by RM Glasse and Shirley Glasse and recent census data. Lancet 1965; 68:1138–1141.

[25] Mathews JD, Glasse RM, Lindenbaum S. Kuru and cannibalism. Lancet 1968; 2(7565):449–452.

[26] Rivers TM. Letter to Gajdusek. In: Farquhar J, Gajdusek DC, editors. Kuru – early letters and field notes from the collection of D. Carleton Gajdusek. Raven Press, New York, 1981:99–100.

[27] Klatzo I. Letter to Gajdusek. In: Farquhar J, Gajdusek DC, editors. Kuru – early letters and field notes from the collection of D. Carleton Gajdusek. Raven Press, N.Y., 1981:155–156.

[28] Poser CM. Notes on the history of the prion diseases. Part I. Clin Neurol Neurosurg 2002; 104(1):1–9.
[29] Klatzo I, Gajdusek DC, Zigas V. Pathology of kuru. Lab Invest 1959; 8:799–847.
[30] Miyakawa T, Watanabe K, Katsuragi S. Ultrastructure of amyloid fibrils in Alzheimer's disease and Down's syndrome. Virchows Arch (B) 1986; 52:99–106.
[31] Berndt RM. Kamano, Jate, Usurufa, and Fore kinship of the Eastern Highlands of New Guinea: a preliminary account. Oceania 1954; 25:156–187.
[32] Berndt RM. Kamano, Jate, Usurufa, and Fore kinship of the Eastern Highlands of New Guinea: a preliminary account. Oceania 1954; 25:23–53.
[33] Gajdusek DC and Gibbs CJ Jr. Attempts to demonstrate a transmissible agent in kuru, amyotropic lateral sclerosis, and other sub-acute and chronic nervous system degenerations of man. Nature 1964; 204:257–258.
[34] Alpers MP. Kuru: implications of its transmissibility for the interpretation of its changing epidemiological pattern. In: Bailey OT, Smith DE, editors. The central nervous system, some experimental models of neurologic diseases. International Academy of Pathology, Monograph No 9. Proceedings of the fifty-sixth annual meeting of the International Academy of Pathology, Washington, DC, 12–15 Mar 1967. Williams and Wilkins, Baltimore, 1968:234–251.
[35] Klatzo I. Report to Smadel. In: Farquhar J, Gajdusek DC, editors. Kuru – early letters and field notes from the collection of D. Carleton Gajdusek. Raven Press, New York, 1981:112–113.
[36] Hamilton L. Letter to Smadel. In: Farquhar J, Gajdusek DC, editors. Kuru – early letters and field notes from the collection of D. Carleton Gajdusek. Raven Press, N.Y., 1981:226–227.
[37] Gajdusek DC. Letters to Price. In: Farquhar J, Gajdusek DC, editors. Kuru – early letters and field notes from the collection of D. Carleton Gajdusek. Raven Press, New York, 1981:106–109.
[38] Gajdusek DC. Letter to Simmons. In: Farquhar J, Gajdusek DC, editors. Kuru – early letters and field notes from the collection of D. Carleton Gajdusek. Raven Press, New York, 1981:162–163.
[39] Imus HA. Letter to Smadel. In: Farquhar J, Gajdusek DC, editors. Kuru – early letters and field notes from the collection of D. Carleton Gajdusek. Raven Press, New York, 1981:108–109.
[40] Gajdusek DC. Letter to Burnet and Anderson. In: Farquhar J, Gajdusek DC, editors. Kuru – early letters and field notes from the collection of D. Carleton Gajdusek. Raven Press, New York, 1981:71–74.
[41] Bennett JH, Rhodes FA, Robson HN. A possible genetic basis for kuru. Am J Hum Genet 1959; 2:169–187.
[42] Kurland L. Letter to Imus. In: Farquhar J, Gajdusek DC, editors. Kuru – early letters and field notes from the collection of D. Carleton Gajdusek. Raven Press, New York, 1981:117–118.
[43] Glasse RM. Fieldwork in the South Fore: the process of ethnographic inquiry. In: Prusiner SB, Collinge J, Powell J, et al., editors. Prion diseases of humans and animals. Ellis Horwood, New York, 1992:77–91.
[44] Williams GR, Fischer A, Fischer JL, et al. An evaluation of the kuru genetic hypothesis. J génét hum 1964; 13:11–21.
[45] Klitzman RL. The trembling mountain: a personal account of kuru, cannibals and mad cow disease. Plenum Trade, New York, 1998.
[46] Rentchnick P. Pathographies (104). Médecine et Hygiène, Genève, 1998; 47:2985–3080.
[47] Gajdusek DC. Infectious amyloids: subacute spongiform encephalopathies as transmissible cerebral amyloidoses. In: Fields DN, Knipe DM, Howley PM, et al., editors. Fields Virology. Lippincott Raven Publisher, Philadelphia, 1996: 2851–2900.
[48] Fischer JL and Fischer A. Studies doubt genetic etiology of kuru. Public Health Reports 1963; 118–119.
[49] Alpers MP and Gajdusek DC. Changing patterns of kuru: epidemiological changes in the period of increasing contact of the Fore people with Western civilization. Am J Trop Med Hyg 1965; 14:852–879.
[50] Gajdusek DC, Gibbs CJ Jr, Alpers M. Experimental transmission of a kuru-like syndrome to chimpanzees. Nature 1966; 209:794–796.
[51] Glasse RM. Cannibalism in the kuru region of New Guinea. Trans N Y Acad Sci (Series 2) 1967; 29:748–754.
[52] Lindenbaum S. Kuru sorcery: disease and danger in the New Guinea highlands. Mayfield, Palo Alto, 1979.
[53] Prusiner SB, Cochran SP, Alpers MP. Transmission of scrapie in hamsters. J Infect Dis 1985; 152:971–978.
[54] Berndt RM and Berndt CH. The first Australians. 2nd ed. Ure Smith, Sydney, 1967.
[55] Helmolt HF. Weltgeschichte. 2. Auflage ed. Bibliographisches Institut, Leipzig und Wien, 1914.

3 History of Prion Research[1]

Stanley B. Prusiner

3.1 Introduction

In 1982 I evoked a good deal of skepticism when I proposed that the infectious agents causing certain degenerative disorders of the central nervous system in animals and, more rarely, in humans might consist of protein and nothing else. At the time, the notion was heretical. Dogma held that the conveyers of transmissible diseases required genetic material, composed of nucleic acid (DNA or RNA), in order to establish an infection in a host. Even viruses, among the simplest microbes, rely on such material to direct synthesis of the proteins needed for survival and replication.

Later, many scientists were similarly dubious when my colleagues and I suggested that these "proteinaceous infectious particles" – or "prions" as I called the disease-causing agents – could underlie inherited, as well as communicable, diseases. Such dual behavior was then unknown to medical science. And we met resistance again when we concluded that prions (pronounced "pree-ons") multiply in an incredible way; they convert normal protein molecules into dangerous ones simply by inducing the benign molecules to change their shape.

Today, however, a wealth of experimental and clinical data has made a convincing case that we are correct on all three counts. Prions are indeed responsible for transmissible and inherited disorders of protein conformation. They can also cause sporadic disease, in which neither transmission between individuals nor inheritance is evident. Moreover, there are hints that the prions causing the diseases explored thus far may not be the only ones. Prions made of rather different proteins may contribute to other neurodegenerative diseases that are quite prevalent in humans. They might even participate in illnesses that attack muscles.

The known prion diseases, all fatal, are sometimes referred to as spongiform encephalopathies. They are so named because they frequently cause the brain to become riddled with holes. These ills, which can brew for years (or even for decades in humans) are widespread in animals.

3.2 Animals and humans affected

The most common form is scrapie, found in sheep and goats. Afflicted animals lose coordination and eventually become so incapacitated that they cannot stand. They also become irritable and, in some cases, develop an intense itch that leads them to scrape off their wool or hair (hence the name "scrapie"). The other prion diseases of animals go by such names as transmissible mink encephalopathy, chronic wasting disease of mule deer and elk, feline spongiform encephalopathy and bovine spongiform encephalopathy. The last, often called mad cow disease is the most worrisome.

Gerald A. H. Wells and John W. Wilesmith of the Central Veterinary Laboratory in Weybridge, England, identified the condition in

---

[1]
Editors' note: This chapter is a reproduction of an article which Stanley B. Prusiner published in 1995. It therefore represents the knowledge on prions at that time. The text is unchanged; only the title has been changed and a few references to other chapters and figures within this book were added by the editors. Some figures were not reprinted due to limited space. Stanley B. Prusiner won the Nobel Prize in Physiology or Medicine in 1997.

1986, after it began striking cows in Great Britain, causing them to become uncoordinated and unusually apprehensive. The source of the emerging epidemic was soon traced to a food supplement that included meat and bone meal from dead sheep. The methods for processing sheep carcasses had been changed in the late 1970s. Where once they would have eliminated the scrapie agent in the supplement, now they apparently did not. The British government banned the use of animal-derived feed supplements in 1988, and the epidemic has probably peaked. Nevertheless, many people continue to worry that they will eventually fall ill as a result

Table 3.1: Prion diseases of humans, which may incubate for 30 years or more, can all cause progressive decline in cognition and motor function; hence, the distinctions among them are sometimes blurry. As the genetic mutations underlying familial forms of the diseases are found, those disorders are likely to be identified by their associated mutation alone. Editors' note: * number of cases etc. up to 1995 (according to original article of S. B. Prusiner).

Disease	Typical symptoms	Route of acquisition	Distribution	Span of overt illness
Kuru	Loss of coordination, often followed by dementia	Infection (through endocannibalism, which stopped by 1958)	Known only in the highlands of Papua New Guinea; some 2,600* cases have been identified since 1957	Three months to one year
Creutzfeldt–Jakob disease	Dementia, followed by loss of coordination, although sometimes the sequence is reversed	Usually unknown (in "sporadic" disease)	*Sporadic form:* 1 person per million worldwide	Typically about one year: range is one month to more than 10 years
		Sometimes (in 10 percent of cases) inheritance of a mutation in the gene coding for the prion protein (PrP)	*Inherited form:* some 100* extended families have been identified	
		Rarely, infection (as an inadvertent consequence of a medical procedure)	*Infectious form:* about 80* cases have been identified	
Gerstmann–Sträussler–Scheinker disease	Loss of coordination, often followed by dementia	Inheritance of a mutation in the PrP gene	Some 50* extended families have been identified	Typically two to six years
Fatal familial insomnia	Trouble sleeping and disturbance of autonomic nervous system, followed by insomnia and dementia	Inheritance of a mutation in the PrP gene	Nine* extended families have been identified	Typically about one year

of having consumed tainted meat (see Chapter 15.9).

The human prion diseases are more obscure. Kuru has been seen only among the Fore highlanders of Papua New Guinea (see Chapters 2, 13, 38). They call it the "laughing death". Vincent Zigas of the Australian Public Health Service and D. Carleton Gajdusek of the U.S. National Institutes of Health described it in 1957, noting that many highlanders became afflicted with a strange, fatal disease marked by loss of coordination (ataxia) and often later by dementia. The affected individuals probably acquired kuru through ritual cannibalism: the Fore tribe reportedly honored the dead by eating their brains. The practice has since stopped, and kuru has virtually disappeared.

Creutzfeldt–Jakob disease, in contrast, occurs worldwide and usually becomes evident as dementia. Most of the time it appears sporadically, striking one person in a million, typically around age 60. About 10 to 15 percent of cases are inherited, and a small number are, sadly, iatrogenic – spread inadvertently by the attempt to treat some other medical problem. Iatrogenic Creutzfeldt–Jakob disease has apparently been transmitted by corneal transplantation, implantation of dura mater or electrodes in the brain, use of contaminated surgical instruments, and injection of growth hormone derived from human pituitaries (before recombinant growth hormone became available).

The two remaining human disorders are Gerstmann–Sträussler–Scheinker disease, (which is manifest as ataxia and other signs of damage to the cerebellum) and fatal familial insomnia (in which dementia follows difficulty sleeping). Both these conditions are usually inherited and typically appear in midlife. Fatal familial insomnia was discovered by Elio Lugaresi and Rossella Medori of the University of Bologna and Pierluigi Gambetti of Case Western Reserve University.

3.3 In search of the cause

I first became intrigued by the prion diseases in 1972, when as a resident in neurology at the University of California School of Medicine at San Francisco, I lost a patient to Creutzfeldt–Jakob disease. As I reviewed the scientific literature on that and related conditions, I learned that scrapie, Creutzfeldt–Jakob disease and kuru had all been shown to be transmissible by injecting extracts of diseased brains into the brains of healthy animals. The infections were thought to be caused by a slow-acting virus, yet no one had managed to isolate the culprit.

In the course of reading, I came across an astonishing report in which Tikvah Alper and her colleagues at the Hammersmith Hospital in London suggested that the scrapie agent might lack nucleic acid, which usually can be degraded by ultraviolet or ionizing radiation. When the nucleic acid in extracts of scrapie-infected brains was presumably destroyed by those treatments, the extracts retained their ability to transmit scrapie. If the organism did lack DNA and RNA, the finding would mean that it was not a virus or any other known type of infectious agent, all of which contain genetic material. What, then, was it? Investigators had many ideas – including, jokingly, linoleum and kryptonite – but no hard answers.

I immediately began trying to solve this mystery when I set up a laboratory at U.C.S.F. in 1974. The first step had to be a mechanical one – purifying the infectious material in scrapie-infected brains so that its composition could be analyzed. The task was daunting; many investigators had tried and failed in the past. But with the optimism of youth, I forged ahead [see "Prions," by Stanley B. Prusiner; Scientific American, October 1984]. By 1982 my colleagues and I had made good progress, producing extracts of hamster brains consisting almost exclusively of infectious material. We had, furthermore, subjected the extracts to a range of tests designed to reveal the composition of the disease-causing component.

3.4 Amazing discovery

All our results pointed toward one startling conclusion: the infectious agent in scrapie (and presumably in the related diseases) did indeed lack nucleic acid and consisted mainly, if not exclusively, of protein. We deduced that DNA and

RNA were absent because, like Alper, we saw that procedures known to damage nucleic acid did not reduce infectivity. And we knew protein was an essential component because procedures that denature (unfold) or degrade protein reduced infectivity. I thus introduced the term "prion" to distinguish this class of disease conveyer from viruses, bacteria, fungi and other known pathogens. Not long afterward, we determined that scrapie prions contained a single protein that we called PrP, for "prion protein".

Now the major question became, where did the instructions specifying the sequence of amino acids in PrP reside? Were they carried by an undetected piece of DNA that traveled with PrP, or were they, perhaps, contained in a gene housed in the chromosomes of cells? The key to this riddle was the identification in 1984 of some 15 amino acids at one end of the PrP protein. My group identified this short amino acid sequence in collaboration with Leroy E. Hood and his co-workers at the California Institute of Technology. Knowledge of the sequence allowed us and others to construct molecular probes, or detectors, able to indicate whether mammalian cells carried the PrP gene. With probes produced by Hood's team, Bruno Oesch, working in the laboratory of Charles Weissmann at the University of Zürich, showed that hamster cells do contain a gene for PrP. At about the same time, Bruce Chesebro of the NIH Rocky Mountain Laboratories made his own probes and established that mouse cells harbor the gene as well. That work made it possible to isolate the gene and to establish that it resides not in prions but in the chromosomes of hamsters, mice, humans and all other mammals that have been examined. What is more, most of the time, these animals make PrP without getting sick.

One interpretation of such findings was that we had made a terrible mistake: PrP had nothing to do with prion diseases. Another possibility was that PrP could be produced in two forms, one that generated disease and one that did not. We soon showed the latter interpretation to be correct.

The critical clue was the fact that the PrP found in infected brains resisted breakdown by cellular enzymes called proteases. Most proteins in cells are degraded fairly easily. I therefore suspected that if a normal, non-threatening form of PrP existed, it too would be susceptible to degradation. Ronald A. Barry in my laboratory then identified this hypothetical protease-sensitive form. It thus became clear that scrapie-causing PrP is a variant of a normal protein. We therefore called the normal protein "cellular PrP" (PrP^{C}) and the infectious (protease-resistant) form "scrapie PrP" (PrP^{Sc}). The latter term is now used to refer to the protein molecules that constitute the prions causing all scrapie-like diseases of animals and humans.

3.5 Prion diseases can be inherited

Early on we had hoped to use the PrP gene to generate pure copies of PrP. Next, we would inject the protein molecules into animals, secure in the knowledge that no elusive virus was clinging to them. If the injections caused scrapie in the animals, we would have shown that protein molecules could, as we had proposed, transmit disease. By 1986, however, we knew the plan would not work. For one thing, it proved very difficult to induce the gene to make the high levels of PrP needed for conducting studies. For another thing, the protein that was produced was the normal, cellular form. Fortunately, work on a different problem led us to an alternative approach for demonstrating that prions could transmit scrapie without the help of any accompanying nucleic acid.

In many cases, the scrapie-like illnesses of humans seemed to occur without having been spread from one host to another, and in some families they appeared to be inherited. (Today researchers know that about 10 percent of human prion diseases are familial, felling half of the members of the affected families.) It was this last pattern that drew our attention. Could it be that prions were more unusual than we originally thought? Were they responsible for the appearance of both hereditary and transmissible illnesses?

In 1988 Karen Hsiao in my laboratory and I uncovered some of the earliest data showing that human prion diseases can certainly be in-

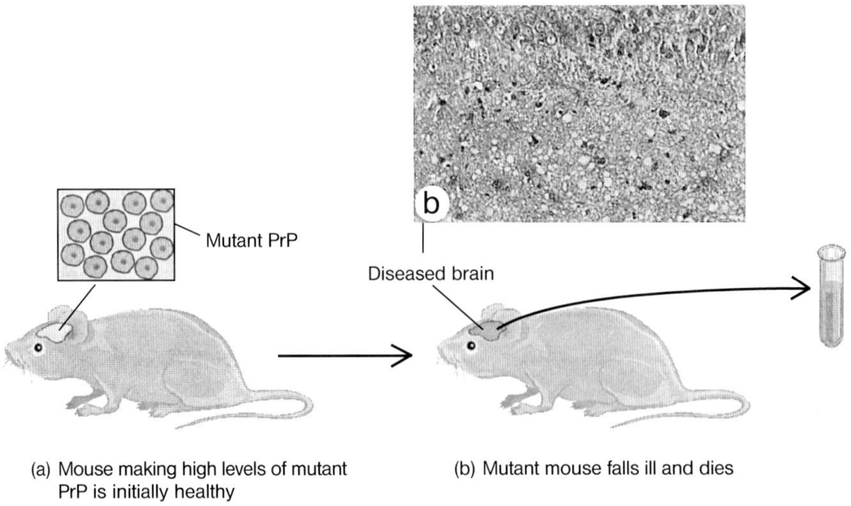

Fig. 3.1: A persuasive experiment. Several studies have shown that prions composed only of PrP are able to convey infection from one animal to another. In one such experiment, the author and his colleagues created mice carrying many copies of a mutant PrP gene **(a)**; these animals made high levels of mutant PrP, some of which appear to adopt the scrapie conformation. Eventually all the mice displayed signs of brain damage and died **(b)**. Then the workers injected brain tissue from the diseased animals into genetically altered mice making low levels of the same mutant PrP protein. (Such mice were chosen as

herited. We acquired clones of a PrP gene obtained from a man who had Gerstmann–Sträussler–Scheinker disease in his family and was dying of it himself. Then we compared his gene with PrP genes obtained from a healthy population and found a tiny abnormality known as a point mutation.

To grasp the nature of this mutation, it helps to know something about the organization of genes. Genes consist of two strands of the DNA building blocks called nucleotides, which differ from one another in the bases they carry. The bases on one strand combine with the bases on the other strand to form base pairs: the "rungs" on the familiar DNA "ladder." In addition to holding the DNA ladder together, these pairs spell out the sequence of amino acids that must be strung together to make a particular protein. Three base pairs together – a unit called a codon – specify a single amino acid. In our dying patient, just one base pair (out of more than 750) had been exchanged for a different pair. The change, in turn, had altered the information carried by codon 102, causing the amino acid leucine to be substituted for the amino acid proline in the man's PrP protein.

With the help of Tim J. Crow of Northwick Park Hospital in London and Jürg Ott of Columbia University and their colleagues, we discovered the same mutation in genes from a large number of patients with Gerstmann–Sträussler–Scheinker disease, and we showed that the high incidence in the affected families was statistically significant. In other words, we established genetic linkage between the mutation and the disease – a finding that strongly implies the mutation is the cause. Until 1994 work by many investigators has uncovered 18 mutations in families with inherited prion diseases; for five of these mutations, enough cases have been collected to demonstrate genetic linkage.

The discovery of mutations gave us a way to eliminate the possibility that a nucleic acid was traveling with prion proteins and directing their

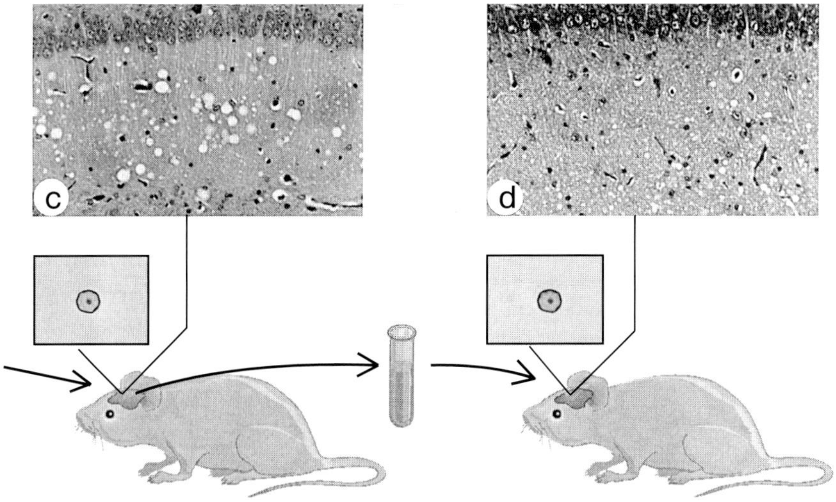

(c) Mouse making low levels of mutant PrP becomes ill after inoculation

(d) Identical mouse becomes ill after receiving inoculation from first recipient

recipients because scrapie PrP is most attracted to PrP molecules having the same composition.) Uninoculated mice did not become ill (indicating that making low levels of the aberrant protein was safe), but many of the treated ones did **(c)**. Moreover, brain tissue transferred from the diseased recipients to their healthy counterparts caused illness once again **(d)**. If the aberrant protein were unable to transmit infection, none of the inoculated animals would have sickened. Original figure by S. B. Prusiner. ©: Scientific American/Spektrum der Wissenschaften. Prionen und Prionkrankheiten, deGruyter, 2001.

multiplication. We could now create genetically altered mice carrying a mutated PrP gene. If the presence of the altered gene in these "transgenic" animals led by itself to scrapie, and if the brain tissue of the transgenic animals then caused scrapie in healthy animals, we would have solid evidence that the protein encoded by the mutated gene had been solely responsible for the transfer of disease. Studies I conducted with Karen K. Hsiao, Darlene Groth in my group and Stephen J. DeArmond, head of a separate laboratory at U.C.S.F., have now shown that scrapie can be generated and transmitted in this way (Fig. 3.1).

These results in animals resemble those obtained in 1981, when Gajdusek, Colin L. Masters and Clarence J. Gibbs, Jr., all at the National Institutes of Health, transmitted apparently inherited Gerstmann–Sträussler–Scheinker disease to monkeys. They also resemble the findings of Jun Tateishi and Tetsuyuki Kitamoto of Kyushu University, Japan, who transmitted inherited Creutzfeldt–Jakob disease to mice. Together the collected transmission studies persuasively argue that prions do, after all, represent an unprecedented class of infectious agents, composed only of a modified mammalian protein. And the conclusion is strengthened by the fact that assiduous searching for a scrapie-specific nucleic acid (especially by Detlev H. Riesner of Heinrich Heine University in Düsseldorf) has produced no evidence that such genetic material is attached to prions (see Chapters 4; 8).

Scientists, who continue to favor the virus theory, might say that we still have not proved our case. If the PrP gene coded for a protein that, when mutated, facilitated infection by a ubiquitous virus, the mutation would lead to viral infection of the brain. Then injection of brain extracts from the mutant animal would spread the infection to another host. Yet in the absence of any evidence of a virus, this hypothesis looks to be untenable.

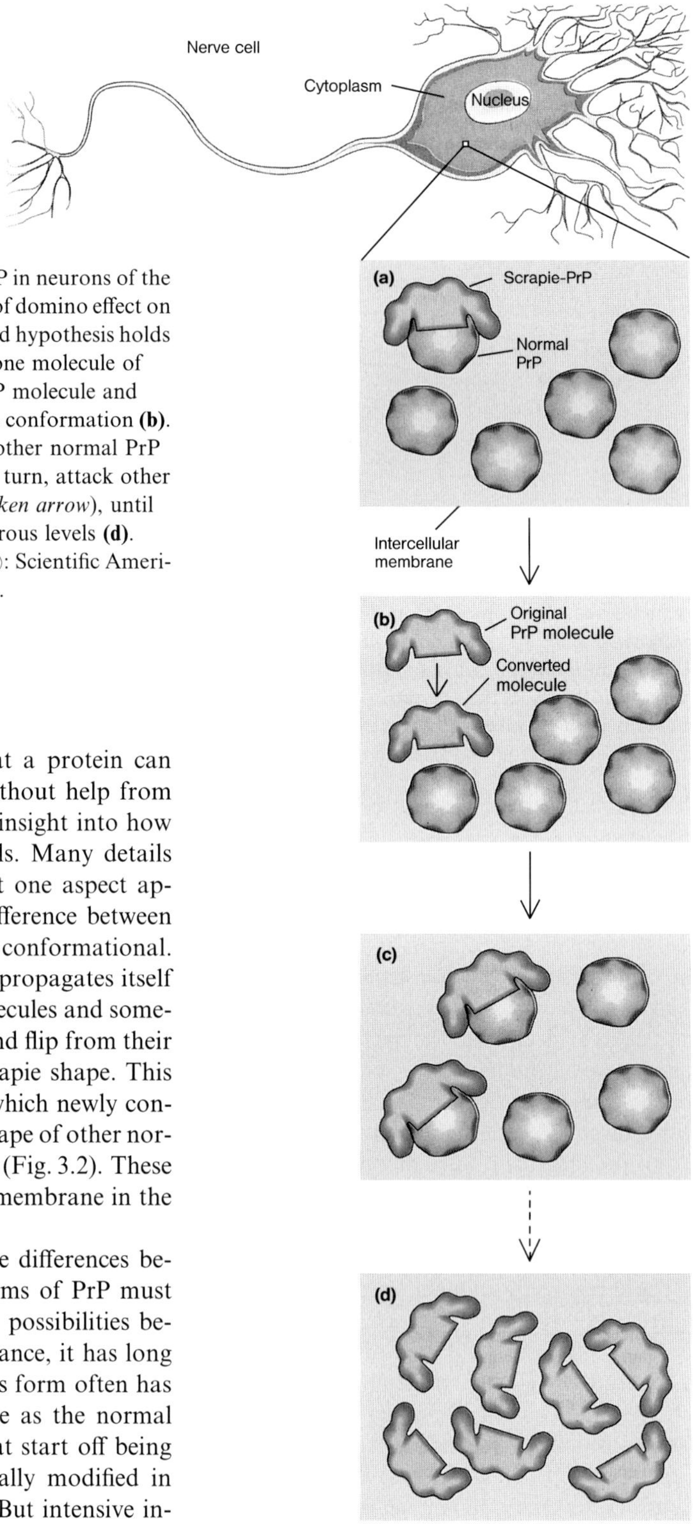

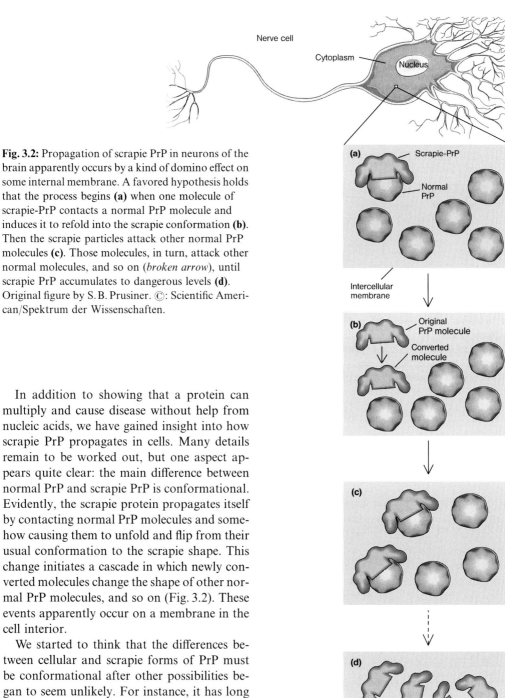

Fig. 3.2: Propagation of scrapie PrP in neurons of the brain apparently occurs by a kind of domino effect on some internal membrane. A favored hypothesis holds that the process begins **(a)** when one molecule of scrapie-PrP contacts a normal PrP molecule and induces it to refold into the scrapie conformation **(b)**. Then the scrapie particles attack other normal PrP molecules **(c)**. Those molecules, in turn, attack other normal molecules, and so on (*broken arrow*), until scrapie PrP accumulates to dangerous levels **(d)**. Original figure by S. B. Prusiner. ©: Scientific American/Spektrum der Wissenschaften.

In addition to showing that a protein can multiply and cause disease without help from nucleic acids, we have gained insight into how scrapie PrP propagates in cells. Many details remain to be worked out, but one aspect appears quite clear: the main difference between normal PrP and scrapie PrP is conformational. Evidently, the scrapie protein propagates itself by contacting normal PrP molecules and somehow causing them to unfold and flip from their usual conformation to the scrapie shape. This change initiates a cascade in which newly converted molecules change the shape of other normal PrP molecules, and so on (Fig. 3.2). These events apparently occur on a membrane in the cell interior.

We started to think that the differences between cellular and scrapie forms of PrP must be conformational after other possibilities began to seem unlikely. For instance, it has long been known that the infectious form often has the same amino acid sequence as the normal type. Of course, molecules that start off being identical can later be chemically modified in ways that alter their activity. But intensive investigations by Neil Stahl and Michael A. Baldwin in my laboratory have turned up no differences of this kind.

3.6 One protein, two shapes

How, exactly, do the structures of normal and scrapie forms of PrP differ? Studies by Keh-Ming Pan in our group indicate that the normal protein consists primarily of alpha-helices, regions in which the protein backbone twists into a specific kind of spiral; the scrapie form, however, contains beta-strands, regions in which the backbone is fully extended. Collections of these strands form beta-sheets. Fred E. Cohen, who directs another laboratory at U.C.S.F., has used molecular modeling to try to predict the structure of the normal protein based on its amino acid sequence. His calculations imply that the protein probably folds into a compact structure having four helices in its core (Fig. 3.3). Less is known about the structure, or structures, adopted by scrapie PrP.

The evidence supporting the proposition that scrapie PrP can induce an alpha-helical PrP molecule to switch to a beta-sheet form comes primarily from two important studies by investigators in my group. Maria Gasset learned that synthetic peptides (short strings of amino acids) corresponding to three of the four putative alpha-helical regions of PrP can fold into beta-sheets. And Jack Nguyen has shown that in their beta-sheet conformation, such peptides can impose a beta-sheet structure on helical PrP peptides. More recently Byron W. Caughey of the Rocky Mountain Laboratories and Peter T. Lansbury of the Massachusetts Institute of Technology have reported that cellular PrP can be converted into scrapie PrP in a test tube by mixing the two proteins together.

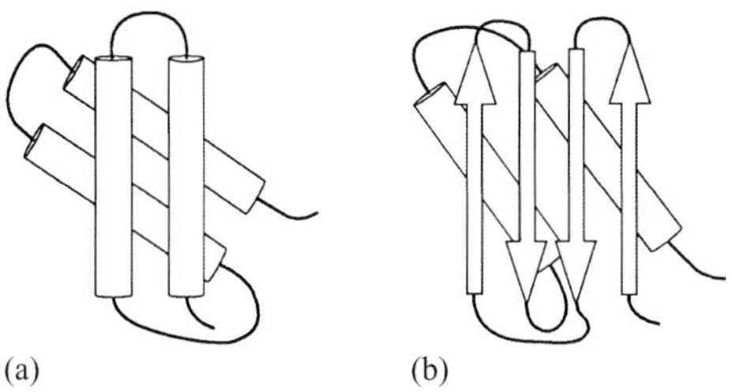

(a) (b)

Fig. 3.3: Prion protein (PrP) is usually harmless. In its benign stage, its backbone twists into multiple helices (shown as cylinders in **(a)**). PrP becomes the infectious, "scrapie" form – a prion – when much of the backbone streches out, forming so-called beta-strands (arrows in the hypothetical structure in **(b)**). Original figure by Fred E. Cohen. ©: Scientific American/Spektrum der Wissenschaften.

PrP molecules arising from mutated genes probably do not adopt the scrapie conformation as soon as they are synthesized. Otherwise, people carrying mutant genes would become sick in early childhood. We suspect that mutations in the PrP gene render the resulting proteins susceptible to flipping from an alpha-helical to a beta-sheet shape. Presumably, it takes time until one of the molecules spontaneously flips over and still more time for scrapie PrP to accumulate and damage the brain enough to cause symptoms.

Fred Cohen and I think we might be able to explain why the various mutations that have been noted in PrP genes could facilitate folding into the beta-sheet form. Many of the human mutations give rise to the substitution of one amino acid for another within the four putative helices or at their borders. Insertion of incorrect amino acids at those positions might destabilize a helix, thus increasing the likelihood that the affected helix and its neighbors will refold into a beta-sheet conformation. Conversely, Hermann Schätzl in my laboratory finds that the harmless differences distinguishing the PrP gene of humans from those of apes and monkeys affect amino acids lying outside of the proposed helical domains – where the divergent amino acids probably would not profoundly influence the stability of the helical regions.

3.7 Treatment ideas emerge

No one knows exactly how propagation of scrapie PrP damages cells. In cell cultures, the conversion of normal PrP to the scrapie form occurs inside neurons, after which scrapie PrP accumulates in intracellular vesicles known as lysosomes. In the brain, filled lysosomes could conceivably burst and damage cells. As the diseased cells died, creating holes in the brain (vacuoles; Fig. 1.1), their prions would be released to attack other cells.

We do know with certainty that cleavage of scrapie PrP is what produces PrP fragments that accumulate as plaques in the brains of some patients. Those aggregates resemble plaques seen in Alzheimer's disease, although the Alzheimer's clumps consist of a different protein. The PrP plaques are a useful sign of prion infection, but they seem not to be a major cause of impairment. In many people and animals with prion disease, the plaques do not arise at all.

Even though we do not yet know much about how PrP scrapie harms brain tissue, we can foresee that an understanding of the three-dimensional structure of the PrP protein will lead to therapies. If, for example, the four-helix-bundle model of PrP is correct, drug developers might be able to design a compound that would bind to a central pocket that could be formed by the four helices. So bound, the drug would stabilize these helices and prevent their conversion into beta-sheets.

Another idea for therapy is inspired by research in which Weissmann and his colleagues applied gene-targeting technology to create mice that lacked the PrP gene and so could not make PrP. By knocking out a gene and noting the consequences of its loss, one can often deduce the usual functions of the gene's protein product. In this case, however, the animals missing PrP displayed no detectable abnormalities. If it turns out that PrP is truly inessential, then physicians might one day consider delivering so-called antisense or antigene therapies to the brains of patients with prion diseases. Such therapies aim to block genes from giving rise to unwanted proteins and could potentially shut down production of cellular PrP. [see "The New Genetic Medicines," by Jack S. Cohen and Michael E. Hogan; Scientific American, December 1994]. They would thereby block PrP from propagating itself.

It is worth noting that the knockout mice provided a welcomed opportunity to challenge the prion hypothesis. If the animals became ill after inoculation with prions, their sickness would have indicated that prions could multiply even in the absence of a preexisting pool of PrP molecules. As I expected, inoculation with prions did not produce scrapie, and no evidence of prion replication could be detected.

The enigma of how scrapie PrP multiplies and causes disease is not the only puzzle starting to be solved. Another long-standing question – the mystery of how prions consisting of a single kind of protein can vary markedly in their effects – is beginning to be answered as well. Iain H. Pattison of the Agriculture Research Council in Compton, England, initially called attention to this phenomenon. Years ago he obtained prions from two separate sets of goats. One isolate made inoculated animals drowsy, whereas the second made them hyperactive. Similarly, it is now evident that some prions cause disease quickly, whereas others do so slowly.

3.8 The mystery of "strains"

Alan G. Dickinson, Hugh Fraser and Moira E. Bruce of the Institute for Animal Health in Edinburgh, who have examined the differential effects of varied isolates in mice, are among those who note that only pathogens containing nucleic acids are known to occur in multiple strains. Hence, they and others assert, the existence of prion "strains" indicates the prion hypothesis must be incorrect; viruses must be at the root of scrapie and its relatives. Yet because efforts to find viral nucleic acids have been unrewarding, the explanation for the differences must lie elsewhere.

One possibility is that prions can adopt multiple conformations. Folded in one way, a prion might convert normal PrP to the scrapie form highly efficiently, giving rise to short incubation times. Folded another way, it might work less efficiently. Similarly, one "conformer" might be attracted to neuronal populations in one part of the brain, whereas another might be attracted to neurons elsewhere, thus producing different symptoms. Considering that PrP can fold in at least two ways, it would not be surprising to find it can collapse into other structures as well.

Since the mid-1980s we have also sought insight into a phenomenon known as the species barrier. This concept refers to the fact that something makes it difficult for prions made by one species to cause disease in animals of another species. The cause of this difficulty is of considerable interest today because of the epidemic of mad cow disease in Britain. We and others have been trying to find out whether the species barrier is strong enough to prevent the spread of prion disease from cows to humans.

3.9 Breaking the barrier

The barrier was discovered by Pattison, who in the 1960s found it hard to transmit scrapie between sheep and rodents. To determine the cause of the trouble, my colleague Michael R. Scott and I later generated transgenic mice expressing the PrP gene of the Syrian hamster – that is, making the hamster PrP protein. The mouse gene differs from that of the hamster gene at 16 codons out of 254. Normal mice inoculated with hamster prions rarely acquire scrapie, but the transgenic mice became ill within about two months.

We thus concluded that we had broken the species barrier by inserting the hamster genes into the mice. Moreover, on the basis of this and other experiments, we realized that the barrier resides in the amino acid sequence of PrP: the more the sequence of a scrapie PrP molecule resembles the PrP sequence of its host, the more likely it is that the host will acquire prion disease. In one of those other experiments, for example, we examined transgenic mice carrying the Syrian hamster PrP gene in addition to their own mouse gene. Those mice make normal forms of both hamster and mouse PrP. When we inoculated the animals with mouse prions, they made more mouse prions. When we inoculated them with hamster prions, they made hamster prions. From this behavior, we learned that prions preferentially interact with cellular PrP of homologous, or like, composition.

The attraction of scrapie PrP for cellular PrP having the same sequence probably explains why scrapie managed to spread to cows in England from food consisting of sheep tissue: sheep and bovine PrP differ only at seven positions. In contrast, the sequence difference between human and bovine PrP is large: the molecules diverge at more than 30 positions. Because the variance is great, the likelihood of transmission from cows to people would seem to be low. Consistent with this assessment are epidemiological studies by W. Bryan Matthews, a professor emeritus at the University of Oxford. Matthews found no link between scrapie in sheep and the occurrence of Creutzfeldt–Jakob disease in sheep-farming countries.

On the other hand, two farmers who had "mad cows" in their herds have recently died of Creutzfeldt–Jakob disease. Their deaths may have nothing to do with the bovine epidemic, but the situation bears watching. It may turn out that certain parts of the PrP molecule are more important than others for breaking the species barrier. If that is the case, and if cow PrP closely resembles human PrP in the critical regions, then the likelihood of danger might turn out to be higher than a simple comparison of the complete amino acid sequences would suggest.

We began to consider the possibility that some parts of the PrP molecule might be particularly important to the species barrier after a study related to this blockade took an odd turn. My colleague Glenn C. Telling had created transgenic mice carrying a hybrid PrP gene that consisted of human codes flanked on either side by mouse codes; this gene gave rise to a hybrid protein. Then he introduced brain tissue from patients who had died of Creutzfeldt–Jakob disease or Gerstmann–Sträussler–Scheinker disease into the transgenic animals. Oddly enough, the animals became ill much more frequently and faster than did mice carrying a full human PrP gene, which diverges from mouse PrP at 28 positions. This outcome implied that similarity in the central region of the PrP molecule may be more critical than it is in the other segments.

The result also lent support to earlier indications – uncovered by Shu-Lian Yang in DeArmond's laboratory and Albert Taraboulos in my group – that molecules made by the host can influence the behavior of scrapie PrP. We speculate that in the hybrid-gene study, a mouse protein, possibly a "chaperone" normally involved in folding nascent protein chains, recognized one of the two mouse-derived regions of the hybrid PrP protein. This chaperone bound to that region and helped to refold the hybrid molecule into the scrapie conformation. The chaperone did not provide similar help in mice making a totally human PrP protein, presumably because the human protein lacked a binding site for the mouse factor.

3.10 The list may grow

An unforeseen story has recently emerged from studies of transgenic mice making unusually high amounts of normal PrP proteins. DeArmond, David Westaway in our group and George A. Carlson of the McLaughlin Laboratory in Great Falls, Montana, became perplexed when they noted that some older transgenic mice developed an illness characterized by rigidity and diminished grooming. When we pursued the cause, we found that making excessive amounts of PrP can eventually lead to neurodegeneration and, surprisingly, to destruction of both muscles and peripheral nerves. These discoveries widen the spectrum of prion diseases and are prompting a search for human prion diseases that affect the peripheral nervous system and muscles.

Investigations of animals that overproduce PrP have yielded another benefit as well. They offer a clue as to how the sporadic form of Creutzfeldt–Jakob disease might arise. For a time I suspected that sporadic disease might begin when the wear and tear of living led to a mutation of the PrP gene in at least one cell in the body. Eventually, the mutated protein might switch to the scrapie form and gradually propagate itself, until the buildup of scrapie PrP crossed the threshold to overt disease. The mouse studies suggest that at some point in the lives of the one in a million individuals who acquire sporadic Creutzfeldt–Jakob disease, cellular PrP may spontaneously convert to the scrapie form. The experiments also raise the possibility that people who become afflicted with sporadic Creutzfeldt–Jakob disease overproduce PrP, but we do not yet know if, in fact, they do.

All the known prion diseases in humans have now been modeled in mice. With our most recent work we have inadvertently developed an animal model for sporadic prion disease. Mice inoculated with brain extracts from scrapie-infected animals and from humans afflicted with Creutzfeldt–Jakob disease have long provided a model for the infectious forms of prion disorders. And the inherited prion diseases have been modeled in transgenic mice carrying mutant PrP genes. These murine representations of the human prion afflictions should not only extend understanding of how prions cause brain degeneration, they should also create opportunities to evaluate therapies for these devastating maladies.

3.11 Striking similarities

Ongoing research may also help determine whether prions consisting of other proteins play a part in more common neurodegenerative conditions, including Alzheimer's disease, Parkinson's disease and amyotrophic lateral sclerosis. There are some marked similarities in all these disorders. As is true of the known prion diseases, the more widespread ills mostly occur sporadically but sometimes "run" in families. All are also usually diseases of middle to later life and are marked by similar pathology: neurons degenerate, protein deposits can accumulate as plaques, and glial cells (which support and nourish nerve cells) grow larger in reaction to damage to neurons. Strikingly, in none of these disorders do white blood cells – those ever present warriors of the immune system – infiltrate the brain. If a virus were involved in these illnesses, white cells would be expected to appear.

Recent findings in yeast encourage speculation that prions unrelated in amino acid sequence to the PrP protein could exist. Reed B. Wickner of the NIH reports that a protein called Ure2p might sometimes change its conformation, thereby affecting its activity in the cell. In one shape, the protein is active; in the other, it is silent.

The collected studies described here argue persuasively that the prion is an entirely new class of infectious pathogen and that prion diseases result from aberrations of protein conformation. Whether changes in protein shape are responsible for common neurodegenerative diseases, such as Alzheimer's, remains unknown, but it is a possibility that should not be ignored.

References

[1] Prusiner SB. The prion diseases. Sci Am 1995; 272:48–57.
[2] Prusiner SB. Prionen-Erkrankungen. Spektrum der Wissenschaften 1995; 3:44–52.

Topic II: Molecular Biology and Genetics

4 The Physical Nature of the Prion

Detlev Riesner

4.1 Introduction

The prion model has become increasingly accepted over the years, and synthetic mammalian prions have recently been established experimentally. Nevertheless, we also briefly review the virus and virino hypothesis here. This is done not only for historical reasons but also for explanatory reasons – to describe the intellectual jump from the classical knowledge of the virus-type infectious agent to the protein-type agent of prions. One should recall that up to the late 1960s the virus-type infection model was the only thinkable model for an undetected infectious agent. We will first describe the general features of the prion model and then briefly consider it in comparison with the virus and virino models.

4.2 The prion model and its nomenclature

In Chapter 3, Prusiner described a very personal experience of how he developed the prion model. The detailed molecular mechanisms are outlined in Chapter 8.6. Here, we will introduce a general nomenclature. The original prion hypothesis claimed that the infectious agent was of a proteinaceous character, i. e., one or several proteins were postulated as major component(s) of the agent. Nucleic acids could not be completely excluded, although none have been detected to date. A simplified form of the prion hypothesis with the advantage of better clarity was the protein-only hypothesis, or even simpler the prion protein-only hypothesis. According to the last formulation, the infectious isoform of the prion protein is the only component of the infectious agent.

The prion is the infectious agent, i. e., the biological phenomenon of a nucleic acid-free agent, whereas the term prion protein describes the major, or only, biochemical component of prions. Since the prion protein was identified as a product of the host genome and was also found to be expressed in the noninfected organism, one had to differentiate between the normal or cellular form, designated as PrP^{C}, and the scrapie form of the agent, PrP^{Sc}. The superscript Sc denotes the historical prion disease scrapie. The biological and functional difference between these two forms is confirmed by the biochemical finding that PrP^{C} is rather sensitive to digestion with proteinase K (PK), whereas PrP^{Sc} is truncated N-terminally; the remainder stays stable and infectious for several hours against PK attack. Because this resistance against PK digestion was the first biochemical property found to be characteristic of PrP^{Sc}, both features i. e., infectivity and PK resistance, were often used synonymously. Later, results showed that this correlation does not hold in all cases, and therefore we adhere to a nomenclature in which PrP^{Sc} is always characterized by infectivity, and any PK-resistant form of PrP is designated as PrPres, which may or may not be infectious.

A clear definition of infectious molecules is difficult. Experimentally, it was found that in the hamster system prions only contain PrPres. Does this mean that all PrPres molecules are identical, and the same as PrP^{Sc}? Since 10^5–10^6 PrP^{Sc} molecules are found within one infectious unit, then either the individual PrP^{Sc} molecules exhibit very low infectivity, or only a very small number represents the infectious unit designated as PrP^* among the many PrP^{Sc} molecules. At present, however, the hypothetical PrP^* molecules cannot be separated experimentally from PrP^{Sc} molecules. The designation PrP^* is not optimal, since other authors use this term to describe the transition state of PrP^{C}–PrP^{Sc}. In a recent publication it was shown that only 500–5,000 PrP^{Sc} molecules are needed to induce an infection, because only that number of molecules remain in the brain, whereas around 96% are washed away immediately after inoculation and, therefore, cannot contribute to infectivity [1].

During the PK digestion of PrP^{Sc} the N-terminal amino acids from amino acid 23 (1–22 are split as a signal peptide) to amino acid 89 are truncated. Only the remaining 141 amino acids are resistant against further digestion and retain the full infectivity. According to a reduction of 33–35,000 Dalton to 27–30,000 Dalton, the truncated product is called PrP27–30. When PrP27–30 is prepared in the presence of detergents, very rigid structures are formed that appear in the electron microscope as rigid rods and exhibit the amyloid-characteristic fluorescence birefringence after Congo Red staining. These structures are called prion rods. Unaware of the prion model, the same or similar structures were already detected in 1981 in brain samples of scrapie-afflicted sheep and were designated as scrapie-associated fibrils, SAF [2] (Fig. 1.5b).

In Fig. 4.1, the prion model is presented in its simple form, the prion protein-only hypothesis. It is shown that PrP^C is expressed from a nuclear host gene. PrP^C is found in many organs, and its cellular function is still being researched (see Chapter 7). If it comes into contact with the invading PrP^{Sc} it is forced to switch from the PrP^C to the PrP^{Sc} conformation. The scheme in Fig. 4.1 does not infer any detail of the PrP^C–PrP^{Sc} transition mechanism (see Chapter 8). It is not indicated whether the PrP^C–PrP^{Sc} contact is direct or indirect, and whether the transformation occurs outside, at the surface, or inside the cell. However, a catalytic cycle is obviously started and the newly formed PrP^{Sc} acts again as an inducer for the transformation of more PrP^C. As such, a prion infection can be described as an amplification of the PrP^{Sc}-specific conformation.

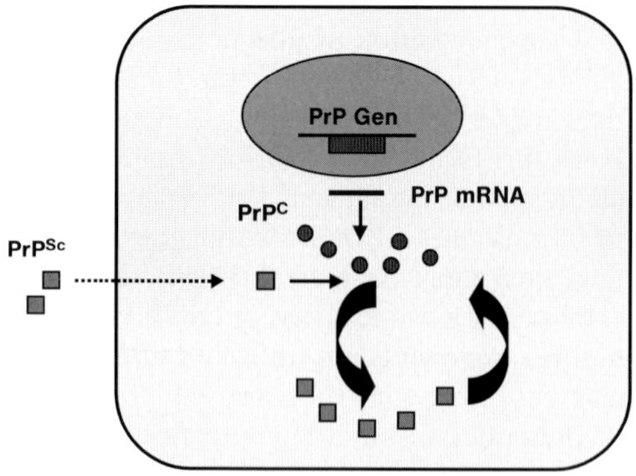

Fig. 4.1: The prion model in its simple form: the prion protein-only hypothesis. Figure modified from S. B. Prusiner (Chapter 3).

4.3 The virus hypothesis

In the 1950s, Björn Sigurdsson classified scrapie as a slow virus disease [3], and a few years later D. Carleton Gajdusek similarly assigned the kuru disease to this category [4]. Nothing else was thinkable at the time, and *slow* described its very long incubation period. Different types of viruses with known and unknown features were discussed as potential agents, which, however, will not be reported here. Until recently, Heino Diringer and Laura Manuelidis [5; 6], respectively, were the most prominent advocates of the virus hypothesis, of which some of its central points are summarized here.

According to Diringer et al., a yet-to-be identified virus-like particle was assumed to be a scrapie agent; it was also called an unconventional virus, suggesting that some of the features of the scrapie agent were atypical for a virus [5]. The agent was believed to contain a scrapie-specific nucleic acid that is protected against the well-known chemical or physical attack by an impenetrable capsid, which explains its exceptional resistance. All attempts, however, to identify a scrapie-specific nucleic acid have failed. The basic question of how to find or exclude an essential nucleic acid will be dealt with in Section 4.5. However, a virus or a virus-like particle needs by definition an information carrier, and only DNA or RNA is possible. In particular, the strain specificity must be encoded in a nucleic acid.

Prion rods or SAF were interpreted as secondary products of the afflicted host; other authors suggested a nucleic acid hidden in these very stable protein aggregates. If one assumes that the prion protein functions as an essential receptor molecule for the infection process, then results with knockout mice (*Prnp*$^{o/o}$ mice) could also be reconciled with the virus hypothesis, because the lack of an essential receptor for a virus would also prevent infection by this virus.

Scrapie, CJD, and other typical prion diseases belong to the amyloidosis diseases [7] as do Alzheimer's, Parkinson's, and Huntington's disease. The only basic difference between the first and the second group is the infectious or noninfectious etiology, respectively.

Diringer emphasized the similarity of the scrapie agent with neurotrophic viruses concerning the track of the infection in the host. After oral uptake, the agent is first multiplied outside the CNS without clinical symptoms during the early phase. Afterwards, the infection spreads via the peripheral nervous system to the CNS, where it is heavily multiplied and leads to a lethal disease [8].

The virus hypothesis is a theoretical construct that can consistently interpret the biological phenomena of the scrapie disease. However, the unconventional properties of the agent were derived from the phenotypic features of the disease and the infectious samples, without delivering, up to now, the biochemical correlate in spite of extensive research. Up to 2004, the only chemical or physical support was in fact, the lack of conclusive evidence for the prion hypothesis. As outlined in Chapter 8, the infectivity of recombinant prion protein can be shown thereby ruling out an essential role of a nucleic acid in infectivity [9]. Also, strain specificity is retained without nucleic acids. This was concluded when residual nucleic acids that were still detected in natural prion samples were destroyed by UV light, which also makes nucleic acids as strain determinants extremely improbable [1]. Thus, the virus hypothesis has in fact stimulated fruitful discussions over the course of many years, but in the end has ultimately failed.

4.4 The virino hypothesis

The virino hypothesis is a special formulation of the virus hypothesis, which has been subject to little research activity in recent years [10]. A very small nucleic acid is assumed to be a scrapie-specific genome. The plant pathogenic viroids [11] with no more than 250–400 nt genome size were the paradigm for virinos. In contrast to viroids it is assumed, however, that virino nucleic acid is virus-like and encapsidated, which can escape detection because of its small size and thorough encapsidation, possibly in aggregates. Primarily, the virino nucleic acids have to encode strain specificity, whereas most other functions of virino multiplication are carried out by host factors, in the sense that they are very similar to viroids. However, experimental evidence is missing, and a recent report by Safar and colleagues strongly argues against functional nucleic acids [1].

4.5 The nucleic acid problem

Nearly 40 years ago, Tikvah Alper and colleagues first proposed the heretical idea that the causative agent of scrapie does not depend on an intrinsic nucleic acid moiety for replication [12]. Although this problem is settled today in the sense that Alper was right, it appears essential for this book to critically review three kinds of experimental approaches that illustrate the vexing nature of the problem of definitively proving that the agent does or does not contain a nucleic acid. It is an intellectually challenging story, which describes the eventful history of scrapie research, and which summarizes many exceptional properties of the agent that are still of great interest today. The three approaches to the search for an agent-specific nucleic acid include (i) infectivity studies of the causative agent after physical and chemical treatment; (ii) search for differences in the nucleic acid patterns of infected and noninfected material; and (iii) direct analysis of nucleic acids in purified infectious material.

4.5.1 Physical, chemical, and biochemical inactivation

Alper's studies with ionizing and UV radiation first indicated that the scrapie genome might be of an unusual size and type. Qualitative differences in resistance to irradiation between the scrapie agent, phages, bacteria, and genetic markers in eukaryotic cells were obvious in one of the early dose-response curves depicted in Fig. 4.2. The target size theory led to the conclusion that the M_r of the putative genome was in the range of $1.0–1.5 \times 10^5$ by ionizing radiation or even smaller in the case of UV radiation. Similar results were obtained by Gibbs and colleagues [13] for the agents of kuru and CJD. In addition to its small size, qualitative differences in comparison to a conventional nucleic acid genome were noticed; the action spectrum from UV irradiation did not have a nucleic acid-specific peak at 254 nm, and the presence of oxygen in dilute suspensions, which normally protects nucleic acids and proteins against radiation damage, nearly sensitized the inactivation of the scrapie agent by an order of magnitude [14]. Inactivation studies are indirect studies. The size estimation of an unknown agent can only be carried out if it is assumed that a comparison with known agents is appropriate. The influence of experimental conditions such as the degree of purity, heterogeneity etc. can hardly be accounted for in a quantitative manner.

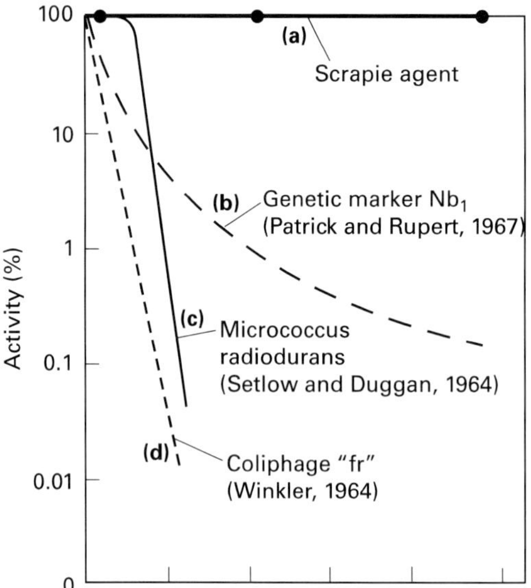

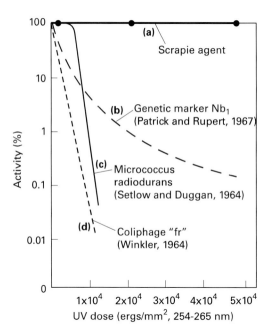

Fig. 4.2: Effects of ultraviolet light in the range of 254–265 nm on biological systems. **(a)** Infectivity of the scrapie agent. **(b)**Transforming activity of the DNA of *H. influenza* for resistance to novobiocin after repair by a recipient cell. The dashed line is an extrapolation of data according to the dose–response relationship given by Patrick and Rupert [36]. **(c)** Survival curve of *M. radiodurans*, an organism with effective intracellular repair mechanisms [37]. **(d)** Inactivation of single-stranded RNA coliphage "fr". The dashed line is an extrapolation of Winkler's [38] data according to the exponential dose response relationship. Figure reproduced in modified form from Alper et al. [12].

The interpretation of the ionizing data has been questioned by Rohwer, who compared the dependence of the inactivation rate and molecular weight of several irradiated viruses and phages [15; 16]. Extrapolating these data to the inactivation dose for the scrapie agent leads to a much larger estimate of the size of a putative nucleic acid genome of 0.75×10^6 in the case of a single-stranded nucleic acid and 1.6×10^6 for a double-stranded one. The controversy between Alper and Rowher focused particularly on the questions of whether the extrapolation procedure and inclusion of data taken under different experimental conditions are at least as doubtful as the application of target size formalism to an agent whose molecular nature is unknown [16–18]. Prusiner and his colleagues took up the issue employing highly purified scrapie material and applying advanced methods of dosage adjustment, and essentially confirmed Alper's early data. If a hypothetical nucleic acid component exists, these studies estimate its size to be a 4- to 5-nucleotide single-stranded or 30- to 45-bp double-stranded nucleic acid. The target size for an infectious protein genome, determined with homogenates of infected brain microsomes, purified prion rods, and prions dispersed in liposomes in these studies, was in all of the cases approximately 55 kDa [19; 20]. Recently, UV irradiation was applied to different scrapie strains, in which it was shown that UV inactivation does not affect the strain properties (see below).

Table 4.1: Stabilities of the scrapie agent and viroids after chemical and enzymatic treatment.

A	Chemical Treatment	Concentration	PSTVd	Scrapie agent
	Et_2PC	10–20 mM	(−)	+
	NH_2OH	0.1–0.5 M	+	−
	Psoralen (AMT)	10–500 μg ml^{-1}	+	−
	Phenol	Saturated	−	+
	SDS	1–10%	−	+
	Zn^{2+}	2 mM	+	−
	Urea	3–8 M	−	+
	Alkali	pH 10	(−)	+
	KSCN	1 M	−	+
B	**Enzymatic treatment**	**Concentration**	**PSTVd**	**Scrapie agent**
	RNase A	0.1–100 μg ml^{-1}	+	−
	DNase	100 μg ml^{-1}	−	−
	Proteinase K	100 μg ml^{-1}	−	+
	Trypsin	100 μg	−	+

+ inactivation; − no change; (−) small change; PSTVd: potato spindle tuber viroid. Reproduced in modified form from Diener et al. [21]

Scrapie infectious material in its highly purified form has been subjected to procedures that distinguish proteins from nucleic acids in their response to hydrolysis, chemical modification, denaturation, or shearing. The influence of these treatments on scrapie activity was then compared to that of viroid infectivity as the prototype of a small pathogen of pure nucleic acid [11]. In similar studies, virus and phages served as the controls. The results of the scrapie–viroid comparison, summarized in Table 4.1 [21], indicate that scrapie-infectious material – in its purified form, the so-called prions, in particular – resist inactivation by procedures that hydrolyze, modify, or shear nucleic acids, but are inactivated by some treatments that denature or modify proteins.

Scrapie infectivity is also resistant to DNase or digestion with RNase, but sensitive to prolonged digestion by proteinase K and trypsin, an observation that originally led to the prion hypothesis (see Chapter 3). These observations have been confirmed in crude homogenates, purified prions, prion-containing liposomes, and in scrapie material propagated *in vitro* in neuroblastoma cells.

4.5.2 Differences in the patterns of nucleic acids from infected and uninfected tissue

A second experimental strategy is based on the premise that an infection with an agent with a nucleic acid component introduces exogenous nucleic acid into cells. Therefore, differences in the nucleic acid profiles of infected *vis-à-vis* uninfected cells or tissues should be distinguishable. Quite different approaches evolved from this strategy.

Marsh and coworkers separated membrane vesicles with high titers of infectivity and treated them with DNase and RNase. After 3'-end labeling of nucleic acids from the vesicles, the samples were analyzed by gel electrophoresis [22]. These experiments documented small RNAs ranging in size up to approximately 100 nucleotides in infected as well as in uninfected tissue. Initially, one 4.3 S RNA was detected only in the samples from infected animals, but in subsequent work with modified procedures, a significant difference could not be confirmed.

Another way to search for a scrapie-specific nucleic acid is to construct a cDNA library from RNA extracted from the brains of infected animals. Then, the library is differentially screened by way of hybridization in order to identify genes whose expression is modulated by a scrapie infection or to find an exogenous scrapie-specific nucleic acid. Several RNAs that increase during infection were identified in this way, but all of them have turned out to be transcripts of host genes [23; 24]. Interestingly, some mitochondrial sequences were found with a high preference. Aiken and colleagues [25] identified ssDNA of about 450 nucleotides from the displacement loop of mitochondrial DNA in purified prion preparations and estimated it to amount to 10 molecules per infectious unit. In any case, scrapie-induced, but not scrapie-infectious nucleic acids were identified. In other words, these nucleic acids were the consequence, not the cause of a scrapie infection.

Much effort has been devoted to searching directly for a scrapie-specific nucleic acid in highly purified infectious material. In some instances one of the ancillary aims of the experiment has been to show that infectivity is or is not associated with nucleic acids. Manuelidis and coworkers [26] found in one such analysis of the sedimentation properties of disaggregated CJD infectious material that infectivity correlated better with cosedimenting nucleic acids than with PrP, but a smaller portion of PrP that cosedimented with infectivity may have contained the fraction that had, after the disaggregation procedure, the critical conformation required for infectivity.

The nucleic acids cosedimenting with infectivity were as large as several kb. Characterizing these nucleic acids in more detail showed some similarity to retroviral RNA and led to the suggestion that the CJD agent resembles retroviruses with transforming activity. However, degrading the large nucleic acids did not lead to a loss of infectivity, and in noninfected tissue those nucleic acids were also found. Thus, the large nucleic acids found in infected samples could not be confirmed as scrapie specific [28].

Attempts to clone a scrapie-specific nucleic acid were carried out starting with highly purified prion preparations in order to minimize contamination by normal hamster nucleic acid. Oesch and colleagues [29] added 10^{11} molecules of globin mRNA as an internal reference to 10^8 infectious units of the purified scrapie sample and produced approximately 10^6 independent recombinants in a bacteriophage λ vector. Assuming that the nucleic acids from the 10^8 infectious units (i. e., at least 10^8 conjectured scrapie-specific nucleic acid molecules) were cloned with the same efficiency as the β-globin mRNA, the ratio of scrapie-specific to β-globin-related clones were expected to be at least $1{:}10^3$. A number of 30,000 plaques were screened by way of hybridization with the globin probe, and all clones with inserts were identified as β-globin with the exception of four clones originating most probably from impurities of the sample. Scrapie-specific nucleic acids have also failed to be identified by this approach using improved cloning procedures and prion preparations of even greater purity (M. Fischer and Ch. Weissmann, personal communication).

4.5.3 Direct analysis of nucleic acids in purified infectious material by physicochemical methods

All nucleic acid analyses described above were designed to detect nucleic acids if they existed. In the following, an approach is described in some detail, which allows finding or excluding a scrapie-specific nucleic acid [30]. The principle of the experiments is depicted schematically in Fig. 4.3. Two types of data were determined for highly purified infectious material: (i) the number of nucleic acid molecules (left branch Fig. 4.3), and (ii) the number of infectious units (right branch Fig. 4.3). Both types of analyses were performed after the destruction of all possible host nucleic acids that were still present in the preparation. From the two results the particle-to-infectivity ratio (P/I) was calculated (Fig. 4.3). It was essential that all types of nucleic acids were quantitatively registered, i. e., DNA and RNA, double-stranded and single-stranded, linear and circular, and chemically modified and, what turned out experimentally

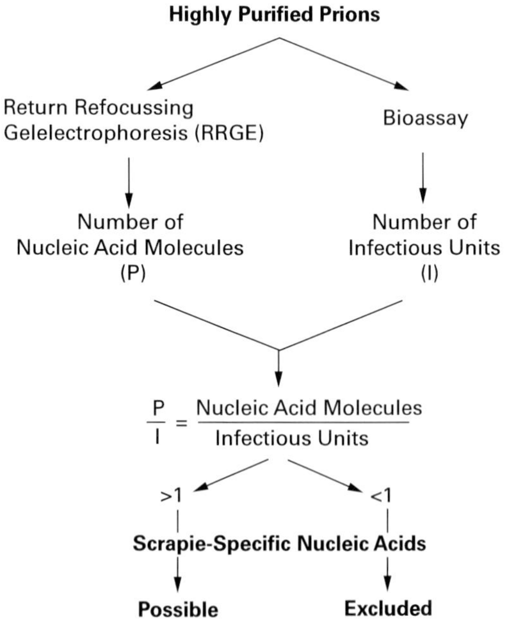

Fig. 4.3: Principle of the biophysicochemical analysis of potential scrapie-specific nucleic acids.

to be most difficult, nucleic acids of heterogeneous size.

In order to safely detect at least semiquantitatively all types of nucleic acids, only gel electrophoresis with silver staining was applicable. Although radioactive labeling and mass spectrometry are more sensitive, both methods depend too heavily on the type of nucleic acid and would not yield quantitative results. The nucleic acids of heterogeneous size were most difficult to analyze quantitatively. They do not show up in sharp bands but in a diffuse and weakly stainable smear. A particular technique, known as return refocusing gel electrophoresis (RRGE), had to be developed for the quantitative analysis [30; 31]. In this technique the gel slot is cut into segments that contain nucleic acids of a well-defined size range. Typical cutting edges were 4, 22, 40, 58, 81, 104, and 127 nucleotides as measured from the start site of the slot. The segments were located at the bottom of a second gel and polymerized into a gel matrix of the same composition as in the first run, typically 9 or 15% polyacrylamide. The second gel electrophoresis was carried out in the opposite direction, i. e., bottom to top so that all nucleic acids of one and the same gel segment are focused in one sharp band that can be stained by silver. This band represents the sum of all nucleic acids in that segment, for example, all nucleic acids between 40 and 58 nucleotides in length. Favorably, the return run was carried out in the presence of 1% SDS, which does not prevent the focusing of nucleic acids but denatures the background proteins and separates them well from the nucleic acids.

Several protocols were developed to prepare prion samples that were still highly infectious but depleted from nucleic acids as much as possible. These protocols contained different combinations of sucrose gradient centrifugation, ultrafiltration, and dispersion in liposomes to obtain highly purified prions. The nucleic acid degradation steps contained hydrolysis by Zn^{2+}, and treatment by several cocktails of DNase and RNase. In these experiments the scrapie infectivity was monitored by an incubation time interval procedure at all of the steps of the preparation. Based on the amount of nucleic acid estimated from RRGE and the titers of the prion fractions prior to boiling in SDS, the ratio of nucleic acid molecules to ID_{50} units (P/I) was calculated.

What type of results can be expected? Two possibilities are shown schematically in Fig. 4.3. If a P/I ratio greater than unity (P/I > 1) is found for a particular nucleic acid, at least one nucleic acid molecule is present for every infectious unit, and the nucleic acid might be essential for infectivity. If, however, a ratio of P/I < 1, infectious units exist without a single nucleic acid molecule, and this type of nucleic acid is, therefore, definitely excluded from being essential for infectivity. If the nucleic acids detected were related to scrapie infectivity, one of two alternative paradigms would be correct. First a putative scrapie-specific nucleic acid of uniform length might be hidden among an ensemble of background nucleic acids. Conventional PAGE would not have detected such a scrapie-specific nucleic acid, even if it were present in sufficient amounts. Second, one must consider the possibility of a scrapie-specific nucleic acid, which is heterogeneous in length. In Fig. 4.4a the number of nucleic acid molecules per ID_{50} unit are plotted as a function of their length as estimated from the individual gel sections. In this plot, the calculation was based upon the first paradigm, i. e., a well-defined scrapie-specific nucleic acid among the heterogeneous background nucleic acids. Data from several published experiments [30; 31] are presented in the plot, and the straight line in the double-logarithmic plot is a mere empirical interpolation of the experimental data without a functional relationship.

A significant decrease in the number of nucleic acid molecules per ID_{50} unit was found as the size of the nucleic acid increased. For small nucleic acid molecules (e. g., 20 nt), about 10 to several 100 molecules/ID_{50} were estimated. If the scrapie-specific nuclei acid were longer (> 76 nt), the P/I ratio would fall below unity. Above 200 nt, the P/I ratio drops several orders of magnitude below unity. If a heterogeneous scrapie-specific nucleic acid is assumed, the P/I ratio has to be integrated over the size range of the putative scrapie genome. Even in this un-

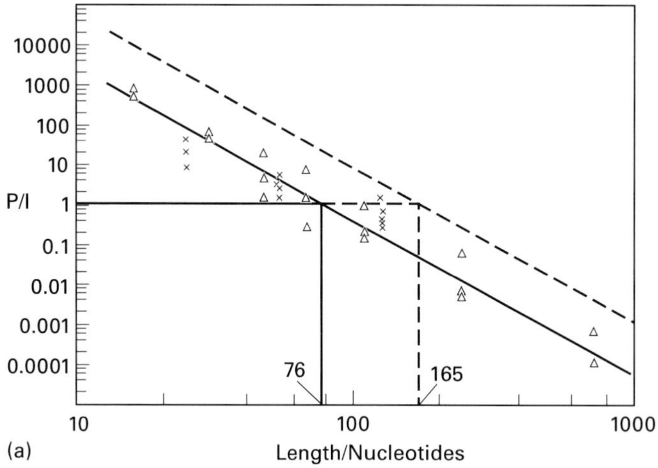

Fig. 4.4: Nucleic acid analysis of highly purified prion samples. Relationship of the particle-to-infectivity (P/I) ratio to the average length of a nucleic acid species from the determinations of residual nucleic acids by RRGE (see text) on five independent prion samples.

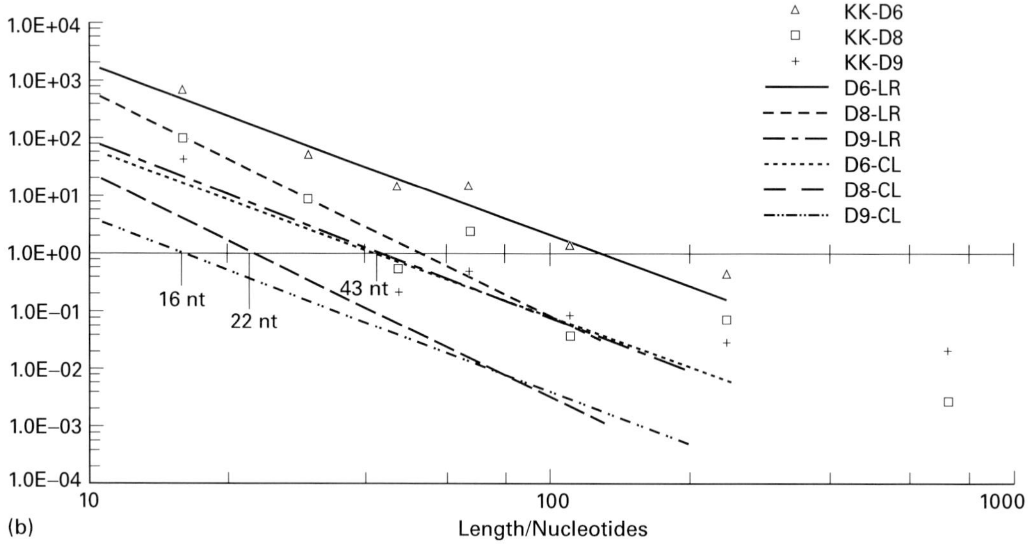

(a) The relationship is linear over a size range of 10–1,000 nucleotides with an intercept of approx. 76 nt for a P/I of unity. Only small nucleic acids < 80–100 nt have P/I > 1. Data (x) were obtained from Meyer et al. [30]; (Δ) from Kellings et al. [31]. The relationships were calculated as follows, using a gel fragment of RRGE as an example: if the fragment contained 450 pg of nucleic acids in the size range of 54–79 nt, and assuming a continuous distribution of different sizes, the 26 species in this class will have an average MW of 22×10^3 and there will be 4.7×10^8 molecules (mol):

$$450 \times 10^{-12}\,\text{g} \times \frac{6 \times 10^{23}\,\text{mol}}{26 \times 22 \times 10^3\,\text{g}} = 4.7 \times 10^8\,\text{mol}$$

of a particular size in this ensemble. Since the starting sample contained $10^{8.7}$ (5×10^8) ID_{50}, the P/I is approximately 1 for a hypothetical discrete scrapie specific nucleic acid in the ensemble. The dashed lines represent the maximum error, i. e. one log ID_{50} for the bioassay and a factor of two in the nucleic acid content.

(b) P/I values of samples prepared according to Safar et al. [1] and the effect of clearance on the P/I ratio with dependence on the average length of the residual nucleic acid. Symbols △, □, and + represent the experimental values from the experiments KK-D6, D8, D9, respectively; straight lines (D6, D8, D9-LR) refer to the corresponding linear regressions, and straight lines (D6, D8, D9-CL) include the correction from clearance. According to 96% clearance, all original data (see text) are reduced to 4% and redrawn to determine the length at P/I = 1. Figure from Safar et al. [1].

precedented case, one may calculate from the plot that all molecules > 240 nt taken together do not add up to a P/I ratio > 1 [31].

In a more recent study [1] protocols for purifying prions and for depleting the highly infectious samples from nucleic acids were further developed, however, with only moderate success. When the amount of residual nucleic acids of several independent experiments was plotted against length as shown in Fig. 4.4a the interpolation yielded a ratio of P/I = 1 at about 50 nucleotides in length (Fig. 4.4b). Consequently, only nucleic acids < 50 nucleotides could not be excluded as essential components of scrapie infectivity. In the same study, the clearance effect of PrPres and infectivity after inoculation was determined (see Chapter 8). It was found that 1 day after inoculation only 4% of PrPres and only 1% of infectivity were left in the brain, whereas the major part was washed away. If it is assumed that only the noncleared portion of PrPres and its coaggregated residual nucleic acids might be essential for infectivity, the data as in Fig. 4.4b (lines with experimental points) have to be reduced to 4% of the nucleic acids (Fig. 4.4b lines without experimental points) and the limit of P/I = 1 is determined around 20 nucleotides.

In summary, all nucleic acids larger than 20–30 nucleotides were excluded by direct biophysical methods. This conclusion was confirmed in its entirety by recent evidence of synthetic prions consisting of only PrP, which proved that no nucleic acids were required for infectivity [9].

For the sake of intellectual clarity, small nucleic acids are not required for infectivity, but this does not exclude them as strain determinants (see Chapter 12). The size of the residual nucleic acids of 20–30 nucleotides might suggest small interfering or microRNAs; in that speculation the synthetic prion might be a new strain without a small RNA as a strain determinant and most other strains have one. A small RNA of host origin as a strain determinant was actually proposed by Weissmann in his prion-propagation theory [32]. In recent nucleic acid analyses (1), two different scrapie strains Sc263 and H139 were inactivated by UV irradiation. It was shown that during inactivation their strain properties, such as strain-specific incubation time, do not change significantly in the sense that their typical differences would be destroyed. It was concluded that the strain-specific properties are also not encoded by small nucleic acids, and that residual nucleic acids that are still found in highly purified prions must be impurities of a host origin.

Finally, it is noteworthy that the clear conclusion drawn above does not contradict other reports in the literature [33–35] that nucleic acids might be involved in the transition of PrP^{C} into PrP^{Sc}. Those nucleic acids might act as host factors in the transition process, but they are definitely not components of an infectious agent. This chapter solely described the nature of the agent, whereas the details of the transition process are described in Chapter 8.

References

[1] Safar JG, Kellings K, Serban H, et al. Search for a prion-specific nucleic acid. J Virol 2005; 79:10796–10806.

[2] Merz PA, Somerville RA, Wisniewski HM, et al. Abnormal fibrils from scrapie infected brain. Acta Neuropathol 1981; 54:63–74.

[3] Sigurdsson B. Rida – a chronic encephalitis of sheep: with general remarks on infections which develop slowly and some of their special characteristics. Br Vet J 1954; 110:341–354.

[4] Gajdusek DC and Zigas V. Degenerative disease of the central nervous system in New Guinea – the epidemic occurrence of kuru in the native population. N Engl J Med 1957; 257:974–978.

[5] Diringer H, Beekes M, Oberdieck U. The nature of the scrapie agent: the virus theory. Ann NY Acad Sci 1994; 724:246–258.

[6] Manuelidis L. Dementias, neurodegeneration, and viral mechanisms of disease from the perspective of human transmissible encephalopathies. Ann NY Acad Sci 1994; 724:259–281.

[7] Safar JG. Infectious amyloid, prions, unconventional viruses, and disease. Neurobiol Aging 1994; 15:279–281.

[8] Baldauf E, Beekes M, Diringer H. Evidence for an alternative direct route of access for the scrapie agent to the brain bypassing the spinal cord. J Gen Virol 1997; 78:1187–1197.

[9] Legname G, Baskakov IV, Nguyen H-OB, et al. Synthetic mammalian prions. Science 2004; 305:673–676.

[10] Dickinson AG and Outram GW. Genetic aspects of unconventional virus infections: the basis of the virino hypothesis. Ciba Found Symp 1988; 135:63–83.
[11] Riesner D and Gross HJ. Viroids. Annu Rev Biochem 1985; 54:531–564.
[12] Alper T, Cramp WA, Haig DA, et al. Does the agent of scrapie replicate without nucleic acid? Nature 1967; 214:764–766.
[13] Gibbs CJ Jr, Gajdusek DC, Latarjet R. Unusual resistance to ionizing radiation of the virus of kuru, Creutzfeldt–Jakob disease, and scrapie. Proc Natl Acad Sci USA 1978; 75:6368–6270.
[14] Alper T, Haig DA, Clarke MC. The scrapie agent: evidence against its dependence for replication of intrinsic nucleic acid. J Gen Virol 1978; 41:503–516.
[15] Rohwer RG. Virus-like sensitivity of the scrapie agent to heat inactivation. Science 1984; 223:600–602.
[16] Rohwer RG. Scrapie infectious agent is virus-like in size and susceptibility to inactivation. Nature 1984; 308:658–662.
[17] Alper T. Scrapie agent unlike viruses in size and susceptibility to inactivation by ionizing or ultraviolet radiation (letter). Nature 1985; 317:750.
[18] Rohwer RG. Estimation of scrapie nucleic acid MW from standard curves for virus sensitivity to ionizing radiation (letter). Nature 1986; 320:381.
[19] Bellinger Kawahara C, Diener TO, McKinley MP, et al. Purified scrapie prion resists inactivation by procedures that hydrolyze, modify, or shear nucleic acids. Virology 1987; 160:271–274.
[20] Bellinger Kawahara CG, Kempner E, Groth D, et al. Scrapie prion liposomes and rod exhibit target sizes of 55,000 Da. Virology 1988; 164:537–541.
[21] Diener TO, McKinley MP, Prusiner SB. Viroids and prions. Proc Natl Acad Sci USA 1982; 79:5220–5224.
[22] German TL, McMillan BC, Castle BE, et al. Comparison of RNA from healthy and scrapie-infected hamster brain. J Gen Virol 1985; 66:839–844.
[23] Duguid JR, Rohwer RG, Seed B. Isolation of cDNAs of scrapie-modulated RNAs by subtractive hybridization of a cDNA library. Proc Natl Acad Sci USA 1988; 85:5738–5742.
[24] Duguid JR, Bohmont CW, Liu NG, et al. Library subtraction of *in vitro* cDNA libraries to identify differentially expressed genes in scrapie infection. Nucleic Acids Res 1989; 18:2789–2792.
[25] Aiken JM, Williamson JL, Borchardt LM, et al. Presence of mitochondrial D-loop DNA in scrapie-infected brain preparation enriched for the prion protein. J Virol 1990; 64:3265–3268.
[26] Sklaviadis TK, Manuelidis L, Manuelidis EE. Physical properties of the Creutzfeldt–Jakob Disease agent. J Virol 1989; 63:1212–1222.
[27] Akowitz A, Sklaviadis T, Manuelidis EE, et al. Nuclease-resistant polyadenylated RNAs of significant size are detected by PCR in highly purified Creutzfeldt–Jakob disease preparations. Microb Pathog 1990; 9:33–45.
[28] Murdoch GH, Sklaviadis T, Manuelidis EE, et al. Potential retroviral RNAs in Creutzfeldt–Jakob disease J Virol 1990; 64:1477–1486.
[29] Oesch B, Groth DF, Prusiner SB, et al. Search for a scrapie-specific nucleic acid: a progress report. In: Bock G, Marsch J, editors. Novel infectious agents and the central nervous system, Ciba Foundation Symposium 1988:209–223.
[30] Meyer N, Rosenbaum V, Schmidt B, et al. Search for a putative scrapie genome in purified prion fractions reveals a paucity of nucleic acids. J Gen Virol 1991; 72:37–49.
[31] Kellings K, Meyer N, Mirenda C, et al. Further analysis of nucleic acids in purified scrapie prion preparations by improved return refocusing gel electrophoresis (RRGE). J Gen Virol 1992; 73:1025–1029.
[32] Weissmann C. A "unified therory" of prion propagation. Nature 1991; 352:679–683.
[33] Cordeiro Y, Machado F, Juliano L, et al. DNA converts cellular prion protein into the beta-sheet conformation and inhibits prion peptide aggregation. J Biol Chem 2001; 276:49400–49409.
[34] Gabus C, Derrington E, Leblance P, et al. The prion protein has RNA binding and chaperoning properties characteristic of nucleocapsid protein NCP7 of HIV-1. J Biol Chem 2001; 276:19301–19309.
[35] Supattapone S. Prion protein converson *in vitro*. J Mol Med 2004; 82:348–356.
[36] Patrick MH and Rupert CS. The effects of host-cell reactivation of assay of UV-irrediated *Haemophilus influenzae* tramsforming DNA. Photochem Photobiol 1967; 6:1.
[37] Setlow JK and Duggan DE. The resistance of *Micrococcus radiodurans* to UV-radiation. Biochem Biophys Acta 1964; 87:664.
[38] Winkler U. Über die fehlende Photo- und Wirtszellreaktivierbarkeit des UV-inaktivierten RNA-Phagen FR. Photochem Photobiol 1964; 3:37.

5 Folding of the Recombinant Prion Protein

Rudi Glockshuber, Jan Stöhr, and Detlev Riesner

5.1 Introduction

Structural and biophysical studies of the cellular isoform of the prion protein PrP^{C} and the scrapie isoform PrP^{Sc} are of central interest in prion research, as they are expected to provide the basis for understanding the molecular mechanism of PrP^{C} conversion to PrP^{Sc}. Prion proteins are secretory cell surface proteins of approximately 210 amino acids. All known mammalian PrP sequences are strikingly similar and pairs of sequences are generally more than 90% identical [1; 2]. Further characteristic features of PrPs are their posttranslational modifications, which include two N-glycosylation sites at Asn181 and Asn196, a single disulfide bond between Cys179 and Cys214, and a glycosylphosphatidyl inositol (GPI) membrane anchor at the C-terminal residue 231 (residue 23 is the N-terminal amino acid of mature PrP; amino acid numbering according to human PrP [1]). Although PrP^{C} and PrP^{Sc} possess identical covalent structures [3–5], they differ considerably in their biochemical and biophysical properties. PrP^{C} is monomeric, soluble in non-denaturing detergents, and sensitive to proteases, whereas PrP^{Sc} forms detergent-insoluble aggregates which are partially proteinase K-resistant. The fact that most of the subunits of PrP^{Sc} are uniformly degraded to an N-terminally truncated form termed PrP27–30 that retains infectivity and ranges from approximately residue 90 to 231, provides strong evidence that PrP^{Sc} is an ordered oligomer [6–8]. Circular dichroism (CD) and infrared spectroscopy data showed that PrP^{C} is rich in α-helical structure [9–11] while PrP^{Sc} has high β-sheet content [10; 12–14]. The protein-only hypothesis states that the prion is identical to PrP^{Sc}, and that PrP^{Sc} propagates through recruitment of PrP^{C} from newly infected cells [15–17]. A more detailed report on the features of PrP^{Sc} is given in Chapter 8.

Studies of the structure of PrP^{C} and alternative conformations of the protein are crucial for understanding the molecular events underlying the formation of PrP^{Sc}. However, difficulties in purification of large amounts of PrP^{C} from its natural source and the insolubility of PrP^{Sc} have long prevented high-resolution structure analysis for both isoforms. In 1996 and 1997, methods became available for the production of large quantities of structurally homogeneous, monomeric recombinant PrP and PrP fragments in *Escherichia coli*, either by secretory expression in the bacterial periplasm [18], or by oxidative refolding from cytoplasmic inclusion bodies [19–23]. Since then, the three-dimensional structures of fragments of recombinant prion proteins from mice [24; 25], hamsters [20; 22], humans [26], and cattle [27] have been determined in solution by nuclear magnetic resonance (NMR) spectroscopy. As expected from the high degree of sequence identity, all structures proved to be very similar. PrP is composed of two structurally distinct moieties. The C-terminal residues 125–231 form a globular domain with a unique fold consisting of three α-helices and a short antiparallel β-sheet, while the N-terminal segment (residues 23–124) is flexibly disordered. The details of the NMR structures are outlined in Chapter 6. The three-dimensional structures of the recombinant prion proteins are in full agreement with all previous physical data on the secondary structure of mammalian PrP^{C}. As a result, it is now generally accepted that solution structures of recombinant proteins, which lack all posttranslational modifications of PrP^{C} except for the disulfide bond Cys179–Cys214, are very similar. A comparison of 1D NMR data of natural, bovine full-length PrP^C with its recombinant produced in *E. coli* reveals very similar structural data [28]. Thus, it can be assumed that posttranslational modifications do not have a major effect on the three-dimensional structure of PrP^C.

This chapter focuses on biophysical studies on folding, stability, and alternative conformations of recombinant prion proteins produced in *E. coli*. It should be noted that this chapter is restricted to mammalian prion proteins – prion proteins from yeast and other species are not discussed. The mammalian prion protein constructs discussed in this chapter include the fragments PrP90–231 from hamster and human PrP, which represent the protease-resistant core of PrP^{Sc}, the full-length murine prion protein PrP23–231, and the structured C-terminal domain PrP121–231 of murine PrP^C. This domain was initially identified by its resistance to proteolytic degradation during periplasmic expression of longer PrP fragments in *E. coli* [18] and was the first PrP segment from which a three-dimensional structure could be obtained [24].

5.2 Folding of recombinant PrP^C

In this section, the folding and the presently known conformational states of disulfide-intact (oxidized) recombinant prion proteins are discussed. Section 5.3 will deal with biophysical studies on folding and aggregation of recombinant PrP with a reduced, i. e., opened, disulfide bond. On the basis of known three-dimensional structures of prion proteins in solution, one would expect that folding of disulfide-intact full-length PrP23–231 and its N-terminally truncated fragment PrP90–231 is restricted to the structured C-terminal segment 125–231. Indeed, folding studies on the disulfide-intact proteins murine PrP23–231, murine PrP121–231, and human PrP90–231 yielded very similar results that are in agreement with the working model that folding of the structured C-terminal domain is essentially independent of the flexibly disordered N-terminal tail. In those studies, mostly fluorescence emission and circular dichroism (CD) spectra were analyzed. Figure 5.1A shows that the CD-spectra of full-length murine PrP and its C-terminal segment, 121–231, are very similar. The process of folding was followed by molar ellipticity at 222 nm (θ_{222}) as shown in Fig. 5.2.

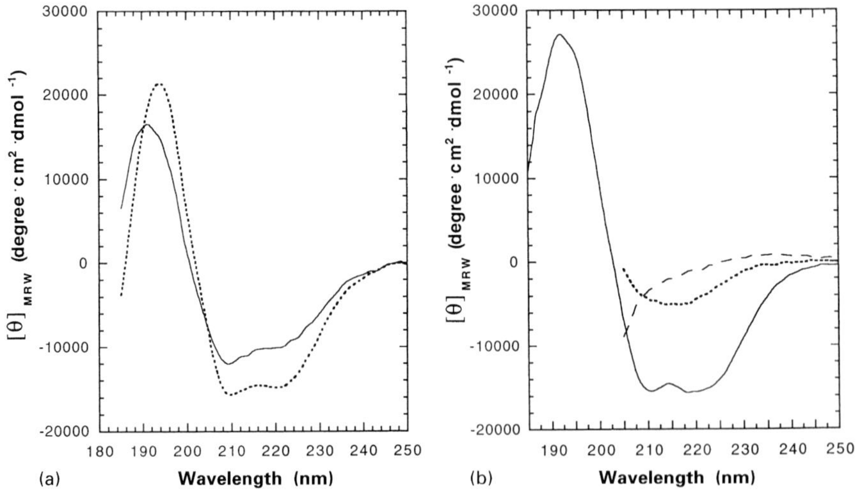

Fig. 5.1 CD spectra of recombinant mouse PrP. **(a)** Folded states of full-length PrP23–231 (———) and the C-terminal domain, 121–231 (- - -). **(b)** Equilibrium folding intermediates of PrP121–231 at pH 4.0; native state in the absence of urea (—), unfolding intermediate at 3.5 M urea (- - -), and denatured state at pH 2.0 and 3.5 M urea (—— —— ——). Figure from R. Glockshuber [64].

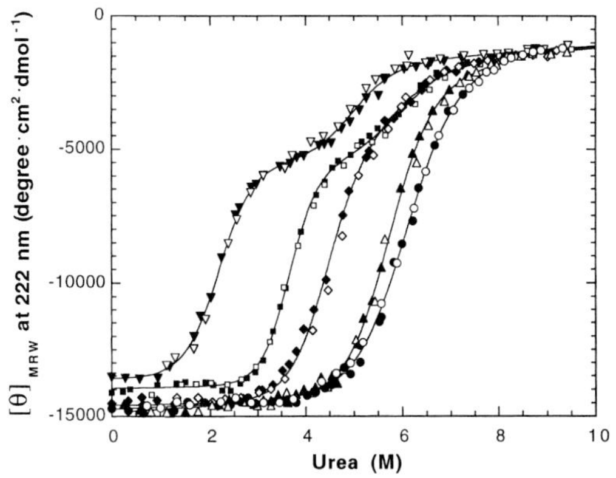

Fig. 5.2: Urea-induced equilibrium unfolding of murine PrP121–231 followed by CD signal at 222 nm. Unfolding experiments were performed at 22 °C and a constant ionic strength of 88 mM. Different unfolding curves refer to pH 4.0 (▼), 4.5 (■), 5.0 (◆), 6.0 (▲), and 7.0 (●). Open symbols refer to the corresponding refolding transitions, which coincide with the unfolding experiments and demonstrate reversibility. Figure from R. Glockshuber [64].

The folding of oxidized, recombinant prion proteins can be summarized as follows. At pH range 4.0–8.0 (in low ionic strength at pH 4–5), unfolding experiments induced by guanidinium chloride (GdmCl) and urea revealed that murine PrP121–231, human PrP90–231, and murine PrP23–231 show completely reversible transitions which are consistent with the two-state model of folding, i. e., completely unfolded and completely folded molecules are in equilibrium at any given denaturant concentration [18; 29–33]. Transitions according to the two-state model are shown in Fig. 5.2 at 88 mM ionic strength and pH 5.0, 6.0, and 7.0. A comparison between the urea-induced unfolding transitions of full-length murine PrP23–231 and its isolated C-terminal domain PrP121–231 showed that the full-length protein was reproducibly slightly less stable: $1G_{fold} = -25.5$ kJ mol^{-1} compared to –29.7 kJ mol^{-1} [32]. Both proteins have identical transition midpoints (6.2–6.3 M urea), but the cooperativity of folding is slightly lower in the case of the full-length protein. As no interactions between the C-terminal domain and the unstructured N-terminal segment were observed in PrP23–231 [25], the lower stability of the full-length protein may result from contacts between residues from the N- and C-terminal parts of PrP23–231 in the unfolded state. Such a residual structure in the unfolded state would reduce the difference in the solvent-accessible surface area between the folded and unfolded state and decrease folding cooperativity [34].

The complete reversibility of unfolding of PrP121–231, PrP90–231, and PrP23–231 has principal implications within the framework of the protein-only hypothesis if one extrapolates the results from recombinant proteins to natural PrP^C. If PrP^C and PrP^{Sc} do indeed have identical covalent structures [4], both PrP^C and PrP^{Sc} will yield identical unfolded forms in the presence of high concentrations of the denaturants urea and GdmCl. This implies that, after reconstitution in vitro, one would always obtain folded PrP^C, independent of whether the experiment was started from PrP^C or PrP^{Sc}. This would explain why all attempts to reconstitute infectivity after complete solubilization of infectious PrP^{Sc} with high concentrations of GdmCl or urea have failed [35]. Reversibility implies, however, that no long-term (days or weeks) aggregation processes of PrP are involved. It was shown later [36] that infectivity could be induced from recombinant PrP if inoculated with amyloid fibers from recombinant murine PrP89–230 (compare Chapter 8).

Further studies fully confirmed the two-state character of the folding of oxidized recombinant PrP. Analysis of the kinetics of the folding of the tryptophane variant F175W of murine PrP121–231 at pH 7.0 revealed extremely dynamic unfolding equilibria, which could only be observed with stopped-flow fluorescence at 4 °C [37]. The kinetics of unfolding and refolding of this mutant domain were in complete agreement with the two-state model and the equilibrium unfolding data. No kinetic intermediates were observed during refolding of the urea-denatured variant F175W of PrP121–231. Indeed, folding of the structured C-terminal PrP domain is one of the fastest protein folding reactions described so far, with an extrapolated folding half-life of 170 μs, whereas unfolding at pH 7.0 and 4 °C takes 4.6 min. This appears to exclude the recruitment of transiently populated PrP folding intermediates as precursors of PrP^{Sc}, which would be expected to cause significantly slower folding. However, a kinetic folding intermediate was detected in the refolding of human PrP90–231 [38]. Under other conditions the α-helical structure was observed as a kinetic intermediate when murine PrP89–230 was refolded from the fully denatured state in 10 M urea; a monomeric α-helical structure was formed very fast, which was transformed very slowly into β-sheet rich oligomers and fibrils. These experiments also showed a hystereis between the unfolding and refolding process of PrP. Only a model with two independent transitions, one between the folded (alpha) and unfolded state and the other between the beta and unfolded state could fit the obtained data [39] (compare Chapter 8). Kinetic folding intermediates have also been detected as source in amyloid formation of human lysozyme mutants [40]. The recruitment of kinetic folding intermediates requires complete unfolding of PrP. The same conclusion was drawn from the analysis of backbone amide hydrogen exchange in human PrP91–231 by NMR spectroscopy [31]. These studies fully confirmed the two-state character of PrP folding at pH 5.5 and low ionic strength, and revealed that only a small portion of residual structure within about 10 residues around the disulfide bond is retained in the unfolded state [31].

Overall, the C-terminal domain of PrP^{C} is a comparatively stable protein, showing perfect two-state behavior of folding at low ionic strength, and is as 'normal and well-behaved' as a protein can be. No property of the domain at physiological pH indicates that it is capable of adopting an entirely different conformation at a lower pH. Such an alternative conformation was first observed as a plateau phase in the GdmCl-induced unfolding transition of hamster PrP90–231 at pH 5.0 [41]. Thereafter, studies on the pH dependence of PrP folding showed that human PrP90–231 at pH 3.6–5.0 in the presence of GdmCl [29; 42], and murine PrP121–231 and PrP23–231 at pH 4.0–4.5 in the presence of urea [43], populate an acid-induced unfolding intermediate at medium denaturant concentration. All these intermediates showed a strong content of β-sheet structure in CD measurements. In the case of murine PrP121–231, the apparent pK_a of the transition from two-state to three-state folding characteristics in the presence of urea is 4.5, indicating that protonation of acidic side chains is responsible for stabilization of the intermediate [43]. Those two-step unfolding curves are shown in Fig. 5.2 for pH 4.0 and pH 4.5. Formation of the intermediate is an intrinsic property of the C-terminal PrP domain 121–231 and independent of the flexible N-terminal segment 23–120. In the case of human PrP91–231, the apparent molecular mass of a dimer was obtained for the intermediate when populated in 1 M GdmCl at pH 4.0 [33].

Importantly, prolonged incubation of the acid-induced unfolding intermediate of oxidized human PrP90–231 in 1 M GdmCl at pH 3.6–5.0 yielded specific fibrillar aggregates that showed increased proteinase K resistance [44]. Experiments under different conditions (neutral pH, presence of monovalent ions) revealed that folding into the more stable PrP^{Sc}-like isoform requires prolonged incubation at destabilizing conditions [39; 45]. In long-term incubation studies, it could be demonstrated that infectivity was generated [36] (compare Chapter 8). In other studies, with oxidized human PrP91–231, it was found that the protein could be converted from the α-helical- to the β-sheet-rich conformation after unfolding and cooling [42].

In summary, the folding of oxidized, recombinant PrP can be described by the following diagram.

pH 4–8, low ionic strength at pH 4–5	$N \rightleftharpoons U$
pH 4, ionic strength > 50 mM	$nN \rightleftharpoons I_n \rightleftharpoons nU$

N, I, and U correspond to the native, intermediate, and unfolded state, respectively. Data on human PrP91–231 indicate that the acid-induced equilibrium intermediate I_n (populated in the presence of GdmCl) is dimeric with $n = 2$ [33].

In summary, this chapter discusses the reversible folding process of recombinant PrP as a model for PrP^C. It is emphasized that knowledge of this process is a prerequisite for understanding the transition from PrP^C to the pathogenic form PrP^{Sc}. In most experiments acidic conditions were applied. As PrP^{Sc} accumulates in endosomes of scrapie-infected cells where acidic pH values of 4.0–6.0 are prevalent, it is tempting to speculate that a change from physiological to acidic pH during endocytosis of PrP^C triggers its conversion to PrP^{Sc} [43; 47]. As an extension of the studies described in this chapter, the refolding process of a PrP^C-like structure into a PrP^{Sc}-like structure will be analyzed in Chapter 8 under a wide variety of conditions including: pH, ionic strength, detergent, and type of PrP.

5.3 The role of the single disulfide bond of PrP

The hypothesis that mammalian prion diseases are only caused by an alternative conformation of the prion protein is essentially based on the fact that no differences between the covalent structures of PrP^C and PrP^{Sc} has been detected so far [4]. PrP^{Sc} subunits thus bear all posttranslational modifications that are found in PrP^C, even though fractions of the unglycosylated, mono- and doubly glycosylated forms of PrP differ in a prion strain-specific manner [7; 46]. The single intramolecular disulfide bond Cys179–Cys214 of PrP is also assumed to be quantitatively formed in both PrP^C and PrP^{Sc}, as denaturant-solubilized PrP^{Sc} only shows reactivity with thiol-specific reagents after treatment with reducing agents [48]. As mentioned above, disulfide bond formation in vivo or oxidative refolding of PrP in vitro has been essential for producing recombinant PrP^C for structure determination. In addition, the oxidized state of PrP was required for the PrP^{Sc}-mediated formation of protease-resistant PrP from PrP^C in vitro [49; 50], and incubation of mouse prions with 2-mercaptoethanol in the presence of SDS decreased scrapie infectivity [51]. This contrasts with the experiments carried out by Turk et al. [48] in which it was not easy to differentiate between inter- and intramolecular disulfide bonds in PrP^{Sc}. In a more recent analysis including gel electrophoresis, Caughey and his coworkers [52] showed that the disulfide bound is formed in PrP^{Sc} subunits. Furthermore, any potential reshuffling of the disulfide bound was suppressed during the cell-free conversion of PrP^C into a PK-resistant, i. e., PrP^{Sc}-like isoform, so that the PK-resistant isoform is generated without temporary breakage and subsequent re-formation of the disulfide bond. The formation of synthetic mammalian prions [36] also occurred under conditions where reshuffling reduction and formation of disulfide bonds was excluded. All these data strongly indicate that the disulfide bond is also permanently present during the conversion of PrP^C into PrP^{Sc} in vivo.

How can these results be interpreted in the light of data showing that reduced recombinant hamster PrP90–231 favors β-sheet-like structure and oligomerization [19; 41]? Furthermore, at low pH (pH 4.0) and low ionic strength, reduced recombinant human PrP91–231 is soluble, shows β-sheet-like CD spectra, possesses a significant degree of tertiary structure and shows the apparent molecular mass of the monomer in gel filtration experiments [33]. Increasing the ionic strength to > 100 mM at pH 4.0 triggers specific aggregation of reduced PrP91–231, leading to amyloid-like fibers with limited proteinase K resistance. Overall, the protease resistance of fibers of reduced PrP91–231 was significantly lower than that of PrP^{Sc}. Nevertheless, these fibers, produced in vitro from reduced recombinant protein, exhibited many similarities with PrP^{Sc}, but infectivity was not reported.

In the known NMR structures of murine, hamster and human recombinant PrP^C, the invariant disulfide bond Cys179–Cys214 is entirely buried in the interior of the structured, C-terminal domain and inaccessible to external reductants. This raises the question of how PrP^C would become reduced in vivo; the question would be even more relevant in the acidic environment of endosomes where disulfide exchange reactions are extremely slow and PrP^{Sc} appears to accumulate.

In summary, the oxidized and reduced forms of the recombinant proteins murine PrP23–231, human PrP91–231, and hamster PrP90–231 adopt entirely different, stable tertiary structures. Some of the properties of the reduced forms are reminiscent of PrP^{Sc}, but experimental evidence clearly argues against a transient reduction of the disulfide bond of PrP during the conversion process in vivo. A remote possibility, that cannot be excluded completely, would be that a small oligomer of reduced PrP serves as a nucleus for growth of large, oxidized PrP^{Sc} oligomers [33; 53]. It would be in accordance with the apparent absence of free thiol groups in PrP^{Sc}, because the fraction of reduced PrP in the oxidized PrP^{Sc} oligomer may be simply too small for detection.

At present, it appears much more probable that the ß-sheet rich, aggregated forms of reduced PrP represent aggregates of denatured protein as known for many proteins [54] and that the stable and buried, intramolecular disulfide bound prevents formation of those aggregates of denatured PrP. Moreover, the disulfide bond may contribute to a constant, structural core, which remains unchanged during the PrP^{C}–PrP^{Sc} conversion.

5.4 Influence of point mutations linked with inherited human prion diseases on the thermodynamic stability of recombinant PrP^{C}

All forms of inherited human prion diseases known so far are linked with dominant mutations in the gene encoding human PrP. This fact has been used as a strong argument in favor of the protein-only hypothesis, as the infectious prion agent spontaneously develops in affected individuals [for reviews, see 17; 55]. Three different phenotypes of inherited human prion diseases are known: the Gerstmann–Sträussler–Scheincker disease (GSS), familial Creutzfeldt–Jakob disease (CJD), and fatal familial insomnia (FFI); the pathology and genetics of these diseases are described in detail in Chapter 26. In addition to point mutations, there are also insertions of additional octapeptide repeats, which cause inherited prion diseases in humans. Octapeptide repeat segments are not required for generation and propagation of prions [56] and are not contained in the infectious, protease-resistant core of PrP^{Sc}. Therefore, the mechanism by which additional octapeptides lead to spontaneous prion formation is most probably quite different from all single amino acid replacements that cause prion diseases and is located within the minimum infectious unit of PrP^{Sc}, i. e., within segment 90–231. A possible mechanism would involve single amino acid replacements either decreasing the thermodynamic stability of PrP^{C} and thereby facilitating its conversion to PrP^{Sc} [57; 58] and/or increasing the stability of PrP^{Sc} [59]. A second model suggests that the mutations accelerate the formation of PrP^{Sc} by decreasing the activation energy barrier for the conformational transition [59], or accelerate the assembly of PrP^{Sc} monomers [32]. As PrP^{Sc} formation is favored at high concentrations of PrP^{C} [60], it could also be the case that the intracellular turnover of PrP is affected. Finally, the interaction between PrP and the postulated host "factor X", which is supposed to promote prion propagation in a species-specific manner [61; 62], might be improved.

Both the thermodynamic and the kinetic models of facilitated, spontaneous prion propagation in inherited prion diseases are in accordance with the observation that sporadic CJD has a very late onset (>65 years; Fig. 15.2), whereas inherited prion diseases are characterized by a comparably early onset – at the age of 30–40 years. The thermodynamic hypothesis of PrP^{C} destabilization can only apply to the structured domain PrP121–231. Consequently, the three GSS-related mutations P102L, P105L, and A117V cannot be interpreted a priori, and indeed, the thermodynamic stability of the variant P102L of human PrP90–231 has been found to be indistinguishable, within experimental error, from that of the wild type protein [30].

The eight disease-related replacements within the structured domain are located in the segment containing α-helices 2 and 3, which form the scaffold of the structured PrP^{C} domain [24; 25] (compare Chapter 6). A detailed structural analysis indicated that only four out of the eight amino acid replacements are likely to be destabilizing, namely D178N, T183A, F198S, and Q217R. Mainly loss of electrostatic interactions, loss of hydrogen bridges, and unfavorable space filling were identified [for details compare 63; 64].

To experimentally test the influence of the eight disease-related replacements on the stability of the structured C-terminal domain of human PrP, murine PrP121–231 may be used as a good model. In addition to its high structural similarity and 94% sequence identity to human PrP121–231, all eight replaced amino acids are identical in wild-type human and murine PrP, and the side chains which form direct contacts with these residues in the structures of murine and human PrP121–231 are also identical [63; 26]. The eight amino acid replacements were introduced individually into murine PrP121–231 and all variants showed the same CD spectra as the wild-type domain, demonstrating that none of these mutations induces a PrP^{Sc}-like conformation a priori, with increased β-sheet and decreased α-helix content [32]. However, exactly those variants predicted to be destabilized (D178N, T183A, F198S, and Q217R) showed a strong tendency to aggregate during secretory expression in *E. coli* and formed periplasmic inclusion bodies [32], which, however, had no characteristic similarities to PrP^{Sc} (G. Cereghetti and R. Glockshuber, unpublished results). Urea-induced equilibrium unfolding experiments proved that all eight variants of PrP121–231 showed completely reversible one-step transitions at pH 7.0 and indeed only the replacements D178N, T183A, F198S, and Q217R were destabilizing; the exchanges V180I, E220K, and V210I, and the human polymorphism replacement M129V essentially had no influence on the thermodynamic stability of PrP121–231. Quantitative numbers of $\Delta\Delta G_{fold}$ are given by Liemann and Glockshuber [32].

Overall, the model of thermodynamic destabilization of PrP^{C} as a cause of spontaneous prion generation in inherited prion diseases might well apply to the mutations D178N, T183A, F198S, and Q217R, but other mechanisms are likely to underlie prion generation in the case of the replacements V180I, E220K, and V210I.

Importantly, there is no correlation between the stabilities of the PrP121–231 variants and the disease phenotypes. This is particularly obvious in the case of the mutation D178N, which, in conjunction with the polymorphism at residue 129 (Met or Val), determines the phenotype of familial prion diseases associated with the D178N mutation [65] (FFI in the case of M129/N178 or inherited CJD for V129/N178). The replacement M129V alone had no effect on protein stability, and the same was observed for the polymorphism replacement in the D178N variant.

When human PrP genes harboring the Dl78N mutation [66] and the Q217R mutation [67] were expressed in human neuroblastoma cells, PrP with D178N exchange was unstable in vivo [66] and temperature-dependent protein misfolding was observed in the case of the Q217R variant [67]. When the murine PrP variants P102L, D178N/M129, T183A, and E200K were expressed in Chinese hamster ovary (CHO) cells, a number of PrP^{Sc}-like properties such as detergent insolubility and protease resistance were observed, and the amino acid exchange T183A blocked the delivery of the protein to the cell surface, probably due to inactivation of the N-glycosylation site at Asn181 [68–72].

In summary, some features of mutant PrPs expressed in eukaryotic cells can be interpreted in terms of thermodynamic stabilities of the corresponding PrP^{C} variants, but it is currently impossible to deduce a clear-cut molecular mechanism of spontaneous prion generation in inherited prion diseases from the thermodynamic stabilities of mutant PrP and data on their expression in cultured cells.

Finally, an important feature of prions in familial forms of the disease may be the stoichiometry and subunit composition of PrP^{Sc} isolated from affected individuals. In the case of the mutations at positions 102, 178, 198, 200, and 217, only the mutant PrP forms protease-resistant PrP^{Sc} [73–77]. In contrast, the mutation at position 210 [78] and also the insertion of five or six additional octapeptide repeats [77] led to heterooligomeric, protease-resistant PrP^{Sc} deposits in the brains of the patients. Consequently, there is no correlation between the thermodynamic stability of PrP^{C} and the tendency of mutated PrPs to form PrP^{Sc} homo- or heterooligomers.

5.5 Outlook

This chapter concentrated on the thermodynamic stability of PrP^C. Experimentally, this task was best performed with recombinant PrP121–231 which represents the structured domain of full-length PrP^C. The reduction to model PrP was derived not only from thermodynamic results, but also, and most convincingly, from the three-dimensional structure as analyzed by high-resolution NMR (see Chapter 6 for details). Thus, the two chapters are closely interrelated. Although understanding the structure and dynamics of PrP^C is essential, a complete understanding of prion function can only be achieved if the structures of both states, i. e., PrP^C and PrP^{Sc}, are known in atomic detail. Some features of PrP^C have been discussed in respect to its capability to be converted into PrP^{Sc}: This topic will be covered further in Chapter 8, which reveals that there is still a long way to go towards a complete molecular understanding of prion propagation.

References

[1] Schätzl HM, Da Costa M, Taylor L, et al. Prion protein gene variation among primates. J Mol Biol 1995; 245:362–374.

[2] Wopfner F, Weidenhofer G, Schneider R, et al. Analysis of 27 mammalian and 9 avian PrPs reveals high conservation of flexible regions of the prion protein. J Mol Biol 1999; 289:1163–1178.

[3] Hope J, Morton LJ, Farquhar CF, et al. The major polypeptide of scrapie-associated fibrils (SAF) has the same size, charge distribution and N-terminal protein sequence as predicted for the normal brain protein (PrP). EMBO J 1986; 5:2591–2597.

[4] Stahl N and Prusiner SB. Prions and prion proteins. FASEB J 1991; 5:2799–2807.

[5] Stahl N, Baldwin MA, Teplow DB, et al. Structural studies of the scrapie prion protein using mass spectrometry and amino acid sequencing. Biochemistry 1993; 32:1991–2002.

[6] Weissmann C, Fischer M, Raeber A, et al. The role of PrP in pathogenesis of experimental scrapie. Cold Spring Harbor Symp Quant Biol 1996; 61:511–522.

[7] Collinge J, Sidle KCL, Meads J, et al. Molecular analysis of prion strain variation and the aetiology of 'new variant' CJD. Nature 1996; 383:685–690.

[8] Raymond GJ, Hope J, Kocisko DA, et al. Molecular assessment of the potential transmissibilities of BSE and scrapie to humans. Nature 1997; 388:285–288.

[9] Baldwin MA, Pan KM, Nguyen J, et al. Spectroscopic characterization of conformational differences between PrPC and PrPSc: an alpha-helix to beta-sheet transition Philos Trans R Soc Lond B Biol Sci 1994; 343:435–441.

[10] Pan KM, Baldwin M, Nguyen J, et al. Conversion of alpha-helices into beta-sheets features in the formation of the scrapie prion proteins. Proc Natl Acad Sci USA. 1993; 90:10962–10966.

[11] Pergami P, Jaffe H, Safar J. Semipreparative chromatographic method to purify the normal cellular isoform of the prion protein in nondenatured form. Anal Biochem 1996; 236:63–73.

[12] Caughey B and Raymond GJ. The scrapie-associated form of PrP is made from a cell surface precursor that is both protease- and phospholipase-sensitive. J Biol Chem 1991; 266: 18217–18223.

[13] Gasset M, Baldwin MA, Fletterick RJ, et al. Perturbation of the secondary structure of the scrapie prion protein under conditions that alter infectivity. Proc Natl Acad Sci U S A 1993; 90:1–5.

[14] Safar J, Roller P P, Gajdusek D C, et al. Conformational transitions, dissociation, and unfolding of scrapie amyloid (prion) protein. J Biol Chem 1993; 268:20276–20284.

[15] Griffith JS. Self-replication and scrapie. Nature 1967; 215:1043–1044.

[16] Prusiner SB. Novel proteinaceous infectious particles cause scrapie. Science 1982; 216:136–144.

[17] Prusiner SB. Prion diseases and the BSE crisis. Science 1997; 278:245–251.

[18] Hornemann S and Glockshuber R. Autonomous and reversible folding of a soluble amino-terminally truncated segment of the mouse prion protein. J Mol Biol 1996; 261:614–619.

[19] Mehlhorn I, Groth D, Stockel J, et al. High-level expression and characterization of a purified 142-residue polypeptide of the prion protein. Biochemistry 1996; 35:5528–5537.

[20] Donne DG, Viles JH, Groth D, et al. Structure of the recombinant full-length hamster prion protein PrP(29–231): the N terminus is highly flexible. Proc Natl Acad Sci USA 1997; 94:13452–13457.

[21] Hornemann S, Korth C, Oesch B, et al. Recombinant full-length murine prion protein, mPrP(23–231): purification and spectroscopic characterization. FEBS Lett 1997; 413:277–281.

[22] James TL, Liu H, Ulyanov NB, et al. Solution structure of a 142-residue recombinant prion protein corresponding to the infectious fragment of the scrapie isoform. Proc Natl Acad Sci USA 1997; 94:10086–10091.
[23] Zahn R, von Schroetter C, Wüthrich K. Human prion proteins expressed in Escherichia coli and purified by high-affinity column refolding. FEBS Lett 1997; 417:400–404.
[24] Riek R, Hornemann S, Wider G, et al. NMR structure of the mouse prion protein domain PrP(121–321). Nature 1996; 382:180–182.
[25] Riek R, Hornemann S, Wider G, et al. NMR characterization of the full-length recombinant murine prion protein, mPrP(23–231). FEBS Lett 1997; 413:282–288.
[26] Zahn R, Liu A, Lührs T, et al. NMR solution structure of the human prion protein. Proc Natl Acad Sci USA 2000; 97:145–150.
[27] Lopez Garcia F, Zahn R, Riek R, et al. NMR structure of the bovine prion protein. Proc Natl Acad Sci USA 2000; 97:8334–8339.
[28] Hornemann S, Schorn C, Wüthrich K. NMR structure of the bovine prion protein isolated from healthy calf brains. EMBO Rep 2004:1159–1164.
[29] Swietnicki W, Petersen R, Gambetti P, et al. PH-dependent stability and conformation of the recombinant human prion protein PrP(90–231). J Biol Chem 1997; 272:27517–27520.
[30] Swietnicki W, Petersen RB, Gambetti P, et al. Familial mutations and the thermodynamic stability of the recombinant human prion protein. J Biol Chem 1998; 273:31048–31052.
[31] Hosszu LL, Baxter NJ, Jackson GS, et al. Structural mobility of the human prion protein probed by backbone hydrogen exchange. Nat Struct Biol 1999; 6:740–743.
[32] Liemann S and Glockshuber R. Influence of amino acid substitutions related to inherited human prion diseases on the thermodynamic stability of the cellular prion protein. Biochemistry 1999; 38:3258–3267.
[33] Jackson GS, Hosszu LL, Power A, et al. Reversible conversion of monomeric human prion protein between native and fibrilogenic conformations. Science 1999; 283:1935–1937.
[34] Myers JK, Pace CN, Scholtz JM. Trifluoroethanol effects on helix propensity and electrostatic interactions in the helical peptide from ribonuclease T1. Protein Sci. 1995; 4:2138–2148.
[35] Prusiner SB, Groth D, Serban A, et al. Attempts to restore scrapie prion infectivity after exposure to protein denaturants. Proc Natl Acad Sci USA 1993; 90:2793–2797.
[36] Legname G, Baskakov IV, Nguyen HO, et al. Synthetic mammalian prions. Science 2004; 305:589.
[37] Wildegger G, Liemann S, Glockshuber R. Extremely rapid folding of the C-terminal domain of the prion protein without kinetic intermediates. Nat Struct Biol 1999; 6:550–553.
[38] Apetri AC and Surewicz WK. Kinetic intermediate in the folding of human prion protein, J Biol Chem 2002; 277:44589–44593
[39] Baskakov JV, Legname G, Prusiner SB, et al. Folding of prion protein to its native α-helical conformation is under kinetic control. J Biol Chem 2001; 276:19687–19690.
[40] Booth DR, Sunde M, Bellotti V, et al. Instability, unfolding and aggregation of human lysozyme variants underlying amyloid fibrillogenesis. Nature 1997; 385:787–793.
[41] Zhang H, Stockel J, Mehlhorn I, et al. Physical studies of conformational plasticity in a recombinant prion protein. Biochemistry 1997; 36:3543–3553.
[42] Jackson GS, Hill AF, Joseph C, et al. Multiple folding pathways for heterologously expressed human prion protein. Biochim Biophys Acta 1999; 1431:1–13.
[43] Hornemann S and Glockshuber R. A scrapie-like unfolding intermediate of the prion protein domain PrP(121–231) induced by acidic pH. Proc Natl Acad Sci USA 1998; 95:6010–6014.
[44] Swietnicki W, Morillas M, Chen SG, et al. Aggregation and fibrillization of the recombinant human prion protein huPrP90–231. Biochemistry 2000; 39:424–431.
[45] Baskakov IV, Legname G, Baldwin MA, et al. Pathway complexity of prion protein assembly into amyloid. J Biol Chem 2002; 277(24):21140–21148.
[46] Bessen RA, Kocisko DA, Raymond, et al. Non-genetic propagation of strain-specific properties of scrapie prion protein. Nature 1995; 375:698–700.
[47] Kelly JW. The environmental dependency of protein folding best explains prion and amyloid diseases. Proc Natl Acad Sci U S A 1998; 95:930–932.
[48] Turk E, Teplow DB, Hood LE, et al. Purification and properties of the cellular and scrapie hamster prion proteins. Eur J Biochem 1988; 176:21–30.
[49] Caughey B, Horiuchi M, Demaimay R, et al. Assays of protease-resistant prion protein and its formation. Methods Enzymol 1999; 309:122–133.
[50] Herrmann LM and Caughey B. The importance of the disulfide bond in prion protein conversion. Neuroreport 1998; 9:2457–2461.

[51] Somerville RA, Millson GC, Kimberlin RH. Sensitivity of scrapie infectivity to detergents and 2-mercaptoethanol. Intervirology 1980; 13:126–129.

[52] Welker E, Raymond LC, Scheraga HA, et al. Intramolecular versus intermolecular disulfide bonds in prion proteins. J Biol Chem 2002; 277:33477–33481.

[53] Ma J and Lindquist S. De novo generation of a PrPSc-like conformation in living cells. Nat Cell Biol 1999; 1:358–361.

[54] Dobson CM. Protein folding and misfolding. Nature 2003; 426:884–890.

[55] Prusiner SB, Scott MR, DeArmond SJ, et al. Prion protein biology. Cell 1998; 93:337–348.

[56] Fischer M, Rulicke T, Raeber A, et al. Prion protein (PrP) with amino-proximal deletions restoring susceptibility of PrP knockout mice to scrapie. EMBO J 1996; 15:1255–1264.

[57] Cohen FE, Pan KM, Huang Z, et al. Structural clues to prion replication. Science 1994; 264:530–531.

[58] Huang Z, Prusiner SB, Cohen FE. Scrapie prions: a three-dimensional model of an infectious fragment. Fold Design 1995; 1:13–19.

[59] Cohen FE. Protein misfolding and prion diseases. J Mol Biol 1999; 293:313–320.

[60] Büeler H, Aguzzi A, Sailer A, et al. Mice devoid of PrP are resistant to scrapie. Cell 1993; 73:1339–1347.

[61] Telling GC, Scott M, Mastrianni J, et al. Prion propagation in mice expressing human and chimeric PrP transgenes implicates the interaction of cellular PrP with another protein. Cell 1995; 83:79–90.

[62] Kaneko K, Zulianello L, Scott M, et al. Evidence for protein X binding to a discontinuous epitope on the cellular prion protein during scrapie prion propagation. Proc Natl Acad Sci USA. 1997; 94:10069–10074.

[63] Riek R, Wider G, Billeter M, et al. Prion protein NMR structure and familial human spongiform encephalopathies. Proc Natl Acad Sci USA 1998; 95:11667–11672.

[64] Glockshuber R. Folding dynamics and energetics of recombinant prion proteins. Adv Protein Chem 2001; 57:83–105.

[65] Goldfarb LG, Petersen RB, Tabaton M, et al. Fatal familial insomnia and familial Creutzfeldt–Jakob disease: disease phenotype determined by a DNA polymorphism. Science 1992; 258:806–808.

[66] Petersen RB, Parchi P, Richardson SL, et al. Effect of the D178N mutation and the codon 129 polymorphism on the metabolism of the prion protein. J Biol Chem 1996; 271:12661–12668.

[67] Singh N, Zanusso G, Chen SG, et al. Prion protein aggregation reverted by low temperature in transfected cells carrying a prion protein gene mutation. J Biol Chem 1997; 272:28461–28470.

[68] Lehmann S and Harris DA. A mutant prion protein displays an aberrant membrane association when expressed in cultured cells. J Biol Chem 1995:270, 24589–24597.

[69] Lehmann S and Harris DA. Mutant and infectious prion proteins display common biochemical properties in cultured cells. J Biol Chem 1996; 271:1633–1637.

[70] Lehmann S and Harris DA. Two mutant prion proteins expressed in cultured cells acquire biochemical properties reminiscent of the scrapie isoform. Proc Natl Acad Sci USA 1996; 93:5610–5614.

[71] Daude N, Lehmann S, Harris DA. Identification of intermediate steps in the conversion of a mutant prion protein to a scrapie-like form in cultured cells. J Biol Chem 1997; 272:11604–11612.

[72] Lehmann S and Harris DA. Blockade of glycosylation promotes acquisition of scrapie-like properties by the prion protein in cultured cells. J Biol Chem 1997; 272:21479–21487.

[73] Kitamoto T, Yamaguchi K, Dohura K, et al. A prion protein missense variant is integrated in kuru plaque cores in patients with Gerstmann–Sträussler syndrome. Neurology 1991; 41:306–310.

[74] Tagliavini F, Prelli F, Porro M, et al. Amyloid fibrils in Gerstmann–Sträussler–Scheinker disease (Indiana and Swedish kindreds) express only PrP peptides encoded by the mutant allele. Cell 1994; 79:695–703.

[75] Barbanti P, Fabbrini G, Salvatore M, et al. Polymorphism at codon 129 or codon 219 of PRNP and clinical heterogeneity in a previously unreported family with Gerstmann–Sträussler–Scheinker disease (PrP-P102L mutation). Neurology 1996; 47:734–741.

[76] Gabizon R, Telling G, Meiner Z, et al. Insoluble wild-type and protease-resistant mutant prion protein in brains of patients with inherited prion disease. Nat Med 1996; 2:59–64.

[77] Chen SG, Parchi P, Brown P, et al. Allelic origin of the abnormal prion protein isoform in familial prion diseases. Nat Med 1997; 3:1009–1015.

[78] Silvestrini MC, Cardone F, Maras B, et al. Identification of the prion protein allotypes which accumulate in the brain of sporadic and familial Creutzfeldt–Jakob disease patients. Nat Med 1997; 3:521–525.

6 Structural Studies of Prion Proteins

Stephan Schwarzinger, Dieter Willbold, and Jan Ziegler

6.1 Introduction

The previous chapter dealt with the biophysical characterization of recombinant mouse prion protein. This included a first characterization of the conformation of the cellular form of prion protein (PrP^C) by circular dichroism (CD) spectroscopy, which indicates a significant amount of helical structure in the cellular isoform. CD has also been successfully applied to study conformational changes, for example, introduced by the addition of small amounts of the denaturant sodium-dodecylsulfate (SDS) resulting in changes of the oligomeric state as well as the secondary structure content depending on the amount of SDS (see Chapter 8). However, techniques such as CD do not allow precise localization of where certain secondary structure elements are positioned or are involved in structural changes. Ultimately, for the understanding of basic principles underlying the conversion of PrP^C into its scrapie isoform (PrP^{Sc}), knowledge of both conformations and eventually of intermediate conformations at atomic detail is required. The situation is shown schematically in Figure 6.1. In this chapter we review the current state of high-resolution structural studies on prion proteins. Besides the wealth of experimental data, which are available at present on the structure of the globular part of PrP, we will outline new approaches on fragments of PrP, interacting ligands, hydrogen exchange and insoluble, PrP^{Sc}-like conformations or PrP, where we can expect further progress in the future.

At present there are only two methods available that provide true resolution at atomic detail, namely, nuclear magnetic resonance (NMR) spectroscopy and X-ray crystal diffraction. In the following we shall briefly review the principles underlying these two techniques. X-ray crystallography utilizes the fact that an electromagnetic beam is scattered by the electron clouds in a single crystal of a given compound. In a single-crystal, molecules are arranged into a so-called asymmetric unit, which is periodically repeated in the three directions of space. If the crystal is hit by an X-ray beam the asymmetric unit produces a particular pattern of scattered rays, typical for the space group in

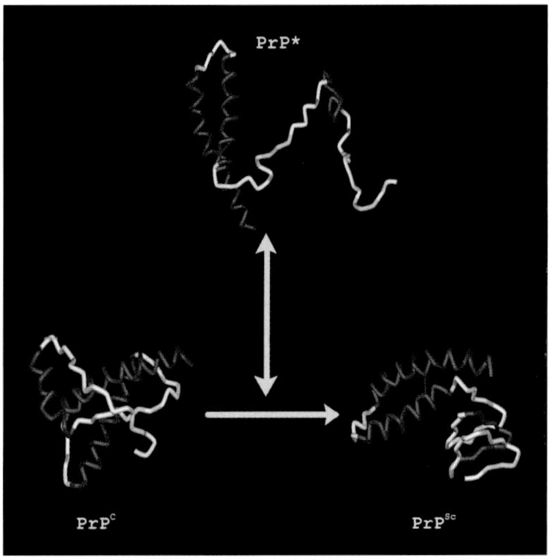

Fig. 6.1: Schematic representation of the conformational changes related to the conversion of cellular prion protein (PrP^C) into the pathogenic form (PrP^{Sc}). Knowledge of the conformations of PrP^C, PrP^{Sc}, and intermediate states called PrP* will provide important insights into the infectious mechanism of transconformation. The structure of PrP^C has been experimentally determined by NMR [Protein Data Bank (PDB) code 1QM2]. PrP^{Sc} has been modeled as described [64], and the schematic PrP* has been obtained from a molecular dynamics (MD) unfolding simulation. ©: Figure by S. Schwarzinger with WebLabViewer Lite, Accelrys, San Diego.

which the crystal grew. Structural information is extracted from the phases and intensities of the individual spots from patterns obtained from numerous different orientations of the crystal in the X-ray beam. The caveats of this technique are that (i) a protein crystal of suitable size and quality has to be grown, and (ii) that the phases of the signals are a priori unknown. Crystallization usually requires screening of large numbers of crystallization conditions, and today is often achieved by utilizing crystallization robots. Phases are routinely obtained by isomorphous heavy-metal replacement, molecular replacement, or more recently, by multiwavelength anomalous diffraction (MAD) techniques. Protein crystallography usually yields positions of the heavy atoms of a protein and can be applied to rather large molecular assemblies, such as the proteasome. However, crystallography of proteins with rather large flexible parts is problematic, either because no crystals can be grown at all or because of large conformational flexibility in the crystal, prohibiting recording of suitable reflections from the flexible parts.

In contrast, high resolution NMR spectroscopy traditionally works in aqueous solutions. It makes use of the radio frequency response that is obtained when a magnetically active nucleus is put into a strong external magnetic field and perturbed by a radio frequency pulse. The radio frequency obtained depends on the local magnetic field of a nucleus, which in turn depends on its chemical environment. Usually, each nucleus can be assigned to a particular resonance frequency. Modern NMR spectroscopy makes use of the magnetically active, stable isotopes ^{1}H (99 % natural abundance), ^{13}C, and ^{15}N, the latter two of which can be conveniently incorporated into the protein via bacterial expression systems. This way in combination with ultra-high field magnets it is now possible to routinely assign the NMR resonances to their respective nuclei and resolve the three-dimensional structures of proteins of up to approximately 25 kDa. This size limit has long been a major obstacle of solution NMR, which, however, has recently been overcome by the development of the so-called TROSY technique. The TROSY technique allows structural studies of proteins and protein complexes of hundreds of kDa [1]. NMR structures have typically been calculated from pairwise interproton distances, available from nuclear Overhauser-type experiments, and torsion angles extracted from scalar coupling constants [2]. Nuclear Overhauser effects (NOEs) can be observed between protons belonging to amino acid residues that may be close or distant to each other within the protein sequence. Observation of such NOEs for a given proton pair indicates them to be close together in space, ultimately yielding an experimentally observed pairwise interproton distance. In general, thousands of these pairwise interproton distances are required to define a high-resolution 3D protein structure. More recently, residual dipolar coupling constants resulting from partial alignment of the protein relative to the external magnetic field have become available as additional structural parameters, which allow the determination of bond vector orientations in space. For example, the relative orientations of amide bond vectors relative to each other are often used as experimental constraints in protein 3D structure determination. In contrast to protein crystallography, solution NMR typically yields an ensemble of converging structures, which already include to some extent information about dynamics within the protein. Typically, these structures are essentially identical for the core and for well-folded parts of the proteins. However, they display structural variation in segments with increased chain flexibility providing information about the extent of motion and alternative conformations or structural disorder in the respective part of the protein. A particular strength of solution NMR spectroscopy is the capability to deal with entirely disordered polypeptide segments, which is of special interest in the context of the intrinsically disordered N-terminus of the PrP, and to extract information about residual or preferred structures in such sequences, for example, from chemical shift deviations [3]. As the only experimental technique available, NMR spectroscopy allows monitoring of internal motions in a molecule on a timescale ranging from nanoseconds to hours. Such studies often reveal important dynamical and conformational features of both folded and disordered polypeptide chains. Solution NMR has also gained importance when it comes to screening for ligands binding to a particular protein and for mapping binding interfaces, for example, in complexes of biological macromolecules.

Both methods have a common disadvantage, namely that they require extremely pure protein with solubility in the mg/ml range (i. e., high µM to mM concentrations). Therefore, production of substantial amounts of properly folded, soluble PrP is of crucial importance for structural studies. As explained in the previous chapter, PrPs are highly conserved among mammals. PrP^{C} is translated as a 253-residue protein with a 22-amino-acid N-terminal leader sequence that is cleaved off during transport to the cell surface. There, the protein is attached to the membrane via a C-terminally linked glycolphosphatidylinositol (GPI) anchor. The protein consists of two domains, the N-terminal (residues 23–120) of which is intrinsically disordered. This domain contains an octarepeat segment (51–91, five times PQGG[T/G/S]WGQ) and is capable of binding copper ions [4]. The C-terminal domain (residues 121–231) adopts a predominantly helical globular fold. It contains a disulfide bridge (Cys179 and Cys214), two sites for N-glycosylation (Asn181 and Asn197), and the GPI anchor is attached to Ser231. Initial attempts to express the PrP were hampered by protease activity cleaving the protein preferentially at residues 112, 118, and 120. Therefore, the first expression system that yielded protein in amounts suitable for structural studies included only the globular C-terminal domain, i. e., residues 121–231 [5]. However, highly efficient transcription systems enabled expression of longer, protease-sensitive constructs as insoluble inclusion bodies, which are insensitive to proteolysis and which could be resolubilized and properly refolded. This way it was possible to produce large quantities of full-length PrP (23–231), a PrP fragment lacking the N-terminal part and the octarepeat region (90–231), as well as the fragments solely consisting of the globular domain 121–231 [6; 7]. In the following, we will initially focus on the structural properties of the globular domain of the PrP, the comparison of structures from different species as well as structures from variants that are associated with increased pathogenic potential, before we briefly review studies on protein dynamics, studies of isolated structural features, investigations on molecules binding to PrP^{C}, intermediate structures, solid state NMR studies, as well as first structural investigations of a scrapie conformer of PrP.

6.2 Structure of the globular domain of PrP^{C}

The first prion structure made available was the C-terminal fragment of mouse prion protein (moPrP), which was solved by NMR spectroscopy at the Eidgenössische Technische Hochschule in Zurich, Switzerland, in 1996 [8]. A little later, the structure of the fragment 90–231 of Syrian hamster prion (shPrP) was solved by NMR at the University of California, San Francisco [9]. As expected from CD data, both structures displayed a high content of α-helices. As seen in Figure 6.2, the globular domain consists of three α-helices, helix 1 (residues 144–154), helix 2 (residues 175–193), and helix 3 (residues 200–219) with a helix-like extension from resi-

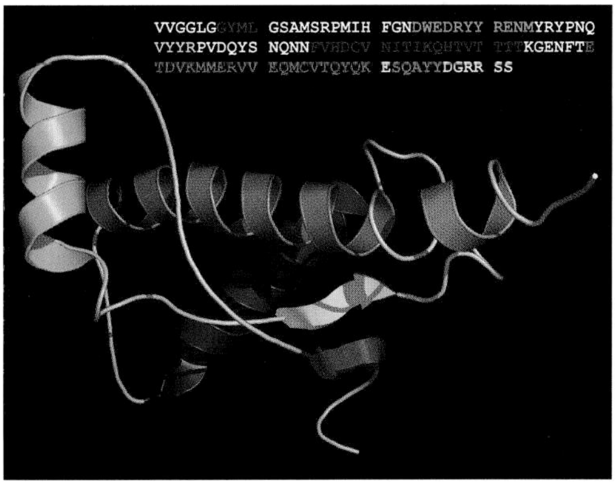

Fig. 6.2: NMR structure of the globular domain of recombinant mouse prion protein (PDB code 1AG2). Elements of regular secondary structure and their corresponding part in the sequence are color coded: strand 1, red; helix 1, green; strand 2, yellow; helix 2, blue; helix 3, magenta; helix 3″, orange. ©: Figure by S. Schwarzinger with PyMOL, DeLano Scientific, San Francisco.

due 222 to 226, as well as a short antiparallel β-sheet with strand 1 extending from residue 128 to 131 and strand 2 from residue 161 to 164. The ensemble of NMR structures converges very well for these elements of regular secondary structure indicating a stable fold. Helices 2 and 3, which do not have a high intrinsic propensity to form helices, are linked and thus stabilized by the disulfide bond. However, increased structural diversity indicating a certain degree of disorder is found in the terminal regions of the polypeptide chain (residues 121–125 and 220–231), but increased backbone flexibility is also detected for the loop between strand 2 and helix 2 (residues 166–171). The N-glycosylation sites are located on the solvent-exposed side of helix 2 and in the loop region between helices 2 and 3.

A central question when analyzing structures of post-translationally modified mammalian proteins that have been obtained by heterologous expression in bacterial expression systems is, whether or not the modifications have a significant impact on the structure. In case of the PrP, which carries a GPI anchor as well as two N-glycan moieties, it was shown that the one-dimensional proton NMR spectra of recombinant bovine PrP and of PrP extracted from calf brains are essentially identical in their most important structural features [10]. Initially, structures have been determined at slightly acidic pH values. Studies of PrP^C at neutral pH, however, revealed an essentially identical structure over the pH range from 4.5 to 7.0. The most pronounced change is a C-terminal extension of helix 1 by about one turn; 3_{10}-helix from residues 153 to 156 as depicted in Figure 6.3 [11]. By comparison of the PrP structure with structures of proteins with known functions [12], a possible signal-peptidase function was predicted as one of several suggestions for the physiological function of PrP^C (see Chapter 7).

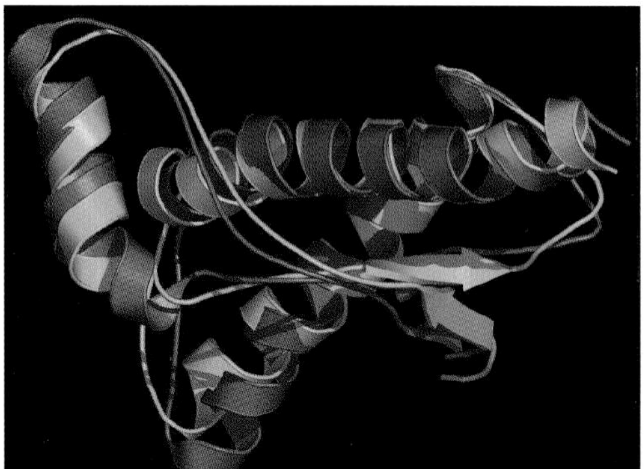

Fig. 6.3: Effect of pH on the structure of human PrP, huPrP(121–231). The NMR structures obtained at pH 4.5 (PDB code 1QM2, shown in green) and pH 7.0 (1HJM, shown in red), respectively, have been overlaid by root-mean-square deviation (RMSD) fitting. The most pronounced changes are observed for helix 1, which is extended at its C-terminus by a turn of a 3_{10}-helix. This causes displacement of the entire helix 1 and the C-terminus of helix 2, which is in contact with helix 1. ©: Figure by S. Schwarzinger with PyMOL, DeLano Scientific, San Francisco.

Based on the known PrP structure, it is possible to locate the amino acid sequence positions for pathogenic mutations and map binding interfaces of other proteins interacting with the PrP. The binding site of the so-called protein X, which is thought to be involved in the pathogenic conformational change to PrP^{Sc} in vivo, could be mapped by biochemical experiments to residues 168, 172, 215, and 219 of the human prion protein, huPrP [13]. From the three-dimensional structure, it is clear that these residues are located in close spatial proximity (Fig. 6.4). Also, the binding epitope of the antibody 6H4, which is capable of preventing scrapie formation, was found to coincide almost exactly with helix 1 [14]. Helix 1 itself is remarkable, since it is the most polar helix in the entire structure database. It shares only very few tertiary contacts with the remainder of the protein. Studies on peptide segments yielded a very high tendency to form stable α-helices [15], also discussed in Section 6.4.

Structure determination of PrPs from different species might provide insights into susceptibility for interspecies prion transmission and into the nature of the species barrier. Thus, structures of the PrPs of mouse [8], hamster [9], humans [16], cattle [17], cat, dog, pig, sheep [18], elk [19], chicken, frog, and turtle [20] have been studied by NMR spectroscopy. The structures of sheep PrP [21] as well as huPrP [22] have also been determined by X-ray crystallography.

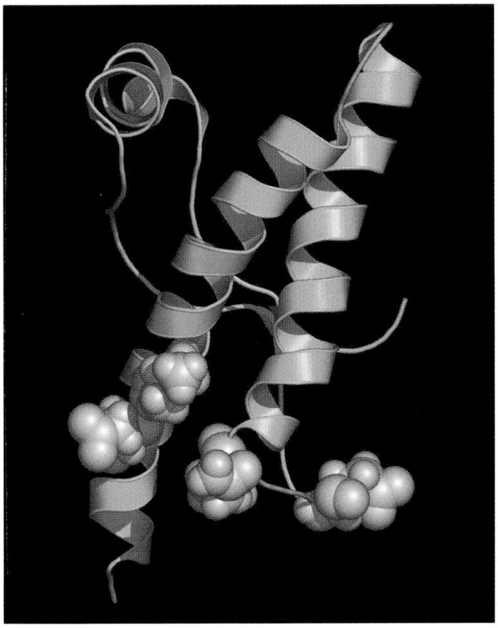

Fig. 6.4: Location of the putative binding interface of "protein X". Residues 168, 172, 215, and 219, shown as cyan spacefill models, represent the proposed binding interface for protein X. ©: Figure by S. Schwarzinger with PyMOL, DeLano Scientific, San Francisco.

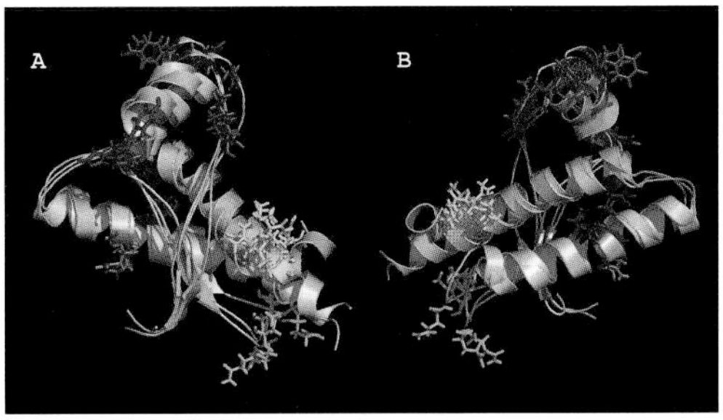

Fig. 6.5: Differences between the structures of the PrP of mouse (PDB code 1AG2), man (PDB code 1QM2), and cattle (PDB code 1DWY). Although the overall fold is well conserved, local differences arise between structures. Positions with differences in the sequence between the three species are colored. A cluster of variations in the sequence is located near or within helix 1 (red), and another cluster is located in the loop between strand 2 and helix 2 (green). Clusters are also found at the C-terminus of helix 3 (yellow) and at the interface between helices 2 and 3 (blue). Another position with differences in the sequence is positioned in the center of helix 2. Figures **(a)** and **(b)** are rotated by 180° for better visibility. Interestingly, the structure similarity between cattle and man is higher that for mouse PrP. ©: Figure by S. Schwarzinger with PyMOL, DeLano Scientific, San Francisco.

In essence, all PrPs studied so far share the same fold. An exception is represented by an X-ray structure of a domain-swap dimer of the huPrP, which will be discussed in more detail below.

An overlay of the structures of PrP^C from mouse, cattle, and man along with differences in sequence is depicted in Figure 6.5. Despite only few differences in sequence and the same overall fold, some deviations in the structures can be seen. In particular, differences in the alignment of helix 1 with the other helices are evident, as well as differences in the orientation of the carboxy-terminus of helix 3. It can also be seen that the structures of bovine and $huPrP^C$ are structurally more closely related than they are to the structure of mouse PrP^C. This actually may indicate why BSE can be transmitted to humans. The recent report on five additional structures of mammalian PrP provides further insights into the possible role of certain structural elements in prion diseases [18]. In particular, it appears that amino acid substitutions in the structurally flexible loop region linking strand 2 and helix 2, which has also previously been assigned as part of the putative protein X binding interface [13], may play an important role in susceptibility to prion diseases. Especially amino acid substitutions which alter charges in this region have been shown to have pronounced effects. For example, sheep carrying the positively charged arginine residue at position 168 of PrP are highly resistant to TSE, while sheep with glutamine or histidine residues at this position display high to medium-high susceptibility for scrapie infection. Based on a comparison of residues varying among different species, the important role of surface charge distribution on the species barrier has already been previously suggested [23].

Amino acid substitutions in the strand 2–helix 2 loop region are reported to have a strong effect on backbone flexibility. Exchange of

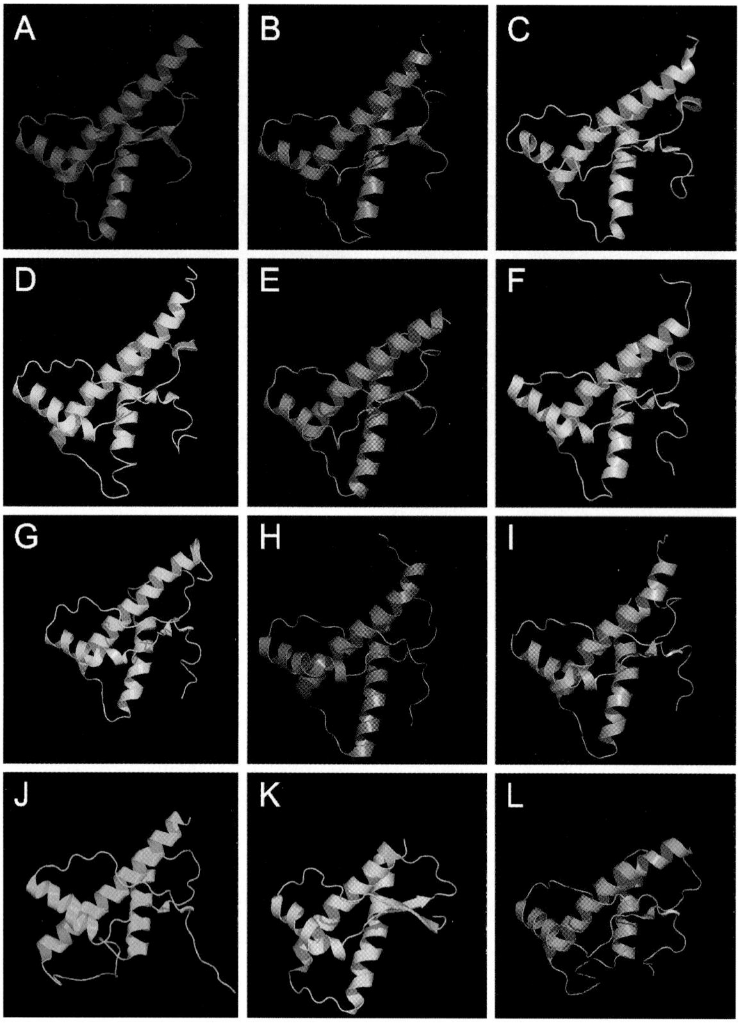

Fig. 6.6: Schematic representation of known PrP structures. The panel shows the structures of the PrP of **(a)** man (PDB code 1QLZ), **(b)** cattle (PDB code 1DWZ), **(c)** sheep (PDB code 1Y2S), **(d)** mouse (PDB code 1XYX), **(e)** Syrian hamster (PDB code 1B10), **(f)** elk (PDB code 1XYW), **(g)** cat (PDB code 1XYJ), **(h)** dog (PDB code 1XYK), **(i)** pig (PDB code 1XYQ), **(j)** chicken (PDB code 1U3M), **(k)** frog *Xenopus laevis* (PDB code 1XU0), and **(l)** turtle *Trachemys scripta* (PDB code 1U5L). While the global fold is well conserved, differences can be found especially in the loop parts and in the positioning of helix 1 relative to helices 2 and 3. In some cases (chicken and turtle), an additional short 3_{10} helix is formed within the globular domain. Despite the low sequence identity of about 30 %, even in these cases the global fold is nearly identical to mammalian prions. Sequence variation within mammalian PrP tends to cluster within the loop 166–172 and at the carboxyterminus of helix 3. Apart from small local structural deviation, these sequence differences do not lead to major structural consequences. However, as these regions are mostly surface exposed, they alter the charge distribution on possible interaction sites of the protein. ©: Figure by J. Ziegler with PyMOL, DeLano Scientific, San Francisco.

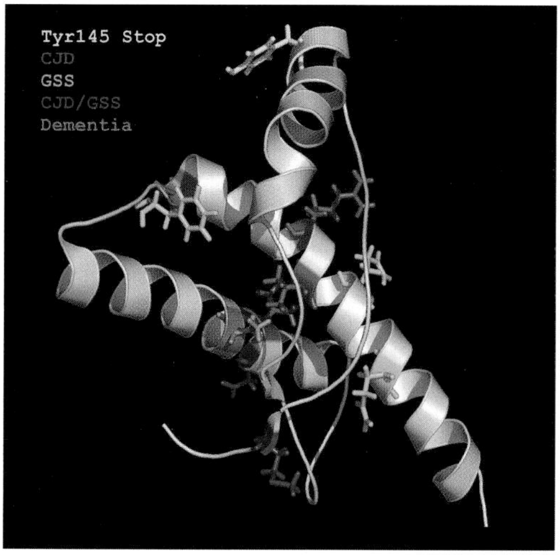

Fig. 6.7: Location of disease-related mutations in the PrP structure. Mutations are mapped on the NMR structure of huPrP(121–231) and have been color coded. Residues shown in red are associated with CJD, those shown in green have been found with GSS. Residue Asp178 is found in either CJD or GSS, depending on whether residue 129 is methionine or valine (both residues shown in blue). Thr183 (magenta) is linked to dementia when changed. Mutation of Tyr145 (yellow) into a stop codon also causes disease. Mutations Pro102Leu, Pro105Leu, Ala117Val, and Met232Arg lie outside the globular domain of PrP and are therefore not shown. ©: Figure by S. Schwarzinger with PyMOL, DeLano Scientific, San Francisco.

Asn173 to Ser in pig significantly reduces the structural variability in this region. Similarly in elk PrP^C, this part of the PrP is very stable [19]. Based on studies of mouse/elk hybrid PrPs, it has been shown that residues Asn170 and Thr174 are decisive for introducing rigidity in the strand 2–helix 2 loop.

Recent studies of non-mammalian PrP with a sequence identity of approximately 30% to mammalian PrP [29] revealed that the architecture of PrP is obviously well conserved over large parts of the animal kingdom. However, significant perturbations in local structure were observed (Fig. 6.6).

Another interesting feature of PrP^C structure is that most PrP mutations related to TSE pathogenesis can be found buried in the interior of the protein, while only a relatively small number of affected residues are surface exposed (Fig. 6.7). However, the effect of such mutations cannot be explained exclusively by a destabilization of the globular domain of PrP^C [24].

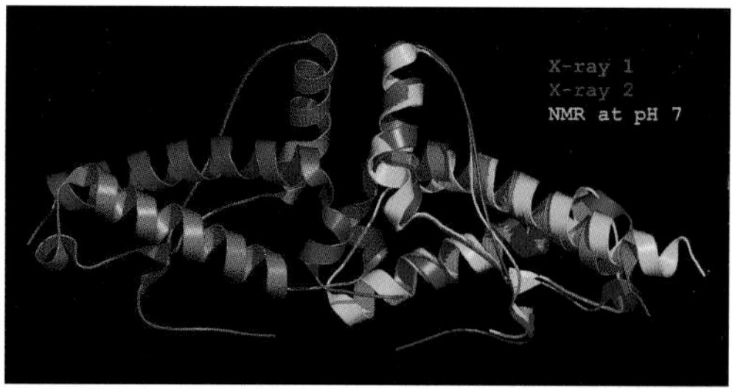

Fig. 6.8: X-ray structure of the domain-swap dimer of huPrP(121–231) (pdb code 1I4M). The dimer is created by rotation along the helix 2–helix 3 loop, so that helix 2 of either monomer (colored red and blue) contacts helix 3 of the other one, and vice versa. The subdomain formed by the β-sheet essentially keeps its orientation relative to the (now domain-swapped) helices 2 and 3. Interestingly, excellent conservation of the native prion fold is revealed when overlaid with the structure of huPrP at pH 7. Structural deviations are most pronounced at the C-terminus of helix 2 and its connecting loop with helix 3. A minor deviation is also observed for the orientation of the C-terminus of helix 3. ©: Figure by S. Schwarzinger with PyMOL, DeLano Scientific, San Francisco.

An X-ray crystallographic study of the huPrP revealed a domain-swapped dimer of PrP [22]. The two monomers orient such that helix 3 of one monomer interacts in a native-like manner with the second monomer forming native-like, although intermolecular disulfide bonds (Fig. 6.8). Formation of the dimer implies opening of the monomeric fold of PrP^C and involves a rearrangement of the native intramolecular disulfide bond. Interestingly, the interface between both monomers contains many "pathogenic" mutations. Thus, it may be speculated, whether a structural rearrangement as observed in the reported dimer plays a role in the conversion of PrP^C to PrP^{Sc}. Up to now, however, a physiological role of a dimer with intermolecular disulfide bonds has not been shown.

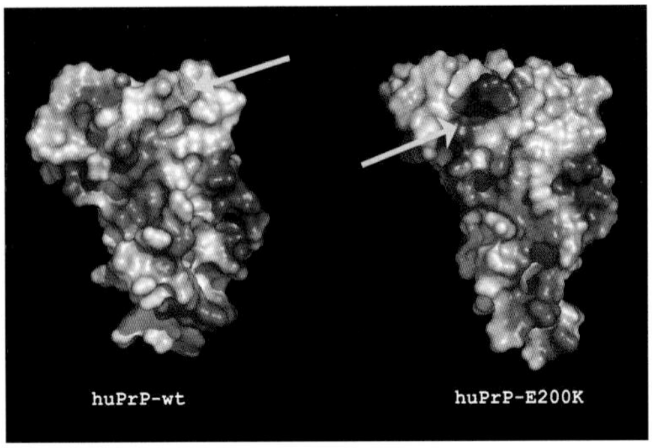

Fig. 6.9: Effect of the "pathogenic" mutation Glu200Lys on the surface charge distribution of PrP. Wild-type huPrP (PDB code 1QM2) and huPrPGlu200Lys (PDB code 1FO7) have been aligned and their vacuum electrostatics have been estimated with PyMOL. Residue 200 is located at the N-terminal part of helix 3. The mutation Glu200Lys leads to a drastic change in the surface charge around this position (marked by arrows). ©: Figure by S. Schwarzinger with PyMOL, DeLano Scientific, San Francisco.

A comparison of the structures of the disease-related mutants Glu200Lys [25] and Met129Val [26] revealed no significant differences. This polymorphism is associated with cases of new variant Creutzfeldt–Jakob disease (vCJD), where patients were shown to be homozygous for methionine at position 129. Furthermore, the mutation Asp178 to Asn is associated with fatal familial insomnia (FFI) when residue 129 is methionine, and with CJD when residue 129 is valine. No differences in structure, dynamics, or stability could be detected between mutant and wild-type protein. This suggests that Met129Val does not act via changes in the structure of PrP^C rather than on the folding or conversion intermediates.

The mutation Glu200Lys involved in familial CJD, is located in the loop connecting helices 2 and 3 at the N-terminus of helix 2. Interestingly, again no structural deviations could be observed between mutant and wild-type protein. The mutation alters the charge distribution at the proteins surface, indicating that surface charge distribution plays an important role in prion diseases (Fig. 6.9).

6.3 Structural studies of full-length PrP

Structural investigations of the flexible N-terminal tail are particularly difficult, as rapid conformational averaging prevents crystallization as well as calculation of structural ensembles from NMR studies. Thus, NMR studies of the structure of full-length PrP only resulted in three-dimensional structures for the globular domain of the protein studies, which are essentially identical to the structures obtained from the isolated C-terminal domains as shown in Fig. 6.10 [16; 17; 27; 28]. However, measurable differences in the structure of the globular domain of huPrP could be detected for the carboxy-terminal parts of helices 2 (187–193) and 3 (219–226). Because these structural variations do not depend on the concentration of the protein over a range from 0.1 mM to 1 mM it was concluded that these structural variations arise from transient contacts to the flexible N-terminus [16; 29].

NMR can reveal differences in the dynamics of individual residues. Measurements of the heteronuclear ^1H-^{15}N Overhauser effect (hetNOE) and of the longitudinal (R_1) and transverse (R_2)

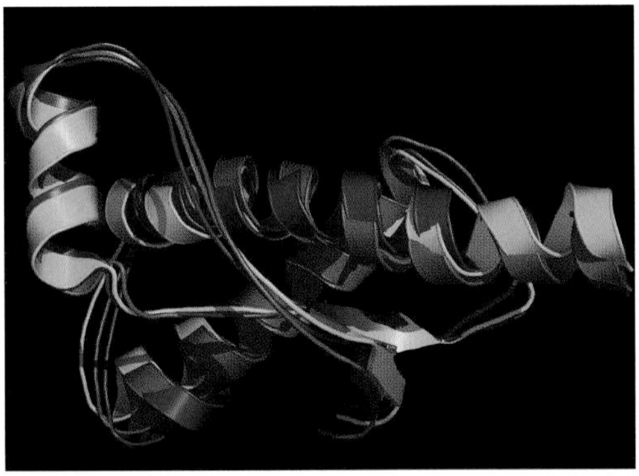

Fig. 6.10: Effect of absence or presence of PrP N-terminal parts on the structure of the globular domain. HuPrP(23–231) blue, PDB code 1QLX; huPrP(90–231) red, PDB code 1QM0; and huPrP(121–231) green, PDB code 1QM2, have been aligned by RMSD minimization. Small structural variations, caused by transient contacts of the flexible N-terminal tail with the globular domain, are centered at the C-terminus of helices 2 and 3. ©: Figure by S. Schwarzinger with PyMOL, DeLano Scientific, San Francisco.

relaxation rates of the amide nitrogen nuclei revealed that the most rigid part of the globular domain is the interface between helices 2 and 3 near the interhelical disulfide bond [30]. Parts of the protein that are not in contact with this core exhibit higher backbone flexibility. Helix 1, which has only a few tertiary contacts with helices 2 and 3, shows increased backbone dynamics in comparison to the core. These data are in line with the structural convergence observed in the NMR-ensembles as well as with amide-hydrogen protection factors (see Section 6.4.3). Methionine 131 in β-strand 1 of sheep PrP undergoes complex motions on the μs to ms timescale that are typical for conformational exchange [30]. Such dynamic behavior in the β-sheet of PrP could be of importance for prion aggregation.

Relaxation measurements have confirmed that the N-terminal region (residues 23–125) does not adopt a compact and stably folded structure. Instead, it samples a large conformational space with a high frequency [30]. Nevertheless, this does not mean that this part of the polypeptide chain moves randomly and does not have residual or preferred structure elements. As previously mentioned, the N-terminal domain appears to be in transient contact with the globular domain [29]. Moreover, the N-terminal domain contains the copper-binding octarepeat unit as well as a copper-binding site involving residues His96 and His111 [31]. Copper, being a paramagnetic nucleus, has so far prohibited high-resolution NMR structures of the copper loaded N-terminus. However, studies of isolated fragments of the octarepeat region reveal presence of a well-defined structure depending on the pH value [32]. Dynamic data derived from hetNOE studies at pH 4.5 and pH 6.2 support this finding by indicating loss of backbone mobility in the central part of the octarepeat region upon pH increase [32]. Similarly, a well-ordered structure could be obtained from NMR studies of octarepeat peptides in micellar environments [33]. It could also be shown that the octarepeat region represents a pH-dependent oligomerization site. Copper binding may modulate this aggregation [32]. For instance, binding of copper to the octarepeat region has recently been shown to induce β-sheet formation in the unstructured domain [34].

Nevertheless, it has been possible to gather some structural insights into the copper-bound states of the N-terminal region of PrP using combinations of NMR, CD, electron paramagnetic resonance (EPR) techniques, and extended X-ray absorption fine structure (EXAFS) spectroscopy [35; 36]. In particular, the latter technique permits investigation of the coordination of copper in the complex with octarepeat peptides. Furthermore, it is possible to obtain an X-ray structure of a peptide HGGGW-Cu^{2+} [37]. The NMR structures [32; 33] display reasonable agreement with the backbone geometry of the crystal structure of the copper-loaded pentapeptide. It appears that β-turn structures are formed from segments encompassing GWGQ [32], and that WGQPH also forms a loop-like core structure [33]. However, the side-chain orientation of the tryptophane residue differs about 180° from the structures obtained from X-ray and NMR studies. Recent EXAFS and EPR studies proposed a three-dimensional structure of a single octarepeat peptide involving a planar coordination of copper with three nitrogen atoms and one oxygen atom [36]. Ap-

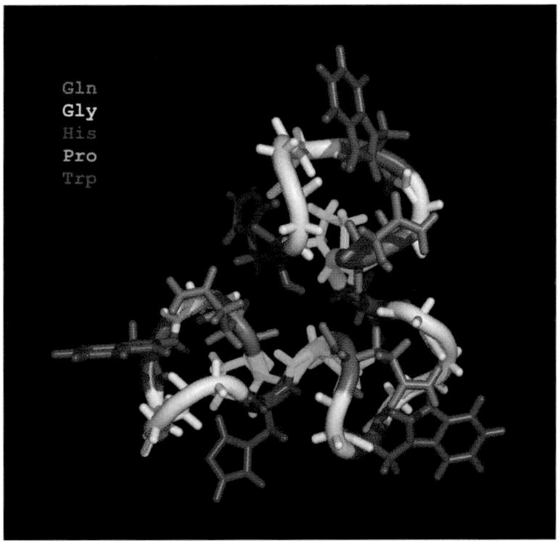

Fig. 6.11: Structure of the copper-free octarepeat residues 61–84 (PDB code 1OEI). ©: Figure by S. Schwarzinger with PyMOL, DeLano Scientific, San Francisco.

parently, a picture of the octarepeat is emerging, in which the nonglycine residues (WGQPH) form a well-structured core linked by three glycine residues (Fig. 6.11). As a result of these studies, it seems that free octarepeat and copper-loaded forms share a high degree of structural similarity. A promising route for obtaining structural insights into the copper-loaded form at high resolution was recently taken by using diamagnetic nickel ions in a ^{1}H-NMR investigation to show that the six copper ions can be bound in the flexible N-terminus of PrP [38]. In particular, this study suggests that each of His96 and His111 are involved in the binding of one copper ion, independent of each other (see also Section 6.4.2).

6.4 NMR studies on isolated structural features of PrP

6.4.1 Peptide studies

While NMR studies on full-length proteins yield a wealth of information on structure and dynamics of a protein in its native context, examining the solution behavior of smaller fragments can provide valuable insights on the behavior and structural preference of smaller parts under deviation from the native environment. This is especially interesting in the case of PrP, as this particular protein undergoes massive structural changes during the conformational transition from PrP^{C} to PrP^{Sc}, in which parts of the protein are excluded from their native environment, possibly populating structural ensembles of unfolded or at least highly dynamic species. The highly dynamic behavior of short peptides in solution might therefore provide a valuable model system for intermediates in prion conversion. Such studies, however, are usually not able to yield structures or even structural models, as there are usually no well-defined distance constraints in such systems. Nevertheless, analysis of chemical shifts and their deviation from random coil values can give information on the existence or propensity of secondary structure elements and their population, even in only partially folded proteins [3; 39]. Additionally, the existence of regular $i, i+3$ and $i, i+4$ NOE connectivity patterns can point out the existence of helices in such peptide systems even at low abundance averaged over time.

Of special interest in this regard are peptide studies on regions which are supposed to take part in the conformational conversion of the PrP. Several studies focused on the region containing the β-sheet, helix 1, and the connecting loops (residues 125–170), which has been suggested to form an initiation site during the conformational transition [8].

Contrary to the expected intrinsic tendency to form sheet-like structures, it was shown that peptides huPrP(125–170), huPrP(142–170), and huPrP(154–170) adopted mainly helical conformers in aqueous solution [40]. The strongest tendency to form helices was observed for the sequence corresponding to helix 1 (145–154). Interestingly, both β-strands did not show any tendency to adopt extended conformations. Additional peptide studies focused on helix 1 in murine [15] and human [41] PrP, and further corroborated the high intrinsic tendency of this sequence to populate helical conformers. Because of its exceptionally high helix propensity helix 1 has been proposed as an energetic barrier in the pathogenic transformation [41].

Other peptide studies on PrP focused on the unstructured amino terminal domain, for which conformational information is largely inaccessible in full-length PrP. The amino terminal sequence of bovine PrP, boPrP(1–30), was shown to populate helical conformers in solutions containing DHPC micelles, thereby simulating a membrane environment [42]. In combination with deuterium exchange data and NMR studies on magnetically aligned micelles, this study points to the insertion of a helical part of this peptide in the membrane in a stable and rigid way. The function of this sequence as a cell-penetrating peptide might play a role in the cell surface accumulation of PrP, facilitating membrane translocation and possibly pathogenesis.

6.4.2 Interaction studies

Observation of chemical shift changes in a protein upon ligand binding is a sensitive method for measuring the strength of an interaction and for defining the protein's interaction surface. Especially useful and sensitive are, for example, heteronuclear single quantum correlation (HSQC) spectra [43]. A ^{1}H-^{15}N-HSQC experiment correlates the chemical shift of a ^{15}N-nitrogen nucleus of an NH_x group with the chemical shift of a directly attached protons. Each resonance signal in the HSQC spectrum thus represents a proton that is directly bound to a ^{15}N-nitrogen atom (H^N proton). Therefore, the spectrum contains the signals of the H^N protons and ^{15}N-nitrogens in the protein backbone. Since there is one backbone H^N per amino acid (except for proline), each HSQC signal represents one single amino acid. Because the chemical shifts of the nuclei whose resonances appear in the HSQC are sensitive to their chemical environment, any binding of a ligand molecule in the vicinity induces changes in chemical shifts ("chemical shift perturbation") of the HSQC cross resonances of the respective amide protons and nitrogens. Therefore, one can conclude, that those amino acid residues that show changes in the chemical shifts of their resonances upon addition of a ligand are somehow affected by the ligand binding. One may further conclude, that these residues build up the ligand binding site, although this is not exactly the same statement. Other experiments to map ligand binding sites are, for example, cross-saturation experiments.

Using such methods, details for the interaction of murine PrP with the SH3 domain of the signal transduction protein Grb2 could be elucidated. The existence of the interaction itself had been known before [44]. NMR spectroscopy, however, revealed the structural and dynamical nature of the interaction site on PrP. Differences in hetNOEs, indicating a change in backbone flexibility upon ligand binding, identified the segment moPrP(101–105) as the interaction site [45]. This sequence fulfils the requirements of a classical SH3 binding motif consisting of at least two proline residues separated by exactly two residues (PxxP), and replacement of either of the two proline residues abolishes the binding. This is especially interesting in the light of the Pro102Leu mutation, a common cause for several familiar TSEs in humans [46], implicating the loss of the Grb2-modulated signal transduction pathway from PrP as a possible cause of the neuropathology of TSEs in humans.

Apart from interactions with other proteins, NMR can also reveal details about the interaction of proteins with smaller ligands. By monitoring the ^{1}H -^{15}N heteronuclear correlation spectra of ^{15}N-labeled huPrP in the absence and presence of the tricyclic aromatic molecule quinacrine, it was found that quinacrine binds to a region formed by residues Tyr225, Tyr226, and Gln227 in helix 3 of PrP [47]. Quinacrine, long known to be an anti-malarial medication, was previously shown to cure cultured cells from PrP^{Sc} infection [48]. As the carboxyterminal region of helix 3 is considered a putative binding site for protein X, quinacrine might interfere with the PrP–protein X interaction, thereby preventing the conformational transition of PrP to its pathogenic isoform [47].

Further interaction studies on PrP have concentrated on copper binding to PrP, which has been proposed to play a significant role in maintaining and regulating synaptic copper concentrations [49]. In addition to the known copper-binding properties of the octarepeat region (see Section 6.3), NMR spectroscopy revealed interactions of the sequences huPrP(106–113) and huPrP(106–126) with copper ions. Copper binding is modulated by the residues Lys106, Lys110, and His111 [50]. For the binding of manganese ions to huPrP(106–126), similar effects were observed [51]. Interestingly, ion coordination and peptide conformation differ for Cu^{2+} and Mn^{2+} ions, which might have implications for the mechanism of neurotoxicity this peptide exhibits in the presence of some bivalent metal ions.

6.4.3 Hydrogen exchange studies of protein dynamics

NMR experiments that monitor hydrogen exchange of proteins form, by some means, a middle ground between structural and dynamical information. The key principle is based on the loss of signal intensity occurring after exchange of protons with deuterons in proteins solvated in heavy water (D_2O). Most experiments focus on the amide protons of the protein backbone. Apart from environmental conditions, such as pH and temperature, hydrogen exchange is predominantly governed by the solvent accessibility of the relevant protons. Therefore, it is a method for the direct detection of solvent accessibility, as well as an indirect method to measure backbone flexibility; backbone motions can render amide protons, which seem to be protected from solvent interaction in the average structure accessible to solvent. Additionally, hydrogen bonding has a significant influence on hydrogen exchange, lowering the observed exchange rates.

Not surprisingly, hydrogen exchange experiments performed on Syrian hamster PrP (shPrP) showed most protected residues located within the regular secondary structure elements of the folded carboxyterminal domain. The strongest protection from hydrogen exchange is observed for helices 2 and 3, whereas helix 1 and both strands of the β-sheet exhibit only intermediate protection, indicating elevated dynamic behavior of those regions [52]. The carboxyterminal ends of helices 2 and 3, however, exhibit lower protection factors than the remaining parts of those helices, which points to the existence of an equilibrium between helical and unfolded states in these regions. Similar observations were reported in experiments on the huPrP [16]. However, it should be pointed out that the protection factors observed for the PrP are comparatively small compared to other well-structured proteins. This may indicate a certain degree of structural breathing, even in the core of the protein. A recent crystallographic study revealed the existence of water molecules buried inside the folded core [53]. The authors could also establish a possible role of these structural water molecules for the stability of the protein.

Furthermore, hydrogen exchange studies are suited to identify regions of the hydrogen bonded backbone which remain stable until the protein is fully unfolded. Kinetic analysis identifies those residues whose protection factors are characteristic for exchange from the unfolded form, rather than from the folded state or from unfolding intermediates. By this approach, regions of the protein which remain folded during the whole unfolding process can be identified. Application of this method to the PrP revealed a core of hyperstable residues, which basically coincide with the core residues of the natively folded protein [54]. As this indicates the persistence of the basic fold throughout the whole unfolding process, the protein has necessarily to go through a nearly completely unfolded state in order to adopt an alternative, β-rich conformation.

Recent methodical advancements [55] allow the investigation of hydrogen-exchange rates in the fibrillar state. The method is based on the very slow exchange rates of protein amide protons in mixtures of organic solvents, such as DMSO, with D_2O. Fibril preparations are exposed to D_2O and are subsequently transferred to a DMSO/D_2O mixture to quench the exchange process and to dissolve the fibrils into their monomers. The latter are then investigated by solution NMR studies to determine which amide protons had exchanged during D_2O exposure. Application of this method to fibrils of the amyloidogenic peptide huPrP(106–126) yielded interesting information on possible conformational behavior of the peptide in its fibrillar state [56]. Hydrogen exchange data under these conditions support models derived from molecular dynamics simulation, depicting the fibrillar structure as a stack of four stranded parallel sheets, shown by the existence of single-exponential exchange kinetics in the central part of the peptide.

6.5 High pressure NMR

Whereas the abovementioned methods mostly focused on the native state of PrP, high pressure NMR presents an approach to the characterization of alternative conformations. Perturbing the native state by application of high pressures allows observation of rare conformers, which become stable under such thermodynamic conditions [57]. Under high pressure, the solvated protein can principally react in two ways: (i) reduction of effective volume by general compression of the structure (indicated by linear change of chemical shifts with pressure increase), or by reduction of effective volume by shifting the equilibrium between at least two different conformations (indicated by nonlinear changes in spectral parameters with increasing pressure).

High pressure studies on the PrP revealed the existence of a metastable, only slightly populated alternative conformer of PrP, which has been hypothesized to be identical to the postulated conversion intermediate PrP^{*} [58]. This intermediate is characterized by transient helix formation in the amino terminal region and within the hydrophobic cluster. Anomalous, nonlinear shift deviations could be observed for the strand 1–helix 1 region, as well as for the loops between helix 1 and strand 2, and the loop between helices 2 and 3. Such alterations can be indicative of hinge motions altering the global structure. Interestingly, regions of conformational instability coincide with regions in which most pathogenic mutations of huPrP cluster together. In a more recent study high-pressure NMR measurements were combined with NMR relaxation measurements [59]. This study reveals regions of complex conformational mobility that cluster in the short antiparallel β-sheet and parts of helices 2 and 3. They are in agreement with the earlier observations [30] that suggest a conformational exchange in strand 1. The same study indicates that the helix 2–helix 3 interface, where many pathogenic mutations cluster, undergoes slow motions, which is somewhat in contrast to the deuterium experiments reported above.

6.6 Structural studies on PrP^{Sc}

In contrast to the soluble and crystallizable PrP^{C}, PrP^{Sc} is not directly accessible for structure determination by NMR or X-ray. The insolubility and strong tendency to form fibrils and rather amorphous aggregates renders this isoform inaccessible to classical methods of structure determination. For this reason, no high-resolution structure of PrP^{Sc} is available up to now.

It is, however, possible to gain insight into several structural features of PrP^{Sc} by various biophysical methods. Solid-state NMR is able to retrieve information on the structural behavior of prion peptide aggregates. While it is not yet possible to obtain high-resolution structures from such experiments, they nevertheless can yield information of secondary structure distributions in the aggregated conformation. Significant differences in secondary structure distribution between the fibrillar and the randomly aggregated, amorphous form of moPrP(89–143) were found for several PrP variants. Especially the Pro102Leu variant exerted significant long-range effects on the conformation of the peptide, showing a strong shift of the conformational distribution towards the mostly extended structures. Similarly, introduction of several alanine to valine exchanges in the palindromic AGAAAAGA-region of shPrP, shifted the conformational distribution of PrP from helical to extended [60]. Interestingly, the mere presence of extended conformers seems not to be sufficient to stimulate disease in animal models. This indicates the necessity of very specific, highly defined conformations for the interaction with PrP^{C} and the development of dis-

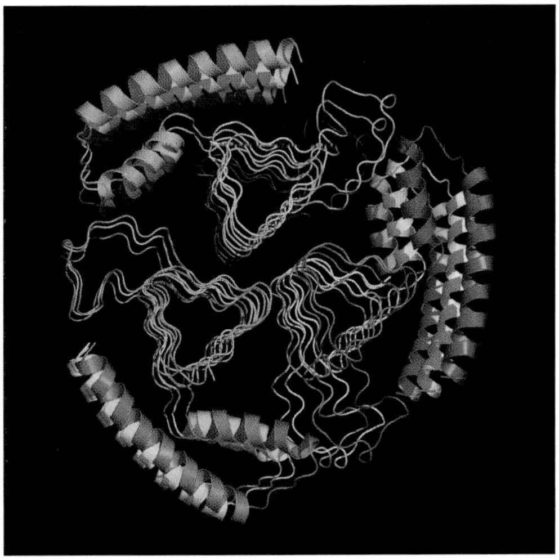

Fig. 6.12: Schematic model of the proposed β-helical conformation of PrP^{Sc}. PrP molecules are arranged in trimeric discs that can staple thereby forming fibrils. The model was constructed as described [64]. Coordinates for Figure 6.12 were kindly provided by M. Stork, LMU Munich. ©: Figure by S. Schwarzinger with PyMOL, DeLano Scientific, San Francisco.

ease. Further evidence for high conformational flexibility of the fibrillar form of the huPrP(109–122) region were concluded from the contents of secondary structure elements in aggregates of this peptide prepared from different solution conditions [61].

A global picture of PrP^{Sc}, albeit with low resolution was obtained using negative stain electron microscopy on two-dimensional crystals of PrP^{Sc} [62]. The method yielded rough structures with a resolution of 0.7 nm, which subsequently could be fitted to a right-handed β-helical model structure based on the structure of the *Methanosarcina thermophila* carbonic anhydrase. Interestingly, the models show helices 2 and 3, which are linked by the intramolecular disulfide bond, retaining their helical conformation. In the remaining part of the protein, consisting of the unstructured amino terminal domain, the β-sheet and helix 1 are incorporated into the β-helix. This underscores the abovementioned importance of the β-strand 1–helix 1–β-strand 2 region for the conformational transition, as this region has to undergo a complete structural rearrangement during conversion to the pathological isoform. Further evidence for the arrangement of such β-helical subunits into a trimeric superstructure could subsequently be obtained by low-resolution fiber diffraction, electron microscopy (EM), and atomic force microscopy (AFM) [63], as well as from molecular modeling studies, as shown in Fig. 6.12 [64].

References

[1] Pervushin K, Riek R, Wider G, et al. Attenuated T2 relaxation by mutual cancellation of dipole–dipole coupling and chemical shift anisotropy indicates an avenue to NMR structures of very large biological macromolecules in solution. Proc Natl Acad Sci USA 1997; 94:12366–12371.

[2] Wüthrich, K. NMR of proteins and nucleic acids. John Wiley & Sons, New York, 1986.

[3] Wishart DS, Sykes BD, Richards FM. The chemical shift index: a fast and simple method for the assignment of protein secondary structure through NMR spectroscopy. Biochemistry 1992; 31:1647–1651.

[4] Brown DR, Qin K, Herms JW, et al. The cellular prion protein binds copper in vivo. Nature 1997; 390:684–687.

[5] Glockshuber R, Hornemann S, Riek R, et al. Three-dimensional NMR structure of a self-folding domain of the prion protein PrP(121–231) [letter; comment]. Trends Biochem Sci 1997; 22:241–242.

[6] Hornemann S, Korth C, Oesch B, et al. Recombinant full-length murine prion protein, mPrP(23–231): purification and spectroscopic characterization. FEBS Lett 1997; 413:277–281.

[7] Zahn R, von Schroetter C, Wuthrich K. Human prion proteins expressed in *Escherichia coli* and purified by high-affinity column refolding. FEBS Lett 1997; 417:400–404.

[8] Riek R, Hornemann S, Wider G, et al. NMR structure of the mouse prion protein domain PrP(121–321). Nature 1996; 382:180–182.

[9] James TL, Liu H, Ulyanov NB, et al. Solution structure of a 142-residue recombinant prion protein corresponding to the infectious fragment of the scrapie isoform. Proc Natl Acad Sci USA 1997; 94:10086–10091.

[10] Hornemann S, Schorn C, Wuthrich K. NMR structure of the bovine prion protein isolated from healthy calf brains. EMBO Rep 2004; 5:1159–1164.

[11] Calzolai L and Zahn R. Influence of pH on NMR structure and stability of the human prion protein globular domain. J Biol Chem 2003; 278:35592–35596.

[12] Glockshuber R, Hornemann S, Billeter M, et al. Prion protein structural features indicate possible relations to signal peptidases [published erratum appears in FEBS Lett 1998 Jul 10; 431(1):130]. FEBS Lett 1998; 426:291–296.

[13] Kaneko K, Zulianello L, Scott M, et al. Evidence for protein X binding to a discontinuous epitope on the cellular prion protein during scrapie prion propagation. Proc Natl Acad Sci USA 1997; 94:10069–10074.

[14] Korth C, Stierli B, Streit P, et al. Prion (PrP^{Sc})-specific epitope defined by a monoclonal antibody. Nature 1997; 390:74–77.

[15] Liu A, Riek R, Zahn R, et al. Peptides and proteins in neurodegenerative disease: helix propensity of a polypeptide containing helix 1 of the mouse prion protein studied by NMR and CD spectroscopy. Biopolymers 1999; 51:145–152.

[16] Zahn R, Liu A, Luhrs T, et al. NMR solution structure of the human prion protein. Proc Natl Acad Sci USA 2000; 97:145–150.

[17] Lopez Garcia F, Zahn R, Riek R, et al. NMR structure of the bovine prion protein. Proc Natl Acad Sci USA 2000; 97:8334–8339.
[18] Lysek DA, Schorn C, Nivon LG, et al. Prion protein NMR structures of cats, dogs, pigs, and sheep. Proc Natl Acad Sci USA 2005; 102:640–645.
[19] Gossert AD, Bonjour S, Lysek DA, et al. Prion protein NMR structures of elk and of mouse/elk hybrids. Proc Natl Acad Sci USA 2005; 102:646–650.
[20] Calzolai L, Lysek DA, Perez DR, et al. Prion protein NMR structures of chickens, turtles, and frogs. Proc Natl Acad Sci USA 2005; 102:651–655.
[21] Haire LF, Whyte SM, Vasisht N, et al. The crystal structure of the globular domain of sheep prion protein. J Mol Biol 2004; 336:1175–1183.
[22] Knaus KJ, Morillas M, Swietnicki W, et al. Crystal structure of the human prion protein reveals a mechanism for oligomerization. Nat Struct Biol 2001; 8:770–774.
[23] Billeter M, Riek R, Wider G, et al. Prion protein NMR structure and species barrier for prion diseases. Proc Natl Acad Sci USA 1997; 94:7281–7285.
[24] Liemann S and Glockshuber R. Influence of amino acid substitutions related to inherited human prion diseases on the thermodynamic stability of the cellular prion protein. Biochemistry 1999; 38:3258–3267.
[25] Zhang Y, Swietnicki W, Zagorski MG, et al. Solution structure of the E200K variant of human prion protein. Implications for the mechanism of pathogenesis in familial prion diseases. J Biol Chem 2000; 275:33650–33654.
[26] Hosszu LL, Jackson GS, Trevitt CR, et al. The residue 129 polymorphism in human prion protein does not confer susceptibility to Creutzfeldt–Jakob disease by altering the structure or global stability of PrP^{C}. J Biol Chem 2004; 279:28515–28521.
[27] Donne DG, Viles JH, Groth D, et al. Structure of the recombinant full-length hamster prion protein PrP(29–231): the N terminus is highly flexible. Proc Natl Acad Sci USA 1997; 94:13452–13457.
[28] Riek R, Hornemann S, Wider G, et al. NMR characterization of the full-length recombinant murine prion protein, mPrP(23–231). FEBS Lett 1997; 413:282–288.
[29] Zahn R. Prion propagation and molecular chaperones. Q Rev Biophys 1999; 32:309–370.
[30] Viles JH, Donne D, Kroon G, et al. Local structural plasticity of the prion protein. Analysis of NMR relaxation dynamics. Biochemistry 2001; 40:2743–2753.
[31] Jackson GS, Murray I, Hosszu LL, et al. Location and properties of metal-binding sites on the human prion protein. Proc Natl Acad Sci USA 2001; 98:8531–8535.
[32] Zahn R. The octapeptide repeats in mammalian prion protein constitute a pH-dependent folding and aggregation site. J Mol Biol 2003; 334:477–488.
[33] Renner C, Fiori S, Fiorino F, et al. Micellar environments induce structuring of the N-terminal tail of the prion protein. Biopolymers 2004; 73:421–433.
[34] Jones CE, Abdelraheim SR, Brown DR, et al. Preferential Cu^{2+} coordination by His96 and His111 induces beta-sheet formation in the unstructured amyloidogenic region of the prion protein. J Biol Chem 2004; 279:32018–32027.
[35] Viles JH, Cohen FE, Prusiner SB, et al. Copper binding to the prion protein: structural implications of four identical cooperative binding sites. Proc Natl Acad Sci USA 1999; 96:2042–2047.
[36] Mentler M, Weiss A, Grantner K, et al. A new method to determine the structure of the metal environment in metalloproteins: investigation of the prion protein octapeptide repeat $Cu^{(2+)}$ complex. Eur Biophys J 2005; 34:97–112.
[37] Burns CS, Aronoff-Spencer E, Legname G, et al. Copper coordination in the full-length, recombinant prion protein. Biochemistry 2003; 42:6794–6803.
[38] Jones CE, Klewpatinond M, Abdelraheim SR, et al. Probing $copper^{2+}$ binding to the prion protein using diamagnetic $nickel^{2+}$ and 1H NMR: the unstructured N terminus facilitates the coordination of six $copper^{2+}$ ions at physiological concentrations. J Mol Biol 2005; 346:1393–1407.
[39] Schwarzinger S, Kroon GJ, Foss TR, et al. Sequence-dependent correction of random coil NMR chemical shifts. J Am Chem Soc 2001; 123:2970–2978.
[40] Sharman GJ, Kenward N, Williams HE, et al. Prion protein fragments spanning helix 1 and both strands of beta sheet (residues 125–170) show evidence for predominantly helical propensity by CD and NMR. Fold Des 1998; 3:313–320.
[41] Ziegler J, Sticht H, Marx UC, et al. CD and NMR studies of prion protein (PrP) helix 1. Novel implications for its role in the PrP^{C}–PrP^{Sc} conversion process. J Biol Chem 2003; 278:50175–50181.
[42] Biverstahl H, Andersson A, Graslund A, et al. NMR solution structure and membrane interac-

tion of the N-terminal sequence (1–30) of the bovine prion protein. Biochemistry 2004; 43:14940–14947.

[43] Görlach M, Wittekind M, Beckman RA, et al. Interaction of the RNA-binding domain of the hnRNP C proteins with RNA. Embo J 1992; 11:3289–3295.

[44] Spielhaupter C and Schatzl H.M. PrP^C directly interacts with proteins involved in signaling pathways. J Biol Chem 2001; 276:44604–44612.

[45] Lysek DA and Wuthrich K. Prion protein interaction with the C-terminal SH3 domain of Grb2 studied using NMR and optical spectroscopy. Biochemistry 2004; 43:10393–10399.

[46] Telling GC, Haga T, Torchia M, et al. Interactions between wild-type and mutant prion proteins modulate neurodegeneration in transgenic mice. Genes Dev 1996; 10:1736–1750.

[47] Vogtherr M, Grimme S, Elshorst B, et al. Antimalarial drug quinacrine binds to C-terminal helix of cellular prion protein. J Med Chem 2003; 46:3563–3564.

[48] Doh-Ura K, Iwaki T, Caughey B. Lysosomotropic agents and cysteine protease inhibitors inhibit scrapie-associated prion protein accumulation. J Virol 2000; 74:4894–4897.

[49] Brown DR and Sassoon J. Copper-dependent functions for the prion protein. Mol Biotechnol 2002; 22:165–178.

[50] Belosi B, Gagelli R, Guerrini R, et al. Copper binding to the neurotoxic peptide PrP106–126: thermodynamic and structural studies. Chembiochem 2004; 5:349–359.

[51] Gaggelli E, Bernardi F, Molteni E, et al. Interaction of the human prion PrP(106–126) sequence with copper(II), manganese(II), and zinc(II): NMR and EPR studies. J Am Chem Soc 2005; 127:996–1006.

[52] Liu H, Farr-Jones S, Ulyanov B, et al. Solution structure of Syrian hamster prion protein rPrP(90–231). Biochemistry 1999; 38:5362–5377.

[53] De Simone A, Dodson GG, Verma CS, et al. Prion and water: tight and dynamical hydration sites have a key role in structural stability. Proc Natl Acad Sci USA 2005; 102:7535–7540.

[54] Hosszu LL, Baxter NJ, Jackson GS, et al. Structural mobility of the human prion protein probed by backbone hydrogen exchange. Nat Struct Biol 1999; 6:740–743.

[55] Ippel JH, Olofsson A, Schleucher J, et al. Probing solvent accessibility of amyloid fibrils by solution NMR spectroscopy. Proc Natl Acad Sci USA 2002; 99:8648–8653.

[56] Kuwata K, Matumoto T, Cheng H, et al. NMR-detected hydrogen exchange and molecular dynamics simulations provide structural insight into fibril formation of prion protein fragment 106–126. Proc Natl Acad Sci USA 2003; 100:14790–14795.

[57] Akasaka K. Highly fluctuating protein structures revealed by variable-pressure nuclear magnetic resonance. Biochemistry 2003; 42:10875–10885.

[58] Kuwata K, Li H, Yamada H, et al. Locally disordered conformer of the hamster prion protein: a crucial intermediate to PrP^{Sc}? Biochemistry 2002; 41:12277–12283.

[59] Kuwata K, Kamatari YO, Akasaka K, et al. Slow conformational dynamics in the hamster prion protein. Biochemistry 2004; 43:4439–4446.

[60] Laws DD, Bitter HM, Liu K, et al. Solid-state NMR studies of the secondary structure of a mutant prion protein fragment of 55 residues that induces neurodegeneration. Proc Natl Acad Sci USA 2001; 98:11686–11690.

[61] Heller J, Kolbert AC, Larsen R, et al. Solid-state NMR studies of the prion protein H1 fragment. Protein Sci 1996; 5:1655–1661.

[62] Wille H, Michelitsch MD, Guenebaut V, et al. Structural studies of the scrapie prion protein by electron crystallography. Proc Natl Acad Sci USA 2002; 99:3563–3568.

[63] Govaerts C, Wille H, Prusiner SB, et al. Evidence for assembly of prions with left-handed beta-helices into trimers. Proc Natl Acad Sci USA 2004; 101:8342–8347.

[64] Stork M, Giese A, Kretzschmar HA, et al. Molecular dynamics simulations indicate a possible role of parallel beta-helices in seeded aggregation of poly-Gln. Biophys J 2005; 88:2442–2451.

7 Function of Cellular Prion Protein (PrP^C) in Copper Homeostasis and Redox Signaling at the Synapse

Jochen Herms and Hans Kretzschmar

7.1 Introduction

The cellular prion protein (PrP^C) is predominantly expressed in the central nervous system (CNS) and is a prerequisite for the pathogenesis of prion diseases. While the exact physiological function of PrP^C remains unclear, increasing evidence indicates a role of PrP^C in oxidative homeostasis in neurons, modulating synaptic transmission and neuronal calcium homeostasis. There is a tentative hypothesis of basic prion function based on the copper-binding ability of PrP^C, i. e., that PrP^C acts as a sensor for free radical stimuli, thereby triggering intracellular calcium-dependent signals that finally translate into modulation of synaptic function and maintenance of neuronal integrity.

7.2 Cellular location of PrP^C

The prion protein is a sialoglycoprotein that is attached to the outer leaflet of the plasma membrane via a C-terminal glycosyl-phosphatidylinositol (GPI) anchor [1]. Although it is most abundant in the CNS, it is also expressed in many non-neuronal tissues, including blood lymphocytes, gastroepithelial cells, heart, kidney, and muscle [2–5]. In the brain, PrP^C is particularly localized at synaptic membranes, most likely the presynaptic domain as indicated by immunohistochemical and immunoelectron microscopic studies [6]. This is also supported by the finding of a PrP^C enrichment in synaptic plasma membrane fractions, shown in Figure 7.1 [7]. A punctuate appearance of plasma membrane staining has suggested that PrP^C is organized in clusters, representing membrane microdomains [8]. Indeed, as other GPI-anchored proteins, PrP^C is enriched in specialized sphingolipid cholesterol compartments called rafts. The predominant presence and location on the plasma membrane of neurons suggests a function that involves the prion protein in synaptic transmission and neuronal excitability [7].

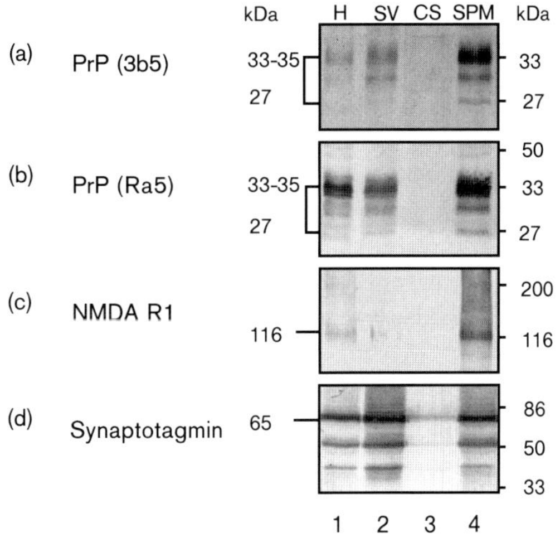

Fig. 7.1: Enrichment of PrP^C in the synaptic plasma membrane fraction. Equal amounts (100 µg/lane) of subcellular fractions from wild-type mouse brain were analyzed by immunoblotting with anti-PrP antibody 3B5 **(a)** and antiserum Ra5 **(b)**. The immunostaining for the NMDA receptor subunit NMDA-R1 **(c)** and synaptotagmin **(d)** was used as a control for post-synaptic membrane proteins and synaptic vesicle proteins, respectively. Subcellular fractions are designated as follows: H, homogenate; SV, crude synaptic vesicle fraction; CS, cytosolic synaptosomal fraction; SPM, synaptic plasma membranes. Figure by J. Herms ©.

7.3 Protein interactions with PrP^C

It has been suggested that PrP^C might function through an interaction with other proteins. A number of proteins at the cell membrane, in endocytic compartments and in the secretory pathway, have been described to be associated with PrP^C. Cell surface proteins associated with PrP^C include N-Cam [9] and Aplp1 [10]. The physiological role of PrP^C binding to these proteins has not been elucidated. The M_r 67000 laminin receptor, a transmembrane protein, reported to associate with PrP^C at the cell surface as well, has been shown to mediate PrP^C internalization [11]. It is, however, intriguing that the binding site for the laminin receptor lies in a domain comprising amino acid residues 144–179 of human PrP^C [11] that is distinct from the domain believed to participate in internalization [12; 13]. An additional PrP^C-associated membrane protein is the stress-inducible protein (STI1) that interacts through the highly conserved and hydrophobic domain comprising amino acids 113–129 of PrP^C [14]. This interaction has been linked to the neuroprotective role of PrP^C through a cAMP/PKA signaling pathway. Other PrP^C-associated proteins present in intracellular vesicles or caveola-like domains such as synapsin, Grb2, and Pint1 [15], p75 [16], caveolin [17–19], and casein kinase 2 [20] have also been identified, but their role is still unknown. BiP, a protein in the endoplasmic reticulum has also been shown to bind to PrP^C. This complex is degraded by the proteasome pathway, thus preventing the formation of aggregates, suggesting that BiP may play a chaperone role in the quality control of the pathway of PrP^C maturation [21]. Another chaperone identified as interacting with PrP^C is the heat shock protein Hsp60. This interaction is believed to catalyze the aggregation of PrP^C [22; 23]. Finally, other cellular proteins such as GFAP [24] and Bcl-2 [25] have also been described as binding to PrP^C, but the cellular compartments and functions of these interactions are as yet unclear.

7.4 PrP^C binds copper(II) ions

In contrast to the heterogeneous group of proteins which are believed, though not proven, to interact with PrP^C, one of the most clearly documented and perhaps one of the most salient features of the prion protein is its ability to bind copper(II) [Cu(II)] ions. Electronspray ionization mass spectroscopy and tryptophan fluorescence spectroscopy studies have demonstrated that among divalent metal ions, PrP^C selectively binds Cu(II) ions [26; 27]. The major Cu(II)-binding site has been identified as the N-terminal region (encompassing residues 60–91 in human PrP^C), specifically an octarepeat domain consisting of four sequential repeats of the sequence PHGGGWGQ [27–29]. The relevance of the Cu(II) binding to the physiological function of PrP^C is reflected in the high degree of conservation of the octarepeat region among prion proteins in mammalian species; a related hexapeptide has evolved in birds [30]. Most studies have concluded that the octarepeat region of $humPrP_{60-91}$ binds four Cu(II) ions [28; 29; 31–34]. There is, however, disagreement as to how copper binds to PrP^C. Investigation of copper binding to the full-length prion protein at a physiological pH of 7.0 indicates positive cooperative binding [34]. In other models, each octarepeat independently binds Cu(II), the minimum binding unit being the HGGGW segment [32; 33]. The octarepeat domain is remarkably selective for $Cu2^+$, with affinity for other metal ion species weak or nonexistent [26; 27; 29]. Moreover, copper binding to PrP^C is highly pH dependent. Binding is most favored at a physiological pH of ~ 7.4, and falls sharply under mildly acidic ($pH < 6.0$) conditions [31; 32; 34]. Affinity data for full-length PrP^C showed that the protein is saturated at about 5.0 μM Cu(II) [34]. Similar dissociation constant (K_d) values were reported for shorter fragments spanning residues 23–98 (5.9 μM) [28] and residues 60–91 (6.7 μM) [29]. These values are compatible with the physiological Cu(II) concentration reached within the synaptic cleft (15.0 μM) during synaptic vesicle release [35].

7.5 Functional relevance of copper binding of PrP^{C} at the synapse

The ability of PrP^{C} to bind copper may not be incidental as first proposed by Brown et al. 1997 [28]. Synaptosomal preparations of *Prnp*$^{o/o}$ reveal a 50 % reduction in copper content with respect to wild-type, as shown in Figure 7.2 [7]. This 50 % reduction is much too strong to be solely attributed to the loss of the copper bound to PrP^{C}. A straightforward explanation would be that PrP^{C} has a direct role in brain copper metabolism, with the protein transporting Cu(II) ions from the extracellular milieu to the acidified cellular vacuoles [36]. However, no differences were observed in the uptake of radioactive copper between *Prnp*$^{o/o}$ and wild-type synaptosomes [37]. Moreover, total brain copper content in *Prnp*$^{o/o}$ and Tga20 mice, which express ~ 10 times the normal level of PrP [38] was not found to be significantly different from that in wild-type mice [39]. Similarly, copper–zinc superoxide dismutase (CuZnSOD) and cytochrome c oxidase activities (both cuproenzymes) in brain extracts from *Prnp*$^{o/o}$, wild-type, and Tga20 mice did not reveal any significant differences [39]. Given these results, it appears unlikely that PrP^{C} is primarily responsible for uptake of copper into neurons or that its chief role is in specialized trafficking pathways for delivery of the metal to CuZnSOD or other cuproenzymes.

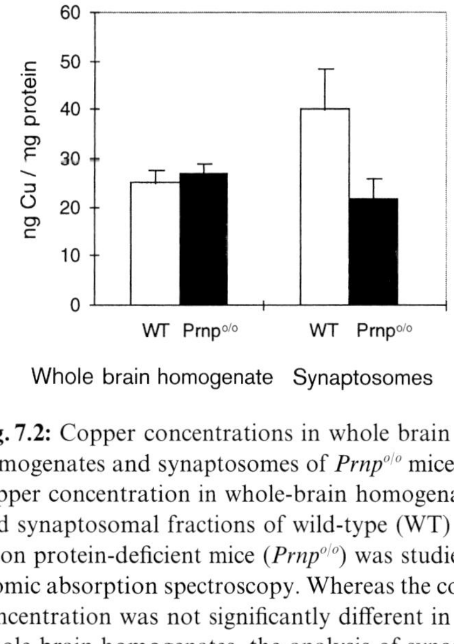

Fig. 7.2: Copper concentrations in whole brain homogenates and synaptosomes of *Prnp*$^{o/o}$ mice. The copper concentration in whole-brain homogenates and synaptosomal fractions of wild-type (WT) and prion protein-deficient mice (*Prnp*$^{o/o}$) was studied by atomic absorption spectroscopy. Whereas the copper concentration was not significantly different in the whole-brain homogenates, the analysis of synaptosomal preparations revealed a significantly reduced copper concentration in *Prnp*$^{o/o}$ synaptosomes compared with wild-type (Students *t*-test; $p < 0.05$). Figure by J. Herms ©.

Another, more likely hypothesis is that PrP^{C} acts to buffer Cu(II) levels in the synaptic cleft, following release of copper ions as a result of synaptic vesicle fusion [35]. Thus, the micromolar K_d of PrP^{C} together with its localization at the presynaptic membrane would allow the prion protein to effectively buffer copper at the site from where it is released. Furthermore, PrP^{C} may be involved in redistribution of copper back into the presynaptic cytosol although it does not seem to directly transport copper across the membrane. The maintenance of copper levels in the presynaptic cytosol is of physiological importance as nerve endings release copper into the synaptic cleft upon depolarization [35; 40]. Importantly, such a hypothesis would explain why the total copper content in brain lysates from *Prnp*$^{o/o}$ mice remains unchanged, because PrP^{C} may alter the distribution but not the overall amount of copper within the brain. Electrophysiological studies in *Prnp*$^{o/o}$ mice have shown abnormalities that can be explained on the basis of increased synaptic copper concentrations as a result of decreased buffering by PrP^{C}. For instance, hippocampal slices from *Prnp*$^{o/o}$ mice exhibited weakened $GABA_A$ receptor-mediated inhibition and impaired long-term potentiation (LTP) [41]. Cu(II) ions are known to be potent inhibitors of $GABA_A$ receptor-mediated currents at low micromolar concentrations [42] and long-term potentiation [43]. Inhibitory synaptic transmission in *Prnp*$^{o/o}$ mice brain slice preparations is significantly affected by the application of low copper concentrations that do not affect synaptic transmission in wild-type control brain slices, as shown in Figure 7.3. [28].

Taken together, the experimental data considered so far seems to be fairly supportive of a hypothesis whereby the location of PrP^{C} in the synaptic membrane allows the protein to re-

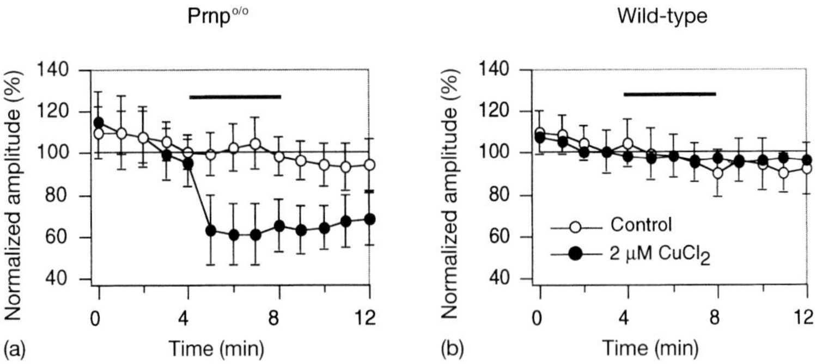

Fig. 7.3: Effect of copper on spontaneous inhibitory synaptic currents in cerebellar Purkinje cells in *Prnp*$^{o/o}$ and wild-type mice. **(a)** In *Prnp*$^{o/o}$ Purkinje cells the sum of IPSC is significantly reduced by 2 μM $CuSO_4$ (filled circles). **(b)** Application of 2 μM $CuSO_4$ (filled circles) to wild-type Purkinje cells does not change the sum of IPSC amplitudes in comparison to untreated control (open circles). Each point represents the mean ± SEM of sum of amplitudes of all synaptic events > 20 pA during 60-s intervals normalized to the values before application of copper. Bar represents time of $CuSO_4$ application. Mean and SE are shown for five experiments. Figure by J. Herms ©.

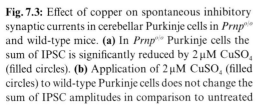

plenish copper content in the presynaptic cytosol, and simultaneously buffer against toxic levels of Cu(II) ions in the synaptic cleft, for example, during high synaptic activity.

7.6 Neuroprotective role of the prion protein in response to copper and oxidative stress

There are several other experimental findings related to prion protein function that cannot be explained on the basis of copper buffering and uptake alone. In particular, a considerable amount of data has been accumulated implicating PrP^C as a neuroprotective molecule, especially in response to pro-oxidative insults. Thus, prion protein-deficient cerebellar neurons showed around 30 % increased sensitivity to O_2^- generated by xanthine/xanthine oxidase [44]. Sensitivity was even greater to H_2O_2, with up to 60 % decreased viability of *Prnp*$^{o/o}$ cerebellar granule neurons (CGN) compared to wild-type [45]. Importantly, *Prnp*$^{o/o}$ CGNs were found to have lower glutathione reductase (GR) activity [45]. Since GR functions in the regeneration of cellular glutathione (GSH), lower GR activity would decrease the breakdown of H_2O_2 by glutathione peroxidase, and hence be related to the increased sensitivity of *Prnp*$^{o/o}$ CGN neurons to H_2O_2. Evidence for a role of PrP^C in cell survival was also demonstrated by Kuwahara [46], who showed that immortalized hippocampal neurons prepared from *Prnp*$^{o/o}$ mice were more susceptible to the apoptosis-inducing serum deprivation insult. Moreover, PrP^C was found to protect human neurons in primary culture against Bax-mediated cell death [47], a phenomenon that was eliminated by depletion of the octapeptide repeats of PrP^C.

Despite such strong evidence favoring the neuroprotective role of cellular prions, particularly against oxidative stress, the actual biological activity of the protein that mediates the neuroprotection is as yet unknown. One has to presume that it would involve Cu(II) binding/release, given the fundamental importance of the N-terminal octarepeat region in defining the biochemical, molecular, and cellular activities of PrP^C. It is in this context that the concept of PrP^C as a synaptic superoxide dismutase (SOD) was proposed [48]. In vitro experiments using recombinant mouse and chicken PrP^C refolded to incorporate Cu(II), showed that PrP^C may have SOD activity [49]. The rate constant for the dismutation reaction was estimated to be $4 \times 10^8\ M^{-1}s^{-1}$, i.e., a magnitude lower than that of CuZnSOD ($6.4 \times 10^9\ M^{-1}s^{-1}$) but similar to MnSOD ($6 \times 10^8\ M^{-1}s^{-1}$). Prion protein immunoprecipitated from wild-type mouse brain was reported to contribute 10–15 % of total SOD activity in vivo [50]. In addition, in those experiments SOD activity of PrP^C was dependent on the level of Cu(II) incorporated into the molecule; thus, incorporation of three or four copper atoms resulted in higher activity than two, while no activity at all was detected with only one copper ion bound to PrP^C [51]. The mechanism of the proposed dismutation reaction is unknown. It is presumed to involve reduction of Cu(II) to Cu(I), with the formation of H_2O_2.

Notwithstanding its attractiveness, there are several arguments that do not favor the hypothesis of synaptic SOD activity of PrP^C. Firstly, it is uncharacteristic of a dismutating enzyme to exhibit pH-dependent binding to its catalytic metal, as does PrP^C. Secondly, affinity of the apoprotein to its metal is normally much higher than the 5–7 μM estimated for Cu(II)-binding to PrP^C. CuZnSOD, for example, has an affinity of about 10^{-14} M for Cu(II) [52]. Thirdly, recent analysis of genetically defined crosses of mice lacking the SOD1 gene with mice lacking or overexpressing PrP^C failed to show any influence of the PrP genotype and gene dosage on SOD1 or SOD2 activity in heart, spleen, brain, and synaptosome-enriched brain fractions [53].

7.7 Redox signaling by PrP^C modulates intracellular calcium homeostasis and synaptic function

At this point it can be confidently concluded that PrP^C maintains synaptic integrity and neurophysiological function by going beyond mere copper buffering (Fig. 7.4). It remains to be elucidated how PrP^C can influence resistance to a wide range of insults. If Cu(II)-bound PrP^C were to activate an intracellular signaling pathway, then such an effect would not be unexpected. The most likely candidate for the intracellular messenger would be calcium. Indeed, there is strong experimental evidence that PrP^C influences the neuronal calcium homeostasis. The most consistent electrophysiological alteration found in PrP-deficient mice is a marked reduction in outward late after-hyperpolarization currents, I_{AHP} [54–56]. Patch-clamp studies on cerebellar

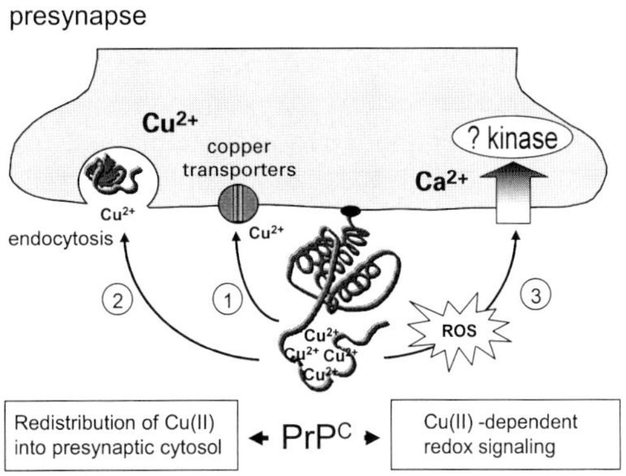

Fig. 7.4: Hypothetical model showing a possible function of PrP^C in copper homeostasis and redox signaling at the synaptic plasma membrane. Cu^{2+} ions released during neurotransmitter vesicle release are buffered by PrP^C and subsequently returned to the pre-synaptic cytosol. This can occur either by copper transport proteins (1) (CTP) or via PrP^C-mediated endocytosis (2). Alternatively, copper bound to PrP^C may interact with ROS, including H_2O_2 and O_2^-, triggering redox signaling and subsequent activation of Ca^{2+}-dependent signaling cascades. (3) Changes in intracellular free Ca^{2+} level lead to modulation of synaptic function and neuroprotection. Figure by J. Herms ©.

Purkinje cells of *Prnp*$^{o/o}$ mice revealed that altered AHPs are most likely due to impaired activation of Ca^{2+}-dependent K^+ channels [55]. Moreover, cerebellar granule neurons from *Prnp*$^{o/o}$ mice were found to exhibit low resting $[Ca^{2+}]_i$ levels, as well as a decreased response of $[Ca^{2+}]_i$ to high K^+-induced depolarization [57].

This disruption of the neuronal calcium homeostasis could explain more complex phenomena observed in *Prnp*$^{o/o}$ mice including alterations in long term potentiation, abnormalities in circadian rhythm and sleep [58; 59], impaired performance in latent learning and disturbed long-term memory [60], and increased susceptibility to epileptic seizures [61]. Therefore, a disturbance of calcium homeostasis is conceivably the basic physiological alteration in prion-deficient neuronal cells.

What, then, is the connection between the physiological Cu(II)-binding/release activity of PrP^C and the regulation of calcium homeostasis? One potential scenario is that PrP^C acts as a modulator of calcium flux in response to Cu(II) and/or reactive oxygen species (ROS), with copper enabling redox signaling and triggering metal-dependent biochemical and cellular responses [62]. Thus, exposure to raised copper levels (e. g., during high synaptic activity) would cause several Cu(II) ions to bind the prion protein and the initiation of Ca^{2+}-activated signaling cascades [63]. Moreover, PrP^C attached to Cu(II) allows the protein to participate in redox reactions [64; 65]. PrP^C might function as a redox signaling molecule, activating an, as yet undefined, membrane protein kinase. We recently obtained data that strongly indicate a functional link between PrP^C expression and phosphatidylinositol 3-kinase (PI 3-kinase) activation, a protein kinase that plays a pivotal role in cell survival [66].

PrP^C can, in addition, react with H_2O_2 [64; 65]. In the presence of physiological concentrations (as low as 10 μM) of H_2O_2 and Cu(II), PrP^C was cleaved into the octapeptide repeat region within each of the metal-binding repeats. Interestingly, O_2^- also induced a site-specific cleavage of the prion amino terminus [64]. Moreover, H_2O_2 cleavage of PrP^C was Cu(II)-dependent (a copper(I)-chelator had no effect) and pH-dependent, thereby further illustrating the participation of prion-bound Cu(II) in the reaction. Thus, PrP^C is cleaved, yielding a 28.5 kDa protein, and Cu(II) ions are released by the following mechanism:

$$\mathrm{PrP{-}Cu^{2+} + H_2O_2 \rightarrow PrP{-}Cu^{+} + 2\,H^{+} + O_2^{-}} \qquad (1)$$

$$\mathrm{PrP{-}Cu^{+} + H_2O_2 \rightarrow PrP{-}OH^{\cdot} + OH^{-} + Cu^{2+}} \qquad (2)$$

It can be concluded that PrP^C does interact with O_2^- in a copper-dependent manner, but unlike with CuZnSOD, this interaction leads to the cleavage of the protein. Such physiological ROS processing may deliver a redox signal that induces protective responses mediated by an increase of the intracellular free calcium concentration.

Alternatively, the oxidative amino-proximally truncated PrP^C molecules may form multimeric, membrane-spanning channel complexes [67], which may be calcium permeable. This hypothesis could explain the neurotoxic effects of N-terminal deleted PrP^C [38] or the neurotoxic effect of Doppel expression [68; 69]. Doppel (Dpl) encodes a 179-residue protein that has sequence similarities to the C-terminus of PrP^C [70]. The artificial expression of truncated PrP^C or Dpl in neurons may induce the formation of calcium-permeable pores within the plasma membrane [67]. This would cause an increase of the basal free calcium concentration, a phenomenon that is known to modulate apoptotic nerve cell death in cerebellar granule cells [71]. The fact that the neurotoxic expression of those proteins disappears when full-length PrP^C is co-expressed [68; 69] may be due to the incorporation of the full-length protein into the channel complex. The intact N-terminus of the PrP^C molecule and the Cu(II) ions bound to it, might block the channel pore.

To conclude, we propose that the physiological role of PrP^C is due to its localization at the plasma membrane, where the prion protein acts as a sensor for ROS stimuli. Here copper may generate a signal through redox chemistry, thereby switching on Ca^{2+}-dependent signaling pathways leading to downstream effects that modulate synaptic transmission and maintain neuronal integrity. Identifying the signaling pathways in which PrP^C is involved is thus of prime importance for the further elucidation of the enigmatic biological function of prion proteins [63].

References

[1] Stahl N, Baldwin MA, Burlingame AL, et al. Identification of glycoinositol phospholipid linked and truncated forms of the scrapie prion protein. Biochemistry 1990; 29:8879–8885.

[2] Bendheim PE, Brown HR, Rudelli RD, et al. Nearly ubiquitous tissue distribution of the scrapie agent precursor protein. Neurology 1992; 42:149–156.

[3] Horiuchi M, Yamazaki N, Ikeda T, et al. A cellular form of prion protein (PrP^C) exists in many non-neuronal tissues of sheep. J Gen Virol 1995; 76:2583–2587.

[4] Fournier J-G, Escaig-Haye F, De Villemeur TB, et al. Distribution and submicroscopic immunogold localization of cellular prion protein

(PrP^C) in extracerebral tissues. Cell Tissue Res 1998; 292:77–84.
[5] Pammer J, Cross HS, Frobert Y, et al. The pattern of prion-related protein expression in the gastrointestinal tract. Virchows Arch 2000; 436(5):466–472.
[6] Fournier JG, Escaig-Haye F, Grigoriev V. Ultrastructural localization of prion proteins: physiological and pathological implications. Microsc Res Tech 2000; 50(1):76–88.
[7] Herms J, Tings T, Gall S, et al. Evidence of presynaptic location and function of the prion protein. J Neurosci 1999; 19:8866–8875.
[8] Laine J, Marc ME, Sy MS, et al. Cellular and subcellular morphological localization of normal prion protein in rodent cerebellum. Eur J Neurosci 2001; 14(1):47–56.
[9] Schmitt-Ulms G, Legname G, Baldwin MA, et al. Binding of neural cell adhesion molecules (N-CAMs) to the cellular prion protein. J Mol Biol 2001; 314(5):1209–1225.
[10] Yehiely F, Bamborough P, Costa MD, et al. Identification of candidate proteins binding to prion protein. Neurobiol Dis 1997; 3:339–355.
[11] Gauczynski S, Hundt C, Leucht C, et al. Interaction of prion proteins with cell surface receptors, molecular chaperones, and other molecules. Adv Protein Chem 2001; 57:229–272.
[12] Shyng S-L, Moulder KL, Lesko A, et al. The N-terminal domain of a glycolipid-anchored prion protein is essential for its endocytosis via clathrin-coated pits. J Biol Chem 1995; 270(24):14793–14800.
[13] Lee KS, Magalhaes AC, Zanata SM, et al. Internalization of mammalian fluorescent cellular prion protein and N-terminal deletion mutants in living cells. J Neurochem 2001; 79(1):79–87.
[14] Zanata SM, Lopes MH, Mercadante AF, et al. Stress-inducible protein 1 is a cell surface ligand for cellular prion that triggers neuroprotection. EMBO J 2002; 21(13):3307–3316.
[15] Spielhaupter C and Schatzl HM. PrP^C directly interacts with proteins involved in signaling pathways. J Biol Chem 2001; 276(48):44604–12.
[16] Della-Bianca V, Rossi F, Armato U, et al. Neurotrophin p75 receptor is involved in neuronal damage by prion peptide 106–126. J Biol Chem 2001; 276(42):38929–33.
[17] Gorodinsky A and Harris DA. Glycolipid-anchored proteins in neuroblastoma cells form detergent-resistant complexes without caveolin. J Cell Biol 1995; 129(3):619–627.
[18] Harmey JH, Doyle D, Brown V, et al. The cellular isoform of the prion protein, PrP^C, is associated with caveolae in mouse neuroblastoma (N_2a)cells. Biochem Biophys Res Commun 1995; 210(3):753–759.
[19] Mouillet-Richard S, Ermonval M, Chebassier C, et al. Signal transduction through prion protein. Science 2000; 289(5486):1925–1928.
[20] Meggio F, Negro A, Sarno S, et al. Bovine prion protein as a modulator of protein kinase CK2. Biochem J 2000; 352(1):191–196.
[21] Jin T, Gu Y, Zanusso G, et al. The chaperone protein BiP binds to a mutant prion protein and mediates its degradation by the proteasome. J Biol Chem 2000; 275(49):38699–38704.
[22] Edenhofer F, Rieger R, Famulok M, et al. Prion protein PrP^C interacts with molecular chaperones of the Hsp60 family. J Virol 1996; 70(7):4724–4728.
[23] Stockel J and Hartl FU. Chaperonin-mediated de novo generation of prion protein aggregates. J Mol Biol 2001; 313(4):861–872.
[24] Oesch B, Teplow TB, Stahl N, et al. Identification of cellular proteins binding to the scrapie prion protein. Biochemistry 1990; 29:5848–5855.
[25] Kurschner C and Morgan JI. The cellular prion protein (PrP) selectively binds to Bcl-2 in the yeast two-hybrid system. Mol Brain Res 1995; 30:165–167.
[26] Stöckel J, Safar J, Wallace AC, et al. Prion protein selectively binds copper (II) ions. Biochem 1998; 37:7185–7193.
[27] Whittal RM, Ball HL, Cohen FE, et al. Copper binding to octarepeat peptides of the prion protein monitored by mass spectrometry. Protein Sci 2000; 9(2):332–343.
[28] Brown DR, Qin K, Herms J, et al. The cellular prion protein binds copper in vivo. Nature 1997; 390:684–687.
[29] Hornshaw MP, McDermott JR, Candy JM. Copper binding to the N-terminal tandem repeat regions of mammalian and avian prion protein. Biochem Biophys Res Commun 1995; 207(2): 621–629.
[30] Gabriel JM, Oesch B, Kretzschmar HA, et al. Molecular cloning of a candidate chicken prion protein. Proc Natl Acad Sci USA 1992; 89:9097–9101.
[31] Viles JH, Cohen FE, Prusiner SB, et al. Copper binding to the prion protein: structural implications of four identical cooperative binding sites. Proc Natl Acad Sci USA 1999; 96:2042–2047.
[32] Miura T, Hori-i A, Mototani H, et al. Raman spectroscopic study on the copper(II) binding mode of prion octapeptide and its pH dependence. Biochemistry USA 1999; 38:11560–11569.

[33] Aronoff-Spencer E, Burns CS, Avdievich NI, et al. Identification of the Cu(2+) binding sites in the N-terminal domain of the prion protein by EPR and CD spectroscopy. Biochemistry 2000; 39(45):13760–13771.
[34] Kramer ML, Kratzin HD, Schmidt B, et al. Prion protein binds copper within the physiological concentration range. J Biol Chem 2001; 276(20):16711–16719.
[35] Hopt A, Korte S, Fink H, et al. Methods for studying synaptosomal copper release. J Neurosci Methods 2003; 128(1–2):159–172.
[36] Pauly PC and Harris DA. Copper stimulates endocytosis of the prion protein. J Biol Chem 1998; 273:33107–33111.
[37] Buchholz M, Giese A, Herms J, et al. Mouse brain synaptosomes efficiently accumulate copper-67 by two distinct prosesses independent of cellular prion protein. J Mol Neurosci 2005; 27(3):347–54.
[38] Shmerling D, Hegyi I, Fischer M, et al. Expression of amino-terminally truncated PrP in the mouse leading to ataxia and specific cerebellar lesions. Cell 1998; 93:203–214.
[39] Waggoner DJ, Drisaldi B, Bartnikas TB, et al. Brain copper content and cuproenzyme activity do not vary with prion protein expression level. J Biol Chem 2000; 275(11):7455–7458.
[40] Kardos J, Kovács I, Hajós F, et al. Nerve endings from rat brain tissue release copper upon depolarization. A possible role in regulating neuronal excitability. Neurosci Lett 1989; 103:139–144.
[41] Collinge J, Whittington MA, Sidle KCL, et al. Prion protein is necessary for normal synaptic function. Nature 1994; 370:295–297.
[42] Sharonova IN, Vorobjev VS, Haas HL. High-affinity copper-block of GABAA-receptor mediated currents in acutely isolated cerebellar Purkinje cells of the rat. Eur J Neurosci 1998; 10:522–528.
[43] Doreulee N, Yanovsky Y, Haas HL. Suppression of long-term potentiation in hippocampal slices by copper. Hippocampus 1997; 7:666–669.
[44] Brown DR, Schulz-Schaeffer WJ, Schmidt B, et al. Prion protein-deficient cells show altered response to oxidative stress due to decreased SOD-1 activity. Exp Neurol 1997; 146:104–112.
[45] White AR, Collins SJ, Maher F, et al. Prion protein-deficient neurons reveal lower glutathione reductase activity and increased susceptibility to hydrogen peroxide toxicity. Am J Pathol 1999; 155:1723–1730.
[46] Kuwahara C, Takeuchi AM, Nishimura T, et al. Prions prevent neuronal cell-line death. Nature 1999; 400:225–226.
[47] Bounhar Y, Zhang Y, Goodyer CG, et al. Prion protein protects human neurons against Bax-mediated apoptosis. J Biol Chem 2001; 19:39145–39149.
[48] Daniels M and Brown DR. Purification and preparation of prion protein: synaptic superoxide dismutase 6. Methods Enzymol 2002; 349:258–267.
[49] Brown DR, Wong BS, Hafiz F, et al. Normal prion protein has an activity like that of superoxide dismutase. Biochem J 1999; 344(1):1–5.
[50] Wong BS, Pan T, Liu T, et al. Differential contribution of superoxide dismutase activity by prion protein in vivo. Biochem Biophys Res Commun 2000; 273(1):136–139.
[51] Brown DR, Clive C, Haswell SJ. Antioxidant activity related to copper binding of native prion protein. J Neurochem 2001; 76(1):69–76.
[52] Rae TD, Schmidt PJ, Pufahl RA, et al. Undetectable intracellular free copper: the requirement of a copper chaperone for superoxide dismutase. Science 1999; 284:805–808.
[53] Hutter G, Heppner FL, Aguzzi A. No superoxide dismutase activity of cellular prion protein in vivo. Biol Chem 2003; 384(9):1279–1285.
[54] Colling SB, Collinge J, Jefferys JGR. Hippocampal slices from prion protein null mice: disrupted Ca^{2+} activated K^+ currents. Neurosci Lett 1996; 209(1):49–52.
[55] Herms J, Tings T, Dunker S, et al. Prion protein affects Ca(2+)-activated K(+) currents in cerebellar Purkinje cells. Neurobiol Dis 2001; 8(2):324–330.
[56] Mallucci GR, Ratte S, Asante EA, et al. Postnatal knockout of prion protein alters hippocampal CA1 properties, but does not result in neurodegeneration. EMBO J 2002; 21(3):202–210.
[57] Herms J, Korte S, Gall S, et al. Altered intracellular calcium homeostasis in cerebellar granule cells of prion protein-deficient mice. J Neurochem 2000; 75(4):1487–1492.
[58] Tobler I, Gaus SE, Deboer T, et al. Altered circadian activity rhythms and sleep in mice devoid of prion protein. Nature 1996; 380:639–642.
[59] Tobler I, Deboer T, Fischer M. Sleep and sleep regulation in normal and prion protein-deficient mice. J Neurosci 1997; 17(5):1869–1879.
[60] Nishida N, Katamine S, Shigematsu K, et al. Prion protein is necessary for latent learning and long-term memory retention. Cell Mol Neurobiol 1997; 17:537–545.

[61] Walz W, Amaral OB, Rockenback IC, et al. Increased sensitivity to seizures in mice lacking cellular prion protein. Epilepsis 1999; 40:1679–1682.

[62] Korte S, Vassallo N, Kramer ML, et al. Modulation of L-type voltage-gated calcium channels by recombinant prion protein. J Neurochem 2003; 87(4):1037–1042.

[63] Vassallo N and Herms J. Cellular prion protein function in copper homeostasis and redox signalling at the synapse. J Neurochem 2003; 86(3):538–544.

[64] McMahon HE, Mange A, Nishida N, et al. Cleavage of the amino terminus of the prion protein by reactive oxygen species. J Biol Chem 2001; 276(3):2286–2291.

[65] Watt NT, Taylor DR, Gillott A, et al. Reactive oxygen species-mediated beta-cleavage of the prion protein in the cellular response to oxidative stress. J Biol Chem 2005; 280:35914–35921.

[66] Vassallo N, Herms J, Behrens C, et al. Activation of phosphatidylinositol 3-kinase by cellular prion protein (PrP^C) and its role in cell survival. Biochem Biophys Res Commun 2005; 332(1):75–82.

[67] Aguzzi A and Polymenidou M. Mammalian prion biology: one century of evolving concepts. Cell 2004; 116(2):313–327.

[68] Rossi D, Cozzio A, Flechsig E, et al. Onset of ataxia and Purkinje cell loss in PrP null mice inversely correlated with Dpl level in brain. EMBO J 2001; 20(4):694–702.

[69] Moore RC, Mastrangelo P, Bouzamondo E, et al. Doppel-induced cerebellar degeneration in transgenic mice. Proc Natl Acad Sci USA 2001; 98:15288–15293.

[70] Moore RC, Lee IY, Silverman GL, et al. Ataxia in prion protein (PrP)-deficient mice is associated with upregulation of the novel PrP-like protein doppel. J Mol Biol 1999; 292:797–818.

[71] Toescu EC. Activity of voltage-operated calcium channels in rat cerebellar granule neurons and neuronal survival. Neuroscience 1999; 94(2): 561–570.

8 The Scrapie Isoform of the Prion Protein PrP^{Sc} Compared to the Cellular Isoform PrP^{C}

Detlev Riesner

8.1 Introduction

Comparison of the properties of the scrapie isoform PrP^{Sc} and the cellular isoform PrP^{C} is of central interest regarding the prion model. Some of these scrapie isoform properties have already been discussed in other chapters, but it is here that we will concentrate on those features of PrP^{Sc} that are acquired only during the transition of PrP^{C} to PrP^{Sc}. These characteristics might also form the basis for approaching the diagnosis and therapy of prion diseases.

8.2 Biological and immunological properties of PrP^{Sc}

8.2.1 Occurrence in the organism

The cellular product PrP^{C} is apathogenic and is expressed in many organs of a healthy organism such as the brain, spinal cord, lungs, spleen, muscles, inter alia, etc. [1–4]. For those organ systems where PrP^{C} has been claimed not to be expressed, the limited sensitivity of the detection system must be carefully considered. Cloning the PrP gene (*Prnp*) has demonstrated that this gene is a host gene and it is not found in highly purified prion samples.

The scrapie isoform exhibits a pattern of occurrence in the infected organism that is somewhat different from the cellular isoform. Accumulation occurs primarily in the central nervous system (CNS), in particular brain tissue and the lymphoreticular tissue. After oral uptake, PrP^{Sc} first shows up in the Peyer's patches of the gastrointestinal section [5]. Low levels of PrP^{Sc} have been found, depending on the species, in other tissue and body fluids, particularly in the amnion [6; 7], and placenta of scrapie-diseased sheep and goats. Until recently, no infectivity of PrP^{Sc} had been detected in muscle, urine, or other organs. This has changed markedly with the increasing sensitivity of tests. Infectivity has been definite in particular muscles of mice [4] (see Chapter 19.7), hamsters [8], and sheep [9], whereas the occurrence in urine is still under debate (see WHO $\text{Tables}_{\text{Appendix}}$ IA-IC).

8.2.2 Lack of immunogenic activity

During the incubation period and clinical phase, neither inflammatory nor immunological responses of the infected organism can be detected. This is not surprising when keeping in mind that both isoforms of PrP, i. e., PrP^{C} and PrP^{Sc} have exactly the same amino acid sequence coded by the host and, therefore, are well known to the host's immune system.

The host immune system recognizes PrP^{Sc} as an integral part of itself. Consequently, in spite of the highly pathogenic potential of PrP^{Sc} and in spite of an intact immune defense system of the host, prions benefit from a type of immunotolerance. Immunotolerance of prion diseases is characterized by:

1. no immune response,
2. no inflammation,
3. no interferon production,
4. no cytotoxic effect within *in vitro*-infected cells, and
5. no qualitative or quantitative change of the level of T and B lymphocytes.

However, anti-PrP-antibodies can be produced after the inoculation of PrP^{Sc} into a foreign host, e. g., mouse PrP^{Sc} into a rabbit (see Chapter 35).

8.3 Chemical, biochemical, and physical properties of PrP^{Sc}

As mentioned above, no differences in chemical composition can be found between PrP^{C} and PrP^{Sc} [10] (see Chapter 5). This means that the amino acid sequence, glycosylation, and lipid anchor of both of the isoforms are identical. There are, however, two restrictions to this statement. First, the glycosylations at Asn181 and Asn197 are heterogeneous per se, and it is impossible to quantitatively compare two heterogeneous compositions. Second, protein analysis would detect heterogeneities only if that amounted to a few percent. Since 10^5–10^6 PrP molecules are found in one infectious unit, it is unclear as to whether all of the molecules contribute to infectivity in the same way or whether only a few PrP molecules designated PrP* are truly infectious among the many PrPres molecules. A few PrP* molecules among the 10^5–10^6 PrPres molecules would be undetectable with existing chemical analytical methods.

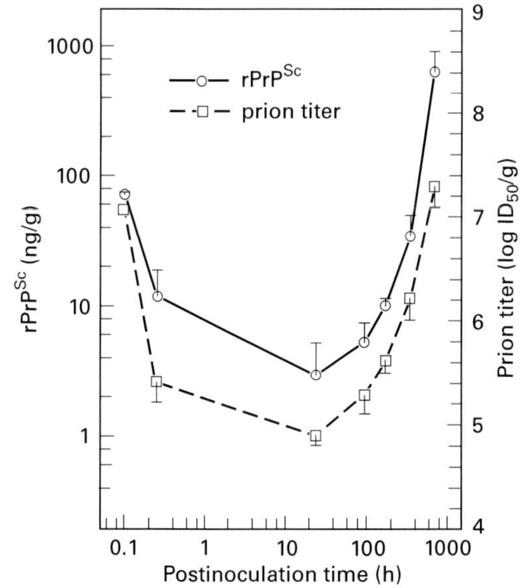

Fig. 8.1: Clearance of scrapie infectivity and PK-resistant PrP ($rPrP^{Sc}$) after intracerebral inoculation. Hamsters were inoculated with prions Sc263 and brains were prepared after various time intervals. Infectivity was determined by bioassay in the next passage and amount of $rPrP^{Sc}$ by Western blotting. Figure from Safar et al. [11].

The number of 10^5–10^6 PrP molecules was determined independently in different laboratories but always included a number of mechanistic concerns. Recently, it was shown that not only the low infectivity of PrP^{Sc} molecules but also a high clearance rate contributed to that number [11]. As shown in Fig. 8.1 the amount of PrPres and infectivity of hamster brain samples were determined in short and long intervals after inoculation. It was found that in the first 24 h after inoculation, infectivity and amount of PrPres in the brain decreased drastically, i. e., to only a few percent, prior to amplification being detected. Obviously most of the inoculated PrP^{Sc} was washed away and only the remaining few percent started the new amplification. If the infectious dose is related to the amount of PrP molecules left over after clearance, 500–5,000 PrPres molecules are estimated within one infectious unit. It is noteworthy that the clearance rates are characteristic not only for the agent and the host but also for the strain of the agent [20].

Purified prions, either in the form of full-length PrP^{Sc} or as PrP27–30, are insoluble even in mild detergents. They have withstood all attempts to prepare soluble infectious prion samples. So far chemical aggregation and prion infectivity is inseparably interrelated. In electron micrographs fibrillar structures, also called prion rods, are visible (Fig. 8.2; compare Fig. 1.5b). After staining with Congo Red, they show the typical fluorescence birefringence of amyloids [12]. In the presence of thioflavin, the fibrillar structures exhibit highly increased fluorescence intensity. Similar structures that were detected in a thin section of the brain of infected animals were originally called scrapie-associated fibrils, SAF [13] (Fig. 1.5b). It should be noted that in the brains of CJD or kuru victims, PrP deposits can be detected as diffuse deposits, amyloid fibers (Fig. 13.2) condensed plaques, or florid plaques (see Chapters 26; 27).

The high tendency to aggregate correlates with a PrP^{Sc}-specific resistance against digestion with proteinase K. In Fig. 8.3 the Western blot of a SDS gel electrophoresis of PrP^{Sc} without

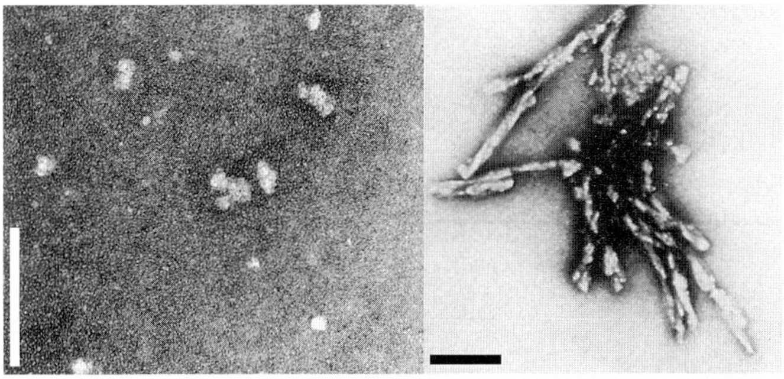

Fig. 8.2: Electron microscopy of prion rods, right panel (23), and of spherical particles of 4–6 PrP27–30 molecules after solubilizaton in SDS buffer, left panel [28]. The bar represents 100 nm. Figure from Riesner and colleagues [28].

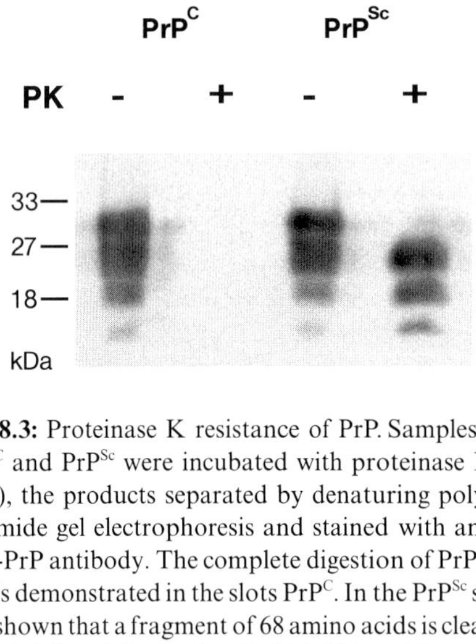

Fig. 8.3: Proteinase K resistance of PrP. Samples of PrP^{C} and PrP^{Sc} were incubated with proteinase K (PK), the products separated by denaturing polyacrylamide gel electrophoresis and stained with an anti-PrP antibody. The complete digestion of PrP^{C} by PK is demonstrated in the slots PrP^{C}. In the PrP^{Sc} slots it is shown that a fragment of 68 amino acids is cleaved from PrP^{Sc} and the product PrP27–30 is stable against further digestion. Lower band, unglycosylated form of PrP; middle band, monoglycosylated form of PrP; upper band, biglycosylated form of PrP.

and with proteinase K digestion, respectively, is depicted. The three characteristic bands of PrP (i. e., without, with one, and with two glycosyl groups) are visible; they disappear completely after the proteinase K digestion of PrP^{C}, which results in small peptides. In the case of PrP^{Sc}, however, the bands remain nearly undiminished in intensity although shifted to a lower molecular weight. These are the N-terminally truncated forms of PrP^{Sc}, generally referred to as PrP27–30. The full-length PrP is cleaved around amino acid 90. The cleavage product is also the product of the most effective purification procedure for prions [14]. Furthermore, it should be noted that most of the presently available routine tests for BSE and test for scrapie are based on the proteinase K resistance of PrP27–30.

PrP27–30 appears as the simpler but still fully infectious molecule as compared to the full-length PrP^{Sc}. It is particularly appropriate for *in vitro* studies of the infectious form. Transgenic animals that only express the truncated form of PrP corresponding to PrP27–30 can be infected and show clinical symptoms, and the disease can also be transmitted to other animals [15]. Consequently, the full length of PrP is not required for replication, and all of the steps can proceed well with PrP27–30, which is confirmed by a report of synthetic mammalian prions [16]. That does not mean, however, that the N-terminus does not influence prion replication – familial prion diseases with mutations in the N-terminus are known (see Chapters 9, 26).

The band pattern of PK-digested PrP as shown in Fig. 8.3 is characteristic but not invariable. It can be used in some cases for strain determination (see Chapter 12). Parchi and his colleagues [17] could differentiate between four types of the CJD disease (see Chapter 26). The group of Prusiner could show that the band pattern, together with the histological lesion profiles in the brain of CJD or FFI patients, respectively, were transmitted even to transgenic mice [18]. Collinge and coworkers also detected a very similar band pattern in BSE-infected cattle and in patients with variant CJD supporting the transmissibility of BSE to humans [19].

The prion model might well be explained on the basis of the conformational changes of the prion protein, which are induced directly or indirectly by the invading PrP^{Sc} (see below). The phenomenon of prion strains, however, is difficult to explain in a similar manner. Although different physicochemical properties were found with different prion strains [20], these could not be attributed to different conformations of single PrP molecules but only to the highly aggregated and insoluble PrP. Even if the principal replication features of prions did not depend on nucleic acids, it was argued that at least the strain specificity might be determined by nucleic acid molecules; this, however, was invalidated in studies described in Chapter 4 [11]. In a model study on yeast prions it could be shown that strain characteristics can be explained by way of different intermolecular contacts of otherwise identical molecules [21].

The lipophilic nature of highly purified prions suggested that not only is the glycolipid anchor linked to PrP but, in addition, lipids might be associated noncovalently with PrP. A chemical analysis revealed specific lipids that amounted to approximately 1 % of purified prions [22]. These are sphingomyelin, α-hydroxy-cerebroside, and cholesterol depending on the method of purification. These lipids are known to be components of the cell membrane exterior in caveolae-like sites where PrP^{C} also accumulates. Although lipids are not essential for infectivity, their presence in natural prions might indicate the origin of prions, namely the site of PrP^{C} accumulation on the outside of the cell membrane.

Early experiments in which prion rods were digested extensively with proteinase K showed that the rod-like structure was maintained in electron micrographs even if PrP was digested by more than 99 % [23]. This result pointed to an additional structural component. It was identified much later as a polymeric sugar consisting of α-1,4-linked and 1,4,6-branched polyglucose [24; 25]. This sugar component is clearly different from the glycosyl groups, which are attached covalently to PrP. A schematic presentation is shown in Figure 8.4. Since the polysaccharide amounted up to 10 % (w/w) of highly purified prions, it might be regarded as a structural scaffold contributing to the high chemical and physical stability of prions. It is noteworthy that in this context the synthetic prions that exclusively consist of mouse recPrP (89–230) exhibit very little PK resistance.

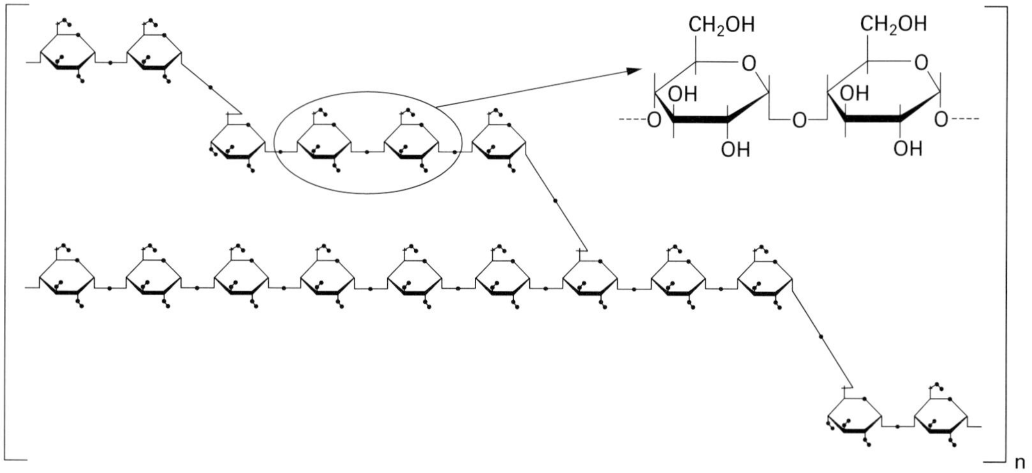

Fig. 8.4: Schematic presentation of the polyglucose scaffold as found in prion rods [24]. The estimated molecular mass is above 200,000 kD. The 1,4 linkage is of the α-type [25]. The letter n indicates repetition of the structural subunit. Figure modified from Dumpitak et al. [25].

8.4 Structure of PrP^{Sc}

Studies on the structure and structural transitions that have been carried out on many proteins have always been interpreted according to their functional aspects. However, there is no precedent for the case where structural transition is directly correlated with the principal step of an infection mechanism. Thus, the structure and structural transitions are of utmost importance for the entire prion model. However, structural studies suffer from a serious problem, i. e., the insolubility of the infectious isoform of PrP, whether it is full-length PrP^{Sc} or PrP27–30. Many attempts to apply a large number of detergents have been undertaken to solubilize PrP27–30. Systematic studies were carried out with guanidinium-hydrochlorid [26; 27], sarkosyl, and sodium-dodecylsulphate (SDS) [28]. Sonication in 0.2% SDS partially disperses prion rods. Prion rods of a smaller size that are still infectious and more prone to spectroscopic studies are produced and in addition a fraction of a variable yield (10–40%) of PrP oligomers of 4–6 PrP units, which have lost infectivity (Fig. 8.2). Some spectroscopic methods that could be applied to soluble recPrP or even PrP^{C}, were also used in particular cases in order to determine the secondary structure of PrP27–30 [28] or PrP^{Sc}. Infrared spectroscopy was used in the microbeam mode on single large aggregates of PrP27–30 [29] and circular dichroism (CD) on SDS-dispersed PrP27–30 or thin films of PrP^{Sc} [30]. From those measurements α-helix and β-sheet contents could be determined. The numbers that are shown in Table 8.1 are not highly accurate, since the analysis on PrP^{Sc} had to be carried out on insoluble samples; however, they clearly show that PrP^{C} is dominated by α-helices and has only little β-sheet content, whereas PrP^{Sc}, both full-length and PrP27–30 are characterized by similar amounts of α-helices and β-sheets.

Table 8.1: Secondary structure of PrP in different isoforms [28–30; 59].

	α-helix (%)	β-sheet (%)
PrP^{C}	43	–
PrP^{Sc}	20	34
PrP27–30	29	31

The three-dimensional structure of recPrP (120–231) was determined for several species and is discussed in Chapter 6. From those and related studies on full-length PrP, even natural PrP, it was concluded that the structure known from the recPrP represents the structure of PrP^{C} fairly well. Thus, a well-defined tertiary structure is restricted to the C-terminal part (120–231), whereas the N-terminal part is flexible. This conclusion still suffers from the fact that natural PrP^{C} is anchored with the glycolipid anchor on the outside of the cell membrane, whereas all of the structural measurements were carried out free in solution.

As mentioned above, PrP^{Sc} and PrP27–30 are inaccessible to structural analysis by NMR or X-ray because of their insolubility. Attempts were made to use the structure described above as a starting model, changing α-helices into β-sheets and in this way developing a model for PrP^{Sc}. These models assume that helices 2 and 3 are unchanged in accordance with the antibody-binding data, but they are incomplete in the sense that they are models for isolated molecules, whereas PrP^{Sc} as well as PrP27–30 were only found in aggregated forms. Thus, one has to assume that the PrP^{Sc} structure is stabilized by intermolecular interactions.

A new approach was followed by Wille from Prusiner's laboratory, who was able to prepare two-dimensional crystalline-like arrays of PrP^{Sc}- or PrP^{Sc}-like molecules [31]. Those samples were studied by way of electron microscopy, and because of the crystalline-like arrangement, images could be reconstructed from the repetitive unit with fairly high resolution. Hexagonal symmetry was visible, but it could not be determined whether one unit was built from three or six molecules. A later study favored three molecules in the unit [32]. The electron density map could be fitted best if a β-helix was assumed instead of β-sheets. The structure of a β-helix consists of a helical arrangement of short parallel β-sheets, and it is known from other fibrillar proteins; spectroscopically that β-sheets and β-helices cannot be differentiated. Therefore, the new model would not contradict

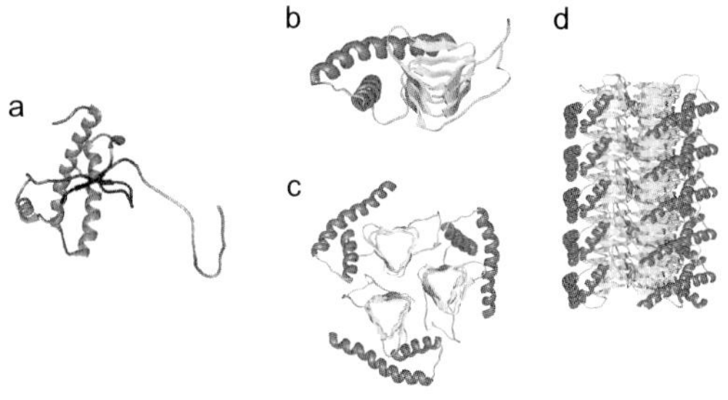

Fig. 8.5: Structure of PrP^{Sc}-like oligomers in comparison with PrP^{C}. Reconstruction of the oligomeric structure from electron microscopic analysis of two-dimensional crystals. PrP^{C} **(a)**, monomeric subunit of PrP^{Sc} **(b)**, trimeric subunit **(c)**, and fibril **(d)**. Figure modified from Govaerts et al. [32].

earlier spectroscopic studies. The model is depicted in Fig. 8.5. In summary, the β-helical N-terminus is located in the inner part of the hexagonal unit with helices 2 and 3 located at the outer part, and the glycosyl groups pointing into the space between the hexagonal units. The model shown in Fig. 8.5 is not a final description of the structure of PrP^{Sc}, but is rather the best model currently available that takes into account the electron microscopic and spectroscopic data as well as the intermolecular stabilization of the PrP^{Sc} structure. Although yeast prions are not in the scope of this monograph, it is noteworthy that the details of the aggregate structure of different yeast proteins with a prion character have been elucidated. These aggregate structures are dominated by cross-β-sheets without a major contribution of α-helices. In that sense they are quite different from the model structure of PrP^{Sc} [21; 33; 34].

8.5 In vitro conversion of PrP and the generation of infectivity

Studies of the in vitro conversion of PrP have had the well-defined aim of generating infectivity from cellular PrP and thereby yielding unequivocal proof of the prion model. Very early on it became clear that this is no easy task and that one might better approach the goal with a step-by-step study of the conformational transitions of PrP. Conformational transitions might be either denaturation processes of PrP^{C} or PrP^{Sc}, transitions between PrP^{C} and PrP^{Sc} induced by varying solvent conditions, or induction of the PrP^{C} to PrP^{Sc} transition by an existing seed of PrP^{Sc}.

The denaturation of the globular domain of recPrP (121–231) by the addition of up to 8 M urea was analyzed quantitatively and is described in Chapter 5. From the reversibility of the denaturation process it was concluded that the PrP^{C}-like conformation is the state of the lowest free energy in buffer without detergent. As described below, this conclusion might be restricted, however, to the fragment PrP (121–231). Similar experiments were carried out, while also taking into account the mechanism of refolding by kinetic analysis, on recombinant PrP (89–230) from mice, which is the recombinant equivalent of PrP27–30 [35]. In this fragment, β-sheet-rich oligomers and fibrils were formed at pH 3.6. However, after switching from the fully denaturing conditions of 10 M urea to native conditions without urea, first the monomeric α-helical conformation was detected and only later (in a very slow process of hours or days) β-sheet-rich oligomers and fibrillar structures showed up. The presence of urea during incubation accelerates the formation of the β-conformation, and if the urea concentration is as high as 5 M, β-sheet-rich oligomers are formed directly, i. e., without running through the α-helical state. Consequently, at an acidic pH, folding of PrP to an α-helical state is under kinetic control and the thermodynamically stable state is the β-sheet-rich state.

The experiments described above might resemble the situation of PrP^{C} when it is folded after synthesis. The PrP^{C}–PrP^{Sc} transition was studied under quite different conditions, i. e., the PrP^{C}-like, α-helical state was first established as in the noninfected organism [36]. Then the transition to the PrP^{Sc}-like conformation was induced by slightly denaturing conditions, e. g., 1 M guanidinium hydrochloride. These experiments were carried out with recombinant PrP at acidic pH (4.0), and the conversion process could be induced by a wide variety of conditions combining mild denaturants and different salts. In accordance with renaturation experiments (see above), it was found that β-sheet formation is always connected with aggregation and that the most stable state, at least at an acidic pH, is the β-sheet-rich aggregated state. An exception was shown under conditions reducing the disulphide bridge; it was reported that acidic and reducing conditions could induce a β-sheet-rich and monomeric state [37]. It is unknown whether this finding is relevant to PrP^{C}–PrP^{Sc} conversion in nature because the intramolecular disulphide bridge is present in both states, and in all other conversion experiments the disulphide bridge was not opened transiently [38].

The *in vitro* conversion was also studied with natural PrP and at neutral pH [28]. Infectious PrP27–30 was converted to an α-helical, oligomeric, and noninfectious form by the addition of 0.3% sodium-dodecylsulphate (SDS) and sonication. Dilution of SDS to 0.01% or lower re-established a β-sheet-rich aggregated and partially proteinase K-resistant conformation. Although natural PrP was used, which was infectious before conversion, infectivity could not be re-established. These experiments were closest to natural conditions if the low concentrations of SDS were regarded as a membrane-like environment. It was also shown that the conversion occurs in steps: first the fast formation of β-sheets concomitant with forming small oligomers, then larger oligomers in minutes to an hour, and finally, large insoluble aggregates in hours up to days. In Fig. 8.6 the kinetics of aggregation was followed by fluorescence correlation spectroscopy (FCS). In this method a few fluorescent particles, in this case fluorescence-labeled PrP aggregates diffuse through a laser beam, resulting in a fluctuating fluorescence signal. The diffusion time can be determined from the fluctuation of the fluorescence intensity and, thereby, the molecular mass. Since a single measurement takes only a few seconds, the kinetics of an aggregation process, i. e., an increase of molecular mass, can be studied well. In Fig. 8.6a the formation of oligomers occurs in approximately 20 min and larger aggregates are formed in a period of just hours or days (Fig. 8.6b).

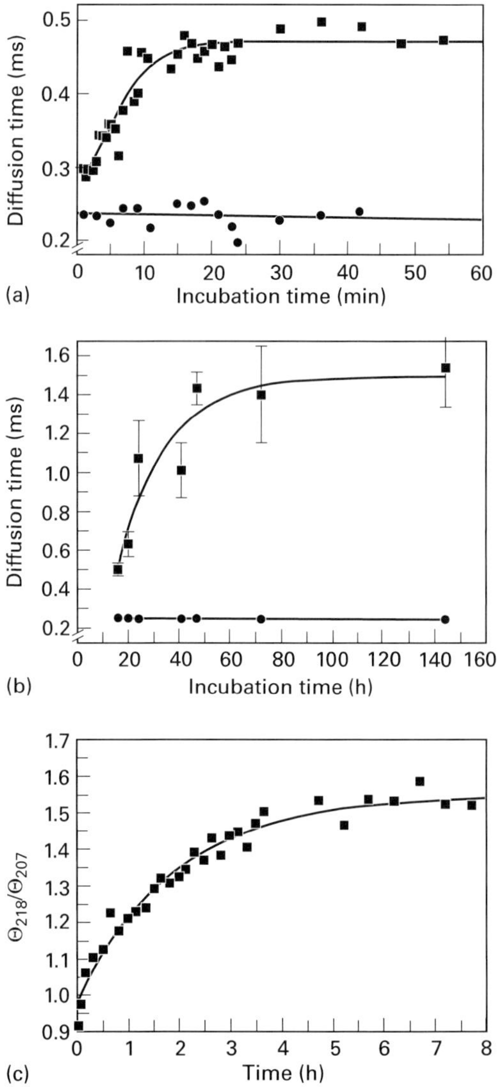

Fig. 8.6: Kinetics of PrP aggregation and conformational transition. **(a)** and **(b)**: aggregation of PrP27–30 (solubilized in 0.2% SDS) was induced by dilution of SDS to 0.01%; aggregation was followed by analyzing the diffusion time from FCS measurement. The spheres represent the constant signal in 0.2% SDS, the squares represent the time-dependent diffusion times after dilution of SDS. Figure from Post et al. [56]. **(c)** Kinetics of the structural transition of recombinant PrP90–231 from α-helical dimer to β-sheet-rich oligomer as followed by CD spectroscopy. High values of ellipticity ratio at two different wavelengths $\Theta_{218}/\Theta_{207}$ indicate a β-sheet-rich structure, low values indicate α-helices. The transition was induced by dilution of SDS to 0.0275% (w/v). Figure from Leffers et al. [40].

Studies in which the SDS concentration was varied in small steps were carried out with recombinant PrP (90–231) with the hamster sequence. Several intermediate states could be identified as depicted in Fig. 8.7 [39; 40]. In 0.06–0.1 % SDS an α-helical dimer is present as a thermodynamically stable state, which is converted in a cooperative manner to a β-sheet-rich oligomeric state if the SDS concentration is lowered from 0.06 to 0.04 %. As recently determined [40], the oligomeric state contains 12–16 molecules and is of particular interest for further biophysical studies since it is stable in solution. In low SDS concentrations (< 0.02 %), large insoluble aggregates (see above) are formed that also remain stable after the SDS has been washed out completely. In buffer without detergent at pH 7, the PrP^{C}-like as well as the PrP^{Sc}-like conformations can be established, with the stable state the PrP^{Sc}-like conformation. The conversion can be induced by very low concentrations of different detergents [see also 41]. As mentioned above, the formation of β-sheet-rich oligomers occurs within 20 min (Fig. 8.6a) if the oligomers are formed from the partially denatured state, the so-called α-helical, random coil state (Fig. 8.7). It is much slower when the equilibrium is shifted between the α-dimer and the β-oligomer (Fig. 8.7), which is actually shown by time-dependent CD measurements in Fig. 8.6c. Consequently, a high activation barrier exists between the α-helical state and β-sheet-rich state, and the transition is always connected with oligomerization [40]. The time needed to form very large aggregates is one day or slightly longer (Fig. 8.6b).

The binding of SDS to PrP was determined quantitatively in terms of numbers of SDS molecules binding to different sites of PrP. The influence of the bacterial chaperonin GroEL and GroEL/ES as a model system on the conversion process was studied with the same method as described above and its action compared with that of SDS [40]. It was concluded that the strongest binding of SDS (five binding sites on PrP) and the binding to GroEL are similar in the sense that hydrophobic sites of PrP are unmasked and a transition state between the α-helical and the β-sheet-rich state is stabilized. Release from GroEL into the water phase leads to aggregation, whereas binding to the GroEL/ES system enables the PrP to refold into the α-helical state without oligomerization. In this conformation, PrP can be released into the water phase after the addition of ATP. The effect of lowering the activation barrier between PrP^{C} and PrP^{Sc} by chaperonins was found in several independent studies, but one should not infer a biological role as a helper for prion formation. The experiments described above point more so to a role of chaperonins – if they meet PrP^{C} and/or PrP^{Sc} at all in the organism – in

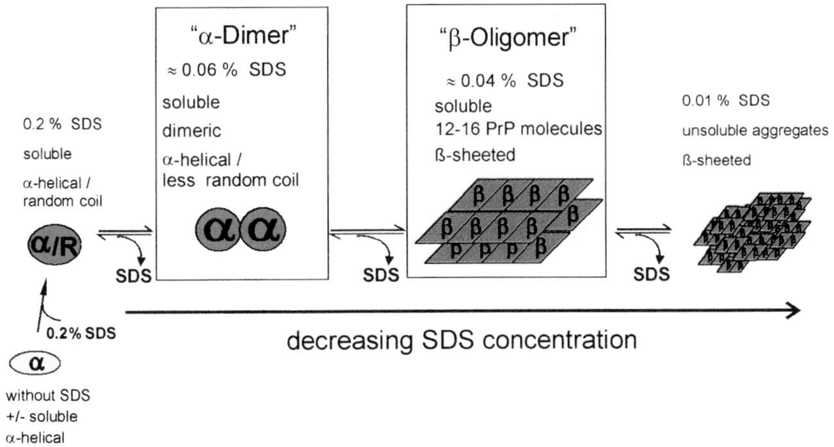

Fig. 8.7: Intermediates of the *in vitro* conversion of hamster recombinant PrP90–231. Data from Leffers et al. [40] and figure from Riesner [57].

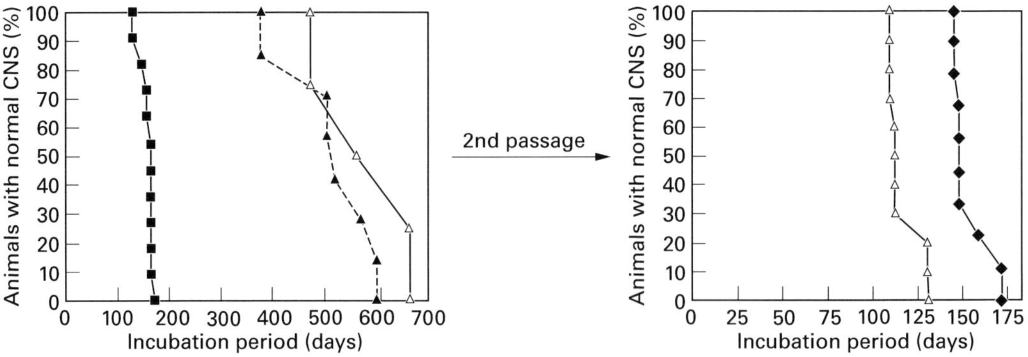

Fig. 8.8: Survival curves after inoculation with synthetic prions. Left: Tg 9949 mice were inoculated with RML prions, ■; seeded amyloid rec Mo PrP89–230, ▲; and unseeded amyloid PrP89–230, △. Right: FVB mice were inoculated with prions RML, △; and brain homogenate from Tg 9949 inoculated with seeded synthetic prions ◆ (see left panel), ▲. Figure modified from Legname et al. [16].

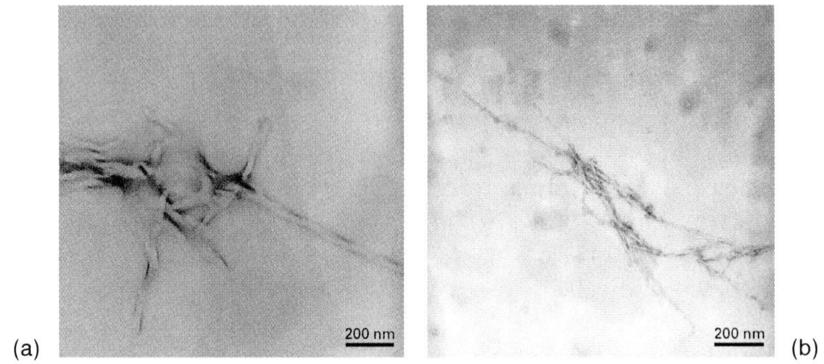

Fig. 8.9: Fibrils reconstructed from hamster recombinant PrP 90–231. Recombinant PrP90–231 was adjusted to 10 mM HEPES and 0.03 % SDS. 250-mM NaCl was subsequently added and the samples were incubated up to 25 days at 37 °C. After centrifugation at 100,000 × g for 1 h, the pellet revealed the formation of fibrils and amorphous aggregates in the absence **(a)** and presence **(b)** of lipids (i. e., sphingomyelin, galactosyl cerebroside, and cholesterol). Bars represent 200 nm. Figure from Leffers et al. [46].

assisting prevention of prion formation by helping to refold to the α-helical structure, rather than favoring prion formation.

Caughey and colleagues have attempted to simulate the infection process *in vitro* by incubating radio-labeled PrP from cell cultures together with PrP^{Sc} and showing that the labeled PrP^{C} acquired PK resistance [26]. The activation barrier was lowered by establishing partially denaturing conditions, including chaperonins [42]. The experiments could be refined in a way that the acquisition of PK resistance showed some characteristics of strain- and species-specificity [43]. Acquisition of infectivity, however, could not be shown, and sophisticated experiments applying hamster PrP^{Sc} and mouse PrP showed that the generation of infectivity was highly unlikely [44].

All or most of the experiments described above attempted to generate infectivity. PK resistance and β-sheet-rich aggregates, in some cases even fibrils were produced, but no infectivity. Therefore, in 2004, when Prusiner and his colleagues reported on "synthetic mammalian prions", a breakthrough was realised [16]. Fibrils were formed from mouse recombinant PrP (89–230) and inoculated into transgenic mice overexpressing PrP (89–230) in a $PrP^{0/0}$ background. As seen from the survival curves (Fig. 8.8), the mice developed clinical signs between 380 and 660 days after inoculation, and PrPres was found in their brain extracts. When the brain extract was transmitted to wild-type mice, the incubation time was only 150 days, i. e., not drastically longer compared to the situation after infection with the natural prion strain RML. What distinguished these experiments in which de novo generation of infectivity suddenly worked when it had not for many years in many laboratories? It was the combination of three features: (i) correct structure of the converted PrP, (ii) exact adaptation of inoculated PrP and expressed PrP in the animal, and (iii) sensitization of the animal to infection by overexpressing PrP. With these experiments the principle of the prion model was proven; even more so in the sense that recombinant PrP without glycosylation and glycolipid anchor was also found to be infectious. Many more experiments are needed to understand the exact mechanisms and determine which additional factors, such as lipids, sugars, factor X, even cellular nucleic acids, etc. will increase the infectivity of prions *in vivo*. For example, it is known that fibril formation is not restricted to recombinant PrP. That is, fibrils could be formed with a similar structure as those with established infectivity from full-length recombinant PrP, from natural denatured PrP27–30, i. e., with the posttranslational modifications, and from PrP^C isolated from noninfected hamsters, as shown in Fig. 8.9 [45; 46].

8.6 Models of prion replication

The models of prion replication most frequently discussed in the literature have been developed by Prusiner [47] and by Lansbury [48]. These are based on the conformational transition of PrP^C into PrP^{Sc} in terms of the protein only-hypothesis, but are hypothetical in the sense that the conformational transitions occur free in solution. More realistic assumptions which locate PrP in the membrane are inaccessible to quantitative consideration. Since Eigen undertook a quantitative comparison of different models, it is appropriate to discuss the models together with the kinetic evaluation of Eigen [49].

The model in Fig. 8.10 was proposed by Prusiner and colleagues [47]. It is referred to as the heterodimer model, because the critical step is the formation of a heterodimer of PrP^C and PrP^{Sc}, which transforms PrP^C to PrP^{Sc}, similar to the enzymatic mechanism of an induced fit. More recently, the model was specified in the respect that protein X has to be involved as a cellular factor in the complex [50]. However, to date, protein X or more generally a factor X could not be identified. The reaction scheme represents a linear autocatalysis with an exponential formation rate of PrP^{Sc} (=B). Regarding an equilibrium PrP^C (=A) $\leftrightarrow$ PrP^{Sc} (=B), B would be the favored state, otherwise there would be no driving force for the catalytic turnover, which at equilibrium is reversible. Eigen's calculations showed that the spontaneous transformation of A into B has to be extremely low ($< 10^{-22} s^{-1}$), otherwise B will grow spontaneously even without infection. However, to make an infection possible in a life span, one would have to assume an extreme enhancement of the $PrP^C \rightarrow PrP^{Sc}$ transition of more than 10^{15} due to heterodimer formation, which is unknown in other enzymatic processes.

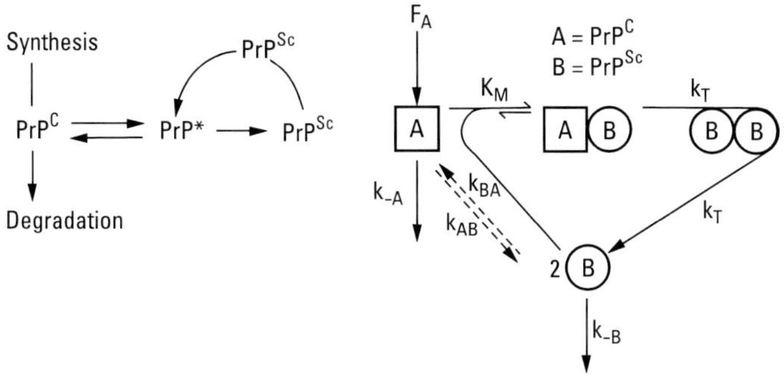

Fig. 8.10: Heterodimer model according to Prusiner and Cohen [58]. Presentation according to Prusiner and Cohen, left; modified according to Eigen [49], right.

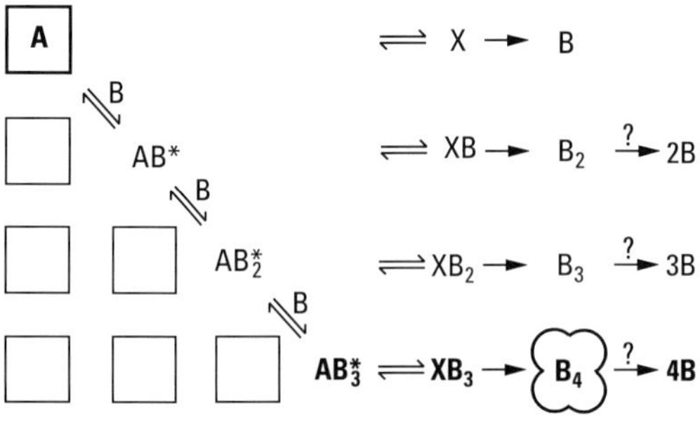

Fig. 8.11: Allosteric or cooperative Prusiner model according to Eigen [49].

Eigen circumvented this apparent inconsistency by proposing an extension of the mechanism, the so-called cooperative Prusiner mechanism, in which several B molecules have to cooperate to transform one molecule A. Thus, the catalytic effect of every single molecule B would be smaller than in the heterodimer model. The scheme of this mechanism is shown in Fig. 8.11 according to an allosteric enzymatic mechanism with three ligand-binding sites. Similar to the mechanism of Fig. 8.10, thermodynamic equilibrium favors state B over A, and the noncatalytic production of B is slow enough compared to metabolic removal, so that the concentration of B cannot reach any appreciable level in the absence of infection. In contrast to linear catalysis, however, the higher order catalysis leads to a threshold concentration for B corresponding to an infection event, which switches the system to the steady state of catalytic amplification of B.

Another mechanism was proposed by Lansbury and colleagues [48; 51] in which fibril formation and generation of infectivity are closely connected. Fibril formation is known for actin, β-amyloid etc., or the so-called linear crystals. As shown in the scheme of Fig. 8.12, species A and B are in a rapid equilibrium, with A being the favorable state. B can form aggregates B_i with decreasing concentrations down to a nucleus B_n. If the nucleus has been formed, the growth of the aggregates is faster than dissociation, and indefinite aggregates will be formed. In contrast, the molecule in state B does not represent a pathogenic form; only the nucleus B_n represents an infectious entity. It was estimated that a very limited range of rate constants, as defined in Fig. 8.12, must exist for Lansbury's mechanism to work [49].

All of the models discussed above have pros and cons when interpreted together with available experimental data. Dimerization of PrP could indeed be observed but as a homodimer of α-helical, PrP^C-like molecules; nevertheless, it shows that PrP^C is prone to an interaction with other PrP molecules [39; 52]. In other words, a heterodimer has not yet been detected experimentally. However, if it was only an intermediate state within the process of forming larger aggregates it would be difficult to detect. Therefore, it cannot be completely excluded.

The experimental evidence that the infectious state is multimeric is overwhelming. As mentioned above, infectivity is always connected with insolubility. A graphical example was produced recently with synthetic prions, when the infectious samples clearly had a fibrillary structure [16]. This feature is in accordance with Lansbury's model. Also in agreement with that model is the finding that the β-sheet-rich aggregates must have a minimum size depending upon the solution conditions between 8 and 14 PrP units [40]. Under particular conditions, these oligomers exhibit the physicochemical features of the nucleus in the Lansbury model, however, infectivity is probably not acquired on this level. Furthermore, one has to assume that the large aggregates have to break into pieces in order to explain the spreading of the agent in the organism.

As mentioned above, the models do not take into account that PrP^C is membrane-anchored

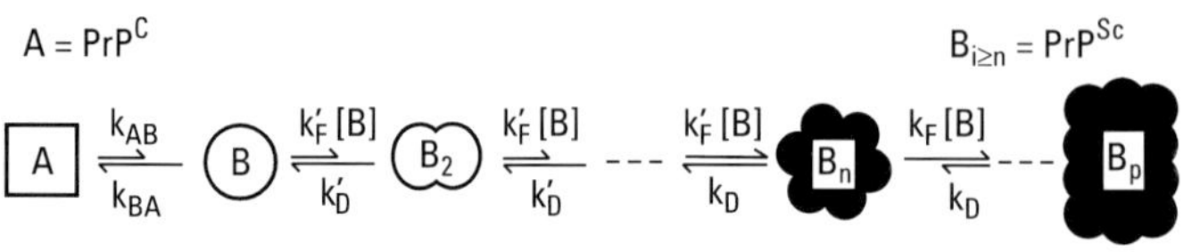

Fig. 8.12: Linear crystallization model according to Lansbury et al. [48].

and more or less homogeneously spread over the surface. PrP^{Sc} approaches from the extracellular compartment, and it is unclear whether contact between PrP^{C} and PrP^{Sc} occurs in the liquid phase or on the membrane. Some results implicate caviolae-like parts of the outer membrane as the site of the beginning of the PrP^{C} transformation [53; 54]. Some of these considerations, together with features of the PrP^{C}–PrP^{Sc} transition free in solution, led to a mechanistic model in which both the membrane and the aqueous phase were essential. This is depicted in Fig. 8.13. PrP^{C} is dispersed on the surface of the membrane, thereby stabilized in its conformation and prevented from aggregation in a β-sheet-rich structure. The process of insertion of PrP^{C} in the membrane and its release into the aqueous phase was studied recently with a model system of PrP^{C} expressed in CHO cells interacting with a membrane of raft-like lipid composition [55]. From those studies, the concentration of PrP^{C} free in solution was found to be between 10^{-10} and 10^{-8} M. If PrP^{C} molecules in this low concentration would meet other PrP^{C} molecules in sufficient numbers, they could then undergo a spontaneous transition into β-sheet-rich aggregates and probably later into infectious fibrils. All of these transitions were shown experimentally in a test tube. Since the concentration of PrP^{C} in the aqueous phase is very low, the spontaneous transition is highly improbable, which is in good correlation with the extremely rare event of spontaneous CJD. The situation might change if a prion particle invades the aqueous phase. If in such a case PrP^{C} molecules were free in solution, either in the form of an α-helical dimer or a β-sheet-rich oligomer, they might become attached to the prion particle. As concluded from recent studies on fibril formation [46], an intermediary of the α-dimer and the β-oligomer is probably prone to attach to invading prion particles or to spontaneously form fibrils. The equilibrium between PrP^{C} bound on the membrane and PrP^{C} in the aqueous phase would be shifted to the latter state with the consequence that the prion particle would grow and new infectivity would be generated.

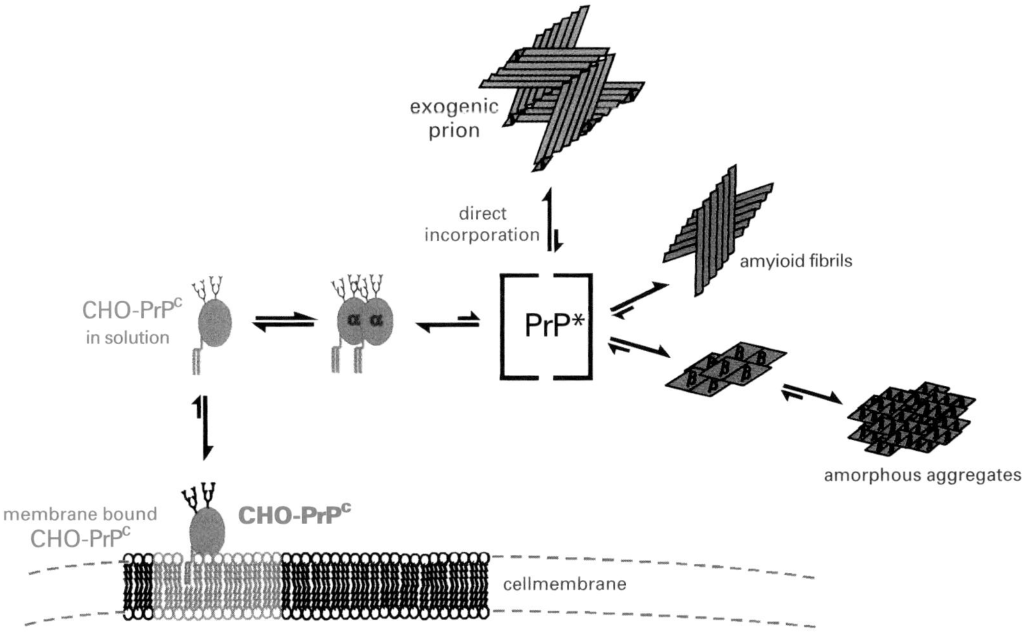

Fig. 8.13: Two-phase model. Figure from Elfrink et al. [55].

8.7 Infectious, sporadic, and familial etiology of prion diseases

Prion diseases are exceptional in that they can be of infectious, sporadic, or of familial etiology. A great achievement, which also lends support to the prion model, is the finding that diseases of quite different etiology can be deduced from a unique phenomenon that is a conformational transition of a host-encoded protein. A conformational transition that is catalyzed by contact with other isoforms of the same protein (i. e., infectious manifestation) should also occur spontaneously according to the laws of thermodynamics, although the probability and the rate may be very low. If only a single or a few PrP^{C} molecules undergo the transition, they could start the autocatalytic cycle, and from the cell where it occurred an infectious-like process could in turn spread through the entire organ or body. This might be an explanation for the sporadic manifestation of CJD and SFI. Close to 90 % of all CJD patients are afflicted by the sporadic form (see Chapter 14). If a protein has two or more structural alternatives, both of which are stable, or one stable and the other metastable etc., one should expect germline mutations that favor PrP^{Sc} over PrP^{C}. Favoring does not have to be exclusive for PrP^{Sc} but may also have a higher incidence for the spontaneous transition of PrP^{C} into PrP^{Sc}. Indeed, many prion diseases with hereditary manifestations are known, and all are connected with mutations in the PrP gene, namely, familial CJD, Gerstmann–Sträussler–Scheinker disease, and familial fatal insomnia (see Chapters 16, 17).

References

[1] Basler K, Oesch B, Scott M, et al. Scrapie and cellular PrP isoforms are encoded by the same chromosomal gene. Cell 1986; 46:417–428.

[2] Bendheim PE. Nearly ubiquitous tissue distribution of the scrapie agent precursor protein. Neurology 1992; 42:149–156.

[3] Oesch B, Westaway D, Walchli M, et al. (1985) A cellular gene encodes scrapie PrP(27–30) protein. Cell 40:735–746.

[4] Bosque PJ, Ryon C, Telling G, et al. Prions in skeletal muscle. Proc Natl Acad Sci USA 2002; 99:3812–3817.

[5] Bons N, Mestre-Frances N, Guiraud I, et al. Prion immunoreactivity in brain, tonsil, gastrointestinal epithelial cells, and blood and lymph vessels in lumerian zoo primates with spongiform encephalopathy. CR Acad Sci III 1979; 320:971–979.

[6] Pattison IH, Hoare MN, Jebbett JN, et al. Spread of scrapie to sheep and goats by oral dosing with foetal membranes from scrapie-affected sheep. Vet Rec 1972; 90:465–468.

[7] Pattison IH, Hoare MN, Jebbett JN, et al. Further observations on the production of scrapie in sheep by oral dosing with foetal membranes from scrapie-affected sheep. Br Vet J 1974; 130:65–67.

[8] Thomzig A, Kratzel Ch, Lenz G, et al. Widespread PrP^{Sc} accumulation in muscles of hamsters orally infected with scrapie. EMBO reports 2003; 4:530–533.

[9] Andreoletti O, Simon S, Lacroux C, et al. PrP^{Sc} accumulation in myocytes from sheep incubating natural scrapie. Nat Med 2004; 10:591–593.

[10] Stahl N, Baldwin MA, Teplow DB, et al. Structural studies of the scrapie prion protein using mass spectrometry and amino acid sequencing. Biochemistry 1993; 32:1991–2002.

[11] Safar JG, Kellings K, Serban H, et al. Search for a prion-specific nucleic acid. J Virol 2005; 79:10796–10806.

[12] Prusiner SB, McKinley MP, Bowman KA, et al. Scrapie prions aggregate to form amyloid-like birefringent rods. Cell 1983; 35:349–358.

[13] Merz PA, Somerville RA, Wisniewski HM, et al. Abnormal fibrils from scrapie-infected brain. Acta Neuropathol 1981; 54:63–74.

[14] McKinley MP, Meyer RK, Kenaga L, et al Scrapie prion rod formation *in-vitro* requires both detergent extraction and limited proteolysis. J Virol 1991; 65:1340–1351.

[15] Fischer M, RulickeT, Raeber A, et al. Prion protein (PrP) with amino-proximal deletions restoring susceptibility of PrP knock-out mice to scrapie. EMBO J 1996; 15:1255–1264.

[16] Legname G, Baskakov IV, Nguyen HOB, et al. Synthetic mammalian prions. Science 2004; 305:673–676.

[17] Parchi P, Castellani R, Capellari S, et al. Molecular basis of phenotypic variability in sporadic Creutzfeldt–Jakob Disease. Ann Neurol 1996; 39:767–778.

[18] Telling GC, Parchi P, DeArmond SJ, et al. Evidence for the conformation of the pathologic iso-

form of the prion protein enciphering and propagating prion diversity. Science 1996; 274:2079–2082.

[19] Collinge J, Sidle KC, Meads J, et al. Molecular analysis of prion strain variation and the aetiology of 'new variant' CJD. Nature 1996; 383:685–690.

[20] Safar J, Wille H, Itri F, et al. Eight prion strains have PrP^{Sc} molecules with different conformations. Nat Med 1998; 4:1157–1165.

[21] Krishman R and Lindquist S. Structural insights into a yeast prion illuminate nucleation and strain diversity. Nature 2005; 435:765–772.

[22] Klein TR, Kirsch D, Kaufmann R, et al. Prion rods contain small amounts of two host sphingolipids as revealed by thin-layer chromatography and mass spectrometry. Biol Chem 1998; 379:655–666.

[23] McKinley MP, Braunfeld MB, Bellinger CG, et al. Molecular characteristics of prion rods purified from scrapie-infected hamster brains. J Infect Dis 1986; 154:110–120.

[24] Appel TR, Dumpitak Ch, Mathiesen U, et al. Prion rods contain an inert polysaccharide scaffold. Biol Chem 1999; 380:1295–1306.

[25] Dumpitak C, Beekes M, Weinmann N, et al. The polysacchride scaffold of PrP27–30 is a common compound of natural prions and consists of α-linked polyglucose. Biol Chem 2005; 386:1149–1155.

[26] Kocisko DA, Come JH, Priola SA, et al. Cell-free formation of protease-resistant prion protein. Nature 1994; 370:471–474.

[27] Prusiner SB, Groth D, Serban A, et al. Attempts to restore scrapie prion infectivity after exposure to protein denaturants. Proc Natl Acad Sci USA 1993; 90:2793–2797.

[28] Riesner D, Kellings K, Post K, et al. Disruption of prion rods generates spherical particles composed of four to six PrP27–30 molecules that have a high α-helical content and are non-infectious. J Virol 1996; 70:1714–1722.

[29] Pan KM, Baldwin M, Nguyen J et al. Conversion of alpha-helices into beta-sheets features in the formation of the scrapie prion protein. Proc Natl Acad Sci USA 1993; 90:10962–10966.

[30] Safar J, Roller PP, Gajdusek DC, et al. Conformational transitions, dissociation, and unfolding of scrapie amyloid (prion) protein. J Biol Chem 1993; 268:20276–20284.

[31] Wille H, Michelitsch MD, Guénebaut V, et al. Structural studies of the scrapie prion protein by electron crystallography. Proc Natl Acad Sci USA 2002; 99:3563–3568.

[32] Govaerts C, Wille H, Prusiner SB, et al. Evidence for assembly of prions with left-handed beta-helices into trimers. Proc Natl Acad Sci USA 2004; 101:8342–8347.

[33] Nelson R, Sawaya MR, Balbirnie M, et al. Structure of the cross-β spine of amyloid-like fibrils. Nature 2005; 435:773–778.

[34] Ritter C, Maddelein ML, Siemer AB, et al. Correlation of structural elements and infectivity of the HET-s prion. Nature 2005; 435:844–848.

[35] Baskakov IV, Legname G, Prusiner SB, et al. Folding of prion protein to its native α-helical conformation is under kinetic control. J Biol Chem 2001; 276:19687–19690.

[36] Swietnicki W, Morillas M, Chen SG, et al. Aggregation and fibrillization of the recombinant human prion protein huPrP90–231. Biochemistry 2000; 39:424–431.

[37] Jackson GS, Hosszu LLP, Power A, et al. Reversible conversion of monomeric human prion protein between native and fibrilogenic conformations. Science 1999; 283:1935–1937.

[38] Welker E, Raymond LD, Scheraga HA, et al. Intramolecular versus intermolecular disulfide bonds in prion proteins. J Biol Chem 2002; 277:33477–33481.

[39] Jansen K, Schäfer O, Birkmann E, et al. Structural intermediates in the putative pathway from the cellular prion protein to the pathogenic form. J Biol Chem 2001; 382:683–691.

[40] Leffers KW, Schell J, Jansen K, et al. The structural transition of PrP into its pathogenic conformation is induced by unmasking hydropholic sites. J Mol Biol 2004; 344:839–853.

[41] Xiong LW, Raymond LD, Hayes S, et al. Conformational change, aggregation and fibril formation induced by detergent treatments of cellular prion protein. J Neurochem 2001; 79:669–678.

[42] DebBurman SK, Raymond GJ, Caughey B, et al. Chaperone-supervised conversion of prion protein to its protease-resistant form. Proc Natl Acad Sci USA 1997; 94:3938–13943.

[43] Raymond G, Hope J, Kocisko DA, et al. Molecular assessment of the transmissibilities of BSE and Scrapie to humans. Nature 1997; 388:285–288.

[44] Hill AF, Antoniou M, Collinge J. Protease-resistant prion protein produced *in vitro* lacks detectable infectivity. J Gen Virol 1999; 80:11–14.

[45] Bocharova OV, Breydo L, Parfenov AS, et al. *In vitro* Conversion of full-length mammalian prion protein produces amyloid form with physical properties of PrP^{Sc}. J Mol Biol 2004; 346:645–659.

[46] Leffers KW, Wille H, Stöhr J, et al. Assembly of natural and recombinant prion protein into fibrils. J Biol Chem. 2005; 386:569–580.
[47] Prusiner SB. Molecular biology of prion diseases. Science 1991; 252:1515–1522.
[48] Come JH, Fraser PE, Lansbury Jr, PT. A kinetic model for amyloid formation in the prion diseases: importance of seeding. Proc Natl Acad Sci USA 1993; 90:5959–5963.
[49] Eigen M. Prionics or the kinetic basis of prion diseases. Biophys Chem 1996; 63:A1–18.
[50] Telling GC, Scott M, Mastianni J, et al. Prion propagation in mice expressing human and chimeric PrP transgenes implicates the interaction of cellular PrP with another protein. Cell 1995; 83:79–90.
[51] Caughey B, Kocisko DA, Raymond GJ, et al. Aggregates of scrapie-associated prion protein induce the cell free conversion of protease-sensitive prion protein to the protease-resistant stage. Chem Biol 1995; 2:807–818.
[52] Mayer RK, Lustig A, Oesch B, et al. A monomer-dimer equilibrium of a cellular prion protein (PrP^{C}) not observed recombinant PrP. J Biol Chem 2000; 275:38081–38087.
[53] Kaneko K, Vey M, Scott M, et al. COOH-terminal sequence of the cellular prion protein directs subcellular trafficking and controls conversion into the scrapie isoform. Proc Natl Acad Sci USA 1997; 94:2333–2338.
[54] Taraboulos A, Scott M, Semenov A, et al. Cholesterol depletion and modification of COOH-terminal targeting sequence of the prion protein inhibit formation of the scrapie isoform. J Cell Biol 1995; 129:121–132.
[55] Elfrink K, Nagel-Steger L, Riesner D. Interaction of PrP^{C} with raft-like lipid membranes. Biol Chem, in press.
[56] Post K, Pitschke M, Schäfer O, et al. Rapid acquisition of β-sheet structure in the prion protein prior to multimer formation. Biol Chem 1998; 379:1307–1317.
[57] Riesner D. Biochemistry and structure of PrP^{C} and PrP^{Sc}. In: Weissmann C, editor. Prions for physicians. Br Med Bull 2003; 66:21–33.
[58] Cohen FE, Pan KM, Huang Z, et al. Structural clues to prion replication. Science 1994; 264:530–531
[59] Caughey BW, Dong A, Bhat K. S, et al. Secondary structure analysis of the scrapie-associted protein PrP (27–30) in water by infrared spectroscopy. Biochemistry 1991; 30:7672–7680.

9 The Phylogeny of Mammalian and Nonmammalian Prion Proteins

Hermann M. Schätzl

9.1 Introduction

With the development of early structural models for prion protein, the comparative analysis of prion protein genes (PrP genes, *Prnp*) of various species turned out to be instrumental. Discussions about structural and functional aspects of prion protein are often based on interspecies sequence comparisons. The PrP genes of more than 70 mammalian species have been analyzed thus far, mainly from primates and ungulates (Fig. 9.1). Until recently, very little was known about PrP genes of other species, such as marsupials and birds. It has become clear that an evolutionary comparison of PrP genes of more phylogenetically distant species will allow further insights into potential functional and structural features. Indeed, the very recent characterization of PrP genes of turtles, *Xenopus laevis*, and several fish species have fulfilled this task. The comparative analysis of PrPs with respect to species barrier scenarios, in which the main determinant seems to be the similarity of the prion proteins of the species involved, is of special interest. In light of BSE and vCJD, it is particularly desirable to be able to interpret and predict possible susceptibilities for the transmission of a given prion strain between certain species.

9.2 The organization of the PrP gene

All known mammalian PrP genes have 5' ends comprising one or two very short exons, which are separated by a long intron (approx. 10 kilobases, kb) from the exon located at the 3' end. The 3' exon harbors the entire open reading frame (ORF) of PrP, which encodes for a protein of approximately 250 amino acids. Therefore, alternative splicing of mRNA should not have a direct influence on PrP. Exact data for exon–intron organization are available for selected PrP genes such as that of Syrian hamster [1], mouse [2; 3], sheep [4], and human [5]. The promoter region does not contain a bona fide TATA-box and reveals similarities to housekeeping genes. Potential binding sites for transcription factors like SP1 and AP1 are present, but a detailed analysis of the transcriptional regulation is often missing.

The murine PrP gene exists in two allelic forms (mouse PrP-A and PrP-B genes). These forms vary by only two amino acid positions and are, nevertheless, responsible for the different incubation times of prion diseases. The murine alleles also have a different exon–intron organization. For example, the PrP-B gene carries an insertion of ~7 kb within the long intron, derived from the integration of a retroelement [3]. The human PrP gene is located on chromosome 20 and the murine PrP gene on chromosome 2. The mRNA transcribed from the PrP gene is usually 2.1–2.5 kb in length. The highest mRNA expression with approximately 50 copies/cell was found in neurons [6]. Other cell types such as astrocytes and lymphocytes (but not granulocytes) also express PrP [7]. A long antisense reading frame is contained on the complementary strand that exists in most species [8]. Only the PrP genes of some rodents, mink, and most birds possess a stop codon in this antisense reading frame. However, no mRNA derived from this putative antisense ORF has been found thus far.

The primary translation product of the PrP gene contains short N- and C-terminal signal peptides which are removed during maturation.

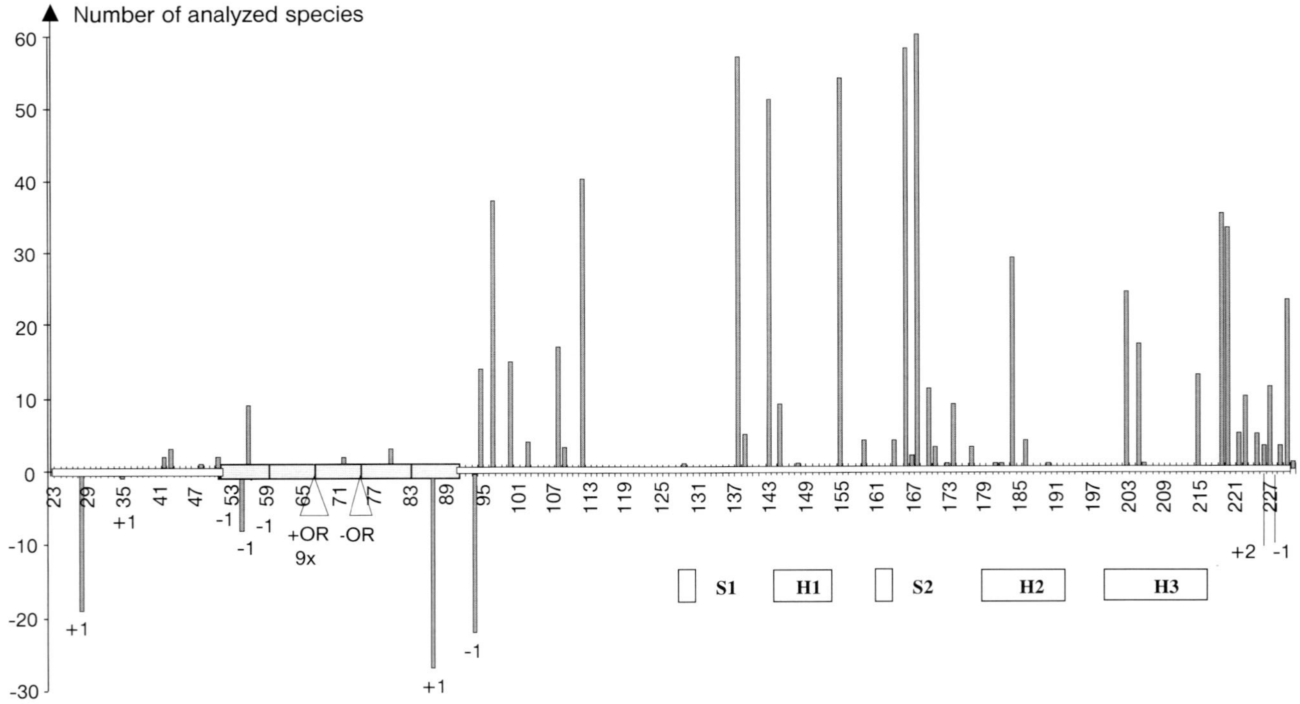

Fig. 9.1: Interspecies variations (mammalian PrP). Matrix of the number of variations in mammals referring to the mature human PrP (aa23–231; 71 mammalian species are included in the analysis). The upper bars denote amino acid exchanges (substitutions, positive axis, number of animal species with exchanges); the lower bars denominate insertions and deletions (negative axis). Triangles mark octarepeat insertions (+OR) and deletions (−OR), respectively (not to scale). Structural aspects are plotted beneath (β-sheets S1 and S2, α-helices H1, H2, and H3, as deduced from NMR). Remarkably, the N-terminal part (aa23–90), i. e., the PK-sensitive part harbors mainly insertions and deletions, whereas in the PK-resistant part (aa90–231) substitutions are found almost exclusively. Figure by H. M. Schätzl ©.

The mature PrP usually harbors a 5-fold repetition of eight amino acids (mainly proline and glycine residues; octarepeat structure) in the N-terminal part and contains two short β-sheeted and three α-helical segments within the C-terminal half [9; 10]. Following translation, the protein is heavily modified by two N-linked glycosylation sites, one disulfide bond, and a glycosylphosphatidylinositol (GPI) moiety, which is attached to the C-terminus of the protein in order to facilitate anchoring at the outer leaflet of the plasma membrane.

9.3 Comparative analysis of PrP genes and prion proteins

The comparative analysis of PrP genes from different species has been proven very helpful in the investigation of species barrier issues and for the initial structure prediction analysis [11–13]. Notably, the initial structure prediction for PrP by F. E. Cohen and S. B. Prusiner was mainly based on the comparison of the primary structures of certain mammalian PrPs and chicken PrP, which were known at that time [11]. In addition, functional aspects of single amino acid positions and of more complex amino acid stretches, called protein domains, have been deduced from such sequence comparisons. It is generally assumed that the degree of conservation at certain positions or entire domains correlates with their biological function and relevance. During evolution, the essentials with regard to structure and function of a protein are often conserved. Other regions of a protein are much more likely to be subjected to mutation and variations, which is reflected, for example, by interspecies variations. Furthermore, certain amino acid positions or domains can be linked to certain biochemical properties like attachment of carbohydrate chains or functions as recognition sequences. Eventually, the comparison of conserved protein domains, known as consensus sequences, to other proteins with known functions can reveal important insights into the function of a given protein. Of special interest for the prion protein is which domains are important for the determination of the species barrier [13].

First of all, the use of the terms "variation", "mutation", and "polymorphism" has to be defined in the special context of PrPs. Amino acid variation designates the exchange of residues at a certain position (codon), when PrPs of two species are compared. The term mutation, in contrast, refers to those amino acid alterations within one species, e. g., in human PrP, which are causally associated with the genetic forms (see Chapter 26) of prion diseases. A polymorphism, however, indicates an amino acid position where different residues, usually two, can occur within one species. This is not pathological *per se* but can be a modulating factor with respect to the susceptibility to endogenous or exogenous prion diseases. Examples are known in human (see Chapter 26), sheep (see Chapter 56) and mouse PrP.

What is currently the best way to compare PrPs? Significant results are usually obtained by the amino acid alignment of two mature PrP molecules, e. g., residues 23–231 of mouse PrP. Such a comparison considers only nucleic acid alterations which result in an amino acid exchange in the mature PrP; N- and C-terminal signal peptides are usually excluded. It is of note that the analysis of nucleic acid exchanges over the entire ORF can be of relevance for fundamental phylogenetic studies. In the following passages, we always refer to amino acid exchanges in the mature PrP.

9.3.1 Variations between the species in relation to structural elements

The tertiary structure of PrP^{Sc} (or more precisely PrP27–30) is now more clearly defined, although is still hotly debated. Due to its unfavorable biophysical properties, i. e., tendency for aggregation and insolubility, it could not be fully investigated by classical X-ray structure analysis or NMR technology (see Chapters 6, 8). In contrast, the tertiary structures of a variety of cellular prion proteins (e. g., Fig. 6.2) or the secondary structure of PrP^{Sc} have been determined. Which domains of PrP^{C} are important for the structural conversion of PrP^{C} into PrP^{Sc}?

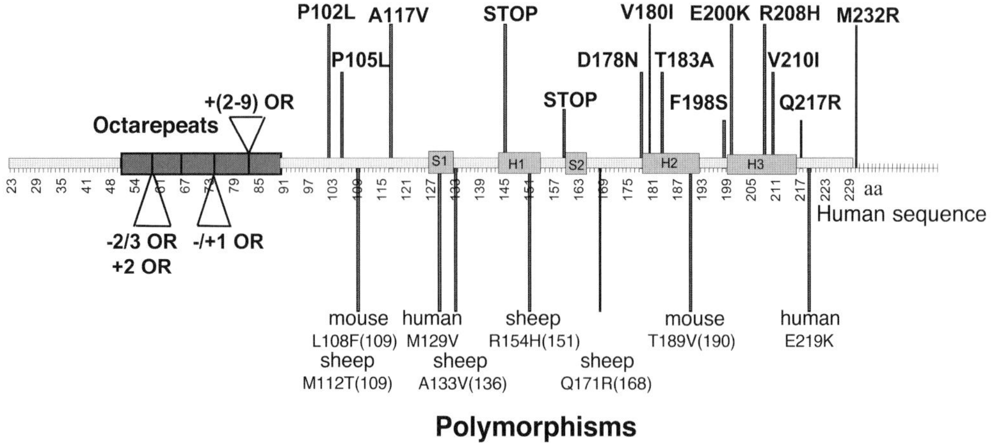

Fig. 9.2: Mutations and polymorphisms in the PrP gene. The pathogenic human mutations (not complete) are shown at the top in topological reference to the structural elements of PrP^C (octarepeats, β-sheets and α-helices derived from NMR). Beneath, selected polymorphisms in human, sheep, and mouse PrP are depicted, of which some can modulate prion disease. Compared to the variations shown in Fig. 9.1 no topological analogy is given between interspecies variations and mutations or polymorphisms. It is obvious that variations seem to exclude the structural elements, whereas mutations and polymorphisms are closely related to them. Figure by H.M. Schätzl ©.

In this context, the comparison of the variations between species with regard to the structural elements of PrP^C, i. e., the β-sheeted and α-helical domains, is of interest (Figs. 9.1–9.3). To study this, we analyzed PrP genes of 52 mammalian and 9 bird species [14]. In total, ~ 70 sequences were available at the time. With regard to amino acid (aa) homologies for mature PrP (without considering polymorphisms), within the following species we found:

1. Primates: 92.9–99.6 % aa identity.
2. Ungulates: 84.7–94.8 % aa identity when compared to human PrP, and 86.6–99.5 % aa identity compared to bovine PrP.
3. Ruminants: ~ 92–93 % aa identity when compared to human PrP.
4. Rodents: ~ 85–92 % aa identity compared to human PrP [2; 3].
5. Cats and dogs: approximately 85 % aa identity compared to human PrP.
6. All examined mammals: aa homologies were in the range of 84.7–99.6 %.

In many cases an important question arises: In exactly how many amino acid positions do the PrPs of two given species vary? With respect to mature PrP (again without considering polymorphisms), the following examples are given between:

1. Sheep and bovines: 6 aa exchanges.
2. Cattle and humans: 13 aa exchanges.
3. Sheep and humans: 14 aa exchanges.

The analyzed avian species share an aa homology of about 90 % (Fig. 9.3). When comparing birds with mammals only ~ 30 % aa identity is found [15; 16].

As already shown in primates [13], our extended study revealed that the structural regions of PrP^C are highly conserved within mammals (Fig. 9.1). Interestingly, the regions between aa187–202 and aa113–137, which are partially located outside the known structured parts, showed no alterations (substitutions) in all examined mammalian PrPs (Fig. 9.1) with the exception of position 129 for Wapiti deer [17]. These regions were conserved even between mammals and birds. An indication for the putative biological relevance of the region aa113–137 was presented in 1998 by D. Shmerling from the research group led by Charles Weissmann. Interestingly, no interspecies variations were

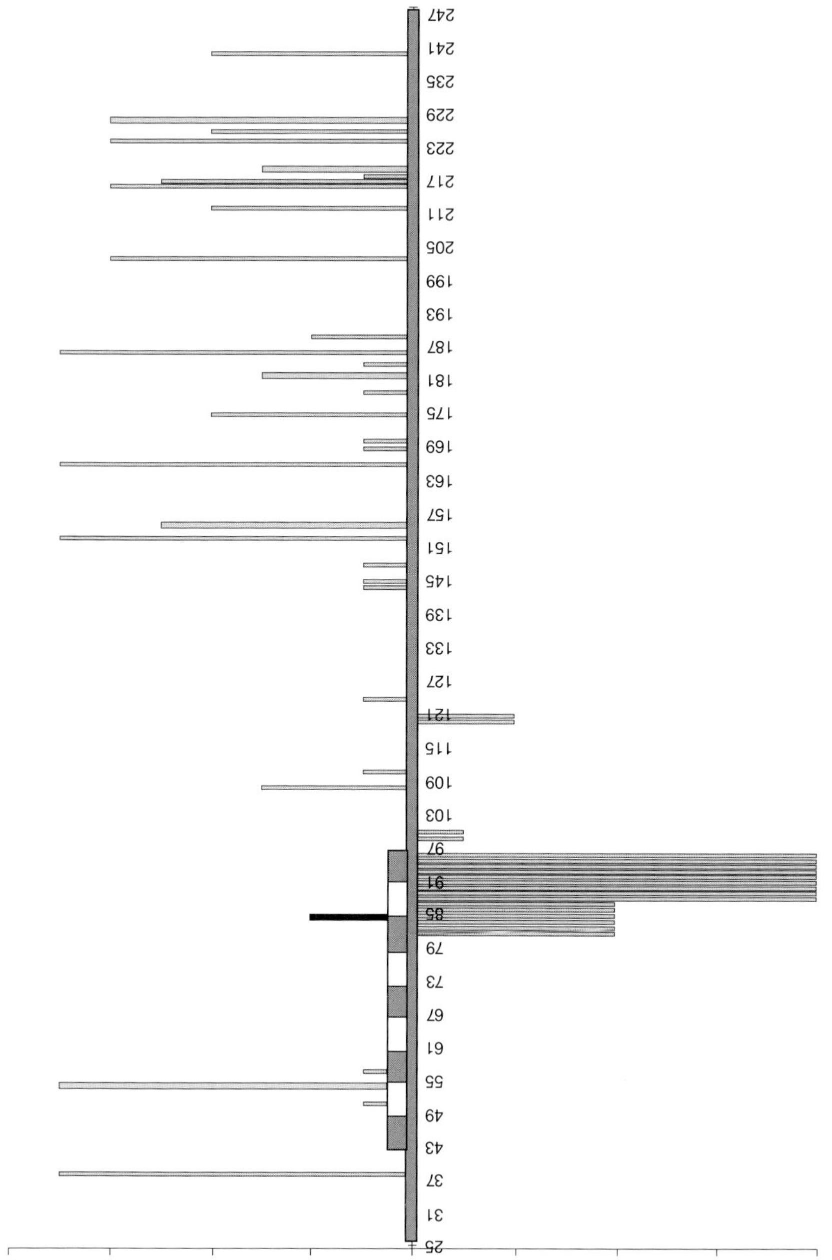

Fig. 9.3: Interspecies variations (avian PrPs). See Fig. 9.1 for set-up; number and topology of variations within avian PrPs is shown in comparison to chicken (aa26–245; eight bird species analyzed). Again, within the N-terminal half mainly insertions and deletions are found (negative axis) and in the C-terminal part substitutions are found (positive axis). Figure by H.M. Schätzl ©.

found at the positions which encode pathological mutations in humans or the polymorphisms of humans, sheep, and mice (Figs. 9.1, 9.2). The same was true for functional positions such as glycosylation sites, the disulfide bond, and serine 230/231, to which the GPI anchor is attached. In summary, we found that the interspecies variations were highly consistent with the structure models of PrP^C and that the potentially structured regions were conserved within mammals.

The first PrP gene to be sequenced after mammals was that of chickens [15; 16]. At the University of Munich we investigated whether the relationships or the degree of conservation, respectively, of avian PrPs corresponded to mammalian PrPs. Another question posed was whether it is possible to predict the architecture and structure of avian PrP based on known data for mammals. Therefore, sequence data from nine bird species were analyzed. It is of note that within bird species the region aa113–128, lying within aa113–137, was again completely conserved. On the other hand, the domain aa187–202, which is conserved within mammals, was highly variable in birds.

The first β-sheet of mammals, consisting of amino acids tyrosine, methionine, leucine, and glycine (sequence YMLG), might have corresponded to the sequence YAMG (tyrosine, alanine, methionine, and glycine) of birds. However, this region was poorly conserved within birds. In contrast, the second β-sheet (VYYR) was completely conserved in both orders. Unfortunately, it was not possible by sequence comparison to deduce the three α-helical domains, which were characterized by NMR analysis of mammalian PrP^C, corresponding to regions in bird PrP. Nevertheless, in order to compare the "architecture" of bird and mammalian PrP, the CONRAD program, which allows the comparison of conserved and variable clusters, was used. As expected, the structural elements (β-sheets and α-helices) were organized in conserved clusters. The arrangement and series of the variable and conserved areas were highly similar between mammals and birds. The highly variable region between aa90–100 was followed by the most conserved area – an area that was frequently linked to specific functions such as prion conversion processes [18] or the function of the species barrier [13].

9.3.2 The primary structure of PrP and the species barrier

It is generally accepted that the amino acid similarity of prion proteins (i. e., the PrP amino acid sequence homology) of donor and recipient species is a major determinant of the species barrier [13; 18–20]. As a rough guideline it can be estimated that the species barrier is small if the PrPs of two species are highly similar, e. g., sheep and cattle. In the case of very different PrP sequences, e. g., in Syrian hamsters and mice, there is usually a very significant species barrier. This has been clearly shown in transgenic mice expressing chimeric PrPs [19–22]. As a caveat, from these studies it cannot be clearly concluded which PrP domains are mainly responsible for the species barrier. The simple numeric comparison of two PrPs is not sufficient, whereas the exact topology of certain amino acid differences seems to be a more decisive factor [23]. Based on earlier analyses of prion infection in rodents and primates, the region around aa112 appeared to be important [13]. Later, studies with transgenic animals pointed to the C-terminal part. In addition, the postulated "factor X" clearly binding to the C-terminal part might have an influence [21; 22] (protein X). A binding site for this factor was mapped as being composed by residues 168, 172, 215, and 219 [24]. The NMR model shows that these four amino acids are in close proximity and might in fact form a kind of epitope [25]. Further important positions seem to be amino acids 184, 186, 203, and 205 [25]. Human and rodent PrPs are identical at these four positions. The exchange at position 186 (Gln–Glu) exists only in cattle and could influence intermolecular PrP–PrP interactions [26].

The theoretical prediction of susceptibility to prion infection from one species to another as deduced from amino acid alignments is highly desirable. In particular, animal species and their products entering the human food chain are of great interest, as there is overwhelming evidence that vCJD is caused by the BSE agent (see Chapter 50). In this respect, cattle, sheep, goats, and deer are very interesting, but also, for example, horses, pigs, chicken, and fish. For example, chickens cannot be infected orally with BSE prions of cattle [27]. The prion proteins of deer and horses are even more remote from cattle than the PrP of pigs [28]. From these two facts it may be concluded that oral transmission of BSE-derived prions from cattle to deer and horses is highly unlikely. Direct experiments are still missing, in particular for horses. For the situation in deer (CWD) see Chapter 21. In this context, it is also important to mention zoological gardens in the UK (Tab. 1.3, 19.1), where ungulates, especially antelope species, and big cats were in contact with BSE-contaminated feed and were obviously infected [29] (see Chapter 20). The detailed comparison of the affected antelope species with cattle and sheep revealed a significant similarity between PrPs, and hence a small species barrier.

9.3.3 The conundrum of cats and dogs

For domestic and wild cats, feline spongiform encephalopathy (FSE) is well documented (see Chapter 23). In contrast, dogs did not appear to be affected, although they were fed with similar or identical pet food [29]. If one disregards the possibility that dogs developed prion diseases but the disease was not recognized, then a comparison of canine and feline PrP would allow a better molecular understanding of the species barrier. In Munich, PrP genes of domestic dogs, dingos, and wolves were investigated (Fig. 9.4). The elucidation of cat PrP was very frustrating and yielded highly variable sequences from various laboratories (with some most definitely incorrect). Importantly, a *bona fide* correct PrP sequence in cats was only very recently published [30]. The research group, led by Flanagan in Liverpool, characterized 124 amino acids of cat PrP several years ago (gene bank no. Y13698). The common stretch of 124 amino acids showed an aa identity of 98.4% with only two amino acid exchanges between dogs and cats (aa187 and aa229). These two exchanges might indeed give some information about the species barrier. The exchange at position 187 (His–Arg) seemed to be specific for cats, because it was not found in any other species. It is located directly beside the bovine specific exchange at position 186 (Gln–Glu). This exchange is assumed to be crucial for PrP–PrP interactions and, therefore, important for the species barrier [26]. Similarly, the PrP of mink (see Chapter 22) and ferrets differed by only two amino acids (aa175, Phe-Lys and aa220, Arg–Gln). Nevertheless, these species seemed to vary significantly in their susceptibility to PrP^{Sc} [31]. Considering this, even the slightest amino acid differences could have a significant impact on the development of prion diseases. The differences between cats and dogs concern substitutions at codon 187 and 229 (histidine or glycine to arginine). Together with the His–Arg exchange at codon 177, which is found only in cats and dogs, these exchanges would alter the positive surface charge of the molecule. This could influence the postulated PrP–PrP interactions that are critical for the initial transmission of prions between species. The molecular differences in PrP could therefore explain the difference between cats and dogs regarding the susceptibility to BSE.

9.3.4 Fundamental consequences of the polymorphism at position 129 in humans

In human PrP, a polymorphism exists at position 129 (Met or Val; see Chapter 26), which has profound effects on the susceptibility to exogenous and endogenous prion diseases. Caucasian populations are ~ 50% Met/Val heterozygous; ~ 40% homozygous for Met, and ~ 10% homozygous for Val. Homozygosity is a clear risk factor for sporadic CJD and CJD acquired by infection [32]. This is especially true for the known cases of vCJD (see Chapter 15). In combination with the mutation at codon 178, this polymorphism dictates whether a familial form of CJD (fCJD) or FFI arises (see Chapter 17). The basic importance of position 129 was initially explained by a possible stabilization of the conformation of PrP^{C} [18]. A publication from the group led by J. Collinge provides very convincing data favoring a balancing selection

```
            1               15              30             45             60                      75
Human       MANL--GCWMLVLFVATWSDLGLCKKRPKP-GGWNTGGSRYPGQGSPGGNRYPPQGGGGWGQPHGGGWGQ--------PHGGGWGQPHGGGWG   82
Cattle      .VKSHI.S.I......M...V.........G.......................................PHGGGWGQ...............   92
Sheep       .VKSHI.S.I......M...V.........G.......................................--------...............   85
Mink        .VKSHI.S.L..........I.F.......G.......................................--------...............   84
Ferret      .VKSHI.S.L..........I.F.......G.......................................--------...............   84
Dog                                                  .............................--------...............
Dingo                                                .............................--------...............
Wolf                                                       .......................--------...............
Cat
                    90             105            120            135            150             165
Human       QPHGG-GWGQGGGTHSQWNKPSKPKTNMKHMAGAAAAGAVVGGLGGYMLGSAMSRPIIHFGSDYEDRYYRENMHRYPNQVVYYRPMDEYSNQNN  174
Cattle      .....G......-..G...............V.........................L...........................V.Q.....  185
Sheep       .....G......-S.................V.........................L....N...........Y..........V.R.....  177
Mink        .....G.......S.G..G............V.........................L....N...........Y........K.V.Q.....  177
Ferret      .....G.......S.G..G............V.........................L....N...........Y........K.V.Q.....  177
Dog         .....G.......S.G..G..N.........V.........................L....N...........Y...D......V.Q.....
Dingo       .....G.......S.G..G............V.........................L....N...........Y...D......V.Q.....
Wolf        .....G.......S...G..N..........V.........................L....N...........Y...D......V.Q.....
Cat                                     ...V.........................L....N...........Y...D......V.Q.....
NMR-MoPrP                                                 ββββ            ααααααααααα      ββββ

                 180            195            210             225            240       250
Human       FVHDCVNITIKQHTVTTTTKGENFTETDVKMMERVVEQMCITQYERESQAYYQRGSSMVLFSSPPVILLISFLIFLIVG  253
Cattle      .........V.E.................I...............Q..........A.VI...................  264
Sheep       .........V...................I.I.............Q..........A.VI...................  256
Mink        .........V...................M.I..........V..Q...E......A.AI...P........L..L...  256
Ferret      L........V...................M.I..........V..QQ..E......A.AI...P........L..L...  256
Dog         ..R......V...................M.I..........V..QK..E......A.AI...P........L..L..
Dingo       ..R......V...................M.I..........V..QK..E......A.AI...P........L..L.
Wolf        ..R......V...................M.I..........V..QK..E......A.A
Cat         ..R......V..R................M.I..........V..QK..E.....RA.A
NMR-MoPrP       ααααααααααααααα      αααααααααααααααααα
```

Fig. 9.4: Amino acid alignment of prion proteins of humans, selected ungulates, and carnivores. The deduced amino acid sequences of carnivores were compared to PrPs of human, cattle, and sheep. Identical amino acids are indicated by dots. Position 97 is polymorphic in dogs (Ser–Gly). The localization of the α-helices and β-sheets of murine PrP as calculated by NMR is depicted. Figure by H. M. Schätzl ©.

at position 129 in human PrP consistent with frequent prehistoric kuru-like epidemics in human evolution [33]. Our analysis of more than 150 samples from over 50 nonhuman species revealed that Met/Met is located at codon 129 [17]. The only exception was one deer species (*Cervus elaphus canadensis;* Wapiti deer), which encoded a leucine on both alleles [17]. Further investigations are trying to elucidate whether this relates to the finding that deer, e. g., *Cervus elaphus canadensis* can develop chronic wasting disease (CWD; see Chapter 21). Interestingly, recent data seem to support this view. Further known polymorphisms that can profoundly modulate the susceptibility to prion infection are present at codons 136 and 171 in sheep [4] and codons 108 and 189 in mice (Fig. 9.2). The latter defines PrP-A and PrP-B genotypes of mice discussed earlier.

9.3.5 Is the phylogeny of PrP genes as expected?

Phylogenetic trees of genes or proteins can be used for the characterization of evolutionary relationships of species. Phylogenetic trees established for PrP are slightly different from those for other proteins; e. g., within primates [13]. Gibbon PrP is clearly closer to human PrP than orangutan PrP, although the orangutan is generally more closely related to humans. The PrPs of New World monkeys are approximately as remote as the PrPs of Old World monkeys from humans. Interestingly, Old World monkeys are phylogenetically more closely related to humans. This shows that expected phylogenetic analogies are not always as expected for PrP. On the other hand, additional phylogenetic PrP investigations within mammals in general did not reveal further unexpected features [14]. Figure 9.5 depicts a representative split-tree phylogenetic analysis of mammalian PrPs.

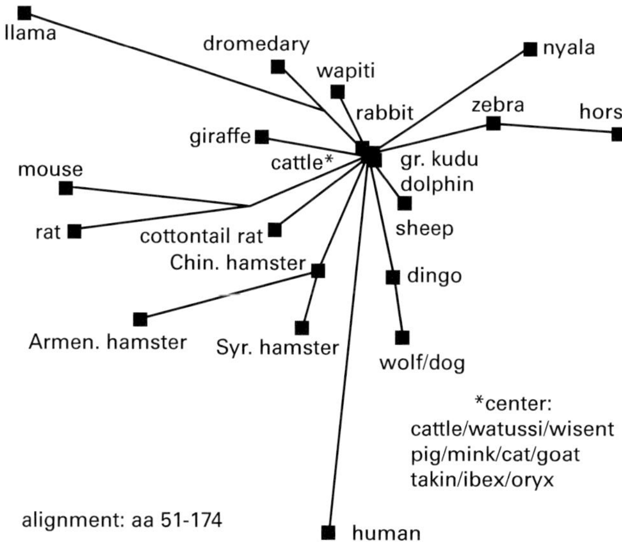

Fig. 9.5: Phylogenetic analysis of PrP by a "split-tree" examination. Amino acid alignment (aa51–174) of the PrPs of 37 mammalian species and delineation of the phylogenetic relationships in the form of a split-tree. PrPs with the highest homology are located in the center of the tree, while more divergent PrPs are located in the periphery. Because most of the analyzed species were ungulates from the family Bovidae, they form the center of the picture (e. g., cattle, watussi, bison, banteng, takin, addax antelope, ibex antelope, oryx antelope, musk-ox, etc.). This kind of analysis (aa51–174) also groups pig and goat, as well as mink and cat (partly sequence) in the center. Representative examples of rodents, carnivores, and ungulates PrPs can be detected as a cluster. Examples of close relationships are Syrian, Chinese, and Armenian hamster, mouse and rat, llama and dromedary, zebra and horse, wild dog, and wolf and dog. Human PrP is significantly more remote. The illustration of various nonhuman primates is not included for the sake of clarity. Figure by H. M. Schätzl [13] ©.

The knowledge of PrP gene regions, which are evolutionarily highly conserved, might provide information about the functional and structural aspects of PrP. It was assumed that species far more remote from mammals, such as fish and reptiles, harbor a PrP gene that shares little similarity (homology) with mammalian PrP genes. Therefore, the report of a cellular PrP (PrP^C) in the brain of salmon was remarkable. However, these data were only based on immunoblot experiments using the species-specific monoclonal antibody 3F4 [34]. This antibody recognizes the PrP of humans and Syrian hamster, but not of mice or cattle. In addition, data from nucleic acid analysis were missing and, therefore, these results are not completely accurate. An even earlier study reporting a PrP gene of the fruit fly *Drosophila* was based on nucleic acid data under conditions with low stringency and turned out later to be false positive. The PrP gene most different from human PrP was, until recently, the PrP gene of birds [15] (Fig. 6.6J).

9.3.6 PrP and PrP-related genes in nonmammalian species

The long-awaited PrPs of turtle (Fig. 6.6L), *Xenopus laevis*, and some fish species were successively characterized and yielded very interesting evolutionary insights [35–40]. Interestingly, in fish, up to five different PrPs or PrP-like gene classes have been identified: a Japanese group described a PrP-like gene in fugu that was

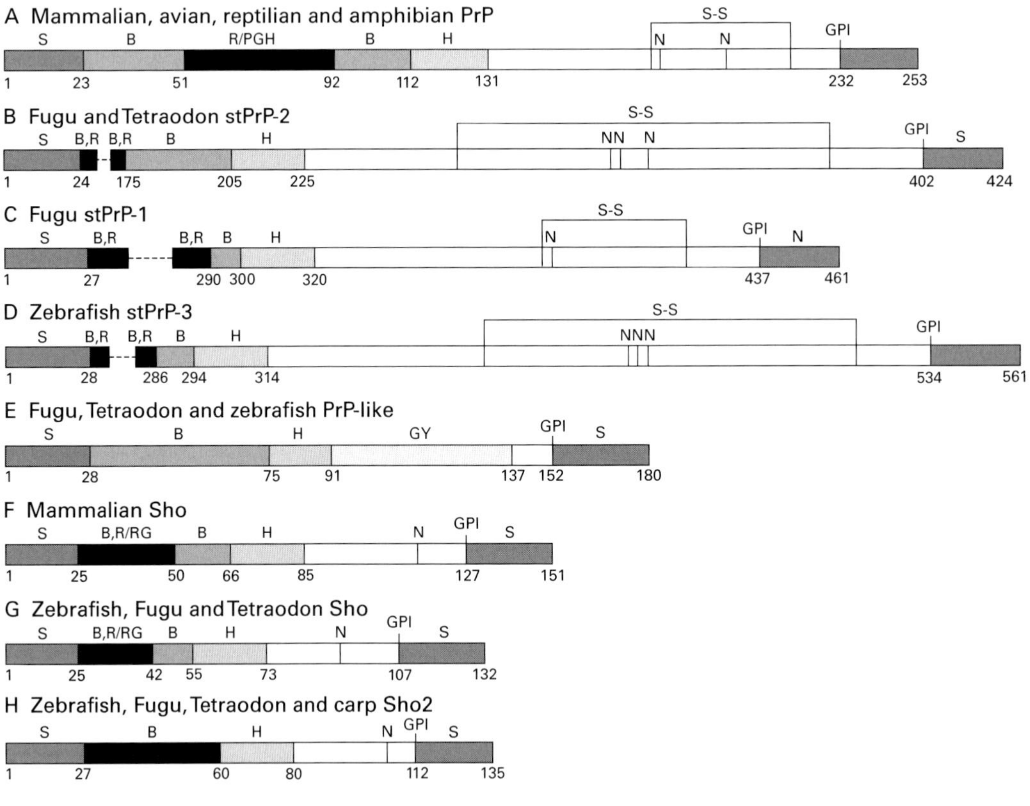

Fig. 9.6: Overall structures of PrPs, PrPs (stPrP-2, st-PrP-1), and PrP-related proteins from fish and shadoo (Sho) proteins. Numbers indicate first residues of each section and last residue of each protein. S, signal sequence; B, basic region; R, repeat region; PGH, Arg–Gly-rich region; H, hydrophobic region; N, N-glycosylation site; S–S, disulphide bond; GPI, glycophosphatidyl inositol anchor. Figure from Premzl et al., 2004 [47] with permission of the authors ©.

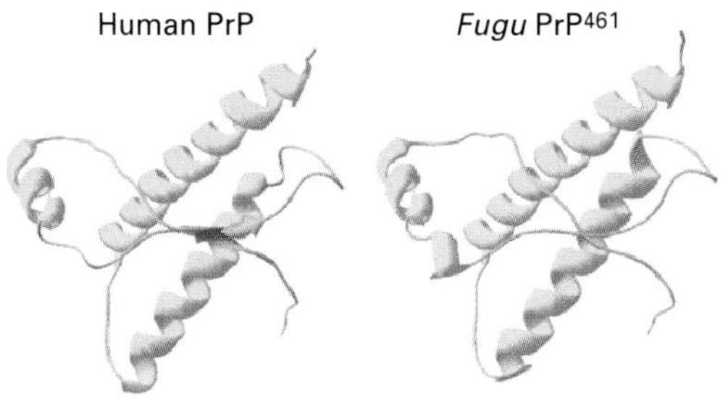

Fig. 9.7: Ribbon diagram showing the conservation of the tertiary structure at the C-terminal region of human and fugu PrPs. Secondary and tertiary protein structures were modeled using the ProModII program at the Swiss Model Automated Protein Modeling Server, based on various mammalian Protein Data Bank structure files. Figure from Rivera-Milla et al., 2003 [38], with permission of the authors.

shorter than mammalian PrP, it did not encode an obvious repeat structure at the N-terminus, and had significant differences from mammalian PrPs [39] (Fig. 9.6 E). In 2003, two German groups described, almost simultaneously, further putative fish PrPs named PrP^{461} or stPrPs (from fugu and salmon; Fig. 9.6 B, C). These fish PrPs were unexpectedly long, but contained almost all the known features of mammalian PrPs, including signal peptides, a putative GPI anchor, a basic region and repeat structures in the N-terminal part, a putative hydrophobic region, N-glycosylation sites, and an S–S bridge. Apart from their unexpected length (i. e., 461 residues in fugu) these fish PrPs were likely to represent the long-awaited PrP homologue of mammalian PrPs [37; 38]. Even a third class of PrP or PrP-like genes named Shadoo or Sho was described, which was again much shorter and only harbored some of the known features of mammalian PrPs [40] (Fig. 9.6 G, H). Fig. 9.7 shows a putative NMR structure of fugu PrP^{461}, in comparison to that proposed for human PrP. Fig. 9.8 shows a phylogenetic tree for selected mammalian, bird, reptile, amphibian, and fish PrPs. Interestingly, the knockout of at least one of these fish PrP genes seems to lead to a very specific phenotype in zebrafish (E. Malaga-Trillo, personal communication).

9.3.7 The N-terminal prion protein: dispensable and important at the same time?

The N-terminal part of mammalian PrP comprising residues ~ 23–90 represents an enigma. Clearly, this part is dispensable for prion infectivity. During usual large-scale enrichment of PrP^{Sc} (i. e., PrP27–30) from infected brains these residues are removed by the PK-digestion step without reducing prion infectivity. Some familial forms of CJD, in contrast, segregate with the insertion of two or more octarepeats in the N-terminal part [41]. Therefore, on the one hand, for some forms of prion diseases this region is dispensable with respect to prion infectivity, while on the other hand, insertions in this part are the cause for familial prion diseases and for the generation of infectivity in such situations (see Chapter 8).

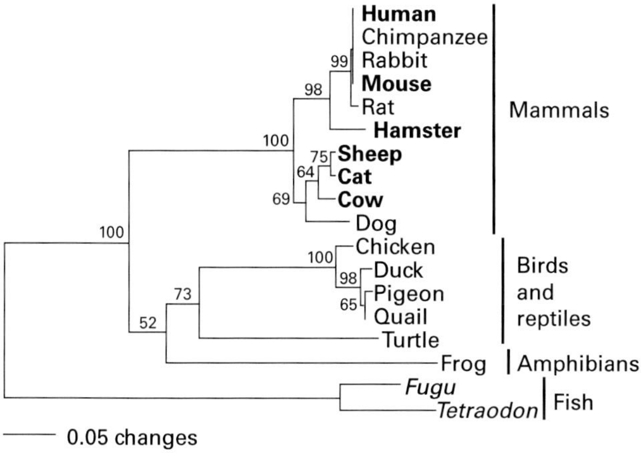

Fig. 9.8: The phylogenetic relationships among vertebrate prion proteins were studied using the parsimony and distance methods provided in the PAUP and MEGA software packages. The neighbor-joining tree is based on genetic distances between amino acid sequences of globular domains. The numbers at the internodes represent bootstrap confidence values (1,000 replications). The horizontal scale bar indicates genetic distances. Taxa known to develop prion diseases are shown in red. Figure from Rivera-Milla et al., 2003 [38], with permission of the authors.

It is of note that NMR studies could not define a tertiary structure for the N-terminal PrP. It was speculated that this lack of structure is significant in allowing a certain flexibility, which might play an important role in the conversion process of PrP^{C} into pathological PrP^{Sc}. It has been shown in in vitro and in vivo experiments that copper ions bind to the repeat structure of PrP^{C} [42] (see Chapter 7). This binding appears to stabilize the structure of the N-terminal PrP [43].

From sequence comparison it is striking that mutations and variations concerning the N-terminal part mainly consist of insertions. In contrast, mutations and variations in the C-terminal part are almost exclusively single amino acid substitutions. This is found for both mammalian and avian PrPs. One outstanding part of the N-terminal part is the repeat region (aa ~ 51–90). Although this region is well conserved within mammals, marsupials, and birds, there are marked differences existing between these classes. In mammals, usually five repeats of clusters consisting of eight (nine) amino acids are found, mainly glycine- and proline-rich (Table 9.1). Interestingly, mammals show significant variability in this region (Table 9.2). In contrast, marsupials, birds, reptiles, and fish have a significantly different architecture of their repeat structures (Table 9.1). Here, the length of the amino acid clusters as well as the number of repetitions varies. In birds, there is a high variability in the number of repetitions at the C-terminal end of the repeat structure. These differences point to a common evolutionary precursor PrP gene, also without duplications in the repeat structure that developed differently within vertebrates, reptiles, fish, and birds. Parts of the repeats seem in fact conserved between birds and mammals, e. g., the marked progression of proline–histidine. Interestingly, in vitro copper binding has initially been shown for mammalian as well as for bird peptides derived from the repeat structure [42] (see Chapter 7). When analyzing the newly discovered fish PrPs in this respect a new classification consisting of two types of four residue repeats was recently suggested [38] (Fig. 9.9).

In summary, the high evolutionary conservation, which exists within mammals and birds, and also between mammals, birds, and to some degree other classes, indicates that the N-terminal part of PrP also has a definite biological function. The binding of copper has already been mentioned [44–46]. Nevertheless, for the entire prion protein, which is characterized by a high degree of phylogenetic conservation, the question remains – which biological functions can be assigned to this remarkable and enigmatic protein?

Table 9.1: Repeat elements in the N-terminal part of PrP.

	Number of repeats[2]	Length (residues)
Mammals	5 [2–7]	8 [9]
Marsupials	4	9–10
Birds	6.5–9	6
Turtles[1]	10	6

[1] Source: [36].
[2] The PrP of *Xenopus laevis* contains no repeat structures [35].

Table 9.2: Variations in octarepeats of mammalian PrP[4].

Species	Alleles	Type	Disease-association
Human[1]	1	Deletion	No
Human[1]	2	Insertion [2-9]	Yes
Orangutan[1]	1	Deletion	Unknown
African green monkey	2	Deletion	Unknown
Squirrel monkey	2	Insertion	Unknown
Spider monkey	2	Deletion	Unknown
Lemur	1	Deletion	Unknown
Cattle[1,2]	1/2	Insertion	Unlikely
Watussi	2	Insertion	Unknown
Wisent	2	Insertion	Unknown
Goat[1,3]	1	Deletion/ insertion	Unlikely
Gazella subgut	2	Insertion	Unknown
Giraffe	2	Insertion	Unknown
Llama	2	Insertion	Unknown

[1] Polymorphism
[2] Normally 6 octarepeats present
[3] In goats and sheep polymorphic alleles ranging from 3 to 7 octarepeats were described.
[4] Source: [Table 9.2 data compiled from 13; 14; 48–51]

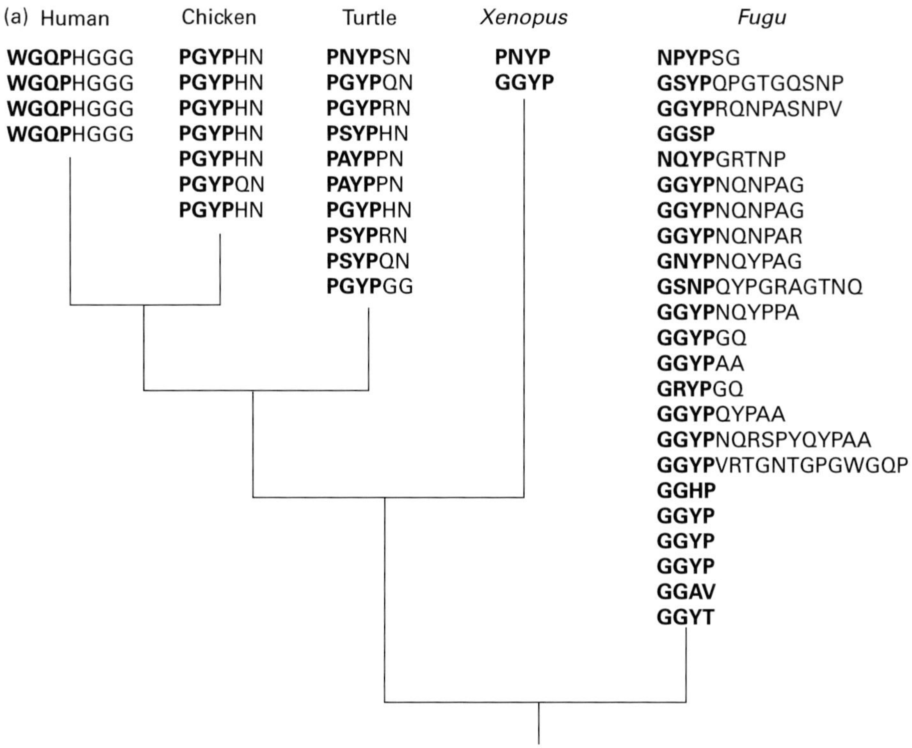

(b)

	132	138	139	142	146	147	160	166	174	223
Human	S	I	I	G	E	D	Q	M	N	S
Mouse	S	M	I	G	E	D	Q	V	N	S
Cow	S	L	I	G	E	D	Q	V	N	S
Sheep	S	L	I	G	E	D	Q	V	N	S
Chicken	R	Y	H	S	Y	R	R	–	V	Y
Duck	R	Y	R	S	Y	R	R	–	V	Y
Pigeon	R	Y	R	S	Y	R	R	–	V	Y
Turtle	S	M	N	R	R	Q	Q	–	R	–
Xenopus	N	Y	Q	N	S	R	R	M	R	–
Fugu	L	H	F	R	E	E	T	K	N	Y
	Type C							Type A		

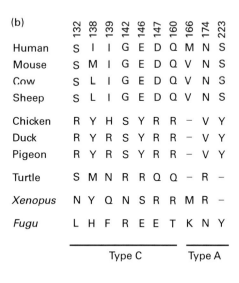

Fig. 9.9: Molecular evolutionary trends in repeat domains of PrP. **(a)** Phylogram indicating the evolution in length, number, and sequence variability of the N-terminal repeats in representative vertebrate PrPs. A comparative alignment of the repeats within a single PrP is shown for each species. The basic tetrapeptide unit shared by all vertebrate classes is highlighted in bold. **(b)** Patterns of amino acid variability at selected A- and C-type residues. Numbers relate to mouse PrP. Dashes indicate deleted residues. Figure from Rivera-Milla et al., 2003 [38], with permission of the authors.

References

[1] Basler K, Oesch B, Scott M, et al. Scrapie and cellular PrP isoforms are encoded by the same chromosomal gene. Cell 1986; 46:417–428.

[2] Oesch B, Westaway D, Walchli M, et al. A cellular gene encodes scrapie PrP 27–30 protein. Cell 1985; 40:735–746.

[3] Westaway D, Cooper C, Turner S, et al. Structure and polymorphism of the mouse prion protein gene. Proc Natl Acad Sci USA 1994; 91:6418–6422.

[4] Westaway D, Zuliani V, Cooper CM, et al. Homozygosity for prion protein alleles encoding glutamine-171 renders sheep susceptible to natural scrapie. Genes Dev 1994; 8:959–969.

[5] Puckett C, Concannon P, Casey C, et al. Genomic structure of the human prion protein gene. Am J Hum Genet 1991; 49:320–329.

[6] Kretzschmar HA, Prusiner SB, Stowring LE, et al. Scrapie prion proteins are synthesized in neurons. Am J Pathol 1986; 122:1–5.

[7] Moser M, Colello RJ, Pott U, et al. Developmental expression of the prion protein gene in glial cells. Neuron 1995; 14:509–517.

[8] Rother KI, Clay OK, Bourquin JP, et al. Long non-stop reading frames on the antisense strand of heat shock protein 70 genes and prion protein (PrP) genes are conserved between species. Biol Chem 1997; 378:1521–1530.

[9] Riek R, Hornemann S, Wider G, et al. NMR structure of the mouse prion protein domain PrP(121–231). Nature 1996; 382:180–182.

[10] Donne DG, Viles JH, Groth D, et al. Structure of the recombinant full-length hamster prion protein PrP(29–231): The N terminus is highly flexible. Proc Natl Acad Sci U S A 1997; 94:13452–13457.

[11] Huang Z, Gabriel JM, Baldwin MA, et al. Proposed three-dimensional structure for the cellular prion protein. Proc Natl Acad Sci USA 1994; 91:7139–7143.

[12] Huang Z, Prusiner SB, Cohen FE. Scrapie prions: a three-dimensional model of an infectious fragment. Folding Design 1995; 1:13–19.

[13] Schätzl HM, Da Costa M, Taylor L, et al. Prion protein gene variation among primates. J Mol Biol 1995; 245:362–374.

[14] Wopfner F, Weidenhöfer G, Schneider R, et al. Analysis of 27 mammalian and 9 avian PrPs reveals high conservation of flexible regions of the prion protein. J Mol Biol 1999; 289:1163–1178.

[15] Harris DA, Falls DL, Johnson FA, et al. A prion-like protein from chicken brain copurifies with an acetylcholine receptor-inducing activity. Proc Natl Acad Sci U S A 1991; 88:7664–7668.

[16] Gabriel JM, Oesch B, Kretzschmar HA, et al. Molecular cloning of a candidate chicken prion protein. Proc Natl Acad Sci USA 1992; 89:9097–9101.

[17] Schätzl HM, Wopfner F, Gilch S, et al. Why is codon 129 of prion protein polymorphic in human beings but not in animals? Lancet 1997; 349(9065):1603–1604.

[18] Prusiner SB. Prion diseases and the BSE crisis. Science 1997; 278:245–251.

[19] Scott M, Foster D, Mirenda C, et al. Transgenic mice expressing hamster prion protein produce species-specific scrapie infectivity and amyloid plaques. Cell 1989; 59:847–857.

[20] Scott M, Groth D, Foster D, et al. Propagation of prions with artificial properties in transgenic mice expressing chimeric PrP genes. Cell 1993; 73:979–988.

[21] Telling GC, Scott M, Hsiao KK, et al. Transmission of Creutzfeldt–Jakob disease from humans to transgenic mice expressing chimeric human-mouse prion protein. Proc Natl Acad Sci USA 1994; 91:9936–9940.

[22] Telling GC, Scott M, Mastrianni J, et al. Prion propagation in mice expressing human and chimeric PrP transgenes implicates the interaction of cellular PrP with another protein. Cell 1995; 83:79–90.

[23] Krakauer DC, Pagel M, Southwood TR, et al. Phylogenesis of prion protein (letter). Nature 1996; 380:675.

[24] Kaneko K, Zulianello L, Scott M, et al. Evidence for protein X binding to a discontinuous epitope on the cellular prion protein during scrapie prion propagation. Proc Natl Acad Sci USA 1997; 94:10069–10074.

[25] Scott MR, Safar J, Telling GC, et al. Identification of a prion protein epitope modulating transmission of bovine spongiform encephalopathy prions to transgenic mice. Proc Natl Acad Sci USA 1997; 94:14279–14284.

[26] Billeter M, Riek R, Wider G, et al. Prion protein NMR structure and species barrier for prion diseases. Proc Natl Acad Sci USA 1997; 84:7281–7285.

[27] Collee JG and Bradley R. BSE: a decade on – part 1. [Review]. Lancet 1997; 349:636–641.

[28] Martin T, Hughes S, Hughes K, et al. Direct sequencing of PCR amplified pig PrP genes. Biochim Biophys Acta 1995; 1270:211–214.

[29] Kirkwood JK and Cunningham AA. Epidemiological observations on spongiform en-

cephalopathies in captive wild animals in the British Isles. Vet Rec 1994; 135:296–303.
[30] Lysek DA, Nivon LG, Wüthrich K. Amino acid sequence of the *Felis catus* prion protein. Gene 2004; 341:249–253.
[31] Bartz JC, McKenzie DI, Bessen RA, et al. Transmissible mink encephalopathy species barrier effect between ferret and mink: PrP gene and protein analysis. J Gen Virol 1994; 75:2947–2953.
[32] Collinge J, Palmer MS, Dryden AJ. Genetic predisposition to iatrogenic Creutzfeldt–Jakob disease. Lancet 1991; 337:1441–1442.
[33] Mead S, Stumpf MP, Whitfield J, et al. Balancing selection at the prion protein gene consistent with prehistoric kurulike epidemics. Science 2003; 300(5619):640–643.
[34] Gibbs CJ Jr and Bolis CL. Normal isoform of amyloid protein (PrP) in brains of spawning salmon. Mol Psychiatr 1997; 2(2):146–147.
[35] Strumbo B, Ronchi S, Bolis LC, et al. Molecular cloning of the cDNA for *Xenopus laevis* prion protein. FEBS Lett 2001; 508:170–174.
[36] Simonic T, Duga S, Strumbo B, et al. cDNA cloning of turtle prion protein. FEBS Lett 2000; 469:33–38.
[37] Oidtmann B, Simon D, Holtkamp N, et al. Identification of cDNAs from Japanese pufferfish (*Fugu rubripes*) and Atlantic salmon (*Salmo salar*) coding for homologues to tetrapod prion protein. FEBS Lett 2003; 538:96–100.
[38] Rivera-Milla E, Stuermer CA, Malaga-Trillo E. An evolutionary basis for scrapie disease: identification of a fish prion mRNA. Trends Genet 2003; 19:72–75.
[39] Suzuki T, Kurokawa T, Hashimoto H, et al. cDNA sequence and tissue expression of *Fugu rubripes* prion protein-like: a candidate for the teleost orthologue of tetrapod PrPs. Biochem Biophys Res Commun 2002; 294:912–917.
[40] Premzl M, Sangiorgio L, Strumbo B, et al. Shadoo, a new protein highly conserved from fish to mammals and with similarity to prion protein. Gene 2003; 314:89–102.
[41] Owen F, Poulter M, Lofthouse R, et al. Insertion in prion protein gene in familial Creutzfeldt–Jakob disease (letter). Lancet 1989; 1:51–52.
[42] Hornshaw MP, McDermott JR, Candy JM. Copper binding to the N-terminal tandem repeat regions of mammalian and avian prion protein. Biochem Biophys Res Commun 1995; 207:621–629.
[43] Miura T, Hori iA, Takeuchi H. Metal-dependent alpha-helix formation promoted by the glycine-rich octapeptide region of prion protein. FEBS Lett 1996; 396:248–252.
[44] Brown DR, Qin K, Herms JW, et al. The cellular prion protein binds copper *in vivo*. Nature 1997; 390:684–687.
[45] Stöckel J, Safar J, Wallace AC, et al. Prion protein selectively binds copper (II) ions. Biochemistry 1998; 37:7185–7193.
[46] Viles JH, Cohen FE, Prusiner SB, et al. Copper binding to the prion protein: structural implications of four identical cooperative binding sites. Proc Natl Acad Sci U S A 1999; 96:2042–2047.
[47] Premzl M, Gready JE, Jermiin LS, et al. Evolution of vertebrate genes related to prion and shadoo proteins: Clues from comparative genomic analysis. Mol Biol Evol 2004; 21:2210–2231.
[48] Palmer MS and Collinge J. Mutations and polymorphisms in the prion protein gene. Hum Mutat 1993; 2:168–173.
[49] Prusiner SB, Fuzi M, Scott M, et al. Immunologic and molecular biologic studies of prion proteins in bovine spongiform encephalopathy. J Infect Dis 1993; 167:602–613.
[50] Gilch S, Spielhaupter C, Schatzl HM. Shortest known prion protein allele in highly BSE-susceptible lemurs. Biol Chem 2000; 381:512–523.
[51] Goldmann W, Chong A, Foster J, et al. The shortest known prion protein allele occurs in goats, has only three octarepeats and is non-pathogenic. J Gen Virol 1998; 79:3173–3176.

10 Knockouts and Transgenic Mice in Prion Research

Eckhard Flechsig, Ivan Hegyi, Alex J. Raeber, Antonio Cozzio, Adriano Aguzzi, and Charles Weissmann

10.1 Introduction

Compelling linkages between the infectious scrapie agent designated prion and PrP were established by biochemical and genetic data, leading to the prediction that animals devoid of PrP should be resistant to experimental scrapie and fail to propagate infectivity. This prediction was indeed borne out, adding substantial support to the "protein only" hypothesis. Since the proposal of the "protein only" hypothesis more than three decades ago, the cloning of the PrP gene, and studies on PrP knockout mice and on mice transgenic for mutant PrP genes, deep insights into prion biology have resulted. Reverse genetics on PrP knockout mice carrying modified PrP transgenes was used to address a variety of problems: mapping PrP regions required for prion replication, studying PrP mutations affecting the "species barrier", modeling familial forms of human prion diseases, analyzing the cell specificity of prion propagation, and investigating the physiological role of PrP by structure–function studies. Transgenic approaches have been used to assess novel intervention strategies against prion diseases. Studies on the mechanism of prion spread in an infected organism have been addressed by transplantation of neuroectodermal or hematopoietic tissue expressing PrP into PrP knockout mice and vice versa. Many questions regarding the role of PrP in susceptibility to prions have been elucidated; however, the physiological role of PrP and the pathological mechanisms of neurodegeneration in prion diseases remain elusive.

10.2 Generation and properties of PrP knockout mice

10.2.1 Generation of PrP knockout mice

Mouse PrP is encoded by a single-copy gene (*Prnp*) that comprises three exons located on chromosome 2 [1; 2]. Several lines of mice devoid of PrP^C have been generated by homologous recombination in embryonic stem (ES) cells using either of two strategies (Fig. 10.1). The "conservative strategy" involves deletions or replacements restricted to the open reading frame (ORF). Mice homozygous for the disrupted gene, such as $Prnp^{o/o}$ [Zürich I] [3] or $Prnp^{-/-}$ [Edinburgh] [4], develop normally, and show no striking phenotype. Although truncated PrP mRNA was expressed in the brain of $Prnp^{o/o}$ [Zürich I] mice, no PrP-related protein was found in either line. PrP knockouts by the "radical strategy" involve deletion of not only the ORF, but also of its flanking regions, in particular the splice acceptor site of the third exon. This type of PrP knockout mouse, such as $Prnp^{-/-}$ [Nagasaki] [5], Rcm0 [6; 7], and $Prnp^{-/-}$ [Zürich II] [8], also developed normally but exhibited severe ataxia and Purkinje cell loss in later life.

Two lines of "conditional PrP knockout" mice have been generated by the conservative strategy. A PrP-expressing transgene under the control of a doxycycline-repressible promoter was introduced into $Prnp^{o/o}$ [Zürich I] mice, however repression was incomplete and resulted in basal PrP^C levels about 5–15% of wild-type. No histopathological changes appeared after repression and no clinical disease after prion inoculation was noted, but prions accumulated nonetheless, albeit at a low level [9]. Mallucci et al. generated $Prnp^{o/o}$ mice that were trans-

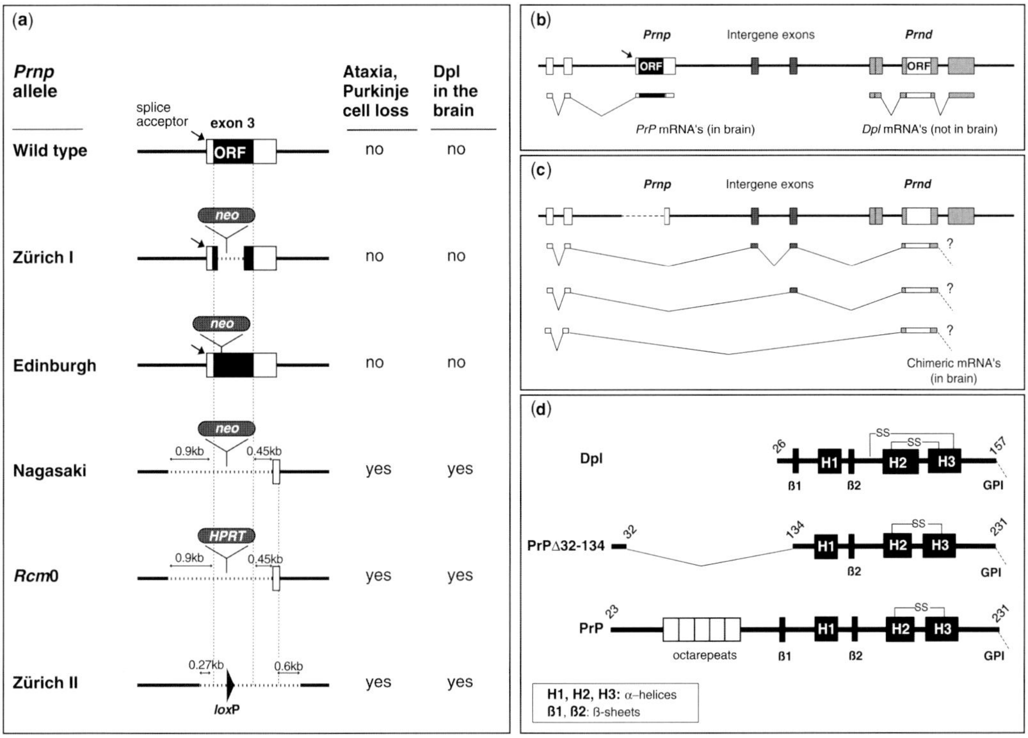

Fig. 10.1: PrP knockout strategies and their consequences. **(a)** Two strategies have been used to target the open reading frame (ORF) of PrP in the third *Prnp* exon by homologous recombination. Gene disruptions restricted to the ORF that retain the indicated splice site do not cause a pathological phenotype (conservative strategy). Gene deletions that extend beyond the ORF and remove the indicated splice site elicit upregulation of Dpl and cause a pathological phenotype (radical strategy). Knockouts were performed by insertion of *HPRT* (hypoxanthine phosphoribosyl transferase) into the ORF or replacement of ORF-containing gene segments by *neo* (neomycin phosphotransferase) or *loxP*, a 34-bp recombination site from phage P1 (black arrowhead). Deleted sequences are shown by dotted lines. Modified from Rossi et al. 2001 [8]. **(b)** Coding and noncoding exons of the PrP gene (*Prnp)*, the Dpl gene (*Prnd*), and intergenic exons of unknown function. In the brain of wild-type mice, the promoter of *Prnp* is active, but the promoter of *Prnd* is silent. **(c)** Exon skipping leads to expression of Dpl under the control of the *Prnp* promoter in the brain. Deletion of the splice site caused formation of chimeric mRNA transcripts comprising the first two exons of *Prnp,* noncoding exons, and the Dpl-encoding exon. **(d)** Comparison of protein domains of Dpl with full-length PrP and with PrP lacking its flexible region, PrPΔ32–134. Figure 10.1 b–c modified from [28].

genic for both a *loxP*-flanked cassette containing the PrP ORF under control of a *Prnp* promoter element and cre recombinase under the control of the neurofilament heavy chain (NFH) promoter [10]. Expression of the recombinase resulted in the ablation of PrP in neurons at 10–12 weeks of age without entailing histopathological changes, but expression of PrP was still detectable in nonneuronal cells. No clinical disease was observed for up to 400 days after prion inoculation [11] (see Section 10.5.1 for more details).

10.2.2 The phenotype of PrP knockout mice

The hope that the generation of PrP null mice would cast light on the physiological role of PrP^C has not been fully realized. The physiological function of PrP is still elusive (see Chapter 7). *Prnp*$^{o/o}$ [Zürich I] and *Prnp*$^{-/-}$[Edinburgh] mice developed and reproduced normally [3; 4], but aging mice show demyelination in the sciatic nerves, albeit without clinical symptoms [12]. Behavioral studies revealed no significant differences to wild-type mice [3; 13], except for alterations in both circadian activity rhythms and sleep patterns that were abolished by the introduction of PrP transgenes [14; 15]. Furthermore, *Prnp*$^{o/o}$ [Zürich I] mice were found to be more susceptible than wild-type mice to acute seizures induced by drugs such as kainic acid [16]. Electrophysiological studies showed that GABA-A receptor-mediated fast inhibition was weakened, long-term potentiation (LTP) was impaired [17–19], afterhyperpolarization potentials in hippocampal neurons were reduced [10], and Ca^{2+}-activated K^+ currents were disrupted in hippocampal cells [20] and cerebellar Purkinje cells [21]. Introduction of high copy numbers of the human [18] or murine PrP gene into PrP null mice restored the LTP response to that seen in wild-type [20] (see Chapter 7). Biochemical changes reported for Zürich-I-type knockout mice suggest impairment of enzymatic activity required for antioxidant defense [22–25]. These data suggest that the absence of PrP^C alters oxidative stress homeostasis.

In contrast, PrP knockout lines with extensive deletions in the *Prnp* gene exhibit severe ataxia and Purkinje cell loss in later life [5–8] as well as abnormal myelination in the spinal cord and sciatic nerves [12]. This phenotype was attributed to the absence of PrP^C, because it was rescued by introduction of a PrP transgene [12]. Because the phenotype was not observed in two lines generated by the conservative strategy, it was suggested that the ataxic phenotype was likely caused by the deletion of sequences flanking the PrP ORF and not by the absence of PrP [26]. As illustrated in Figure 10.1, this puzzle was solved 3 years later with the discovery of a novel gene, designated *Prnd*. It was found that the phenotype is associated with ectopic expression of the *Prnd* product, Doppel (Dpl), in the brains of all three ataxic PrP knockout lines [6; 8; 27; 28].

Dpl is encoded by *Prnd*, located 16 kb downstream of *Prnp* and is not transcribed in brain of wild-type mice [6] (Fig. 10.1). As a consequence of the radical PrP knockout strategy, the acceptor splice site of the PrP-encoding exon 3 is lost and chimeric transcripts containg the first two noncoding exons of the *Prnp* locus linked to the Dpl-encoding *Prnd* exon are formed. This places Dpl expression under control of the *Prnp* promoter and Dpl transcripts have been preferentially found in Purkinje cells [6; 27], explaining the Purkinje cell degeneration. Time to appearance of Purkinje cell loss and ataxia was inversely correlated with the expression level of Dpl in brain and the phenotype was rescued by coexpression of PrP^C [8; 12]. Thus, ectopic expression of Dpl in the absence of PrP^C, rather than absence of PrP^C itself, causes ataxia and Purkinje cell loss in all three ataxic PrP knockout lines. This interpretation is supported by the finding that transgenic *Prnp*$^{o/o}$ mice expressing Dpl ubiquitously in brain developed severe ataxia associated with loss of both granule and Purkinje cells in the cerebellum [29]. As in the case of the ataxic PrP knockout lines, age of onset was inversely related to the level of Dpl expression, and introduction of a hamster-PrP-encoding transgene resulted either in complete abrogation of the cerebellar syndrome in mice expressing moderate Dpl levels or partial abrogation in animals expressing high levels of Dpl [29]. Similarly, Dpl targeted to Purkinje cells of *Prnp*$^{o/o}$ [Zürich I] mice caused Purkinje cell loss and ataxia. Introduction of PrP was sufficient to abrogate the onset of disease caused by moderate levels of Dpl [30].

How is brain damage caused by Dpl expression and why does PrP prevent it? Dpl is an N-glycosylated, GPI-anchored protein normally expressed in a variety of tissues but not in postnatal brain [6; 7; 31]. Its physiological function is still elusive, but its absence causes male sterility in mice [32; 33]. Dpl is unlikely to be involved in prion pathogenesis [29; 34; 35]. While Dpl and PrP show about 25% sequence similarity and share similar globular domains [7; 31; 36], Dpl lacks a counterpart to the flexible N-terminal half of PrP (Fig. 10.1). It has been proposed that ectopic expression of Dpl elicits an increase of heme oxygenase 1 and both neuronal and inducible NO synthase, causing oxidative stress deleterious to sensitive neurons; this effect would be counteracted by the antioxidant properties of PrP^C [37]. However, further studies did not find evidence that Dpl elicits enhanced oxidative damage to brain proteins [38] or reduced activities of enzymes involved in oxidative stress, such as superoxide dismutase (SOD) or glutathione reductase activities [39].

Another explanation is based on the structural similarity between Dpl and the truncated PrP lacking the flexible N-terminal half [28] and is discussed in Section 10.4.1. Indeed, expression of this truncated PrP targeted to Purkinje cells of $Prnp^{o/o}$ [Zürich I] mice caused Purkinje cell loss and ataxia (as in ataxic PrP knockout lines), which are suppressed in the presence of wild-type PrP [40]. Thus, Dpl and truncated PrP might cause Purkinje cell degeneration by the same mechanism.

Mice devoid of PrP developed normally and did not show any obvious behavioral differences. Minor electrophysiological changes and alterations in circadian activity rhythms and sleep pattern were found. However, the physiological function of the highly conserved protein is still unknown. The phenotype found in three PrP knockout lines is caused by ectopic expression of a novel gene Dpl and not by the absence of PrP. As a consequence of the radical PrP knockout strategy, the acceptor splice site of the PrP-encoding exon is lost, placing Dpl under control of the PrP promoter. Dpl and PrP share similar globular domains, but Dpl lacks the flexible N-terminal part of PrP. This structural similarity may explain why PrP rescues the phenotype caused by Dpl.

10.2.3 Resistance of PrP knockout mice to scrapie

The finding that mice devoid of PrP ($Prnp^{o/o}$) are resistant to scrapie provides central support for the "protein only" hypothesis. $Prnp^{o/o}$ [Zürich] mice remained free of clinical symptoms for at least 2 years and showed no scrapie-specific pathology as late as 1 year after inoculation [41]. In contrast, wild-type C57BL/6 mice ($Prnp^{+/+}$) inoculated intracerebrally with mouse-adapted prion strain [42] developed clinical symptoms at about 160 days and died about 10 days later (Fig. 10.2). Moreover, $Prnp^{o/o}$[Zürich I] mice challenged with scrapie prions failed to propagate prions in brain and spleen, whereas prion levels in brain and spleen of wild-type mice increased to about 8.6 and 6.9 log LD_{50} units/ml of 10% homogenate respectively, by 140 days post infection [41]. Similar results were reported for the $Prnp^{o/o}$[Zürich] mice bred in San Francisco and inoculated with the Chandler isolate [43], $Prnp^{-/-}$[Edinburgh] mice inoculated with the ME7 mouse strain [44], and for $Prnp^{-/-}$ [Nagasaki] mice challenged with the mouse-adapted Fukuoka-1 strain of CJD prions [45]. All three lines of PrP null mice showed a complete lack of scrapie-typical neuropathology following intracerebral inoculation with prions and there is no evidence for net generation of scrapie prions; the occasional low-level infectivity detected in the brains of PrP null mice after intracerebral inoculation may be due to residual inoculum or, less likely, to contamination. It has however been speculated that in the absence of PrP, a temperature-sensitive or otherwise labile form of scrapie agent might be generated [45; 46].

10.2.4 Properties of mice hemizygous for the $Prnp^o$ allele

Mice carrying a single *Prnp* allele ($Prnp^{o/+}$ mice) showed no abnormal phenotype related to behavior and development [3; 4]. Surprisingly, they showed prolonged incubation times of about 290 days to appearance of disease, as compared to 160 days in the case of $Prnp^{+/+}$ mice. $Prnp^{o/+}$ mice harbored high levels of infectivity and PrP^{Sc} at 140 days after inoculation, as did wild-type controls, but survived thereafter for at least another 140 days without showing severe clinical symptoms [47]. Prolonged incubation times of 259 ± 27 days in $Prnp^{o/+}$ [Nagasaki] mice as compared to 138 ± 13 days

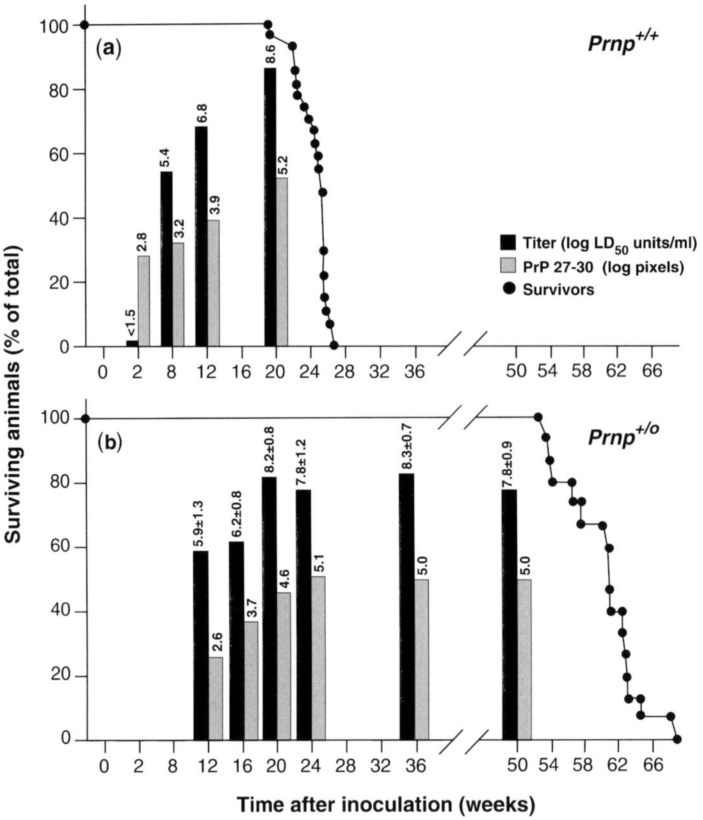

Fig. 10.2: Survival, prion titers, and PrP^{Sc} in brains of scrapie-inoculated mice. **(a)** Wild-type (*Prnp*$^{+/+}$) and **(b)** hemizygous (*Prnp*$^{o/+}$) mice at various times after inoculation with mouse prions. Although both types of mice have similar levels of prions and PrP^{Sc} at 20 weeks post infection, the wild-type mice succumb within the following 8 weeks, while the hemizygous knockout mice survive for yet another 30 weeks or more without clinical symptoms. Data from Büeler et al. 1994 [47].

in *Prnp*$^{+/+}$ mice were also reported for the mouse-adapted Fukuoka-1 strain of CJD prions [45]. *Prnp*$^{o/+}$ [Edinburgh] mice challenged either with the ME7 mouse strain or the mouse-adapted BSE strains 301V and 301C also developed clinical symptoms with a delay of 80–150 days as compared with wild-type mice [44]. There appeared, however, to be no difference in the severity of the clinical signs and the pathology in the brain at the terminal stages of disease. PrP gene dosage appears therefore to affect the timing of disease but not the final pathology.

In summary, whereas in wild-type mice inoculated with mouse RML prions the increase in prion titer and PrP^{Sc} levels is followed within weeks by scrapie symptoms and death, *Prnp*$^{o/+}$ [Zürich] mice remained free of symptoms for many months despite similar levels of scrapie infectivity and PrP^{Sc} (Fig. 10.2). These findings suggest that clinical symptoms are not necessarily correlated with the overall accumulation of PrP^{Sc} in brain; it has been suggested that the pathological processes must extend to a so-called clinical target area before disease and death ensue [48]. There are in fact several reports in which prion disease leads to death without substantial accumulation or even detectable levels of PrP^{Sc} in the brain [49–53]. Probably, discrete changes in the postulated target areas, which might be in the brain stem or upper spinal cord [50] and which have not been routinely searched for, might control the onset of the disease. It is of practical interest that under certain conditions clinically healthy mice can harbor high levels of infectivity for long periods of time [54–58]. That is, it suggests that cattle or humans, which are also in apparent good health, could contain infectious agent in their central nervous and lymphoreticular systems and perhaps fail to show clinical symptoms. Recently, a patient developed vCJD after receiving a blood transfusion derived from an unrecognized vCJD patient [59]. Another transfusion recipient died of unrelated causes, but was found at autopsy to harbor PrP^{Sc} in his lymphoid system [60]. These are points to consider when animal or human tissues are used for preparing pharmaceuticals (see Chapter 46), grafts (see Chapter 51), or when blood is transfused.

10.2.5 Restoration of susceptibility of PrP knockout mice to prions by PrP transgenes

The protein only hypothesis requires not only the demonstration that resistance to scrapie is due to the absence of PrP, but also that restoration of PrP expression in PrP knockout mice restores scrapie susceptibility. Therefore, murine *Prnp* genes were introduced by transgenesis into *Prnp*$^{o/o}$ mice. Two lines of mice transgenic for the half-genomic *Prnp* construct (see Section 10.3.1 for details), expressing PrP-A at about 3–4 and 6–7 times the level of wild-type mice were challenged with mouse RML prions. As shown in Figure 10.3, they developed disease after 2 and 3 months, respectively, compared to about 5 months in the case of wild-type (C57BL/6) mice [61]. This confirmed that incubation times are inversely related to PrP expression levels, as reported earlier in a different context [62]. The mouse line Tg*a20* was rendered homozygous and, because of its short incubation time, used for a rapid scrapie infectivity assay [61; 63]. Taken together, these experiments support the conclusion that scrapie-resis-

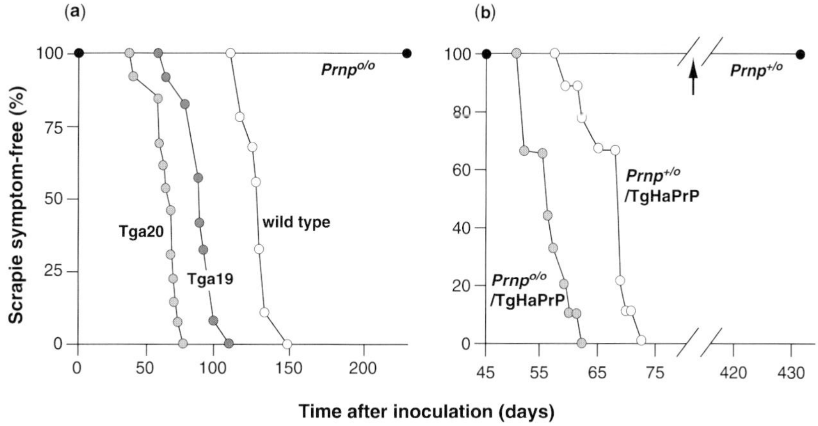

Fig. 10.3: Susceptibility to scrapie of PrP knockout mice carrying various transgenes encoding full-length or truncated PrP. **(a)** *Prnp*$^{o/o}$ mice were rendered transgenic for *Prnp* genes. Tg*a19*/+ mice had 3–4 times the normal PrP^{C} level; Tg*a20*/+ mice had 6–7 times the normal level. Figure 10.3a from Fischer et al. 1996 [61]. **(b)** *Prnp*$^{o/o}$ and *Prnp*$^{o/+}$ mice with hamster PrP transgenes were inoculated with the Sc237 isolate of hamster prions. Arrow: One animal died spontaneously without scrapie symptoms and one was killed because of a tumor. Mice with a mouse PrP allele are completely resistant to hamster prions while the presence of hamster PrP in a PrP knockout mouse renders them highly susceptible. The additional presence of a mouse PrP allele reduces susceptibility to hamster prions, as evidenced by increased incubation times. Figure 10.3b modified from Büeler et al. 1993 [41].

tant phenotype of the Zürich I PrP knockout mice is indeed due to ablation of PrP.

10.3 Transgenesis and gene replacement

When a certain phenotype is generated by gene ablation – such as ataxia in some PrP knockout lines or resistance to scrapie – it is important to show that this is indeed the consequence of the targeted modification and not to some unintended event, such as elimination of enhancers regulating another gene, disruption of an unidentified gene, or gain of function as described for Dpl [28] (see Section 10.2.2). The most effective, albeit (as shown above) not infallible strategy to link a phenotype to the knockout of a specific protein is to abrogate the effect of the deletion by introducing into the knockout animal a transgenic vector encoding the protein in question. Such gene insertion is usually performed by nuclear injection into a one-cell embryo of a DNA segment containing the coding sequence under the control of a promoter aimed at reproducing the expression pattern of the wild-type gene as closely as possible. Following injection, the DNA segment containing the cloned gene usually concatenates prior to integration, resulting in random insertion of multiple gene copies into one or few sites of the genome; therefore, each resulting transgenic mouse line is unique. Depending on the site of insertion, expression of the transgene can be silenced or modulated by regulatory elements in the neighboring regions. Perhaps these effects can be mitigated by flanking the transgene with so-called locus control regions [64] (LCRs). Moreover, by using mice carrying *loxP* sequences in a characteristic favorable locus of their genome, direct insertion of one or a few gene copies into that site by transient expression of Cre recombinase is possible to [65; 66].

Alternatively, modifications of the genome may be achieved by in situ gene replacement using homologous recombination in ES cells (see Section 10.3.2). In situ gene replacement is based on homologous recombination rather than random integration and allows modification of a resident gene in a variety of fashions. It has the advantage that the normal copy number is preserved and that the modified gene, because it remains in its physiological environment, will not be influenced by alien enhancers or by position-dependent silencing. However, protein expression levels depend not only on the properties of the promoter, but also on the stability of the mRNA and the protein, both of which may be profoundly affected by sequence changes, so that the cellular levels of a protein encoded by the normal gene and by its modified counterpart need not be the same.

10.3.1 Transgene vectors for PrP expression

Three vector types have been generated and are commonly used to express the wild-type PrP gene: the PrP cosmid, the PrP minigene, designated "half-genomic" PrP vector and the "cos-Tet" vector. The 40-kb mouse cosmid, derived from the murine *Prnp*^{b} allele (see Section 10.4.4 for details) contains 6 kb of the promoter sequence, the three exons and two introns and approximately 18 kb of 3' downstream sequence that includes *Prnd* [6; 8; 67]. Because such large cosmid vectors are laborious to work with, the half-genomic PrP vector was constructed by deleting the large intron 2 and all but 2.2 kb of 3' flanking sequence. Thus, the half-genomic PrP vector contains the same promoter sequence as the cosmid, but the coding sequence is derived from the *Prnp*^{a} locus and *Prnd* is deleted [8; 61]. The expression pattern elicited by the half-genomic construct is similar to wild type, except that cerebellar Purkinje cells express neither PrP nor PrP mRNA at detectable levels [61]. Perhaps the half-genomic PrP vector lacks a Purkinje-cell-specific enhancer, which could be located in the large intron or in the distal part of the 3' noncoding region, both of which are absent in the half-genomic construct. The lack of Purkinje-cell-specific PrP expression has also been found in mice expressing mutated PrP [68; 69] under control of MoPrP vector. XhoI [70], a construct similar to the half-genomic vector. The third vector, called "cos-Tet", encompasses the two exons and one intron of the hamster *Prnp* gene and approximately 24 and 6 kb of the 5' and 3' flanking region, respectively, and lacks the Dpl ORF [71].

Overexpression of the cosmid transgenes, such as hamster PrP and PrP-B (encoded by the *Prnp^b* allele), leads to a profound necrotizing myopathy involving skeletal muscle, demyelinating polyneuropathy, and focal vacuolation of the central nervous system as the mice age [72]. However, no impairments were observed with mice transgenic for the half-genomic construct, even though PrP-A expression was equally high [61] (i. e., up to seven times the wild-type PrP level). Because a similar phenotype occurred in mice harboring hamster PrP cosmid transgenes lacking *Prnd* [72] and because mice expressing Dpl from the "cosTet" do not exhibit myopathic changes [29], the myopathy is not due to expression of Dpl. Because expression of PrP-A (encoded by the *Prnp^a* allele) at levels similar to those of hamster PrP or PrP-B is not pathogenic [61; 73], the neuromyopathy could be due to overexpression of PrP-B or some unknown feature of the cosmid.

10.3.2 Replacements of *Prnp*

A double replacement gene targeting strategy was used to introduce point mutations into the *Prnp* locus of a 129/Ola derived ES line lacking the enzyme hypoxanthine phosphoribosyl transferase (HPRT). In the first step, the entire PrP ORF was replaced with an HPRT minigene by homologous recombination, and successfully targeted clones were isolated by selection in a gancyclovir-containing medium without HAT that is lethal for cells lacking HPRT. In the second step, the HPRT minigene was replaced with a mutated PrP coding sequence, again by homologous recombination. Cells devoid of HPRT, selected by virtue of their resistance to 6-thioguanine, contained the altered *Prnp* allele that was indistinguishable from the wild-type allele, except for the desired mutation [53; 74; 75]. Selected ES cell clones were injected into blastocysts from C57BL/6 mice, and chimaeric offspring derived from both 129 and C57BL/6 cells, identified by the presence of two coat colors, were bred with 129/Ola mice to detect germline transmission of the transgene and establish the line.

A different "knockin" technology was used by Kitamoto et al. to replace the murine PrP ORF by a mutated human counterpart [76]. In the first step, a murine *Prnp* DNA segment containing a cassette with the human PrP ORF and a neomycin resistance gene flanked by *loxP* sites was introduced into the *Prnp* locus of ES cells by homologous recombination. The neomycin resistance gene allowed the selection of positive clones from which mice carrying the cassette were generated. In a second step, a *Cre* expression construct was injected into fertilized oocytes carrying the modified allele, which led to elimination of the neomycin resistance gene, leaving a single *loxP* site in the 3' noncoding region of the human gene [76]. This type of approach can be simplified by eliminating the resistance gene in the modified ES cells prior to generating mice [77]. Elimination of selection genes is recommended, because it has been reported that in at least one instance a neomycin resistance cassette caused embryonal lethality [78], and that the full-length, albeit not the truncated [79], herpes simplex virus thymidine kinase (HSV-TK) gene may lead to male sterility [80].

10.4 Reverse genetics: studies on the structure–function relationship of PrP

The ability to restore susceptibility to scrapie in PrP knockout mice by introduction of PrP-encoding transgenes or to replace the wild-type gene by modified counterparts opened the way for reverse genetics on PrP. Reverse genetics was first established as an approach to the elucidation of structure–function relationships of the genome of the phage Qβ, a classical model in the early days of molecular biology [81]. To determine the function of a known gene, a mutation is introduced in the gene of interest and the resulting effect or phenotype is analyzed. In contrast, classical genetics involves screening for a particular phenotype followed by the identification of the responsible gene. Here we focus on several questions that are addressed by reverse genetics: Which parts of PrP are essential for sustaining prion propagation? Is the species barrier affected by PrP mutations? Is the mutated expression of PrP, which causes familial prion diseases in humans, sufficient to elicit transmissible disease in mice? Is expression of PrP sufficient to enable a cell to propagate prions?

10.4.1. Exploring the physiological role of PrP by transgenic studies.

Despite the nearly normal phenotypes of Zürich-I *Prnp*$^{o/o}$ mice, it cannot be excluded that PrP might have important functions, since the absence of PrP might be compensated during embryogenesis or subsequently by some functionally homologous protein(s). For instance, knockout mice lacking either APP or the APP-like protein APLP-2 remain healthy, whereas the knockout of both genes is lethal, suggesting functional redundancy of the two genes [82]. If indeed a functional homologue of PrP exists, and if this homologue and PrP have a common ligand, it might be possible to interfere with their function by a mutated version of PrP, which binds and blocks the ligand thereby impeding function. A candidate for such a PrP mutant was discovered when screening a series of amino proximal deletions [83] (Fig. 10.4).

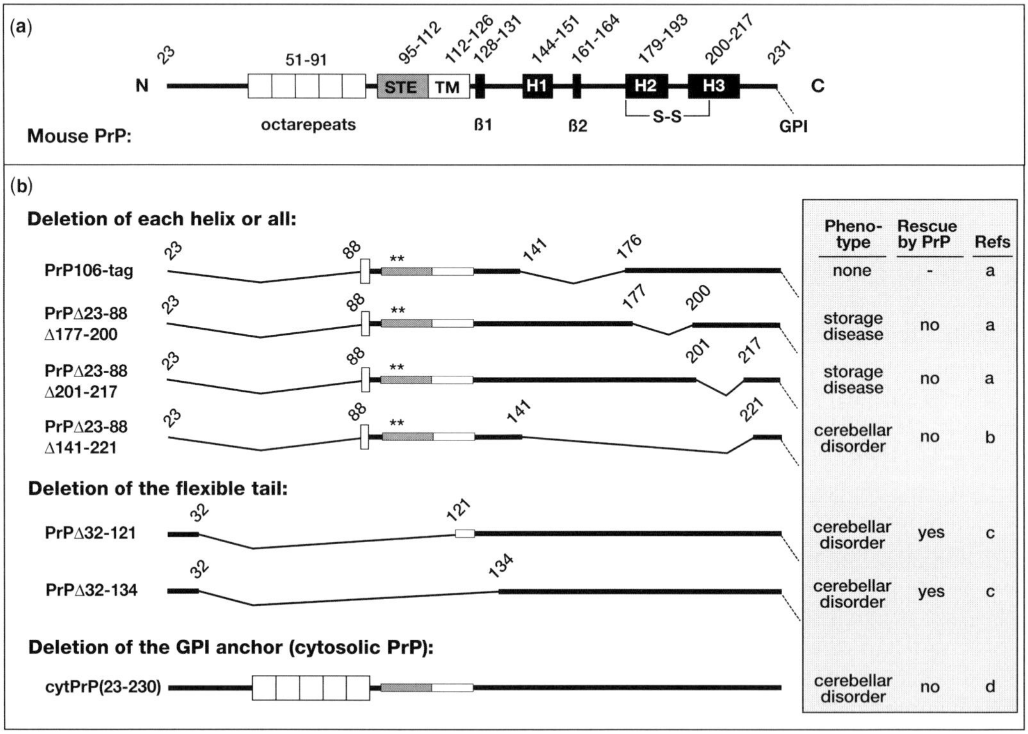

Fig. 10.4: Structure and function analysis of mouse PrP. **(a)** Scheme of the murine PrP sequence. The globular part of PrP comprises three α-helices (H1–H3) and a short antiparallel β-sheet composed of β1 and β2 [194; 195]. The five octarepeats, the stop-transfer effector site (STE), and the optional transmembrane (TM) region containing a highly conserved sequence motif are located in the flexible amino terminal region. The single disulfide bond (S–S) and the glycolipid (GPI) attachment site are shown. **(b)** Mice expressing various PrP transgenes on a Zürich-I PrP knockout background as shown in Fig. 10.1 have been analyzed for the appearance of phenotypic abnormalities. The ability of wild-type PrP to abolish the pathological phenotype is only found in transgenic mice expressing amino-proximal PrP deletions extending to residues 121 and 134. References: (a) Muramoto et al. 1997 [90]; (b) Supattapone et al. 2001a [91]; (c) Shmerling et al. 1998 [83]; (d) Ma et al. 2002 [69]. Figure from Flechsig and Weissmann 2004 [87].

Deletions of the flexible part of PrP causing a cerebellar syndrome in PrP knockout mice that is suppressed by wild-type PrP

Transgenic *Prnp*$^{o/o}$ mice expressing PrP with amino proximal deletions were generated by using the half-genomic vector and analyzed for phenotypic abnormalities. Expression of PrP lacking amino acids 32–93 (octarepeats) or 32–106 did not result in an abnormal phenotype [83]. However, *Prnp*$^{o/o}$ mice expressing PrP with

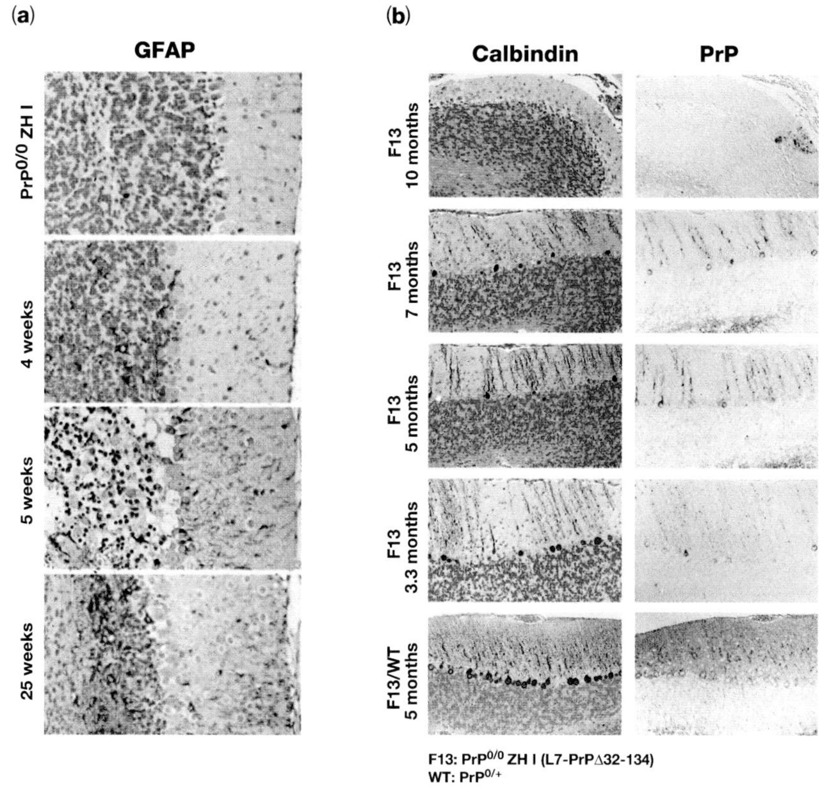

Fig. 10.5: Cerebellar sections from *Prnp*$^{o/o}$ mice expressing truncated PrPΔ32–134 under the control of different promoters. Cerebellum is normally developed until early postnatal life leading to formation of molecular layer, Purkinje cell layer, and granule cell layer. **(a)** Under control of the half-genomic PrP promoter, truncated PrP^{C} is expressed ubiquitously in brain except for the Purkinje cells. Massive degeneration of cerebellar granule cell layer is apparent beginning at 5 weeks. Note strong gliosis also affecting the molecular layer as shown by immunostaining for glial fibrillary acidic protein (GFAP). At the end stage of disease, mice suffer from a profound cerebellar syndrome and the thickness of the granule cell layer is considerably reduced, but Purkinje cells appear intact throughout this process [83]. **(b)** Under control of the Purkinje-cell-specific L7 promoter, truncated PrP^{C} is expressed in Purkinje cells, but nowhere else in the brain. Three-week-old mice show normal cerebellar structure with an intact Purkinje cell layer, as revealed by immunostaining for calbindin. After onset of the cerebellar syndrome, 14-week-old mice present with Purkinje cell loss and moderate gliosis. At the end stage of the disease, there is an almost complete Purkinje cell loss and shrinkage of the molecular layer; a single, remaining Purkinje cell is shown positive for PrP. A slight decrease of granule cells may be secondary to the extensive Purkinje cell loss. Age-matched *Prnp*$^{o/o}$ and *Prnp*$^{+/+}$ mice show an intact Purkinje cell layer. The syndrome is abrogated in the presence of full-length PrP^{C} [40]. Nagasaki-type PrP knockout mice (Fig. 10.1) exhibit similar Purkinje cell loss and ataxia, caused by ectopic expression of Dpl. The phenotype caused by Dpl is also abrogated by full-length PrP^{C}.

deletions of the flexible tail extending to amino acid 121 or 134 developed severe ataxia and apoptosis of the cerebellar granule cell layer as early as 1–3 months of age [83] (Fig. 10.5). Neurons in the cortex and elsewhere expressed truncated PrP at similar levels as granule cells but did not undergo cell death, arguing against an unspecific toxic effect. Purkinje cells did not express the truncated PrP, for reasons explained in Section 10.3.1, and were spared the disease. Strikingly, the pathological phenotype was completely abolished by the introduction of a single wild-type *Prnp* allele, although the level of truncated PrP remained unchanged, exceeding that of the wild-type counterpart [83]. To map the PrP region that suppresses the defect, mice expressing PrP with the largest, pathogenic deletion (PrPΔ32–134) were crossed with mice expressing PrP with shorter deletions. PrP lacking the octarepeats (PrPΔ32–93) still completely suppressed the syndrome, whereas PrPΔ32–106 could not abolish the defect anymore. However, the duration of the disorder to the terminal state was prolonged by 2 months [84]. When the truncated PrP was specifically targeted to Purkinje cells of *Prnp*$^{o/o}$[Zürich I] mice using the L7 promoter, ataxia and Purkinje cell degeneration developed, while the cerebellar granule layer remained unaffected [40] (Fig. 10.5). The phenotype resembled that of the ataxic PrP knockout lines with upregulated Dpl in brain (see Section 10.2.2), which was also abrogated by PrP. Similarly, transgenic *Prnp*$^{o/o}$ mice expressing Dpl targeted to Purkinje cells also developed ataxia associated with loss of Purkinje cells and introduction of PrP resulted in abrogation of the phenotype in mice expressing moderate levels of Dpl [30]. Because the overall structure of Dpl is remarkably similar to that of the globular domain of PrP lacking the flexible N-terminus, the mechanism of pathogenesis might be the same in both cases [8; 28; 30; 40].

These findings have been explained by a model in which truncated PrP^C or Dpl acts as dominant inhibitor of a putative functional homologue of PrP^C, with both competing for the same putative PrP^C ligand [83]. We proposed a two-step mechanism according to the address–message concept [85], in which binding of the globular part of PrP to its ligand (address) is the first step and activation of responses with the N-terminal tail of PrP (message) is the second step [86; 87]. Dpl and truncated PrP would bind to the conjectured ligand and prevent its activation, but because wild-type PrP would have a higher binding affinity, it overcomes the inhibition. Examples of such bipartite function of molecules include several hormones [85] or bacterial flagellin when acting as an elicitor in plants [88]. Alternative models to explain the pathological effect of Dpl and the truncated PrP are discussed elsewhere [89].

Internal deletions of α-helices causing a storage disease

Mice expressing PrP with the four octarepeats (PrPΔ23–88) deleted or even, in addition, lacking the first α-helix and one β-sheet (Δ141–176, "PrP106") remained healthy [90]. However, deletion of either the second α-helix (Δ177–202) or third α-helix (Δ201–217) strongly affected correct PrP folding and caused a lethal illness resembling neuronal storage disease [90]. Figure 10.4 summarizes the effect of PrP with deletion of three α-helices singly or jointly. *Prnp*$^{o/o}$ mice expressing low levels of a 61-amino acid PrP protein (PrP61) that contains only the amino acid sequence 88–141 and the GPI-attachment site (residues 221–231) developed a neurological disorder at a few months of age and died a few days later. PrP61-mediated neurotoxicity was not prevented by coexpression of wild-type PrP and the disease was not transmissible to mice expressing PrP106 [91].

Expression of cytosolic PrP and effects

In cultured cells, mature PrP^C is subject to retrograde transport to the cytosol and degradation by proteasomes. Based on the observation that accumulation of even small amounts of cytosolic PrP was strongly neurotoxic to cultured cells [92], Lindquist and colleagues suggested that formation of cytosolic PrP may contribute to neuronal degeneration in prion disease [69]. Interestingly, wild-type mice transgenic for murine PrP23–230, which lacks the signal sequence, accumulated cytosolic PrP and developed severe ataxia with cerebellar degeneration and gliosis. Although the mechanism for the formation of cytosolic PrP species (retrotranslocation or inefficient translocation of PrP due to overexpression) is debated [93], it appears that PrP^C located partly or entirely in the cytosol is extremely toxic. However, cytosolic PrP has also been found in the cytosol in subpopulations of neurons in the hippocampus, neocortex, and thalamus, but not in the cerebellum of uninfected wild-type mice [94]. Thus, the role of cytosolic PrP in prion disease remains unknown.

Although overexpression of a variety of PrP mutations gives rise to nontransmissible clinical disease, usually associated with accumulation of insoluble and/or toxic forms of PrP, in no case was the phenotype abrogated by the coexpression of wild-type PrP other than in that of the Shmerling syndrome described above (Fig. 10.4).

10.4.2 Mapping of PrP regions required for prion propagation

Treatment of prion preparations with protease cleaves off about 60 amino terminal residues of PrP^{Sc} [95] without abrogating infectivity [96]. This raised the question as to whether amino-proximally truncated PrP^C could serve as substrate for the conversion to PrP^{Sc} and sustain susceptibility to scrapie in mice. Thus, a number of transgenic mouse lines expressing PrP with deletions have been generated in order to delineate the minimal sequence of PrP required to sustain prion replication (Fig. 10.6). *Prnp*$^{o/o}$ mice expressing PrP with deletions from position 32–80 (Δ32–80) [61] or 32–93 (Δ32–93) [50; 83] were normal and after intracerebral inocu-

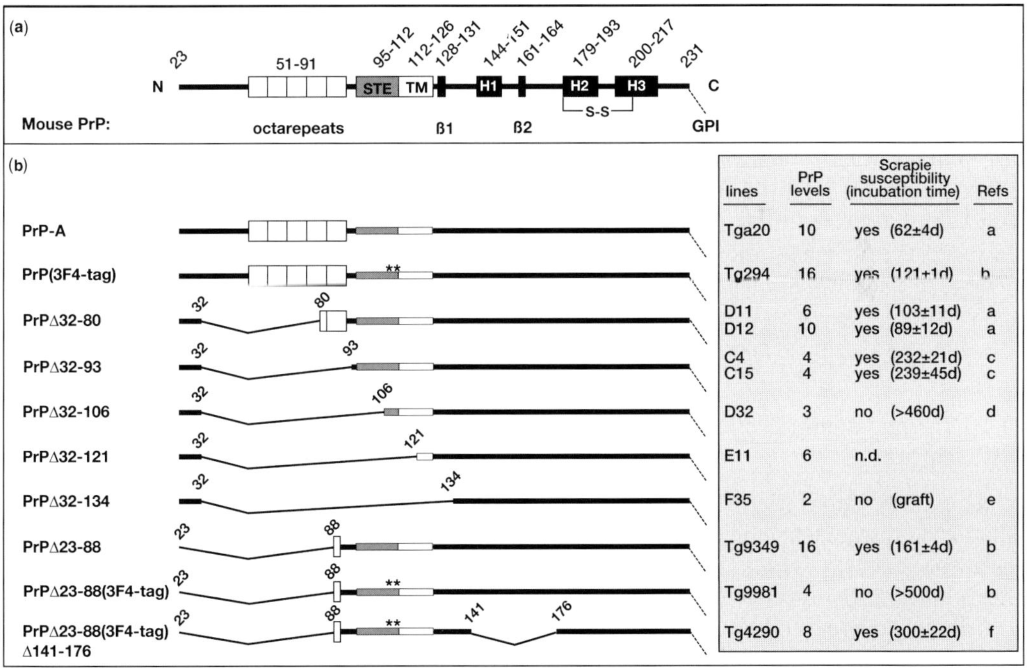

Fig. 10.6: Delineation of PrP regions expendable for prion propagation in PrP knockout mice. **(a)** The scheme of mouse PrP sequence is described in Fig. 10.4. **(b)** Susceptibility of PrP knockout mice expressing various PrP transgenes after intracerebral inoculation with mouse prion (RML). Tag: 3F4 epitope (**). References: (a) Fischer et al. 1996 [61]; (b) Supattapone et al. 2001b [97]; (c) Flechsig et al. 2000 [50] (d) E. F, I. H., AA and CW, unpublished results; (e) I. H., EF, AA and CW, unpublished results; (f) Supattapone et al. 1999 [101]. Figure from Flechsig and Weissmann, 2004 [87].

lation with mouse prions (RML) developed disease, propagated prions, and accumulated protease-resistant truncated PrP^{Sc}. Removal of the five octarepeats (PrPΔ32–93), resulted in a longer incubation time and a lower level of both infectivity and PrP^{Sc} [50]. The longer incubation time might be due to the reduced conversion efficiency of the truncated PrP [50; 97]. Surprisingly, brains of terminally sick mice expressing PrPΔ32–93 failed to show histopathology typical of mouse scrapie; however, in the spinal cord, gliosis and motor neuron loss were as in terminally sick wild-type controls. PrP with deletions extending to position 106 or 134 was unable to restore susceptibility to prions (E. F., I. H., A. A and C. W., unpublished data; Fig. 10.6). Remarkably, at least 60 residues of the amino proximal region of mature PrP^{C}, which contain the entire octarepeat region, are dispensable. This is remarkable because amplification of the octarepeat region is associated with familial CJD and GSS [98; 99] and appears to affect the type of cerebellar amyloid deposits [100]. Amazingly, PrP with a partial deletion of the flexible tail (Δ23–88), lacking the first α-helix and one β-sheet (Δ141–176) still confers prion susceptibility to $Prnp^{o/o}$ mice [101]. The resulting PrP106 contains only 106 amino acids, compared to the 208 residues in full-length PrP. However, $PrP106^{Sc}$ miniprions are only transmissible to mice expressing the same PrP deletion (PrP106), but not to wild-type mice, probably due to an artificial species barrier.

10.4.3 PrP mutations affecting the species barrier

In many cases, prions derived from one species fail to elicit disease when inoculated into a different species or do so only inefficiently and/or after a long incubation time. This phenomenon is attributed to a species barrier [102]; however, at least in the case of mice inoculated intracerebrally with hamster prions, PrP^{Sc} and infectivity may accumulate in the brain despite the absence of clinical symptoms [56; 57] so that the barrier is not absolute. Introduction into mice of PrP transgenes containing all or part of the PrP sequence of the prion donor can overcome the species barrier. Thus, $Prnp^{+/+}$ mice transgenic for hamster PrP transgenes are susceptible to both mouse and hamster prions [62; 103], while $Prnp^{o/o}$ mice containing Syrian hamster PrP transgenes are susceptible to hamster but not mouse-derived prions; interestingly, the presence of a mouse PrP allele diminished the susceptibility to hamster prions [41] (Fig. 10.3). Similarly, $Prnp^{o/o}$ mice expressing human transgenes were susceptible to human CJD prions, while $Prnp^{+/+}$ mice with the same transgene cluster were as resistant to human prions as wild-type mice [104; 105]. MHu2M, a chimeric PrP that contains nine human PrP-specific residues between positions 96–167 in a murine framework, was almost as effective in rendering mice susceptible to human prions on a $Prnp^{+/+}$ background as it was on a $Prnp^{o/o}$ background. This showed that the critical region for "human species specificity" lay in the region 96–168 of human PrP and, along with other data, led to the suggestion that a species-specific "factor X" that bound to residues 168, 172, 215, and 219 near the C terminus participated in prion formation [104; 106]. $Prnp^{o/o}$ mice expressing MHu2M have been used as recipients for a variety of CJD prions [107]. Surprisingly, transgenic mice overexpressing human PrP are less susceptible to human variant CJD (vCJD) than wild-type mice [108]. $Prnp^{o/o}$ mice transgenic for bovine PrP genes are susceptible to bovine [109–111] as well as to vCJD and sheep scrapie prions while mice containing ovine $Prnp^{VRQ}$ or $Prnp^{ARQ}$ transgenes showed vastly decreased incubation time for sheep scrapie prions as compared to wild-type mice [112; 113]. Inasmuch as a decrease in incubation time is viewed as a lowering of the species barrier, it is noteworthy that this can be achieved by mutations in murine PrP that do not increase the similarity to the PrP of the prion donor. Mice homozygous for the targeted PrP mutation P101L ($Prnp^{a[P101L,P101L]}$) had shorter incubation times with both sheep (SSBP1) and hamster (263K) prions but extended incubation times with murine ME7 and human vCJD prions, as compared with wild-type mice. Thus, the P101L mutation can alter incubation times across three species barriers in a strain-dependent manner [114]. Interestingly, this mutation is located in the flexible region and not in the domain postulated to bind the conjectural protein X [104; 106]. Polymorphisms in the PrP gene have been shown to alter incubation time (see Chapter 10.4.4) and susceptibility to prion disease in mice [75], sheep [115], and man [116; 117]. These polymorphisms may control the rate of replication of the infectious agent [118], but the mechanism by which this is achieved is not known.

10.4.4 PrP mutations linked to incubation time of prion strains

As described above, concordance between the sequence of the PrP gene of the prion donor and the recipient as well as the expression level of PrP^C, both have a major influence on susceptibility and incubation time. Moreover, two different prion strains, even when derived from the same inbred mouse line and thus associated with the same PrP^{Sc} sequence, may exhibit different incubation times and patterns of histopathology in the same mouse strain. This shows that features other than PrP sequence concordance influence incubation time. It has previously been suggested that an essential nucleic acid [119] ("virino" hypothesis) or a nonessential nucleic acid [120] ("co-prion" hypothesis) is the crucial feature; currently the prevalent idea is that different conformations of the PrP^{Sc} molecule, imposed on the same sequence (or even on a different one) specify strain properties [121–123].

The *Sinc* gene has been shown to be the major gene controlling survival time of mice exposed to scrapie strains [124; 125] (see Chapter 12). Animals homozygous for the *Sinc s7* allele had short incubation times when infected with Chandler isolate of mouse prions and those with the *Sinc p7* allele had long incubation times. With a different agent, the mouse-adapted BSE strain 301V, incubation times were long for the *Sinc s7* allele and short for the *Sinc p7* allele [126]. The *Sinc s7* allele was genetically linked to the $Prnp^a$ allele characterized by the residues 108 Leu and 189 Thr, whereas the *Sinc p7* allele was linked to the $Prnp^b$ allele with 108 Phe and 189 Val [127]. The close linkage of the *Sinc* and PrP genes suggested that they might be congruent [127–129], however analysis using transgenes of the two alleles were inconclusive because the effects of variable gene copy numbers and PrP^C levels were superimposed on those of the alleles [73; 130].

Using the double replacement gene targeting technique, PrP codons 108 and 189 in ES cells derived from a *Sinc s7* ($Prnp^a$ allele) mouse were altered to 108Phe and 189Val, which are characteristic for the $Prnp^b$ allele [74]. ES clones with the mutant PrP gene yielded chimeras that were bred with 129/Ola mice to generate an inbred line of mice carrying a PrP gene with 108Phe189Val ($Prnp^{a[L108F,T189V]}$). Mice homozygous for the $Prnp^a$ allele or the mutated allele were inoculated with the 301V agent. The incubation time in the mutant mice carrying the ($Prnp^{a[L108F,T189V]}$) allele was 133 days, about 110 days shorter than that of their wild-type littermates carrying the $Prnp^a$ allele [75]. As mentioned above, mice homozygous for the targeted PrP mutation P101L had extended incubation times with murine ME7 as compared with wild type [114]. The finding that point mutations in the PrP gene in mice with otherwise identical genetic background caused a dramatic change in incubation time supports the view that *Sinc* and *Prnp* are indeed congruent [75]. However, it is also clear that loci other than *Prnp* contribute importantly to incubation time [131–133].

10.4.5 Modeling familial forms of human prion disease in mice

More than 20 different mutations of the PrP gene have been identified in families suffering from familial human prion diseases, such as GSS and familial CJD (see Chapter 26). It has been speculated that mutated PrP genes give rise to unstable PrP^C that can spontaneously convert into abnormal PrP^{Sc} conformers, whereas sporadic forms of the disease are attributed to rare spontaneous transitions of PrP^C into PrP^{Sc} or to somatic mutations in the PrP gene. However, studies on thermodynamic stability of the C-proximal globular domain of PrP with those mutations do not generally support the concept of mutation-induced PrP^C destabilization [134; 135]. Several attempts to elicit transmissible prion disease in mice by expressing mutated PrPs associated with inherited prion diseases have failed, with the possible exception of PrP^{P101L}. Mice overexpressing murine PrP^{P101L} [136] (i. e., the counterpart of human PrP^{P102L} linked to familial GSS) 8-fold, on a *Prnp*$^{o/o}$ background, spontaneously developed neurodegeneration at 140 days of age, but little if any PrP^{Sc} could be detected [137]. The disease could not be transmitted to wild-type mice, but only to transgenic mice that expressed the same mutation at low levels (2-fold) and spontaneously developed neurodegeneration only late in life [137–139]. It was suggested that the amino acid difference constituted an artificial transmission barrier to wild-type mice. However, human P102L-linked GSS can be transmitted to both monkeys and wild-type mice [140; 141]. Thus, it has been argued that the P101L mutation may be an important susceptibility factor rather than a direct cause of GSS [114; 142].

In addition to the P102L-linked GSS mutation, two different mutations associated with inherited CJD in human were analyzed. *Prnp*$^{o/o}$ mice expressing PrP with the counterparts of the human CJD-linked mutations at T183A or E200K did not develop any pathological signs [137; 143]. Furthermore, expression of a mouse PrP version of a nine-octapeptide insertion associated with CJD in humans produced a slowly progressive cerebellar disorder and progressive myopathy [68; 144]. Because the disease was not transmissible, it qualifies as a "proteinopathy" rather than a prion disease. In a further attempt to generate spontaneous prion disease in transgenic mice, Ala was replaced by Val at positions 113, 115, and 118 of PrP ("AV3") to promote de novo β-sheet formation in the flexible tail [145]. Although founders developed a fatal neurological disorder [146], this was not transmissible and no protease-resistant PrP was detected [145]. Thus, with one possible exception, human familial prion diseases have not been modeled successfully in the mouse.

Many PrP mutations are pathogenic in mice, mostly or perhaps only when overexpressed. It is not surprising that neurons accumulating abnormal forms of protein may suffer damage. With the possible exception of the disease caused by overexpression of PrP^{P101L} [137–139], none of the PrP-linked diseases in mice were shown to be transmissible. Thus, it is appropriate to distinguish between prion diseases, which are transmissible, and "nontransmissible conformational diseases", or "nontransmissible proteinopathies", which are not [140]. In the case of human prion diseases, experimental transmission to animals has been achieved for some but not all familial cases [140]. Doubtless, in any particular instance lack of transmission may be due to inadequate recipients, but in view of the results with mice, it is possible that at least some of the familial human spongiform encephalopathies may be truly nontransmissible conformational diseases and therefore, in view of the definition of prions as transmissible agents, not prion diseases. Beyond the nine-octapeptide insertion and the triple mutation (AV3) that cause nontransmissible neurodegeneration in the mouse, further examples of this type of mutation have been described in Section 10.4.1. [69; 90; 91].

10.5 Transgenic approaches to study intervention strategies against prion diseases

The central role of PrP in prion propagation suggested several potential strategies for treatment or prophylaxis of prion diseases. Ablation of the PrP gene was suggested early on as a possible approach to generate prion-resistant farm animals [41] and several attempts to produce PrP knockout animals have been performed by sequential targeting of both PrP alleles in fibroblasts derived from cattle [147] or sheep [148]. Because sheep, and particularly cattle, have been used to produce biomedical products such as gelatin, collagen, and, increasingly, human proteins after genetic modification, PrP knockout animals would help to ensure that these recombinant products are free of prion infectivity.

10.5.1 Depletion of neuronal PrP prevents disease and reverses spongiosis

To determine whether depletion of PrP^C could alleviate disease, Mallucci et al. have studied a conditional PrP knockout model [10] (see Section 10.2.1). *Prnp*$^{o/o}$ mice transgenic for both a *loxP*-flanked cassette containing the PrP ORF under control of a *Prnp* promoter element and cre recombinase under the control of the neurofilament heavy chain (NFH) promoter [10], were intracerebrally inoculated with mouse prions at 4 weeks of age [11]. In this line cre is expressed at 10–12 weeks of age and causes ablation of PrP. Surprisingly, depletion of neuronal PrP^C about 8 weeks after infection reversed early spongiform change and prevented neuronal loss and onset of clinical disease. This occurred despite the accumulation of PrP^{Sc} to levels seen in terminally ill wild-type animals. These findings provide a rationale for targeting PrP^C as a therapeutic intervention in prion disease. Recently, introduction of small interfering RNAs (RNAi) against PrP was shown to diminish levels of PrP^C and PrP^{Sc} in cultured cells [149; 150]. Whether RNAi can be used as a pharmacotherapeutic remains to be determined.

10.5.2 Dominant-negative inhibitors of prion formation confers protection against prion disease

The finding that susceptibility and resistance to natural scrapie in sheep is largely controlled by polymorphisms of the PrP gene led to the concept of selective breeding of disease-resistant genotypes. Certain configurations of polymorphic residues of PrP in sheep (positions 136, 154, and 171) afford protection against prion disease [151]. For example, Cheviot sheep with two 171Q alleles (171Q/Q) are susceptible to scrapie while the configuration 171Q/R and 171R/R confers protection [152]; thus, the 171R allele may be considered as dominant negative. Indeed, amino acid substitutions at the C-terminal residues 167 (which corresponds to 171 in sheep), 171, 214, or 218 of mouse PrP^C act as dominant-negative inhibitors of PrP^{Sc} formation in cultured cells [106]. Interestingly, amino acid replacements at multiple sites were less effective than single-residue substitutions and an intact N-terminal region is required to sustain the dominant-negative effect [153]. To investigate the dominant-negative effect, transgenic mice expressing murine PrP^{Q167R} at about wild-type level were generated [154]. *Prnp*$^{o/o}$ mice containing the murine PrP^{Q167R} transgene remained healthy and failed to accumulate PrP^{Sc} after inoculation with mouse scrapie prions, whereas *Prnp*$^{+/+}$ mice with the same transgene had an extended life span after inoculation, but developed atypical neurological symptoms at about 450 days (wild-type controls had an incubation time of 127 days) and showed accumulation of PrP^{Sc} and neuropathological changes typical for scrapie no later than 300 days. Thus, PrP^{Q167R} is not converted to the scrapie form by RML prions and affords partial protection to micc containing wild-type PrP. The mechanism of dominant-negative inhibitors is still unknown, but has been attributed to sequestration of the conjectural protein X [154]. Recently, inoculation studies of sheep have shown that a genotype associated with the highest resistance can still be infected with BSE by intracerebral inoculation [155]. Moreover, two atypical scrapie cases have been reported among sheep homozygous for the resistant polymorphism. These two cases did not show any clinical symptoms, but had low levels of proteinase K-resistant PrP; however, both the glycosylation pattern of PrP^{Sc} and its sensitivity to limited proteinase K digestion differed from those of typical scrapie cases [156]. Although the relevance of these findings remains to be determined, it may have important implications for disease eradication strategies.

10.5.3 Anti-PrP-antibodies prevent the onset of prion disease upon peripheral inoculation

Several studies have suggested that anti-PrP antibodies could provide a promising approach to prophylaxis and therapy. *In vitro* preincubation with anti-PrP antisera was reported to reduce the prion titer of infectious hamster brain homogenates by up to 2 log units [157] and an anti-PrP antibody inhibited formation of PrP^{Sc} in a cell-free system [158]. Further studies have shown that antibodies [159] or Fab fragments against PrP [160] abrogate PrP^{Sc} and infectivity in prion-infected cultured cells. Although active immunization is hampered by tolerance to PrP in wild-type mice, auto-antibodies can be induced in wild-type mice albeit with limited therapeutic effect in mice [161–163]. Tolerance has been circumvented by transgenic expression of a monoclonal antibody against PrP. Heppner et al. [164] prepared transgenic mice expressing the variable V(D)J region of the heavy chain of the monoclonal PrP antibody 6H4 [165]. At the age of 150 days transgenic $Prnp^{o/o}$, $Prnp^{o/+}$, and $Prnp^{+/+}$ mice expressed similar titers of anti-PrP antibodies, indicating that if deletion of autoreactive B cells occurs, it is not sufficient to prevent anti-PrP immunity. The avidity of anti-PrP antibody serum consisting of the transgenic μ heavy chain and a large repertoire of endogenous κ and λ chains was approximately 100-fold lower than that of the original 6H4 antibody. Expression of the 6H4μ heavy chain conferred protection from scrapie upon i.p. inoculation of mouse RML prions [164]. This delivers proof-of-principle that a protective humoral response against prions can be mounted by the immune system and suggests that B cells are not intrinsically tolerant to PrP^C. Further studies have shown that passive immunization could also be effective against peripheral infection [166]. Mice were infected by the i.p. route and a monoclonal antibody was applied repeatedly, twice per week, at a dose of 2 mg starting from 1 or 4 weeks post infection. This antibody treatment was able to prevent the onset of the disease for more than 500 days.

Importantly, the protective effect of the monoclonal antibodies was only observed when prions were administered peripherally and not when they were injected intracerebrally, probably because the blood–brain barrier prevented the antibodies from reaching an effective concentration in the brain. In this connection, it is noteworthy that extensive neuronal apoptosis in the hippocampus resulted following intracranial delivery of a monoclonal antibody against a specific epitope of PrP [167]. Although these experiments warrant a cautionary note in regard to passive intracerebral immunization, the use of Fab fragments might circumvent the problem, because they have a therapeutic effect in tissue culture [160], but are not toxic when injected intracerebrally [167]. In summary, many problems have to be addressed in the quest for effective vaccines in prophylaxis or in the therapy of prion diseases in animals and humans.

10.6 Investigating the mechanism of prion propagation by ectopic expression of PrP

Several questions regarding the cell specificity of prion propagation and the spread of prions throughout the body have been addressed by generating mice that express PrP in only one type of tissue or organ, rather than almost ubiquitously, as wild-type mice do (Table 10.1). Because PrP is expressed both in neurons [168] and astrocytes [169], the question arose whether PrP expression in each cell type alone is sufficient for prion propagation. Expression of hamster PrP under control of the neuron-specific enolase (NSE) promoter, which is considered to be neuron specific, rendered transgenic $Prnp^{o/o}$ mice susceptible to hamster prions [170]. Thus, PrP expression in neurons is sufficient for prion propagation. It was surprising to find that expression of PrP targeted to astrocytes of $Prnp^{o/o}$ mice by the glial fibrillary acid protein (GFAP) promoter also rendered mice susceptible to intracerebrally administered prions and led to clinical disease and prion propagation [54]. Thus,

Table 10.1: Ectopic expression of PrP in PrP null mice and cell specificity of prion replication.*

Ectopic expression of PrP	Promoter	Prion propagation	References
Neurons	NSE	Yes	[170]
Astrocytes	GFAP	Yes	[54]
T lymphocytes	Lck	No	[177]
B lymphocytes	CD19	No	[178]
Hepatocytes	transthyretin	Yes	[191]
Muscle cells	α-actin	Yes	[191]
Oligodendrocytes	MBP	No	[190]

* Ectopic expression of PrP in the indicated cell types of PrP knockout mice was accomplished by using cell-specific promoter elements, such as neuron-specific enolase (NSE) for neurons, glial fibrillary acid protein (GFAP) for astrocytes, lck for T cells, CD19 for B cells, and myelin basic protein (MBP) for oligodendrocytes.

not only do astrocytes appear to be competent for prion replication, but the brain pathology exhibited by scrapie-infected terminally ill transgenic mice and wild-type mice is also highly similar.

In most mouse scrapie models, infectivity appears in the spleen within days after intracerebral inoculation and reaches a plateau at 6–8 weeks, whereas infectivity in the brain is detected only 3–4 few weeks after inoculation and reaches titers two orders of magnitude higher than in the spleen [41; 171; 172]. In spleen, PrP^{Sc} is found mainly in follicular dendritic cells [173; 174] (FDCs). Further studies found mouse prions also associated with spleen-derived B and T cells and with the FDC-containing stromal fraction, but not with neutrophils. No infectivity was detected in circulating lymphocytes in this model [175]. To address the question of whether B and T cells were able to propagate prions or whether they acquired them from another source, PrP was specifically targeted to B or T cells: *Prnp*$^{o/o}$ mice expressing PrP under the control of the T-cell-specific lck promoter [176] showed high levels of PrP in T cells, both in thymus and spleen, but not in brain [177]. No clinical symptoms became apparent and no infectivity was detected in spleen, thymus, or brain of these mice up to 1 year after intraperitoneal inoculation. *Prnp*$^{o/o}$ mice expressing PrP under control of the B-cell-specific CD19 promoter had 10- to 20-fold higher PrP levels on B cells compared to wild-type mice. Yet, neither pathology nor prion propagation was observed after intraperitoneal inoculation with scrapie prions [178]. Thus, the presence of PrP on T or B cells does not suffice to support prion replication and prions associated with splenic T and B lymphocytes must stem from another source, most likely FDCs [179; 180].

From the spleen and likely from other sites, prions proceed along the peripheral nervous system to finally reach the brain, either directly via the vagus nerve [181; 182] or via the spinal cord, under involvement of the sympathetic nervous system [183; 184]. The finding that *Prnp*$^{o/o}$ mice with a PrP-expressing brain graft (see Chapter 11) developed scrapie pathology in the graft after intracerebral [63], but not after intraperitoneal infection [185], shows that PrP is required for the spread of prions from the periphery to the brain. After reconstitution of the hematopoietic system of those graft-bearing *Prnp*$^{o/o}$ mice with PrP-expressing cells, prion accumulation in the spleen was the same as in wild-type animals. Nonetheless, intraperitoneal inoculation failed to produce scrapie pathology in the neurografts [186]. Thus, transfer of infectivity from spleen to the CNS is crucially dependent on the expression of PrP in a tissue compartment that cannot be reconstituted by bone marrow transfer, likely the peripheral nervous system [183; 184]. This was indeed shown to be the case using *Prnp*$^{o/o}$ mice transgenic for hamster PrP under control of the NSE promoter, which express PrP in brain and nerves but not in spleen, lymph nodes, or bone marrow [170].

To determine whether the spread of prions from the periphery to the CNS is dependent on hematogenous or lymphoreticular tissue (LRS), NSE-hamster PrP-transgenic mice [170] were inoculated with hamster prions either orally or intraperitoneally. In both cases, scrapie disease ensued with the same incubation time as in mice expressing hamster PrP under the control of the PrP promoter [187]. Thus, at least after peripheral exposure to high doses of hamster prions, PrP^{C}-expressing LRS was not required for prion amplification or transport. Although PrP expression on neurons is sufficient for the spread of prions from the periphery into the brain, the underlying mechanism is still unknown, because the velocity of the propagation (approx. 1 mm/day) corresponds neither to fast nor to slow axonal transport [183; 188]. Schwann cells might be involved, which expressed PrP in wild-type mice and were able to propagate prions in culture [189]. To assess the role of oligodendrocytes and Schwann cells in prion propagation, PrP under the control of the myelin basic protein (MBP) promoter was introduced in *Prnp*$^{o/o}$ mice [190]. PrP was detected in oligodendrocytes and Schwann cells but not in neurons and astrocytes. After intracerebral, intraperitoneal, or intraocular challenge with RML mouse prions, MBP-PrP transgenic mice remained resistant and did not contain prion infectivity in the brain. Moreover, intraocular prion inoculation of MBP-PrP transgenic mice carrying PrP-expressing grafts in the brain (see Chapter 11) failed to elicit spongiform encephalopathy or prion infectivity in the graft. Thus, oligodendrocytes do not support cell-autonomous prion replication and neural spread of mouse prions [190]. Further transgenic studies have shown that expressing PrP under the control of the α-actin promoter in *Prnp*$^{o/o}$ mice allows prion replication in muscle [191]. Inoculation of mice expressing PrP under the direction of the liver-specific transthyretin promoter/enhancer resulted in low levels of prion infectivity in the liver, but also in the brain [191]. From the experiments with B and T lymphocytes, it can be concluded that the presence of PrP on the cell surface does not suffice to support prion replication; perhaps location within a particular membrane region is required [192; 193] as well as other components, such as the postulated factor X [104; 106] or a receptor.

References

[1] Basler K, Oesch B, Scott M, et al. Scrapie and cellular PrP isoforms are encoded by the same chromosomal gene. Cell 1986; 46(3):417–428.
[2] Sparkes RS, Simon M, Cohn VH, et al. Assignment of the human and mouse prion protein genes to homologous chromosomes. Proc Natl Acad Sci USA 1986; 83(19):7358–7362.
[3] Büeler H, Fischer M, Lang Y, et al. Normal development and behaviour of mice lacking the neuronal cell-surface PrP protein. Nature 1992; 356(6370):577–582.
[4] Manson JC, Clarke AR, Hooper ML, et al. 129/Ola mice carrying a null mutation in PrP that abolishes mRNA production are developmentally normal. Mol Neurobiol 1994; 8(2–3):121–127.
[5] Sakaguchi S, Katamine S, Nishida N, et al. Loss of cerebellar Purkinje Cells in aged mice homozygous for a disrupted PrP gene. Nature 1996; 380:528–531.
[6] Moore RC, Lee IY, Silverman GL, et al. Ataxia in prion protein (PrP)-deficient mice is associated with upregulation of the novel PrP-like protein doppel. J Mol Biol 1999; 292(4):797–817.
[7] Silverman GL, Qin K, Moore RC, et al. Doppel is an N-glycosylated, glycosylphosphatidylinositol-anchored protein. Expression in testis and ectopic production in the brains of Prnp(0/0) mice predisposed to Purkinje cell loss. J Biol Chem 2000; 275(35):26834–26841.
[8] Rossi D, Cozzio A, Flechsig E, et al. Onset of ataxia and Purkinje cell loss in PrP null mice inversely correlated with Dpl level in brain. Embo J 2001; 20(4):694–702.
[9] Tremblay P, Meiner Z, Galou M, et al. Doxycycline control of prion protein transgene expression modulates prion disease in mice. Proc Natl Acad Sci USA 1998; 95(21):12580–12585.
[10] Mallucci GR, Ratte S, Asante EA, et al. Postnatal knockout of prion protein alters hippocampal CA1 properties, but does not result in neurodegeneration. Embo J 2002; 21(3):202–210.
[11] Mallucci G, Dickinson A, Linehan J, et al. Depleting neuronal PrP in prion infection prevents disease and reverses spongiosis. Science 2003; 302(5646):871–874.
[12] Nishida N, Tremblay P, Sugimoto T, et al. A mouse prion protein transgene rescues mice deficient for the prion protein gene from purkinje cell degeneration and demyelination. Lab Invest 1999; 79(6):689–697.
[13] Lipp HP, Stagliar-Bozicevic M, Fischer M, et al. A 2-year longitudinal study of swimming navigation in mice devoid of the prion protein: no evidence for neurological anomalies or spatial learning impairments. Behav Brain Res 1998; 95(1):47–54.
[14] Tobler I, Gaus SE, Deboer T, et al. Altered circadian activity rhythms and sleep in mice devoid of prion protein. Nature 1996; 380(6575):639–642.
[15] Tobler I, Deboer T, Fischer M. Sleep and sleep regulation in normal and prion protein-deficient mice. J Neurosci 1997; 17(5):1869–1879.
[16] Walz R, Amaral OB, Rockenbach IC, et al. Increased sensitivity to seizures in mice lacking cellular prion protein. Epilepsia 1999; 40(12):1679–1682.
[17] Collinge J, Whittington MA, Sidle KC, et al. Prion protein is necessary for normal synaptic function. Nature 1994; 370(6487):295–297.
[18] Whittington MA, Sidle KC, Gowland I, et al. Rescue of neurophysiological phenotype seen in PrP null mice by transgene encoding human prion protein. Nat Genet 1995; 9(2):197–201.

[19] Manson JC, Hope J, Clarke AR, et al. PrP gene dosage and long term potentiation. Neurodegeneration 1995; 4(1):113–114.
[20] Colling SB, Collinge J, Jefferys JG. Hippocampal slices from prion protein null mice: disrupted Ca(2+)-activated K+ currents. Neurosci Lett 1996; 209(1):49–52.
[21] Herms JW, Tings T, Dunker S, et al. Prion protein affects Ca2+-activated K+ currents in cerebellar purkinje cells. Neurobiol Dis 2001; 8(2):324–330.
[22] Keshet GI, Bar-Peled O, Yaffe D, et al. The cellular prion protein colocalizes with the dystroglycan complex in the brain. J Neurochem 2000; 75(5):1889–1897.
[23] Miele G, Jeffrey M, Turnbull D, et al. Ablation of cellular prion protein expression affects mitochondrial numbers and morphology. Biochem Biophys Res Commun 2002; 291(2):372–377.
[24] Brown DR, Nicholas RS, Canevari L. Lack of prion protein expression results in a neuronal phenotype sensitive to stress. J Neurosci Res 2002; 67(2):211–224.
[25] Klamt F, Dal-Pizzol F, Conte da Frota MJ, et al. Imbalance of antioxidant defense in mice lacking cellular prion protein. Free Radic Biol Med 2001; 30(10):1137–1144.
[26] Weissmann C. PrP effects clarified. Curr Biol 1996; 6(11):1359.
[27] Li A, Sakaguchi S, Atarashi R, et al. Identification of a novel gene encoding a PrP-like protein expressed as chimeric transcripts fused to PrP exon 1/2 in ataxic mouse line with a disrupted PrP gene. Cell Mol Neurobiol 2000; 20(5):553–567.
[28] Weissmann C and Aguzzi A. PrP's double causes trouble. Science 1999; 286(5441):914–915.
[29] Moore RC, Mastrangelo P, Bouzamondo E, et al. Doppel-induced cerebellar degeneration in transgenic mice. Proc Natl Acad Sci USA 2001; 98(26):15288–15293.
[30] Anderson L, Rossi D, Linehan J, et al. Transgene-driven expression of the Doppel protein in Purkinje cells causes Purkinje cell degeneration and motor impairment. Proc Natl Acad Sci USA 2004; 101(10):3644–3649.
[31] Lu K, Wang W, Xie Z, et al. Expression and structural characterization of the recombinant human doppel protein. Biochemistry 2000; 39(44):13575–13583.
[32] Behrens A, Genoud N, Naumann H, et al. Absence of the prion protein homologue Doppel causes male sterility. Embo J 2002; 21(14):3652–3658.
[33] Paisley D, Banks S, Selfridge J, et al. Male infertility and DNA damage in Doppel knockout and prion protein/Doppel double-knockout mice. Am J Pathol 2004; 164(6):2279–2288.
[34] Behrens A, Brandner S, Genoud N, et al. Normal neurogenesis and scrapie pathogenesis in neural grafts lacking the prion protein homologue Doppel. EMBO Rep 2001; 2(4):347–352.
[35] Tuzi NL, Gall E, Melton D, et al. Expression of doppel in the CNS of mice does not modulate transmissible spongiform encephalopathy disease. J Gen Virol 2002; 83(3):705–711.
[36] Mo H, Moore RC, Cohen FE, et al. Two different neurodegenerative diseases caused by proteins with similar structures. Proc Natl Acad Sci USA 2001; 98(5):2352–2357.
[37] Wong BS, Liu T, Paisley D, et al. Induction of HO-1 and NOS in doppel-expressing mice devoid of PrP: implications for doppel function. Mol Cell Neurosci 2001; 17(4):768–775.
[38] Qin K, Coomaraswamy J, Mastrangelo P, et al. The PrP-like protein Doppel binds copper. J Biol Chem 2003; 278(11):8888–8896.
[39] Atarashi R, Nishida N, Shigematsu K, et al. Deletion of N-terminal residues 23–88 from prion protein (PrP) abrogates the potential to rescue PrP-deficient mice from PrP-like protein/Doppel-induced neurodegeneration. J Biol Chem 2003; 278(31):28944–28949.
[40] Flechsig E, Hegyi I, Leimeroth R, et al. Expression of truncated PrP targeted to Purkinje cells of PrP knockout mice causes Purkinje cell death and ataxia. Embo J 2003; 22(12):3095–3101.
[41] Büeler H, Aguzzi A, Sailer A, et al. Mice devoid of PrP are resistant to scrapie. Cell 1993; 73(7):1339–1347.
[42] Chandler RL. Encephalopathy in mice produced with scrapie brain material. Lancet 1961; 1:1378–1379.
[43] Prusiner SB, Groth D, Serban A, et al. Ablation of the prion protein (PrP) gene in mice prevents scrapie and facilitates production of anti-PrP antibodies. Proc Natl Acad Sci USA 1993; 90(22):10608–10612.
[44] Manson JC, Clarke AR, McBride PA, et al. PrP gene dosage determines the timing but not the final intensity or distribution of lesions in scrapie pathology. Neurodegeneration 1994; 3(4):331–340.
[45] Sakaguchi S, Katamine S, Shigematsu K, et al. Accumulation of proteinase K-resistant prion protein (PrP) is restricted by the expression level

of normal PrP in mice inoculated with a mouse-adapted strain of the Creutzfeldt–Jakob disease agent. J Virol 1995; 69(12):7586–7592.
[46] Chesebro B and Caughey B. Scrapie agent replication without the prion protein? Curr Biol 1993; 3:696–698.
[47] Büeler H, Raeber A, Sailer A, et al. High prion and PrPSc levels but delayed onset of disease in scrapie-inoculated mice heterozygous for a disrupted PrP gene. Mol Med 1994; 1(1):19–30.
[48] Kimberlin RH, Cole S, Walker CA. Pathogenesis of scrapie is faster when infection is intraspinal instead of intracerebral. Microb Pathog 1987; 2(6):405–415.
[49] Collinge J, Palmer MS, Sidle KCL, et al. Transmission of fatal familial insomnia to laboratory animals. Lancet 1995; 346:569–570.
[50] Flechsig E, Shmerling D, Hegyi I, et al. Prion protein devoid of the octapeptide repeat region restores susceptibility to scrapie in PrP knockout mice. Neuron 2000; 27(2):399–408.
[51] Lasmezas CI, Deslys JP, Robain O, et al. Transmission of the BSE agent to mice in the absence of detectable abnormal prion protein. Science 1997; 275:402–405.
[52] Manuelidis L, Fritch W, Xi YG. Evolution of a strain of CJD that induces BSE-like plaques. Science 1997; 277(5322):94–98.
[53] Manson JC, Jamieson E, Baybutt H, et al. A single amino acid alteration (101L) introduced into murine PrP dramatically alters incubation time of transmissible spongiform encephalopathy. Embo J 1999; 18(23):6855–6864.
[54] Raeber AJ, Race RE, Brandner S, et al. Astrocyte-specific expression of hamster prion protein (PrP) renders PrP knockout mice susceptible to hamster scrapie. Embo J 1997; 16(20):6057–6065.
[55] Frigg R, Klein MA, Hegyi I, et al. Scrapie pathogenesis in subclinically infected B-cell-deficient mice. J Virol 1999; 73(11):9584–9588.
[56] Hill AF, Joiner S, Linehan J, et al. Species-barrier-independent prion replication in apparently resistant species. Proc Natl Acad Sci USA 2000; 97(18):10248–10253.
[57] Race R, Raines A, Raymond GJ, et al. Long-term subclinical carrier state precedes scrapie replication and adaptation in a resistant species: analogies to bovine spongiform encephalopathy and variant Creutzfeldt–Jakob disease in humans. J Virol 2001; 75(21):10106–10112.
[58] Thackray AM, Klein MA, Aguzzi A, et al. Chronic subclinical prion disease induced by low-dose inoculum. J Virol 2002; 76(5):2510–2517.
[59] Knight RS and Will RG. Prion diseases. J Neurol Neurosurg Psychiatry 2004; 75(Suppl 1): i36–42.
[60] Llewelyn CA, Hewitt PE, Knight RS, et al. Possible transmission of variant Creutzfeldt–Jakob disease by blood transfusion. Lancet 2004; 363(9407):417–421.
[61] Fischer M, Rülicke T, Raeber A, et al. Prion protein (PrP) with amino-proximal deletions restoring susceptibility of PrP knockout mice to scrapie. EMBO J 1996; 15(6):1255–1264.
[62] Scott M, Foster D, Mirenda C, et al. Transgenic mice expressing hamster prion protein produce species-specific scrapie infectivity and amyloid plaques. Cell 1989; 59:847–857.
[63] Brandner S, Isenmann S, Raeber A, et al. Normal host prion protein necessary for scrapie-induced neurotoxicity. Nature 1996; 379(6563): 339–343.
[64] Li Q, Zhang M, Han H, et al. Evidence that DNase I hypersensitive site 5 of the human beta-globin locus control region functions as a chromosomal insulator in transgenic mice. Nucleic Acids Res 2002; 30(11):2484–2491.
[65] Araki K, Araki M, Yamamura K. Targeted integration of DNA using mutant lox sites in embryonic stem cells. Nucleic Acids Res 1997; 25(4):868–872.
[66] Soukharev S, Miller JL, Sauer B. Segmental genomic replacement in embryonic stem cells by double lox targeting. Nucleic Acids Res 1999; 27(18):e21.
[67] Westaway D, Mirenda CA, Foster D, et al. Paradoxical shortening of scrapie incubation times by expression of prion protein transgenes derived from long incubation period mice. Neuron 1991; 7(1):59–68.
[68] Chiesa R, Piccardo P, Ghetti B, et al. Neurological illness in transgenic mice expressing a prion protein with an insertional mutation. Neuron 1998; 21(6):1339–1351.
[69] Ma J, Wollmann R, Lindquist S. Neurotoxicity and neurodegeneration when PrP accumulates in the cytosol. Science 2002; 298(5599): 1781–1785.
[70] Borchelt DR, Davis J, Fischer M, et al. A vector for expressing foreign genes in the brains and hearts of transgenic mice. Genet Anal 1996; 13(6):159–163.
[71] Scott MR, Kohler R, Foster D, et al. Chimeric prion protein expression in cultured cells and transgenic mice. Protein Sci 1992; 1(8):986–997.
[72] Westaway D, DeArmond SJ, Cayetano Canlas J, et al. Degeneration of skeletal muscle, periph-

eral nerves, and the central nervous system in transgenic mice overexpressing wild-type prion proteins. Cell 1994; 76(1):117–129.

[73] Carlson GA, Ebeling C, Yang SL, et al. Prion isolate specified allotypic interactions between the cellular and scrapie prion proteins in congenic and transgenic mice. Proc Natl Acad Sci USA 1994; 91(12):5690–5694.

[74] Moore RC, Redhead NJ, Selfridge J, et al. Double replacement gene targeting for the production of a series of mouse strains with different prion protein gene alterations. Biotechnology (N Y) 1995; 13(9):999–1004.

[75] Moore RC, Hope J, McBride PA, et al. Mice with gene targetted prion protein alterations show that Prnp, Sinc and Prni are congruent. Nat Genet 1998; 18(2):118–125.

[76] Kitamoto T, Nakamura K, Nakao K, et al. Humanized prion protein knock-in by cre-induced site-specific recombination in the mouse. Biochem Biophys Res Comm 1996; 222(3):742–747.

[77] Luo JL, Yang Q, Tong WM, et al. Knock-in mice with a chimeric human/murine p53 gene develop normally and show wild-type p53 responses to DNA damaging agents: a new biomedical research tool. Oncogene 2001; 20(3):320–328.

[78] Fiering S, Epner E, Robinson K, et al. Targeted deletion of 5'HS2 of the murine beta-globin LCR reveals that it is not essential for proper regulation of the beta-globin locus. Genes Dev 1995; 9(18):2203–2213.

[79] Cohen JL, Boyer O, Salomon B, et al. Fertile homozygous transgenic mice expressing a functional truncated herpes simplex thymidine kinase delta TK gene. Transgenic Res 1998; 7(5):321–330.

[80] Braun RE, Lo D, Pinkert CA, et al. Infertility in male transgenic mice: disruption of sperm development by HSV-tk expression in postmeiotic germ cells. Biol Reprod 1990; 43(4):684–693.

[81] Weissmann C. Reverse genetics: a new approach to the elucidation of structure-function relationships. Trends Biochem Sci 1978; 3:N109–111.

[82] von Koch CS, Zheng H, Chen H, et al. Generation of APLP2 KO mice and early postnatal lethality in APLP2/APP double KO mice. Neurobiol Aging 1997; 18(6):661–669.

[83] Shmerling D, Hegyi I, Fischer M, et al. Expression of amino-terminally truncated PrP in the mouse leading to ataxia and specific cerebellar lesions. Cell 1998; 93(2):203–214.

[84] Flechsig E, Hegyi I, Schmerling D, et al. PrP lacking the flexible tail causes a severe cerebellar syndrome that is suppressed in the presence of wild-type PrP or PrP devoid of all five octarepeats. In: 1 ELSO conference in Geneva, 2–6. 9. 2000.

[85] Schwyzer R. ACTH: a short introductory review. Ann N Y Acad Sci 1977; 297:3–26.

[86] Weissmann C, Shmerling D, Rossi D, et al. Structure-function analysis of prion protein. In: Smith GL, Irving W, McCauley J, et al., editors. New Challenges to health: the treat of virus infection: 60 SGM symposium. Cambridge University Press, Cambridge, 2001:179–194.

[87] Flechsig E and Weissmann C. The role of PrP in health and disease. Curr Mol Med 2004; 4:337–353.

[88] Meindl T, Boller T, Felix G. The bacterial elicitor flagellin activates its receptor in tomato cells according to the address-message concept. Plant Cell 2000; 12(9):1783–1794.

[89] Behrens A and Aguzzi A. Small is not beautiful: antagonizing functions for the prion protein PrP(C) and its homologue Dpl. Trends Neurosci 2002b; 25(3):150–154.

[90] Muramoto T, DeArmond SJ, Scott M, et al. Heritable disorder resembling neuronal storage disease in mice expressing prion protein with deletion of an alpha-helix. Nat Med 1997; 3(7):750–755.

[91] Supattapone S, Bouzamondo E, Ball HL, et al. A protease-resistant 61-residue prion peptide causes neurodegeneration in transgenic mice. Mol Cell Biol 2001; 21(7):2608–2616.

[92] Ma J and Lindquist S. Wild-type PrP and a mutant associated with prion disease are subject to retrograde transport and proteasome degradation. Proc Natl Acad Sci USA 2001; 98(26): 14955–14960.

[93] Priola SA, Chesebro B, Caughey B. A view from the top – prion diseases from 10,000 feet. Science 2003; 300(5621):917–919.

[94] Mironov A, Jr., Latawiec D, Wille H, et al. Cytosolic prion protein in neurons. J Neurosci 2003; 23(18):7183–7193.

[95] Hope J, Multhaup G, Reekie LJ, et al. Molecular pathology of scrapie-associated fibril protein (PrP) in mouse brain affected by the ME7 strain of scrapie. Eur J Biochem 1988; 172(2): 271–277.

[96] McKinley MP, Bolton DC, Prusiner SB. A protease-resistant protein is a structural component of the scrapie prion. Cell 1983; 35(1):57–62.

[97] Supattapone S, Muramoto T, Legname G, et al. Identification of two prion protein regions that modify scrapie incubation time. J Virol 2001b; 75(3):1408–1413.
[98] Goldfarb LG, Brown P, Gajdusek DC. The molecular genetics of human transmissible spongiform encephalopathy. In: Prusiner SB, editor. Prion diseases of humans and animals. Ellis Horwood, London, 1992:139–153.
[99] Collinge J. Prion diseases of humans and animals: their causes and molecular basis. Annu Rev Neurosci 2001; 24(369):519–550.
[100] Vital C, Gray F, Vital A, et al. Prion encephalopathy with insertion of octapeptide repeats: the number of repeats determines the type of cerebellar deposits. Neuropathol Appl Neurobiol 1998; 24(2):125–130.
[101] Supattapone S, Bosque P, Muramoto T, et al. Prion protein of 106 residues creates an artifical transmission barrier for prion replication in transgenic mice. Cell 1999; 96(6):869–878.
[102] Pattison JH. Experiments with scrapie with special reference to the nature of the agent and the pathology of the disease. US. Governement Printing Office, Washington D.C., 1965.
[103] Prusiner SB, Scott M, Foster D, et al. Transgenetic studies implicate interactions between homologous PrP isoforms in scrapie prion replication. Cell 1990; 63(4):673–686.
[104] Telling GC, Scott M, Mastrianni J, et al. Prion propagation in mice expressing human and chimeric PrP transgenes implicates the interaction of cellular PrP with another protein. Cell 1995; 83(1):79–90.
[105] Telling GC, Scott M, Hsiao KK, et al. Transmission of Creutzfeldt–Jakob disease from humans to transgenic mice expressing chimeric human–mouse prion protein. Proc Natl Acad Sci USA 1994; 91(21):9936–9940.
[106] Kaneko K, Zulianello L, Scott M, et al. Evidence for protein X binding to a discontinuous epitope on the cellular prion protein during scrapie prion propagation. Proc Natl Acad Sci USA 1997; 94(19):10069–10074.
[107] Mastrianni JA, Capellari S, Telling GC, et al. Inherited prion disease caused by the V210I mutation: transmission to transgenic mice. Neurology 2001; 57(12):2198–2205.
[108] Hill AF, Desbruslais M, Joiner S, et al. The same prion strain causes vCJD and BSE. Nature 1997; 389(6650):448–450.
[109] Buschmann A, Pfaff E, Reifenberg K, et al. Detection of cattle-derived BSE prions using transgenic mice overexpressing bovine PrP(C). Arch Virol Suppl 2000; 16:75–86.
[110] Scott MR, Safar J, Telling G, et al. Identification of a prion protein epitope modulating transmission of bovine spongiform encephalopathy prions to transgenic mice. Proc Natl Acad Sci USA 1997; 94(26):14279–14284.
[111] Scott MR, Will R, Ironside J, et al. Compelling transgenetic evidence for transmission of bovine spongiform encephalopathy prions to humans. Proc Natl Acad Sci USA 1999; 96(26): 15137–15142.
[112] Crozet C, Flamant F, Bencsik A, et al. Efficient transmission of two different sheep scrapie isolates in transgenic mice expressing the ovine PrP gene. J Virol 2001; 75(11):5328–5334.
[113] Vilotte JL, Soulier S, Essalmani R, et al. Markedly increased susceptibility to natural sheep scrapie of transgenic mice expressing ovine PrP. J Virol 2001; 75(13):5977–5984.
[114] Barron RM, Thomson V, Jamieson E, et al. Changing a single amino acid in the N-terminus of murine PrP alters TSE incubation time across three species barriers. Embo J 2001; 20(18):5070–5078.
[115] Goldmann W, Hunter N, Smith G, et al. PrP genotype and agent effects in scrapie: change in allelic interaction with different isolates of agent in sheep, a natural host of scrapie. J Gen Virol 1994; 75(5):989–995.
[116] Goldfarb LG, Petersen RB, Tabaton M, et al. Fatal familial insomnia and familial Creutzfeldt–Jakob disease: disease phenotype determined by a DNA polymorphism. Science 1992b; 258(5083):806–808.
[117] Palmer MS, Dryden AJ, Hughes JT, et al. Homozygous prion protein genotype predisposes to sporadic Creutzfeldt–Jakob disease. Nature 1991; 352(6333):340–342.
[118] Dickinson AG and Outram GW. The scrapie replication-site hypothesis and its implications for pathogenesis. In: Prusiner SB, editor. Slow transmissible diseases of the nervous system. Academic Press, New York, 1979.
[119] Dickinson AG and Outram GW. Genetic aspects of unconventional virus infections: the basis of the virino hypothesis. Ciba Found Symp 1988; 135:63–83.
[120] Weissmann C. A "unified theory" of prion propagation. Nature 1991; 352(6337):679–683.
[121] Bessen RA and Marsh RF. Biochemical and physical properties of the prion protein from two strains of the transmissible mink encephalopathy agent. J Virol 1992; 66(4):2096–2101.

[122] Safar J, Wille H, Itri V, et al. Eight prion strains have PrP(Sc) molecules with different conformations. Nat Med 1998; 4(10):1157–1165.
[123] Telling GC, Parchi P, DeArmond SJ, et al. Evidence for the conformation of the pathologic isoform of the prion protein enciphering and propagating prion diversity. Science 1996; 274(5295):2079–2082.
[124] Carlson GA, Kingsbury DT, Goodman PA, et al. Linkage of prion protein and scrapie incubation time genes. Cell 1986; 46(4):503–511.
[125] Dickinson AG, Meikle VM, Fraser H. Identification of a gene which controls the incubation period of some strains of scrapie agent in mice. J Comp Pathol 1968; 78(3):293–299.
[126] Bruce M, Chree A, McConnell I, et al. Transmission of bovine spongiform encephalopathy and scrapie to mice: strain variation and the species barrier. Philos Trans R Soc Lond B Biol Sci 1994; 343(1306):405–411.
[127] Westaway D, Goodman PA, Mirenda CA, et al. Distinct prion proteins in short and long scrapie incubation period mice. Cell 1987; 51(4):651–662.
[128] Hunter N, Dann JC, Bennett AD, et al. Are Sinc and the PrP gene congruent? Evidence from PrP gene analysis in Sinc congenic mice. J Gen Virol 1992; 73(10):2751–2755.
[129] Hunter N, Hope J, McConnell I, et al. Linkage of the scrapie-associated fibril protein (PrP) gene and Sinc using congenic mice and restriction fragment length polymorphism analysis. J Gen Virol 1987; 68(10):2711–2716.
[130] Carlson GA, Ebeling C, Torchia M, et al. Delimiting the location of the scrapie prion incubation time gene on chromosome 2 of the mouse. Genetics 1993; 133(4):979–988.
[131] Lloyd SE, Onwuazor ON, Beck JA, et al. Identification of multiple quantitative trait loci linked to prion disease incubation period in mice. Proc Natl Acad Sci U S A 2001; 98(11):6279–6283.
[132] Manolakou K, Beaton J, McConnell I, et al. Genetic and environmental factors modify bovine spongiform encephalopathy incubation period in mice. Proc Natl Acad Sci USA 2001; 98(13):7402–7407.
[133] Stephenson DA, Chiotti K, Ebeling C, et al. Quantitative trait loci affecting prion incubation time in mice. Genomics 2000; 69(1):47–53.
[134] Liemann S and Glockshuber R. Influence of amino acid substitutions related to inherited human prion diseases on the thermodynamic stability of the cellular prion protein. Biochemistry 1999; 38(11):3258–3267.
[135] Swietnicki W, Petersen RB, Gambetti P, et al. Familial mutations and the thermodynamic stability of the recombinant human prion protein. J Biol Chem 1998; 273(47):31048–31052.
[136] Hsiao K, Baker HF, Crow TJ, et al. Linkage of a prion protein missense variant to Gerstmann–Sträussler syndrome. Nature 1989; 338(6213):342–345.
[137] Telling GC, Haga T, Torchia M, et al. Interactions between wild-type and mutant prion proteins modulate neurodegeneration in transgenic mice. Genes Dev 1996b; 10(14):1736–1750.
[138] Hsiao KK, Groth D, Scott M, et al. Serial transmission in rodents of neurodegeneration from transgenic mice expressing mutant prion protein. Proc Natl Acad Sci USA 1994; 91:9126–9130.
[139] Kaneko K, Ball HL, Wille H, et al. A synthetic peptide initiates Gerstmann–Sträussler–Scheinker (GSS) disease in transgenic mice. J Mol Biol 2000; 295(4):997–1007.
[140] Tateishi J, Kitamoto T, Hoque MZ, et al. Experimental transmission of Creutzfeldt–Jakob disease and related diseases to rodents. Neurology 1996; 46(2):532–537.
[141] Brown P, Gibbs CJ Jr, Rodgers Johnson P, et al. Human spongiform encephalopathy: the National Institutes of Health series of 300 cases of experimentally transmitted disease. Ann Neurol 1994; 35(5):513–529.
[142] Chesebro B. Prion protein and the transmissible spongiform encephalopathy diseases. Neuron 1999; 24(3):503–506.
[143] DeArmond SJ, Sanchez H, Yehiely F, et al. Selective neuronal targeting in prion disease. Neuron 1997; 19(6):1337–1348.
[144] Chiesa R, Drisaldi B, Quaglio E, et al. Accumulation of protease-resistant prion protein (PrP) and apoptosis of cerebellar granule cells in transgenic mice expressing a PrP insertional mutation. Proc Natl Acad Sci USA 2000; 97(10):5574–5579.
[145] Prusiner SB and Scott MR. Genetics of prions. Annu Rev Genet 1997; 31:139–175.
[146] Hegde RS, Mastrianni JA, Scott MR, et al. A transmembrane form of the prion protein in neurodegenerative disease. Science 1998; 279(5352):827–834.
[147] Kuroiwa Y, Kasinathan P, Matsushita H, et al. Sequential targeting of the genes encoding immunoglobulin-mu and prion protein in cattle. Nat Genet 2004; 36(7):775–780.
[148] Denning C, Burl S, Ainslie A. Deletion of the alpha(1,3)galactosyl transferase (GGTA1) gene

and the prion protein (PrP) gene in sheep. Nat Biotechnol 2001; 19(6):559–562.
[149] Tilly G, Chapuis J, Vilette D, et al. Efficient and specific down-regulation of prion protein expression by RNAi. Biochem Biophys Res Commun 2003; 305(3):548–551.
[150] Daude N, Marella M, Chabry J. Specific inhibition of pathological prion protein accumulation by small interfering RNAs. J Cell Sci 2003; 116(13):2775–2779.
[151] Hunter N, Moore L, Hosie BD, et al. Association between natural scrapie and PrP genotype in a flock of Suffolk sheep in Scotland. Vet Rec 1997; 140(3):59–63.
[152] Hunter N. Genotyping and susceptibility of sheep to scrapie. In: Baker HF, et al., editors. Prion diseases. Humana Press, Totowa, New Jersey, 1996:211–221.
[153] Zulianello L, Kaneko K, Scott M, et al. Dominant-negative inhibition of prion formation diminished by deletion mutagenesis of the prion protein. J Virol 2000; 74(9):4351–4360.
[154] Perrier V, Kaneko K, Safar J, et al. Dominant-negative inhibition of prion replication in transgenic mice. Proc Natl Acad Sci USA 2002; 99(20):13079–13084.
[155] Houston F, Goldmann W, Chong A, et al. Prion diseases: BSE in sheep bred for resistance to infection. Nature 2003; 423(6939):498.
[156] Buschmann A, Luhken G, Schultz J, et al. Neuronal accumulation of abnormal prion protein in sheep carrying a scrapie-resistant genotype (PrPARR/ARR). J Gen Virol 2004; 85(9):2727–2733.
[157] Gabizon R, McKinley MP, Groth D, et al. Immunoaffinity purification and neutralization of scrapie prion infectivity. Proc Natl Acad Sci USA 1988; 85:6617–6621.
[158] Horiuchi M and Caughey B. Specific binding of normal prion protein to the scrapie form via a localized domain initiates its conversion to the protease-resistant state. Embo J 1999; 18(12):3193–3203.
[159] Enari M, Flechsig E, Weissmann C. Scrapie prion protein accumulation by scrapie-infected neuroblastoma cells abrogated by exposure to a prion protein antibody. Proc Natl Acad Sci USA 2001; 98(16):9295–9299.
[160] Peretz D, Williamson RA, Kaneko K, et al. Antibodies inhibit prion propagation and clear cell cultures of prion infectivity. Nature 2001; 412(6848):739–743.
[161] Polymenidou M, Heppner FL, Pellicioli EC, et al. Humoral immune response to native eukaryotic prion protein correlates with antiprion protection. Proc Natl Acad Sci USA 2004; 101(Suppl2):14670–14676.
[162] Sigurdsson EM, Brown DR, Daniels M, et al. Immunization delays the onset of prion disease in mice. Am J Pathol 2002; 161(1):13–17.
[163] Gilch S, Wopfner F, Renner-Muller I, et al. Polyclonal anti-PrP auto-antibodies induced with dimeric PrP interfere efficiently with PrPSc propagation in prion-infected cells. J Biol Chem 2003; 278(20):18524–18531.
[164] Heppner FL, Musahl C, Arrighi I, et al. Prevention of scrapie pathogenesis by transgenic expression of anti-prion protein antibodies. Science 2001b; 294(5540):178–182.
[165] Korth C, Stierli B, Streit P, et al. Prion (PrPSc)-specific epitope defined by a monoclonal antibody. Nature 1997; 390(6655):74–77.
[166] White AR, Enever P, Tayebi M, et al. Monoclonal antibodies inhibit prion replication and delay the development of prion disease. Nature 2003; 422(6927):80–83.
[167] Solforosi L, Criado JR, McGavern DB, et al. Cross-linking cellular prion protein triggers neuronal apoptosis in vivo. Science 2004; 303(5663):1514–1516.
[168] Kretzschmar HA, Prusiner SB, Stowring LE, et al. Scrapie prion proteins are synthesized in neurons. Am J Pathol 1986; 122(1):1–5.
[169] Moser M, Colello RJ, Pott U, et al. Developmental expression of the prion protein gene in glial cells. Neuron 1995; 14(3):509–517.
[170] Race RE, Priola SA, Bessen RA, et al. Neuron-specific expression of a hamster prion protein minigene in transgenic mice induces susceptibility to hamster scrapie agent. Neuron 1995; 15(5):1183–1191.
[171] Eklund CM, Kennedy RC, Hadlow WJ. Pathogenesis of scrapie virus infection in the mouse. J Infect Dis 1967; 117(1):15–22.
[172] Kimberlin RH and Walker CA. Pathogenesis of mouse scrapie: dynamics of agent replication in spleen, spinal cord and brain after infection by different routes. J Comp Pathol 1979; 89(4):551–562.
[173] McBride PA, Eikelenboom P, Kraal G, et al. PrP protein is associated with follicular dendritic cells of spleens and lymph nodes in uninfected and scrapie-infected mice. J Pathol 1992; 168(4):413–418.
[174] Kitamoto T, Muramoto T, Mohri S, et al. Abnormal isoform of prion protein accumulates in follicular dendritic cells in mice with Creutzfeldt–Jakob disease. J Virol 1991; 65:6292–6295.

[175] Raeber AJ, Klein MA, Frigg R, et al. PrP-dependent association of prions with splenic but not circulating lymphocytes of scrapie-infected mice. Embo J 1999; 18(10):2702–2706.

[176] Chaffin KE, Beals CR, Wilkie TM, et al. Dissection of thymocyte signaling pathways by in vivo expression of pertussis toxin ADP-ribosyltransferase. Embo J 1990; 9(12):3821–3829.

[177] Raeber AJ, Sailer A, Hegyi I, et al. Ectopic expression of prion protein (PrP) in T lymphocytes or hepatocytes of PrP knockout mice is insufficient to sustain prion replication. Proc Natl Acad Sci USA 1999; 96(7):3987–3992.

[178] Montrasio F, Cozzio A, Flechsig E, et al. B lymphocyte-restricted expression of prion protein does not enable prion replication in prion protein knockout mice. Proc Natl Acad Sci USA 2001; 98(7):4034–4037.

[179] Montrasio F, Frigg R, Glatzel M, et al. Impaired prion replication in spleens of mice lacking functional follicular dendritic cells. Science 2000; 288(5469):1257–1259.

[180] Mabbott NA, Mackay F, Minns F, et al. Temporary inactivation of follicular dendritic cells delays neuroinvasion of scrapie. Nat Med 2000; 6(7):719–720.

[181] Beekes M, McBride PA, Baldauf E. Cerebral targeting indicates vagal spread of infection in hamsters fed with scrapie. J Gen Virol 1998; 79(3):601–607.

[182] Baldauf E, Beekes M, Diringer H. Evidence for an alternative direct route of access for the scrapie agent to the brain bypassing the spinal cord. J Gen Virol 1997; 78(5):1187–1197.

[183] Kimberlin RH and Walker CA. Pathogenesis of mouse scrapie: evidence for neural spread of infection to the CNS. J Gen Virol 1980; 51(1):183–187.

[184] Beekes M, Baldauf E, Diringer H. Sequential appearance and accumulation of pathognomonic markers in the central nervous system of hamsters orally infected with scrapie. J Gen Virol 1996; 77(8):1925–1934.

[185] Brandner S, Raeber A, Sailer A, et al. Normal host prion protein (PrPC) is required for scrapie spread within the central nervous system. Proc Natl Acad Sci U S A 1996; 93(23):13148–13151.

[186] Blättler T, Brandner S, Raeber AJ, et al. PrP-expressing tissue required for transfer of scrapie infectivity from spleen to brain. Nature 1997; 389(6646):69–73.

[187] Race R, Oldstone M, Chesebro B. Entry versus blockade of brain infection following oral or intraperitoneal scrapie administration: role of prion protein expression in peripheral nerves and spleen. J Virol 2000; 74(2):828–833.

[188] Glatzel M and Aguzzi A. PrP(C) expression in the peripheral nervous system is a determinant of prion neuroinvasion. J Gen Virol 2000; 81(11):2813–2821.

[189] Follet J, Lemaire-Vieillc C, Blanquet-Grossard F, et al. PrP expression and replication by Schwann cells: implications in prion spreading. J Virol 2002; 76(5):2434–2439.

[190] Prinz M, Montrasio F, Furukawa H, et al. Intrinsic resistance of oligodendrocytes to prion infection. J Neurosci 2004; 24(26):5974–5981.

[191] Bosque PJ, Ryou C, Telling G, et al. Prions in skeletal muscle. Proc Natl Acad Sci USA 2002; 99(6):3812–3817.

[192] Taraboulos A, Scott M, Semenov A, et al. Cholesterol depletion and modification of COOH-terminal targeting sequence of the prion protein inhibit formation of the scrapie isoform. JCell Biol 1995; 129:121–132.

[193] Vey M, Pilkuhn S, Wille H, et al. Subcellular colocalization of the cellular and scrapie prion proteins in caveolae-like membranous domains. Proc Natl Acad Sci USA 1996; 93(25): 14945–14949.

[194] Riek R, Hornemann S, Wider G, et al. NMR characterization of the full-length recombinant murine prion protein, mPrP(23–231). FEBS Lett 1997; 413(2):282–288.

[195] Riek R, Hornemann S, Wider G, et al. NMR structure of the mouse prion protein domain PrP(121–321). Nature 1996; 382(6587):180–182.

11 Transplantation as a Tool in Prion Research

Eckhard Flechsig, Ivan Hegyi, Alex J. Raeber, Antonio Cozzio, Adriano Aguzzi, and Charles Weissmann

11.1. Introduction

Among the many unresolved questions of prion diseases are those concerning the molecular mechanisms of pathogenesis in the CNS and the mechanism of prion spread in the infected organism. A useful approach to address some of these problems is based on the transplantation of tissue expressing PrP into *Prnp*$^{o/o}$ mice and vice versa, particularly in neuroectodermal or hematopoietic tissues. The neurografting procedure is straightforward: neuroectoderm is derived from embryos and injected into the caudoputamen or lateral ventricles of recipient mice using a stereotaxic frame [1; 2]. Thereby, tissue from a donor mouse, which would usually die young due to lethal disease, can be kept alive in a healthy recipient. Another important technique is transplantation of hematopoietic tissue, either bone marrow or fetal liver, into lethally irradiated mice.

11.2 Prion-infected neurografts fail to cause neuropathological changes in PrP knockout mice

An interesting question concerns the molecular mechanism underlying neuropathological changes, in particular cell death, resulting from prion disease. Depletion of PrP^{C} is an unlikely cause in view of the finding that abrogation of PrP does not cause scrapie-like neuropathological changes [3] even when elicited postnatally [4]. More likely, toxicity of PrP^{Sc} or some other PrP^{C} conformer is responsible. To address the question of neurotoxicity, brain tissue of *Prnp*$^{o/o}$ mice was exposed to a continuous source of PrP^{Sc}. To this end, neuroectodermal tissue from transgenic mice overexpressing PrP [5] was transplanted into the forebrain of *Prnp*$^{o/o}$ mice and the "pseudochimeric" brains were inoculated with scrapie prions. All grafted and scrapie-inoculated mice remained free of scrapie symptoms for at least 70 weeks; this exceeded the survival time of scrapie-infected donor mice at least 7-fold. Therefore, the presence of a continuous source of PrP^{Sc} and of scrapie prions does not exert any clinically detectable adverse effects on a mouse devoid of PrP^{C}. On the other hand, the infected grafts developed characteristic histopathological features of scrapie (Fig. 11.1). The course of the disease in the graft was very similar to that observed in the brain of scrapie-inoculated wild-type mice. Importantly, grafts had extensive contact with the recipient brain, and prions could navigate between the two compartments, evidenced by the fact that inoculation of wild-type animals carrying PrP-expressing neurografts resulted in scrapie pathology in both graft and host tissue. Nonetheless, histopathological changes never extended into host tissue, even at the latest stages (> 450 days), although PrP^{Sc} was detected in both grafts and recipient brain, and immunohistochemistry revealed PrP deposits in the hippocampus, and occasionally in the parietal cortex, of all animals [6]. Thus, prions moved from the grafts to some regions of the PrP-deficient host brain without causing pathological changes or clinical disease. The distribution of PrP^{Sc} in the white matter tracts of the host brain suggests diffusion within the extracellular space [7] rather than axonal transport (S. Brandner and A. Aguzzi, unpublished results).

These findings suggest that the expression of PrP^{C} by an infected cell, rather than the extracellular deposition of PrP^{Sc}, is the critical pre-

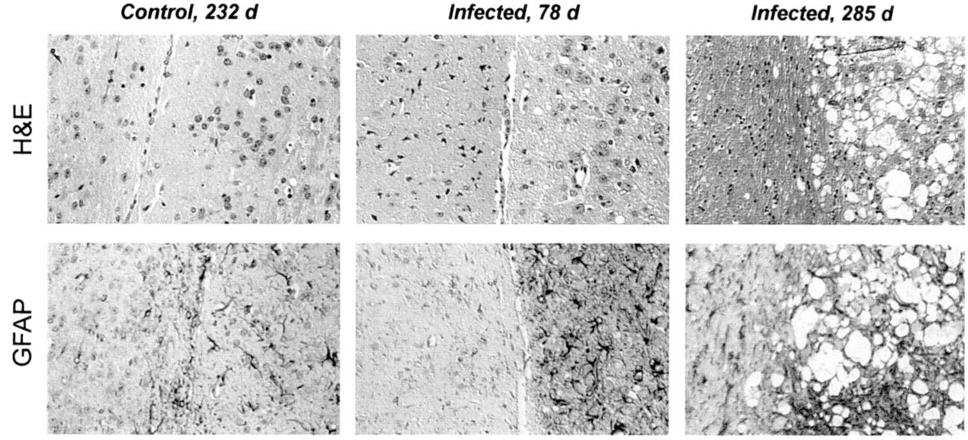

Fig. 11.1: Typical appearance of noninfected and scrapie-infected PrP-expressing neurografts in the brain of PrP knockout mice. PrP knockout mice were grafted with neuroectodermal tissue from PrP-expressing mouse Tg*a20* embryos and inoculated with the RML mouse scrapie strain where indicated. Animals were sacrificed at indicated time points and brain sections stained either with hematoxylin-eosin (H&E) or an antibody against glial fibrillary acidic protein (GFAP). The graft in the host brain is shown in each panel on the right side. The infected graft shows mild spongiform changes and proliferation of glial cells early after inoculation (middle panel). Typical spongiform changes and gliosis in a chronically infected graft are shown 285 days post inoculation, whereas the surrounding host tissue is not affected (right panel). The noninfected graft shows no spongiosis and only little gliosis. For further details see Brandner et al. 1996 [6]. Figure by A. Aguzzi.

requisite for the development of scrapie pathology. Perhaps PrP^{Sc} is inherently nontoxic and PrP^{Sc} plaques found in prion diseases are an epiphenomenon rather than a cause of neuronal damage. Therefore, one may propose that availability of PrP^{C} for some intracellular processes elicited by the infectious agent (perhaps the formation of a toxic form of PrP, indicated as PrP* [8]; differing from the bulk of protease-resistant, aggregated material designated PrP^{Sc}) is responsible for spongiosis, gliosis, and neuronal death. This would be in agreement with the fact that in several instances, and especially in fatal familial insomnia, spongiform pathology is detectable although very little PrP^{Sc} is present [9; 10].

11.3 Spread of prions in the CNS requires PrP^{C}-expressing tissue

Intracerebral (i. c.) inoculation of tissue homogenate into suitable recipients is the most effective method for transmission of prion diseases and may even facilitate circumvention of the species barrier. However, prion diseases can also be initiated ocularly by conjunctival instillation [11], corneal grafts [12], and intraocular (i.o.) injection [13]. The latter method has proved particularly useful to study neural spread of the agent, since the retina is a part of the central nervous system (CNS) and i.o. injection does not produce direct physical trauma to the brain. Diachronic spongiform changes along the retinal pathway following i.o. infection argue in favor of axonal spread [13; 14].

As discussed above, expression of PrP^{C} is required for prion replication and also for neurodegenerative changes to occur. This raises the question as to whether spread of prions within the CNS is also dependent on PrP^{C} expression in neural pathways. Again, grafts in the brain served as an indicator for infectivity in an otherwise scrapie-resistant host. After inoculation of prions into the eyes of engrafted *Prnp*$^{o/o}$ mice, no signs of spongiform encephalopathy or deposition of PrP^{Sc} were detected in the grafts. In control experiments, unilateral i.o. inoculation led to progressive appearance of scrapie pathology along the optic nerve and optic tract, which then extended into the rest of the brain as previously described [13; 15]. Thus, infectivity administered to the eye of PrP-deficient hosts does not induce scrapie in a PrP-expressing brain graft [14].

Because *Prnp*$^{o/o}$ mice are not exposed to PrP during maturation of their immune system, engraftment with PrP^{C}-producing tissue might lead to an immune response to PrP and possibly to neutralization of infectivity. Indeed, analysis of sera from engrafted *Prnp*$^{o/o}$ mice revealed significant anti-PrP antibody titers. Because even mock-inoculated and uninoculated engrafted *Prnp*$^{o/o}$ mice showed an immune response to PrP, while i.c. inoculation of nonengrafted *Prnp*$^{o/o}$ mice did not [16], PrP^{C} presented by the intracerebral graft (rather than the inoculum or graft-borne PrP^{Sc}) was clearly the offending antigen. To test whether grafts would develop scrapie if infectivity were administered before they elicited a potentially neutralizing immune response, mice were inoculated 24 h after grafting. Again, no disease was detected in the graft of two mice inoculated intraocularly. In order to definitively rule out the possibility that prion transport was disabled by a neutralizing immune response, the experiments were repeated in PrP-tolerant mice. Transgenic *Prnp*$^{o/o}$ mice overexpressing PrP in T lymphocytes (Tg33) are resistant to scrapie and do not contain scrapie infectivity in brain or spleen after inoculation with scrapie prions [17]. When engrafted with PrP-overexpressing Tg*a20* neuroectoderm, these mice did not develop antibodies to PrP after i.c. or i.o. inoculation. As previously, i.o. inoculation with prions did not provoke scrapie in the graft, supporting the conclusion that lack of PrP^{C}, rather than an immune response to PrP, prevented spread [14].

The prion itself is surprisingly sessile and spread might proceed along a PrP^{C}-paved chain of cells. Perhaps prions require PrP^{C} for propagation across synapses; PrP^{C} is present in the synaptic region [18] and, as discussed above, certain synaptic properties are altered in *Prnp*$^{o/o}$ mice [19–21]. Perhaps transport of prions within (or on the surface of) neuronal processes is PrP^{C} dependent. Within the framework of the protein-only hypothesis, these findings may be accommodated by a "domino effect" model in which spreading of scrapie prions in the CNS occurs per continuitatem through conversion of PrP^{C} by adjacent PrP^{Sc} [22].

11.4 Spread of prions from extracerebral sites to the CNS

As discussed above, PrP^{C} seems indispensable for prion spread within the CNS. But prions need not be delivered directly to the CNS. Intraspinal, intraperitoneal, intramuscular, intravenous (i.v.), and subcutaneous injections [15; 23–25] or scarification [26; 27] are effective, albeit less efficient. Oral infection has been demonstrated in many animal species [28–35]. In mice and sheep [36], and probably also in humans in the case of vCJD [37], prions accumulate in the spleen before doing so in the brain, even when infectivity is administered i.c. Prions can accumulate in all components of the lymphoreticular system (LRS), including lymph nodes and intestinal Peyer's patches, where they are found almost immediately following oral infection [38]. In immunodeficient mice, i.p. infection does not lead to replication of prions in spleen or brain; however, transfer of wild-type spleen cells to these mice restores susceptibility of the CNS to i.p. inoculation [39–42]. This suggests that components of the immune system are required for efficient transfer of prions from the site of peripheral infection to the CNS. Perhaps prions injected i.p. or ingested are first brought to lymphatic organs, specifically to germinal centers, by mobile immune cells; dendritic cells are currently being discussed for this role [43; 44], as well as the involvement of other cells [45]. Opsonization by complement system components is likely to be relevant, since mice genetically engineered to lack complement factors [46], or mice deficient in the C3 complement component by administration of cobra venom [47], exhibit a remarkable resistance to peripheral prion inoculation. It is unlikely that lymphocytes transport prions to the CNS [48].

Rather, it would seem that in the spleen, B-cell-dependent, mature FDCs allow prions to multiply to a relatively high titer [49–51], facilitating transfer to the peripheral nervous system. However, prion amplification in lymph nodes can occur in cells other than FDCs [45; 52]. Following invasion of peripheral nerve endings in the LRS, the CNS is reached and further spread occurs transsynaptically and along fiber tracts. Prions are first found in the CNS segments to which the sites of peripheral inoculation project [15; 53].

Because PrP^C is crucial for prion spread within the CNS [14], it may also be required for the spread of prions from peripheral sites to the CNS. Indeed, PrP-expressing neurografts in *Prnp*$^{o/o}$ mice did not develop scrapie histopathology after i.p. or i.v. inoculation with scrapie prions. After reconstitution of the hematopoietic system of *Prnp*$^{o/o}$ mice with PrP-expressing cells, prion accumulation in the spleen was as in wild-type animals. Surprisingly, however, i.p. or i.v. inoculation failed to produce scrapie pathology in the neurografts in 27 out of 28 animals, in contrast to i.c. inoculation. Thus, transfer of infectivity from spleen to the CNS is crucially dependent on the expression of PrP in a tissue compartment that cannot be reconstituted by bone marrow transfer [54], likely the peripheral nervous system [15; 53]. In particular, the autonomic nervous system has been held responsible for transport from lymphoid organs to the CNS [55–58], in a PrP^C-dependent manner [54; 59; 60]. Chemical sympathectomy and immunosympathectomy significantly delayed the development of scrapie, and mice with sympathetic hyperinnervation of immune organs had a shortened incubation time and a 100-fold increase in prion titer in the spleen [61].

It is unclear how prions are actually transported by peripheral nerves. Axonal or nonaxonal transport mechanisms may be involved, and nonneuronal cells (such as Schwann cells) may play a role. While PrP expression on neurons is required and sufficient for prion transport along peripheral nerves [60], PrP expression exclusively on Schwann cells and oligodendrocytes failed to restore propagation and transport of prions in *Prnp*$^{o/o}$ mice [62]. The "domino effect" mechanism implies that incoming PrP^{Sc} converts resident PrP^C on the axolemmal surface, thereby propagating the infection along the axon. While speculative, this model supports findings that the velocity of neural prion spread is much slower than expected from fast axonal transport [63] and that PrP^{Sc} is deposited periaxonally [64; 65].

References

[1] Aguzzi A, Kleihues P, Heckl K, et al. Cell type-specific tumor induction in neural transplants by retrovirus-mediated oncogene transfer. Oncogene 1991; 6:113–118.

[2] Isenmann S, Brandner S, Aguzzi A. Neuroectodermal grafting: a new tool for the study of neurodegenerative diseases. Histol Histopathol 1996; 11(4):1063–1073.

[3] Büeler H, Fischer M, Lang Y, et al. Normal development and behaviour of mice lacking the neuronal cell-surface PrP protein. Nature 1992; 356(6370):577–582.

[4] Mallucci GR, Ratte S, Asante EA, et al. Postnatal knockout of prion protein alters hippocampal CA1 properties, but does not result in neurodegeneration. Embo J 2002; 21(3):202–210.

[5] Fischer M, Rülicke T, Raeber A, et al. Prion protein (PrP) with amino-proximal deletions restoring susceptibility of PrP knockout mice to scrapie. EMBO J 1996; 15(6):1255–1264.

[6] Brandner S, Isenmann S, Raeber A, et al. Normal host prion protein necessary for scrapie-induced neurotoxicity. Nature 1996; 379(6563): 339–343.

[7] Jeffrey M, Goodsir CM, Bruce ME, et al. Murine scrapie-infected neurons in vivo release excess prion protein into the extracellular space. Neurosci Lett 1994; 174(1):39–42.

[8] Weissmann C. A 'unified theory' of prion propagation. Nature 1991; 352(6337):679–683.

[9] Collinge J, Palmer MS, Sidle KCL, et al. Transmission of fatal familial insomnia to laboratory animals. Lancet 1995; 346:569–570.

[10] Aguzzi A and Weissmann C. Sleepless in Bologna: transmission of fatal familial insomnia. Trends Microbiol 1996; 4(4):129–131.

[11] Scott JR, Foster JD, Fraser H. Conjunctival instillation of scrapie in mice can produce disease. Vet Microbiol 1993; 34(4):305–309.

[12] Duffy P, Wolf J, Collins G, et al. Possible person-to-person transmission of Creutzfeldt–Jakob disease. N Engl J Med 1974; 290(12):692–693.
[13] Fraser H. Neuronal spread of scrapie agent and targeting of lesions within the retino-tectal pathway. Nature 1982; 295(5845):149–150.
[14] Brandner S, Raeber A, Sailer A, et al. Normal host prion protein (PrPC) is required for scrapie spread within the central nervous system. Proc Natl Acad Sci USA 1996; 93(23):13148–13151.
[15] Kimberlin RH and Walker CA. Pathogenesis of scrapie (strain 263K) in hamsters infected intracerebrally, intraperitoneally or intraocularly. J Gen Virol 1986; 67(2):255–263.
[16] Büeler H, Aguzzi A, Sailer A, et al. Mice devoid of PrP are resistant to scrapie. Cell 1993; 73(7):1339–1347.
[17] Raeber AJ, Sailer A, Hegyi I, et al. Ectopic expression of prion protein (PrP) in T lymphocytes or hepatocytes of PrP knockout mice is insufficient to sustain prion replication. Proc Natl Acad Sci USA 1999; 96(7):3987–3992.
[18] Fournier JG, Escaig Haye F, Billette de Villemeur T, et al. Ultrastructural localization of cellular prion protein (PrPc) in synaptic boutons of normal hamster hippocampus. C R Acad Sci III 1995; 318(3):339–344.
[19] Collinge J, Whittington MA, Sidle KC, et al. Prion protein is necessary for normal synaptic function. Nature 1994; 370(6487):295–297.
[20] Whittington MA, Sidle KC, Gowland I, et al. Rescue of neurophysiological phenotype seen in PrP null mice by transgene encoding human prion protein. Nat Genet 1995; 9(2):197–201.
[21] Manson JC, Hope J, Clarke AR, et al. PrP gene dosage and long term potentiation. Neurodegeneration 1995; 4(1):113–114.
[22] Aguzzi A. Neuro-immune connection in spread of prions in the body? The Lancet 1997; 349:742–743.
[23] Buchanan CR, Preece MA, Milner RD. Mortality, neoplasia, and Creutzfeldt–Jakob disease in patients treated with human pituitary growth hormone in the United Kingdom. Bmj 1991; 302(6780):824–828.
[24] Kimberlin RH and Walker CA. Pathogenesis of mouse scrapie: dynamics of agent replication in spleen, spinal cord and brain after infection by different routes. J Comp Pathol 1979; 89(4):551–562.
[25] Kimberlin RH, Cole S, Walker CA. Pathogenesis of scrapie is faster when infection is intraspinal instead of intracerebral. Microb Pathog 1987; 2(6):405–415.
[26] Taylor DM, McConnell I, Fraser H. Scrapie Infection Can Be Established Readily Through Skin Scarification in Immunocompetent But Not Immunodeficient Mice. Journal of General Virology 1996; 77(7):1595–1599.
[27] Carp RI. Transmission of scrapie by oral route: effect of gingival scarification. Lancet 1982; 1(8264):170–171.
[28] Anderson RM, Donnelly CA, Ferguson NM, et al. Transmission dynamics and epidemiology of BSE in British cattle. Nature 1996; 382(6594):779–788.
[29] Bradley R. Bovine spongiform encephalopathy-distribution and update on some transmission and decontamination studies. In: Gibbs CJ Jr, editor. Bovine spongiform encephalopathy: the BSE dilemma. Springer Verlag, New York, 1996:11.
[30] Foster JD, Hope J, Fraser H. Transmission of bovine spongiform encephalopathy to sheep and goats. Vet Rec 1993; 133(14):339–341.
[31] Prusiner SB, Cochran SP, Alpers MP. Transmission of scrapie in hamsters. J Infect Dis 1985; 152(5):971–978.
[32] Bons N, Mestre-Frances N, Belli P, et al. Natural and experimental oral infection of nonhuman primates by bovine spongiform encephalopathy agents. Proc Natl Acad Sci USA 1999; 96(7):4046–4051.
[33] Jeffrey M, Martin S, Gonzalez L, et al. Differential diagnosis of infections with the bovine spongiform encephalopathy (BSE) and scrapie agents in sheep. J Comp Pathol 2001; 125(4):271–284.
[34] Maignien T, Lasme Zas CI, Beringue V, et al. Pathogenesis of the oral route of infection of mice with scrapie and bovine spongiform encephalopathy agents. J Gen Virol 1999; 80(11):3035–3042.
[35] Ridley RM and Baker HF. Oral transmission of BSE to primates. Lancet 1996; 348(9035):1174.
[36] Kimberlin RH and Walker CA. Pathogenesis of experimental scrapie. Ciba Found Symp 1988; 135:37–62.
[37] Hill AF, Zeidler M, Ironside J, et al. Diagnosis of new variant Creutzfeldt–Jakob disease by tonsil biopsy. Lancet 1997; 349:99.
[38] Kimberlin RH and Walker CA. Pathogenesis of scrapie in mice after intragastric infection. Virus Res 1989; 12(3):213–220.
[39] Lasmezas CI, Cesbron JY, Deslys JP, et al. Immune system-dependent and -independent replication of the scrapie agent. J Virol 1996; 70(2):1292–1295.

[40] Klein MA, Frigg R, Raeber AJ, et al. PrP expression in B lymphocytes is not required for prion neuroinvasion. Nat Med 1998; 4(12):1429–1433.

[41] Klein MA, Frigg R, Flechsig E, et al. A crucial role for B cells in neuroinvasive scrapie. Nature 1997; 390(6661):687–690.

[42] O'Rourke KI, Huff TP, Leathers CW, et al. SCID mouse spleen does not support scrapie agent replication. J Gen Virol 1994; 75:1511–1514.

[43] Aucouturier P, Geissmann F, Damotte D, et al. Infected splenic dendritic cells are sufficient for prion transmission to the CNS in mouse scrapie. J Clin Invest 2001; 108(5):703–708.

[44] Huang FP, Farquhar CF, Mabbott NA, et al. Migrating intestinal dendritic cells transport PrP(Sc) from the gut. J Gen Virol 2002; 83(1): 267–271.

[45] Oldstone MB, Race R, Thomas D, et al. Lymphotoxin-alpha- and lymphotoxin-beta-deficient mice differ in susceptibility to scrapie: evidence against dendritic cell involvement in neuroinvasion. J Virol 2002; 76(9):4357–4363.

[46] Klein MA, Kaeser PS, Schwarz P, et al. Complement facilitates early prion pathogenesis. Nat Med 2001; 7(4):488–492.

[47] Mabbott NA, Bruce ME, Botto M, et al. Temporary depletion of complement component C3 or genetic deficiency of C1q significantly delays onset of scrapie. Nat Med 2001; 7(4):485–487.

[48] Raeber AJ, Klein MA, Frigg R, et al. PrP-dependent association of prions with splenic but not circulating lymphocytes of scrapie-infected mice. Embo J 1999; 18(10):2702–2706.

[49] Montrasio F, Frigg R, Glatzel M, et al. Impaired prion replication in spleens of mice lacking functional follicular dendritic cells. Science 2000; 288(5469):1257–1259.

[50] Mabbott NA, Mackay F, Minns F, et al. Temporary inactivation of follicular dendritic cells delays neuroinvasion of scrapie. Nat Med 2000; 6(7):719–720.

[51] Brown KL, Stewart K, Ritchie DL, et al. Scrapie replication in lymphoid tissues depends on prion protein- expressing follicular dendritic cells. Nat Med 1999; 5(11):1308–1312.

[52] Prinz M, Montrasio F, Klein MA, et al. Lymph nodal prion replication and neuroinvasion in mice devoid of follicular dendritic cells. Proc Natl Acad Sci USA 2002; 99(2):919–924.

[53] Beekes M, Baldauf E, Diringer H. Sequential appearance and accumulation of pathognomonic markers in the central nervous system of hamsters orally infected with scrapie. J Gen Virol 1996; 77(8):1925–1934.

[54] Blättler T, Brandner S, Raeber AJ, et al. PrP-expressing tissue required for transfer of scrapie infectivity from spleen to brain. Nature 1997; 389(6646):69–73.

[55] Beekes M, McBride PA, Baldauf E. Cerebral targeting indicates vagal spread of infection in hamsters fed with scrapie. J Gen Virol 1998; 79(3):601–607.

[56] Clarke MC and Kimberlin RH. Pathogenesis of mouse scrapie: distribution of agent in the pulp and stroma of infected spleens. Vet Microbiol 1984; 9(3):215–225.

[57] Cole S and Kimberlin RH. Pathogenesis of mouse scrapie: dynamics of vacuolation in brain and spinal cord after intraperitoneal infection. Neuropathol Appl Neurobiol 1985; 11(3):213–227.

[58] McBride PA and Beekes M. Pathological PrP is abundant in sympathetic and sensory ganglia of hamsters fed with scrapie. Neurosci Lett 1999; 265(2):135–138.

[59] Glatzel M and Aguzzi A. PrP(C) expression in the peripheral nervous system is a determinant of prion neuroinvasion. J Gen Virol 2000; 81(11):2813–2821.

[60] Race R, Oldstone M, Chesebro B. Entry versus blockade of brain infection following oral or intraperitoneal scrapie administration: role of prion protein expression in peripheral nerves and spleen. J Virol 2000; 74(2):828–833.

[61] Glatzel M, Heppner FL, Albers KM, et al. Sympathetic innervation of lymphoreticular organs is rate limiting for prion neuroinvasion. Neuron 2001; 31(1):25–34.

[62] Prinz M, Montrasio F, Furukawa H, et al. Intrinsic resistance of oligodendrocytes to prion infection. J Neurosci 2004; 24(26):5974–5981.

[63] Kimberlin RH, Hall SM, Walker CA. Pathogenesis of mouse scrapie. Evidence for direct neural spread of infection to the CNS after injection of sciatic nerve. J Neurol Sci 1983; 61(3):315–325.

[64] Glatzel M and Aguzzi A. Peripheral pathogenesis of prion diseases. Microbes Infect 2000b; 2(6):613–619.

[65] Hainfellner JA and Budka H. Disease associated prion protein may deposit in the peripheral nervous system in human transmissible spongiform encephalopathies. Acta Neuropathol (Berl) 1999; 98(5):458–460.

12 Prion Strains

Martin H. Groschup, Anja Gretzschel, and Thorsten Kuczius

12.1 Introduction

Based on the present state of knowledge, there is no doubt that the prion protein (PrP) plays a central role in the transmission and pathogenesis of transmissible spongiform encephalopathies (TSE). Much of the scientific data even indicate that the pathological prion protein (PrP^{Sc}) is the infectious agent itself. The prion model relates to a completely new class of pathogens that is clearly distinct, for example, from viruses and bacteria. However, irrefutable evidence for this model is still lacking. An important argument against such a model is the existence of a variety of different prion strains. As shown over the past four decades, prion strains are characterized by a number of different biological and biochemical features: their host spectrum and incubation periods, the neuropathological changes, courses of the disease, and the signs and symptoms they cause in experimental transmissions, their resistance to physical and chemical inactivation procedures, and last but not least the biochemical characteristics of the corresponding PrP^{Sc}. Prion strains are able to change their determining characteristics – especially when they change their host species [1]. However, the molecular basis for prion strain characteristics is still an enigma.

12.2 Definition of the term "prion strain"

The definition of the term prion strain has developed over the years depending on the methods available for determining the strain characteristics. In 1961, Pattison and Millson were the first to present clinical evidence that the clinical picture of scrapie can vary considerably in small ruminants [2]. In subsequent years, comparative studies on the transmission to other species were carried out so that the pathogens could be differentiated.

In addition, it was possible to differentiate between individual scrapie strains based on their incubation periods within the same rodent line. A detailed description of the neuropathological findings (lesion profile) of different scrapie strains was published by Fraser and Dickinson in the late 1960s [3]. The criteria for establishing the lesion profiles defined at that time are still valid today. Incubation period and lesion profile are the main criteria for differentiating the strains [4; 5]. Additionally, numerous working groups have described differences in clinical symptoms [6–12].

In the 1980s and 1990s, the development of improved immunochemical detection methods finally permitted a detailed analysis of PrP^{Sc} and the finding that the latter is glycosylated to a greater or lesser extent in the different strains [13]. One should emphasize, however, that the amino acid sequence of PrP^{Sc} is identical for different strains in the same host, since this is a host encoded protein. It has not yet been conclusively explained how a single protein can encode for such a variety of strain characteristics. As various prion strains can be replicated in the same host species, these characteristics must be pathogen-specific and not host-specific. In other words, scrapie strains must carry their own information pertaining to the development of their specific characteristics. Therefore, a number of scientists have put much effort into answering the question of whether, in addition to PrP^{Sc}, the TSE agents contain a so far undetected genetic substance in the form of a short nucleic acid (see Chapter 4). Other scientists have put forward the theory that the information on strain-specific characteristics may be contained in different conformation types of PrP^{Sc} or even in the differences in the prion protein glycosylation.

12.3 Characteristics of prion strains

Prion strains can be differentiated based on seven criteria: i) clinical symptoms; ii) incubation period; iii) transmissibility; iv) histopathological lesion profiles; v) inactivation behavior; vi) PK resistance and cleavage site of PrP^{Sc}; vii) glycoprofile of PrP^{Sc}.

12.3.1 Clinical symptoms

In small ruminants, scrapie may be associated with different clinical traits, as described by Pattison and Millson in 1961. They infected goats experimentally with scrapie by intracerebral (i.c.) injection of brain homogenates of a scrapie-infected sheep [2]. After an incubation period of 43 months, 7 out of the 10 animals showed clinical signs of scrapie. Interestingly, however, two clearly different clinical pictures were observed: the "drowsy" syndrome and the "scratching" syndrome. Pattison and Millson then used brain samples from the 7 goats suffering from scrapie to inoculate other goats. Surprisingly, the differences in the signs observed in the seven goats were also observed in the newly infected animals depending on whether they had been infected with tissue of the scratching or of the drowsy goats. As these different characteristics were still observed in goats as many as 10 passages later, it was concluded that there must be two different scrapie strains.

Characteristic clinical symptoms are also observed in the passage of established scrapie strains in mice and hamsters, which have been observed since the work done by Pattison and Millson. In mice, scrapie infection with the ME7 strain is characterized by a considerable weight gain (obesity), whereas other pathogens such as 139A or 22L have no effect on body weight [14–17]. Also, the glucose tolerance of mice may be impaired as observed after the transmission of ME7 or 22L strains. Hamsters infected with the 139H and 22CH strains showed weight gain, hyperglycemia, hyperinsulinemia [18], and an enlargement of liver, kidneys, and endocrinal organs. In contrast, these parameters remained unchanged after infection with strain 263K.

Two clearly distinguishable clinical traits can be observed after transmission of the two TME strains isolated from mink to hamsters [19; 20]. While the hyper (HY) strain causes overexcitement and cerebellar ataxia, which causes the animals to stagger, the drowsy (DY) strain is characterized by lethargy and ataxia, while hyperesthesia is not observed [21].

12.3.2 Incubation period

In the 1940s and 1950s, English scientists conducted extensive scrapie infection trials in sheep. Almost all trials were carried out using subpassages of a pool of animals suffering from scrapie (scrapie sheep brain pool, SSBP/1). These experiments were not always as successful and reproducible as desired. Very often only a small percentage (30 %) of the inoculated sheep became ill [22]. In addition, long incubation periods of several years complicated the evaluation of the results. Improvements were seen when goats were chosen as study animals instead of sheep. In this species, the infection rate was close to 100 %. However, with a 3- to 5-year incubation period, it still took an extended amount of time before results could be obtained. A significant breakthrough was achieved in 1961 by the British scientist R. Chandler who transmitted the isolates (scrapie strains) – thus far only used in sheep and goats – to mice [23]. In mice, the incubation period was reduced considerably over the first three passages to sometimes less than 1 year. Surprisingly, however, no uniform incubation period developed. In contrast, nine different scrapie isolates were obtained from SSBP/1 and its subpassages in sheep, goats, and mice, which differed with regard to their respective incubation period in the different mouse lines VM, C57BL, RIII and CD1-white Swiss (Table 12.1).

The incubation period of different prion strains is determined after serial passages in different mouse lines and is defined as the time interval between infection and onset of the clinical signs of the disease. Due to the species barrier, the incubation period in a new species can only be exactly determined after two or three passages. If the prion strain is not highly variable, the incubation period in one species remains extremely consistent and reproducible (see Chapter 50) – at least as long as the transmission conditions do not change. A prolongation of the incubation period can occur with the first transmission of a strain from one mouse line to another. In contrast, mouse BSE strain 301V shows a significant shortening (to approximately 120 days) of the incubation period after serial mouse-to-mouse passage in VM mice. This is the shortest incubation period ever observed for a TSE disease in nontransgenic mice [27]. Therefore, defined incubation periods can be indicated for a large number of murine scrapie models. In the majority of cases, the standard deviation from the mean incubation periods is less than 2 % [1]. Therefore, the incubation period is considered to be both a simple and important criterion for the characterization of a prion strain.

Table 12.1: Incubation time of different TSE strains in C57BL and VM mice as well as in hamsters. Source: [24–26].

TSE strains	Incubation time (days)	
	C57BL mice	VM mice Primarily isolated in C57BL mice (*Sinc*^{s7})
ME7	168	346
D	360	620
→ 7D	180	360
22C	170	447
22L	157	210
79A	161	309
139A	166	208
301C	198	345
		Primarily isolated in VM mice (*Sinc*^{p7})
87V	>700	290
22A	474	199
→22F[1]	295	–
22H	191	327
79V	approx. 240	approx. 280
111A	>600	>750
201V	203	308
301V	274	116

→ mutant variants

[1] The 22F strain developed from 22A after having been passaged in C57BL mice.

Hamster-passaged TSE strains

TSE strains in hamster	Incubation time (days) After adaptation
263K[2]	65
79AH	175
79VH	64
139H[2]	122
22LH	128
22AH	151
22CH	140
ME7H	251
TME Hyper[2]	65
TME Drowsy[2]	168

[2] Currently known TME strains and the scrapie strain. 263K can only be replicated in hamster, not in mice. The 139H strain (from 139A) is another hamster-adapted scrapie strain.

In order to compare the incubation periods of individual strains in mice, all animals must be infected according to the same protocol. Different routes of infection and inoculation doses cause varying results. Thus, as sort of general rule, 30 µl of a 10 % brain homogenate of an animal suffering from scrapie is inoculated i.c. into a group of 10–25 mice. Scrapie strains which replicate especially well in the brain of a donor mouse lead to shorter incubation periods, as the infection dose is especially high. In contrast, strains that only cause low pathogen titers in the brain appear to have longer incubation periods, independent of their specific strain characteristics. However, this effect could only be ruled out by using the same number of inoculated prions in all trials. Unfortunately, the prion concentration in the brain of diseased mice is only known for a small number of scrapie strains.

The incubation periods of prion strains are not only strain specific, but also species specific, i. e., they depend on innate, genetically determined factors of the recipient species. Evidence of this was first found in rodent models for scrapie. Here, defined strains had different incubation periods depending on the mouse line used. Based on this observation, it was assumed that the mouse lines possess a scrapie incubation period (*sinc*) gene [8; 28; 29]. Thus, in homozygous s7s7 (s = short; 7 = ME7) mouse lines such as RIII or C57BL, the ME7 strain has a short incubation period, whereas in homozygous p7p7 (p = prolonged) mouse lines such as VM, it has a long incubation period. The situation, however, might be just the opposite in other prion strains such as 22A. Strain 22A has a short incubation period in p7p7 mice and a long incubation period in s7s7 mice. According to the present state of knowledge, the *sinc* gene corresponds to the PrP gene. Homozygous s7s7 mice are also known as Prn^a and homozygous p7p7 mice are known as Prn^b. Consequently, PrP is structured differently in s7s7 mice and in p7p7 mice: it differs in 2 of the 254 amino acids [30–32]. In Prn^a mice, leucine instead of phenylalanine is found at amino acid position 108 and threonine instead of valine is found at position 189. Analogous to the *sinc* gene in mice, the existence of what is known as a scrapie incubation period (*sip*) gene in sheep has also been observed [33]. We also now know that *sip* genes code for different PrP amino acid sequences. Also, in the PrP gene of sheep, different amino acids are known at at least nine positions. This is known as amino acid polymorphism (see Chapter 9). Three of these polymorphisms are of major importance for incubation time and progression of disease and can lead to 12 possible combinations. However, only five alleles seem to appear with any frequency. These five alleles are VRQ, ARH, AHQ, ARQ, and ARR. Their combination leads to 15 genotypes that are commonly found in sheep. The incidence of these genotypes is different in distinct breeds and these 15 genotypes clearly differ in their susceptibility to scrapie. The ARR/ARR genotype seems to be highly resistant, whereas the VRQ/VRQ genotype seems to be highly susceptible to scrapie. Therefore, these genotypes provide a basis for risk classification in the establishment of scrapie-resistance breeding programs (see Chapters 18; 56).

Published data suggest that there are also PrP gene polymorphisms in cattle, in addition to the well-known octapeptide repeat polymorphism [34].

Although the influence of PrP gene polymorphisms on incubation time is well known, other evidence indicates that PrP amino acid differences are not the only genetic influence. There may be other factors, such as additional genetic loci, that contribute to variation in incubation period when inbred mice with the same PrP genotype reveal major differences in incubation times in response to a defined prion strain [35].

12.3.3 Transmissibility

The experimental transmissibility of prion diseases is described in Chapter 43. It is well known

Table 12.2: Experimental transmission of scrapie to different mouse strains and other species.

Strains		Passages in	TSE strains
A	Primarily isolated in $Sinc^{s7}$ mice		
	22C (uncloned)	$Sinc^{p7}$	22H
	22C (cloned)	$Sinc^{p7}$	22C
	ME7 (cloned)	$Sinc^{p7}$	ME7
	139A (uncloned)	$Sinc^{p7}$	139A
	87A	$Sinc^{s7}$	87A + ME7
B	Primarily isolated in $Sinc^{s7}$ mice		
	22A	$Sinc^{s7}$	22F
C	Cloned, mouse-passaged strains		
	ME7	Hamster → mouse	ME7
	22A	Hamster → mouse	22A
	22C	Hamster → mouse	22C-H/M (not 22C)
	139A	Hamster → mouse	139A-H/M (not 139A)
	139A	Rat → mouse	139A

that the transmissibility depends on both, the donor and the recipient species (Tab. 12.2). In addition, there are differences in the stability of transmissibility on the level of the prion strain, for example the mouse-adapted scrapie strain 139A can be transmitted from mice to hamsters and retransmitted back to mice while the scrapie strain 263K is no longer virulent for mice after several passages in hamsters [10].

12.3.4 Histopathological lesion profiles

In the 1960s and 1970s, the scrapie strains, which had been differentiated based on their incubation periods, were transmitted systematically to experimental animals (mice) in order to develop criteria for the evaluation of the histopathological changes found in the brain of the diseased host [3]. The prion strains differed with regard to their localization, their severity, and the kind of spongiform changes as well as the kind, amount, and distribution of plaques within the brain. The histopathological changes were independent of the infection route and dose [36].

The criteria listed in the following text constitute the basis of a semiquantitative method to differentiate the individual prion strains. This involves the histological examination and evaluation of the intensity, quality, and localization of spongiform lesions in nine defined regions of the gray and in three defined regions of the white matter [3]. The degree of vacuolation is expressed on a scale from 0 to 5 and shown graphically in the form of lesion profiles to permit a clear and reproducible differenti-

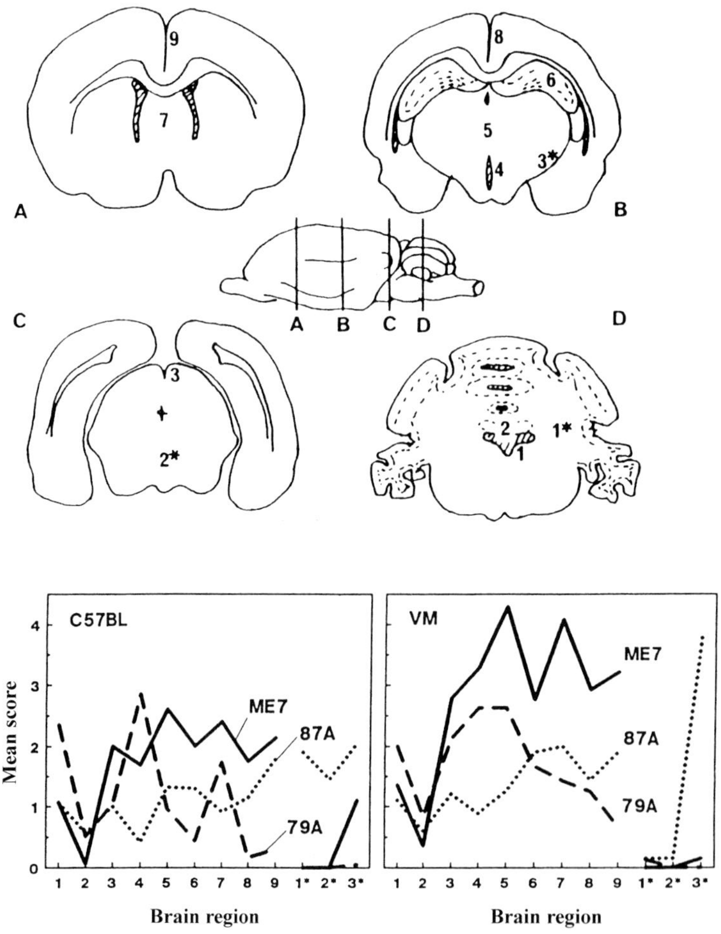

Fig. 12.1: Schematic presentation of the mouse brain sections and target areas used for lesion profile scoring (above). Lesion profiles of three scrapie strains (ME7, 87V and 79A) in C57BL and VM mice (below). Figure adapted from [4].

ation of the prion strains [7] (Fig. 12.1). In addition to these lesion profiles, other criteria such as presence and distribution of amyloid-like plaques [37], degeneration of the neurons in the hippocampus [38], and loss of retinal photoreceptor cells [39] are used to describe prion strains. The pattern of PrP^{Sc} deposition, PrP^{Sc} labeling of astrocytes, localization and level of astrocytosis as well as the exact localization of the vacuolation – intracellularly in the nerve cells or extracellularly in the neuropil – are also taken into account.

Although the distribution of the spongiform changes in the brain (lesion profile) is mainly influenced by the prion strain, the recipient mouse line with its individual *sinc*-gene also has an influence [7; 40]. For example, the lesion profile of strain 22A in p7p7 mice (e. g., VM mice) differs considerably from that in s7s7 mice (e. g., C57BL mice) in regions 5–9. Differences between the lesion profiles of different mouse lines can also be found for scrapie strains ME7, 22L, 22C, 139A, and 79A [7].

In order to clarify whether a causal association between scrapie, BSE, and vCJD can be established, the formation of vacuoles after transmission of infected ovine, bovine, or vCJD brains to RIII mice and other inbred mice was analyzed (see below). Histological examinations of BSE and vCJD inoculated mice revealed similar degenerative changes of the brain, which differed clearly from the lesion profile of the examined scrapie strains to date; they also differed clearly from the lesion profile of sCJD [4; 27; 41; 42]. This approach confirms a link between BSE and vCJD (see Chapter 50).

12.3.5 Inactivation behavior

In some cases, prion strains differ considerably with regard to their resistance to chemical and physical inactivation methods. For example, scrapie strain 22A has a significantly higher heat resistance in porous load autoclaving compared to the other strains [43]. In 1994, David Taylor and his colleagues showed that scrapie strains 263K and ME7 are more susceptible to sodium hydroxide (see Chapter 44) than BSE [44].

12.3.6 PK resistance and cleavage-site of PrP^{Sc}

Originally, strains were characterized and defined largely by biological parameters. Since the 1980s, the central role of the PrP in pathogenesis has been established [45–49]. The transformation of the cellular prion protein (PrP^{C}) to its pathological isoform (PrP^{Sc}) is considered to be the key event (see Chapter 8) in TSE. PrP^{Sc} has a strongly increased tendency to aggregate [50; 51] compared to that of PrP^{C} and is partially resistant to proteinase K cleavage [45; 52; 53]. While PrP^{C} is cleaved rapidly and completely by proteinase K, only part of the PrP^{Sc} amino acid chain is cleaved at the N-terminus under the same conditions.

Different strains differ significantly in their resistance to proteolytic degradation [46; 54; 55]. The PrP^{Sc} of scrapie strain 139A, for example, proved to be considerably more sensitive to proteinase K than the PrP^{Sc} of the 87V and ME7 scrapie strains, whereas BSE prions are more sensitive than all known scrapie strains [56]. In addition, PrP^{Sc} of atypical scrapie strains (see Chapter 54) seems to be more sensitive to proteinase K than that of classical scrapie strains.

Moreover, it is not only the general extent of resistance but also the exact location of the PK cleavage site by which PrP^{Sc} derived from infections with different prion strains can be distinguished. PrP^{Sc} shows three characteristic protein bands on Western blots which result from the different degree of usage of the glycosylation sites, i. e., there may be two, one or no glycosylation site in use (see below) [13; 57–60]. After PK proteolysis, the non-glycosylated N-terminally truncated fragment of PrP^{Sc} does not have the same molecular size (approx. 17–21 kDa) in all prion strains [54; 61]. This effect can be made even more visible when all PrP^{Sc} is artificially deglycosylated enzymatically. The nonuniform molecular sizes are caused by more or less amino terminal truncations of PrP^{Sc} by PK in the different prion strains. Moreover, even the PK concentration can to a certain degree have an effect on the penultimate cleavage site. With a lower concentration of PK, the terminal amino acid of BSE PrP^{Sc} was at amino acid 96 (N), while that of scrapie PrP^{Sc} was at amino acid 81 or 86 (both G). However, with an increase in the PK concentration, the N-terminus of PrP^{Sc} shifts towards amino acid 89 (G) in most scrapie strains but remains stable in BSE PrP^{Sc} [62]. The molecular mass of the fragment of strain 87V is lower by approximately 1 kDa. It has been shown that the nonglycosylated band of PrP^{Sc} from BSE in cattle or experimental BSE in sheep shows lower values than the nonglycosylated band of classical scrapie cases [56; 63; 64]. Ovine BSE gave even smaller molecular sizes than BSE in cattle. These particular characteristics can be used to discriminate between BSE and scrapie in sheep. For this purpose, monoclonal antibodies are used, which are directed against epitopes that are located directly at or in the vicinity of the proteinase K cleavage site of the PrP^{Sc} fragment (depending on the particular prion strain). For example one such prototypic antibody designated mab P4 targets exactly the overhanging end of scrapie PrP^{Sc}, but cannot detect the further truncated BSE PrP^{Sc}, as shown in Figure 12.2 [56; 63; 64].

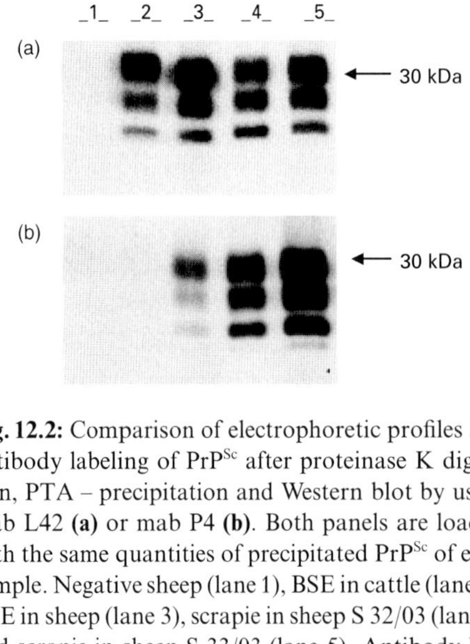

Fig. 12.2: Comparison of electrophoretic profiles and antibody labeling of PrP^{Sc} after proteinase K digestion, PTA – precipitation and Western blot by using mab L42 **(a)** or mab P4 **(b)**. Both panels are loaded with the same quantities of precipitated PrP^{Sc} of each sample. Negative sheep (lane 1), BSE in cattle (lane 2), BSE in sheep (lane 3), scrapie in sheep S 32/03 (lane 4) and scrapie in sheep S 33/03 (lane 5). Antibody L42 detected all BSE and scrapie samples **(a)**. Antibody P4 produced no signal with cattle BSE and only a weak signal with ovine BSE, but an even stronger signal with all scrapie samples than produced with antibody L42 **(b)**. Figure by M.H. Groschup.

12.3.7 Glycoprofile of PrP^{Sc}

As already indicated, characteristic differences in the glycosylation of PrP^{Sc} are also found between the different prion strains. On immunoblots this can be deduced from the different intensities of the nonglycosylated, the mono- and the diglycosylated PrP^{Sc} bands [54; 55; 65–67]. In order to define these differences, the term "glycoprofile" has come into use.

In the ME7, 87V, and 236K strains, the diglycosylated band represents the main fraction of the total intensity, whereas in the Chandler and 79A scrapie strains the monoglycosylated band displays maximum intensity. The genotype of the recipient may be of minor significance for the glycoprofile of the strains [68], as identical glycosylation patterns were also observed after passages in mice with different *sinc* genes [13; 66]. In diseased experimental animals, mouse-passaged BSE isolates and isolates of the vCJD induced the same characteristic glycoprofile of PrP^{Sc}. It was this similarity, together with other factors, that supported the conclusion that BSE and vCJD are caused by the same pathogen [65] (see Chapter 53).

Surprisingly, a diversity of prion strains can also be detected after passage of a given field isolate in mice (see below). For example, the ME7 and 87V strains were derived from the same sheep suffering from scrapie [5], but differ significantly with regard to their glycoprofile. Vice versa BSE strains, 301C and 301V, which were selected after passage in genotypically different mice and which have remarkably different biological characteristics, can be distinguished from all known scrapie strains based on their common glycoprofile. BSE PrP^{Sc} is particularly unique since the diglycosylated band has the maximum intensity, and the nonglycosylated band is particularly weak in almost all species infected to date [66; 67].

These features of prion strains can be used to discriminate between BSE and scrapie in sheep (Fig. 12.3). Interestingly, recent data suggest that there is one scrapie strain, CH 1641,

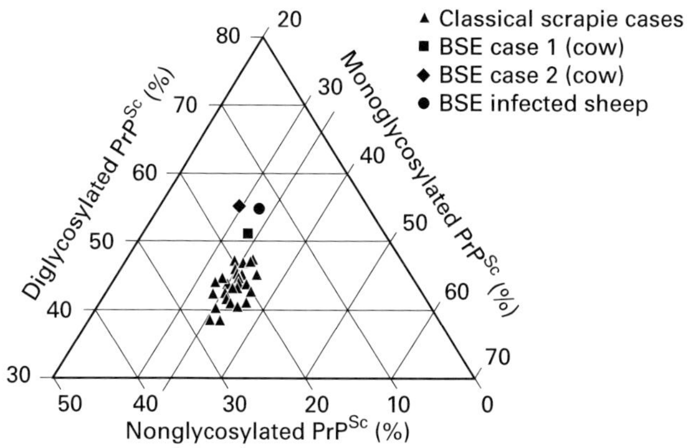

Fig. 12.3: PrP^{Sc} glycosylation of 33 classical scrapie cases, two BSE cases and one case of BSE in sheep, depicted in the form of an equilateral triangle. The percentage of the diglycosylated band is reflected on the upper left side of the triangle, that of the monoglycosylated bands at the upper right side and that of the nonglycosylated fraction at the bottom. Figure by M. H. Groschup.

that has a very similar, but not identical, glycoprofile to BSE [69]. In addition, there may be more than one BSE strain [70–72] (see Chapters 1.9.2; 19.2; atypical BSE).

12.4 Adaptation and selection

Stable scrapie strains are able to change their characteristics after transmission to other species (e. g., mice), while instable strains can change their characteristics also in their original host. Due to the complex indirect interactions between the genotype of the recipient species and the parameters determining the strain characteristics, the mechanisms involved remain unknown. In the following paragraph, two possible explanations are given:

1) Adaptation: One possibility is that the strain characteristics of TSE pathogens are determined on the molecular level in a currently unknown way. A sudden or continuous modification of these molecular characteristics due to adaptation might lead to the development of new strains. It has been suggested that adaptation may include the conversion of host PrP^{C} to host PrP^{Sc} by a donor PrP^{Sc} molecule having a different primary structure and the formation of multiple PrP^{Sc} conformations from a single donor PrP^{Sc} conformation that is responsible for generating new strains [73].

2) Selection: Often, two distinguishable scrapie strains have been identified from field isolates of animals suffering from scrapie, if different recipient mouse lines are chosen. An explanation for this phenomenon might be that the donor animals were indeed infected with a mixture of strains from which one strain was eventually selected. Recent data provide the first evidence for the selection of TSE strains at the molecular level [73]. During the transmission of TME into Syrian hamsters, the strains were selected according to their dose and rate of agent replication. It is also possible that adaptation and selection processes occur in the prion strain development. In certain strains, modifications occur more easily and more drastically. It is possible that adaptation gives them a selective advantage if they are transmitted to a suitable recipient species, i. e., genotype.

12.4.1 Murine scrapie strain class I

In mice, scrapie strains can be divided into three different classes based on the stability of their biological properties. Class I strains are completely stable, even if passaged in mouse lines of different genotypes. Class II strains are stable in the mouse genotype used for their isolation, but gradually modify their features during successive passages in other genotypes. In contrast, class III strains are unstable, even in the mouse genotype used for their isolation, and sometimes change their biological characteristics during just one mouse passage.

The ME7 strain is the prototype of a class I strain [1; 5; 11]. Its biological characteristics do not change, even during passage in different mouse genotypes or in other species such as sheep or hamsters. After retransmission to the original species, the incubation period returns to that of the original pathogen and stabilizes. The retransmitted ME7 strain also produces the same histopathological lesion profile as the original strain.

12.4.2 Murine scrapie strain class II

A typical class II strain is scrapie strain 22A. 22A shows stable biological characteristics as long as it is passaged in VM mice, the mouse line used for its isolation. However, if 22A is passaged in C57BL mice, for example, the incubation periods and lesion profiles may slowly change. After four to five passages, the strain develops stable characteristics again and is then designated 22F [5]. 22F has almost the same characteristics as strain 22C, which was also isolated from SSBP/1, but by means of C57BL mice. This slow transformation from 22A to 22F/22C might be due to a combination of adaptation and selection.

12.4.3 Murine scrapie strain class III

Strain 87A is regarded as the prototype of a class III strain. Isolates with the characteristics of 87A were derived from six different sheep from four different breeds. If 87A is passaged in C57BL mice in low concentrations, it shows stable characteristics. However, if the concentration of the prion strain used for inoculation within the same mouse genotype is increased, a sudden reduction in the incubation period and stabilization on the new level may occur. This newly developed strain was designated 7D (Table 12.1); it produces a different histopathological profile from 87A. In contrast, 7D always seems to be identical with the ME7 scrapie strain [74]. Thus, it may be assumed that ME7 is a modified strain and derived from 87A.

The question has arisen whether scrapie strains, which have been repeatedly passaged and selected in rodents, remain infectious in sheep. In 1988, Foster and Dickinson showed that the mouse scrapie strains 79A, 22A, 22C, and ME7 are still highly infectious in sheep [75].

12.5 Known prion strains in different species

In order to be able to evaluate the significance of the prion strains, their respective history of isolation or generation should be taken into consideration. Indeed, most strains can be traced back to a small number of sources, while the exact origin of a number of strains remains unclear.

12.5.1 Human prion strains

For a long time, human prion diseases could only very infrequently be transmitted to mice, and only a small number of isolates were obtained. A breakthrough was finally achieved in the early 1990s when improved inoculation protocols and transgenic mouse models (see Chapter 10) were used. At this time the polymorphism at position 129 of human PrP was discovered. Heterozygous Met/Val strongly reduces prion susceptibility, whereas homozygous Met/Met or Val/Val strongly increases susceptibility. Finally, examinations of proteinase resistance and cleavage site, glycoprofile, and genotype of the human PrP^{Sc} clearly indicate the presence of distinct types of CJD [65; 76]. These glycotypes are colloquially referred to as strains although the before-mentioned characteristics have never been proven.

The greater majority of the CJD represent the classic CJD phenotype, which is characterized by a PrP^{Sc} type 1 (higher molecular size after PK cleavage), while a smaller number of the CJD cases are associated with PrP^{Sc} type 2 (lower molecular size after PK cleavage). PrP^{Sc} type 1 and 2 consist largely of monoglycosylated prion protein molecules. Based on the clinical and neuropathological findings in these patients and the respective PrP genotype at codon 129 (homozygocity encoding for either valine or methionine, heterozygosity), these two types can be further subdivided. Most recent studies have shown that PrP^{Sc} type 1 and type 2 are heterogeneous species which can be further divided into five molecular subtypes that fit the current histopathological classification of sCJD variants [77]. PrP^{Sc} (representing type 3) is different from type 2 PrP^{Sc} in that the diglycosylated moiety is the most abundantly stained band on immunoblots, which is similar to the BSE PrP^{Sc} (see above) [65].

12.5.2 Scrapie strains from small ruminants

From the 1940s to the 1970s, research on scrapie was mainly conducted in Great Britain. For more than 30 years, virtually all the material used by British scientists to infect sheep and goats came from the scrapie subpassaged brain pool, (known as SSBP/1-material) and its subpassages [78]. SSBP/1 was a pool of brain homogenates of three different sheep suffering from scrapie: one Cheviot sheep and two Cheviot × Border-Leicester crossbred animals. This was quite a mix of starting material, containing at least three potentially different strains of scrapie. Subsequently, SSBP/1 was passaged in sheep and goats up to 20 times. Today, little is known (and was probably the case at that time) about the exact scrapie status of these animals, which were often purchased from another sheep flock. Considering the high incidence of scrapie in Great Britain – in 1990, a questionnaire survey found that 30% of the shepherds had seen sheep with clinical signs reminiscent of scrapie in their flock [79] – it cannot be excluded that subclinically scrapie-infected animals were used in the studies. These animals then developed scrapie because of a natural rather than an experimental infection. Therefore, the number of SSBP/1 (/1 = number of passages) derived scrapie strains may have increased artificially several times over the years. For example, it is very likely that the so-called drowsy goat strain resulted from a contamination with field isolates [78].

Therefore, it is not surprising that to date eight different murine scrapie strains have been isolated from sheep and goats infected with SSBP/1 or its subpassages: 22C, 22A, 22M, 22L, 80V, 79A, 79V, and 139A. Today, strain 139A is also used under the synonymous designation Chandler (named after its discoverer) as well as RML (RML: subpassage replicated by R. Marsh at the Rock Mountain Laboratory, Montana, and then passed on to S. Prusiner). Due to their likely contamination, these isolates are now divided into the SSBP/1 family (22C, 22A, 22M, 22L) in sheep, the hamster-passaged isolates 302K and 263K, the scratching strain (22C, 80V), and the drowsy strain (79A, 79V, 139A) in goats, shown in Figure 12.4. There are three hypotheses why the drowsy strain differs significantly from the scratching strain [78]:

1. The drowsy strains are a contamination with field isolates caused by using a goat already infected with scrapie.
2. The strains 79A and 139A were contained in the SSBP/1 homogenate, but in contrast to 22A and 22C only in low doses. If this assumption is true, no distinct formation of vacuoles should occur in the white substance during the first C57BL mouse passage.
3. The pathogens changed due to an adaptation to certain goats of the drowsy strain.

Some of the scrapie strains were already known during the first transmissions of scrapie from sheep and goat to mice. Among these were strains 79A and 79V, which were isolated from the same starting material by inoculation in s7s7 mice (C57BL) or p7p7 mice (VM). These showed stable characteristics after the seventh passage. As a rule, however, the strains used today were developed by multiple passages in mice. Thus, the strain designated 139A (Chandler, RML) was passaged in mice at least 30 times, scrapie strain 79A at least 9 times, and strains 22A, 22C, 22L, 22H, and 22F at least 25 times (Fig. 12.4). In this process, the strains were cloned frequently and repeatedly, i. e., successive infection trials in mice were carried out using pathogen concentrations containing only one infectious unit, e. g., within the framework of an endpoint titration.

Furthermore, transmissions and successive passages from SSBP/1 or drowsy goat isolates to animal species other than mice were carried out. By adaptation and/or selection, further defined strains were obtained such as the mouse-adapted strains 302K and 431K after 17 subpassages (for each of the two) in rats and after two (302K) or five (431K) passages in hamsters. The hamster strain 263K used in many laboratories today was also generated in these transmission studies. Similar to strain 139A in mice,

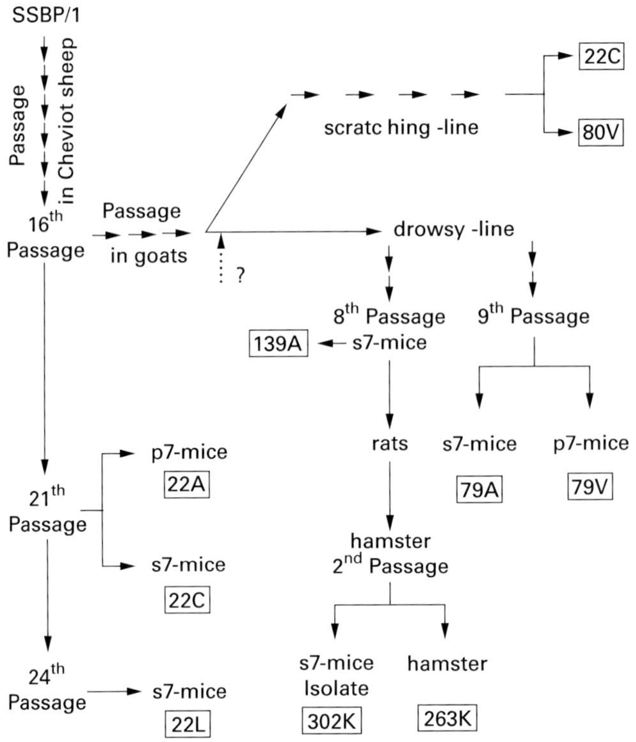

Fig. 12.4: Historical overview of the numerous passages of the scrapie pool SSBP/1 (scrapie sheep brain pool) derived from several sheep brain homogenates in different animal lines (Figure adapted from [74]). Cheviot sheep passages led to the mouse-adapted scrapie strains 22A, 22C and 22L. Goat passages led to two lines, namely the "scratching" line, which led to the strains 22C and 80V, and the "drowsy" line leading to the mouse scrapie strains 139A, 79A and 79V, and the hamster-passaged scrapie strain 263K, and mouse scrapie strain 302K.
In contrast to the "scratching" line, the mouse-adapted scrapie strains 139A, 79A and 79V are characterized by large vacuoles in the white substance of C57BL mice. There are three hypotheses with which the enormous difference between "drowsy" and "scratching" line can be explained. (1) The "drowsy" strains are the result of a field contamination resulting from the use of a scrapie-incubating goat. (2) The 79A and 139A strains were present in SSBP/1 homogenate, but, in contrast to 22A or 22C, only in small doses. This assumption would lead to the clear suppression of vacuolation in the white substance in the first C57BL mouse passages. (3) The pathogens underwent changes due to the goat adaptation to initiate the drowsy line.

strain 263K is designated Chandler scrapie in hamsters in older publications (Table 12.3).

Since the 1970s, new isolates have been obtained from sheep suffering from scrapie. These have been passaged in different mouse lines and characterized in mouse typing experiments. From the majority of British scrapie cases, pathogens with the phenotypical characteristics of strain ME7 were obtained, which was probably derived from 87A. Examples are isolates 58A and 201C [41]. Thus, at least in Great Britain, infections with strain ME7 seem to represent the majority of endemic scrapie cases. In addition, however, strains were found that differed completely from ME7 and the SSBP/1 or drowsy goat isolates. Among these are the two phenotypically distinguishable strains 87A and 87V, which were isolated from the brain of a Cheviot × Border-Leicester crossbred animal suffering from scrapie (while interestingly, strain 7D was also isolated from the latter) using mouse lines VM and C57BL [74]. The charac-

Table 12.3: Scrapie strains which were isolated from diseased sheep and goats.

A	Isolated strains	Scrapie origin	Number
	87A, 87V, ME7	Dorset Down	1
	87A, 87V, ME7	Suffolk	5
	87A, 87V, ME7	Cheviot × Border-Leicester	1
	87A, 87V, ME7	Blackface[1)]	2
	87A, 87V, ME7	Southdown	1
	87A, 87V, ME7	Welsh Mountain	1
	87A, 87V, ME7	Cheviot[2)]	1
	22A, 177A	Icelandic sheep	3
	111A, ME7	Colbred × Suffolk	1
	124A	goats[2)]	1
B	**Untyped isolates**	**Scrapie origin**	**Number**
	–	Cheviot	1
	–	Border-Leicester	1
C	**No first passage in mice**	**Scrapie origin**	**Number**
	–	Swaledale	1
	–	Blackface[1)]	1
	–	Dorset Horn	1
	–	Cheviot	2
	–	Icelandic sheep	2
	–	Finnish land race	1

[1)] Animals had been in contact with diseased Suffolk sheep. [2)] Preliminary strain identity.

teristics of other field isolates mentioned in the literature designated 31A, 51C, 125A, 138A, and 153A correspond to strain 87A [5]. Isolate 124A derived from goats has thus far not been characterized in detail, but it also seems to be very similar to strain 87A.

Other known scrapie strains from Great Britain are strains 201V and 111A derived from sheep. In addition to the British scrapie isolates, two Icelandic scrapie isolates were characterized. While one of them corresponded with strain 22A, the second one, designated 177A, had a phenotype similar to that of strain 111A.

Additionally, in 1991, Carp and Callahan examined scrapie isolates from five diseased American Suffolk sheep by inoculating VM and C57BL mice [80]. They obtained five pairs of isolates designated C600 and C601, C602 and C603, C604 and C605, C606 and C607, and C608 and C609, all of which partially differed with regard to their symptoms and/or incubation periods.

Interestingly, in 1999 an experimental scrapie isolate of natural scrapie, CH1641, was found. The glycoprofile and PK cleavage site of the nonglycosylated PrP^{Sc} fragment were similar but not identical to BSE; it differed from BSE in that it did not transmit easily to mice, whereas BSE transmits without difficulty [69].

12.5.3 The BSE strains

Since the discovery of BSE in Great Britain in the mid 1980s, the central question has been whether it was originally caused by transmission of a scrapie pathogen to cattle, or whether it was caused by an independent pathogen (see Chapter 19.8–9). This question remains unanswered. Due to the complex indirect interaction between the strain and the PrP genotype of the recipient species, it is unlikely that this question will ever be answered conclusively. Within the framework of studies carried out in order to answer this question, M. Bruce and her colleagues transmitted BSE material to different mouse lines (RIII, C57BL, VM, etc.) shown in Figure 12.5. Surprisingly, the incubation periods for all BSE field isolates were highly con-

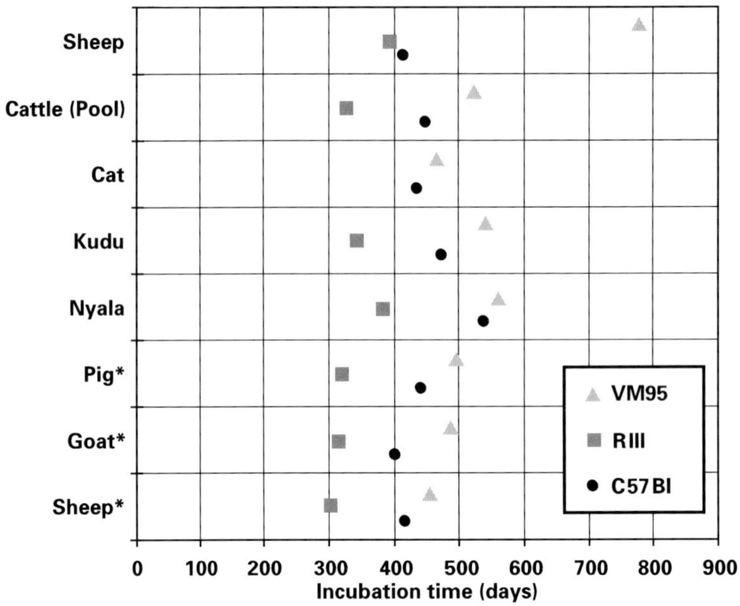

Fig. 12.5: Incubation time of C57BL (●), RIII (■) and VM95 mice (▲) after intracerebral injection of a 1 % brain homogenate derived from a scrapie-infected sheep, BSE-infected cattle (mean values of a pool of seven cattle), an FSE-infected cat, a BSE-infected kudu and nyala and a vCJD-infected person. *BSE brain homogenate was inoculated into pigs, goats, and sheep and subsequently passaged in mice. Figure adapted from [38].

sistent in the respective mouse lines (Fig. 12.5). The histopathology of all BSE field isolates from cattle is largely uniform and shows a characteristic lesion profile (Fig. 12.6 bottom), which differs from all scrapie strains described so far (Fig. 12.6 top). BSE infected exotic ruminants such as kudu as well as cats suffering from FSE were also analyzed, yielding the same results (see Chapters 20; 23). Neuropathology and lesion profiles remain similar in different mouse lines after the first passage. The results obtained in these studies [41] are described in detail in Chapter 50. All of these observations indicate that a single BSE strain was in circulation in the field and responsible for the BSE epidemic in Europe. However, during further passages in genotypically highly different mouse lines, the selection of two distinct BSE strains was observed. In s7s7 mouse passages (C57BL) known as strain 301C and in p7p7 mouse lines (VM), strain 301V developed. Compared to strain 301C, strain 301V shows significantly shorter incubation periods in p7p7 mice and slightly longer incubation periods in s7s7 mice. It is of note that the histopathological lesion profiles and the glycosylation patterns of these two strains of laboratory isolates are identical.

Recent data from different countries suggest that more BSE strains may exist in cattle [70; 71; 81]: atypical BSE phenotypes have recently been described in France and Germany (also designated H-type) and Italy and Germany (also designated L-type or BASE; see Chapter 19.2.2). Both types were first found in older animals, one having a significantly higher molecular mass, but a conventional glycopattern of the proteinase K treated abnormal prion protein (PrP^{Sc}) (H-type) and the other being characterized by a slightly lower molecular mass and a distinctly different glycopattern (L-type). These deviant BSE types indicate the possible existence of sporadic BSE cases in bovines. At the time of writing, the infectious nature of the German atypical BSE case of the L-type has been verified, as it was successfully transmitted into bovinized transgenic mice [81]. Similar to kuru

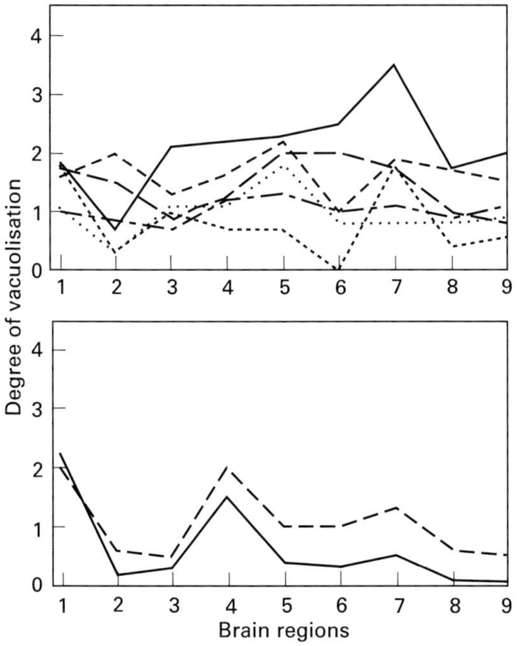

Fig. 12.6: Comparison of the lesion profiles in RIII mice after intracerebral injection. (Above) Six different sheep scrapie isolates are presented. (Below) Presentation of British BSE field isolates as pool (-) and one Swiss BSE field isolate (- - -). Figure by M. H. Groschup.

which may have originated from the cannibalistic recycling of a case of sporadic Creutzfeldt–Jakob disease one century or so ago, the intraspecies recycling of a sporadic bovine BSE case could have also caused the BSE epidemic in cattle in the UK, apart from the widely postulated interspecies transmission of small ruminant scrapie to bovines.

12.5.4 Prion strains of other animals

After passages to Syrian hamsters [19; 20], two distinguishable strains were isolated from TME-infected mink, which were designated hyper (HY) and drowsy (DY) strains [21]. They differed in the electrophoretic mobility of the PK-resistant, nonglycosylated protein fragment: for strain HY the fragment was approximately 21 kDa, for strain DY approximately 19 kDa [19]. In the field, only one TME strain has been detected (see Chapter 22.2).

Chronic wasting disease (CWD) is a prion disease of Rocky Mountain elk, mule deer, and white-tailed deer (Table 1.3). In the sequence of the PrP gene of elk, one amino acid exchange (Met to Leu at codon 132) was found. Elk that are homozygous Met/Met seem to be overrepresented in CWD-infected elk when compared to groups of uninfected elk [82–84]. Published data suggest that elk and deer are possibly infected by different CWD strains [85–88].

12.6 Explanatory approaches with regard to the development and existence of prion strains

Based on the indicated differentiation criteria, more than twenty-four different strains have been identified and differentiated thus far. This diversity is used as an argument against the protein-only hypothesis (see Chapter 4). The existence of strains with very different phenotypical characteristics might indicate that the pathogen consists of a second strain-specific component in addition to the PrP.

As the amino acid sequence of PrP is determined in the genome of the host species, it is hard to imagine how strain characteristics might be coded according to the prion model. The following hypotheses have been proposed. The order in which they are listed is not significant, as none have been conclusively proven to date.

1. A regulatory pathogen DNA or a viral nucleic acid might be responsible for the difference in characteristics (see Chapter 4).
2. The binding of another regulatory factor, e. g., a protein X, might lead to the development of strain characteristics.
3. The glycosylation of PrP^{Sc} might allow differentiation of strains, specifically due to an involvement of different brain cells (neurons) in its formation, i. e., target-cell theory.
4. The folding of PrP^{Sc} might allow the differentiation of strains in a specific manner, i. e., strain-specific conformation.

As the search for viral nucleic acids has thus far been unsuccessful, S. Prusiner and other scientists have postulated that the existence of different strains might be due to the existence of several different PrP^{Sc} conformations. One should keep in mind, that in this case not only the tertiary structure of PrP^{Sc} but also the quaternary structure of the highly aggregated PrP^{Sc} molecules like SAF (Fig. 1.5b), prion rods, or amyloid plaques have to be considered. If this is the case, one can assume that a specific conformation can transform PrP^{C} highly efficiently into the PrP^{Sc} isoform, resulting in a short incubation period. Another PrP^{Sc} conformation might not be as efficient, resulting in a longer incubation period. Different conformations of PrP^{Sc} might also have an affinity for different brain sections. This could possibly also explain the difference in the clinical signs or symptoms. However, although this is a very attractive concept, no proof exists for this hypothesis of prion strains to date.

References

[1] Adams DH. Relation between Carp (multiple-sclerosis associated) agent and multiple sclerosis. Lancet 1976; 1:444–446.

[2] Pattison IH and Millson GC. Scrapie produced experimentally in goats with special reference to the clinical syndrome. J Comp Pathol 1961; 71:101–108.

[3] Fraser H and Dickinson AG. The sequential development of the brain lesions of scrapie in three strains of mice. J Comp Pathol 1968; 78:301–311.

[4] Bruce ME. Strain typing studies of scrapie and BSE. In: Baker HF, Ridley RM, editors. Methods in molecular medicine: Prion diseases. Humana Press, Totowa, NJ, 1996: 223–238.

[5] Bruce ME and Dickinson AG. Biological stability of different classes of scrapie agent. In: Prusiner SB, Hadlow WJ, editors. Slow transmissible diseases of the nervous system. Academic Press, New York, Vol 2, 1979: 71–86.

[6] Mori S, Hamada C, Kumanishi T, et al. A Creutzfeldt–Jakob disease agent (Echigo-1 strain) recovered from brain tissue showing the 'panencephalopathic type' disease. Neurology 1989; 39:1337–1342.

[7] Bruce ME, McConnell I, Fraser H, et al. The disease characteristics of different strains of scrapie in Sinc congenic mouse lines: implications for the nature of the agent and host control of pathogenesis. J Gen Virol 1991; 72:595–603.

[8] Dickinson AG and Meikle VM. Host-genotype and agent effects in scrapie incubation: change in allelic interaction with different strains of agent. Mol Gen Genet 1971; 112:73–79.

[9] Kimberlin RH, Cole S, Walker C. Transmissible mink encephalopathy (TME) in Chinese hamsters: identification of two strains of TME and comparisons with scrapie. Neuropathol Appl Neurobiol 1986; 12:197–206.

[10] Kimberlin RH and Walker C. Evidence that the transmission of one source of scrapie agent to hamsters involves separation of agent strains from a mixture. J Gen Virol 1978; 39:487–496.

[11] DeArmond SJ, Mobley WC, DeMott DL, et al. Changes in the localization of brain prion proteins during scrapie infection (published erratum appears in Neurology 1987 Nov; 37(11):1770). Neurology 1987; 37:1271–1280.

[12] Anonym. Obese protein slims mice (news). Science 1995; 269:475–476, 540, 543, 546.

[13] Kascsak RJ, Rubenstein R, Merz PA, et al. Immunological comparison of scrapie-associated fibrils isolated from animals infected with four different scrapie strains. J Virol 1986; 59:676–683.

[14] Carp RI, Callahan SM, Sersen EA, et al. Preclinical changes in weight of scrapie-infected mice as a function of scrapie agent-mouse strain combination. Intervirology 1984; 21:61–69.

[15] Kim YS, Carp RI, Callahan SM, et al. Scrapie-induced obesity in mice. J Infect Dis 1987; 156:402–405.

[16] Kaneko K, Wille H, Mehlhorn I, et al. Molecular properties of complexes formed between the prion protein and synthetic peptides. J Mol Biol 1997; 270(4):574–586.

[17] Outram GW. Changes in drinking and feeding habits of mice with experimental scrapie. J Comp Pathol 1972; 82:415–427.

[18] Carp RI, Kim YS, Callahan SM. Scrapie-induced alterations in glucose tolerance in mice. J Gen Virol 1989; 70:827–835.

[19] Bessen RA and Marsh RF. Biochemical and physical properties of the prion protein from two strains of the transmissible mink encephalopathy agent. J Virol 1992; 66:2096–2101.

[20] Bessen RA and Marsh RF. Distinct PrP properties suggest the molecular basis of strain variation in transmissible mink encephalopathy. J Virol 1994; 68:7859–7868.

[21] Bessen RA and Marsh RF. Identification of two biologically distinct strains of transmissible mink encephalopathy in hamsters. J Gen Virol 1992; 73:329–334.

[22] Pattison IH. The relative susceptibility of sheep, goats and mice to two types of the goat scrapie agent. Res Vet Sci 1966; 7:207–212.

[23] Chandler RL. Encephalopathy in mice produced by inoculation with scrapie brain material. Lancet 1961; 1:1378–1379.

[24] Bruce ME, and Fraser H. Scrapie strain variation and its implications. Curr Top Microbiol Immunol 1991; 172:125–138.

[25] Bruce ME, McConnell I, Fraser H, Dickinson AG. The disease characteristics of different strains of scrapie in Sinc congenic mouse lines: implications for the nature of the agent and host control of pathogenesis. J Gen Virol 1991; 72: 595–603.

[26] Fraser HM, Bruce ME, McConnell I. Murine scrapie strains, BSE models and genetics. In: Bradley R, Savey M, Marchant B, editors. Subacute Spongiform Encephalopathies. Commission of European Communities. Kluwer Academic, Netherlands, 1991:131–136.

[27] Bruce ME, Boyle A, Cousens S, et al. Strain characterization of natural sheep scrapie and comparison with BSE. J Gen Virol 2002; 83(Pt 3):695–704.

[28] Dickinson AG and Fraser H. Scrapie pathogenesis in inbred mice: an assessment of host control and response involving many strains of agent. In: Ter Meulen V, Katz M, editors. Slow virus infections of the central nervous system. Springer, Berlin, 1977:3–14.

[29] Dickinson AG, Meikle VM, Fraser H. Identification of a gene which controls the incubation period of some strains of scrapie in mice. J Comp Pathol 1968; 78:293–299.

[30] Carlson GA, Kingsbury DT, Goodman PA, et al. Linkage of prion protein and scrapie incubation time genes. Cell 1986; 46:503–511.

[31] Hunter N, Hope J, McConnell I, et al. Linkage of the scrapie-associated fibril protein (PrP) gene and Sinc using congenic mice and restriction fragment length polymorphism analysis. J Gen Virol 1987; 68:2711–2716.

[32] European Community 1990. Council Directive 90/667/EEC of 27 November 1990 laying down the veterinary rules for the disposal and processing of animal waste, for its placing on the market and for the prevention of pathogens in feedstuffs of animal or fish origin and amending Directive 90/425/EEC.

[33] Hunter N. Genotyping and susceptibility of sheep to scrapie. In: Baker HF, Ridley RM, editors. Methods in molecular medicine: Prion diseases. Humana Press, Totowa, NJ, 1996:211–221.

[34] Sander P, Hamann H, Pfeiffer I, et al. Analysis of the sequence variability of the bovine prion protein gene (PRNP) in German cattle breeds. Neurogenetics 2005; 5:19–25.

[35] Lloyd SE, Onwuazor ON, Beck JA, et al. Identification of multiple quantitative trait loci linked to prion disease incubation period in mice. Proc Natl Acad Sci U S A 2001; 98:6279–6283.

[36] Carp RI, Meeker H, Sersen E. Scrapie strains retain their distinctive characteristics following passages of homogenates from different brain regions and spleen. J Gen Virol 1997; 78(Part 1):283–290.

[37] Bruce ME, Dickinson AG, Fraser H. Cerebral amyloidosis in scrapie in the mouse: effect of agent strain and mouse genotype. Neuropathol Appl Neurobiol 1976; 2:471–478.

[38] Scott JR and Fraser H. Degenerative hippocampal pathology in mice infected with scrapie. Acta Neuropathol (Berl) 1984; 65:62–68.

[39] Foster JD, Fraser H, Bruce ME. Retinopathy in mice with experimental scrapie. Neuropathol Appl Neurobiol 1986; 12:185–196.

[40] Fraser H and Dickinson AG. Scrapie in mice. Agent-strain differences in the distribution and intensity of grey matter vacuolation. J Comp Pathol 1973; 83:29–40.

[41] Bruce ME, Chree A, McConnell I, et al. Transmission of bovine spongiform encephalopathy and scrapie to mice: strain variation and the species barrier. Philos Trans R Soc Lond B 1994; 343:405–411.

[42] Brown DA, Bruce ME, Fraser JR, et al. Comparison of the neuropathological characteristics of bovine spongiform encephalopathy (BSE) and variant Creutzfeldt–Jakob disease (vCJD) in mice. Neuropathol Appl Neurobiol 2003; 29:262–272.

[43] Kimberlin RH, Walker C, Millson GC et al. Disinfection studies with two strains of mouse-passaged scrapie agent. Guidelines for Creutzfeldt–Jakob and related agents. J Neurol Sci 1983; 59:355–369.

[44] Taylor DM, Fraser H, McConnell I, et al. Decontamination studies with the agents of bovine spongiform encephalopathy and scrapie. Arch Virol 1994; 139:313–326.

[45] McKinley MP, Bolton DC, Prusiner SB. A protease-resistant protein is a structural component of the scrapie prion. Cell 1983; 35:57–62.

[46] Merz PA, Kascsak RJ, Rubenstein R, et al. Antisera to scrapie-associated fibril protein and prion protein decorate scrapie-associated fibrils. J Virol 1987; 61:42–49.

[47] Prusiner SB. Molecular biology of prion diseases. Science 1991; 252:1515–1522.

[48] Prusiner SB, Bolton DC, Groth DF, et al. Further purification and characterization of scrapie prions. Biochemistry 1982; 21:6942–6950.
[49] Prusiner SB, McKinley MP, Bowman KA, et al. Scrapie prions aggregate to form amyloid-like birefringent rods. Cell 1983; 35:349–358.
[50] Merz PA, Rohwer RG, Kascsak R, et al. Infection-specific particle from the unconventional slow virus diseases. Science 1984; 225:437–440.
[51] Merz PA, Somerville RA, Wisniewski HM, et al. Abnormal fibrils from scrapie-infected brain. Acta Neuropathol 1981; 54:63–74.
[52] Bolton DC, McKinley MP, Prusiner SB. Identification of a protein that purifies with the scrapie prion. Science 1982; 218:1309–1311.
[53] Diringer H, Gelderblom H, Hilmert H, et al. Scrapie infectivity, fibrils and low molecular weight protein. Nature 1983; 306:476–478.
[54] Adams JM. Persistent or slow viral infections and related diseases. West J Med 1975; 122:380–393.
[55] Kascsak RJ, Rubenstein R, Merz PA, et al. Biochemical differences among scrapie-associated fibrils support the biological diversity of scrapie agents. J Gen Virol 1985; 66:1715–1722.
[56] Lezmi S, Martin S, Simon S, et al. Comparative molecular analysis of the abnormal prion protein in field scrapie cases and experimental bovine spongiform encephalopathy in sheep by use of western blotting and immunohistochemical methods. J Gen Virol 2004; 78:3654–3662.
[57] Hope J, Morton LJ, Farquhar CF, et al. The major polypeptide of scrapie-associated fibrils (SAF) has the same size, charge distribution and N-terminal protein sequence as predicted for the normal brain protein (PrP). EMBO J 1986; 5:2591–2597.
[58] Hope J, Multhaup G, Reekie LJ, et al. Molecular pathology of scrapie-associated fibril protein (PrP) in mouse brain affected by the ME7 strain of scrapie. Eur J Biochem 1988; 172:271–277.
[59] Manuelidis L, Valley S, Manuelidis EE. Specific proteins associated with Creutzfeldt–Jakob disease and scrapie share antigenic and carbohydrate determinants. Proc Natl Acad Sci U S A 1985; 82:4263–4267.
[60] Oesch B, Westaway D, Walchli M, et al. A cellular gene encodes scrapie PrP 27–30 protein. Cell 1985; 40:735–746.
[61] Rubenstein R, Merz PA, Kascsak RJ, et al. Detection of scrapie-associated fibrils (SAF) and SAF proteins from scrapie-affected sheep. J Infect Dis 1987; 156:36–42.
[62] Hayashi HK, Yokoyama T, Takata M, et al. The N-terminal cleavage site of PrP(Sc) from BSE differs from that of PrP(Sc) from scrapie. Biochem Biophys Res Commun 2005; 328:1024–1027.
[63] Nonno R, Esposito E, Vaccari G, et al. Molecular analysis of Italian sheep scrapie and comparison with cases of bovine spongiform encephalopathy (BSE) and experimental BSE in sheep. J Clin Microbiol 2003; 41:4127–4133.
[64] Thuring CMA, Erkens JH, Jacobs JG, et al. Discrimination between scrapie and bovine spongiform encephalopathy in sheep by molecular size, immunoreactivity and glycoprofile of prion protein. J Clin Microbiol 2004; 42:972–980.
[65] Collinge J, Sidle KC, Meads J, et al. Molecular analysis of prion strain variation and the aetiology of 'new variant' CJD. Nature 1996; 383:685–690.
[66] Kuczius T, Haist I, Groschup MH. Molecular analysis of BSE and scrapie strain variation. J Infect Dis 1998; 178:693–699.
[67] Somerville RA, Chong A, Mulqueen OU, et al. Biochemical typing of scrapie strains. Nature 1997; 386:564.
[68] Telling GC, Parchi P, DeArmond SJ, et al. Evidence for the conformation of the pathologic isoform of the prion protein enciphering and propagating prion diversity. Science 1996; 274:2079–2082.
[69] Hope J, Wood SC, Birkett CR, et al. Molecular analysis of ovine prion protein identifies similarities between BSE and an experimental isolate of natural scrapie, CH 1641. J Gen Virol 1999; 80:1–4.
[70] Casalone C, Zanusso G, Acutis P, et al. Identification of a second bovine amyloidotic spongiform encephalopathy: molecular similarities with sporadic Creutzfeldt–Jakob disease. Proc Natl Adad Sci USA 2004; 101:3065–3070.
[71] Biacabe AG, Laplanche JL, Ryder S, et al. Distinct molecular phenotypes in bovine prion diseases. EMBO Rep 2004; 5:110–115.
[72] Lloyd SE, Linehan JM, Desbruslais M, et al. Characterization of two distinct prion strains derived from bovine spongiform encephalopathy transmissions to inbred mice. J Gen Virol 2004; 85:2471–2478.
[73] Bartz JC, Bessen RA, McKenzie D, et al. Adaptation and selction of prion protein strain conformations following interspecies transmission of transmissible mink encephalopathy. J Gen Microbiol 2000; 74:5542–5547.
[74] Bruce ME and Dickinson AG. Biological evidence that scrapie agent has an independent genome. J Gen Virol 1987; 68:79–89.

[75] Foster JD and Dickinson AG. The unusual properties of CH1641, a sheep-passaged isolate of scrapie. Vet Rec 1988; 123:5–8.
[76] Parchi P, Giese A, Capellari S, et al. Classification of sporadic Creutzfeldt–Jakob disease based on molecular and phenotypic analysis of 300 subjects. Ann Neurol 1999; 46(2):224–233.
[77] Notari S, Capellari S, Giese A, et al. Effects of different experimental conditions on the PrPSc core generated by protease digestion: implications for strain typing and molecular classification of CJD. J Biol Chem 2004; 279(16):16797–16804.
[78] Dickinson AG. Scrapie in sheep and goats. Front Biol 1976; 44:209–241.
[79] Morgan KL, Nicholas K, Glover MJ, et al. A questionnaire survey of the prevalence of scrapie in sheep in Britain. Vet Rec 1990; 127:373–376.
[80] Anonym. Case records of the Massachusetts General Hospital. Weekly clinicopathological exercises. Case 43 –1977. N Engl J Med 1977; 297:930–937.
[81] Buschmann A, Gretzschel A, Biacabe AG, et al. Atypical BSE cases in Germany – proof of transmissibility and biochemical characterization. Vet Microbiol 2006; submitted.
[82] O'Rourke KI, Spraker TR, Hamburg LK, et al. Polymorphisms in the prion precursor functional gene but not the pseudogene are associated with susceptebility to chronic wasting disease in white-tailed deer. J Gen Virol 2004; 85:1339–1346.
[83] O'Rourke KI, Besser TE, Miller MW, et al. PrP genotypes of captive and free-ranging Rocky Mountain elk (Cervus elaphus nelsoni) with chronic wasting disease. J Gen Microbiol 1999; 80:2765–2779.
[84] Salman MD. Chronic wasting disease in deer and elk: scientific facts and findings. J Vet Med Sci 2003; 65:761–768.
[85] Bruce ME, Will RG, Ironside JW, et al. Transmissions to mice indicate that new variant CJD is caused by the BSE agent. Nature 1997; 389:498–501.
[86] Williams ES and Young S. Spongiform encephalopathy of Rocky Mountain elk. J Wildl Dis 1982; 18:465–471.
[87] Williams ES and Young S. Chronic wasting disease of captive mule deer: a spongiform encephalopathy. J Wildl Dis 1980; 16:89–98.
[88] Race RE, Raines A, Baron TG, et al. Comparison of abnormal prion protein glycoform patterns from transmissible spongiform encephalopathy agent-infected deer, elk, sheep and cattle. J Gen Virol 2002; 76:12365–12368.

Topic III: Portraits of Prion Diseases

Each chapter within the topic of *Portraits of Prion Diseases* is divided into the following sections:

1. History
2. Forms or variants
3. Incubation periods, transmissibility, and susceptibility
4. Clinical signs and course of disease
5. Differential diagnoses
6. Epidemiology
7. Pathology
8. Is ... a new disease?
9. Risk factors
10. Surveillance, prevention and control
11. Editor's note/further information

The goal of each chapter within this topic is to help readers gain a systematic overview in a relatively short period of time.

Editor's note: In addition to the information given – particularly that found in Section 10 of Chapters 13–25 – as well as in all other sections of the book that relate to surveillance, prevention, and control of prion diseases, we highly recommend that the current legislation of any the country of interest, and the pertinent recommendations of international organizations such as the WHO (Public Health) or the OIE (Animal Health) are taken into account. This is particularly important for decision makers and their advisors.

13 Portrait of Kuru

Beat Hörnlimann and Michael P. Alpers

13.1 History

Kuru first appeared among the people of the Fore linguistic group and their neighboring groups in the Eastern Highlands of Papua New Guinea (PNG) [1]. In addition to the name kuru, the local people also called this incurable syndrome uruna [2] or guzigli [3]. The linguistic groups Fore, Gimi, Keiagana, Kanite, Yate, Yagaria, Kamano, Usurufa, and Auyana all suffered from kuru; later immigrant cases also occurred among the Awa. The people believed that this disease and its effects were the result of a new form of sorcery (kuru sorcery; see Chapter 2).

Western medicine was not cognizant of this disease until 1956, when V. Zigas reported the first case in writing [4]. It is now well established that the disease is connected to endocannibalism-related activities in the preparation of ritualistic meals [4–7]. Based on reliable accounts of older members of the community recollecting early cases of kuru, the onset of the epidemic can be dated back to approximately the beginning of the 20th century. From this evidence, the epidemic must have lasted throughout the entire 20th century and the true total number of cases may therefore have been about twice as high as the reported 2,700 cases since kuru surveillance began in 1957 (Fig. 38.1).

13.2 Forms or variants

With a few exceptions there was only one clinical form of kuru. It was characterized by a surprisingly homogeneous clinical picture and course of disease. There was probably only one infectious agent (prion strain; compare Chapter 12) involved in the kuru epidemic but this is not known and will most likely not be determined retrospectively.

13.3 Incubation period, transmissibility, and susceptibility

The incubation period under field conditions lasts from 4 years to several decades [7; 8]. This wide range of incubation periods is consistent with transmission occurring as a consequence of eating various amounts (per person) of the infected brains of deceased relatives (endocannibalism or transumption). In addition, parenteral transmission could have occurred by manual and sometimes indirect contact with mucous membranes while removing the brain from the skull or through contamination of cuts and scratches (Fig. 2.6). Parenteral transmission is characterized by higher infective efficiency than oral transmission, reflected in a shorter incubation period given the same load of infective tissue (Fig. 1.6). Furthermore a new genetic study has shown that there is increased susceptibility to kuru of carriers of *PRNP* 129 methionine/methionine genotype [9] (see Chapter 38), leading to a shorter incubation period than that of carriers of a heterozygotic codon 129.

Transmission of the disease by natural, direct horizontal transfer from one person to another has not been described. As is normally the case for prion diseases – except for scrapie and chronic wasting disease (CWD) – kuru-infected subjects do not excrete prions (see Chapter 42). They are also not passed on by breast-feeding [5; 10]. Based on epidemiological evidence kuru was never transmitted vertically [11]: none of the children born since 1959 that were breastfed by mothers who, at the time of nursing or already during pregnancy and birth, showed clinical symptoms of kuru, ever contracted the disease.

The incubation period in an infected ape after the first experimental transmission of kuru was almost 2 years after intracerebral inoculation

[12]. It was later determined that the experimental incubation period for kuru was consistently longer than for Creutzfeldt–Jakob disease (CJD) [13]; (see Chapter 41).

13.4 Clinical signs and course of disease

A surprisingly uniform picture of clinical symptoms and disease progression characterized kuru. On average, the time from clinical onset until death was 12 months for adults and slightly shorter for children [1] (Table 30.1), perhaps because relative to adult body mass children consumed comparatively higher amounts of infective brain tissue (see Chapter 38.7).

While in 66% of all cases of sporadic CJD (sCJD) the clinical duration is less than 6 months [14], this is the case for only a small proportion of all cases of kuru. Characteristic symptoms during the initial phase of the clinical progression were headaches and joint and limb pain. In general, impaired movement predominated. After just a few weeks, unsteadiness in gait and general clumsiness were always detectable (ataxia due to cerebellar lesions; Fig. 13.1a). With increasing progression, a rough rhythmical shaking of the whole body developed, especially when standing. This postural tremor was also called kuru tremor (Fig. 13.1a) and led to the name "kuru" in the Fore language, meaning "trembling", "shaking" [15], or "shivering" [16].

Starting usually at an early stage grotesque laughter of patients, often accompanied by actual euphoria (Fig. 13.1b), was also a commonly observed sign and gave rise to the name, "the laughing death" [4], despite the fact that it was inappropriate and incorrect (in the late stage of the disease this sign disappeared). Some patients displayed intermittent strabismus starting at the intermediate stage (Fig. 13.1c). Several months after the beginning of the disease (though sometimes occurring earlier) signs such as dysmetria, dysarthria, hypotonia, etc. appeared. The clinical progression of the disease was marked by progressive cerebellar ataxia. During the later stages very strong disturbances in all motor coordination (Fig. 13.1d) formed a dominant part of the clinical picture. Typically, impaired sensory modalities and myoclonus (as occurs in CJD) were not detected. Electromyographical examinations as well as an EEG showed no signs of deviation for kuru [17]. Dementia, or the deterioration of mental capabilities, occurred relatively rarely and usually not until the terminal stage (whereas dementia is a main early component of sCJD); there were only very few known cases of kuru that displayed early signs of dementia.

Fig. 13.1: Clinical signs of kuru. All figures are reproduced with permission of D.C. Gajdusek and Elsevier Publishers [18]. ►
(a) Six kuru victims, two girls and four boys, in front of the kuru hospital in Okapa. The smallest boy in the photo (centre) is 5 years old and is in a clinically early stage of kuru. The symptoms began to show 6 months before the picture was taken. Without assistance, he is only able to walk with his legs wide apart, and manifests a tremor as well as ataxia (incoordinated way of standing or walking). The adult man (on the left) who is supporting a girl suffering from kuru, and a taller boy in the center who is holding another kuru victim in an upright position, are the only persons in the photo not suffering from kuru. The taller boy on the right-hand side who is preventing a sick girl from collapsing, claimed that he had also begun to suffer from kuru. This could, however, not be confirmed by clinical examination at that time. Indeed, the boy died of kuru within a period of 1 year. **(b)** A girl in an advanced state of kuru. She was only able to walk with the help of others. A smile observed from the beginning of the disease developed into real euphoria, combined with giggling and laughter. **(c)** Fluctuating strabismus and ocular ataxia in a young boy in the intermediate stage of the clinical course of kuru. **(d)** The boy from the Gimi linguistic group (Kimi; Fig. 2.2) is in an advanced state of kuru. His father has to support him as he is no longer able to walk upright without help. He cannot touch his finger tips because of extreme coordination disturbances and cerebellar lesions caused by kuru.

(a)

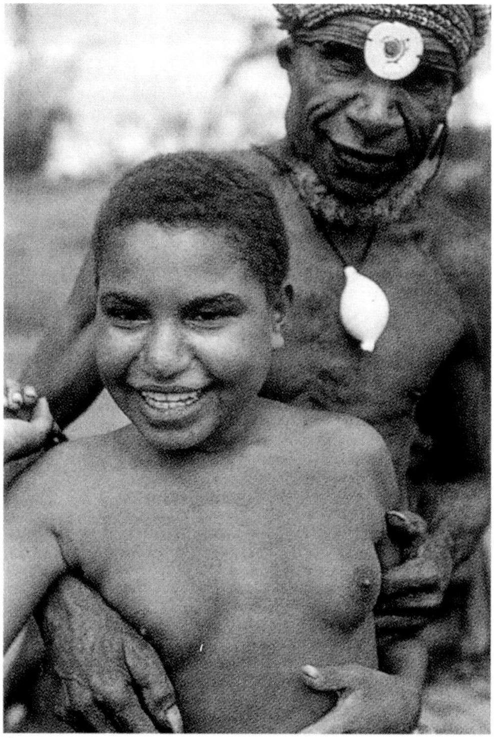
(b)

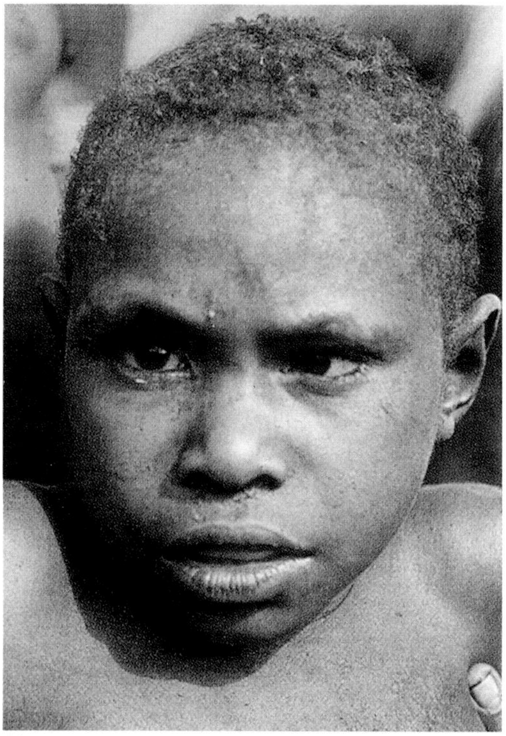
(c)

(d)

13.5 Differential diagnoses

The differential diagnoses playing a role for prion diseases in humans are summarized in Table 30.2. Kinsmen of kuru patients did not confuse kuru with other diseases [19–20], making it possible to trace cases from previous generations (see Chapters 2, 38).

13.6 Epidemiology

The epidemiology of this disease is more extensively described and discussed in Chapter 38, but to summarize, some basic epidemiological facts are mentioned here.

Approximately 30,000–40,000 people lived in the kuru region, which remained limited to parts of the Eastern Highlands of PNG. The group most affected by kuru was the Fore with a population of approximately 11,000 at the time [19]. Since 1957, when systematic kuru research began, a total of approximately 2,700 patients have contracted kuru.

It is estimated that the epidemic started at the beginning of the 20th century. It peaked between 1957 and 1960 when more than 200 individuals per year fell ill to kuru (Fig. 38.1). In 1960, for the first time, a decline to below 200 cases per year was registered after endocannibalism had largely stopped in the mid-1950s [11] (see Chapter 2).

The resulting break in the spread of infections, as seen first by the decline of age-specific incidences in the youngest affected group (Fig. 38.2), was not detectable until later.

In 1967, it was possible for the first time to connect the decline in kuru cases with the cessation of endocannibalism in terms of a hypothesis based on scientific grounds [5]. Later more evidence was found [7] based on thorough questioning of many affected relatives and taking into account the transmissibility established in 1966 [12]. Since 1967, there have been no further cases among children under the age of 10 [11] (Table 38.3; Fig. 38.2).

Both the distribution among the sexes and the pronounced wide range in age at the time of death, varying from 4 to over 60 years of age, were very unusual (Table 38.2), and reflect the wide range of incubation periods of 4–40 years or more. These phenomena can be explained by various factors, such as (i) the varying time of exposure, (ii) varying amounts of infectious agent transmitted, and (iii) host genetics – methionine/methionine homozygosity of codon 129 of the prion protein gene (*PRNP*) is associated with a shorter incubation period [9] or, in other words, an increased susceptibility to kuru. These parameters are discussed in detail in Chapter 38. It should be mentioned here, however, that the observation of sociocultural

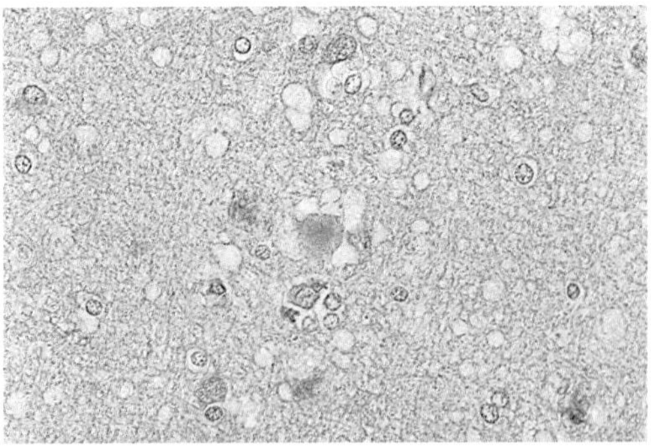

(a)

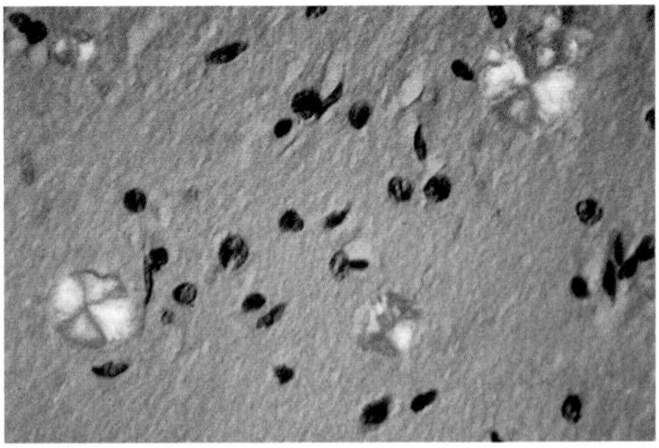

(b)

Fig. 13.2: (a) The central amyloid plaque or kuru plaque (blue) is surrounded by vacuoles. Dyes: Alcian blue and nuclear fast red × 125 (original magnification). Slide kindly provided by H. Budka. **(b)** Amyloid plaques (often called kuru plaques) as found in the majority of prion diseases: their number varies depending on the species and agent strain (see Chapter 12). Birefringence in polarizing light. Slide kindly provided by C. Weissmann.

differences between the lifestyle of women and girls compared to that of men and boys helped to establish the interrelation between the endocannibalistic rites and the epidemiological patterns (sex and age distribution) of kuru (see Chapters 2, 38).

13.7 Pathology

In brain sections of kuru patients microscopically small holes (vacuoles, spongiform changes) typical of prion diseases or transmissible spongiform encephalopathies are detectable. Kuru plaques (Fig. 13.2a), i.e., deposits characterized by birefringence in polarizing light (Fig. 13.2b), are also observed. In the case of kuru, pathological changes of the parencephalon predominate and manifest themselves as accentuated cerebellar signs and symptoms (especially disruptions in coordination). The pathology of kuru is described in more detail in Chapter 26 and the pertinent literature [17; 21–25].

13.8 Is kuru a new disease?

No, kuru probably started about 100 years ago and it should soon be a disease of the past. The topic, however, is still of current interest for reasons mentioned in other parts of this book (see Chapters 1–3, 38). The disease probably originated as a case of sporadic CJD occurring among the Fore or their neighbors [26].

13.9 Risk factors

The transmission of kuru-infected brain or spinal cord tissue is the primary risk factor. Not only ingestion of such infectious tissue represents a risk, but also intensive physical contact with it, for example by skin lesions or mucous membranes of the eye. For prion diseases, the latter, parenteral mode of transmission is generally more efficient than the oral mode, i.e., eating infectious or cross-contaminated tissue parts [27]. Nevertheless, the facts described in Chapter 2.11–14 on endocannibalism and the wide range of incubation periods in kuru strongly supports the consumption of (various amounts of) infective tissue of dead relatives as being the main mode of transmission of kuru, particularly when taking into account that host genetics is a cofactor with respect to the length of the incubation period [9].

Endocannibalism alone could not have been the cause for kuru. It was, however, a potent means of intraspecies recycling of the kuru agent, once it was present in the population (compare intraspecies recycling of BSE agents among the British cattle population through feeding recycled meat-and-bone-meal [28]).

In any case, many years, decades, or centuries ago numerous other ethnic groups throughout the world practiced cannibalism [29] without – according to current knowledge – kuru or any related disease ever appearing, for example in the Caribbean/Antilles [30; 31], Mexico [31], Peru [31], Australia (also endocannibalism in part) [31; 32], New Zealand (31), Fiji Islands [31], Africa (within the Congo Basin partly also endocannibalism) [31; 33], Sumatra [31], approximately 1200 B.C. in Egypt following a famine [31], approximately 800 B.C. in Samaria [34], America [35–37] including endocannibalism [38], as well as in Europe [31][1].

The initial very slow increase in the slope of the incidence or epidemic curve (Fig. 38.1) and the tight geographical confines (Figs. 2.1, 2.2) of the epidemic principally allow the conclusion that all cases of kuru originated from a single source. However, the true origins of kuru can only be speculated upon: the first case – approximately 100 years ago (Fig. 1.3) – could have been what is known as a sporadic prion disease which may have arisen spontaneously, as is believed to be the case for sCJD [11; 41; 42] (see Chapters 3, 8; Table 1.2). It is conceivable that the brain of a clinically or subclinically sCJD-afflicted person initiated the kuru epi-

[1] A critical comment on cannibalism, which is worth reading, was written by anthropologist Margaret Mackenzie [39]. Recent genetic studies [40] have shown that the balancing selection found in the Fore to minimize the transmission of kuru is found in all human populations, suggesting that similar endocannibalistic practices might have been widespread in the remote human past.

demic as the index case within the context of intense endocannibalism [11; 26]. The incidence of sCJD seems to be the same in PNG [42] as it is in the rest of the world (1–2 cases per 1 million inhabitants/year). The probability of such a case occurring among the Fore is low, but not zero, and once it had occurred the intense endocannibalistic mortuary practices (transumption) would have provided the necessary intraspecies recycling, which is the only way the epidemic could have been sustained.

13.10 Surveillance, prevention, and control

Even today after the epidemic has subsided, monitoring is required. This should be done to track "spontaneously developing" sporadic cases of kuru, which theoretically cannot be ruled out (particularly if there is a local genetic susceptibility). It will also allow the monitoring of additional acquired cases, which have yet to manifest themselves because of their long incubation times.

The most important measures for prevention and control of kuru are to cease the practice of endocannibalism and to protect oneself from parenteral transmission of infectious brain or spinal cord material (e. g., by not handling such materials with bare hands; especially in diagnostics and research laboratories; see Chapters 48–49).

In the mid-1950s, before the beginnings of systematic kuru research (see Chapter 2) and therefore long before the relevant correlation between endocannibalism and this disease was known [5], cannibalism was stopped in the Okapa district. Hence, the decline of the kuru epidemic was a direct but unintended result of the prohibition of this practice imposed by the Australian administration. The effectiveness of this measure is evidenced retrospectively by the fact that there have rarely been any cases of kuru among individuals born after 1956 and none among those born after 1959 [7; 11], and that there has been a continuous decline in cases since 1959.

13.11 Editor's note

Looking deeper than a purely scientific perspective, the chief editor wishes to add that the practice of endocannibalism must not be judged or condemned from a purely western viewpoint. Endocannibalism has – at least when taken from a historical and ethnological perspective – a right to be viewed from the eyes of the Fore people, i. e., on the basis of their own cultural background.

References

[1] Alpers MP. Epidemiology and clinical aspects of kuru. In: Prusiner SB, McKinley MP, editors. Prions. Academic Press, San Diego, 1987:451–465.

[2] Gajdusek DC. Letter to Burnet and Anderson. In: Farquhar J, Gajdusek DC, editors. Kuru – Early letters and field notes from the collection of D. Carleton Gajdusek. Raven Press, New York, 1981:71–74.

[3] Berndt RM. A "devastating disease syndrome" – Kuru sorcery in the Eastern Central Highlands of New Guinea. Sociologus 1958; 8:4–28.

[4] Zigas V. Laughing death – The untold story of kuru. Humana Press, Clifton, New Jersey, 1990.

[5] Alpers MP. Kuru: implications of its transmissibility for the interpretation of its changing epidemiologic pattern. In: Bailey OT, Smith DE, editors. The central nervous system, some experimental models of neurological diseases. International Academy of Pathology, Monograph No 9. Proceedings of the fifty-sixth annual meeting of the International Academy of Pathology, Washington, DC, 12–15 Mar 1967. Williams and Wilkins, Baltimore, 1968:234–251.

[6] Klitzman RL. The trembling mountain: a personal account of kuru, cannibals and mad cow disease. Plenum Trade, New York and London, 1998.

[7] Klitzman RL, Alpers MP, Gajdusek DC. The natural incubation period of kuru and the episodes of transmission in three clusters of patients. Neuroepidemiology 1984; 3:3–20.

[8] Prusiner SB, Gajdusek DC, Alpers MP. Kuru with incubation period exceeding two decades. Ann Neurol 1982; 12:1–9.

[9] Lee HS, Brown P, Cervenakova L, et al. Increased susceptibility to kuru of carriers of the PRNP 129 methionine/methionine genotype. J Infect Dis 2001; 183:192–196.

[10] EU. Report from the scientific veterinary committee on the risk analysis for colostrum, milk and milk products. Leg Vet Zoot 1996; VI/8197/96 Version J:1–26.

[11] Alpers MP. Epidemiology and ecology of kuru. In: Prusiner SB, Hadlow WJ, editors. Slow transmissible diseases of the nervous system. Academic Press, New York, London, Sydney, Toronto, San Francisco. Vol 1, 1979:67–90.

[12] Gajdusek DC, Gibbs CJ Jr, Alpers M. Experimental transmission of a kuru-like syndrome to chimpanzees. Nature 1966; 209:794–796.

[13] Gibbs CJ Jr, Gajdusek DC, Asher DM, et al. Creutzfeldt–Jakob disease (spongiform encephalopathy): transmission to the chimpanzee. Science 1968; 161:388–389.

[14] Brown P, Cathala F, Sadowsky D, et al. Creutzfeldt–Jakob disease in France: II. Clinical characteristics of 124 consecutive verified cases during the decade 1968–1977. Ann Neurol 1979; 6:430–447.

[15] Farquhar J and Gajdusek DC. Kuru – Early letters and field notes from the collection of D. Carleton Gajdusek. Raven Press, New York, 1981.

[16] Newby J. Cannibals – Interview with Dr. Mike Alpers. Webpage 2003: www.abc.net.au/catalyst/stories/s938896.htm.

[17] Scrimgeour EM, Masters CL, Alpers MP, et al. A clinicopathological study of a case of kuru. J Neurol Sci 1983; 59:265–275.

[18] Gajdusek DC. Kuru in childhood: implications for the problem of whether bovine spongiform encephalopathy affects humans. In: Court L, Dodet B, editors. Transmissible subacute spongiform encephalopathies: Prion diseases. Elsevier, Paris, 1996:15–26.

[19] Gajdusek DC and Zigas V. Studies on the pathogenesis of kuru (Untersuchungen über die Pathogenese von Kuru). Klin Wochenschr 1958; 36(10):445–459.

[20] Glasse RM. The spread of kuru among the Fore. A preliminary report. Department of Public Health, Territory of Papua and New Guinea. Typed script 1962; 1–9.

[21] Adams H, Beck E, Shenkin AM. Creutzfeldt–Jakob disease: further similarities with kuru. J Neurol Neurosurg Psychiatry 1974; 37:195–200.

[22] Beck E, Daniel PM, Alpers M, et al. Neuropathological comparisons of experimental kuru in chimpanzees with human kuru. Int Arch Allergy Immunol 1969; 36:553–562.

[23] Klatzo I, Gajdusek DC, Zigas V. Pathology of kuru. Lab Invest 1959; 8:799–847.

[24] Krucke W, Beck E, Vitzthum HG. Creutzfeldt–Jakob disease. Some unusual morphological features reminiscent of kuru. Z Neurol 1973; 206:1–24.

[25] Lantos PL, Bhatia K, Doey LJ et al. Is the neuropathology of new variant Creutzfeldt–Jakob disease and kuru similar? Lancet 1997; 350:187–188.

[26] Alpers MP and Rail L. Kuru and Creutzfeldt–Jakob disease: clinical and aetiological aspects. Proc Aust Assoc Neurol 1971; 8:7–16.

[27] Prusiner SB, Cochran SP, Alpers MP. Transmission of scrapie in hamsters. J Infect Dis 1985; 152:971–978.

[28] Wilesmith JW, Ryan JB, Atkinson MJ. Bovine spongiform encephalopathy: epidemiological studies on the origin. Vet Rec 1991; 128(9):199–203.

[29] Spiel C. Menschen essen Menschen (people eat people). C. Bertelsmann Verlag, 1972.

[30] Meyers. Konversations-Lexikon (Nachschlagewerk des allgemeinen Wissens). Stichwort Kannibalen. Bibliographisches Institut, Leipzig und Wien, 1896:850.

[31] Meyers. Konversations-Lexikon (Nachschlagewerk des allgemeinen Wissens). Stichwort Anthropophagie. Bibliographisches Institut, Leipzig und Wien, 1896:665–666.

[32] Berndt RM and Berndt CH. The first Australians. 2nd ed. Ure Smith, Sydney, 1967.

[33] Helmolt HF. Weltgeschichte. 2. Auflage ed. Bibliographisches Institut, Leipzig und Wien, 1914.

[34] Kings 2. bilingual-bible.com (Chapter 6, Verse 28–29). Webpage 2005: www.bilingual-bible.com/browse/2-kings/chapter6.php.

[35] de Montellano BO. Counting skulls: comment on the Aztec cannibalism theory of Harner and Harris. Am Anthropol 1983; 85(2):403–406.

[36] Chagnon N and Hames R. Protein deficiency and tribal warfare in Amazonia: new data. Science 1979; 203:910–913.

[37] Harner M. The ecological basis for Aztec sacrifice. Am Ethnol 1977; 4:117–135.

[38] Dole GE. Endocannibalism among the Amahuaca Indians. Trans NY Acad Sci 1962; 24(2):567–573.

[39] Mackenzie M. New Zealand and cannibalism: the brainpot. Webpage 2006: http://wais.stanford.edu/NewZealand/newzealand—cannibalismbrainpot20301.html.

[40] Mead S, Stumpf MP, Whitfield J, et al. Balancing selection of the prion protein gene consistent with prehistoric kurulike epidemics. Science 2005; 300:640–643.

[41] Brown P and Gajdusek DC. The human spongiform encephalopathies: kuru, Creutzfeldt–Jakob disease, and the Gerstmann–Sträussler–Scheinker syndrome. Curr Top Microbiol Immunol 1991; 172:1–20.

[42] Hornabrook RW and Wagner F. Creutzfeldt–Jakob disease. P N G Med J 1975; 18:226–228.

14 Portrait of Creutzfeldt–Jakob Disease

Herbert Budka

14.1 History

A peculiar progressive neurological disease termed spastic pseudosclerosis and its neuropathology were first described by Hans Gerhard Creutzfeld [1] and Alfons Maria Jakob [2–5]. What Creutzfeldt described in 1920 does not seem to correspond with what we today consider to be Creutzfeldt-Jakob disease (CJD) [6]. Nevertheless, Walter Spielmeyer suggested in 1922 to name the disease after its first two observers, his students Creutzfeldt and Jakob, in the order that their publications appeared. This sequence of eponyms has been repeatedly criticized but remains deeply entrenched [7]. Later reexamination by Masters of five cases originally reported by Jakob in 1921 [2–4] and 1923 [5] clarified that at least his third and fifth case had indeed the spongiform neuropathology of CJD [8]. In 1995, researchers succeeded in identifying a D178N mutation of the prion protein (PrP) gene (*PRNP*) in brain samples [9; 10] from two members of the Backer family that had originated from Jakob's laboratory and been stored for more than 70 years [9; 10]. Thus Jakob is also linked with familial CJD, today also called genetic CJD (fCJD/gCJD).[1]

In the following decades, CJD cases were reported in many countries, proving the worldwide appearance of the disease (Table 36.1). Epidemiology failed to provide evidence of any spread by transmission or of an environmental source in the large majority of cases that were categorized as sporadic (idiopathic) CJD (sCJD) [11]. A causal relation with other transmissible spongiform encephalopathies (TSEs) of animals or humans is not known and can only be hypothesized. In medical history, molecular biology and biochemistry have played an important role in gaining deeper insights into the clinical-pathological spectrum of the disease [6; 12], including the subtypes of sCJD (see Chapter 26, 30.9). It was not until the 1960s that CJD, like kuru, was transmitted experimentally for the first time [13]. In addition to sCJD and fCJD/gCJD, almost 400 iatrogenic CJD cases (iCJD) have become known – the results of inadvertent transmission of CJD between humans by "medical accidents" [14–16]. A new variant of CJD (vCJD) was described in 1996 [17], and it is virtually certain that this variant is the result of the transmission of BSE from cattle to humans [18] (see Chapter 50). CJD is geno- and phenotypically distinct from other human TSEs such as classical cases of Gerstmann–Sträussler–Scheinker disease (GSS; see Chapter 16), fatal familial insomnia (FFI; see Chapter 17), and kuru (see Chapter 13).

14.2 Forms or variants

Sporadic CJD, familial/genetic CJD, and forms of the disease that have been acquired by infection need to be distinguished [19; 20]. All forms of the disease can be transmitted experimentally and are linked with "signature" patterns of the proteinase-resistant PrP^{Sc} after proteolysis (PrPres) seen on Western blots of diseased central nervous system (CNS) tissue [21; 22]. Two different nomenclatures have been in use for these molecular PrP^{Sc}-banding types, distinguishing mainly two [21] or three [22] distinct types in sCJD and iCJD according to the molecular mass of the unglycosylated PrP^{Sc} fragment, although recent data suggest considerable

[1] The term "genetic CJD", abbreviated "gCJD", seems more appropriate in the context of the clinical history of a patient or affected family, as there is a family history only in about 50% of all genetic CJD cases.

heterogeneity [23] and the regular coexistence of multiple PrP^{Sc} types in CJD brains [24].

14.2.1 Sporadic (idiopathic, probably spontaneously occurring) CJD (sCJD)

Sporadic CJD cases make up the large majority of all known CJD cases. As there is no evidence of an external source, it is generally assumed that the disease is endogenously caused by a somatic *PRNP* mutation, or a spontaneous conformational change of PrP, which would then, in a chain reaction, lead to the conversion of normal PrP^{C} into the abnormal, disease-associated, and infectious PrP^{Sc} configuration.

14.2.2 Familial (genetic, hereditary) CJD (fCJD/gCJD)

fCJD/gCJD occurs in an autosomal dominant mode of inheritance, but a family history is lacking in 47% of all genetic TSEs [25]. Many different *PRNP* point mutations and insertions (Table 26.5) have been identified [26].

14.2.3 Iatrogenic CJD (iCJD)

This infectious form of CJD comprises cases that were caused by medical accidents (accidental transmission; Table 14.1) [14; 15]. The overwhelming majority of iCJD cases have been caused either by the parenteral administration of contaminated growth hormone preparations from the pituitary gland of deceased people, or by contaminated grafts of dura mater. A few cases occurred after neurosurgery including insertion of deep brain electrodes, and corneal transplantations. Further information on this topic is provided in Chapter 36.

14.2.4 Variant CJD (vCJD)

This form of CJD [17] results from human exposure to BSE prions [18] and is described in detail in Chapter 15.

14.3 Incubation period, transmissibility, and susceptibility

Information on the incubation period of CJD is naturally only available for CJD forms that are acquired by infection. In the case of iCJD, the mean incubation period is 1.5–1.6 years following direct intracerebral contact of infectious material by neurosurgery or depth electrodes; 9 years after dura mater transplantations; and 12–13 years after peripheral inoculation with hormone injections [14].

Table 14.1: Iatrogenic Creutzfeldt–Jakob disease: risk factors, number of cases, mode of transmission (entrance), and incubation period (updated numbers of cases from [61]).

Mode of transmission	Number of cases	Entrance	Mean incubation period (range)
Instrumentation			
• neurosurgery	5	Intracerebral	17 months (12–28)
• stereotactic EEG	2	Intracerebral	18 months (16–20)
Tissue transplantation			
• cornea transplant	4 1)	Nervus opticus	1)
• dura mater graft	169 2)	Brain surface	9 years (1.5–23)
Tissue extracts, injected			
• growth hormone	191 3)	Hematogenic	13 years (4–27)
• sex hormone (gonadotrophin, FSH)	4	Hematogenic	13 years (12–16)
All	375		

1) Two cases with definite CJD in the donors were diagnosed: one in the USA with an incubation period of 18 months [47], and the other in Germany with that of 30 years [48] (see Chapter 51). Two possible transmissions were reported from Japan [49] and the USA [50] but could not definitely be traced back to a donor suffering from CJD; in these cases, the incubation period would have been 15 months, and 15 to 24 years, respectively.

2) Of these, 113 cases in Japan.

3) Of these, 106 cases in France.

For vCJD, see Chapter 15. Transmissibility and susceptibility are discussed in Chapters 26, 36, and 42.

14.4 Clinical signs and course of the disease

Typical signs and symptoms of "classical" CJD feature a *rapidly progressive dementia* that is associated with several neurological symptoms (Tables 30.1 and 30.4), in particular *myoclonus*; *pyramidal* or *extrapyramidal symptoms* such as spasticity, hyperreflexia, or tremor and rigor; *ataxia* or *visual symptoms*; or *akinetic mutism*. A clinical diagnosis of probable CJD is supported by laboratory examinations such as conspicuous alterations in the EEG (periodic sharp wave complexes, PSWC) (sensitivity 66 %, specificity 74 %; Fig. 30.2) and/or detectability of the 14–3–3 protein in the cerebrospinal fluid (sensitivity 94 %, specificity 84 %; Table 31.5-9) [27]. A combination of such clinical signs and symptoms with laboratory support forms the basis of currently valid clinical-diagnostic CJD criteria for surveillance purposes, as developed by EuroCJD and supported by WHO [11; 19; 20; 28] (Table 31.1). In addition to laboratory support by EEG and 14–3–3 protein in the CSF, the MRI usually shows, aside from potential atrophy, bilateral symmetric areas of increased signal intensity in the caudate nuclei and putamina on long repetition time images [19; 29], in particular in T2 diffusion-weighted and/or FLAIR images. Nevertheless, MRI features have not yet been proven to be of consistently high sensitivity and specificity, and thus have not been included in the surveillance diagnostic criteria. Many iCJD cases, in particular those following treatment with growth hormone, feature a predominantly ataxic clinical syndrome [15]. For the complete case definitions, see Table 31.1 (and Table 29.1).

In summary: the clinical diagnosis of *probable* CJD is based on the combined aforementioned alterations that are typical of CJD. *Possible* CJD has the same signs and symptoms as probable CJD but lacks laboratory support and has a duration of less than 2 years. *Definite* CJD is diagnosed only by neuropathology. Aside from bioassays, the definite diagnosis is currently performed by histopathology in addition to immunohistochemical tests (Fig. 26.1e,f) and/or Western blotting for PrP^{Sc}, complemented by genetic analyses of the *PRNP* if fCJD/gCJD is suspected [19].

Analysis of the age at death reveals differences in the age distribution between the four subtypes: vCJD and iCJD cases occur mainly in the age class of younger than 39 years, fCJD/gCJD is mostly distributed in the age group of 50 to 69, and sCJD predominates in the age group of 60 to 79 [20].

sCJD, fCJD/gCJD, and iCJD caused by dural grafting have the shortest median survival times of less than 6 months; iCJD caused by cadaveric growth hormone and vCJD have median survival times of about 1 year. Only 15 % of sCJD patients are alive after 1 year. In sCJD, longer survival was correlated with younger age at onset of illness, female gender, codon 129 heterozygosity, CSF 14–3–3 protein, and PrP^{Sc} type 2A [30].

The genotype at the polymorphic *PRNP* codon 129, as well as the different PrP^{Sc} types seen on immunoblots, have been shown to significantly influence the phenotype of sCJD; at least six subtypes with distinct clinical and neuropathological presentation have been distinguished [11; 21]. The MM1 and MV1 types are the most frequent and correspond to the classical CJD presentation described above, whereas the VV2 and MV2 types correspond to a predominantly ataxic/cerebellar presentation (see Table 26.3).

14.5 Differential diagnoses

Table 30.2 shows the large number of diseases that have to be taken into consideration when neurological disease symptoms are observed. The most important differential diagnoses concern other dementing processes, in particular Alzheimer's disease followed by vascular dementia and Lewy body dementia [11]. Several centers have reported cases in which the symptoms suggested that the disease was "most likely CJD"; this also included alterations in the EEG and the detection of 14–3–3 protein in the cerebrospinal fluid, which were rated as typical CJD signs. Nevertheless, subsequent autopsy clearly identified the disease as Alzheimer's in many cases. This emphasizes the importance of postmortem investigations, particularly in all dementing disorders. iCJD often reveals distinctive cerebellar symptoms, but care has to be taken to clearly distinguish it from spino-cerebellar diseases such as Friedreich's ataxia.

14.6 Epidemiology

14.6.1 sCJD

The EuroCJD database of 4,441 human TSE cases contains 84% sCJD [20]. The incidence and mortality rates for sCJD are similar because of the short mean duration of disease. The overall annual mortality rate from sCJD (definite and probable sCJD cases) in the period 1999 to 2002 was 1.39 cases per million (pooled data from all EuroCJD countries). The rates in individual countries range from 0.48 in Slovakia to 2.23 in Switzerland, while most other countries showed similar rates [20]. There is a relatively consistent incidence rate of sCJD around the world [11] (Tables 36.1–2).

14.6.2 fCJD/gCJD

The EuroCJD database contains 10% genetic TSEs (also including GSS and FFI, which are geno- and phenotypically distinct from fCJD/gCJD; [20] (see Chapters 16; 17; 26). More than 30 *PRNP* aberrations, in particular different point mutations and insertions, have been linked with the development of fCJD/gCJD [26] (Table 26.5). In Slovakia and in Libyan immigrants in Israel, an excess of CJD has been observed that results from local clusters of fCJD/gCJD due to the E200K mutation [31].

14.6.3 iCJD

The EuroCJD database contains 3% iCJD (20). Worldwide, almost 400 iCJD cases have been described [14] (Table 14.1).

14.6.4 vCJD

The EuroCJD database contains 3% vCJD (20). By May 2006, 191 vCJD cases had occurred worldwide (161 in the UK). Further data are given in Chapter 15.

14.7 Pathology

The macroscopic investigation of the brain during autopsy might not reveal any obvious sign of abnormality. However, in the majority of cases, a generalized, occasionally also focal atrophy can be observed, in either the cerebrum or the cerebellum. The classical triad of spongiform change with loss of nerve cells and gliosis is the neuropathological hallmark of CJD, and has more recently been supplemented by the additional criterion of deposition of the disease-associated PrP in CNS tissue [11; 32]. Immunohistochemistry (IHC) is of utmost diagnostic importance in visualizing these deposits (Fig. 26.1-2); therefore, appropriate techniques and antibodies must be used [33]. The type and location of PrP^{Sc} deposits [34] depend on the *PRNP* sequence, the prion strain (see Chapter 12), and the molecular characteristics of the prion protein [PrP^{Sc} types 1 and 2 according to the Western blotting classification of Parchi et al. [21] (see Chapter 26.4.1)]. Diagnosis in atypical cases usually relies on demonstration of PrP^{Sc} in tissue [34]. In addition to the most frequent PrP^{Sc} deposits at synapses and as extracellular plaque-like deposits, neuronal cell bodies, dendrites, axons, astrocytes, and microglia harbor granular PrP^{Sc} deposits that testify to intraaxonal transport and spread, intracellular production, processing and degradation of prions in the brain [35]. Extra-CNS deposition of PrP^{Sc} in sCJD has been observed in peripheral nervous tissue [36; 37], olfactory epithelium [38], dendritic cells and macrophages in vessel walls [39], spleens and skeletal muscle [40], with particularly high amounts in muscle tissue with inclusion body myositis (IBM) [41].

14.8 Is CJD a new disease?

No, because sporadic and familial/genetic forms of the disease have been known since the

1920s [1–5]. In contrast, iCJD and vCJD are more recent forms of CJD that are associated with environmental and other risk factors (see Chapter 36) linked with modern life and medicine.

14.9 Risk factors

14.9.1 sCJD

Specific genetic factors affect the probability of developing sCJD [42] or iCJD [43] (see Chapter 26). Approximately 72 % of all sCJD patients are homozygote for methionine (MM) at *PRNP* codon 129, whereas only 37 % of the unaffected population is MM homozygous [42]. Only about 11 % of sCJD patients are heterozygous for the amino acids methionine and valine (MV) at codon 129; in contrast, approximately 51 % of the normal population are MV heterozygotes. The rest is VV homozygous, which corresponds to approximately 17 % of sCJD patients and approximately 12 % of the normal population. Thus *PRNP* codon 129 is an important indicator of CJD susceptibility.

In the EuroCJD database, about 68 % of cases are MM, with 67 % of sCJD cases and 100 % of vCJD cases exhibiting this genotype. In iCJD, the relative excess of MM is less marked, and the distribution of codon 129 genotypes is different between France (61 % MM) and the UK (only 5 % MM). There is no significant difference in the distribution of codon 129 genotypes by country in sCJD [20].

14.9.2 fCJD/gCJD

It occurs in families but also in those without a family history. The penetrance of most *PRNP* mutations seems to be close to 100 %. In the E200K mutation, the penetrance has been indicated to be as low as 59.5 % in Slovakia [44], but it is nearly 100 % by age 85 in fCJD/gCJD of Libyan immigrants in Israel [45].

14.9.3 iCJD

The most frequent cause of infection relates to the treatment of people with hormones derived from the pituitary glands of deceased people (growth hormone, gonadotrophins) [14]. The second risk factor is the transplantation of dura mater grafts, particularly in Japan [46] (Table 14.1). A few patients have acquired iCJD by way of cornea transplantations [47–50] (see Chapter 51) or through the use of contaminated surgical instruments [15].

14.9.4 vCJD (formerly nvCJD)

Epidemiological and, in particular, experimental findings (see Chapter 50) suggest that vCJD can be traced back to the transmission of BSE from cattle to humans [14; 51; 52]. Further details on the mechanism of transmission, likely incubation period, and other aspects are discussed in Chapter 15.3.

With regard to the issue of blood as a risk factor in the development of CJD, no experimental or epidemiological evidence has so far become available that suggests that sCJD is transmitted by way of blood transfusions or through the use of blood products. A higher frequency of CJD cases was not observed in people who are at a higher risk of contracting the disease as a result of their higher exposure to blood products (e. g., hemophiliacs). Although more recently very small amounts of PrP^{Sc} have become detectable in peripheral tissues such as muscle and spleen in sCJD [40], their transmission risks are unclear at present.

This relatively favorable situation with sCJD has dramatically changed with the appearance of vCJD: in vCJD, PrP^{Sc} and infectivity are not restricted to the CNS as in other CJDs but spread in substantial amounts to lymphatic tissue such as the tonsils and the appendix [53; 54]. For this reason, the possibility of the infectious agent existing in blood has not been ruled out (see Chapter 51). Indeed, recently two cases of iatrogenic vCJD were reported as secondary human-to-human transmissions via blood transfusions [55; 56].

14.10 Surveillance, prevention, and control

Surveillance of human TSEs has become sufficiently established primarily in the developed world. In Europe and some allied countries, surveillance is carried out within the EuroCJD network. Surveillance should also be implemented in developing countries [28].

Prevention and control measures must focus on infection control in medical and hospital settings, as well as in the food sector. Basically, tissue and body fluids of patients and people suspected of suffering from a prion disease must not be used to produce therapeutic and other products.

In general, measures must be taken at all levels in order to prevent the transmission of the disease. Three potential areas of risk have been identified:

1. The possibility of the patient being infected with CJD when undergoing invasive medical interventions (see Chapter 47).
2. The risk of infection for people working in medicine, pathology, or research laboratories who handle infectious tissue (see Chapters 47–49).
3. The risk of infection from certain animal products or certain tissues or body fluids, either by way of food, medicine, or cosmetics (see Chapters 46, 47, and 53).

14.10.1 Protection of the patient

Iatrogenic transmission of CJD can best be prevented by destroying the instruments that have been used for surgical operations, in particular operations on the eye or CNS, of people that are suspected of suffering from CJD. If economic considerations do not allow the destruction of used, potentially contaminated instruments, then the instruments must be treated according to the guidelines described in Chapter 47.

Nowadays, it is possible to produce pituitary gland hormones through genetic engineering. The WHO has stated that only genetically engineered hormones should be used, and that dura mater grafts (see Chapter 51) from diseased people should not be used [28]. Similar restrictions should also apply to the selection of blood donors (see Chapter 51).

14.10.2 Protection of people working in clinical medicine, neuropathology, or in research

The measures described in Chapters 44 and 47–49 have to be taken into consideration when handling potentially prion-contaminated material [57].

14.10.3 Protection of the consumer from BSE-contaminated food, medicines, and cosmetics

The most important measure that was introduced in order to prevent humans from consuming potentially BSE-/TSE-contaminated tissues or from coming into contact with such tissues is the careful and controlled exclusion of certain animal tissues (specified risk material, SRM) from the food production. The preventive measures are described in Chapter 53.3.1.

Preventive measures concerning the selection and decontamination of raw animal tissue, body fluid, and organs for use in the production of medicines and cosmetics are equally important; as described in detail in Chapter 46.

14.10.4 Treatment

While there are some promising strategies and already some interesting in vitro and experimental in vivo results on the potential treatment of CJD [59], there is currently no effective therapy established. Current top priorities for research [60] are, aside from the development of new therapeutic agents, in particular the development of tests for a reliable diagnosis in vivo (live test) and for screening blood and blood products from pre-symptomatic donors.

References

[1] Creutzfeldt HG. Über eine eigenartige herdförmige Erkrankung des Zentralnervensystems. Z Ges Neurol Psychiat 1920; 57:1–18.

[2] Jakob A. Über eigenartige Erkrankungen des Zentralnervensystems mit bemerkenswertem anatomischem Befunde (spastische Pseudosklerose-Encephalomyelopathie mit disseminier-

ten Degenerationsherden). Dtsch Z Nervenheilk 1921; 70:132–146.

[3] Jakob A. Über eigenartige Erkrankungen des Zentralnervensystems mit bemerkenswertem anatomischen Befunde (Spastische Pseudosklerose-Encephalomyelopathie mit disseminierten Degenerationsherden). Z Ges Neurol Psychiat 1921; 64:147–228.

[4] Jakob A. Über eine der multiplen Sklerose klinisch nahestehende Erkrankung des Centralnervensystems (spastische Pseudosklerose) mit bemerkenswertem anatomischem Befunde. Mitteilung eines vierten Falles. Med Klin 1921; 17:382–386.

[5] Jakob A. Spastische Pseudosklerose. In: Foerster O, Willmanns K, editors. Die extrapyramidalen Erkrankungen mit besonderer Berücksichtigung der pathologischen Anatomie und Histologie und der Pathophysiologie der Bewegungsstörungen. Monographien aus dem Gesamtgebiete der Neurologie und Psychiatrie. Julius Springer, Berlin, Vol 37, 1923:215–345.

[6] Richardson EP Jr and Masters CL. The nosology of Creutzfeldt–Jakob disease and conditions related to the accumulation of PrP^{CJD} in the nervous system. Brain Pathol 1995; 5(1):33–41.

[7] Poser CM and Bruyn GW. Creutzfeldt–Jakob disease. In: Koehler PJ, Bruyn GW, Pearce JMS, editors. Neurological eponyms. Oxford University Press, Oxford/New York, 2000:283–290.

[8] Masters CL and Gajdusek DC. The spectrum of Creutzfeldt–Jakob disease and the virus-induced subacute spongiform encephalopathies. In: Smith WT, Cavanagh JB, editors. Recent advances in neuropathology. Churchill Livingstone, New York, 1982:139–163.

[9] Kretzschmar HA, Neumann M, Stavrou D. Codon 178 mutation of the human prion protein gene in a German family (Backer family): sequencing data from 72-year-old celloidin-embedded brain tissue. Acta Neuropathol 1995; 89(1):96–98.

[10] Brown P, Cervenakova L, Boellaard JW, et al. Identification of a PRNP gene mutation in Jakob's original Creutzfeldt–Jakob disease family [letter]. Lancet 1994; 344(8915):130–131.

[11] Budka H, Head MW, Ironside JW, et al. Sporadic Creutzfeldt–Jakob disease. In: Dickson D, editor. Neurodegeneration. The molecular pathology of dementia and movement disorders. ISN Neuropath Press, Basel, 2003:287–297.

[12] Liberski PP and Budka H. An overview of neuropathology of the slow unconventional virus infections. In: Liberski PP, editor. Light and electron microscopic neuropathology of slow virus disorders. CRC Press, Boca Raton, FL, 1993: 111–149.

[13] Gajdusek DC, Gibbs CJ Jr, Alpers M. Experimental transmission of kuru-like syndrome to chimpanzees. Nature 1966; 209:794-796.

[14] Will RG. Acquired prion disease: iatrogenic CJD, variant CJD, kuru. Br Med Bull 2003; 66(1):255–265.

[15] Brown P, Preece M, Brandel J-P, et al. Iatrogenic Creutzfeldt–Jakob disease at the millennium. Neurology 2000; 55:1075–1081.

[16] Brown P, Preece MA, Will RG. "Friendly fire" in medicine: hormones, homografts, and Creutzfeldt–Jakob disease. Lancet 1992; 340:24–27.

[17] Will RG, Ironside JW, Zeidler M, et al. A new variant of Creutzfeldt–Jakob disease in the UK. Lancet 1996; 347:921–925.

[18] Budka H, Dormont D, Kretzschmar H, et al. BSE and variant Creutzfeldt–Jakob disease: never say never. Acta Neuropathol 2002; 103: 627–8.

[19] Kovács GG, Voigtländer T, Gelpi E, et al. Rationale for diagnosing human prion disease. World J Biol Psychiat 2004; 5:83–91.

[20] Ladogana A, Puopolo M, Croes EA, et al. Mortality from Creutzfeldt–Jakob disease and related disorders in Europe, Australia, and Canada. Neurology 2005; 64(9):1586–1591.

[21] Parchi P, Giese A, Capellari S, et al. Classification of sporadic Creutzfeldt–Jakob disease based on molecular and phenotypic analysis of 300 subjects. Ann Neurol 1999; 46:224–233.

[22] Hill AF, Joiner S, Wadsworth JDF, et al. Molecular classification of sporadic Creutzfeldt–Jakob disease. Brain 2003; 126:1333–1346.

[23] Head MW, Bunn TJR, Bishop MT, et al. Prion protein heterogeneity in sporadic but not variant Creutzfeldt–Jakob disease: UK cases 1991–2002. Ann Neurol 2004; 55(6):851–859.

[24] Polymenidou M, Stoeck K, Glatzel M, et al. Coexistence of multiple PrPSc types in individuals with CJD. Lancet Neurol 2005; 4:805–814.

[25] Kovács GG, Puopolo M, Ladogana A, et al. Genetic prion disease: the EUROCJD experience. Hum Genet 2005; 118(2):166–174.

[26] Kovács GG, Trabattoni G, Hainfellner JA, et al. Mutations of the prion protein gene: phenotypic spectrum. J Neurol 2002; 249(11):1567–1582.

[27] Zerr I, Pocchiari M, Collins S, et al. Analysis of EEG and CSF 14–3–3 proteins as aids to the diagnosis of Creutzfeldt–Jakob disease. Neurology 2000; 55:811–815.

[28] WHO. WHO manual for surveillance of human transmissible spongiform encephalopathies including variant Creutzfeldt–Jakob disease. WHO Communicable Disease Surveillance and Response, Geneva, 2003.

[29] Collie DA, Sellar RJ, Zeidler M, et al. MRI of Creutzfeldt–Jakob disease: imaging features and recommended MRI protocol. Clin Radiol 2001; 56:726–739.

[30] Pocchiari M, Puopolo M, Croes EA, et al. Predictors of survival in sporadic Creutzfeldt–Jakob disease and other human transmissible spongiform encephalopathies. Brain 2004; 127(10):2348–2359.

[31] Lee HS, Sambuughin N, Cervenakova L, et al. Ancestral origins and worldwide distribution of the PRNP E200K mutation causing familial Creutzfeldt–Jakob disease. Am J Hum Genet 1999; 64:1063–1070.

[32] Budka H, Aguzzi A, Brown P, et al. Neuropathological diagnostic criteria for Creutzfeldt–Jakob disease (CJD) and other human spongiform encephalopathies (prion diseases). Brain Pathol 1995; 5:459–466.

[33] Kovács GG, Head MW, Hegyi I, et al. Immunohistochemistry for the prion protein: comparison of different monoclonal antibodies in human prion disease subtypes. Brain Pathol 2002; 12:1–11.

[34] Budka H. Neuropathology of prion diseases. Br Med Bull 2003; 66(1):121–30.

[35] Kovacs GG, Preusser M, Strohschneider M, et al. Subcellular localization of disease-associated prion protein in the human brain. Am J Pathol 2005; 166(1):287–294.

[36] Ishida C, Okino S, Kitamoto T, et al. Involvement of the peripheral nervous system in human prion diseases including dural graft associated Creutzfeldt–Jakob disease. J Neurol Neurosurg Psychiat 2005; 76(3):325–329.

[37] Hainfellner JA and Budka H. Disease associated prion protein may deposit in the peripheral nervous system in human transmissible spongiform encephalopathies. Acta Neuropathol (Berl) 1999; 98:458–460.

[38] Zanusso G, Ferrari S, Cardone F, et al. Detection of pathologic prion protein in the olfactory epithelium in sporadic Creutzfeldt–Jakob disease. N Engl J Med 2003; 348(8):711–719.

[39] Koperek O, Kovacs GG, Ritchie D, et al. Disease-associated prion protein in vessel walls. Am J Pathol 2002; 161(6):1979–1984.

[40] Glatzel M, Abela E, Maissen M, et al. Extraneural pathologic prion protein in sporadic Creutzfeldt–Jakob disease. N Engl J Med 2003; 349(19):1812–1820.

[41] Kovács GG, Lindeck-Pozza E, Chimelli L, et al. Creutzfeldt–Jakob disease and inclusion body myositis: abundant disease-associated prion protein in muscle. Ann Neurol 2004; 55(1):121–125.

[42] Palmer MS, Dryden AJ, Hughes JT, et al. Homozygous prion protein genotype predisposes to sporadic Creutzfeldt–Jakob disease. Nature 1991; 352:340–342.

[43] Collinge J, Palmer MS, Dryden AJ. Genetic predisposition to iatrogenic Creutzfeldt–Jakob disease. Lancet 1991; 337:1441–1442.

[44] Mitrova E and Belay G. Creutzfeldt–Jakob disease with E200K mutation in Slovakia: characterization and development. Acta Virol 2002; 46:31–39.

[45] Meiner Z, Gabizon R, Prusiner SB. Familial Creutzfeldt–Jakob disease. Codon 200 prion disease in Libyan Jews. Medicine (Baltimore) 1997; 76(4):227–237.

[46] Centers for Disease Control and Prevention (CDC). Update: Creutzfeldt–Jakob disease associated with cadaveric dura mater grafts: Japan, 1979 to 2003. MMWR 2003; 52(48):1179–1181.

[47] Duffy P, Wolf J, Collins G, et al. Possible person-to-person transmission of Creutzfeldt–Jakob disease. N Engl J Med 1974; 290(12):692–693.

[48] Heckmann JG, Lang CJ, Petruch F, et al. Transmission of Creutzfeldt–Jakob disease via a corneal transplant. J Neurol Neurosurg Psychiat 1997; 63(3):388–390.

[49] Uchiyama S, Ishida C, Yago S, et al. An autopsy case of Creutzfeldt–Jakob disease associated with corneal transplantation [in Japanese]. Dementia 1994; 8:466–473.

[50] Hammersmith K, Cohen EJ, Rapuano CJ, et al. Creutzfeldt–Jakob disease following corneal transplantation. Cornea 2004; 23(4):406–408.

[51] Collinge J, Sidle KCL, Meads J, et al. Molecular analysis of prion strain variation and the aetiology of "new variant" CJD. Nature 1996; 383:685–690.

[52] Bruce ME, Will RG, Ironside J, et al. Transmissions to mice indicate that "new variant" CJD is caused by the BSE agent. Nature 1997; 389:498–501.

[53] Hilton DA, Ghani AC, Conyers L, et al. Prevalence of lymphoreticular prion protein accumulation in UK tissue samples. J Pathol 2004; 203:733–739.

[54] Ironside JW, Head MW, Bell JE, et al. Laboratory diagnosis of variant Creutzfeldt–Jakob disease. Histopathology 2000; 37:1–9.

[55] Peden AH, Head MW, Ritchie DL, et al. Preclinical vCJD after blood transfusion in a PRNP codon 129 heterozygous patient. Lancet 2004; 264:527–529.

[56] Llewelyn CA, Hewitt PE, Knight RSG, et al. Possible transmission of variant Creutzfeldt–Jakob disease by blood transfusion. Lancet 2004; 363:417–421.

[57] Budka H, Aguzzi A, Brown P, et al. Tissue handling in suspected Creutzfeldt–Jakob disease (CJD) and other human spongiform encephalopathies (prion diseases). Brain Pathol 1995; 5:319–322.

[58] Advisory Committee on Dangerous Pathogens. Precautions for work with human and animal transmissible spongiform encephalopathies, 1st ed. HMSO, London, 1994.

[59] Mallucci G and Collinge J. Rational targeting for prion therapeutics. Nat Rev Neurosci 2005; 6(1):23–34.

[60] Erdtmann R and Sivitz LB. Advancing prion science. Guidance for the National Prion Research Program. The National Academies Press, Washington, DC, 2003.

[61] Brown P. Transmissible spongiform encephalopathy. In: Goetz C, editor. Textbook of clinical neurology, chapter 43, 3rd ed. W. B. Saunders, Philadelphia, 2006: in press.

15 Portrait of Variant Creutzfeldt–Jakob Disease

Robert G. Will and James W. Ironside

15.1 History

Surveillance of Creutzfeldt–Jakob disease (CJD) was instituted in the United Kingdom in 1990 following the occurrence of bovine spongiform encephalopathy (BSE). The primary aim was to identify any change in the epidemiology or other characteristics of CJD. In 1993, a collaborative surveillance program for CJD was established in six European countries, including France, Germany, Italy, the Netherlands, Slovakia, and the UK. In the early 1990s, there was an increase in the incidence of CJD in the UK, but this was paralleled by a similar change in the incidence in other countries, which had not reported cases of BSE [1]. In late 1995 and early 1996, a small number of cases of CJD with a remarkably young age at death were identified in the UK and further investigation indicated that these cases had an unusual clinical phenotype [2], with early psychiatric symptoms and a prolonged duration of illness in comparison with sporadic CJD. Crucially, the neuropathology in these cases was distinct from previous cases of CJD. By March 1996, 10 cases of what was believed to be a new variant of CJD had been identified. Importantly, no similar case had, at that stage, been identified in any other country in the European surveillance system. This evidence raised the possibility that this new disease, now known as variant CJD (vCJD), might be causally linked to BSE exposure in the UK [3].

15.2 Forms or variants

By May 2006, 161 cases of vCJD had been identified in the UK, 17 cases in France, 4 in Ireland, 2 in the USA, and single cases in 7 other countries (see below). There has been remarkable consistency in the clinical and pathological features in all these cases, and all tested cases have been methionine homozygotes at codon 129 of the prion protein gene. On current evidence, there is only one form of vCJD. However, it is possible that in individuals who are either valine homozygotes or heterozygotes there may be a different clinicopathological phenotype. Such cases have not, as yet, been identified.

15.3 Incubation period, transmissibility, and susceptibility

The current hypothesis is that variant CJD was caused by exposure to high-titer bovine tissue in the human food chain in the UK in the 1980s. During this period, bovine spinal cord may have contaminated mechanically recovered meat, and there may have been fairly extensive population exposure to the BSE agent. Legislative measures introduced in the UK, including the specified bovine offals (SBO) ban in 1989 and the "Over 30 Months Rule" (OTMS) in 1996, have significantly reduced such exposure.

The incubation period in sporadic CJD is not known. In iatrogenic CJD with a peripheral route of infection, the minimum incubation period is about 4.5 years, similar to that in kuru. By analogy, human infection with the BSE agent is likely to have a minimum incubation period of at least 4.5 years and probably longer, as the incubation period is usually prolonged after cross-species infection. The average incubation period in human growth hormone related CJD is about 12 years (Table 14.1), but the average incubation period in vCJD is unknown and might be longer; some estimates have been derived from mathematical models (see Chapter 1.9.3).

Transmission studies in inbred strains of mice have demonstrated that the BSE agent has remarkably consistent transmission characteristics, which are independent of the source of the BSE agent [4]. The incubation periods (Fig. 50.1) and the pathological "lesion profiles" (Fig. 50.2) in mice are very similar following inoculation of BSE itself, and other conditions that are thought to be BSE related, such as feline spongiform encephalopathy and prion diseases of exotic zoo ungulates in the UK (see Chapter 20). The transmission characteristics of vCJD in the same inbred lines of mice are indistinguishable from previous BSE transmissions, providing strong evidence that the agent causing vCJD is the same as the BSE agent [5–7] (see Chapter 50). Transmission studies in bovine transgenic mice have supported these findings, and have shown very similar transmission characteristics and brain pathology following inoculation with BSE and vCJD. These findings are quite different to those following inoculation with scrapie and sporadic CJD. Biochemical studies of the abnormal prion protein in the central nervous system (CNS) in vCJD have shown a similar glycosylation profile to that seen in BSE and other BSE-related conditions, including a macaque model of BSE that also shows similar neuropathology to vCJD.

15.4 Clinical signs and course of disease

vCJD is characterized clinically by a prolonged prodromal phase dominated by psychiatric symptoms, followed by a terminal phase that is similar to classical CJD [8; 9]. Painful sensory symptoms occur in about half the cases, and involuntary movements are a consistent feature, although chorea and dystonia occur as well as myoclonus. All patients develop ataxia, which is followed by a progressive dementing process, which terminates in a state of helplessness and sometimes with akinetic mutism. The mean survival is about 14 months from the first symptom, which contrasts with sporadic CJD, in which the mean survival averages 4 months.

The electroencephalogram usually shows slow activity, but may be normal even after neurological signs develop. The cerebrospinal fluid is acellular with a normal or slightly elevated protein content. The 14–3–3 immunoassay is positive in about half the cases (see Chapter 31). The most helpful investigation is MRI brain scanning, as the great majority of cases of vCJD exhibit symmetrical posterior thalamic high signal, which is relatively specific for this condition (compare Fig. 31.6). A definite diagnosis of vCJD depends on the histological examination of brain material; some patients have undergone brain biopsy, although this procedure has risks.

15.5 Differential diagnoses

Since March 1996, there have been approximately twice as many referrals with suspect vCJD as cases with confirmation of this diagnosis. In the majority of these cases, the patients have presented with an encephalopathic illness with subsequent recovery. In patients who have died, the major differential diagnosis is sporadic CJD and, in particular, cases of sporadic CJD associated with a valine homozygosity or heterozygosity at codon 129 of the PrP gene (Table 26.3). Other alternative diagnoses included SSPE, herpes simplex encephalitis, Wilson's disease, and hypoxic encephalopathy.

15.6 Epidemiology

vCJD was first identified in the UK and there had been 161 cases until May 2006 (Fig. 15.1). Thirty cases have been identified outside the UK, including 17 in France, 4 in Ireland, 2 in the USA, and single cases in Italy, Canada, Japan, Saudi Arabia, Spain, the Netherlands, and Portugal. The Canadian and US cases, two Irish cases, and possibly the Japanese case, are likely to have been exposed to BSE while resident in the UK in the 1980s or early 1990s. Systematic surveillance for CJD is carried out in a number of countries. Within the UK, cases of vCJD are widely distributed geographically, and there is no good evidence of clustering of cases, although mortality rates are approximately double in the northern half of the UK. The number of deaths per annum from vCJD peaked in the UK in 2000 and has subsequently

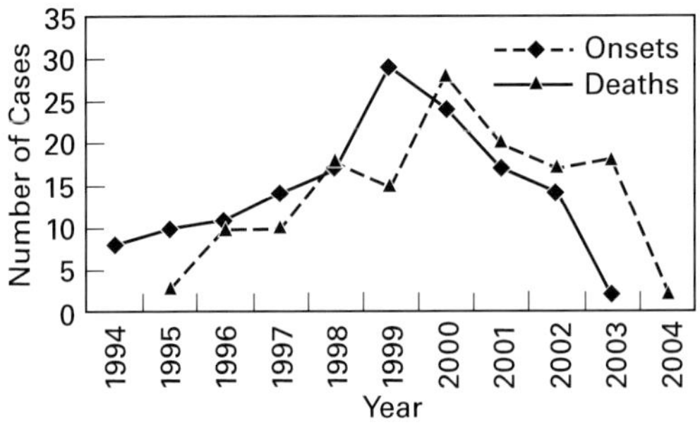

Fig. 15.1: Annual numbers of onset and deaths per annum for vCJD in the UK, 1994–2004. Figure by R. G. Will ©.

declined (Fig. 15.1). Recent mathematical analysis of deaths and clinical onsets indicates that the epidemic of vCJD in the UK may be in decline.

The mean age of death in vCJD is approximately 29 years, which compares with a mean age at death in sporadic CJD of about 66 years (Fig. 15.2). The reason for the relatively early age at death in vCJD is unknown, but may relate to distinct patterns of dietary exposure in different age groups or age-related susceptibility. The youngest patient was aged 14 years at death and the oldest was aged 74 years.

15.7 Pathology

Neuropathologically, vCJD is characterized by large numbers of florid plaques (Fig. 15.3a) in the cerebral and cerebellar cortex. These plaques are comprised of an eosinophilic core and a peripheral region composed of radiating fibrils, surrounded by a rim of spongiform change. These plaques stain intensely on immunohistochemistry for PrP (Fig. 15.3b), which also demonstrates large numbers of smaller plaques and amorphous PrP deposits that are not evident on routine stains. Other key neuropathological features include severe spongiform change in the caudate nucleus and marked neuronal loss and gliosis in the posterior thalamus, particularly the pulvinar (Fig. 15.3c). An important pathological feature of vCJD is the presence of abnormal PrP in non-CNS tissues, particularly in lymphoid tissues throughout the body [tonsil (Fig. 15.3d), lymph node, spleen, thymus, and gut] and sensory ganglia, including dorsal root and trigeminal ganglia. Infectivity has been confirmed in the tonsil and spleen by transmission to inbred lines of mice, where the levels of infectivity were estimated to be around 1.5–2 logs lower than in the CNS. The biochemical profile of abnormal PrP in Western blots from the brain in vCJD shows a uniform pattern that is generally distinct from cases of

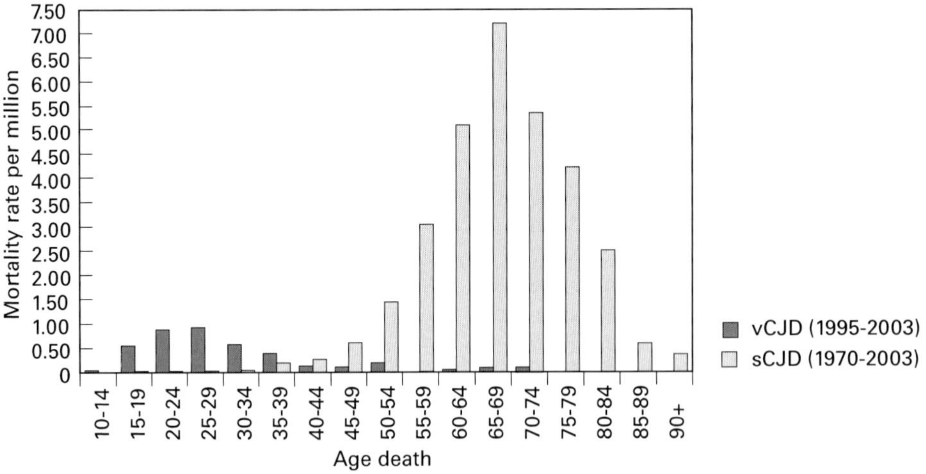

Fig. 15.2: Age distribution and mortality rates of sCJD and vCJD patients in the UK, 1970–2003. Figure by R. G. Will ©.

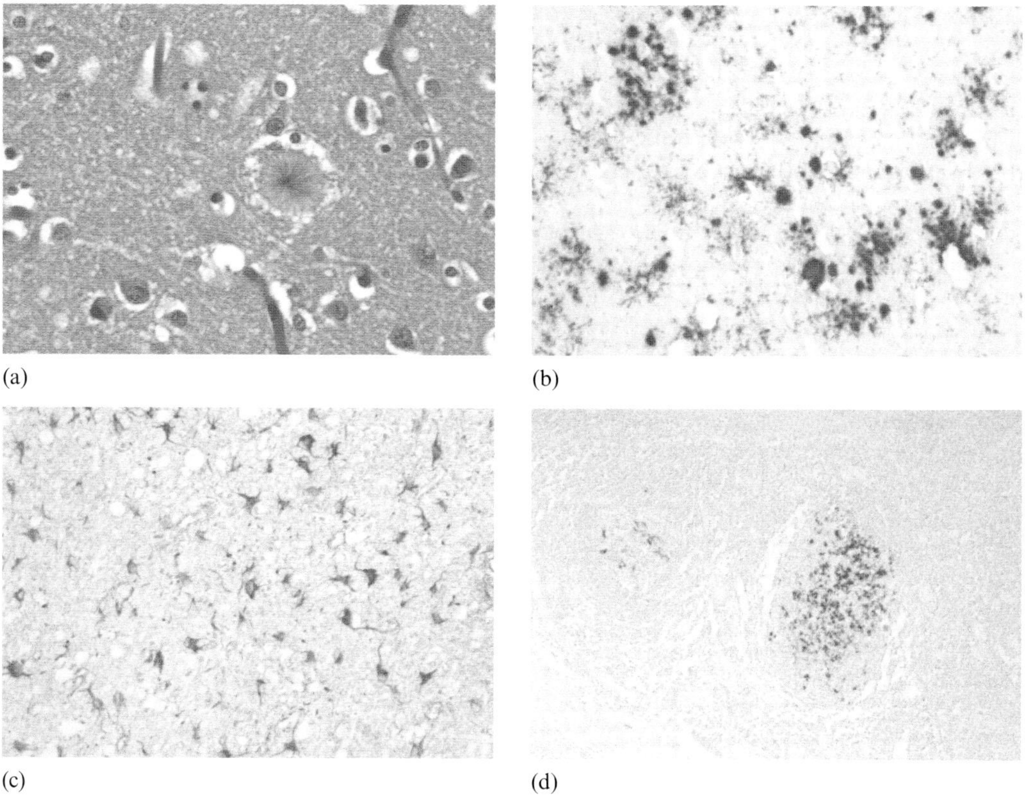

Fig. 15.3: **(a)** A florid plaque in the frontal cortex in a case of variant CJD (vCJD). This comprises a dense core within a fibrillary pale-stain in peripheral region in the plaque, which is surrounded by spongiform change. The adjacent neuropil is intact. Hematoxylin and eosin, original magnification ×20. Figure by James Ironside ©. **(b)** Immunohistochemistry (IHC) for PrP shows strong staining of the florid plaques in the cerebral cortex in vCJD, and in addition demonstrates multiple smaller plaques and amorphous PrP deposits that cannot be visualized on routine stains. 3F4 antibody, original magnification ×10. Figure by James Ironside ©. **(c)** IHC for glial fibrillary acidic protein (GFAP) demonstrates severe astrocytosis in the pulvinar in vCJD. No plaques are present but there is patchy spongiform change and almost total neural loss. GFAP antibody, original magnification ×10. Figure by James Ironside ©. **(d)** IHC for PrP in the tonsil in vCJD shows labeling of follicular dendritic cells within germinal centers. The intervening lymphoid tissue is unlabelled. 3F4 antibody, original magnification ×4. Figure by James Ironside ©.

sporadic CJD and other forms of human prion disease in terms of the pattern of glycosylation and the molecular size of the unglycosylated fragment, although a few reports exist of partially similar biochemical findings in a small number of cases. A similar glycosylation pattern is found in abnormal PrP from non-CNS tissues in vCJD, although tissue-specific variations in glycosylation have been identified.

15.8 Is vCJD a new disease?

vCJD was first suspected to be a new disease because of the young age at death. Cases of sporadic CJD dying at an early age had been previously described in other countries, but these reports were exceptional and only 4 cases of CJD in teenagers had been reported in the world literature. By March 1996, 10 cases of CJD had been identified with an average age at death of 29 years. There have now been 161 cases of vCJD in the UK, of which 51 were aged less than 25 years at death.

The clinical phenotype of vCJD is relatively distinct from sporadic CJD (Table 30.3). Psychiatric features, painful sensory symptoms, prolonged duration of illness, and an atypical EEG have all been described in sporadic CJD, but as a group, the vCJD cases are remarkably similar and exhibit clinical features that are distinct from those usually seen in sporadic CJD.

When vCJD was first described, the neuropathological features, including florid plaque formation (Fig. 15.3b), were crucial to the hypothesis that this was a new disease. Since the "specific" neuropathological features were published, contemporary and archive material in other countries has been examined. However, to date, no past case with these pathological features has been identified. The balance of this evidence strongly suggests that vCJD is indeed a new disease.

The hypothesis that BSE and vCJD were causally linked was based, in part, on the colocalization of this new human disease and a putative new risk factor, BSE [10]. There have now been 161 cases of vCJD in the UK, and, with the exception of the case in Saudi Arabia, the other cases of vCJD identified outside the UK have spent periods of residence either in the UK [those in Ireland (2/4), Japan, Canada, and the USA] or in countries in which BSE cases have been identified in indigenous animals. The continuing occurrence of vCJD in individuals residing in the UK and other countries with cases of BSE supports the hypothesis of a causal link with BSE and also suggests that vCJD has not been identified solely as a result of improved case ascertainment (see Chapter 50). Furthermore, the experimental evidence summarized above on the relationship of the BSE agent to that of the agent in vCJD strongly supports a causative relationship between these disorders.

15.9 Risk factors

All tested cases of vCJD, to date, have a methionine homozygote genotype (Table 26.3) at codon 129 of *PRNP*, and this genetic factor, together with residence in the UK, appears to be the main risk factor for the development of disease. However, the genotype at codon 129 may influence the incubation period, and it is possible that cases with other genotypes at this locus may be identified in the future.

A case control study, using hospital-based controls, has been undertaken in order to identify risk factors for the development of disease. Putative risk factors that have been studied include past medical exposure, occupation, contact with animals, and dietary history. An apparent association between vCJD and dietary consumption of bovine-derived meat products has been identified, but these results should be interpreted with caution because of recall bias and because evidence on risk factors is obtained from a surrogate witness. Furthermore, identifying specific dietary risk factors may be very difficult if the relevant dietary exposure took place many years in the past and if products containing high-titer bovine tissue were intermittently contaminated.

15.10 Surveillance, prevention, and control

If vCJD is caused by the BSE agent, the cases that have been identified to date may reflect exposure to BSE in the 1980s and perhaps before BSE was even recognized. A range of measures was introduced in the UK in the late 1980s in order to minimize human exposure to the BSE agent, for example through the food chain. Subsequently, additional measures have been introduced in the light of scientific findings in the UK and many other countries (Tables 39.2–3), and the current human risks from BSE are probably low, on the assumption that appropriate measures have both been introduced and enforced. However, with the potentially prolonged incubation period in vCJD, it is possible that there will be many more such cases in the UK, and perhaps in other countries, even if human exposure to the BSE agent was minimized many years ago.

The occurrence of iatrogenic CJD indicates that there is a potential for case-to-case transmission of human prion disease via medical procedures. The risks of onward transmission of a vCJD may be greater than sporadic CJD because vCJD is caused by a novel infectious agent for humans, and there is evidence that the peripheral pathogenesis in vCJD is distinct from sporadic CJD (compare Chapter 28). Although epidemiological evidence does not suggest that sporadic CJD is transmitted iatrogenically through blood or blood products, this evidence may not be relevant to vCJD [11], and recent reports have raised the possibility of transfusion transmission of vCJD [12; 13]. A range of measures, including the leukodepletion of all blood donations and external sourcing of plasma for the production of blood derivatives, have been introduced in the UK in order to minimize the risks of onward transmission of vCJD (see Chapter 51). Plasma products to be used in children born after 1995 are now imported into the UK, and recently a ban has been introduced on any individual becoming a blood donor in the UK if they have received a blood transfusion there since 1980. Deferral of blood donors who have been resident in the UK for defined periods has also been introduced in many other countries.

References

[1] Alperovitch A, Brown P, Weber T, et al. Incidence of Creutzfeldt–Jakob disease in Europe in 1993 (letter). Lancet 1994; 343:918.

[2] Will RG, Zeidler M, Stewart GE, et al. Diagnosis of new variant Creutzfeldt–Jakob disease. Ann Neurol 2000; 47:575–582.

[3] Will RG, Ironside JW, Zeidler M, et al. A new variant of Creutzfeldt–Jakob disease in the UK. Lancet 1996; 347:921–925.

[4] Bruce ME, Chree A, McConnell I, et al. Transmission of bovine spongiform encephalopathy and scrapie to mice: strain variation and species barrier. Philos Trans R Soc Lond Biol 1994; 343:405–411.

[5] Hill AF, Desbruslais M, Joiner S, et al. The same prion strain causes vCJD and BSE. Nature 1997; 389:448–449.

[6] Bruce ME, Will RG, Ironside JW, et al. Transmissions to mice indicate that new variant CJD is caused by the BSE agent. Nature 1997; 389:498–501.

[7] Scott MR, Will R, Ironside J, et al. Compelling transgenetic evidence for transmission of bovine spongiform encephalopathy prions to humans. Proc Natl Acad Sci USA 1999; 96(26):15137–15142.

[8] Zeidler M, Stewart GE, Barraclough CR, et al. New variant Creutzfeldt–Jakob disease: neurological features and diagnostic tests. Lancet 1997; 350(9082):903–907.

[9] Zeidler M, Johnstone EC, Bamber RWK, et al. New variant Creutzfeldt–Jakob disease: psychiatric features. Lancet 1997; 350:908–910.

[10] Will RG. New variant Creutzfeldt–Jakob disease. In: Brown F, Griffiths E, Horaud F, et al., editors. Safety of biological products prepared from mammalian cell culture. Karger, Basel, 1998:79–84.

[11] Will RG and Kimberlin RH. Creutzfeldt–Jakob disease and the risk from blood or blood products. Vox Sanguinis 1998; 75:178–180.

[12] Llewelyn CA. Possible transmission of variant Creutzfeldt–Jakob disease by blood transfusion. Lancet 2004; 363(9407):417–421.

[13] Peden AH, Head MW, Ritchie DL, et al. Preclinical vCJD after blood transfusion in a PRNP codon 129 heterozygous patient. Lancet 2004; 364(9433):527–529.

16 Portrait of Gerstmann-Sträussler-Scheinker Disease[1]

Herbert Budka

16.1 History

In 1936, Josef Gerstmann, Ernst Sträussler, and Isaac Scheinker[2], who, at that time, all worked in Vienna, described "a rare hereditary familial disease of the central nervous system" in a young woman [1]. Previously, in 1928, Gerstmann had published a preliminary description of the clinical picture of the disease [2]. The family in question was well known among Viennese neurologists in the early decades of the 20th century. As early as 1912, at a meeting of the Viennese Association for Neurology and Psychiatry, Dimitz presented the clinical picture of a young family member [3]. After World War II, four members of this family died and the neuropathological picture was described by Braunmühl [4] and Seitelberger [5]. In 1962, Seitelberger was the first to realize the great clinical and neuropathological similarities of this disease with kuru (see Chapter 13). In the following years, Seitelberger lost contact with the family. A major reason for this was that several family GPs frequently diagnosed their disease as "Tabes dorsalis" (a form of tertiary syphilis), which progressively destroys the structures of the dorsal column of the spinal cord. The doctors, unaware of the far less common Gerstmann–Sträussler–Scheinker disease (GSS), were unable to treat and help the family members. The fact that the family was diagnosed as a "luetic family" did not lend confidence in these doctors, particularly in the small village settings where the family used to live. In the late 1980s, the original Austrian GSS family was rediscovered: a case that resembled Creutzfeld–Jakob disease (CJD) ultimately turned out to belong to the "H" part of the original GSS family [6]. The pedigree was updated to include now 221 family members from nine generations (from the late 18th century onward). Of these, at least 20 patients suffered from definite GSS [7] (Fig. 16.1). After Seitelberger's report [5], additional families were found that had similar clinical and neuropathological features [8; 9].

The successful experimental transmission of GSS finally led the scientists to include the disease in the group of spongiform encephalopathies (TSEs or prion diseases) [8]. The classical *PRNP* mutation P102L of GSS was the first genetic aberration discovered in a human TSE [10] and also identified in the original GSS family [6]. On the basis of these findings, GSS can be classified as a familial or genetic prion disease [11].

The paramount feature of GSS brain pathology is the multicentric plaque (Fig. 16.2), which has been proposed as the definition criterion of GSS (Table 29.3). On the basis of the examination of several families and the typical features associated with the disease, it was recommended that GSS only be diagnosed in families with progressive ataxia and/or dementia, the presence of many multicentric PrP plaques in brain tissue, and a GSS-associated *PRNP* mutation [12].

---

[1] Historically and in the literature often called "... syndrome".

[2] In the original 1936 article, Scheinker's first name (Isaac) was only initialised (I.). For political reasons, it was, at that time, not appropriate for an author of an article in a German journal to have a Jewish first name. From 1938 onward, all three authors became victims of the Nazis' persecution and lost their medical and university positions. Both Gerstmann and Scheinker managed to emigrate to the USA – Scheinker fled first to Paris and worked for one year in the Salpêtrière. Later, in the USA, as Mark I. Scheinker, he wrote several textbooks on medical and neurosurgical neuropathology (1947–1951). Sträussler, protected by his "Aryan" wife, survived the Nazi period in Vienna.

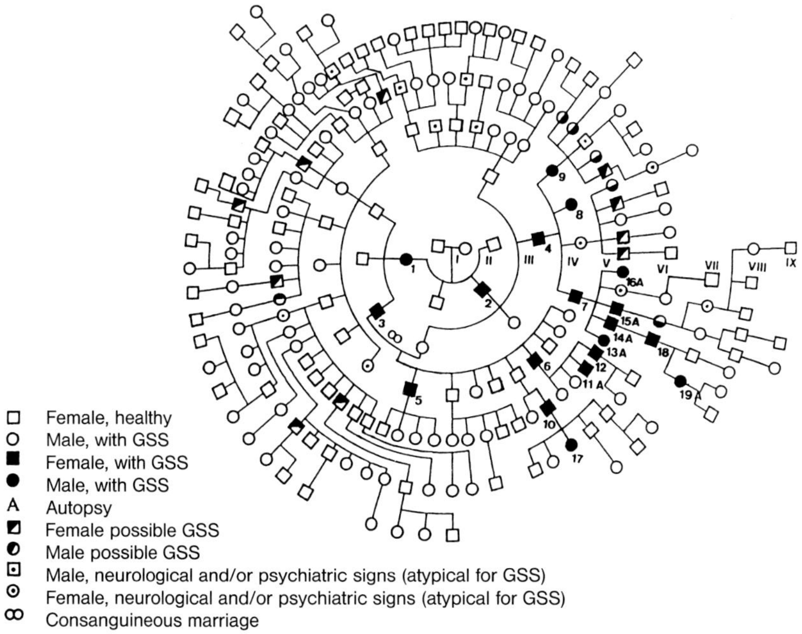

Fig. 16.1: Family tree of the original Austrian GSS family. The family tree includes 221 members over nine generations. Sixteen family members were diagnosed with potential and 20 with definite GSS. Two examined patients were homozygote for methionine (MM; at codon 129 of the *PRNP*). J. Gerstmann and his colleagues described patient 16A in 1936. Figure by H. Budka; source: [7].

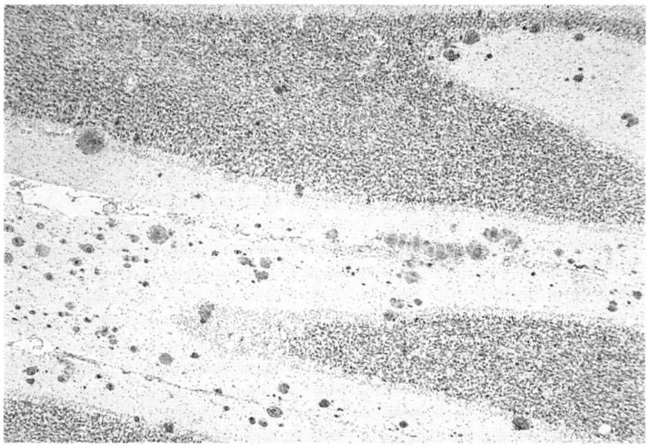

Fig. 16.2: Cerebellar cortex of a member of the original Austrian GSS family. Numerous distinctive, multicentric brown PrP plaques are located in the molecular layer. Immunohistochemistry involving an anti-PrP-antibody (brown staining substrate). Counter-staining with hematoxylin. Slide by H. Budka ©.

16.2 Forms or variants

While there is significant clinical heterogeneity, there are two major clinical forms of GSS: "classical" cases in which ataxia dominates, and a rarer, primarily dementing form [13]. Biochemically, Western blots show different patterns of prion protein fragments that correlate with distinct clinical phenotypes in P102L GSS [14] as well as with presence or absence of spongiform change in brain tissue [15]. A 7 kDa prion protein fragment – an integral component of the PrP region required for infectivity – is the major amyloid protein also in A117V GSS [16].

16.3 Incubation period, transmissibility, and susceptibility

As GSS is not an acquired prion disease, no natural incubation period can be observed.

It has been shown that GSS can be transmitted experimentally to rats, mice [17], and apes

[8; 18]. Nevertheless, the success of the transmission varies considerably, even within a single "family" [8; 18]. Kitamoto and Tateishi only succeeded in transmitting the disease caused by a *PRNP* 102 mutation to rodents in five out of nine cases. It was impossible to transmit one case of the original Austrian GSS family [7]. Successful experimental transmission had an incubation period from 20 to 32 weeks. The disease could not be transmitted when mutations other than P102L were involved. One isolate from Japan, the Fujisaki or Fukuoka strain, was passaged to mice, rats, guinea pigs, and squirrel monkeys, and seems to be the only well-characterized nontransgenic experimental model for human TSEs [19].

Carriers of a predisposing mutation are "susceptible" to the disease, and the penetrance seems to be near 100% [20]. Noteworthy, however, is a report on monozygotic twins, one of whom was healthy 8 years after the other became diseased [21].

16.4 Clinical signs and course of disease

GSS is the prototype of a familial or genetic TSE with an autosomal dominant mode of inheritance that is classically associated with a mutation in codon 102 of *PRNP* [6; 7]. Other GSS-associated mutations affect codons 105, 117, 131, 145, 187, 198, 202, 212, 217, and 232 (see Chapters 26.5.3 and 26.6). The mutation in codon 102 of *PRNP* was originally associated with the classical GSS form in which ataxia dominates [10; 11; 13]. The clinical picture begins very slowly, showing symptoms of progressive spinal ataxia and pyramidal signs. Higher-order brain activities are affected at a much later stage of the disease; manifestations range from forgetfulness, reduced intellectual performance, up to severe dementia. Overall, 82% of GSS patients ultimately have dementia, whereas myoclonus is rare (15%) [22]. The duration of illness is longer than 5 years in 31% of cases [22], with a median of 39 months [23]. GSS manifests usually in people around the mean age of 45, but this can vary considerably (Table 30.1). A positive family history has been found in 70% of cases [24].

The age of members of the original Austrian GSS family ranged between 25 and 59 at the time of disease onset, with a mean age of 44.3 years and a median of 45 years. The age at death ranged between 31 and 62 years (mean: 52.8; median: 54.5 years). The duration of disease ranged from 2 to 17 years (mean: 6.6; median 6.5) (Table 30.1).

One of the members of the original Austrian GSS family suffered from a primary dementing form of GSS rather than from the classical ataxic phenotype. Nevertheless, the patient had a PrP genotype that was identical to that of a member suffering from ataxic GSS [7], i.e., a point mutation in codon 102 leading to the substitution of proline through leucine, and homozygosity for methionine at codon 129 [P102L M129M].

Families with a consistent primary dementing phenotype usually possess different *PRNP* mutations [13; 25; 26] (Table 26.5) but in exceptional cases may have P102L [27]. The primarily dementing form of the disease progresses much more rapidly, as in CJD.

The *PRNP* codon 129 on the mutated allele usually encodes methionine. There is only one record on P102L 129V with an atypical presentation of seizures, lower limb paresthesias, and bilateral deafness [28].

The L105V mutation has spastic paraparesis as a peculiar presenting feature. In a Hungarian family with the A117V mutation, lower motor neuron deficits with loss of anterior horn motor neurons and chronic neurogenic muscle atrophy, and parkinsonism occurred in addition to intellectual decline and ataxia [29]. Another A117V patient clinically mimicked frontotemporal dementia [30].

Thus considerable clinical heterogeneity is found both within and between families with the same mutation. As Austrian [7] and Italian [31] families show, multiple phenotypes may be observed within one family. However, haplotype-specific patterns of clinical presentation have been described [20].

MRI (compare Fig. 31.6) has revealed progressive brain atrophy in GSS [32]. EEG alterations that were typical for CJD (see Chapter 30) were documented in one member of the original Austrian GSS family who carried the P102L mutation [7]. A detailed description of the clinical symptoms of GSS is given in Chapter 30.5.

16.5 Differential diagnoses

Cerebellar and spinal signs and symptoms occurring in the classical form of GSS in which ataxia dominates must be distinguished from other spinocerebellar degenerations or multisystem atrophy. The dementing phenotype is to be distinguished from CJD, Alzheimer's disease (AD), and other dementias. Whenever GSS is suspected, it is important to examine *PRNP* (see Chapter 29). Pathologically, other amyloidoses must be distinguished, including CJD and vCJD. Immunoblotting might be particularly helpful in such situations.

It is difficult to distinguish the dementing form of GSS, which is mainly characterized by pathological alterations of the telencephalon/cerebrum, from CJD. This can be resolved by genetic analysis of *PRNP*.

16.6 Epidemiology

Although exact epidemiological data are not available for GSS due to the scarcity of the disease, the most commonly cited incidence is 2 to 5 per 100 million; at least 56 families have been reported worldwide [20]. In the EuroCJD database of 455 genetic cases out of a TSE total of 4,441, 52 (11.4%) were classified as GSS [24].

16.7 Pathology

Depending on the clinical phenotype (form in which ataxia dominates, or the dementing form) [13], either the cerebellar cortex or the cerebral cortex are predominantly affected by neuronal loss and gliosis. Spongiform change of the neuropil differs in extent from one case to another [8]. However, all forms of GSS have in common abundant and prominent amyloid deposits in the form of large, often multicentric plaques composed of PrP fragments [33; 34]. The size and extent of these plaques can vary considerably. The deposited abnormal PrP was found to be of exclusively mutant allelic origin [35].

The majority of distinctive plaques are found in the molecular layer of the cerebellum (Fig. 16.2). This confirms the highly informative role of PrP immunostaining in the cerebellum for all types of genetic TSEs [22]. In GSS forms characterized by a P105L mutation of *PRNP*, the majority of plaque deposits are found in the cerebrum, hardly any in the cerebellum [36]. Dystrophic neurites that resemble those in AD have been observed around amyloid plaques [37]. Exceptionally, PrP deposits may extend to the peripheral nervous system [38].

An American family living in the state of Indiana (Indiana kindred) with the F198S *PRNP* mutation (see Chapters 26.4.2; 26.5.3) reveals distinctive neurofibrillary tangles like those of AD [26]. Such neurofibrillary tangles were also found in a member of a French GSS family [39], in addition to βA4 (Alzheimer amyloid) depositions [40]. PrP can be detected in Western blots, particularly in the basal ganglia and in the cerebellum [41]. The 145STOP mutation is characterized by a PrP cerebral amyloid angiopathy.

As multicentric plaques may be observed also in *PRNP* insertions, particularly in the 192 base-pair insertion, these disorders are usually grouped with GSS although their clinical presentation differs from classical GSS by earlier age at onset, shorter duration of illness, and higher frequency of psychiatric symptoms and myoclonus [22].

16.8 Is GSS a new disease?

No, it is not. The original Austrian GSS family had already been known to Viennese neurologists prior to Gerstmann's publications. The family tree of this family traces as far back as to the 18th century [7] (Fig. 16.1).

16.9 Risk factors

Apart from the genetic factors (Table 26.5), no other risk factors are known.

16.10 Surveillance, prevention, and control

Generally, GSS does not differ from CJD in respect of surveillance (see Chapter 14.10).

Because of the genetic predisposition to GSS, no preventive measures can be taken. If predisposing genotypes are identified, families need to be informed about the potential consequences by geneticists or neurologists (for ethical considerations, see Chapter 29.3.3).

Recently, similar levels of infectivity were experimentally detected in the blood of mice infected with human-derived vCJD and GSS prion strains [42]. As vCJD has become very likely to transmit via blood, such experimental data suggest intensifying research on blood infectivity in GSS as well, in order to prevent potential secondary transmissions.

References

[1] Gerstmann J, Sträussler E, Scheinker I. Über eine eigenartige hereditär-familiäre Erkrankung des Zentralnervensystems. Zugleich ein Beitrag zur Frage des vorzeitigen lokalen Alterns. Z ges Neurol Psychiat 1936; 154:736–762.

[2] Gerstmann J. Über ein noch nicht beschriebenes Reflexphänomen bei einer Erkrankung des zerebellären Systems. Wien Medizin Wochenschr 1928; 78:906–908.

[3] Dimitz L. Bericht des Vereines für Psychiatrie und Neurologie in Wien. (Vereinsjahr 1912/13). Sitzung vom 11. Juni 1912. Jahrb Psychiatr Neurol 1913; 34:384.

[4] von Braunmühl A. Über eine eigenartige hereditär-familiäre Erkrankung des Zentralnervensystems. Arch Psychiat Z Neurol 1954; 191:419–449.

[5] Seitelberger F. Eigenartige familär-hereditäre Krankheit des Zentralnervensystems in einer niederösterreichischen Sippe. Wien klin Wochenschr 1962; 74:687–691.

[6] Kretzschmar HA, Honold G, Seitelberger F, et al. Prion protein mutation in family first reported by Gerstmann, Sträussler, and Scheinker. Lancet 1991; 337:1160.

[7] Hainfellner JA, Brantner-Inthaler S, Cervenakova L, et al. The original Gerstmann–Sträussler–Scheinker family of Austria: divergent clinicopathological phenotypes but constant PrP genotype. Brain Pathol 1995; 5:201–211.

[8] Masters CL, Gajdusek DC, Gibbs CJ Jr. Creutzfeldt–Jakob disease virus isolations from the Gerstmann–Sträussler syndrome. With an analysis of the various forms of amyloid plaque deposition in the virus-induced spongiform encephalopathies. Brain 1981; 104:559–588.

[9] Schumm F, Boellard JW, Schlote W, et al. Morbus Gerstmann–Sträussler–Scheinker. Familie Sch. – Ein Bericht über drei Kranke. Arch Psychiatr Nervenkr 1981; 230:179–196.

[10] Hsiao K, Baker HF, Crow TJ, et al. Linkage of a prion protein missense variant to Gerstmann-Sträussler syndrome. Nature 1989; 338:342–345.

[11] Doh-ura K, Tateishi J, Sasaki H, et al. Pro-leu change at position 102 of prion protein is the most common but not the sole mutation related to Gerstmann-Sträussler syndrome. Biochem Biophys Res Commun 1989; 163:974–979.

[12] Farlow MR, Tagliavini F, Bugiani O, et al. Gerstmann–Sträussler–Scheinker disease. In: Vinken PJ, Bruyn GW, Klawans HL, editors. Handbook of Clinical Neurology. Elsevier, Amsterdam, 1991:619–633.

[13] Hsiao K and Prusiner SB. Inherited human prion diseases. Neurology 1990; 40:1820–1827.

[14] Parchi P, Chen SG, Brown P, et al. Different patterns of truncated prion protein fragments correlate with distinct phenotypes in P102L Gerstmann–Sträussler–Scheinker disease. Proc Natl Acad Sci USA 1998; 95:8322–8327.

[15] Piccardo P, Dlouhy SR, Lievens PMJ, et al. Phenotypic variability of Gerstmann–Sträussler–Scheinker disease is associated with prion protein heterogeneity. J Neuropathol Exp Neurol 1998; 57:979–988.

[16] Tagliavini F, Lievens PM-J, Tranchant C, et al. A 7 kDa prion protein fragment – an integral component of the PrP region required for infectivity – is the major amyloid protein in Gerstmann–Sträussler–Scheinker disease A117V. J Biol Chem 2001; 276:6009–6015.

[17] Tateishi J, Ohta M, Koga M, et al. Transmission of chronic spongiform encephalopathy with kuru plaques from humans to small rodents. Ann Neurol 1979; 5:581–584.

[18] Baker HF, Duchen LW, Jacobs JM, et al. Spongiform encephalopathy transmitted experimentally from Creutzfeldt–Jakob and familial Gerstmann–Sträussler–Scheinker diseases. Brain 1990; 113:1891–1909.

[19] Kordek R, Hainfellner JA, Liberski PP, et al. Deposition of the prion protein (PrP) during the evolution of experimental Creutzfeldt–Jakob disease. Acta Neuropathol 1999; 98:597–602.

[20] Ghetti B, Bugiani O, Tagliavini F, et al. Gerstmann–Sträussler–Scheinker disease. In: Dickson D, editor. Neurodegeneration: the molecular pa-

thology of dementia and movement disorders. ISN Neuropath Press, Basel, 2003:318–325.

[21] Hamasaki S, Shirabe S, Tsuda R, et al. Discordant Gerstmann–Sträussler–Scheinker disease in monozygotic twins. Lancet 1998; 352(9137): 1358.

[22] Kovács GG, Trabattoni G, Hainfellner JA, et al. Mutations of the prion protein gene: phenotypic spectrum. J Neurol 2002; 249(11):1567–1582.

[23] Pocchiari M, Puopolo M, Croes EA, et al. Predictors of survival in sporadic Creutzfeldt–Jakob disease and other human transmissible spongiform encephalopathies. Brain 2004; 127(10): 2348–2359.

[24] Kovács GG, Puopolo M, Ladogana A, et al. Genetic prion disease: the EUROCJD experience. Hum Genet 2005; 118(2):166–174.

[25] Hsiao KK, Cass C, Schellenberg GD, et al. A prion protein variant in a family with the telencephalic form of Gerstmann–Sträussler–Scheinker syndrome. Neurology 1991; 41:681–684.

[26] Ghetti B, Dlouhy SR, Giaccone G, et al. Gerstmann–Sträussler–Scheinker disease and the Indiana kindred. Brain Pathol 1995; 5(1):61–75.

[27] Majtenyi C, Brown P, Cervenakova L, et al. A three-sister sibship of Gerstmann–Sträussler–Scheinker disease with a CJD phenotype. Neurology 2000; 54:2133–2137.

[28] Young K, Clark HB, Piccardo P, et al. Gerstmann–Sträussler–Scheinker disease with the PRNP P102L mutation and valine at codon 129. Brain Res Mol Brain Res 1997; 44(1):147–150.

[29] Kovács GG, Ertsey C, Majtényi C, et al. Inherited prion disease with A117V mutation of the prion protein gene: a novel Hungarian family. J Neurol Neurosurg Psychiat 2001; 70:802–805.

[30] Woulfe J, Kertesz A, Frohn I, et al. Gerstmann–Sträussler–Scheinker disease with the Q217R mutation mimicking frontotemporal dementia. Acta Neuropathol 2005; 110:317–319.

[31] Barbanti P, Fabbrini G, Salvatore M, et al. Polymorphism at codon-129 or codon-219 of PRNP and clinical heterogeneity in a previously unreported family with Gerstmann–Sträussler–Scheinker disease (PrP-P102L mutation). Neurology 1996; 47:734–741.

[32] Wimberger D, Uranitsch K, Schindler E, et al. Gerstmann–Sträussler–Scheinker syndrome: MR findings. J Comput Assist Tomogr 1993; 17:326–327.

[33] Prusiner SB, Hsiao KK, Bredesen DE, et al. Prion disease. In: McKendall RR, editor. Handbook of clinical neurology. Elsevier Science, Amsterdam, 1989:543–580.

[34] Brown P. The phenotypic expression of different mutations in transmissible human spongiform encephalopathy. Rev Neurol (Paris) 1992; 148:317–327.

[35] Muramoto T, Tanaka T, Kitamoto N, et al. Analyses of GSS with 102Leu219Lys using monoclonal antibodies that specifically detect human prion protein with 219Glu. Neurosci Lett 2000; 288:179–182.

[36] Kitamoto T, Amano N, Terao Y, et al. A new inherited prion disease (PrP-P105L mutation) showing spastic paraparesis. Ann Neurol 1993; 34:808–813.

[37] Liberski PP, Kloszewska I, Boellaard J, et al. Dystrophic neurites of Alzheimer's disease and Gerstmann–Sträussler–Scheinker disease dissociate from the formation of paired helical filaments. Alzheimer's Res 1996; 1:89–93.

[38] Hainfellner JA and Budka H. Disease associated prion protein may deposit in the peripheral nervous system in human transmissible spongiform encephalopathies. Acta Neuropathol (Berl) 1999; 98:458–460.

[39] Tranchant C, Sergeant N, Wattez A, et al. Neurofibrillary tangles in Gerstmann–Sträussler–Scheinker syndrome with the A117V prion gene mutation. J Neurol Neurosurg Psychiatry 1997; 63(2):240–246.

[40] Bugiani O, Giaccone G, Verga L, et al. Beta PP participates in PrP-amyloid plaques of Gerstmann–Sträussler–Scheinker disease, Indiana kindred. J Neuropathol Exp Neurol 1993; 52(1):64–70.

[41] Brown P, Kenney K, Little B, et al. Intracerebral distribution of infectious amyloid protein in spongiform encephalopathy. Ann Neurol 1995; 38:245–253.

[42] Cervenakova L, Yakovleva O, McKenzie C, et al. Similar levels of infectivity in the blood of mice infected with human-derived vCJD and GSS strains of transmissible spongiform encephalopathy. Transfusion 2003; 43(12):1687–1694.

17 Portrait of Fatal Familial Insomnia and Sporadic Fatal Insomnia

Herbert Budka and Ellen Gelpi

17.1 History

Fatal familial insomnia (FFI) is a rare disease, first described in Italy in 1986. This neurodegenerative disorder has an autosomal dominant mode of inheritance and is characterized by progressive, uncontrollable alterations of the sleep–wake cycle, dysautonomy, motor signs, and prominent thalamic pathology [1; 2]. FFI belongs to the group of hereditary prion diseases as a result of a specific *PNRP* mutation at codon 178 [3; 4]. In addition, the polymorphic codon 129 of the mutated allele encodes for methionine (Met). Thus FFI is peculiar in having two genetic determinants for phenotypic expression. A microsatellite analysis suggested that the D178N mutations had independent origins in each affected pedigree or apparently sporadic case; a de novo spontaneous *PRNP* mutation was observed [5]. Thus evidence was provided that FFI has originated independently over the world from multiple recurrent mutational events, in contrast to what was found for genetic Creutzfeldt–Jakob disease (CJD) with the *PRNP* 200 mutation [5].

In 1995, it was possible to transmit the disease experimentally, and FFI was subsequently included among the transmissible spongiform encephalopathies (TSEs) or prion diseases [6].

Although the disease has only been known for 20 years and remains relatively rare, FFI is distinct from all other TSEs in many aspects and thus has become an interesting model for researching the correlation between genotype, clinical or neuropathological phenotype, pathogenesis of brain tissue damage, and transmissibility of TSEs [7].

More recently, a very rare sporadic form of "fatal insomnia" (sporadic fatal insomnia, SFI) has become recognized as phenotypically indistinguishable from FFI, but without a *PRNP* mutation [8]. However, SFI is genetically homogenous in that all patients carry homozygosity for methionine at *PRNP* codon 129. Earlier reports of "thalamic dementia" or "thalamic form of Creutzfeldt–Jakob disease" are likely to have described the same entity [7] (see Chapter 26.5.4).

17.2 Forms or variants

Despite its variability in signs and symptoms (see below), FFI has no peculiar subtypes or variants. The same is true for SFI. However, it has become recognized that the classical FFI genotype (D178N, M129M) may associate also with a CJD-like phenotype [9].

Clinically, the duration of FFI and its progression depend on whether the patients are Met/Met homozygous or Met/Val heterozygous at codon 129 of *PRNP* (see below). On a molecular level, only one FFI form with a PrP^{Sc} type 2 pattern [10] on Western blots exists.

In SFI, PrP^{Sc} type 2 is present in all subjects in amount and distribution similar to FFI. However, its PrP^{Sc} does not show the striking under-representation of the unglycosylated isoform of the protein that is characteristic of FFI [8].

17.3 Incubation period, transmissibility, and susceptibility

Experiments on the transmissibility of FFI to apes and mice were initially unsuccessful; the first success in transmitting FFI experimentally was obtained in 1995 [6]. Experimental transmission of SFI to mice was achieved in 1999 [11]. As FFI and SFI are not acquired TSEs, no natural incubation period can be observed.

Carriers of the predisposing mutation are susceptible to the disease; the penetrance seems to be high. In all described cases of SFI, homozygosity for methionine at *PRNP* codon 129 has been found and thus might be considered to be a susceptibility factor.

17.4 Clinical signs and course of disease

This is similar if not identical in both FFI and SFI. There is a higher male affection in FFI (ratio: 1.46) [12]. Age of disease manifestation, duration of the disease, and the clinical and neuropathological symptoms are very heterogeneous. As the name of the disease implies, often a characteristic disturbance in the wake–sleep rhythm (insomnia) can be observed along with other progressive neurological deficiencies. If insomnia does not manifest clinically, the sleep disorder might be identified by way of polysomnography [13]. Cognitive decline has been observed in 78 % of FFI patients [12]. PET examinations reveal severe hypometabolism in the thalamus and a mild hypometabolism in the cortex [13]. Similar changes are observed in SFI: thalamic hypoperfusion or hypometabolism on cerebral blood flow SPECT or [18F]2-fluoro-2-deoxy-d-glucose PET [14].

In a review on FFI [15], 14 FFI patients from five different families were presented. The investigation showed that progression of the disease was either rapid (9.1 months) or slow (30.8 months) (Table 30.1). The observed length depended on whether the patients were homozygous (Met/Met at *PRNP* codon 129; rapid progression) or heterozygous (Met/Val at *PRNP* codon 129; slow progression) [13]. In addition, at the beginning of the disease, homozygous patients experienced marked dream-like (oneiroid) episodes, insomnia, and dysautonomy, whereas heterozygous patients suffered from ataxia, dysarthria, early onset of sphincter dysfunction and grand-mal attacks, as well as pronounced cortical damage [13]. Exceptions were also observed [16; 17].

More details are available for six members of the Austrian FFI family first described in 1999 [18] (see updated pedigree in Fig. 17.1). In these patients, age at clinical manifestation of disease was between 20 and 63 years. Duration of disease was between 6 and 20 months. The progression of the disease was relatively uniform and followed the following pattern: early and very marked weight loss, dysarthria, sleep disturbances, forgetfulness, perioral myo-

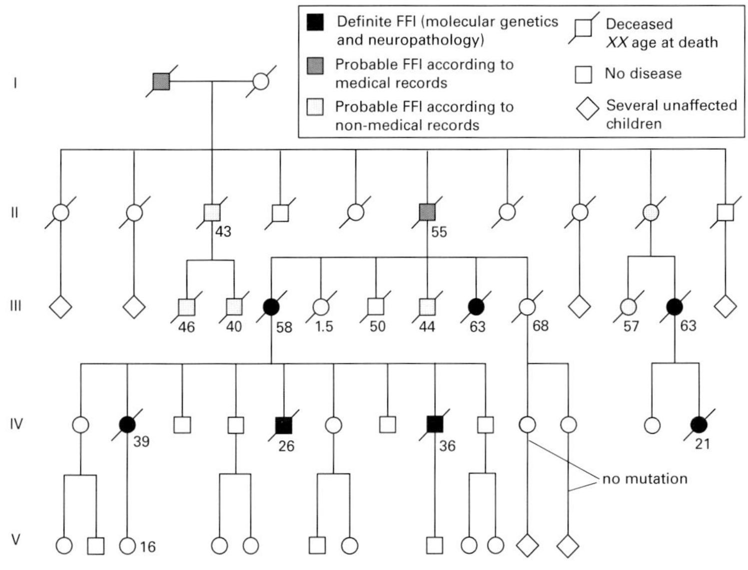

Fig. 17.1: Updated pedigree of the Austrian FFI family first described in 1999 [18]. The family comprises more than 50 members over five generations. Figure by H. Budka ©.

clonus, diplopia, gait ataxia, vegetative disturbances, and reduced vigilance (in 5/6 patients); apathy and marked fatigue (in 2/6 patients); tremor, dizziness, dementia, and hallucinations (in 1/6 patients). A detailed description of the clinical symptoms of FFI is given in Chapter 30.6 (SFI: see Chapter 30.7).

17.5 Differential diagnoses

Although the clinical picture of the two diseases has certain characteristics, in particular when combined with polysomnography, it is difficult to distinguish the diseases from other neurodegenerative diseases affecting the thalamus and hypothalamus. Absence of a family history does not rule out FFI, as this is found only in 50 % of cases [19]. Molecular genetic methods need to be used to get a definitive diagnosis: FFI should only be diagnosed in individuals possessing the D178N mutation (exchange of aspartate by asparagine in codon 178) [4] in combination with a methionine-encoding triplet in the polymorphic codon 129 [3; 20] of the mutated allele. It is interesting to note that the same *PRNP* mutation in codon 178 does not lead to FFI but to familial CJD (fCJD) if codon 129 of the mutated allele codes for valine instead of methionine (s. Chapt. 26.6). However, this rule of thumb has increasingly become blurred, as considerable clinical and pathological overlapping was observed among homozygous 129MM patients especially in the Basque country, favoring the view that FFI and CJD178 are the extremes of a spectrum rather than two distinct and separate entities [9]. Rarely, another *PRNP* mutation such as E200K-129 M may mimic clinical and pathological features of FFI [21]. On the other hand, the typical genetic FFI constellation occasionally may not feature the typical clinical or pathological presentation [22; 23].

SFI can be distinguished from FFI not only by the absence of a *PRNP* mutation, but also by its PrP^{Sc} pattern on Western blots [8].

17.6 Epidemiology

Because of the rarity of the diseases, accurate epidemiological data are not available for FFI and SFI. Therefore, it is impossible to calculate exact statistical parameters. However, FFI is the most frequent genetic prion disease in Germany [24], but is much rarer in Italy [25], for example. An overview of the global occurrence of FFI [15] found three FFI cases in France, seven in Germany [26], two in Great Britain, four in Italy, three in the USA, two in Australia, and one affected family in Japan. A more recent review cites 26 presumably unrelated kindreds with 55 FFI cases [1].

An Austrian FFI family comprised 50 family members from five generations; FFI was most probably or definitely diagnosed in 15 patients from four generations (Fig. 17.1). The definite diagnoses were based on neuropathological and molecular genetic findings [18].

SFI occurs even more rarely than FFI. Again, exact data are not available, in particular because – before the recognition of SFI – it was confused with CJD as its "thalamic form". A recent publication cites nine well-characterized SFI cases [1].

17.7 Pathology

Histopathological findings include loss of nerve cells and gliosis, in particular in the thalamus (anterior ventral, medio-dorsal, and pulvinar nuclei; Fig. 17.2a and b) and the inferior olivary nucleus of the medulla oblongata (Fig. 17.2e and f). Occasionally, mild spongiform alterations have been observed in the cerebral cortex including the pre-subiculum [27]. It is very puzzling that in FFI, little or no PrP^{Sc} can be detected by immunohistochemistry (IHC) despite marked local brain damage [18]. This finding is very important for the diagnosis of the disease. Negative immunohistochemical results do not, therefore, exclude FFI. In our experience, it is the cerebellar cortex that, even with minimal or absent morphological changes (Fig. 17.2c), has the highest chance to contain focal PrP deposits (Fig. 17.2d) when many tissue blocks are studied. The paraffin-embedded tissue (PET) blot may be more sensitive than IHC to demonstrate PrP in FFI [28]. Also, the amount of PrPres detectable by Western blotting is much lower than in other human TSEs

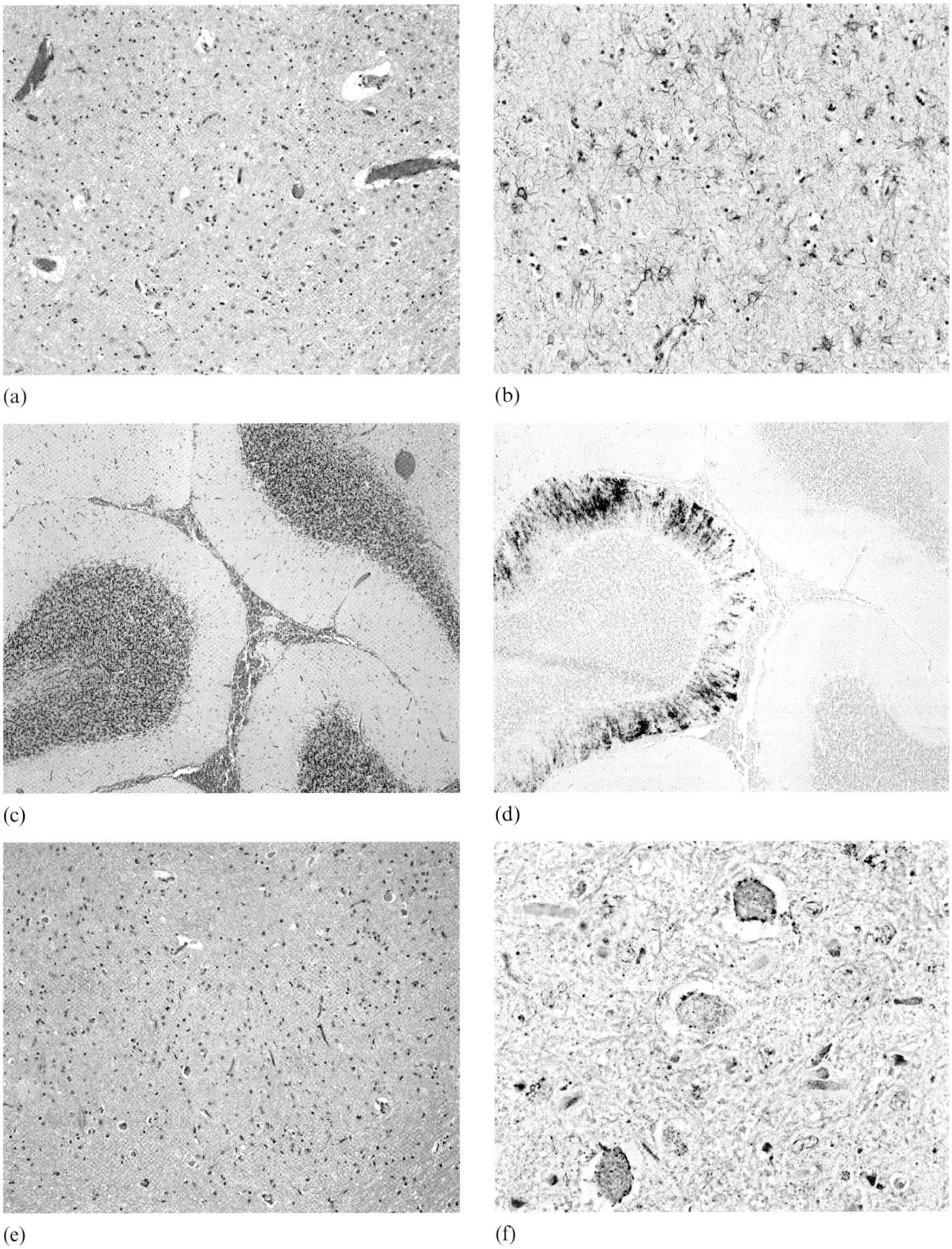

Fig. 17.2: Histopathology of a recent Austrian FFI case (female of generation III in Fig. 17.1, who died at the age of 63). **(a)** and **(b)**: Thalamus with severe degeneration – neuronal loss in (a), astrogliosis in (b). **(c)** and **(d)**: Cerebellar cortex without prominent morphology (c) but with focally prominent linear PrP deposition (d). **(e)** and **(f)**: Inferior olivary nucleus featuring prominent degeneration (nerve cell loss and gliosis in (e)) and granular, sometimes perineuronal PrP deposits (f). (a), (c), and (e): hematoxylin & eosin stain; (b), (d), and (f): immunocytochemistry for glial fibrillary acidic protein (GFAP, b) or PrP (monoclonal antibody 12F10, (d) and (f)). Figure by H. Budka ©.

[1]. Immunopositivity in the form of granular PrP deposition in vacuoles of inferior olive neurons in FFI is rare but characteristic [18]. The dissociation between local tissue damage and detection of PrP^{Sc}, as particularly evident in the thalamus, supports a pathogenetic loss of function model rather than the neurotoxicity of PrP^{Sc} as the cause of the tissue damage [18]. Apoptosis is frequently observed in the major damaged brain regions and seems to be an important mechanism that is responsible for the loss of nerve cells; it does, however, not correlate with the local deposition of PrP^{Sc} [29]. While this is similar to other human TSEs, FFI differs markedly from them in terms of selective neuronal vulnerability [30]. An alteration of the serotonergic system of the brainstem might represent the functional substrate of some typical symptoms of FFI [31].

In the Austrian FFI family [18], the brain damage of five of the six patients on whom a post-mortem examination was carried out was relatively uniform. Apart from the known degeneration of thalamus and inferior olives, diffuse cortical astrocytosis, marked degeneration of nuclei and tracts in the brain stem, and pseudohypertrophy of the olivary nucleus of the medulla oblongata were observed. The latter is a sequel of transsynaptic degeneration following damage to the central tegmental tract. In the molecular layer of the cerebellar cortex of two patients, characteristic linear PrP deposits perpendicular to the surface in the molecular layer were observed (Fig. 17.2d), which are occasionally also found in other familial TSEs.

Pathological changes in SFI do not significantly differ from FFI [1]. However, occasional observations appeared to lie halfway between CJD and FFI with regard to neuropathology [32].

17.8 Are FFI and SFI new diseases?

No, despite the fact that FFI was not described until 1986. As in other familial TSEs, FFI can be traced back several generations by way of family trees (Fig. 17.1), indicating that the disease must have been around for at least several decades.

SFI has become recognized only in the late 1990s. However, a "thalamic" form of CJD has been known for decades and likely corresponds to SFI. Thus SFI, like FFI, does not appear to be a new disease.

17.9 Risk factors

FFI is caused by mutation of *PRNP*. Risk factors other than the D178N mutation and the occurrence of methionine at the polymorphic codon 129 of the mutated allele are unknown.

SFI is found only in 129 MM patients with PrP^{Sc} type 2. Hence, the only known susceptibility factor is the genetic background of homozygosity for Met at *PRNP* codon 129.

17.10 Surveillance, prevention, and control

Generally, FFI and SFI do not differ from CJD in respect to surveillance (see Chapter 14.10).

In FFI, no measure is known that would prevent the manifestation because of the genetic predisposition. If predisposing genotypes are identified, the families need to be informed about the potential consequences by geneticists or neurologists (for ethical considerations, see Chapter 29.3.3).

References

[1] Gambetti P, Parchi P, Chen SG, et al. Fatal insomnia: familial and sporadic. In: Dickson D, editor. Neurodegeneration: the molecular pathology of dementia and movement disorders. ISN Neuropath, Basel, 2003:326–332.

[2] Lugaresi E, Medori R, Montagna P, et al. Fatal familial insomnia and dysautonomia with selective degeneration of thalamic nuclei. N Engl J Med 1986; 315(16):997–1003.

[3] Goldfarb LG, Petersen RB, Tabaton M, et al. Fatal familial insomnia and familial Creutzfeldt–Jakob disease: disease phenotype determined by a DNA polymorphism. Science 1992; 258:806–808.

[4] Medori R, Tritschler HJ, LeBlanc A, et al. Fatal familial insomnia, a prion disease with a mutation at codon 178 of the prion gene. N Engl J Med 1992; 326:444–449.

[5] Dagvadorj A, Petersen RB, Lee HS, et al. Spontaneous mutations in the prion protein gene causing transmissible spongiform encephalopathy. Ann Neurol 2002; 52:355–359.

[6] Tateishi J, Brown P, Kitamoto T, et al. First experimental transmission of fatal familial insomnia. Nature 1995; 376(6539):434–435.

[7] Budka H. Fatal familial insomnia around the world. Introduction. Brain Pathol 1998; 8:553.

[8] Parchi P, Capellari S, Chin S, et al. A subtype of sporadic prion disease mimicking fatal familial insomnia. Neurology 1999; 52(9):1757–1763.

[9] Zarranz JJ, Digon A, Atares B, et al. Phenotypic variability in familial prion diseases due to the D178N mutation. J Neurol Neurosurg Psychiatry 2005; 76(11):1491–1496.

[10] Parchi P, Giese A, Capellari S, et al. Classification of sporadic Creutzfeldt–Jakob disease based on molecular and phenotypic analysis of 300 subjects. Ann Neurol 1999; 46:224–233.

[11] Mastrianni JA, Nixon R, Layzer R, et al. Prion protein conformation in a patient with sporadic fatal insomnia. N Engl J Med 1999; 340:1630–1638.

[12] Kovács GG, Trabattoni G, Hainfellner JA, et al. Mutations of the prion protein gene: phenotypic spectrum. J Neurol 2002; 249(11):1567–1582.

[13] Montagna P, Cortelli P, Avoni P, et al. Clinical features of fatal familial insomnia: phenotypic variability in relation to a polymorphism at codon 129 of the prion protein gene. Brain Pathol 1998; 8:515–520.

[14] Hamaguchi T, Kitamoto T, Sato T, et al. Clinical diagnosis of MM2-type sporadic Creutzfeldt–Jakob disease. Neurology 2005; 64(4):643–648.

[15] Budka H. Fatal familial insomnia around the world. Brain Pathol 1998; 8:553–570.

[16] McLean CA, Storey E, Gardner RJ, et al. The D178N (cis-129M) "fatal familial insomnia" mutation associated with diverse clinicopathologic phenotypes in an Australian kindred. Neurology 1997; 49(2):552–558.

[17] Rossi G, Macchi G, Porro M, et al. Fatal familial insomnia: genetic, neuropathologic, and biochemical study of a patient from a new Italian kindred. Neurology 1998; 50:688–692.

[18] Almer G, Hainfellner JA, Brücke T, et al. Fatal familial insomnia: a new Austrian family. Brain 1999; 122(1):5–16.

[19] Kovács GG, Puopolo M, Ladogana A, et al. Genetic prion disease: the EUROCJD experience. Hum Genet 2005; 118(2):166–174.

[20] Monari L, Chen SG, Brown P, et al. Fatal familial insomnia and familial Creutzfeldt–Jakob disease: different prion proteins determined by a DNA polymorphism. Proc Natl Acad Sci USA 1994; 91(7):2839–2842.

[21] Taratuto AL, Piccardo P, Reich EG, et al. Insomnia associated with thalamic involvement in E200K Creutzfeldt–Jakob disease. Neurology 2002; 58(3):362–367.

[22] Sasaki K, Doh-ura K, Wakisaka Y, et al. Fatal familial insomnia with an unusual prion protein deposition pattern: an autopsy report with an experimental transmission study. Neuropathol Appl Neurobiol 2005; 31(1):80–87.

[23] Taniwaki Y, Hara H, Doh-Ura K, et al. Familial Creutzfeldt–Jakob disease with D178N-129M mutation of *PRNP* presenting as cerebellar ataxia without insomnia [letter]. J Neurol Neurosurg Psychiat 2000; 68:388.

[24] Windl O, Giese A, Schulz-Schaeffer W, et al. Molecular genetics of human prion diseases in Germany. Hum Genet 1999; 105:244–252.

[25] Ladogana A, Puopolo M, Poleggi A, et al. High incidence of genetic human transmissible spongiform encephalopathies in Italy. Neurology 2005; 64(9):1592–1597.

[26] Kretzschmar H, Giese A, Zerr I, et al. The German FFI cases. Brain Pathol 1998; 8(3):559–561.

[27] Gambetti P, Parchi P, Petersen RB, et al. Fatal familial insomnia and familial Creutzfeldt–Jakob disease: clinical, pathological and molecular features. Brain Pathol 1995; 5:43–51.

[28] Scaravilli F, Cordery RJ, Kretzschmar H, et al. Sporadic fatal insomnia: a case study. Ann Neurol 2000; 48:665–668.

[29] Dorandeu A, Wingertsmann L, Chrétien F, et al. Neuronal apoptosis in fatal familial insomnia. Brain Pathol 1998; 8:531–537.

[30] Guentchev M, Wanschitz J, Voigtlaender T, et al. Selective neuronal vulnerability in human prion diseases. Fatal familial insomnia differs from other types of prion diseases. Am J Pathol 1999; 155:1453–1457.

[31] Wanschitz J, Klöppel S, Jarius C, et al. Alteration of the serotonergic nervous system in fatal familial insomnia. Ann Neurol 2000; 48:788–791.

[32] Piao Y-S, Kakita A, Watanabe H, et al. Sporadic fatal insomnia with spongiform degeneration in the thalamus and widespread PrPSc deposits in the brain. Neuropathology 2005; 25(2):144–149.

18 Portrait of Scrapie in Sheep and Goat

Beat Hörnlimann, Lucien van Keulen, Martha J. Ulvund, and Ray Bradley

18.1 History

Scrapie is a clinical term applied to sheep, goats, and moufflon affected by a transmissible spongiform encephalopathy (TSE) agent, also called prions. Confirmation of scrapie requires examination of the brain to determine the presence of pathognomic spongiform encephalopathy and/or detection of the misfolded form of prion protein PrP^{Sc} by a range of methods (see Chapter 32).

With few exceptions, most sheep-producing countries worldwide are, or may be, affected by scrapie. Australia, New Zealand, and Argentina are countries generally regarded as devoid of animal TSE. The scrapie status of sheep and goats in many countries is not known with certainty due to absent or inadequate surveillance for the disease (see Chapter 54–57).

The clinical signs of scrapie were accurately recorded by Leopoldt (Fig. 1.2) in Germany in the 18th century, though the disease may have existed in Roman times [1]. One of the confounding problems is the various names ascribed to the disease in different countries; for example, murrain, goggles, rickets, shewcroft, shakers, scratchie, rubbers, cuddie trot, and scrapie in Great Britain (GB), *la tremblante* and *la prurigo lombaire* in France, *Traberkrankheit* in Germany, *la trambladera* in Spain, and *riđa* in Iceland. The first well-documented record of scrapie in GB dates back to the year 1732 [2]. Further reports about a scrapie-like disease among the English breeds Dorset Horn, Wiltshire Horn, and Norfolk Horn go back to the period between 1750 and the beginning of the 1800s. Parry records the crucial importance of the export of fine-wooled Spanish Merinos after 1700 to several countries in northern Europe, following which severe and epidemic forms of scrapie occurred in some sheep populations in Germany, France, and the Danube valley but not in others in Britain, some parts of Germany, and Scandinavia [3; 4]. These different experiences are probably explained by differences in the prevalence of scrapie or susceptible genotypes in Merinos of different breeds. According to reports from Spain, the spread of scrapie is today restricted to the region of Aragon in the northeastern part of the country [5; 6] (Badiola, personal communication).

Although scrapie was not officially verified in Norway, for instance, before 1981, there are several reports of clinical cases of *gnave-og travesjuke* (itching and trotting disease) in 1890, 1916, and 1938, all associated with imports of British Cheviot sheep and descendants of these [7].

Between 1920 and 1958, scrapie became a major problem in the Suffolk breed, causing considerable financial loss in some flocks. International trade with English Suffolk sheep seems to have then been responsible for the introduction of scrapie into numerous countries including Australia and New Zealand (where it was eliminated while sheep were under quarantine restrictions), the USA [8], and also Norway (1958). Concern about the extent of the disease has led to an increase in scrapie research since the 1930s, especially in England and France [3].

Since 1950, there have been numerous reports worldwide of an increase in scrapie and an extension in the breeds affected. This appears to be associated with a massive increase in sheep movements, particularly in the post-World War II era. Most sheep breeds have been affected by scrapie, although it is more common in some than in others. Following the introduction of harmonized active surveillance for scrapie in sheep and goats throughout the EU in 2003, new knowledge about the incidence of the disease has come forward. In 2004, 2,663 cases were reported in sheep and 398 in goats, the majority by active surveillance. Cyprus had the most cases (1,208 in sheep and 354 in goats), followed by France (459 and 27), the UK (331 and 0), Italy (139 and 2), the Netherlands (105 and 0), Ireland (101 and 0), Germany (100 and 0), and Greece (71 and 15). Other countries had fewer than 50 cases in sheep each and none in goats [9].

The discovery of bovine spongiform encephalopathy (BSE) and variant Creutzfeldt–Jakob disease (vCJD) in man, which is derived from BSE, has resulted in a resurgence of scrapie (and BSE) research and monitoring worldwide. In 1993, scrapie became a notifiable disease in the EU.

18.2 Forms or variants (Scrapie agent strains)

More than 20 agent strains of scrapie have been distinguished using laboratory mice [10]; they are described in detail in Chapter 12.5.2. Some have been cloned after passage at limiting dilution and are regarded as murine strains of scrapie agent.

Biologically, scrapie strains can be distinguished by the length of the incubation period and the distribution patterns of pathological changes in the brain (the lesion profile) determined after challenge of certain in-bred lines of laboratory mice with different PrP genotypes. A restricting factor, however, is that not all isolates of confirmed sheep or goat scrapie transmit to mice. It is also important to recognize that strain identification in mice may not reflect the ovine strain from which it is derived, especially as few studies have challenged sheep or goats with cloned murine strains. Murine strains of scrapie derived from British sheep and goats have been studied extensively, but little has been reported about the murine strains derived from sheep or goats in other countries. In some isolates of naturally infected sheep, more than one scrapie strain can be found [11]. More work of this kind is in progress as a result of finding cases of atypical scrapie (Nor98) in the European flock in recent years, and as a result of the confirmation by strain-typing that BSE has occurred in one French goat [12; 13].

Following the discovery of BSE in cattle, the experimental transmission of BSE to both sheep and goats by the oral route with clinical signs similar to scrapie, and the epidemiological evidence that these small ruminant species also were exposed to the BSE agent in feed, it has become important to ensure that the correct etiological diagnosis is reached when investigating a case of scrapie [14]. It is important to determine that the disease is caused by a scrapie strain, and not by the BSE agent. This is because the BSE agent is a human pathogen responsible for vCJD in man, whereas scrapie strains appear harmless to man. The importance of this is emphasized following the confirmation of BSE in a single goat in France and possibly in a British goat [12; 15]. Methods, not based on strain-typing in mice, have been developed to differentiate between scrapie and BSE in sheep [16; 17], and reviews of diagnostic methods have been published [18; 19]. Biochemical methods for the differentiation of TSE strains are still being developed and evaluated. Two methods with potential for practical implementation are the use of proteinase K (PK) digestion and Western blot for the analysis of the molecular mass and glycoform profile [20; 21].

18.3 Incubation period, transmissibility, and susceptibility

Scrapie is most often found among sheep between 2 and 5 years of age, and gender does not appear to have an effect [22; 23]. For establishing epidemiological models, usually an average age of 3.5 years, at which the animals fall ill, is assumed [3; 24]. Since it is thought that most animals are infected at birth or shortly thereafter, the age at clinical onset and period of incubation are coincident.

Historically, the etiology of scrapie has been disputed. Initially, discussion centered upon whether scrapie was a genetic or an infectious disease. H.B. (James) Parry postulated that scrapie was an autosomal recessive genetic disease, but at the same time conceded that the affected animals harbored a transmissible agent that was infectious by artificial routes [25]. He did not find any hints pointing to a natural transmission pathway.[1] Today it is generally accepted that scrapie is a transmissible disease with a pronounced genetic component (see Chapter 56). However, the role that the genetic components play is not yet completely understood. It is known that the *Sip* gene (*s*crapie incubation period gene) in sheep and the *Sinc* gene (*s*crapie incubation gene) in mice have an effect on the length of the incubation period; these designations are identical to the PrP gene (*Prnp*) [26].

The murine *Sinc* gene now known as *Prnp* consists of two alleles, s7 and p7, which differ in their PrP amino acid sequence encoded by codons 108 and 189. Experimental infections, e. g., with the ME7-scrapie strain, result in a short incubation period in the case of s7-homozygous mice, a prolonged incubation period in p7-homozygous mice, and a medium-length incubation period in heterozygous mice (s7p7). However, inoculation of these mice with the 22A-scrapie strain results in a prolonged incubation period in the s7-homozygous mice and a short incubation period in the p7 homozygous mice [27]. Thus the incubation period is determined both by the genotype of the host and the strain of the agent as well as the dose and the route by which it is administered.

In a flock of experimental Cheviot sheep, two genotypes of the *Sip* gene were detected. The genotypes sA (short incubation allele) and pA (prolonged incubation allele) correspond to mutations in the amino acid sequence encoded by codon 136 of the PrP gene [28]. In these scrapie-susceptible Cheviot sheep, the incubation period length, which is primarily determined by these two genotypes, is additionally influenced by sequence changes in codons 154 and 171. Further polymorphisms have been reported but seem to play little or no role in naturally occurring scrapie infections as compared to the polymorphisms in codons 136, 154, and 171 [29] (see Chapter 56). A similar reversal in incubation period length as described with ME7 and 22A in mice occurs in Cheviot sheep after inoculation with different scrapie isolates. Inoculation of SSBP1 scrapie (derived from a pool of brains from sheep with scrapie) into sA-homozygous Cheviot sheep results in a short incubation period, whereas inoculation in pA-homozygous animals results in a prolonged incubation period. When sA and pA sheep are inoculated with a different scrapie isolate, CH1641, the incubation period lengths are reversed [30].

Experimental transmission of scrapie was first accomplished in sheep by Cuillé and Chelle in France in 1936/1938 [31; 32]. In the early 1960s, Chandler added to this discovery by experimentally infecting mice with scrapie. Iatrogenic scrapie occurred in GB in the 1930s [33–36] and in Italy in the 1990s [37], both episodes being due to contamination of non-commercially prepared vaccines with the scrapie agent.

In the 1970s, evidence was presented in several studies [8; 38; 39] demonstrating that scrapie was transmitted naturally between sheep. Important among these was the study by Pattison showing that the placenta from infected ewes could transmit disease to scrapie-free sheep and goats by the oral route [40]. Subsequently the detection of infectivity or PrP^{Sc} in the placenta of scrapie-susceptible adult offspring from scrapie-infected ewes was reported by other groups [41–45].

Maternal transmission could be defined as transmission of the infectious agent from infected ewes to their offspring in utero, during parturition, or in the immediate post parturient period. Pattison's studies [40] showed that the last of these is feasible and is generally regarded as likely in practice. In addition, of course, any susceptible sheep (and perhaps goats) might become infected from an environment (like a lambing pen) contaminated with infected products of the conceptus (like placenta) [8; 38–40; 46]. The precise age at which lambs are infected is not known. If it occurs several days or weeks after birth, even if infected from the dam, it might be more correctly described as horizontal transmission. Thus, in practice, a clear distinc-

[1] Contrary to observations and hypotheses from the 18th and 19th centuries (see Chapter 1.3).

tion between maternal, vertical, and horizontal transmissions of scrapie is blurred.

It is believed that the route of exposure is oral [40; 46–48], but entry via the conjunctiva [49] or broken skin [50; 51] cannot be excluded. Direct transmission via hay mites has also been proposed [52]. Some more details concerning the transmissibility of scrapie and other prion diseases are described in Chapter 42.

Owing to their complexity, the naturally occurring pathways of scrapie transmission are not yet completely elucidated. Important questions remain unanswered, for example:

- What is the minimum oral infectious dose for sheep and goats?
- Do non-pregnant sheep excrete the agent?
- Is excretion of the infectious agent a continuous process or intermittent?
- Apart from the rather poorly understood genetic influences, do goats follow the same pattern as sheep?
- Do healthy carriers exist [53]? If so, are they a source of infection for other sheep, goats, and the environment [53; 54]? (Remark: We know that pre- or subclinical cases exist, but we do not know whether clinically healthy, genetically resistant sheep exposed to infection could act as carriers and be a source of infection for other sheep or the environment).

18.4 Clinical signs and course of disease

After oral infection, accumulation and perhaps replication of the scrapie agent first occur in gut-associated lymphoid tissue (GALT) such as Peyer's patches and the mesenteric lymph nodes. After invasion of the GALT, dissemination occurs to all other lymphoid tissues [55; 56] and notably the spleen. In this preclinical stage, PrP^{Sc} can be detected in biopsies of easily accessible lymphoid tissues (tonsils, third eyelid, rectal lymphoid tissue), thus enabling ante mortem tests for scrapie infection to be developed [57–60]. However, there have been a few reports of scrapie cases in which no involvement of the lymphoid system could be demonstrated [59; 61].

Transmission to the central nervous system (CNS) starts with invasion of the enteric nervous system of the gut, possibly facilitated by the infection of the Peyer's patches. From there, the brain is invaded through an ascending infection along the parasympathetic and sympathetic efferent nerves innervating the gut (n. vagus and n. splanchnicus). The portal of entry into the CNS is thus the dorsal motor nucleus of the vagus in the brain stem and the intermediolateral column in the spinal cord, from where the scrapie agent eventually spreads to involve the entire neuraxis [62]. In clinical disease, infectivity is widely dispersed throughout the body (WHO $\text{Tables}_{\text{Appendix}}$ IA–IC). Infectivity can, for example, be detected in the pituitary gland, adrenal gland, bone marrow, pancreas, thymus, liver, and in the peripheral nervous system. In pregnant ewes, infectivity may be present in the placenta [40; 46], though the titers of infectivity in these tissues do not usually reach those found in the brain of sheep with terminal disease.

There is no diagnostic blood test for scrapie available, although blood transfusion studies have shown that in natural scrapie and experimental BSE in sheep [63; 64], the blood is infected both during incubation and in the clinical phase of disease. Previous studies of ovine blood from sheep with scrapie have shown no detectable infectivity to be present. There are three reasons for this apparent conflict of results. The first is that, in the recent studies, sheep and not mice were used for the transfusion studies, thus eliminating the species barrier. Secondly, the PrP genotypes of donor and recipient were matched, thus maximizing the chance of infection being established. Thirdly, large volumes (500 ml) were used by a route only 10 % less efficient than the intracerebral route.

The clinical signs of scrapie vary considerably, especially during the initial clinical phase. Since the knowledge of the clinical signs is extremely important for monitoring the disease (see Chapter 32), this topic is discussed separately in Chapter 34. However, the most important facts are summarized here [see also 65].

Lesions in the CNS cause functional changes that are responsible for changes in behavior, mental status, and abnormalities of posture and gait. Associated with these, especially in the later stages of clinical disease, are general changes such as loss of weight and poor bodily condition despite an unchanging appetite (Fig. 34.1f). In lactating sheep, milk yield may be reduced.

The clinical phase can last from 2 weeks to 6 months [66; 67], rarely longer [3]. Presenting signs may be so mild as to go unnoticed or may only be obvious to the shepherd, e. g., the animals may separate themselves from the rest of the flock. Subsequently, more obvious signs develop and sheep may become nervous and sometimes aggressive. Some animals "stargaze" (Fig. 34.1e) or press their head or entire body (Fig. 34.1b) against fixed objects.

Pruritus is a common sign but is absent in the traditional form of scrapie (*riđa*) in Icelandic sheep. This leads to rubbing, scraping (hence scrapie), loss of wool, and economic loss (Fig. 34.1). The skin may sometimes be rubbed raw. Some sheep bite their legs or pull wool from their sides with their teeth. A "nibble reflex" (Fig. 34.1c) can often be evoked by scratching the lumbar area of the back. Horned animals (including goats) may use the horns to rub irritating sites with great precision.

Hypersensitivity, irritability, and trembling (hence *la tremblante* in French) are other common signs.

If left undisturbed at rest, an affected animal may appear normal. However, when stimulated by a sudden noise, excessive movement, or the stress of handling, tremor may become excessive or the animal may even fall down in a convulsive-like state.

Gait abnormalities include trotting (hence *Traberkrankheit* in German) and bunny-hopping especially when stimulated to run. Severe ataxia may occur during further progression of the disease, causing the animal to sway, support its hindquarters against fences or walls when standing, and have difficulty rising. These signs may progress to recumbency.

Not all animals show all signs, but usually more than one sign is present and, more often than not, several. Scrapie is, as all prion diseases, always fatal.

18.5 Differential diagnoses

Especially in the early stages of clinical scrapie, the signs may be attributable to other diseases, including external parasite infestation, rabies, Aujeszky's disease, listeriosis, ketosis, poisoning. Table 34.2 lists other differential diagnoses that should be considered.

As the disease progresses toward termination and before recumbency and death, the signs are distinctive enough in most cases to make a confident clinical diagnosis. Nevertheless, it is always necessary to confirm the diagnosis post mortem.

18.6 Epidemiology

The epidemiology of scrapie was reviewed in detail about 10 years ago [68] with the aim of summarizing and evaluating the data on the etiology of natural scrapie, especially that concerning the transmission of infection between related animals. The main conclusion at that time was that scrapie is an infectious disease with a genetic influence on incubation period. The increased risk of infection in offspring is resultant upon increased genetic susceptibility, and a large proportion of cases in high-incidence flocks results from horizontal transmission. Scrapie agents (excreted by the placenta of affected ewes) may remain infectious for a long time in the environment, contributing potentially to an increasing pressure of infection in the affected regions or flocks.

In 1990 two British epidemiological approaches tried to estimate the prevalence of scrapie in GB [69; 70]; years before Hoinfille's review [68]. Further informative epidemiological studies have been conducted since 1999 [71–74]. However, the true prevalence and incidence rate of scrapie is still unknown in many countries, but knowledge is improving in the EU and some other countries that have a passive and targeted active surveillance program. Even here the figures may be underestimated [24; 69; 75]. As long as there is no practical, rapid, preclinical test for scrapie infection in the live animal (biopsy of lymph nodes, tonsil, or third eyelid, though otherwise effective, is not regarded as practical on a large scale), determining the true incidence will be thwarted. Reliance on reporting of clinical signs of scrapie to the State Veterinary Service, although mandatory in the EU and some other countries, is unwise. This is because there are commercial incentives to conceal the disease to aid the selling of valuable breeding stock that would have no value if blighted by scrapie in the flock. Since there is no internationally validated test system available for the diagnosis of infection or disease in live animals, one must rely on post mortem examination of the brain by an approved laboratory using an approved method (see Chapters 32 and 35).

In some countries, the scrapie incidence is limited to certain regions, ways of keeping the flocks, or specific breeds. For example, in the United States more than 87% of all diagnosed cases occur in Suffolk sheep [24]. Another example is Iceland, where the disease remained limited to a region in the North for 70 years after its introduction, and only spread to other regions in the middle of the 20th century [23].

The export of scrapie-infected sheep from affected areas or countries is one reason for the existence of the disease in many countries today. The correlation between the novel occurrence of scrapie in a region and the preceding animal trade are illustrated in studies undertaken by Hourrigan and his colleagues [8].

Australia and New Zealand have been able to remain unaffected by scrapie owing to their rigorous quarantine measures [76; 77] and the slaughtering of the affected as well as all in-contact sheep after an imported scrapie occurrence. They take very stringent quarantine and monitoring precautions if sheep genes have to be imported for breed improvement or other purposes.

Following introduction of infection into a flock, scrapie is on average detected some 3.5 years later; in the meantime, the disease may have been spread to other sheep. Clinical cases in such sheep may appear at 4–5 years, decreasing to 1.5–2 years during the course of an epidemic [23; 78]. The following explanation may in part account for this: the level of infectious agent and the pressure of infection most likely increase slowly over time after the initial introduction into the flock – due to contamination of the "environment" by offspring-rearing scrapie-infected suckling ewes [78]. It is known for all prion diseases that – within certain limits – an increase in the infective dose results in a reduced incubation period.

The number of clinically affected animals within a scrapie-infected flock may vary considerably, with mortality rates ranging from 3 to 20% [23; 79; 80]. While the annual mortality rate of most affected flocks is around 3–5% [23], reports of losses from 10–20% are not uncommon [23; 79; 80]. One large farm in Iceland reported an annual mortality rate of 50% [23], but the disease is virtually eliminated there now due a rigorous control program.

18.7 Pathology

The neuropathology of scrapie is described in Chapter 27.

18.8 Is scrapie a new disease?

No (see Chapter 1.3 and Section 18.1).

18.9 Risk factors

Scrapie-risk factors and risks of transmission are described in detail in Chapters 42 and 56. The most important known risk factors of scrapie are summarized here:

1. (i) Contact with the placenta; (ii) contact with an infected environment including pasture and lambing pens; (iii) consumption of feed containing TSE-infected meat-and-bone-meal (MBM).

   Although transmission via feed has not been formally proven in sheep and goats, it is clearly responsible for the transmission of BSE to cattle and therefore remains a possible route. At least in BSE-affected countries, feed bans of various kinds are designed to protect all ruminant species from feed-borne transmission of TSE.
2. Distinct genetic predispositions present a risk factor for an increased susceptibility to scrapie (see Chapter 56). The susceptibility of sheep to scrapie infection is strongly in-

fluenced by particular alleles of the PrP gene and associated with polymorphisms at codons 136, 154, and 171 [81]. With very few exceptions, naturally infected sheep of a number of breeds in the USA, UK, and the European mainland often display the genotypes valine/valine (VV) and valine/alanine (VA) at codon 136 or glutamine/glutamine (QQ) at codon 171 (for references and further details, see Chapter 56). For instance, Suffolk, a British terminal sire breed has historically (before molecular genotyping was available) been associated with the introduction of scrapie into other countries of flocks. Using genotyping the frequent scrapie occurrence in Suffolk was found to be correlated with homozygosity QQ at codon 171. This situation has been improved by genetic selection (see Chapter 56). A recent paper [82] has analyzed PrP gene data from officially confirmed cases of scrapie and controls in GB and calculated odds ratios for the occurrence of disease compared with that in wild-type ARQ/ARQ genotypes. The results are similar to those of previous studies and show the VRQ/VRQ has odds ratio point estimates greater than 20, ARQ/VRQ and ARH/VRQ between 5 and 20, and other genotypes all less than one. Sheep homozygous for alanine, arginine, and arginine, respectively (ARR), show high resistance to natural and experimental scrapie (which may be overcome by challenge by the intracerebral route) [83]. National scrapie plans in the UK and some other European countries have involved genetic testing and selection of rams for breeding that have high resistance to scrapie with a view to increasing the number of sheep with resistance alleles and thus reducing the incidence of scrapie in the national flock.

Despite all currently available knowledge, many questions about the role of genetics in relationship to scrapie susceptibility still remain unanswered, for example:

- Do certain genotypes fully prevent natural scrapie infection, or can these "resistant" sheep also incur an infection without clinical manifestation of the disease (carrier state) [53] (see Chapter 55)?
- When a flock is selected for scrapie resistance as part of a breeding program, but unintentionally only resistant against a specific, currently circulating scrapie strain, is there a risk that this breed is more susceptible to a different scrapie strain (or the BSE strain), and, as a consequence, would proportionally more animals succumb to the disease after the introduction of a new scrapie strain (or the BSE strain) (see Chapter 56)?

18.10 Surveillance, prevention, and control

A number of features of scrapie (mainly ignorance or incomplete knowledge of key facts) militate against eliminating scrapie once it has occurred in a country or flock. Thus the first line of approach is to prevent introduction even if scrapie already exists, as this could result in the introduction of new strains of agent, atypical scrapie, or even BSE that could propagate in the sheep or goat population if undetected.

Other than Australia and New Zealand, no country has yet been successful in completely eliminating the disease if it has been introduced. Iceland, which has historically had regions of the country with endemic scrapie, has come the nearest to eliminating the disease by using draconian methods to succeed (compare also Norway; see Chapter 57). These include flock slaughter, leaving vacant for a number of years, removing top soil from around contaminated buildings, and burning anything wooden.

Australia and New Zealand were fortunate that original imports were presumably devoid of infection and subsequent imports that did carry infection were restricted to sheep under quarantine restrictions [76; 77]. If a country is devoid of scrapie, then it is wisest to adopt stringent import controls and lifetime quarantine and monitoring of imported animals to reduce the likelihood of occurrence. Importation of genetically resistant sheep or embryos derived from highly monitored flocks both before and after introduction and over a long period assist in this process.

Care should also be taken over the importation of sheep products including those for human consumption but particularly those that may find their way back to sheep like feed or edible material. What applies to sheep should also be applied to goats and moufflon, as all are susceptible to scrapie.

The nature of scrapie prevents identification of infected but clinically normal animals, vaccination or treatment to eliminate infection. Clinical examination and biopsy of lymphoreticular tissues are likely to be inefficient and incompletely successful, especially when applied to genetically resistant sheep.

Before attempting to control scrapie on a national or flock basis, several things are required. Shepherds must be trained to detect clinical scrapie and be aware of the signs that lead to suspicion of the disease, which must be notified to the responsible authority without delay for further examination. If pregnant or recently lambed, the animal should be isolated immediately. All individual animals must be permanently identified, and flock records should indicate the breeding line. These must be kept for a period of 10 years or so. Tracing of animals into, out of, and within the flock is needed to identify possible source and destination flocks and at-risk families.

In the EU, two alternatives are given when a case of scrapie is detected. Either the whole flock is destroyed or relevant animals are genotyped and only resistant ones kept for breeding.

An epidemiological investigation system should be in place to attempt to identify the source of infection and when it was introduced. This should include checking feed supplies and fertilizer containing MBM, as well as the more obvious importation of live animals. Since some outbreaks of scrapie have resulted from the use of locally manufactured vaccines, such origins should be fully investigated. To date, commercially prepared vaccines have never been implicated in scrapie transmission.

Whether or not depopulation is necessary or employed, attention should be paid to the cleaning and then disinfection of equipment and buildings exposed to scrapie infectivity, particularly at the time of lambing. Pastureland cannot easily be effectively decontaminated but should be fenced off and taken out of use for sheep and goats for two or more seasons (for details see Chapters 44 (disinfection) and 57 (Norway); and also compare with experiences with measures against scrapie in Iceland (Literature and Internet)). If repopulation of the flock is envisaged, then scrapie-free supervised flocks should be the source, and selecting only genetically resistant animals should be considered (see Chapter 56). Future attention should be directed to the design and use of lambing facilities, and effective management of parturition. Particularly important is the collection and safe disposal of placentas. Collectively these procedures will reduce, but may not eliminate risk and continuous monitoring for scrapie should be maintained.

From the point of view of the controlling authority, movement of animals from a scrapie-infected flock should be restricted. All sheep movements should be recorded in detail to enable tracing. Clinical inspection of flocks and auditing of the records may facilitate control. Furthermore, targeted surveillance on risk animals and older slaughter animals and especially those of non resistant genotypes is necessary.

References

[1] Gaiger SH. Scrapie. J Comp Path 1924; 37:259–277.

[2] McGowan JP. Scrapie in sheep. Scottish J Agric 1922; 5:365–375.

[3] Parry HB. Scrapie disease in sheep. In: Oppenheimer DR, editor. London: Academic Press, 1983.

[4] Greig JR. Scrapie. Transcript, Highland and Agriculture Society, Scotland 1940; 52:71–90.

[5] Perote M. BSE risk assessment and surveillance in Spain. In: Bradley R, Marchant B, editors. Transmissible spongiform encephalopathies. Proceedings of a consultation on BSE with the Scientific Veterinary Committee of the European Communities, September 14–15, 1993. Brussels: EU: VI/4131/94-EN document ref. F. II.3 – JC/0003, 1994:69–76.

[6] Garcia de Jalon JA, De las Heras M, Ferber L. Evolutión de la enfermedad del prurigo lumbar (scrapie) en Aragón. Med Vet 1991; 8:241–144.

[7] Hopp P, Bratberg B, Ulvund MJ. Skrapesjuke hos sau i Norge. Historikk og epidemiologi.

[Scrapie in sheep in Norway. History and epidemiology]. Norsk veterinærtidsskrift 2000; 112:368–375.
[8] Hourrigan J, Klingsporn A, Clark WW, et al. Epidemiology of scrapie in the United States. In: Prusiner SB, Hadlow WJ, editors. Slow transmissible diseases of the nervous system. Academic Press, New York, Vol 1. 1979:331–356.
[9] European Commission. Report on the monitoring and testing of ruminants for the presence of transmissible spongiform encephalopathy (TSE) in the EU in 2004. 31 May 2005. Health and Consumer Protection Directorate General. European Commission, Brussels, 2005:1–93.
[10] Bruce ME, McConnell I, Fraser H, et al. The disease characteristics of different strains of scrapie in Sinc congenic mouse lines: implications for the nature of the agent and host control of pathogenesis. J Gen Virol 1991; 72:595–603.
[11] Fraser H. A survey of primary transmission of Icelandic scrapie (Rida) in mice. In: Court LA, Cathala F, editors. Virus non-conventionnels et affections du systeme nerveux central. Masson, Paris, 1993:34–36.
[12] European Commission. BSE agent in goat tissue: first known naturally occurring case confirmed. Euro Surveillance 2005; 10: E050203.1.
[13] Eloit M, Adjou K, Coulpier M, et al. BSE agent signatures in a goat. Vet Rec 2005; 156(16):523–524.
[14] Houston EF and Gravenor MB. Clinical signs in sheep experimentally infected with scrapie and BSE. Vet Rec 2003; 152(11):333–334.
[15] Jeffrey M, Martin S, Gonzalez L, et al. Immunohistochemical features of PrP(d) accumulation in natural and experimental goat transmissible spongiform encephalopathies. J Comp Pathol 2006; 134(2/3):171–181.
[16] Jeffrey M, Martin S, Gonzalez L, et al. Differential diagnosis of infections with the bovine spongiform encephalopathy (BSE) and scrapie agents in sheep. J Comp Pathol 2001; 125(4):271–284.
[17] Stack MJ, Chaplin MJ, Clark J. Differentiation of prion protein glycoforms from naturally occurring sheep scrapie, sheep-passaged scrapie strains (CH1641 and SSBP1), bovine spongiform encephalopathy (BSE) cases and Romney and Cheviot breed sheep experimentally inoculated with BSE using two monoclonal antibodies. Acta Neuropathol (Berl) 2002; 104(3):279–286.
[18] Stack MJ. Western immunoblotting techniques for the study of transmissible spongiform encephalopathies. In: Lehmann S, Grassi J, editors. Techniques in prion research. Birkhauser, Basel, 2004:97–116.
[19] Gavier-Widen D, Stack MJ, Baron T. Diagnosis of transmissible spongiform encephalopathies in animals: a review. J Vet Diagn Invest 2005; 17(6):509–527.
[20] Collinge J, Sidle KC, Meads J, et al. Molecular analysis of prion strain variation and the aetiology of 'new variant' CJD. Nature 1996; 383:685–690.
[21] Bessen RA and Marsh RF. Biochemical and physical properties of the prion protein from two strains of the transmissible mink encephalopathy agent. J Virol 1992; 66:2096–2101.
[22] Dickinson AG. Scrapie in sheep and goats. Front Biol 1976; 44:209–241.
[23] Sigurdarson S. Epidemiology of scrapie in Iceland and experience with control measures. In: Bradley R, Savey M, Marchant B, editors. Subacute spongiform encephalopathies. Commission of European Communities. Kluwer Academic, Amsterdam, 1991:233–242.
[24] Wineland NE, Detwiler LA, Salman MD. Epidemiologic analysis of reported scrapie in sheep in the United States: 1,117 cases (1947–1992). J Am Vet Med Assoc 1998; 212(5):713–718.
[25] Parry HB. Natural scrapie in sheep – I. Clinical manifestation and general incidence, treatment, and related syndromes [Report]. Scrapie Seminar, ARS 91–53. USDA, Washington, DC, January 27–30, 1964.
[26] Moore RC, Hope J, McBride PA, et al. Mice with gene targeted prion protein alterations show that *Prnp*, *Sinc* and *Prni* are congruent. Nat Genet 1998; 18:118–125.
[27] Dickinson AG and Meikle VM. A comparison of some biological characteristics of the mouse-passaged scrapie agents 22A and ME7. Genetic Res 1969; 13:213–225.
[28] Hunter N, Foster JD, Dickinson AG, et al. Linkage of the gene for the scrapie-associated fibril protein (PrP) to the *Sip* gene in Cheviot sheep. Vet Rec 1989; 124:364–366.
[29] Hunter N, Foster JD, Goldmann W, et al. Natural scrapie in a closed flock of Cheviot sheep occurs only in specific PrP genotypes. Arch Virol 1996; 141:809–824.
[30] Foster JD and Dickinson AG. The unusual properties of CH1641, a sheep-passaged isolate of scrapie. Vet Rec 1988; 123:5–8.
[31] Cuillé J and Chelle PI. La maladie dite tremblante du mouton est-elle inoculable? C R Acad Sci Paris 1936; 203:1552–1554.

[32] Cuillé J and Chelle PI. La tremblante du mouton est bien inoculable. C R Acad Sci Paris 1938; 206:78–79.
[33] Gordon WS, Brownlee A, Wilson DR. Looping-ill, tick-borne fever and scrapie. In: Dawson MH, editor. Third International Congress for Microbiology, New York, September 2–9, 1939. International Association of Microbiologists, New York, 1940:262–263.
[34] Gordon WS. Advances in veterinary research. Looping-ill, tick born fever and scrapie. Vet Rec 1946; 58:516.
[35] Greig JR. Scrapie in sheep. J Comp Pathol 1950; 60:263–266.
[36] Taylor DM. Understanding the difficulty in inactivating TSE agents. A comprehensive update on the latest TSE research developments [Proceedings]. IIR Conference, London, March 9, 1998.
[37] Caramelli M, Ru G, Casalone C, et al. Evidence for the transmission of scrapie to sheep and goats from a vaccine against *Mycoplasma agalactiae*. Vet Rec 2001; 148(17):531–536.
[38] Brotherston JG, Renwick CC, Stamp JT, et al. Spread of scrapie by contact to goats and sheep. J Comp Path 1968; 78:9–17.
[39] Dickinson AG, Stamp JT, Renwick CC. Maternal and lateral transmission of scrapie in sheep. J Comp Pathol 1974; 84:19–25.
[40] Pattison IH, Hoare MN, Jebbett JN, et al. Spread of scrapie to sheep and goats by oral dosing with foetal membranes from scrapie-affected sheep. Vet Rec 1972; 90:465–468.
[41] Race R, Jenny A, Sutton D. Scrapie infectivity and proteinase K-resistant prion protein in sheep placenta, brain, spleen and lymph node – implications for transmission and antemortem diagnosis. J Infect Dis 1998; 178:949–953.
[42] Onodera T, Ikeda T, Muramatsu Y, et al. Isolation of scrapie agent from the placenta of sheep with natural scrapie in Japan. Microbiol Immunol 1993; 37:311–316.
[43] Andreoletti O, Lacroux C, Chabert A, et al. PrP(Sc) accumulation in placentas of ewes exposed to natural scrapie: influence of foetal PrP genotype and effect on ewe-to-lamb transmission. J Gen Virol 2002; 83(10):2607–2616.
[44] Tuo W, O'Rourke KI, Zhuang D, et al. Pregnancy status and fetal prion genetics determine PrPSc accumulation in placentomes of scrapie-infected sheep. Proc Natl Acad Sci U S A 2002; 99(9):6310–6315.
[45] Tuo W, Zhuang D, Knowles DP, et al. Prp-c and Prp-Sc at the fetal-maternal interface. J Biol Chem 2001; 276(21):18229–18234.
[46] Pattison IH, Hoare MN, Jebbett JN, et al. Further observations on the production of scrapie in sheep by oral dosing with foetal membranes from scrapie-affected sheep. Br Vet J 1974; 130:65–67.
[47] Hadlow WJ, Kennedy RC, Race RE. Natural infection of Suffolk sheep with scrapie virus. J Infect Dis 1982; 146:657–664.
[48] Pattison IH and Millson GC. Experimental transmission of scrapie to goats and sheep by the oral route. J Comp Pathol 1961; 71(2):171.
[49] Haralambiev H, Ivanov I, Vesselinova A, et al. An attempt to induce scrapie in local sheep in Bulgaria. Zbl Vet Med (B) 1973; 20:701–709.
[50] Stamp JT, Brotherston JG, Zlotnik I, et al. Further studies on scrapie. J Comp Pathol 1959; 69:268–280.
[51] Taylor DM, McConnell I, Fraser H. Scrapie infection can be established readily through skin scarification in immunocompetent but not immunodeficient mice. J Gen Virol 1996; 77:1595–1599.
[52] Wisniewski HM, Sigurdarson S, Rubenstein R, et al. Mites as vectors for scrapie [letter]. Lancet 1996; 347:1114.
[53] Race R and Chesebro B. Scrapie infectivity found in resistant species. Nature 1998; 392:770.
[54] Madec JY, Simon S, Lezmi S, et al. Abnormal prion protein in genetically resistant sheep from a scrapie-infected flock. J Gen Virol 2004; 85(11): 3483–3486.
[55] van Keulen LJ, Vromans ME, van Zijderveld FG. Early and late pathogenesis of natural scrapie infection in sheep. APMIS 2002; 110(1):23–32.
[56] Andreoletti O, Berthon P, Marc D, et al. Early accumulation of PrP(Sc) in gut-associated lymphoid and nervous tissues of susceptible sheep from a Romanov flock with natural scrapie. J Gen Virol 2000; 81(12):3115–3126.
[57] Schreuder BE, van Keulen LJ, Vromans ME, et al. Preclinical test for prion diseases [letter]. Nature 1996; 381:563.
[58] Schreuder BE, van Keulen LJ, Vromans ME, et al. Tonsillar biopsy and PrPSc detection in the preclinical diagnosis of scrapie. Vet Rec 1998; 142:564–568.
[59] O'Rourke KI, Baszler TV, Besser TE, et al. Preclinical diagnosis of scrapie by immunohistochemistry of third eyelid lymphoid tissue. J Clin Microbiol 2000; 38(9):3254–3259.
[60] Gonzalez L, Dagleish MP, Bellworthy SJ, et al. Postmortem diagnosis of preclinical and clinical scrapie in sheep by the detection of disease-as-

sociated PrP in their rectal mucosa. Vet Rec 2006; 158(10):325–331.
[61] van Keulen LJ, Schreuder BE, Meloen RH, et al. Immunohistochemical detection of prion protein in lymphoid tissues of sheep with natural scrapie. J Clin Microbiol 1996; 34:1228–1231.
[62] van Keulen LJ, Schreuder BE, Vromans ME, et al. Pathogenesis of natural scrapie in sheep. Arch Virol Suppl 2000; 16:57–71.
[63] Houston F, Foster JD, Chong A, et al. Transmission of BSE by blood transfusion in sheep. Lancet 2000; 356(9234):999–1000.
[64] Hunter N, Foster J, Chong A, et al. Transmission of prion diseases by blood transfusion. J Gen Virol 2002; 83(11):2897–2905.
[65] Cockcroft PD and Clark AM. The Shetland Islands scrapie monitoring and control programme: analysis of the clinical data collected from 772 scrapie suspects 1985–1997. Res Vet Sci 2006; 80(1):33–44.
[66] Bossers A, Schreuder BE, Muileman IH, et al. PrP genotype contributes to determining survival times of sheep with natural scrapie. J Gen Virol 1996; 77:2669–2673.
[67] Radostits OM, Blood DC, Gay CC. Veterinary medicine – a textbook of the diseases of cattle, sheep, pigs, goats and horses. Baillière Tindall, London, 1994.
[68] Hoinville LJ. A review of the epidemiology of scrapie in sheep. Rev Sci Tech Off Int Epiz 1996; 15:827–852.
[69] Morgan KL, Nicholas K, Glover MJ, et al. A questionnaire survey of the prevalence of scrapie in sheep in Britain. Vet Rec 1990; 127:373–376.
[70] Martin WB. Prevalence of scrapie in British sheep flocks [letter]. Vet Rec 1990; 126:409.
[71] Hoinville L, McLean AR, Hoek A, et al. Scrapie occurrence in Great Britain. Vet Rec 1999; 145(14):405–406.
[72] Hoinville LJ, Hoek A, Gravenor MB, et al. Descriptive epidemiology of scrapie in Great Britain: results of a postal survey. Vet Rec 2000; 146(16):455–461.
[73] Elliott H, Gubbins S, Ryan J, et al. Prevalence of scrapie in sheep in Great Britain estimated from abattoir surveys during 2002 and 2003. Vet Rec 2005; 157(14):418–419.
[74] Del RV, Ryan J, Elliott HG, et al. Prevalence of scrapie in sheep: results from fallen stock surveys in Great Britain in 2002 and 2003. Vet Rec 2005; 157(23):744–745.
[75] Dickinson AG. Scrapie in sheep and goats. In: Kimberlin RH, editor. Slow virus diseases of animals and man. North-Holland, Amsterdam, 1976:209–241.
[76] AQIS. Requirements for the approval and operation of scrapie freedom assurance programs for imported sheep and goats. Australian Quarantine and Inspection Service, April 10, 1995:1–25.
[77] MacDiarmid SC. Scrapie: the risk of its introduction and effects on trade. Aust Vet J 1996; 73:161–164.
[78] Foster JD and Dickinson AG. Age at death from natural scrapie in a flock of Suffolk sheep. Vet Rec 1989; 125:415–417.
[79] Pattison IH. Scrapie in the Welsh Mountain breed of sheep and its experimental transmission to goats. Vet Rec 1965; 77(47):1388–1390.
[80] Young GB, Stamp JT, Renwick CC, et al. Field observations on scrapie incidence [report]. USDA scrapie seminar, ARS 91–93, Washington DC, January 27–30, 1964:199–206.
[81] Dawson M, Hoinville LJ, Hosie BD, et al. Guidance of the use of PrP genotyping as an aid to the control of clinical scrapie. Vet Rec 1998; 142:623–625.
[82] Tongue SC, Pfeiffer DU, Warner R, et al. Estimation of the relative risk of developing clinical scrapie: the role of prion protein (PrP) genotype and selection bias. Vet Rec 2006; 158(2):43–50.
[83] Houston F, Goldmann W, Chong A, et al. Prion diseases: BSE in sheep bred for resistance to infection. Nature 2003; 423(6939):498.

19 Portrait of Bovine Spongiform Encephalopathy in Cattle and Other Ungulates

Beat Hörnlimann, Jutta Bachmann, and Ray Bradley

19.1 History

Late in 1986 a diagnosis was made at United Kingdom MAFF's Central Veterinary Laboratory (CVL) that would change forever the cattle livestock industry. CVL's Pathology Department had a surveillance function for animal diseases and provided expert pathology services as back-up, especially in neuropathology, to the Veterinary Investigation Service (VIS). The VIS was MAFF's eyes and ears for animal disease in the field through its laboratory service to practicing veterinarians. In November 1986, the brains of two cows from southern England presenting with progressive neurological signs were submitted to CVL by the VIS. Two veterinary neuropathologists, G.A.H. Wells and M. Jeffrey, diagnosed both as having a hitherto unrecognized scrapie-like spongiform encephalopathy [1][1], and the new cattle disease was named bovine spongiform encephalopathy (BSE).

At the time it was not known whether these were merely rare scientific curiosities or the first cases of an epidemic. However, as more BSE cases were identified, it soon became apparent that the second was correct. By the end of 1986 senior staff in MAFF were alerted to the possible consequences of an epidemic and, once the seriousness of the disease was recognized, an enormous research program was implemented.

By mid-1987 the clinical and pathological definition of the disease had been refined and what was probably the largest veterinary epidemiological investigation ever commenced, led by John W. Wilesmith. By the end of the year a hypothesis for the vehicle of transmission was established; namely that BSE was a feed-borne disease with meat-and-bone-meal (MBM) being the vehicle [3].

This led in July 1988 to a ban on feeding ruminant protein to ruminants (with the exception of dairy products). This ban is referred to hereafter as 'The Ruminant Feed Ban of 18 July 1988' or just 'The first feed ban'.

Wilesmith and colleagues summarized the results of the first investigation as follows: "The form of the epidemic was typical of an extended common source in which all affected animals were index cases. The use of therapeutic[2] or agricultural chemicals on affected farms presented no common factors. Specific genetic analyses eliminated BSE from being exclusively determined by simple Mendelian inheritance. Neither was there any evidence that it was introduced into Great Britain by imported cattle or semen. The study supported previous evidence of etiological similarities between BSE and scrapie of sheep. The findings were consistent with exposure of cattle to a scrapie-like agent, via cattle feedstuffs containing ruminant-derived protein. It is suggested that exposure began in 1981/82 and that the majority of affected animals became infected in calfhood" [3].

Further studies examined changes which had occurred in rendering (a method of cooking abattoir and other animal waste) in the recent past. Rendering is a process using heat to remove water, separate tallow (fat) from greaves (pro-

[1] Five months earlier, in June 1986, the brain of a captive nyala (*Tragelaphus angasi*) had been submitted to the same laboratory and lesions identified as a unique scrapie-like spongiform encephalopathy were observed [2]. Years later, it was determined that this disease was almost certainly caused by the BSE agent (see Footnote 1 of Table 1.3 and Chapter 50.3).

[2] Compare "BSE and organophosphates" in an internet search, for example through www.google.com.

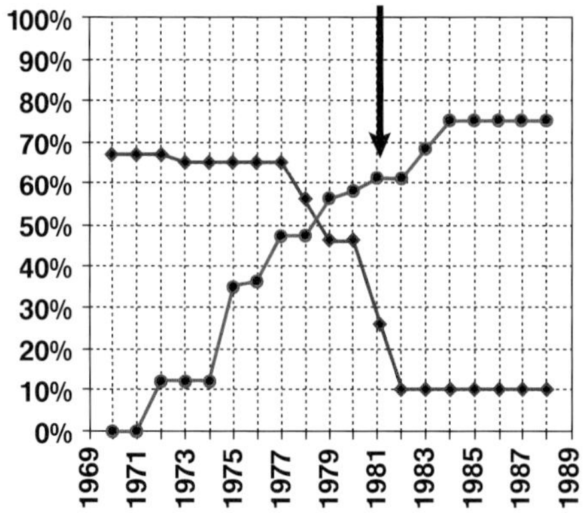

Fig. 19.1: The degree of prion inactivation achieved through rendering before the 1970s seems to have been mainly due to the application of sufficient wet heat (steam; ~ 120 °C) to remove residual solvents. The changes in rendering practices occurred in the UK from 1971 to 1988. Solvent extraction (blue curve) was largely discontinued – particularly between 1977 and 1984 – and replaced by a batch system working on lower temperature (< 100 °C) and a shorter total heat exposure time. In 1975, about 65 % of the MBM manufactured in the UK had been subjected to solvent extraction, but by 1982 the proportion had dropped to 10 % [4]. The arrow pointing to the situation in 1981 indicates the point in time when the first cattle may have been exposed to an effective oral dose according to J. Wilesmith [3]. Source of figure: *USDA:APHIS:VS* AHI1.01/91, modified by B. Hörnlimann.

tein), and produce MBM for use in animal feed (details see Section 19.9) or fertilizer by grinding the greaves.

Wilesmith and colleagues summarized the results of the extended investigations as follows: "The onset of this exposure was related to the cessation, in all but two[3] rendering plants, of the hydrocarbon solvent extraction of fat from meat-and-bone-meal. A further possible explanation, related to the geographical variation in the reprocessing of greaves to produce meat-and-bone-meal, was identified for the geographical variation in the incidence of BSE" [quoted from 4]. Figure 19.1 demonstrates the cessation of the traditional rendering process between 1971 and 1988.

A retrospective examination of adult cattle brains in pathological archives revealed a single case of BSE that had previously been overlooked. That case had occurred in April 1985 [5] but at the time was not sufficiently distinctive to evoke a definitive diagnosis of a new disease.

The origin of BSE is not known, and may never be (details see Section 19.9). It may have arisen from a cattle-adapted scrapie-like agent from sheep. At the time of the probable initial exposures of cattle (1981–1982)[4], scrapie of sheep and goats [6] was the only naturally occurring animal prion disease recognized in the UK and clearly is a strong candidate suspect source [7]. Scrapie had been recognized since at least 1732 [8] (see Chapter 1.3). However, subsequent biological and molecular strain typing of the BSE agent have not been able to align it with any known scrapie strain (compare Fig. 12.3). Thus, the mystery of its origin remains unsolved (Fig. 19.2).

19.2 Forms or variants

19.2.1 The BSE strain in domestic UK cattle

Although similar to scrapie, BSE differs in having a stereotypic lesion profile (Fig. 27.4). Furthermore, although there are polymorphisms in *Prnp* of cattle, they do not appear to influence the occurrence of BSE. This is in contrast to the situation in sheep where other polymorphisms significantly affect the incubation period, resistance, and susceptibility to prions (TSE agents; see Section 19.3). These features, and the consistency of the biological characteristics of the causal agents in mice ("agent strains" see below), distinguish BSE from all other prion diseases [9].

[3] See Section 19.9

[4] According to Anderson et al. [7] and information used for Figure 19.1, some animals, which were born in the 1970s and later developed BSE, might already have been infected in the second half of the 70s. However, this assumption cannot be confirmed and thus remains pure hypothesis. In this context, it is of note that the hypothesis would have been supported by the multiple GBR assessments that suggested that the UK was exporting BSE infectivity long before the disease was first reported in the UK.

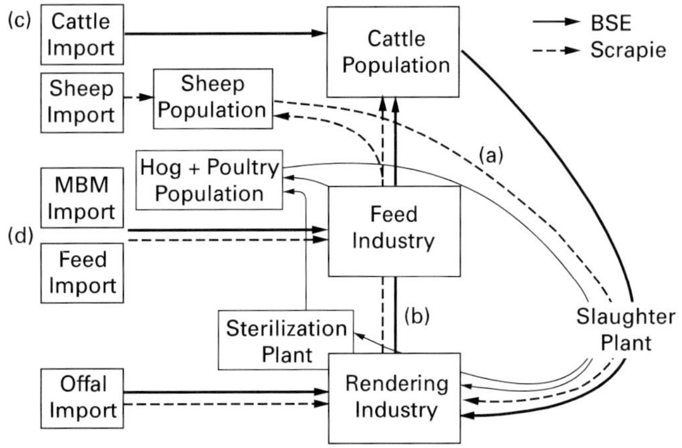

Fig. 19.2: The factors *increased incidence of scrapie in UK sheep* and an *increasing sheep population* have – in combination – contributed to higher quantities of scrapie infectious animal waste (sheep heads) that entered the rendering system **(a)** and subsequently the feed cycle, flowing back to ruminants (and to pigs and poultry → cross-contamination). The most probable, although unprovable, explanation for the BSE epidemic is that scrapie prions have been transmitted across the species barrier between sheep and cattle by infectious MBM. Any contaminating scrapie (or BSE) agent would no longer have been fully inactivated after major changes in rendering (Fig. 19.1 and see Section 19.9) leading to a striking increase of infectivity (PrP^{Sc}) in MBM **(b)**, which served as the vehicle of transmission. In other countries BSE cases were traced back to the import of live cattle, some of which were already infected before they left the country of origin **(c)**; and in some it was possible to prove that British MBM was imported **(d)**. MBM was also shipped indirectly along comlicated trading routes of the global market to the final countries of destination [65]. Figure by B. Hörnlimann ©.

Moira E. Bruce and colleagues at the Institute for Animal Health's Neuropathogenesis Unit in Edinburgh have traditionally used a panel of four inbred mice strains (and their crosses) to type TSE strains from field isolates by determining the incubation period and lesion profile in the mouse brain. A pre-requisite is, of course, that an isolate transmits to mice in the first place, which the BSE agent does. However, not all isolates of scrapie do.

When brain from BSE cases was inoculated into C57 black mice (*Sinc* genotype s7s7) further passage resulted in the isolation of strain 301C. When inoculated into VM mice (*Sinc* genotype p7p7) and sub-passaged, 301V was isolated. These murine prion strains were quite different from all known strains of scrapie from sheep (compare Fig. 12.3) and goats. Bruce and colleagues concluded that a single strain of agent was responsible for BSE and that it was biologically unlike any other TSE agent known at the time [10; 11].

Thus, while scrapie in sheep [10] is caused by several different prion strains (see Chapter 12), BSE appears to be caused by a single major strain of agent (10); at least according to studies to date on isolates from the UK, Switzerland, and France[5] [11]. However, while scrapie strains have been circulating for more than 270 years, BSE has only been recognized for about 20 years. So, if BSE is not eradicated, from a biological point of view, it is probable that strain variation of the BSE agent will emerge over time. However, the BSE epidemic in the UK is declining rapidly and forecasts suggest that it will die out in the next decade (see Chapter 39), unless idiopathic atypical BSE cases (see below), comparable to sporadic Creutzfeldt–Jakob disease (sCJD) in humans, are found to be occurring spontaneously.

[5] It is recommended that other countries affected by BSE undertake similar studies to assess the possibility of strain variation of the BSE agent.

19.2.2 Atypical BSE

For a few "atypical" BSE cases in several countries it is unclear whether they are caused by one (or several) different unknown strain(s) of the BSE agent. The possible natural occurrence of phenotypic variation in BSE, or novel expression of a TSE of cattle, has been raised by reports from Japan [12], Italy [13], France [14], Denmark, Germany, and the Netherlands, in which a small number of cattle with molecular and/or pathological features atypical of BSE have been described. While they all have unusual electrophoretic profiles of the PrP^{Sc} accumulating in the brain, only the report from Italy describes the immunohistochemical differences in the form and distribution of disease-specific PrP in the brain (see Chapter 27.5). The Italian group of Cristina Casalone and others [13] demonstrated that the disorder was pathologically characterized by the presence of PrP-immunopositive amyloid plaques, as opposed to the lack of amyloid deposition in typical BSE cases, and by a different pattern of regional distribution and topology of brain PrP^{Sc} accumulation. Therefore, these cases – in animals older[6] than most of those with typical BSE – have been called bovine amyloidotic spongiform encephalopathy (BASE). In addition, Western blot analysis showed a PrP^{Sc}-type with predominance of the low molecular mass glycoform and a protease-resistant fragment of lower molecular mass than BSE-PrP^{Sc} isolated from UK BSE cases. Strikingly, the molecular signature of atypical bovine PrP^{Sc} was reported to be similar to that encountered in a distinct subtype of sCJD [13].

19.2.3 The BSE strain in nyala and greater kudu (zoo ungulates)

Early in the BSE epidemic, new spongiform encephalopathies of ruminant species in zoos and wildlife parks (Table 19.1; see Chapter 20) and a variety of domestic and captive wild cats were identified (Table 1.3; see Chapters 20, 23). Most, if not all, were presumed to have been infected through feed; MBM in the case of the ruminants, probably contaminated skeletal meat from heads and vertebral columns from infected cattle fed to the large cats (see Chapter 20), and contaminated[7] pet food fed to the domestic cats.

Transmission experiments carried out by Bruce and colleagues showed that the affected *Tragelaphus spp.*, nyala and greater kudu (Table 19.1), were infected by the same strain of agent that caused BSE in cattle, despite differences in the sequence of the PrP in these species (references see Chapter 50 and Table 1.3). It is assumed that the diseases occurring in the subfamilies Bovinae (*Taurotragus oryx*), Hippotraginae (*Oryx spp.*) (Table 19.1), and other ungulates (Table 1.3) are also attributable to the BSE prion strain. This assumption is based on observations related to geographical distribution (mainly in UK zoos; Table 19.1; compare to Fig. 39.2), temporal distribution (period of BSE epidemic in the UK; compare Fig. 20.1 with Fig. 39.1) and similar feeding histories (see Chapter 20).

19.3 Incubation time, transmissibility, and susceptibility

19.3.1 Incubation time

The incubation time (or period) depends on a number of factors and is quite variable. Time of infection and age at clinical onset (Fig. 39.3) of disease are closely linked by the incubation time, as illustrated in Figure 39.4.

A number of studies [7; 15–20] have estimated mean values and distributions of incubation time based, among other factors, on the wide age distribution observed at the time of onset of disease (Fig. 39.3) and on the results of a case-control study (see Chapter 40). According to this study, approximately 60% of the British animals were infected in their first months of life (compare Fig. 39.4). A mean incubation time of approximately 5 years was deduced, given that the animals were reared under field

[6] The Japanese atypical BSE cases, conversely, were reported in very young animals (20 and 23 months old), and were detected by rapid tests at the slaughterhouse.

[7] Containing infectious MBM or cross-contaminated with infectious spinal cord and brain.

Tab. 19.1: BSE in the *Tragelaphus spp.* "nyala" and "greater kudu" or related prion diseases (TSE) in *Oryx spp.* and a *Taurotragus sp.* kept in zoos. Incomplete list of individual animals; for further affected ungulates compare Table 1.3. Year and country of birth and deaths indicated if available. Source: [59].

BSE/TSE	Genus	Species	Subspecies	Year of birth	Country of origin (zoo)	Year of death	Country of death
+	*Tragelaphus*	*angasi*	Nyala*	1983	UK	1986	UK
+	*Oryx*	*gazella*	Gemsbok	1983	UK	1986	UK
p	*Tragelaphus*	*strepsiceros*	Greater kudu	1985	UK	1987	UK
+	*Tragelaphus*	*strepsiceros*	Greater kudu	1987	UK	1989	UK
+	*Tragelaphus*	*strepsiceros*	Greater kudu	1988	UK	1990	UK
+	*Tragelaphus*	*strepsiceros*	Greater kudu	1988	UK	1991	UK
p	*Tragelaphus*	*strepsiceros*	Greater kudu	1988	UK	1991	UK
+	*Tragelaphus*	*strepsiceros*	Greater kudu	1989	UK	1992	UK
p	*Tragelaphus*	*strepsiceros*	Greater kudu	1990	UK	1992	UK
+	*Taurotragus*	*oryx*	Eland	1987	UK	1989	UK
+	*Taurotragus*	*oryx*	Eland	1989	UK	1991	UK
+	*Taurotragus*	*oryx*	Eland	1989	UK	1991	UK
p	*Taurotragus*	*oryx*	Eland	1988	UK	1990	UK
+	*Taurotragus*	*oryx*	Eland	1990	UK	1992	UK
+	*Taurotragus*	*oryx*	Eland	1991	UK	1994	UK
+	*Taurotragus*	*oryx*	Eland	1993	UK	1995	UK
+	*Oryx*	*leucoryx*	Arabian oryx	1986	Switzerland	1991	UK
+	*Oryx*	*dammah*	Scimitar-horned oryx	1990	na	1993	UK

* First case, published by M. Jeffrey et al. as a scrapie-like spongiform encephalopathy; it was a unique occurrence at that time [2].
\+ Diagnosis confirmed post mortem.
p Indicates probable / suspicious case.
na Not available.

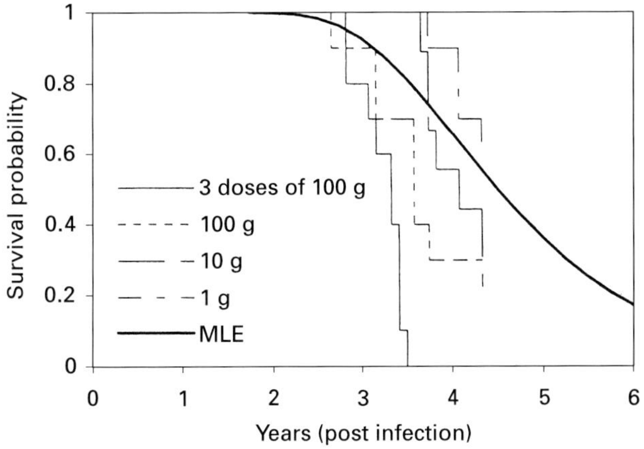

Fig. 19.3: Survival curves and modelled distribution of the "experimental BSE incubation time": Results of an attack-rate-study using infective material of a BSE cattle brain pool, which was orally (1 g, 10 g, 100 g or 100 g per day on three consecutive days) administered to calves. The mean "experimental incubation time" is approximately 36 ± 2 months. Source: [20].

conditions. There are differences in experimental infections in cattle, such as shorter incubation time (mean: 36 ± 2 months) (Fig. 19.3) [19; 20].

19.3.2 Transmissibility

The BSE agent is transmissible to cattle but BSE is not a contagious disease. The vehicle for the indirect transmission is inappropriately prepared MBM[&,8]. Hence, BSE is mainly, if not entirely, initiated by exposure to feed contaminated by MBM[&] [21]. Today we know from ongoing studies in the UK that BSE can be transmitted orally with as little as 1 mg of infectious brain tissue [22] (Wells, personal communication, see Section 19.7).

The feeding of contaminated feed is the only proven source of BSE in the UK. Transmission via tallow (fat; see Chapter 46.4.2) in milk replacers has not been proven. However, one recent statistical analysis outside the UK suggests that contaminated milk replacers (containing bovine tallow) may have played a minor role as a significantly higher proportion of the BSE-affected farms had fed proprietary concentrates and/or milk replacers to calves [23].

Horizontal transmission of BSE in cattle appears unlikely. This is based on epidemiological analysis which has revealed no or only weak evidence for transmission between cattle [7]. That is, direct horizontal transmission ratios below 1 % would have remained undetected [24]. The mean within-herd incidence in the UK was 3.5% in 1992 (at the peak of the epidemic). Given that this reflects the real situation at that time, experts concluded that direct horizontal transmission from cow to cow does not occur [25] or, if it occurs, it is insufficient to maintain an epidemic [26]. Consequently, there is no evidence that BSE is a contagious disease in cattle, which contrasts with scrapie in small ruminants (see Chapter 18) and CWD in certain North American cervid species (see Chapter 21).

Similarly, vertical transmission of BSE in cattle appears unlikely. If it occurs at all, maternal transmission seems to be insignificant in the UK epidemic. That is, if maternal transmission occurs, it is at a rate insufficient to maintain an epidemic.

By examining the current stage of the UK epidemic (Fig. 39.1), it is evident that the former interpretation of the data by experts, in which they concluded that neither horizontal (including lateral) nor vertical (including maternal) transmission would play a sustaining role in the epidemic, is supported by the facts, that is, the low number of cases occurring nowadays in the UK (Table 19.2; Fig. 39.1).

19.3.3 Susceptibility of the host

Experimental models of prion diseases have demonstrated that the combination of the agent strain (see Chapter 12) and the genotype of the host PrP gene play a major role in determining relative incubation times between model systems (Fig. 50.1). Together agent strain and PrP genotype also affect the targeting of infection to different organs and to different parts of the brain (Fig. 12.1; also Fig. 50.2). The size of the dose required to infect the host is also affected by these two factors.

The use of the words "susceptible" and "resistant" in what follows requires careful definition. They should be seen as relative terms in a continuum of susceptibility, not as absolute statements. By "more susceptible" it is implied that animals can be infected by a relatively small amount of infectivity and by a relatively inefficient route (e. g., the oral route). By contrast "more resistant" implies that a larger dose of infective material is required to infect the animal and possibly by a more efficient route (e. g. intracerebral injection). Although it is often the case that more susceptible models have relatively short incubation times, susceptibility and resistance should not be confused with length of incubation time, since in some cases highly susceptible animals can have long incubation times.

---

[8] MBM (→) marked with an '&' symbol as superscript indicates that the reader can obtain comprehensive information about MBM and other important residual feed-related BSE risks or BSE risks in bovine-derived products in the document [22].

Cattle. There is no evidence of genetic differences in susceptibility of cattle to BSE. *Therefore, all cattle breeds and individuals must be considered to be susceptible.* There is a recent report on a racial predisposition for the breed "BS" in Germany [27]; but it is based on statistical significance only and remains an unproven hypothesis.

Goats. Less is known about TSEs in goats than in sheep. Classical scrapie in goats has, for example, been described by Hadlow et al. [28; 29]. Atypical scrapie (Nor 98) in goats has been diagnosed a few times recently, for instance in Norway on 6 June 2006 (S. Benestad, personal communication; compare Chapter 54).

In the light of Eloit's publication [30] about BSE in a French goat in 2005, risk managers are challenged to define a goat-specific approach to avoid more cases of BSE in goats. The following is relevant in this context:

- Scrapie occurs less frequently in goats than in sheep.
- It is likely that all goats are susceptible to BSE or scrapie by the oral route under certain conditions and to a variable degree. It is of note that the dimorphism in codon 142 of the caprine *Prnp* appears to be associated with different incubation times in experimental infections with BSE or scrapie. Recent research has shown that goats have similarly complex PrP-genetics as sheep. However, the relationships between breed, *Prnp*-polymorphisms and susceptibility are not yet understood and therefore a breeding program towards TSE resistance in goats is not currently feasible.

Sheep. In contrast to goats, BSE under natural field conditions has not yet been diagnosed in sheep. As far as the genetic susceptibility of sheep to experimental BSE is concerned [compare 31–33], the effects of sheep PrP genotypes on incubation time and pathogenesis are complex and not yet fully understood (compare Chapter 56). Current knowledge is based on a few experiments [31–33] carried out on small numbers of animals involving only a small proportion of sheep breeds. The results to date are difficult to interpret and it would be unwise to make generalizations on host susceptibility to BSE in sheep at present.

- Based on preliminary observations of TSEs in sheep, it is assumed, as a worst case, that after infection, there is a rapid increase in the amount of infectivity in lymphoid and some other tissues (WHO Tables$_{\text{Appendix}}$IA-IC) in both susceptible and semi-resistant sheep (genotypes). In contrast, resistant sheep (genotypes) may just harbor less infectivity that is ineffective to produce clinical signs (onset of disease) throughout life time.

19.4 Clinical signs and course of disease

Since knowledge of the clinical signs of BSE is of utmost importance for veterinarians, farmers, and other professionals involved with clinical or passive[9] surveillance, the clinical manifestations are described in detail in Chapter 33. A complete list of signs is shown in Table 33.1.

The diagnosis "BSE" must be considered in adult cattle, particularly those exhibiting abnormal behavior (Fig. 33.2) or locomotor disorders (downer cows; see Chapter 53.3.3). The mean age of affected cows at onset of disease is approximately 5 years (Fig. 39.3) [34; 35]. The maximum reported age of a BSE case has been 22 years and 7 months and the youngest has been 20 months [25].

BSE can only be confirmed post mortem but in order to decrease the risk of potentially infected cattle entering the normal slaughter process, ante mortem inspections to identify suspects on presentation at the slaughter plant remain an important component of BSE surveillance (see Chapters 32.2, 35, 53.3.3).

19.5 Differential diagnoses

Other diseases, which are accompanied by changes in behavior and locomotion, must be

[9] The importance of a high-quality passive surveillance is emphasized in Chapter 32. The passive surveillance of BSE in the UK has proven the importance of this instrument and of its potential to lead – combined with a control program – to a successful decrease in the incidence rate of BSE, as long as it was high or relatively high.

taken into consideration. These include metabolic disorders such as (chronic) hypomagnesemia, nervous acetonemia, parturient paresis (hypocalcemia), the bacterial infection cerebral listeriosis, the viral infections rabies [36] and Aujeszky's disease, poisoning due to consumption of material containing soluble lead salts or botulinum toxin, traumatic injuries to the head, vertebral column, and spinal cord [34], and other diseases for which the reader has to consult country-specific veterinary literature, since there may be differences between countries and continents.

19.6 Epidemiology

Retrospective epidemiological considerations on the course of the BSE epidemic are summarized in detail in Chapter 39. Some conclusions from analytical data which led to the support of the MBM-hypothesis are covered in detail in Chapter 40. In this section, only a few basic facts are listed:

19.6.1 Number of cases

In the UK, the average national incidence was 0.97 % of adult cattle in 1992 when the epidemic reached its peak; this corresponded to about 970 BSE cases per 100,000 adult breeding cattle. By 2005 the annual incidence had dropped to about four cases per 100,000, demonstrating that the epidemic is now waning. The number of cases reported each week is now lower than when the disease was made notifiable in 1988.

19.6.2 Geographical spread of BSE

Countries which have experienced BSE cases, and to what extent, are listed in Table 19.2.

International and national up-to-date data on BSE occurrence (number of cases, incidence, prevalence, etc.) can be found on the Internet, particularly the web site of the World Organisation for Animal Health (OIE) and on the "UK Defra Animal Health and Welfare" web site [25; 37–39]. The competent authorities of countries other than the UK report their own national situations, so relevant data can be found on corresponding web sites and – partly – in the multiple GBR assessments (references in Table 53.1). For these reasons, and particularly because the data change monthly, it is not the objective of this text to comment on the latest number of cases for all the countries involved (Table 19.2).

19.6.3 Age at the time of infection

The transmission of the BSE agent occurred mainly in the first year of cattle life [40–42] (Fig. 39.4). Detailed explanations of this are described in Chapters 39 and 40.

19.7 Pathology and pathogenesis

BSE presents the classical neuropathological features of prion diseases; they are described in detail by Gerald Wells in Chapter 27. Therefore, the following is restricted to some basic observations on BSE pathogenesis.

When BSE occurred in the mid-1980s, some of the questions needing urgent answers were: i) "how is BSE – *experimentally and naturally* – transmissible to other cattle?" (see Section 19.3), ii) is BSE – *experimentally and naturally* – transmissible to other species?" (see Chapters 18, 20, 41, 42) and particularly iii) "could BSE be transmitted *naturally* to humans?" (see Chapters 15, 50). To answer the third question – at least in parts – it was important to study BSE pathogenesis as well as body tissue and liquid distribution of the agent (WHO $\text{Tables}_{\text{Appendix}}$ IA-IC) in animal models, and to undertake surveillance for new variants of CJD (see Chapters 15; 29). Milk, for instance, from cows clinically affected with BSE, showed no detectable infectivity [21; 43]. In contrast, mechanically recovered meat (MRM), for instance used in hamburger [44], could contain BSE infectivity.

19.7.1 General remarks on the pathogenesis studies

An understanding of the pathogenesis of BSE as well as tissue distribution of infectivity at different ages and incubation stages is important

Table 19.2: BSE cases in the UK [38] and in non-UK countries [39] as of 1 January 2006. Total number of cases from active and passive surveillance according to official publications by the World Organisation for Animal Health, Paris (OIE). Furthermore the following countries or regions are affected: Azores (1 case), Falkland Islands (1 case), Sultanate of Oman (2 cases), and Sweden (1 case).

	1987 & before	1988	1989	1990	1991	1992	1993	1994	1995	1996	1997	1998	1999	2000	2001	2002	2003	2004	2005	Total on 1 Jan. 2006
Austria	0	0	0	0	0	0	0	0	0	0	0	0	0	0	1	0	0	0	1	
Belgium	0	0	0	0	0	0	0	0	0	0	1	6	3	9	46	38	15	11	2	
Canada	0	0	0	0	0	0	1	0	0	0	0	0	0	0	0	0	2	1	1	
Czech Republic	0		0	0	0	0	0	0	0	0	0	0	0	0	2	2	4	7	8	
Denmark	0	0	0	0	0	1	0	0	0	0	0	0	0	1	6	3	2	1	na	
Finland	0	0	0	0	0	0	0	0	0	0	0	0	0	0	1	0	0	0	0	
France	0	0	0	0	5	0	1	4	3	12	6	18	31	161	274	239	137	54	18	
Germany	0	0	0	0	0	1	0	3	0	0	2	0	0	7	125	106	54	65	na	
Greece	0	0	0	0	0	0	0	0	0	0	0	0	0	0	1	0	0	0	na	
Ireland (Republic of)	0	0	15	14	17	18	16	19	16	73	80	83	91	149	246	333	183	126	69	
Israel	0	0	0	0	0	0	0	0	0	0	0	0	0	0	0	1	0	0	0	
Italy	0	0	0	0	0	0	0	2	0	0	0	0	0	0	48	38	29	7	7	
Japan	0	0	0	0	0	0	0	0	0	0	0	0	0	0	3	2	4	5	7	
Liechtenstein	0	0	0	0	0	0	0	0	0	0	0	2	0	0	0	0	0	0	na	
Luxembourg	0	0	0	0	0	0	0	0	0	0	1	0	0	0	0	1	0	0	1	
Netherlands	0	0	0	0	0	0	0	0	0	0	2	2	2	2	20	24	19	6	na	
Poland	0	0	0	0	0	0	0	0	0	0	0	0	0	0	0	4	5	11	18	
Portugal	0	0	0	1	1	1	3	12	15	31	30	127	159	149	110	86	133	92	37	
Slovakia	0	0	0	0	0	0	0	0	0	0	0	0	0	0	5	6	2	7	na	
Slovenia	0	0	0	0	0	0	0	0	0	0	0	0	0	0	1	1	1	2	1	
Spain	0	0	0	0	0	0	0	0	0	0	0	0	0	2	82	127	167	137	75	
Switzerland	0	0	0	1	9	15	29	64	68	45	38	14	50	33	42	24	21	3	3	
USA	0	0	0	0	0	0	0	0	0	0	0	0	0	0	0	0	0	0	1	
Total (Non-UK countries)			15	17	31	36	50	104	102	161	160	252	336	513	1013	1035	778	535	249	5 387
Total (BSE cases in UK)	446	2514	7228	14407	25359	37280	35090	24438	14562	8149	4393	3235	2301	1443	1202	1144	611	343	151	184 296

Important note by the editors: All the footnotes added by the OIE are not printed in this table; they and special explanations to individual countries and years must be checked on the Internet / source: [38; 39].

in formulating measures to protect human and animal health. The following provides a summary of some information used as a basis to define SRM-related age limits (cut-off points) as indicated in Chapter 53.3.2. To investigate the pathogenesis of BSE and to determine whether tissue infectivity in cattle differs from that of scrapie in sheep tissues (upon which the original UK 1989[10] specified offal ban was based; SBO ban), experimental studies were undertaken [45; 46].

In other words, the pathogenesis studies were undertaken in order to determine whether the 1989 classification of tissues posing a risk to humans (see SBO ban; later SRM ban), based primarily on scrapie data [28], was correct, and to identify edible tissues that are uninfected throughout the incubation time and into the clinical phase of BSE. SRM-relevant results[11] of mouse bioassays are shown in Figure 53.2 [45; 46]. Bioassays in cattle and mice [47] demonstrated that spleen and lymph nodes, for instance, contain at least one million times less infectivity than brain tissue from clinically affected cattle.

In natural and experimental [46] BSE, the distribution of infectivity during the incubation time and in the clinical stage of disease is much more restricted than in scrapie of sheep [48] and goats [28]. In cattle, BSE infectivity is confined to the distal ileum from about month 6 to 18 [49] (Fig. 53.2) of the incubation time. It is also present in the palatine tonsil early in the incubation time [45]. In the CNS, infectivity can be detected by rapid BSE test between 6 and 3 months before the onset of clinical signs [44].

19.7.2 Results and details of the pathogenesis studies

The first experimental oral exposure study of BSE in cattle was initiated in 1991 using 30 calves, each dosed with 100 g of BSE-contaminated brain titrating $10^{3.5}$ ID_{50} /g[12] in RIII mice. Animals were sequentially killed from two months up to 40 months after exposure and tissues taken for infectivity assays in RIII or C57BL mice. In another set of experiments, the infectivity of a BSE brainstem pool (from five cases of BSE) was compared in RIII mice and cattle to evaluate the species barrier. In RIII mice the brainstem pool contained $10^{3.3}$ mouse [intracerebral (i.c.) and intraperitoneal (i.p.)] ID_{50}/g while in cattle the same material titrated $10^{6.0}$ i.c. ID_{50} /g; thus, the underestimate of infectivity titer in mice was 500-fold (2.7 $\log_{10}$). The limit of detection of infectivity by the cattle assay was approximately 10^{-1} cattle i.c. ID_{50} units or $10^{-3.7}$ mouse i.c. + i.p. ID_{50} units/g [52].

In the first UK pathogenesis study, infectivity was first detected in a pool of distal ileum from three exposed animals six months after exposure. PrP^{Sc} was detected in Peyer's patches. It is noteworthy that in the UK study PrP^{Sc} was not detected in the ileum of cows with natural BSE, which contrasts the finding with Tg(bov) mice in the German pathogenesis study (see below). One pool of lymph nodes (retropharyngeal, mesenteric, popliteal and prescapular) and one pool of spleens from five BSE cases were negative on assay in mice and cattle, suggesting that infectivity, if present, must have been at a titer less than 0.1 cattle i.c. ID_{50} /g. One pool of tonsils from three calves in the first experimental oral exposure study was positive by cattle bioassay at ten months post inoculation (p.i.). Also, a pool of nictitating membranes (lymphoid tissue from ten clinically suspect cases of BSE, including nine later confirmed cases) was positive (incubation period 31 months) implying an approximate infectious titer of 101–102 cattle i.c. ID_{50} /g (Wells, unpublished). Assays of tissues from naturally infected clinical cases in RIII mice showed infectivity only in central nervous system ([53] and Wells, unpublished).

A second UK BSE pathogenesis study initiated in 1997 used 300 cattle divided into three

[10] Measures taken in GB, the EU, and the USA for instance – from 1988–2000 – to prevent the spread of BSE to animals and humans are comprehensively summarized in two tables in [50; 51].

[11] Further study results related to tissue distribution of TSE infectivity in vCJD, BSE, scrapie, and other prion diseases are comprehensively summarized in the WHO $\text{Tables}_{\text{Appendix}}$IA-IC, re-printed in the appendix of this book.

[12] ID_{50}, 50 % infectious dose, is a term preferred by some investigators to LD50, 50 % lethal dose. The terms are essentially equivalent for prion diseases and both are used in this book and in the literature.

groups receiving oral doses of 100 g or 1 g of BSE-infected brain, or remaining as undosed controls. Tissues were sampled from six exposed cattle and three unexposed controls killed sequentially at three-month intervals, starting three months p.i. Sampling was concluded at 45 months (100-g-dose group) or at 78 months p.i. (1-g-dose group) [52]. Further study of the collected material is in progress (Wells, personal communication).

Bioassays of infectivity have inherent limitations. Assays conducted in wild-type mice underestimate infectivity because of the species barrier effect. This problem may be overcome by assay in cattle, but the cost and protracted timescale of such studies are prohibitive for routine application. Assays in mice expressing multiple copies of the bovine *Prnp* also overcome the species barrier and offer high sensitivity (see below: the German pathogenesis study). All assays for TSE infectivity have limitations in sensitivity due to the non-uniform distribution of infectivity in tissues and the consequent variation resulting from focal sampling and small starting volumes of inocula; these problems are compounded when assays are conducted on tissue pools [52].

In a third pathogenesis study, in Germany, 56 calves were dosed orally with 100 g of BSE brain while 18 control calves received normal brain. Samples of blood, urine, and CSF were collected at two-month or four-month intervals. Serial kills of about four BSE exposed and one control animal each were performed every four months. More than 150 body tissues and fluids were collected during each necropsy. Conventional immunohistochemistry and immunoblotting were used to detect abnormal PrP, and Tg(bov) mice overexpressing bovine PrP^C were inoculated to detect infectivity in the samples. The sensitivity of the Tg(bov)-mouse bioassay model was verified by comparative infectivity titrations of a single BSE brain stem pool in both conventional wild-type RIII mice and in the Tg(bov) mice; infectivity titers of the BSE pool were $10^{7.7}$ i.c. ID_{50} /g in Tg(bov) mice and $10^{3.3}$ i.c. ID_{50} /g in RIII mice, i.e., assay in Tg(bov) mice was more than 10,000-fold more sensitive than assay in RIII mice and about tenfold more sensitive than reported in the UK pathogenesis study for cattle (see above) inoculated with BSE agent i.c. [52, 54].

The German BSE pathogenesis studies included an investigation of a late-stage pregnant cow with naturally acquired BSE. Tissues from that animal were assayed in RIII and Tg(bov) mice; no infectivity was detected in either spleen or lymph nodes, and all samples from the reproductive tract were also free from detectable infectivity. However, infectivity was detected in the Peyer's patches of the distal ileum by assay with Tg(bov) mice (mean incubation time 540 days) while the RIII mice remained negative even after a subpassage (see above). Infectivity was also found in some peripheral nerves (but not radial nerve) assayed in Tg(bov) mice, with incubation times of 438 and 538 days after inoculations with sciatic and facial nerves respectively. One of ten Tg(bov) mice inoculated with one specific muscle (semitendinosus) was positive 520 days p.i. [52; 54].

19.8 Is BSE a new disease?

Since the origin of BSE is not known, and probably never will be known, the question of whether BSE is a new disease can not be answered.

19.9 Risk factors

Studies have not revealed a means of BSE transmission which is capable of maintaining an epidemic (under field conditions), other than the feed-borne route. The following summarizes the risk factors that led to the emergence of BSE[13].

19.9.1 Historical aspects of rendering and feeding of cattle with MBM[&]

In the early 1900s it was realized [55] that the solid material that remains after the tallow (fat) has been extracted from rendered animal tissues

[13] Parts of Section 19.9 are extracts from a 2003 review by D.M. Taylor et al. [63] in which the authors discuss comprehensively the role of rendered MBM as a dietary supplement in propagating the UK epidemic of BSE, together with the role of MBM in spreading BSE outside the UK.

is rich in protein, and might usefully be fed back to animals as a dietary supplement.

The practice of feeding such animal proteins to other animals became common world-wide after the Second World War. Animal by-products are rendered by cooking to remove water and inactivate (conventional) micro-organisms, followed by separation of fat from the solid residue. The residue is ground to produce MBM which provided a valuable source of protein for use in animal feed, especially for omnivores (i. e. as a component of pig and poultry feed) and – much later – carnivores (i. e. as a component of pet food). It was also used as an ingredient in ruminant feed. MBM was used in the European Union (since 1993 only for feeding non-ruminant species) until its complete prohibition in animal feeds in 2001 (referred to as "the total feed ban"). Authors' note: The practice of feeding MBM to herbivores has been heavily criticized – because it is "unnatural" –, and we recommend that animal-derived protein ($MBM^{\&}$) should not be fed back to the same species, even among omnivores and carnivores, in the future.

Changes in rendering procedures in the 1970s and 80s (Fig. 19.1) contributed to scrapie-like agents surviving some of the changed rendering steps. Thus, any contaminating scrapie or BSE agent would no longer have been fully inactivated (see Chapter 45) leading to a striking increase of infectivity (prions, PrP^{Sc}) in MBM, which served as the vehicle of transmission. Whatever the original source, BSE was amplified in UK cattle through the change in established rendering and calf/cattle feeding practices. In other words, the epidemiological data are consistent with an origin of BSE in an increased exposure via feed of the BSE agent from cattle or to a cattle-adapted scrapie-like agent from sheep [compare 4]. BSE has, in any case, resulted from an increase in exposure to feed/MBM contaminated with prions.

In the context of the ongoing debate about the origin of BSE [56], the uniform pathology of the disease (stereotypic lesion profile; Fig. 27.4) over the course of the epidemic [9; 57] and the fact that only one major BSE strain has been isolated under UK field conditions [11] are impressive facts, contrasting with the many different scrapie strains in sheep (Fig. 12.3). This does not rule out the scrapie-origin hypothesis, but the British BSE Inquiry [5] and SSC reports [58] did not come to the conclusion that scrapie is the origin of BSE. It remains uncertain which of the two hypotheses is correct and it is unlikely that this question will ever be answered with certainty.

19.9.2 BSE risk factors

BSE first occurred in the South of England during the 1980s, after $MBM^{\&}$ had been fed to cattle for about 70 years. Several hypotheses to the question "why did BSE start in England?" were proposed relating to the following: (i) an increasing incidence of scrapie in UK sheep; (ii) an increasing UK sheep population – relative to the UK cattle population; (iii) the changes in rendering practices [3; 60], e. g. the abandonment of the solvent extraction processes; (iv) the use of ruminant-derived MBM, even in calf rations in the UK.

The third point is based on the observation that, in the past, rendered tissues in the UK had commonly been subjected to solvent extraction, which enhanced the yield of tallow and produced a low-fat MBM that had attracted premium prices. The practice was, however, largely discontinued between 1977 and 1984 (Fig 19.1). The process had involved exposure of rendered materials to hot solvents and then to wet heat to remove remaining solvents. Given that the mean incubation time of BSE is approximately 5 years and that the disease was first observed in spring 1985 [5; compare 1], the emergence of the BSE epidemic appears to have been associated with the discontinuation of the use of solvent extraction by the rendering industry. In 1975, about 65 % (Fig. 19.1) of the MBM manufactured in the UK had been subjected to solvent extraction, but by 1982 the proportion had dropped to 10 % [4]. With hindsight it was hypothesized that the application of wet heat during the second step of solvent extraction, i. e. the removal of solvent residuals, to already rendered tissues might have provided sufficient additional inactivation of BSE or scrapie infectivity to prevent these agents being present in MBM at titers that would constitute an effective oral dose for cattle [3]. However, not much was known at the time about the (scrapie) agent-inactivating potential of the solvent extraction process used by the rendering industry.

In contrast, it was well known since the 1960s that scrapie agents are much more resistant to the environment [61] and chemical inactivation (Table 44.1) than conventional micro-organisms, so that it should have been logical that the organic solvents per se could not generally have a significant effect to inactivate scrapie agents in crude tissue preparations [62].

By the time of the earliest effective dietary exposure of cattle to prions that caused BSE (probably 1981/1982), most of the UK rendering industry had abandoned the use of solvent extraction. However, two renderers in Scotland continued the practice using heptane and hexane. These two plants are known [63] to have manufactured a large proportion of the MBM used, at least in 1988, in Scotland [4]. Given the low incidence of BSE in Scotland at that time (Fig. 39.2), and the fact that the incidence later increased through the acquisition of subclinically BSE-infected cattle from England, it has been suggested that the solvent extraction process had at least a partial capacity to inactivate prions and consequently to reduce the titers of BSE or scrapie infectivity [3].

D.M. Taylor and colleagues [63] reported on experiments in which BSE- and scrapie-infectious tissues were exposed to hot solvents (heptane, hexane, perchlorethylene and petroleum spirit), followed by exposure to dry heat and wet heat (steam). Taylor considered in [63] that on average, the traditional solvent extraction systems achieved about a ten-fold reduction in the titer of the BSE or scrapie agents tested. His data suggest also that the hot solvents per se had relatively little effect on infectivity, and that the degree of inactivation achieved was mainly due to the application of wet heat to remove residual solvents (the second heating step lasted usually several hours).

It thus appears that the abandonment of solvent extraction (Fig. 39.1) by the UK rendering industry was not the single key factor that permitted the emergence of BSE. However, if – as seems likely – a number of factors conspired to allow BSE to emerge, then the abandonment of solvent extraction (Fig. 19.1) was one of several key risk factors [63] (compare Fig. 19.2).

In summary, the BSE epidemic probably occurred in the British Isles due to the combination of several critical risk factors, mentioned above, including a growing sheep population in which scrapie occurred increasingly, and which was – relative to the British cattle population – much larger. The other risk factors that have to be taken into account are that in the UK, at the time, rendering practices were changed and the British cattle industry used more and more ruminant-derived MBM[&] in calf rations [3; 60]. The changed rendering practices included, for example, the abandonment of the solvent extraction [other examples in 63].

Many other countries rendered and recycled ruminant proteins to varying extents but, in the absence of a high scrapie prevalence and taking into account differences in animal husbandry (such as the age at which calves were first fed concentrates containing rendered ruminant materials), similar BSE epidemics did not emerge elsewhere [64]. An international comparison of risk factors[14] suggests that an epidemic of similar magnitude outside the British Isles is unlikely.

Wilesmith's extensive epidemiological studies [3; 4] indicated clearly the dietary origin of BSE, and the important role of the feeding of MBM to cattle. That BSE was spread through contaminated MBM is a reflection of the fact that prions are extremely resistant to inactivation (compare Chapters 45; 52) by procedures that readily inactivate conventional micro-organisms. In some countries, this problem was minimized decades before the emergence of the BSE epidemic by existing legislation. For example, in Germany the rendering industry had been required by law since 1939 to render at high temperatures with the aim of inactivating anthrax spores; in that country it had been suggested almost a century ago that the raw material (abattoir waste) entering the rendering system

[14] Compare Figure 41.1 of [65] and Table 41.3 of [65].

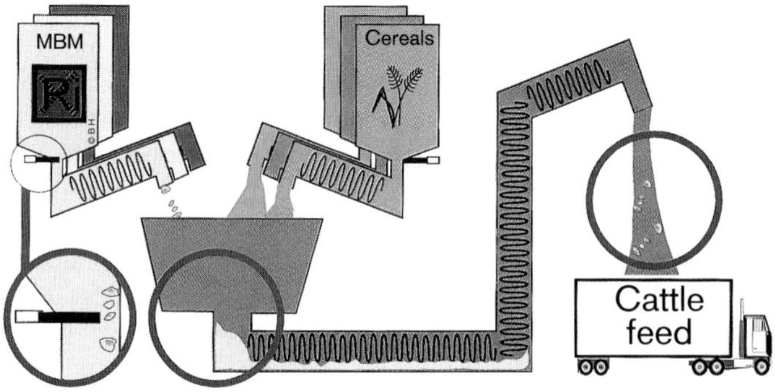

Fig. 19.4: In animal feed production plants (mills) cross-contamination happens mainly due to gravity or adhesion. Cross-contamination could also occur in silos, silo lorries or storage silos of mills and farms. Figure by B. Hörnlimann.

should be exposed to temperatures as high as 132°C for several hours [55].

19.9.3 The permanent risk of cross-contamination in feed mills

A simple but utmost important message that was learned from the BSE crisis – and this applies world-wide – is that the introduction of regulations is not sufficient unless adequate inspection and enforcement systems are implemented to ensure compliance. For instance, the introduction of the UK ban in the late 1980s on the use of specified bovine offals (SBO; later referred to as specified risk materials (SRM; see Chapter 53) for any food or feed purposes should have immediately ended this practice. Nevertheless, it was later acknowledged that such materials were subsequently entering rendering plants producing MBM for monogastric [5; 65] species (i. e., pig, poultry, dogs, and cats; see above and Chapters 23–25). In theory, this should not have presented any problems because pigs, poultry, and dogs (see Chapter 9) are not susceptible to BSE by dietary challenge; but cats are, which was known since 1990 [66] (see Chapters 20; 23). In contrast, the fact that pig and poultry are resistant to oral BSE transmission, was known only much later (see Chapters 24, 25).

Furthermore, the fact that feed for pigs, poultry, and cattle was commonly prepared in the same feed mills, and that cattle feed was consequently becoming cross-contaminated with the BSE agent present in MBM that was legally fed to non-ruminant species between 1988 and 1996, was not recognized as another major problem for many years, when it led to the occurrence of BSE cases in cattle born after the ban (BABs) and born after the re-inforced ban (BARBs; see Chapter 39.3.5).

19.10 Surveillance, prevention, and control

The basic principles of surveillance are described in Chapter 32, and the technical details of the post mortem BSE tests in Chapter 35.

Internationally recommended approaches for the prevention and control of BSE, including the importance of the assessment of the GBR, are described by Stuart C. MacDiarmid et al. in Chapter 53.

References

[1] Wells GAH, Scott AC, Johnson CT, et al. A novel progressive spongiform encephalopathy in cattle. Vet Rec 1987; 121:419–420.

[2] Jeffrey M and Wells GAH. Spongiform encephalopathy in a nyala (*Tragelaphus angasi*). Vet Pathol 1988; 25:398–399.

[3] Wilesmith JW, Wells GAH, Cranwell MP, et al. Bovine spongiform encephalopathy: epidemiological studies. Vet Rec 1988; 123(25):638–644.

[4] Wilesmith JW, Ryan JB, Atkinson MJ. Bovine spongiform encephalopathy: epidemiological studies on the origin. Vet Rec 1991; 128(9):199–203.

[5] Anonym. The BSE inquiry. Information from the UK Government. Webpage 2006: www.bseinquiry.gov.uk.

[6] Hörnlimann B. Evaluation of BSE risk factors among European countries. In: Gibbs CJ Jr, editor. Bovine spongiform encephalopathy – the BSE dilemma. Springer, New York 1996: 384–394. See also: Scrapie occurrence and BSE risk factors among European countries, published on the Internet on 28 March 1996. Webpage 2006: www.priondata.org/data/A_scrapieex.html # European.

[7] Anderson RM, Donnelly CA, Ferguson NM, et al. Transmission dynamics and epidemiology of BSE in British cattle. Nature 1996; 382(6594):779–788.

[8] Hörnlimann B. Prions – a challenge of Hygeia [in German: Prionen – eine Herausforderung der Hygieia]. Schweiz Arch Tierheilkd 2005; 147(1): 25–27.

[9] Wells GAH and Wilesmith JW. The neuropathology and epidemiology of bovine spongiform encephalopathy. Brain Pathol 1995; 5:91–103.

[10] Bruce ME, Boyle A, Cousens S, et al. Strain characterization of natural sheep scrapie and comparison with BSE. J Gen Virol 2002; 83(3):695–704.

[11] Bruce ME, Chree A, McConnell I, et al. Transmission of bovine spongiform encephalopathy and scrapie to mice: strain variation and the species barrier. Philos Trans R Soc Lond B 1994; 343:405–411.

[12] Yamakawa Y, Hagiwara K, Nohtomi K, et al. Atypical proteinase K-resistant prion protein (PrPres) observed in an apparently healthy 23-month-old Holstein steer. Jpn J Infect Dis 2003; 56(5-6):221–222.

[13] Casalone C, Zanusso G, Acutis P, et al. Identification of a second bovine amyloidotic spongiform encephalopathy: molecular similarities with sporadic Creutzfeldt–Jakob disease. Proc Natl Acad Sci U S A 2004; 101(9):3065–3070.

[14] Biacabe AG, Laplanche JL, Ryder S, et al. Distinct molecular phenotypes in bovine prion diseases. EMBO Rep 2004; 5(1):110–115.

[15] Donnelly CA, Ghani AC, Ferguson NM, et al. Analysis of the bovine spongiform encephalopathy maternal cohort study: evidence for direct maternal transmission. Appl Statist 1997; 46(3):321–344.

[16] Bradley R. Bovine spongiform encephalopathy – distribution and update on some transmission and decontamination studies. In: Gibbs CJ Jr, editor. Bovine spongiform encephalopathy – the BSE dilemma. Springer, New York 1996:11–27.

[17] Ferguson NM, Donnelly CA, Woolhouse MEJ, et al. Genetic interpretation of heightened risk of BSE in offspring of affected dams. Proc R Soc Lond Biol 1997; 264(1387):1445–1455.

[18] Supervie V and Costagliola D. The unrecognised French BSE epidemic. Vet Res 2004; 35(3):349–362.

[19] Wells GAH, Dawson M, Hawkins SA, et al. Preliminary observations on the pathogenesis of experimental bovine spongiform encephalopathy. In: Gibbs CJ Jr, editor. Bovine spongiform encephalopathy – the BSE dilemma. Springer, New York 1996:28–44.

[20] Donnelly CA, Ghani AC, Ferguson NM, et al. Recent trends in the BSE epidemic. Nature 1997; 389(6654):903.

[21] Anonym. Overview of the BSE risk assessments of the European Commission's Scientific Steering Committee (SSC) and its TSE/BSE ad hoc group adopted between September 1997 and April 2003. Prepared under the scientific secretariat of P. Vossen, J. Kreysa and M. Goll. (Subject to editorial changes). Document of 5 June 2003. Webpage 2006: ec.europa.eu/comm/food/fs/sc/ssc/out364_en.pdf.

[22] EFSA. Quantitative assessment of the residual BSE risk in bovine-derived products. The EFSA Journal 2005; 307:1–135. Webpage 2006: www.efsa.eu.int/science/biohaz/biohaz_documents/1280/efsaqrareport2004_final20dec051.pdf.

[23] Clauss M, Sauter-Louis C, Chaher E, et al. Investigations of the potential risk factors associated with cases of bovine spongiform encephalopathy in Bavaria, Germany. Vet Rec 2006; 158(15):509–513.

[24] Wilesmith JW. Creutzfeldt–Jakob disease and bovine spongiform encephalopathy – cohort study of cows is in progress (Letter). BMJ 1996; 312:843.

[25] General Q&A – Section 1: BSE epidemic. Webpage 2006: www.defra.gov.uk/animalh/bse/general/qa/section1.htmlq9.

[26] Wilesmith JW. Recent observations on the epidemiology of bovine spongiform encephalopathy. In: Gibbs CJ Jr, editor. Bovine spongiform encephalopathy – the BSE dilemma. Springer, New York 1996:45–55.

[27] Sauter-Louis C, Clauss M, Chaher E, et al. Breed predisposition for BSE: epidemiological evidence in Bavarian cattle. Schweiz Arch Tierheilk 2006; 148:245–250.

[28] Hadlow WJ, Kennedy RC, Race RE, et al. Virologic and neurohistologic findings in dairy goats affected with natural scrapie. Vet Pathol 1980; 17:187–199.

[29] Hadlow WJ, Prusiner SB, Kennedy RC, et al. Brain tissue from persons dying of Creutzfeldt–Jakob disease causes scrapie-like encephalopathy in goats. Ann Neurol 1980; 8:628–32.

[30] Eloit M, Adjou K, Coulpier M, et al. BSE agent signatures in a goat. Vet Rec 2005; 156(16):523–524.

[31] Foster JD, Bruce ME, McConnell I, et al. Detection of BSE infectivity in brain and spleen of experimentally infected sheep. Vet Rec 1996; 138:546–548.

[32] Foster JD, Hope J, Fraser H. Transmission of bovine spongiform encephalopathy to sheep and goats. Vet Rec 1993; 133:339–341.

[33] Foster JD, Hope J, McConnell I, et al. Transmission of bovine spongiform encephalopathy to sheep, goats, and mice. Ann N Y Acad Sci 1994; 724:300–303.

[34] Wilesmith JW, Hoinville LJ, Ryan JB, et al. Bovine spongiform encephalopathy: aspects of the clinical picture and analyses of possible changes 1986–1990. Vet Rec 1992; 130:197–201.

[35] Wilesmith JW and Ryan JB. Bovine spongiform encephalopathy: observations on the incidence during 1992. Vet Rec 1993; 132:300–301.

[36] Foley GL and Zachary JF. Rabies-induced spongiform change and encephalitis in a heifer. Vet Pathol 1995; 32:309–311.

[37] Defra. Department for Environment, Food and Rural Affair. Webpage 2006: www.defra.gov.uk.

[38] World Organisation for Animal Health (OIE). Number of cases of bovine spongiform encephalopathy (BSE) reported in the United Kingdom. Webpage 2006: www.oie.int/eng/info/en_esbru.htm.

[39] World Organisation for Animal Health (OIE). Number of reported cases of bovine spongiform encephalopathy (BSE) in farmed cattle worldwide (excluding the United Kingdom). Webpage 2006: www.oie.int/eng/info/en_esbmonde.htm.

[40] Bradley R. Experimental transmission of bovine spongiform encephalopathy. In: Court L, Dodet B, editors. Transmissible subacute spongiform encephalopathies: Prion diseases. Elsevier, Paris 1996:51–56.

[41] Wilesmith JW, Ryan JB, Hueston WD. Bovine spongiform encephalopathy: case-control studies of calf feeding practices and meat and bonemeal inclusion in proprietary concentrates. Res Vet Sci 1992; 52:325–331.

[42] Wilesmith JW. BSE: epidemiological approaches, trials and tribulations. Prev Vet Med 1993; 18:33–42.

[43] European Community 1996. Report from the scientific veterinary committee on the risk analysis for colostrum, milk and milk products, Brussels 1996:1–26.

[44] Bradley R, Collee JG, Liberski PP. Variant CJD (vCJD) and bovine spongiform encephalopathy (BSE): 10 and 20 years on: part 1. Folia Neuropathol 2006; 44(2):93–101.

[45] Wells GAH, Spiropoulos J, Hawkins SA, et al. Pathogenesis of experimental bovine spongiform encephalopathy: preclinical infectivity in tonsil and observations on the distribution of lingual tonsil in slaughtered cattle. Vet Rec 2005; 156(13):401–407.

[46] Wells GAH, Hawkins SA, Green RB, et al. Preliminary observations on the pathogenesis of experimental bovine spongiform encephalopathy (BSE): an update. Vet Rec 1998; 142:103–106.

[47] Wells GAH. Pathogenesis of BSE in bovines. Abstract of a paper presented at the Joint WHO/FAO/OIE Technical Consultation on BSE: Public Health, Animal Health and Trade, 2001.

[48] Hadlow WJ, Kennedy RC, Race RE. Natural infection of Suffolk sheep with scrapie virus. J Infect Dis 1982; 146:657–664.

[49] Bradley R. Bovine spongiform encephalopathy – Update. Acta Neuropathol Exp 2002; 62:183–195.

[50] Brown P, Will RG, Bradley R, et al. Bovine spongiform encephalopathy and variant Creutzfeldt–Jakob disease: background, evolution, and current concerns. Table A. Measures taken to prevent the spread of bovine spongiform encephalopathy (BSE) to animals. Webpage 2006: www.cdc.gov/ncidod/EID/vol7no1/brown.htmAppendix.

[51] Brown P, Will RG, Bradley R, et al. Bovine spongiform encephalopathy and variant Creutzfeldt–Jakob disease: background, evolution, and current concerns. Table B. Measures taken to prevent the spread of bovine spongiform encephalopathy (BSE) to humans. Webpage 2006: www.cdc.gov/ncidod/EID/vol7no1/brown.htmAppendix.

[52] WHO. Guidelines on tissue infectivity distribution in transmissible spongiform encephalopathies. Report of the WHO Consultation on 14–16 Sept. 2005. Geneva 2006:1–72.

[53] Foster JD, Hope J, McConnell I, et al. Transmission of bovine spongiform encephalopathy to sheep, goats, and mice. Ann N Y Acad Sci 1994; 724:300–303.

[54] Buschmann A and Groschup MH. Highly bovine spongiform encephalopathy-sensitive trans-

genic mice confirm the essential restriction of infectivity to the nervous system in clinically diseased cattle. J Infect Dis 2005; 192(5):934–942.
[55] Klimmer M. Die Futtermittelkunde – Animalische Futtermittel. [Translation: The practice of feeding – Animal-derived feed compounds; Chapter B.-X., particularly pages 194–195]. In: Veterinärhygiene – Grundriss der Gesundheitspflege und Fütterungslehre der landwirtschaftlichen Haussäugetiere; Kapitel B.-X. Paul Parey Verlag, Berlin 1914:192–196.
[56] EC. Hypotheses on the origin and transmission of BSE. Adopted by the scientific steering committee at its meeting of 29–30 November 2001. Webpage 2006: ec.europa.eu/comm/food/fs/sc/ssc/out236_en.pdf.
[57] Wells GAH and Wilesmith JW. Bovine spongiform encephalopathy and related diseases. In: Prusiner SB, editor. Prion biology and diseases. Cold Spring Harbor Laboratory Press, New York 2004:595–628.
[58] SSC. Opinions and information. Webpage 2006: www.priondata.org/data/A_ECSSC.html.
[59] Anonym. Zoo prion disease: review of scientific literature. Webpage 2006: www.mad-cow.org/zoo_cites_annotated.html.
[60] Taylor DM. Bovine spongiform encephalopathy and its association with the feeding of ruminant-derived protein. Dev Biol Stand 1993; 80:215–224.
[61] Johnson CJ, Phillips KE, Schramm PT, et al. Prions adhere to soil minerals and remain infectious. PLoS Pathog 2006; 2(4):e32.
[62] Taylor DM. Transmissible degenerative encephalopathies. Inactivation of the causal agents. In: Russel AD, Hugo WB, Ayliffe GAJ, editors. Principles and practice of disinfection preservation and sterilisation. Blackwell, Oxford, 1999: 222–236.
[63] Taylor DM and Woodgate SL. Rendering practices and inactivation of transmissible spongiform encephalopathy agents. Rev Sci Tech Off Int Epiz 2003; 22(1):297–310.
[64] Defra. Animal feeding practices before BSE. Webpage 2006: www.defra.gov.uk/animalh/bse/controls-eradication/causes.html.
[65] Hörnlimann B. BSE in non-UK countries: spread by live cattle and MBM imports [in German: Bovine Spongiforme Enzephalopathie (BSE): BAB-Fälle in der Schweiz (original publication of summary of Master of Public Health thesis; submitted at ISPM University of Zürich, 1999]. In: Hörnlimann B, Riesner D, Kretzschmar H, editors. Prionen und Prionkrankheiten. de Gruyter, Berlin and New York 2001:337–358.
[66] Wyatt JM, Pearson GR, Smerdon TN, et al. Spongiform encephalopathy in a cat. Vet Rec 1990; 126:513.

20 Portrait of Prion Diseases in Zoo Animals

James K. Kirkwood and Andrew A. Cunningham

20.1 History

In 1986, a scrapie-like spongiform encephalopathy was diagnosed post mortem in a nyala *Tragelaphus angasi*, which showed clinical signs similar to scrapie [1]. Since that time, transmissible spongiform encephalopathies (TSEs; prion diseases) have been diagnosed in other zoo species. Most of these animals inhabited, or were exported from, zoos in Great Britain [2–4] (Tab. 19.1), but some cases that have been reported more recently were in animals at or from zoos elsewhere in Europe [5–7]. As of 2005, all of the animals in which the diagnosis was confirmed were from the families Bovidae (→ BSE) and Felidae (→ FSE), and included, in addition to nyala: scimitar-horned oryx *Oryx dammah* [2; 8], eland *Taurotragus oryx* [2; 3; 9], greater kudu *Tragelaphus strepsiceros* [2; 10], gemsbok *Oryx gazella* [1], Arabian oryx *Oryx leucoryx* [2; 8], American bison *Bison bison* [4], cheetah *Acinonyx jubatus* [2; 7; 11–13], puma *Felis concolor* [3; 14], ocelot *Felis pardalis* [3], Asian golden cat *Catopuma temminki* [5], leopard cat *Prionailurus bengalis* [15], tiger *Panthera tigris* [3], and lion *Panthera leo* [16] (Tab. 1.3).

Single cases of chronic wasting disease (CWD), which occurred in mule deer *Odocoileus hemionus* in a Canadian zoo, are not included in this chapter (see Chapter 21). In addition, this list does not include cases of BSE in domesticated species in zoos, i. e., BSE in Ankole or other domestic cattle (Tab. 1.3), or spongiform encephalopathies (SEs) assumed to be scrapie in moufflon sheep *Ovis musimon* [17]. A form of SE which was thought probably to have a different etiology has also been reported in ostriches *Struthio camelus* in Germany [18]. These cases are dealt with in Chapter 25. Cases, thought to be caused by the BSE agent have also been reported in a rhesus macaque *Macacca mulatta* and in lemurs *Eulemur species* at zoos in France [6], but there is some doubt as to whether or not the presence of an SE was confirmed in these cases [16]. Of interest in relation to the occurrence of FSE in zoo felids, is the small epidemic of TSE in domestic cats since 1990 (see Chapter 23). These cases have also been attributed to exposure to the BSE agent [19].

20.2 Forms or variants

The cases of scrapie-like SEs in zoo animals have been investigated to varying degrees. Only in the case of the nyala and the greater kudu have attempts been made to type the strain of agent by transmission in a panel of mouse strains (see Chapter 50). In these cases, the pattern of incubation periods and the neuropathological profiles in the panel of mouse strains were very similar to those cases induced by inoculation of brain material from cattle with BSE [20; 21]. This finding, together with the temporal and geographic coincidence of the zoo animal cases with the BSE epidemic in domestic cattle and the knowledge of routes by which the affected zoo animals could have been exposed to the BSE agent, strongly suggests that the BSE agent was the cause of cases that occurred in UK zoos [2; 3]. This is depicted in Figure 50.1. Although not reported for zoo felids with FSE, evidence from mouse transmissions [22] and Western blotting analysis indicate a BSE origin for FSE in domestic cats (see Chapters 23; 50).

20.3 Incubation period, transmissibility, and susceptibility

Almost all species of zoo animals in the UK were potentially exposed to the BSE agent through feed (see below) or feed supplements such as vitamin and mineral preparations containing meat-and-bone-meal (MBM). The pattern of cases of BSE in zoo animals, however, has shown taxonomic patchiness, with cases having been reported only in members of the families Bovidae and Felidae.

The incubation periods of cases in zoo animals can only be estimated because dates of exposure are unknown. Data on ages at death, which are the only clues available to estimate incubation periods were summarized in 1999 [3]. These data strongly suggest that incubation periods are longer in cat species (Felidae) than in bovids (Bovidae). Age at death of most zoo felids with TSE was around 60–80 months (and much longer in some cases); while age of death of most zoo bovids with TSE was approximately 30 months [3]. The shortest incubation period observed was 19 months in a greater kudu [10].

Ruminant-derived protein (i. e., protein recovered from cattle carcasses; MBM) was widely incorporated into proprietary feeds for ruminants (and other animals) during the 1980s. Many zoo ruminants were fed on proprietary feeds (see Chapter 19.2) that may have been contaminated in this way with BSE agent. It is thought that the etiology of many, if not all, zoo bovid cases can be explained through exposure in this way. However, there has been concern that some of the cases that occurred, particularly in the greater kudu, may have arisen through transmission between animals; feeds offered to some of the animals that became affected were thought to be at very low risk of including ruminant-derived protein [2; 3].

Zoo felids were presumed to have been exposed to the BSE agent through consumption of raw cattle tissues characterized by a high infectivity titer. Most of these animals were known to have been exposed to parts of cattle carcasses deemed unfit for human consumption, including vertebral columns containing spinal cord [2; 3]. To date, there is no evidence for transmission of the disease between animals in zoo felids.

20.4 Clinical signs and course of disease

The clinical signs reflect progressive loss of aspects of central nervous system (CNS) function. Typically, signs in both bovids and felids have an insidious onset and relatively slow progression (over weeks). The clinical signs include: ataxia, abnormal postures including head and ear carriage, fine muscle tremors and myoclonus (a sudden nervous twitch of individual muscles), abnormal activity patterns, wasting, hyperesthesia (hypersensitivity towards touch), and anxiety [2; 3; 13]. In some bovid cases, the clinical signs appeared to have a sudden onset and a rapid progression, with death or euthanasia occurring within only a few days [1; 2]. Nibbling of the tail base was observed in the nyala [1], and head and flank rubbing was observed in greater kudu [2] as in scrapie (Fig. 34.1b). In some animals, excessive lip movements (lip trembling, or nibbling movements of the lips as in scrapie; Fig. 34.1d), reduced rumination, and emaciation were observed. However, the picture is variable and, for example, wasting was not a consistent finding. Early clinical signs reported in some felids included apparent lameness, senile behavior, and signs of cerebellar ataxia when trying to eat. Ataxia of gait is the most consistent sign reported in zoo felids [7; 12; 14; 23] and may be accompanied by tremors.

20.5 Differential diagnoses

Prion diseases cannot be differentiated on the basis of clinical signs from some other diseases of the CNS. Diseases of bovids in which similar signs occur include: metabolic disorders (ketosis, copper deficiency), cerebrocortical necrosis, CNS neoplasia, congenital CNS dysplasias, hepatic encephalopathy, cranial trauma, and CNS infections, e. g., rabies, listeriosis, and toxoplasmosis [24].

The differential diagnosis in felids is dealt with in Chapter 23.5 (there might be differences in domestic and wild cats). In particular, differential diagnosis in zoo felids should include se-

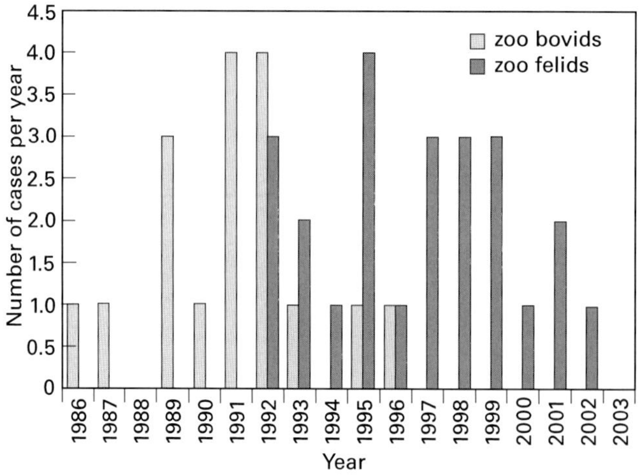

Fig. 20.1: The annual number of cases of scrapie-like spongiform encephalopathy in wild bovids or wild felids inhabiting or originating from zoos in Great Britain from 1985 to February 2004. Figure provided by J.K. Kirkwood; to Defra, 2004. Crown copyright ©.

quelae to feline parvovirus and feline herpes virus 1 infections [13; 25].

20.6 Epidemiology

The number of cases of TSE each year in zoo bovids reached a peak in 1991/2 and has since declined (Fig. 20.1). The first case of FSE in zoo felids was diagnosed in 1991 in Great Britain and the number of cases reached a peak in 1995 and has since declined (Fig. 20.1).

All of the cases of TSEs diagnosed in zoo bovids to date have been in animals bred and kept in Great Britain. Cases of FSE in zoo felids that have been diagnosed in other countries, i. e., cheetah in Australia, Republic of Ireland, and France [12; 13; 16], and Asian golden cat in Australia [5], were in animals exported from the UK, with two exceptions, one in a cheetah born and kept in France [7] and the other in an Asian golden cat [5], which originated in Germany. In these two incidents the countries from which they were reported are not regarded as the countries of TSE origin and infection is presumed to have occurred in continental Europe. Cases in animals exported from the UK were probably infected at a young age in the country of origin. The rhesus macaque in France reported to have SE [6] was also exported from the UK. The lemurs diagnosed with TSE in France could not have been exposed in the UK [6]. It is known that MBM was exported from the UK to continental Europe for feed manufacture and this may have been the source of exposure for these animals, but it is noteworthy that no cases of TSEs have been reported in nonhuman primates in the UK where dietary exposure to contaminated feed was even more likely to have occurred.

Thus, with the exception of scrapie in sheep, moufflon, and goats, and of CWD in deer, the prevalence of prion diseases in zoo animals bred or kept outside of Great Britain appears to have been extremely low. However, it is possible that a very low and sporadic incidence could have been overlooked.

A very wide range of species of wild animals is kept in Britain but cases have only occurred (or been detected) in bovids and felids (Tab. 1.3). It is likely that animals of other families, e. g., Carnivora: Procyonidae, Ursidae, Mustelidae[1], Canidae (see Chapter 9); and Artiodactyla: Cervidae, Giraffidae, and others were also exposed to contaminated food.

One of the most interesting questions concerning prion diseases in zoo animals is why particular species from certain, non-closely related

[1] TSEs have never been diagnosed in zoo mustelids; this is in contrast to mink (*Mustela vison*) kept on farms (see Chapter 22).

taxa succumbed to the disease, in particular cats (see Chapter 9). Furthermore, within the families Bovidae and Felidae, the distribution of cases among species appears also to have been non-random [2; 3], i. e., no proportionality to the size of the animal populations can be observed. For example, the number of cases that have occurred in greater kudu was surprisingly large in view of the small overall population size [2], while among some other bovid species kept in much larger numbers no cases were observed. This pattern suggests that genetic factors may influence variation in susceptibility between species (see Chapter 9). However, some caution is needed in drawing this conclusion since (i) little is known about possible variation in intensity of exposure between individuals and species, and because (ii) it is possible that cases are yet to emerge in some species in which the disease has a longer incubation period. (A database of prion diseases in primates and other zoo animals can be found in the internet [26]).

20.7 Pathology

In affected zoo ruminants there are no macroscopic lesions except, occasionally, wasting. Self mutilation from biting was observed in the affected nyala [1]. Histopathological changes are restricted to the CNS and closely resemble those of scrapie and BSE. They include vacuolation of neurons, evident as spongiform change, due mainly to vacuoles within neurites and as vacuoles within neuronal perikarya. There is also degeneration and loss of neurons and gliosis, notably astrocytosis [27]. There are differences in the lesion profile (see Chapter 12.3.4) between host species [27].

Few cases of FSE in non-domestic cats have been described in detail, but, in general the lesions are similar to those of FSE in domestic cats. Most prominent is a vacuolation of the gray matter, consisting of small empty holes in the neuropil [28; 29] (see Chapter 27). Vacuolar changes may be widespread, extending throughout the neuraxis. Vacuolation and PrP immunohistochemical staining of the cerebellar cortex, the cerebral cortex, corpus striatum, thalamus, brainstem, and spinal cord has been variably recorded in puma, cheetah, and Asian golden cat [5; 7; 14; 23]. Additional alterations of neurons are rare, but a microgliosis is prominent in most cases, as is an astrocytic reaction.

20.8 Are prion diseases of zoo animals new?

The temporal cluster of cases of scrapie-like SEs that has occurred in zoo animals at or from zoos in Great Britain since 1986, appears to be due to the spread of BSE (i. e., in domestic cattle), via animal feed, to zoo animals [2; 3].

Cases of SE were reported in white tigers *Panthera tigris* at the Bristol Zoo between 1970 and 1977 [30] but these were considered to be unlike the cases seen after 1986 because of differences in histopathological appearance and because attempts to transmit the condition to laboratory animals were unsuccessful [3].

We cannot conclude that prion diseases never occurred in zoo animals in earlier times as cases could have been missed or misdiagnosed (see Chapters 9, 21, 23 and 25). As discussed elsewhere [2] considering that scrapie-like SEs were recognized prior to 1980 in 6 of the relatively small number of mammal species (man, sheep, goat, mink, mule deer, and elk; compare with the situation of today: 26 species are affected by TSE (Table 1.3)) whose diseases have been studied closely, it would be surprising if other prion diseases did not occur in some of the other 4,050 or so species of mammals whose biology is far less well known (nor perhaps should the possibility of prion diseases occurring in birds or other classes be ruled out).

20.9 Risk factors

The cases of prion disease in British zoo animals since 1986 indicate that there is a risk associated with feeding some BSE-infected cattle tissues (or protein products derived or contaminated from them), to various species of wild animals. The pattern of incidence of the disease among taxa has been uneven, suggesting that there may be differences in (genetic) susceptibility between species. But there are no grounds, at this stage, for assuming that any species of zoo animal is resistant to prion diseases.

BSE in exotic Bovidae species. The pattern of cases in greater kudu suggested that maternal and horizontal transmission may have occurred. However, contaminated feeds may have included traces of ruminant-derived protein from cattle (see Chapter 39.3.5; BABs) [2; 3]. Since transmission of scrapie among domestic small ruminants (sheep and goat) is possible, and it can be assumed also that CWD is horizontally transmissible in deer (see Chapter 21) [4], it is possible that maternal infection or infection of in-contact animals might be risk factors in zoo animal incidents.

It has been found that the infectious agent is present in a wider range of tissues in greater kudu with BSE than in cattle with BSE, and the tissue distribution is more like that seen in cases of scrapie in genetically susceptible sheep [16]. In addition to the tissues found to be positive for scrapie infectivity in such sheep, however, infectivity in greater kudu has also been detected in skin, conjunctiva, and salivary glands [16]. These findings may be relevant to the possibility of transmission between kudu and show that a wider range of tissues from kudu (and perhaps other zoo ruminants) might be a source of infection; e. g. to carnivores, than just those tissues found to contain infectivity in cattle with BSE.

Scrapie in exotic species. Since scrapie has been recognized in domestic sheep *Ovis aries* and goats *Capra hircus* for many years and it has been reported in moufflon *Ovis musimon* [17], it is possible that the disease occurs in other species within the group *Caprini*. It seems likely that zoo carnivores may have been exposed to scrapie through eating infected sheep or goat carcasses but that no such cases of prion disease have been recognized as related to this origin (although caution should be exercised in assuming that this means that they could not have occurred).

It is thought that Creutzfeldt–Jakob disease (CJD) can arise spontaneously at a low incidence (about 1–2 per million) in humans. It is possible that prion diseases may occasionally arise spontaneously in other species. No such cases have been observed in captive or free-living wild animals but their detection would require extraordinarily intense surveillance.

20.10 Surveillance, prevention, and control

Detection of TSEs in zoo animals requires extraordinary vigilance and high-quality monitoring. A number of the reported cases were not found to show classic clinical signs but were detected only because brain histology was undertaken.

The most important factor in preventing spread of BSE of cattle to zoo animals is preventing infected tissues from cattle entering zoo animal feeds. Although not all prion diseases may present a threat to other species (e. g., there is no evidence that ingestion of any tissue from scrapie-infected sheep causes disease in humans or carnivores), clearly some do pose a threat, and in the zoo context it is probably simplest and safest to recommend that no tissues from any animal at risk of having a prion disease should be used as food or as an ingredient in the feed of any animal [2; 3]. In the UK, the European Community, and in some other countries legislation exists to prevent the use of potentially infected tissues from cattle being used as feed for any animals [31; 32]. It is apparent from the decline in the number of cases occurring in Britain (Fig. 20.1), that these measures have been effective in controlling the BSE-related TSE epidemic in zoos.

If TSEs are detected in new host species in captive or free-living wild animals, from the disease control perspective it would be prudent to assume that CNS and possibly other tissues (see Chapter 53.2; Fig. 53.2) might contain infectivity and could be a source of disease through ingestion or iatrogenic inoculation to other animals. The possibility that the disease could be naturally transmissible should also be seriously considered. The occurrence of vertical or horizontal transmission could result in the disease becoming endemic in the population, as occurs in scrapie and as is thought to occur in CWD.

It has been suggested that, in order to prevent the risk of BSE spreading from captive to free-living populations, no zoo animals that may have been exposed to the BSE agent or their (direct) offspring or in-contacts should be used for reintroductions to the wild. Furthermore, in view of the great importance of some zoo animals for conservation programs, translocations of such animals between zoos should not be undertaken without careful consideration of the risks [2; 33]. It has been recommended that this level of caution should be extended in the control of any prion disease in zoo animals unless detailed knowledge is available on the 'natural' transmissibility (see Chapter 42) of prion diseases in zoo or wild animals [2].

References

[1] Jeffrey M and Wells GAH. Spongiform encephalopathy in a nyala (*Tragelaphus angasi*). Vet Pathol 1988; 25:398–399.

[2] Kirkwood JK and Cunningham AA. Epidemiological observations on spongiform encephalopathies in captive wild animals in the British Isles. Vet Rec 1994; 135:296–303.

[3] Kirkwood JK and Cunningham AA. Scrapie-like spongiform encephalopathies (prion diseases) in non-domesticated species. In: Fowler ME, Miller RE, editors. Zoo and wild animal medicine: current therapy 4. W.B. Saunders Co, Philadelphia, 1999:662–668.

[4] Williams ET, Kirkwood JK, Miller M. Spongiform encephalopathies. In: Williams ES, Barker IK, editors. Infectious diseases of wild mammals. Iowa State University Press, Ames, Iowa, 1999:292–301.

[5] Young S and Slocombe RF. Prion-associated spongiform encephalopathy in an imported Asiatic golden cat (*Catopuma temmincki*). Aust Vet J 2003; 81:295–296.

[6] Bons N, Mestre-Frances N, Belli P, et al. Natural and experimental oral infection of non-human primates by bovine spongiform encephalopathy agents. Proc Natl Acad Sci U S A 1999; 96:4046–4051.

[7] Lezmi S, Bencsik A, Monks E, et al. First case of feline spongiform encephalopathy in a captive cheetah born in France: PrPSc analysis in various tissues revealed unexpected targeting of kidney and adrenal gland. Cell Biol 2003; 119:415–422.

[8] Kirkwood JK, Wells GAH, Wilesmith JW, et al. Spongiform encephalopathy in an Arabian oryx (*Oryx leucoryx*) and a greater kudu (*Tragelaphus strepsiceros*). Vet Rec 1990; 127:418–420.

[9] Fleetwood AJ and Furley CW. Spongiform encephalopathy in an eland (letter). Vet Rec 1990; 126:408–409.

[10] Kirkwood JK, Cunningham AA, Wells GAH. Spongiform encephalopathy in a herd of greater kudu (*Tragelaphus strepsiceros*): epidemiological observations. Vet Rec 1993; 133:360–364.

[11] Kirkwood JK, Cunningham AA, Flach EJ, et al. Spongiform encephalopathy in another captive cheetah (*Acinonyx jubatus*): evidence for variation in susceptibility or incubation periods between species? J Zoo Wildlife Med 1995; 26:577–582.

[12] Peet R and Curran JM. Spongiform encephalopathy in an imported cheetah (*Acinonyx jubatus*). Aust Vet J 1992; 69:117.

[13] Vitaud C, Flach EJ, Thornton SM, et al. Clinical observations on four cases of feline spongiform encephalopathy in cheetahs (Acinonyx jubatus). Proceedings of the European Association of Zoo and Wildlife Veterinarians. Chester, UK, 21st-24th May, 1998:133–138.

[14] Willoughby K, Kelly DF, Lyon DG, et al. Spongiform encephalopathy in a captive puma (*Felis concolor*). Vet Rec 1992; 131:431–434.

[15] Defra. BSE Statistics. Department for Environment, Food & Rural Affairs. 2005.

[16] Cunningham AA, Kirkwood JK, Dawson M, et al. Bovine spongiform encephalopathy infectivity in greater kudu (*Tragelaphus strepsiceros*). Emerg Infect Dis 2004; 10:1044–1049.

[17] Wood JL, Lund LJ, Done SH. The natural occurrence of scrapie in moufflon. Vet Rec 1992; 130:25–27.

[18] Schoon HA, Brunckhorst D, Pohlenz J. Spongiforme Enzephalopathie beim Rothalsstrauss (*Struthio camelus*). Ein kasuistischer Beitrag. Tierärztl Prax 1991; 19:263–265.

[19] Wells GAH and Wilesmith JW. Bovine spongiform encephalopathy and related diseases. In: Prusiner SB, editor. Prion biology and diseases. Cold Spring Harbor Laboratory Press, New York, 2004:595–628.

[20] Bruce ME, Chree A, McConnell I, et al. Transmission of bovine spongiform encephalopathy and scrapie to mice: strain variation and the species barrier. Philos Trans R Soc Lond B 1994; 343:405–411.

[21] Jeffrey M, Scott JR, Williams A, et al. Ultrastructural features of spongiform encephalopathy transmitted to mice from three species of bovidae. Acta Neuropathol 1992; 84:559–569.

[22] Fraser H, Pearson GR, McConnell I, et al. Transmission of feline spongiform encephalopathy to mice. Vet Rec 1994; 134:449.

[23] Baron T, Belli P, Madec JY, et al. Spongiform encephalopathy in an imported cheetah in France. Vet Rec 1997; 141:270–271.
[24] Jeffrey M, Simmons MM, Wells GAH. Observations on the differential diagnosis of bovine spongiform encephalopathy in Great Britain. In: Bradley R, Marchant B, editors. Transmissible spongiform encephalopathies. Proceedings of a consultation on BSE with the Scientific Veterinary Committee of the European Communities, 14–15 September 1993. EU: VI/4131/94-EN Document Ref. F. II.3 – JC/0003, Brüssel 1994:347–358.
[25] Walzer C, Kübber-Heiss A, Gelbmann W, et al. Acute hind limb paresis in cheetah (*Acinonyx jubatus*) cubs. Proceedings of the European Association of Zoo and Wildlife Veterinarians. Chester, UK, 21st-24th May, 1998:267–273.
[26] Zoo prion disease: review of scientific literature. Webpage 2006: www.mad-cow.org/zoo_cites_annotated.html.
[27] Wells GAH and McGill IS. Recently described scrapie-like encephalopathies of animals: case definitions. Res Vet Sci 1992; 53:1-10.
[28] Pearson GR, Wyatt JM, Henderson JP, et al. Feline spongiform encephalopathy: a review. In: Raw ME, Parkinson TJ, editors. The Veterinary Annual. Blackwell Scientific Publications, London, 1993:1-10.
[29] Wyatt JM, Pearson GR, Gruffyd-Jones TJ. Feline spongiform encephalopathy. Feline Pract 1993; 21:7-9.
[30] Kelly DF, Pearson H, Wright AI, et al. Morbidity in captive white tigers. In: Montali RJ, Migaki G, editors. The comparative pathology of zoo animals. Smithsonian Institution Press, Front Royal, Virginia, 1980:183–188.
[31] Amendment Order. Bovine Spongiform Encephalopathy (No 2) Amendment Order 1990. Statutory Instrument No. 1930. Her Majesty's Stationery Office, London 1990.
[32] Order. The Bovine Spongiform Encephalopathy Order 1988, Statutory Instrument No 1039. Her Majesty's Stationery Office, London 1988.
[33] Cunningham AA. Bovine spongiform encephalopathy and British zoos. J Zoo Wildlife Med 1991; 11:605–634.

21 Portrait of Chronic Wasting Disease in Deer Species

Elizabeth S. Williams* and Michael W. Miller

21.1 History

Chronic wasting disease (CWD) of certain North American cervid (deer family) species was first recognized as a syndrome in captive mule deer (*Odocoileus hemionus hemionus*; Fig. 21.1). In the late 1960s, biologists conducting physiological and nutritional studies on this species noticed that their adult research animals often died of a syndrome of progressive weight loss and behavioral changes. Because of the clinical picture of the syndrome, the biologists thought perhaps these animals were suffering from a nutritional deficiency, possible exposure to toxins, or stresses due to captivity.

It was not until 1978 that this "chronic wasting" condition was identified as a transmissible spongiform encephalopathy (TSE or prion disease) by histological examination of brains of affected deer from wildlife research facilities in Colorado and Wyoming, USA [1]. Within a few years a similar TSE was diagnosed in wapiti (*Cervus elaphus nelsoni*, also called Rocky Mountain elk) [2] held in the same facilities where mule deer previously had been diagnosed with the condition. During this time, a case of CWD also was diagnosed in a black-tailed deer (*O. hemionus columbianus*) held in the wildlife research facility in Wyoming where some of the mule deer cases had been observed. Chronic wasting disease was also found at approximately this time in a zoo in Ontario, Canada, in several mule deer that had originated from zoos in Colorado and elsewhere.

Chronic wasting disease was first detected in commercially propagated wapiti on a game farm in Saskatchewan, Canada, in 1996 [3]. Subsequent investigations soon led to detection of several infected game farms in South Dakota, USA, as well as others in Saskatchewan [3; 4]. Development and implementation of state, provincial, and federal surveillance programs for captive cervids in the late 1990s and early 2000s in both the USA and Canada have thus far led to the detection of CWD in farmed cervids in nine states and two provinces; most of the affected facilities have held captive wapiti, but some have held captive white-tailed deer (*O. virginianus*). The movement of captive wapiti in commerce also appears responsible for the emergence of CWD in Korea [5].

It was not until 1981 that CWD was first diagnosed in a free-ranging wapiti [6], although there are unpublished records of suspected cases occurring in free-ranging cervids prior to that time. More formal surveillance and routine histological examination of suspected clinical cases in Colorado and Wyoming beginning in the 1990s led to the recognition that free-ranging mule deer and white-tailed deer also were infected with CWD in some areas [6]. The recognition of established CWD foci in the wild in Colorado and Wyoming led to development of techniques for large-scale surveillance of free-ranging populations in the 1990s [7]. Surveys undertaken by state and provincial wildlife management agencies throughout North America beginning in the late 1990s and early 2000s have thus far revealed foci of CWD in free-ranging cervids in 10 states and 2 provinces.

21.2 Forms or variants

To date there is only one recognized form of CWD. However, the relatively wide natural host range (four species – of one with two af-

* This chapter is dedicated to the memory of Elizabeth S. Williams†.

fected subspecies – in three different genera) and polymorphisms in the protein coding region of the prion gene of each host species (Table 1.3) provide a potential biological mechanism for CWD-associated prion strain variation to arise. Data from several studies suggest the possibility of CWD strain variation, including differences in cross-species native prion protein conversion efficiencies [8], variation in glycoform ratios in CWD-associated prion within and among host species [9], and variable incubation and neuroanatomical distribution of CWD-associated prion in transgenic mice expressing a generic coding region for cervid prion protein [10]. Recently developed transgenic mouse lines susceptible to CWD should afford models for more complete assessment of potential strain variation within and among affected host species.

21.3 Incubation period, transmissibility, and susceptibility

The natural incubation period of CWD is not specifically known. Both yearling (< 2-year-old) deer and wapiti have been diagnosed with the disease, suggesting a minimum incubation period of approximately 1.5 years. This is also supported by the incubation periods observed in experimentally inoculated mule deer [4]. Alternatively, first signs of CWD have occurred in animals as old as 15 years; however, in such cases it has not been possible to determine the age when exposure to the agent occurred.

The exact mechanism of natural transmission of CWD is not known with certainty.

Nonetheless, some form of direct or indirect horizontal transmission apparently sustains epidemics [11]; maternal transmission (from mother to young), if it occurs, is of limited importance. Live, infected animals are probably the primary source of new infections and geographic spread. However, environments contaminated with excreta or decomposed carcass remains may harbor some infectivity for years [12].

So far, there is no evidence that CWD is associated with a feed-borne pathogen as is the case with bovine spongiform encephalopathy (BSE). Deer and wapiti in game farms are not known to have been fed rendered ruminant proteins, e. g., meat-and-bone-meal (MBM). Free-ranging deer and wapiti feed on natural forage and have no known contact with MBM or similar slaughter by-products (see Chapters 19, 40).

Chronic wasting disease has been observed naturally in four species in three genera, all North American members of the family Cervidae. Infections in mule deer and white-tailed deer are encountered more often than in wapiti; a single natural case has been reported recently in moose (*Alces alces*). To date, CWD has not been observed in other free-ranging North American ruminant species including bighorn sheep (*Ovis canadensis*) and pronghorn antelope (*Antilocapra americana*) that have had ample opportunities for natural exposure. Similarly, TSE has not been diagnosed in cattle living on ranges overlapping those of infected deer and wapiti [13], and there is no indication that domestic animals such as cattle, sheep, and goats are naturally susceptible to CWD [8].

As with other TSEs, CWD can be transmitted to a variety of species by experimental means. It has been transmitted on first passage by intracerebral (i.c.) inoculation to domestic mink (*Mustela vison*; see Chapter 22), domestic ferrets (*Mustela putorius furo*), and squirrel monkeys (*Saimiri sciureus*; R. Marsh et al., personal communication) [14], to mule deer and a domestic goat [14], and to cattle [15]. Additional studies of susceptibility of domestic livestock and cervids by i.c. or oral exposure are currently underway in order to investigate the susceptibility of farm animals and deer species thus far not affected. Data from i.c. inoculation studies may be of limited utility in predicting natural host range: for example, CWD can be transmitted (with limited success) to cattle via i.c. exposure, but cattle exposed by oral inoculation or direct contact with infected deer have remained healthy for over 8 years.

Knowledge about possible genetic influences on CWD susceptibility is less extensive than for scrapie in sheep, but has improved in recent years [16]. For the three main natural host species, one or more substitution polymorphisms have been identified in the protein-coding region of the prion gene as follows: in wapiti, at codon 132 (methionine [M] or leucine [L]) [17]; in mule deer, at codons 20 (aspartate [D] or glycine [G]; removed during processing) and 225 (serine [S] or phenylalanine [F] [18]; and in white-tailed deer, at codons 95 (glutamine [Q] or histidine [H]), 96 (G or S), and 116 (alanine [A] or G) [19]. A processed pseudogene with high sequence identity to the functional prion gene occurs in all mule deer examined to date, as well as in about 25% of white-tailed deer studied [18–20]; this pseudogene appears inconsequential in CWD pathognesis, but complicated earlier studies attempting to sequence the deer prion protein. Data from natural exposure and experimental infections indicate that deer and wapiti of all common PrP genotypes are susceptible to CWD; however, in each host species individuals of the less common genotypes tend to be underrepresented among infected subpopulations [17; 20; 21], and onset of clinical disease appears to be delayed [16; 20] (M. Miller et al., unpublished data; K. O'Rourke et al., personal communication).

21.4 Clinical signs and course of disease

The most prominent clinical signs of end-stage CWD are behavioral alterations and loss of body condition in adult deer (Fig. 21.1a) and wapiti [1; 2]. As in any disease, clinical signs vary among individuals and not all signs are shown by all animals; moreover, consistent cli-

Fig. 21.1: (a) Female mule deer (*Odocoileus hemionus hemionus*) in the terminal stage of clinical chronic wasting disease (CWD). The animal displays excessive salivation, "depression", and wasting or emaciation Source: [1]. **(b)** Healthy mule deer. Photographs courtesy of E. Williams ©.

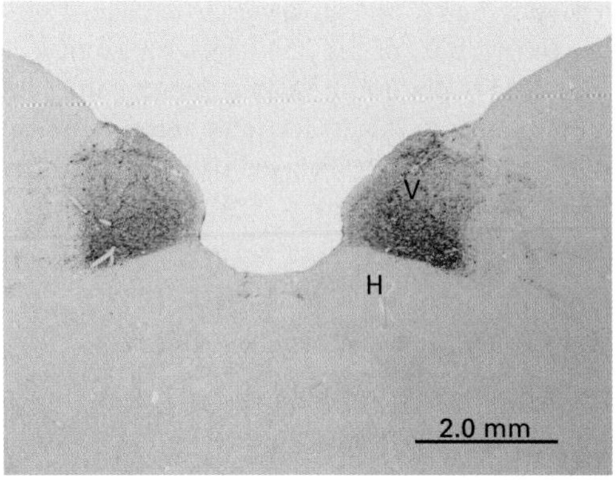

(a)

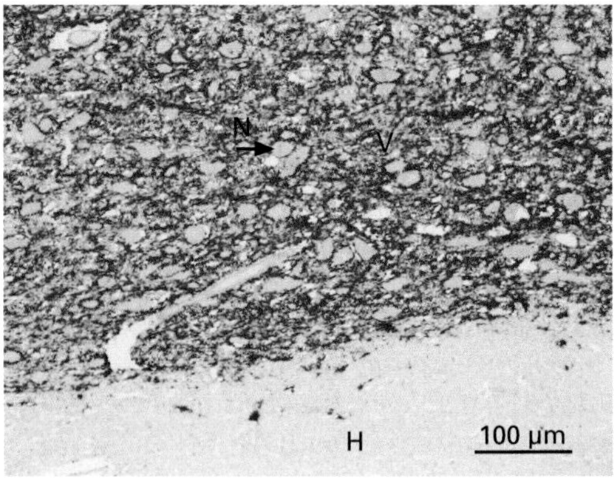

(b)

Fig. 21.2: Immunohistochemical labeled tissue section of the medulla oblongata, at the obex, of a mule deer affected by CWD. **(a)** PrP^{Sc} labeling in the dorsal motor nucleus of the vagus nerve (V). Note the absence of PrP^{Sc} labeling in the hypoglossal nucleus (H). **(b)** Higher magnification of PrP^{Sc} labeling in the dorsal motor nucleus of the vagus nerve (V) and absence of labeling in the hypoglossal nucleus (H). Neuronal perikarya (N) in the dorsal motor nucleus of the vagus nerve are not labeled. Primary antibody: Anti-PrP99/Streptavidin-alkaline phosphatase method; Hematosylin/bluing counterstain. Ventana Medical Systems, Inc. Photomicrograph by Terry Spraker, College of Veterinary Medicine, Colorado State University, Fort Collins ©.

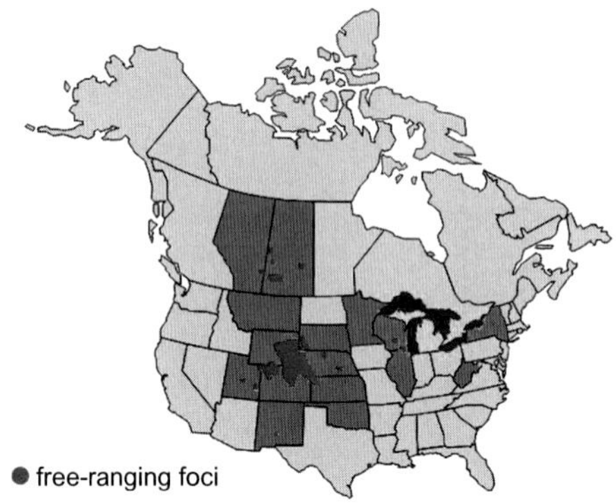

(a)

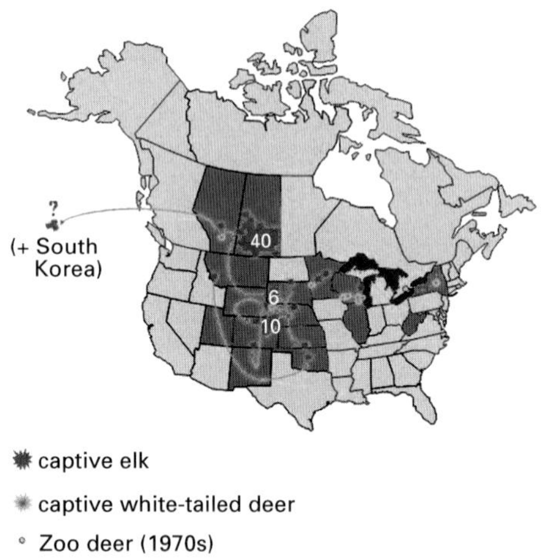

(b)

Fig. 21.3: Geographic distribution of CWD in cervids (2006). All states and provinces where ≥ 1 CWD case has been diagnosed since 1990 are shaded.
(a) Multiple foci of CWD in free-ranging cervids in 11 states and 2 provinces through various surveillance approaches; these foci do not all appear to be epidemiologically connected. **(b)** Cases of CWD have been detected in captive cervid facilities in 9 states and 2 provinces since 1996, but likely were occurring undetected prior to that time; known or suspected epidemiological connections are shown as arrows, but sources remain undetermined for many outbreaks. In addition to North American, an undetermined number of CWD cases have occurred in South Korea among captive wapiti imported from North America. Data sources: respective state and provincial wildlife and natural resource management agency reports, US Department of Agriculture, and Canadian Food Inspection Agency. Figures by Mike Miller ©.

nical signs of CWD occur relatively late in the overall course of disease. In captivity, the earliest signs are subtle changes in individual behavior, usually detected by caretakers most familiar with the individual animal. These may include changes in the way the animal interacts with caretakers or other animals in the herd, repetitive walking of a path in its pen, and possibly periods of somnolence. Affected animals lose weight even though they continue to eat. In animals that live to the terminal stage of disease, emaciation (wasting) is typical (see Fig. 21.2b). Itching, as in scrapie, has never been observed. Clinical signs that may be observed include: polydypsia; polyuria; ataxia (stumbling and incoordination); head tremor; hyperexcitability; dilated, spastic or flaccid oesophagus; abnormal movements of the tongue; sialorrhea (excessive salivation); odontoprisis (tooth grinding); and difficulties swallowing.

The clinical course of CWD is variable and typically lasts from a few weeks to perhaps a year or more. In part, the apparent duration of clinical disease will depend upon astuteness of observers in recognizing the earliest clinical signs of disease. In a few white-tailed deer, the clinical course was very short, lasting only a few days, and one animal died acutely without displaying typical clinical signs [22]. Emerging data suggest that the amino acid sequence comprising the native prion protein gene (*Prnp*) may influence the overall length of the disease course in both deer and wapiti (see Section 21.3).

21.5 Differential diagnoses

Behavioral alterations and central nervous system signs, as well as loss of weight may also be due to a variety of causes including meningeal worm (*Parelaphostrongylus tenuis*) infection, bacterial or viral meningitis, meningoencephalitis, trauma, intoxication, etc.

Emaciation of cervids may occur for many reasons, only one of which is CWD. Thus, diseases and conditions leading to loss of weight should be considered as differential diagnoses. Some possibilities include malnutrition or starvation, dental attrition, or musculoskeletal problems preventing normal foraging behavior. Some infectious diseases leading to poor body condition include parasitism, paratuberculosis, tuberculosis, and chronic bronchopneumonia. Difficulties in swallowing caused by CWD sometimes lead to fatal aspiration pneumonia. Death due to pneumonia is not uncommon, particularly in captive cervids, and CWD always should be suspected and ruled out when unusual cases of pneumonia are encountered in captive North American deer, wapiti, or closely related species, especially when loss of body condition or behavioral changes also are apparent.

21.6 Epidemiology

Cervids may develop CWD any time of the year; however, in the wild more cases are observed in the fall and winter months. This may have more to do with increased likelihood of detection (because wild cervids are hunted and seen more readily during the fall and winter months) rather than any particular seasonality of the disease [6], although a similar pattern also occurs in captive cervids. Among mule deer, males and older individuals are more likely to be infected [23].

Considerable effort is currently being directed at determination of the distribution and prevalence of CWD in free-ranging and captive cervids in North America and elsewhere (see Section 21.10). Surveillance and detection in free-ranging populations is challenging because disease distribution appears to be patchy and uniform sampling is rarely achieved. Overall prevalence of sub- or preclinical CWD cases in endemic areas can reach surprisingly high levels in free-ranging populations. In portions of southeast Wyoming and northeast Colorado, prevalence was about 5% in mule deer and 0.5% in sympatric wapiti [7]; in southcentral Wisconsin, USA, overall prevalence was approximately 3% in an endemic focus detected more recently in free-ranging white-tailed deer [24]. Although there is some perception that CWD has "spread" dramatically since 2000, the growing number of incidences are likely attributed – at least in part – to more extensive and intensive monitoring and surveillance undertaken during the last few years.

Prevalence of CWD in captive cervids can be even higher than observed in the wild; in captive mule and white-tailed deer, for example, entire cohorts have become infected and succumbed to CWD over the course of several years [1; 11; 14; 22]. The factors contributing to the remarkably high prevalence sometimes seen in captive deer and wapiti remain to be determined.

21.7 Pathology

The neuropathology of CWD is described in Chapter 27, and also has been covered in a recent review article [16].

21.8 Is CWD a new prion disease?

The origin of CWD is not known, but two possible sources are considered most likely. Chronic wasting disease may be a new or a reemerging disease of cervids (its relatively limited geographic distribution supports this hypothesis). It is also possible that CWD is the result of scrapie infection in cervids.

Arguments against scrapie as a source of CWD include the apparently limited distribution of CWD compared to scrapie and its potential geographic overlap with cervid species worldwide, the relatively long incubation seen after i.c. inoculation of CWD into goats and cattle [14; 15], and the apparent strain differences between CWD and other recognized prion diseases [8; 25]. Arguments supporting scrapie as a possible source of CWD include (i) moderate ability of prions derived from CWD affected deer and wapiti to convert cellular prion protein from sheep to the abnormal isoform and, conversely, for prions derived from scrapie affected sheep to convert cellular prion protein from deer or wapiti to abnormal isoforms [8]; (ii) the similarities in glycoform patterns on Western blots of disease-associated prion protein derived from CWD-affected deer and wapiti and sheep scrapie [9]; (iii) the occurrence of CWD-like clinical disease and lesions in wapiti inoculated i.c. with scrapie agent [26]; and (iv) the lack of cohesive epidemiological explanations for all of the CWD foci detected in North America in recent years. Experimental studies are needed to more clearly define the relationship between CWD and scrapie and the potential for scrapie exposure to lead to the occurrence of new CWD foci in North America and elsewhere.

Although there is some uncertainty about the origin of the CWD prion, strain typing and epidemiological data have shown more clearly that the CWD strain is not identical with the BSE strain [25] (compare Chapter 50). Important evidence for this is that cervids in North America do not have access to recycled ruminant protein (major BSE risk factor; see Chapter 40). On the other hand, cervids in zoos and game farms in the UK were undoubtedly exposed to the BSE agent in feed, yet none have been diagnosed with the disease [27]. This is in contrast to other wild ruminant species that were also exposed to BSE-contaminated feed and subsequently became infected with BSE (see Chapter 20).

21.9 Risk factors

Without detailed knowledge on the origin(s) and relative importance of various transmission routes, it is difficult to determine what risk factors are associated with the geographic spread and establishment of CWD among captive and free-ranging cervid populations.

Direct contact with deer or wapiti with CWD or from known infected or exposed herds should be considered a risk factor for susceptible animals [3–5; 11; 14]. And because of the possibility of environmental contamination with the CWD agent [4; 12; 14; 27a], housing cervids on pastures or in facilities or regions that have previously held animals with CWD also should be considered a possible risk. Scrapie exposure should also be considered as a potential risk factor [4; 8; 9; 26], particularly in cases where other more likely risk factors cannot be identified or are relatively implausible.

21.10 Surveillance, prevention, and control

Mortalities of adult cervids on game farms should be examined for lesions and other diagnostic evidence of CWD (compare Chapter

32.2.3), not only in North America but in other places where prion disease occurs in domestic or wild ruminant species. With the recognition of CWD in the commercial cervid industry in the USA and Canada, herd monitoring and certification programs are being developed and implemented.

Surveys designed to detect CWD-infected free-ranging cervid populations are now being conducted in most states and provinces in North America. Several approaches have been used separately and in combination to detect new foci of CWD. Common approaches include collection and testing of deer, wapiti, or moose showing clinical signs compatible with CWD, sampling of hunter-harvested cervids during annual hunting seasons for CWD examination (see Section 21.6), and sampling of vehicle-killed cervids [7; 28]. Surveys focused on either symptomatic or vehicle-killed animals tend to bias sampling toward detecting infected individuals, and are thus better strategies for monitoring CWD distribution than for estimating prevalence. Live-animal testing via tonsil biopsy also has been used to augment CWD surveillance in urban and exurban areas in Colorado [29–31].

Once detected, captive wapiti or deer on most infected game farms have been depopulated in Canada, the USA, and Korea. Quarantine or monitoring of CWD-infected herds for research purposes is currently practiced in a few situations. Attempts to eliminate CWD from commercial facilities via partial depopulation or other control approaches have not yet been fully evaluated.

Cervid depopulation and decontamination were attempted at wildlife research facilities in Colorado and Wyoming but were not successful [14; 32]. The reasons for these failures were unclear. It was not known with certainty if recurrence of CWD in these facilities was due to residual environmental contamination or reintroduction of infected animals into the facilities, although the former seems most likely.

Experiences have shown that control of CWD in free-ranging cervids is problematic. Attempts to either control or eliminate CWD in free-ranging populations have been undertaken only recently, and it is too early to assess the effectiveness of these attempts.

Preventive measures intended to reduce opportunities for further spreading and establishing new foci of CWD include (i) surveillance programs for both captive and free-ranging populations, (ii) regulations preventing translocation or movements in commerce of live animals and carcass parts from endemic regions or directly neighboring wild populations to other areas (see Chapter 20.10), (iii) regulations prohibiting release of animals from captive herds into free-ranging populations, (iv) recommendations to double fence established game farms in CWD-endemic areas, and (v) discouraging the establishment of new captive facilities in such areas.

References

[1] Williams ES and Young S. Chronic wasting disease of captive mule deer: a spongiform encephalopathy. J Wildl Dis 1980; 16:89–98.

[2] Williams ES and Young S. Spongiform encephalopathy of Rocky Mountain elk. J Wildl Dis 1982; 18:465–471.

[3] Kahn S, Dube C, Bates L, et al. Chronic wasting disease in Canada: part 1. Can Vet J 2004; 45(5):397–404.

[4] Williams ES and Miller MW. Chronic wasting disease in deer and elk in North America. Rev Sci Tech 2002; 21(2):305–316.

[5] Sohn HJ, Kim JH, Choi KS, et al. A case of chronic wasting disease in an elk imported to Korea from Canada. J Vet Med Sci 2002; 64(9):855–858.

[6] Spraker TR, Miller MW, Williams ES, et al. Spongiform encephalopathy in free-ranging mule deer (*Odocoileus hemionus*), white-tailed deer (*Odocoileus virginianus*), and Rocky Mountain elk (*Cervus elaphus nelsoni*) in northcentral Colorado. J Wildl Dis 1997; 33:1–6.

[7] Miller MW, Williams ES, McCarty CW, et al. Epizootiology of chronic wasting disease in free-ranging cervids in Colorado and Wyoming. J Wildl Dis 2000; 36(4):676–690.

[8] Raymond GJ, Bossers A, Raymond LD, et al. Evidence of a molecular barrier limiting susceptibility of humans, cattle and sheep to chronic wasting disease. EMBO J 2000; 19(17):4425–4430.

[9] Race RE, Raines A, Baron TG, et al. Comparison of abnormal prion protein glycoform patterns from transmissible spongiform encephalopathy agent-infected deer, elk, sheep, and cattle. J Virol 2002; 76(23):12365–12368.
[10] Browning SR, Mason GL, Seward T, et al. Transmission of prions from mule deer and elk with chronic wasting disease to transgenic mice expressing cervid PrP. J Virol 2004; 78(23):13345–13350.
[11] Miller MW and Williams ES. Prion disease: horizontal prion transmission in mule deer. Nature 2003; 425(6953):35–36.
[12] Miller MW, Williams ES, Hobbs NT, et al. Environmental sources of prion transmission in mule deer. Emerg Infect Dis 2004; 10(6):1003–1006.
[13] Gould DH, Voss JL, Miller MW, et al. Survey of cattle in northeast Colorado for evidence of chronic wasting disease: geographical and high-risk targeted sample. J Vet Diagn Invest 2003; 15(3):274–277.
[14] Williams ES and Young S. Spongiform encephalopathies in Cervidae. Rev Sci Tech Off Int Epiz 1992; 11:551–567.
[15] Hamir AN, Kunkle RA, Cutlip RC, et al. Experimental transmission of chronic wasting disease agent from mule deer to cattle by the intracerebral route. J Vet Diagn Invest 2005; 17(3):276–281.
[16] Williams ES. Chronic wasting disease. Vet Pathol 2005; 42(5):530–549.
[17] O'Rourke KI, Besser TE, Miller MW, et al. PrP genotypes of captive and free-ranging Rocky Mountain elk (*Cervus elaphus nelsoni*) with chronic wasting disease. J Gen Virol 1999; 80:2765–2779.
[18] Brayton KA, O'Rourke KI, Lyda AK, et al. A processed pseudogene contributes to apparent mule deer prion gene heterogeneity. Gene 2004; 326:167–173.
[19] O'Rourke KI, Spraker TR, Hamburg LK, et al. Polymorphisms in the prion precursor functional gene but not the pseudogene are associated with susceptibility to chronic wasting disease in white-tailed deer. J Gen Virol 2004; 85(5):1339–1346.
[20] Jewell JE, Conner MM, Wolfe LL, et al. Low frequency of PrP genotype 225SF among free-ranging mule deer (*Odocoileus hemionus*) with chronic wasting disease. J Gen Virol 2005; 86(8):2127–2134.
[21] Johnson C, Johnson J, Clayton M, et al. Prion protein gene heterogeneity in free-ranging white-tailed deer within the chronic wasting disease affected region of Wisconsin. J Wildl Dis 2003; 39(3):576–581.
[22] Miller MW and Wild MA. Epidemiology of chronic wasting disease in captive white-tailed and mule deer. J Wildl Dis 2004; 40(2):320–327.
[23] Miller MW and Conner MM. Epidemiology of chronic wasting disease in free-ranging mule deer: spatial, temporal, and demographic influences on observed prevalence patterns. J Wildl Dis 2005; 41(2):275–290.
[24] Joly DO, Ribic CA, Langenberg JA, et al. Chronic wasting disease in free-ranging Wisconsin white-tailed deer. Emerg Infect Dis 2003; 9(5):599–601.
[25] Bruce ME, Will RG, Ironside JW, et al. Transmissions to mice indicate that new variant CJD is caused by the BSE agent. Nature 1997; 389:498–501.
[26] Hamir AN, Miller JM, Cutlip RC, et al. Transmission of sheep scrapie to elk (*Cervus elaphus nelsoni*) by intracerebral inoculation: final outcome of the experiment. J Vet Diagn Invest 2004; 16(4):316–321.
[27] Williams ES, Kirkwood JK, Miller MW. Transmissible spongiform encephalopathies. In: Williams ES, Barker IK, editors. Infectious diseases of wild mammals. Ames, Iowa: Iowa State University Press, 2001:292–301.
[27a] Johnson CJ, Phillips KE, Schramm PT, et al. Prions adhere to soil, minerals and remain infectious. PLoS. Pathog. 2, e32. 2006.
[28] Krumm CE, Conner MM, Miller MW. Relative vulnerability of chronic wasting disease infected mule deer to vehicle collisions. J Wildl Dis 2005; 41(3):503–511.
[29] Wolfe LL, Conner MM, Baker TH, et al. Evaluation of antemortem sampling to estimate chronic wasting disease prevalence in free-ranging mule deer. J Wildl Manage 2002; 66:564–573.
[30] Wolfe LL, Miller MW, Williams ES. Feasibility of "test-and-cull" for managing chronic wasting disease in urban mule deer. Wildl Soc Bull 2004; 32:500–505.
[31] Farnsworth ML, Wolfe LL, Hobbs NT, et al. Human land use influences chronic wasting disease prevalence in mule deer. Ecol Appl 2005; 15:119–126.
[32] Miller MW, Wild MA, Williams ES. Epidemiology of chronic wasting disease in captive Rocky Mountain elk. J Wildl Dis 1998; 34:532–538.

22 Portrait of Transmissible Mink Encephalopathy

William J. Hadlow

22.1 History

Transmissible mink encephalopathy (TME) is a rare, noncontagious, invariably fatal neurological disease of ranch-raised mink (*Mustela vison*) caused by a feed-borne pathogen indistinguishable from the scrapie agent [1]. It was identified in 1947 as a new disease of mink in Wisconsin, USA [2]. At that time, the characteristic degenerative changes seen in the brains of affected mink were regarded as those of a neurotoxicosis, presumably the result of exposure to an unknown noxious substance on the farm. The disease was not recognized again until it appeared in 1961 on five farms in Wisconsin [2]. Because all were provided with the same feed, it was considered the source of the noxious substance responsible for the disease on each farm. Two years later, TME occurred on a farm in Idaho, USA, when its neurohistologic resemblance to scrapie was realized [3; 4]. This prompted studies showing that it, like scrapie, could be transmitted experimentally [3; 5]. The same year it occurred on a farm in Ontario, Canada [6], and on two farms in Wisconsin that used feed from the same supplier [2]. The most recent incident of TME in the United States was in 1985, again in Wisconsin – 22 years after its last recorded occurrence in North America [7]. Elsewhere, it occurred in Finland in 1966 [1] and in the former German Democratic Republic in 1967 [8]. The disease was reported in the Soviet Union in 1979 [9] where it reappeared several times in the early 1980s [10; 11].

22.2 Forms or variants

Wherever TME has occurred in North America, its clinicopathologic expression has been the same [2; 3; 5–7]. Only the morbidity/mortality it causes in a herd has varied. This seems so as well when the disease has appeared elsewhere, though slight differences in its neuropathologic pattern have been reported [10; 12]. Because of the uncertainty about the identity and source of the causal agent, variations in its biologic behavior are expected, and they could account for variations in the expression of TME from one outbreak to another [13; 14]. This may be most apparent in its neuropathologic pattern, as in the disease caused by some strains of the scrapie agent in mink that was otherwise indistinguishable from typical TME [14; 15].

22.3 Incubation period, transmissibility, and susceptibility

TME is not naturally transmissible, neither horizontally nor vertically [16]. (The disease has occurred in a few kits that had cannibalized a dead or dying affected mother [5]). Ostensibly arising from a dead-end infection with an agent not native to mink, TME has no way of sustaining itself in a population of mink [13; 16]. Each occurrence of it represents a new chance or accidental infection with an exogenous agent contaminating the mink's feed now and again. Eating the feed may not be the only way mink become infected. Intradermal inoculation is also likely, as when unweaned littermate kits bite one another during the fighting that commonly occurs at feeding time [14]. Whatever the main route of natural exposure is, the incubation period of the disease has been 7–12 months [2]. So even though mink may become infected as 3-month-old kits, clinical disease does not supervene until they are adults (yearlings and older), which are the typical victims.

Experimental transmission of TME is readily achieved by all parenteral routes – intradermal,

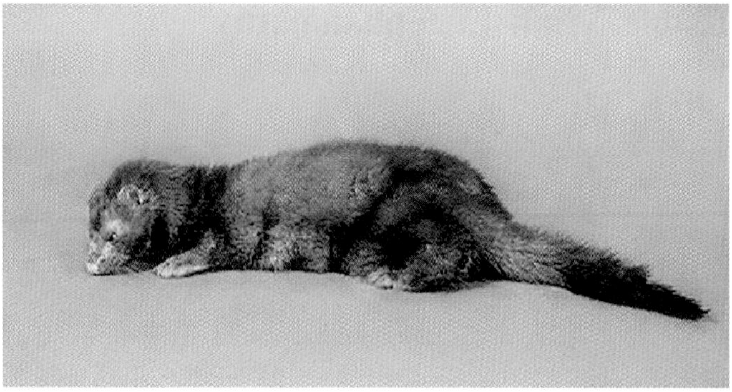

Fig. 22.1: Typical somnolent attitude of a pastel mink in advanced stage of TME. Photograph by W.J. Hadlow ©.

subcutaneous, intramuscular, intraperitoneal, and intracerebral [3; 5; 8; 10; 17; 18]. Depending largely on the concentration of the causal agent in the inoculum, the incubation period has varied from 6 to 12 months but may be longer. The intracerebral route is the most effective one, with incubation periods as short as 4 months [17]. After oral exposure, accomplished simply by adding the inoculum to the feed, the incubation period has been 7–8 months [5; 7; 8; 10; 17]. In all instances, the experimental disease has been indistinguishable from the natural one.

Apart from mink, the TME agent has a wide experimental host range that includes the domestic ferret; striped skunk; pine and beech martens; raccoon; Syrian and Chinese hamsters; squirrel, rhesus, and stumptail macaque monkeys; sheep; goat; and cattle [18–25]. After intracerebral inoculation, the incubation period of the experimental disease has varied from 5 months in skunks to 65 or more months in sheep. The neurohistologic changes and their topographic distribution also have differed widely among the host species. In the skunk and the raccoon they were remarkably similar to those in the mink [21]. Unlike the scrapie agent, the TME agent in both the United States and the Soviet Union has not been pathogenic for the laboratory mouse [6; 15; 19; 26].

22.4 Clinical signs and course of disease

The clinical picture of TME is distinctive, usually making its diagnosis obvious [1; 2]. Onset is insidious with behavioral changes that include hyperexcitability, hyperesthesia, and increased aggressiveness. Almost as though frenzied, the mink vigorously attacks objects that are moved along the sides of the pen. Its responses to sound and touch are exaggerated. It defecates at random sites instead of normally at a single site. Appetite is maintained for a while, though the mink has difficulty eating and tramples its feed. Within a few days to a week after onset, unsteadiness of the hindquarters supervenes, causing the mink to fall to the side when forced to move rapidly. As the hyperexcitability wanes, it is often found in a drowsy state from which it can be aroused only briefly. The mink's tail often curls over its back like that of a squirrel. Occasionally whole body tremor occurs. Convulsions are rare. Some mink circle intermittently. Vision often becomes greatly impaired.

As the disease progresses, coordination of the hind limbs worsens, making forward movement almost impossible. Although the mink is able to move its hind limbs, they are held flexed close to the body. In advanced disease, compulsive biting of self and objects dominates the mink's behavior; this behavior causes severe self-mutilation that usually results in death when the tail is partially amputated. Eventually, the mink becomes less aware of its surroundings and spends much of the time in deep somnolence from which it is not easily aroused (Fig. 22.1). In the end it becomes stuporous and is often found dead with its nose pressed into a corner of the pen or with its jaws firmly clenching the wire mesh. Typically, the disease evolves slowly and relentlessly over a period of weeks. In a few mink, however, the course is either rapidly progressive leading to death in about 1 week, or is prolonged for several months. But most mink die in an unkempt, debilitated state 2–7 weeks after onset of clinical signs. TME is always fatal.

22.5 Differential diagnoses

The early behavioral changes in mink affected with TME might be mistaken for the effects of a neurotoxin or an acute infection of the central nervous system, Aujeszky's disease, for example [27]. But given about 1 week to evolve, TME is readily recognized by its characteristic clinical signs that distinguish it from other known naturally occurring neurological diseases of ranch-raised mink. Moreover, in a typical outbreak the many affected mink will most likely be in different stages of the disease, thereby showing an array of clinical signs that makes its identification more certain. Microscopic examination of the brain will confirm the diagnosis.

22.6 Epidemiology

The simultaneous occurrence of TME on several farms sharing the same feed firmly established it as the source of the infectious agent [2]. From early on the neuropathologic likeness of TME to scrapie pointed to their having an etiologic relationship. Presumably, the incriminated feed (typically a wet mixture of cereal and raw animal tissues) included tissues of sheep infected with the scrapie agent. But no certain evidence of this was ever obtained from the outbreaks in North America [14]. Even so, the scrapie agent might sometimes be responsible, for certain wild American strains of it caused an encephalopathy indistinguishable from TME in mink 12–24 months after intracerebral, but not after oral, inoculation (14; 15; 28; unpublished observations). Moreover, a neurological disease essentially identical with scrapie occurred in sheep and goats inoculated intracerebrally with a brain homogenate from a TME-affected mink [23]. Quite possibly, a strain of the scrapie agent exists that when fed to mink will give rise to TME in 7 to 12 months, as in the natural disease [1]. The occurrence of TME in the Soviet Union is said to have been traced to feeding carcasses of sheep affected with scrapie [11].

In the United States, tissues of cattle, mainly "downer" dairy cows commonly used in preparing mink feed, are now considered an even more likely source of the TME agent [7]. Reportedly, cattle in the United States are infected, mostly subclinically, with an as yet unidentified encephalopathic agent responsible for the sporadic occurrence of TME [7; 29]. Providing support for this still unproved notion is the finding that mink brain homogenates from three separate American outbreaks of TME were pathogenic for cattle when inoculated intracerebrally [24]. Each caused a neurological disease with some histologic features of bovine spongiform encephalopathy (BSE). Then, too, brain homogenates from cows affected with BSE were pathogenic for mink when injected intracerebrally or given orally, causing an encephalopathy somewhat resembling TME clinically and neuropathologically [30]. Yet despite these and other suggestive findings, uncertainty remains about the source and identity of the TME agent and about its epidemiologic relation to other extant infectious agents that cause TSE in man and animals.

As a sporadic disease, TME has occurred in the US notably in Wisconsin where a third of the country's mink pelts are produced [1]. For unknown reasons it has not occurred in two other major mink-producing states, Utah and Minnesota. Its 1963 occurrence in Idaho was an isolated event unrelated to the outbreaks in Wisconsin the same year. Of the other countries that have large populations of ranch-raised mink, only Canada, Finland, and the Soviet Union have recorded outbreaks of the disease. Presumably, the occurrence of TME in the former German Democratic Republic in 1967 was another isolated event.

Significantly, only adult mink are affected, males and females equally. In the United States the morbidity/mortality in each outbreak has been high – from 10–30 % in some herds and 60–100 % in others [2] (Tab. 22.1). The variable outcome might be explained in part by differences in the virulence of the causal agent from outbreak to another. If eating the contaminated

Table 22.1: Some examples of TME outbreaks in North America. At the time 4,400 mink were kept on average per mink farm. (The number of mink farms decreased from about 1,000 in 1988 to 415 in 1996. In the US, a total of 813,800 farmed mink were kept in 1996.)

Year	State/country of outbreak	Number of affected farms	Number of mink kept in farm of outbreak	Number of affected animals per outbreak	Proportion of affected animals per outbreak (within-herd incidence %)
1947*	Wisconsin, USA	2	1,250	1,250	100
1961	Wisconsin, USA	5	na	na	10–30
1963	Wisconsin, USA	2	1,178	1,178	100
1963	Idaho, USA	1	700	700	100
1963	Canada	1	1,700	na	na
1985	Wisconsin, USA	1	4,400	≈ 2,600	≈ 60

* second affected farm in Minnesota [1].
na = not available

feed is the main mode of natural exposure, it has been highly effective in those herds in which most or all adult mink (2,600 in one herd) succumbed to the disease.

Although all common color phases (autosomal recessive traits) of mink are susceptible to TME [17], nothing is known about the relation of the mink's genetic constitution to the occurrence of the disease. A study of the mink's prion protein (PrP) gene did not identify polymorphisms that might influence not only susceptibility but also the incubation period, clinical course, and neuropathologic pattern of the disease [31]. The many mink affected in each sporadic occurrence of TME rules out its ever having arisen de novo from a mutation of the PrP gene [32].

22.7 Pathology

The neuropathology of TME is described in Chapter 27.3.

22.8 Is TME a new disease?

Even though its cause and very nature were a mystery when first observed in 1947, TME was readily identified as a new disease of ranch-raised mink by persons long experienced with mink and their diseases [2]. Whether the causal agent is truly new is another matter. Its physicochemical and biological properties, including those of strains identified by passage in hamsters, have yet to clearly distinguish it from closely related pathogens, e. g., scrapie agent, that cause similar neurodegenerative diseases in man and animals [22; 33; 34]. Whatever its identity and source (sheep, cattle, or other) may be, it is not indigenous to the mink, which is an aberrant host inadvertently and rarely exposed to it from the outside.

22.9 Risk factors

In the US, TME has been extremely rare in view of the many mink at risk – since 1947, five outbreaks involving 11 farms provide little evidence supporting an increasing risk of the disease [35]. Yet its origin elsewhere is unclear. The husbandry practices and general farm conditions under which it sporadically occurs need defining. Sheep may be a risk factor in areas where scrapie is prevalent [35]. And the recent emphasis on downer dairy cows as the source of the causal agent cannot be ignored, however vague this evidence may be at present [35; 36]. Wildlife reservoirs and environmental contamination have been suggested as other possible sources [35].

22.10 Surveillance, prevention, and control

If the scrapie agent, or a variant thereof, causes TME, then controlling scrapie in sheep would help prevent the occurrence of the disease in mink [1]. However, if it results from an unrecognized infection in cattle with a similar agent, then preventive measures would be less straightforward [37]. In any event, the potential risk of feeding tissues of sheep and cattle to mink should be brought to the attention of mink farmers [1]. Efforts to prevent or control TME should be made with the understanding that it does not represent a great threat to the mink industry [1; 4].

References

[1] Marsh RF and Hadlow WJ. Transmissible mink encephalopathy. Rev Sci Tech Off Epiz 1992; 11:539–550.

[2] Hartsough GR and Burger D. Encephalopathy of mink (I) – Epizootiologic and clinical observations. J Infect Dis 1965; 115:387–392.

[3] Hadlow WJ. Discussion of paper by D Burger and GR Hartsough. In: Gajdusek DC, Gibbs CJ Jr, Alpers M, editors. Slow, latent, and temperate virus infections, NINDB Monograph No 2. U.S. Government Printing Office, Washington, D.C., 1965:303–305.

[4] Marsh RF. Slow virus diseases of the central nervous system. In: Brandly CA, Cornelius CE, editors. Advances in veterinary science and comparative medicine, Vol 18. Academic Press, New York, 1974:155–178.

[5] Burger D and Hartsough GR. Encephalopathy of mink (II) – experimental and natural transmission. J Infect Dis 1965; 115:393–399.

[6] Hadlow WJ and Karstad L. Transmissible encephalopathy of mink in Ontario. Can Vet J 1968; 9:193–196.

[7] Marsh RF, Bessen RA, Lehmann S, et al. Epidemiological and experimental studies on a new incident of transmissible mink encephalopathy. J Gen Virol 1991; 72:589–594.

[8] Hartung J, Zimmermann H, Johannsen H. Infektiöse Enzephalopathie beim Nerz 1. Mitteilung: Klinisch-epizootiologische und experimentelle Untersuchungen. Monatsh Veterinärmed 1970; 25:385–388.

[9] Danilov EP, Bukina NS, Akulova BP. Encephalopathy in mink (in Russian). Krolikovod Zverovod 1974; 17:34.

[10] Duker II, Geller VI, Chizhov VA, et al. Clinical and morphological investigation of transmissible mink encephalopathy (in Russian). Vopr Virusol 1986; 31:220–225.

[11] Gorham J. Viral and bacterial diseases of mink in Soviet Union. Fur Rancher 1991; 71:3,10–11.

[12] Johannsen U and Hartung J. Infektiöse Enzephalopathie beim Nerz. 2. Mitteilung: Pathologisch-morphologische Untersuchungen. Monatsh Veterinärmed 1970; 25:389–395.

[13] Hadlow WJ, Race RE, Kennedy RG. Temporal distribution of transmissible mink encephalopathy virus in mink inoculated subcutaneously. J Virol 1987; 61:3235–3240.

[14] Marsh RF and Hanson RP. On the origin of transmissible mink encephalopathy. In: Prusiner SB, Hadlow WJ, editors. slow transmissible diseases of the nervous system, Vol 1. Academic Press, New York, 1979:451–460.

[15] Roikhel V, Fokina G, Sobolev S, et al. Relationship among causative agents of subacute transmissible encephalopathies (in Russian). Vopr Virusol 1992; 37:149–153.

[16] Hanson RP and Marsh RF. Biology of transmissible mink encephalopathy and scrapie. In: Zeman W, Lennette EH, editors. Slow virus diseases. Williams & Wilkins, Baltimore, 1973:10–15.

[17] Marsh RF, Burger D, Hanson RP. Transmissible mink encephalopathy: behavior of the disease agent in mink. Am J Vet Res 1969; 30:1637–1642.

[18] Hartung J, Johannsen U, Zimmermann H. Infektiöse Enzephalopathie beim Nerz. 3. Mitteilung: Ergebnisse weiterer Praxiserhebungen und Ubertragungsversuche. Monatsh Veterinärmed 1975; 30:23–27.

[19] Marsh RF, Burger D, Eckroade R, et al. A preliminary report on the experimental host range of the transmissible mink encephalopathy agent. J Infect Dis 1969; 120:713–719.

[20] Eckroade RJ, ZuRhein GM, Marsh RF, et al. Transmissible mink encephalopathy: experimental transmission to the squirrel monkey. Science 1970; 169:1088–1090.

[21] Eckroade RJ, ZuRhein GM, Hanson RP. Transmissible mink encephalopathy in carnivores: clinical, light and electron microscopic studies in raccoons, skunks and ferrets. J Wildl Dis 1973; 9:229–240.

[22] Kimberlin RH, Cole S, Walker CA. Transmissible mink encephalopathy (TME) in Chinese hamsters: identification of two strains of TME and comparisons with scrapie. Neuropathol Appl Neurobiol 1986; 12:197–206.

[23] Hadlow WJ, Race RE, Kennedy RC. Experimental infection of sheep and goats with transmissible mink encephalopathy virus. Can J Vet Res 1987; 51:135–144.

[24] Robinson MM, Hadlow WJ, Knowles DP, et al. Experimental infection of cattle with the agents of transmissible mink encephalopathy and scrapie. J Comp Pathol 1995; 133:241–251.
[25] Hamir AN, Kunkle RA, Miller JM, et al. First and second cattle passage of transmissible mink encephalopathy by intracerebral inoculation. Vet Pathol 2006; 43(2):118–126.
[26] Taylor DM, Dickinson AG, Fraser H, et al. Evidence that transmissible mink encephalopathy agent is biologically inactive in mice. Neuropathol Appl Neurobiol 1986; 12:207–215.
[27] Löliger H-C: Pelztierkrankheiten. Gustav Fisher Verlag, Stuttgart, 1970:36–38.
[28] Hanson RP, Eckroade RJ, Marsh RF, et al. Susceptibility of mink to sheep scrapie. Science 1971; 172:859–861.
[29] Marsh RF and Hartsough GR. Evidence that transmissible mink encephalopathy results from feeding infected cattle. In: Murphy BD, Hunter DB, editors. Proc 4th Int Sci Congr Fur Anim Prod. Canada Mink Breeders Assoc, Toronto, 1988:204–207.
[30] Robinson MM, Hadlow WJ, Huff TP, et al. Experimental infection of mink with bovine spongiform encephalopathy. J Gen Virol 1994; 75:2151–2155.
[31] Kretzschmar HA, Neumann M, Riethmüller G, et al. Molecular cloning of a mink prion protein gene. J Gen Virol 1992; 73:2757–2761.
[32] Ridley RM and Baker HF. The paradox of prion disease. In: Baker HF, Ridley RM, editors. Prion diseases. Humana Press, Totowa, New Jersey, 1966:1–13.
[33] Marsh RF and Hanson RP. Physical and chemical properties of the transmissible mink encephalopathy agent. J Virol 1969; 3:176–180.
[34] McKenzie D, Bartz JC, Marsh RF. Transmissible mink encephalopathy. Semin Virol 1996; 7:201–206.
[35] Bridges V, Bleem A, Walker K. Risk of transmissible mink encephalopathy in the U.S. In: Animal health insight. Fort Collins, Colorado: USDA-APHIS-VS Animal Health Information. Fall 1991:7–13.
[36] Robinson MM. An assessment of transmissible mink encephalopathy as an indicator of bovine scrapie in U.S. cattle. In: Gibbs CJ Jr, editor. Bovine spongiform encephalopathy. The BSE dilemma. Springer-Verlag, New York, 1996:97–107.
[37] Walker KD, Hueston WD, Hurd HS, et al. Comparison of bovine spongiform encephalopathy risk factors in the United States and Great Britain. J Am Vet Med Assoc 1991; 199:1554–1561.

23 Portrait of Transmissible Feline Spongiform Encephalopathy

Marion Hewicker-Trautwein and Ray Bradley

23.1 History

The first reported occurrence of feline spongiform encephalopathy (FSE) in a British 5-year-old Siamese domestic cat (*Felis cattus*) was in 1990 [1], 4 years after the first case of BSE was recognized. For further information on the history of FSE in wild Felidae kept in zoological gardens see Chapter 20. Long before transmissible spongiform encephalopathies (TSE or prion diseases) had been described in Felidae, the only carnivore species in which natural TSE was reported was the farmed mink [2] (see Chapter 22) though the Creutzfeldt–Jakob disease agent had been experimentally transmitted to domestic cats in the USA [3] and in Slovakia [4].

23.2 Forms or variants

To date the single strain of TSE agent identified and responsible for natural FSE in domestic cats is biologically [5] and molecularly [6] indistinguishable from the BSE agent. It is presumed from epidemiological evidence – although inconclusively – that the same agent strain is responsible for FSE in captive wild cats.

23.3 Incubation period, transmissibility, and susceptibility

Transmission studies have shown that FSE can be transmitted experimentally to susceptible mice from formalin-fixed or unfixed brain tissue originating from domestic cats with FSE [5; 7]. In a panel of five inbred strains of mice the transmissions gave the same ranking of incubation periods as BSE from cattle (Fig. 50.1). Moreover, the lesion profile of FSE was closely similar to that of BSE when transmitted to RIII mice, confirming that the cats were infected with the BSE agent (Fig. 50.1).

23.4 Clinical signs and course of disease

The mean age of 11 domestic cats affected by FSE, which originated from different geographical areas of the UK, was 6 years with a range of 2–10 years [8–10].

In all cases, the clinical signs were gradual in onset over several weeks or months. The main clinical signs and their frequency are (i) behavioral changes (aggression, timidity), (ii) hyperesthesia to tactile and auditory stimuli, polyphagia, decreased or increased grooming, (iii) locomotor dysfunction such as ataxia (most frequently), hypermetria, or crouching gait, and (iv) hypersalivation and polydipsia [9–11].

In many cases, behavioral changes were the first signs noted by owners. Some animals, without being threatened or attacked, showed aggressive behavior directed at the owner or at other household pets. Other individuals showed timidity, manifested as fear of contact with other cats or humans. They showed hiding behavior or seemed afraid of going outdoors. In some cases, the animals had an episodic staring gaze or made circling movements within a certain area of the room.

In all cats, progressive locomotor dysfunction occurred. In most cases the owners sought veterinary advice because they had noted hind-limb ataxia and/or inability to judge distances such as when jumping. In advanced cases, fore- and hind-limb ataxia was observed, frequently occurring with a rapid, crouching, hypermetric gait. Because of incoordination some animals showed aberrant defecation.

A further important sign, noted in almost all cases, was tactile hyperesthesia. This often showed as a fine continuous tremor, particularly of the head and ears. Many animals showed an exaggerated response to auditory stimuli. In 7 of 11 cases, grooming either decreased or increased. In several animals, hypersalivation, polydipsia, and polyphagia were observed. In addition, dilated, responsive or nonresponsive pupils were reported [8; 12].

For a number of cats affected by FSE, clinical findings and outcome have been summarized in reviews [8–10]; all were euthanized due to an unfavorable prognosis.

23.5 Differential diagnoses

Neuropathological signs may occur in a number of feline diseases [13]. The most relevant differential diagnoses include: infections with feline leukemia virus, feline immunodeficiency virus, feline infectious peritonitis virus, rabies virus, toxoplasmosis, neoplasms such as lymphosarcoma and brain tumors, congenital diseases such as cerebellar hypoplasia, storage diseases, and hepatic encephalopathy associated with congenital portal-systemic shunts, trauma, thiamine deficiency, and intoxication [9; 10].

23.6 Epidemiology

Between April 1990 and May 2004, 90 cases of FSE in domestic cats have been recognized in the UK, one of which was in Northern Ireland. From the rest of Europe four cases have been reported: one in a 6-year-old female domestic shorthaired cat in Norway, in 1995 [14], one in a cat about 9 years old from the Principality of Liechtenstein, in 1996 [15], one in a 6-year-old female Birman cat in Switzerland, in 2001 [16], and a second case in a cat in Switzerland, in 2003 [17].

Furthermore, FSE cases have been recognized in captive wild felids in zoological gardens or wildlife parks in the UK, Australia, Ireland and France. The species affected have been: cheetah (*Acinonyx jubatus*), puma (*Felis concolor*), ocelot (*Felis pardalis*), leopard cat (*Prionailurus bengalis*), lion (*Panthera leo*), and tiger (*Panthera tigris*) all of which were in, or originated from, Great Britain [15; 18–22] (see Chapter 20). A case of FSE has also been reported in an Asian golden cat (*Felis temmincki*) in an Australian zoo [23; 24]. The cat was born in a German zoo, moved to the Netherlands, and succumbed to disease presumably following exposure in Europe.

23.7 Pathology

In most cases, spongiform alterations can be observed in all regions of the brain but the extent of lesions may vary among individual cases. The most intense lesions are present in the medial corpus geniculatum and in the corpus striatum. Spongiform changes are frequent in the cerebral and cerebellar cortex and in the cerebellar cortex it is the deeper layers that are predominantly affected. The spongy change comprises vacuolation of the gray matter, consisting of small empty holes in the neuropil [9; 10].

While not as prominent as spongiform change there is also vacuolation in the cytoplasm of neurons. Predilection sites for neuronal vacuolation are the nucleus ruber in the mesencephalon, the dorsal nucleus of the nervus vagus, and the nuclei vestibulares in the medulla oblongata. Also in the spinal cord, single vacuoles may be found in the cytoplasm of neurons. Additional alterations of neurons are rare. In all cats, there is gliosis involving proliferation of both astrocytes and microglial cells. Pearson points out the diagnostic value of the vacuoles appearing in the neuropil of gray matter and in nerve cells [9]. With the exception of those FSE cases, which have been diagnosed since 1990, there is no information in the "historical literature" or in archival samples providing evidence of histopathological changes in cat brains consistent with FSE prior to 1990 [25].

23.8 Is FSE a new disease?

Retrospective studies of the post mortem diagnoses of a total of 286 domestic cats with neurological signs presented to the University of Bristol Veterinary School shows that FSE is a new disease [8–10; 25]. The studies revealed that during the period from January 1975 to March 1990 no case with comparable pathological findings had occurred. The results of biological strain typing studies [5] in mice using inocula from brain tissue of different species with natural TSE or experimental BSE, indicates a close similarity between experimental murine FSE and experimental murine BSE i. e., the responsible biological agent strain is the same. In particular, the lesion profile and mean incubation period (Fig. 50.1) observed in these studies closely resemble one another. The presence of scrapie-associated fibrils (SAF) and of prion protein (PrP^{Sc}) was confirmed in the brains of several cats affected by FSE [7; 26; 27].

23.9 Risk factors

Several factors point to the possibility of FSE originating from exposure to bovine BSE: (i) the BSE-like features of the FSE agent in experimental studies (see Chapter 50), (ii) the fact that FSE, like BSE is regarded as a new disease, and (iii) the first cases of reported FSE followed the first cases of BSE in cattle by approximately 5 years. Like cattle, cats might have been exposed to similar infected feed sources [9]. Anamnestic data reveal that cats, which later developed FSE, had received different types of food, i. e., commercial cat food products such as canned and dry foods [10; 12]. Some types of cat food could have contained bovine-derived meat-and-bone-meal that is considered the major vehicle carrying the BSE agent to cattle in feed.

All exotic felids (other than the single Asian golden cat), which died spontaneously or were euthanized because of FSE, were born in zoological gardens in the UK. Epidemiological data available for cheetahs and pumas reveal that these animals had been fed with bovine material that could have been infected with the BSE agent, e. g., vertebral column containing spinal cord [18; 19].

According to Kirkwood (see Chapter 20), there is no definitive answer to the question as to whether FSE is horizontally transmissible from cat to cat under natural conditions.

23.10 Surveillance, prevention, and control

Prevention of FSE is best approached by avoidance of feed material potentially containing the BSE agent, e. g., cooked or uncooked parts of brain or spinal cord from species susceptible to TSE (especially cattle), animal by-products derived from such material, or indeed any material now defined as specified risk material (see Chapter 52.2; Fig. 52.2).

References

[1] Wyatt JM, Pearson GR, Smerdon TN, et al. Spongiform encephalopathy in a cat. Vet Rec 1990; 126:513.

[2] Jervis G. Sheep, minks, savages and presenile dementia. The story of slow viruses (2). Psychiatr Q Suppl 1968; 42:371–435.

[3] Gibbs CJ Jr and Gajdusek DC. Experimental subacute spongiform virus encephalopathies in primates and other laboratory animals. Science 1973; 182:67–68.

[4] Mayer V, Mitrova E, Orolin D. Creutzfeldt–Jakob disease in Czechoslovakia and a working concept of its surveillance. In: Prusiner SB, et al., editors. Slow transmissible diseases of the nervous system. Academic Press, New York, 1979:287–303.

[5] Bruce ME, Chree A, McConnell I, et al. Transmission of bovine spongiform encephalopathy and scrapie to mice: strain variation and the species barrier. Philos Trans R Soc Lond B 1994; 343:405–411.

[6] Collinge J, Sidle KC, Meads J, et al. Molecular analysis of prion strain variation and the aetiology of 'new variant' CJD. Nature 1996; 383:685–690.

[7] Fraser H, Pearson GR, McConnell I, et al. Transmission of feline spongiform encephalopathy to mice. Vet Rec 1994; 134:449.

[8] Gruffyd-Jones TJ, Galloway PE, Pearson GR. Feline spongiform encephalopathy. J Small Anim Pract 1991; 33:471–476.

[9] Pearson GR, Wyatt JM, Henderson JP, et al. Feline spongiform encephalopathy: a review. In: Raw ME, Parkinson TJ, editors. The veterinary annual. Blackwell Scientific Publications, London, 1993:1–10.

[10] Wyatt JM, Pearson GR, Gruffyd-Jones TJ. Feline spongiform encephalopathy. Feline Pract 1993; 21:7–9.

[11] Wyatt JM, Pearson GR, Smerdon TN, et al. Naturally occurring scrapie-like spongiform en-

cephalopathy in five domestic cats. Vet Rec 1991; 129:233–236.
[12] Leggett MM, Dukes J, Pirie HM. A spongiform encephalopathy in a cat. Vet Rec 1990; 127:586–588.
[13] Kelly DF, Wells GAH, Haritani M, et al. Neuropathological findings in cats with clinically suspect but histologically unconfirmed feline spongiform encephalopathy. Vet Rec 2005; 156(15):472–477.
[14] Bratberg B, Ueland K, Wells GAH. Feline spongiform encephalopathy in a cat in Norway. Vet Rec 1995; 136:444.
[15] MAFF. Bovine spongiform encephalopathy in Great Britain. A progress report – December 1999. MAFF, 3 Whitehall Place, London SW1A 2HH, 1999.
[16] Demierre S, Botteron C, Cizinauskas S, et al. Feline spongiforme Enzephalopathie: Erster klinischer Fall in der Schweiz. Schweiz Arch Tierheilk 2002; 144:550–557.
[17] Anonym. Swiss FSE case raises questions. Anim Pharm 2005; 525:4.
[18] Baron T, Belli P, Madec JY, et al. Spongiform encephalopathy in an imported cheetah in France. Vet Rec 1997; 141:270–271.
[19] Kirkwood JK and Cunningham AA. Epidemiological observations on spongiform encephalopathies in captive wild animals in the British Isles. Vet Rec 1994; 135:296–303.
[20] Kirkwood JK, Cunningham AA, Flach EJ, et al. Spongiform encephalopathy in another captive cheetah (*Acinonyx jubatus*): evidence for variation in susceptibility or incubation periods between species? J Zoo Wildl Med 1995; 26:577–582.
[21] Peet R and Curran JM. Spongiform encephalopathy in an imported cheetah (*Acinonyx jubatus*). Aust Vet J 1992; 69:117.
[22] Willoughby K, Kelly DF, Lyon DG, et al. Spongiform encephalopathy in a captive puma (*Felis concolor*). Vet Rec 1992; 131:431–434.
[23] Anonym. Imported zoo cat falls victim to rare disease. Aust Vet J 2003; 80:445.
[24] Young S and Slocombe RF. Prion-associated spongiform encephalopathy in an imported Asiatic golden cat (*Catopuma temmincki*). Aust Vet J 2003; 81:295–296.
[25] Bradshaw JM, Pearson GR, Gruffydd-Jones TJ. A retrospective study of 286 cases of neurological disorders of the cat. J Comp Pathol 2004; 131:112–120.
[26] Bruce ME, McBride PA, Jeffrey M, et al. PrP in pathology and pathogenesis in scrapie-infected mice. Mol Neurobiol 1994; 8:105–112.
[27] Pearson GR, Wyatt JM, Gruffyd-Jones TJ, et al. Feline spongiform encephalopathy: fibril and PrP studies. Vet Rec 1992; 131:307–310.

24 Portrait of Experimental BSE in Pigs

Gerald A.H. Wells, Stephen A.C. Hawkins, Joachim Pohlenz*, and Danny Matthews

24.1 History

There are no reports of a naturally occurring transmissible spongiform encephalopathy (TSE) or prion disease in pigs. Prior to studies of the transmissibility of the bovine spongiform encephalopathy (BSE) agent, information on the susceptibility of the pig to TSE infection, either from experimental or natural exposure, was solely confined to an unsuccessful attempt to transmit the agent of kuru to pigs. Eight pigs were each inoculated with a different individual source of kuru agent and observed for periods of 52–76 months [1]. Because, in 1987–1988, at the start of the UK research program on BSE, it was clear that pigs had been exposed to ruminant derived meat-and-bone-meal (MBM), studies were conducted to examine the transmissibility of the BSE agent to pigs by parenteral routes of inoculation and by feeding. Successful transmission occurred after simultaneous inoculations by intracerebral, intravenous, and intraperitoneal routes, each pig receiving a total of 1 g of brain homogenate from BSE-affected cows. There was no evidence of transmission after feeding pigs large amounts of infected brain material [2–5].

24.2 Forms or variants

In the absence of a naturally occurring TSE or prion disease in pigs, the issue of different forms does not arise. The pathological phenotype of BSE in pigs after transmission by multiple parenteral routes was consistent among the recipients that developed disease [4], and on transmission of brain material from an affected pig to mice, the biological characteristics of the disease were similar to those seen on direct transmission of the BSE agent from cattle to mice [6]. Retention of this biological identity after a single passage through other mammalian species has been a constant feature of the BSE agent.

24.3 Incubation period, transmission, and susceptibility

After the simultaneous triple-route parenteral exposure of ten piglets to the BSE agent, clinical disease was produced in five, with an incubation period range of 69–150 weeks. Two piglets died of intercurrent disorders shortly after inoculation, and preclinical spongiform encephalopathy was detected in two pigs euthanized electively at 105 and 106 weeks post-inoculation (p.i.). The remaining clinically normal test animal, euthanized electively at 107 weeks p.i., had no evidence of pathological changes or infectivity in the brain. Infectivity, detected by bioassay in inbred mice, was found in the central nervous system (CNS) of all parenterally inoculated pigs that had brain lesions and in the stomach, jejunum, distal ileum, and pancreas of those terminally affected [2; 4; 5].

In contrast to the parenteral inoculation study, pigs fed brain tissue from cattle infected with BSE on each of three successive occasions, at 1–2 week intervals, did not develop disease throughout a post-exposure observation period of 7 years [5]. Each pig was exposed to a total of 1.2 kg of the infected brain tissue, and the amount fed to them on each occasion was equiv-

* This chapter is dedicated to the memory of Joachim Pohlenz†.

alent to the maximum daily intake of MBM in rations for pigs aged 8 weeks. This provided an exposure to the bovine CNS of 50,000 times more than that estimated to have resulted from dietary exposure of pigs to such tissues in MBM in the field, suggesting the existence of a substantial cattle–pig species barrier to infection with the BSE agent by the oral route.

A similarly designed but, as of yet, incomplete study of the transmissibility of scrapie agent to pigs by feeding them affected sheep brain tissue has so far not found any evidence of transmission [7].

24.4 Clinical signs and course of disease

The first pig to become affected after parenteral exposure to the BSE agent developed clinical signs and was euthanized at 74 weeks post-exposure [2]. The other four pigs that developed clinical disease did so much later and were euthanized 139–163 weeks after inoculation. The clinical course, prior to euthanasia, ranged from 5 weeks (in the first affected pig) to 13 weeks [5]. The clinical signs were similar in all of the affected pigs, with initial behavioral changes that began with apparent agonistic reactions to attendants and apprehension. Later on, the behavior suggested apparent confusion or depression. The behavioral changes were accompanied by an initial mild hind-limb ataxia and then a progressive locomotor disability with generalized ataxia of gait, weakness, and adventitial movement disorders. Finally, there was persistent recumbency with difficulty in rising. Some pigs showed aimless biting activities or, after initial aggressive signs, would become quiet, timid, and easily frightened, or persistently approached attendants with continual vigorous vocalizations. A low carriage of the head and the ears was noted, and adventitial movements included tremor in the shoulder regions and flank and of the ears. Occasionally there was myoclonus.

24.5 Differential diagnoses

The distinctive, slowly progressive clinical signs observed in the experimentally affected pigs after parenteral inoculation provides a basis on which a potential naturally occurring TSE in the species might be recognized. The neurological nature of the signs is likely to alert clinicians initially to the possibility of differential diagnoses that would include notifiable diseases, particularly classical swine fever (hog cholera) and African swine fever. Pseudorabies (Aujeszky's disease), rabies, Japanese encephalitis, and Teschen–Talfan disease might also present with some clinical similarities. Additionally, bacterial meningoencephalitides may result in similar clinical signs. Further differential diagnoses could include cerebrospinal angiopathy, thiamin deficiency, water deprivation/salt intoxication, and selenium poisoning. As in other species, a TSE of pigs is likely to be present only in mature animals. Careful neuropathological examination would enable a TSE in pigs to be distinguished easily from other neurological diseases.

24.6 Epidemiology

As a naturally occurring TSE is not recorded in pigs, there are no epidemiological data.

24.7 Pathology

Essential to interpreting the neuropathological examination for evidence of a TSE in any species is the knowledge that vacuolation of neurons, occurring localized and at low frequency, is an incidental, nonspecific finding in most, if not all, mammalian species including the domestic pig [8; 9]. Neuronal vacuolation [8; 10] and the other morphological changes present in the TSE may occur in a variety of other diseases. From the experiment to transmit the BSE agent to pigs by injection [4], vacuolation of the perikaryon of occasional neurons in the dorsal nucleus of the vagus nerve occurred in the age-matched healthy control pigs examined. In addition, these and all pigs in the unsuccessful experiment to transmit the BSE agent by feeding [5] had varying severity of neuropil vacuolation of the rostral colliculus, but none had evidence of the presence of disease-specific PrP in the CNS [4]. Subsequent examination of archived porcine brain material from other parts of the world supports this view [7]. In the seven pigs that developed disease after experimental infection, the lesions comprised a typical spongiform encephalopathy, with severe spongiosis of gray matter, affecting most brain regions, but especially the forebrain, and an astrocytic response. Disease-specific PrP configurations were detected by immunohistochemistry. The density of immunolabeling was often mild and did not correspond obviously to the severity of vacuolar change. Characteristic SAFs [11] were visualized in tissue extracts from the CNS of the first animal to develop disease in the experimental study [2], but not in subsequently affected animals [5].

24.8 Is there a TSE of pigs?

Unsurprisingly, inasmuch as pigs can be infected with BSE agent by parenteral inoculation, they are susceptible to TSEs. But, given that, as stated previously, no naturally occurring TSE of pigs has been reported, that the circumstances of dietary exposures of pigs to the BSE agent in the UK almost certainly did not ever represent an effective exposure dose, and that the clinical signs of BSE in the pig are probably distinctive, it seems unlikely a TSE occurs in this species with a readily detectable prevalence. If, nevertheless, hypothetically, sporadic or genetic forms of prion disease, as recorded in man, occur in animals, cases could be so rare as to escape any form of farm-animal disease surveillance. Until such forms of TSE are demonstrated in nonhuman species, this possibility remains entirely theoretical.

24.9 Risk factors

The feeding of animal proteins of mixed-species origin to pigs in the UK has its origins in the early part of the 20th century, providing the possibility of exposure of pigs to scrapie-infected tissues for almost a century [7] without apparent adverse consequences. In 1996, in the UK, BSE control measures were extended to a ban on the feeding of mammalian MBM to pigs and poultry. Since January 2001, use of all processed mammalian protein in feeds for farmed animals has been banned throughout the European Union (EU). Prior to this in the UK, there was ample opportunity for repeated primary exposures of commercial pigs to BSE infection, including the considerable potential for pig-to-pig recycling of infection. The fact that this did not result in natural cases of TSE in pigs suggests that pigs did not become infected with BSE. However, these materials continue to be used in pig rations in some other parts of the world, raising a risk of cross-contamination in the preparation of feed. In addition, by analogy with rodent models of TSEs, the possibility must be considered of inducing a subclinical carrier state in pigs, whereby they become infected but do not develop disease within their lifetime. While the failure to detect infectivity in the intestine and lymphoid tissues of pigs fed BSE-infected material would argue against this situation, it nevertheless poses a theoretical risk of within-species recycling and eventual amplification of the agent if pigs consume MBM of pig origin. Furthermore, the inclusion of MBM in ruminant feed, where it is still practiced, has the potential to perpetuate the recycling of ruminant protein and, therefore, possibly the BSE agent if intestinal contents from pigs are included in the product.

24.10 Surveillance, prevention, and control

Surveillance for the detection of BSE in pigs has been reported only in the Republic of Ireland with negative results on a sample of 1,107 adult cull pigs [12]. In the light of current assessments of the geographical risk of occurrence of BSE worldwide, it is recommended not to use animal protein in pig feed. Where such use is continued, it is advocated that it should be closely monitored [7]. The practice should be subject to local risk assessments for the use of animal by-products in feed, including the possible application of the removal of certain tissues (specified risk materials) where the product is of cattle origin. Risk-reduction measures should also be implemented, including those related to the conditions of rendering.

Molecular genetic investigations of porcine *Prnp* [13] may be of value in determining factors that influence the species barrier between bovine and porcine species, but, to date, have shown only homogeneity among different breeds of pig [13; 14].

References

[1] Gibbs CJ Jr, Gajdusek DC, Amyx H. Strain variation in the viruses of Creutzfeldt–Jakob disease and kuru. In: Prusiner SB, Hadlow WJ, editors. Slow transmissible diseases of the nervous system, Vol 2. Academic Press, New York, 1979:87–110.

[2] Dawson M, Wells GA, Parker BN, et al. Primary parenteral transmission of bovine spongiform encephalopathy to the pig [letter]. Vet Rec 1990; 127:338.

[3] Dawson M, Wells GA, Parker BNJ, et al. Transmission studies of BSE in cattle, pigs and domestic fowl. In: Bradley R, Marchant B, editors. Transmissible spongiform encephalopathies. Proceedings of a consultation on BSE with the Scientific Veterinary Committee of the European Communities, September 14–15, 1993. Brussels: EU: VI/4131/94-EN document ref. F. II.3 – JC/0003, 1994:161–167.

[4] Ryder SJ, Hawkins SA, Dawson M, et al. The neuropathology of experimental bovine spongiform encephalopathy in the pig. J Comp Pathol 2000; 122:131–143.

[5] Wells GA, Hawkins SA, Austin AR, et al. Studies of the transmissibility of the agent of bovine spongiform encephalopathy to pigs. J Gen Virol 2003; 84:1021–1031.

[6] Bruce ME, Chree A, McConnell I, et al. Transmission of bovine spongiform encephalopathy and scrapie to mice: strain variation and the species barrier. Philos Trans R Soc Lond B 1994; 343:405–411.

[7] Matthews D and Cooke BC. The potential for transmissible spongiform encephalopathies in non-ruminant livestock and fish. Rev Sci Tech Off Int Epiz 2003; 22:283–296.

[8] Wells GA and McGill IS. Recently described scrapie-like encephalopathies of animals: case definitions. Res Vet Sci 1992; 53:1–10.

[9] Pumarola M, Juanola B, Fatzer R. Neuronal vacuoles in the canine brain. Schweiz Arch Tierheilk 1995; 137:62–64.

[10] Kortz GD, Meier WA, Higgins RJ, et al. Neuronal vacuolation and spinocerebellar degeneration in young Rottweiler dogs. Vet Pathol 1997; 34(4):296–302.

[11] Merz PA, Rohwer RG, Kascsak R, et al. Infection-specific particle from the unconventional slow virus diseases. Science 1984; 225:437–440.

[12] Onnasch H, Bassett H, Callanan JJ, et al. Neurosurveillance for TSE in at-risk pigs [abstract]. Res Vet Sci 2004; 76(Suppl A):22.

[13] Martin T, Hughes S, Hughes K, et al. Direct sequencing of PCR amplified pig PrP genes. Biochim Biophys Acta 1995; 1270:211–214.

[14] Lipp O, Ritzmann M, Kixmöller M, et al. Homogeneity of the prion protein gene in various European and Asian pig breeds. J Vet Med B 2004; 51:97–98.

25 Portrait of a Spongiform Encephalopathy in Birds and the Transmissibility of Mammalian Prion Diseases to Birds

Gerald A.H. Wells, Joachim Pohlenz*, Stephen A.C. Hawkins, and Danny Matthews

25.1 History

In 1986, a disease associated with progressive ataxia and other central nervous system (CNS) disturbances was observed in an adult male red-necked ostrich (*Struthio camelus*) from a zoo in Northern Germany. Neuropathological investigations of a further case from the same zoo, in an adult female, showing identical clinical signs, revealed spongiform alterations of the brain stem similar to those seen in prion diseases [1] (Fig. 25.1). Since then, four further cases with a similar disease from two additional premises have been observed [2] (J. Pohlenz, personal observations). The temporal coincidence of these cases with the recognition of bovine spongiform encephalopathy (BSE) in the United Kingdom (UK) and the subsequent recording of BSE occurrence throughout much of Europe raises the possibility of a causal link between the diseases, but none has ever been established thus far. Separate concerns regarding the potential exposure of other domesticated animal species, including poultry, to feed contaminated with the BSE agent gave rise in 1990 to studies of the transmissibility of BSE to the domestic fowl [3].

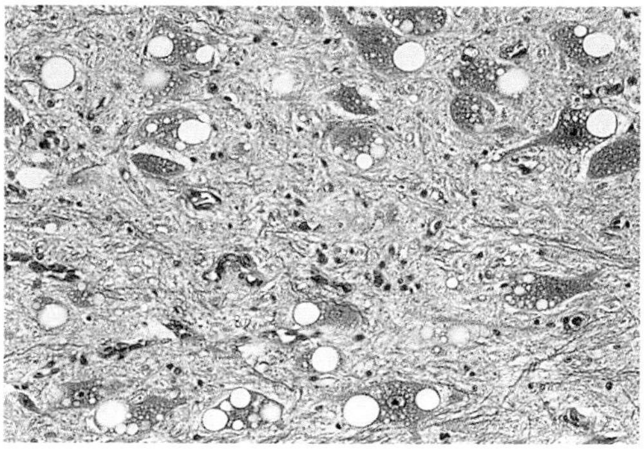

Fig. 25.1: Histologic finding in the brain of an ostrich suffering from spongiform encephalopathy. There are optically empty, ovoid to spheric vacuoles of varying size in the neuropil and intraneuronally. Slide by J. Pohlenz ©.

* This chapter is dedicated to the memory of Joachim Pohlenz†.

25.2 Avian prion diseases?

The occurrence of prion diseases in avian species has never been confirmed. Furthermore, there is very little information on neuropathological entities in birds that would satisfy the morphological definition of a mammalian spongiform encephalopathy (SE; degenerative brain changes, characterized by vacuolation, principally in gray matter). Perhaps prion diseases of birds, if they occur, would not present such pathology. Attempts to transmit transmissible mink encephalopathy (TME) [4], or BSE, to the domestic fowl [3] have been unsuccessful (described below), so there is no experimental model of a prion disease in birds. There is therefore no known pathomorphological basis for their diagnosis. Only partial homology between mammalian PrP and the product of a putative equivalent gene in birds [5; 6] also presents difficulties for immunological surveillance of prion diseases in the order Aves.

A report of a prion disease case in a 30-month-old domestic fowl showing behavioral abnormalities and neurological signs, from a nonveterinary source [7], was unconfirmed upon further investigation [8]. The report referred to a single affected chicken; brain sections submitted by the author and examined by scientists at both animal and human transmissible spongiform encephalopathy (TSE) reference centers showed evidence of viral encephalitis but no pathology suggestive of a TSE. The alleged PrP immunostaining reported was considered to be nonspecific, and the unsatisfactory nature of the material submitted for diagnosis prevented further study.

25.3 Transmission studies

Experiments to examine the transmissibility of the ostrich SE to mice and hamsters were unsuccessful, as were studies to demonstrate PrP^{Sc} in the brain tissue (H. Diringer, personal communication).

In experiments to examine the host range of TME (4), two chickens were inoculated intravenously with mink-passaged TME agent. They were euthanized, when asymptomatic, 1 and 5 months post-inoculation. By assay in mink, infectivity was demonstrated in lymphoid tissues of both birds, but not in their brains. It is unclear whether the TME agent detected was the result of infection or persistence and transport of inoculum in the birds.

Studies of the transmissibility of the BSE agent to the domestic fowl [3; S.A.C. Hawkins and others, unpublished] have included both *parenteral* inoculations (i.c. injection of day-old chicks, followed by i.p. injection at 2 weeks old) and per os exposures (deposition of inoculum into the esophagus and crop on three occasions at 4, 5, and 6 weeks of age). The birds were retained up to 5 years after exposure; tissues were sampled for histopathological examination, assays of tissue infectivity, and blind subpassage. There was no evidence of transmission during the study. No pathology indicative of a TSE was identified in the brain or spinal cord of any of the birds in this primary transmission experiment. A blind pass of nervous system tissue pools (frontal cortex and sciatic nerve) from selected birds of each of the primary studies was conducted in a panel of inbred mice, with negative results. The same two tissue pools were each inoculated intracerebrally into groups of ten day-old chicks. These birds were again retained for 5 years; no evidence of transmission of a TSE was detected. Interpretation of the negative results of immunohistochemical examinations of brain material for evidence of PrP using a number of antibodies to mammalian PrP was compromised in the absence of an avian positive control model.

There are no reports of attempts to transmit other TSEs to birds.

25.4 Clinical signs and course of the SE of ostriches

Typical signs in all of the five birds (Table 25.1) were a slowly progressive ataxia, including disturbances of balance and inappetance [1]. Signs in the two ostriches affected in 1992/1993, which were observed in more detail for a period of 6–8 weeks, also included an unusual twisting of the neck, and increasingly, sternal recumbency [9]. In these birds, the disease had a remissive progressive course with an apparent paresis of the neck and, with deteriorating general condition, eventual permanent recumbency. The age and gender of the animals are listed in Table 25.1.

Table 25.1: Data on the described outbreaks of spongiform encephalopathy in ostriches.

Animal/ Zoo	Age	Gender	Weight (kg)	Year [1]	Reference
1 / A	adult	female	150	1986	[1; 2]
2 / B	adult	female	80	1988	[2]
3 / B	juvenile	male	60	1989	[2]
4 / C	adult	female	76	1992/ 1993	[*]
5 / C	adult	male	78	1992/ 1993	[*]

[1] Year of diagnosis
* J. Pohlenz, unpublished

25.5 Differential diagnoses of the SE of ostriches

Neurological signs in ostriches have been associated with Newcastle disease [2; 10] and bacterial infections, including listeriosis (associated with silage feeding) [9]. Other general possible differential causes include – at least in theory – nutritional deficiency diseases, metabolic disorders, and intoxications.

25.6 Epidemiology of the SE of ostriches

The ostriches developed the disease in three different zoological gardens of Northern Germany. Cases occurring in premises A and B were not investigated epidemiologically. Both animals from premises C developed disease in the winter of 1992/1993 within a 3-week period. A third animal in the same group remained healthy and was investigated at a later date, with no significant findings. These ostriches originated from Namibia and were imported as chicks to Germany. They were reared in another location and then transferred to premises C approximately 9 months prior to the first appearance of clinical signs. No similar cases have been reported outside of Germany.

25.7 Pathology of the SE of ostriches

The neuropathological changes gave rise to the descriptive diagnosis of an SE in these birds. These morphologic lesions, distributed mainly in the brainstem (confined to the medulla oblongata in case 2; Table 25.1) were bilaterally symmetrical in all of the affected animals [2; 11] (J. Pohlenz, personal observations). The vacuolar changes that varied greatly in severity among the affected birds comprised a spongiosis of the neuropil and optically empty ovoid to spherical vacuoles in the perikarya of neurons. Strikingly, the vacuoles in the neuronal cytoplasm varied greatly in size, from a fine foamy appearance to single, or multiple, large vacuoles distending the neuron. Populations of large neurons in the nucleus ruber, the nucleus vestibularis, and the formatio reticularis were particularly affected. Lipofuscin pigment, identified from its histological staining characteristics, was present in many neurons and was most apparent in non-vacuolated neurons [2]. A mild gliosis was observed in some areas of the brainstem in some of the cases.

25.8 What is the significance of the SE of ostriches?

The neurodegenerative changes described in the ostriches appear not to have been reported previously in the literature. They differ in some details from typical TSE lesions in mammals. The very wide size range of intraneuronal vacuoles is not seen in mammalian TSE. Also, the spongiosis of the neuropil in the ostrich SE extends sometimes to involve white matter (G. A. H. Wells, personal observation), providing an appearance resembling intramyelinic vacuolation, attributable in mammals to a wide range of metabolic and toxic causes. However, white matter changes may also feature in mammalian TSE, and the possibility, or probability, that an avian TSE would present differently from such a disease in mammals cannot be discounted. The background of an incomplete etiological investigation, in particular the absence of an investigation of toxicological or nutritional causes and the absence of any evidence of the disease being transmissible or having any association with a pathogenesis implicating the accumulation in the central nervous system of a misfolding, host-encoded protein, calls into question the inclusion of this disease in a book on prion diseases. It is the unique nature of the observation in birds that, nevertheless, leaves the possibility of a prion-like causation.

25.9 Risk factors and prevention

The diets of the ostriches included commercial poultry feed and reportedly also uncooked waste products from animals that had been the subject of emergency slaughter. Speculation with regard to the latter aside, it cannot be excluded therefore that all of these birds had access to feed containing meat-and-bone-meal (MBM) from commercial sources in Europe that might have contained the BSE agent.

Domestic poultry in the UK were undoubtedly exposed to the BSE agent [3]. They consumed the same ruminant protein that gave rise to the BSE epidemic in cattle, but there has been no evidence of an epidemic in these species, although it might be argued that few would survive to an age at which they might be expected to show clinical signs. Early in the course of the epidemic, the possibility that BSE could be transmitted to poultry and that the use of poultry offal in poultry feed could have propagated infectivity within the poultry industry was considered. As a result, some feed companies discontinued using poultry offal and feather meal in poultry feeds from the late 1980s onward. Others continued to use these ingredients until the use of mammalian MBM in livestock feed was banned in 1996.

Despite the 1996 ban in the UK, the feeding of mammalian MBM to pigs and poultry remained legal in other countries of Europe. Since January 2001 [3], use of all processed mammalian protein in feeds for farmed animals has been banned throughout the European Union, with periodic adjustments, but these materials continue to be used in pig, poultry, and indeed ruminant feeds in many other parts of the world. While specific legislation is not applicable to zoo animals, voluntary precautions will have been taken in many countries with a geographical risk for occurrence of BSE.

In mammals, to varying degrees, a significant, but not sole, risk factor associated with susceptibility to a TSE is the occurrence of certain polymorphisms of the PrP gene. The existence of a prion protein in birds has been demonstrated, although research has shown that it has a very low level of homology, of approximately 30 %, to that found in mammals [5; 6], but this work is based on a small number of tested birds (see Chapter 9).

There is no evidence that poultry under field conditions [3], or other birds, particularly necrophageous species [12], could, after oral exposure, act as healthy carriers in the spread of TSE agents, but the hypothesis has been considered.

The European Commission [12] concluded from a review of the world literature and other available data that there was no evidence of the existence of a naturally occurring TSE in poultry.

25.10 Surveillance

With the current state of knowledge as to the possibility of birds having, or acquiring, prion disease, it would seem impractical and unwarranted to conduct any specific monitoring of bird populations. It would also be advisable, as in mammals, through routine veterinary diagnostic approaches to investigate unusual nervous disease presentations and/or laboratory-diagnosed encephalopathies of uncertain etiology in domestic and zoo avian species. A study of the brains of a sample of healthy birds and mortalities of farmed ostriches in Italy did not reveal any significant changes [11].

References

[1] Schoon HA, Brunckhorst D, Pohlenz J. Spongiforme Enzephalopathie beim Rothalsstrauss (*Struthio camelus*). Ein kasuistischer Beitrag. Tierärztl Prax 1991; 19:263–265.

[2] Schoon HA, Brunckhorst D, Pohlenz J. Beitrag zur Neuropathologie beim Rothalsstrauss (*Struthio camelus*) – Spongiforme Enzephalopathie. Verh Ber Erkrg Zootiere (Akademischer Verlag) 1991; 33:309.

[3] Matthews D and Cooke BC. The potential for transmissible spongiform encephalopathies in non-ruminant livestock and fish. Rev Sci Tech Off Int Epiz 2003; 22:283–296.

[4] Marsh RF, Burger D, Eckroade R, et al. A preliminary report on the experimental host range of the transmissible mink encephalopathy agent. J Infect Dis 1969; 120:713–719.

[5] Gabriel JM, Oesch B, Kretzschmar HA, et al. Molecular cloning of a candidate chicken prion protein. Proc Natl Acad Sci U S A 1992; 89:9097–9101.

[6] Wopfner F, Weidenhöfer G, Schneider R, et al. Analysis of 27 mammalian and 9 avian PrPs reveals high conservation of flexible regions of the prion protein. J Mol Biol 1999; 289:1163–1178.

[7] Narang H. Failure to confirm a TSE [transmissible spongiform encephalopathy] in chickens. Vet Rec 1997; 141:255–256.

[8] Cawthorne RJG. Failure to confirm a TSE [transmissible spongiform encephalopathy] in chickens. Vet Rec 1997; 141:203.

[9] Klomburg S. Ungeklärte Todesfälle bei Antilopen und Straussen – Fallberichte. Tagungsbericht 14. Arbeitstagung der Zootierärzte im

deutschsprachigen Raum. November 4–6, Bochum, 1994:30–31.

[10] Samberg Y, Hadash DU, Perelman B, et al. Newcastle disease in ostriches (*Struthio camelus*): field case and experimental infection. Avian Pathology 1989; 18:221–226.

[11] Bozzetta E, Casalone C, Caramelli M, et al. Neuropathological findings in ostriches with specific reference to transmissible spongiform encephalopathies in Italy. XXXVIII Covegno della societa Italiana di Patologia Aviare. Forli, Italy, September 30 – October 1, 1999. Selezione-Veterinaria 2000:783.

[12] European Commission. Opinion on necrophagous birds as possible transmitters of TSE/BSE. Adopted by the Scientific Steering Committee at its meeting of November 7–8, 2002. Webpage 2006: www.europa.eu.int./comm/food/fs/sc/ssc/out295_en.pdf.

Topic IV: Pathology

26 Pathology and Genetics of Human Prion Diseases

Hans Kretzschmar and Piero Parchi

26.1 Introduction

Human prion diseases are grouped into three etiologic categories (Table 26.1). The largest one is generally referred to as "sporadic" and includes sporadic Creutzfeldt–Jakob disease (sCJD) and sporadic fatal insomnia (SFI). As the etiology of this group is not clear, it should be designated "idiopathic". It has been hypothesized that sporadic CJD may be a spontaneous disease, a concept that can be explained by the prion theory, but is not proven. Acquired prion diseases are accidentally transmitted from infected humans as in iatrogenic CJD (iCJD), by ritual cannibalism as in kuru, or from BSE-infected animals as in variant CJD or vCJD. The third group encompasses genetic or familial diseases, which are categorized according to their predominant clinical and pathological features as familial CJD (fCJD), Gerstmann–Sträussler–Scheinker disease (GSS), and fatal familial insomnia (FFI). These are associated with a large number of mutations of the prion protein gene (*PRNP*) and may be clinically and pathologically indistinguishable from idiopathic CJD or may have distinct clinical and neuropathological characteristics.

Historically it took decades for the concept of human prion diseases to develop. The first cases were described as "spastic pseudosclerosis" [1]. Walter Spielmeyer later suggested the name of "Creutzfeldt–Jakob Krankheit (disease)" for this group of diseases [2]. In the early 1920s they were interpreted as degenerative [1; 3; 4] or hereditary disorders [5]. The conceptual link with infectious diseases came about in the 1960s, when kuru was shown to be a transmissible disease similar to scrapie [6] and Igor Klatzo recognized similarities between kuru and CJD [7]. Today the clinical, pathological, and bio-

Table 26.1: Human prion diseases.

	Disease	Etiology
Idiopathic	Sporadic Creutzfeldt–Jakob disease (sCJD)	Unknown. Possibly spontaneous conversion of PrP^{C} into PrP^{Sc} or somatic *PRNP* mutation
	Sporadic fatal insomnia (SFI)	Unknown. Possibly spontaneous conversion of PrP^{C} into PrP^{Sc} or somatic *PRNP* mutation. Found only in 129 MM patients with PrP^{Sc} type 2
Hereditary	Familial CJD (fCJD) / Genetic CJD (gCJD)	Various *PRNP* mutations
	Gerstmann–Sträussler–Scheinker disease (GSS)	Various *PRNP* mutations
	Fatal familial insomnia (FFI)	*PRNP* mutation D178N with M129
Acquired	Iatrogenic CJD (iCJD)	Accidental transmission through treatment with prion-contaminated preparations of human growth hormone, dura mater grafts etc.
	(new) Variant CJD, (n)vCJD	Infection by bovine prions (BSE-contaminated food or other products)
	Kuru	Infection through ritualistic endocannibalism in the Fore population in Papua New Guinea (historical)

chemical changes, in particular the various forms of human PrP^{Sc} in human prion diseases, have been characterized extensively and a number of distinct subgroups have emerged that will be the basis of future epidemiology and surveillance [8]. At present, most phenotypic variants of idiopathic CJD have been associated with a particular form of PrP^{Sc} [9].

Pathologically, the common denominator of all human prion diseases is the presence of spongiform change as well as protease-resistant PrP. Spongiform changes may be widespread and severe but in some cases, in particular FFI, may be very subtle or may even be absent.

26.2 Neuropathologic features of prion diseases in humans

26.2.1 Light microscopic morphologic features

Macroscopic findings in human prion diseases range from inconspicuous in some cases to moderate atrophy of the cerebral cortex, the basal ganglia, and the cerebellum, to the most severe cerebral and cerebellar atrophy in other cases, in particular those with a long clinical course.

Neuronal loss, spongiform change, and astrocytic gliosis are the defining features of prion diseases observed in light microscopy. Neuronal loss may at times be difficult to assess without systematic morphometry or may be most severe with practically no discernible neurons left in the cerebral cortex and basal ganglia as well as extreme loss of cells of the internal granular layer of the cerebellum. Although the hippocampus proper, areas CA1–CA4, and the fascia dentata are spared from neuronal loss and spongiform change in the majority of cases, they are affected to some degree in CJD brains of VV1, VV2, and MV2 patients (see below). The thalamus shows variable degrees of atrophy and is most severely affected in FFI and SFI, in particular the ventral anterior and dorsomedial nuclei. In the cerebellum, Purkinje cells appear normal in the vast majority of CJD brains even with severe atrophy of the internal granular layer, but they sometimes show shrinkage and axonal swellings (torpedoes). In H&E-stained section, brainstem pathology is inconspicuous in most cases but shows spongiform change, nerve cell loss, and gliosis in the midbrain tectum and substantia nigra in VV2, MV2, and to a lesser degree in MM2 cases (see below).

Vacuolar changes comprise a number of descriptive forms. Spongiform change consists of small, round, sometimes opaque vacuoles of 2–5 μm in diameter (Fig. 26.1). Spongiform change may be found in the cerebral and cerebellar cortex, in the basal ganglia, thalamus, and brain stem. It is found in almost all cases of human prion diseases, but may be restricted to very small areas such as the entorhinal cortex in FFI and SFI and may be practically absent in cases with severe status spongiosus (see below). It should be noted that changes indistinguishable from those seen widely distributed in CJD can also be found restricted to the temporal lobe in cases of dementia with Lewy bodies (DLB). Typical spongiform change must be distinguished from "confluent vacuoles" that are seen in MM2-cortical cases (MM2-C) or MM cases with a mixture of PrP^{Sc} type 1 and 2. Confluent vacuoles are observed in the cerebral neocortex and basal ganglia; they are absent from the cerebellum and brainstem as well as from the allocortex. *Status spongiosus* was defined by Masters and Richardson in 1978 and was described as being "... characterized by the appearance of cavitation of the neuropil in the presence of a dense glial meshwork. The cavitation was situated between glial fibers, and the cavities were of irregular size and shape. This lesion corresponds to what was originally described by Probst (1903), Fischer (1911) and Spielmeyer (1922)" [10]. Status spongiosus is not specific for CJD, it is also occasionally observed in other degenerative diseases, in particular Pick's disease and Alzheimer's disease, and is seen in end-stage metabolic diseases including Leigh syndrome, and mitochondrial disorders, in particular MELAS. It should be distinguished from the laminar necrosis seen in anoxic-ischemic encephalopathies and boundary zone infarcts.

Spongiform change, confluent vacuoles, and status spongiosus must be distinguished from other types of vacuolar changes, in particular spongy degeneration of the 2nd cortical layer in

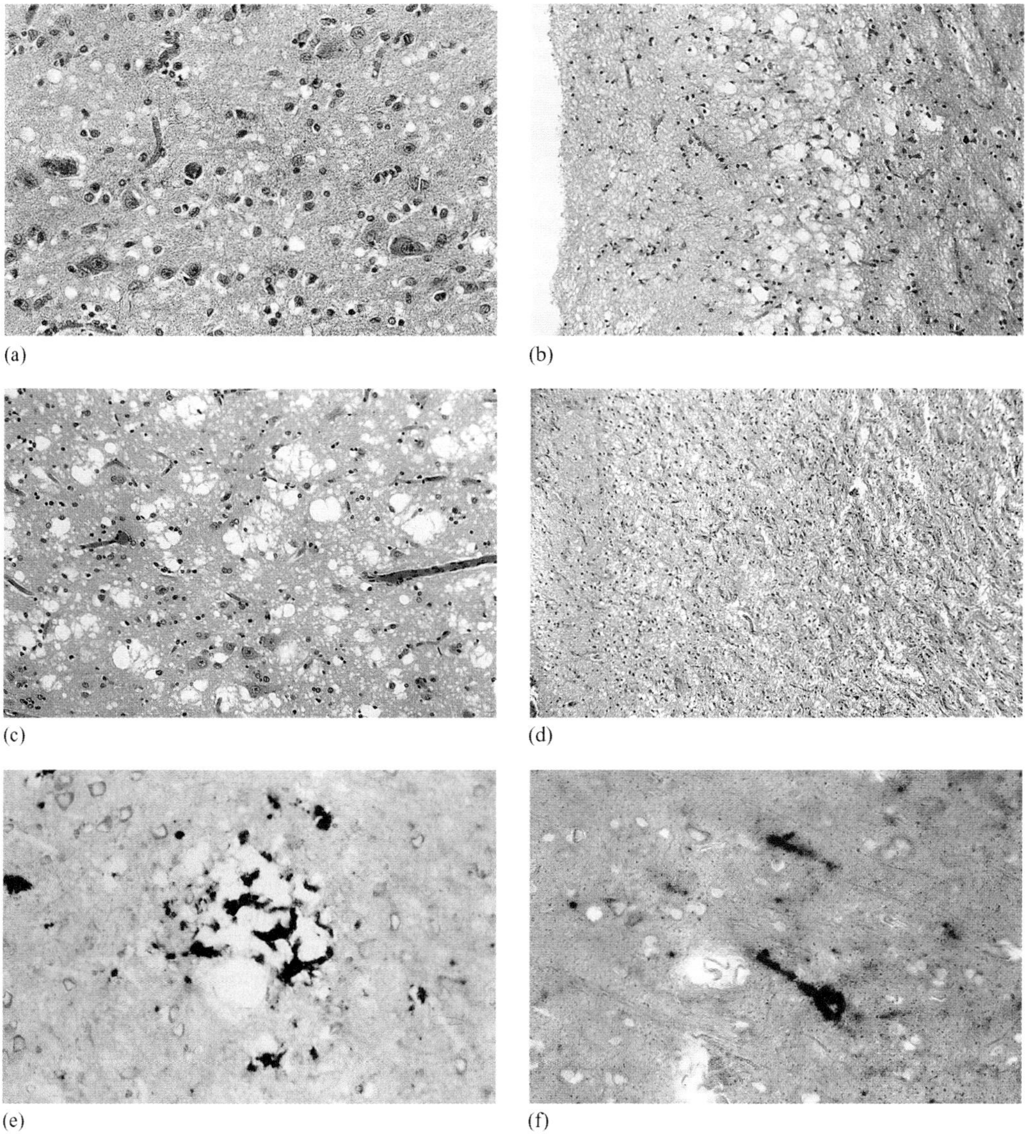

Fig. 26.1: Light microscopic changes in human prion diseases. **(a)** Spongiform change in the cortex of a sporadic CJD (sCJD) case (MM1). There are fine, partially opaque vacuoles in the neuropil. H&E stain, × 40 (original magnification). **(b)** For comparison the spongy degeneration in the second cortical layer in a case of Alzheimer's disease is shown. This type of vacuolar change in the upper cortical layers is seen in neurodegenerative diseases with severe nerve cell loss in the cortex. H&E stain, × 20 (original magnification). **(c)** Confluent vacuoles in a case of sCJD. This type of confluent vacuolation is most often observed in MM2 cases and is also seen in MM1 cases with focal PrP^{Sc} type 2. H&E stain, × 40 (original magnification). **(d)** Status spongiosus. There is almost complete neuronal loss with severe astrocytic gliosis and numerous shrinkage clefts in the remaining neuropil of the cortex of a CJD patient. H&E stain, × 20 (original magnification). **(e)** Confluent vacuoles with perivacuolar PrP^{Sc} staining. Typically dense deposits of PrP^{Sc} are seen surrounding confluent vacuoles. IHC with antibody L42, × 40 (original magnification). **(f)** Perineuronal PrP^{Sc} staining, which often includes neuronal staining, is typically observed in VV2 cases. IHC with antibody L42, × 40 (original magnification). Figure by H. Kretzschmar.

Table 26.2: Neuropathology of spongy change. The nomenclature shown here is used by most neuropathologists with minor variations.

Spongy change	Morphology	Observed in
Spongiform change	Small (2–10 μm), round to oval, vacuoles, located predominantly in neuronal processes	Prion diseases (all subtypes)
Confluent vacuoles	Clusters of coalescing large (10–50 μm) vacuoles	CJD, in particular MM2 (Table 26.3)
Status spongiosus	Almost complete nerve cell loss, severe astrocytic gliosis, tissue rarefaction, large pericellular clefts	Prion diseases, end-stage neurodegenerative and metabolic diseases
Spongy degeneration of the 2nd cortical layer	Vacuolar clefts in the 2nd cortical layer due to tissue shrinkage	Common in late stages of neurodegenerative diseases with severe astrocytic gliosis (Pick disease, Alzheimer's disease, MND with dementia, etc.)
Other spongy changes	Irregular tissue rarefaction with vacuoles, edematous vacuoles	Encephalitis, post-hypoxic states

late stages of a variety of neurodegenerative diseases with severe cortical atrophy, including Alzheimer's disease (AD), Pick disease, motor neuron disease (MND) with dementia, and prion diseases (Table 26.2).

26.2.2 Pathogenesis

There is now detailed knowledge on the phenotypes of human prion diseases and the correlation with genetics and the form of PrP^{Sc} (see below). In a large number of human disease cases there is quite good correlation of the clinical appearance and the pathological changes in various areas of the brain. For example, the predilection for the neocortex and in particular the occipital cortex in the Heidenhain variant, tallies with the clinically observed dementia and cortical blindness. Although the wide phenotypic variability of human prion diseases has been related to the existence of different prion strains, capable of transmitting diseases to syngenic animals that differ in topography, type of lesion and intracerebral distribution of the protease-resistant PrP (PrP^{Sc}), there is still little understanding as to why certain areas of the brain or certain neuronal cell types should be more susceptible to damage than others and why this should correlate with the particular molecular form of PrP^{Sc}.

A disturbance of inhibitory GABAergic mechanisms in prion diseases has been postulated by various authors who noted a marked reduction in parvalbumin-positive (PV+) cells and morphological changes of the remaining PV+ neurons in the cerebral cortex and hippocampus [11; 12]. Parvalbumin is a cytosolic calcium-binding protein implicated in the buffering and transport of Ca^{++} as well as in the regulation of various enzyme systems. In the isocortex and hippocampus it is predominantly expressed in a subset of fast-spiking inhibitory GABAergic interneurons.

However, the pathogenesis underlying some of the key features in CJD such as periodic sharp wave complexes (PSWC) in electroencephalogram (EEG) tracings and myoclonus are not well understood. Since there is electrophysiological evidence that the thalamus, in particular the reticular thalamic nucleus, serves as a pacemaker generating highly synchronous electric activity, an attempt was made to correlate damage to PV+ neurons in various thalamic nuclei with clinical changes. PV+ neurons were significantly reduced in the ventral tier thalamic nuclei, and the mediodorsal and

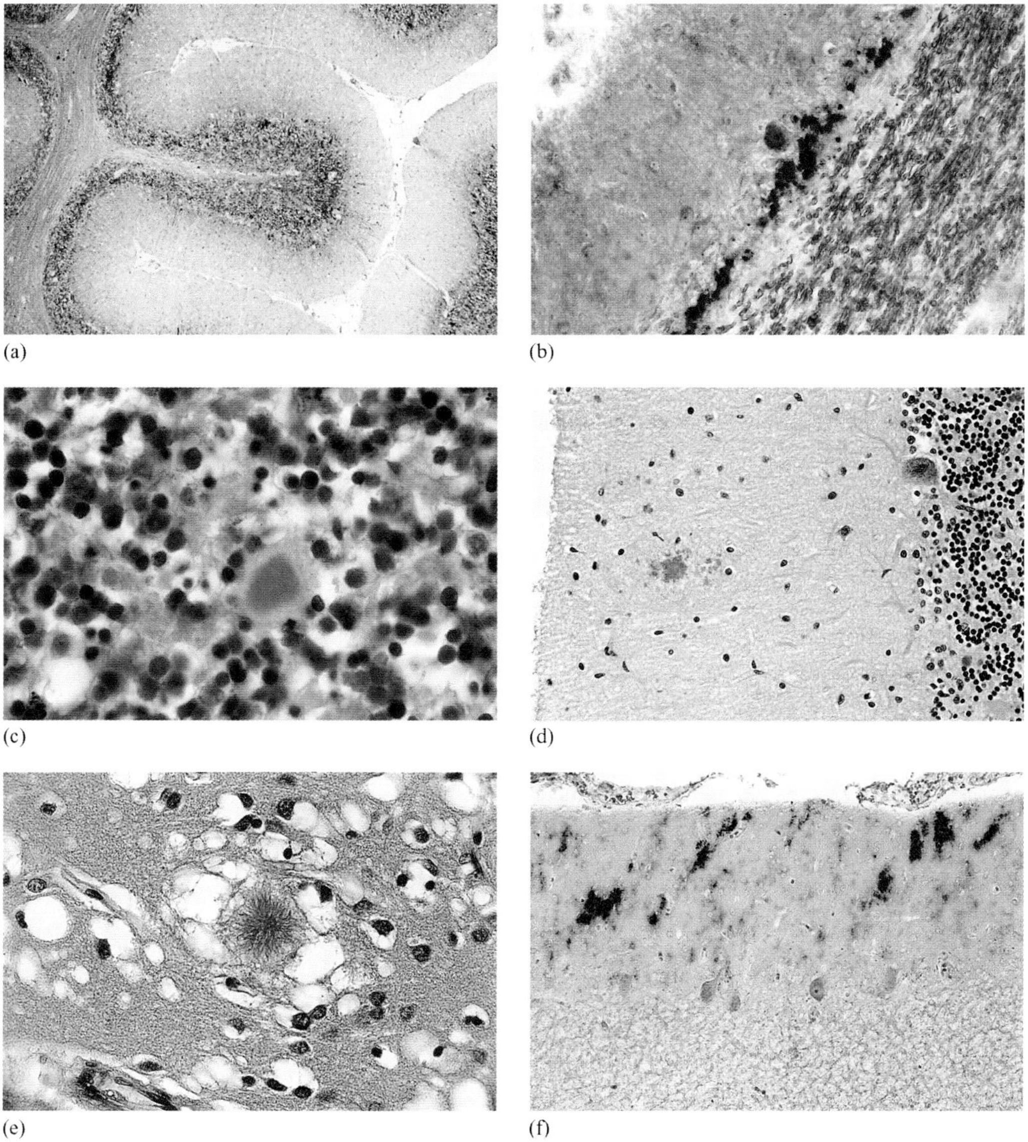

Fig. 26.2: Immunohistochemistry in human prion diseases. **(a)** Synaptic PrP^{Sc} staining. Delicate PrP^{Sc} staining is seen in the molecular and granular cell layers in the cerebellum. IHC with antibody L42, × 10 (original magnification). **(b)** Plaque-like deposits in the cerebellum of a VV2 sCJD case. Plaque-like deposits are not visible in routine H&E- or PAS-stained sections. IHC with antibody L42, × 20 (original magnification). **(c)** A kuru plaque is shown in the cerebellar granular cell layer in a sCJD case (MV2). H&E stain, × 63 (original magnification). **(d)** Multicentric plaque in the cerebellar molecular layer of a GSS case. In humans this type of PrP plaque is only seen in genetic disease and mutations of *PRNP*. H&E stain, × 20 (original magnification). **(e)** A florid plaque in a case of vCJD (tissue provided by J. Ironside, University of Edinburgh). These plaques consist of an eosinophilic, sometimes fibrillar core surrounded by a region of radiating fibrils. This core structure is encircled by a rim of spongiform change. In human prion disease this form of PrP plaque is seen in large numbers in all vCJD cases and has been observed in small numbers in some cases of iCJD after dura mater implantation. H&E stain, × 63 (original magnification). **(f)** PrP^{Sc} IHC in the cerebellum of a genetic CJD case (insert mutation) showing rare patchy staining in the molecular layer. IHC with antibody L42, × 20 (original magnification). Figures by H. Kretzschmar ©.

reticular nuclei. Patients with typical EEG changes (PSWC) and myoclonus had a predominant loss of PV+ cells in the reticular thalamic nucleus [13]. Assuming that the reticular thalamic nucleus controls the genuine electric activity of the thalamus by the action of PV+ inhibitory neurons [14] damage to PV+ cells would reduce inhibition, which in turn would contribute to the generation of PSWC.

PrP^{Sc} depositions: Light microscopy of H&E-stained sections shows tissue deposits of PrP^{Sc} only in kuru plaques and multicentric plaques. Various other forms of PrP^{Sc} depositions can be visualized using immunohistochemistry (IHC) and paraffin-embedded tissue (PET) blotting.

Kuru plaques. Compact PrP^{Sc} depositions that are visible in routine H&E- or PAS-stained sections (Fig. 26.2) are termed kuru plaques, whereas plaque-like deposits are much smaller and are visible only on IHC staining with antibodies against PrP. Kuru plaques are pathognomonic for prion diseases but are seen only in a minority of sCJD cases, where they are almost exclusively associated with heterozygosity for methionine and valine at codon 129 of *PRNP* and PrP^{Sc} type 2 [8]. In addition they are present in GSS, vCJD, and kuru. They are most commonly seen in the internal granule cell layer in the cerebellum and are difficult to find in the neocortex without IHC staining.

Multicentric plaques. In GSS, multicentric plaques are regularly found. Most often they consist of a central large plaque that is surrounded by smaller "satellites" (Fig. 26.2). Other forms of multicentric plaques consist of small groups of tiny plaques or cloudy accumulations of densely packed PrP^{Sc}. These have also been termed cluster plaques.

Florid plaques. A central core surrounded by a ring of spongiform changes (Fig. 26.2) characterizes florid plaques. In human prion diseases they are encountered in high density in vCJD but are also seen in small numbers in a minority of iCJD cases after dura mater implantation.

Kuru plaques, multicentric plaques, and florid plaques can be seen in H&E-stained sections and have been shown to contain PrP^{Sc} by using IHC with antibodies against PrP. In addition a number of quite characteristic PrP^{Sc} deposits are only visible using IHC, i. e., plaque-like deposits, "synaptic" PrP staining, and perivacuolar as well as perineuronal staining. Immunohistochemical techniques for PrP^{Sc} do not differ significantly from other IHC applications. However, it is of paramount importance to use denaturing tissue pretreatments such as immersion in guanidium hydrochloride, autoclaving in H_2O, and hydrolytic autoclaving in a solution of 10 mM HCl, or a combination of these pretreatments. Many groups use pretreatment with formic acid for enhancement of PrP^{Sc} staining [15] and tissue decontamination [16] at the same time. Pretreatment with proteinase K (PK) is not regularly used. It often leads to enhanced PrP staining but the tissue dissolves in the process. PrP^{C} is not seen after the described tissue pretreatment, therefore PK pretreatment is not necessary for a distinction from PrP^{Sc}. Immunohistochemically the following forms of PrP^{Sc} can be observed:

1. Plaque-like depositions consist of small accumulations of PrP^{Sc} not visible in H&E-stained sections and are primarily seen in VV2 cases (see below; Fig. 26.2).
2. Synaptic staining is a form of delicate and often hazy PrP^{Sc} deposition that was first described by Kitamoto [17]. It is commonly found in the molecular and granular cell layers in the cerebellum and in the isocortex in MM1 and MV1 cases (Fig. 26.2).
3. Perivacuolar staining is associated with large confluent vacuoles (Fig. 26.1) typical of MM2 sCJD and in many MM1 and MV1 cases in addition to synaptic staining.
4. Perineuronal PrP^{Sc} accumulations are typically found in VV2 sCJD cases (Fig. 26.1).

26.2.3 IHC and PET blotting features

In diagnostic neuropathology, IHC with antibodies against PrP is used in conjunction with a number of pretreatment protocols that in one way or another abolish PrP^{C} immunostaining. Tissue pretreatments include immersion in 4 M guanidium isothiocyanate at 4°C for 2 h, hydrated autoclaving at 121 °C for 10 min, immersion in formic acid for 5 min and others. Immunohistochemistry with various antibodies against PrP and varying pretreatment regimes is satisfactory for prion disease diagnosis in the vast majority of human prion diseases.

That antibodies do not stain PrP^C is clearly shown by the fact that staining in control cases is negative. But what do PrP antibodies really stain in human prion disease cases? This is by no means clear. PK treatment of formalin-fixed tissue sections is possible. A great deal less PrP is seen in tissue sections after PK treatment than after the regimes described above – which might be either due to the fact that only a part of the PrP seen in usual IHC is PK resistant or the fact that PK treatment leads to generalized loss of tissue on the glass slides, including partial loss of PrP of any kind.

To circumvent this problem, Steve DeArmond invented a technology that makes possible the visualization of PK resistant PrP after transferring tissue sections to a nitrocellulose membrane, the histoblot technique. This technique is very sensitive, depicting only PK-resistant PrP, and has been used extensively in animal research [18]. It has two disadvantages, i. e., it requires unfixed tissue and the optic resolution is inferior to IHC.

We further developed the histoblot technique for the use of formalin-fixed and paraffin-embedded tissues which had hitherto been considered impossible. The PET blot was found to be easy to perform and reliable [19]. In an experimental animal system we showed that it is as sensitive as the Western blot and allows satisfactory anatomic resolution, while the latter is inferior to conventional light microscopic IHC. Also, the PET blot shows no particular advantage over IHC in cases that have plaques or plaque-like PrP deposits or strong synaptic-type staining.

Given its relatively high sensitivity, the PET blot often shows positive results in two situations in which IHC is often very weak or sometimes absent, i. e., FFI (SFI) and VV1 cases of sCJD.

26.2.4 Electron microscopy

Electron microscopic studies using human tissues are impeded by post mortem changes. High quality ultrastructural investigations have been performed with tissues from experimentally infected rodents and nonhuman primates [20]. In murine scrapie models, ultrastructural vacuoles are primarily found in neurites. They are less frequently found in axons, axon terminals and neuronal perikarya, in astrocytes, oligodendrocytes, and in myelin. The vacuoles may be surrounded by a membrane, a double membrane, or may be without a membranous boundary. The pathogenesis of vacuolar structures in prion diseases is still unknown; they may originate in the smooth endoplasmic reticulum or other subcellular organelles such as mitochondria. The pathogenic significance of vacuolar change in prion diseases is also unknown. A number of studies have shown that prion replication and infectivity may precede spongiform change, that areas without spongiform change may be infectious, and that vacuolar change is not seen in certain experimental systems [21–23]. Vacuoles indistinguishable from those seen in prion diseases in animals may be observed in prematurely aging mice, in rabies in skunks and foxes [24], and in retroviral infections of mice. Interestingly, depletion of endogenous neuronal PrP^C by conditional PrP^C knockout in mice with established neuroinvasive prion infection reversed early spongiform change [25]. In human neuropathology, spongiform change may be practically absent in some FFI cases and other familial prion diseases.

Tubulovesicular structures were first described in scrapie-infected mice in 1968 [26] and were also later observed in CJD [27]. They measure 30–35 nm in diameter and are found in axon terminals and preterminals and dendrites. According to the published data, tubulovesicular structures appear to be disease specific, however, their relevance has remained elusive. They do not seem to contain PrP [28].

26.3 PrP^{Sc} in non-neuronal tissues

In human prion disease, PrP^{Sc} has been shown to be present in the CNS including the retina [29] by IHC, Western blotting, and the PET blot. Reports on PrP^{Sc} or infectivity in various other tissues must be considered in more detail, particularly in respect to findings in vCJD or sCJD.

In vCJD, PrP^{Sc} was noted in neurons of sympathetic ganglia [30]. This observation may indicate the involvement of the sympathetic nervous system in prion propagation in humans after oral uptake. The presence of PrP^{Sc} in tonsils [31] and other lymphoreticular organs including lymph follicles in the appendix [32] has been used for the in vivo diagnosis and epidemiology [33] of vCJD. It is very likely that the blood in vCJD patients may carry infectivity [34], however, it has not been possible to detect PrP^{Sc} in the blood of humans.

By using differential precipitation with sodium phosphotungstic acid, which increases the sensitivity of Western blot analysis for PrP^{Sc} by up to three orders of magnitude, PrP^{Sc} was detected in 10 of 28 spleen specimens and in 8 of 32 muscle samples in sCJD cases from Switzerland [35].

PrP^{Sc} was also identified in intracranial blood vessel walls in vCJD and sCJD by IHC and PET blotting [36]. It was located primarily in the media of deep perforating arteries and occasionally of meningeal and basal arteries. PrP^{Sc}-bearing cells were HLA-DR and S-100 immunoreactive cells in the intima that are components of the vascular-associated dendritic cell network as well as HLA-DR and CD-68 immunoreactive macrophages of the intima and media. Vascular PrP^{Sc} was found only in a minority of sCJD and vCJD cases while the authors of that study could not show its presence in four iCJD cases. Delicate granular PrP immunoreactivity was also found in the extracranial carotid artery and ascending aorta in a small number of vCJD patients. The authors of the study suggest that PrP^{Sc} deposits in intracranial vessel walls may represent PrP^{Sc} derived from the periarterial interstitial fluid, which joins the cerebrospinal fluid (CSF) that may harbor PrP^{Sc}. This may be similar to AD, where β-amyloid, often seen in perivascular locations, has been suggested to follow interstitial fluid drainage along the periarterial space of intracerebral leptomeningeal arteries and the intracranial carotid to join deep cervical lymph nodes. It is suggested that the presence of PrP^{Sc} in vascular-associated dendritic cells might possibly play a role in neuroinvasion of prion disease. At present it seems difficult to know the importance of these findings for prion pathogenesis.

Paul Brown reported transmission of disease to nonhuman primates from extraneuronal tissues of patients with sCJD, iCJD, and kuru including occasional transmissions from the lungs, liver, kidneys, spleen, and lymph nodes [37].

26.4 The human prion protein and prion protein gene (*PRNP*)

Human PrP^C is a glycoprotein of 253 amino acids before cellular processing [38]. There is an 85–90 % homology to PrP of other mammalian species. PrP^C is a membrane protein expressed mainly in neurons, but also in astrocytes and a number of other cells [39; 40]. It has an N-terminal signal sequence of 22 amino acids, which is cleaved off the translation product. Twenty-three terminal amino acids are removed when glycosylphosphatidylinositol (GPI) is attached to serine residue 230. Mature PrP^C is attached to the cell surface by this GPI anchor and undergoes endocytosis and recycling. It seems, however, that PrP may exist in alternative membrane topologies (ctmPrP and ntmPrP), whose implications for the function of PrP^C and its role in pathogenesis are just beginning to be elucidated. There are two N-glycosylation sites that are glycosylated differently in different human CJD variants (see below). The N-terminal moiety of the protein contains an octapeptide repeat [(PHGGGWGQ) × 4], which has been suggested to function in copper binding [41–44].

Whereas PrP^C is found on the surface of many cell types in all mammals and birds studied so far, PrP^{Sc} is generated from PrP^C in a post-translational process and is closely associated with infectivity. In terms of the prion hypothesis, it is part and parcel of the infectious agent, the

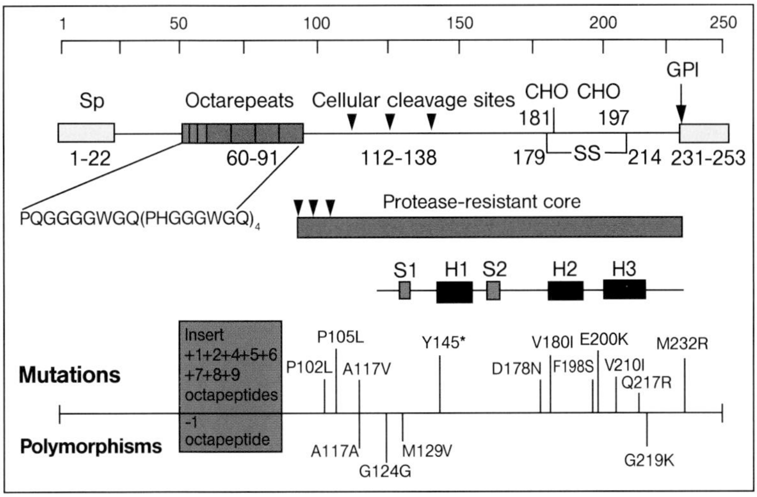

Fig. 26.3: Structural features of human prion protein (PrP) and the human prion protein gene (*PRNP*). Shown are the cellular isoform of PrP (PrP^C), the protease-resistant core of PrP^{Sc}, features of the secondary structure of PrP^C, and some of the known mutations and polymorphisms of *PRNP*. Abbreviations: SP signal peptide; GPI glycosyl-phosphatidylinositol; CHO glycosylation sites; H1, H2, H3 the three α-helices (Fig. 6.6A); S1, S2 β-pleated sheets. Figure by H. Kretzschmar.

prion. NMR structural studies have shown that the C-terminal half of PrP^C contains a two-stranded antiparallel β-sheet (S1 and S2) and three α-helices, whereas the N-terminal moiety is thought to have no definite structure in aqueous solution. Association with membranes seem to impart a specific structure on PrP^C [45]. PrP^C and PrP^{Sc} seem to differ mainly in their folded structures. PrP^{Sc} purified from hamster brain consisted of 42 % α-helical and only 3 % β-sheet structure, whereas PrP^{Sc} purified from scrapie-infected hamster brain is composed of 30 % α-helix and 43 % β-sheet. PrP^{Sc} shows increased protease resistance, is insoluble in aqueous solution, and tends to form fibrils that show birefringence after binding of Congo red.

The human PrP gene (*PRNP*) is located on the short arm of chromosome 20. It has a simple genomic structure and consists of two exons and a single intron 13 kb in length. The entire protein-coding region is located in exon 2. In families with inherited prion diseases, different point mutations and insertion mutations have been described in the open reading frame of *PRNP* (Fig. 26.3). The insertional mutations are situated in the N-terminal half of the protein in an octapeptide repeat region, whereas the point mutations cluster in the central and C-terminal regions of the protein.

The common polymorphism at amino acid position 129 of the prion protein, where humans carry a methionine (M) or valine (V) allele, clearly influences susceptibility to the sporadic and iatrogenic types of prion diseases and, furthermore, determines in part, the phenotype of the sporadic, as well as of some of the inherited prion diseases. Several studies have revealed a marked over-representation of homozygotes (mainly for methionine) at this position in cases of sCJD compared to the normal population. CJD homozygotes at codon 129 also show a shorter incubation time in iCJD. There is also a strong correlation of codon 129 genotype and clinicopathological phenotype (see below).

26.4.1 Western blotting in the diagnosis of human prion diseases

Several lines of evidence indicate that human PrP^{Sc} exists in a number of molecular subtypes that show differences in their glycosylation pattern, conformation, and degree of protease resistance. However, the true nature of this phenomenon is still unknown. In particular, the question of whether the PrP^{Sc} varieties encipher the human prion strains by themselves or rather reflect differing PrP molecular interactions, remains elusive.

Whatever the case, there is wide agreement that each particular PrP^{Sc} variety correlates with the clinical and neuropathological features of the disease [8]. It has been convincingly shown that in human prion disease there is a very strong correlation between the particular physicochemical properties of PrP^{Sc} and the pathological and clinical disease phenotypes. It has further been shown that PrP^{Sc} heterogeneity involves both the site of protein cleavage by proteases and the degree of protein glycosylation.

Based on the electrophoretic mobility of the core fragments of PrP^{Sc} after PK digestion, two major human molecular PrP^{Sc} types can be distinguished [46; 47]. PrP^{Sc} type 1 has a relative electrophoretic mobility of 21 kDa and a primary cleavage site at residue 82 while PrP^{Sc} type 2 has a relative molecular mass of 19 kDa and a primary cleavage site at residue 97 [48] (Fig. 26.4). PrP^{Sc} types 1 and 2 characterize all subtypes of CJD, as well as kuru, independent of the apparent etiology of the disease, i. e., sporadic, inherited or acquired by infection [46; 48], thus suggesting a common mechanism of PrP^{Sc} formation in all forms of prion disease.

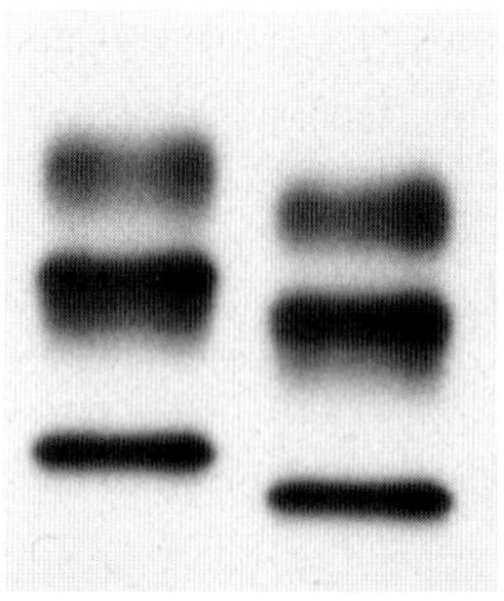

Fig. 26.4: PrP^{Sc} types 1 and 2 in sporadic CJD. Western blot of proteinase K-treated brain homogenates stained with the 3F4 monoclonal antibody, which recognizes the PrP residues 109–111. The three bands are composed of (from top to bottom) diglycosylated PrP, monoglycosylated PrP, and unglycosylated PrP. The figure shows two distinct patterns of human PrP^{Sc}, which differ with respect to the electrophoretic mobility of the protein core. Type 1 (lane 1) has a relative molecular mass of 21 kDa, whereas type 2 (lane 2) has a mass of 19 kDa. The mobility of the core varies between the two PrP^{Sc} types because different numbers of amino acids are removed from the N-terminal fragment of PrP by proteinase K digestion. Figure by P. Parchi.

Based on glycosylation, both PrP^{Sc} types 1 and 2 can be further distinguished in at least two subtypes. In the large majority of cases PrP^{Sc} glycosylation is characterized by an overrepresentation of the monoglycosylated form. This type of glycosylation has been named pattern A to be distinguished from pattern B, which is found in familial CJD and FFI linked to the E200K or D178N mutations [48; 49] or in variant CJD [46; 50] and is characterized by a predominance of the fully glycosylated form. Since glycosylation is a co-translational process that is known to differ between cell types, the different glycoform ratios may reflect the involvement or the targeting of distinct neuronal populations. Alternatively, different prion strains may preferentially convert certain PrP^{Sc} glycoforms. However, in inherited prion diseases, the glycoform ratio of PrP^{Sc} is also reflected by altered PrP^{C} processing induced by specific *PRNP* mutations, and therefore, it varies independent of the agent strain [51; 52]. The E200K and the D178M mutations represent 2 examples of such an influence on the PrP^{C} metabolism [51; 52].

Since the original proposal by Parchi et al. [46; 47], two alternative classifications of human PrP^{Sc} types have been published [50; 53; 54]. Both of them indicated that the sCJD MM1 subtype identified by Parchi et al. [46; 47] is heterogeneous and includes two distinct PrP^{Sc} isoforms and two phenotypes. The lack of a uniform classification of sCJD and a proper nomenclature has generated significant confusion in the field. However, recent insights into the biochemistry of PrP^{Sc} cleavage by proteases have solved these problems of diverging nomenclatures and molecular classification [9]. The most significant finding in this regard has been that pH variations among CJD brain homogenates in standard buffers influence PK activity as well as the size of PrP^{Sc} after PK cleavage. Thus, it has been convincingly shown that the PrP^{Sc} heterogeneity that had been used to identify putative subtypes of PrP^{Sc} type 1 in some studies [50; 53; 54] simply represents technical variations related to the lack of homogeneous experimental conditions.

On the other hand, the study of PrP^{Sc} in rigorous experimental conditions as well as the use of a more sensitive gel electropheresis technique demonstrated that PrP^{Sc} types 1 and 2 are indeed heterogeneous species, which can be further distinguished into molecular subtypes that fit the current histopathological classification of sporadic CJD in 6 subtypes. Using a standardized high buffer strength for brain homogenization, protease K digestion at pH 6.9 with a high PK concentration, and long running gels Notari et al. [9] have shown that specific PrP^{Sc} properties are associated with most of the six sCJD phenotypes as well as vCJD. For example, PrP^{Sc} from MV cases shows a unique doublet or triplet band that differs from other type 2 proteins and type 1 PrP^{Sc} from 129 valine homozygotes migrates faster than type 1 PrP^{Sc} from MM1 and MV1 samples when PK digestion is performed at a pH under 7.2 (Fig. 26.5).

Another interesting aspect of the molecular pathology of CJD concerns the finding of PrP^{Sc} types 1 and 2 associated in the same brain. Originally reported by Parchi et al. [8], the existence of CJD subjects with mixed molecular and pathological phenotypes has been confirmed in at least three studies [55–57]. Nevertheless, difficulties remain in both the understanding of the biological basis of this phenomenon as well as its relevance e. g., how many CJD cases have both PrP^{Sc} types. The issue has become particularly intriguing after two recent studies [58; 59] have claimed that each patient classified as CJD type 2 also shows at least some associated PrP^{Sc} type 1, thus pointing to a regular coexistence of multiple PrP^{Sc} types in patients with CJD. Although of interest, this interpretation must be taken with caution because, once again, the possibility that a PrP^{Sc} type 1-like signal due to an incomplete type 2 protease digestion is being erroneously interpreted as a real PrP^{Sc} type 1 signal has not been convincingly ruled out.

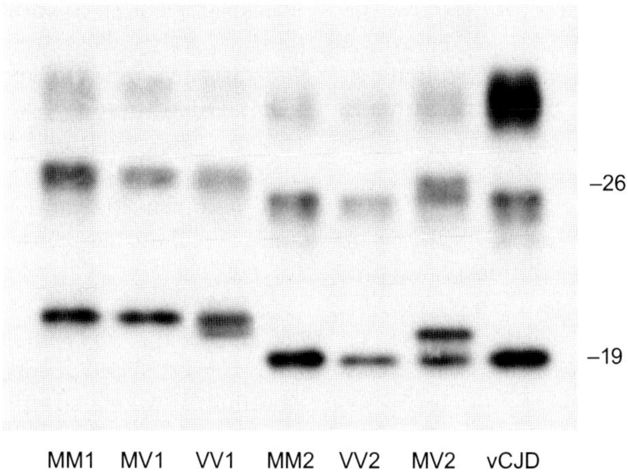

Fig. 26.5: Distinction of subtypes of PrP^{Sc} type 1 and type 2. Immunoblot analysis of frontal cortex homogenates from MM1, MV1, VV1, MM2-cortical, VV2, MV2, and vCJD subjects is shown. Homogenates were prepared in lysis buffer with 100 mM Tris at pH 6.9. Aliquots were digested with 2000 µg/ml of PK for 1 hour at 37 °C. Samples were run in a Tris-glycine PAGE 15% gel (15 cm long). Approximate molecular masses are in kilodaltons. Based on the analysis of PrP^{Sc} electrophoretic mobility it is possible to distinguish 4 out of 5 sporadic CJD pathological subtypes. Reproduced with permission from Notari et al. J Biol Chem 2004; 279:16797–16804 ©.

Finally, novel truncated PrP^{Sc} fragments of smaller size than types 1 and 2 have recently been identified in CJD [60]. Although more studies are needed to fully characterize the presence, characteristics, and biological significance of these peptides, preliminary results indicate that their search and characterization may be useful in the molecular distinction of CJD subtypes.

26.4.2 Polymorphisms of *PRNP*

A number of polymorphisms of *PRNP* have been described. There are silent polymorphisms with no recognizable influence on disease susceptibility or disease phenotype. Among these are the A117A (GCA→GCG) and G124G (GGC→GGG) silent changes [61; 62]. Loss of one octarepeat has been reported with no recognizable influence on disease susceptibility [63; 64], while the loss of two octarepeats has been reported in CJD [65; 66].

The polymorphism for methionine (ATG) or valine (GTG) at codon 129 of *PRNP* is a determining factor for susceptibility to acquired and idiopathic prion disease. In addition, the pathological and clinical features of idiopathic and genetic prion diseases are severely influenced by this polymorphism (see below). A number of studies have demonstrated that methionine homozygotes are more susceptible to sCJD, whereas heterozygotes are less susceptible [67–70]. In a large study of 300 sCJD cases and 544 controls in Europe and the USA in the normal population, 37 % were MM, 51 % were MV, and 12 % were VV. In contrast, among the sCJD cases 71.6 % were MM, 11.7 % were MV, and 16.7 % were VV [8]. All vCJD cases reported to date have been methionine homozygotes. PrP^{Sc} deposition in lymphoreticular organs has been reported in one MV patient who died of a ruptured aneurysm and who never showed clinical signs of vCJD. Five years before his death he had received a blood transfusion from a person who later had developed vCJD [71].

The phenotypes of sCJD, fCJD, and GSS are strongly influenced by this polymorphism. As an example, in sCJD plaque-like deposits and kuru plaques are only found in cases with at least one valine codon (for further details see Section 26.5.1). The D178N mutation of *PRNP* is associated with two different clinical and pathological phenotypes of familial prion disease dependent on the polymorphism at codon 129. D178N in coupling with methionine is associated with FFI whereas in coupling with valine (D178D-129V) this mutation causes fCJD. Codon 129 also influences the age of presentation in GSS associated with the F198S mutation [72; 73].

Heterozygosity at another polymorphic *PRNP* codon, E219K, may be associated with resistance to sCJD. This notion is based on a Japanese study in which 6 % of the population were found to show the polymorphism [74] and an additional study of 20 definite and 65 probable sCJD cases that were all homozygous EE [75]. However, a probable case of iCJD with the E219K polymorphism was identified in that study.

Heterozygosity is thought to confer resistance to disease transmission by inhibition of homologous protein interaction [68]. In a study on the survivors of the kuru epidemic who had had multiple exposures to endocannibalistic feasts, Mead and coworkers found predominantly *PRNP* 129 heterozygotes, in marked contrast to younger unexposed Fore [76]. Kuru seems to have imposed a strong balancing selection on the Fore and may have eliminated 129 homozygotes in the Fore population that had partaken in cannibalistic rites. To the authors of that study the worldwide *PRNP* haplotype diversity and coding allele frequencies suggest that strong balancing selections occurred during the evolution of modern humans. They suggest also that repeated episodes of endocannibalism-related prion disease epidemics in ancient human populations may have made coding heterozygosity at *PRNP* a significant selective advantage, leading to the selection observed today.

Unexpectedly, the codon 129 polymorphism may play an as yet enigmatic role in the pathogenesis of Alzheimer's disease and/or cognitive impairment in the elderly. An association between Val homozygosity and cognitive impairment in the elderly or early cognitive decline was described in two studies [77; 78]. Significant associations between homozygosity at codon 129 and early onset Alzheimer's disease were reported in two different investigations [79; 80]. This may be related to the postulated function of PrP^C in oxidative stress and neuroprotection. However, the association with codon 129 homozygosity is unclear.

26.5 Neuropathological phenotypes of human prion diseases

26.5.1 Creutzfeldt–Jakob disease

In the majority of cases CJD is encountered as a sporadic disease (80–90 %) with no recognizable transmission and of unknown origin (idiopathic CJD). sCJD is often thought to originate de novo and in terms of the prion hypothesis may be a spontaneous disease. Pathological and clinical features of sCJD are determined by a codon 129 polymorphism and the specific PrP^{Sc} type (see above). Among the acquired diseases, the new variant of CJD (vCJD), which is caused by BSE prions, is characterized by a peculiar and very uniform neuropathological appearance. Accidentally or iatrogenically transmitted CJD (iCJD) is neuropathologically indistinguishable from sCJD with the notable exception of a small number of dura mater-related transmissions that show low numbers of florid plaques. Ten to fifteen percent of all CJD cases are genetic or familial CJD (fCJD) with neuropathological changes closely related to particular *PRNP* mutations.

Sporadic CJD (idiopathic CJD): phenotypes

sCJD most often affects patients in their 60s. It usually presents with dementia and various neurological signs and runs a relentless course leading to death, usually within 6 months. Clinical diagnostic criteria established by Masters have been modified (Table 29.1). The definite diagnosis can only be made by neuropathological or biochemical examination of the brain. Well-documented phenotypic heterogeneity of sCJD in the absence of a genome of the infectious agent has been a puzzling observation defying the prion hypothesis. This apparent contradiction now seems to be resolved. Parchi and coworkers described two different molecular types of PrP^{Sc} associated with distinct clinical and pathological characteristics of sCJD [47; 81]. On transmission to laboratory animals the properties of PrP^{Sc} types 1 and 2 have been shown to propagate faithfully [82]. The two types of PrP^{Sc} are distinguished by their different physicochemical properties, in particular their appearance on Western blot transfers after digestion with proteinase K. The unglycosylated forms of PrP^{Sc} are seen as proteins of approximately 21 kDa (type 1) and 19 kDa (type 2) relative molecular mass (Fig. 26.4). As recently shown, proteinase K has two preferential cleavage sites at codons 97 and 82 of PrP^{Sc}, most likely related to two different conformations of PrP^{Sc} [48]. In addition, the common methionine and valine polymorphism at codon 129 of *PRNP* was shown to modify the phenotype of the disease. The full spectrum of sCJD variants was defined in a large series of 300 CJD patients and molecular and pathological characteristics were compared. Six distinct types of sCJD were described by molecular analysis. Almost 90 % of sCJD patients were homozygous at codon 129, the vast majority being MM homozygotes, whereas only half of the normal population in Europe and the US are homozygous. PrP^{Sc} types 1 and 2 were found associated with all *PRNP* genotypes. However, there was a strong association between PrP^{Sc} type 1 and MM homozygosity and between PrP^{Sc} type 2 and VV or MV patients. A more detailed study using standardized high buffer strength for brain homogenization, protease K digestion at pH 6.9 with a high PK concentration, and long running gels has shown that specific PrP^{Sc} properties are associated with most, if not all of the six phenotypes of sCJD as well as vCJD [9].

Analyses of the pathological lesion pattern (lesion profile) and PrP IHC defined six pathological variants of sCJD that are by and large associated with the six groups of sCJD as defined by molecular analysis (Table 26.3). Among the MM2 patients there is one subgroup, termed MM2-cortical (MM2-C), whose pathologic phenotype closely resembles MM1 patients with the exception of the presence of confluent vacuoles and the absence of cerebellar pathology, whereas another subgroup, termed MM2-thalmic (MM2-T) shows striking resemblance to the FFI phenotype. IHC for PrP^{Sc} is positive in all subgroups of sCJD, but may be extremely weak and hardly recognizable and therefore not diagnostically reliable in VV1 and MM2-T cases.

As to the clinical signs at onset, the absence of cognitive impairment found in a large group of VV2 patients and the absence of ataxia in the VV1 and MM2-C subgroups is remarkable. During the course of the disease, dementia was found in all subjects except in a small group of patients who rapidly lapsed into a stupor after a short initial phase with exclusively neurological signs. Ataxia was always more common in VV2 and MV2 subjects. Visual signs and myoclonus were prominent in the MM1 and MV1 subgroups.

The EEG shows typical PSWC in MM1 and MV1 subjects while these are rare in MV2 and VV2 patients, and absent in VV1, MM2-C, and MM2-T patients. As a consequence, these patients are classified as "possible CJD" according to classical clinical diagnostic criteria. The 14–3–3 proteins in the CSF have proven to be good surrogate markers for most variants of sCJD with the exception of MV2 subjects [83].

MM1 and MV1 subjects comprise 70 % of sCJD cases exhibiting features that have been recognized as typical of CJD, i. e., these subjects show dementia, myoclonus, and PSWC. On histological examination they show delicate spongiform changes in the cerebral cortex, particularly the occipital cortex, neostriatum, thalamus, and cerebellum. Brain stem nuclei and the hippocampus are usually spared. Immunohistochemistry shows synaptic PrP^{Sc} deposition in the cortex and cerebellum. The second most common subgroup are VV2 patients (16 %), who are clinically characterized by prominent ataxia and lack of PSWC in the EEG. Histology is characterized by spongiform degeneration of the deep cortical laminae of the neocortex, the limbic cortex, and subcortical nuclei. Immunohistochemically plaque-like deposits of PrP^{Sc} and prominent perineuronal staining are found. With 9 % of all cases MV2 patients are the third most common group and are clinically similar to the VV2 type. These patients show no typical EEG and in most cases the proteins 14–3–3 are also negative in the CSF. Histologically there is involvement of cortical and subcortical structures. The most prominent feature is the presence of cerebellar kuru plaques, which, in contrast to plaque-like deposits, are visible in routine H&E-stained sections as homogeneous eosinophilic structures. The rare MM2-T phenotype (2 %) is clinically and neuropathologically indistinguishable from the FFI phenotype. There is prominent atrophy of the thalamus and the inferior olivary nucleus, minimal and focal cortical spongiform change, and minimal and diagnostically unreliable IHC staining for PrP^{Sc}. Progressive insomnia, absence of PSWC, and visual hallucinations are found in the MM2-T group, which has also been termed sporadic fatal insomnia (SFI). The cortical MM2 phenotype is also rare (2 %). It shows progressive dementia with no PSWCs. Histologically there are large confluent vacuoles with coarse perivacuolar PrP^{Sc} deposits. Cases with a mixture of MM1 and MM2 have been reported. VV1 patients are the smallest subgroup of sCJD (1 %). There are very young patients in this group. Clinically they show progressive dementia, with no typical EEG. Histologically in these cases there is massive spongiform degeneration with numerous small vacuoles in the neocortex and striatum. Immunohistochemical detection of PrP^{Sc} is minimal and sometimes diagnostically unreliable.

Panencephalopathic CJD, a disease with rampant destruction of the entire cerebrum including massive myelin loss [84] and activation of microglia and macrophages may be an unspecific end-stage condition that can be associated with most if not all CJD variants in individuals with a prolonged course of disease [85].

The detailed analysis of molecular and phenotypic variants of sCJD thus forms the basis for the recognition of unusual subtypes, which in the past may not have been diagnosed as CJD because of their early age at onset, their

Table 26.3: Molecular and phenotypic features of sCJD variants.

sCJD variant	Other classification	Percent of cases	Duration (months)	Age at onset (years)	Clinical features	Neuropathological features
MM1 or MV1	Myoclonic, Heidenhain variant	70	3.9 (1–18)	65.5 (42–91)	Rapidly progressive dementia, early and prominent myoclonus, typical EEG; visual impairment or unilateral signs at onset in 40 % of cases	"Classic" CJD pathology: spongiform change with small (2–10 μm) round to oval vacuoles, predominantly located to neuronal processes; often predominant involvement of occipital cortex; "synaptic"-type PrP staining; in addition, one third of cases shows confluent vacuoles and perivacuolar PrP staining
VV2	Ataxic variant, Brownell-Oppenheimer variant	16	6.5 (3–8)	61.3 (41–80)	Ataxia at onset, dementia late in the disease, no typical EEG in most cases	Prominent involvement of subcortical nuclei including brain stem; in neocortex, spongiosis is often confined to deep layers; PrP staining shows plaque-like focal deposits as well as prominent perineuronal staining
MV2	Kuru-plaque variant	9	17.1 (5–72)	59.4 (40–81)	Ataxia and progressive dementia, no typical EEG, long duration (> 2 years) in some cases	Cerebellar kuru plaques, other pathological changes similar to VV2 with more consistent plaque-like deposits
MM2 thalamic	Thalamic variant, sporadic fatal insomnia (SFI)	2	15.6 (8–24)	52.3 (36–71)	Insomnia and psychomotor hyperactivity in most cases, in addition to ataxia and cognitive impairment, no typical EEG	Prominent atrophy of the thalamus and inferior olive with little pathology in other areas; spongiosis may be absent or focal, minimal PrP^{Sc} deposition
MM2 cortical	-	2	15.7 (9–36)	64.3 (49–77)	Progressive dementia, no typical EEG	Large confluent vacuoles with perivacuolar PrP^{Sc} staining in all cortical layers; cerebellum is relatively spared
VV1	-	1	15.3 (14–16)	39.3 (24–49)	Progressive dementia, no typical EEG	Severe pathology in the cerebral cortex and striatum with sparing of brain stem nuclei and cerebellum; no large confluent vacuoles, and very faint synaptic PrP^{Sc} staining

unusual clinical course, and even their atypical pathology. Therefore there should be much greater clinical awareness of sCJD cases in young patients with a long clinical course and progressive dementia but no PSWC as represented by MM2-C and VV1 cases.

Accidentally transmitted CJD (iatrogenic CJD, iCJD)

iCJD after treatment with pituitary hormones, implantation of human dura mater, the use of contaminated surgical instruments, and cornea transplantation (see Chapter 42) cannot in general be distinguished from sCJD either clinically or pathologically. A study of 51 iCJD cases after treatment with pituitary-derived human growth hormone [74] found the following genetic distribution at codon 129 of *PRNP*: MM 63 %, MV 12 %, and VV25 %. The mean incubation time of the MM and VV groups was 8 years and in the MV group the earliest case was identified 5 years after the first patient in the homozygous groups.

In contrast to the classically held opinion that iCJD is indistinguishable from sCJD clinically and pathologically, a small number of iatrogenic cases after dura implantation have been reported from Japan, France, and Germany that may diverge from this view (Table 26.4). The common denominator of these cases is the presence of plaques that are indistinguishable from florid plaques seen in vCJD. However, these plaques are rare in iCJD as opposed to vCJD and there is much less widespread PrP^{Sc} deposition in iCJD. In addition, the banding pattern of PrP^{Sc} on Western blots shows PrP^{Sc} that is clearly distinguishable from PrP^{Sc} seen in vCJD. Thus there may be a separate type of iCJD with florid plaques that is distinguishable from vCJD pathologically and biochemically. Clinically these cases have been reported to have incubation times between 9 and 19 years, clinical disease ranges from 5 to 18 months, and in most cases the EEG shows only unspecific changes. Thus no clear clinical syndrome is recognizable for this subgroup at present. A number of similar cases have been mentioned in abstracts of scientific meetings, but it is difficult to be sure of the identity of these cases. It is not possible at present to obtain reliable data on the prevalence of this possible variant of iCJD. Some researchers assume that this subtype may constitute 10 % of iCJD after dura mater implantation.

Variant CJD (vCJD)

A new variant of CJD was first described in 10 patients in the UK in 1996 [87]. These patients

Table 26.4: Summary of iCJD cases with florid plaques.

Year of dura mater graft	Onset of disease	Latency (years)	Duration of disease (months)	Age at onset (years)	EEG (PSWC)	Myoclonus	Gender	Codon 129	Ref.
1981	1999	19	18	58	Slowing, no PSWC	Yes	Male	MM	[131]
1984	1995	11	5	52	Diffuse flattening, no characteristic EEG	No	Female	MM	[132; 133]
1985	1994	9	17	47	Slow activity, no PSWC	Yes	Female	MM	[134]
1985	1996	11	8	68	Slow background activity, no periodic discharges	No	Male	MM	[135] Case 1
1986	1996	10	15–17	68	Slow background activity, periodic synchronous discharges on the day before death	No	Female	MM	[135] Case 2

had a mean age of 29 years and presented with psychiatric disturbances, whereas signs more typical of CJD developed later in the course of disease. Neuropathology at autopsy was exceptional, showing extensive depositions of PrP^{Sc} in various areas of the brain in a fashion that had only been described in hereditary diseases before 1996. In addition, there were florid plaques with a dense eosinophilic core staining positive for PrP surrounded by a pale region of radiating fibrils, which were encircled by a rim of microvacuolar spongiform change (see Fig. 26.2). Florid plaques had not previously been seen in human prion disease.

A detailed neuropathological description of the first 89 cases [88] shows that in these cases spongiform change is widespread in a patchy distribution within the cerebral cortex, but no spongiform change is detectable in the hippocampus. Confluent spongiform change is present in the basal ganglia and is disproportionally severe compared with the small number of plaques in that region. Focal spongiform change is seen in many thalamic nuclei, but the posterior thalamus including the pulvinar is relatively spared. In the cerebellar cortex there is prominent spongiform change; confluent vacuoles are observed in the molecular layer in a patchy distribution often associated with florid plaques. Neuronal loss is widespread and in the medial and posterior nuclei of the thalamus there is extensive neuronal loss with marked astrocytosis that is most severe in the pulvinar. Florid plaques are easily seen on H&E stains measuring up to 150 μm and are particularly well visualized using Congo red stain. They are seen in largest numbers in the occipital and cerebellar cortex. In addition, kuru plaques are seen. IHC for PrP^{Sc} shows strong staining of florid plaques and smaller plaques often arranged in irregular clusters ("cluster plaques"). In addition there is widespread pericellular deposition of PrP around small neurons. A synaptic pattern of immunoreactivity is observed in the thalamus, hypothalamus, midbrain, pons, and medulla. Peripheral nerves contain no detectable PrP on IHC. PrP staining is identified in follicular dendritic cells and macrophages within germinal centers in the pharyngeal, lingual, and palatine tonsil, with a more restricted pattern in germinal centers in the appendix, Peyer's patches, spleen, and lymph nodes from cervical, mediastinal, and mesenteric regions.

By May 2006 161 cases of vCJD had been identified in the UK, 17 cases in France, 4 in Ireland, 2 in the USA, and single cases in 7 other countries. Information on the clinical appearance has been compiled, and a definition for suspect vCJD cases is now available (Table 29.1). All vCJD patients to date have been methionine homozygotes at codon 129 of the prion protein gene; the youngest patient was 12 years old and the oldest was 74 years old at onset. PrP^{Sc} deposition in lymphoreticular organs has been reported in one MV patient who died of a ruptured aneurysm and who never showed clinical signs of vCJD. Five years before his death he had received a blood transfusion from a person who later had developed vCJD [71].

It is not possible at present to define the expected clinical and pathological features in codon 129 valine homozygotes or in heterozygotes that are likely to be observed in the future. vCJD has shown significantly new features in the pattern of extracerebral deposition of PrP^{Sc} in the tonsils, lymph nodes, spleen, and appendix. This has raised concern that blood cells may also harbor the infectious agent and that the disease might spread by blood transfusions. Indeed, one patient died of vCJD at the age of 69 having received red blood cells from a young donor who developed vCJD 3.5 years after donating blood [34]. There is no proof that the infectious agent spread via blood in this case, but given the odds of contracting vCJD via blood or the food chain and in the light of animal experiments that have shown that BSE transmitted to sheep can be transmitted by using blood of the infected sheep [89], this appears to be the most likely scenario.

It is hypothesized that primary transmission to humans causing vCJD has occurred by the consumption of food or other products containing large amounts of the BSE agent (BSE prions). This hypothesis is strengthened by epidemiological and experimental findings. The appearance of vCJD 10 years after BSE in the country with the highest incidence of BSE is highly consistent with this hypothesis. The banding pattern of PrP^{Sc} in vCJD has been shown to differ from sCJD and resembles the pattern in BSE (Fig. 26.5). Although BSE and vCJD show significant differences pathologically, upon transmission to genetically homogeneous animals (inbred mice) they elicit practically identical patterns, whereas, in these strain-typing experiments, all tested scrapie strains and sCJD cases were different [90]. These findings have been confirmed in transgenic animals expressing bovine PrP [91].

The available epidemiological data appear to be insufficient to predict the number of cases that must be expected to occur in the future, because neither the mean incubation time nor the shape of the epidemiological curve, the exact mode of transmission, nor any other parameter that might possibly be important is known at present.

Familial Creutzfeldt–Jakob disease

This disease is described in detail in Section 26.6.

26.5.2 Kuru

Kuru was first described in the late 1950s as a fatal neurodegenerative disease affecting the Fore people in the eastern highlands of Papua New Guinea. This disease was mainly characterized by ataxia and predominantly affected women and children. On the suggestion of W. J. Hadlow, a veterinary pathologist, D.C. Gajdusek and C.J. Jr. Gibbs successfully transmitted the disease to chimpanzees in 1966 [92]. Later, ritualistic cannibalism of deceased clan members was identified as the mode of transmission in the Fore population. After cannibalistic practices had ceased, the disease disappeared. Kuru incubation times of more than three decades have been reported and a diminishing number of exceptional cases of this disease are still reported every year.

All neuropathological changes of CJD may also be observed in kuru. Thus, pathologically, single kuru cases are indistinguishable from CJD cases [7; 93; 94]. However, kuru plaques and cerebellar pathology, which is related to the predominant ataxic clinical appearance of kuru, seem to be more common in kuru than in sCJD. It is not quite clear how the clinical and pathological changes observed in kuru fit into the well-established classification scheme of sCJD [8]. Among nine historical kuru cases that have recently been re-examined genetically and neuropathologically, kuru plaques were found in four, one case of which was a heterozygote and three were methionine homozygotes [95]. The significance of these findings is not clear at present. The association of kuru plaques and methionine homozygosity is exceptional in CJD [96]. Reports on cases with florid plaques in methionine homozygotes after dura mater implants may imply that PrP plaques in methionine homozygotes are associated with acquired rather than sporadic disease. Although this interpretation would agree with the findings from kuru, at present there is not sufficient data to make this argument conclusive.

26.5.3 Gerstmann–Sträussler–Scheinker disease

Gerstmann–Sträussler–Scheinker disease (GSS) was first described in 1928 and 1936 in an Austrian family [97; 98]. GSS is a historic term that has been used to describe a predominantly ataxic hereditary syndrome in patients with multicentric PrP plaques. It has been found in association with *PRNP* mutations at codons 102, 105, 117, 145, 198, 202, and 217 (Table 26.5). Since in some affected individuals and families the predominant clinical feature is dementia and, on the other hand, some CJD cases show dense depositions of PrP plaques, the distinction from CJD may be artificial and may be replaced by another terminology in the future. Clinically the majority of affected patients seem to be mainly ataxic, they develop dysphagia, dysarthria, hyporeflexia, and dementia. Dementia can be severe or it may develop late and can be minor compared to other clinical signs. Subjects carrying the same *PRNP* mutation, sometimes belonging to same family, have been reported to develop typical GSS or CJD [99–101], and these divergent phenotypic expressions have been correlated to the PrP^{Sc}

forms accumulating in the brain. Neuropathologically, GSS is a distinct entity characterized by large multicentric PrP^{Sc}-containing plaques (Fig. 26.2). Other features show more variation between different mutations. There may be marked spongiform change as in the P102L mutation, neurofibrillary tangles associated with the Y145*, F198S, and Q217R mutations, or amyloid angiopathy, which is found in cases with the 145* mutation.

26.5.4 Fatal familial insomnia (FFI) and sporadic fatal insomnia (SFI)

FFI is associated with a D178N mutation of *PRNP* in coupling with methionine at codon 129. It is one of the most common genetic prion diseases worldwide and together with SFI most likely represents what was described as "thalamic dementia" in earlier literature. Both FFI and SFI have been successfully transmitted experimentally to animals [82; 102–105].

Clinically, FFI frequently begins with disturbances of vigilance and sleep that are often neglected. The patients go on to develop hallucinatory states, autonomic disturbances such as high blood pressure and hyperthermia, disturbances of gait, disequilibrium, myoclonus, and pyramidal involvement and finally often die in a vegetative state. The onset of disease is between 36 and 62 years, and the disease duration varies from 8 to 72 months.

Thc EEG activity in FFI is generally slowed. In most FFI patients typical PSWC are not found in the EEG, but may be present in patients with a long clinical duration of disease. The 14–3–3 protein examination is often negative in the CSF in FFI and thus is of no diagnostic value [106].

The neuropathology of FFI shows (i) severe neuronal loss and gliosis in the thalamus, in particular the ventral anterior and the dorsomedial nuclei, (ii) severe neuronal loss and gliosis in the inferior olive, and (iii) cortical involvement, frequently limited to the entorhinal cortex in cases of short duration. Variable astrocytic gliosis and spongiform changes are found in other areas of the cortex in cases of long clinical duration. The amount of PrP^{Sc} found in FFI is much lower than in sCJD, and is often difficult to detect using IHC and may only be discovered using Western blotting or the PET blot technique in some cases. In addition to clinically and pathologically typical FFI cases there is a small number that appears more similar to CJD [107; 108].

A clinical syndrome similar to FFI has been reported in at least twelve patients with sporadic disease and with no *PRNP* mutation. This disease has been termed sporadic fatal insomnia (SFI). The neuropathologic changes seen in these cases are indistinguishable from FFI [103; 109; 110]. All reported patients were homozygous for methionine at codon 129 and they had PrP^{Sc} type 2. This disease has been systematically described as the MM2-T subtype of sCJD [8].

26.5.5 "Atypical" prion diseases

Rare or uncommon combinations of clinical or pathological changes in human prion diseases have been termed "atypical". The majority of those cases are genetic diseases. In particular, rare cases with insert mutations have shown pathological changes that are not observed in more common types of sCJD or GSS [111]. It took years for FFI to be recognized as an "atypical prion disease". Now a well-described entity, it seems to be one of the most common familial prion diseases [112; 113]. Some of the atypical cases can only be diagnosed by molecular genetic investigation of *PRNP* since they show practically no spongiform change or PrP^{Sc} on IHC investigation. It is not known if there are sporadic cases of human prion disease that show practically no spongiform change or IHC PrP^{Sc} similar to these atypical genetic cases.

26.6 Genetics of human prion diseases: phenotypes of familial (genetic) prion diseases

One of the first descriptions of CJD, still called spastic pseudosclerosis at that time, referred to a familial case, which later was identified as carrying the D178N mutation [5; 114]. Thus it was known very early on that spongiform encephalopathies could be familial. It has taken many years for the recognition of GSS as a transmissible disease, which is similar to CJD [115] as well as the fact that both can be caused by *PRNP* mutations [116]. Some familial human prion diseases are clinically and neuropathologically indistinguishable from sporadic disease while others are associated with pathological changes that are uncommon or unknown in sporadic disease. Thus, multicentric plaques have only been described in familial prion disease (GSS). The clinical features may also differ significantly from sporadic disease, in particular the clinical course may exceed 2 years. While the 14–3–3 protein is detected in the majority of sCJD cases, in fCJD the sensitivity is only around 50 %. The sensitivity may even be lower in FFI [106]. Experience has shown that by prescreening suspect cases with the 14–3–3 test a large number of familial cases are missed. Sequencing of *PRNP* is necessary in all suspect cases of CJD and in all cases of dementia in which the diagnosis is unclear.

The most important *PRNP* mutations and their phenotypes are described in some detail in this chapter, while a rough outline of the clinical and neuropathological features is shown in Table 26.5.

E200K. This is the most common genetic prion disease. Clinically and neuropathologically this disease cannot be distinguished from sCJD. The largest known group with this mutation are Libyan immigrants to Israel who have a 100-fold higher incidence of CJD [117]. Originally it was believed that the increased incidence in this group was caused by the consumption of brains and eyes of scrapie-infected sheep [118]. However, in practically all CJD cases in this group a E200K *PRNP* mutation was shown to be associated with CJD [117; 119]. The increased incidence of CJD that has been observed in certain regions of Slovakia is also familial and caused by the E200K mutation [120]. The penetrance of this mutation has been debated and has been estimated to be slightly higher than 50 % by some, while others assume it is closer to 100 % [121].

D178N. The D178N mutation can be the cause of two different clinically and neuropathologically defined diseases. In coupling with methionine at codon 129 it is associated with FFI and in coupling with valine at codon 129 it causes fCJD (see above). FFI (D178N-129M) is one of the most common genetic prion diseases.

P102L. This mutation is the most common cause of GSS. The family first reported by Gerstmann, Sträussler, and Scheinker [98] also had this mutation [122]. Clinically and neuropathologically this mutation is associated with the very typical characteristics of GSS with prominent ataxia, dementia in late stages of disease, and numerous kuru plaques and multicentric plaques. However, there is quite a large variation of features associated with the P102L mutation; even members of the same family can develop different disease phenotypes. This does not seem to be related to the codon 129 polymorphism [99–101], but rather to the physicochemical properties of the PrP^{Sc} forms accumulating in the brain. Brain tissue of P102L cases has been shown to cause spongiform encephalopathy after experimental transmission to primates and rodents [123; 124]. Thus it was shown for the first time that GSS is a genetic and transmissible disease. Transgenic mice overexpressing a murine PrP gene with a corresponding P101L mutation spontaneously developed a spongiform encephalopathy indistinguishable from scrapie in mice. The disease was also reported to be transmissible to transgenic mice expressing the P196L mutation at low levels and to hamsters but not to CD-1 Swiss mice [125; 126]. However, when the P102L mutation was introduced into the murine PrP gene by gene targeting, these mice did not develop spontaneous disease but showed significant changes in incubation of prion disease of various sources [127].

Insert mutations. Between amino acid residues 51 and 91 the wild-type PrP shows five incomplete repeats of the four octapeptides PHGGGWGQ and one preceding nonapeptide PQGGGTWGQ. In families with insert muta-

Table 26.5: Genotypes and phenotypes of genetic human prion diseases. Other mutations that have to date been observed in a very small number of families or individuals include the deletion of two octarepeats [65], R148H [157], T188A [158], E211Q [159; 160], Q160Stop [161], E196K [159], V203 I [159], T188K [159].

Genotype	Disease	Clinical features	Pathology	References
Insertion mutations	CJD	Progressive dementia in the majority of cases. In addition a large number of changes can be observed, some of which are rarely seen in other prion diseases such as dysphasia, apraxia, and personality disorders	Spongiform change. Other changes vary widely. PrP plaques are often noted in the cerebellum.	[111; 128; 129; 136]
P102L	GSS	Progressive cerebellar syndrome with ataxia, dementia often later in the course	Multicentric plaques, predominantly in the cerebellum	[116; 122]
P105L	GSS	Spastic paraparesis	PrP^{Sc} deposition in the cerebral cortex	[137]
A117V	GSS	Progressive ataxia and dementia	Multicentric plaques in the cortex or cerebellum	[138; 139]
Y145*	GSS	Slowly progressive dementia with a clinical course exceeding 20 years	PrP in cerebral blood vessels; neurofibrillary tangles (NFT)	[140; 141]
N171S		Psychiatric disturbances	Unknown	[142]
D178N-129M	FFI	Progressive insomnia, disturbances of autonomic and endocrine systems, progressive dementia, and other CJD-typical features often appear late	Nerve cell loss in the thalamus and inferior olive, spongiform change and PrP deposition may be minimal	[73; 113; 143]
D178N-129V	fCJD	Progressive dementia	Spongiform change, nerve cell loss, astrocytic gliosis	[73; 114; 143]
V180I	fCJD	Progressive dementia	Spongiform change, nerve cell loss, astrocytic gliosis	[144; 145]
T183A	fCJD	Progressive dementia with frontotemporal clinical features and long duration	Severe spongiform change with minimal gliosis	[146; 147]
F198S	GSS	Progressive ataxia and dementia	Multicentric plaques, NFT in the cortex and subcortical nuclei	[62; 148]
E200K	fCJD	Prominent progressive dementia in most cases, cerebellar and thalamic symptoms in a minority	Spongiform change, nerve cell loss, astrocytic gliosis	[119; 120; 149]
R208H	fCJD	Progressive dementia	Spongiform change, nerve cell loss, astrocytic gliosis	[150; 151]
V210I	fCJD	Progressive dementia	Spongiform change, nerve cell loss, astrocytic gliosis	[152–154)
Q212P	GSS	Ataxia	"mild amyloid deposition"	[155]
Q217R	GSS	Progressive dementia, ataxia later in the disease	Multicentric plaques and NFT in the cortex	[62; 72]
M232R	fCJD	Progressive dementia	Spongiform change, nerve cell loss, astrocytic gliosis	[145; 156]

* = STOP

tions between one and nine additional octarepeats have been identified. An unequal crossover and recombination are thought to be the cause of extra-repeat formation [128]. According to this hypothesis more than one round of unequal crossover must have taken place in cases of more than four extra repeats. Clinical and neuropathological changes associated with insert mutations show considerable variations among the more than 30 families described to date. These include patients with classical CJD or GSS and others with no morphologically recognizable phenotype that on clinical examination may only show progressive dementia [111; 129]. In cases with one to four repeats the mean age at onset is described as relatively high (mean 60 years, range 52–82 years), the duration of disease is described as short (mean 6 months, range 2–14 months), patients often present with progressive dementia, ataxia, visual disturbances, myoclonus, and PSWC. Histopathologically they are indistinguishable from CJD. In contrast, patients with five or more repeats are described as being relatively young at the onset of disease (mean 32 years, range 21–61 years), having a long clinical course of disease (mean 6 years, ranging from 3 months to 19 years), presenting with a slowly progressive mental deterioration, cerebellar and extrapyramidal signs, and often lacking PSWC. Neuropathologically they are a heterogeneous group which may show changes similar to CJD or GSS or changes that cannot easily be classified [130].

References

[1] Jakob A. Über eigenartige Erkrankungen des Zentralnervensystems mit bemerkenswertem anatomischem Befunde (spastische Pseudosklerose-Encephalomyelopathie mit disseminierten Degenerationsherden). Dtsch Z Nervenheilkd 1921; 70:132–146.

[2] Spielmeyer W. Histopathologie des Nervensystems. Springer-Verlag, Berlin, 1922.

[3] Creutzfeldt HG. Über eine eigenartige herdförmige Erkrankung des Zentralnervensystems. Z Ges Neurol Psychiatr 1920; 57:1–18.

[4] Jakob A. Spastische Pseudosklerose. In: Jakob A, editor. Die extrapyramidalen Erkrankungen. Springer-Verlag, Berlin, 1923:215–245.

[5] Meggendorfer F. Klinische und genealogische Beobachtungen bei einem Fall von spastischer Pseudosklerose. Z Ges Neurol Psychiatr 1930; 128:337–341.

[6] Hadlow WJ. Scrapie and kuru. Lancet 1959; 2:289–290.

[7] Klatzo I, Gajdusek DC, Zigas V. Pathology of the kuru. Lab Invest 1959; 8:799–847.

[8] Parchi P, Giese A, Capellari S, et al. Classification of sporadic Creutzfeldt–Jakob disease based on molecular and phenotypic analysis of 300 subjects. Ann Neurol 1999; 46(2):224–233.

[9] Notari S, Capellari S, Giese A, et al. Effects of different experimental conditions on the PrPSc core generated by protease digestion: implications for strain typing and molecular classification of CJD. J Biol Chem 2004; 279(16):16797–16804.

[10] Masters CL, Richardson EP Jr. Subacute spongiform encephalopathy (Creutzfeldt–Jakob disease). The nature and progression of spongiform change. Brain 1978; 101:333–344.

[11] Ferrer I, Casas R, Rivera R. Parvalbumin-immunoreactive cortical neurons in Creutzfeldt–Jakob disease. Ann Neurol 1993; 34:864–866.

[12] Guentchev M, Hainfellner J-A, Trabattoni GR, et al. Distribution of parvalbumin-immunoreactive neurons in brain correlates with hippocampal and temporal cortical pathology in Creutzfeldt–Jakob disease. J Neuropathol Exp Neurol 1997; 56:1119–1124.

[13] Tschampa HJ, Herms JW, Schulz-Schaeffer WJ, et al. Clinical findings in sporadic Creutzfeldt–Jakob disease correlate with thalamic pathology. Brain 2002; 125(11):2558–2566.

[14] Steriade M and Contreras D. Spike-wave complexes and fast components of cortically generated seizures. I. Role of neocortex and thalamus. J Neurophysiol 1998; 80(3):1439–1455.

[15] Kitamoto T, Ogomori K, Tateishi J, et al. Formic acid pretreatment enhances immunostaining of cerebral and systemic amyloids. Lab Invest 1987; 57:230–236.

[16] Brown P, Wolff A, Gajdusek DC. A simple and effective method for inactivating virus infectivity in formalin-fixed tissue samples from patients with Creutzfeldt–Jakob disease. Neurology 1990; 40:887–890.

[17] Kitamoto T, Shin RW, Doh-ura K, et al. Abnormal isoform of prion proteins accumulates in the synaptic structures of the central nervous system in patients with Creutzfeldt–Jakob disease. Am J Pathol 1992; 140:1285–1294.

[18] Taraboulos A, Jendroska K, Serban D, et al. Regional mapping of prion proteins in brains. Proc Natl Acad Sci U S A 1992; 89:7620–7624.

[19] Schulz-Schaeffer WJ, Tschoke S, Kranefuss N, et al. The paraffin-embedded tissue blot detects PrP(Sc) early in the incubation time in prion diseases. Am J Pathol 2000; 156(1):51–56.

[20] Jeffrey M, Goodbrand IA, Goodsir CM. Pathology of the transmissible spongiform encephalopathies with special emphasis on ultrastructure. Micron 1995; 26(3):277–298.

[21] Marsh RF, Sipe JC, Morse SS, et al. Transmissible mink encephalopathy. Reduced spongiform degeneration in aged mink of the Chediak-Higashi genotype. Lab Invest 1976; 34:381–386.

[22] Baringer JR, Bowman KA, Prusiner SB. Replication of the scrapie agent in hamster brain precedes neuronal vacuolation. J Neuropathol Exp Neurol 1983; 42:539–547.

[23] Bruce ME. Agent replication dynamics in a long incubation period model of mouse scrapie. J Gen Virol 1985; 66:2517–2522.

[24] Bundza A and Charlton KM. Comparison of spongiform lesions in experimental scrapie and rabies in skunks. Acta Neuropathol 1988; 76:275–280.

[25] Mallucci G, Dickinson A, Linehan J, et al. Depleting neuronal PrP in prion infection prevents disease and reverses spongiosis. Science 2003; 302(5646):871–874.

[26] David-Ferreira JF, David-Ferreira KL, Gibbs CJ Jr, et al. Scrapie in mice: ultrastructural observations in the cerebral cortex. Proc Soc Exp Biol Med 1968; 127:313–320.

[27] Liberski PP, Budka H, Sluga E, et al. Tubulovesicular structures in human and experimental Creutzfeldt–Jakob disease. Eur J Epidemiol 1991; 7:551–555.

[28] Liberski PP, Jeffrey M, Goodsir C. Tubulovesicular structures are not labeled using antibodies to prion protein (PrP) with the immunogold electron microscopy techniques. Acta Neuropathol 1997; 93(3):260–264.

[29] Head MW, Northcott V, Rennison K, et al. Prion protein accumulation in eyes of patients with sporadic and variant Creutzfeldt–Jakob disease. Invest Ophthalmol Vis Sci 2003; 44(1):342–346.

[30] Haik S, Faucheux BA, Sazdovitch V, et al. The sympathetic nervous system is involved in variant Creutzfeldt–Jakob disease. Nat Med 2003; 9(9):1121–1123.

[31] Hill AF, Zeidler M, Ironside J, et al. Diagnosis of new variant Creutzfeldt–Jakob disease by tonsil biopsy. Lancet 1997; 349:99–100.

[32] Wadsworth JD, Joiner S, Hill AF, et al. Tissue distribution of protease resistant prion protein in variant Creutzfeldt–Jakob disease using a highly sensitive immunoblotting assay. Lancet 2001; 358(9277):171–180.

[33] Ghani AC, Donnelly CA, Ferguson NM, et al. Assessment of the prevalence of vCJD through testing tonsils and appendices for abnormal prion protein. Proc R Soc Lond B Biol Sci 2000; 267(1438):23–29.

[34] Llewelyn CA, Hewitt PE, Knight RS, et al. Possible transmission of variant Creutzfeldt–Jakob disease by blood transfusion. Lancet 2004; 363(9407):417–421.

[35] Glatzel M, Abela E, Maissen M, et al. Extraneural pathologic prion protein in sporadic Creutzfeldt–Jakob disease. N Engl J Med 2003; 349(19):1812–1820.

[36] Koperek O, Kovacs GG, Ritchie D, et al. Disease-associated prion protein in vessel walls. Am J Pathol 2002; 161(6):1979–1984.

[37] Brown P, Gibbs CJ Jr, Rodgers Johnson P, et al. Human spongiform encephalopathy: the National Institutes of Health series of 300 cases of experimentally transmitted disease. Ann Neurol 1994; 35:513–529.

[38] Kretzschmar HA, Stowring LE, Westaway D, et al. Molecular cloning of a human prion protein cDNA. DNA 1986; 5:315–324.

[39] Kretzschmar HA, Prusiner SB, Stowring LE, et al. Scrapie prion proteins are synthesized in neurons. Am J Pathol 1986; 122:1–5.

[40] Moser M, Colello RJ, Pott U, et al. Developmental expression of the prion protein gene in glial cells. Neuron 1995; 14:509–517.

[41] Hornshaw MP, McDermott JR, Candy JM, et al. Copper binding to the N-terminal tandem repeat region of mammalian and avian prion protein: structural studies using synthetic peptides. Biochem Biophys Res Commun 1995; 214(3):993–999.

[42] Brown DR, Qin K, Herms JW, et al. The cellular prion protein binds copper in vivo. Nature 1997; 390:684–687.

[43] Herms JW, Tings T, Madlung A, et al. The prion protein is predominantly localized in presynaptic membranes. Neurobiol Aging 1998; 19:299.

[44] Herms J, Tings T, Gall S, et al. Evidence of presynaptic location and function of the prion protein. J Neurosci 1999; 19(20):8866–8875.

[45] Renner C, Fiori S, Fiorino F, et al. Micellar environments induce structuring of the N-terminal tail of the prion protein. Biopolymers 2004; 73(4):421–433.

[46] Parchi P, Capellari S, Chen SG et al. Typing prion isoforms. Nature 1997; 386(6622):232–234.

[47] Parchi P, Castellani R, Capellari S, et al. Molecular basis of phenotypic variability in sporadic Creutzfeldt–Jakob disease. Ann Neurol 1996; 39(6):767–778.

[48] Parchi P, Zou W, Wang W, et al. Genetic influence on the structural variations of the abnormal prion protein. Proc Natl Acad Sci USA 2000; 97(18):10168–10172.

[49] Monari L, Chen SG, Brown P, et al. Fatal familial insomnia and familial Creutzfeldt–Jakob disease: different prion proteins determined by a DNA polymorphism. Proc Natl Acad Sci USA 1994; 91(7):2839–2842.

[50] Collinge J, Sidle KC, Meads J, et al. Molecular analysis of prion strain variation and the aetiology of 'new variant' CJD. Nature 1996; 383:685–690.

[51] Petersen RB, Parchi P, Richardson SL, et al. Effect of the D178N mutation and the codon 129 polymorphism on the metabolism of the prion protein. J Biol Chem 1996; 271:12661–12668.

[52] Capellari S, Parchi P, Russo CM, et al. Effect of the E200K mutation on prion protein metabolism. Comparative study of a cell model and human brain. Am J Pathol 2000; 157(2):613–622.

[53] Wadsworth JD, Hill AF, Joiner S, et al. Strain-specific prion-protein conformation determined by metal ions. Nat Cell Biol 1999; 1(1):55–59.

[54] Zanusso G, Farinazzo A, Fiorini M, et al. pH-dependent prion protein conformation in classical Creutzfeldt–Jakob disease. J Biol Chem 2001; 276(44):40377–40380.

[55] Puoti G, Giaccone G, Rossi G, et al. Sporadic Creutzfeldt–Jakob disease: co-occurrence of different types of PrP(Sc) in the same brain. Neurology 1999; 53(9):2173–2176.

[56] Kovacs GG, Head MW, Hegyi I, et al. Immunohistochemistry for the prion protein: comparison of different monoclonal antibodies in human prion disease subtypes. Brain Pathol 2002; 12(1):1–11.

[57] Head MW, Bunn TJ, Bishop MT, et al. Prion protein heterogeneity in sporadic but not variant Creutzfeldt–Jakob disease: UK cases 1991–2002. Ann Neurol 2004; 55(6):851–859.

[58] Polymenidou M, Stoeck K, Glatzel M, et al. Coexistence of multiple PrPSc types in individuals with Creutzfeldt–Jakob disease. Lancet Neurol 2005; 4(12):805–814.

[59] Yull HM, Ritchie DL, Langeveld JP, et al. Detection of type 1 prion protein in variant Creutzfeldt–Jakob disease. Am J Pathol 2006; 168(1):151–157.

[60] Zou WQ, Capellari S, Parchi P, et al. Identification of novel proteinase K-resistant C-terminal fragments of PrP in Creutzfeldt–Jakob disease. J Biol Chem 2003; 278(42):40429–40436.

[61] Wu Y, Brown WT, Robakis NK, et al. A PvuII RFLP detected in the human prion protein (PrP) gene. Nucleic Acids Res 198715:3191.

[62] Hsiao K, Dlouhy SR, Farlow MR, et al. Mutant prion proteins in Gerstmann–Sträussler–Scheinker disease with neurofibrillary tangles. Nat Genet 1992; 1:68–71.

[63] Jeong BH, Nam JH, Lee YJ, et al. Polymorphisms of the prion protein gene (*PRNP*) in a Korean population. J Hum Genet 2004; 49:319–324.

[64] Palmer MS, Mahal SP, Campbell TA, et al. Deletions in the prion protein gene are not associated with CJD. Hum Mol Genet 1993; 2:541–544.

[65] Beck JA, Mead S, Campbell TA, et al. Two-octapeptide repeat deletion of prion protein associated with rapidly progressive dementia. Neurology 2001; 57(2):354–356.

[66] Capellari S, Parchi P, Wolff BD, et al. Creutzfeldt–Jakob disease associated with a deletion of two repeats in the prion protein gene. Neurology 2002; 59(10):1628–1630.

[67] Laplanche JL, Delasnerie-Lauprêtre N, Brandel JP, et al. Molecular genetics of prion diseases in France. Neurology 1994; 44:2347–2351.

[68] Palmer MS, Dryden AJ, Hughes JT, et al. Homozygous prion protein genotype predisposes to sporadic Creutzfeldt–Jakob disease. Nature 1991; 352:340–342.

[69] Salvatore M, Genuardi M, Petraroli R, et al. Polymorphisms of the prion protein gene in Italian patients with Creutzfeldt–Jakob disease. Hum Genet 1994; 94:375–379.

[70] Windl O, Dempster M, Estibeiro JP, et al. Genetic basis of Creutzfeldt–Jakob disease in the United Kingdom: a systematic analysis of predisposing mutations and allelic variations in the *PRNP* gene. Hum Genet 1996; 98:259–264.

[71] Peden AH, Head MW, Ritchie DL, et al. Preclinical vCJD after blood transfusion in a

PRNP codon 129 heterozygous patient. Lancet 2004; 364(9433):527–529.

[72] Dlouhy SR, Hsiao K, Farlow MR, et al. Linkage of Indiana kindred of Gerstmann–Sträussler–Scheinker disease to prion protein gene. Nat Genet 1992; 1:64–67.

[73] Goldfarb LG, Petersen RB, Tabaton M, et al. Fatal familial insomnia and familial Creutzfeldt–Jakob disease: disease phenotype determined by a DNA polymorphism. Science 1992; 258:806–808.

[74] Kitamoto T and Tateishi J. Human prion diseases with variant prion protein. Philos Trans R Soc Lond B Biol Sci 1994; 343(1306):391–398.

[75] Shibuya S, Higuchi J, Shin R-W, et al. Protective prion protein polymorphisms against sporadic Creutzfeldt–Jakob disease. Lancet 1998; 351:419.

[76] Mead S, Stumpf MP, Whitfield J, et al. Balancing selection at the prion protein gene consistent with prehistoric kurulike epidemics. Science 2003; 300(5619):640–643.

[77] Berr C, Richard F, Dufouil C, et al. Polymorphism of the prion protein is associated with cognitive impairment in the elderly: The EVA study. Neurology 1998; 51:734–737.

[78] Croes EA, Dermaut B, Houwing-Duistermaat JJ, et al. Early cognitive decline is associated with prion protein codon 129 polymorphism. Ann Neurol 2003; 54(2):275–276.

[79] Riemenschneider M, Klopp N, Xiang W, et al. Prion protein codon 129 polymorphism and risk of Alzheimer's disease. Neurology 2004; 63(2):364–366.

[80] Dermaut B, Croes EA, Rademakers R, et al. *PRNP* Val129 homozygosity increases risk for early-onset Alzheimer's disease. Ann Neurol 2003; 53(3):409–412.

[81] Parchi P, Giese A, Capellari S, et al. The molecular and clinico-pathologic spectrum of phenotypes of sporadic Creutzfeldt–Jakob disease (sCJD). Neurology 1998; 50:A336.

[82] Telling GC, Parchi P, DeArmond SJ, et al. Evidence for the conformation of the pathologic isoform of the prion protein enciphering and propagating prion diversity. Science 1996; 274:2079–2082.

[83] Zerr I, Pocchiari M, Collins S, et al. Analysis of EEG and CSF 14–3–3 proteins as aids to the diagnosis of Creutzfeldt–Jakob disease. Neurology 2000; 55(6):811–815.

[84] Mizutani T, Okumura A, Oda M, et al. Panencephalopathic type of Creutzfeldt–Jakob disease: primary involvement of the cerebral white matter. J Neurol 1981; 44:103–115.

[85] Ghorayeb I, Series C, Parchi P, et al. Creutzfeldt–Jakob disease with long duration and panencephalopathic lesions: Molecular analysis of one case. Neurology 1998; 51:271–274.

[86] Deslys J-P, Jaegly A, d'Aignaux JH, et al. Genotype at codon 129 and susceptibility to Creutzfeldt–Jakob disease. Lancet 1998; 351:1251.

[87] Will RG, Ironside JW, Zeidler M, et al. A new variant of Creutzfeldt–Jakob disease in the UK. Lancet 1996; 347:921–925.

[88] Ironside JW, McCardle L, Horsburgh A, et al. Pathological diagnosis of variant Creutzfeldt–Jakob disease. APMIS 2002; 110(1):79–87.

[89] Houston F, Foster JD, Chong A, et al. Transmission of BSE by blood transfusion in sheep. Lancet 2000; 356(9234):999–1000.

[90] Bruce ME, Will RG, Ironside JW, et al. Transmissions to mice indicate that 'new variant' CJD is caused by the BSE agent. Nature 1997; 389:489–501.

[91] Scott MR, Will R, Ironside J, et al. Compelling transgenetic evidence for transmission of bovine spongiform encephalopathy prions to humans. Proc Natl Acad Sci U S A 1999; 96(26):15137–15142.

[92] Gajdusek DC, Gibbs CJ Jr, Alpers M. Experimental transmission of a kuru-like syndrome to chimpanzees. Nature 1966; 209:794–796.

[93] Krucke W, Beck E, Vitzthum HG. Creutzfeldt–Jakob disease. Some unusual morphological features reminiscent of kuru. Z Neurol 1973; 206(1):1–24.

[94] Beck E, Daniel PM, Alpers M, et al. Neuropathological comparisons of experimental kuru in chimpanzees with human kuru. Int Arch Allergy Immunol 1969; 36:553–562.

[95] Cervenáková L, Goldfarb LG, Garruto R, et al. Phenotype-genotype studies in kuru: Implications for new variant Creutzfeldt–Jakob disease. Proc Natl Acad Sci USA 1998; 95:13239–13241.

[96] Puoti G, Limido L, Cotrufo R, et al. Sporadic Creutzfeldt–Jakob disease with MM1-type prion protein and plaques. Neurology 2004; 62(7):1239.

[97] Gerstmann J. Über ein noch nicht beschriebenes Reflexphänomen bei einer Erkrankung des zerebellären Systems. Wien Med Wochenschr 1928; 78:906–908.

[98] Gerstmann J, Sträussler E, Scheinker I. Über eine eigenartige hereditär-familiäre Erkrankung des Zentralnervensystems. Zugleich ein

Beitrag zur Frage des vorzeitigen lokalen Alterns. Z Neurol 1936; 154:736–762.
[99] De Michele G, Pocchiari M, Petraroli R, et al. Variable phenotype in a P102L Gerstmann–Sträussler–Scheinker Italian family. Can J Neurol Sci 2003; 30(3):233–236.
[100] Barbanti P, Fabbrini G, Salvatore M, et al. Polymorphism at codon 129 or codon 219 of *PRNP* and clinical heterogeneity in a previously unreported family with Gerstmann–Sträussler–Scheinker disease (PrP-P102L mutation). Neurology 1996; 47(3):734–741.
[101] Tanaka Y, Minematsu K, Moriyasu H, et al. A Japanese family with a variant of Gerstmann–Sträussler–Scheinker disease. J Neurol Neurosurg Psychiatry 1997; 62(5):454–457.
[102] Tateishi J, Brown P, Kitamoto T, et al. First experimental transmission of fatal familial insomnia. Nature 1995; 376:434–435.
[103] Mastrianni JA, Nixon R, Layzer R, et al. Brief report: Prion protein conformation in a patient with sporadic fatal insomnia. N Engl J Med 1999; 340(21):1630–1638.
[104] Collinge J, Palmer MS, Sidle KCL, et al. Transmission of fatal familial insomnia to laboratory animals. The Lancet 1995; 346:569–570.
[105] Scott MRD, Telling GC, Prusiner SB. Transgenetics and gene targeting in studies of prion diseases. In: Prusiner SB, editor. Prions. Prions. Prions. Springer-Verlag, Berlin, 1996:95–123.
[106] Zerr I, Bodemer M, Gefeller O, et al. Detection of 14–3–3 protein in the cerebrospinal fluid supports the diagnosis of Creutzfeldt–Jakob disease. Ann Neurol 1998; 43:32–38.
[107] Kretzschmar H, Giese A, Zerr I, et al. The German FFI cases. Brain Pathol 1998; 8:559–561.
[108] Zerr I, Giese A, Windl O, et al. Phenotypic variability in fatal familial insomnia (D178N-129M) genotype. Neurology 1998; 51:1398–1405.
[109] Parchi P, Capellari S, Chin S, et al. A subtype of sporadic prion disease mimicking fatal familial insomnia. Neurolgy 1999; 52:1757–1763.
[110] Scaravilli F, Cordery RJ, Kretzschmar H, et al. Sporadic fatal insomnia: a case study. Ann Neurol 2000; 48(4):665–668.
[111] Collinge J, Owen F, Poulter M, et al. Prion dementia without characteristic pathology. Lancet 1990; 336:7–9.
[112] Lugaresi E, Medori R, Montagna P, et al. Fatal familial insomnia and dysautonomia with selective degeneration of thalamic nuclei. N Engl J Med 1986; 315:997–1003.
[113] Medori R, Tritschler H-J, LeBlanc A, et al. Fatal familial insomnia, a prion disease with a mutation at codon 178 of the prion protein gene. N Engl J Med 1992; 326:444–449.
[114] Kretzschmar HA, Neumann M, Stavrou D. Codon 178 mutation of the human prion protein gene in a German family (Backer family): sequencing data from 72-year-old celloidin-embedded brain tissue. Acta Neuropathol (Berl) 1995; 89:96–98.
[115] Masters CL and Gajdusek DC. The spectrum of Creutzfeldt–Jakob disease and the virus-induced subacute spongiform encephalopathies. Recent Adv Neuropathol 1982; 2:139–163.
[116] Hsiao K, Baker HF, Crow TJ, et al. Linkage of a prion protein missense variant to Gerstmann–Sträussler syndrome. Nature 1989; 338:342–345.
[117] Gabizon R, Rosenmann H, Meiner Z, et al. Mutation and polymorphism of the prion protein gene in Libyan Jews with Creutzfeldt–Jakob disease (CJD). American Journal of Human Genetics 1993; 53:828–835.
[118] Kahana E, Alter M, Braham J, et al. Creutzfeldt–Jakob disease: focus among Libyan Jews in Israel. Science 1974; 183:90–91.
[119] Goldfarb LG, Korczyn AD, Brown P, et al. Mutation in codon 200 of scrapie amyloid precursor gene linked to Creutzfeldt–Jakob disease in Sephardic Jews of Libyan and non-Libyan origin. Lancet 1990; 336:637–638.
[120] Goldfarb LG, Mitrová E, Brown P, et al. Mutation in codon 200 of scrapie amyloid protein gene in two clusters of Creutzfeldt–Jakob disease in Slovakia. Lancet 1990; 336:514–515.
[121] Goldfarb LG, Brown P, Mitrowá E, et al. Creutzfeldt–Jakob disease associated with the *PRNP* codon 200 Lys mutation. An analysis of 45 families. Eur J Epidemiol 1991; 7:477–486.
[122] Kretzschmar HA, Honold G, Seitelberger F, et al. Prion protein mutation in family first reported by Gerstmann, Sträussler, and Scheinker. Lancet 1991; 337:1160.
[123] Masters CL, Gajdusek DC, Gibbs CJ Jr. Creutzfeldt–Jakob disease virus isolations from the Gerstmann–Sträussler syndrome. With an analysis of the various forms of amyloid plaque deposition in the virus-induced spongiform encephalopathies. Brain 1981; 104:559–588.
[124] Tateishi J, Sato Y, Nagara H, et al. Experimental transmission of human subacute spongiform encephalopathy to small rodents. IV. positive transmission from a typical case of Gerstmann–Sträussler–Scheinker disease. Acta Neuropathol (Berl) 1984; 64:85–88.

[125] Hsiao K, Scott M, Foster D, et al. Spontaneous neurodegeneration in transgenic mice with mutant prion protein. Science 1990; 250:1587–1590.

[126] Hsiao K, Groth D, Scott M, et al. Serial transmission in rodents of neurodegeneration from transgenic mice expressing mutant prion protein. Proc Natl Acad Sci USA 1994; 91:9126–9130.

[127] Manson JC, Jamieson E, Baybutt H, et al. A single amino acid alteration (101L) introduced into murine PrP dramatically alters incubation time of transmissible spongiform encephalopathy. EMBO J 1999; 18(23):6855–6864.

[128] Goldfarb LG, Brown P, McCombie WR, et al. Transmissible familial Creutzfeldt–Jakob disease associated with five, seven, and eight extra octapeptide coding repeats in the *PRNP* gene. Proc Natl Acad Sci USA 1991; 88:10926–10930.

[129] Collinge J, Brown J, Hardy J, et al. Inherited prion disease with 144 base pair gene insertion. 2. Clinical and pathological features. Brain 1992; 115:687–710.

[130] Gambetti P, Kong Q, Zou W, et al. Sporadic and familial CJD: classification and characterisation. Br Med Bull 2003; 66:213–239.

[131] Kretzschmar HA, Sethi S, Foldvari Z, et al. Iatrogenic Creutzfeldt–Jakob disease with florid plaques. Brain Pathol 2003; 13(3):245–249.

[132] Kopp N, Streichenberger N, Deslys JP, et al. Creutzfeldt–Jakob disease in a 52-year-old woman with florid plaques. Lancet 1996; 348:1239–1240.

[133] Deslys JP, Lasmézas CI, Streichenberger N, et al. New variant Creutzfeldt–Jakob disease in France. Lancet 1997; 349:30–31.

[134] Takashima S, Tateishi J, Taguchi Y, et al. Creutzfeldt–Jakob disease with florid plaques after cadaveric dural graft in a Japanese woman. Lancet 1997; 350:865–866.

[135] Shimizu S, Hoshi K, Muramoto T, et al. Creutzfeldt–Jakob disease with florid-type plaques after cadaveric dura mater grafting. Arch Neurol 1999; 56:357–362.

[136] Skworc KH, Windl O, Schulz-Schaeffer WJ, et al. Familial Creutzfeldt–Jakob disease with a novel 120-bp insertion in the prion protein gene. Ann Neurol 1999; 46(5):693–700.

[137] Itoh Y, Yamada M, Hayakawa M, et al. A variant of Gerstmann–Sträussler–Scheinker disease carrying codon 105 mutation with codon 129 polymorphism of the prion protein gene: a clinicopathological study. J Neurol Sci 1994; 127:77–86.

[138] Mastrianni JA, Curtis MT, Oberholtzer JC, et al. Prion disease (PrP-A117V) presenting with ataxia instead of dementia. Neurology 1995; 45:2042–2050.

[139] Tranchant C, Sergeant N, Wattez A, et al. Neurofibrillary tangles in Gerstmann–Sträussler–Scheinker syndrome with the A117V prion gene mutation. J Neurol Neurosurg Psychiatry 1997; 63:240–246.

[140] Ghetti B, Piccardo P, Spillantini MG, et al. Vascular variant of prion protein cerebral amyloidosis with tau-positive neurofibrillary tangles: The phenotype of the stop codon 145 mutation in *PRNP*. Proc Natl Acad Sci USA 1996; 93:744–748.

[141] Kitamoto T, Iizuka R, Tateishi J. An amber mutation of prion protein in Gerstmann–Sträussler syndrome with mutant PrP plaques. Biochem Biophys Res Commun 1993; 192:525–531.

[142] Samaia HB, Mari JD, Vallada HP, et al. A prion-linked psychiatric disorder. Nature 1997; 390:241.

[143] Gambetti P, Parchi P, Petersen RB, et al. Fatal familial insomnia and familial Creutzfeldt–Jakob disease: clinical, pathological and molecular features. Brain Pathol 1995; 5:43–51.

[144] Ishida S, Sugino M, Koizumi N, et al. Serial MRI in early Creutzfeldt–Jakob disease with a point mutation of prion protein at codon 180. Neuroradiology 1995; 37:531–534.

[145] Kitamoto T, Ohta M, Dohura K, et al. Novel missense variants of prion protein in Creutzfeldt–Jakob disease or Gerstmann–Sträussler syndrome. Biochem Biophys Res Commun 1993; 191:709–714.

[146] Nitrini R, Rosemberg S, Passos-Bueno MR, et al. Familial spongiform encephalopathy associated with a novel prion protein gene mutation. Ann Neurol 1997; 42:138–146.

[147] Grasbon-Frodl EM, Lorenz H, Mann U, et al. Loss of glycosylation associated with the T183A mutation in human prion disease. Acta Neuropath 2004; 108(6):476–484.

[148] Ghetti B, Dlouhy SR, Giaccone G, et al. Gerstmann–Sträussler–Scheinker disease and the Indiana kindred. Brain Pathol 1995; 5:61–75.

[149] Chapman J, Arlazoroff A, Goldfarb LG, et al. Fatal insomnia in a case of familial Creutzfeldt–Jakob disease with the codon 200Lys mutation. Neurology 1996; 46:758–761.

[150] Mastrianni JA, Iannicola C, Myers RM, et al. Mutation of the prion protein gene at codon 208 in familial Creutzfeldt–Jakob disease. Neurology 1996; 47:1305–1312.

[151] Roeber S, Krebs B, Neumann M, et al. Creutzfeldt–Jakob disease in a patient with an R208H mutation of the prion protein gene (*PRNP*) and a 17-kDa prion protein fragment. Acta Neuropathol (Berl) 2005; 109(4):443–448.

[152] Pocchiari M, Salvatore M, Cutruzzolá F, et al. A new point mutation of the prion protein gene in Creutzfeldt–Jakob disease. Ann Neurol 1993; 34:802–807.

[153] Ripoll L, Laplanche JL, Salzmann M, et al. A new point mutation in the prion protein gene at codon 210 in Creutzfeldt–Jakob disease. Neurology 1993; 43:1934–1937.

[154] Mastrianni JA, Capellari S, Telling GC, et al. Inherited prion disease caused by the V210I mutation: Transmission to transgenic mice. Neurology 2001; 57(12):2198–2205.

[155] Young K, Piccardo P, Kish SJ, et al. Gerstmann–Sträussler–Scheinker disease (GSS) with a mutation at prion protein (PrP) residue 212. J Neuropathol Exp Neurol 1998; 57:518.

[156] Hoque MZ, Kitamoto T, Furukawa H, et al. Mutation in the prion protein gene at codon 232 in Japanese patients with Creutzfeldt–Jakob disease: a clinicopathological, immunohistochemical and transmission study. Acta Neuropathol 1996; 92(5):441–446.

[157] Pastore M, Castellani R, Chin S, et al. Creutzfeldt–Jakob disease associated with the novel R148H prion protein gene mutation. J Neuropathol Exp Neurol 2002; 61:491.

[158] Collins S, Boyd A, Fletcher A, et al. Novel prion protein gene mutation in an octogenarian with Creutzfeldt–Jakob disease. Arch Neurol 2000; 57(7):1058–1063.

[159] Peoc'h K, Manivet P, Beaudry P, et al. Identification of three novel mutations (E196K, V203I, E211Q) in the prion protein gene (*PRNP*) in inherited prion diseases with Creutzfeldt–Jakob disease phenotype. Hum Mutat 2000; 15(5):482.

[160] Ladogana A, Almonti S, Petraroli R, et al. Mutation of the *PRNP* gene at codon 211 in familial Creutzfeldt–Jakob disease. Am J Med Genet 2001; 103(2):133–137.

[161] Finckh U, Muller-Thomsen T, Mann U, et al. High prevalence of pathogenic mutations in patients with early-onset dementia detected by sequence analyses of four different genes. Am J Hum Genet 2000; 66(1):110–117.

27 The Pathology of Prion Diseases in Animals

Gerald A.H. Wells, Stephen J. Ryder, and William J. Hadlow

27.1 Introduction

This chapter summarizes the morphological features of the naturally occurring transmissible spongiform encephalopathies (TSE) or prion diseases of animal species affected (Table 1.3). Each disease entity is considered in the chronological order of its original description. By definition, these diseases resemble, in all the major pathological features, the prion diseases of humans (see Chapter 26), but in each species clear differences arise in the phenotypic expression of the diseases. In particular the relative emphasis of the component changes and the patterns of their distribution vary. Over the few decades of documentation of animal TSE pathology the focus has extended from the initial descriptions of the light microscopic morphological changes in the central nervous system (CNS), comprising the triad of vacuolar changes, neuronal degeneration, and glial responses, to the immunohistochemical demonstration of the disease-specific form of PrP and its accumulation patterns throughout the body. Given the etiological uncertainties surrounding TSE, the central role of microscopic pathological features in defining each disease phenotype remains paramount. There are no specific macroscopic changes in prion diseases of animals. Important to the interpretation of the microscopic changes is a knowledge of the incidental background vacuolar changes that appear to be features of the normal CNS of most species.

27.2 Pathology of scrapie in sheep and goats

Scrapie is a naturally occurring transmissible spongiform encephalopathy of sheep, goats, and moufflon [1–4] that has been known as a clinical entity in sheep in Europe for over two centuries [5]. In this respect it is the earliest known prion disease, although its characterization pathologically began only in the twentieth century. The clinical disease is characterized by severe pruritus and wasting, in addition to neurological signs after a very long incubation period. Histopathological changes are those associated with all prion diseases; vacuolation of gray matter of the CNS, gliosis, neuronal degeneration, and accumulation of the disease-specific form of the prion protein, PrP^{Sc}. Accumulation of PrP^{Sc} is the only detectable disease-related change outside of the nervous system. PrP^{Sc} is initially detected in lymphoid tissues of the gastrointestinal tract, in some cases many years prior to the onset of clinical signs [6]. Immunohistochemistry has shown disease-specific PrP to be associated with follicular dendritic cells and macrophages within the germinal centers of lymphoid follicles (Fig. 27.1a). There is a progressive build up of PrP^{Sc} in all lymphoid tissues until at end stage disease most, if not all, lymph nodes show abundant PrP^{Sc} accumulation [7; 8]. Despite the quantities of abnormal protein that accumulate in germinal centers, there is no evidence of any functional abnormality of the lymphoid system. The extent of lymphoid tissue involvement has been shown to vary; in some sheep, in particular those of the VRQ/ARR PrP genotype, there is little [7] or no [8] detectable PrP^{Sc} in lymphoid tissue in the clinical disease. It is unclear if this is a result of a low level accumulation or complete absence of accumulation but it shows that widespread and abundant accumulation of PrP^{Sc} in lymphoid tissue is not a prerequisite for involvement of the nervous system or the development of clinical disease. The role of abundant PrP^{Sc}

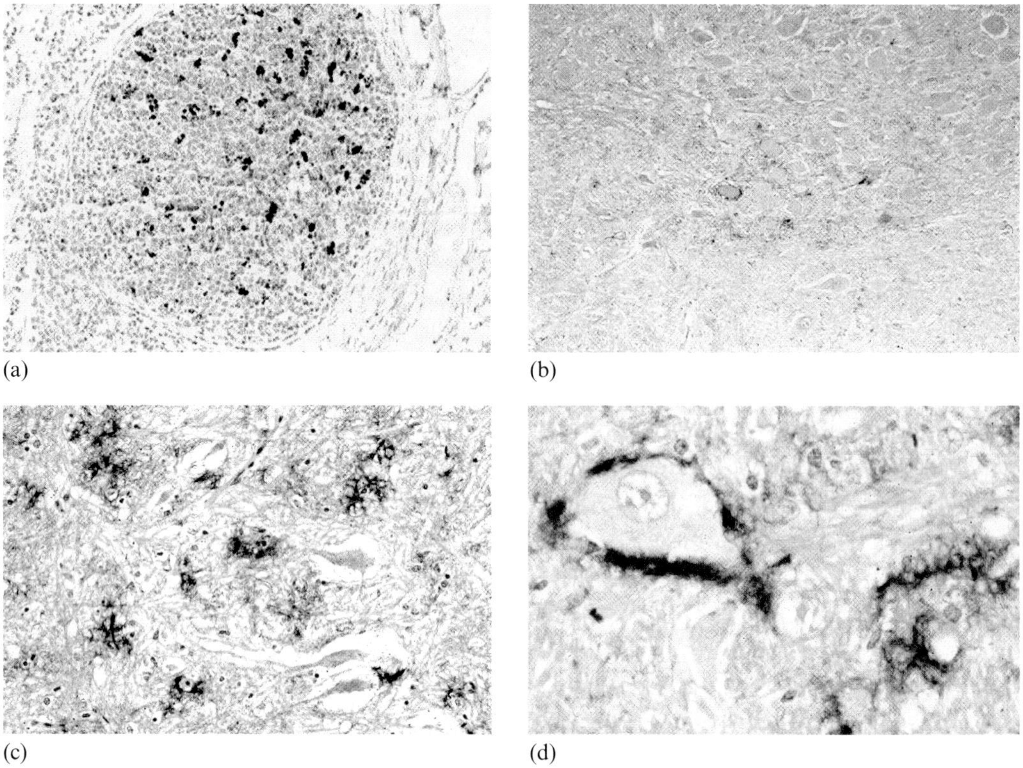

Fig. 27.1: (a) Disease-specific PrP immunolabeling of macrophages (course, intense multifocal labelling) and of follicular dendritic cells (faint, reticulated labelling) in lymph node of sheep with scrapie (PrP-IHC). **(b)** Disease-specific PrP immunolabeling confined mainly to the ventral border of the dorsal parasympathetic nucleus of the vagus nerve of a pre-clinically scrapie infected sheep (PrP-IHC). **(c)** Multifocal stellate periglial PrP immunolabeling in scrapie affected sheep brain (PrP-IHC). **(d)** Perineuronal PrP immunolabeling in scrapie affected sheep brain (PrP-IHC). PrP = Prion protein. IHC = Immunohistochemistry. Crown copyright ©.

accumulation in many sheep therefore remains enigmatic, but it may be involved in transmission of the disease, as systematic distribution of infection increases the potential for shedding of agent via excretory or secretory routes.

Disease-specific PrP accumulates in the autonomic peripheral nervous system (PNS) and CNS of all scrapie affected sheep, in advance of any morphological changes [9]. Distribution is influenced initially by the stage of disease. In pre-clinical stages of the disease, once the CNS is involved, PrP^{Sc} is found predominantly in the brain stem, in the dorsal parasympathetic nucleus of the vagus nerve, consistent with entry via the vagus nerve innervating the gastrointestinal tract [10; 11] (Fig. 27.1b). From these early bilaterally symmetrical deposits, PrP^{Sc} accumulation progressively involves other parts of the CNS and, in many sheep, the entire neuraxis. The morphology and distribution of PrP^{Sc} deposits vary widely between affected sheep [11–13]. A number of different types of deposits have been described. In the neuropil, deposits include fine particulate or granular forms and larger, apparently unstructured aggregates. Large aggregates are also found underlying the pia, ependyma, and surrounding blood vessels (associated with the glia limitans). Labeling of glia takes two forms; stellate periglial deposits (Fig. 27.1c) and dense punctate intra-glial deposits. Amyloid plaques are rare in natural scrapie but intra-mural vascular accumulations of amyloid are observed. Neurons show punctate or granular deposits in the cytoplasm and around the membrane of perikarya (Fig. 27.1d) and neurites.

Gliosis and cell loss have historically been described as hallmark lesions of scrapie. No studies have been carried out to objectively quantify neuronal numbers in natural scrapie, but significant cell loss is not evident on routine histopathological examination. Evidence for neuronal degeneration in the form of chromatolysis, neuronophagia, and dark, shrunken neurons are occasionally seen [3; 4]. Gliosis, in the form of astrocyte pleomorphism, hypertrophy, and hyperplasia is seen in some cases of scrapie, often in gray matter areas of the diencephalon, but can be detected in all areas of the brain. In many cases there is no obvious astrocytic response compared to normal sheep, but this is complicated by the apparent heterogeneity of astrocytes and high levels of the astrocyte protein GFAP in the brains of normal sheep [14]. Vacuolar pathology seen in routine histological preparations of brain tissue is the hallmark lesion of sheep scrapie. Vacuolation of grey matter, generally bilaterally symmetrical, takes two forms; intra-cytoplasmic vacuolation of neuronal perikarya and vacuolation of the neuropil (Fig. 27.2). Both types are commonly seen in scrapie, either separately or in combination, but the diagnostic value of intraneuronal vacuolation is complicated by the common occurrence of incidental intra-cytoplasmic vacuolation of neuronal perikayra, particularly in the dorsal parasympathetic nucleus of the vagus nerve in normal sheep [15]. Artifactual spaces, resembling vacuolation, in the neuropil may also arise from autolysis and technical factors.

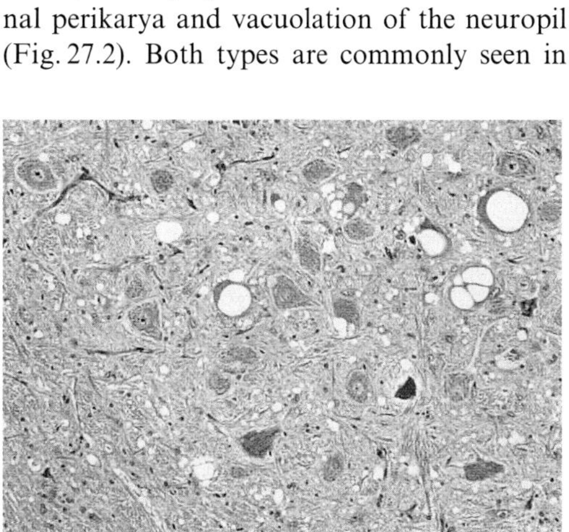

Fig. 27.2: Intracytoplasmic vacuolation of neuronal perikarya and vacuolation of neuropil in the dorsal parasympathetic nucleus of the vagus nerve in a scrapie affected sheep (HE). Crown copyright ©.

Considerable variation in the distribution of vacuolar pathology in natural scrapie has been reported. A review of the pathology in 222 natural cases of scrapie diagnosed over approximately a 10-year period classified cases into 7 patterns based on lesion distribution [3]. In all cases vacuolation was present in the medulla oblongata in the dorsal parasympathetic nucleus of the vagus nerve. Lesions at this level of the brainstem have considerable diagnostic significance and this is commonly the only part of the brain examined for scrapie diagnosis. The patterns varied in the extent of vacuolation in more rostral areas of the brain [3]. Despite its considerable diagnostic significance, vacuolation is not invariably present in the brains of scrapie affected sheep. Experimental scrapie in lambs, succumbing to disease at the relatively young age (for natural scrapie) of approximately seven months showed minimal or no vacuolar pathology.

An unusual distribution of vacuolar pathology has been described in a, seemingly, novel scrapie phenotype, termed the Nor98 variant, from Norway [16]. In these cases there is vacuolation of the cerebellar and cerebral cortices, but not of the medulla at the level of the obex. Similar "atypical" presentations of scrapie have since been found elsewhere, many in sheep with PrP genotypes not usually associated with classical scrapie, including the ARR/ARR genotype in which natural clinical scrapie is rare [17]. The Nor98 cases, the prototypes of "atypical" scrapie, have little or no abnormal PrP at the obex, but in most cases exhibit an intense PrP^{Sc} deposition/accumulation in the cerebellar cortex and are characterized at a molecular level by a smaller and less stable protease-resistant core of PrP^{Sc}. The extent to which such cases are etiologically atypical is uncertain. None of the evidence gathered so far suggests that "atypical" cases of scrapie are caused by the BSE agent. Scrapie is a highly variable disease, thought to be caused by a number of strains of infectious agent. Whether these "atypical" cases represent part of that spectrum of disease phenotypes, or a new disease entity, remains to be established. In the meantime, the differences between such cases and the forms of scrapie that have up to now been termed "classical scrapie" have practical implications for diagnosis of TSE in sheep.

The variability in type and distribution of lesions, whether vacuolar changes or PrP^{Sc} accumulation in the brains of scrapie affected sheep, is a striking feature in all reports of the disease. The significance of this is unclear. In mice the distribution and the relative severity of lesions (and the incubation period of the disease) have been shown to be affected by a number of host factors, in particular, but not exclusively, the PrP genotype and the route of challenge [18]. However, agent strain has a major influence on the distribution of lesions (the lesion profile). It is tempting to attribute at least some of the variation seen in natural scrapie to agent strain diversity, but this has not yet been conclusively shown.

27.3 Pathology of transmissible mink encephalopathy

Spongiform alterations in the neuropil of the gray matter, neuronal degeneration, and astrocytosis [19–21] characterize the essential histopathological changes in transmissible mink encephalopathy (TME). Generally the most obvious changes are the spongiform alterations which vary from localized to diffuse neuropil vacuolation. The intensity of vacuolation is remarkably less in older mink homozygous for the "Aleutian" gene, which is responsible for the metallic grey color of the fur [22]. Dark, shrunken, angular neuronal perikarya are the main expression of neuronal degeneration, which results in loss of nerve cells. Much less commonly, neurons, mainly those in the brain stem, have large vacuoles in the perikaryon. Astrocytosis (hypertrophy and hyperplasia of astrocytes in gray matter) is also prominent. Most reactive astrocytes appear as enlarged, pale, naked nuclei. Those in the molecular layer of the cerebral cortex, notably at depths of sulci, typically have wisps of cytoplasm, but obvious gemistocytes are unusual. Amyloid plaques are not found in TME [19]. Neuropathological alterations are characterized by a bilateral symmetrical pattern, especially in the telencephalon [19; 23]. Early alterations occur in the dorsomedial gyri of the frontal lobes of the cerebral neocortex [19; 24]. Caudally the degree of degenerative change is decreased and the occipital lobes are spared. The septal area, the corpus striatum, as well as the diencephalon are commonly involved. Severe degenerative changes are present in the thalamus, especially in the caudal dorsal part and in the medial geniculate nucleus. More diffuse alterations are seen in the hypothalamus. In the hippocampal formation the lesions are generally moderate. Those in the amygdala are always severe. The lesions in the brain stem are generally less intense and more variable, compared with more rostral changes. In the midbrain the caudal colliculi and the central gray matter are especially involved. In the pons and medulla oblongata only limited changes occur. The cerebellum is not affected. There are no published accounts of the extraneural pathogenesis of TME in the natural host, nor of the distribution patterns of disease-related PrP.

27.4 Pathology of chronic wasting disease

Chronic wasting disease (CWD) affects North American species of deer; elk (*Cervus elaphus nelsoni*), white tailed deer (*Odocoileus virginianus*), and mule deer (*Odocoileus hemionus*; Table 1.3) [25]. The disease is characterized by progressive weight loss, generally in the absence of neurological signs, but microscopic lesions are confined to the CNS. Like other TSE there are no macroscopic lesions. Spongiform lesions with intracytoplasmic vacuoles, especially in the neurons of the brainstem, are the predominant characteristics [26]. Neuronal degeneration and increased numbers of astrocytes (astrocytosis) occur, but are not prominent changes. The deposition of amyloid is relatively common and can be detected in haematoxylin and eosin (HE) stained sections, or by PrP immunohistochemistry [27–30]. Amyloid plaques of different sizes, often surrounded by vacuolated neurons, are found in all species, but with decreasing frequency in white-tailed deer, elk, and mule deer [25]. These are comparable to florid plaques, which occur in the variant form of CJD (vCJD). The distribution of lesions in the brains of affected animals is very uniform, with only minor differences seen between the three affected species. The most severely affected areas are the dorsal nucleus of the vagus nerve in the medulla oblongata, the hypothalamus, many nuclei in the thalamus, and in the olfactory cortex [26]. Alterations in the medulla oblongata at the level of the obex are invariably present, a feature that has enabled the histological investigation of hunter-killed animals in surveillance programs to be based on this brain area alone. While several nuclei in the medulla oblongata, pons, mesencephalon and telencephalon show changes in clinically affected animals [26; 31], the parasympathetic nucleus of the vagus nerve in the dorsal portion of the medulla oblongata at the obex is the most important site to be examined for diagnosis of CWD in apparently clinically normal animals [30; 32]. The disease-associated prion protein, PrP^{Sc}, as demonstrated by immunohistochemistry, is found in the brain, palatine tonsils, visceral and regional lymph nodes, Peyer's patches, other lymphoid tissue of small and large intestine, and spleen of affected deer [33]. In the brain, the disease-specific PrP accumulation is seen initially in the dorsal motor nucleus of the vagus nerve, prior to the occurrence of spongiform change [26; 34]. Sigurdson et al [35] found PrP^{Sc} in pituitary gland, adrenal medulla and pancreatic islets and in the PNS involving the vagosympathetic trunk, sympathetic trunk, nodose ganglion, myenteric plexus, brachial plexus, and sciatic nerve, but not the trigeminal (Gasserian) ganglion, coeliac ganglion, cranial cervical ganglion, or spinal nerve roots (dorsal and ventral) in a small sample of mule deer naturally affected with CWD. These findings suggest that there is, at least in the clinical disease, extensive involvement of multiple organ systems, including central and peripheral nervous tissues, endocrine organs and alimentary tract, the last suggesting a possible means of agent shedding. Lesions caused by CWD are ultrastructurally similar to those seen in other prion diseases [36; 37]. Electronmicroscopically, scrapie-associated fibrils (SAF; Fig. 1.5b) are found in brain extracts and are morphologically identical to those found in brain extracts from other TSE-affected species [31; 38]. The pathogenesis of CWD has been studied in experimentally infected mule deer after oral exposure to brain homogenate from a clinical case of CWD [33]. PrP^{Sc} was detected in alimentary-tract-associated lymphoid tissues (involving one or more of the following: retropharyngeal lymph node, tonsil, Peyer's patches, and ileocaecal lymph node) as early as 42 days post-inoculation (p.i.) and in all fawns examined thereafter (53 to 80 days p.i. when the study was terminated). No PrP^{Sc} was detectable in neural tissue in any fawn. In other, ongoing, pathogenesis studies in orally exposed deer [34] PrP^{Sc} was detected in alimentary lymphoid tissue three months before it was consistently present in the brain. Consistent with findings in naturally infected deer, PrP^{Sc} was first detected in the brain in the parasympathetic nucleus of the vagus nerve some 10 months prior to the presence of spongiform changes in this nucleus at 16 months p.i. In similarly infected elk, PrP^{Sc} was also first detected in the parasympathetic nucleus of the vagus nerve, preceding spongiform changes by 6 months.

27.5 Pathology of bovine spongiform encephalopathy

Bovine spongiform encephalopathy (BSE) was first recognized as a scrapie-like disease of domestic cattle in Great Britain in 1986 [39]. In many respects the pathology of BSE resembles that of scrapie. However, differences, particularly in relation to the extraneural pathogenesis of BSE have emerged from studies of both the natural [40–42] and experimental [43–49] disease in cattle.

In contrast to scrapie of sheep and goats and to chronic wasting disease (CWD) of cervids, extraneural agent distribution appears restricted, consistently involving only the distal ileum. In the distal ileum, disease-specific PrP is present in macrophages within only a small proportion of follicles of Peyer's patches throughout much of the disease course from 6 months after experimental oral exposure. In similarly exposed calves killed six months post exposure, no evidence of disease-associated PrP was found in mesenteric lymph nodes or at other levels of intestine [43].

This primary lymphoid tissue involvement, unlike the situation in the other species, does not extend, over the course of the incubation period, to systematic spread throughout the lymphoreticular system, even in cattle exposed orally to a large infective dose of agent [47; 48]. Investigations of the pathogenesis of experimental scrapie in laboratory rodents [50–60] and natural scrapie of sheep [9; 53; 55; 60] and, indeed, experimental BSE infection of sheep [61], indicate that after oral intake the agents replicate in both the lymphatic and/or neural tissue of the intestine. The subsequent spread to the brain and spinal cord is, at least in experimental models, considered to be via the tracts of the autonomic nervous system. The only evidence to date of infectivity in the autonomic nervous system of cattle with BSE is that of inconsistently detectable disease-specific PrP accumulation in the ganglion cells of the myenteric plexus (but not the submucosal plexus) of the distal ileum of naturally and experimentally affected cattle [43]. Disease-specific PrP accumulation has been reported in the vagus nerve of a naturally occurring case of BSE in Japan [62]. From results of sequential time point pathogenesis studies following oral exposure, infectivity in the PNS of cattle has been found only in dorsal root ganglia and trigeminal ganglion, concurrent with CNS infectivity [48]. Other PNS tissues (cranial cervical ganglion, stellate ganglion, sciatic, facial, and phrenic nerves) and skeletal muscles (triceps, masseter, sternocephalicus, and longissimus dorsi) examined by mouse assay have proved negative [47; 48]. Similarly, tissue pools of sciatic with radial nerve and of masseter, semitendinosis, and longissimus dorsi muscles, from animals at selected time points of the oral exposure study in cattle and inoculated into cattle, have not transmitted [47].

While muscle homogenates have not transmitted disease to cattle from cattle infected experimentally with BSE, there is a single recent report of a trace of infectivity (detected by assay in mice overexpressing the bovine PrP gene) in skeletal muscle tissue (M. semitendinosus) in a clinical case of BSE in Germany [40]. In the same case and in a clinical case of BSE in Japan [62], infectivity or disease-specific PrP, respectively, have been detected in certain somatic peripheral nerve trunks (including the sciatic nerve), raising questions as to the precise cellular localization of infectivity in the muscle tissue. Retrograde spread of agent via peripheral nerves after CNS infection may explain involvement of sensory ganglia and the potential for spread to sensory and somatic motor nerve trunks. Bioassays of tissue infectivity in mice or cattle have also revealed traces of infection in bone marrow (Fig. 53.2), only at the time of clinical onset [49] and in tonsil, only early in the incubation period [47] of experimentally exposed cattle.

The pathologic changes of BSE are confined to the CNS and in the clinical disease are dominated histologically by spongiform change or vacuolation of the neuropil (Fig. 27.3a). Ultrastructural studies of this vacuolation show membrane-bounded vacuoles within neuronal processes [63]. Vacuolation of neuronal perikarya is most prominent in certain neuroanatomic locations, particularly the vestibular nuclei. Evidence of neuronal degeneration is seen occasionally and includes necrotic neurons, basophilic shrunken neurons, neurophagia, and dystrophic neurites. Substantial neuronal loss has been demonstrated by morphometric studies of the vestibular nuclei [64; 65]. Astrocytic hypertrophy is regularly seen, although usually it is not as marked as in sheep scrapie. Astrocytosis is evident using GFAP immunohistochemical labeling. Consistent with the light microscopic appearances, ultrastructural changes are mainly seen in gray matter

[63] and resemble closely those of other TSEs. In addition to the vacuoles within neurites there are hypertrophic astrocytes and dystrophic cells containing accumulations of neurofilaments, mitochondria, lamellar bodies, and other electron-dense profiles. Immunohistochemical examination for disease-specific PrP demonstrates widespread immunolabeling which generally corresponds to the distribution of vacuolar changes [66]. Disease-associated immunolabeling is mostly particulate in form (Fig. 27.3b) and diffuse within affected areas, but also appears perineuronally (so-called synaptic type), in a linear form (following the course of neurites; Fig. 27.3c) and as a stellate form (periglial). In addition to these forms in the neuropil, a coarse granular form of labeling occurs in the perikarya of some neurons (Fig. 27.3c). Amyloidosis, based on demonstration of apple green birefringence in Congo Red-stained sections, a feature of some PrP aggregations in scrapie, is very infrequent and localized in BSE [66].

From the first preliminary identification of BSE as a new disease in cattle, it was characterized by a distinctive pathological presentation. Subsequent studies using semi-quantitative methods previously applied to mouse models of scrapie [67], assessed the severity of the vacuolar changes relative to their distribution – the lesion profile [44; 66; 68–70]. These studies established that BSE produced a consistent neuropathological pattern of changes, a feature which was in contrast to the variable lesion patterns reported in scrapie of sheep. Neuropil vacuolation in the solitary tract nucleus or the nucleus of the spinal tract of the trigeminal nerve occurred in 99.6% of clinically suspect cases confirmed by a more extensive histopathological examination of the brainstem [71]. In the 1980's the application of methods of detection of the prion protein (disease-spe-

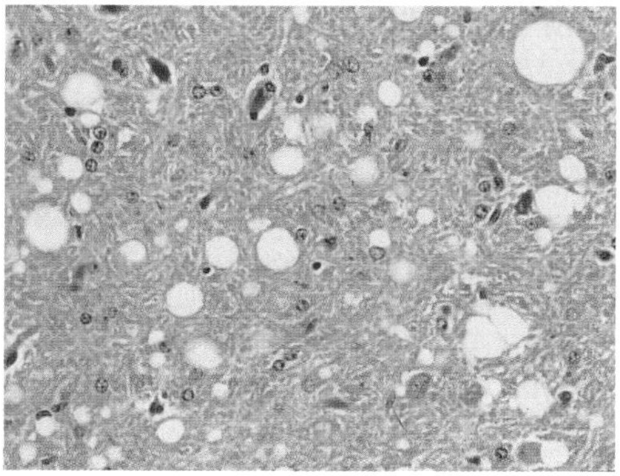
(a)

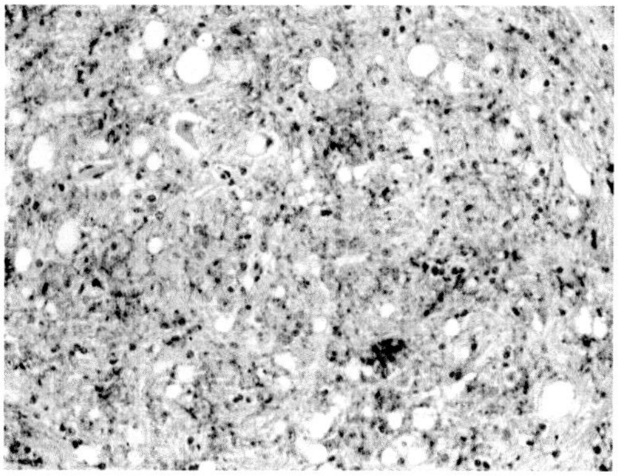
(b)

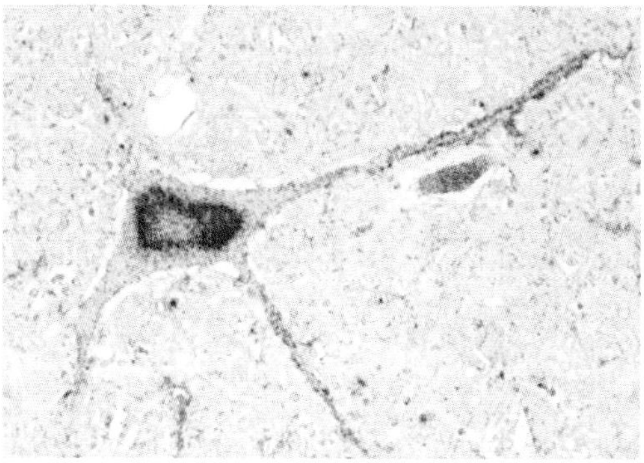
(c)

Fig. 27.3: (a) Spongiform change or vacuolation of neuropil in the medulla oblongata of a BSE affected cow (HE). **(b)** Particulate PrP immunolabeling in the solitary tract nucleus of a BSE affected cow (PrP-IHC). **(c)** Perineuronal PrP immunolabeling forming linear deposits on the course of neurites and intense granular PrP immunolabeling in a neuronal perikaryon (PrP-IHC). PrP = Prion protein. IHC = Immunohistochemistry. Crown copyright ©.

cific PrP) was not a routine approach to diagnosis of TSE; histopathological examination of the brain was the standard. The stereotypical pattern of the brain lesions was therefore important because it had the potential to simplify the histopathologic diagnosis, enabling diagnosis of BSE by the examination of a single brainstem section (medulla oblongata at the level of the obex).

This simplified diagnostic protocol was crucial to the management of the BSE epidemic in the UK, where at peak incidence (1992/93), some 1000 suspected cases were reported in a week. The lesion profile approach has been used to monitor the pattern of the vacuolar changes in the brains of cases sampled sequentially through the epidemic. The lesion profile of BSE has remained constant over the course of the epidemic in Britain (Fig. 27.4). This has provided crucial evidence of an apparent biological stability of the disease phenotype in cattle with BSE and has been important in supporting the notion, further proven by strain typing studies in mice, that the epidemic was sustained by a single, or major strain of TSE agent [72; 73]. Areas with the most severe lesion scores are in midbrain (central gray matter), pons and medulla (nucleus of the spinal tract of the trigeminal nerve), diencephalon (hypothalamic nuclei), and hippocampus. The severity of changes in the brain invariably diminishes rostrally [44]. Lesions in the spinal cord are present throughout cervical, thoracic, and lumbar regions, with the most marked changes in the substantia gelatinosa of the dorsal gray horns.

Based on experience of field cases of BSE and studies of the distribution and relative severity of vacuolar changes, it has been speculated that the solitary tract nucleus and the spinal tract nucleus of the trigeminal nerve are possible primary sites of brain lesions in BSE [74]. Subsequent study of early disease-specific PrP distribution in the brains of cattle infected with BSE agent by the oral route [10] indicated PrP accumulation initially in these same medullary nuclei. This is in contrast to natural scrapie and experimental BSE in sheep [10], in which preclinical accumulation of PrP was found solely in the dorsal nucleus of the vagus nerve. The possible significance of these observations in terms of differences in entry of the agent into the CNS between BSE in cattle and scrapie and BSE in sheep remains unclear.

The possible natural occurrence of phenotypic variation in BSE or novel expression of a TSE of cattle has been raised by reports from Japan [75], Italy [76], and France [77] and since, in several other European countries, of a small number of cattle with molecular and/or pathological features atypical of BSE as defined in

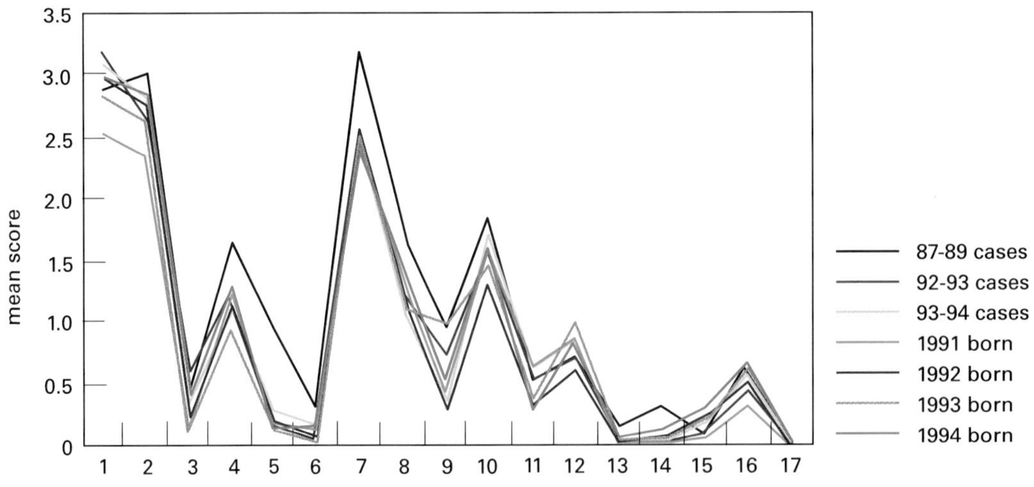

Fig. 27.4: The constant lesion profile of BSE through the major period of the UK epidemic [48], 2004. 1-17, major neuroanatomical nuclei from medulla oblongata through to forebrain. Crown copyright ©.

the UK. These "atypical" cases were detected through the results of rapid tests (Table 35.3) applied to slaughtered cattle in active surveillance programs of the respective countries and are of undetermined clinical status. While they all have unusual electrophoretic profiles of the protease-resistant prion protein accumulating in the brain, only the report from Italy [76] describes immunohistochemical differences in form and distribution of disease-specific PrP in the brains compared to typical BSE. In particular these show a widespread distribution of small PrP plaques which, notably, occur in some areas, in white matter.

27.6 Pathology of BSE in non domestic captive ungulate species

The basic features of the pathology in the zoo ungulate species attributed to BSE are described in Chapter 20. Detailed studies of the distribution of CNS changes in affected species (see references in Table 1.3) have not been published, but Simmons et al [78] examined the profile of vacuolar changes in the brains of eland, oryx, and greater kudu and found them to be different from the lesion profile of BSE and to differ among the three species. In contrast to the limited distribution of infectivity in tissues of cattle with BSE, infectivity in tissues in affected kudu appears widespread [79], similar to that seen in CWD and scrapie.

27.7 Pathology of feline spongiform encephalopathy

Feline spongiform encephalopathy (FSE) was first reported in domestic cats in the UK in 1990 [80; 81]. The disease occurred contemporaneously with the BSE epidemic in cattle, and strain typing of the infectious agent isolated from cats in mice has shown that it is caused by the same strain of agent [72]. The disease has essentially the same features as other spongiform encephalopathies of domestic animals and is described in Chapter 23. However, the distribution and the severity of the brain lesions seen in cats is strikingly different from those in ruminants, and similar to those of TME of ranch-reared mink. Although lesions are widespread, the brainstem is relatively spared compared to the diencephalon, in particular the medial geniculate nucleus (Fig. 27.5a), and the corpus striatum. The predominant lesion in all areas is vacuolation of the neuropil, with only occasional vacuoles of neuronal perikarya observed. The intensity of spongiform change in many areas is greater than that seen in ruminant species. In some cases neuronal loss may be evident in routine HE sections, particularly in the cerebellar cortex (Fig. 27.5b). Immunohistochemically, fine particuate accumulations of abnormal PrP are detected throughout the CNS. Amyloid plaques have not been observed in cats.

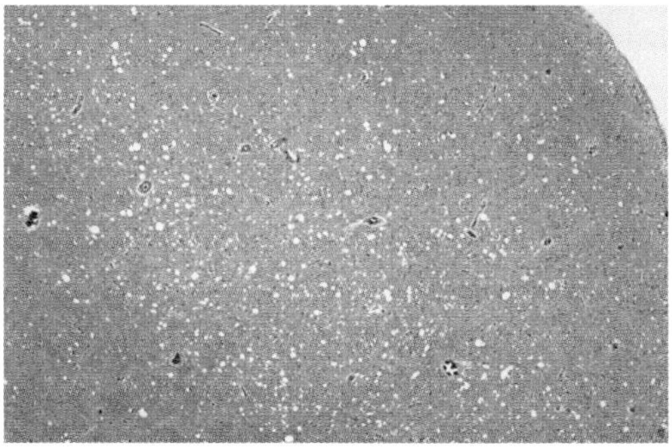

(a)

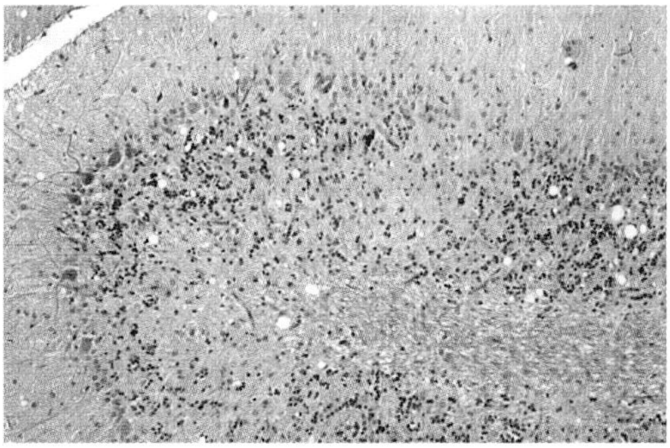

(b)

Fig. 27.5: (a) Severe spongiform change in the medial geniculate nucleus of a domestic cat with FSE. (HE). **(b)** Neuronal loss in the cerebellar cortex, with marked depletion of internal granule layer cells in a cat with FSE. (HE). Crown copyright ©.

Only limited accumulations of abnormal PrP are detectable in lymphoid tissues. In a small sample of cases PrP was detected inconsistently in Peyer's patches and in the spleen [82]. A similar disease has been reported in a variety of non-domesticated cats (Table 1.3), predominantly, though not exclusively, in British zoological collections (see Chapter 20).

References

[1] Wood JL and Done SH. Natural scrapie in goats: neuropathology. Vet Rec 1992; 131:93–96.

[2] Wood JL, Lund LJ, Done SH. The natural occurrence of scrapie in moufflon. Vet Rec 1992; 130:25–27.

[3] Wood JL, McGill IS, Done SH, et al. Neuropathology of scrapie: a study of the distribution patterns of brain lesions in 222 cases of natural scrapie in sheep, 1982– 1991. Vet Rec 1997; 140(7):167–174.

[4] Jeffrey M and Gonzales L. Pathology and pathogenesis of bovine spongiform encephalopathy and scrapie. In: Harris D, editor. Mad cow disease and related spongiform encephalopathies. Springer-Verlag, Berlin, 2004:65–97.

[5] Parry HB. Scrapie disease in sheep. In: Oppenheimer DR, editor. Academic Press, London, 1983.

[6] van Keulen LJ, Schreuder BE, Vromans ME, et al. Scrapie-associated prion protein in the gastrointestinal tract of sheep with natural scrapie. J Comp Pathol 1999; 121(1):55–63.

[7] Andreoletti O, Berthon P, Marc D, et al. Early accumulation of PrP(Sc) in gut-associated lymphoid and nervous tissues of susceptible sheep from a Romanov flock with natural scrapie. J Gen Virol 2000; 81(Pt 12):3115–3126.

[8] van Keulen LJ, Schreuder BE, Meloen RH, et al. Immunohistochemical detection of prion protein in lymphoid tissues of sheep with natural scrapie. J Clin Microbiol 1996; 34:1228–1231.

[9] van Keulen LJ, Schreuder BE, Vromans ME, et al. Pathogenesis of natural scrapie in sheep. Arch Virol Suppl 2000; 16:57–71.

[10] Ryder SJ, Bellworthy S, Wells GAH. A comparison of early PrP distribution in the CNS in natural scrapie and experimental BSE in both cattle and sheep (abstract). Symposium "Characterisation and Diagnosis of Prion Diseases of Animals and Man", Tubingen, Germany 23–25 September, 1999.

[11] Ryder SJ, Spencer YI, Bellerby PJ, et al. Immunohistochemical detection of PrP in the medulla oblongata of sheep: the spectrum of staining in normal and scrapie-affected sheep. Vet Rec 2001; 148(1):7–13.

[12] Gonzalez L, Martin S, Begara-McGorum I, et al. Effects of agent strain and host genotype on PrP accumulation in the brain of sheep naturally and experimentally affected with scrapie. J Comp Pathol 2002; 126(1):17–29.

[13] van Keulen LJ, Schreuder BE, Meloen RH, et al. Immunohistochemical detection and localization of prion protein in brain tissue of sheep with natural scrapie. Vet Pathol 1995; 32:299–308.

[14] Mackenzie A. Immunohistochemical demonstration of glial fibrillary acidic protein in scrapie. J Comp Pathol 1983; 93:251–259.

[15] Zlotnik I and Rennie JC. A comparative study of the incidence of vacuolated neurons in the medulla from apparently healthy sheep of various breeds. J Comp Pathol 1958; 68:411–415.

[16] Benestad SL, Sarradin P, Thu B, et al. Cases of scrapie with unusual features in Norway and designation of a new type, Nor98. Vet Rec 2003; 153(7):202–208.

[17] European Food Safety Authority (EFSA). Opinion of the scientific panel on biological hazards on the request from the European Commission on classification of atypical transmissible spongiform encephalopathy (TSE) cases in small ruminants. The EFSA Journal 2005; 276:1–30.

[18] Bruce ME, Boyle A, McConnell I. TSE strain typing in mice. In: Lehmann S, Grassi J, editors. Techniques in prion research. Birkhäuser Verlag, Switzerland, 2004:132–146.

[19] Eckroade RJ, ZuRhein GM, Hanson RP. Experimental transmissible mink encephalopathy: brain lesions and their sequential development in mink. In: Prusiner SB, Hadlow WJ, editors. Slow transmissible diseases of the nervous system, Vol 1. Academic Press, New York, 1979: 409–449.

[20] Duker II, Geller VI, Chizhov VA, et al. Clinical and morphological investigation of transmissible mink encephalopathy (in Russian). Vopr Virusol 1986; 31:220–225.

[21] Marsh RF and Hadlow WJ. Transmissible mink encephalopathy. Rev Sci Tech Off Int Epiz 1992; 11:539–550.

[22] Marsh RF, Sipe JC, Morse SS, et al. Transmissible mink encephalopathy. Reduced spongiform degeneration in aged mink of the Chediak-Higashi genotype. Lab Invest 1976; 34:381–386.

[23] Johannsen U and Hartung J. Infectious encephalopathy in the mink. 2. Pathological-morphological studies (in German). Monatsh Veterinarmed 1970; 25(10):389–395.

[24] Hadlow WJ, Race RE, Kennedy RC. Temporal distribution of transmissible mink encephalopathy virus in mink inoculated subcutaneously. J Virol 1987; 61:3235–3240.

[25] Williams ES. Chronic wasting disease. Vet Pathol 2005; 42(5):530–549.

[26] Williams ES and Young S. Neuropathology of chronic wasting disease of mule deer (Odocoileus hemionus) and elk (Cervus elaphus nelsoni). Vet Pathol 1993; 30:36–45.

[27] Guiroy DC, Williams ES, Yanagihara R, et al. Topographic distribution of scrapie amyloid-immunoreactive plaques in chronic wasting disease in captive mule deer (Odocoileus hemionus hemionus). Acta Neuropathol (Berl) 1991; 81:475–478.

[28] Guiroy DC, Williams ES, Yanagihara R, et al. Immunolocalization of scrapie amyloid (PrP27–30) in chronic wasting disease of Rocky Mountain elk and hybrids of captive mule deer and white-tailed deer. Neurosci Lett 1991; 126:195–198.

[29] O'Rourke KI, Baszler TV, Miller JM, et al. Monoclonal antibody F89/160.1.5 defines a conserved epitope on the ruminant prion protein. J Clin Microbiol 1998; 36(6):1750–1755.

[30] Spraker TR, Zink RR, Cummings BA, et al. Distribution of protease-resistant prion protein and spongiform encephalopathy in free-ranging mule deer (Odocoileus hemionus) with chronic wasting disease. Vet Pathol 2002; 39(5):546–556.

[31] Spraker TR, Miller MW, Williams ES, et al. Spongiform encephalopathy in free-ranging mule deer (Odocoileus hemionus), white-tailed deer (Odocoileus virginianus), and Rocky Mountain elk (Cervus elaphus nelsoni) in north-central Colorado. J Wildl Dis 1997; 33:1–6.

[32] Peters J, Miller JM, Jenny AL, et al. Immunohistochemical diagnosis of chronic wasting disease in preclinically affected elk from a captive herd. J Vet Diagn Invest 2000; 12(6):579–582.

[33] Sigurdson CJ, Williams ES, Miller MW, et al. Oral transmission and early lymphoid tropism of chronic wasting disease PrPres in mule deer fawns (Odocoileus hemionus). J Gen Virol 1999; 80:2757–2764.

[34] Williams ES and Miller MW. Chronic wasting disease in deer and elk in North America. Rev Sci Tech Off Int Epiz 2002; 21(2):305–316.

[35] Sigurdson CJ, Spraker TR, Miller MW, et al. PrP(CWD) in the myenteric plexus, vagosympathetic trunk and endocrine glands of deer with chronic wasting disease. J Gen Virol 2001; 82(10):2327–2334.

[36] Guiroy DC, Williams ES, Liberski PP, et al. Ultrastructural neuropathology of chronic wasting disease in captive mule deer. Acta Neuropathol 1993; 85:437–444.

[37] Guiroy DC, Liberski PP, Williams ES, et al. Electron microscopic findings in brain of Rocky Mountain elk with chronic wasting disease. Folia Neuropathol 1994; 32:171–173.

[38] Guiroy DC, Williams ES, Song KJ, et al. Fibrils in brain of Rocky Mountain elk with chronic wasting disease contain scrapie amyloid. Acta Neuropathol 1993; 86:77–80.

[39] Wells GAH, Scott AC, Johnson CT, et al. A novel progressive spongiform encephalopathy in cattle. Vet Rec 1987; 121:419–420.

[40] Buschmann A and Groschup MH. Highly bovine spongiform encephalopathy-sensitive transgenic mice confirm the essential restriction of infectivity to the nervous system in clinically diseased cattle. J Infect Dis 2005; 192(5):934–942.

[41] Fraser H and Foster JD. Transmission to mice, sheep and goats and bioassay of bovine tissues. In: Bradley R, Marchant B, editors. Transmissible spongiform encephalopathies. Proceedings of a consultation on BSE with the Scientific Veterinary Committee of the European Communities, 14–15 September 1993. EU: VI/4131/94-EN Document Ref. F. II.3 – JC/0003, Brussels, 1994:145–159.

[42] Middleton DJ and Barlow RM. Failure to transmit bovine spongiform encephalopathy to mice by feeding them with extraneural tissues of affected cattle. Vet Rec 1993; 132:545–547.

[43] Terry LA, Marsh S, Ryder SJ, et al. Detection of disease-specific PrP in the distal ileum of cattle exposed orally to the agent of bovine spongiform encephalopathy. Vet Rec 2003; 152(13):387–392.

[44] Wells GAH, Hawkins SA, Cunningham AA, et al. Comparative pathology of the new transmissible spongiform encephalopathies. In: Bradley R, Marchant B, editors. Transmissible spongiform encephalopathies. Proceedings of a consultation on BSE with the Scientific Veterinary Committee of the European Communities, 14–15 September 1993. EU: VI/4131/94-EN Document Ref. F. II.3 – JC/0003, Brussels, 1994:327–346.

[45] Wells GAH, Dawson M, Hawkins SA, et al. Preliminary observations on the pathogenesis of experimental bovine spongiform encephalopathy. In: Gibbs CJ Jr, editor. Bovine spongiform en-

cephalopathy – the BSE dilemma. Springer, New York, 1996:28–44.
[46] Wells GAH and Wilesmith JW. Bovine spongiform encephalopathy and related diseases. In: Prusiner SB, editor. Prion biology and diseases. Cold Spring Harbor Laboratory Press, New York, 2004: 595–628.
[47] Wells GAH, Spiropoulos J, Hawkins SA, et al. Pathogenesis of experimental bovine spongiform encephalopathy: preclinical infectivity in tonsil and observations on the distribution of lingual tonsil in slaughtered cattle. Vet Rec 2005; 156(13):401–407.
[48] Wells GAH, Hawkins SA, Green RB, et al. Preliminary observations on the pathogenesis of experimental bovine spongiform encephalopathy (BSE): an update. Vet Rec 1998; 142:103–106.
[49] Wells GAH, Hawkins SAC, Green RB, et al. Limited detection of sternal bone marrow infectivity in the clinical phase of experimental bovine spongiform encephalopathy (BSE). Vet Rec 1999; 144:292–294.
[50] Beekes M, Baldauf E, Diringer H. Sequential appearance and accumulation of pathognomonic markers in the central nervous system of hamsters orally infected with scrapie. J Gen Virol 1996; 77(8):1925–1934.
[51] Baldauf E, Beekes M, Diringer H. Evidence for an alternative direct route of access for the scrapie agent to the brain bypassing the spinal cord. J Gen Virol 1997; 78(5):1187–1197.
[52] Beekes M, McBride PA, Baldauf E. Cerebral targeting indicates vagal spread of infection in hamsters fed with scrapie. J Gen Virol 1998; 79(3): 601–607.
[53] Beekes M and McBride PA. Early accumulation of pathological PrP in the enteric nervous system and gut-associated lymphoid tissue of hamsters orally infected with scrapie. Neurosci Lett 2000; 278(3):181–184.
[54] Blattler T, Brandner S, Raeber AJ, et al. PrP-expressing tissue required for transfer of scrapie infectivity from spleen to brain. Nature 1997; 389(6646):69–73.
[55] Glatzel M and Aguzzi A. Peripheral pathogenesis of prion diseases. Microbes Infect 2000; 2(6):613–619.
[56] Kimberlin RH and Walker C. Pathogenesis of mouse scrapie: evidence for neural spread of infection to the CNS. J Gen Virol 1980; 51:183–187.
[57] Kimberlin RH and Walker C. Pathogenesis of scrapie in mice after intragastric infection. Virus Res 1989; 12:213–220.
[58] Kimberlin RH, Field HJ, Walker C. Pathogenesis of mouse scrapie: evidence for spread of infection from central to peripheral nervous system. J Gen Virol 1983; 64:713–716.
[59] McBride PA, Schulz-Schaeffer WJ, Donaldson M, et al. Early spread of scrapie from the gastrointestinal tract to the central nervous system involves autonomic fibers of the splanchnic and vagus nerves. J Virol 2001; 75(19):9320–9327.
[60] McBride PA and Beekes M. Pathological PrP is abundant in sympathetic and sensory ganglia of hamsters fed with scrapie. Neurosci Lett 1999; 265(2):135–138.
[61] Jeffrey M, Ryder S, Martin S, et al. Oral inoculation of sheep with the agent of bovine spongiform encephalopathy (BSE). 1. Onset and distribution of disease-specific PrP accumulation in brain and viscera. J Comp Pathol 2001; 124(4):280–289.
[62] Iwamaru Y, Okubo Y, Ikeda T, et al. PrPSc distribution of a natural case of bovine spongiform encephalopathy. In: Kitamoto T, editor. Prions. food and drug safety. Springer Verlag, New York, 2006:179.
[63] Liberski PP, Yanagihara R, Wells GAH, et al. Comparative ultrastructural neuropathology of naturally occurring bovine spongiform encephalopathy and experimentally induced scrapie and Creutzfeldt-Jakob disease. J Comp Pathol 1992; 106:361–381.
[64] Jeffrey M, Halliday WG, Goodsir CM. A morphometric and immunohistochemical study of the vestibular nuclear complex in bovine spongiform encephalopathy. Acta Neuropathol 1992; 84:651–657.
[65] Jeffrey M and Halliday WG. Numbers of neurons in vacuolated and non-vacuolated neuroanatomical nuclei in bovine spongiform encephalopathy-affected brains. J Comp Pathol 1994; 110:287–293.
[66] Wells GAH and Wilesmith JW. The neuropathology and epidemiology of bovine spongiform encephalopathy. Brain Pathol 1995; 5:91–103.
[67] Fraser H and Dickinson AG. The sequential development of the brain lesions of scrapie in three strains of mice. J Comp Pathol 1968; 78:301–311.
[68] Hawkins SAC, Wells GAH, Simmons MM, et al. The topographic distribution pattern of vacuolation in the central nervous system of cattle infected orally with bovine spongiform encephalopathy. British Cattle Veterinary Association, Edinburgh 8–12 July, 1996: 431.
[69] Simmons MM, Harris P, Jeffrey M, et al. BSE in Great Britain: consistency of the neurohis-

topathological findings in two random annual samples of clinically suspect cases. Vet Rec 1996; 138:175–177.

[70] Wells GAH, Wilesmith JW, McGill IS. Bovine spongiform encephalopathy: a neuropathological perspective. Brain Pathol 1991; 1:69–78.

[71] Wells GAH, Hancock RD, Cooley WA, et al. Bovine spongiform encephalopathy: diagnostic significance of vacuolar changes in selected nuclei of the medulla oblongata. Vet Rec 1989; 125:521–524.

[72] Bruce ME, Chree A, McConnell I, et al. Transmission of bovine spongiform encephalopathy and scrapie to mice: strain variation and the species barrier. Philos Trans R Soc Lond B 1994; 343:405–411.

[73] Green R, Horrocks C, Wilkinson A, et al. Primary isolation of the bovine spongiform encephalopathy agent in mice: agent definition based on a review of 150 transmissions. J Comp Pathol 2005; 132(2–3):117–131.

[74] Wells GAH. Pathology of nonhuman spongiform encephalopathies: variations and their implications for pathogenesis. In: Brown F, editor. Transmissible spongiform encephalopathies – Impact on animal and human health. Dev Biol Stand Vol. 80. Karger, Basel, 1993:61–69.

[75] Yamakawa Y, Hagiwara K, Nohtomi K, et al. Atypical proteinase K-resistant prion protein (PrPres) observed in an apparently healthy 23-month-old Holstein steer. Jpn J Infect Dis 2003; 56(5–6):221–222.

[76] Casalone C, Zanusso G, Acutis P, et al. Identification of a second bovine amyloidotic spongiform encephalopathy: molecular similarities with sporadic Creutzfeldt-Jakob disease. Proc Natl Adad Sci USA 2004; 101:3065–3070.

[77] Biacabe AG, Laplanche JL, Ryder S, et al. Distinct molecular phenotypes in bovine prion diseases. EMBO Rep 2004; 5(1):110–115.

[78] Simmons MM, Hickey TE, Cummingham AA, et al. Lesion profiling in naturally occurring transmissible spongiform encephalopathies (abstract). Symposium on Prion and Lentiviral Diseases, Reykjavík, Iceland. August 20–22, 1998: 94.

[79] Cunningham AA, Kirkwood JK, Dawson M, et al. Bovine spongiform encephalopathy infectivity in greater kudu (Tragelaphus strepsiceros). Emerg Infect Dis 2004; 10:1044–1049.

[80] Wyatt JM, Pearson GR, Smerdon TN, et al. Spongiform encephalopathy in a cat. Vet Rec 1990; 126:513.

[81] Wyatt JM, Pearson GR, Smerdon TN, et al. Naturally occurring scrapie-like spongiform encephalopathy in five domestic cats. Vet Rec 1991; 129(11):233–236.

[82] Ryder SJ, Wells GAH, Bradshaw JM, et al. Inconsistent detection of PrP in extraneural tissues of cats with feline spongiform encephalopathy. Vet Rec 2001;148(14):437–441.

28 Pathophysiology of Prion Diseases Following Peripheral Infection

Walter J. Schulz-Schaeffer, Hans Kretzschmar, and Michael Beekes

28.1 Introduction

Prion diseases are transmissible within and between species with only a few exceptions. As early as in the 1930s, Cuillé and Chelle were able to demonstrate the experimental transmissibility of scrapie [1]. In the 1960s Gajdusek showed that Creutzfeldt–Jakob disease (CJD) was an experimentally transferable disease [2]. Later the transmissibility of genetically determined prion diseases was also demonstrated [3; 4]. The spread of kuru among the Fore population of Papua New Guinea through ritual cannibalism as well as the spread of bovine spongiform encephalopathy (BSE) among cattle herds of Great Britain by way of insufficiently decontaminated animal feed, demonstrated that oral intake of the infectious agent can be responsible for the propagation of these diseases.

28.2 Administration of the pathogen in animal experiments

In animal experiments, a single oral administration of 10 mg of brain tissue from terminally diseased hamsters was sufficient to transfer the hamster-adapted scrapie strain 263K to healthy animals [5]. As shown by Büeler et al. [6] and Manson et al. [7], the development of a prion disease requires the presence of the physiological, cellular host prion protein (PrP^C). Mice in which the prion protein gene is not functional ($Prnp^{o/o}$) cannot be infected experimentally. When embryonic telencephalic tissue from PrP-competent mice ($Prnp^{+/+}$) is implanted into $Prnp^{o/o}$ mice, nervous tissue with differentiated nerve and glia cells develop in the transplanted tissue [8]. When the transplanted animals are infected intercerebrally with a mouse-adapted scrapie strain, pathological changes of spongiformity, gliosis and nerve cell loss occur in the graft only. Interestingly, the scrapie isoform of the prion protein (PrP^{Sc}) is also detectable in the surrounding brain tissue of the $Prnp^{o/o}$ mice and along the projections of the corpus callosum without any sign of tissue destruction [9]. This indicates that the accumulation of PrP^{Sc} apparently only has a toxic effect on PrP^C-expressing nervous tissue.

28.3 Cell culture experiments

Cell culture experiments with the neurotoxic fragment that represents the amino acid sequence 106–126 of the prion protein produced comparable results [10]. The peptide 106–126 only had a toxic effect on nerve cells expressing PrP^C. Moreover, it was shown that the neurotoxic effect is mediated through microglial cells. The neurotoxic effect of the peptide 106–126 on PrP^C-expressing neuronal cells can be offset through selective elimination of microglia in a mixed cerebellar culture with L-leucine-methylester [11]. Furthermore, microglia in vivo seem to play an important role in the pathogenesis of prion diseases. Experimentally infected mice examined at 30-day intervals after intracerebral infection using different mouse-adapted scrapie strains showed a microglial activation that precedes the loss of nerve cells [12].

28.4 Peripheral paths of infection

The most efficient route of infection is intracerebral inoculation, followed by intravenous, intraperitoneal, and intramuscular administration of infectious agent. Provided that in the case of intraspecies transmission one lethal dose of the agent is necessary to intracerebrally infect and subsequently kill a recipient with a probability of 50% (1 $LD_{50i.c.}$), an approximately ten-fold amount of that infectivity (i.e., 10 $LD_{50i.c.}$) is needed for achieving the same effect intravenously; for intraperitoneal, subcutaneous and oral infection 100, 10^4 and 10^5 $LD_{50i.c.}$, respectively, is required [13; 14]. Spread of the infectious agent of TSE from the peripheral tissues to the central nervous system (CNS) is apparently mediated by PrP^C-expressing tissues. *Prnp*$^{o/o}$ mice, in which a PrP-competent nervous tissue graft has been implanted, cannot be infected via a peripheral route, i.e., intraperitoneally. This result can be explained by the production of anti-PrP-autoantibodies in reaction to the scrapie agent administration for only a portion of the animals in this experiment [15]. In peripherally challenged wild-type mice, an increasing amount of PrP^{Sc} can usually be found in the spleen before it is detectable in central nervous tissues. In contrast, in peripherally infected *Prnp*$^{o/o}$ mice with PrP-competent cerebral grafts PrP^{Sc} replication in lymphoreticular tissues is not traceable [15].

28.5 Significance of the hematopoetic system, in particular the spleen

There is evidence that the cellular prion protein is also physiologically important for cells of the hematopoietic system. PrP^C is expressed on the cell surface of lymphocytes [16–20] and monocytes [21]. With the activation of lymphocytes, PrP^C expression is upregulated and increased [22]. T cells, NK cells, and monocytes exhibit similar PrP^C levels, whereas PrP^C surface staining on B cells is significantly lower and PrP^C is virtually absent on granulocytes. Within the T-cell compartment, CD8+ cells show a significantly higher PrP^C expression than CD4+ cells. Similarly, CD3+ cells co-expressing the activation marker CD56 (N-CAM) exhibit significantly higher PrP^C expression levels than their CD56-counterparts. Culture of CD14+ pB monocytes for 12–48 h in the presence of interferon gamma (IFN-gamma) results in a significant increase in PrP^C expression in a time- and concentration-dependent manner. This effect is partially abrogated by the addition of the metabolic inhibitor cycloheximide, indicating the role of protein synthesis in this process. These results show that PrP^C expression on human hemopoietic cells correlates with the activation and developmental status of these cells, suggesting an important functional role of PrP^C in the hemopoietic system [23]. A failure in PrP^C expression apparently leads to a reduction in lymphocyte activation, since the concavalin A-induced proliferation of "splenocytes" (presumably T lymphocytes) that were extracted from spleen tissue of *Prnp*$^{o/o}$ mice is reduced in vitro as compared to PrP-competent splenocytes [24]. The physiological significance of the increased PrP^C expression accompanying cell activation remains unexplained.

Nevertheless, the spread of the scrapie agent after peripheral inoculation apparently does not only depend on the physiological expression of PrP^C within the cells of the lymphatic system. Transgenic mice that carry a genetic defect, which disrupts the differentiation of B lymphocytes (Rag 2, Rag 1, Agr deficiency) or the immunological B cell response (μMT, SCID) cannot (or can only very inefficiently, i.e., by high doses of infectivity) be infected with the scrapie agent through peripheral (intraperitoneal) inoculation, whereas intracerebral infection is largely successful [25–27]. After cerebral inoculation, the course of disease and the pathological alterations in the CNS cannot be distinguished from the cerebral infection in control animals; hence this genetic defect does not seem to influence the development of the pathological changes in the CNS. The control animals display infectivity in the spleen found in neither the peripherally nor the cerebrally infected transgenic animals with a deficiency of competent B cells. If the immune deficiency is cured by administration of pluripotent hemapoietic cells, PrP^{Sc} replication again takes place in the spleen after infection with the scrapie agent [15]. Peripheral replication in the spleen and peripheral infectibility are also possible if reconstitution is through hematopoietic stem cells from *Prnp*$^{o/o}$ mice [28].

Altogether, PrP^{Sc} replication in the lymphoreticular cells of mouse models seems to influence susceptibility to peripheral scrapie infection with the strains 139A and ME7. Genetic asplenia or a splenectomy, before an experimental scrapie infection, extends the incubation time, whereas a thymectomy does not [29; 30]. On the other hand, a splenectomy does not seem to have any effect on the transmission of the CJD agent to mice [31] or after peroral administration of scrapie agent [32].

Neonatally infected mice are slightly more resistant to scrapie infection than adult animals [33]. Immunosuppression with corticosteroids or amphotericin B reduces susceptibility [34; 35], as does administration of polyanions [36; 37]. Total-body radiation apparently does not lead to a lengthening of the incubation time after intraperitoneal infection [38]. From this it was concluded that fast-dividing cells do not play a significant role in peripheral infection.

Presuming that PrP^{C}-expressing cells are the target cells for the spread or replication of the scrapie agent in the lymphoreticular system, these could be the long-living, radiation-resistant follicular dendritic cells (FCD) in spleen and lymph tissue [39]. FDCs express PrP^{C} [40]. These cells accumulate PrP^{Sc} in intraperitoneally infected mice [41] and in the preclinical stage of patients with vCJD as shown by immunohistochemical studies on appendix tissue [42]. Their sensitivity to steroids [34] might explain the longer incubation time for peripheral infection under administration of corticosteroids. On the other hand, TNF receptor 1 knock-out mice, which do not form germinal centers in the lymphatic organs and in which FDCs are not detectable because TNFR 1 is required for differentiation [43], can be infected either intracerebrally or intraperitoneally with the scrapie agent [25]. One possible explanation of these results is that FDCs are involved in the replication of peripheral PrP^{Sc}, while B cell-mediated processes are necessary for the neuroinvasion. How the infectious agent accesses peripheral nerves from lymphatic tissue is not exactly known. Recently, it has been shown that the distance between FDCs and sympathetic nerves affects the speed of RML scrapie neuroinvasion in a mouse model. In mice deficient for the chemokine receptor CXCR5 ($Blr1^{-/-}$ mice), FDCs and B lymphocytes lie adjacent to perivascular nerves in the architecture of splenic lymphoid follicles, whereas in wild-type mice FDCs reside near the periphery of the white pulp. The FDC translocation accelerates the transfer of intraperitoneally administered RML scrapie into the spinal cord [44; 45].

Recent results indicate that cells of the lymphoreticular system, other than FDCs or B cells, may be involved in PrP^{Sc} replication or translocation for neuroinvasion. CD11+ lymphoid dendritic cells are able to propagate prion infectivity [46] and it is hypothesized that other cells may also be involved [47].

28.6 Species-specific differences

Peripheral replication of the scrapie agent apparently varies in different species. In experimentally infected mice, replication can be shown in the lymphoreticular system before the corresponding titer can be detected in the CNS [48–49], but the pathogenetic involvement of the spleen appears to be less pronounced in hamsters [50–51]. Similar findings as for mice have also been reported for field cases of scrapie in sheep [52]. In contrast, in cattle that were experimentally orally infected with BSE, infectivity was only found in the distal ileum (Peyer's patches) and not in the spleen or lymph nodes, from the sixth month on after oral infection [53]. Evidence for organ infectivity was based on the experimental infection of RIII mice, the mouse strain that reacts with the highest sensitivity to the BSE agent of those investigated. When VM/Dk mice are infected intraperitoneally with a mouse-passaged BSE strain (301V), PrP^{Sc} is first detected by Western blotting in the brain tissue but not in the spleen or lymph nodes [54]. In natural sheep scrapie, accumulation of PrP^{Sc} can be detected immunohistochemically in lymph follicles of tonsil tissue [55]. However, there have been a few reports of scrapie cases in which no involvement of the lymphoid system could be demonstrated [55a] (see Chapter 18.4). With immunohistochemistry and Western blotting, PrP^{Sc} could also be detected in the tonsil tissue of patients with vCJD [56], as well as in the appendix that had been removed months before the onset of clinical symptoms of a later verified vCJD patient [42]. However, detection of PrP^{Sc} in lymphatic tissue in cases of sporadic CJD, using either immunohistochemistry or Western blot, has so far not been possible [57] with one exception [58].

28.7 Neuronal spread of infection to the central nervous system

Next to replication in the lymphoreticular system, the neuroinvasive behavior of the infectious agent is of critical importance for the pathogenesis of TSEs. Numerous findings indicate that after peripheral exposure, the peripheral nervous system plays an essential role in the spread of the agent to the brain and spinal cord. This holds true for scrapie in sheep [59; 60], experimental scrapie in mice [61–64] and in hamsters [50; 51], as well as for experimental forms of CJD [65] and BSE [53] in animals.

Earlier studies in which the spatial–temporal course of replication of the infectious agent in parenterally infected mice and hamsters was investigated, showed that the infectious process spreads to the spinal cord at the thoracic level, and from there moves upward into the brain [31; 62; 63; 66–69]. Analogous findings were also reported in mice, even after intragastric infection [48].

The uptake of TSE agents in food apparently plays a decisive role in the oral transmission of TSEs [5]. How the infection reaches the CNS after feeding on the scrapie agent and how it spreads thereafter has been systematically investigated in Syrian hamsters, which were perorally challenged with 263K scrapie. On the basis of a close quantitative association of infectivity and PrP^{Sc} in this experimental model [50; 51], PrP^{Sc} was used as a biochemical marker for tracing the spread of infection through the body in time course studies.

The mapping of the spatial–temporal course of PrP^{Sc} accumulations in the brain and spinal cord of orally infected hamsters proved the existence of an alternative route of the spread of the infectious agent from the periphery directly to the brain through the medulla oblongata, i. e., bypassing the spinal cord [50; 51]. Kimberlin and Walker [63] already mentioned the possibility of an alternative route of spread to the brain other than via the spinal cord, but for a long time there were no experimental data available to support this. In the meantime, investigations on the course of infection in hamsters have shown by immunohistochemical prion protein detection that the first target area in which the scrapie infection in the brain of orally infected animals can be seen is the nucleus dorsalis nervi vagi, which is the parasympathetic nucleus of the vagus nerve [70]. The next target area comprises the nucleus tractus solitarii and the ependymal cells surrounding the central canal of the spinal cord [70; 71]. In the spinal cord, the earliest possible detection of PrP^{Sc} after oral infection coincides with the earliest possible detection of prion proteins in the brain – and is in the intermediolateral column at the Th4–Th9 level. The infectious process spreads from this region in the spinal cord in a rostral and caudal direction. In the C1 to C3 cervical area of the spinal cord, PrP^{Sc} is first immunohistochemically detectable when it is already present in several target areas in the medulla oblongata. This also indicates that the early target areas in the medulla oblongata are not the result of an infectious process ascending the spinal cord [71]. The immunohistochemical targeting of pathological prion protein was confirmed using the paraffin-embedded tissue blot (PET blot) method [72] that enables a highly sensitive topographical detection of the proteinase K-resistant prion protein.

The present data suggest that after oral uptake the scrapie agent spreads from the gastrointestinal tract along the efferent parasympathetic fibers of the vagus nerve, retrograde to the nucleus dorsalis nervi vagi and the nucleus tractus solitarii, thus entering the brain via the medulla oblongata. Parallel to this, after passing the gastrointestinal tract, the infectious agent also reaches the gray matter in the area of the thoracic spinal cord via the splanchnic nerve. Therefore, apparently both the parasympathetic and sympathetic vegetative nervous systems play an important role in the infectious spread to the CNS [71]. Pathomorphological studies have confirmed the experimentally established parasympathetic pathway in naturally occurring scrapie of sheep, in cattle BSE and in chronic wasting disease of deer [73–76].

The current data show that after the nucleus dorsalis nervi vagi and the nucleus tractus solitarii along the known, parasympathetic-related neural circuit is reached, the scrapie agent spreads further to the efferent ganglion nodosum of the vagus nerve, which is located outside of the brain. Regarding the sympathetic route from the spinal cord, an analogous temporal progression of target areas can be postulated: the infectious agent reaches the intermediolateral collum in the area of the spinal segments Th4–Th9 via the afferent prevertebral ganglia (ganglion coeliacum and ganglion mesentericum superius) along the splanchnic nerve, and then spreads further to the corresponding afferent paravertebral spinal ganglia [71].

Moreover, it could be demonstrated that the scrapie agent is able to spread into muscles of experimentally infected mice and hamsters [77; 78] as well as of sheep in experimentally and probable naturally occurring scrapie [79]. Time course studies were able to show that in the well-established hamster model a spread to skeletal muscles and the tongue occurs prior to the onset of clinical symptoms but after reaching the CNS. With a pathomorphological analysis using the PET blot method the spread along motor nerve fibers to the neuromuscular junction of single fibers in a distribution of a motor unit and an accumulation within single nerve fibers was recently demonstrated [80]. Therefore, the scrapie agent may be able to spread from the CNS anterograde to muscle tissues and may replicate in motor nerves as well as in muscle fibers.

References

[1] Cuillé J and Chelle PL. Pathologie animale. La maladie dite tremblante du mouton est-elle inoculable? C R Acad Sci (III) 1936; 26:1552–1554.

[2] Gajdusek DC, Gibbs CJ Jr, Alpers M. Experimental transmission of a kuru-like syndrome to chimpanzees. Nature 1966; 209:794–796.

[3] Masters CL, Gajdusek DC, Gibbs CJ Jr. Creutzfeldt–Jakob disease virus isolations from the Gerstmann–Sträussler syndrome. With an analysis of the various forms of amyloid plaque deposition in the virus-induced spongiform encephalopathies. Brain 1981; 104:559–588.

[4] Tateishi J, Brown P, Kitamoto T, et al. First experimental transmission of fatal familial insomnia. Nature 1995; 376:434–435.

[5] Diringer H, Beekes M, Oberdieck U. The nature of the scrapie agent: the virus theory. Ann N Y Acad Sci 1994; 724:246–258.

[6] Büeler H, Fischer M, Lang Y, et al. Normal development and behaviour of mice lacking the neuronal cell-surface PrP protein. Nature 1992; 356:577–582.

[7] Manson JC, Clarke AR, Hooper ML, et al. 129/Ola mice carrying a null mutation in PrP that abolishes mRNA production are developmentally normal. Mol Neurobiol 1994; 8:121–127.

[8] Isenmann S, Brandner S, Sure U, et al. Telencephalic transplants in mice: characterization of growth and differentiation patterns. Neuropathol Appl Neurobiol 1996; 22:108–117.

[9] Brandner S, Isenmann S, Raeber A, et al. Normal host prion protein necessary for scrapie-induced neurotoxicity. Nature 1996; 379:339–343.

[10] Forloni G, Angeretti N, Chiesa R, et al. Neurotoxicity of a prion protein fragment. Nature 1993; 362:543–546.

[11] Brown DR, Schmidt B, Kretzschmar HA. Role of microglia and host prion protein in neurotoxicity of prion protein fragment. Nature 1996; 380:345–347.

[12] Giese A, Brown DR, Groschup M, et al. Role of microglia in neuronal cell death in prion disease. Brain Pathol 1998; 8:449–457.

[13] Diringer H. Durchbrechen von Speziesbarrieren mit unkonventionellen Viren. Bundesgesundheitsblatt 1990; 33,435–440.

[14] Diringer H, Roehmel J, Beekes M. Effect of repeated oral infection of hamsters with scrapie. J Gen Virol 1998; 79:609–612.

[15] Blättler T, Brandner S, Raeber AJ, et al. PrP-expressing tissue required for transfer of scrapie infectivity from spleen to brain. Nature 1997; 389:69–73.

[16] Lavelle GC, Sturman L, Hadlow WJ. Isolation from mouse spleen of cell populations with high specific infectivity for scrapie virus. Infect Immun 1972; 5:319–323.

[17] Kingsbury DT, Smeltzer DA, Gibbs CJ Jr, et al. Evidence for normal cell-mediated immunity in

scrapie-infected mice. Infect Immun 1981; 32: 1176–1180.
[18] Kuroda Y, Gibbs CJ Jr, Amyx HL, et al. Creutzfeldt–Jakob disease in mice: persistent viremia and preferential replication of virus in low-density lymphocytes. Infect Immun 1983; 41:154–161.
[19] Oesch B, Westaway D, Wälchli M, et al. A cellular gene encodes scrapie PrP 27–30 protein. Cell 1985; 40:735–746.
[20] Manson J, West JD, Thomson V, et al. The prion protein gene: a role in mouse embryogenesis? Development 1992; 115:117–122.
[21] Dodelet VC and Cashman NR. Prion protein expression in human leukocyte differentiation. Blood 1998; 91:1556–1561.
[22] Cashman NR, Loertscher R, Nalbantoglu J, et al. Cellular isoform of the scrapie agent protein participates in lymphocyte activation. Cell 1990; 61:185–192.
[23] Dürig J, Giese A, Schulz-Schaeffer W, et al. Differential constitutive and activation-dependent expression of prion protein in human peripheral blood leucocytes. Br J Haematol 2000; 108(3):488–495.
[24] Mabbott NA, Brown KL, Manson J, et al. T-lymphocyte activation and the cellular form of the prion protein. Immunology 1997; 92:161–165.
[25] Klein MA, Frigg R, Flechsig E, et al. A crucial role for B cells in neuroinvasive scrapie. Nature 1997; 390:687–690.
[26] Lasmézas CI, Cesbron JY, Deslys J-P, et al. Immune system-dependent and -independent replication of the scrapie agent. J Virol 1996; 70:1292–1295.
[27] Fraser H, Brown KL, Stewart K, et al. Replication of scrapie in spleens of SCID mice follows reconstitution with wild-type mouse bone marrow. J Gen Virol 1996; 77:1935–1940.
[28] Klein MA, Frigg R, Raeber AJ, et al. PrP expression in B lymphocytes is not required for prion neuroinvasion. Nat Med 1998; 4:129–1433.
[29] Fraser H and Dickinson AG. Pathogenesis of scrapie in the mouse: the role of the spleen. Nature 1970; 226:462–463.
[30] Fraser H and Dickinson AG. Studies of the lymphoreticular system in the pathogenesis of scrapie: the role of spleen and thymus. J Comp Pathol 1978; 88:563–573.
[31] Mohri S, Handa S, Tateishi J. Lack of effect of thymus and spleen on the incubation period of Creutzfeldt–Jakob disease in mice. J Gen Virol 1987; 68:1187–1189.
[32] Kimberlin RH and Walker CA. The role of the spleen in the neuroinvasion of scrapie in mice. Virus Res 1989; 12:201–212.
[33] Outram GW, Dickinson AG, Fraser H. Developmental maturation of susceptibility to scrapie in mice. Nature 1973; 241:536–537.
[34] Outram GW, Dickinson AG, Fraser H. Reduced susceptibility to scrapie in mice after steroid administration. Nature 1974; 249:855–856.
[35] Pocchiari M, Schmittinger S, Masullo C. Amphotericin B delays the incubation period of scrapie in intracerebrally inoculated hamsters. J Gen Virol 1987; 68:219–223.
[36] Ehlers B and Diringer H. Dextran sulphate 500 delays and prevents mouse-scrapie by impairment of agent replication in spleen. J Gen Virol 1984; 65:1325–1330.
[37] Kimberlin RH and Walker CA. Suppression of scrapie infection in mice by heteropolyanion 23, dextran sulphate and some other polyanions. Antimicrob Agents Chemother 1986; 30:409–413.
[38] Fraser H, Bruce ME, Davies D, et al. The lymphoreticular system in the pathogenesis of scrapie. In: Prusiner SB, Collinge J, Powell J, et al., editors. Prion diseases of humans and animals. Ellis Horwood, New York, 1992:308–317.
[39] Fraser H, Davies D, McConell I. Are radiation-resistant, post-mitotic, long-lived (RRPMLL) cells involved in scrapie replication? In: Court LA, Dormont D, Brown P, et al., editors. Unconventional virus diseases of the central nervous system. Fontenay-aux-Roses: Commissariat á l'Energie Atomique, 1989:563–574.
[40] McBride PA, Eikelenboom P, Kraal G, et al. PrP protein is associated with follicular dendritic cells of spleens and lymph nodes in uninfected and scrapie-infected mice. J Pathol 1992; 168:413–418.
[41] Kitamoto T, Muramoto T, Mohri S, et al. Abnormal isoform of prion protein accumulates in follicular dendritic cells in mice with Creutzfeldt–Jakob disease. J Virol 1991; 65:6292–6295.
[42] Hilton DA, Fathers E, Edwards P, et al. Prion immunoreactivity in appendix before clinical onset of variant Creutzfeldt–Jakob disease. Lancet 1998; 352:703–704.
[43] Le Hir M, Bluethmann H, Kosco-Vilbois MH, et al. Differentiation of follicular dendritic cells and full antibody responses require tumor necrosis factor receptor-1 signaling. J Exp Med 1996; 183:2367–2372.
[44] Prinz M, Heikenwalder M, Junt T, et al. Positioning of follicular dendritic cells within the

spleen controls prion neuroinvasion. Nature 2003; 425:957–62.
[45] Mabbott NA and Bruce ME. Prion disease: bridging the spleen–nerve gap. Nat Med 2003; 9:1463–1464.
[46] Aucouturier P, Feissmann F, Damotte D, et al. Infected splenic dendritic cells are sufficient for prion transmission to the CNS in mouse scrapie. J Clin Invest 2001; 108:703–708
[47] Oldstone MB, Race R, Thomas D, et al. Lymphotoxin-α- and lymphotoxin-β-deficient mice differ in susceptibility to scrapie: evidence against dendritic cell involvement in neuroinvasion. J Virol 2002; 76:4357–4363.
[48] Kimberlin RH and Walker CA. The role of the spleen in the neuroinvasion of scrapie in mice. Virus Res 1989; 12:201–212.
[49] Scott JR. Scrapie pathogenesis. Br Med Bull 1993; 49:778–791.
[50] Beekes M, Baldauf E, Diringer H. Sequential appearance and accumulation of pathognomonic markers in the central nervous system of hamsters orally infected with scrapie. J Gen Virol 1996; 77:1925–1934.
[51] Baldauf E, Beekes M, Diringer H. Evidence for an alternative direct route of access for the scrapie agent to the brain bypassing the spinal cord. J Gen Virol 1997; 78:1187–1197.
[52] Hadlow WJ, Kennedy RC, Race RE. Natural infection of Suffolk sheep with scrapie virus. J Infect Dis 1982; 146:657–664.
[53] Wells GAH, Hawkins SAC, Green RB, et al. Preliminary observations on the pathogenesis of experimental bovine spongiform encephalopathy (BSE): an update. Vet Rec 1998; 142:103–106.
[54] Farquhar CF, Dornan J, Moore RC, et al. Protease-resistant PrP deposition in brain and non-central nervous system tissues of a murine model of bovine spongiform encephalopathy. J Gen Virol 1996; 77:1941–1946.
[55] Schreuder BEC, van Keulen LJM, Vromans MEW, et al. Preclinical test for prion diseases. Nature 1996; 381:563.
[55a] Ligios C, Cancedda MG, Madau L, et al. PrP(Sc) deposition in nervous tissues without lymphoid tissue involvement is frequently found in ARQ/ARQ Sarda breed sheep preclinically affected with natural scrapie. Arch Virol 2006; Apr 20; [Epub ahead of print].
[56] Hill AF, Zeidler M, Ironside J, et al. Diagnosis of new variant Creutzfeldt–Jakob disease by tonsil biopsy. Lancet 1997; 349:99–100.
[57] Head MW, Ritchie D, Smith N, et al. Peripheral tissue involvement in sporadic, iatrogenic, and variant Creutzfeldt–Jakob disease: an immunohistochemical, quantitative, and biochemical study. Am J Pathol 2004; 164:143–53.
[58] Glatzel M, Abela E, Maissen M, et al. Extraneural pathologic prion protein in sporadic Creutzfeldt–Jakob disease. N Engl J Med 2003; 349:1812–20.
[59] van Keulen LJM, Schreuder BEC, Meloen RH, et al. Immunohistochemical detection and localization of prion protein in brain tissue of sheep with natural scrapie. Vet Pathol 1995; 32:299–308.
[60] Groschup MH, Weiland F, Straub OC, et al. Detection of scrapie agent in the peripheral nervous system of a diseased sheep. Neurobiol Dis 1996; 3:191–195.
[61] Fraser H. Neuronal spread of scrapie agent and targeting of lesions within the retino-tectal pathway. Nature 1982; 295:149–150.
[62] Kimberlin RH and Walker CA. Pathogenesis of mouse scrapie: patterns of agent preplication in different parts of the CNS following intraperitoneal infection. J R Soc Med 1982; 75:618–624.
[63] Kimberlin RH and Walker CA. Invasion of the CNS by scrapie agent and its spread to different parts of the brain. In: Court LA, Cathala F, editors. Virus non conventionels et affections du systeme nerveux central. Masson, Paris 1993:17–33.
[64] Inaba K, Inaba M, Romani N, et al. Generation of large numbers of dendritic cells from mouse bone marrow cultures supplemented with granulocyte/macrophage colony-stimulating factor. J Exp Med 1992; 176:1693–1702.
[65] Muramoto T, Kitamoto T, Tateishi J, et al. Accumulation of abnormal prion protein in mice infected with Creutzfeldt–Jakob disease via intraperitoneal route: a sequential study. Am J Pathol 1993; 143:1470–1479.
[66] Kimberlin RH and Walker CA. Pathogenesis of mouse scrapie: evidence for neural spread of infection to the CNS. J Gen Virol 1980; 51:183–187.
[67] Kimberlin RH and Walker CA. Pathogenesis of mouse scrapie. Evidence for direct neural spread of infection to the CNS after injection of sciatic nerve. J Neurol Sci 1983; 61:315–325.
[68] Kimberlin RH and Walker CA. Pathogenesis of scrapie (strain 263K) in hamsters infected intracerebrally, intraperitoneally or intraocularly. J Gen Virol 1986; 67:255–263.
[69] Carp RI, Callahan SM, Patrick BA, et al. Interaction of scrapie agent and cells of the lymphoreticular system. Arch Virol 1994; 136:255–268.

[70] Beekes M, McBride PA, Baldauf E. Cerebral targeting indicates vagal spread of infection in hamsters fed with scrapie. J Gen Virol 1998; 79:601–607.

[71] McBride PA, Schulz-Schaeffer WJ, Donaldson M, et al. Early spread of scrapie from the gastrointestinal tract to the central nervous system involves autonomic fibers of the splanchnic and vagus nerves. J Virol 2001; 75:9320–9327.

[72] Schulz-Schaeffer WJ, Tschöke S, Kranefuss N, et al. The paraffin-embedded tissue blot detects PrP(Sc) early in the incubation time in prion diseases. Am J Pathol 2000; 156:51–56.

[73] van Keulen LJ, Schreuder BE, Vromans ME, et al. Pathogenesis of natural scrapie in sheep. Arch Virol Suppl 2000; 16:57–71.

[74] Schulz-Schaeffer WJ, Fatzer R, Vandevelde M, et al. Detection of PrP(Sc) in subclinical BSE with the paraffin-embedded tissue (PET) blot. Arch Virol Suppl 2000; 16:173–180.

[75] Williams ES and Mille MW. Pathogenesis of chronic wasting disease in orally exposed mule deer (Odocoileus hemionus): preliminary results. Proc 49th Wildl Dis Assoc Conf 2000:29.

[76] Sigurdson CJ, Spraker TR, Miller MW, et al. PrP(CWD) in the myenteric plexus, vagosympathetic trunk and endocrine glands of deer with chronic wasting disease. J Gen Virol 2001; 82:2327–2334.

[77] Bosque PJ, Ryou C, Telling G, et al. Prions in skeletal muscle. Proc Natl Acad Sci USA 2002; 99:3812–3817.

[78] Thomzig A, Kratzel C, Lenz G, et al. Widespread PrPSc accumulation in muscles of hamsters orally infected with scrapie. EMBO Rep 2003; 4:530–533.

[79] Andreoletti O, Simon S, Lacroux C, et al. PrPSc accumulation in myocytes from sheep incubating natural scrapie. Nat Med 2004; 10:591–593.

[80] Thomzig A, Schulz-Schaeffer W, Kratzel C, et al. Preclinical deposition of pathological prion protein PrPSc in muscles of hamsters orally exposed to scrapie. J Clin Invest 2004; 113:1465–1472.

Topic V: Surveillance, Clinical Aspects and Diagnostics

29 Introduction to Surveillance for Human Prion Diseases

Beat Hörnlimann, Hans Kretzschmar, Robert G. Will, Otto Windl and Herbert Budka

29.1 Introduction

Surveillance of human prion disease is carried out in order to provide accurate data on the incidence and characteristics of these conditions. Surveillance involves registration of suspected cases of human prion disease and assessment of clinical and investigative features to allow reliable case classification. It is important to distinguish between sporadically occurring (idiopathic) cases, genetic forms of human prion disease and infectious/acquired forms such as variant Creutzfeldt–Jakob disease (vCJD) or iatrogenic prion disease. An important objective is the reliable identification of newly-occuring variants of prion diseases. Identification of cases of vCJD requires careful monitoring of clinical, pathological, and biochemical disease phenotypes. Risk factors for the development of these conditions may also be assessed by systematic epidemiological surveillance through methodologies such as case-control studies.

Hospitals and pathological institutes should cooperate closely in order to achieve area-wide surveillance. Patients displaying symptoms that suggest a prion disease should be admitted to a hospital and examined neurologically. Should a patient potentially suffering from a prion disease die, neuropathological examination of the brain is essential to establish a definite diagnosis.

29.2 Surveillance of patients on an out-patient and in-patient basis

CJD or other prion diseases should be diagnosed as early as possible. It is recommended that patients suspected of suffering from a prion disease should be promptly referred to a neurological center in order to undergo detailed assessment (see Chapter 30), including neurological examination, and specialist investigations such as EEG, examination of the cerebrospinal fluid (CSF; see below), and MRI brain scan (see Chapter 31). vCJD poses a particular problem in early diagnosis because the initial symptoms are mainly psychiatric [1].

In many patients, the initial clinical suspicion of a prion disease can be discounted on the basis of the clinical features and the results of specialist investigation (differential diagnoses: Tab. 30.2). Post mortem neuropathological examinations should be carried out on patients suspected of suffering from a prion disease.

29.2.1 Case definitions based on clinical features and diagnostic investigations

Case definitions or diagnostic criteria for human prion disease, including CJD, are summarized in Tables 29.1, 29.3, 31.1 and in reference [2]. These (WHO) criteria are useful for comparisons of surveillance data, but are not rules for diagnosis as 'atypical' signs not fitting the diagnostic criteria may still be the result of prion disease. Clinical characteristics of Gerstmann–Sträussler–Scheinker Syndrome (GSS) and fatal familial insomnia (FFI) / sporadic fatal insomnia (SFI) are listed below. The signs, symptoms, and differential diagnosis of human prion diseases are described in greater detail in Chapter 30.

Diagnostic criteria for CJD include the categories definite, probable, and possible, and a further category 'not CJD' (Table 29.1 and reference [3]). Most scientific groups and published international data on CJD include cases classi-

Table 29.1: Definition of terms: [3]. For case definition / diagnostic criteria (by the WHO) which are for surveillance purposes only, see Table 31.1 or [2].

Sporadic CJD (sCJD): This is the most common form of CJD, with a worldwide incidence of approx. 1.5 per million of the population per year. It is predominantly a disease of late middle age and the cause is unknown. In a typical case, the invariably progressive, rapid clinical evolution and the relatively short illness duration, readily distinguishes sporadic CJD from most other dementing illnesses.	**Definite sCJD:** Definite cases pathologically confirmed, in most cases by post mortem examination of brain tissue (rarely by brain biopsy). Sporadic cases account for approx. 85% of all cases. **Probable sCJD:** Cases with a history of rapidly progressive dementia, typical EEG and at least two of the following clinical features: myoclonus, visual or cerebellar signs, pyramidal/extrapyramidal signs or akinetic mutism OR cases with a history of rapidly progressive dementia, at least two of the above clinical features, duration of less than 2 years and a positive 14–3–3 test. Probable cases have not been confirmed pathologically; some cases are never confirmed pathologically because a post mortem examination does not take place (for instance where the relatives of the patient refuse consent) and these cases remain permanently in the probable category. **Possible sCJD:** Cases include patients showing the same clinical symptoms with a disease duration of less than 2 years. However, the EEG of these patients have not revealed PSWC and the 14–3–3 protein has not been detected in their CSF. **"no case" / neither Probable nor Possible sCJD:** Cases do not fulfill the clinical criteria for probable or possible CJD. Upon the death of such a patient, post mortem examinations should in all cases lead to a definitive answer. Occasionally, CJD can be confirmed. However, this group of patients often includes hereditary (familial) prion diseases.
Variant CJD (vCJD): Characterized clinically by a progressive neuropsychiatric disorder leading to ataxia, dementia, and myoclonus (or chorea) without the typical EEG appearance of CJD. Neuropathology shows marked spongiform change and extensive florid plaques throughout the brain	**Definite vCJD:** see case definition in Table 31.1d **Probable vCJD:** see case definition in Table 31.1d. **Possible vCJD:** see case definition in Table 31.1d
Genetic or familial prion diseases	**Familial CJD or Genetic CJD:** Cases occurring in families associated with mutations in the PrP gene or cases with definite or probable CJD in a first degree relative. (Approx. 10% of all human prion diseases). **GSS and FFI/SFI:** The exact case definitions are summarized in Table 31.1. Clinical characteristics for GSS (see also Chapters 16; 30) and FFI (see also Chapters 17, 30) are described in Table 29.3.
Accidentally transmitted prion disease or **iatrogenic CJD (iCJD)** Relevant iatrogenic risk factors for the classification as iatrogenic CJD: It is of note, that the relevance of any exposure to disease causation must take into account the timing of the exposure in relation to disease onset. The worldwide number of reported iCJD cases and the range of incubation periods are summarized in Table 14.1.	**Definite iCJD:** see "Definite sCJD" with a recognized iatrogenic risk factor **Probable iCJD:** see "Probable sCJD" with a recognized iatrogenic risk factor (see below) OR progressive predominant cerebellar syndrome in human pituitary hormone recipient. **Iatrogenic risk factors:** – Treatment with human pituitary growth hormone, human pituitary gonadotrophin or human dura mater graft OR – Corneal graft in which the corneal donor has been classified as definite or probable human prion disease OR – Exposure to neurosurgical instruments previously used in a case of definite or probable human prion disease.

fied as 'definite' or 'probable', but do not report 'possible' cases because of significant diagnostic uncertainty in this group.

Biopsy of brain and tonsils

If CJD or another prion disease is suspected, brain biopsy may be undertaken, allowing histopathological, immunohistochemical, and biochemical examination of the central nervous system (CNS). Frozen unfixed tissue should be stored and used for Western blotting. In many cases brain biopsy may allow a definitive opinion on whether the suspected patient suffers from a prion disease. However, when the typical pathological changes are not identified in a brain biopsy, the possibility of a prion disease cannot be excluded as pathological and biochemical abnormalities may not be present in all areas of the brain. As brain biopsy is associated with risks (see Chapter 47), and does not allow the reliable exclusion of prion diseases, this procedure is only justified in exceptional cases. As PrP^{Sc} has been found in tonsillar tissue, lymph nodes, and spleen of vCJD patients, tonsil biopsy may be considered for diagnosis when vCJD is suspected (Fig. 15.3d).

Although PrP^{Sc} had not previously been detected in the lymphatic and/or other tissues of patients diagnosed with idiopathic (sporadic) CJD (sCJD), a recent study using a very sensitive detection method, has demonstrated PrP^{Sc} in spleen and/or muscle of about one third of sCJD patients at autopsy, particularly in those cases with a long duration of disease [4].

Cerebrospinal fluid tests

The reliable detection of PrP^{Sc} in CSF is not yet possible and currently no disease-specific CSF test is available (Fig. 31.4). The CSF of patients with prion diseases contains a number of proteins that are elevated in CJD and are used as *surrogate markers* for disease, in particular the 14–3–3 protein (see Chapter 31).

Subsequent steps

If the neurological examination and specialist investigations substantiate the diagnosis of a prion disease, the neurologist should inform the relatives of the diagnosis. In the event of the patient's death, the treating general practitioner or neurologist is advised to ask the family for consent for a post mortem examination in order to obtain a neuropathological diagnosis.

29.3 Surveillance in neuropathological laboratories and diagnostic laboratories

In cases of suspected prion disease, CNS samples should be sent to specialized neuropathological laboratories. Many countries have specific, national reference laboratories that are prerequisites for standardized diagnosis following international criteria. Centralized laboratories may also perform investigations essential to surveillance of human prion diseases, including PrP^{Sc} typing and sequencing of the PrP gene or *PRNP*. The identification of unusual and new disease variants is of particular importance.

29.3.1 Recommendations for the removal, pretreatment, and shipment of tissue samples

The definite diagnosis of CJD/prion disease can only be made by neuropathological examination. Therefore, it is necessary to carry out an autopsy on cases of suspected prion disease. Ideally, the entire procedure, from the time of tissue removal up to the examination in the laboratory is carried out in close cooperation with the reference laboratory. It is vital that internationally accepted recommendations regarding precautionary measures are observed in autopsies of suspect cases [5] (see Chapters 47–49).

Table 29.2 provides details on the tissue samples that should be sent to diagnostic laboratories carrying out histopathological and immunohistochemical examinations, PrP gene analysis, biochemical detection of PrP^{Sc} protein, and any additional examinations that may be necessary.

Table 29.2: Organs and tissues used for the diagnosis of human prion diseases; as proposed by the authors [see also 18].

Sample	Procedure
Frontal pole of the cerebrum, a brain slice at the level of the mamillary bodies, an occipital pole and a cerebellar hemisphere	Freeze at −70°C
Remainder of brain / tissue blocks	Fix in buffered 4% formalin for 3 weeks prior to further analysis (for practical information on autopsies see Chapter 48). Cut slices 3 mm thick and decontaminate in concentrated formic acid for 1 h. Fix tissue blocks again in formalin for 48 h and embed in paraffin (Paraplast®) for histological examination.
a) Brain: – neocortex (frontal, parietal, temporal, occipital) – basal ganglia – hippocampus – thalamus – cerebellum (vermis and hemisphere) – midbrain – pons – medulla oblongata – pituitary gland b) Spinal cord (optional): – cervical section – thoracic section – lumbar section c) Lymphatic tissue is sampled in suspect vCJD cases: – tonsils [19] – lymph nodes – spleen	Histological and immunohistochemical examination
All other organs	Decontaminate in formic acid (see Chapters 44; 48) prior to embedding in paraffin.

Autopsies and the shipment of tissue to a specialized neuropathological reference laboratory must be in accordance with valid legislation. Since legislation differs from one country to another, it is mandatory to coordinate the procedure with the respective authorities and reference centers.

When shipping tissue samples taken from patients suspected of suffering from a prion disease, care must be taken that suitable packaging is used and the international guidelines for shipping hazardous goods are observed. According to the International Air Transport Association (IATA) Dangerous Goods Regulations 2005 [6] the tissue samples may be transported as "diagnostic specimens" when they are transported for diagnostic or investigational purposes and must be classified in Division 6.2 under UN 3373. Tissue samples classified as Diagnostic Specimens UN 3373 have to be packed in compliance with the IATA packing instruction P650. Suitable packaging includes commercially available containers for the shipment of infectious organs and tissues.

29.3.2 Methods of examination

Histopathology

Histopathological examination of the brain allows for relatively rapid, specific, and highly sensitive diagnosis of prion diseases. The examination is based on the detection of spongiform alterations of the neuropil (Fig. 26.1). The morphological alterations differ in their extent in the different brain regions. In some cases kuru plaques (Fig. 26.2) can be detected using conventional light microscopy. In patients suffering from GSS (Tab. 29.3), both monocentric kuru plaques and multicentric plaques are detected [7–9] (see Chapter 26). vCJD is characterized by widespread typical florid plaques (see Fig. 15.3a). Rare cases of human prion disease show minimal or no spongiform changes.

Immunohistochemistry (IHC)

Currently, standard diagnostics involve the detection of PrP^{Sc} in the brain, which can be carried out by IHC using formalin-fixed and paraffin-embedded tissue. The PET blot is a convenient complementary technique. As prion diseases are comparatively rare, it is recommended that such examinations be carried out in, or at least confirmed by, a reference laboratory.

The detection of prion protein depositions (PrP^{Sc}) in the brain with antibodies against PrP is a reliable method for the diagnosis of prion diseases [10–12]. The sensitivity of this method can be improved by pretreating the tissue; steam autoclaving and the use of formic acid or guanidinium thiocyanate (GdnSCN) have proven most effective. The immunohistochemical reactions (Fig. 15.b–d) should be assessed by an experienced neuropathologist who will take into consideration the different forms of PrP^{Sc} depositions (see Chapter 26).

Table 29.3: Diagnostic clinical characteristics for GSS, FFI and SFI (see also Table 30.5: listing of possible signs in greater detail).

Disease	Diagnostic clinical characteristics
Gerstmann–Sträussler–Scheinker disease (GSS) see Chapter 16	Chronic and progressive ataxia
	Dementia often only occurring in the terminal stage of disease
	A clinical duration of 2–27 years with a mean of 7 +/-5 years and a median of 6.5 years (see Chapter 30.5)
	Histological features include multicentric plaques in the cerebrum and cerebellum, kuru plaques, and rarely spongiform change.
	Association with a number of mutations of *PRNP* (P102L, P105L, A117V, F198S, Q217R) (Tab. 26.5)
Fatal familial insomnia (FFI) see Chapter 17	Insomnia[1)]
	Vegetative signs and symptoms
	Progressive dementia with severe personality disorders[2)]
	Clinical duration: see Chapter 30.6
	Histological examinations only reveal minor spongiform change in the temporal cortex. Atrophy and astrocytic gliosis are observed in the thalamus and olives.
	On the molecular level, FFI is characterized by a D178N mutation and the presence of the amino acid methionine at codon 129 of *PRNP*.
Sporadic fatal insomnia (SFI) see Chapters 17; 30.7	Clinical features similar if not identical to those of FFI

1) Frequently insomnia is not clinically apparent, but a disordered sleep can be recognized by polysomnography.
2) Frequently late and relatively inconspicuous.

Western blot

The immunostaining of PrP^{Sc} following electrophoretic separation and membrane transfer of the proteins in brain homogenates of patients (Western blotting) is of increasing importance in the differentiation of subtypes of sporadic CJD [13]. Different migration patterns of PrP^{Sc}-specific signals also help to discriminate between vCJD and sporadic CJD [14]. The Western blot analysis requires fresh-frozen material sampled during autopsy and material kept at –80 °C prior to analysis. Western blots and other diagnostic methods are described in detail in Chapters 31 and 35.

Analysis of the PrP gene (PRNP)

Modern prion disease surveillance programs require analysis of the PrP gene (*PRNP*). For several reasons the analysis of *PRNP* is recommended in all cases in which prion disease is suspected:

1. The analysis of *PRNP* is necessary in order to identify hereditary prion diseases. Since most families are unaware of their genetic predisposition, it is insufficient to base the identification of hereditary cases on a known family history of disease.
2. Strictly speaking, a diagnosis of sporadic prion diseases can only be made with confidence if *PRNP* mutations have been excluded.
3. Atypical forms of prion diseases (see Chapter 26.5.5) are often the result of *PRNP* mutations and can only be identified by *PRNP* sequencing.[1]
4. The analysis of *PRNP* also allows distinction from other hereditary neurodegenerative diseases, such as the hereditary forms of Alzheimer's disease.
5. In sporadic cases, it is useful to analyze the polymorphism at codon 129 of the *PRNP* as the clinical and pathological picture of the disease is significantly influenced by this polymorphism (Tab. 26.3).
6. When vCJD is clinically diagnosed, pathogenic mutations that might lead to a clinical picture similar to that of vCJD should be excluded.

Molecular analyses are carried out using genomic DNA of the patient suspected of suffering from the disease. Genomic DNA is usually isolated from white blood cells (EDTA blood). In cases in which it becomes necessary to isolate DNA from brain material, special safety measures must be observed similar to those used in PrP^{Sc} diagnostics involving Western blots. The protein-coding region (open reading frame) of the *PRNP* is amplified using PCR. The PCR product is either sequenced directly or following its cloning in a bacterial vector.

29.3.3 Ethical considerations (genetic analysis)

Genetic analysis involves a number of ethical considerations. In order for such an analysis to be carried out informed consent must be obtained in writing. In addition, the patients and/or their relatives may require genetic counseling. Results of the analysis should be communicated by genetic counseling physicians. If a disease-specific mutation is detected in the PrP gene, the relatives of the patient concerned must[2] be given the choice of whether they want to be informed of the result. Genetic analysis of relatives should only be carried out following genetic counseling.

29.4 International cooperation and the importance of national health authorities

National health authorities coordinate the activities of the different institutions involved in the surveillance of prion diseases and also provide information on these diseases. These authorities process data on notifiable diseases in collaboration with the leading European reference centers and WHO.

The WHO, based in Geneva, has initiated a program for global improvement in prion disease surveillance with the aim of worldwide har-

[1] Compare sporadic CJD / "no case" in Table 29.1.

[2] National regulations might differ and should be followed.

monization. Since 1993, approximately 100 laboratories in European Union (EU) Member States and other countries focusing on neuropathological and basic research, have been participating in a research program for the improvement of neuropathological diagnostics in the field of human prion diseases.

The BIOMED 1 program – *Surveillance of CJD in the European Community* (Ref. No. PL 9209988) – integrated research groups from Germany, the United Kingdom, France, Italy, the Netherlands, and Slovakia (see Chapters 36 and 37). Early in 1997, the study expanded into the BIOMED 2 program; in which close cooperative activities were initiated between Austria, Spain, Canada, Australia, Switzerland [15], and further countries.

29.5 Epidemiological surveillance and case control studies

The EU initiated the systematic recording of prevalence and geographic distribution of human prion diseases in 1993. Notification of CJD has been mandatory in many countries including Switzerland since 1987, Germany since 1994, and Austria since 1996[3].

Epidemiological analyses are carried out in order to assess the prevalence, incidence, mortality, etc. of the diseases. Prerequisites for the successful assessment are the cooperation of a large number of hospitals, standardized diagnostics, and close cooperation between clinicians and neuropathological laboratories. The EU CJD surveillance program helped identify vCJD (nvCJD) in the UK in 1996.

Investigations aimed at identifying risk factors are carried out in some countries by epidemiological research groups. It is difficult to identify potential risk factors as prion diseases are rare. In addition, if caused through infection, they are characterized by particularly long incubation periods (problem of recall bias). Case control studies have proved the most suitable type of study for this kind of investigation. These involve a comparison of the responses from patients' relatives with those of at least one healthy control person or his or her relatives, using a standardized questionnaire on putative risk factors. A Europe-wide coordination of the search for risk factor has been completed for sporadic CJD (EuroCJD project of the EU, see Chapters 36; 37). As prion diseases are rare, epidemiological data obtained in several countries must be pooled to achieve sufficient power for statistical analysis. The harmonization of diagnostic criteria and data acquisition are necessary prerequisites for these studies.

[3] Only confirmed deaths are notified.

29.6 The problem of possible phenotypic variation of disease caused by BSE prions in humans

All tested clinical cases of vCJD have been 129MM homozygotes at codon 129 of *PRNP*. However, findings from experiments in transgenic animals indicate that the phenotype of the disease caused by bovine spongiform encephalopathy (BSE) prions in humans may vary according to the genetic background [16], for example transgenic mice expressing human PrP with valine at codon 129 and infected with BSE prions show a distinct phenotype. On the other hand, transgenic animals expressing the human PrP gene with 129MM exhibit a phenotype similar to that of sporadic CJD [17]. This implies that it may be impossible to distinguish sporadic CJD from vCJD on clinical and pathological grounds. Although we cannot be sure that humans will express the same phenotypes as transgenic mice, these findings should be taken as a serious incentive to continue surveillance programs to monitor the overall mortality of prion diseases, possible changes in phenotypes, and alterations in the frequency of affected genotypes.

References

[1] Zeidler M, Johnstone EC, Bamber RWK, et al. New variant Creutzfeldt–Jakob disease: psychiatric features. Lancet 1997; 350:908–910.

[2] The National Creutzfeldt–Jakob Disease Surveillance Unit. National Creutzfeldt–Jakob disease surveillance – diagnostic criteria. Webpage 2006: www.cjd.ed.ac.uk/criteria.htm.

[3] EuroCJD. Definition of terms. Webpage 2006: www.eurocjd.ed.ac.uk/def.htm.
[4] Glatzel M, Abela E, Maissen M, et al. Extraneural pathologic prion protein in sporadic Creutzfeldt–Jakob disease. N Engl J Med 2003; 349(19):1812–1820.
[5] Budka H, Aguzzi A, Brown P, et al. Tissue handling in suspected Creutzfeldt–Jakob disease (CJD) and other human spongiform encephalopathies (prion diseases). Brain Pathol 1995; 5:319–322.
[6] International Air Transport Association (IATA). Dangerous Goods Regulations. Webpage 2006: www.iata.org/whatwedo/dangerous_goods/download.htm.
[7] Barcikowska M, Liberski PP, Boellaard JW, et al. Microglia is a component of the prion protein amyloid plaque in the Gerstmann–Sträussler–Scheinker syndrome. Acta Neuropathol 1993; 85:623–627.
[8] Doh-ura K, Tateishi J, Kitamoto T, et al. Creutzfeldt–Jakob disease patients with congophilic kuru plaques have the missense variant prion protein common to Gerstmann–Sträussler syndrome. Ann Neurol 1990; 27:121–126.
[9] Nochlin D, Sumi SM, Bird TD, et al. Familial dementia with PrP-positive amyloid plaques: a variant of Gerstmann–Sträussler syndrome. Neurology 1989; 39:910–918.
[10] Hainfellner JA, Liberski PP, Guiroy DC, et al. Pathology and immunocytochemistry of a kuru brain. Brain Pathology 1997; 7(1):547–553.
[11] Kitamoto T, Tateishi J, Sato Y. Immunohistochemical verification of senile and kuru plaques in Creutzfeldt–Jakob disease and the allied disease. Ann Neurol 1988; 24:537–542.
[12] Tateishi J, Kitamoto T, Hashiguchi H, et al. Gerstmann–Sträussler–Scheinker disease: immunohistological and experimental studies. Ann Neurol 1988; 24:35–40.
[13] Parchi P, Castellani R, Capellari S, et al. Molecular basis of phenotypic variability in sporadic Creutzfeldt–Jakob disease. Ann Neurol 1996; 39(6):767–778.
[14] Collinge J, Sidle KC, Meads J, et al. Molecular analysis of prion strain variation and the aetiology of 'new variant' CJD. Nature 1996; 383:685–690.
[15] Hörnlimann B, Aguzzi A, Raeber P-A, et al. Überwachung transmissibler spongiformer Enzephalopathien (TSE) – Konsensbericht zur Harmonisierung der CJD-Überwachung. Bull BAG/OFSP Bern 1997; No. 50:9–11.
[16] Wadsworth JD, Asante EA, Desbruslais M, et al. Human prion protein with valine 129 prevents expression of variant CJD phenotype. Science 2004; 306(5702):1793–1796.
[17] Asante EA, Linehan JM, Desbruslais M, et al. BSE prions propagate as either variant CJD-like or sporadic CJD-like prion strains in transgenic mice expressing human prion protein. EMBO J 2002; 21(23):6358–6366.
[18] Ironside JW, Head MW, Bell JE, et al. Laboratory diagnosis of variant Creutzfeldt–Jakob disease. Histopathology 2000; 37(1):1–9.
[19] Ironside JW, Hilton DA, Ghani A, et al. Retrospective study of prion-protein accumulation in tonsil and appendix tissues. Lancet 2000; 355(9216):1693–1694.

30 Clinical Findings in Human Prion Diseases

Matthias Sturzenegger and Robert G. Will

30.1 Introduction

Human prion diseases are characterized by a broad spectrum of symptoms and clinical signs. As pathognomonic clinical signs and reliable (test)-criteria are not available, the disease cannot be reliably diagnosed in patients while they are alive (intra vitam) but only suspected – with different degrees of certainty [1; 2]. Absolute certainty can only be gained following examination of the brain (Table 29.2).

The recognition of a possible prion disease depends on (i) the experience and specialist knowledge of the examining physician, (ii) a comprehensive history (in particular family history), (iii) the recording of the course of the disease, and (iv) the results of additional investigations.

Because of their similarity to other diseases, the differential diagnosis of prion diseases must include additional paraclinical examinations. These are especially important in the initial stage of the disease when nonspecific clinical symptoms and signs prevail (Table 30.2) in order to exclude other diseases with similar symptoms. These kinds of investigations are best carried out at a neurological department and rely on imaging techniques (MRI), EEG, the examination of cerebrospinal fluid, blood tests and biopsies. At stages during which the typical symptoms and signs are fully expressed, the diagnosis is much easier. The terminal phase, characterized by akinetic mutism or coma, however, is once again nonspecific. A definite diagnosis is only possible following histopathological examination, usually after autopsy, which is highly recommended if a prion disease is suspected. A brain biopsy during life is only rarely indicated, e. g., upon tumor suspicion, or if there is a reasonable possibility of an alternative, potentially treatable disease. Currently, no proven treatment is available [3].

Chapters 29, 30 and 31 address the general practitioner and neurologist and discuss prerequisites and methods of adequate clinical assessment, correct diagnosis, and monitoring of prion diseases.

30.1.1 Initial examination and history

The earlier in the course of a prion disease the physician is contacted by a patient, the more nonspecific are the symptoms presented. In particular, general practitioners, psychiatrists and neurologists are challenged to be able to recognize a prion disease in its early stages (see Chapter 29). As variant Creutzfeldt–Jakob disease (vCJD) is initially characterized by depression and other psychiatric symptoms, psychiatrists assume an important role in the early detection of a neurological disorder [4; 5]. The spectrum of neurological features in vCJD is more restricted than the more extended spectrum of symptoms and signs in sporadic CJD (sCJD). The possibility of such a disease must be taken into consideration when there is a combination of psychiatric and neurological symptoms (see Fig. 30.1). In familial forms of CJD, the clinical picture of the disease can vary considerably, even within the members of one family [6]. A detailed family history together with genetic analysis is particularly important in the identification of familial prion diseases (fCJD, GSS, FFI; Tables 26.5, 29.3). The occurrence of cases of dementia in the families concerned is another factor which may prompt genetic analysis of the prion protein gene (*PRNP*) [7].

Clinical features which can be observed in all human prion diseases, and those which should

	Possible symptoms (nonspecific)	**Possible findings (upon clinical examination)**
Psychiatric	Apathy Dysphoria Irritability Jumpiness Eating disorder Sleep disorder	Adynamia Isolation Depression Delusions Stupor
Neurological		
- Cognitive	Memory impairment Poor Concentration Lacking in drive (apathy)	Reduced short term memory Reduced attentiveness Orientation disorder
- Visual	Unspecific visual disorders Diplopia Hallucinations	Hemianopia Agnosia Nystagmus
- Cerebellar	Speech disturbance Difficulties in swallowing Coordination disorder Unsteadiness of gait	Dysarthria Dysphagia Dysmetria of limbs Intention tremor
- Extra-/pyramidal	Weakness Muscular rigidity Muscular atrophy Muscular twitching	Ataxia of stance and gait Rigidity Spasticity, hyperreflexia Muscular atrophy Myoclonus (spontaneous, triggered)
- Unspecific	Paresthesias Epileptic fits	

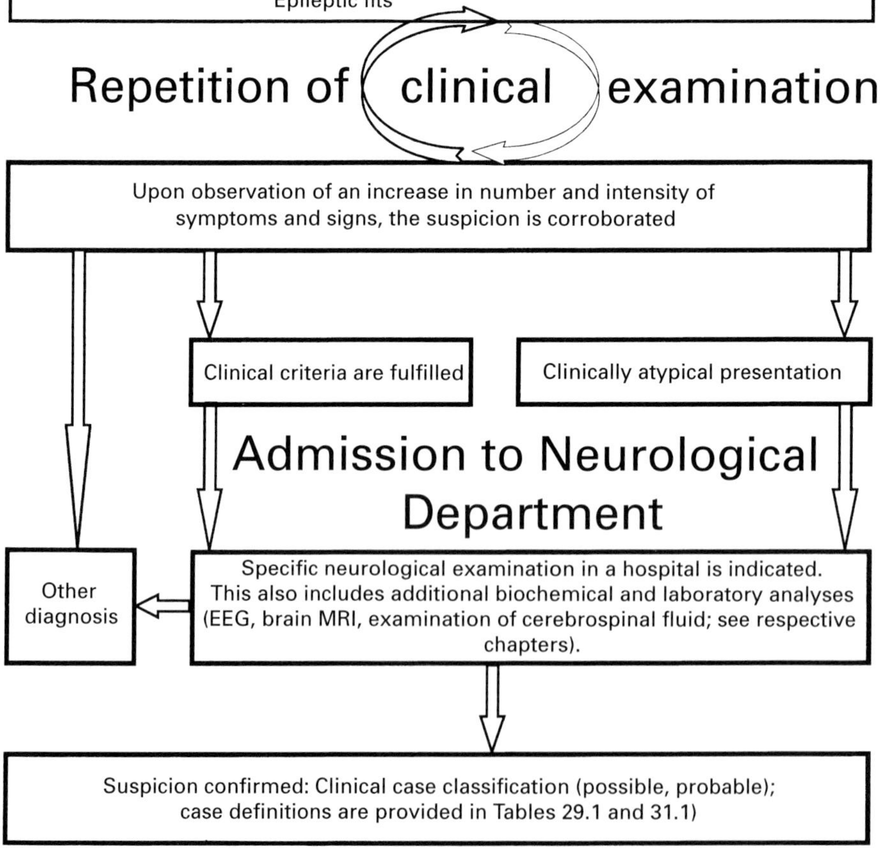
Repetition of clinical examination
Upon observation of an increase in number and intensity of symptoms and signs, the suspicion is corroborated
Clinical criteria are fulfilled
Clinically atypical presentation
Admission to Neurological Department
Other diagnosis
Specific neurological examination in a hospital is indicated. This also includes additional biochemical and laboratory analyses (EEG, brain MRI, examination of cerebrospinal fluid; see respective chapters).
Suspicion confirmed: Clinical case classification (possible, probable); case definitions are provided in Tables 29.1 and 31.1)

raise the suspicion of an underlying prion disease, are shown in Figure 30.1.

30.1.2 Course of the disease

The suspicion of a prion disease is strengthened if initial symptoms or signs become more intense rapidly, i. e., within a few weeks after onset, or if they accumulate quickly during the course of the disease. Although the variability of the phenotypic spectrum remains broad for a relatively long period of time, this variability diminishes in the later stages. Few cases remain absolutely atypical in their phenomenology until death (see Chapter 26.5.5).

Upon the suspicion of a prion disease, it is important to re-examine the patient at short intervals. The rapid progression of the disease within a few weeks, is one of the major features of the majority of human prion diseases. Exceptions are Gerstmann–Sträussler–Scheinker disease (GSS) and occasionally sCJD. One GSS patient lived for 17 years after the disease had been diagnosed (Table 30.1).

When rapid clinical progression of the disease and an accumulation of new symptoms or signs are observed, it is imperative that the patient is seen by a neurologist. The patients concerned should be admitted to a neurological clinic.

The clinical features of the individual prion diseases overlap to varying degrees. As shown in Table 30.1, the diagnostically relevant differences in the clinical picture of the different prion diseases depend in particular on (i) the age of the affected patients, (ii) the duration of the disease and the speed of its progression, (iii) the chronological sequence of the main symptoms (initial versus later stages of the disease), (iv) the expression of the symptoms, and (v) the combination of the observed clinical signs.

◀ **Fig. 30.1:** Possible symptoms and findings upon clinical examination which might suggest the presence of a human prion disease; schematic guide on what to do in order to corroborate the initial diagnostic suspicion. Figure by M. Sturzenegger and B. Hörnlimann ©.

30.2 Differential diagnosis and additional paraclinical investigations

Differential diagnosis has to cope with a broad spectrum of degenerative, inflammatory, metabolic, toxic, and post-traumatic brain disorders (Table 30.2). Initially, when no details are known on the further progression of the symptoms, the spectrum is particularly broad. As shown in Table 30.2, there are, however, some diseases, which can be specifically treated and cured if diagnosed at an early stage. Therefore, their early identification is particularly important. Three investigations are of particular diagnostic utility: electroencephalogram (EEG), cerebrospinal fluid examination (CSF) and magnetic resonance imaging (MRI). Diagnostic evaluation should always include cerebral imaging, if possible MRI. With this technique there are not only early findings corroborating underlying prion disease such as high signal in putamen and caudate or posterior thalamus (see Chapter 31.9), but also it is possible to exclude hydrocephalus or focal or multifocal lesions, such as encephalitis, ischemic brain infarctions or intracranial space occupying processes. Among electrophysiological examination techniques, EEG is of greatest value, especially in sCJD (Fig. 30.2). In 50–90 % of all cases, the EEG can reveal characteristic, though nonspecific, findings [8]. In particular, on repeated examinations, EEGs can show a typical chronological progression. In the early stages of the disease, only nonspecific abnormalities [9] are found such as disorganization and diffuse slowing of the normal background rhythms. Later, repetitive, generalized, synchronous, pseudoperiodic 0.5–2/s bi- or triphasic complexes (so called periodic sharp wave complexes, PSWCs) become more and more frequent. These complexes may first appear unilaterally and then spread diffusely. They are nearly pathognomonic and can be observed in about two thirds of all cases. Their appearance can, however, fluctuate which makes it necessary to repeat the EEG examination several times at intervals of a few weeks (Fig. 30.2). Their absence does not exclude the diagnosis [10]. Furthermore PSWCs may also present in other conditions such as, e. g., hepatic encephalopathy or drug toxicity.

Table 30.1: Overview of the signs and symptoms occurring in human prion diseases. Clusters of signs are shown, as wcll as indicators regarding the course of the disease and particular findings, which determine the suspicion of a specific form of human prion disease.

Form	Age at the onset of the disease	Main symptoms in the initial stage	Main symptoms in the late stage	Duration of disease	Specifics
sCJD	Average: 65 years	• Rapidly progressive dementia and myoclonus	• Multifocal neurological disturbance, often with ataxia and rigidity. • Akinetic mutism terminally	1–130 months; average: 4.5–7 months	Atypical features in about 20 % of cases e. g., long duration, insidious onset of dementia, cerebellar onset
fCJD	Average: 50 years	• Usually similar to sCJD, depends on specific mutation	• Usually similar to sCJD, depends on specific mutation	Average: 21 months	Familial-genetic; autosomal dominant inheritance
iCJD	Depends on the mechanism of transmission and age at the time of infection	• In the case of central transmission, the same symptoms as in sCJD • If peripherally transmitted, symptoms similar to kuru	• In the case of central transmission, the same symptoms as in sCJD • If peripherally transmitted, symptoms similar to kuru	10* months – 30 years (depending on the mechanism of transmission)	Central transmission by brain surgery, intracerebral EEG recordings, etc.; peripheral transmission through pituitary hormones, etc. (iatrogenic)
vCJD	14–74 years; average: 28 years	• Psychiatric disorder (apathy, depression) • Gait ataxia • Early, persistent, painful paraesthesia	• Progressive ataxia and dementia	7-40 months; average: 14 months	Variant CJD, first described in 1996
GSS	25–61 years; average: 44 ± 4 years	Slowly progressive instability in walking and standing; frequent falls Sometimes also • Dysarthria • Dysmetria of the limbs • Tremor	After 1–3 years: • Spasticity • Myoclonus • Progressive, finally very severe dementia	2–17 years; average: 7 ± 5 years	Familial-genetic (autosomal dominant inheritance)
Kuru (Fig.13.1)	Bimodality: approx. 4–20 years; approx. 20–60 years or older	• Rapidly progressive gait ataxia and trunk ataxia	• Dysarthria • Tremor • Choreoathetosis • Emotional instability	Average: 12 months; less in children	A few new cases annually still identified today
FFI	20–71 years; average: 51 years	• Progressive sleeping problems • Intermittent dream-like phases (during the day) • Dysautonomia expressed as chronic stress syndrome	• Impairment of memory and orientation ability • Gait ataxia • Myoclonus	31 ± 1 months or 9 ± 1 months (see Chapter 17.4)	Familial-genetic (autosomal dominant inheritance)

* following FSH injection, see Chapter 36.4.2 and [51; 52].

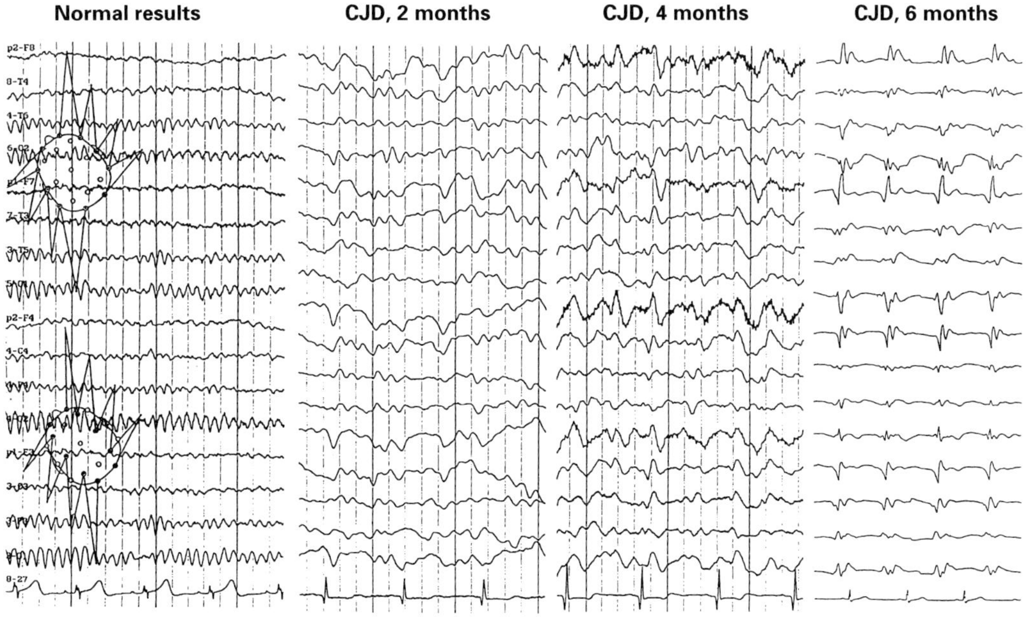

Fig. 30.2: Typical electroencephalographic findings (EEG) in sCJD patients. A 64-year-old patient with severe dementia, generalized myoclonus and spasticity. CJD was later confirmed at autopsy. The curves were recorded 2, 4, and 6 months after the first symptoms were observed. These were characterized by dysphoria and rapidly worsening memory impairments. After two months EEG demonstrates generalized diffuse slowing of normal background rhythm. After four months, repetitive, generalised, synchronous pseudoperiodic bi- or triphasic complexes can be observed (= periodic sharp wave complexes, PSWC). After a period of six months, these complexes dominate and normal background rhythms are almost completely missing (EEG criteria in CJD according to Steinhoff: (1) periodicity (most important criterion), (2) PSWC, (3) frequency 0.5–2/sec, (4) duration of 200–600 msec, (5) amplitude > 150–300 μV, (6) generalized with frontal maximum; partially regionally lateralized and (7) subsequent "burst supression". Figure by M. Sturzenegger ©.

With further progression of the disease, the EEG tracing becomes flatter with generalized low voltage. Frequently, hyper-responsiveness to external, i. e., acoustic, tactile, or visual stimuli with activation of the sharp waves can be observed.

Polysomnographic examination has proved particularly helpful upon suspicion of fatal familial insomnia (FFI; see Chapter 17). Electroneurographic, electromyographic, and evoked potential studies assist in identifying the subclinical affection of the peripheral nervous system, the spinal tracts, or the muscles. Muscle tissue is generally not affected by prion diseases, at least clinically. Useful application of these diagnostic tools requires plenty of experience, particularly in the interpretation of the results. It should be emphasized that the needle electrodes used *must* be disposable needles whenever a dementia patient is examined, if there is a possibility of a prion disease. Of great importance in differential diagnosis is the examination of CSF. Inflammatory disorders (encephalitis, meningoencephalitis) can easily be recognized because of cerebrospinal fluid pleocytosis; serological examinations and PCR of CSF may lead to the identification of the infective agent. Cell count, total protein content, and glucose concentration, as well as CSF pressure typically display normal values in patients suffering from prion diseases. In contrast, the identification of increased levels of neuronal marker proteins in

Table 30.2: Differential diagnosis of Creutzfeldt–Jakob disease (listed according to frequency).

Disease group	Disease
Degenerative brain diseases	Alzheimer's disease DLB (Dementia with Lewy bodies) Chorea (e. g., Huntington's chorea) Progressive supranuclear palsy (PSP) Amyotrophic lateral sclerosis (ALS) Pick disease
Inflammatory encephalopathies	Viral Encephalitis (HSV, etc.) Progressive multifocal leukencephalopathy (PML)
Metabolic encephalopathies	Hepatopathy Uremia Hypo-, hyperthyreosis Hypo-, hypercalcemia Hypo-, hyperglycemia Korsakow syndrome Wernicke encephalopathy
Toxic encephalopathies	Intoxication with metals (bromine, bismuth, mercury, lithium) or drugs
Paraneoplastic encephalopathies	Limbic encephalomyelitis Paraneoplastic cerebellar degeneration
Miscellaneous (partially post-traumatic)	Postanoxic encephalopathy Chronic subdural haematoma Hydrocephalus Binswanger's encephalopathy

the CSF such as neuron-specific enolase, protein 14–3–3 [11; 12] and protein S100b are helpful in the differential diagnosis of CJD. An elevated 14–3–3 protein may be found in sCJD cases in the relatively early stages (see Chapter 31.3). On their own, these proteins are not specific for CJD. However, other CNS disorders, in which these changes can also be identified, show differences in other cerebrospinal fluid parameters, in disease duration or findings on brain imaging, which generally allow the differentiation of the disease [13]. The CSF 14–3–3 assay should not be used as a screening test in unselected cases. Routine chemical examinations of blood and urine generally show normal values. It is important to adequately classify the significance of these paraclincal examinations and interpret their possible results. They are discussed in detail in Chapter 31.

30.3 Clinical features of Creutzfeldt–Jakob disease

30.3.1 Epidemiological parameters

Sporadic CJD is by far the most frequent human prion disease, but with an incidence of 1 case/1 million people/year, it is still rather rare. 80–90 % of all human prion diseases occur as sporadic forms (sCJD). External risk factors for this form of the disease have not been identified, i. e., no "natural" route of acquisition has been found for sCJD (see Chapters 36; 42). Yet, that sCJD could be an acquired illness has not been excluded; two recent case control studies in the UK and Australia found a relative excess of previous surgery in the sCJD cases.

Familial CJD (fCJD) is very rare. This disease is characterized by an autosomal, dominant pattern of inheritance and an association with mutations of *PRNP* on chromosome 20 [14]. Of all patients diagnosed with CJD, 4–15 % are classified as familial prion diseases (depending on the study and whether GSS and FFI are included or not).

Iatrogenic CJD (iCJD) is also extremely rare [15]. Apart from iCJD, only kuru and vCJD, (believed to be caused by BSE) are aquired forms of human prion disease (see Chapter 1). iCJD can, for example, be transmitted by a number of mechanisms (Table 14.1): (i) surgical interventions on the brain with contaminated surgical instruments or brain electrodes for deep intracerebral EEG recordings (see Chapter 47), (ii) implantation of contaminated dura mater and cornea (see Chapter 51), (iii) therapeutic use of pituitary gland hormones (growth and sex hormones) taken from cadavers [16].

Over the past 30 years more than 350 cases of CJD are reported to have been caused by the transmission of infection from person to person through medical/surgical treatments. Dura mater and pituitary growth hormone account for most cases (Table 14.1). Further details will be provided in Chapter 36.4.2.

30.3.2 Factors determining the manifestation of the disease

Disease manifestation mainly depends on: (i) the mechanism of transmission or entrance of the infective agent, (ii) the PrP genotype of the host (in the case of GSS, FFI and fCJD) (see Chapters 26.4–6), and (iii) the prion strain (see Chapters 12; 50).

Mechanism of transmission. An important factor influencing the clinical phenotype of iatrogenic CJD is the route through which the infective agent enters the body [17]. iCJD cases that are transmitted peripherally by i.v. injection, e.g., of contaminated of pituitary hormones [15] are clinically very similar to kuru (see Chapter 13.4). Both iCJD and kuru are characterized by an initial cerebellar syndrome and only late onset and minor dementia. These two forms of prion disease, as well as vCJD, follow a different clinical course from the iCJD patients that were infected centrally through neurosurgical interventions (see Chapter 47). The latter evolves very similar to (probably) spontaneously-arising sCJD with rapidly progressive dementia and myoclonus (definition of spontaneous: see Chapters 3; 8) [18]. Relating to the variable phenotype of iCJD, there seem to be genetic influences, but the evidence indicates that irrespective of codon 129 genotype, the route of exposure influences the clinical phenotype of iCJD. The titer of inoculum and site of inoculation influence the incubation period. Direct intracerebral inoculation may be associated with incubation times of only 16–28 months. The longest delays (up to 30 years) are associated with subcutaneous injections of pituitary hormones. Susceptibility to transmssion of iCJD is influenced by variation at codon 129 of *PRNP*.

PrP genotype. The familial forms of the prion diseases (fCJD, GSS, and FFI) are associated with mutations of *PRNP* (see Chapter 26.6). Even within a single family with identical mutations, different clinical phenotypes can be observed. These differences include the duration of the disease, the type and expression of the symptoms, as well as differences in histopathology. Regarding the variations in the clinical phenotype related to the different underlying mutations, each may display characteristic features. These are described in detail in Chapter 26.

Although an overall clinical consistency can be observed, there is phenotypic variability in fCJD, even among patients who have the same mutation. This suggests the involvement of other regulatory genes or other factors in the expression of the disease. Sequencing of *PRNP*, IHC examination, and Western blot analysis of PrP^{Sc} demonstrate that the clinical phenotypes of sCJD cases mainly correlate with the methionine/valine polymorphism at *PRNP* codon 129 and the PrP^{Sc} banding pattern on the Western blot (PrP glycoform type) [19–21] (see Chapter 26). It should be stressed however, that many cases of genetic prion disease cannot be clinically distinguished form non-genetic forms; a family history of genetic prion disease may not be present; the penetrance is typically high, but age dependent and variable. Therefore, to exclude or identify genetic prion disease, *PRNP* mutation screening has to be undertaken.

Prion strain. The specific prion strain also has an influence on the clinical phenotype (see Chapter 12).

30.3.3 Age at the onset of CJD symptoms and duration of the disease

The onset of sCJD and familial CJD forms occurs typically in older people with an average age of 65 years and a peak incidence in the seventh decade for sporadic (Fig. 15.2) and in the fifth decade for familial forms. A gradual development of the symptoms is much more frequent than a sudden, stroke-like onset. The duration of the disease is estimated, depending on the study, at an average of 4.5 – 7 months (1-130 months) for sporadic CJD and 21 months for the familial CJD forms. The disease is always fatal, and in 90% of all patients, death occurs within one year and in two thirds within six months. Fourteen percent have a disease duration of one year and 5% of two years or more. The most important difference between sCJD and fCJD is thus the early age of disease onset and the longer duration of the disease in the familial cases (Table 30.1).

vCJD is observed in patients with an average age at onset of approximately 29 years (Table 30.1).

30.3.4 Prodromal stage of sCJD

Weeks, sometimes months before the first neurological and cognitive deficits are observed, nearly half of the patients complain about nonspecific symptoms such as fatigue, lassitude, dizziness, headache, difficulties in concentration, loss of weight, sleep disorders, and less frequently about abnormal eating behavior or abnormal sweating [13]. Relatives may report dysphoric or depressive moods, change of personality, irritability, jumpiness or tics.

In about one third of the patients, the actual disease starts with mental disturbances such as memory impairment, confusion, behavioral disorders accompanied by depression and agitation. In another third of patients the disease starts with physico-neurological symptoms (cerebellar, oculomotor, visual), and in the final third with a combination of the two [13; 22].

30.3.5 Major clinical sCJD signs

The major feature of sCJD is a rapidly progressive dementia [2; 19]. The degradation of mental functions dominates the initial phase of the disease (in 69% of all cases) as well as its later course (in 97–100% of all cases, depending on the study; Table 30.4). The dementia exhibits cortical characteristics such as aphasia,

Table 30.3: Differences between sporadic Creutzfeldt–Jakob disease (sCJD) and variant CJD (vCJD).

	Sporadic CJD (sCJD)	Variant CJD (vCJD) [4; 26]
Age at onset of disease, average	65 years	29 years
Initial signs	• Dementia • Myoclonus	• Psychiatric symptoms (depression, anxiety, delusions) • Paraesthesia (painful) • Ataxia (after 6 months)
Periodic complexes in EEG	Up to 70%	Typical PSWC only < 1%
14–3–3 proteins in CSF	90–95%	50%
Course of disease	Rapid progression (average 4.5–7 months) 1–130 months	Prolonged (average 14 months) 7–40 months
PRNP genotype (codon 129 polymorphism)	Excess of methionine homozygotes, no mutation	So far, only methionine homozygotes; no mutations
Histology	No plaques or small numbers of kuru plaques	Many florid plaques in the cerebrum and cerebellum
PrP^{Sc} banding pattern (Western blot)	Type 1 and 2A	Type 2B (as in BSE)

Table 30.4: Sporadic Creutzfeldt–Jakob disease (sCJD): Clinical signs observed in 232 patients – modified from [22].

Signs	Initial stages (%)	Later stages (%)
Mental decline	69	97–100
• Impairment of memory	48	100
• Behavioral disorders	29	57
• Degeneration of higher cortical functions	16	73
Cerebellar disorders	33	71–86
• Dysarthria	na	85
• Ataxia of gait and trunk	na	60
Visual impairments	17	42
• Oculomotor disorders	6	16
Vertigo	13	19
Headache	11	18
Sensation disturbance	6	11
Vegetative disorders	3	7
Pyramidal signs	2	52–62
Extrapyramidal signs	2	56–73
• Rigidity	0	51
Amyotrophy	0.5	12
Epilepsy	0.5	19
Motor disturbances	0.4	91
• Myoclonus	5	78–89
Periodic EEG	0	60
• Triphasic complexes 1/sec	0	48

agraphia, agnosia, acalculia, memory impairment, and poor concentration (Table 30.1). In the late stage when akinetic mutism may be observed, evidence of dementia or previous demential development might prove difficult or even impossible to establish. In contrast, the definitive clinical signs of dementia are sometimes not apparent during the very early phases of the disease, for example in cases with an early cerebellar syndrome.

The second most frequent feature is multifocal myoclonus, which, in combination with dementia, is highly suggestive of a prion disease. Initially myoclonus occurs in only approximately 5% of cases, but becomes more and more frequent during the further progression of the disease (78–89%). However, the myoclonus may be intermittent and can also disappear. Therefore, patients must be repeatedly examined to identify myoclonic movements. They can be evoked with a variety of external stimuli or detected by monitoring with long-term video recordings. Other involuntary choreiform or athetoid movements can occur, although less frequently.

Symptoms of cerebellar disturbance can be recognized at the beginning of the disease in approximately one third of all patients, in the later stages of the disease in 71–86%. These disturbances include dysarthria (approximately 86%), ataxia of gait and trunk (approximately 60%), dysmetria of the limbs and nystagmus.

Extrapyramidal signs such as rigidity, tremor, dystonia, chorea, or athetosis are initially rare (0–2%) but become more frequent during the progression of the disease (51–73%). Pyramidal signs such as spastic hypertonia, hyperreflexia, and Babinski's sign occur initially in only about 2% of the cases, later in 52–62%. Visual symptoms such as double vision, hemianopia, visual agnosia, cortical blindness, or optical hallucinations manifest themselves initially in 17% of all patients, later on this cluster of symptoms becomes part of the main symptoms of the disease (42% of all patients) (Table 30.1). Epileptic fits can be observed in one fifth of the patients. Muscular atrophy is rather rare (12% in the course of the disease) [23].

In individual cases, nearly any combination of the above-mentioned symptoms and findings is possible.

Editor's note: Section 30.9 is an Annex containing a "Historical classification" (since readers might find it helpful for clarification when they are reading corresponding papers).

30.4 Clinical features of variant CJD

The most important features of vCJD are young age at onset, conspicuous initial psychiatric symptoms, a relatively protracted clinical course, lack of periodic EEG activity and specific findings on histopathology (Fig. 15.3) [24].

The onset of the disease is on average at the age of 29 years (Table 30.3). The initial symptoms involve behavioral abnormalities such as isolation and social withdrawal, as well as psychiatric features such as apathy, depression, and anxiety. Although these psychiatric symptoms are initial symptoms in the majority of cases of vCJD [5], they are nonspecific. It must be emphasized that psychiatric symptoms such as depression, sleep disorders, and behavioral abnormalities are also quite frequent in sCJD and can sometimes be observed in the prodromal stages. On the basis of clinical presentation alone, the sporadic form and the vCJD in the initial stage of the disease may be difficult to distinguish. Retrospective data analysis of vCJD patients examined in detail neuropsychologically suggests early fairly generalized cognitive impairment, particularly affecting memory, executive functions, attention, and visuopreceptual reasoning [25].

vCJD must be included in the differential diagnosis when neurological deficits such as ataxia, involuntary movements and cognitive deficiencies are observed in patients with preceding psychiatric features [4] (see Chapter 15.4; Table 31.1d). Further characteristic symptoms of vCJD at a relatively early stage are gait ataxia and impaired sensory function in the form of persistent painful paraesthesia [26] which are reported in about 35% of cases at presentation [5]. A central, probably thalamic origin of these symptoms is assumed. Myoclonus has been observed in many patients, but only during the later stages of the disease; some patients displayed choreoathetotic involuntary movements. By two months, nearly 60% are reporting neurological symptoms, but it is generally more than 4 months before clearcut neurological signs such as gait disturbance, slurred speech, or tremor are evident, and it usually takes more than 6 months before involuntary movements and cognitive impairment are manifest [5]. All patients finally developed progressive dementia. However, memory impairment is rarely an initial symptom. Although there is an overlap of clinical features, the vCJD cases as a group are clinically relatively distinct from sCJD and are also remarkably homogeneous in comparison to the disparate clinical presentations in sCJD [17]. The course of vCJD compared to sCJD is prolonged, with an average duration 14 months (7–40) until death. EEG readings may be normal, but can show progressive slowing of the normal background rhythms. However, with the exception of recordings carried out late in the clinical course in two cases, the periodic activity which is characteristic of sCJD has not been observed in vCJD. MRI is very helpful and reveals high signal in the posterior thalamus (the "pulvinar sign") on T2 and especially FLAIR and DWI sequences in about 90% of cases (Fig. 31.6). These are located posteriorly (dorsomedial nuclei and pulvinar of the thalamus), whereas they can be found anteriorly in sCJD (caudate and putamen). CSF 14–3–3 protein is detected in only about half the cases and CSF tau protein may be a more useful marker [27]. Tonsil biopsy may allow the identification of the disease associated prion protein, and thus ante mortem diagnostic confirmation. The variant form, contrary to the sporadic form, is characterized by high amounts of PrP^{Sc} in lymphoreticular tissues such as tonsils and spleen, and, to a lesser degree, lymph nodes in all patients [28].

The diagnosis can only be confirmed on histopathological examination [26] and laboratory analyses [29]. The currently valid case definition can be found in Tables 29.1 and 31.1 as well as reference [30].

30.5 Clinical features of Gerstmann–Sträussler–Scheinker disease

Gerstmann-Sträussler-Scheinker disease (GSS) [31; 32] is very rare and is characterized by autosomal dominant inheritance (see Chapter 16). The major clinical signs involve slowly developing cerebellar ataxia with chronic progressive course and late onset of dementia. The disease usually starts between 25 and 61 years of age, with the first symptoms being difficulty in walking or standing, causing frequent falls (Table 30.1). Subsequently, the cerebellar symptoms increase and involve dysarthria, dysphagia, and dysphonia, then nystagmus and saccadic eye movements [33], increased hand

and arm dysmetria with intention tremor; and eventually a pancerebellar syndrome with increasing disability. Psycho-organic alterations such as irritability, impairment of memory, and learning disturbances often develop in the course of the disease. In addition, in the later stages of the disease, the following symptoms can also be observed: myoclonus, spasticity, abnormal dystonic or athetoid movements, finally very severe dementia, and mutism. Immobility, confinement in bed, and dysphagia lead to cachexia. After a duration of approximately five years (2.5–11) the disease leads to death.

MRI and CT of the brain often only reveal no more than a nonspecific brain atrophy predominantly affecting the cerebellum. EEG shows a progressive diffuse slowing of the normal background rhythms but without periodic activity.

Genetic background of GSS. Different *PRNP* mutations can lead to GSS, however, with different clinical phenotypes (see Chapter 26). The most frequent variant is the CCG»CTG single-basepair mutation which leads to the substitution of proline through leucine at codon 102 (102Leu). This mutation was observed in non-related families in England, Japan, USA, Canada, Austria, Italy, Israel, and Germany. The different mutations are also associated with age difference at the time of onset of the symptoms and different duration times of the disease. The 117Val mutation, which was observed in families in France and America, causes an early onset of the disease (< 40 years) and is frequently associated with a pseudobulbar syndrome. The 198Ser mutation, identified in families from America, is clinically analogous to the 102Leu mutation but may display a dopamine-responsive parkinsonian syndrome [2; 34].

Even in patients of the same family, considerable variability among clinical and histopathological features have been observed (see Chapter 26.5.3). None of the listed clinical symptoms or signs have been observed in all patients. This suggests that the *PRNP* mutation is not the only factor which determines the clinical and pathological phenotype [32; 35] of GSS. Discordance within families points to additional genetic and environmental modifying factors, including, again, codon 129 status.

30.6 Clinical features of fatal familial insomnia

Like fCJD and GSS, fatal familial insomnia (FFI) also has autosomal dominant inheritance (see Chapter 17). Apart from the *PRNP* gene mutation, other genetic factors influence the clinical phenotype. Notably, the DI178 mutation of the *PRNP* gene gives rise to fCJD when associated with 129V on the mutant allele and results in FFI when associated with 129M. There are considerable differences between FFI and CJD regarding clinical, histopathological and molecular findings. In brief, the characteristic clinical features of FFI are: sleep disturbances, behavior disorder as well as conspicuous vegetative, motor and cognitive disturbances (Table 30.5) [34].

The disease onset is between ages of 20 and 71 years, with the average being 51. Disease duration is 31 $\pm$ 1 month or 9 $\pm$ 1 month (depending on the sequence at codon 129 of *PRNP*) [36; 37]. The initial symptoms of the disease are variable, but usually the main feature is progressive insomnia. Sometimes, insomnia can only be determined with multimodal sleep analysis (polysomnography, PSG) [38]. PSG reveals a disruption of the normal sleep-wake cycle that becomes gradually worse and finally leads to the complete cessation of the sleep-wake rhythm with the destruction of the physiological cyclical sleep pattern. There is reduced total sleep time, virtual absence of REM periods and deeper sleep phases characterized by K-complexes, spindles, and slow waves. The loss of sleep activity, in particular of the recovering slow-wave sleep, is directly attributed to thalamic degeneration. Another main feature of the disease is progressive dysautonomy with unbalanced sympathetic overactivity and a chronic stress syndrome, which is characterized as follows: loss of the physiological fall of blood pressure at night, tachycardia, tachypnea, apnea, hyperhidrosis, fever, increased lacrimation, and saliva secretion, as well as chronic secondary arterial hypertension. There are also progressive cognitive disturbances initially presenting with reduc-

Table 30.5: Fatal familial insomnia (FFI): Clinical signs observed in 24 patients – modified from [36].

Parameter	Feature/Clinical sign
Age at onset of disease	51 years (20–71 years)
Disease duration	14 months (8-32 months) (bimodal)
Sleep and vigilance	Progressive insomnia (polysomnography) Episodic dreamy states Hallucinations Stupor
Cognitive functions	Memory impairment (long-term memory) Orientation disorders Attention disorders Confusion
Autonomous functions	Hyperhidrosis Tachycardia Hyperthermia Arterial hypertension Tachypnea Apnea
Motor functions	Dysarthria Dysphagia Ataxia Myoclonus (spontaneous, evoked) Hyperreflexia Babinski's signs
Endocrine functions	Lower ACTH levels Higher cortisol and catecholamine levels Abnormal circadian rhythm of the following hormones: – Growth hormone (GH) – Melatonin – Prolactin
EEG (during sleep)	Loss of delta activity, sleep spindles and K-complexes Abnormal REM phases
EEG (when awake)	Progressive flattening and slowing of the basal rhythm Missing drug-induced sleep activity
Positron emission tomography (PET)	Hypometabolism in the anterior thalamus (less frequently in the fronto-parietal cortex)

ed attention and vigilance. Fluctuating vigilance with rapid change between wakefulness and oneiric states – with the polygraphic equivalent of a true status dissociatus – can be observed early in the disease. At a later stage of the disease, memory impairment (in particular long-term memory) and disturbances in orientation (in particular chronological sequences), confusion and dream-like states are observed [39]. The latter are characterized by acting out dream contents (oneiric states).

Endocrinologically, a gradual attenuation of the circadian oscillations of somatotropin and later also prolactin secretion can be observed. Adrenocortical overactivity is indicated by hypercortisolism and persistently elevated catecholamine levels. The secretion pattern of melatonin reveals declining plasma levels and a loss of rhythmic secretion. In the final stages of the disease, a complete loss of all circadian autonomous or neuroendocrine rhythmicity is observed (Table 30.5). These features are related to the neuropathologically characteristic restricted degeneration confined to mediodorsal and anteroventral thalamic nuclei with important functions within the limbic system and central autonomic network.

Motor disturbances observed in FFI include myoclonus, cerebellar ataxia with gait ataxia, dysarthria, dysmetria, dysphagia and pyramidal signs. In the later stages of the disease, the patients become apathetic, mutistic, somnolent, and comatose. PET studies with radiolabeled fluorodeoxyglucose reveal a characteristic thalamic hypometabolism, whereas conventional neuroimaging (CT, MRI) is usually normal or may show nonspecific cerebral and cerebellar atrophy. The EEG reveals a gradually progressive nonspecific slowing of the normal background rhythms (Table 30.5). Haematological and biochemical examinations usually reveal normal results. Protein 14–3–3 detection in CSF is usually absent (see Chapter 31).

Genetic background of FFI. There is a mutation on *PRNP* codon 178 which causes a replacement of aspartate by asparagine (178Asn) and a methionine polymorphism at codon 129. Several families from Italy, France, Germany, and the USA have been found to carry this mutation [36; 40; 41]. Rare cases of sporadic fatal insomnia (SFI) have been reported with quite similar clinical and neuropathological findings but no family history and no mutation in *PRNP* (see Chapter 17) [36; 42]. These cases are also referred to as the MM2 thalamic variant of CJD [43; 44].

30.7 Clinical features of sporadic fatal insomnia

The clinical features of this disease (see Chapter 17) are similar if not identical to those of FFI [36; 42; 43; 44]. Sporadic fatal insomnia (SFI) is of unknown etiology. In 1999, P. Parchi and colleagues described this very rare sporadic form of "fatal insomnia" which occurs without a known *PRNP* mutation [44]. Possibly it is caused by a spontaneous conversion of PrP^{C} into PrP^{Sc} or by a somatic *PRNP* mutation. Until this disease was recognized as a distinct entity, it is highly likely that similar cases were previously labeled as "the thalamic variant of CJD" or "thalamic dementia" (see Chapter 26.5.4).

30.8 Clinical features of kuru

In the 1950s, kuru was observed as an epidemic in the Fore tribe and in neighboring tribes living in the eastern highlands of Papua New Guinea [45; 46] (Fig. 2.1, 2.2). Over 2,700 cases of kuru have been documented since 1957 in a total population of 36,000 people within the kuru region.

(Editors' note: Kuru is still a disease: between mid-1996 and mid-2004 there were 11 new cases of kuru recorded. Therefore it appeared appropriate to describe the clinical signs in detail; here and in Chapter 13.4.)

The spectrum of the kuru symptoms is similar to those of CJD, but in contrast, kuru has a remarkably homogeneous picture of clinical signs and course [17]. The time until death from first observation of disease signs is on average 12 months (6–36, and hence longer compared to CJD) in adults and slightly less in children [45] (Table 30.1).

Characteristic prodromal symptoms are headache as well as arthralgia and melalgia. After a few weeks, the signs of paleocerebellar disease with midline ataxia of the trunk can be observed: problems in standing and walking (Fig. 13.1a), unclear language and possibly also double vision (Fig. 13.1c). Instability of the trunk worsens and leads to titubation, i. e., rhythmic coarse tremor of the entire body when standing upright. After a few weeks neocerebellar affection develops with coarse postural tremor, dysmetria, dysdiadochokinesis, dysarthria, hypotension and finally the ability to stand and walk is lost. Later, a transient phase can be observed characterized by muscular hypertonus, hyperreflexia and clonic spasms. Choreoathetotic and dystonic movement disorders can often be observed. Frequently, emotional instability, characterised by pathological crying or 'laughing' (Fig. 13.1b) is observed. Typically, sensory functions remain intact and myoclonus is absent.

EEG and electromyographic examinations are normal [47]. In the final stage of this rapidly progressive disease, dysphagia associated with cachexia, incontinency and then death occurs which is often induced by pneumonia and infected decubital ulcers.

The major difference from sCJD is the early, progressive cerebellar ataxia (always seen in kuru cases) and coordination disturbances (Fig. 13.1d). Dementia, the main early symptom of CJD, is rare in kuru. If it does occur, it does so in the terminal stage of the disease. A further difference is the broad age spectrum of the kuru patients.

30.9 Annex: historical classification

(This annex relates to section 30.3.5 "Major clinical sCJD signs").

In the early literature, five clinical CJD variants or types were described, corresponding to the combination of specific symptoms and signs. As the historical classification might stimulate the consideration of the possibility of CJD or other prion disease through the observation of certain clinical symptoms, the syndromes are outlined below.

1. The most frequent syndrome, observed in at least 70 % of cases, comprises dementia and myoclonus together with cerebellar and pyramidal symptoms. This was previously referred to as the *diffuse-cerebral type of CJD*.
2. Cases with pronounced fronto-cortical and pyramidal-extrapyramidal motor symptoms were, referred to as the *Jakob variant* or *spastic pseudosclerosis* [48].
3. sCJD mainly characterized by occipito-parietal disturbances and prevalence of visual, agnostic, and dyskinetic symptoms are referred to as the *Heidenhain's variant* [49]. There may be relatively isolated visual disturbances culminating in cortical blindness.
4. Cases with predominant early cerebellar symptoms are categorized as *cerebellar CJD* or the *Brownell-Oppenheimer variant* (see Chapter 26). As in kuru, dementia is only observed in the later stages of the disease [50].
5. The *amyotrophic form of CJD* characterized by muscular atrophy is very rare and is probably not a prion disease [23].
6. The thalamic form of CJD is probably identical with sporadic fatal insomnia (SFI).

There is a good correlation between these clinical types of sporadic CJD (and SFI) and the molecular classification of sporadic human prion diseases (see Chapter 26). One has to keep in mind that the correct diagnosis of CJD is not only based on individual symptoms or combinations thereof. In particular, the time course with usually rapid deterioration obvious even after a few weeks, is most suggestive of sCJD. The clinical suspicion of a prion disease must be followed by *PRNP* genotyping (provided there is informed consent), in order to identify potentially hereditary cases.

In the terminal phase of the disease, patients become mute, do not react to stimuli (akinetic mutism), are somnolent, incontinent, cachectic and finally die of intercurrent illness, often bronchopneumonia.

References

[1] Knight RS and Will RG. Prion diseases. J Neurol Neurosurg Psychiatry 2004; 75 Suppl 1:i36–i42.

[2] Collins SJ, Lawson VA, Masters CL. Transmissible spongiform encephalopathies. Lancet 2004; 363(9402):51–61.

[3] Knight R. Therapeutic possibilities in CJD: patents 1996–1999. Exp Opin Ther Patents 2000; 10:49–57.

[4] Zeidler M, Johnstone EC, Bamber RWK, et al. New variant Creutzfeldt–Jakob disease: psychiatric features. Lancet 1997; 350:908–910.

[5] Spencer MD, Knight RS, Will RG. First hundred cases of variant Creutzfeldt–Jakob disease: retrospective case note review of early psychiatric and neurological features. BMJ 2002; 324(7352): 1479–1482.

[6] Howard RS. Creutzfeldt–Jakob disease in a young woman. Lancet 1996; 347:945–948.

[7] van Duijn CM, Delasnerie-Lauprêtre N, Masullo C, et al. Case-control study of risk factors of Creutzfeldt–Jakob disease in Europe during 1993–95. European Union (EU) Collaborative Study Group of Creutzfeldt–Jakob disease (CJD). Lancet 1998; 351(9109):1081–1085.

[8] Steinhoff BJ, Racker S, Herrendorf G, et al. Accuracy and reliability of periodic sharp wave complexes in Creutzfeldt–Jakob disease. Arch Neurol 1996; 53(2):162–166.

[9] Donnet, A, Farnarier G, Gambarelli D, et al. Sleep electroencephalogram at the early stage of Creutzfeldt–Jakob disease. Clin Electroencephalogr 1992; 23(3):118–125.

[10] Chiofalo N, Fuentes A, Galvez S. Serial EEG findings in 27 cases of Creutzfeldt–Jakob disease. Arch Neurol 1980; 37:143–145.

[11] Zeidler M. 14–3–3 cerebrospinal fluid protein and Creutzfeldt–Jakob disease. Ann Neurol 2000; 47(5):683–684.

[12] Green A, Thompson E, Zeidler M, et al. Raised concentrations of brain specific proteins in patients with sporadic and variant CJD (Abstract). J Neurol Neurosurg Psychiatry 2000; 69:419.

[13] Poser S, Zerr I, Schulz-Schaeffer WJ, et al. Creutzfeldt–Jakob disease, a sphinx of current neurobiology. Dtsch Med Wochenschr 1997; 122(37):1099–1105.

[14] Brown P. The phenotypic expression of different mutations in transmissible human spongiform encephalopathy. Rev Neurol (Paris) 1992; 148:317–327.

[15] Brown P, Preece M, Brandel JP, et al. Iatrogenic Creutzfeldt–Jakob disease at the millennium. Neurology 2000; 55(8):1075–1081.

[16] Will RG. Epidemiology of Creutzfeldt–Jakob disease. Br Med Bull 1993; 49:960–970.

[17] Will RG. Acquired prion disease: iatrogenic CJD, variant CJD, kuru. Br Med Bull 2003; 66:255–265.
[18] Brown P, Preece MA, Will RG. "Friendly fire" in medicine: hormones, homografts, and Creutzfeldt–Jakob disease. Lancet 1992; 340:24–27.
[19] Gambetti P, Kong Q, Zou W, et al. Sporadic and familial CJD: classification and characterisation. Br Med Bull 2003; 66:213–239.
[20] Notari S, Capellari S, Giese A, et al. Effects of different experimental conditions on the PrPSc core generated by protease digestion: implications for strain typing and molecular classification of CJD. J Biol Chem 2004; 279(16):16797–16804.
[21] Parchi P, Castellani R, Capellari S, et al. Molecular basis of phenotypic variability in sporadic Creutzfeldt–Jakob disease. Ann Neurol 1996; 39:767–778.
[22] Brown P, Gibbs CJ Jr, Rodgers-Johnson P, et al. Human spongiform encephalopathy: the National Institutes of Health series of 300 cases of experimentally transmitted disease. Ann Neurol 1994; 35(5):513–529.
[23] de Silva R. Human spongiform encephalopathy. In: Baker HF, Ridley RM, editors. Methods in molecular medicine: prion diseases. Humana Press, Totowa, NJ. 1996:15–33.
[24] Will RG, Zeidler M, Stewart GE, et al. Diagnosis of new variant Creutzfeldt–Jakob disease. Ann Neurol 2000; 47(5):575–582.
[25] Kapur N, Abbott P, Lowman A, et al. The neuropsychological profile associated with variant Creutzfeldt–Jakob disease. Brain 2003; 126(12): 2693–2702.
[26] Zeidler M, Stewart GE, Barraclough CR, et al. New variant Creutzfeldt–Jakob disease: neurological features and diagnostic tests. Lancet 1997; 350(9082):903–907.
[27] Green AJ, Thompson EJ, Stewart GE, et al. Use of 14–3–3 and other brain-specific proteins in CSF in the diagnosis of variant Creutzfeldt–Jakob disease. J Neurol Neurosurg Psychiatry 2001; 70(6):744–748.
[28] Wadsworth JD, Joiner S, Hill AF et al. Tissue distribution of protease resistant prion protein in variant Creutzfeldt–Jakob disease using a highly sensitive immunoblotting assay. Lancet 2001; 358(9277):171–180.
[29] Ironside JW, Head MW, Bell JE, et al. Laboratory diagnosis of variant Creutzfeldt–Jakob disease. Histopathology 2000; 37(1):1–9.
[30] Verity CM, Nicoll A, Will RG, et al. Variant Creutzfeldt–Jakob disease in UK children: a national surveillance study. Lancet 2000; 356 (9237):1224–1227.
[31] Gerstmann J, Sträussler E, Scheinker I. Über eine eigenartige hereditär-familiäre Erkrankung des Zentralnervensystems. Zugleich ein Beitrag zur Frage des vorzeitigen lokalen Alterns. Z Ges Neurol Psychiat 1936; 154:736–762.
[32] Ghetti B, Dlouhy SR, Giaccone G, et al. Gerstmann–Sträussler–Scheinker disease and the Indiana kindred. Brain Pathol 1995; 5:61–75.
[33] Grant MP, Cohen M, Petersen RB, et al. Abnormal eye movements in Creutzfeldt–Jakob disease. Ann Neurol 1993; 34:192–197.
[34] Collins S, McLean CA, Masters CL. Gerstmann–Sträussler–Scheinker syndrome, fatal familial insomnia, and kuru: a review of these less common human transmissible spongiform encephalopathies. J Clin Neurosci 2001; 8(5): 387–397.
[35] Brown P, Goldfarb LG, Brown WT, et al. Clinical and molecular genetic study of a large German kindred with Gerstmann–Sträussler–Scheinker syndrome. Neurology 1991; 41(3): 375–379.
[36] Gambetti P, Petersen R, Monari L, et al. Fatal familial insomnia and the widening spectrum of prion diseases. Br Med Bull 1993; 49:980–994.
[37] Lugaresi E, Medori R, Montagna P, et al. Fatal familial insomnia and dysautonomia with selective degeneration of thalamic nuclei. N Engl J Med 1986; 315:997–1003.
[38] Silburn P, Cervenakova L, Varghese P, et al. Fatal familial insomnia: a seventh family. Neurology 1996; 47:1326–1328.
[39] Gallassi R, Morreale A, Montagna P, et al. Fatal familial insomnia: behavioral and cognitive features. Neurology 1996; 46:935–939.
[40] Seilhean D, Duyckaerts C, Hauw JJ. Insomnie fatale familiale et maladie a prion. Rev Neurol 1995; 151:225–230.
[41] Kretzschmar H, Giese A, Zerr I, et al. The German FFI cases. Brain Pathol 1998; 8(3):559–561.
[42] Montagna P, Gambetti P, Cortelli P, et al. Familial and sporadic fatal insomnia. Lancet Neurol 2003; 2(3):167–176.
[43] Reder AT, Mednick AS, Brown P, et al. Clinical and genetic studies of fatal familial insomnia. Neurology 1995; 45(6):1068–1075.
[44] Parchi P, Capellari S, Chin S, et al. A subtype of sporadic prion disease mimicking fatal familial insomnia. Neurology 1999; 52(9):1757–1763.
[45] Alpers MP. Epidemiology and clinical aspects of kuru. In: Prusiner SB, McKinley MP, editors.

Prions. Academic Press, San Diego, 1987:451–465.
[46] Gajdusek DC, Gibbs CJ Jr, Alpers M. Experimental transmission of a kuru-like syndrome to chimpanzees. Nature 1966; 209:794–796.
[47] Scrimgeour EM, Masters CL, Alpers MP, et al. A clinicopathological study of cases of kuru. J Neurol Sci 1983; 59:265–275.
[48] Jakob AM. Über eigenartige Erkrankungen des Zentralnervensystems mit bemerkenswertem anatomischem Befunde (Spastische Pseudosklerose – Encephalomyelopathie mit disseminierten Degenerationsherden). Dtsch Z Nervenheilk 1921; 70:132–146.
[49] Heidenhain A. Klinische und anatomische Untersuchungen über eine eigenartige organische Erkrankung des Zentralnervensystems im Praesenium. Z Ges Neurol Psychiat 1929; 118:49–114.
[50] Gomori AJ, Partnow MJ, Horoupian DS, et al. The ataxic form of Creutzfeldt–Jakob disease. Arch Neurol 1997; 29:318–323.
[51] Cochius JI, Burns RJ, Blumbergs PC, et al. Creutzfeldt–Jakob disease in a recipient of human pituitary-derived gonadotrophin. Aust N Z J Med 1990;20(4):592–593.
[52] Cochius JI, Hyman N, Esiri MM. Creutzfeldt–Jakob disease in a recipient of human pituitary-derived gonadotrophin: a second case: J Neurol Neurosurg Psychiatry 1992;55(11):1094–1095.

31 Methods for the Clinical Diagnosis of Human Prion Diseases

Inga Zerr

31.1 Introduction

The clinical signs and symptoms of patients with Creutzfeldt–Jakob disease (CJD) are heterogeneous. This heterogeneity is the result of the involvement of various brain structures and a number of different disease variants. Since the currently available diagnostic options are not sufficient to establish the clinical diagnosis while the patient is still alive, clinical criteria are needed to establish the clinical diagnosis. The classification criteria are related to the etiology of the disease, which can be divided into three categories: sporadic, iatrogenic and familial or genetic CJD (Table 31.1a–c). A new variant of CJD, now termed vCJD, was first described in 1996. In the meantime, clinical criteria for vCJD have been established and include a variety of neurological signs and symptoms as well as magnetic resonance imaging and tonsil biopsy [1].

There are some problems associated with the use of clinical classification criteria which can be related to the genetic background. In some patients with sporadic CJD, the disease progresses rapidly. However, family members and physicians may not notice the development of dementia, despite this being one of the most important features of CJD diagnosis. Although all other neurological signs may be present, these patients cannot be classified as probable or possible CJD cases according to the criteria [2; 3].

In such patients, disease often starts with rapid progressive ataxia. These patients are frequently homozygous for valine at codon 129 of the prion protein gene and type 2 PrP^{Sc} PrP (VV2 molecular subtype) [4; 5] (see Chapter 26). In contrast, patients with more common clinical signs and periodic bi- and triphasic sharp wave EEG complexes are methionine homozygotes and have PrP^{Sc} type 1 [5; 6].

The established classification criteria also include some symptoms and signs of the disease (e. g., myoclonus; Table 31.1a) that do not occur until the disease has progressed. Akinetic mutism, another clinical criterion, is usually not evident until end stage CJD [7]. The current criteria may be observed in patients with the fully developed disease, but might not be sensitive enough to detect the disease at an early stage. Therefore, clinical diagnosis of CJD always requires follow-up examinations and should not be made on the basis of one examination alone.

Distinct molecular subtypes of sporadic CJD (sCJD) have been described, leading to the implementation of a molecular classification of the disease into MM1, MM2, MV1, MV2, VV1 and VV2 subtypes (see Chapter 26). Distinct phenotypes of the disease are associated with certain molecular subtypes. The differences between phenotypes are based on neuropathological lesion profiles and clinical syndromes. However, the sensitivity of clinical tests varies among phenotypes/molecular subtypes [8] (see below). As the currently used clinical criteria only include the electroencephalogram (EEG) and 14–3–3 detection in the cerebrospinal fluid (CSF), some atypical cases such as MV2 subtypes may not be diagnosed correctly using the available clinical criteria.

The diagnostic process for members of families in which a familial form of CJD is known to have occurred previously should be especially careful and comprehensive to rule out a curable disease. The diagnosis should definitely not be based on molecular genetic findings alone [9].

Table 31.1: Current diagnostic criteria for the surveillance of sporadic (a), iatrogenic (b), familial (c), and variant CJD (d); as developed by Euro-CJD and adopted by WHO [10; 33; 86].

(a) Diagnostic criteria for sporadic CJD

Definite	Neuropathologically confirmed
Probable	Rapid progressive dementia and at least two of the following clinical features: • Myoclonus • Visual or cerebelar signs • Pyramidal or extrapyramidal features • Akinetic mutism and typical EEG (periodic sharp and slow wave complexes, PSWCs) or 14–3–3 proteins in CSF and duration of < 2 years
Possible	Rapid progressive dementia and at least two of the clinical features as above, and • duration of < 2 years, • but not PSWCs in EEG and no 14–3–3 detection in CSF

(b) Diagnostic criteria for iatrogenic CJD (accidentally transmitted CJD)

Definite	Definite CJD with a recognized iatrogenic risk factor (see below)
Probable	Progressive predominant cerebellar syndrome in human pituitary hormone recipients or probable CJD with a recognised iatrogenic risk factor: • Treatment with human pituitary growth hormone, human pituitary gonadotrophin or human dura mater graft, or • Corneal graft in which the corneal donor has been classified as definite or probable human prion disease, or • Exposure to neurosurgical instruments previously used in a case of definite or probable human prion disease

(c) Diagnostic criteria for familial CJD or genetic CJD

Definite	Definite TSE and definite or probable TSE in 1st degree relative or definite TSE with a pathogenic *PRNP* mutation
Probable	Progressive neuropsychiatric disorder and definite or probable TSE in 1st degree relative or progressive neuropsychiatric disorder and pathogenic *PRNP* mutation (see box in "Genetic TSE" of www.cjd.ed.ac.uk/criteria.htm – or Table 26.5).

(d) Diagnostic criteria for variant CJD

Criteria	**I**	**A**	Progressive neuropsychiatric disorder
		B	Duration of illness > 6 months
		C	Routine investigations do not suggest an alternative diagnosis
		D	No history of potential iatrogenic exposure
		E	No evidence of a familial form of TSE
	II	**A**	Early psychiatric symptoms[a]
		B	Persistent painful sensory symptoms[b]
		C	Ataxia
		D	Myoclonus or chorea or dystonia
		E	Dementia
	III	**A**	EEG does not show the typical appearance of sporadic CJD[c] (or no EEG performed)
		B	Bilateral pulvinar high signal on MRI brain scan
	IV	**A**	Positive tonsil biopsy[d]
Definite	**I**	**A**	and neuropathological confirmation of vCJD[e]
Probable	**I** and 4/5 of **II** and **III A** and **III B** or **I** and **IV A**[d]		
Possible	**I** and 4/5 of **II** and **III A**		

[a] Depression, anxiety, apathy, withdrawal, delusions.
[b] This includes both frank pain and/or dysaesthesia.
[c] Generalised triphasic periodic complexes at approximately one per second.
[d] Tonsil biopsy is **not** recommended routinely, nor in cases with EEG appearances typical of sporadic CJD, but may be useful in suspect cases in which the clinical features are compatible with vCJD and MRI does not show bilateral pulvinar high signal.
[e] Spongiform change and extensive PrP deposition with florid plaques throughout the cerebrum and cerebellum.

31.2 Electroencephalogram

Characteristic EEG changes in CJD were first described in the middle of the twentieth century [11]. Since then, periodic bi- and triphasic sharp wave complexes (PSWCs) have been considered as an electroencephalographic pattern that may be an important clue to a positive diagnosis (Fig. 31.1). The EEG may show bilateral alpha activity with generalized irregular theta and delta waves in the early stage of the disease [11; 12]. In addition, intermittent high-amplitude rhythmic delta activity is seen on the EEG. Changes caused by hyperventilation or photostimulation are not present. Sometimes PSWCs can be detected as early as three weeks after the onset of CJD. More often, however, PSWCs do not occur until about week 12 as shown in Figure 30.2, in a few patients even later. In some cases their presence can only be detected after audio and tactile stimulation. Early in the disease process, there is no correlation between PSWCs and the severity of the disease. However, there is a correlation between early occurrence and shorter survival [11] (Fig. 31.2). PSWCs occur episodically at the onset of the disease, but later in the disease process the EEG shows continuous periodic sharp waves of 100–600 ms duration at a frequency of 0.5–2 Hz (Table 31.2). The background activity decreases during disease progression and is gradually replaced by theta and delta activity. In end stage CJD the EEG shows an isoelectric line [11]. As revealed by simultaneous EEG recordings, the occurrence of myoclonus does not always correlate with PSWCs [11].

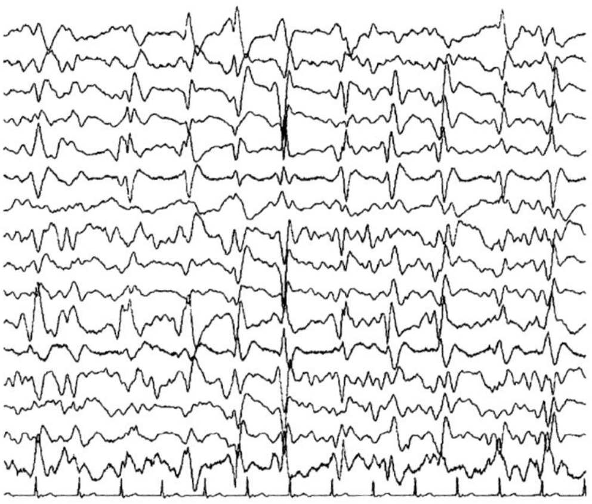

Fig. 31.1: Detection of periodic sharp and slow wave complexes (PSWCs) in the EEG. Patient with sporadic CJD. Figure by I. Zerr ©.

The EEG is one of the two diagnostic tests that were included in the set of diagnostic criteria (Table 31.1). The frequency of characteristic EEG changes varies in the literature (sensitivity 60–80 %). Since it is exactly these changes that frequently lead to the clinical diagnosis, patients with PSWCs are often overrepresented in some studies, resulting in the risk

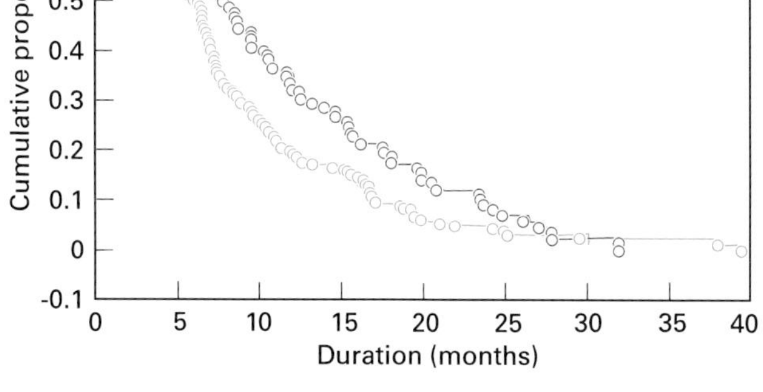

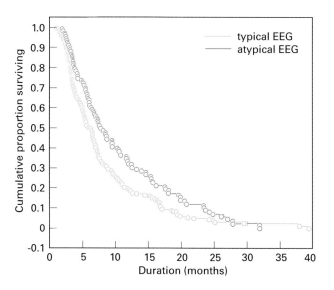

Fig. 31.2: Early detection of PSWCs in the EEG correlates with shorter survival. Survival rate: 188 cases with and 99 without typical EEG. Figure by I. Zerr ©.

Table 31.2: Diagnostic criteria for periodic sharp and slow wave complexes in the EEG in CJD (according to [14]). See also Figure 30.2.

Diagnostic criteria PSWCs
Periodicity (most important criterium)
"Sharp-wave" complexes
At least five complexes
Frequency 0.5–2/sec
Duration 200–600 msec
Amplitude > 150–300 μV

that the sensitivity of the EEG may be overestimated. Prospective studies using standardized criteria show that an average of 67% of the patients have PSWCs on at least one EEG [13; 14]. Several EEG recordings may increase the diagnostic sensitivity [15], but EEG changes are not pathognomonic for CJD [15]. Their diagnostic specificity is up to 86% [16].

On the other hand, the EEG is a valuable tool for the classification of more rapidly progressive prion diseases. CJD cases in which the disease duration is unusually long may pose diagnostic problems because PSWCs are often not detected. Here, too, there seems to be a link with the molecular PrP^{Sc} type [5; 11; 17; 18]. With the exception of patients with a mutation at codon 200 (E200K), PSWCs are also absent in most cases of fCJD and in cases of Gerstmann–Sträussler–Scheinker disease (GSS), fatal familial insomnia (FFI) and vCJD.

The list of possible differential diagnoses for diseases associated with PSWCs is comprehensive and includes a variety of encephalitides (e. g., subacute sclerosing panencephalitis), metabolic encephalopathies, lithium intoxication, hepatic encephalopathy, hyperammonemia, hypoglycemia, multiple cerebral abscesses, mianserin intoxication, electrolyte imbalance and some isolated cases of intracerebral lymphoma [19].

31.3 Analysis of cerebrospinal fluid

31.3.1 General information

Among the diagnostic markers available to date, the analysis of CSF is currently the most sensitive method to support the clinical suspicion of CJD (Table 31.3). Typical clinical symptoms together with periodic bi- and triphasic EEG complexes and detection of 14–3–3 proteins in the CSF almost always suggest a positive diagnosis of CJD. However, the EEG is less effective than biochemical markers (Table 31.3) as characteristic EEG changes are only present in 67% of the cases. Abnormally increased levels of the 14–3–3, tau, γ-enolase, S100b proteins, on the other hand, may indicate acute neuronal damage or astrocyte activation and, therefore, should not be used to diagnose CJD without other supporting evidence.

Table 31.3: Clinical diagnostic tests [major references].

Test	n	Method	Sensitivity	Specificity*
MRI [75]	238		63%	88%
EEG [16]	150		67%	86%
Biochemical markers in CSF				
14–3–3 [3]	1136	Western blot	95%	93%
γ-Enolase ≥ 35 ng/ml [102]	1276	ELISA	81%	92%
S100 ≥ 8 ng/ml [44]	135	ELISA	84%	91%
tau ≥ 1400 pg/ml [45]	290	ELISA	93%	91%
PrP^{Sc} [57]	37	FCS	20%	100%

* in the relevant differential diagnosis

These biochemical markers, which are of great value in the differential diagnosis of dementia, should be used exclusively in the clinical context. In Alzheimer's disease (AD), the major differential diagnosis of CJD, abnormal levels of 14–3–3 proteins, were found in very few CSF samples (Table 31.4). Tau levels are high in AD, but levels in CJD exceed those in AD, therefore, another cut-off level must be used (1400 pg/ml). Some abnormal CSF marker profiles might be used for differentiation of various types of rapid progressive dementia. In CJD, all markers tested so far in larger groups (14–3–3, tau, γ-enolase, S100b) are elevated. In AD, 14–3–3 is usually negative and the levels of γ-enolase and S100b are only slightly elevated. In other diseases with extensive neuronal damage, such as

Table 31.4: Overview of the diagnostic value of biochemical parameters in the diagnosis of CJD.

	14–3–3	tau	phosphorylated tau	γ-enolase	S100b
Creutzfeldt–Jakob disease	↑↑↑	↑↑↑	↑	↑↑↑	↑↑↑
Alzheimer's disease	↑*	↑	↑↑	(↑)	↑
Lewy body dementia	–	↑	↑	(↑)	↑
Vascular dementia	–	↑	↑	?	?
Fronto-temporal dementia	–	↑	(↑)	?	↑

* according to Hans Kretzschmar

hemorrhage, meningoencephalitis, epileptic fit, 14–3–3 levels might be elevated in the CSF, but other neuronal markers might be normal. On the other hand, CSF marker proteins may be negative in 5–10 % of the CJD cases, depending on the laboratory.

31.3.2 Routine parameters

Routine analyses of CSF samples from patients with CJD or GSS usually reveal normal results (20). Only one CJD patient out of 165 confirmed and probable CJD cases included in our statistics had mild pleocytosis of 14 cells/mm^3. A slight unspecific increase in total protein was observed in about one-third of the cases. Oligoclonal bands were found in the CSF of some patients using isoelectric focusing [20–23]. However, the presence of such bands is not particularly common in CJD patients.

31.3.3 Surrogate markers in sporadic CJD

p130 and p131 proteins

In 1986, the research team of M. Harrington characterized two proteins by using two-dimensional gel electrophoresis [24]. These proteins were detectable in the CSF of patients with CJD, and were found in about 50 % of patients with herpes simplex encephalitis. The p130 protein has a molecular mass of 26 kDa and an isoelectric point of 5.2. The molecular mass of the p131 protein is 29 kDa and the isoelectric point is 5.1.

While these two proteins were detected in all 21 patients examined in the original study [24], this degree of sensitivity could not be achieved in our study, in which the proteins were detected in 84 % of the sporadic CJD cases [25]. No evidence of p130 and p131 has been found so far in some genetic forms of human prion diseases. While this diagnostic test is positive in early stage CJD, when loss of neurological function and cognitive impairment are rarely present, it is not suitable for the use in presymptomatic or predictive testing. In cases of AD, the most important differential diagnosis of CJD, these proteins have not been found in the CSF, in particular not in the CSF of AD patients with very short disease duration and with PSWCs [3].

14–3–3 proteins

Amino acid sequence analyses showed that the proteins p130 and p131 belong to the family of 14–3–3 proteins [26]. While p130 and p131 can only be detected by using the time consuming

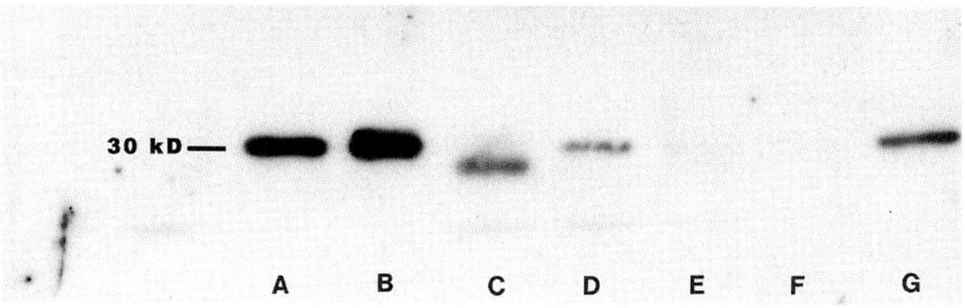

Fig. 31.3: Detection of the 14–3–3 protein by Western blot analysis. A, B, D: patients with CJD; C: female patient with neurosyphilis; E: patient with Alzheimer's disease; F: negative control; G: positive control. Figure by I. Zerr ©.

and costly 2D gel electrophoresis technique, the 14–3–3 proteins can be identified by Western blot analysis (Fig. 31.3). Antibodies against various 14–3–3 isoforms are commercially available. This test is more sensitive than the detection of p130 and p131 using gel electrophoresis. However, as was shown in a prospective study, the enhanced sensitivity of this test is accompanied by a decrease in specificity, leading to false-positive results in Western blot analysis. The amount of 14–3–3 in CSF seems to be linked to the degree of neuronal destruction and to the stage of the disease. Experimental studies on chimpanzees showed that the 14–3–3 proteins are detectable as early as the onset of the disease or shortly before [26]. The test might produce negative results during the terminal stages. In patients with CJD, the proteins are found at the onset of the first neurological symptoms [3]. At present, 14–3–3 is the most sensitive marker of CJD [3; 8; 27–33]. It is found in 95 % of patients with definite and probable CJD [3]. Its specificity in the differential diagnosis between CJD and other forms of dementia is 95 %, but it is not suited for general screening of patients with dementia [8; 30]. There are suggestions that the detection of some 14–3–3 isoforms might be useful to support the diagnosis [31; 34]. The origin of 14–3–3 in the CSF is still unknown. It has been suggested that it could be the product of neuronal damage. Alternatively, 14–3–3 could be involved in the disease pathogenesis, since an upregulation could be recently shown in glial cells in patients with sporadic CJD [35].

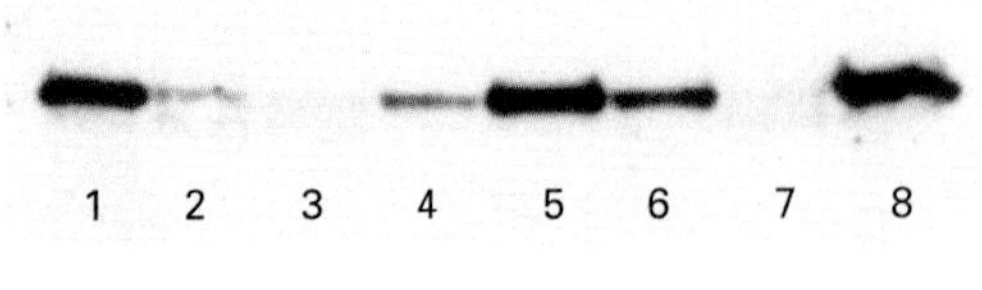

Fig. 31.4: Detection of the 14–3–3 protein during disease course in various conditions (sporadic CJD and epileptic fit). lane 1: "false-positive" of 14–3–3 levels in CSF in a patient with epileptic fit; lanes 2 and 3: same patient, lumbar punctures in the follow-up; lane 4: sporadic CJD, early in the disease course; lane 5: same patient as in lane 4, 4 weeks later; lane 6: sporadic CJD; lane 7: negative control; lane 8: positive control. Figure by I. Zerr ©.

The sensitivity is high in sporadic and genetic CJD as well as in iatrogenic cases [9; 36] (Table 31.5). In vCJD, only 44 % test positive [29; 37]. Patients with FFI and GSS usually test negative. False-positive results are observed in cases of acute destructive neurological diseases, which, however, are clinically easily distinguishable from CJD (Table 31.6). In some other diseases that are relevant to the differential diag-

Table 31.5: 14–3–3 increase in prion diseases.

	n	positive
Sporadic CJD	1136	95 %
Variant CJD	93	44 %
Iatrogenic CJD	20	(75 %)
Genetic	120	86 %
FFI	30	(11 %)
GSS	7	(29 %)

Table 31.6: 14–3–3 increase in other diseases.

	n	positive
Neoplasia	72	38 %
Ischemia	46	31 %
Inflammatory	170	29 %
Neurodegenerative	550	13 %
Epileptic fits	22	18 %

Table 31.7: 14–3–3 in the follow-up: CJD (Lp = Lumbar puncture).

1st Lp	2nd Lp	n	days between Lps
+	+	21	21 (6–180)
–	–	4	111 (9–350)
–	+	2	40, 167
+	–	2	12, 41

Table 31.8: 14–3–3 in the follow-up: other diseases (Lp = Lumbar puncture).

Disease	1st Lp	2nd Lp	n	days between Lps
Epileptic fit	+	–	3	16 (14–25)
Inflammatory	+	–	3	24
	+	+	2	10, 28
Metabolic	+	–	1	37

nosis of CJD, such as psychiatric diseases, Parkinson's disease, disseminating encephalomyelitis, Pick's disease, Hashimoto's encephalitis and other neurological diseases, the 14–3–3 protein is not found. Table 31.6 shows the rate of false-positive 14–3–3 findings in a set of patients with various neuronal destruction diseases.

Repeated lumbar punctures might be useful in cases of doubt. In CJD, there is usually a further increase in 14–3–3 levels as the disease progresses (Fig. 31.4, Tab. 31.7). In contrast, in patients with an acute neuronal damage, the levels return to normal during the recovery (Table 31.8).

14–3–3 during the course of the disease

To prove the hypothesis that 14–3–3 levels increase in CJD and decrease in other neurological conditions, where the elevation was caused by a single event, we have analyzed lumbar punctures in patients with CJD and controls (n = 38).

Lumbar punctures were conducted for diagnostic reasons in the participating hospitals. In CJD patients, there was almost always an increase or at least a persistence of the 14–3–3 levels (Table 31.7). In 21 CJD patients, 14–3–3 was detectable in both lumbar punctures. In 2 patients, 14–3–3 was negative in the first lumbar puncture but positive in the second one six months later. Four patients remained negative, two patients were initially positive but tested negative two weeks and one month later.

In other diseases, 14–3–3 levels declined in patients who recovered (Table 31.8). All three patients suffering from epileptic fits had normal 14–3–3 tests in the second lumbar puncture, as well as a patient with a metabolic encephalopathy. Similar results were obtained in inflammatory diseases – three patients who had recovered from meningoencephalitis had normal 14–3–3 concentrations in the second lumbar puncture. Those patients in whom elevated 14–3–3 levels in the CSF persisted, still had inflammatory CSF findings.

We were able to show that 14–3–3 levels usually increase in CJD patients as the disease progresses. A decrease in 14–3–3 levels in other diseases, where 14–3–3 elevation was caused by a single event such as hypoxia or meningoencephalitis, was also observed.

A multi-center European study has analyzed the value of 14–3–3 detection in cases with differential diagnosis of CJD. The conclusion was that the sensitivity of 14–3–3 was 89–95% in this setting, depending on the laboratory and method used [33].

14–3–3 is a surrogate marker that can be used to support the clinical diagnosis of CJD (Table 31.9). Furthermore, it is of great diagnostic value when CJD is considered in the differential diagnosis of patients whose diagnosis is uncertain. For this reason 14–3–3 was included as a biochemical marker in the set of clinical criteria for probable CJD [33] (Table 31.1a).

γ-Enolase (neuron-specific enolase)

γ-Enolase is mainly synthesized in neurons and neuroendocrine cells and has a molecular weight of 78 kDa. Immunoassays such as enzyme-linked immunosorbent assays (ELISAs),

Table 31.9: Reliability of the clinical methods for the diagnosis of CJD.

	Typical clinical picture			
	—	and positive EEG findings	and detection of 14–3–3	and positive EEG, or 14–3–3
Confirmed CJD (n = 90)	99% (89/90)	70% (63/90)	96% (86/90)	99% (89/90)
No CJD (n = 124)	13% (16/124)	3% (4/124)	1% (1/124)	3% (4/124)

Definitions: *Confirmed CJD*: neuropathologically confirmed cases. *No CJD*: patients with an initial clinical diagnosis of CJD whose diagnosis was not confirmed. *Typical clinical picture*: the clinical criteria of *possible CJD* are fulfilled.

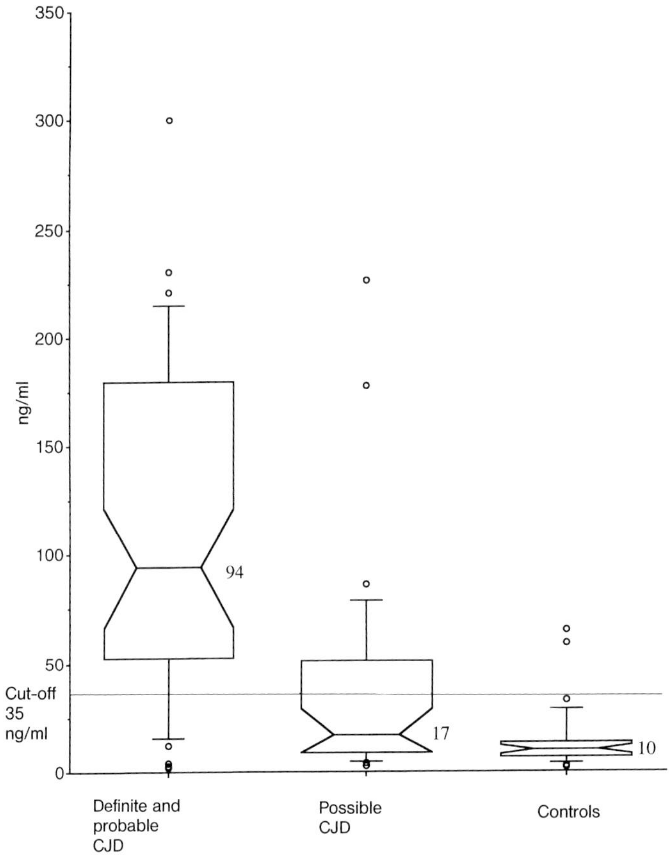

Fig. 31.5: Neuron-specific enolase (NSE or γ-enolase) levels in the CSF of patients with CJD and in the CSF of controls (cut-off = optimum threshold value). Figure by I. Zerr ©.

radioimmunoassays (RIAs) or dissociation-enhanced lanthanide fluorescent immunoassays (DELFIAs) are commercially available. High levels of γ-enolase in CSF and serum may be a useful prognostic marker of acute neuronal damage (e. g., hypoxia, apoplexy and intracerebral hemorrhage) and disease progression. Abnormally high γ-enolase concentrations were detected in early stage CJD [38–41]. Parallel to the development of brain atrophy visible on CT scans, γ-enolase levels decrease during disease progression and return to normal in end stage CJD, according to previous studies [39; 41–43]. After neuroradiological exclusion of brain tumors, cerebral infarction and intracerebral hemorrhage, high levels of γ-enolase in CSF can support the clinical suspicion of CJD (sensitivity 81 %; Fig. 31.5). In spite of an increase in γ-enolase levels in acute disease processes, such as hypoxia or hemorrhage, this diagnostic marker may help to discriminate CJD from other forms of dementia; in this context its specificity is 92 %, and the optimum cut-off point is 35 ng/ml (Youden index).

S100b protein

S100b is an acidic calcium-binding protein that is mainly synthesized in glial cells of the central and peripheral nervous system and in melanocytes. At present, S100b serves primarily as a marker for prognostic assessment of cerebral lesions, for example in patients who have had cardiac surgery. A detection of this protein in CSF may be an indicator of CJD. CSF analyses revealed that 84 % of the patients with confirmed or probable CJD had abnormally high levels of S100b (> 8 ng/ml). Its specificity with respect to the diagnostic discrimination between CJD and other forms of dementia was 91 % [44].

Tau protein

Tau protein, which is a constituent of neurofibrillary tangles, was first described to be elevated in CSF of patients with Alzheimer's disease. Meanwhile, the data obtained in other diseases indicate abnormally high levels of this protein in other dementia (Table 31.4). Levels in CJD are extremely high and are therefore useful to support the diagnosis in addition to 14–3–3 detection. The sensitivity of tau detection is around 93% [45; 46]. A phospho-tau/total tau ratio in CSF might be a useful marker to discriminate CJD from other dementia and this parameter should be studied in a larger group of patients [47; 48].

Other biochemical parameters

Other proteins such as the brain-specific isoform of creatine kinase (CK-BB) and G0, a protein that is mainly present in the neuropil of the cortex, are also abnormally high in the CSF of CJD patients [42]. In addition, there is a substantial increase in ubiquitin in CSF (107.6–377.0 ng/ml) in the early and middle stages of the disease compared to healthy controls (7.3–21.0 ng/ml) [49]. Elevated levels of some cytokines were also described [50–52]. Two studies have shown increased levels of prostaglandin E2 in the CSF of patients with sporadic and variant CJD and higher levels correlated with shorter survival time [53; 54]. Increased lipid peroxidation in the CSF and plasma in CJD patients may support the hypothesis that oxidative mechanisms are involved in the pathogenesis of CJD [55]. Cystatin C has been suggested recently as a potential marker for CJD in CSF, but also has to be evaluated further [56].

Detection of the abnormal prion protein in CSF

A detection of the abnormal PK-resistant form of the prion protein (PrP^{Sc} or PrPres) in living organisms could lead to a clinical diagnostic test. To date, it has only been possible to identify the PK-sensitive PrP in CSF, leukocytes and thrombocytes [57–60]. Although still under development, ultrasensitive diagnostic tests, such as fluorescence-correlated spectroscopy (FCS), could facilitate the detection of PrP^{Sc} in CSF [61]. Advances in the PrP^{Sc} detection methodologies, i. e., increased test sensitivity or/and novel amplification techniques, such as cyclic amplification, might allow development of a preclinical or clinical CSF test in the future [62; 63].

31.3.4 CSF analysis in the diagnosis of new variant CJD (vCJD)

Normal CSF parameters are observed in vCJD patients [64]. While the 14–3–3 proteins are detected in most sCJD cases, they are only found in 44% of vCJD cases [65]. Abnormally high levels of surrogate markers are also observed in patients with vCJD, although these are lower than in patients with sCJD. This difference might be due to either a different localization of the disease process or to a protracted disease course [29; 37].

31.3.5 CSF analysis in the diagnosis of genetic (familial) forms of prion diseases

According to current information, the changes in the CSF of patients with familial forms of CJD are comparable to those found in sCJD samples. 14–3–3 proteins are detectable in patients with an E200K and V210I mutation [3; 9]. However, in cases of GSS, the 14–3–3 proteins are detectable in only 29% of patients [3]. The use of CSF analysis in the diagnosis of patients with FFI may be problematic. CSF samples from 30 patients examined were almost always 14–3–3 negative [3; 66].

31.4 Imaging techniques

31.4.1 Computed tomography

A number of other diseases relevant to the differential diagnosis of CJD, among them potentially reversible conditions such as normal pressure hydrocephalus, chronic subdural hematoma or space-occupying lesions, can be identified using computed tomography (CT).

However, neither a single computed tomography (CT) scan nor follow-up CT scans can substantiate the diagnosis of CJD.

Normal examination results are frequently seen in the prodromal stage. CT may reveal increasing brain atrophy during the progression of the disease, which may be most prominent in the end stage of akinetic mutism [67–69]. To date, there has only been one case in which post mortem examination revealed that white matter changes identified by CT as hypodense areas corresponded to CJD-specific focal changes [70].

31.4.2 Magnetic resonance imaging

Sporadic CJD

Magnetic resonance imaging (MRI) is used for the differential diagnosis of rapidly progressive dementia. Using this imaging technique it is possible to identify changes that may either corroborate or rule out the clinical suspicion of CJD. In addition to non-specific cerebral atrophy detectable by MRI, T2-, proton-weighted (PD), FLAIR and diffusion-weighted imaging can also detect symmetric hyperintense signals in the basal ganglia [71]. In isolated cases, these MRI changes may be observed as early as a few weeks or months after the onset of CJD, sometimes even before the first clinical signs become evident [39; 72] T1-weighted images reveal no abnormalities.

When compared with histopathological changes, MRI signal abnormalities correlate significantly with those brain areas that are most affected by astrocytosis and vacuolization [71; 73].

Symmetric hyperintense signals in putamen and caudate nucleus were observed in CJD patients (Fig. 31.6a-b) [71; 74; 75]. Follow-up examinations revealed rapidly progressive atrophy and striatal atrophy [76]. The degree of atrophy correlates with the period of time between the first symptoms and the MRI examination. Furthermore, MRI shows marked atrophy in all cases in which the disease duration is longer than four months [76]. Since the changes in basal ganglia are mainly symmetric, they are not always diagnosed as pathological – especially not when the changes are mild. It was reported that radiologists spontaneously identified them in only one out of eight iatrogenic CJD (iCJD) cases and in three out of 84 sCJD cases [77].

As well as the changes detectable by T2- and PD-weighted imaging, FLAIR imaging can also detect abnormal signal intensity in basal ganglia (Fig. 31.6c). Compared to other imaging techniques, FLAIR images may also show the involvement of the cortex in the form of increased signal intensity or atrophy more clearly [78–81].

There is growing evidence that diffusion-weighted imaging might be the most sensitive technique to detect changes in cases of CJD. As with FLAIR imaging, the advantage of diffusion-weighted imaging is that it can reveal changes that are not visible on T2- and PD-weighted images [81–83]. Furthermore, the short examination time (approximately 400 ms per sequence) reduces the incidence of movement artifacts due to restlessness in the CJD patients. A clear advantage of the technique is that increased signal intensity is evident, not only in the basal ganglia, but also in the cortex

Fig. 31.6: Imaging techniques for the diagnosis of human prion diseases: Magnetic resonance imaging (MRI) scans of patients with sporadic CJD (sCJD). **(a)** This proton-weighted MRI scan at the level of the basal ganglia shows signal increase of both caudate heads and putamina. Fifty-six-year-old male patient, 6 months after the onset of CJD and 2 months prior to his death. **(b)** This T2-weighted image at the level of the basal ganglia shows increased signal intensity in both caudate heads and putamina. Forty-nine-year-old female patient, 9 months after the onset of CJD and 7 months prior to her death. **(c)** This fluid-attenuated inversion recovery (FLAIR)-weighted image shows a symmetric increased signal of the basal ganglia (arrows). Forty-nine-year-old female patient, 7 month after the onset of CJD and 16 months prior to her death. **(d)** This diffusion-weighted image (DWI) shows signal increase of both caudate heads and putamina, and also slight signal increase of both thalami. Sixty-three-year old female patient, 5 months after the onset of CJD, 1 month prior to her death. Figures by I. Zerr.

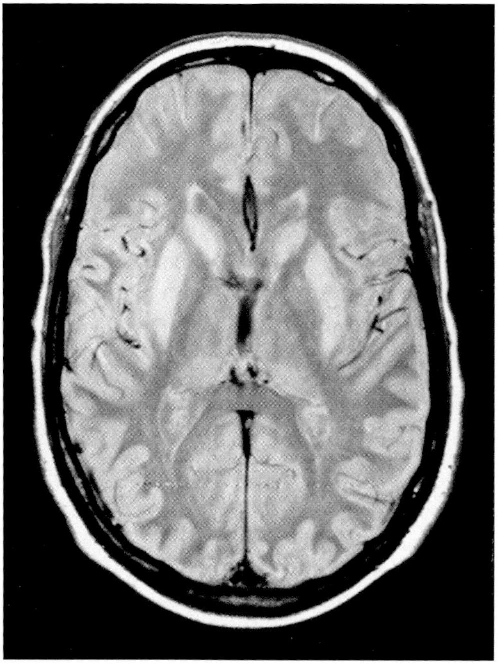

(a)

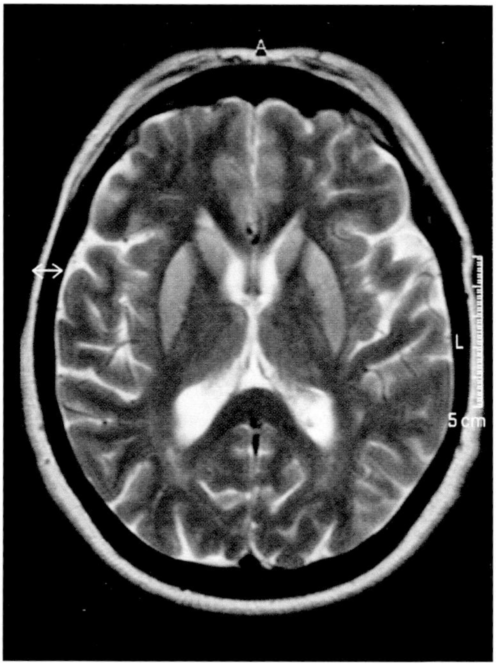

(b)

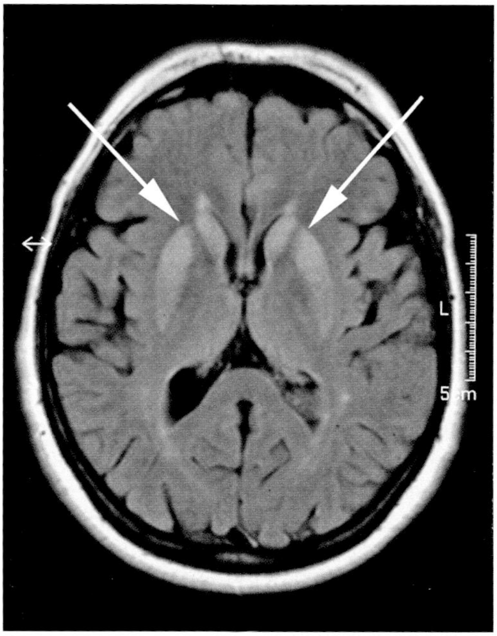

(c)

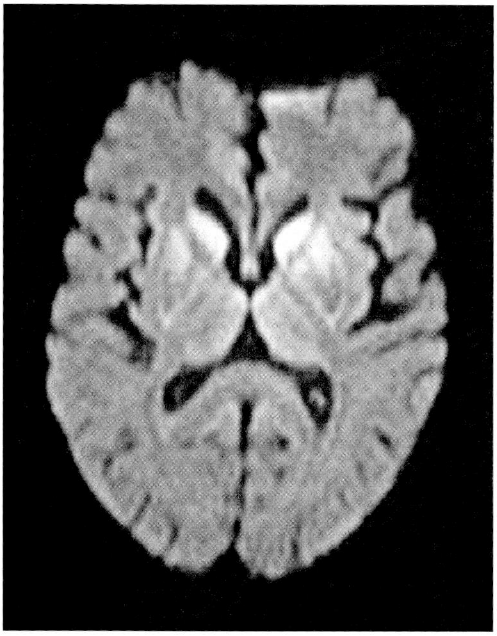

(d)

Table 31.10: Hyperintensities in the MRI in sporadic CJD (n = 163) [84].

MRI findings	%
Nucleus caudatus, Putamen	67 %
Thalamus	7 %
Occipital cortex	4 %
Cerebellum	3 %

(Fig. 31.6d). Other brain areas might also be affected (Table 31.10).

In sporadic CJD, patients with hyperintense basal ganglia were shown to have a poorer prognosis [75]. There are some indications that suggest that some atypical molecular subtypes of sporadic CJD, such as patients with MV2 subtype (prolonged disease duration, dementia and ataxia at onset, no typical EEG, kuru-plaque depositions in the brain), could be better identified by means of the MRI than other clinical tests [8; 75].

Variant CJD

The distribution patterns of MRI changes in sCJD differ from those observed in vCJD (Table 31.11). In patients with sCJD the changes are most prominent in putamen and caudate nucleus. In contrast, in patients with vCJD, the increase in signal intensity is most marked in the posterior thalamus on T2-weighted images [85]. The characteristic distribution pattern of the lesions on MRI is an important diagnostic indicator of vCJD and was therefore included in the set of clinical diagnostic criteria (Table 31.1d). Hyperintensities in the posterior thalamus are seen in 80 % of patients with vCJD [1; 87].

Table 31.11: Comparison of appearance of MRI in vCJD and sCJD (according to [86]).

MRI findings	Variant CJD	Sporadic CJD
Caudate head hyperintense	++	+++
Putamen hyperintense	+	+++
Posterior ('pulvinar') and dorsomedial thalamus hyperintense	+++	+
Periaqueductal grey matter hyperintense	++	+
Cerebral atrophy	+	+++
Cerebral cortex	+	++

GSS

MRI has revealed progressive brain atrophy in GSS [88]. No other abnormalities have been reported so far.

31.4.3 Other imaging techniques

Neuroimaging with specific ligand techniques or specific tracers is able to visualize metabolic processes in the central nervous system. A reduction in glucose metabolism can be shown by positron emission tomography (PET). In neurodegenerative diseases, a reduced metabolism was found in cerebral regions which are involved in the early stages of the pathological process. Because of the diminished metabolism of the brain tissue, there is also a reduction in the regional utilization of oxygen and of the regional cerebral blood flow. This hypo-perfusion can be detected by single photon emission computed tomography (SPECT).

SPECT

In several dementing diseases, typical deficits of cerebral perfusion, detected by SPECT, were described [89–91]. Bilateral hypo-perfusion of the temporal and parietal lobe is typically seen in Alzheimer's disease. Similar results were found in Parkinson's disease with dementia and dementia of Lewy body type. Reduction of tracer uptake of the frontal lobe was also seen in Pick's disease and amyotrophic lateral sclerosis (ALS) with dementia.

In CJD, there is a striking correlation between the involvement of certain parts of the brain and the affected site [92–94]. The most frequent hypo-perfusion is found in the parietal cortex. Most patients have reduced tracer uptake at least on one side of this location. The occipital brain is affected more often than the frontal lobes (German CJD surveillance study).

Reduction of regional cerebral blood flow (rCBF) was also seen in the thalamic nucleus, in the cerebellum, and in the basal ganglia in single cases [95; 96]. In contrast to this, we found that in the majority of CJD cases posterior cortical regions and in a smaller number subcortical and cerebellar structures are involved in a marked hypoperfusion. At these sites the typical pathological findings like spongiform changes, astrocytic gliosis and loss of granular cells are also present in CJD.

Positron emission tomography [^{18}F] FDG-PET

A widespread change in the metabolism in brains of patients with CJD is observed using ^{18}F-2-fluoro-2-desoxy-D-glucose (FDG) and positron emission tomography (PET). A reduction of cerebral glucose metabolism in at least one temporal or parietal region [97–100] has been observed in all CJD patients so far investigated. The occipital lobe, cerebellum and basal ganglia are also involved. Changes are widespread, often markedly asymmetric and differ from typical pattern seen in Alzheimer's disease or other neurodegenerative disorders.

The diagnostic utility of both techniques has not yet been evaluated systematically in patients with sporadic CJD, but the results obtained to date suggest that both techniques might be of interest and should be developed further. In addition, recent data on FFI and in vCJD suggest a typical hypo-perfusion/hypometabolism pattern: a reduced tracer uptake is seen in the thalamus in both fatal familial insomnia and sporadic familial insomnia. In variant CJD, a reduced perfusion in cortical structures was described [95].

31.5 Sensitivity of clinical diagnostic tests in distinct molecular CJD subtypes

The molecular classification of sporadic CJD is based on the genotype at polymorphic codon 129 of the prion protein gene *(PRNP)* and the physicochemical properties of the protease-resistant core of the pathological PrP^{Sc} which is designated according to electrophoretic mobility as type 1 or 2 protein. There are six different molecular subtypes, MM1 and MV1, MM2 (thalamic and cortical subtypes), VV1, VV2 and MV2, which are characterized by the size of the protease-resistant fragment of the pathological prion protein, type 1 and 2, and homozygosity (MM or VV) or heterozygosity (MV) for methionine or valine at codon 129 of the prion protein gene [5; 6; 8].

Table 31.12: Sensitivity of technical investigations and molecular CJD subtypes.

CJD-phenotype	PSWCs (EEG) % (n)	14–3–3 in CSF % (n)	hyperintense basal ganglia MRI % (n)
MM1/MV1	80 (62/78)	96 (75/78)	71 (15/21)
MM2	33 (1/3)	100 (3/3)	(0/2)
MV2	0 (0/10)	30 (3/10)	89 (8/9)
VV1	(0/2)	(1/1)	(0/1)
VV2	0 (0/15)	100 (15/15)	70 (7/10)

The MM1/MV1 type (homo-/heterozygosity for methionine and presence of the PrP^{Sc} type [1] is considered as the classical form of CJD with early occurrence of dementia and rapid disease progression (median five months). There are also molecular types associated with a protracted course and a heterogeneous clinical picture, for example, patients with VV1 or MV2 phenotypes [6]. A subsequent study revealed differences in the sensitivity of different diagnostic examinations (EEG, CSF, MRI) in particular molecular types [8; 28] (Table 31.12). In general, the detection of the 14–3–3 protein was the most sensitive test (sensitivity 96% or 100%, respectively). The CSF examination only failed to show good results when the patient had the MV2 molecular type (sensitivity = 30%). However, MRI achieved a better sensitivity, specifically for this subtype of CJD.

The most frequent phenotypes of sporadic CJD are designated as MM1/MV1, VV2 and MV2:

- MM1/MV1: The most common phenotype with comparatively short disease duration (median five months) and a peak incidence is in the seventh decade of life. Most often the onset of the disease is characterized by cortical visual impairment and rapidly progressive dementia, which, in more than two-thirds of the patients, is often already present at the onset of the disease, while ataxia is rarely seen at this stage. The course of the disease is accompanied by severe involvement of the pyramidal and extrapyramidal systems; MM1/MV1 patients often develop myoclonus.

- VV2: These patients are younger (median 62 years), and their disease duration is longer (median eight months). Progressive ataxia is often the first clinical sign, while visual symptoms are less frequent at the onset and are usually manifested as blurred vision or diplopia. Dementia is rarely found at this stage, although all patients become demented during the course of the disease. Neuropathological examination typically shows pronounced involvement of the cerebellum and brain stem and spongiform changes in the deep layers of the cerebral cortex.
- MV2: Patients with the MV2 phenotype belong to the third most frequent group. The long disease duration (median 17 months) and diverse clinical signs are notable features. Dementia is the first clinical sign in about half of the cases. However, the disease can also

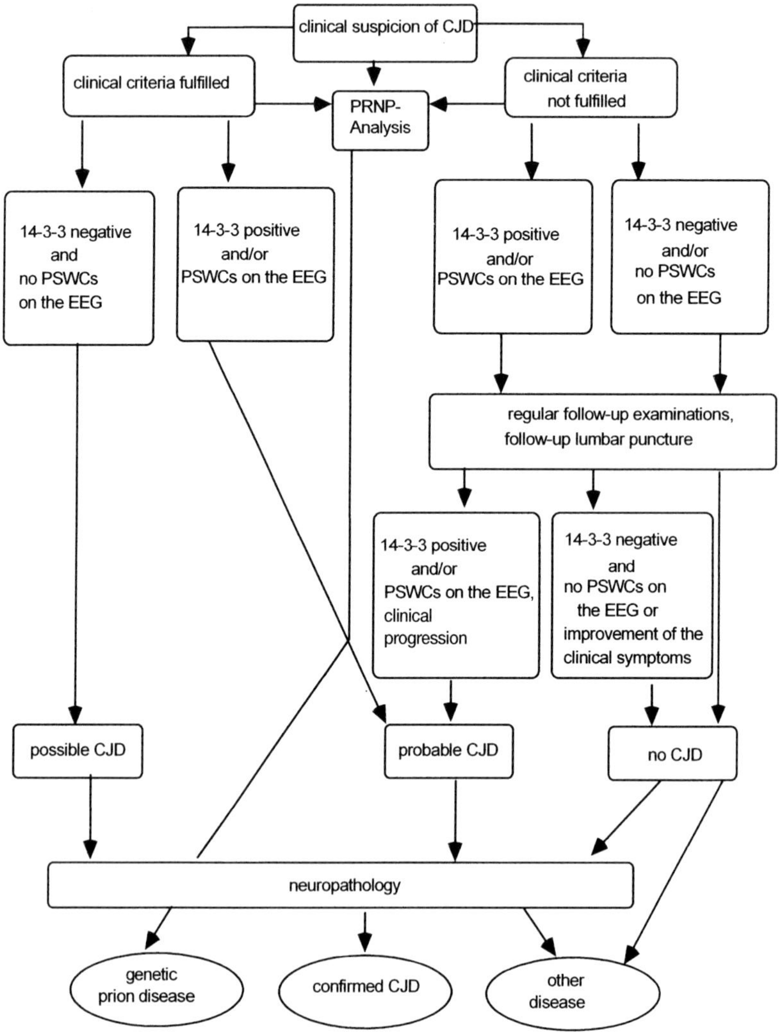

Fig. 31.7: Flowchart: EEG and laboratory examinations for the diagnosis of CJD. Figure by I. Zerr ©.

start with ataxia or extrapyramidal signs. Neuropathologically, this phenotype is characterized by kuru plaques. Focal lesions are distributed in the cortex, although they are also found in the basal ganglia, thalamus and cerebellum.

The three most frequently occurring phenotypes of CJD do not only differ in the clinical presentation and neuropathology. One striking observation is the different sensitivity of the clinical diagnostic tests. Typical EEG changes seem to occur most often in patients with the MM1/MV1 subtype. The detection of the 14–3–3 protein represents the technical investigation that can be used for patients with a VV2 subtype as well as for MM1/MV1 subtype (sensitivity 95%). There are no typical EEG changes in patients with VV2 subtype. The clinical diagnosis of patients with the MV2 subtype is particularly difficult to establish, not only because of the prolonged disease duration. In this case, no PSWCs in the EEG are observed, and the sensitivity of 14–3–3 protein detection in CSF is low. MRI helps to fill the diagnostic gap. According to results obtained so far, almost all patients show hyperintensities in their basal ganglia, which emphasizes the important diagnostic role of MRI in this phenotype [8; 75].

31.6 Diagnostic procedure

Clinical signs are the primary indicators of CJD while the patient is still alive. The suggested clinical diagnostic scheme is shown in Figure 31.7. Although MRI is not yet included in the clinical criteria, it is urgently needed in patients with rapid progressive dementia in order to exclude other diseases, to confirm the clinical diagnosis of CJD, and to differentiate between sporadic and variant disease forms (Fig. 30.2). In addition, repeated EEG should be carried out. The detection of 14–3–3 in the CSF is an important factor in diagnosis and tests should be performed. If there is any doubt, a repeat lumbar puncture is required.

Clinical and final neuropathological classification of sporadic CJD: A retrospective evaluation of 564 autopsy-confirmed CJD cases or suspects showed that only 15/564 (3%; "*no CJD*") patients did not meet the diagnostic criteria of *probable* (496/564 (88%)) or *possible* (59/564 (9%)) *CJD* while they were alive.

In the German CJD surveillance study, only single cases who were initially classified as *probable* CJD, were not confirmed at autopsy [33; 101].

The clinical classification criteria together with the presence of 14–3–3 in CSF are sensitive tests in the diagnosis of CJD. PSWCs in the EEG, abnormally increased concentrations of brain-derived proteins in CSF or detection of hyperintense basal ganglia in the MRI support the clinical CJD diagnosis. However, follow-up examinations are essential in cases of atypical disease presentation.

References

[1] Will RG, Zeidler M, Stewart GE, et al. Diagnosis of new variant Creutzfeldt–Jakob disease. Ann Neurol 2000; 47(5):575–582.

[2] WHO. Human transmissible spongiform encephalopathies. Weekly epidemiological record WHO, Geneva 1998; 47:361–365.

[3] Zerr I, Bodemer M, Gefeller O, et al. Detection of 14–3–3 protein in the cerebrospinal fluid supports the diagnosis of Creutzfeldt–Jakob disease. Ann Neurol 1998; 43(1):32–40.

[4] Giese A, Schulz-Schaeffer WJ, Windl O, et al. A novel classification scheme of sporadic Creutzfeldt–Jakob disease based on molecular analysis. Clin.Neuropathol. 1997; 16:259.

[5] Parchi P, Castellani R, Capellari S, et al. Molecular basis of phenotypic variability in sporadic Creutzfeldt–Jakob disease. Ann Neurol 1996; 39(6):767–778.

[6] Parchi P, Giese A, Capellari S, et al. Classification of sporadic Creutzfeldt–Jakob disease based on molecular and phenotypic analysis of 300 subjects. Ann Neurol 1999; 46(2):224–233.

[7] Otto A, Zerr I, Lantsch M, Weidehaas K, et al. Akinetic mutism as a classification criterion for the diagnosis of Creutzfeldt–Jakob disease. J Neurol Neurosurg Psychiatry 1998; 64(4): 524–528.

[8] Zerr I, Schulz-Schaeffer WJ, Giese A, et al. Current clinical diagnosis in Creutzfeldt–Jakob disease: identification of uncommon variants. Ann Neurol 2000; 48(3):323–329.

[9] Rosenmann H, Meiner Z, Kahana E, et al. Detection of 14–3–3 protein in the CSF of genetic

Creutzfeldt–Jakob disease. Neurology 1997; 49(2):593–595.
[10] The National Creutzfeldt–Jakob Disease Surveillance Unit. National Creutzfeldt–Jakob disease surveillance – diagnostic criteria. Webpage 2006: www.cjd.ed.ac.uk/criteria.htm.
[11] Levy SR, Chiappa KH, Burke CJ, et al. Early evolution and incidence of electroencephalographic abnormalities in Creutzfeldt–Jakob disease. J Clin Neurophysiol 1986; 3(1):1–21.
[12] Hansen HC, Zschocke S, Sturenburg HJ, et al. Clinical changes and EEG patterns preceding the onset of periodic sharp wave complexes in Creutzfeldt–Jakob disease. Acta Neurol Scand 1998; 97(2):99–106.
[13] Donnet A, Farnarier G, Gambarelli D, et al. Sleep electroencephalogram at the early stage of Creutzfeldt–Jakob disease. Clin Electroencephalogr 1992; 23(3):118–125.
[14] Steinhoff BJ, Racker S, Herrendorf G, et al. Accuracy and reliability of periodic sharp wave complexes in Creutzfeldt–Jakob disease. Arch Neurol 1996; 53(2):162–166.
[15] Steinhoff BJ, Herrendorf G, Zerr I, et al. Creutzfeldt–Jakob disease. A sphinx of modern neurobiology. Dtsch Med Wochenschr 1998; 123(6):169.
[16] Steinhoff BJ, Zerr I, Glatting M, et al. Diagnostic value of periodic complexes in Creutzfeldt–Jakob disease. Ann Neurol 2004; 56(5):702–708.
[17] Brown P, Rodgers-Johnson P, Cathala F, et al. Creutzfeldt–Jakob disease of long duration: clinicopathological characteristics, transmissibility, and differential diagnosis. Ann Neurol 1984; 16:295–304.
[18] Notari S, Capellari S, Giese A, et al. Effects of different experimental conditions on the PrP^{Sc} core generated by protease digestion: Implications for strain typing and molecular classification of CJD. J Biol Chem 2004; 279:16797–16804.
[19] Will RG. Epidemiological surveillance of Creutzfeldt–Jakob disease in the United Kingdom. Eur J Epidemiol 1991; 7(5):460–465.
[20] Jacobi C, Arlt S, Reiber H, et al. Immunoglobulins and virus-specific antibodies in patients with Creutzfeldt–Jakob disease. Acta Neurol Scand 2005; 111(3):185–190.
[21] Brown P, Gibbs CJ Jr, Rodgers Johnson P, et al. Human spongiform encephalopathy: the National Institutes of Health series of 300 cases of experimentally transmitted disease. Ann Neurol 1994; 35:513–529.
[22] Jones HR Jr, Hedley-Whyte ET, Freidberg SR, et al. Ataxic Creutzfeldt–Jakob disease: diagnostic techniques and neuropathologic observations in early disease. Neurology 1985; 35(2):254–257.
[23] Olsson JE. Brain and CSF proteins in Creutzfeldt–Jakob disease. Eur Neurol 1980; 19(2): 85–90.
[24] Harrington MG, Merril CR, Asher DM, et al. Abnormal proteins in the cerebrospinal fluid of patients with Creutzfeldt–Jakob disease. N Engl J Med 1986; 315(5):279–283.
[25] Zerr I, Bodemer M, Otto M, et al. Diagnosis of Creutzfeldt–Jakob disease by two-dimensional gel electrophoresis of cerebrospinal fluid. Lancet 1996; 348(9031):846–849.
[26] Hsich G, Kenney K, Gibbs CJ Jr, et al. The 14–3–3 brain protein in cerebrospinal fluid as a marker for transmissible spongiform encephalopathies. N Engl J Med 1996; 335(13):924–930.
[27] Aksamit AJ Jr, Preissner CM, Homburger HA. Quantitation of 14–3–3 and neuron-specific enolase proteins in CSF in Creutzfeldt–Jakob disease. Neurology 2001; 57(4):728–730.
[28] Castellani RJ, Colucci M, Xie Z, et al. Sensitivity of 14–3–3 protein test varies in subtypes of sporadic Creutzfeldt–Jakob disease. Neurology 2004; 63:436–442.
[29] Green AJ, Thompson EJ, Stewart GE, et al. Use of 14–3–3 and other brain-specific proteins in CSF in the diagnosis of variant Creutzfeldt–Jakob disease. J Neurol Neurosurg Psychiatry 2001; 70(6):744–748.
[30] Lemstra AW, van Meegen MT, Vreyling JP, et al. 14–3–3 testing in diagnosing Creutzfeldt–Jakob disease: a prospective study in 112 patients. Neurology 2000; 55(4):514–516.
[31] Sanchez-Valle R, Saiz A, Graus F. 14–3–3 Protein isoforms and atypical patterns of the 14–3–3 assay in the diagnosis of Creutzfeldt–Jakob disease. Neurosci Lett 2002; 320(1-2):69–72.
[32] Zerr I, Bodemer M, Weber T. The 14–3–3 brain protein and transmissible spongiform encephalopathy. N Engl J Med 1997; 336(12):874–875.
[33] Zerr I, Pocchiari M, Collins S, et al. Analysis of EEG and CSF 14–3–3 proteins as aids to the diagnosis of Creutzfeldt–Jakob disease. Neurology 2000; 55(6):811–815.
[34] Wiltfang J, Otto M, Baxter HC, et al. Isoform pattern of 14–3–3 proteins in the cerebrospinal fluid of patients with Creutzfeldt–Jakob disease. J Neurochem 1999; 73(6):2485–2490.
[35] Kawamoto Y, Akiguchi I, Jarius C, et al. Enhanced expression of 14–3–3 proteins in reac-

tive astrocytes in Creutzfeldt–Jakob disease brains. Acta Neuropathol (Berl) 2004; 108(4):302–308.
[36] Brandel JP, Peoc'h K, Beaudry P, et al. 14–3–3 protein cerebrospinal fluid detection in human growth hormone-treated Creutzfeldt–Jakob disease patients. Ann Neurol 2001; 49(2):257–260.
[37] Green AJ, Ramljak S, Muller WE, et al. 14–3–3 in the cerebrospinal fluid of patients with variant and sporadic Creutzfeldt–Jakob disease measured using capture assay able to detect low levels of 14–3–3 protein. Neurosci Lett 2002; 324(1):57–60.
[38] Beaudry P, Cohen P, Brandel JP, et al. 14–3–3 protein, neuron-specific enolase, and S-100 protein in cerebrospinal fluid of patients with Creutzfeldt–Jakob disease. Dement Geriatr Cogn Disord 1999; 10(1):40–46.
[39] Rother J, Schwartz A, Harle M, et al. Magnetic resonance imaging follow-up in Creutzfeldt–Jakob disease. J Neurol 1992; 239(7):404–406.
[40] Kohira I, Tsuji T, Ishizu H et al. Elevation of neuron-specific enolase in serum and cerebrospinal fluid of early stage Creutzfeldt–Jakob disease. Acta Neurol Scand 2000; 102(6):385–387.
[41] Kropp S, Zerr I, Schulz-Schaeffer WJ, et al. Increase of neuron-specific enolase in patients with Creutzfeldt–Jakob disease. Neurosci Lett 1999; 261(1-2):124–126.
[42] Jimi T, Wakayama Y, Shibuya S, et al. High levels of nervous system-specific proteins in cerebrospinal fluid in patients with early stage Creutzfeldt–Jakob disease. Clin Chim Acta 1992; 211(1–2):37–46.
[43] Wakayama Y, Shibuya S, Kawase J, et al. High neuron-specific enolase level of cerebrospinal fluid in the early stage of Creutzfeldt–Jakob disease. Klin Wochenschr 1987; 65(16):798–801.
[44] Otto M, Stein H, Szudra A, et al. S-100 protein concentration in the cerebrospinal fluid of patients with Creutzfeldt–Jakob disease. J Neurol 1997; 244(9):566–570.
[45] Otto M, Wiltfang J, Cepek L, et al. Tau protein and 14–3–3 protein in the differential diagnosis of Creutzfeldt–Jakob disease. Neurology 2002; 58(2):192–197.
[46] Otto M, Wiltfang J, Tumani H, et al. Elevated levels of tau-protein in cerebrospinal fluid of patients with Creutzfeldt–Jakob disease. Neurosci Lett 1997; 225(3):210–212.
[47] Riemenschneider M, Wagenpfeil S, Vanderstichele H, et al. Phospho-tau/total tau ratio in cerebrospinal fluid discriminates Creutzfeldt–Jakob disease from other dementias. Mol Psychiatry 2003; 8(3):343–347.
[48] Van EB, Green AJ, Vanmechelen E, et al. Phosphorylated tau in cerebrospinal fluid as a marker for Creutzfeldt–Jakob disease. J Neurol Neurosurg Psychiatry 2002; 73(1):79–81.
[49] Manaka H, Kato T, Kurita K, et al. Marked increase in cerebrospinal fluid ubiquitin in Creutzfeldt–Jakob disease. Neurosci Lett 1992; 139(1):47–49.
[50] Sharief MK, Green A, Dick JP, et al. Heightened intrathecal release of proinflammatory cytokines in Creutzfeldt–Jakob disease. Neurology 1999; 52:1289–1291.
[51] Stoeck K, Bodemer M, Ciesielczyk B, et al. Interleukin 4 and interleukin 10 levels are elevated in the cerebrospinal fluid of patients with Creutzfeldt–Jakob disease. Arch Neurol 2005; 62(10):1591–1594.
[52] Stoeck K, Bodemer M, Zerr I. Pro- and anti-inflammatory cytokines in the CSF of patients with Creutzfeldt–Jakob disease. J Neuroimmunol 2005; Epub ahead of print.
[53] Minghetti L, Cardone F, Greco A, et al. Increased CSF levels of prostaglandin E(2) in variant Creutzfeldt–Jakob disease. Neurology 2002; 58(1):127–129.
[54] Minghetti L, Greco A, Cardone F, et al. Increased brain synthesis of prostaglandin E2 and F2-isoprostane in human and experimental transmissible spongiform encephalopathies. J Neuropathol Exp Neurol 2000; 59(10):866–871.
[55] Arlt S, Kontush A, Zerr I, et al. Increased lipid peroxidation in cerebrospinal fluid and plasma from patients with Creutzfeldt–Jakob disease. Neurobiol Dis 2002; 10(2):150–156.
[56] Sanchez JC, Guillaume E, Lescuyer P, et al. Cystatin C as a potential cerebrospinal fluid marker for the diagnosis of Creutzfeldt–Jakob disease. Proteomics 2004; 4:2229–2233.
[57] Bieschke J, Giese A, Schulz-Schaeffer W, et al. Ultrasensitive detection of pathological prion protein aggregates by dual-color scanning for intensely fluorescent targets. Proc Natl Acad Sci U S A 2000; 97(10):5468–5473.
[58] Meiner Z, Halimi M, Polakiewicz RD, et al. Presence of prion protein in peripheral tissues of Libyan Jews with Creutzfeldt–Jakob disease. Neurology 1992; 42(7):1355–1360.
[59] Perini F, Frangione B, Prelli F. Prion protein released by platelets. Lancet 1996; 347(9015): 1635–1636.

[60] Tagliavini F, Prelli F, Porro M, et al. A soluble form of prion protein in human cerebrospinal fluid: implications for prion-related encephalopathies. Biochem Biophys Res Commun 1992; 184(3):1398–1404.

[61] Giese A, Bieschke J, Eigen M, et al. Putting prions into focus: application of single molecule detection to the diagnosis of prion diseases. Arch Virol Suppl 2000; 16:161–171.

[62] Castilla J, Saa P, Soto C. Detection of prions in blood. Nat Med 2005; 11(9):982–985.

[63] Saborio GP, Permanne B, Soto C. Sensitive detection of pathological prion protein by cyclic amplification of protein misfolding. Nature 2001; 411:810–813.

[64] Zeidler M, Stewart GE, Barraclough CR, et al. New variant Creutzfeldt–Jakob disease: neurological features and diagnostic tests. Lancet 1997; 350(9082):903–907.

[65] Will RG, Zeidler M, Brown P, et al. Cerebrospinal-fluid test for new-variant Creutzfeldt–Jakob disease. Lancet 1996; 348(9032):955.

[66] Zerr I, Giese A, Windl O, et al. Phenotypic variability in fatal familial insomnia (D178N-129M) genotype. Neurology 1998; 51(5):1398–1405.

[67] Hayashi R, Hanyu N, Kuwabara T, et al. Serial computed tomographic and electroencephalographic studies in Creutzfeldt–Jakob disease. Acta Neurol Scand 1992; 85(3):161–165.

[68] Schlenska GK and Walter GF. Serial computed tomography findings in Creutzfeldt–Jakob disease. Neuroradiology 1989; 31(4):303–306.

[69] Westphal KP and Schachenmayr W. Computed tomography during Creutzfeldt–Jakob disease. Neuroradiology 1985; 27(4):362–364.

[70] Kawata A, Suga M, Oda M, et al. Creutzfeldt–Jakob disease with congophilic kuru plaques: CT and pathological findings of the cerebral white matter. J Neurol Neurosurg Psychiatry 1992; 55(9):849–851.

[71] Finkenstaedt M, Szudra A, Zerr I, et al. MR imaging of Creutzfeldt–Jakob disease. Radiology 1996; 199(3):793–798.

[72] Onofrj M, Fulgente T, Gambi D, et al. Early MRI findings in Creutzfeldt–Jakob disease. J Neurol 1993; 240(7):423–426.

[73] Mittal S, Farmer P, Kalina P, et al. Correlation of diffusion-weighted magnetic resonance imaging with neuropathology in Creutzfeldt–Jakob disease. Arch Neurol 2002; 59(1):128–134.

[74] Collie DA, Sellar RJ, Zeidler M, et al. MRI of Creutzfeldt–Jakob disease: imaging features and recommended MRI protocol. Clin Radiol 2001; 56(9):726–739.

[75] Meissner B, Koertner K, Bartl M, et al. Sporadic Creutzfeldt–Jakob disease: magnetic resonance imaging and clinical findings. Neurology 2004; 63(3):450–456.

[76] Uchino A, Yoshinaga M, Shiokawa O, et al. Serial MR imaging in Creutzfeldt–Jakob disease. Neuroradiology 1991; 33(4):364–367.

[77] Zeidler M, Will RG, Ironside JW, et al. Creutzfeldt–Jakob disease and bovine spongiform encephalopathy. Magnetic resonance imaging is not a sensitive test for Creutzfeldt–Jakob disease. BMJ 1996; 312(7034):844.

[78] Lim CC, Tan K, Verma KK, et al. Combined diffusion-weighted and spectroscopic MR imaging in Creutzfeldt–Jakob disease. Magn Reson Imaging 2004; 22(5):625–629.

[79] Mendez OE, Shang J, Jungreis CA. Diffusion-weighted MRI in Creutzfeldt–Jakob disease: a better diagnostic marker than CSF protein 14-3-3? J Neuroimaging 2003; 13(2):147–151.

[80] Schwaninger M, Winter R, Hacke W, et al. Magnetic resonance imaging in Creutzfeldt–Jakob disease: evidence of focal involvement of the cortex. J Neurol Neurosurg Psychiatry 1997; 63(3):408–409.

[81] Shiga Y, Miyazawa K, Sato S, et al. Diffusion-weighted MRI abnormalities as an early diagnostic marker for Creutzfeldt–Jakob disease. Neurology 2004; 63(3):443–449.

[82] Bahn MM and Parchi P. Abnormal diffusion-weighted magnetic resonance images in Creutzfeldt–Jakob disease. Arch Neurol 1999; 56(5):577–583.

[83] Demaerel P, Sciot R, Robberecht W et al. Accuracy of diffusion-weighted MR imaging in the diagnosis of sporadic Creutzfeldt–Jakob disease. J Neurol 2003; 250(2):222–225.

[84] Schröter A, Zerr I, Henkel K, Tschampa HJ, et al. Magnetic resonance imaging (MRI) in the clinical diagnosis of Creutzfeldt–Jakob disease. Arch Neurol 2000; 57:1751–1757.

[85] Chazot G, Broussolle E, Lapras C, et al. New variant of Creutzfeldt–Jakob disease in a 26-year-old French man. Lancet 1996; 347(9009):1181.

[86] WHO. WHO manual for surveillance of human transmissible spongiform encephalopathies including variant Creutzfeldt–Jakob disease. WHO Library Cataloguing-in-Publication Data 2003.

[87] Collie DA, Summers DM, Sellar RJ, et al. Diagnosing variant Creutzfeldt–Jakob disease

with the pulvinar sign: MR imaging findings in 86 neuropathologically confirmed cases. AJNR Am J Neuroradiol 2003; 24(8):1560–1569.

[88] Prusiner SB, Hsiao KK, Bredesen DE, et al. Prion disease. In: McKendall RR, ed. Handbook of Clinical Neurology. Elsevier Science Publishers, Amsterdam, 1989:543–80.

[89] Collins S, Boyd A, Fletcher A, et al. Recent advances in the pre-mortem diagnosis of Creutzfeldt–Jakob disease. J Clin Neurosci 2000; 7(3):195–202.

[90] Matsuda M, Tabata K, Hattori T, et al. Brain SPECT with 123I-IMP for the early diagnosis of Creutzfeldt–Jakob disease. J Neurol Sci 2001; 183(1):5–12.

[91] Read SL, Miller BL, Mena I, et al. SPECT in dementia: clinical and pathological correlation. J Am Geriatr Soc 1995; 43(11):1243–1247.

[92] Jibiki I, Fukushima T, Kobayashi K, et al. Utility of 123I-IMP SPECT brain scans for the early detection of site-specific abnormalities in Creutzfeldt–Jakob disease (Heidenhain type): a case study. Neuropsychobiology 1994; 29(3):117–119.

[93] Kirk A and Ang LC. Unilateral Creutzfeldt–Jakob disease presenting as rapidly progressive aphasia. Can J Neurol Sci 1994; 21(4):350–352.

[94] Kothbauer-Margreiter I, Baumgartner RW, Bassetti C, et al. Hemisensory deficit in a patient with Creutzfeldt–Jakob disease. Eur Neurol 1996; 36(2):108–109.

[95] de Silva R, Patterson J, Hadley D, et al. Single photon emission computed tomography in the identification of new variant Creutzfeldt–Jakob disease: case reports. BMJ 1998; 316(7131): 593–594.

[96] Frazzitta G, Grampa G, La Spina I, et al. SPECT in the early diagnosis of Creutzfeldt–Jakob disease. Clin Nucl Med 1998; 23(4):238–239.

[97] Bar KJ, Hager F, Nenadic I, et al. Serial positron emission tomographic findings in an atypical presentation of fatal familial insomnia. Arch Neurol 2002; 59(11):1815–1818.

[98] Engler H, Lundberg PO, Ekbom K, et al. Multitracer study with positron emission tomography in Creutzfeldt–Jakob disease. Eur J Nucl Med Mol Imaging 2003; 30(1):85–95.

[99] Henkel K, Zerr I, Hertel A, et al. Positron emission tomography with [(18)F]FDG in the diagnosis of Creutzfeldt–Jakob disease (CJD). J Neurol 2002; 249(6):699–705.

[100] Tsuji Y, Kanamori H, Murakami G, et al. Heidenhain variant of Creutzfeldt–Jakob disease: diffusion-weighted MRI and PET characteristics. J Neuroimaging 2004; 14(1):63–66.

[101] Poser S, Mollenhauer B, Krauss A, et al. How to improve the clinical diagnosis of Creutzfeldt–Jakob disease. Brain 1999; 122:2345–2351.

[102] Zerr I, Bodemer M, Röcker S, et al. Cerebrospinal fluid concentration of neuron-specific enolase in diagnosis of Creutzfeldt–Jakob disease. Lancet 1995; 345:1609–1610.

32 Introduction to Surveillance for Animal Prion Diseases

Beat Hörnlimann, Marcus G. Doherr, Danny Matthews, and Stuart C. MacDiarmid

32.1 Introduction

Programs to survey for animal prion diseases (TSEs) are expected to produce reliable estimates of the incidence, defined as the proportion of new disease cases detected in a risk population over a given period of time, of BSE, scrapie and other TSEs. Surveillance typically describes the combined activities of monitoring (disease detection) and control (implementation of measures once cases are found). Such systems traditionally depend on the cooperative efforts of livestock owners and all other persons who are in contact with animals susceptible to TSEs. In particular, the assistance of practicing veterinarians and people working in animal clinics and diagnostic laboratories is important.

The examination of cattle, sheep and goats displaying neurological disorders, and their subsequent reporting as "clinical suspects" (passive or baseline or clinical surveillance) has led to the detection of BSE and scrapie in many countries. This passive approach often was the only surveillance system in place in a country or region. The introduction of new rapid (screening) test procedures in 1999 (see Chapter 35) has significantly improved the ability of national surveillance programs to detect new cases. This additional active or targeted surveillance provides more accurate estimates of the TSE incidence in the screened populations. When implemented correctly, active surveillance enables the detection of infected animals that leave the population from being slaughtered some months before clear clinical signs indicative of BSE or scrapie develop. This could apply to animals that have been exposed to the same BSE infection risk as the index case within a herd (defined as risk cohorts). However, it has been shown by retrospective investigation of BSE test-positive cattle that the majority had presented some signs that were consistent with BSE. Unfortunately, because (i) the imprecise nature of those signs, (ii) the perception that BSE (or scrapie) does not exist in the country or farm, (iii) insufficient education of the farmers in terms of recognizing the signs of disease and (iv) the often greater age of the animals ("marking them for routine replacement") results in them leaving the population without the suspicion of BSE or scrapie. With an appropriate active surveillance system in place, they might be detected after death.

A major aim of surveillance systems in general is to collect sufficiently precise information on the disease incidence for spatial and temporal trends in epidemics to be followed. The gathering of epidemiological data from farms, and risk assessments to quantify the exposure to the infectious agents via different routes, enable competent national authorities to evaluate the effectiveness of measures for control and eradication and to make adjustments where necessary.

This chapter provides an overview of the minimal requirements demanded of an efficient BSE or scrapie surveillance system. Elementary, internationally accepted recommendations have been established by the World Organisation for Animal Health in Paris (OIE) [1], and have been subject to regular review and modification over the years [2; 3] as BSE has been identified in more countries [4] and the importance of active surveillance in strengthening monitoring programs has become evident.

An internationally acceptable surveillance program depends upon a transparent policy of notification by farmers and veterinarians, based upon satisfactory levels of compensation to farmers following the confirmation of disease. Financial disincentives to reporting should be avoided. For example, the farmer should not be expected to bear greater than normal veterinary or disposal costs from having dead cattle sampled on-farm. The cost of on-farm sampling and carcass disposal needs to be adequately covered by the competent national authority. Reporting of clinical suspects certainly is discouraged if international trade rules adversely penalize countries and farmers following disease confirmation. In such circumstances there is therefore a danger of under-reporting of cases [5].

32.2 Passive surveillance for animal prion diseases

32.2.1 The recognition of clinical signs

Any animal exhibiting clinical signs suggestive of BSE or scrapie should be reported to the appropriate authority as soon as possible. A large proportion of incubating animals display abnormal behavior and paraesthesia in the early stage of clinical manifestation. To maintain rigorous passive surveillance, and to ensure that sick or suspect animals are recognized at all stages of clinical manifestation, all professionals in close contact with live cattle, sheep, and goats should have a comprehensive knowledge of the clinical signs suggestive of prion diseases (see Chapters 33, 34). In addition to farmers and their veterinarians, cattle dealers and drivers of animal transport vehicles, employees in abattoirs and small butchers may recognize clinical signs suggestive of these diseases in the course of their handling and inspection of animals during transportation or prior to slaughter. Most importantly, veterinarians working in abattoirs as official meat inspectors who observe neurological or behavioral abnormalities during ante mortem inspection (see Chapter 33.7) must be aware of their responsibility to notify such cases in order to exclude them from the food chain. In this context it should be mentioned that stress associated with transportation to the abattoir, as well as the process prior to slaughter, have been reported to trigger or enhance the presentation of clinical signs of BSE [6–8]. Recognition and reporting thus depends upon an in-depth knowledge of the typical disease signs (Tables 33.1, 34.1).

32.2.2 The role of veterinarians and animal owners in surveillance for prion diseases

Between 1988 and 1994, most European Union countries, in accordance with EU legislation, introduced mandatory reporting of suspect BSE and scrapie cases. Livestock farmers and practicing veterinarians played a crucial role in the early identification of clinical suspects at that time. International and national regulations require that animals displaying clinical signs suggestive of BSE or scrapie be reported to the national veterinary authorities. If following further veterinary examination, or during ante mortem inspection, the animals are still considered suspect, they must be culled and their brains examined in a certified laboratory in order to verify whether or not they are infected with a TSE. Countries implement different mechanisms for the submission of brain samples, in accordance with their own infrastructure for disposal of livestock and for laboratory testing. Shipment and transportation of samples must comply with the specific packaging and postal guidelines. Information on appropriate preservation measures for the tissue and transport can be obtained from national and international reference laboratories [9]. In cases where the suspicion of disease is confirmed by the examining laboratory, further measures such as decontamination, tracing to the farm of origin, cohort identification, and culling will need to be taken.

32.2.3 The role of diagnostic laboratories

Officially authorized laboratories play a key role in the diagnosis of BSE and scrapie. They are obliged to follow examination procedures that are either defined in national or European law, or required to be compliant with procedures published by the OIE, as the latter is the basis for guarantees that support international law on trade in animals and animal products [1; 9]. National reference laboratories are responsible for the control of nationwide standardized diagnostics and the re-examination of ambiguous cases according to these international criteria.

Various research associations supported financially by the European Commission (such as the European Scrapie and SRTSE Networks, a research project on "Surveillance and Diagnosis of Ruminant TSE" (see Chapter 55) and the Neuroprion project (established in 2003) [10]) have been involved in the continuous improvement, standardization and evaluation of laboratory methods used for the diagnosis of prion diseases. A more formal role was established in the EU in 2001 with the legal designation of the UK National Reference Laboratory as EU Community Reference Laboratory [11]. It is required, through ring trials, to monitor the performance of the EU National Reference Laboratories on a regular basis. It also offers a similar, but more limited, service to non-EU laboratories under its designation as OIE-Reference Laboratory. Other OIE-Reference Laboratories exist in Switzerland, Japan and Canada.

Histopathological, immunohistochemical (IHC) and immunochemical (SAF-immunoblot) assays are used for the verification of prion diseases in suspect animals. The procedures for these confirmatory tests are laid down in the OIE *Manual of Diagnostic Tests and Vaccines for Terrestrial Animals* [2].

Histopathological diagnosis is based on the evidence of bilaterally symmetrical spongiform alterations in the neuropil, the loss of nerve cells and gliosis at key target sites. The pattern of changes, and their severity, vary between different brain regions of the respective host species and with the strain (there are multiple strains of scrapie; see Chapters 12; 27). The severity of lesions also varies from animal to animal, from extensive vacuolation to instances where vacuolation is barely detectable.

IHC or other immunochemical methods identify abnormal prion protein (PrP^{Sc}) and its distribution in the brain. Although used to confirm interpretation of results obtained by histopathological examination [1], the increased sensitivity and specificity of the immunochemical methods relative to histopathology has resulted in their more widespread use as primary tests [2; 12; 13].

In an immunoblot, PrP^{Sc} is extracted from a tissue homogenate, either directly through digestion with an enzyme called proteinase K (PK), or after first being concentrated by ultracentrifugation or other means, and then subjected to PK treatment. Following electrophoresis, the PK-resistant PrP^{Sc} molecules are stained by binding with PrP-specific antibodies. A number of excellent poly- and monoclonal antibodies are available for this method, but their suitability for use in any particular test has to be validated before introduction [14–16].

Other rapid BSE tests, which have been evaluated by the EU and approved by several countries, are used in the context of the active surveillance (see Section 32.3). Accordingly, specialized commercial laboratories (Table 35.1) are responsible for the quality of the tests used, although their performance is overseen by respective National Reference Laboratories. In some countries only official laboratories carry out the rapid testing. Rigorous procedures are required in all testing laboratories. This applies not only in the context of surveillance specifically for BSE or scrapie, but also when considering differential diagnoses (see Chapters 33.3, 34.4) made by animal clinics, institutes for animal pathology and by state veterinarians, i. e., whenever investigations suggest the existence of a prion disease. Given a certain level of disease awareness at these institutions and in accordance to most national TSE legislation, any suspicion of a TSE in any species should be reported to the competent national authority for further examination. Such procedures have led to the identification of prion diseases [17] in zoo animals (see Chapter 20) and FSE in domestic (see Chapter 23) or wild cats (see Chapter 20).

32.2.4 The role of international, national and regional authorities

The international recommendations put forward for the control of BSE and scrapie by the

OIE [1] and the EU [11] are an excellent basis for the effective surveillance of livestock on a national level. They are not identical, but reflect the greater prevalence of BSE in Europe, and the closer political and statutory links within the EU in comparison to the world community. OIE recommendations are subject to greater political pressures before the 167 member countries adopt them, since many of those countries either do not have BSE or scrapie, or do not have the resources to conduct large scale surveillance. National authorities may base their statutory measures for reporting and surveillance on such international guidelines, although at times they may go further or implement intermediate rules depending on the structure of their national livestock industries. Before farmers, veterinarians and all other professionals involved in the handling of live animals can comply with the obligation to report suspect cases, care must be taken to provide these groups with easy-to-understand information on BSE, scrapie or other TSEs. Education should involve, for example, the production and publication of comprehensible, practical videos [18–20], some of which may now be accessible on the Internet [18; 21; 22], together with the description of the clinical signs of BSE (see Chapter 33) and scrapie (see Chapter 34).

In addition to other activities, the competent national authority should (i) coordinate the activities of diagnostic laboratories involved in the surveillance program at all levels (states, counties, districts, Kantons), (ii) design monitoring programs, and (iii) coordinate the activities of the responsible official and practicing veterinarians to ensure that they comply with internationally accepted targets. These targets have varied over recent years, depending on the global incidence of disease, greater awareness of the spread of BSE, and on improvements in scientific understanding and diagnostic methodologies.

To optimize surveillance, national authorities should ensure the standardization of epidemiological data collection and analysis, in accordance with criteria documented internationally on prion diseases. This will insure that the outcome is acceptable to, and understood by, stakeholders at home and abroad. The results, their interpretation and the consequent recommendations should be made public to all those involved in the surveillance program and to the appropriate international organizations such as the OIE [2; 3] and European Commission [4; 23] through journals or websites. In some countries, such as member states of the EU, such information is compulsorily reported to the European Commission, and further analyzed and published on the Commission's website. The following methods of data compilation and studies are important not only for each country but also with regard to international comparison:

1. Publication of BSE and scrapie incidences on the basis of data provided by the surveillance authorities and the statistical offices (number of TSE cases per year, size of the animal population, age structure, etc.).
2. Evaluation studies for the assessment of efficiency and effectiveness of the measures taken.
3. Disclosure of risk factors on the basis of observational and comparative studies. The clarification of the potential risk factors usually takes place (upon the occurrence of an epidemic) only after a case study has been formulated to elaborate hypotheses (see Chapter 40). It is most advantageous to collect detailed epidemiological evidence in the field, i. e., by actually visiting the affected animal breeders on their farms and asking them to provide answers to standardized questionnaires. Although this kind of investigation produces a much better quality of data than those sent by post, it is considerably more time-consuming and requires greater resources.
4. Risk assessments which take into account the quality of monitoring/surveillance within the individual countries are inevitably complex. Geographical BSE risk (GBR) assessments in respective countries are necessary for the documentation of the risk factors and their mutual integration [4].

In 1994, the World Trade Organization (WTO) promulgated the *Agreement on the Application of Sanitary and Phytosanitary Measures*, the so-called "SPS Agreement" [24], which obliges member countries to reduce trade barriers between them. Under the SPS Agreement, trade between countries may only be restricted when the health of humans, animals or plants of the importing country is endangered. To comply with the SPS Agreement, the importing countries are entitled to verify the documented GBR assessment from the exporting countries [4; see also 25; 26].

BSE and scrapie data from individual countries, when published internationally, can only be compared if the quality standards of the national monitoring programs and risk assessments [4] are similar. The Scientific Steering Committee of the European Commission took this requirement into account when the GBR assessments were carried out [4]. Thus, international harmonization supports worldwide consumer safety, and safeguards equal market opportunities for all concerned.

The process of seeking international agreement on appropriate methods of surveillance continues to highlight the importance of surveillance of clinical disease, and the detection of affected animals.

32.3 "Active" surveillance for animal prion diseases

32.3.1 Evaluation of innovative diagnostic test methods (BSE tests)

The major aim of passive surveillance is the identification of as many clinically suspect cases as possible. It depends on the competence of many workers in the field, and is thus subject to many, greatly varying, factors such as education, willingness to report, data protection and degree of compensation. Even though TSE data have been collected from many different countries in this manner, the accuracy with which they can be compared is limited. Against this background, great efforts have been made to establish tests for the active surveillance of animal prion diseases (see Chapter 35). The major aim of active surveillance, which may be carried out either in all parts of a country or in specific target areas or subpopulations, is the additional identification of (i) infected animals that have not yet developed clinical signs (it is estimated that such preclinical infections may be detectable within 3 to 6 months of onset of signs); (ii) clinical cases which may not present the full range of expected clinical signs; and (iii) typical but unrecognized or unreported cases. In this respect, an important step forward was made by Prionics AG in 1999, first in Switzerland, by the introduction of a new post mortem BSE test into comprehensive surveillance programs [27]. The European Commission and more recently the European Food Safety Authority on behalf of the Commission have in past years evaluated a range of commercially available post mortem BSE tests (see Chapter 35; Table 35.1), and more recently similar tests for scrapie in small ruminants. EU legislation lists those that are rated suitable for the purpose of monitoring BSE [27; 29] or scrapie [20; 30; 31].

32.3.2 Scrapie tests (tests on live animals)

In small ruminants it is also possible to use certain peripheral tissues [31], in addition to brain material, for testing for scrapie. However, there have been a few reports of scrapie cases without involvement of the lymphoid system (references see Chapter 18.4). In contrast to cattle, abnormal prion protein (PrP^{Sc}) can – in most cases – be found in a large number of lymphatic organs in certain infected sheep. This is partly determined by their genetic background (see Chapter 56), and also by the scrapie strain (see Chapter 12), route of infection and incubation stage. In recent years this information has led to the development of scrapie tests that can be used on live animals. Lymphatic tissue can be collected under general anaesthesia from the pharyngeal cavity (tonsil test) [32], or part of the third eye lid [33] can be removed under local anaesthesia. Additionally, lymphoid tissue located immediately inside the anus (RAMALT) can be sampled without any anaesthesia and subjected to the same confirmatory tests [30; 31]. It remains to be seen whether the influence of genotype on peripheral distribution of infectivity and abnormal prion protein will limit the extent to which such approaches can be used for large-scale surveillance. They may however still be of value in confirming a clinical diagnosis in vivo, and thus for flock-level freedom certification programs (see Chapter 57) except for animals without involvement of the lymphoid system (see above).

Given the extraordinary importance that scrapie undoubtedly has among all human and animal prion diseases, the quotation of the scrapie researcher S. Gaiger in 1924 seems most apt: "My opinion is that the main problem which confronts the research worker in scrapie is first to find the precise causal agent, and, following this, to find some test by which latent infection may be detected" [quoted from 34].

References

[1] World Organisation for Animal Health (OIE). International animal health code – mammals, birds and bees. Article 2.3.13 "Bovine spongiform encephalopathy". Manual of diagnostic tests and vaccines for terrestrial animals. 1999.

[2] World Organisation for Animal Health (OIE). Manual of diagnostic tests and vaccines for terrestrial animals. 2004.

[3] World Organisation for Animal Health (OIE). Terrestrial animal health code (2005). Webpage 2006: www.oie.int/eng/normes/MCode/a_summry.htm.

[4] European Commission 2003. Opinions on GBR in third countries. Webpage 2006: http://europa.eu.int/comm/food/fs/bse/scientific_advice03_en.html.

[5] Abbott A. Germany uses BSE fears to seek ban on British beef (news). Nature 1994; 368:178.

[6] Bennett AD, Birkett CR, Bostock CJ. Molecular biology of scrapie-like agents. Rev Sci Tech Off Int Epiz 1992; 11:569–603.

[7] Hörnlimann B and Braun U. BSE: Clinical signs in Swiss BSE cases. In: Bradley R, Marchant B, editors. Transmissible spongiform encephalopathies. EU: VI/4131/94-EN Document Ref. F. II.3 – JC/0003, Brussels, 1994:289–299.

[8] Braun U, Schicker E, Hörnlimann B. Diagnostic reliability of clinical signs in cows with suspected bovine spongiform encephalopathies. Vet Rec 1998; 143:101–105.

[9] World Organisation for Animal Health (OIE). Reference Laboratories. Webpage 2006: www.oie.int/eng/OIE/organisation/en_LR.htm.

[10] European Commission. Inventory of national research activities in transmissible spongiform encephalopathies (TSEs) in Europe. Webpage 2006: http://europa.eu.int/comm/research/quality-of-life/pdf/tse-finalreport-april2003_en.pdf.

[11] EU. Main EU legislation on BSE. Webpage 2006: http://europa.eu.int/comm/food/fs/bse/bse19_en.html.

[12] Groschup MH, Beekes M, McBride PA, et al. Deposition of disease-associated prion protein involves the peripheral nervous system in experimental scrapie. Acta Neuropathol (Berl) 1999; 98(5):453–457.

[13] Hardt M, Baron T, Groschup MH. A comparative study of immunohistochemical methods for detecting abnormal prion protein with monoclonal and polyclonal antibodies. J Comp Pathol 2000; 122(1):43–53.

[14] Groschup MH, Harmeyer S, Pfaff E. Antigenic features of scrapie prion protein and cellular prion proteins of various mammalian species. J Immunol Meth 1997; 207:89–101.

[15] Harmeyer S, Pfaff E, Groschup MH. Synthetic peptide vaccines yield monoclonal antibodies to cellular and pathological prion proteins of ruminants. J Gen Virol 1998; 79:937–945.

[16] Korth C, Stierli B, Streit P, et al. Prion (PrPSc)-specific epitope defined by a monoclonal antibody. Nature 1997; 390(6655):74–77.

[17] Anonym. Bovine spongiform encephalopathy in Switzerland in a zoo zebu. Webpage 2006: www.rense.com/general54/dmdmet.htm.

[18] Braun U, Pusterla N, Schicker E. Clinical findings in cattle with bovine spongiform encephalopathy (BSE); 1997. Video, available in different languages on Webpage 2006: www.bse.unizh.ch/english/video/content.htm.

[19] Ulvund MJ. Skrapesjuke – kliniske symptom (Scrapie – clinical symptoms; video in Norwegian). Statens Dyrehelsetilsyn, Oslo, 1987.

[20] Veterinary Laboratories Agency. United Kingdom. Webpage 2006: www.vla.gov.uk.

[21] Veterinary Laboratories Agency. Clinical signs of scrapie in sheep. Webpage 2006: www.defra.gov.uk/corporate/vla/science/documents/science-scrapie-res.pdf.

[22] Veterinary Laboratories Agency. Clinical signs of bovine spongiform encephalopathy in cattle. Webpage 2006: www.defra.gov.uk/corporate/vla/science/documents/science-bse-res.pdf.

[23] European Commission 2005. The TSE Road Map. Webpage 2006: http://europa.eu.int/comm/food/food/biosafety/bse/roadmap_en.pdf.

[24] WTO. The results of the Uruguay round of multilateral trade negotiations: the legal texts. GATT Secretariat, Geneva, 1994:69–84.
[25] MacDiarmid SC. Scrapie: the risk of its introduction and effects on trade. Aust Vet J 1996; 73(5):161–164.
[26] MacDiarmid SC and Pharo HJ. Risk analysis: assessment, management and communication. Rev Sci Tech Off Int Epiz 2003; 22(2):397–408.
[27] Prionics. First results on new Swiss BSE surveillance system presented 25 Sept. 1999 (Prion disease news of historical importance). Webpage 2006: www.prionics.ch/online/news/e-archive.htm.
[28] European Commission – Scientific Steering Commitee. The evaluation of tests for the diaganosis of transmissible spongiform encephalopathy in bovines (includes evaluation protocol). Webpage 2006: www.priondata.org/data/A_ECassess.html.
[29] Moynagh J and Schimmel H. Tests for BSE evaluated. Nature 1999; 400:105.
[30] Gonzalez L, Jeffrey M, Sisó S, et al. Diagnosis of preclinical scrapie in samples of rectal mucosa. Vet Rec 2005; 156(26):846–847.
[31] Gonzalez L, Dagleish MP, Bellworthy SJ, et al. Postmortem diagnosis of preclinical and clinical scrapie in sheep by the detection of disease-associated PrP in their rectal mucosa. Vet Rec 2006; 158(10):325–331.
[32] van Keulen LJ, Schreuder BE, Meloen RH, et al. Immunohistochemical detection of prion protein in lymphoid tissues of sheep with natural scrapie. J Clin Microbiol 1996; 34:1228–1231.
[33] O'Rourke KI, Baszler TV, Parish SM, et al. Preclinical detection of PrPSc in nictitating membrane lymphoid tissue of sheep. Vet Rec 1998; 142:489–491.
[34] Gaiger HS. Scrapie. J Comp Path 1924; 37:259–277.

33 Clinical Findings in Bovine Spongiform Encephalopathy

Ernst Schicker, Ueli Braun, Beat Hörnlimann, and Timm Konold

33.1 Introduction

The clinical manifestation of bovine spongiform encephalopathy (BSE) presents a broad spectrum of signs, various combinations of which are most pronounced in later stages of the disease. To identify BSE susepcts in the early clinical stages it is important to collect a detailed clinical history and to conduct thorough general clinical and neurological examinations [1–6].

The neurological signs can be traced to a lesion in the diencephalon and brainstem, thalamus, and cerebellum [7; 8]. Currently, no pathognomic complex of signs is known that would allow accurate clinical diagnosis of BSE [9]. In contrast to scrapie, however, the pathology of BSE (see Chapter 27) is comparatively uniform. This chapter aims to provide the practicing veterinarian with a guide for the systematic examination of cattle for signs of BSE.

The clinical picture is characterized by unspecific general signs, such as emaciation, reduced activity, reduced appetite, and lowered heart rate as well as by neurological signs. Nearly all animals show behavioral abnormalities characterized by increased sensorimotor irritability, which is manifested in anxiety, restlessness, or nervousness. In virtually all affected animals, hyperresponsiveness to external stimuli is most obvious; the animals overreact to tactile stimuli in particular. Many animals are hypersensitive to sudden noise and some are also sensitive to sudden light changes. The majority of animals also display abnormalities in locomotion such as hind limb ataxia. This often causes difficulty in rising to a standing position. Recumbency is seen in the final stage of the disease. Hematological and chemical analysis of blood does not reveal uniform deviations from normal values and is in most cases unobtrusive. The composition of the cerebrospinal fluid (CSF) is normal. Laboratory analysis only assists in clarification of differential diagnoses.

33.2 Clinical history and course of the disease

Cattle of all breeds can be affected by BSE from approximately 20 months to the age of 22 years and 7 months [10]. Dairy breeds are more frequently affected than beef suckler cows. The age-specific incidence is highest among 4- to 6-year-old animals (Fig. 39.3).

The most frequent initial signs observed involve deterioration in body condition, a decrease in weight and milk production as well as neurological signs. The first signs recognized by farmers that are associated with BSE are changes in behavior. Animals that used to behave normally suddenly begin to kick violently, refuse to enter the milking parlor or to enter doorways, remain distant from the rest of the herd in the field, and are apprehensive. They may exhibit trembling (muscle fasciculations or tremors) and hyperesthesia. The first signs of abnormal locomotion may be mild hind limb ataxia, difficulty in rising or even recumbency [11–15]. Once the farmer has observed the first locomotor disorder in what appears to be the early stage of the disease, the animal may well be in a rather advanced stage. Changes in behavior and sensitivity have most likely existed for some time and have probably escaped the attention of the farmer [1; 4].

Clinical signs can be triggered by stress in the animals [16] and can vary on a daily basis. If the animals are kept in a calm and familiar environment, this will usually lead to the attenuation of clinical signs [13]. After a progressive course of the disease of on average between 40 and 60 days, an individual pattern of clinical signs usually develops: Milk production is reduced in approximately 50%[1]–100%[2] of affected animals as soon as clinical signs have become obvious. In later stages of the disease, the animals often lose weight. The animal's appetite, however, is only noticeably reduced in the final stages of the disease [11; 15; 17].

33.3 Differential diagnosis of BSE

A definite diagnosis of BSE can only be made post mortem [14] (see Chapters 27; 32; 35). BSE must be considered in dairy cows that show normal appetite and rectal temperature despite only moderate or poor nutritional status and particularly, when abnormal behavior or locomotor disorders are observed. Other diseases, which are accompanied by changes in behavior and locomotion, must also be taken into consideration. These include metabolic disorders such as hypomagnesemia, nervous acetonemia, parturient paresis (hypocalcemia), the bacterial infection listeriosis, the viral infections rabies and Aujeszky's disease, poisoning due to consumption of material containing soluble lead salts or botulinum toxin, and traumatic injuries to the head, vertebral column, and spinal cord [11; 18].

In 2003, C. Saegerman and colleagues [19] established a classification of neurological or neurologically expressed disorders that occur in Western European cattle aged 12 months and over on the basis of etiology, frequency and condition of appearance, age and type of animals concerned, and the main clinical signs observed. Neurologically expressed disorders have been classified according to different groups of causes: biological, nonbiological and nonspecific or unknown. Differential diagnosis of neurologically expressed disorders is an essential element in the clinical surveillance of BSE. A growing number of etiologies are described in the scientific literature.

[1] According to British observations in approx. 174,000 BSE cases (Timm Konold, unpublished observation).

[2] In 50 cattle, according to a Swiss study by Schicker [1].

Several studies failed to establish a differential diagnosis in many suspected BSE cases that were negative for BSE or any other disease on examination of the brain [20–22]. Although it is possible that some of these represented metabolic diseases that did not result in pathological findings in the brain it is likely that other nonneurological conditions may produce clinical signs similar to BSE, which may result in reporting as a BSE suspect. However, an examination as detailed below should exclude BSE in most of these cases. In particular, hypersensitivity to external stimuli seems to be more frequent in BSE cases than in BSE suspects without pathological confirmation of the disease [6; 23].

33.4 General clinical examination findings

The main clinical findings include poor nutritional status of the animal despite apparent normal appetite, chronic weight loss, and reduced milk production [11; 17]. The frequency of ruminal contractions is in most cases normal, whereas strength of rumen movements and rumen fill are generally reduced. This indicates reduced food intake although the cow's appetite is usually described as normal. The California Mastitis Test (CMT) may be clearly positive in BSE cases.

BSE cases usually have a normal body temperature; stress-related increase of rectal temperature may be observed in the final phase of the disease or during transportation of sick animals. It is often possible to observe bradycardia, which continues even in a state of extreme discomfort [17; 24].

33.5 Neurological examination findings

33.5.1 Changes in behavior

Almost all BSE cases display abnormal behavior (Fig. 33.1). If possible, it is best to assess the state of the animal in a familiar environment, for example, a cowshed or pasture. Changes in the animal's behavior are characterized by increased sensorimotor distress, which is mainly expressed in defensive behavior patterns such as panic-stricken behavior, anxiety, restlessness, and nervousness and less frequently in aggressive behavior [6; 11]. Panic-stricken behavior is characterized by flinching or startling, shying behavior – often with a slightly anxious facial expression (Fig. 33.1a), staring, restlessness, nervousness, ear twitching, aggressiveness with increased defense readiness, maliciousness, and sudden kicking. Further alterations in behavior include frequent bruxism, salivation, frequent nose licking (Fig. 33.1b), as well as trembling, which is either confined to the head and neck area or extends, for example, to the shoulders and even the entire body. These signs are mainly due to the discomfort the animal experiences during its examination and are less frequent when the animal is kept in the cowshed, unaware of being observed. Some animals are reluctant to go through doorways and appear to jump over the smallest of obstacles.

33.5.2 Changes in sensitivity

BSE cases regularly display hypersensitivity to tactile, auditory, and visual stimuli (Table 33.1). If animals are assessed according to the procedure described below, then it will be possible to observe hypersensitivity to tactile stimuli in virtually all affected animals [1; 4]. When carrying out this examination, the response of the animal to different tactile stimuli is checked on the entire surface of the body. Sometimes severe and dangerous self-defensive reactions may be provoked and to ensure the examiner's and animal's safety, the examination should be ceased. Initially, the response of the animal to manipulations is tested: hand stroking of either side of the head and neck of the animal (Fig. 33.1c). The intensity of the stimulus is increased by touching the lateral surfaces of the neck and head as well as the forehead and the mouth with the tip of a ballpoint pen (Fig. 33.1d). Hypersensitive animals try to get away, try to fend the examiner off with their head or horns (head tossing or bobbing) and become increasingly irritated, which expresses itself in salivation, snorting, and trembling. By touching the skin with a ballpoint pen it is possible to assess the animal's sensitivity over the body. To test for hind limb hypersensitivity, the "broom test" or "stick test" is a most adequate and simple method. This test is based on the observation that BSE-affected cattle kick while straw is spread around the animal with a fork or while the hind limbs are touched with a farm tool or during milking. To exploit this reaction in BSE examinations, the examiner touches the fetlocks of the animal's hind limbs several times with a broom or preferably a flexible stick made of PVC to avoid backlash injuries when the animal suddenly kicks out. Hypersensitive animals usually kick out violently. All BSE cases that exhibit hypersensitivity to tactile stimuli show hyperesthesia of the head and neck area. Towards the rear of the animal, the hypersensitivity is less pronounced and it is only rarely found on the entire body.

BSE cases are very often also hypersensitive to sudden noise. This is tested by producing some noise close to the animal, e. g., by knocking on a brass cup with a rubber hammer (Fig. 33.1e) or by clapping the hands loudly. (Ideally, this should not be done directly in front of the animal to prevent an additional visual stimulus created by the movement of the hand.) Hypersensitivity to external stimuli is usually repeatable in BSE cases. Thus, the reaction to noise must be checked several times to assess if habituation occurs. Hypersensitivity to noise can range from very mild (e. g., the cow flinches slightly, usually only with the head, neck and forelimbs at most) to rather severe responses. In the latter case the entire body jerks and the animal pulls at the rope with which it is bound to a pole (Fig. 33.1d; Table 33.1 remark at bottom), or it may struggle in a squeeze chute; it may also react with trembling and salivation.

Hypersensitivity to sudden light changes can be observed in almost 50% of affected cattle. The reaction of the animals can be assessed by keeping them in a dark room. A halogen lamp is directed towards the head of the animal and switched on and off. Alternatively, the reaction can also be tested by using a camera flash, especially if the examination room cannot be darkened [6; 25]. Hypersensitivity to light is expressed by a repeatable mild to severe response, which is similar to the animal's hypersensitivity to noise.

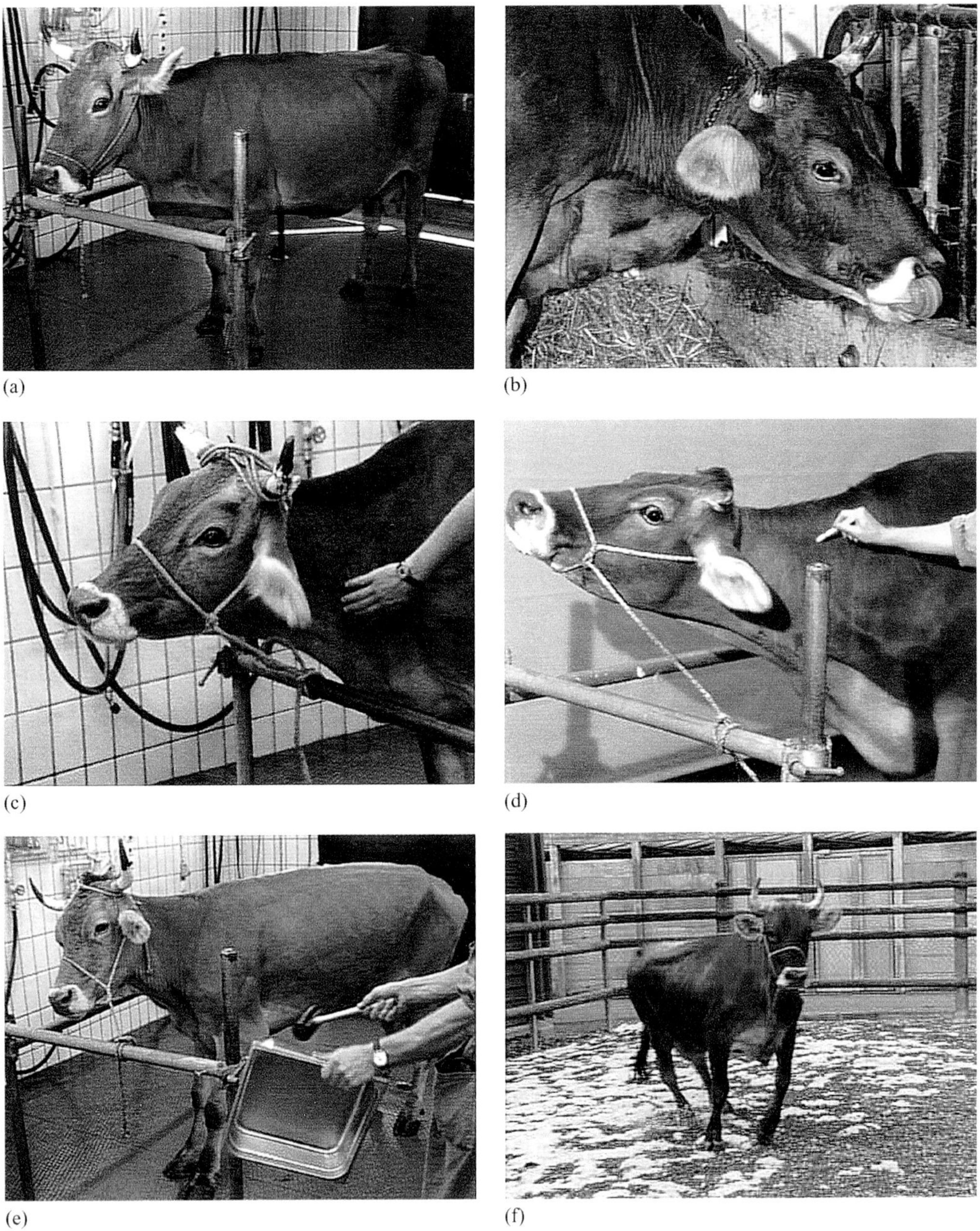

Fig. 33.1: Different neurological signs in BSE-affected cattle: Source: reprints from video material of the University of Zurich, Veterinary Faculty, Department for Farm Animals (E. Schicker and U. Braun ©). **(a)** Attentive cow in the examination room with a nervous and anxious expression. **(b)** Agitated animal with staring eyes and frequent nose licking. **(c)** Strong hypersensitivity to manipulation of the neck; this is expressed in an anxious facial expression, nose wrinkling, shaking of the head. It is impossible to embrace the neck with both arms because of the animal's severe defensive reactions. **(d)** Strong hyperesthesia of the animal when touched with the tip of a ballpoint pen; the animal tries to escape and pulls at the rope; anxious facial expression, nose wrinkling, and salivation. **(e)** Animal with hypersensitivity to noise; this manifests itself in fear and retreat. **(f)** Swaying and staggering when allowed to walk freely (ataxia), a highly conspicuous BSE sign.

In addition, BSE cases usually overreact to sudden movements characterized by startle reactions, panic, or aggressiveness towards the stimulus. This can be assessed directly by testing the menace response or – more remotely – by waving a hand or a clipboard towards the animal [6].

33.5.3 Changes in locomotion

The movements of the animal can be assessed when it is led or when it is allowed to walk freely within a paddock. When the animal is being led from the cowshed to outdoors, attention should be paid to the way the animal crosses the dung trench and how it goes through the doorway. Once outside, the animal should be led to and over small obstacles; variations in its manner of moving should be closely observed. These are even more noticeable if the animal is allowed to walk freely (Fig. 33.1f; Table 33.1).

The majority of affected animals display disorders in locomotion. The most frequent disorder observed is general or hind limb ataxia. Ataxia mostly expresses itself in hypermetria, staggering or swaying [11] (Fig. 33.1f). Difficulty in rising or inability to rise, and subsequent falling due to ataxia are less frequently observed. Some animals may get up in an abnormal way – rising on their forelimbs first. Recumbency can be taken as a sign of a very advanced stage of BSE and is mainly observed in final stages of the disease [11]. Recumbent animals may adopt an abnormal leg posture with one or both hind limbs stretched out behind [26; 27]; "frog sitting" has also been reported [28]. They may also present with traumatic lesions and periarthritis as a result of occasional falling prior to recumbency [29]. Another key sign is "pacing", which can be found in one out of four BSE cases [5].

33.5.4 Additional findings

Changes in kinesthesia are rare. Reflexes and the assessment of the cranial nerves are usually very inconspicuous [1; 4].

33.5.5 Frequency of the neurological findings

In the majority of BSE-affected cattle, changes in behavior, sensitivity, and locomotion can be observed simultaneously; but some BSE cases may only display signs in two of these three categories [1; 4; 6]. Other authors have observed at least one of the three main features: apprehension, hyperesthesia, or ataxia in nearly all affected animals [11]. The pattern of the signs depends on the stage of the disease and also undergoes daily fluctuations, which makes it extremely difficult to define the frequency of certain clinical signs.

The frequency of observed neurological signs depends on the subjective assessment and, as far as abnormal behavior is concerned, also on the stage of the incubation time when the examination is carried out. It could be shown that with increased experience, the examiners were able to diagnose the disease at a much earlier stage. In addition, the way of collecting data seems to be of great importance. Table 33.1 retrospectively summarizes the data provided by groups A, B, and D and includes signs of the disease that were observed during the entire period of its clinical manifestation (observed by veterinarians and farmers). Data obtained by group C, however, was derived from a single standardized examination in a veterinary clinic [1; 4].

There may be considerable variation in the frequency of behavioral signs based on a single standardized examination as demonstrated in Fig. 33.2, which compares the frequency of behavioral abnormalities in two groups of BSE cases observed at the University of Zurich at different times during the epidemic. A change in the clinical presentation of BSE over time is theoretically possible but this is not supported by the uniform pathology (stereotypic lesion profile) over the course of the BSE epidemic [30–31] and only one BSE strain could be isolated under field conditions [32]. It is more likely that differences in the frequency of expressed clinical signs in both groups are attributable to (i) a different clinical stage at the time of the examination (animals in group E had a longer clinical duration as it was decided to euthanize

Table 33.1: Systematic comparison of the frequency of neurological signs in cattle with BSE diagnosed post mortem.

Signs	Frequency (%)			
	Group A n = 36 Switzerland [16]	Group B n = 17,154 UK [11]	Group C n = 50 Switzerland [1; 4]	Group D n = 193 France [41]
Changes in behavior	na	na	96	na
Changes in temperament or character	80	78	na	65
Abnormal behavior	42	63	na	69
Panic-stricken behavior	na	na	66	na
Restlessness or nervousness	75	na	64	82
Apprehension	72	86	66	92
Aggressiveness	47	na	42	36
Sudden kicking	na	na	20	na
Abnormal ear position/ear twitching	25	61	37	46
Staring expression	na	na	14	na
Salivation	na	na	31	na
Bruxism	36	40	47	41
Nose wrinkling and flehmen	na	na	16	na
Licking of nose	31	na	31	25
Nervous of doorways	31	42	13	na
Kicking during milking/ difficulty milking (group D)	44	50	30	40
Muscle tremor	64	68	33	82
Frenzy	11	31	2	8
Changes in sensitivity (Group B: hyperesthesia = hypersensitivity to touch and sound combined; group D: hyperesthesia not specified)	na	75	98	84
Hypersensitive to touch	50	na	98	na
Head shyness	72	37	94	na
Pressing and rubbing the head on objects	17	18	na	na
Impaired vision/blindness	8	2	0	4
Hypersensitivity to noise	50	na	84	na
Hypersensitivity to light	47	na	45	na
Changes in locomotion	na	na	96	na
Abnormal stance	na	na	20	na
Abnormal head carriage	25	48	6	45
Difficulty in rising	19	na	33	na
Recumbency	36	15	10	42
Ataxia	61	77	91	74
Falling (or: collapsing)	47	40	16	69
Fetlock knuckling	19	9	na	na
Paresis	3	8	na	39
Circling	6	3	na	na

n = number of animals examined
na = not available
Editor's note: Examination protocols proposed by different authors may not be directly comparable because of the possible differences in the temperament of breeds and in the housing of cattle (tethered versus loose housing). For example, whilst some (Swiss) cattle breeds may willingly accept being led by a rope or tied to a poll, other breeds, e. g., in France or in the UK, may require restraint in a squeeze chute, and locomotion can only be evaluated when the animal moves freely on pasture or in a yard.

the animals at a later stage, data not shown), (ii) the subjectivity of the examiner (clinician) of certain clinical signs (e. g., apprehension, staring expression), which may be interpreted differently by different observers, as well as better familiarization with the clinical signs of BSE over time and thus greater attention to specific clinical signs, such as nose licking, nose wrinkling, and flehmen.

33.6 Laboratory findings

Hematological and blood chemical findings: Hematology and blood biochemistry are usually unremarkable in BSE cases; deviations from the normal values are generally associated with secondary diseases and unspecific for BSE [33; 34]. A chemical analysis of the blood will allow the diagnostic differentiation of BSE from magnesium deficiency. In the early stage of hypomagnesemia (grass tetany) with serum magnesium concentrations of < 0.8 mmol/l, panic-stricken behavior and increased excitability may be observed [1; 4].

It is of note that the diagnosis of hypomagnesemia or acetonemia does not necessarily rule out the absence of BSE. Acetonemia may be present in BSE cases and a veterinarian may be tempted to diagnose "nervous ketosis" in a BSE case, in particular if BSE is unexpected (T. Konold, unpublished observation). The repeated unsuccessful treatment of a metabolic disease with neurological signs should alert the veterinarian that the underlying disease may be of a different neurological disease. In these situations BSE is highly probable.

CSF findings: The analysis of CSF only allows the differentiation of BSE from inflammation caused by viral and bacterial (also fungal, even if very rare) infections and parasitic infestation of the central nervous system, e. g., listeriosis [35; 36]. However, on rare occasions, BSE may be accompanied by listeriosis or any other type of encephalitis [6], which possibly alter the clinical presentation (e. g., cranial nerve deficits in listeriosis in addition to signs of BSE). An examination of CSF from BSE-affected cattle showed that cytology and protein content was similar to that of healthy cattle [1; 4]. It had been shown that the 14–3–3 test [37] (see Chapter 31) was unable to produce statistically significant differences between normal and BSE-infected animals [38]. Thus, its application in buiatrics cannot be justified.

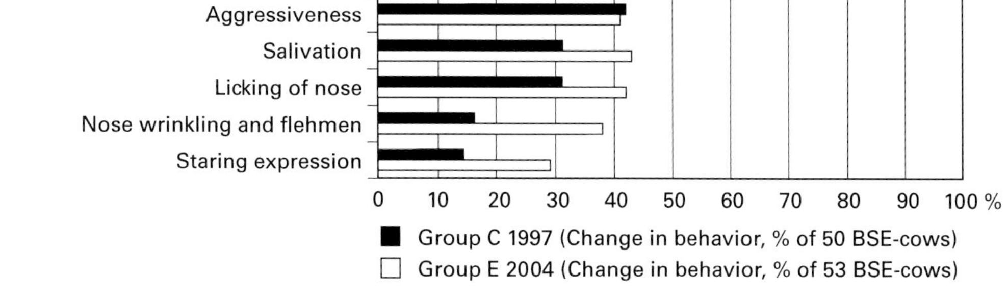

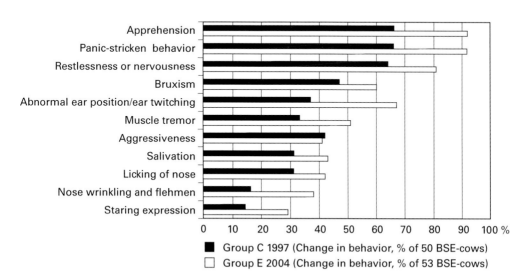

Fig. 33.2: Variation in the frequency of behavioral signs based on a single standardized examination. Group C; n = 50 cows, study by E. Schicker 1997 [1] and group E; n = 53 cows, study by S. Hauri 2004 [42]. Both clinical studies were supervised under the same leading examiner, U. Braun, at the Vetsuisse Faculty of the University of Zurich. Original figure by B. Hörnlimann.

33.7 Examination at abattoirs – a diagnostic challenge

It has been demonstrated that a clinical diagnosis of BSE can usually be made following the specific clinical and neurological examination of cattle for signs of BSE as described above. Only a small percentage of cattle suffering from other conditions, display BSE-like behavioral, sensory and/or locomotor changes with similar frequency and severity [3; 9; 27; 39]. The method has its limitations when animals are examined at abattoirs, where a full clinical and neurological examination is impractical [39], or when animals are recumbent and clinical history is inadequate to confirm the presence of locomotor changes prior to recumbency.

A simplified approach has been proposed for the ante mortem inspection [40]. The presence of four out of the seven following clinical signs is suggestive of BSE:

- Aggressiveness
- Teeth grinding
- Staring eyes
- Reduced rumination
- Recumbency
- Overreactivity
- Difficulty in rising to a standing position.

Casualty slaughter cattle represent the group with the highest BSE risk. Moreover, those animals that display behavioral, sensory, or neurogenous locomotor abnormalities should always be regarded as potential BSE cases, in particular if this is accompanied by poor bodily condition.

All cattle with a high risk to be infected with the BSE agent (see Chapter 53.3.3) should be tested additionally by a rapid BSE test.

References

[1] Schicker E. Klinische Befunde bei Kühen mit boviner spongiformer Enzephalopathie (BSE). DVM thesis. Veterinary Faculty of the University of Zurich, 1997.

[2] Braun U, Pusterla N, Schicker E. Clinical findings in cattle with bovine spongiform encephalopathy (BSE). Video, available in different languages on webpage 2006: www.bse.unizh.ch/english/video/content.htm.

[3] Braun U, Schicker E, Hörnlimann B. Diagnostic reliability of clinical signs in cows with suspected bovine spongiform encephalopathy. Vet Rec 1998; 143:101–105.

[4] Braun U, Schicker E, Pusterla N. Clinical findings in 50 cows with bovine spongiform encephalopathy (BSE). Berl Münch Tierärztl Wochenschr 1998; 111(1):27–32.

[5] Braun U, Gerspach C, Ryhner T, et al. Pacing as a clinical sign in cattle with bovine spongiform encephalopathy. Vet Rec 2004; 155(14):420–422.

[6] Konold T, Bone G, Ryder S, et al. Clinical findings in 78 suspected cases of bovine spongiform encephalopathy in Great Britain. Vet Rec 2004; 155(21):659–666.

[7] Austin AR, Hawkins SAC, Kelay NS, et al. New observations on the clinical signs of BSE and scrapie. In: Bradley R, Marchant B, editors. Transmissible spongiform encephalopathies. EU: VI/4131/94-EN Document Ref. F. II.3 – JC/0003 Brüssel, 1994:277–287.

[8] Scott PR. Detailed neurological examination of suspected cases of bovine spongiform encephalopathy (BSE) – PhD thesis. University of Edinburgh, Edinburgh, 1990.

[9] Braun U, Amrein E, Estermann U, et al. Reliability of a diagnosis of BSE made on the basis of clinical signs. Vet Rec 1999; 145(7):198–200.

[10] Defra. BSE: Statistics – Confirmed cases of BSE in GB by year of birth where known. Webpage 2006: www.defra.gov.uk/animalh/bse/statistics/bse/yrbirth.html.

[11] Wilesmith JW, Hoinville LJ, Ryan JB, et al. Bovine spongiform encephalopathy: aspects of the clinical picture and analyses of possible changes 1986–1990. Vet Rec 1992; 130:197–201.

[12] Wilesmith JW and Ryan JB. Bovine spongiform encephalopathy: recent observations on the age-specific incidences. Vet Rec 1992; 130:491–492.

[13] Wilesmith JW, Wells GAH, Cranwell MP, et al. Bovine spongiform encephalopathy: epidemiological studies. Vet Rec 1988; 123:638–644.

[14] Wells GAH, Scott AC, Johnson CT, et al. A novel progressive spongiform encephalopathy in cattle. Vet Rec 1987; 121:419–420.

[15] Cranwell MP, Hancock RD, Hindson J, et al. Bovine spongiform encephalopathy. Vet Rec 1988; 122:190.

[16] Hörnlimann B and Braun U. BSE: Clinical signs in Swiss BSE cases. In: Bradley R, Marchant B, editors. Transmissible spongiform encephalopathies. EU: VI/4131/94-EN Document Ref. F. II.3 – JC/0003 Brussels, 1994:289–299.

[17] Winter MH, Aldridge BM, Scott PR, et al. Occurrence of 14 cases of bovine spongiform encephalopathy in a closed dairy herd. Br Vet J 1989; 145:191–194.
[18] Foley GL and Zachary JF. Rabies-induced spongiform change and encephalitis in a heifer. Vet Pathol 1995; 32:309–311.
[19] Saegerman C, Claes L, Dewaele A, et al. Differential diagnosis of neurologically expressed disorders in Western European cattle. Rev Sci Tech Off Int Epiz 2003; 22(1):83–102.
[20] McGill IS and Wells GAH. Neuropathological findings in cattle with clinically suspect but histologically unconfirmed bovine spongiform encephalopathy (BSE). J Comp Pathol 1993; 108:241–260.
[21] Agerholm JS, Tegtmeier CL, Nielsen TK. Survey of laboratory findings in suspected cases of bovine spongiform encephalopathy in Denmark from 1990 to 2000. APMIS 2002; 110(1):54–60.
[22] Miyashita M, Stierstorfer B, Schmahl W. Neuropathological findings in brains of Bavarian cattle clinically suspected of bovine spongiform encephalopathy. J Vet Med B Infect Dis Vet Public Health 2004; 51(5):209–215.
[23] Saegerman C, Speybroeck N, Roels S, et al. Decision support tools for clinical diagnosis of disease in cows with suspected bovine spongiform encephalopathy. J Clin Microbiol 2004; 42(1): 172–178.
[24] Austin AR, Pawson L, Meek SC, et al. Abnormalities of heart rate and rhythm in bovine spongiform encephalopathy. Vet Rec 1997; 141:352–357.
[25] Braun U. Clinical signs and diagnosis of BSE. Schweiz Arch Tierheilkd 2002; 144(12):645–652.
[26] McElroy MC and Weavers ED. Clinical presentation of bovine spongiform encephalopathy in the Republic of Ireland. Vet Rec 2001; 149(24): 747–748.
[27] Konold T, Sivam SK, Ryan J, et al. Analysis of clinical signs associated with bovine spongiform encephalopathy in casualty slaughter cattle. Vet J 2006; 171(3):438–444.
[28] Collin E. Troubles nerveux chez les bovins et ESB. Le Point Vet 2002; 33(229):40–41.
[29] van Wuijckhuise L, Vellema P, Terbijhe RJ. BSE: clinical diagnosis and field experience. Tijdschr Diergeneeskd 2001; 126(8):279–281.
[30] Wells GAH and Wilesmith JW. Bovine spongiform encephalopathy and related diseases. In: Stanley Prusiner, editor. Prion biology and diseases, 2nd edn. Cold Spring Harbor Laboratory Press, New York, 2004:595–628.
[31] Wells GAH and Wilesmith JW. The neuropathology and epidemiology of bovine spongiform encephalopathy. Brain Pathol 1995; 5:91–103.
[32] Bruce ME, Chree A, McConnell I, et al. 1994. Transmission of bovine spongiform encephalopathy and scrapie to mice: strain variation and the species barrier. Philos Trans R Soc Lond B Bio Sci 1994; 343:405–411.
[33] Johnson CT and Whitaker CJ. Bovine spongiform encephalopathy. Vet Rec 1988; 122:142.
[34] Scott PR, Aldridge BM, Clarke M, et al. Bovine spongiform encephalopathy in a cow in the United Kingdom. J Am Vet Med Assoc 1989; 195:1745–1747.
[35] Scott PR. The collection and analysis of cerebrospinal fluid as an aid to diagnosis in ruminant neurological disease. Br Vet J 1995; 151:603–614.
[36] Scott PR, Aldridge BM, Clarke M, et al. Cerebrospinal fluid studies in normal cows and cases of bovine spongiform encephalopathy. Br Vet J 1990; 146:88–90.
[37] Hsich G, Kenney K, Gibbs CJ Jr, et al. The 14–3–3 brain protein in cerebrospinal fluid as a marker for transmissible spongiform encephalopathies. N Engl J Med 1996; 335(13):924–930.
[38] Robey WG, Jackson R, Walters RL, et al. Use of cerebrospinal fluid levels of 14–3–3 in predicting neuredegeneration in confirmed BSE symptomatic cattle. Vet Rec 1998; 143:50–51.
[39] Nowotni A, Wendel H, Klee W. Clinical examination of cattle for BSE in a cattle facility and slaughterhouse. Dtsch Tierärztl Wochenschr 2004; 111(1):5–7.
[40] Hett AR, Rufenacht J, Perler L, et al. A method to identify BSE suspects in emergency slaughtered cattle during ante mortem examination. Schweiz Arch Tierheilkd 2002; 144(12):654–662.
[41] Cazeau G, Ducrot C, Morignat E, et al. Surveillance clinique de l'ESB en France – Etude quantitative des caractéristiques des animaux et des signes cliniques observés en cas de suspicion. Rev Med Vet 2002; 153:785–794.
[43] Hauri S. Untersuchungen bei 53 Kühen mit Boviner Spongiformer Enzephalopathie (BSE). DVM thesis. Veterinary Faculty of the University of Zurich 2004. Webpage 2006: www.dissertationen.unizh.ch/2004/hauri/diss.pdf.

34 Clinical Findings in Scrapie

Martha J. Ulvund

34.1 Introduction

This chapter will describe mainly the clinical signs of sheep scrapie, but deals also, briefly and comparatively, with scrapie in goats [1]. The clinical signs of scrapie vary widely among individual sheep. In recent years it has become clear that the natural occurrence of sheep scrapie is highly influenced by polymorphisms in the ovine PrP gene (*Prnp*). Such polymorphisms determine susceptibility to scrapie agents and are among the factors that effect expression of the disease. In Suffolk sheep and other breeds, homozygosity for the *Prnp* allele encoding glutamine-171 (Q_{171}) renders sheep susceptible to natural scrapie [2; 3]. In so-called valine-breeds (encoding valine (V) at codon 136), such as Cheviot and Texel, the $V_{136}R_{154}Q_{171}$ allele is primarily associated with high susceptibility to scrapie [4] (see Chapter 56). Sheep with the AHQ allele are less susceptible to scrapie than ARQ, but more resistant than ARR, which generally renders sheep even more resistant to scrapie, the most resistant sheep being homozygous ARR/ARR [5].

Recently a variant form of scrapie (Nor98) with unusual features was described in PrP genotypes rarely associated with scrapie previously [6]. The Nor98 cases were associated with polymorphisms at codons 141 (phenylalanine) and 154 (histidine); none of the cases encoded the $V_{136}R_{154}Q_{171}$ allele [7]. Similar so-called "atypical" scrapie cases were subsequently found in Germany and elsewhere [8–11]. Differences between the previous forms, now termed classical scrapie, and the Nor98 form include the vacuolation and PrP^{Sc} distribution patterns, the Western blot pattern, and the clinical signs [6; 12]. The influence of the PrP gene on the clinical signs in atypical–Nor98 and in classical scrapie has, however, not been fully elucidated.

Many sheep that suffer from scrapie display signs of changed behavior, ataxia, pruritus, and weight loss [13]. While clinical signs of BSE have been studied extensively (Table 33.1, Fig. 33.1), comprehensive studies of the clinical signs of scrapie in relation to PrP genotypes, are rather limited. Parry [13], who has produced the most comprehensive review of the clinical signs of scrapie to date, carried out his studies before the significance of the PrP genotypes on signs and PrP^{Sc} deposition within the brain were known. The latest detailed clinical information on sheep scrapie relates to studies in Ireland [14].

The variability of the clinical disease according to breed, flock, region, and country presents considerable difficulties to the conduct of definitive studies of sheep with scrapie. Scrapie may, for example, manifest itself in one region as a nervous form, characterized by ataxia, and in another region as a pruritic form. The significance of different strains of the scrapie agent, and their association with PrP genotypes, is largely unknown in sheep. The general importance and influence of scrapie strains on clinical signs are dealt with in Chapter 12. In contrast, in BSE, considered to be caused by one main pathogen strain, clinical expression is far less variable (see Chapter 33). The clinical diagnosis of scrapie is often difficult because none of the clinical signs, singularly or in combination, are definite proof of the disease. Table 34.1 shows the results of a survey of important clinical signs of scrapie in sheep and goats (on initial examination) in order of percentage frequency (%) of occurrence, representing 129 Irish, 64 Norwegian, 24 Spanish and 550 Italian sheep, and

500 Italian goat scrapie cases. As can be seen from the table, hind limb ataxia and head tremor were the initial most prominent signs in Ireland, while altered mental status, pruritus, and wool loss were the initial most prominent signs among Norwegian and Spanish sheep. In Italian sheep, both hind limb ataxia and pruritus were seen frequently, while in goats with scrapie, biting, aggressiveness, difficulty with milking, etc. were seen, in addition to pruritus and ataxia (Table 33.1).

Among 18 cases of Nor98 scrapie, eight of the positive samples originated from clinically healthy, slaughtered animals, three from fallen stock that probably died of causes other than scrapie, and seven from animals with clinical signs dominated by ataxia. None had pruritus [6; 12]. Incoordination, circling, nervousness and a hopping gait were reported in two atypical Irish cases [10].

Although sheep are susceptible to experimental BSE, with clinical signs similar to scrapie [15], confirmed cases of BSE have never to date been detected naturally in sheep. BSE has recently been detected in an asymptomatic French goat [16] in the course of surveillance. There are no clinical signs which clearly differentiate experimental BSE in sheep from natural scrapie. However, there is a tendency that experimental BSE in sheep presents as sudden-onset ataxia in the absence of pruritus [15; 17], although intense pruritus and ataxia typical of scrapie were reported in one sheep [18].

Scrapie in goats [1] may appear similar to scrapie in sheep, the most common signs being hyperesthesia, ataxia, and pruritus, but in Italian milking goats, the first signs were reluctance to be milked and a defensive type of aggression, followed by allotriophagia and biting [19].

Several other diseases may display signs similar to scrapie (Table 34.2) and therefore, if scrapie is suspected on the basis of the clinical examination, the diagnosis must be confirmed by necropsy and pathological examinations including histology, and immunohistochemical and immunochemical methods.

34.2 Case history

The clinical approach to sheep and goats believed to be suffering from scrapie should include careful collection of the clinical history and the conduct of a detailed clinical examination in order to exclude the possibility of other diseases. Whether or not suspect cases of scrapie are reported depends on the awareness and interest of the farmer and his willingness to contact a veterinarian to attend diseased sheep [20].

Signs of the disease usually appear between the age of 2 and 5 years, and very often at about 3 years of age [14; 19; 21–24]. Clinical signs have very rarely been observed in animals over 10 years of age [25; 26] and in a small number of animals, signs have been found as early as seven months of age [22]. Sheep with VRQ homozygosity seem to have the shortest incubation period [27; 28]. The mean age at which scrapie signs appear in Nor98 cases is 6 years [11]. It is particularly difficult to identify scrapie in the early stages of the disease. According to Parry [13], slight restlessness, a fixed stare (Fig. 34.1a) or reluctance to walk forward when the animals are driven, can be regarded as early signs of scrapie. Subtle changes in behavior, or paresthesia, are often only recognized by shepherds or animal attendants who know the animals very well [13]. Video surveillance has proven useful in detecting very early signs of experimental scrapie in sheep and in studying suspect field cases in the laboratory setting [27; 29].

Usually, only one or a very small number of animals within a flock are diagnosed with clinical signs at any one time, but may herald a greater incidence of pre-clinical cases. For example, examination of the distribution of PrP^{Sc} in nervous and lymphoid tissues of sheep in a naturally infected scrapie flock with two clinical cases at the time of slaughter of the whole flock, revealed nine times as many animals (18 sheep) with PrP^{Sc} deposits in nervous and or lymphoid tissues; all cases being of susceptible PrP genotypes (at least one VRQ allele) [29].

(a)

(b)

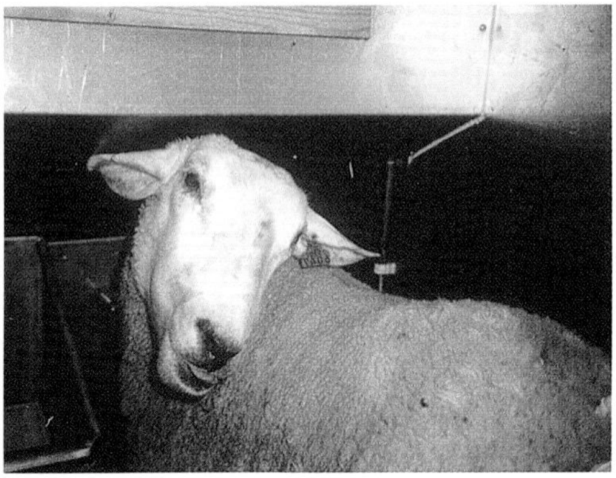

(c)

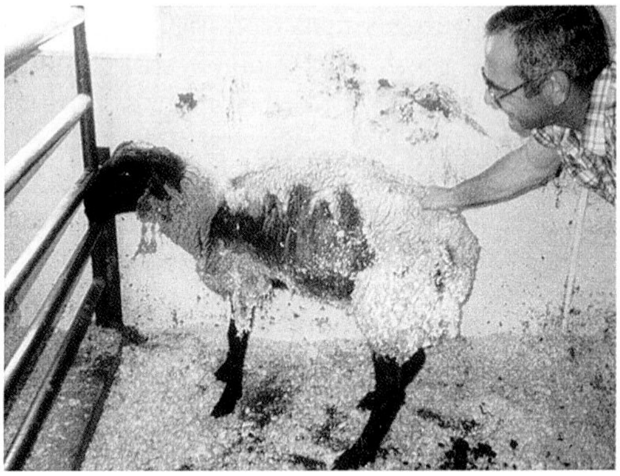

(d)

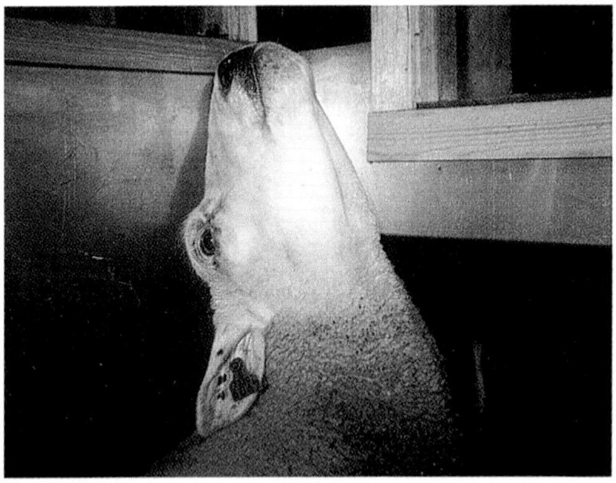

(e)

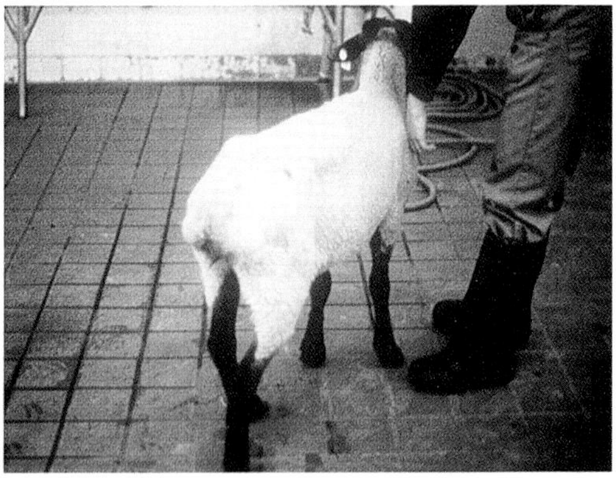

(f)

34.3 Findings of the general clinical examination

Body temperature, respiratory, and heart rates are most often normal [14]. During the examination, the animals may show signs of stress with tachycardia and irregular heartbeat [13]. In the early phase of the disease, appetite and bodily condition are not affected. Rumen activity is often normal, although rumenal tympany is occasionally observed [13]. Sheep may spend less time eating forage, and may regurgitate and masticate less frequently [30; 31]. Some animals have been reported with a distended abdomen resulting from abomasal impaction [32; 33]. In a recent study, one out of six sheep with abomasal distention had PrP^{Sc} deposits [34] in the coeliacomesenteric ganglion, but the authors regarded abomasal distention to be an acquired form of dysautonomia, not necessarily associated with scrapie [35]. Polydipsia and polyuria may also occur [13].

The subtle onset and slow clinical progression of the disease largely prevents an early diagnosis, and thus the implementation of measures to counteract the spread of scrapie. Farmers and veterinarians who notify the authorities of suspected scrapie at the earliest possible opportunity contribute considerably to reducing the spread of the disease. Animals that are infected with scrapie can, seemingly in contrast to BSE infected cattle, excrete the pathogen, contaminate the environment and thus constitute a source of subsequent infections (see Chapter 42). If no measures are taken to intervene, the morbidity or within-flock incident rate will increase – whereas the incubation period may decrease – over the years as a gradually increasing "infection pressure" builds up on an affected farm [29] (see Chapter 18.6). Suspect cases should, if possible, be PrP genotyped, and lymphoid biopsies taken for diagnostic examination (see below).

◀ **Fig. 34.1:** Diverse neurological abnormalities observed in scrapie-infected sheep. **(a)** Highly attentive sheep with anxious and fearful appearance. Even during the prodromal and early stage of the disease, animals may isolate themselves from the rest of the flock. **(b)** Scrapie sheep with pruritus (early or middle stage), manifest as rubbing of the body against objects. **(c)** Nibble reflex; nibbling movements of the lips and nibbling reaction of the tongue seen when the back of the animal is scratched by the examiner. **(d)** Associated with extreme pruritus, animals, if scratched on the back, display not only the nibble reflex but thrust the back towards the examiner, seemingly obtaining pleasure from the experience. This animal, in the medium to late scrapie stage, also shows extensive wool loss. The wool that is still present is soiled and matted. **(e)** Sometimes, scrapie sheep stretch their head upwards, spontaneously, or when scratched on the back. **(f)** In the late stage of the disease, emaciation becomes very noticeable (chronic wasting). Photos by M. J. Ulvund (a,c,e) and L. Detwiler (b,d,f) ©.

34.4 Differential diagnoses

Abnormal behavior can also be one of the early signs of Borna disease, characterized by animals separating themselves from the flock, and sometimes combined with jumpiness. Drowsiness, rocking gait, somnolence or excitability, and tooth grinding may all be signs of neurosis resulting from causes other than scrapie. Isolation from the flock, insecure gait, and inability or difficulty in rising may be caused by botulism. Anxiety may be an initial sign of rabies in sheep or goats. Signs of paresthesia associated with itching and muscle twitching can be observed in Aujeszky's disease, a disease characterized by hypersensitivity and paralysis. Pruritus (Fig. 34.1b), wool loss or skin damage occur in parasitic skin diseases; for example, the infestation with Chorioptes mites leads to extreme pruritus. Pruritus, wool loss, and emaciation (Fig. 34.1f) may indicate Psoroptic mange. Sarcoptic mange is usually only observed on the head. Infestation with biting or sucking lice (*Damalinia ovis, Linognathus spp.*), sheep keds (*Melophagus ovinus*), and occasionally also trichophytia (dermatomycosis) may cause pruritus. Itching may, in rare cases, be observed in dermatophilosis (*Dermatophilus congolensis*) and also photosensitization [36–40].

Changes in posture and gait are observed in many sheep diseases. Stumbling gait, dragging of the hind legs, pareses and paralyses, and emaciation despite regular appetite, may suggest visna. Circling and conjunctivitis may in-

Table 34.1 Clinical signs of scrapie in sheep (on initial examination) in order of frequency (%) from Irish (129 cases), Norwegian (64 cases), Spanish (24 cases) and Italian (550 cases) data and signs of scrapie in 500 Italian goats [14; 19; 24; 56].

Clinical signs	Ireland n = 129	Norway n = 64	Spain n = 24	Italy (sheep) n = 550	Italy (goats) n = 500
General signs					
Depression	15	10	54	–	–
Normal to good appetite	–	12	100	–	–
Reduced appetite	–	4	–	–	–
Dirty wool	–	17	–	–	–
Wool loss	19	29	71	–	–
Abomasal dilatation	–	2	–	–	–
Cardiac arrhythmia	–	–	50	–	–
Regurgitation	–	–	–	23	18
Changes in behavior					
Head tremor	61	23	42	57	40
Altered mental status	57	54	71	–	–
Nibble response (reflex)	51	14	46	10	17
Teeth grinding	44	2	42	–	–
Altered head carriage	41	10	–	–	–
Hyperresponsive	28	29	50	63	66
Abnormal micturition	16	–	–	–	–
Anxious, apprehensive	13	–	–	78	67
Salivation	7	4	–	13	15
Aggressiveness	–	–	–	–	70
Reluctance to be milked	–	–	–	83	85
Changes in sensitivity					
Pruritus	25	47	71	83	85
"Cannibalism"	–	–	–	–	3
Pica*	–	–	–	–	55
Biting	–	–	–	–	65
Changes in locomotion					
Hind limb ataxia	71	27	54	81	64
Dysmetria	50	36	–	–	–
Abnormal posture	51	10	–	–	–
Hind limb weakness	36	27	42	–	–
Circling	–	1	–	–	–
Other signs					
Thin, BCS <1.5	38	31	58	22	85
Labial edema	–	–	–	43	8
Visual impairment	–	–	–	64	53
Pyknolepsy**	–	–	–	72	65
Hypogalactia (reduced milk yield)	–	–	–	75	57

* Allotriophagia / A craving to lick, bite or eat whatever object is around, including nonfood items, such as dirt, paint chips, and clay.

** Brief epileptiform attacks.

Table 34.2: Diseases to be considered in the differential clinical diagnosis of scrapie.

Group of signs	Disease
Behavioral disorders/ abnormal behavior	Borna disease Coenurosis (gid) Botulism Rabies Bacterial meningoencephalitis Ryegrass staggers
Paresthesia	Aujeszky's disease (pseudorabies) Chorioptic mange (limbs) Psoroptic mange (sheep scab) Sarcoptic mange (head, limbs) Louse infestation (pediculosis) Sheep keds (*Malophagus ovinus*) Dermatophilosis Photosensitization Cutaneous myiasis (strike) Forage mites Flea infestation
Abnormal posture and gait	Visna Listeriosis
Diverse combinations of signs	Louping ill Rumen acidosis Hypocalcemia Hypomagnesemia Ketosis (pregnancy toxemia) Polioencephalomalacia (cerebrocortical necrosis, CCN) Brain abscess Hemorrhage
Slowly progressive emaciation	Maedi Pulmonary adenomatosis (Jaagsiekte) Paratuberculosis (Johne's disease) Pseudotuberculosis Distomatosis Helminthiasis Dental disease Protein-energy malnutrition Cobalt deficiency Abomasal impaction Neoplasia

dicate listeriosis. If ataxia, a hopping, or rocking gait, and tremor are observed, louping ill may be suspected. Other differential diagnoses include cerebrocortical necrosis (CCN, polioencephalomalacia), brain abscesses and tumors, meningitides, and other brain lesions. Motor disorders and nervousness associated with depression may occur in combination with rumen acidosis, hypocalcemia, hypomagnesemia, and ketosis [36; 37; 39].

If emaciation and weakness (chronic wasting) continue to progress gradually, maedi disease, pulmonary adenomatosis (Jaagsiekte), paratuberculosis, and pseudotuberculosis must be excluded as potential differential diagnoses. Loss of wool is common in animals suffering from the chronic form of distomatosis; these animals also gradually lose weight despite unaltered feeding habits (Table 34.2).

34.5 Course of the disease

The neurological signs may vary depending on the stage of the clinical disease. The expression of clinical signs may be initially intermittent, or precipitated, or accelerated by stressful situations.

34.5.1 Typical clinical course of disease

The clinical signs usually progress slowly over several weeks to months, and are not alleviated by "symptomatic therapy" [25; 41], but scrapie has also been confirmed in cases with an apparently acute onset, and short duration. Affected sheep may also be found recumbent or dead [14; 21; 42]. In very rare cases, the final phase of the disease lasts more than a year and is associated with intermittent remission of the signs [13]. According to Parry, the progression of the clinical signs can be divided into three typical phases:

1. An early stage characterized by behavioral disorders (Fig. 34.1a), susceptibility to stress, rocking gait, frequent drinking, and slight itchiness (Fig. 34.1b), for example at the base of the tail.
2. An intermediary stage with weakness, ataxia, tongue play and lip tremor (Fig. 34.1c), marked itchiness and loss of wool (Fig. 34.1d) as well as occasional head lifting (Fig. 34.1e).
3. A late stage associated with emaciation (Fig. 34.1f), pruritus, uncoordinated movements, tremor, and recumbency or difficulty in standing.

Although at the initial stage of the disease, central nervous system deficits are the most frequent signs observed, some animals first display signs of pruritus and loss of wool (Table 34.1). While some sick animals display signs of behavioral, sensory, and motor disorders, others display only behavioral abnormalities with, or without, pruritus [39; 40; 43; 44], or just motor disorders. In occasional cases, posterior ataxia may be the only sign observed. In the final stage of the disease, sick animals lose a lot of weight and are finally no longer able to stand.

According to Healy et al. [14], the signs which increased most in frequency over time were pruritus, wool loss, nibble reflex, teeth grinding, head tremor, hyper-responsiveness, ataxia, and reduced body condition score (BCS < 1.5). The presence of a nibble reflex was strongly associated with the *Prnp* alleles ARQ and ARH. Pruritus was negatively associated with ataxia, and pruritus and teeth grinding were associated with a positive nibble reflex [14].

The within-flock incidence rate depends on the genotype; it varies in the majority between 0.5 and 5%. In one report, 50% was observed in Iceland [22]. As in all prion diseases scrapie is invariably lethal in all animals that reach the clinical stage of the disease [38].

34.5.2 Unusual course of disease

According to Parry [13] and Joubert [45], the paralytic form of scrapie associated with rapidly progressive weakness, resembles acute myasthenia. Affected sheep become recumbent early in the course of the disease and are unable to rise, with death occuring within two weeks [46]. Sudden death of scrapie-infected animals, without premonitory signs has been described in Shetland and Japan [21; 47]. Some scrapie-infected animals, in particular older ones, may display a slowly progressive weakeness of long duration [13].

34.5.3 Influence of stress on the course of disease

The clinical signs can be provoked or worsened through stress [19; 48] (see Chapter 33). Transportation, handling, driving, parturition and nutritional stress of late pregnancy, high environmental temperatures, and concurrent illnesses may accelerate the clinical signs. Scrapie-infected animals often display signs of ataxia or paraplegia shortly after lambing [13]. During transport to the slaughterhouse, a combination of stress factors can affect the animals. Consequently, the loading ramp at the slaughterhouse is an ideal site for random clinical scrapie surveillance (see Chapter 32.2.1). If stress is eliminated, certain signs may diminish in intensity. This phenomenon is also true of BSE in cattle [49] (see Chapter 33).

34.6 Findings of the neurological examination

34.6.1 Behavioral disorders

Sheep breeders are often the first to detect certain behavioral disorders. In the field, one may observe that scrapie-infected animals stay back when they are driven. Inexperienced observers will have difficulties identifying subtle signs such as general nervousness, restlessness, slight trembling and stretching of the head. In the form of scrapie associated with apathy, or drowsiness, the head is held pointed downwards, and the ears are held in an abnormal position. Significant behavioral disorders may include nervousness, aggression, teeth grinding, and salivation. Scratching (Fig. 34.1b) or excessive licking of affected body parts are direct causes of pruritus (see below), as are nibbling at the body, biting, or sudden "snapping" movements at their own legs. In some cases, dullness, sleepiness, and apathy are observed [13].

34.6.2 Paresthesia

Paresthesia is often recognized as pruritus with itching, scratching, and bite reflexes which in turn lead to bilateral wool loss at the thighs, flanks (Fig. 34.1d), back, legs, tail base and head. The scratching of body parts against different objects may lead to wool loss, skin lesions and swellings, including hematomas on the ears. Loss of wool may also occur without pruritus.

The "nibble response", or " – reflex" is defined as the induction of rhythmic head movements, lip movements, or both, with or without rubbing of the body in response to vigorous palpation or scratching by the examiner over the back from the lumbar region to the shoulder (Fig. 34.1d). This response is present in 10–50% of cases (Table 34.1). Head tremor is common. Some animals also reveal hyper-responsiveness to touch and manipulation.

Hyper-responsiveness to sudden noise (clapping of hands) may cause the animal to wince or startle [50]. Lactating animals kick when attempts are made to milk them [19].

34.6.3 Posture and movement disorders

Motor disorders often become apparent as a gradually altered, trotting-like gait with a strange lifting of the front legs and a stumbling over objects, a feature by which the disease is called Traberkrankheit in German, from the verb to trot – traben [44]. Affected animals gradually lose the ability to gallop and may display a crouching gait, bunny hopping, hyper- or hypometria, and incoordination. There may also be weakness of the hind legs or the dragging of the limbs, and difficulties in rising. In the final stage of the disease, the animals may suddenly fall and be unable to rise [13]. Videotape recording has been used as an effective tool in assessing gait [14].

34.6.4 Additional findings

In some cases, blindness or a "trembling" vocalization may be observed [13; 19; 22] (Table 34.1)

34.7 Laboratory findings

Scrapie can only be diagnosed with certainty after death (see Chapters 27, 32) by examination of brain material. Diagnostic methods include the histopathological demonstration of spongiform encephalopathy and the immunohistochemical (IHC) and/or immunoblot (Western blot, dot blot) demonstration of PrP^{Sc}. In certain PrP genotypes, biopsies of lymphatic tissues to detect PrP^{Sc} allow the in vivo diagnosis of the disease (live tests). Relevant tissues for biopsies are tonsils, [28; 34; 51], third eyelid [28; 52], and rectum [53].

No serological test is available for the diagnosis because the scrapie agent does not elicit any specific immune response. Detection of abnormal prion protein in the blood of sheep with scrapie by using capillary electrophoresis immunoassay has been reported [54].

Evaluations of non-specific analytes, such as endogenous metabolites, hormones or enzymes as markers for scrapie infection have been performed and significant elevations of plasma 20 β-dihydrocortisol and urine creatinine were found in scrapie affected sheep compared with healthy animals [55].

References

[1] Wood JNL, Done SH, Pritchard GC, et al. Natural scrapie in goats: case histories and clinical signs. Vet Rec 1992; 131:66–68.

[2] Westaway D, Zuliani V, Cooper CM, et al. Homozygosity for prion protein alleles encoding glutamine-171 renders sheep susceptible to natural scrapie. Genes Dev 1994; 8:959–969.

[3] Hunter N, Moore L, Hosie BD, et al. Association between natural scrapie and PrP genotype in a flock of Suffolk sheep in Scotland. Vet Rec 1999; 140:59–63.

[4] Goldmann W, Hunter N, Benson G, et al. Different scrapie-associated fibril proteins (PrP) are encoded by lines of sheep selected for different alleles of the Sip gene. J Gen Virol 1991; 72:2411–2417.

[5] Tranulis MA, Osland A, Bratberg B, et al. Prion protein gene polymorphisms in sheep with natural scrapie and healthy controls in Norway. J Gen Virol 1999; 80:1073–1077.

[6] Benestad SL, Sarradin P, Thu B, et al. Cases of scrapie with unusual features in Norway and designation of a new type, Nor98. Vet Rec 2003; 153:202–208.

[7] Moum T, Olsaker I, Hopp P, et al. Polymorphisms at codons 141 and 154 in the ovine prion protein gene are associated with scrapie Nor98 cases. J Gen Virol 2005; 86:231–235.

[8] Buschmann A, Lühken G, Schultz J, et al. Neuronal accumulation of abnormal prion protein in sheep carrying a scrapie-resistant genotype (PrPARR/ARR). J Gen Virol 2004; 85:2727–2733.

[9] Gavier-Widen D, Noremark M, Benestad SL, et al. Recognition of the Nor98 variant of scrapie in the Swedish sheep population. J Vet Diagn Invest 2004; 16:562–567.
[10] Onnasch H, Gunn HM, Bradshaw BJ, et al. Two Irish cases of scrapie resembling Nor98. Vet Rec 2004; 155:636–637.
[11] Bosschere H, De Roels S, Benestad SL, et al. Scrapie case similar to Nor98 diagnosed in Belgium via active surveillance. Vet Rec 2004; 155:707–708.
[12] Bratberg B, Benestad SL, Sarradin P, et al. An unusual form of scrapie, Nor98. Conference on methods for control of scrapie. 15 & 16 May 2003; Programme & abstracts: page 14, Oslo, Norway.
[13] Parry HB. In: Oppenheimer DR, editor. Scrapie disease in sheep. Academic Press, London, 1983.
[14] Healy AM, Weavers E, McElroy M, et al. The clinical neurology of scrapie in Irish sheep. J Vet Intern Med 2003; 17:908–916.
[15] Foster JD, Parnham D, Chong A, et al. Clinical signs, histopathology and genetics of experimental transmission of BSE and natural scrapie to sheep and goats. Vet Rec 2001; 148:165–171.
[16] Anonym. EU Press Release. IP/05/105. Jan 28, 2005.
[17] Houston EF and Gravenor MB. Clinical signs in sheep experimentally infected with scrapie and BSE. Vet Rec 2003; 152:333–334.
[18] Baron T, Madec JY, Calavas D, et al. Comparison of French natural scrapie isolates with bovine spongiform encephalopathy and experimental scrapie infected sheep. Neurosci Lett 2000; 284:175–178.
[19] Capucchio MT, Guarda F, Pozzato N, et al. Clinical signs and diagnosis of scrapie in Italy: a comparative study in sheep and goats. J Vet Med A 2001; 48:23–31.
[20] Hopp P, Ulvund MJ, Jarp J. A case-control study on scrapie in Norwegian sheep flocks. Prev Vet Med 2001; 51:183–198.
[21] Clark AM and Moar JA. Scrapie: a clinical assessment. Vet Rec 1992; 130:377–378.
[22] Sigurdarson S. Epidemiology of scrapie in Iceland and experience with control measures. In: Bradley R, Savey M, Marchant B, editors. Subacute spongiform encephalopathies. Commission of European Communities. Kluwer Academic, Netherlands, 1991:233–242.
[23] Ulvund MJ, Bratberg B, Tranulis MA. Prion diseases in animals. Norsk vetr tidsskr 1996; 108:455–466.
[24] Ulvund MJ, Ersdal C, Seljeskog E. Scrapie – following up, clinic, autopsy and diagnostics. Husdyrforsøksmøtet 2002, Ås–NLH, Norway 2002; 73–76.
[25] Dickinson AG. Scrapie in sheep and goats. In: Kimberlin RH, editor. Slow virus diseases of animals and man. North-Holland, Amsterdam, 1976:209–241.
[26] Hoinville LJ. A review of the epidemiology of scrapie in sheep. Rev Sci Tech Off Int Epiz 1996; 15:827–852.
[27] Ersdal C, Ulvund MJ, Espenes A, et al. Mapping PrPSc propagation in experimental and natural scrapie in sheep with different PrP genotypes. Vet Pathol 2005; 42:258–274.
[28] Ryder S, Dexter G, Bellworthy S, et al. Demonstration of lateral transmission of scrapie between sheep kept under natural conditions using lymphoid tissue biopsy. Res Vet Sci 2004; 76:211–217.
[29] Ersdal C, Ulvund MJ, Benestad S L, et al. Accumulation of pathogenic prion protein (PrP^{Sc}) in nervous and lymphoid tissues of sheep with subclinical scrapie. Vet Pathol 2003; 40:164–174.
[30] Austin AR and Simmons MM. Reduced rumination in bovine spongiform encephalopathy and scrapie. Vet Rec 1993; 132:324–325.
[31] Healy AM, Hanlon AJ, Weavers E, et al. A behavioural study of scrapie affected sheep. Appl Anim Behav Sci 2002; 79:89–102.
[32] Hovers KA. Scrapie in young texel ewes confused by concurrent hepatic pathology (Ann Proc). Sheep Vet Society 1991; 142–143.
[33] Sharp MW and Collings DF. Ovine abomasal enlargement and scrapie (letter). Vet Rec 1987; 120:215.
[34] van Keulen LJ, Schreuder BE, Meloen RH, et al. Immunohistochemical detection of prion protein in lymphoid tissues of sheep with natural scrapie. J Clin Microbiol 1996; 34:1228–1231.
[35] Pruden SJ, McAllister MM, Schulthess PC, et al. Abomasal emptying defect of sheep may be an acquired form of dysautonomia. Vet Pathol 2004; 41:164–169.
[36] Behrens H. Lehrbuch der Schafkrankheiten. Paul Parey, Berlin, 1979.
[37] Dedié K and Bostedt H. Schafkrankheiten. Eugen Ulmer, Stuttgart, 1985.
[38] Detwiler LA, Jenny AL, Rubenstein R, et al. Scrapie: a review. Sheep & Goat Res J 1996; 12:111–131.
[39] Radostits OM, Blood DC, Gay CC. Veterinary medicine – a textbook of the diseases of cattle, sheep, pigs, goats and horses. Baillière Tindall, London, 1994.

[40] Kimberlin RH. Scrapie. In: Martin WB, et al., editors. Diseases of sheep. Blackwell Science, Oxford, 2000:163–169.
[41] Bossers A, Schreuder BE, Muileman IH, et al. PrP genotype contributes to determining survival times of sheep with natural scrapie. J Gen Virol 1996; 77:2669–2673.
[42] Humphery RW, Clark AM, Begara-McGorum I, et al. Estimation of scrapie prevalence in cull and found-dead sheep on the Shetland Islands. Vet Rec 2004; 154:303–304.
[43] Fraser H. Scrapie in sheep and goats, and related diseases. In: Martin WB, et al., editors. Diseases of sheep. Blackwell Science, London, 2000:207–218.
[44] Kümper H. Scrapie aus klinischer Sicht. Tierärztl Prax 1994; 22:115–120.
[45] Joubert L, Lapras M, Gastellu J, et al. Un foyer de tremblante du mouton en Provence. Bull Soc Sci Vet et Med Comp 1972; 74:165–184.
[46] Ulvund MJ. Scrapie: a clinical challenge. Norwegian School of Veterinary Science, Book of Publications, Oslo, 1998:5–23.
[47] Onodera T and Hayashi T. Diversity of clinical signs in natural scrapie cases occurring in Japan. Japan Agr Res Quarterly 1994; 28:59–61.
[48] Toumazos P. Scrapie in Cyprus. Br Vet J 1991; 147:147–154.
[49] Hörnlimann B and Braun U. BSE: Clinical signs in Swiss BSE cases. In: Bradley R, Marchant B, editors. Transmissible spongiform encephalopathies. EU: VI/4131/94–EN Document Ref. F.II.3 – JC/0003, Brüssel, 1994: 289–299.
[50] Ulvund MJ. Skrapesjuke – kliniske symptom (Scrapie – clinical symptoms; video in Norwegian). Thorud K, Statens Dyrehelsetilsyn, FIM VIDEO, Bergen, 1997.
[51] Schreuder BE, van Keulen LJ, Vromans ME, et al. Tonsillar biopsy and PrP^{Sc} detection in the preclinical diagnosis of scrapie. Vet Rec 1998; 142:564–568.
[52] O'Rourke KI, Baszler TV, Parish SM, et al. Preclinical detection of PrP^{Sc} in nictitating membrane lymphoid tissue of sheep. Vet Rec 1998; 142:489–491.
[53] Espenes A, Press CM, Landsverk T, et al. Detection of (PrP^{Sc}) in rectal biopsy and necropsy samples from sheep with experimental scrapie. J Comp Pathol 2006; 134(2–3):115–125.
[54] Schmerr MJ, Jenny AL, Bulgin MS, et al. Use of capillary electrophoresis and fluorescent labelled peptides to detect abnormal prion protein in the blood of animals that are infected with a transmissible spongiform encephalopathy. J Chromatography 1999; 53:207–214.
[55] Picard-Hagen N, Gayrard V, Laroute V, et al. Discriminant value of blood and urinary corticoids for the diagnosis of scrapie in live sheep. Vet Rec 2002; 150:680–684.
[56] Vargas F, Bolea R, Monleón E, et al. Clin. characterisation of natural scrapie in a native Spanish breed of sheep. Vet Rec 2005; 156:320–322.

35 Diagnosis of Bovine Spongiform Encephalopathy by Immunological Methods

Alex J. Raeber, Markus Moser, and Bruno Oesch

35.1 Introduction

The only reliable molecular marker for prion diseases is PrP^{Sc}, the pathological conformer of the prion protein that accumulates in the central nervous system (CNS) and to a lesser extent in lymphoreticular tissues. For bovine spongiform encephalopathy (BSE), several commercial diagnostic kits based on the post mortem immunochemical detection of PrP^{Sc} in brain tissue are now available. These rapid screening tests have been used in active surveillance of BSE and have greatly improved the detection of infected cattle before their entry into the human food chain. At present, with the exception of scrapie tests (see Chapter 34.7), no diagnostic test exists for the detection of prion diseases in live animals or humans. New diagnostic techniques aimed at increasing sensitivity and specificity of PrP^{Sc} detection in body fluids and at identifying novel surrogate markers are under development (compare Chapter 31). In this chapter, we review the currently approved rapid tests for the post mortem detection of the BSE-associated pathological prion protein, as well as emerging technologies for the ante mortem diagnosis of prion diseases.

35.2 Properties of the normal and disease-associated form of the prion protein

The normal form of the prion protein (PrP^C) can be distinguished from the disease-associated form PrP^{Sc} by the following criteria (see also Chapter 8):

- Different sensitivities to proteolytic digestion by proteases such as proteinase K (PK): whereas PrP^C is totally digested by PK, PrP^{Sc} loses only its N-terminus [1].
- Different physicochemical properties: PrP^{Sc} forms large aggregates, which can be separated from PrP^C by centrifugation [2; 3].
- Both forms of PrP have the same amino acid sequences, but differ in their tertiary structure. While PrP^C is predominantly α-helical (ca. 40 %) with low β-sheet content (< 5 %), PrP^{Sc} has considerably more β-sheet structure (ca. 40 %) and a similar α-helical content (30 %). Most of the antibodies raised against PrP recognize linear sequences on PrP and are therefore unable to differentiate between the two forms of PrP. Several antibodies have been published that recognize a conformational epitope on PrP^{Sc} and are able to discriminate between the two forms of PrP [4–7]. Some of these antibodies bind exclusively to PrP^{Sc} without cross-reacting to PrP^C and thereby allow for a direct one-step detection of PrP^{Sc} [4].

35.3 Rapid tests approved by European authorities

35.3.1 Detection principles of rapid tests

Most diagnostic tests approved by the EU authorities exploit the relative protease resistance of PrP^{Sc} in brain samples to discriminate between PrP^C and PrP^{Sc}, in combination with immunological detection of the protease-resistant part of PrP^{Sc} (PrP27–30). Sample preparation is one crucial part of the diagnostic tests and influences the diagnostic sensitivity and specificity as well as the throughput. Because of the membrane attachment of PrP^C and PrP^{Sc}, tissue solubilization requires detergent extraction. In addition, conditions for PK digestion were optimized in order to allow complete digestion of PrP^C but only removing the N-terminus of PrP^{Sc}, thereby leaving intact the protease-resistant core, PrP27–30, for the subsequent detection by anti-PrP-antibodies. Various other proteases can be used to discriminate between the protease-sensitive PrP^C molecules and the protease-resistant PrP^{Sc} structure. However, while PK digests the whole N-terminus of PrP^{Sc}, other proteases such as trypsin or pronase do not result in such a well-defined removal of the N-terminus – they leave the whole PrP^{Sc} molecule intact or digest away only a few amino acids [8]. For non-size-discriminating immunoassays this issue is of no direct importance since size shifts are not a part of the result output. In Western blot assays a size shift is directly visible and constitutes a specificity criterion for a positive assay [9]. This has the additional advantage of providing a direct in-process control for the performance of the protease. For ELISA-based or similar assays it is of paramount importance that the digestion conditions are solid as even slightly incomplete digestion of PrP^C can lead to false positive results.

In 1999, the European Commission carried out a first evaluation of rapid post mortem tests for BSE [10]. Four tests were evaluated on brain tissue from clinical BSE cases and the EU commission approved the following three rapid tests for BSE monitoring in cattle according to regulation (EC) 999/2001 (Tab. 35.1):

- Prionics®-Check WESTERN (Western blot, Prionics AG, Switzerland): Simple automated one-step sample preparation procedure followed by protease treatment. Detection occurs after the separation of the treated sample by denaturing polyacrylamide gel electrophoresis and transfer to a membrane using a PrP-specific antibody and an alkaline phosphatase-coupled secondary antibody detection system generating chemiluminescence. The presence of a PrP-immunoreactive signal with the additional two criteria of a reduced molecular weight (due to digestion of the N-terminus of PrP^{Sc}) and a typical three-band pattern (due to different glycosylation forms of PrP) result in a TSE-positive diagnosis.
- TeSeE test (ELISA, Bio-Rad, USA): Sample preparation involving protease treatment followed by precipitation and a centrifugation step for enrichment of the analyte. Detection is performed by sandwich ELISA with two different monoclonal antibodies. The capture antibody is immobilized on the microplate and after incubation with the solution containing the analyte. A color-converting enzyme-coupled antibody is used for detection. Cutoff setting with gray zone (leading to repetition of the sample).
- Enfer test (ELISA, Enfer Scientific Ltd., Ireland): One-step sample extraction in a 'stomacher', which is a plastic bag in which the brain tissue is gently dissociated followed by protease treatment. The ELISA plates are coated directly with the digested brain homogenates containing the analyte, followed by detection via polyclonal antibodies and an enzyme-coupled secondary antibody. Detection based on chemiluminescence using a fixed cutoff setting after blank subtraction.

A second call to participate in an evaluation for rapid BSE testing was issued in 2002. This exercise was split into a laboratory evaluation and a field trial to assess the performance of the new tests under field conditions and allow a comparison to already approved tests [11]. The following two tests passed the laboratory evaluation and the field trial, and were approved by the EU commission in 2003 for BSE monitoring in cattle (Table 35.1).

- Prionics®-Check LIA (ELISA, Prionics AG, Switzerland): Simple one-step sample preparation procedure followed by protease treatment. Detection occurs by sandwich ELISA using monoclonal antibodies, one of them enzyme labeled. The digested homogenate is incubated with the detection antibody and the mixture is then transferred to another microtiter plate that is coated with the capture antibody. Detection with an enzyme-coupled second monoclonal antibody is based on chemiluminescence using a plate-specific cutoff setting.
- Conformation-dependent immunoassay CDI-5 (ELISA, Inpro, USA): In addition to a mild protease treatment and a phosphotungstic acid (PTA) precipitation, the CDI uses the differential binding of antibodies to

Table 35.1: Evaluation of rapid post mortem tests for BSE.

	Laboratory evaluation			Field trial				
				Healthy slaughtered			Poor quality	
	Sensitivity in % (95% CI in %)	Specificity in % (95% CI in %)	Detection limit	Sensitivity in % (95% CI in %)	Specificity in % (95% CI in %)	IR	Specificity in % (95% CI in %)	IR
1999								
Prionics®-Check WESTERN	100 (99)	100 (99.7)	1:100[a]	nd	nd	nd	nd	nd
Bio-Rad TeSeE	100 (99)	100 (99.7)	≥1:200	nd	nd	nd	nd	nd
Enfer TSE kit	100 (99)	100 (99.7)	1:30	nd	nd	nd	nd	nd
2003								
Prionics®-Check LIA	100 (93.8)[b]	100 (98)	≥1:200	100 (98.7)	100 (99.97)	0.86	100 (98.5)	
CDI-5	100 (93.8)	100 (98)	≥1:200	100 (98.6)	100 (99.97)	1.45	100 (98.5)	
2004								
CediTect® BSE Test	100 (94)	100 (98)	≥1:200	99.5 (97.7)[c]	100 (99.97)	0	100 (98.6)0	
FRELISA BSE	100 (94)	100 (98)	≥1:200	ip	ip	ip	ip	ip
IDEXX HerdChek® BSE Test	100 (94)	100 (98)	≥1:200	100 (98.5)	99.99 (99.95)[d]	0.39	100 (98.5)	0
Institut Pourquier Speed®it BSE	100 (94)	100 (98)	1:64	100 (98.5)	100 (99.97)	1.88	100 (98.5)0	
PDL Rapid BSE Test	96 (87.4)[e]	100 (98)	1:10					
Prionics®-Check PrioSTRIP	100 (94)	100 (98)	1:100	100 (98.5)	100 (99.97)	0.37	99.6 (98.2)	1.34
Priontype® post mortem ELISA	100 (94)	100 (98)	1:25	ip	ip	ip	ip	ip
Roboscreen Beta Prion BSE Test	100 (94)	100 (98)	≥1:200	100 (98.5)	100 (99.97)	0.67	100 (98.5)0	
Roche PrionScreen	100 (94)	99.3 (96.2)[f]	1:100	100 (98.5)	100 (99.97)	0.2	100 (98.5)	27
Enfer TSE kit v2	100 (94)	100 (98)	≥1:200	100 (98.5)	100 (99.97)	5.81	100 (98.5)	0

95% CI in %: 95% confidence interval in %
IR: initial reactive rate in ‰
nd: not done
ip: in progress
[a] Detection limit in the 1999 EU evaluation was 1:10; the EU evaluation in 2002 showed a detection limit between 1:81–1:243
[b] One sample which was initially a false negative was found positive after retesting of additional tissue and must be considered as a sample with low and/or heterogeneous PrP^{Sc}.
[c] Three samples were initially classified as false negative. A change in the cutoff setting reduced this number to 1 false negative.
[d] Four samples were classified initial reactive; three of them were negative after retesting and one re-tested positive.
[e] Three positive samples were not correctly classified initially and re-tested again negative in a duplicate analysis. Re-testing of these samples starting with new tissue slices were positive for one sample but remained negative for two samples. Based on this weakness in diagnostic sensitivity and the low performance in the analytical sensitivity, the test was not allowed to enter the field trial.
[f] One negative sample out of three initial reactives was positive in re-testing.

native or denatured PrP^{Sc}. The detection antibody recognizes a conformation-dependent epitope, which, while always exposed in the noninfectious form (PrP^{C}), becomes exposed in the infectious form of PrP (PrP^{Sc}) only upon denaturation. Quantification of the binding events leads to a signal difference (signal 'total denatured PrP' minus signal 'native PrP^{C}' equals PrP^{Sc}), which is used as a diagnostic criterion.

A third evaluation was carried out in 2004 following a call for expression of interest in January 2003. This evaluation exercise was managed jointly by the European Commission's Directorate General Joint Research Centre, the Institute for Reference Materials and Measurements (IRMM), and the Directorate General Health and Consumer Protection in collaboration with a working group of the European Food Safety Agency (EFSA). Again, this evaluation was carried out in two phases: a laboratory evaluation [12] and a field trial [13]. Of a total of 10 tests that were assessed in the laboratory evaluation, 9 advanced to enter the field trial. Based on low diagnostic sensitivity, the PDL rapid BSE test (Prion Developmental Laboratories, Inc., USA) was not subject to the field trial. By November 2004, the following 7 tests successfully passed the field trial and received approval by the EU commission (Table 35.1).

- CediTect® BSE test (ELISA, CEDI Diagnostics, the Netherlands): Protease-digested samples are fixed to an activated PVDF membrane by filtration in a 96-well format. One replicate of each sample is further treated with a chaotropic agent resulting in the T-sample; the second replicate (N-sample) as a control is not exposed to that reagent. Detection of the protease-resistant PrP is based on a chemiluminescent readout using an enzyme-conjugated monoclonal antibody. Calculation of the ratio of the T- and N-values is used to determine the status of the homogenate.
- IDEXX HerdCheck® BSE Antigen Test Kit, EIA (ELISA, IDEXX Laboratories Inc., USA): Sample homogenization is carried out with a ribolyser. Detection of PrP^{Sc} is performed by a classical two-sided immunoassay using a chemical polymer (technology licensed from Microsense Biotechnologies, London, UK) for selective capture of PrP^{Sc} – without the need for a protease digestion step – and detection by an enzyme-conjugated monoclonal antibody using a colorimetric substrate. A plate-specific cutoff setting is used in the diagnosis of the sample.
- Institut Pourquier Speed'it BSE (ELISA, Institut Pourquier, France): Tissues are homogenized using a ribolyser followed by treatment of the samples with PK. Detection of PrP27–30 uses a classical two-sided immunoassay with two monoclonal antibodies. Capture and detection of the protease-resistant core of PrP^{Sc} is performed in a single step in a microplate. Detection is based on chemiluminescence using a plate-specific cutoff setting.
- Prionics®-Check PrioSTRIP (Immunochromatographic assay, Prionics AG, Switzerland): Rapid microplate-based chromatographic immunoassay using two monoclonal antibodies to detect the protease-resistant core of PrP^{Sc}. The homogenate is subjected to treatment with PK and then incubated with the conjugate consisting of a colored latex bead-labelled antibody. The PrioSTRIP® – a comb shaped cartridge holding eight strips – is inserted into the sample conjugate mixture in a microplate, and the immunocomplexes are captured by a monoclonal antibody sprayed as a line perpendicular to the long axis of a strip (Fig. 35.1). The resulting immunosandwich appears as a colored line and the reading can be done either visually or automatically allowing a quantitative analysis using a specially designed cartridge holder for a maximum of 12 cartridges (i. e., a total of 96 strips), a flat-bed scanner, and dedicated software (PrioSCAN®).
- Roboscreen Beta Prion BSE EIA Test Kit (ELISA, Roboscreen GmbH, Germany): Tissues are homogenized using a ribolyser followed by treatment of the samples with PK. After protease digestion, the homogenate is subjected to a protein precipitation and delipidation step. The test is a classical two-

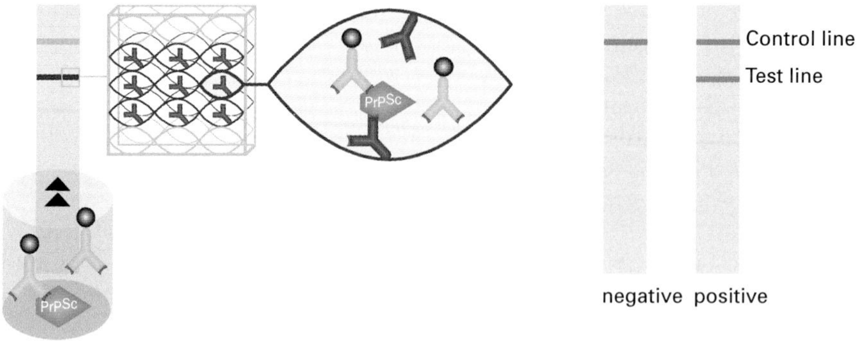

Fig. 35.1: Detection of protease treated PrP^{Sc} in the PrioSTRIP® assay. Medulla oblongata samples are homogenized and reacted with proteinase K to digest cellular PrP. Samples of digested homogenates are then mixed with the PrioSTRIP® conjugate consisting of anti-PrP mAb-bound colored latex particles. Cassettes consisting of 8 individual strips are then inserted into the sample/conjugate mixtures causing the analytes to chromatograph up the strip towards an absorbent waste pad. During the chromatography PrP binds to the latex-bound mAb, thereby forming immuno-complexes. These complexes are captured at the test line position resulting in a visible line across the strip. Excess conjugate is bound at the control line. PrioSTRIP® forces analyte and antibodies into close proximity and concentrates the analyte at the test line. This increases assay sensitivity, accelerates reaction kinetics and minimizes incubation times. Figure by PRIONICS ©.

sided immunoassay using two monoclonal antibodies. Detection is based on a colorimetric enzyme reaction using a fixed cutoff setting with gray zone (leading to repetition of the sample).

- Roche Applied Science PrionScreen (ELISA, Roche Diagnostics GmbH, Germany): Homogenization and protease digestion of the tissue samples is carried out in a 96-well plate homogenizer. Samples are then analyzed in a classical two-sided immunoassay using two monoclonal antibodies to detect the protease-resistant core of PrP^{Sc}. Capture and detection of PrP27–30 is performed in a single step in a streptavidin-coated microplate using a biotin-conjugated capture and an enzyme-conjugated detection antibody. Detection is based on a colorimetric enzyme reaction using a plate-specific cutoff setting.
- Enfer TSE Kit version 2.0, automated sample preparation (ELISA, Enfer Scientific Ltd., Ireland): Due to a change in the homogenization system, the EC approved Enfer TSE test was subjected to a full re-evaluation in accordance with the rules defined by the Community Reference Laboratory for TSE. The new homogenization procedure uses a disposable plastic tube in which a tissue sample is ground between two rough surfaces. The process of homogenization is automated and the samples are then transferred to a liquid handling robot for the subsequent ELISA.

In November 2004, field trials for the following two tests were still ongoing:

- FRELISA BSE (ELISA, Fujirebio, Japan): Tissue homogenization with beads followed by a treatment with DNAse I, collagenase and protease. The homogenate is then subjected to protein precipitation, protein unfolding and sonication steps. Detection of protease-resistant PrP^{Sc} is performed by a one-step ELISA using two monoclonal antibodies, one of them an enzyme conjugate, which allows a colorimetric detection with a plate-specific cutoff setting.
- Priontype® post mortem ELISA (Labor Diagnostik Leipzig GmbH, Germany): Homogenization is performed with ceramic beads

followed by protein denaturation, protein precipitation, resolubilization of the protein pellet and a stringent protein unfolding step. PrP^{Sc} is captured by a monoclonal antibody that differentiates between PrP^{C} and PrP^{Sc}. Detection with an enzyme-coupled second monoclonal antibody is based on a colorimetric reaction using a plate-specific cutoff setting.

By the end of 2004, more than 40 million rapid tests had been performed in Europe within the mandatory rapid testing of risk animals and of slaughtered cattle above the age of 30 months or 24 months in some countries (e. g., Germany). Currently, the Prionics®-Check tests are the most used BSE-tests worldwide followed by the TeSeE test. In total, about 10 million BSE tests are performed in Europe and about 1 million in Japan, annually.

35.3.2 Laboratory evaluation of rapid tests

The hallmark of any diagnostic test is its diagnostic sensitivity (ability to identify infected reference samples as positive) and specificity (ability to identify uninfected reference samples as negative). This will determine the usefulness of a test in the field, i. e., how many true and false positives it will produce. The diagnostic sensitivity and specificity of TSE tests was evaluated by the EU commission as well as in individual countries using 50 positive and 150 negative samples from animals previously diagnosed for TSE with a reference method such as histopathology or immunohistochemistry (IHC).

Table 35.2: Results of BSE testing in France from July 1, 2001 – March 30, 2002.

	Normal slaughter		Risk category	
	Prionics	Bio-Rad	Prionics	Bio-Rad
Negative	2,356,421	753,124	161,308	37,336
Positive	79	26	92	24
Total	2,356,500	753,150	161,400	37,360
χ^2	0.02	0.27		

Numbers of positively and negatively tested cattle from two different populations (normal slaughter and cattle from the risk category) in France using two different assay systems (Prionics®-Check WESTERN and BioRad). The low χ^2values (compared to $\chi^2\alpha = 3.8$) of a χ^2 test indicate no significant difference in the relative populations of positively tested animals between the two assay systems. Source: [16].

In addition, the evaluation protocol included an assessment of the detection limit using serial dilutions of a BSE-positive brain homogenate diluted in negative homogenates. A comparison of three ELISA tests showed very similar analytical sensitivities (Fig. 35.2). Determination of the detection limit for the analyte PrP^{Sc} should give an indication on the test's capability to detect preclinical cases of BSE. However, the analytical sensitivity may or may not correlate with the diagnostic sensitivity depending on the appearance and concentration of the analyte during the course of the disease. In the case of protease-resistant PrP as an analyte of TSE diagnosis, the two parameters differ because PrP^{Sc} does not accumulate linearly but exponentially in the brain of infected animals [14]. Furthermore, PrP^{Sc} may be present in different aggregation states during the incubation phase of the

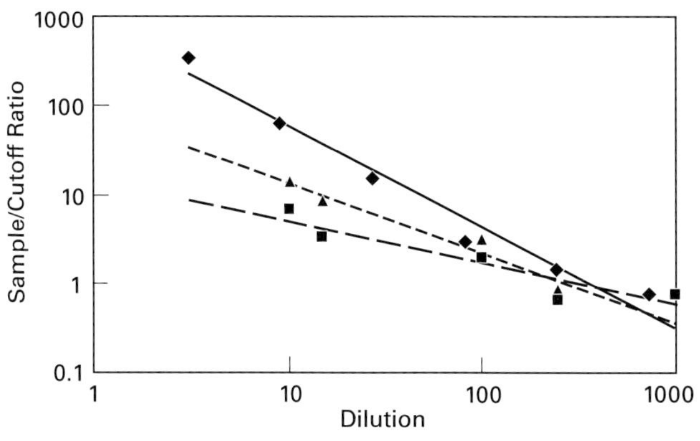

Fig. 35.2: Similar analytical sensitivity of rapid TSE tests. BSE positive brain homogenate was diluted in a negative homogenate in dilutions down to 10^{-3} and analyzed with the Prionics®-Check LIA (-- ◆ --), the CDI (-- ■ --) and the TeSeE (-- ▲ --) ELISA. Measurements are expressed as ratio of mean signals to the cut off value for the LIA and the Platelia. For the CDI the measurements are expressed as ratio of mean signals to negative controls. Data taken from [45]. Figure by PRIONICS.

disease. It has been shown that small PrP^{Sc} aggregates, which may be present in preclinical animals cannot be easily precipitated [15]. Therefore, PrP^{Sc} in a dilution of brain sample of a clinical animal does not necessarily reflect PrP^{Sc} in a preclinical animal. For this reason, slight variations in analytical sensitivity do not necessarily reflect differences in diagnostic sensitivity. This is illustrated by a comparative study on the surveillance data of France [16] (Table 35.2), which shows equal diagnostic sensitivity for two of the most used BSE tests in Europe despite some claims of higher analytical sensitivity for one of the tests.

35.3.3 Field assessment of rapid tests

The rapid BSE tests were further evaluated in a field trial in order to determine their performance under normal routine conditions. This enabled a thorough assessment of parameters like robustness, applicability in different laboratories, expected rate of false initial reactives under high throughput field conditions, and necessary cut off adjustment before market introduction. For this evaluation, over 10,000 tissue samples from healthy slaughtered cattle that were tested negative with an already approved rapid BSE test were analyzed to assess the diagnostic specificity. The diagnostic sensitivity of a new test was likewise assessed using a set of 200 true positive samples that were tested positive using a reference test. In addition the tests were assessed for their performance on 200 poor quality samples. This is important because field samples are often in a deteriorated state due to poor storage conditions of carcass samples over long periods of time. Such samples show extended autolysis and a robust diagnostic test has to give accurate results under such conditions. Previous studies have shown that PrP^{Sc} can be readily detected in autolyzed samples with Western blot and IHC [17–18].

Whereas it has been extensively documented that sample autolysis does not influence PrP^{Sc} detection by Western blot, a recent study carried out by the National Institute of Animal Health of Japan demonstrated that some ELISAs might be unsuitable for screening of fallen stock samples [19]. Hayashi and coworkers found that one of the commonly used ELISA rapid tests showed a significant reduction in optical density (OD) values on deteriorated samples.

35.4 Ante mortem TSE test development

35.4.1 PrP^{Sc} as a disease marker

To date, PrP^{Sc} is used as a marker for the post mortem diagnosis of TSEs because high concentrations are found in the CNS of infected animals and humans. In contrast, PrP^{Sc} concentrations in peripheral tissues and body fluids are several orders of magnitude lower than in brain tissues and hence, ultrasensitive detection methods or enrichment steps for PrP^{Sc} become mandatory for ante mortem test development. Infectivity studies in sheep and in rodent models have shown that the levels of prion infectivity in blood are in the order of 1–10 infectious units (IU) ml^{-1} in the presymptomatic period and that these levels rise to about 100 IU ml^{-1} in the clinical phase [20]. Based on estimations that 100 IU contain about 1 picogram (one trillionth of a gram) of PrP^{Sc} [21], it can be assumed that a sensitive blood test would need to detect PrP^{Sc} in blood at a level of 10 fg ml^{-1} ($\sim 3 \times 10^{-16}$ M) to 1 pg ml^{-1} ($\sim 3 \times 10^{-14}$ M). While current ELISAs approved for the post mortem diagnosis of TSEs have their limit of detection for PrP in the picogram range [22], a number of ultrasensitive detection methods are being developed with the aim to establish a sensitive preclinical test for TSEs. Among these techniques are spectroscopy-based methods such as confocal fluorescence spectroscopy (CFS) [23] and capillary electrophoresis in combination with a competitive immunoassay [24].

In CFS, single fluorescent molecules are excited by a laser beam either in solution or on a glass surface and the emitted light is detected by a confocal microscope equipped with a photon counter. With such a confocal microscope setup, it is possible to detect a single fluorescent molecule within a femtoliter size volume, corresponding to a nanomolar concentration. Bieschke et al. [23] have used fluorescence spectroscopy with correlation analysis (FCS) for the detection of PrP^{Sc} aggregates in solution in the femtomolar range using antibody probes tagged with fluorescent dyes. Furthermore, using this technology they were able to detect PrP^{Sc} in the cerebrospinal fluid of 5 out of 24 CJD patients but in none of the 13 control subjects. CFS has also been studied for the ultrasensitive detection of fluorescent molecules on glass surfaces. By using two anti-PrP-specific monoclonal antibodies – a capture antibody immobilized on the glass surface of a microplate and a detection antibody conjugated with a fluorescent dye – it was possible to detect picomolar concentrations of recombinant PrP spiked into serum [25].

The immunocompetitive capillary electrophoresis (ICCE) assay is based on the extraction of protease-resistant PrP from a biological sample followed by a competitive immunoassay using an anti-PrP-antibody and a fluorescein-labeled PrP peptide that is competitively displaced by PrP27–30 from the sample and detection of the free and bound peptide by laser-induced fluorescence during capillary electrophoresis. Schmerr et al. [24] have reported the detection of protease-resistant PrP in blood of scrapie-infected sheep and CWD-infected elk but not in normal sheep and elk. However, since attempts to detect protease-resistant PrP in blood from human TSE patients and CJD-infected chimpanzees have failed, the applicability of this assay for blood screening has been questioned based on difficulties with the reproducibility of the PrP extraction procedure and the variability in the immunocompetitive assay [26].

An alternative approach to increasing the sensitivity of the detection method for PrP^{Sc} is to use enrichment procedures for PrP^{Sc} in the sample before detection. In recent years, a diverse array of molecules capable of distinguishing between native PrP^{C} and PrP^{Sc} have been developed and could be used in the concentration of PrP^{Sc} from blood. These include plasminogen [27], anti-DNA antibodies and a DNA-binding protein [7], RNA aptamers [28; 29], PrP^{Sc}-specific monoclonal antibodies [4–6] and motif-grafted antibodies containing the replicative interface of PrP^{C} [30]. Most promising among these molecules are antibodies that specifically recognize the disease-associated form of the prion protein and that were shown to be capable of detecting PrP^{Sc} in brain homogenates without the need of PK digestion [4]. An additional advantage of an immunoaffinity purification method is that protease-sensitive forms of PrP^{Sc}, which constitute a significant fraction of disease-specific PrP, are detected as well. Recently, it has been shown that the conformation-specific antibody 15B3 is capable of detecting protease-sensitive forms of PrP^{Sc} in a transgenic mouse model of Gerstmann–Sträussler–Scheinker disease suggesting that this monoclonal antibody might be a valuable tool for the development of a blood-based TSE test [31].

An alternative strategy to increasing the sensitivity of the detection method or to enriching the analyte is to increase the amount of the analyte present in body fluids by using an in vitro amplification method. But since prions are

Table 35.3: Atypical BSE cases identified in France (2004), Italy (2004) and Japan (2003).

Country	Frequency	Age, yr	Unglycosylated PrP^{Sc}	Histology	Test used for initial diagnosis
Italy	2/8	15	Low Mr	Plaques	Prionics®-Check WESTERN
		11	Low Mr	Plaques	Prionics®-Check WESTERN
France	3/58	10	High Mr	n. d.	Platelia
		15	High Mr	n. d.	Prionics®-Check WESTERN
		8	High Mr	n. d.	Prionics®-Check WESTERN
Japan	1/9	2	Low Mr	negative	Platelia

Frequency and characterization of atypical BSE cases that were identified in routine surveillance of cattle in France, Italy and Japan. Frequency refers to the number of atypical cases identified per total BSE positives animals analyzed in the reported study.
n. d. not done.

composed entirely of protein, PCR technology cannot be used for amplification of the agent prior to a detection step. To overcome these problems, cell-free conversion systems were developed that mimic the PrP^{Sc}-templated refolding of PrP^{C} into new PrP^{Sc} molecules. However, early in vitro conversion systems produced only low levels of newly formed PrP^{Sc}, making it difficult to analyze the infectious properties of de novo formed PrP^{Sc} [32]. A technology known as protein misfolding cyclic amplification (PMCA) was shown to mimic the PrP refolding process more closely and allowed indefinite amplification of PrP^{Sc} [33; 34]. In this method, minute quantities of an infectious brain homogenate are incubated for 30 min with an excess of normal brain homogenate followed by a 40 s pulse of sonication to disrupt large PrP^{Sc} aggregates into smaller fragments and allow further growth of the fragments with a reaction kinetic reminiscent of a nucleation-dependent seeding mechanism. However, the practical use of this technology as a diagnostic test for TSEs might be limited due to the requirement for normal brain homogenate as a substrate in the refolding reaction.

35.4.2 Surrogate markers

Alternatively, disease-indicating markers other than PrP^{Sc} present in easily accessible body fluids may be needed for the development of an early preclinical TSE test. Using differential RNA display technologies, it was found that a novel transcript coding for erythroid differentiation-related factor (EDRF) showed a progressive decrease in the expression during the course of the disease in scrapie-infected mice [35]. In a similar approach, cellular nucleic acids were identified in the serum of BSE-infected cattle that appeared to correlate with the development of the disease [36]. If such nucleic acids are indeed present in blood, this might form the basis of a simple PCR-based diagnosis of TSE and related disorders from live organisms.

Another simple, non-invasive test to diagnose TSEs uses a technique known as high-resolution electrocardiogram (ECG) measuring a unique heart rate variability (HRV) signature observed at the early stages of infection [37]. The test was used on more than 90 bovines fed with brain homogenate from BSE field cases and showed a significant difference in the HRV between controls and cattle exposed to BSE between 29 and 42 months after infection.

Instead of using single or multiple biomolecular markers for the diagnosis of a TSE, complex proteomic profiles obtained by fourier transform infrared (FT-IR) spectroscopy and analyzed by computational pattern recognition techniques have been used to classify serum sample spectra based on disease-related features [38]. This technology uses a large number of TSE positive and negative samples to train an artificial neural network with a classification algorithm. Afterwards, unknown samples can be classified as positive or negative based on distinct spectral patterns. In an experimental model of scrapie using hamsters, FT-IR spectroscopy on blood serum allowed the identification of healthy and clinically sick animals with sensitivities and specificities of 97 % and 100 %, respectively [39]. Furthermore, the same technology was applied to the classification of BSE using more than 800 serum samples including BSE positives, healthy controls, and animals with other diseases for the training of the computer algorithm. In two blinded validation studies including more than 350 sera, combined sensitivities and specificities of BSE detection of 96 and 92 % were achieved, respectively [38].

35.5 Identification of atypical BSE strains

Classically, TSE strains have been characterized through inoculation of brain homogenates into a series of mouse lines. Distinct strains are then differentiated based on incubation time and lesion profile within the same mouse line. These are the most important criteria for differentiation of strains. More recently and with the availability of immunochemical diagnostic methods, a more detailed analysis was made possible based on glycosylation patterns of protease-treated PrP^{Sc}. On Western blots, protease-treated PrP^{Sc} is resolved into three bands that correspond to the di-, mono-, and unglycosylated prion protein. These bands can exhibit different mobility patterns, presumably reflecting different conformations of PrP^{Sc} associated with different prion strains. To date, all BSE cases that were analyzed showed the same PrP glycopattern with the highest intensity of the diglycosylated form of PrP suggesting that a single strain of agent is responsible for BSE [40; 41].

New evidence has now been published suggesting the existence of more than one strain of BSE and supporting the notion that the BSE epidemic might have had its origin in a sporadic form of BSE occurring at very low frequency in cattle. The atypical BSE strains identified, e.g., in Italy [42], France [43], and Japan [44], exhibited novel molecular and pathological phenotypes and were detected among routinely diagnosed BSE cases following active surveillance of the disease using rapid TSE tests for the detection of PrP^{Sc} in brain tissue of cattle. Table 35.3 summarizes the atypical cases and their phenotypic presentations.

All atypical BSE cases were found in apparently healthy cattle with no clinical signs suggestive of BSE. Whereas the cases in Italy and France were found in old cattle, the Japanese BSE case was confirmed in an animal less than 2 years old. This underscores the importance of continued active surveillance of cattle at the slaughterhouse and demonstrates that current post mortem rapid tests are capable of detecting BSE in presymptomatic animals. Characterization of the atypical BSE cases was performed by strain typing using Western blot analysis of the protease-treated PrP^{Sc}.

Whereas the Italian and Japanese cases showed a PrP^{Sc} type with predominance and a lower apparent molecular mass of the unglycosylated PrP, the French animals presented with a higher apparent molecular mass of the unglycosylated PrP (Fig. 35.3). Based on these results, it was speculated that the Italian and Japanese cases shared similarities whereas the French cases represented a distinct BSE subtype. However, further investigations on the pathology of the atypical BSE cattle revealed striking differences between the Japanese and Italian cases. In contrast to the Japanese case, which showed no BSE-typical pathology, the Italian cases showed PrP-positive amyloid plaques, a feature that has not been associated with typical BSE. These findings prompted the authors to name this disease bovine amyloidotic spongiform encephalopathy (BASE). The authors then showed that the PrP^{Sc} glycoforms associated with BASE are, surprisingly, very similar to the glycoforms found in some types of sporadic CJD (sCJD). This raises the question whether a similar link between atypical BSE cases and sCJD exists as between typical BSE and vCJD. However, it might be premature for such far-reaching conclusions based on molecular characterization of a small number of novel BSE phenotypes. More solid support for a common etiology between BASE and sCJD has to await characterization of these TSE agents by

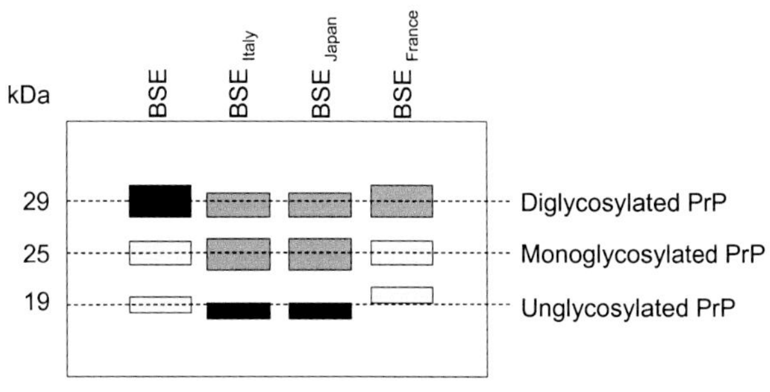

Fig. 35.3: PrP^{Sc} glycotype pattern of atypical BSE cases. Schematic representation of a Western blot of protease-treated PrP^{Sc} from a typical BSE case and from atypical cases as reported in Italy, France and Japan. Gray scales of the bands represent the intensity of the PrP band with black corresponding to the highest and light gray to lowest intensity. Apparent molecular weight markers are depicted on the left in kDa. Figure by PRIONICS ©.

inoculation into mice. Further characterization of the French atypical cases revealed a striking similarity to cattle that developed BSE following experimental infection with sheep scrapie. Thus, the question was raised whether this represents a scrapie strain that was transmitted from sheep to cattle by contact or through environmental contamination.

From the low numbers of cases analyzed so far, it is difficult to predict how frequent these atypical BSE cases in the cattle population are. Nevertheless, assuming a frequency of 1/4 to 1/20 as reported in the three independent studies (Table 35.3), it will be crucial to apply the most stringent testing criteria in order to reliably detect atypical BSE cases in the future. The Italian cases and two of the French cases were identified with the Prionics®-Check WESTERN blot. One of the French cases and the Japanese case were found by the TeSeE ELISA test and the strain characterization was later performed by Western blot analysis. Thus, Western blot-based BSE tests have the advantage that they allow the identification of new BSE strains because detection of PrP^{Sc} is based on several molecular criteria such as immunological reactivity, molecular weight, and glycosylation pattern.

References

[1] Oesch B, Westaway D, Walchli M, et al. A cellular gene encodes scrapie PrP27–30 protein. Cell 1985; 40:735–746.

[2] Bolton DC, McKinley MP, Prusiner SB. Identification of a protein that purifies with the scrapie prion. Science 1982; 218(4579):1309–1311.

[3] Hope J, Morton LJ, Farquhar CF, et al. The major polypeptide of scrapie-associated fibrils (SAF) has the same size, charge distribution and N-terminal protein sequence as predicted for the normal brain protein (PrP). EMBO J 1986; 5:2591–2597.

[4] Korth C, Stierli B, Streit P, et al. Prion (PrP^{Sc})-specific epitope defined by a monoclonal antibody. Nature 1997; 390:74–77.

[5] Paramithiotis E, Pinard M, Lawton T, et al. A prion protein epitope selective for the pathologically misfolded conformation. Nat Med 2003; 9:893–899.

[6] Curin-Serbec V, Bresjanac M, Popovic M, et al. Monoclonal antibody against a peptide of human prion protein discriminates between Creutzfeldt–Jacob's disease-affected and normal brain tissue. J Biol Chem 2004; 279:3694–3698.

[7] Zou W, Zheng J, Gray DM, et al. Antibody to DNA detects scrapie but not normal prion protein. Proc Natl Acad Sci USA 2004; 101:1380–1385.

[8] McKinley MP, Meyer RK, Kenaga L, et al. Scrapie prion rod formation in vitro requires both detergent extraction and limited proteolysis. J Virol 1991; 65:1340–1351.

[9] Schaller O, Fatzer R, Stack M, et al. Validation of a western immunoblotting procedure for bovine PrP(Sc) detection and its use as a rapid surveillance method for the diagnosis of bovine spongiform encephalopathy (BSE). Acta Neuropathol (Berl) 1999; 98:437–443.

[10] Moynagh J and Schimmel H. Tests for BSE evaluated. Nature 1999; 400:105. Webpage: www.europa.eu.int/comm/food/fs/bse/bse12_en.html.

[11] Schimmel H, Catalani P, Le Guern L, et al. The evaluation of five rapid tests for the diagnosis of transmissible spongiform encephalopathy in bovines (2nd study). Webpage 2002: www.europa.eu.int/comm/food/food/biosafety/bse/bse42_en.pdf.

[12] Philipp W, van Iwaarden P, Goll M, et al. The evaluation of 10 rapid post mortem tests for the diagnosis of transmissible spongiform encephalopathy in bovines. Webpage 2004: www.efsa.eu.int/science/tse_assessments/bse_tse/694_en.html, www.irmm.jrc.be.

[13] Philipp W and Vodrazka P. The field trial of seven new rapid post mortem tests for the diagnosis of bovine spongiform encephalopathy in bovines. Webpages 2004: www.efsa.eu.int/science/tse_assessments/bse_tse/694_en.html, www.irmm.jrc.be.

[14] Jendroska K, Heinzel FP, Torchia M, et al. Proteinase-resistant prion protein accumulation in Syrian hamster brain correlates with regional pathology and scrapie infectivity. Neurology 1991; 41:1482–1490.

[15] Gabizon R, McKinley MP, Groth DF, et al. Properties of scrapie prion protein liposomes. J Biol Chem 1988; 263:4950–4955.

[16] Anonym. Direction générale de l'alimentation [French Food Agency]. Results of BSE testing in France. DGAL 2002. Webpage 2006: www.agriculture.gouv.fr/esbinfo/esbinfo.htm.

[17] Debeer SO, Baron TG, Bencsik AA. Immunohistochemistry of PrP^{Sc} within bovine spongiform encephalopathy brain samples with graded

autolysis. J Histochem Cytochem 2001; 49:1519–1524.

[18] Chaplin MJ, Barlow N, Ryder S, et al. Evaluation of the effects of controlled autolysis on the immunodetection of PrP(Sc) by immunoblotting and immunohistochemistry from natural cases of scrapie and BSE. Res Vet Sci 2002; 72:37–43.

[19] Hayashi H, Takata M, Iwamaru Y, et al. Effect of tissue deterioration on post mortem BSE diagnosis by immunobiochemical detection of an abnormal isoform of prion protein. J Vet Med Sci 2004; 66:515–520.

[20] Hunter N, Foster J, Chong A, et al. Transmission of prion diseases by blood transfusion. J Gen Virol 2002; 83:2897–2905.

[21] Brown P, Cervenakova L, Diringer H. Blood infectivity and the prospects for a diagnostic screening test in Creutzfeldt–Jakob disease. J Lab Clin Med 2001; 137:5–13.

[22] Biffiger K, Zwald D, Kaufmann L, et al. Validation of a luminescence immunoassay for the detection of PrP(Sc) in brain homogenate. J Virol Methods 2002; 101:79–84.

[23] Bieschke J, Giese A, Schulz-Schaeffer W, et al. Ultrasensitive detection of pathological prion protein aggregates by dual-color scanning for intensely fluorescent targets. Proc Natl Acad Sci USA 2000; 97:5468–5473.

[24] Schmerr MJ, Jenny AL, Bulgin MS, et al. Use of capillary electrophoresis and fluorescent labeled peptides to detect the abnormal prion protein in the blood of animals that are infected with a transmissible spongiform encephalopathy. J Chromatogr 1999; 853:207–214.

[25] Dietrich A, Bossart K, Oesch B, et al. Evaluation of confocal fluorescence spectroscopy for the detection of pathological prion proteins. Chimia 2005; 59:250–255.

[26] Cervenakova L, Brown P, Soukharev S, et al. Failure of immunocompetitive capillary electrophoresis assay to detect disease-specific prion protein in buffy coat from humans and chimpanzees with Creutzfeldt–Jakob disease. Electrophoresis 2003; 24:853–859.

[27] Fischer MB, Roeckl C, Parizek P, et al. Binding of disease-associated prion protein to plasminogen. Nature 2000; 408:479–483.

[28] Rhie A, Kirby L, Sayer N, et al. Characterization of 2'-fluoro-RNA aptamers that bind preferentially to disease-associated conformations of prion protein and inhibit conversion. J Biol Chem 2003; 278:39697–39705.

[29] Sayer NM, Cubin M, Rhie A, et al. Structural determinants of conformationally selective, prion-binding aptamers. J Biol Chem 2004; 279:13102–13109.

[30] Moroncini G, Kanu N, Solforosi L, et al. Motif-grafted antibodies containing the replicative interface of cellular PrP are specific for PrP^{Sc}. Proc Natl Acad Sci USA 2004; 101:10404–10409.

[31] Nazor K, Kuhn F, Seward T, et al. Immunodetection of disease-associated mutant PrP, which accelerates disease in GSS transgenic mice. EMBO J 2005; 24:2472–2480.

[32] Kocisko DA, Come JH, Priola SA, et al. Cell-free formation of protease-resistant prion protein. Nature 1994; 370:471–474.

[33] Saborio GP, Permanne B, Soto C. Sensitive detection of pathological prion protein by cyclic amplification of protein misfolding. Nature 2001; 411:810–813.

[34] Castilla J, Saa P, Hetz C, et al. In vitro generation of infectious scrapie prions. Cell 2005; 121:195–206.

[35] Miele G, Manson J, Clinton M. A novel erythroid-specific marker of transmissible spongiform encephalopathies. Nat Med 2001; 7:361–364.

[36] Brenig B, Schutz E, Urnovitz H. Cellular nucleic acids in serum and plasma as new diagnostic tools. Berl Muench Tieraerztl Wochenschr 2002; 115:122–124.

[37] Pomfrett CJD, Glover DG, Bollen BG, et al. Perturbation of heart rate variability in cattle fed BSE-infected material. Vet Rec 2004; 154:687–691.

[38] Lasch P, Schmitt J, Beekes M, et al. Antemortem identification of bovine spongiform encephalopathy from serum using infrared spectroscopy. Anal Chem 2003; 75:6673–6678.

[39] Schmitt J, Beekes M, Brauer A, et al. Identification of scrapie infection from blood serum by Fourier transform infrared spectroscopy. Anal Chem 2002; 74:3865–3868.

[40] Collinge J, Sidle KCL, Meads J, et al. Molecular analysis of prion strain variation and the aetiology of "new variant" CJD. Nature 1996; 383:685–690.

[41] Stack J, Chaplin J, Clark J. Differentiation of prion protein glycoforms from naturally occurring sheep scrapie, sheep-passaged scrapie strains (CH1641 and SSBP1), bovine spongiform encephalopathy (BSE) cases and Romney and Cheviot breed sheep experimentally inoculated with BSE using two monoclonal antibodies. Acta Neuropathol (Berl) 2002; 104:279–286.

[42] Casalone C, Zanusso G, Acutis P, et al. Identification of a second bovine amyloidotic spongiform encephalopathy: molecular similarities with

sporadic Creutzfeldt–Jakob disease. Proc Natl Acad Sci USA 2004; 101:3065–3070.

[43] Biacabe A, Laplanche J, Ryder S, et al. Distinct molecular phenotypes in bovine prion diseases. EMBO Reports 2004; 5:110–115.

[44] Yamakawa Y, Hagiwara K, Nohtomi K, et al. Atypical proteinase K-resistant prion protein (PrPres) observed in an apparently healthy 23-month-old holstein steer. Jpn J Infect Dis 2003; 56:221–222.

[45] European Commission. 2002. The evaluation of five rapid tests for the diagnosis of transmissible spongiform encephalopathy in bovines (2nd study). Webpage 2006: http://europa eu int/comm/food/fs/bse/bse42_en pdf.

Topic VI: Epidemiology

36 Epidemiology and Risk Factors of Creutzfeldt–Jakob Disease

Inga Zerr and Sigrid Poser*

36.1 Introduction

Concern about the transmissibility of bovine spongiform encephalopathy (BSE) to humans by consumption of contaminated bovine products prompted epidemiological studies on the occurrence of Creutzfeldt–Jakob disease (CJD) throughout Europe. Previous statistical surveys of the incidence of the disease were almost exclusively limited to death certificate statistics. No differentiation could be made between sporadic, genetic (familial), and iatrogenic CJD.

Retrospective studies have revealed an annual incidence of 0.31–0.47 per million inhabitants in Great Britain (GB) [1; 2] (Table 36.1) for the period between 1970 and 1989. From 1990 onwards, the incidence was 0.70–0.74 in prospective studies. Shortly thereafter, between 1993 and 1996, it was shown that the increase in incidence was due to methodological reasons. Improved diagnosis and surveillance, especially in the age group of 70 years and older, were important factors [3; 4].

This chapter, together with Chapter 37, provides a systematic overview of the epidemiological parameters and their determining factors in the development of CJD (causes, risk factors, and determinants).

36.2 Descriptive epidemiology

36.2.1 Incidence

The annual incidence of CJD is 1–2 cases per million worldwide (Table 36.2). It is calculated as the number of definite and probable CJD cases, because a post mortem examination for

* This chapter is dedicated to the memory of Sigrid Poser†.

Table 36.1: Annual incidence of Creutzfeldt–Jakob disease. For methodological reasons (i. e., improved monitoring of the disease), the incidence rates increased during the periods shown.

Country	Period	Incidence (cases per million people/year)
Australia	1979–1992	0.75
Austria	1994–1995	1.27 [1)]
Chile	1955–1972	0.10
	1973–1977	0.31
	1978–1983	0.69
Czechoslovakia	1972–1986	0.66
France	1968–1977	0.34
	1978–1982	0.58
	1992–1995	0.96 [1)]
Germany	1979–1990	0.31
	1993–1997	0.85 [1)]
Great Britain	1964–1973	0.09
	1973–1979	0.31
	1980–1984	0.47
	1985–1989	0.46
	1990–1994	0.70 [1)]
	1995–1996	0.74 [1)]
Hungary	1960–1986	0.39
Iceland	1960–1990	0.27
Israel	1963–1987	0.91
	1989–1997	0.90
Italy	1958–1971	0.05
	1972–1986	0.09
	1993–1995	0.56 [1)]
Japan	1975–1977	0.45
Netherlands	1993–1995	0.81 [1)]
New Zealand	1980–1989	0.88
Slovakia	1999–2002	0.48
Sweden	1985–1996	1.20
Switzerland	1999–2002	2.23
United States of America	1973–1977	0.26
	1979–1990	0.90

CJD is also reported to occur in Argentina, Belarus, Belgium, Brazil, Canada, China, Columbia, Egypt, Finland, Greece, India, Indonesia, Iran, Mexico, Northern Ireland, Norway, Oman, Papua New Guinea, Peru, Poland, Portugal, Romania, Senegal, South Africa, Spain, Taiwan, Thailand, Tunisia, Uruguay, Venezuela, West Bengal and Yugoslavia.

1) Results from prospective studies.

Table 36.2: Annual mortality rates in Europe, Canada, and Australia attributed to CJD. Source: [8].

Year	Germany	France	Canada	Australia	UK	Italy	Nether-lands	Austria	Switzer-land
1993	0.4	0.6	–	1.13	0.6	0.5	0.8	0.8	1.44
1994	0.9	0.8	0.07	0.67	0.9	0.6	1.2	1.2	1.42
1995	1.0	1.0	0.10	1.16	0.6	0.5	0.5	1.2	1.27
1996	0.9	1.2	0.44	1.53	0.7	0.9	0.9	1.2	1.41
1997	1.3	1.4	0.60	1.30	1.0	0.8	1.2	0.8	1.41
1998	1.4	1.4	0.80	1.44	1.1	1.1	1.2	1.0	1.27
1999	1.2	1.5	1.02	1.48	1.1	1.4	1.1	0.8	1.13
2000	1.3	1.5	1.14	1.67	0.9	1.0	0.7	1.0	1.40
2001	1.5	1.8	0.97	1.18	1.0	1.4	0.9	1.4	2.5
2002	1.2	1.8	1.12	0.87	1.2	1.4	0.9	0.9	2.63
2003	1.4	1.7	0.85	1.06	1.3	1.3	0.8	1.9	2.34
2004	0.9	–	0.38	0.35	0.8	1.2	0.9	1.0	–

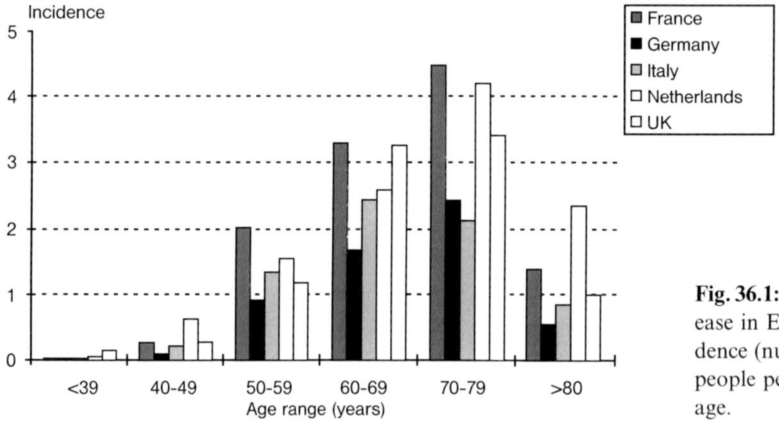

Fig. 36.1: Creutzfeldt–Jakob disease in Europe in 1993–1995. Incidence (number of cases per million people per year), distribution by age.

the diagnosis cannot always be performed. Because of the short disease duration, figures relating to the incidence and mortality are close and mortality figures are used in official statistics.

Between 1993 and 2002, data on 3,720 sporadic CJD patients were analyzed in an EU-funded study (QLRT-1999–31709) [5]. The incidence of sporadic CJD (sCJD) revealed by this study was relatively homogeneous among European countries, Australia, and Canada. The study results show that the average CJD mortality rate was 1.67 per million per annum for all cases and 1.39 for sCJD. Variations between the individual countries can be partly explained by the difference in the number of post mortem examinations. The autopsy rates of CJD suspects vary between countries (e. g., 100% in Slovakia and 49% in Germany) [6]. In recent years, a high sCJD mortality has been reported in Switzerland compared to other countries; probably due to improved surveillance [7]. There have been no cases of variant CJD (vCJD) in Switzerland.

36.2.2 Age-specific incidence and mortality

An earlier study in the years 1993–1995 on age-specific analysis of the incidence shows an identical peak in the age group between 70 and 79 years of age in European countries with the exception of Italy, where the peak of incidence was between 60 and 69. CJD occurs less frequently in patients aged over 80 years (Fig. 36.1), and it is slightly more frequent in women than in men (1.4:1; see Chapter 37). This observation was confirmed in a recent multicenter European study [9].

The distribution of the yearly age- and gender-specific mortality rates showed low rates in people under 50 and a peak mortality in the 60–79 years age group [9]. The analysis of pooled data for three time periods (1993–1995, 1996–1998, and 1999–2002) showed mortality rates from sCJD increasing with time in all age groups. However, this effect was most prominent in older patients [9], and might – for all groups – be due to improved surveillance.

36.2.3 Survival

Predicting factors of survival times have been studied in a large, multicenter study [10]. On the basis of the data obtained from more than 2,300 patients with sCJD or other forms of prion diseases, several potentially disease-modifying factors were studied. Longer survival was associated with female gender in sCJD, genetic CJD in codon 200 (E200K) and 210 (V210I) mutation carriers, and vCJD. Younger age at onset of illness correlated with longer survival in sCJD and vCJD. Codon 129 genotype affected survival in sCJD and fatal familial insomnia (FFI; D178N), and the median disease duration was longer in heterozygous patients [10].

36.2.4 Geographical distribution

The incidence of CJD is low and relatively constant worldwide (Table 36.1). However, the surveillance systems are likely to differ in quality between various countries. In countries with comparable surveillance systems, the figures are very similar.

Clusters or areas of high incidence were observed in the past and suggestions that they could be linked to the codon 200 mutation in the prion protein gene (*PRNP*) have been reported in subsequent studies [11–16]. In addition, several studies have reported time–space clusters of sCJD in various countries [17–22]. Such findings are difficult to interpret. They might occur by chance in a rare disease, might be linked to an increase in disease awareness in some local hospitals, or might be due to an as yet unidentified external risk factor.

36.3 Risk factors for sporadic CJD

36.3.1 Findings from case-control studies

The risk factors for CJD were investigated in several case-control studies. However, the methodologies used were not identical, in particular with respect to control group selection. The results obtained from various case-control studies differed from country to country [9]. Statistically significant findings are summarized in Table 36.3.

In a European case-control study, a total of 405 patients and controls were evaluated. Two risk factors emerged from this evaluation: a family history of dementia and consumption of *meat* [5a; 23]. Other potential risk factors were related to medical risks and the results are described below. Case-control studies in CJD are hampered in general by several factors. Firstly, in a rare disease, only a comparatively small number of patients can be investigated in a study. Therefore, data have to be either pooled from various countries or the investigations

Table 36.3: Investigation of CJD risk factors in case-control studies.

Risk factor	Reference
Surgery	[24; 68–70]
Physical trauma of any kind	[69; 71]
Tonometry 2 years prior to the onset of CJD	[69]
Family history of dementia (after exclusion of familial forms of CJD)	[5a; 72]
Previous herpes zoster infection	[72]
Increased consumption of particular types of meat and brain	[5; 23; 72]
Contact with fish, squirrels, and rabbits	[73]
Contact with leather or organic fertilizers (horn shavings)	[5a]

Table 36.4: Case-control studies; medical risk factors.

n	Medical risk factors	Reference
60 CJD 103 controls	Surgery up to 5 years before onset	[71]
26 CJD 40 controls	Physical trauma or surgery; tonometry	[69]
92 CJD 184 controls	Herpes zoster infection; family history of dementia	[72]
405 CJD 405 controls	Family history of dementia	[5a]
241 CJD 784 controls	Surgery	[70]
326 CJD 326 controls	Surgery	[24]

have to span several years, which might cause methodological problems. Secondly, at the time of CJD diagnosis, patients are usually severely demented. As a result, the information on potential risk factors has to be obtained from a secondary source (proxy interview). Thirdly, the studies attempt to obtain information in retrospect and identify potential risk factors that are likely to have appeared decades before the onset of the disease. The long time period between potential exposure and onset of the disease also contributes to methodological problems of the case-control studies in the field.

36.3.2 Medical risk factors

Several studies have analyzed medical risk factors for sCJD. The findings are given in Table 36.4. Although the design of the studies varied considerably, depending on the types of the controls, surgery of any kind was found to be a risk factor in several of these studies. The most recent study included 326 CJD patients and matched population controls and revealed an increased risk for CJD in patients with a history of medical surgery. However, the risk could not be attributed to a specific kind of surgery [24].

36.3.3 Findings from uncontrolled studies

Several uncontrolled studies have suggested a connection between some factors and CJD (Table 36.5). Given the possibility that medical personnel may be at risk of becoming infected with CJD, the following case reports deserve particular attention. A specialist in internal medicine with no documented exposure to CJD-infected material was diagnosed with CJD. His case history revealed that 30 years earlier he had worked in a department of pathology for 12 months [25]. According to other case reports, one pathologist (26), one neurosurgeon [27], and two histology assistants [28; 29] contracted CJD. Furthermore, an orthopedist was reported to have performed experiments with sheep dura mater 30 years before the onset of CJD [30] (compare Chapters 47–49).

Table 36.5: Investigation of CJD risk factors in uncontrolled studies; no statistical data can be drawn from these studies.

Risk factor	Reference
Craniotomy	[19]
Contact with ferrets and sheep	[74; 75]
Consumption of lamb and mutton	[76]
Medical/paramedical profession	[77]
Family history of dementia	[77]

Controlled case studies, however, showed no increased risk of CJD among medical personnel [5a; 31] (see Chapter 47). The consumption of brain tissue from free-range animals was sometimes associated with an increased risk of sCJD [32; 33].

The observation of an apparently higher level of sCJD among dairy farmers led to the assumption that these cases might be associated with cases of BSE in their herds of cattle [34]. A comparison of the occupation-specific incidence rates between the various European countries revealed an unexpectedly high occurrence of CJD in dairy farmers (four cases per million people per year) but no difference compared with Great Britain [3]. However, considering the advanced age at which the dairy farmers died of CJD (compare Section 36.2.2), these data are not significant. The age-specific incidence among individuals aged between 55 and 74 years without occupational exposure is usually higher than in the overall population [9; 35]. The clinical and neuropathological features observed in dairy farmers – including those in

Great Britain – did not differ from those in sCJD patients. Experimental transmission of the disease to mice showed a similar neuropathological lesion pattern and no significant difference in incubation period compared to sCJD in other patients [36]. In contrast, inoculation of brain tissue homogenates from vCJD patients into the brains of mice resulted in a lesion pattern that differed markedly from that seen in sCJD cases (see Chapter 50).

36.4 Epidemiology and risk factors of acquired forms of CJD

CJD acquired by infection includes iatrogenic CJD (iCJD) and vCJD. Reports of isolated cases of human-to-human transmission had to be revoked in some instances [18; 37; 38]. Recently, Brown reported a case of a married couple in which both partners developed the disease in succession [37], but this isolated case does not permit any conclusions regarding human-to-human transmission.

36.4.1 Variant CJD

By May 2006, 161 vCJD cases had been reported from the UK, 17 cases from France [39], 4 from the Republic of Ireland, 2 from the USA, and 1 each from Italy (Sicily), the Netherlands, Spain, Portugal, Canada, Japan and Saudi Arabia. Recent figures for vCJD in UK and France are shown in Fig. 36.2. Since many factors influence disease susceptibility (age, genetic background, dose of infectious agent, agent strain), it would be speculative to predict the incidence of vCJD based on the currently available data [40].

Results from molecular biological studies and experimental animal models provide evidence of a causal link between vCJD and BSE [36; 41; 42] (see Chapter 50). A recent study linked a higher risk of vCJD with increased beef product consumption [43].

Analyses of *PRNP* revealed no pathogenic mutation in patients with vCJD. However, most patients examined were homozygous for methionine (MM) at codon 129 of *PRNP* (see Chapter 15.9). There is a recent report on a patient heterozygous for methionine and valine (MV) at codon 129 who had received a blood transfusion (see Chapter 51) and died a few years later of an unrelated disease. PrP^{Sc} was identified in lymphoreticular organs but not in the central nervous system (CNS). 129MV heterozygotes might have a longer incubation time than MM homozygotes.

36.4.2 Iatrogenic CJD

On the basis of evidence to date, a human-to-human transmission of the CJD agent can only occur iatrogenically through exposure to infectious tissue [44]. The sources of infection identified in cases of iCJD were dura mater, cornea, and growth or sex hormones extracted from the pituitary glands of human corpses. In some cases, the transmission occurred through the use

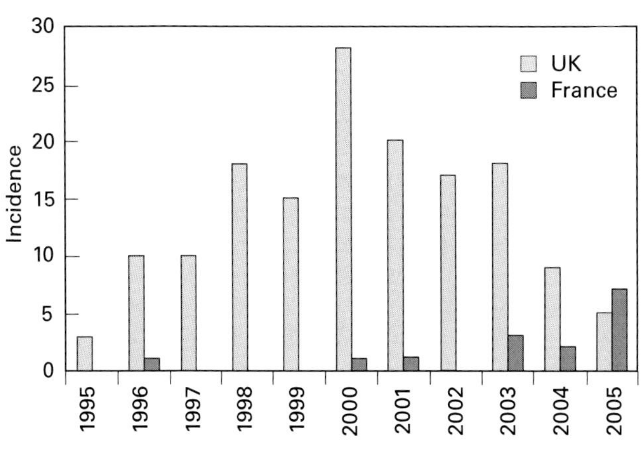

Fig. 36.2: Variant CJD in UK and France.

of insufficiently sterilized surgical instruments during neurosurgical procedures or through implanted EEG electrodes (Table 14.1).

Cornea

The first documented case of iCJD was caused by corneal transplantation. Most probably, it was only identified as iatrogenic in origin because of the short time period of only some months between infection and onset of the disease. In 1974, the 55-year-old female patient was reported to have developed the disease after corneal transplantation [45]. As it turned out later, the donor had died of CJD.

A Japanese research group reported the case of another patient with iCJD, although in this case no data were available on the cornea donor. Therefore, no definite causal link could be established [46]. The third known case of iCJD occurred in Germany. A female patient died of CJD 30 years after receiving corneal transplants from a donor who had also died of CJD [47] (see Chapter 51.3).

Stereotactic EEG and neurosurgical instruments

In 1974, two patients contracted the disease from stereotactic electrodes within 16 and 20 months of the examination. Some weeks earlier, these electrodes had been used on a patient with CJD and then inadequately sterilized [48]. Studies on chimpanzees provided further evidence that the infection is transmissible through contaminated electrodes [49]. Four other cases of iatrogenic CJD were caused by insufficiently sterilized neurosurgical instruments [18; 50].

Dura mater

The transmissibility of CJD through dura mater is well documented. In most cases in which it was possible to trace back the origin of the transplants, the source of infection was lyophilized dura mater from one company. Prior to 1987, dura mater grafts were not subjected to treatment with NaOH, leading to insufficient inactivation of the agent [51; 52]. In more than 97 cases reported from Japan up to 2003, dura mater was the source of infection. Most of these patients acquired the disease through neurosurgical procedures [51; 53]. In Japan, one out of 1250 dura mater recipients developed CJD. No cases have been reported among patients who received the dural graft after 1991 [54]. The sources of infection in most cases were commercial products. Noncommercial dura mater grafts were used in only a few cases [55] (see Chapter 51.4).

Direct contact with the brain surface does not appear to be a prerequisite for infection in dura mater-associated iCJD cases. Dura mater is not only used in neurosurgical operations but also during extracranial operative procedures. In Japan, a case of dura mater-associated iCJD occurred after orthopedic surgery [51]. The following is a summary of other reports of patients who developed iCJD following surgical treatment: a 28-year-old female patient contracted the disease through dura mater within 19 months after cholesteatoma surgery [56]. A 25-year-old patient contracted iCJD after embolization of a nasopharyngeal angiofibroma through the external carotid artery [57]. A third patient developed the disease after embolization of an intercostal artery [58].

Growth hormones (human growth hormone)

Until 1985, patients with primary hypopituitarism were treated with human cadaveric pituitary growth hormone (hGH). Apparently some of the batches were contaminated with the infectious agent and single recipients developed CJD after treatment with cadaveric hGH [59]. The average incubation period in these patients is 12 years (age range: 4.5–30 years) [50].

A total of 103 iCJD cases were registered worldwide until the beginning of 1999. Of these, 61 were reported in France [60], with the remaining cases from GB and the USA. The hormone preparations used in these countries originated from different sources [50]. Some patients in Brazil and New Zealand received hormone preparations from either the USA or Australia [50]. The infectivity of the preparations used was confirmed by animal experiments [61]. According to estimations, about half of the hGH produced in France between January 1984 and March 1985 could have been contaminated [62].

Follicle-stimulating hormone (gonadotropin)

After treatment with human pituitary follicle-stimulating hormone (FSH), five Australian women developed CJD [63]. The first symptoms were noticed within 12 to 16 years after infertility treatment. Two patients were aged 39 and 44 years at the onset of the disease, and they died after about 10 months [64; 65]. As these cases occurred exclusively in Australia, the preparations used most likely came from the same source [63].

Blood transfusion and blood products

There is no evidence for the transmissibility of sCJD by blood or blood products. However, vCJD may very likely be transmissible by blood transfusion [66] (see Chapter 51).

36.4.3 Incubation period and its determining factors

The incubation period of CJD is dependent on dose of infection and route of transmission (central route versus peripheral route). Transmission of the infectious agent through, for example, stereotactic insertion of intracerebral electrodes or through application near the CNS (central route of transmission) results in an incubation period of 18 to 20 months, while transmission via the peripheral route (e. g., by injection of hormones) results in an average incubation period of 12 to 13 years.

The incubation period is additionally determined by genetic factors. Genetic analyses of *PRNP* in normal control samples revealed the following results: 37 % of the control population was homozygous for methionine at codon 129, 12 % was homozygous for valine (VV), and 51 % was heterozygous for methionine and valine (see Chapter 26.4.2). A completely different distribution of the polymorphism at this codon was found in iCJD cases in which the infection occurred via the *central* route: 80 % of these patients were homozygous for methionine. Among the patients who were infected via the peripheral route (i. e., by injection of human growth hormone), the disease occurred particularly often in patients who were homozygous for either methionine or valine at codon 129, while the number of patients with heterozygosity for MV increased after a longer incubation period [67].

References

[1] Cohen CH. Does improvement in case ascertainment explain the increase in sporadic Creutzfeldt–Jakob disease since 1970 in the United Kingdom? Am J Epidemiol 2000; 152(5):474–479.

[2] Will RG, Matthews WB, Smith PG, et al. A retrospective study of Creutzfeldt–Jakob disease in England and Wales 1970–1979. II: Epidemiology. J Neurol Neurosurg Psychiatry 1986; 49(7):749–755.

[3] Cousens SN, Zeidler M, Esmonde TF, et al. Sporadic Creutzfeldt–Jakob disease in the United Kingdom: analysis of epidemiological surveillance data for 1970–96. BMJ 1997; 315 (7105):389–395.

[4] Will RG. Incidence of Creutzfeldt–Jakob disease in the European Community. In: Gibbs CJ Jr, ed. Bovine spongiform encephalopathy. The BSE dilemma. New York: Springer, 1996:364–374.

[5] Pocchiari M, Puopolo M, Croes EA, et al. Predictors of survival in sporadic Creutzfeldt–Jakob disease and other human transmissible spongiform encephalopathies. Brain 2004; 127: 2348–2359.

[5a] Van Duijn CM, Delasnerie-Lauprêtre N, Masullo C, et al. Case-control study of risk factors of Creutzfeldt–Jakob disease in Europe during 1993–95. Lancet 1998; 351(9109):1081–1085.

[6] Will RG, Alpérovitch A, Poser S, et al. Descriptive epidemiology of Creutzfeldt–Jakob disease in six European countries, 1993–1995. Ann Neurol 1998; 43(6):763–767.

[7] Glatzel M, Rogivue C, Ghani A, et al. Incidence of Creutzfeldt–Jakob disease in Switzerland. Lancet 2002; 360(9327):139–141.

[8] EuroCJD. Total cases of CJD/GSS (deaths). Webpage 2006: www.eurocjd.ed.ac.uk./allcjd.htm.

[9] Ladogana A, Puopolo M, Croes EA, et al. Mortality from Creutzfeldt–Jakob disease and re-

lated disorders in Europe, Australia, and Canada. Neurology 2005; 64:1586–1591.
[10] Pocchiari M, Poupolo M, Croes EA, et al. Predictors of survival in sporadic Creutzfeldt–Jakob disease and other human transmissible spongiform encephalopathies. Brain 2004; 10:2348–2359.
[11] Mayer V, Orolin D, Mitrova E. Cluster of Creutzfeldt–Jakob disease and presenile dementia [letter]. Lancet 1977; 2(8031):256.
[12] Kahana E, Alter M, Braham J, et al. Creutzfeldt–Jakob disease: focus among Libyan Jews in Israel. Science 1974; 183(120):90–91.
[13] Galvez S, Masters C, Gajdusek C. Descriptive epidemiology of Creutzfeldt–Jakob disease in Chile. Arch Neurol 1980; 37(1):11–14.
[14] Goldfarb LG, Mitrova E, Brown P, et al. Mutation in codon 200 of scrapie amyloid protein gene in two clusters of Creutzfeldt–Jakob disease in Slovakia [letter]. Lancet 1990; 336(8713):514–515.
[15] Goldfarb LG, Korczyn AD, Brown P, et al. Mutation in codon 200 of scrapie amyloid precursor gene linked to Creutzfeldt–Jakob disease in Sephardic Jews of Libyan and non-Libyan origin. Lancet 1990; 336:637–638.
[16] Brown P, Galvez S, Goldfarb LG, et al. Familial Creutzfeldt–Jakob disease in Chile is associated with the codon 200 mutation of the PRNP amyloid precursor gene on chromosome 20. J Neurol Sci 1992; 112(1-2):65–67.
[17] Collins S, Boyd A, Fletcher A, et al. Creutzfeldt–Jakob disease cluster in an Australian rural city. Ann Neurol 2002; 52(1):115–118.
[18] Matthews WB. Epidemiology of Creutzfeldt–Jakob disease in England and Wales. J Neurol Neurosurg Psychiatry 1975; 38(3):210–213.
[19] Will RG and Matthews WB. Evidence for case-to-case transmission of Creutzfeldt–Jakob disease. J Neurol Neurosurg Psychiatry 1982; 45(3):235–238.
[20] Huillard d'Aignaux JN, Cousens S, Delasnerie-Lauprêtre N, et al. Analysis of the geographical distribution of sporadic Creutzfeldt–Jakob disease in France between 1992 and 1998. Int J Epidemiol 2002; 31:495–496.
[21] Arakawa K, Nagara H, Itoyama Y, et al. Clustering of three cases of Creutzfeldt–Jakob disease near Fukuoka City, Japan. Acta Neurol Scand 1991; 84(5):445–447.
[22] Farmer PM, Kane WC, Hollenberg SJ. Incidence of Creutzfeldt–Jakob disease in Brooklyn and Staten Island [letter]. N Engl J Med 1978; 298(5):283–284.
[23] Davanipour Z, Alter M, Sobel E, et al. A case-control study of Creutzfeldt–Jakob disease. Dietary risk factors. Am J Epidemiol 1985; 122(3): 443–451.
[24] Ward HJT, Everington D, Croes EA, et al. Sporadic Creutzfeldt–Jakob disease and surgery – a case control study using community controls. Neurology 2002; 59(4):543–548.
[25] Berger JR and David NJ. Creutzfeldt–Jakob disease in a physician: a review of the disorder in health care workers. Neurology 1993; 43(1):205–206.
[26] Gorman DG, Benson DF, Vogel DG, et al. Creutzfeldt–Jakob disease in a pathologist. Neurology 1992; 42(2):463.
[27] Schoene WC, Masters CL, Gibbs C Jr, et al. Transmissible spongiform encephalopathy (Creutzfeldt–Jakob disease). Atypical clinical and pathological findings. Arch Neurol 1981; 38(8):473–477.
[28] Miller DC. Creutzfeldt–Jakob disease in histopathology technicians. N Engl J Med 1988; 318: 853–854.
[29] Sitwell L, Lach B, Atack E, et al. Creutzfeldt–Jakob disease in histopathology technicians. New Engl J Med 1988; 318(13):854.
[30] Weber T, Tumani H, Holdorff B, et al. Transmission of Creutzfeldt–Jakob disease by handling of dura mater [letter]. Lancet 1993; 341 (8837):123–124.
[31] Mitrova E and Belay G. Creutzfeldt–Jakob disease in health professionals in Slovakia. Europ J Epidemiol 2000; 16:353–355.
[32] Berger JR, Waisman E, Weisman B. Creutzfeldt–Jakob disease and eating squirrel brains. Lancet 1997; 350(9078):642.
[33] Kamin M and Patten BM. Creutzfeldt–Jakob disease. Possible transmission to humans by consumption of wild animal brains. Am J Med 1984; 76(1):142–145.
[34] Almond JW, Brown P, Gore SM, et al. Creutzfeldt–Jakob disease and bovine spongiform encephalopathy: any connection? BMJ 1995; 311(7017):1415–1421.
[35] Hörnlimann B. Epidemiology of BSE and CJD in Switzerland. AAJM 1998; 7:45–47.
[36] Bruce ME, Will RG, Ironside JW, et al. Transmissions to mice indicate that "new variant" CJD is caused by the BSE agent. Nature 1997; 389(6650):498–501.
[37] Brown P, Cervenakova L, McShane L, et al. Creutzfeldt–Jakob disease in a husband and wife. Neurology 1998; 50(3):684–688.
[38] Hainfellner JA, Jellinger K, Budka H. Testing

for prion protein does not confirm previously reported conjugal CJD. Lancet 1996; 347(9001): 616–617.
[39] Chazot G, Broussolle E, Lapras C, et al. New variant of Creutzfeldt–Jakob disease in a 26-year-old French man. Lancet 1996; 347(9009): 1181.
[40] Cousens SN, Vynnycky E, Zeidler M, et al. Predicting the CJD epidemic in humans. Nature 1997; 385(6613):197–198.
[41] Collinge J, Sidle KC, Meads J, et al. Molecular analysis of prion strain variation and the aetiology of "new variant" CJD. Nature 1996; 383(6602):685–690.
[42] Lasmezas CI, Deslys JP, Demalmay R, et al. BSE transmission to macaques. Nature 1996; 381 (6585):743–744.
[43] Ward H. Variant Creutzfeldt–Jakob disease in the UK 1995–2003: a case-control study of potential risk factors. In: First International Conference of the European Network of Excellence NeuroPrion 2004. Paris, May 25, 2004:40.
[44] Brown P, Preece M, Brandel J-P, et al. Iatrogenic Creutzfeldt–Jakob disease at the millennium. Neurology 2000; 55:1075–1081.
[45] Duffy P, Wolf J, Collins G, et al. Possible person-to-person transmission of Creutzfeldt–Jakob disease. N Engl J Med 1974; 290(12):692–693.
[46] Uchiyama K, Ishida C, Yago S, et al. An autopsy case of Creutzfeldt–Jakob disease associated with corneal transplantation. Dementia 1994; 8:466–473.
[47] Heckmann JG, Lang CJ, Petruch F, et al. Transmission of Creutzfeldt–Jakob disease via a corneal transplant. J Neurol Neurosurg Psychiatry 1997; 63(3):388–390.
[48] Bernoulli C, Siegfried J, Baumgartner G, et al. Danger of accidental person-to-person transmission of Creutzfeldt–Jakob disease by surgery [letter]. Lancet 1977; 1(8009):478–479.
[49] Gibbs C Jr, Asher DM, Kobrine A, et al. Transmission of Creutzfeldt–Jakob disease to a chimpanzee by electrodes contaminated during neurosurgery. J Neurol Neurosurg Psychiatry 1994; 57(6):757–758.
[50] Brown P. Environmental causes of human spongiform encephalopathy. In: Baker HF, Ridley RM, editors. Prion diseases. Humana Press, Totowa, NJ, 1996:139–154.
[51] Centers for Disease Control and Prevention (CDC). Creutzfeldt–Jakob disease associated with cadaveric dura mater grafts in Japan, January 1979–May 1996. Jama 1998; 279:11–12.
[52] Diringer H and Braig HR. Infectivity of unconventional viruses in dura mater [letter]. Lancet 1989; 1(8635):439–440.
[53] Matusi S, Sadaike T, Hamada C, et al. Creutzfeldt–Jakob disease and cadaveric dura mater grafts in Japan: an updated analysis of incubation time. Neuroepidemiology 2004; 24:22–25.
[54] Centers for Disease Control and Prevention (CDC). Update: Creutzfeldt–Jakob disease associated with cadaveric dura mater grafts – Japan, 1979–2003. MMWR Morb Mortal Wkly Rep 2003; 48:1179–1181.
[55] Pocchiari M, Masullo C, Salvatore M, et al. Creutzfeldt–Jakob disease after non-commercial dura mater graft [letter]. Lancet 1992; 340(8819): 614–615.
[56] Thadani V, Penar PL, Partington J, et al. Creutzfeldt–Jakob disease probably acquired from a cadaveric dura mater graft. Case report. J Neurosurg 1988; 69(5):766–769.
[57] Antoine JC, Michel D, Bertholon P, et al. Creutzfeldt–Jakob disease after extracranial dura mater embolization for a nasopharyngeal angiofibroma. Neurology 1997; 48(5):1451–1453.
[58] Defebvre L, Destee A, Caron J, et al. Creutzfeldt–Jakob disease after an embolization of intercostal arteries with cadaveric dura mater suggesting a systemic transmission of the prion agent. Neurology 1997; 48(5):1470–1471.
[59] Koch TK, Berg BO, De AS, et al. Creutzfeldt–Jakob disease in a young adult with idiopathic hypopituitarism. Possible relation to the administration of cadaveric human growth hormone. N Engl J Med 1985; 313(12):731–733.
[60] Goullet D. The effect of non-conventional transmissible agents (prions) on disinfection and sterilization processes. World Symposium on Central Service in Hospitals 1999. Zentr Steril 1999; 7(5):305–318.
[61] Gibbs C Jr, Asher DM, Brown PW, et al. Creutzfeldt–Jakob disease infectivity of growth hormone derived from human pituitary glands [letter]. N Engl J Med 1993; 328(5):358–359.
[62] Huillard d'Aignaux J, Alperovitch A, Maccario J. A statistical model to identify the contaminated lots implicated in iatrogenic transmission of Creutzfeldt–Jakob disease among French human growth hormone recipients. Am J Epidemiol 1998; 147(6):597–604.
[63] Cooke J. Cannibals, cows and the CJD catastrophe. Random House, Sydney, Australia 1998.
[64] Cochius JI, Burns RJ, Blumbergs PC, et al. Creutzfeldt–Jakob disease in a recipient of human pituitary-derived gonadotropin. Aust N Z J Med 1990; 20(4):592–593.

[65] Cochius JI, Hyman N, Esiri MM. Creutzfeldt–Jakob disease in a recipient of human pituitary-derived gonadotrophin: a second case. J Neurol Neurosurg Psychiatry 1992; 55(11):1094–1095.
[66] Klein R and Dumble LJ. Transmission of Creutzfeldt–Jakob disease by blood transfusion [letter]. Lancet 1993; 341(8847):768.
[67] Deslys JP, Jaegly A, d'Aignaux JH, et al. Genotype at codon 129 and susceptibility to Creutzfeldt–Jakob disease [letter]. Lancet 1998; 351 (9111):1251.
[68] Kondo K. Epidemiology of Creutzfeldt–Jakob and related disorders [author's translation]. No To Shinkei 1982; 34(5):451–463.
[69] Davanipour Z, Alter M, Sobel E, et al. Creutzfeldt–Jakob disease: possible medical risk factors. Neurology 1985; 35(10):1483–1486.
[70] Collins S, Law MG, Fletcher A, et al. Surgical treatment and risk of sporadic Creutzfeldt–Jakob disease: a case-control study. Lancet 1999; 353(9154):693–697.
[71] Kondo K and Kuroiwa Y. A case control study of Creutzfeldt–Jakob disease: association with physical injuries. Ann Neurol 1982; 11(4):377–381.
[72] Harries-Jones R, Knight R, Will RG, et al. Creutzfeldt–Jakob disease in England and Wales, 1980–1984: a case-control study of potential risk factors. J Neurol Neurosurg Psychiatry 1988; 51(9):1113–1119.
[73] Davanipour Z, Alter M, Sobel E, et al. Transmissible virus dementia: evaluation of a zoonotic hypothesis. Neuroepidemiology 1986; 5(4):194–206.
[74] Lo Russo F, Neri G, Figà-Talamanca L. Creutzfeldt–Jakob disease and sheep brain. A report from Central and Southern Italy. Ital J Neurol Sci 1980; 1(3):171–174.
[75] Matthews WB, Campbell M, Hughes JT, et al. Creutzfeldt–Jakob disease and ferrets [letter]. Lancet 1979; 1(8120):828.
[76] Alter M, Hoenig E, Pratzon G. Creutzfeldt–Jakob disease: possible association with eating brains [letter]. N Engl J Med 1977; 296(14):820–821.
[77] Masters CL, Harris JO, Gajdusek DC, et al. Creutzfeldt–Jakob disease: patterns of world-wide occurrence and the significance of familial and sporadic clustering. Ann Neurol 1979; 5(2):177–188.

37 Creutzfeldt–Jakob Disease in Germany

Inga Zerr, Sigrid Poser*, and Hans Kretzschmar

37.1 Introduction

Creutzfeldt–Jakob disease (CJD) is a rapidly progressive disease of the central nervous system (CNS). In addition to a dramatic cognitive decline, focal neurological deficits (ataxia, myoclonus) occur early in the disease. Akinetic mutism is seen later in the disease course. In countries with good medical care, patients with CJD, sooner or later, are likely to be examined by a neurologist or psychiatrist. The prerequisite for the following prospective study is that all cases of suspected CJD be referred to the epidemiological research team by a neurologist or psychiatrist while the patient is hospitalized. Other epidemiological studies have shown that the number of notified cases increases as a result of improved registration procedures. In the United Kingdom (UK), the increase in the number of registered patients with CJD aged over 70 years was attributable to improved diagnostic tests and increased awareness of the disease [1].

37.2 German CJD surveillance study

37.2.1 Study design

On the basis of the use of a uniform set of criteria, clinical neurological examinations, post mortem examinations, and genetic analyses have been performed in cases of suspected CJD in Germany since June 1, 1993. The methodology used in this study was adopted from the Creutzfeldt–Jakob Disease Surveillance Unit in Edinburgh. For the first time in 1993, the epidemiological research group in Göttingen, Germany, requested the cooperation of 1,300 neurologists, psychiatrists, and geriatricians in notifying cases of suspected CJD (Fig. 37.1). Since then, these colleagues, who are for the most part hospital physicians, have been regularly contacted at 4-month intervals. This epidemiological study is based on voluntary cooperation by the hospitals. Up to now, 89% of the cases have been reported by neurologists, 8% by psychiatrists, and the other cases by general practitioners, neuropathologists, etc., or in some cases, directly by the family members of the patients, in particular during the first years of the study. In a first step, a physician of the research group performs a physical examination of the patient with suspected CJD in the notifying hospital. Subsequently, the comprehensive case history, physical examinations, and previous medical history of the patient up to the time of the examination by the study physician are evaluated. In addition, copies are made of the electroencephalogram (EEG), the imaging scans (computed CT), and the magnetic resonance tomography (MRT). A cerebrospinal fluid (CSF) investigation is carried out in order to exclude an inflammatory disease and to detect 14–3–3 proteins (see Chapter 31).

The family members of the patients are requested to complete an epidemiological questionnaire, which includes detailed questions on personal and family case history as well as potential risk factors.

Following the physical examination in the patient's hospital, each case is discussed by the research team and classified according to the criteria for CJD [2] (Tab. 29.1 and 31.1). The reporting physicians are informed in writing about the classification of the case into the category of either "probable cases", "possible

* This chapter is dedicated to the memory of Sigrid Poser†.

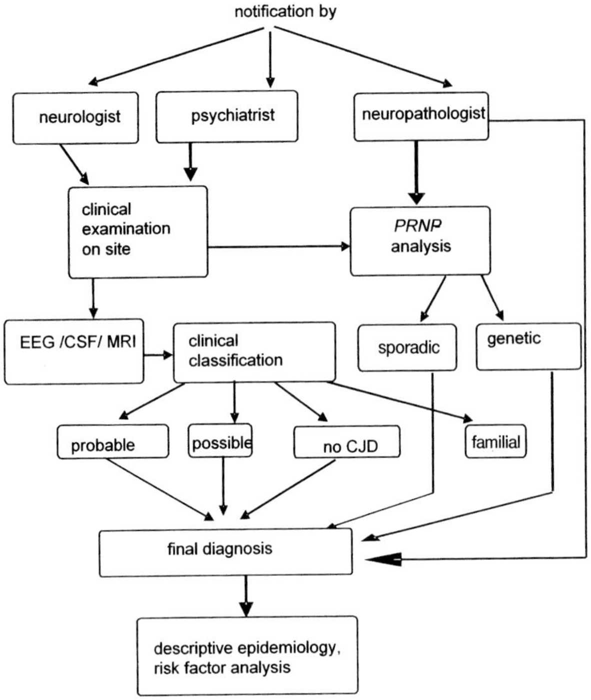

EEG = Electroencephalogram
CSF = Cerebrospinal fluid
MRI = Magnetic resonance imaging

Fig. 37.1: Study design. Figure by Inga Zerr ©.

cases", or "other cases", along with all potential differential diagnoses taken into account by the epidemiological research team. In order to be able to follow the disease progression, the study physicians regularly contact family members of the patient and the treating physicians. It is thus possible to collect new data and to reclassify the patient while s/he is still alive or after post mortem examination.

37.2.2 Diagnosis

Currently, the definite diagnosis of CJD can only be made by neuropathological examination [3]. The clinical diagnostic classification is made while the patient is still alive and is based on standardized criteria. Apart from the clinical symptoms, EEG and detection of the 14–3–3 proteins in CSF are included in the classification [4] (see Chapter 31). A genetic analysis of the prion protein gene (*PRNP*) is always desirable in order to detect a possible genetic predisposition or mutation. Prior to analysis, the family members of the patient decide on what is to be done with the result of the genetic analysis. On the basis of comprehensive counseling by the study physician, all family members together discuss whether or not they wish to be informed about the result. Whenever possible, the family should reach a consensus [5]. A genetic counseling center is involved to inform the relatives of the results of genetic tests.

37.3 CJD epidemiology in Germany

37.3.1 General description

A total of 1,834 cases of suspected CJD were included in the epidemiological study from June 1, 1993 to September 30, 2004. The diagnostic

Table 37.1: Classification of notified suspected cases of human prion diseases in Germany (6/1993–9/2004).

	All	Female	Male	f:m ratio
Definite CJD	623	357	266	1.3:1
Probable sCJD	530	312	218	1.4:1
Possible sCJD	148	80	68	1.2:1
Iatrogenic (iCJD)	6	3	3	1:1
GSS (Gerstmann–Sträussler–Scheinker disease)	10*			
Familial CJD/genetic CJD (fCJD/gCJD)	57*			
FFI (fatal familial insomnia)	26*			
No prion disease	434	271	163	1.7:1
All	1834	1074	760	1.4:1

* The total of 93 cases of inherited prion diseases occurred in 51 female and 42 male patients (f : m ratio = 1.2 : 1).

Table 37.2: The most common genetic prion diseases in Germany (1993–2004).

Mutation	Disease	N
D178N-129M	Fatal familial insomnia (FFI)	26
E200K	Genetic CJD (gCJD/fCJD)	19
V210I	Genetic CJD (gCJD/fCJD)	12
P102L	Gerstmann–Sträussler–Scheinker disease (GSS)	10

classification of the cases into "probable" or "possible" sporadic CJD (sCJD), familial or iatrogenic CJD (iCJD), or "other cases" is shown in Table 37.1. The mutation at codon 178 in conjunction with methionine at codon 129 (D178N-129M), that is associated with fatal familial insomnia (FFI), is currently the most common mutation in Germany and accounts for 2 % of all prion disease cases. It was detected in 26 patients. A family history was often negative or not known to the relatives [6; 7]. The proportion of familial CJD or genetic CJD cases varies among countries in Europe [8]. In Germany, mutations of *PRNP* are found in 10 % of patients tested. The most common mutations are found at codon 200 and at codon 210 [8; 9] (Table 37.2).

A total of six patients suffered from iCJD (Table 37.3). In five of these six cases, the disease was caused by dura mater transplants. The median incubation time between dura mater graft and disease was 13 years (range: 10–20). One case was attributed to an accidental transmission via corneal graft [10].

Since the beginning of the study, the number of patients reported to the CJD Surveillance Unit in Göttingen has continuously increased by 20 % during the past years. The incidence and mortality data are shown in Table 37.4. There may be two major reasons for this observation. First, cooperation has improved, and second, the attitude toward notification has

Table 37.3: Iatrogenic CJD in Germany.

	Pat. 1	Pat. 2	Pat. 3	Pat. 4	Pat. 5	Pat. 6
Year at onset	1994	1995	1997	1999	2000	2002
Age at onset	39	45	57	42	57	65
Duration (months)	3	11	2	4	10	12
Sex	m	f	f	m	m	f
Exposure	dura (1984)	cornea (1965)	dura (1987)	dura (1986)	dura (1980)	dura (1987)
Incubation time (years)	10	30	10	13	20	15
Classification	probable	probable	definite	probable	definite*	probable

* Kretzschmar et al., 2003 [11].

Table 37.4: Incidence and mortality in Germany.

	Incidence°	Mortality°
1993	0.7*	0.4*
1994	0.9	0.9
1995	1.1	1.0
1996	1.1	0.9
1997	1.3	1.3
1998	1.4	1.4
1999	1.3	1.2
2000	1.3	1.3
2001	1.5	1.5
2002	1.2	1.2
2003	1.4	1.1
2004, 30 Sept.	1.3*	0.9*

° per 1 million
* extrapolated

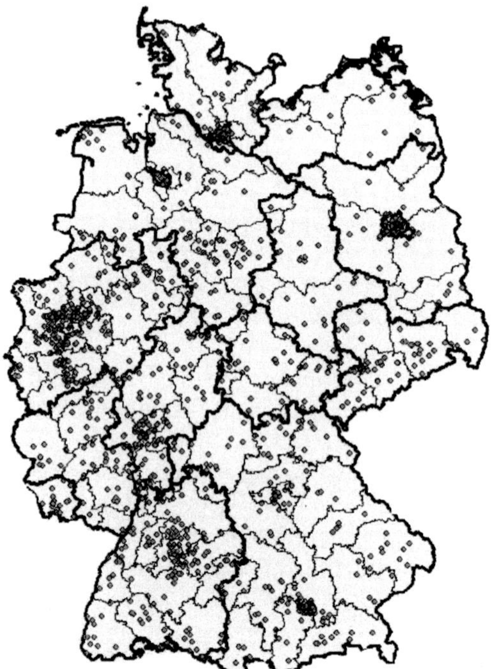

Fig. 37.2: Sporadic CJD in Germany (geographic distribution by place of residence, 6/93–6/04).

Table 37.5: Age at onset of disease and classification.

CJD	Minimum	Median	Maximum
Definite	19	66	88
Probable	35	67	88
Possible	23	68	90
No CJD	14	64	88

changed. It appears that there is a growing tendency to report a case of suspected CJD on the basis of clinical suspicion alone. Furthermore, improved diagnostic tests have made it possible to identify more patients with early-stage CJD. One reliable and helpful diagnostic method in this context is the analysis of the CSF.

37.3.2 Calculation of the incidence

The incidence in Germany is around 0.9–1.5 case/year/million inhabitants for the duration of the observation period [10; 12] (Table 37.4).

In order to allow a comparison of the incidence in Germany with that found in other European studies, the disease mortality rate is also given in Table 37.4.

Figure 37.2 shows a distribution of sCJD in Germany according to the patients' place of residence. No time–space cluster of sCJD can be seen, and the numbers of sporadic cases are proportional to the population density.

37.3.3 Age- and gender-related incidence

Table 37.5 shows the age distribution of the patients and their clinical classification. The youngest CJD patient was 19 years old at the onset of the disease and the oldest patient with definite CJD was 88 years old.

Table 37.6: Age-specific incidence (per 1 million per year) stratified by gender.

	≤ 39	40–49	50–59	60–69	70–79	≥ 80
Female	0.06	0.4	2.41	6.09	5.95	1.28
Male	0.01	0.3	2.02	4.89	6.76	1.85
All	0.08	0.35	2.21	5.52	6.22	1.43

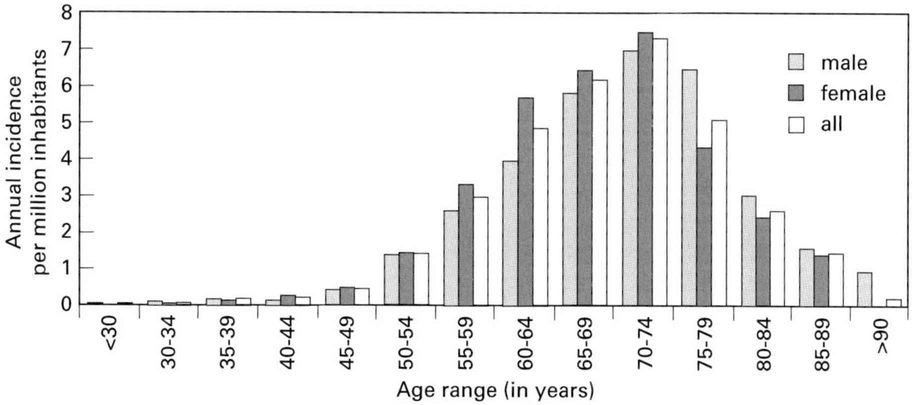

Fig. 37.3: Age-specific incidence stratified by gender and 5-year periods.

Women are more frequently affected than men (Table 37.1), which is particularly evident in the age group between 60 and 69 years. The age-related annual incidence is 4.82 per million women and 3.2 for men (Table 37.6). The age-related incidence is lowest in the group of patients below 39 years of age (0.03/year/million inhabitants). While the age-related incidence increases with age and is highest in the group of patients aged 60–69 years, it decreases in the group of patients aged above 74 years (Fig. 37.3).

37.3.4 Post mortem examination

The consent for post mortem examination is always sought in order to be able to make a definite diagnosis. In a few cases of refusal of consent by family members, no post mortem examination could be carried out. The overall CJD post mortem rate is good, especially in view of the otherwise continuously decreasing willingness in Germany to consent to a post mortem examination [13]. This success is also due to the guidance and encouragement provided to family members by the study physicians. To some extent, the autopsy rate is related to the clinical classification of the patients after inclusion in the study (62 % for "probable", 48 % for "possible" CJD, and 50 % for those who were classified as "other cases").

37.3.5 Young sporadic CJD patients

During the 10 years of the study, 54 patients under the age of 50 years were diagnosed with definite and probable sCJD (Table 37.7). Their clinical course often started with psychiatric abnormalities such as affective disorders, aggression, personality change, depression, hallucinations, and paranoia. However, in contrast to vCJD, dementia and focal neurological signs followed early in the disease. They had a longer disease duration compared with the older CJD patients (Fig. 37.4). Interestingly, only a minor proportion had typical periodic sharp wave complexes (PSWCs) in the EEG (23 % vs. 66 %), and the sensitivity of other tests such as

Table 37.7: Notified and examined suspected CJD with age at onset below 50.

	Notified cases	Definite and probable
1994	17	3
1995	13	2
1996	29	5
1997	11	6
1998	21	8
1999	12	7
2000	6	2
2001	14	6
2002	13	5
2003	16	7
9/2004	6	3
Total	158	54

Table 37.8: Diagnostic tests in CJD patients stratified by age at onset of disease.

Age	PSWCs in EEG	14–3–3 in CSF	Hyperintense basal ganglia in MRI	Codon 129 genotype MM	MV	VV
< 50	23 %	86 %	35 %	52 %	15 %	33 %
> 50	66 %	94 %	63 %	68 %	16 %	16 %

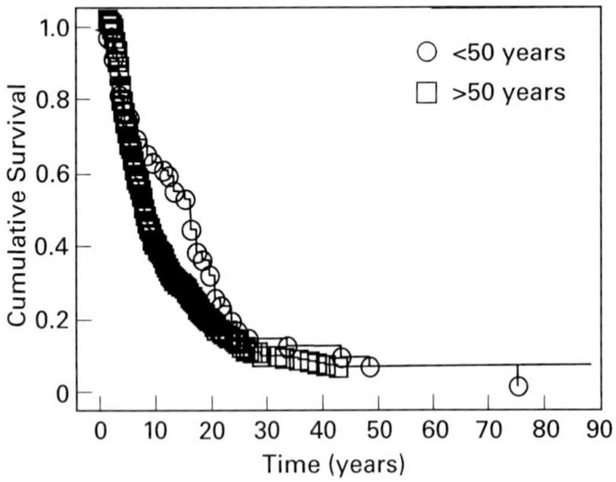

Fig. 37.4: Cumulative survival time stratified by age at onset of disease.

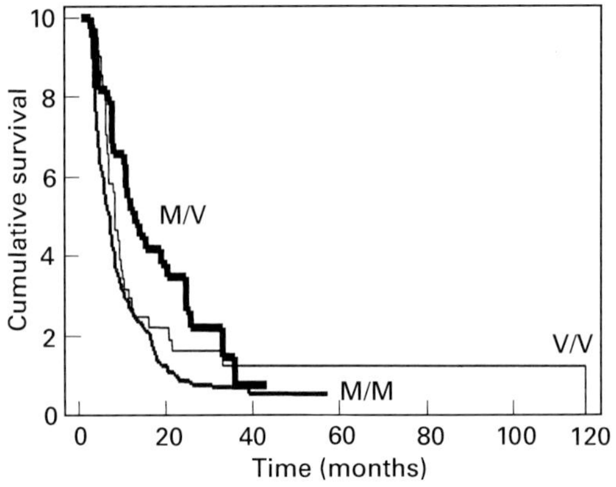

Fig. 37.5: Cumulative survival in sporadic CJD stratified by codon 129 genotype.

14–3–3 and MRI is also lower (Table 37.8). There was no apparent increase in numbers during the years 1993–2004. Six to eight patients below 50 at onset were diagnosed with sCJD per year. Codon 129 distribution in these patients differs from that seen in all CJD groups (Table 37.8); this may explain the divergent EEG, CSF, and MRI findings. The most prominent finding is the high frequency of valine homozygous patients in the group with disease onset below 50.

37.4 Prognostic factors in sporadic CJD

Codon 129 genotype is known to influence disease susceptibility and incubation time in acquired CJD and the disease phenotype in sCJD. No effect on gender distribution was observed in our population (Table 37.9), and the predominance of females was observed in all three codon 129 genotypes. The effect of genotype on survival time in sCJD was more dramatic. Patients homozygous for methionine or valine had a shorter survival time than methione/valine heterozygous patients (Fig. 37.5).

The predictors of survival in sCJD and other prion diseases were recently analyzed in a large European multicenter study [14]. The study population comprised 2,304 sCJD cases. In sCJD, the longer survival correlated with younger age at onset of illness, female gender, codon 129 heterozygosity, and type 2a of the prion protein. One interesting observation concerned a clear decrease in the survival time with increasing age. This observation was made in several studies. The reason for this is not clear and may reflect age-related variation in disease

Table 37.9: Codon 129 genotype and gender in sCJD patients.

	Female	Male	f:m ratio	All
M/M	389	254	1.5:1	643
M/V	127	83	1.5:1	210
V/V	91	68	1.3:1	159

resistance to terminal infection. The influence of female gender on disease survival (longer in females than in males) may indicate the influence of sex-specific factors that influence the clinical phase of the disease, but possibly also hormonal effects, genetic determinants outside *PRNP,* or neuroanatomical factors [14].

References

[1] Cousens SN, Zeidler M, Esmonde TF, et al. Sporadic Creutzfeldt–Jakob disease in the United Kingdom: analysis of epidemiological surveillance data for 1970–1996. Br Med J 1997; 315(7105):389–395.

[2] Zerr I, Pocchiari M, Collins S, et al. Analysis of EEG and CSF 14–3–3 proteins as aids to the diagnosis of Creutzfeldt–Jakob disease. Neurology 2000; 55(6):811–815.

[3] Kretzschmar HA, Ironside JW, DeArmond SJ, et al. Diagnostic criteria for sporadic Creutzfeldt–Jakob disease. Arch Neurol 1996; 53:913–920.

[4] Steinhoff BJ, Zerr I, Glatting M, et al. Diagnostic value of periodic complexes in Creutzfeldt–Jakob disease. Ann Neurol 2004; 56(5):702–708.

[5] Kropp S, Riedemann C, Zerr I, et al. Do relatives of Creutzfeldt–Jakob patients want to become genetically enlightened? [letter]. Dtsch Med Wochenschr 1998; 123(34–35):1023.

[6] Kretzschmar H, Giese A, Zerr I, et al. The German FFI cases. Brain Pathol 1998; 8(3):559–561.

[7] Zerr I, Giese A, Windl O, et al. Phenotypic variability in fatal familial insomnia (D178N-129M) genotype. Neurology 1998; 51:1398–1405.

[8] Krasemann S, Zerr I, Weber T, et al. Prion disease associated with a novel nine octapeptide repeat insertion in the PRNP gene. Brain Res Mol Brain Res 1995; 34:173–176.

[9] Noch S, Zerr I, Poser S, et al. A novel insertion mutant comprising nine extra octapeptide repeats within the human prion gene. Med Genetik 1995; 2:266–267.

[10] Poser S, Mollenhauer B, Krauss A, et al. How to improve the clinical diagnosis of Creutzfeldt–Jakob disease. Brain 1999; 122(Pt 12):2345–2351.

[11] Kretzschmar HA, Sethi S, Foldvari Z, et al. Iatrogenic Creutzfeldt–Jakob disease with florid plaques. Brain Pathol 2003; 13(3):245–249.

[12] Zerr I, Bodemer M, Gefeller O, et al. Detection of 14–3–3 protein in the cerebrospinal fluid supports the diagnosis of Creutzfeldt–Jakob disease. Ann Neurol 1998; 43(1):32–40.

[13] Höpker WW and Wagner S. Die klinische Obduktion. Eine nicht verzichtbare Maßnahme einer Medizin im Wandel. Dtsch Arztebl 1998; 95(25):A1596–A1600.

[14] Pocchiari M, Puopolo M, Croes EA, et al. Predictors of survival in sporadic Creutzfeldt–Jakob disease and other human transmissible spongiform encephalopathies. Brain 2004; 127(10): 2348–2359.

38 The Epidemiology of Kuru

Michael P. Alpers and Beat Hörnlimann

38.1 Introduction

Kuru is an invariably fatal neurological disease with a subacute course of 12 months on average. Epidemiological surveillance of kuru has been continuous from 1957 to the present. According to the many observations and analyses of epidemiological data on kuru, the disease was transmitted by means of oral transmission through ingestion of infected brain matter, though in a few cases parenteral transmission may have occurred [1].

38.2 Frequency of cases and progression of the epidemic

In the middle of the 20th century, the population affected by kuru in the remote southern part of the Eastern Highlands of Papua New Guinea (PNG) numbered approximately 30,000–40,000, of whom 11,000 were Fore [2]. Overall since 1957, when systematic kuru research was begun, approximately 2,700 patients have been affected by kuru.

From 1957 to 1959 more than 200 individuals fell victim to this disease and died from it each year (Fig. 38.1). In 1960, for the first time, there was a decline in numbers to less than 200 cases per year. In the 1960s there were 130 patients dying per year on average and in the 1970s just over 40 patients per year. In addition to the decline in the number of deaths, in the 1960s the average age of those afflicted started to go up. This was due to a decrease in the age-specific mortality among the 0–9 year olds (Fig. 38.2: 0–9; Table 38.1). Already in the mid-1960s the mortality among this age group was almost zero. The next (accelerated) decline in mortality occurred among the age group of 10–19 year olds during the second half of the 1960s. Among this group the age-specific zero-mortality was reached in 1974 (Fig. 38.2: *10–19*). The declines in age groups 20–29 years and 30–39 years were almost parallel but at a higher level; the slower decline in patients dying at 40 years and older started speeding up in 1969 (Fig. 38.2: *20–29*, *30–39* and *40+*).

In the first half of the 1980s approximately 15 people died from kuru each year and in the second half there were about 10 per year. In the first half of the 1990s about 6 people died yearly until 1995, and since then, on average, only between one and two individuals annually have died from kuru (Fig. 38.1). In 1987 it was predicted that kuru cases would still occur for at least another decade [3], based on the epidemiological patterns of decline in villages where kuru no longer occurs. The fact that cases were still being reported as late as 2005 allows the conclusion that an incubation period of more than 45 years is possible. The epidemic curve is most clearly depicted when data are plotted in 5-year periods, which is possible now that the span of kuru investigation exceeds 45 years. Figure 38.3 shows this in a preliminary way, with some points missing, since not all the epidemiological data have yet been extracted for analysis.

Though they were not aware of the interrelation between the disease and endocannibalism, the Australian government of the Trust Territory effectively suppressed cannibalism in the Okapa area of the Eastern Highlands beginning in 1954. Thereafter, endocannibalistic rituals of transumption of dead relatives in the kuru region were aggressively stopped until they were completely abandoned by 1960.

Retrospectively speaking, the age-specific dynamics in the decline in incidence since 1959 (Fig. 38.1) are consistent with the cessation of

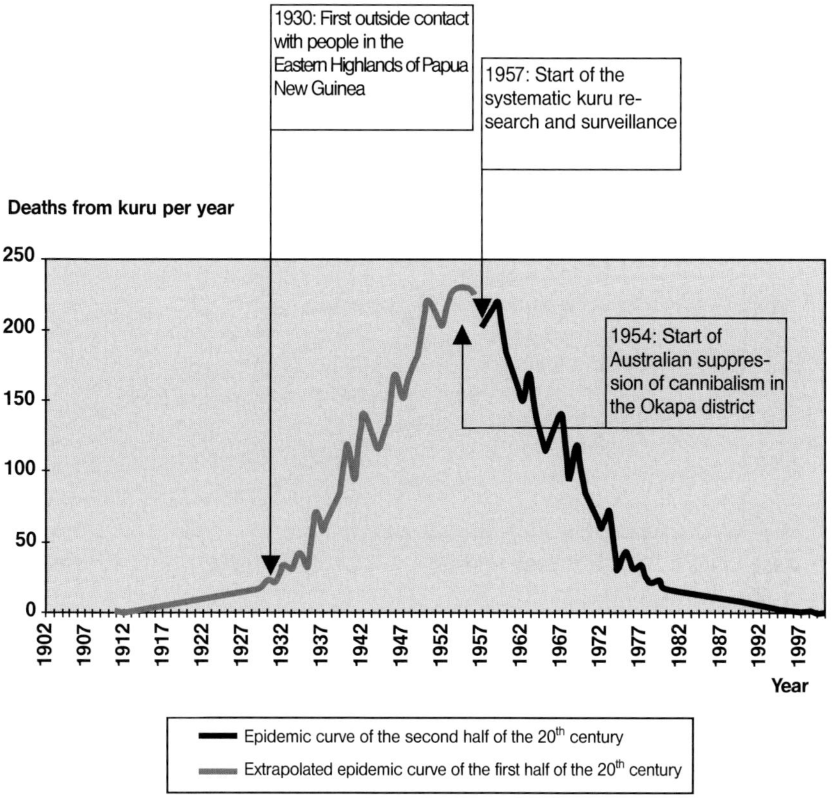

Fig. 38.1: Development of the kuru epidemic during the 20th century. The course observed until 1956 is not based on actual data but on a reconstruction of the course since 1957 by mirroring the curve, though its shape is generally consistent with the oral history given by the local people (see Chapter 2). It can be said with certainty that the peak of the curve was in the 1950s. Original figure by B. Hörnlimann and M.P. Alpers.

Table 38.1: Age distribution of the youngest ever kuru patients (according to age at time of death). These cases were recorded between 1957 and 1966. After 1966, no children of these age groups fell ill with kuru. Data from [1].

Age (years at death)	Number of kuru patients per year									
	1957	1958	1959	1960	1961	1962	1963	1964	1965	1966
4	2	0	0	0	0	0	0	0	0	0
5	3	3	2	1	2	0	0	0	0	0
6	8	5	3	3	0	0	0	0	0	0
7	16	6	8	4	0	2	1	0	0	1
Total	29	14	13	8	2	2	1	0	0	1

endocannibalism – presenting a break in the infectious chain of events.

We note that it is essential to view the mortuary practices involving endocannibalism objectively and not to judge these traditions from an outsider's point of view. The traditions had great strength and meaning to the Fore people but there is not space here to discuss the com-

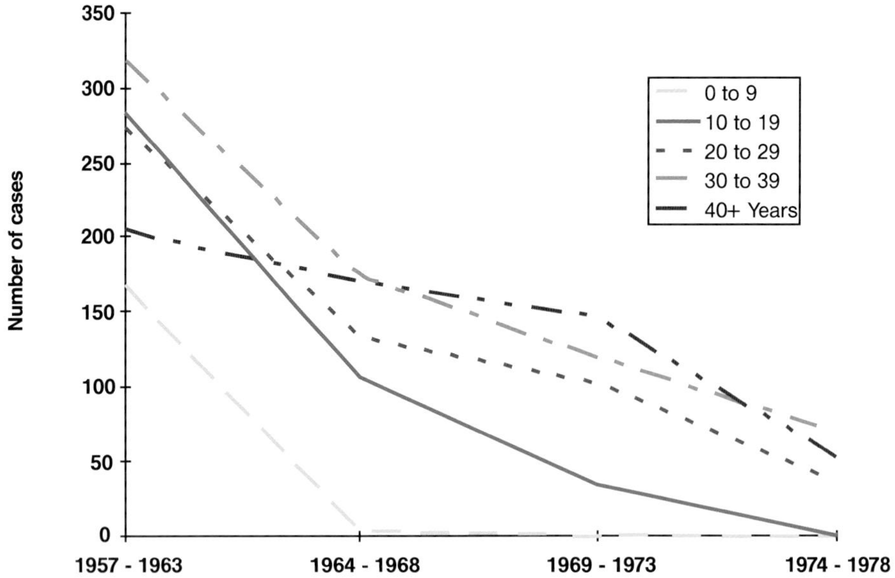

Fig. 38.2: Progressive decrease of kuru mortality from 1957 onwards according to age groups. Some of the 0–9 year-old kuru patients included in the yellow graph are shown in Table 38.1 according to age classes. Source: [1]; Figure by B. Hörnlimann and M.P. Alpers.

plex historical (compare Chapter 2) and ethnological aspects of the mortuary practices of the Fore and their neighbors involving transumption of the dead.

The following epidemiological description of kuru in this chapter is based on data from the literature [1] and on the 2,592 cases of kuru reported from 1957 to 1982 which correspond to approximately 96 % of all cases reported up to the year 2005.

38.3 The geographical spread and related cultural events

The presence of kuru was limited to a relatively small area of land of about 2,500 km^2 in the Eastern Highlands of PNG (Fig. 2.2). There were a few cases of male kuru patients with onset far outside of the known kuru region (Fig. 2.1). However, these men, without exception, originally came from the kuru region where they presumably had contracted the infection; in many cases they had left their home community more than ten years before the disease symptoms began [3; D.C. Gajdusek, personal communication].

The reconstructed origin and the geographical spread of kuru were described in detail by Glasse in 1962 [4] and by Mathews in 1965 [5]. The disease probably originated in Uwami, a village to the northwest of the Fore area within the region of the Keiagana group. Possibly infected tissue of a person dying of sporadic Creutzfeldt–Jakob disease (CJD) was transferred to other people for the first time during a single endocannibalistic meal (see Chapter 13.9). After the seed of the epidemic was putatively planted in this way, the brains of an expanding number of infected individuals constituted a source of infection for new victims. Since almost everybody who died in the community was eaten at a mortuary feast at which all members of the family participated, the disease was able to build up to epidemic proportions over an extended period of only three generations.

From Uwami kuru spread to Awande and then into the northern region (North Fore) and,

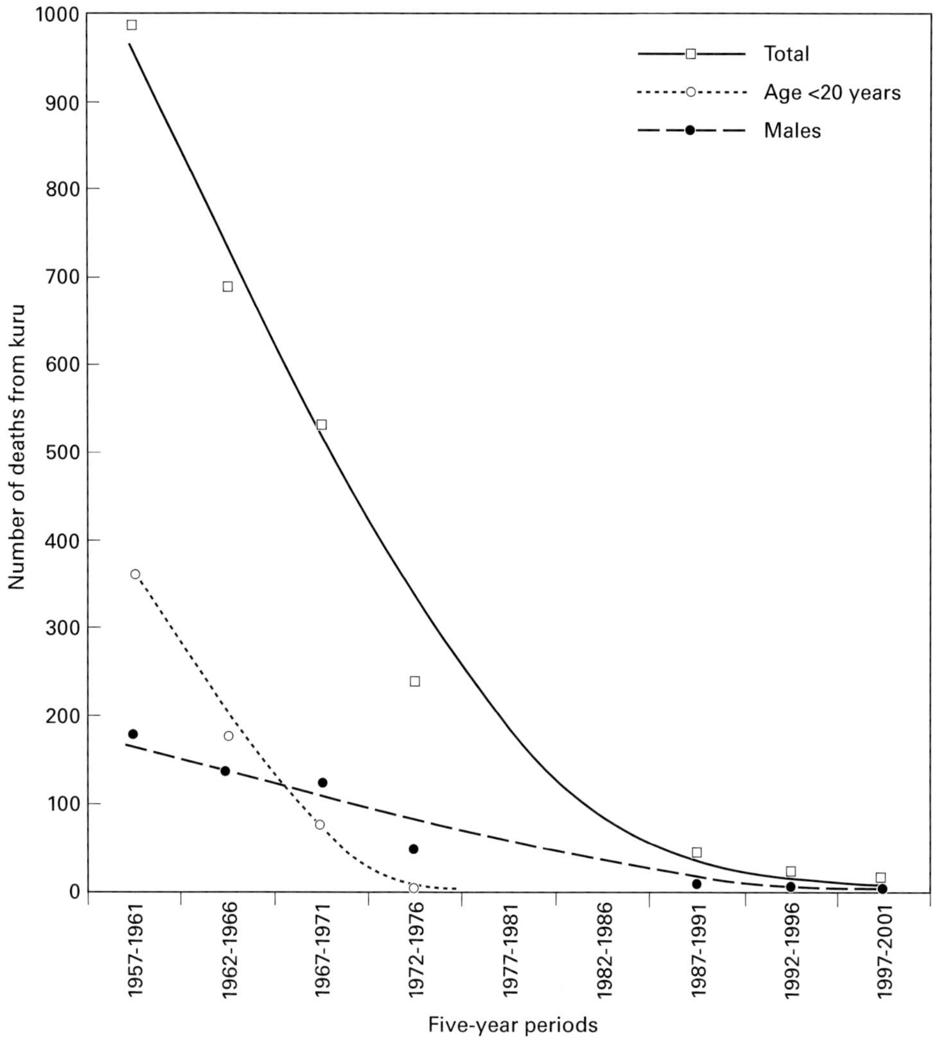

Fig. 38.3: The number of deaths from kuru in five-year periods, 1957–2001, for all patients, those aged under 20 years and male patients. The curves represent the best estimates of the shape of the declining epidemic of kuru, as determined by work in progress. Data are plotted for the periods 1957–1976 [1] and 1987–2001 [17]; the data for the intervening period 1977–1986 are still being extracted from extensive field notes. Data from Alpers, 1979 [1], Alpers and Kuru Surveillance Team, 2005 [17] and J. Collinge et al. (manuscript in preparation); Figure by B. Hörnlimann and M.P. Alpers.

more dominantly, to the South Fore. A possible explanation for the difference in kuru incidence between the South and North Fore is given in Chapter 2.14.

Altogether 62 % of all cases of kuru diagnosed from 1957 to the end of the 1970s occurred among the South Fore and 20 % among the North Fore [1]. The remaining kuru cases were observed among members of nine other linguistic groups in neighboring regions (Fig. 2.2) [1].

Many women who moved to neighboring groups to get married did not contract kuru until they lived in the new community (see Chapters 2.10 and 2.11). This was because they had been infected long before when they had been exposed to the actual risk factor "close contact with infectious brain.'' Thus a nutritional or an environmental rather than a genetic factor was able to explain the family pattern of the disease.

With the benefit of knowledge of the transmissibility of kuru [6], the mode of transmission and the epidemiological changes began to be understood [3; 7–9].

38.4 Distribution according to sex and age

Since the beginning of systematic kuru research in 1957 (see Chapter 2), but especially since Gajdusek and his colleagues realized that kuru was a transmissible disease [6], one of the most pressing questions was the following: "Why is there such a strange distribution according to sex among kuru victims?" (Table 38.2; Fig. 38.4). It was conspicuous that boys and girls contracted kuru at about the same rate of incidence; among adults, however, the disease predominated among women. Among the women of the South Fore, kuru was the most frequent cause of death in the villages with the highest rate of incidence during the late 1950s. In one community, the ratio between adult male and female inhabitants sank to one woman to 3.4 men in 1964. Among the patients 79 % of all 2,592 people afflicted with kuru from 1957 to 1982 were female (87 % of adults aged 20 years or more and 53 % of those under 20 years; Table 38.2).

Of all patients only 10 % were adult males; this proportion was 2% when kuru was first investigated and has gradually increased with the change in epidemiological patterns. Of all cases, 77 % occurred among adults aged 20 or more years with the female to male (F:M) ratio

Table 38.2: Distribution according to sex- and age-group (age at the time of death) of kuru cases between 1957 and 1982.

Age (years at death)	Female	Male	Total	F/M
5–9*	108	65	173	1.7
10–14	110	109	219	1.0
15–19	98	108	206	0.9
20–29	433	140	573	3.1
30–39	649	78	727	8.3
40 and older	655	39	694	16.8
Total	2,053	539	2,592	3.8

* includes two patients who died aged 4 years

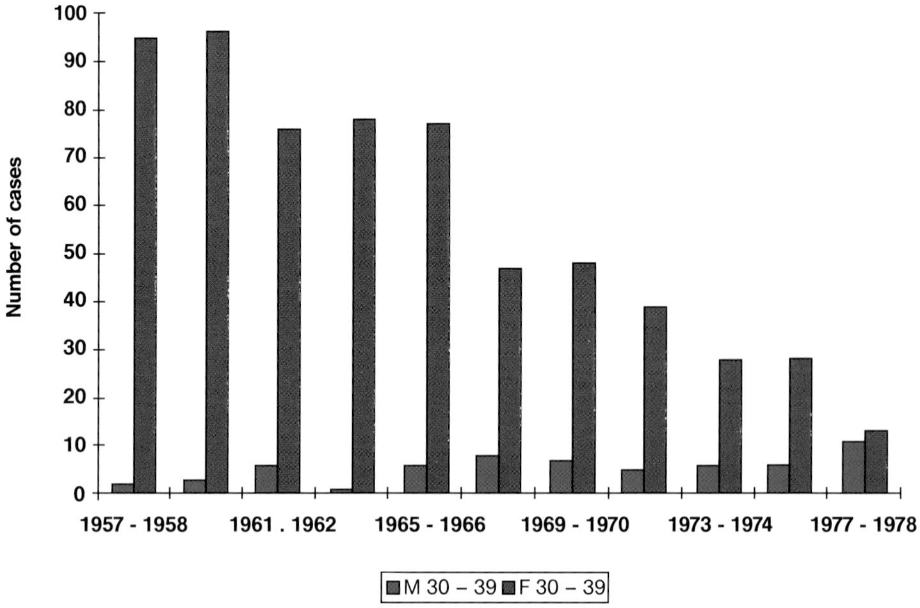

Fig. 38.4: Distribution of the kuru patients according to sex in the age group 30–39 year-olds (M = male, F = female). Source: [1]; Figure by B. Hörnlimann and M.P. Alpers.

being 6.8 (1,737 women to 257 men). Among the 23 % of all cases occurring in children and adolescents under 20 years of age, the F:M ratio was 1.1 (316 girls to 282 boys; Table 38.2). The F:M ratios among more finely subdivided age groups during the time from 1957 to 1982 are shown in Table 38.2.

Based on the unusual observation that both adult women (Table 38.2; Fig. 38.4) and children and adolescents of both sexes were principally affected, the conclusion was drawn that sex- and age-specific differences in exposure to kuru must have been important. Indeed, the retrospective analysis of former socioculturally based differences between the lifestyles of Fore women and girls and the men and boys provided a significant clue to solving the puzzle of kuru.

38.5 Sociocultural background of the sex- and age-specific distribution

As can be seen from the data, the risk of infection for boys dropped significantly after a certain age compared to that of girls. Traditionally, boys only stayed with their mothers and sisters until they reached the age of approximately 6–8 years. Girls, however, always stayed within the surroundings of the women. As they grew, they had to take on the chores and duties of the adult women, which included the preparation of meals.

After the boys became independent of their mother they had to go through a progression of initiation rites before they were accepted as an adult man. Throughout this time they lived in the men's house, and their behavior was determined by the adult men. They did not participate in the preparations necessary for the mortuary feasts nor did they any longer consume the parts of the body given to the women and girls. The men and initiate boys, as far as they participated in the endocannibalistic meals at all, usually ate only fat or muscle (which does not contain infectious prions). The internal organs, especially the brain (which contained most of the infectivity), were eaten by the women and children (see Chapter 2.9). For these reasons men were much less affected by kuru than women.

38.6 Explanation for the survival time curve (time of infection and infectious dose)

Highly variable amounts of the infectious agent taken up during a meal under natural circumstances provide one way of explaining the broad range of incubation periods found in kuru patients. For prion diseases it is generally known that the infectious dose (amount of infectious agent or titer) and the incubation period within a certain range are inversely proportional to one another. In addition, it has been shown by Lee et al. that host genetics is a co-factor with respect to the length of the incubation period. Heterozygosity of codon 129 of the prion protein gene, for example, is associated with a longer incubation period [10].

Some kuru victims may have received very large amounts of infectious doses [11], which is an explanation for those patients who experienced a relatively short incubation period [12]. Also, multiple exposures have an influence on the total amount of infectious agent in the body, since it is possible to accumulate infectious agents taken up during each exposure (meal). Those who were already adults between 1954 and 1959 – the time during which consumption of the dead at traditional mortuary feasts was abandoned – would have been exposed to a higher risk of kuru than those who were children at the time, since it was more likely for adults to have already participated in several endocannibalistic rites before 1954.

On the other hand not all members of one family or kin who had partaken of the same mortuary feasts fell ill; one reason could be because not all participants had actually ingested infectious tissue (especially not the men, since they ate, if at all, predominantly skeletal muscle) or because the infective dose was not sufficiently high in all cases to trigger an infection.

After very complex investigations and detailed interviews with surviving relatives of 65 kuru victims, Klitzman et al. published extremely informative data on this subject in 1984 [12]. The kuru victims "Nonon" (1948), "Tom" (1950), "Nen" (1953 or 1954) and "An" (1954) were devoured. This remarkable study was designed, among other things, to investigate in detail the last mortuary feasts in communities where the participants could be identified and their subsequent fate determined by enquiry and by reference to the epidemiological record [12]. The surviving relatives were always certain of the cause of death, since kuru is characterized by a protracted course and very consistent and recognizable signs and symptoms (Fig. 13.1; see Chapter 2.6). The insight gained from this study [12] confirms the hypothesis that transumption of the dead was the cause of transmission of the disease.

The feasts chosen to be investigated in the study were those in honor of persons who had died of kuru. It was found that most of those who were identified as being participants at the feasts subsequently died of kuru. Of course older participants had been exposed to the infectious agent more than once since they would have participated in other feasts held earlier than the last one in the village, so the incubation period for older people determined by this study was a minimum one; however, for those who were young children at the time the calculated incubation period could only have been the actual one.

The striking outcome of the study was the finding that in some cases several family members, often living in different communities, came down with kuru within months of each other after an incubation period of 30 years. In other cases, in contrast, the incubation periods were very different.

From the study of Klitzman et al. the only persons for whom we have a more or less exact estimate of the incubation period are those for whom the mortuary feast was their first as well as their last exposure to infective brain tissue [12].

In all these investigations it is of note that we cannot allow for the dose of infection, which is as much unknown in the individual as the route of infection (normally oral but in some cases could have been parenteral).

38.7 Explanation of kuru cases among children (modes of infection)

Children of both sexes were infected with their mothers. The women were exposed to infection through handling and consuming the brain, in particular. Children were fed brain in large amounts because it was considered to be a growth food. Males effectively ceased to be exposed when they left their mothers (by about the age of 6-8 years). Females continued to be exposed throughout life, but during adolescence (starting, in traditional times, before puberty, at about 10 years of age) or even earlier, after growth was no longer promoted and before adult status had been reached, the exposure of females lessened significantly. It is of interest that modelling the kuru epidemic mathematically also led to the conclusion of a reduced infection rate during the age period 10–14 years [13].

This could explain the bimodal distribution of kuru with age in females seen in the past [14; 15] (assuming that incubation periods in girls were mostly relatively short, in the order of 4 years, or perhaps even less). In males in the past [14; 15] the number of deaths from kuru in the age group 10–19 years declined to a similar level as in females, before the dramatic fall that occurred after the age of 20 years. More young females aged 5–9 years were affected than males [14; 15] (Table 38.2) and we may postulate relatively longer incubation periods in young males than in young females (possibly related to differences in infective load). These matters are still being intensively studied [16].

The bimodality in females was only seen in the early years of kuru investigation. Once the cohort born since transmission ceased had grown past childhood, no new childhood cases were added to the records; adolescent cases continued to occur for longer until they too finally cut out. The number of adult cases at that time,

Table 38.3: Changes in the age range of kuru patients by decade. Data from Alpers, 1979 [1] and Alpers and Kuru Surveillance Team, 2005 [17].

Age (years)	Last year in which a patient of this age died
< 10	1967
< 20	1973
< 30	1987
< 40	1991

in contrast, remained high (Fig. 38.2). The age composition has changed dramatically by decade (Table 38.3) and all patients since 1992 have been aged 40 years or more.

38.8 Conclusion and present significance of kuru

Throughout their lives, female individuals were repeatedly involved in the preparation of ritualistic meals and the consumption of infectious brain tissue, during both of which kuru can be transmitted. On the other hand, male members of the community for the most part were only exposed to the risk of kuru up to the age of about 6–8 years.

Here it should be mentioned that regarding the modes of transmission and the age at death of kuru-afflicted children and adolescents a certain parallel to the peripherally transmitted cases of the iatrogenic CJD (iCJD) can be drawn. iCJD is a form of CJD which can be caused for example by parenteral administration of contaminated growth hormone to children with stunted growth. It can then clinically manifest itself among young people. With this disease the average incubation period is twelve years (Table 14.1; see Chapter 36.4.2) [18]. Clinically this disease resembles kuru more than sporadic CJD.

Today kuru is still a disease much studied by scientists (Fig. 38.5). Kuru may actually be an important foundation for understanding the variant CJD (vCJD) which has only been around since 1996. It is currently most likely that patients with vCJD have been infected via the oral uptake of BSE-contaminated material (cattle brain or spinal cord). The comparison with kuru is obvious and kuru should be considered in the interpretation and description of the vCJD epidemic curve.

Fig. 38.5: Kuru researcher M.P. Alpers together with the editor of this book, B. Hörnlimann (left). 2006; photo by Deborah Lehmann.

References

[1] Alpers MP. Epidemiology and ecology of kuru. In: Prusiner SB, Hadlow WJ, editors. Slow transmissible diseases of the nervous system. Academic Press, New York, Vol 1, 1979:67–90.

[2] Gajdusek DC and Zigas V. Untersuchungen über die Pathogenese von Kuru. Klin Wochenschr 1958; 36:445–459.

[3] Alpers MP. Epidemiology and clinical aspects of kuru. In: Prusiner SB, McKinley MP, editors. Prions. Academic Press, San Diego, 1987:451–465.

[4] Glasse RM. The spread of kuru among the Fore. A preliminary report. Department of Public Health, Territory of Papua and New Guinea. Typed script 1962:1–9.

[5] Mathews JD. The changing face of kuru – an analysis of pedigrees collected by RM Glasse and Shirley Glasse and recent census data. Lancet 1965; 1:1138–1141.

[6] Gajdusek DC, Gibbs CJ Jr, Alpers M. Experimental transmission of a kuru-like syndrome to chimpanzees. Nature 1966; 209:794–796.

[7] Alpers MP. Kuru: implications of its transmissibility for the interpretation of its changing epidemiologic pattern. In: Bailey OT, Smith DE, editors. The central nervous system, some experimental models of neurological diseases. International Academy of Pathology, Monograph No 9. Proceedings of the fifty-sixth annual meeting of the International Academy of Pathology, Washington, DC, 12–15 Mar 1967. Williams and Wilkins, Baltimore, 1968:234–251.

[8] Mathews JD, Glasse RM, Lindenbaum S. Kuru and cannibalism. Lancet 1968; 2(7565):449–452.

[9] Alpers MP. Reflections and highlights: a life with kuru. In: Prusiner SB, Collinge J, Powell J, et al, editors. Prion diseases of humans and animals. Ellis Horwood, New York, 1992:66–76.

[10] Lee HS, Brown P, Cervenakova L, et al. Increased susceptibility to kuru of carriers of the PRNP 129 methionine/methionine genotype. J Infect Dis 2001; 183:192–196.

[11] Donnelly CA and Ferguson NM. Statistical aspects of BSE and vCJD – models for epidemics. Chapman & Hall/CRC, Boca Raton, 2000.

[12] Klitzman RL, Alpers MP, Gajdusek DC. The natural incubation period of kuru and the episodes of transmission in three clusters of patients. Neuroepidemiology 1984; 3:3–20.

[13] Huillard d'Aignaux JN, Cousens SN, Maccario J, et al. The incubation period of kuru. Epidemiology 2002; 13:402–408.

[14] Alpers MP. Epidemiological changes in kuru, 1957 to 1963. In: Gajdusek DC, Gibbs CJ Jr, Alpers M, editors. Slow, latent, and temperate virus infections. NINDB Monograph No 2. US Government Printing Office, Washington DC, 1965:65–82.

[15] Alpers MP and Gajdusek DC. Changing patterns of kuru: epidemiological changes in the period of increasing contact of the Fore people with Western civilization. Am J Trop Med Hyg 1965; 14:852–879.

[16] Whitfield J. Variations in mortuary practices in the kuru-affected region. Abstract A45 in program and abstracts of the forty-first annual symposium of the Medical Society of Papua New Guinea, Goroka. 5–9 Sep 2005:41.

[17] Alpers MP for the Kuru Surveillance Team. The epidemiology of kuru in the period 1987 to 1995. Commun Dis Intell 2005; 29:391–399.

[18] Brown P, Preece MA, Brandel JP, et al. Iatrogenic Creutzfeldt–Jakob disease at the millennium. Neurology 2000; 55:1075–1081.

39 The Course of the BSE Epidemic – Retrospective Epidemiological Considerations

Beat Hörnlimann, Judith B. Ryan, and Stuart C. MacDiarmid

39.1 Introduction

In this chapter, we summarize basic data on the bovine spongiform encephalopathy (BSE) epidemic in the United Kingdom (UK) and describe the factors affecting its course over the past 20 years. We aim to show the relationships between such factors as the age at infection, incubation time, age at clinical onset, and the time at which countermeasures started to influence the course of the epidemic. The effect of the countermeasures provide indirect evidence in support of the feed-borne hypothesis, as described in detail in Chapter 40.

The occurrence of "born after the ban" cases (BABs) has delayed the attainment of BSE eradication in the UK. The reasons for the occurrence of BABs were, in summary, the incomplete elimination of sources of cross-contamination of cattle rations and insufficient compliance with the first feed ban. These deficiencies were rectified by the reinforced feed ban imposed in 1996.

Up-to-date information on numbers of BSE cases and recent epidemiological observations in the UK can be found on the World Organisation for Animal Health (OIE) and Defra Animal Health and Welfare web sites [1–3]. Reported numbers of cases in all affected countries (until 1 January 2006) are shown in Table 19.2. Up-to-date information on numbers of BSE cases and animals tested using rapid tests can be found on the OIE web site [3; 4]. Additionally, the veterinary authorities in most countries report national BSE statistics on their web sites and, to a certain extent, in the various geographical BSE risk (GBR) assessments (see references in Table 53.1).

39.2 Basic epidemiological data on BSE in the UK

39.2.1 Number of BSE cases

Up to January 1, 2006, a total of 184,296 cases of BSE had been confirmed in the UK [3]; this corresponds to approximately 97.1% of all cases diagnosed worldwide. Globally, the number of BSE cases had reached 189,683 by the beginning of 2006 [3; 4], of which 5,387 were in countries other than the UK [4]. Data in the following part of this chapter, except where otherwise stated, relate to the UK.

The annual incidence of BSE in 1992, when the epidemic peaked, was slightly less than 1% of the population at risk *of developing the disease* (over 4 million animals[1]), most of which were adult dairy cattle. (A total of 37,280 cases in 1992; Table 19.2). By 2005, the annual incidence had dropped to about 4 cases per 100,000, clearly demonstrating a waning epidemic. The number of new cases each week is now lower than when the disease was made notifiable in 1988. By the beginning of 2006, 3 to 4 new suspected clinical cases were being reported each week, compared with over 1,000 per week in early 1993 [1]. About 16% of the suspect-cases in 2006 were confirmed as BSE, in comparison to over 80% of suspects confirmed in 1993.

Since the introduction of rapid post mortem diagnostic tests in 2000/2001, it has become possible to identify infected cattle in an advanced preclinical stage of the BSE incubation period (see Chapter 35). However, in the UK, where the surveillance system has been based

[1] Dairy cattle (Fig. 39.2; bottom-right); in comparison, the total UK sheep population at the time was about 40 million.

Table 39.1: Active UK surveillance 1999–2005; number of cattle tested BSE-positive. Source of data: [1].

	Britain	Northern Ireland	UK total
Pre-1999	0	0	0
1999	18	0	18
2000	44	55	99
2001	332	40	372
2002	594	70	664
2003	375	51	426
2004	227	26	253
2005	164	22	186
Total	1754	264	2018

mainly on the detection of clinical cases, only 2,018 cases had been detected by rapid tests up to January 2006 (Table 39.1);[2] these comprise only 1.09% of the total 184,296 UK cases.

39.2.2 The cattle population at risk

After exposure to the BSE agent, mainly through rations containing contaminated meat-and-bone-meal (MBM[3]; see Chapter 19.9) and mainly in their first year of life, calves destined for milk production were one category at risk *of developing clinical BSE*. The offspring of the beef suckler cows were a second category at risk *of developing clinical disease*. By March 2006, a total of 147,109 cases had been confirmed in UK dairy cattle on 23,434 farms. This represents about 80% of the total number of cases. In beef suckler cows, 22,355 cases (12%) were confirmed in the same period on 10,588 farms. The remaining 8% of confirmed cases occurred in other categories of cattle.

While the clinical cases referred to above are significant, of greater importance when considering the public health aspects of BSE, the epidemiologically important cattle included all animals at risk *of becoming infected* with BSE. That is, exposed directly or indirectly (cross-contamination) to infectious MBM[&] derived from brain and spinal cord tissue of infected cattle or sheep (see Chapters 19.8–9, 39.3.5, 53.2–3).

So, in this context, as well as the approximately 4 million adult dairy cattle and the beef suckler cows, all those calves reared for fattening comprised a third category of animals at risk *of being infected*, even though they were unlikely to live long enough to develop overt signs of BSE. This category was much larger than the other two categories.

Calves reared for fattening, together with preclinically infected dairy and beef suckler cattle, posed a BSE risk to other animals (through recycling of infectivity) as well as to humans *through contaminated meat products*. According to the statistical studies of Donnelly et al. [5] and Anderson et al. [6], the number of beef cattle slaughtered and consumed[4] while preclinically infected was several times higher than the number of clinical BSE cases.

As most cattle bred primarily for beef were slaughtered at an age less than the mean incubation period of about 5 years, most animals exposed to, and infected with, BSE did not survive long enough to develop clinical disease. In the absence of tests that could have measured actual rates of infection before the onset of disease, modellers such as Donnelly, Anderson, and their colleagues had to estimate the rates of infection in the birth cohorts concerned. Using back-calculation models, Donnelly et al. [5] postulated that there had been a substantial under-ascertainment of clinical cases over the course of the UK epidemic: Donnelly and colleagues estimated that, regardless of the stage of incubation, a total of around 3.5 to 4 million animals may have been infected with BSE [5]. In making such calculations, one inevitably has to make a number of assumptions, some of which are somewhat "soft". Hence, despite the excellent work behind the mathematical models, the conclusion that around 20 times more cattle had been infected in the UK than actually detected and confirmed remains theoretical.

---

2 There still are no tests available to detect BSE in live animals.

3 The reader can obtain comprehensive information about MBM and other important residual feed-related BSE risks or BSE risks in bovine-derived products in the document [38].

4 As compliance with the ban on SRM or SBM (specified bovine material) has become higher over time, the public health risk has constantly decreased.

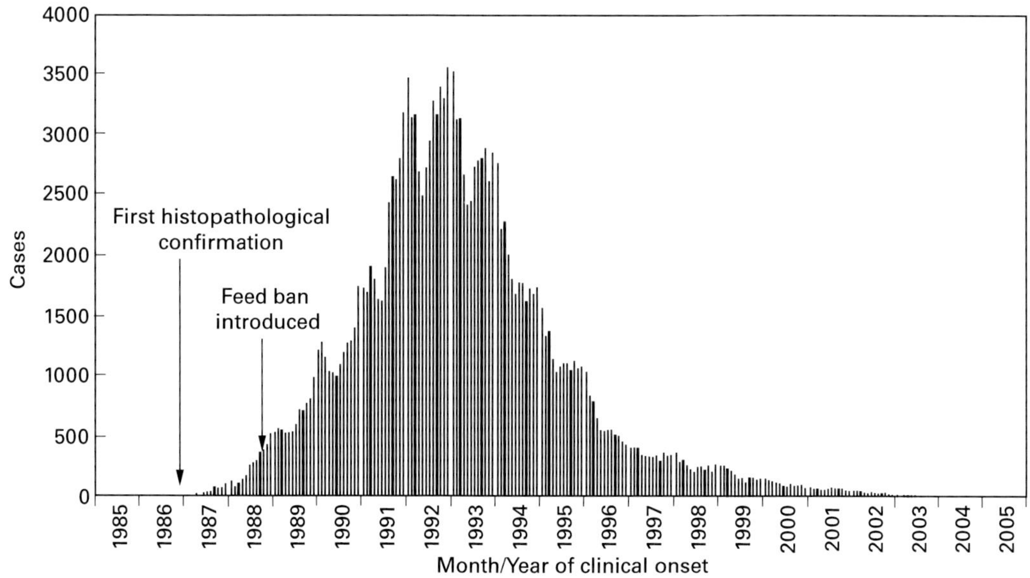

Fig. 39.1: BSE cases confirmed by passive surveillance plotted by month of clinical onset. Figure taken from [35]; Crown copyright ©.

Despite this, the approaches of Donnelly et al. [5] and Anderson et al. [6] are of major importance in assessing the risks of human exposure to BSE. The World Health Organization (WHO) assumed in 1996 [7] that some public health measures against BSE (Tables 39.2, 39.3), such as the specified bovine offal (SBO) ban of September 1990, may not have been strictly complied with initially [7; 8]. Such incomplete compliance, coupled with a lack of rapid BSE tests prior to 2000, meant that tissues harvested from preclinically infected cattle could not be fully excluded from the human food chains at the time. This could have resulted in significant human exposure to the BSE agent in some types of meat product (compare Chapters 1.9.3, 15.9) and, indeed, a link has been established between increased consumption of certain beef product and a higher risk of variant Creutzfeldt–Jakob disease (vCJD) [9]. Nevertheless, as compliance with the SBO and SRM (specified risk material) bans improved over time, the public health risk progressively decreased. This may be part of the reason for the relatively small number of vCJD cases actually observed to date (see Chapter 15).

39.2.3 The temporal distribution of the BSE cases

The BSE epidemic in the UK started in November 1986 [10] and its chronological course is shown in Fig. 39.1 [11]. The small number of probable cases observed between spring 1985 and the end of 1986 (Fig. 40.1), when BSE was first confirmed as a specific disease entity [12; 13], are based on retrospective clinical investigations.

From the beginning of 1988, the epidemic curve rose exponentially [2] (Fig. 39.1), reaching its peak in 1992 with an annual total of 37,280 cases. A decline in incidence was noted from January 1993, although the rate of decline slowed following the occurrence of BAB cases (see Section 39.3.5).

The seasonal pattern of the epidemic curve has been remarkable, with higher BSE incidence rates consistently observed during the winter months.

39.2.4 The geographical distribution of the BSE cases

The wide geographic distribution of BSE cases was noted from early in the course of the epidemic. By the end of 1986, cases had probably occurred in 17 British counties (Fig. 40.1). In 1988, Wilesmith and colleagues summarized the results of their first investigation thus: "The form of the epidemic was typical of an extended common source in which all affected animals were *index cases*..." [14]. Nonetheless, in the following years considerable differences were observed in the regional incidences. Until 1990, the highest annual incidence rates were observed in southern England. In the following years, higher rates were seen in the east of the country.

The number of BSE cases identified in eastern England only began to decrease from 1995; that is, later than in other regions (Fig. 39.2). The annual incidence of BABs (see Section 39.3.5) was higher in that part of the country [8; 15] (Fig. 39.2). By June 1997, the distribution of BSE-affected holdings across the UK was determined largely by factors influencing the amount of contaminated feedstuffs to which they had been exposed [8].

During the whole epidemic, the highest densities of confirmed BSE cases per 100 cattle per square kilometer occurred in the southwest region of England and in Dyfed in the southwest of Wales. In Wales, a small number of holdings experienced large numbers of cases. In the southwest region of England, a large number of holdings experienced small numbers of cases [8].

39.2.5 BSE cases within the affected herds

In more than 50% of affected herds (as of March 31, 2006), the number of BSE cases recorded was relatively small. About 35% experienced a single case only, while about 15% experienced just two cases. There were, however, several large herds with a high incidence but which, when feeding practices were changed, subsequently managed to eradicate the disease and have had no cases for several years [1].

So, why was it that not all animals in a herd or cohort that had been fed the same contaminated feed, and which were retained long enough, developed clinical BSE? (Compare Chapter 41.3.1: comment on pseudo-resistance.)

The differences in within-herd incidence rates may be explained as follows:

1. As highlighted in Section 39.2.2, the majority of infected cattle would have been slaughtered for consumption long before reaching clinical onset, and before the index case in the cohort reached clinical onset.
2. Another reason is that brain and spinal cord tissue fragments from infected cattle would not have been mixed homogeneously in the rendered products, nor would contaminated MBM have been mixed homogeneously with the other ingredients in cattle rations. That is, in practice, BSE infectivity was distributed unevenly in feedstuffs and, therefore, not all animals fed the same feedstuffs were exposed to an equal dose of infectivity.

It follows from these observations that whole-of-herd slaughter, applied in controlling contagious diseases, is unnecessary with BSE and has little effect on the evolution of the epidemic (see Chapter 53.2.2). This assertion is supported by the small number of cases that have been detected among cattle culled as cohorts of BSE cases (D. Matthews, personal communication). An additional consideration is that a whole-of-herd slaughter policy could discourage the reporting of suspects.

39.3 The factors that determined the course of the epidemic

39.3.1 The age of animals at onset of disease

The mean age at onset of disease was approximately 5 years [16; 17]. The maximum observed age of a BSE case was 22 years and 7 months and the youngest 20 months [1]. However, clinical BSE has rarely been seen in animals less than 3 years old. The age distribution of cases is shown in Fig. 39.3. The age at onset of disease depends on (i) the time of infection ("when was the individual animal infected?"), and (ii) the incubation period ("how long was the incuba-

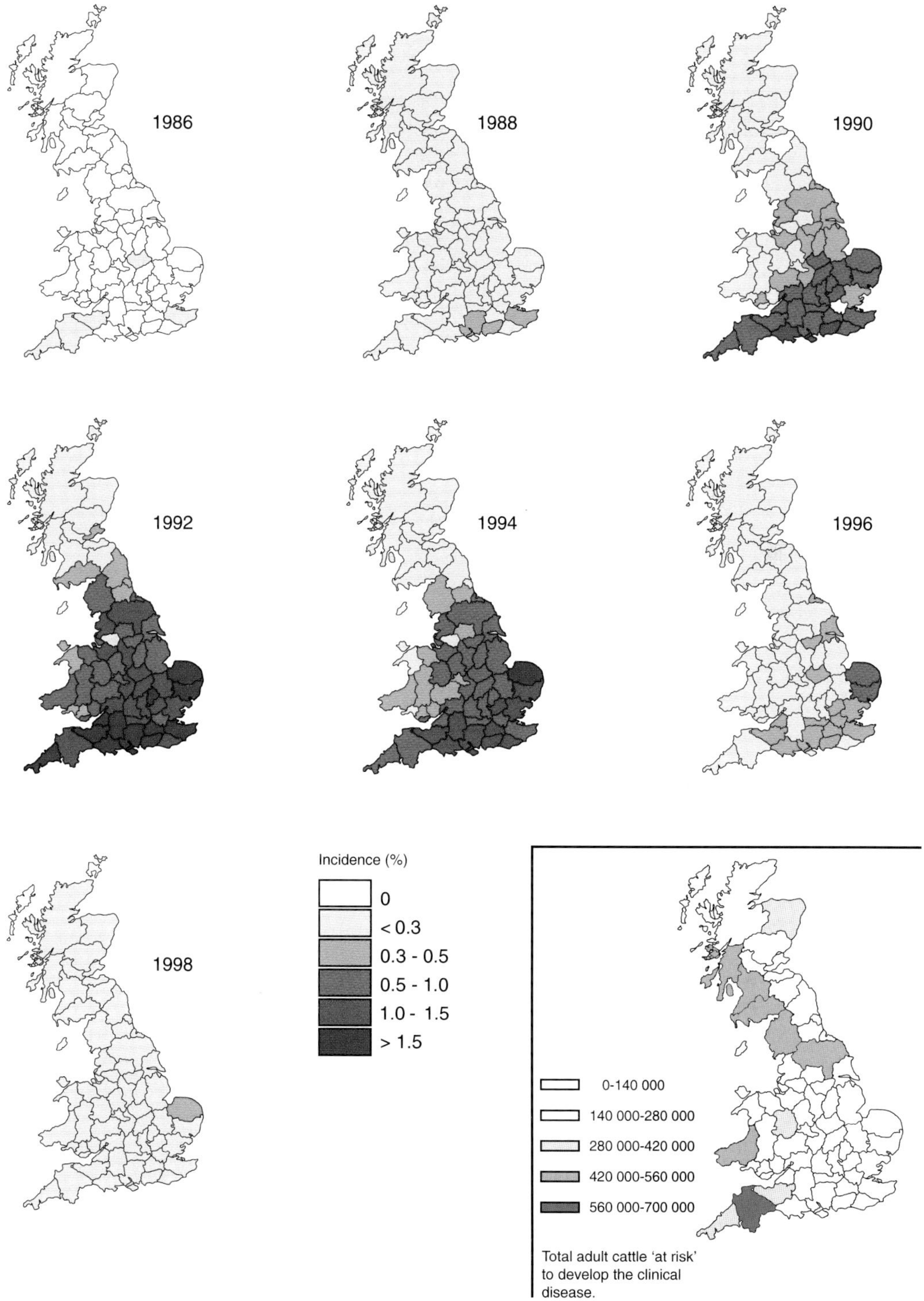

Fig. 39.2: The development of the annual BSE incidences (in %) in the British counties from 1986 to 1998. Adult cattle are the animals "at risk of developing clinical BSE"; the bottom-right map displays the size of that adult cattle population by county in 1994. Figure from [36]; Crown copyright ©.

tion period in the individually affected animal?").

The statistical distribution of the time of infection is closely related to the age of onset of clinical disease, linked by the incubation period, as shown schematically in Fig. 39.4 [based on the 1997 analyses of Donnelly et al.; 18].

39.3.2 The time of infection and incubation period

It is assumed that the transmission of BSE to calves occurred mainly in their first year of life [19; 20] and more rarely in older animals [21] (Fig. 39.4). On the basis of the results of a case control study (see Chapter 40), about 60 % of the animals in the UK epidemic were infected in their first months of life. One assessment estimated a mean age of about 18 weeks (4 months or 0.34 years) for the age at infection [18].

A number of studies have estimated mean value and distribution of incubation period, based, among other things, on the wide age distribution observed at the time of onset of disease [6; 18; 22–24]. On the basis of 175,322 cases (Fig. 39.3), a mean incubation period of ap-

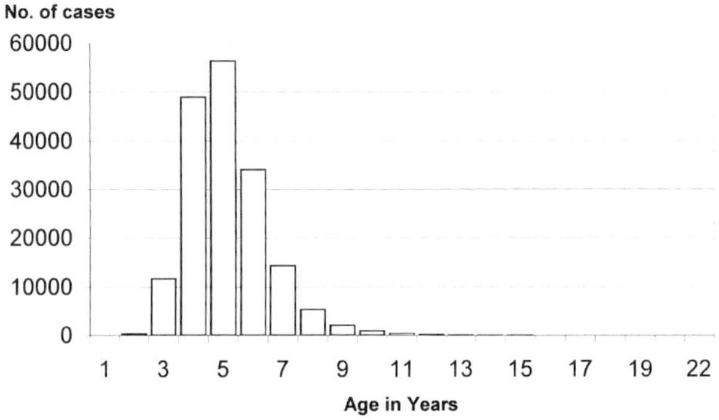

Fig. 39.3: Age distribution of confirmed cases of BSE from 1986 to May 2006 (known age cases with a clinical onset; $N = 175{,}322$). Note: 5 years, for example, means everything from 5 years and 0 days, to 5 years and 364 days. Data/figure by J. B. Ryan, VLA, Weybridge. Crown copyright ©.

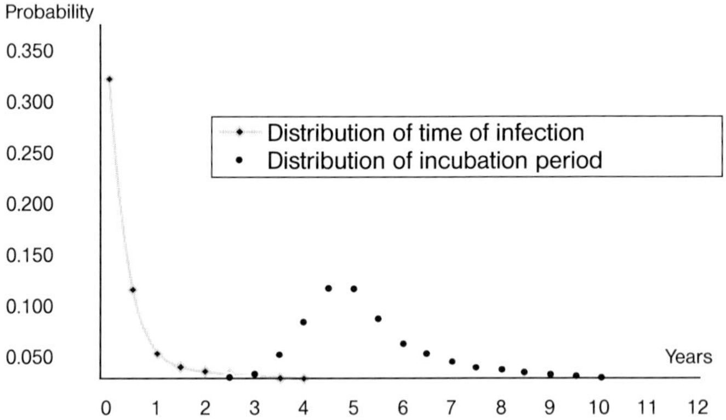

Fig. 39.4: Simplified schematic distribution of the incubation period of BSE (with an open end from 10 years to 22 years; compare Fig. 39.3). Comparison with the distribution of the time of infection. *Comment:* The age distribution shown in Fig. 35.4 is a function of the age distribution at the time of infection and the incubation period. Neither can be measured and so both lines are hypothetical. How they were originally constructed can be found on page 333 of [18]. Schematic diagram based on data from information of reference [18]; modified by B. Hörnlimann ©.

proximately 5 years was deduced for animals reared under field conditions.

In contrast, a considerably shorter incubation period was determined in an experimental attack-rate study (mean: 36 ± 2 months; Figure 19.3). In that study, doses of 1 g, 10 g, 100 g, or 300 g of infectious brain tissue were administered orally to calves at 4 months of age [25; compare 26]. The infectious doses were probably much larger than for animals under field conditions, explaining the shorter incubation period compared to "natural" BSE cases. Large variations in age at exposure and titer of infectivity *under field conditions* explain the wide range of age at onset of disease [1] (Fig. 39.3).

Initially [12] it proved difficult to devise scientifically based intervention measures. However, the first epidemiological study [27] provided evidence that the feeding of calves with ruminant-derived MBM, contaminated with a scrapie-like agent, was the most important risk factor (see Chapter 19.9). In consequence, the British government promulgated the first feed ban on July 18, 1988.

The long incubation period of BSE was an important factor contributing to the large scale of the British BSE epidemic, as BSE infectivity was recycled from preclinically infected cattle back to calves through rations containing contaminated MBM (Fig. 19.2). Infection was recycled and amplified unnoticed from 1981/1982 (the first infections) until the feed-borne hypothesis was accepted. Once the hypothesis was accepted, decision-makers and risk managers put measures into effect (Tables 39.2 and 39.3). The long incubation period also explains why the effect of the 1988 feed ban was not reflected in the epidemic curve until the beginning of 1993, beside the incomplete compliance [7; 8].

39.3.3 The influence of "synchronized births" on the seasonal variation of the epidemic

The seasonal peaks observed in the epidemic curve during winter months (Fig. 39.1) result from complex, interrelated factors, each of which is quite variable. The most obvious is the calving season in the UK, determined by a strong synchronization of mating, with the result that most calving takes place in autumn (Fig. 39.5).

The highest monthly incidences in the UK are observed in the periods between the main calving season and winter. It is possible to relate this to the stress cows experienced during calving or early stages of lactation, either of which may trigger and enhance the clinical manifestation of BSE and thus influence the time of observed onset of the disease, especially when the animals are in an advanced stage of incubation [16; 28]. The time of calving and the first months of lactation are probably the most important stress factors.[5] Furthermore, during this period, the cows are handled more intensively than at other times, and so signs of BSE are more likely to be observed by the farmer (see Chapter 32).

While it could be assumed that the synchronized calving season would manifest itself in the seasonal pattern of the epidemic curve, a comparison of Figs. 39.1 and 39.5 shows that the periodicities of the two curves are not identical. Over the course of the epidemic, the seasonal peak has been mitigated and shifted by several months depending on other factors. For instance, there was a gradual shift in the time of calving during the course of the BSE epidemic.

In contrast with Fig. 39.1, the seasonality seen in Fig. 39.5 is very evident. In this figure, the curve is plotted by month of birth of BSE cases, reflecting clearly the calving season. Apart from regular seasonal fluctuations, the number of calves born that subsequently developed BSE increased significantly up to autumn 1987, before numbers decreased due to the partial effect of the first feed ban imposed on July 18, 1988 (Fig. 39.5).

39.3.4 The effect of time of intervention on the course of the epidemic

The first infection of the UK cattle population probably occurred in 1981 or 1982, according

[5] It is possible that other kinds of stress, such as movement of the animal to another farm, also trigger clinical disease [16; 28] (see Chapter 33.2). Stress can be a factor triggering clinical onset of scrapie [29] (see Chapter 34).

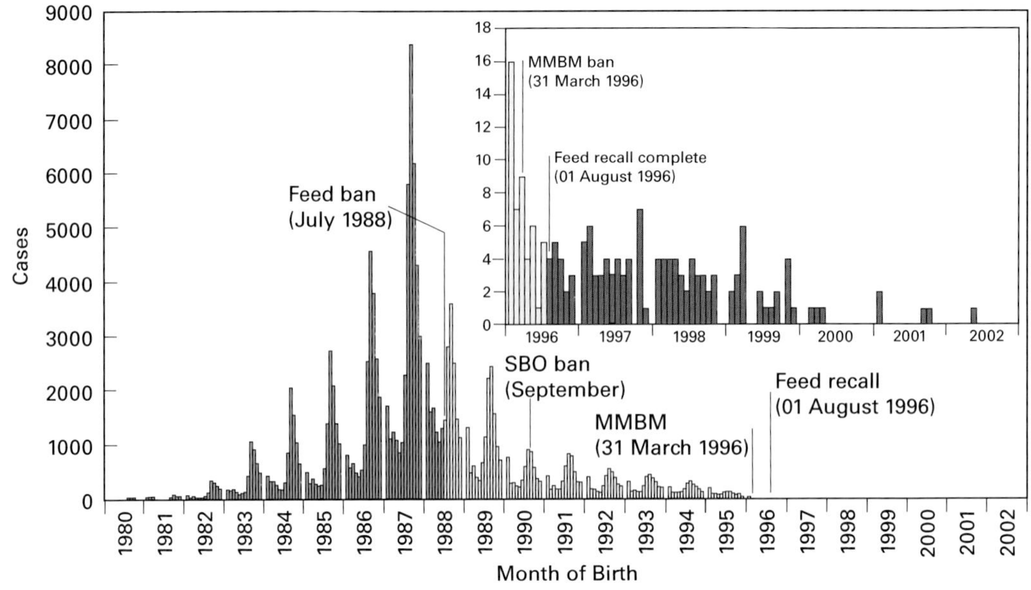

Fig. 39.5: Confirmed cases of BSE with known dates of birth, plotted by month of birth; and reflection of the effect of the feed ban of July 18, 1988. (Animals belonging to the same age cohort in the UK were born between July 1 and June 30 of the following year.) Data valid 01 April 2006. The inset shows 1996 to 2002 on a larger scale as the cases for later years do not show on the scale used in the larger graph. Yellow bars = BABs; gray bars = BARBs; blue bars = Born before the ban (BBBs). MMBM = Mammalian meat-and-bone-meal. Figure taken from [37]; Crown copyright ©.

to Wilesmith [30],[6] but the long incubation period meant that the first BSE cases were not observed until spring 1985 (see Figure 40.1). The delay between the occurrence of these early cases and the first histologic confirmation of BSE at the end of 1986 [12] may have been a result of field veterinarians initially assuming that other diseases, such as chronic hypomagnesemia, were responsible and so did not send appropriate samples to diagnostic laboratories. In addition, there was a lack of obligation on farmers or veterinarians to notify suspect cases until the middle of 1988 [31] (Tables 39.2 and 39.3).

As mentioned above, the incubation period also influenced the interval between the introduction of the initial feed ban in 1988 and the peak of the epidemic at the end of 1992. Despite the substantial effect of the ban in preventing new infections, the number of cases detected increased exponentially until the beginning of 1992 (Table 19.2; Fig. 39.1).

Nevertheless, even before 1992 it had been observed by epidemiologists that the 1988 feed ban was resulting in an increase in the mean age of new cases [17; 32]. Age-specific incidence data are shown in Fig. 39.6. It is evident from these curves that the incidence in 2–3-year-old cattle started to decrease in 1991–1992 (black curve in figure), but this was not statistically significant. This age group comprised only a small proportion of the BSE cases (0.18 % of cases until October 21, 1991).

In the period 1992–1993 the incidence among the 3–4-year-old cattle started to drop (red curve in figure), followed by a significant de-

[6] According to Anderson [6] and information used for Fig. 19.1, some animals, which were born in the 1970s and later developed BSE, might already have been infected in the second half of the 1970s. However, this assumption cannot be confirmed and thus remains hypothetical. In this context, it is of note that the hypothesis would have been supported by the multiple GBR assessments that suggested that the UK was exporting BSE infectivity long before BSE was first recognized in the UK.

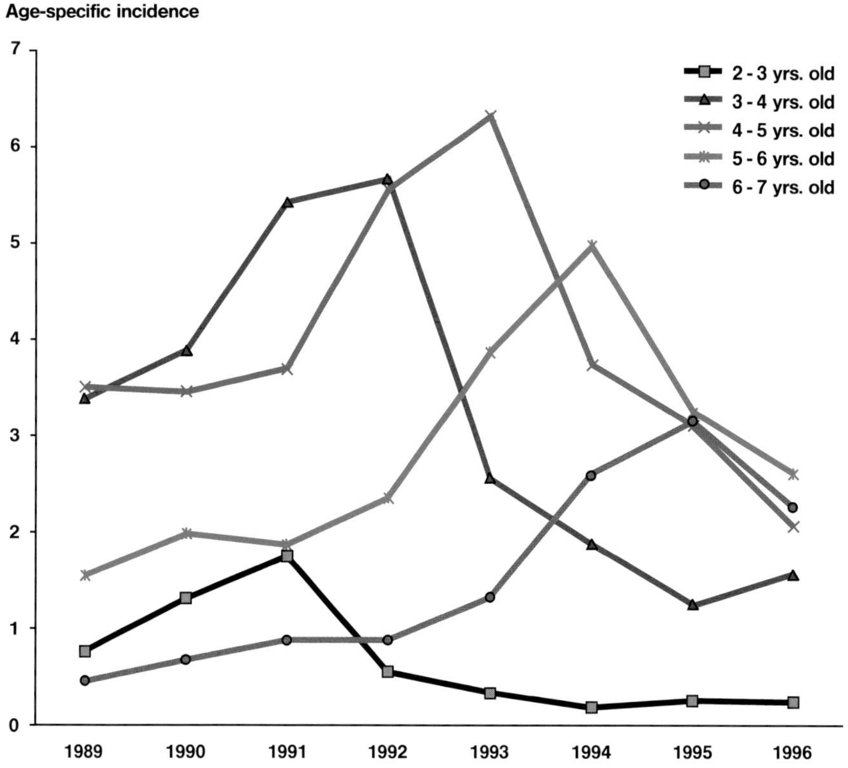

Fig. 39.6: Age-specific BSE incidence in Great Britain from 1989 until 1996 (given in %; grouped according to the age at which the animals showed clinical signs). If the peaks of the curves of the different age groups are connected, this graph also mirrors the course of the epidemic. Source of data: MAFF, reprinted with permission, 2001 in German edition of this book (1st ed. – Prionen und Prionkrankheiten). Figure by B. Hörnlimann ©.

crease among the 4–5-year-old animals (blue curve in figure) in the period 1993–1994, etc.

These interpretations are consistent with later analyses [10] indicating that the control measures imposed in 1988 and 1990 (Tabs 39.2–3) brought the epidemic under control, despite progress being slower in some regions such as eastern England.

39.3.5 The BAB and BARB BSE cases

The length of the incubation period (see Section 39.3.2) was not the only factor responsible for the continuing increase in BSE cases after the imposition of the first feed ban. As mentioned, compliance with the first feed ban was not 100%. Other reasons for the occurrence of BABs and the slow decrease of the epidemic after 1992 are:

- There was some on-farm carry-over of infected feed produced before the ban.
- Additionally, SBO/SRM controls would have allowed some infectivity to enter the MBM and feed production cycle legally up to 1996 because, in hindsight, it can be seen that cross-contamination in feed mills was unavoidable (Fig. 19.4), as cattle skulls contaminated with residual infectious brain could be legally used to produce MBM for nonruminant animal (pig, poultry) feed[7] up to August 1, 1996, when the reinforced feed ban was implemened.

[7] The same kinds of conclusions were reached in other countries, for example Switzerland [33; 34].

Table 39.2: Measures taken in Great Britain, the EU, and the USA to prevent the spread of bovine spongiform encephalopathy (BSE) to animals (1988–2000). Source: [31]. © Copyright from CDC, Atlanta, USA.

Precautions	Great Britain[a]	European Union[a]	United States
BSE made a notifiable disease	Jun 1988	Apr 1990	Nov 1987
BSE surveillance, with histologic examination of brains	Jun 1988	May 1990	May 1990
Ban on ruminant protein in ruminant feed	Jul 1988		
Ban on export of UK cattle born before July 1988 feed ban		Jul 1989	
Ban on import of live ruminants and most ruminant products from all BSE countries			Jul/Nov 1989
Ban on export of UK cattle > 6 months of age		Mar 1990	
Ban on SBO[b] for use in animal nutrition; ban on export of SBO and feed containing SBO to EU[c] countries	Sep 1990		
High-risk waste to be rendered at 133 °C/3 bar/20 min (or other approved procedure)		Nov 1990	
Ban on export of SBO and feed containing SBO to non-EU countries	Jul 1991		
Ban on MBM[d] from SBO in fertilizer	Nov 1991		
After Jan 1, 1995, rendering methods must sterilize BSE		Jun 1994	
Ban on mammalian MBM in ruminant feed		Jul 1994	
BSE surveillance includes immunohistologic features of brains			Oct 1993
Ban on mammalian protein in ruminant feed[e]	Nov 1994		Aug 1997
Ban on import of live ruminants and most ruminant products (including meat products) from all countries of Europe			Dec 1997
Immunologic testing for ruminant protein in animal feed		Jul 1995	
Mammalian MBM prohibited from all animal feed/fertilizer	Mar/Apr 1996		
Slaughtered cattle > 30 months old (except certain beef cattle > 42 months old) ruled unfit for animal use (hides for leather excluded)	Mar 1996		
Mammalian MBM and MBM-containing feed recalled	Jun 1996		
All mammalian waste to be rendered at 133 °C/3 bar/20 min (or other approved procedure)		Jul 1996	
Cattle tracing system improved	Sep 1998		
Quarantine of 3 sheep flocks imported from Europe with possible exposure to BSE (4 animals die with atypical TSE[f])			Oct 1998
BSE surveillance of fallen stock (downer cows) is intensified			Oct 1998
Proposal to eradicate scrapie is rejuvenated			Nov 1999
Allow export of deboned beef from cattle > 30 months old born after July 1996	Aug 1999		
Prohibit use of animal protein, including MBM and blood meal (but excluding milk, or fish meal for nonruminants) in feed for any farmed animal species (effective Jan 1, 2001)		Dec 2000	
Prohibit importation of rendered protein and rendering wastes originating or processed in Europe			Dec 2000

[a] In Northern Ireland and Scotland, dates of implementation sometimes differed from those shown for England and Wales; in addition, individual European Union countries often adopted different measures on different dates.
[b] SBO = specified bovine offals (brain, spinal cord, thymus, tonsil, spleen, and intestines from cattle > 6 months of age)
[c] EU = European Union
[d] MBM = meat and bone meal (protein residue produced by rendering)
[e] Some exemptions, e. g., milk, blood, and gelatin.
[f] TSE = transmissible spongiform encephalopathy

Table 39.3: Measures taken in Great Britain, the EU, and the USA to prevent the spread of bovine spongiform encephalopathy (BSE) to humans (1988–2000). Source: [31]. © Copyright from CDC, Atlanta, USA.

Precautions	Great Britain[a]	European Union[a]	United States
Compulsory slaughter of BSE-affected cattle	Aug 1988		
Destroy milk from affected cattle (except for milk fed to cows' own calves)	Dec 1988		
Ban on import of UK cattle born after July 1988 feed ban		Jul 1989	
Ban on SBO[b] for domestic consumption	Nov 1989		
Ban on export to EU[c] of SBO and certain other tissues, including lymph nodes, pituitaries, and serum	Apr 1990	Apr 1990	
Ban on export of live UK cattle (except calves < 6 months old)	Jun 1990	Jun 1990	
Ban on use of head meat after skull opened	Mar 1992		
FDA recommends use of BSE/scrapie-free sources for materials used in dietary supplements; request for safety plans			Nov 1992
Cell lines used for biologicals should be BSE agent-free			May 1993
FDA requests that bovine source materials (except gelatin) used in manufacture of regulated products be restricted to BSE-free countries			Dec 1993
Bone-in beef only from farms with no BSE for 6 years; if not BSE-free, must be deboned with visible nervous and lymphatic tissue removed		Jul 1994	
FDA requests that bovine-derived materials for animal use or for cosmetics and dietary supplements not be sourced from BSE countries			Aug 1994
Thymus and intestines from calves < 6 months old made SBO	Nov 1994		
Import of beef only from UK cattle (i) > 30 months, or (ii) from herds BSE-free for 6 years, or (iii) if not BSE-free, deboned with visible nervous tissue and specified lymph nodes removed		Jul 1995	
SBO ban broadened to include whole skull (SBM)	Aug 1995		
MRM[d] from bovine vertebral column banned and export prohibited	Dec 1995		
Removal of lymph nodes and visible nervous tissue from bovine meat > 30 months exported to EU	Jan 1996		
Ban on export of all UK cattle and cattle products except milk		Mar 1996	
SBM[e] ban broadened to include entire head (excluding uncontaminated tongue)	Mar 1996		
Slaughtered cattle > 30 months (or certain beef cattle > 42 months) ruled unfit for animal or human use (hides excepted)	Mar 1996		
FDA urges manufacturers of FDA-regulated human products to take steps to assure freedom from BSE agent			May 1996
Partial lifting of export ban on tallow and gelatin		Jun 1996	
SBM ban broadened to include certain sheep and goat heads, spleens, and spinal cords (SRM)	Sep 1996		
FDA recommends withdrawal of plasma and plasma products made from pools to which persons who later died of CJD had contributed			Dec 1996
CNS[f] tissues excluded from cosmetic products for use in EU		Jan 1997	
BSE cohort cattle in UK ordered slaughtered and destroyed	Jan 1997		

Table 39.3 (continued)

Precautions	Great Britain[a]	European Union[a]	United States
Proposed ban on SRM[g] in cosmetics for use in EU (effective Oct 2000)		Jul 1997	
SBM controls for cosmetics and medicinal products	Mar 1997		
FDA request to manufacturers that no bovine gelatin from BSE countries be used in injectable, implantable, or ophthalmic products; and that special precautions be applied to gelatin for oral and topical use			Sep/Dec 1997
Ban on marketing cosmetic products containing SRM prepared before April 1, 1998		Mar 1998	
Allow export of beef and beef products from cattle > 30 months in certified BSE-free herds from Northern Ireland		Mar 1998	
Importation of all plasma and plasma products for use in UK	Aug 1998		
FDA limits plasma product withdrawals to pools at risk for contamination by vCJD donors			Sep 1998
Slaughter and destruction of offspring born to BSE-affected cattle after July 1996	Jan 1999		
FDA guidance to defer blood donors with > 6 months cumulative residence in UK during 1980–1996			Nov 1999
Leukodepletion of whole blood donations from UK residents	Jul/Nov 1999		
Public FDA discussion about possible risk associated with vaccines produced with bovine-derived materials from BSE countries			Jul 2000
Withdrawal and destruction of a potentially tainted 1989 lot of polio vaccine from one manufacturer	Oct 2000		
SRM ban implemented (effective Oct 2000)		Jul 2000	
Ban on slaughter techniques that could contaminate cattle carcasses with brain emboli (e. g., pithing or pneumatic stun guns), effective Jan 2001		Jul 2000	
All cattle > 30 months old must have brain examinations for proteinase-resistant protein (PrP) before entering the food chain (effective Jan–Jun 2001)		Dec 2000	

[a] In Northern Ireland and Scotland, dates of implementation sometimes differed from those shown for England and Wales; in addition, individual European Union countries often adopted different measures on different dates.
[b] SBO = specified bovine offals (brain, spinal cord, thymus, tonsils, spleen, and intestines from cattle > 6 months old)
[c] EU = European Union
[d] MRM = mechanically recovered meat
[e] SBM = specified bovine materials (SBO plus entire head, including eyes but excluding tongue)
[f] CNS = central nervous system
[g] SRM = specified risk materials (SBM plus sheep and goat heads and spleens from animals of any age, and spinal cords from animals > 1 year old)

- Some farmers probably fed pig or poultry rations to cattle by intention, thus also allowing the transmission of infective residuals (author, personal observations).
- There might have been some continued cross-contamination of cattle feed with MBM produced after 1 August 1996 (see below).

The on-going risk of cross-contamination in feed mills is described in Chapter 19.9.3.

The occurrence of BABs also demonstrated that rendering methods after 1988 were still not sufficient to completely inactivate BSE prions (see Chapters 19.9, 45, and 52).

In the UK on March 29, 1996, the 1988 ban on feeding mammalian protein to ruminants was reinforced by a ban on the feeding of mammalian MBM to any farmed livestock (MMBM). This measure was introduced to prevent the cross-contamination of ruminant feeds by feeds intended for other species such as pigs or poultry. Additionally, new EU-wide controls were implemented on August 1, 1996, banning all processed animal proteins (with a few exceptions) from rations intended for any animals used for food production.

Allowing for completion of a feed recall scheme that disposed of feed produced before March 29, 1996, the ban on feeding of mammalian MBM to farmed livestock was considered effective after July 31, 1996. Even so, some cases have been recorded in cattle born after that date. These are referred to as BARBs (born after the reinforced ban; Fig. 39.5; inset). The occurrence of BARBs did not significantly influence the course of the UK epidemic, in contrast to the occurrence of BABs. Fewer than 0.1 % of all UK BSE cases are BARBs. Further measures put into effect after 1996 are comprehensively summarized in Tables 39.2 and 39.3. These have led to the current situation whereby the number of cases in the UK continues to decline rapidly and the latest forecasts suggest that the epidemic will die out over the next few years.

39.3.6 The influence of surveillance

Active surveillance for BSE became an issue in the EU in 2001 after the first program using a rapid diagnostic test, capable of detecting infection several months before the onset of disease, was developed and applied successfully in Switzerland (see Chapter 35). Consequently, BSE was discovered in several countries that had hitherto regarded themselves as free from the disease (see Table 19.2).

As in the case of other diseases, passive surveillance (see Chapter 32.2) may result in an underestimation of the numbers of clinical cases. Furthermore, with BSE, passive or clinical surveillance does not pick up any preclinical cases. The implementation of active or targeted surveillance programs (see Chapter 32.3), using the rapid post mortem BSE tests that have become commercially available since 2000/2001, has been an important advance (see Chapter 35).

These rapid tests are today specifically used in risk groups, such as "downer cows", after having initially been applied generally to cattle over 30 months of age (see Chapter 53.3.1) or, in some countries, over 24 months.

Chapters 32, 35, and 53 describe why and how the quality of surveillance can be improved despite the limitations (compare Chapter 53.3.4) of rapid BSE tests. Targeted surveillance definitely permits the identification of a larger proportion of infected animals compared with passive surveillance alone. However, animals in the early stages of incubation can still not be detected with any of the tests on the market (Table 35.1). This is because in the early stage of incubation the disease-specific marker PrP^{Sc} cannot be detected in the brain or spinal cord with the tests currently available.

It is of note that, in contrast to the UK, in several other countries it was only the implementation of active BSE surveillance that was able to detect BSE cases, probably because passive surveillance was not sensitive enough to detect or report earlier clinical cases.

References

[1] Defra. BSE epidemic (General Q&A – Section 1). Webpage 2006: www.defra.gov.uk/animalh/bse/general/qa/section1.html.

[2] Defra. Department for Environment, Food and Rural Affairs. Webpage 2006: www.defra.gov.uk.

[3] World Organisation for Animal Health (OIE). Number of cases of bovine spongiform encephalopathy (BSE) reported in the United Kingdom. Webpage 2006: www.oie.int/eng/info/en_esbru.htm.

[4] World Organisation for Animal Health (OIE). Number of reported cases of bovine spongiform encephalopathy (BSE) in farmed cattle worldwide (excluding the United Kingdom). Webpage 2006: www.oie.int/eng/info/en_esbmonde.htm.

[5] Donnelly CA, Ferguson NM, Ghani AC, et al. Implications of BSE infection screening data for the scale of the British BSE epidemic and current European infection levels. Proc Biol Sci 2002; 269(1506):2179–2190.

[6] Anderson RM, Donnelly CA, Ferguson NM, et al. Transmission dynamics and epidemiology of BSE in British cattle. Nature 1996; 382(6594): 779–788.

[7] World Health Organization. Zoonoses control – bovine spongiform encephalopathy. Week Epid Rec 1996; 11:83–85.

[8] Stevenson MA, Wilesmith JW, Ryan JB, et al. Descriptive spatial analysis of the epidemic of bovine spongiform encephalopathy in Great Britain to June 1997. Vet Rec 2000; 147(14):379–384.

[9] Ward H. Variant Creutzfeldt–Jakob disease in the UK 1995–2003: a case-control study of potential risk factors. First International Conference of the European Network of Excellence NeuroPrion, May 25, 2004:40.

[10] Wilesmith JW, Ryan JB, Stevenson MA. Temporal aspects of the epidemic of bovine spongiform encephalopathy in Great Britain: holding-associated risk factors for the disease. Vet Rec 2000; 147(12):319–325.

[11] Defra. BSE: Statistics – incidences of BSE. Webpage 2006: www.defra.gov.uk/animalh/bse/statistics/incidence.html.

[12] Wells GAH, Scott AC, Johnson CT, et al. A novel progressive spongiform encephalopathy in cattle. Vet Rec 1987; 121:419–420.

[13] Wells GAH, Hawkins SA, Hadlow WJ, et al. The discovery of bovine spongiform encephalopathy and observations on the vacuolar changes. In: Prusiner SB, Collinge J, Powell J, et al., editors. Prion diseases of humans and animals. Ellis Horwood, New York, 1992:256–274.

[14] Wilesmith JW, Wells GAH, Cranwell MP, et al. Bovine spongiform encephalopathy: epidemiological studies. Vet Rec 1988; 123(25):638–644.

[15] Wilesmith JW. Recent observations on the epidemiology of bovine spongiform encephalopathy. In: Gibbs CJ Jr, editor. Bovine spongiform encephalopathy – the BSE dilemma. Springer, New York, 1996:45–55.

[16] Wilesmith JW, Hoinville LJ, Ryan JB, et al. Bovine spongiform encephalopathy: aspects of the clinical picture and analyses of possible changes 1986–1990. Vet Rec 1992; 130:197–201.

[17] Wilesmith JW and Ryan JB. Bovine spongiform encephalopathy: observations on the incidence during 1992. Vet Rec 1993; 132:300–301.

[18] Donnelly CA, Ghani AC, Ferguson NM, et al. Analysis of the bovine spongiform encephalopathy maternal cohort study: evidence for direct maternal transmission. Appl Statist 1997; 46(3):321–344.

[19] Bradley R. Experimental transmission of bovine spongiform encephalopathy. In: Court L and Dodet B, editors. Transmissible subacute spongiform encephalopathies: prion diseases. Elsevier, Paris, 1996:51–56.

[20] Wilesmith JW, Ryan JB, Hueston WD. Bovine spongiform encephalopathy: case-control studies of calf feeding practices and meat and bonemeal inclusion in proprietary concentrates. Res Vet Sci 1992; 52:325–331.

[21] Wilesmith JW. BSE: epidemiological approaches, trials and tribulations. Prev Vet Med 1993; 18:33–42.

[22] Bradley R. Bovine spongiform encephalopathy – distribution and update on some transmission and decontamination studies. In: Gibbs CJ Jr, editor. Bovine spongiform encephalopathy – the BSE dilemma. New York: Springer, 1996:11–27.

[23] Ferguson NM, Donnelly CA, Woolhouse MEJ, et al. Genetic interpretation of heightened risk of BSE in offspring of affected dams. Proc R Soc Lond Biol 1997; 264(1387):1445–1455.

[24] Medley GFH and Short NRM. A model for the incubation period distribution of transmissible spongiform encephalopathies and predictions of the BSE epidemic in the United Kingdom. Preprint 1996. (Reference according to Donnelly CA, Ghani AC, Ferguson NM, et al. Appl Statist 1997; 46:321–344.)

[25] Wells GAH, Dawson M, Hawkins SA, et al. Preliminary observations on the pathogenesis of experimental bovine spongiform encephalopathy. In: Gibbs CJ Jr, editor. Bovine spongiform encephalopathy – the BSE dilemma. Springer, New York, 1996:28–44.

[26] Donnelly CA, Ghani AC, Ferguson NM, et al. Recent trends in the BSE epidemic. Nature 1997; 389(6654):903.

[27] Wilesmith JW, Wells GAH, Cranwell MP, et al. Bovine spongiform encephalopathy: epidemiological studies. Vet Rec 1988; 123(25):638–644.

[28] Hörnlimann B and Braun U. BSE: Clinical signs in Swiss BSE cases. In: Bradley R, Marchant B, editors. Transmissible spongiform encephalopathies. EU: VI/4131/94-EN document ref. F. II.3 – JC/0003, Brussels, 1994:289–299.

[29] Toumazos P. Scrapie in Cyprus. Br Vet J 1991; 147:147–154.

[30] Wilesmith JW, Ryan JB, Atkinson MJ. Bovine spongiform encephalopathy: epidemiological studies on the origin. Vet Rec 1991; 128:199–203.

[31] Brown P, Will RG, Bradley R, et al. Bovine spongiform encephalopathy and variant Creutzfeldt–Jakob disease: background, evolution, and current concerns. Emerg Infect Dis 2001; 7(1):6–16.

[32] Wilesmith JW and Ryan JB. Bovine spongiform encephalopathy: recent observations on the age-specific incidences. Vet Rec 1992; 130:491–492.

[33] Guidon D. Microscopy as a means of testing compliance with the (Swiss) feed ban [Die Kontrolle der Einhaltung des Tiermehl-Beimischungsverbotes durch Mikroskopie]. In: Hörnlimann B, Riesner D, Kretzschmar H, editors. Prionen und Prionkrankheiten. De Gruyter, Berlin, 2001:494–496.

[34] Hörnlimann B. BSE in Non-UK countries: spread by live cattle and MBM imports [in German; original publication of Hörnlimann B. Bovine Spongiforme Enzephalopathie (BSE): BAB-Fälle in der Schweiz (summary of Master of Public Health thesis)]. ISPM University of Zürich 1999. In: Hörnlimann B, Riesner D, Kretzschmar H, editors. Prionen und Prionkrankheiten. De Gruyter, Berlin, 2001:337–358.

[35] Defra. Graph of BSE cases confirmed by passive surveillance plotted by month and year of clinical onset (Appendix 1a) of webpage document. Webpage 2006: www.defra.gov.uk/animalh/bse/statistics/bse/monthlystats.pdf.

[36] MAFF. Bovine spongiform encephalopathy in Great Britain. A progress report. Ministry of Agriculture, Fisheries and Food, London, 1999.

[37] Defra. Graph of confirmed cases of BSE with known dates of birth, plotted by month of birth (Appendix 2) of webpage document. Webpage 2006: www.defra.gov.uk/animalh/bse/statistics/bse/monthlystats.pdf.

[38] EFSA. Quantitative assessment of the residual BSE risk in bovine-derived products. The EFSA Journal 2005; 307:1–135. See also Webpage 2006: www.efsa.eu.int/science/biohaz/biohaz_documents/1280/efsaqrareport2004_final20dec051.pdf.

40 The Causes of the BSE Epidemic

Susanne Dahms and Beat Hörnlimann

40.1 Introduction

BSE was diagnosed for the first time in Great Britain in 1985/1986 on histological tissue preparations of an infected cow [1]. The number of BSE cases continued to increase, and BSE was soon found all over Great Britain. After some months it became clear that this was indicating the beginning of an epidemic. Therefore, in May 1987, a first epidemiological case-series study was initiated [2] with the aim of systematically collecting and describing available data on observed BSE cases. Results of this investigation suggested that there was only *one single* factor that was common to all herds, namely feedstuffs used for rearing calves that contained animal-waste-derived material like meat-and-bone-meal (MBM). On the basis of this observation, other information provided in this chapter, and the neuropathological similarity of BSE with the scrapie disease in sheep and goats, it was hypothesized that BSE might originate from feeding scrapie-contaminated MBM to cattle (feed-borne hypothesis) [2; 3]. Alternatively, it was hypothesized that BSE had been an extremely rare disease in cattle in its own right, which suddenly began to spread rapidly because of the technical modifications in the production of MBM and similar products from animal-waste-derived material (see Chapter 19.9).

The feed-borne hypothesis was more thoroughly investigated in a two-step case-control study. This study examined whether infected animals took in the BSE pathogen – during the first month of life – via commercially available mixed feedstuffs containing ruminant protein in the form of MBM. It will be shown in this chapter that it was possible to substantiate the feed-borne hypothesis.

40.2 The case-series study: development of the feed-borne hypothesis

At the beginning of the epidemic, numerous hypotheses existed on the possible etiology of the new cattle disease. In a first case-series study, epidemiologists searched for factors that were common to all affected herds or cases in order to clarify where, when, under what animal management conditions, and in which cattle breeds and age groups the disease was found. The main results of this study can be summarized [4] as follows:

1. BSE was a new cattle disease, which was clinically observed for the first time in 1985 (Fig. 40.1).
2. BSE was not introduced to Great Britain from abroad, neither through imported live cattle nor through animal products or cattle sperm.
3. BSE cases were spread over a large geographical area. However, higher incidences were found in the southern regions of Great Britain (Fig. 39.2).
4. In general, BSE cases were more often diagnosed in dairy cow herds than in beef suckler herds.
5. The disease was observed in adult animals only (mainly in 3–7-year-old animals). The majority of diseased animals fell sick at an age of 4–5 years.
6. The within-herd incidence was rather low. In the beginning there was often only one or only a few BSE cases per herd (see Chapter 39.2.5).
7. Epidemiologically, no association could be found with:
   - Breed (thus BSE seemed not to be a genetically determined disease);

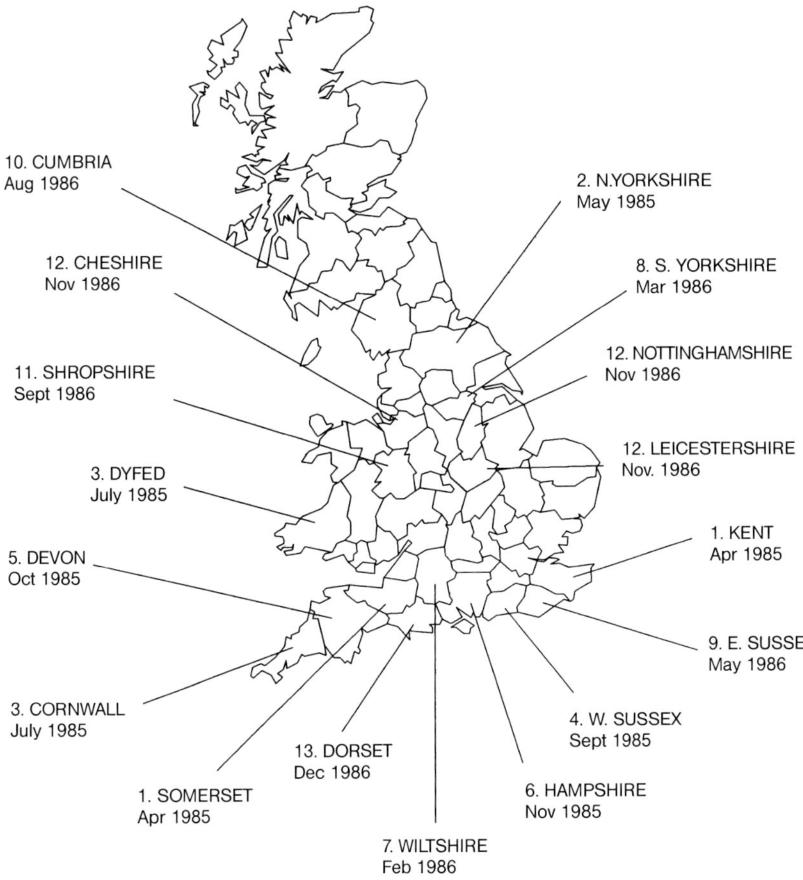

Fig. 40.1: Geographical distribution of the first British BSE cases (place of clinical onset) between May 1985 and the end of 1986, i. e., until the first confirmation of BSE was officially reported after laboratory examination [1]. Figure by J.W. Wilesmith taken from [5].

- Husbandry of sheep or goats on affected farms (this factor was of great interest because of scrapie);
- Administration of hormones, animal drugs, or exposure to agrochemicals used on the farms.

8. All early BSE cases seemed to be *index* cases, i. e., there was no evidence for a direct horizontal, maternal, or other way of direct transmission from animal to animal. Therefore, BSE seemed to be a disease that was widely spread geographically but had a common source of infection (common-source epidemic).
9. The only common factor that could be found in either affected herds or in herds from which infected animals were purchased was feeding of young calves with commercial feedstuff containing MBM (feed-borne hypothesis).

40.3 The case-control study: investigations on the feed-borne hypothesis

Further investigations indicated that in Great Britain a constellation of four risk factors explaining the start of the BSE epidemic coincided geographically some years before 1986 when the first BSE case was histologically confirmed [2;

3] (see Chapter 52). It seemed that scrapie pathogens had crossed the species border from sheep to cattle. The risk factors involved:

1. The frequent, endemic presence of scrapie in the sheep population (source of infection).
2. The high density of sheep in comparison to the cattle population (infection pressure).
3. Feeding of animal meal such as MBM (transmission vector), which contained recycled sheep tissue, to cattle of all age groups.
4. The technical modifications in the production of animal meal such as MBM from animal-waste-derived material: the heating temperatures were decreased and fat was no longer extracted with solvents (missing or inadequate inactivation of the infectious agent).

After it had been discovered that BSE could be traced back to these risk factors, attempts were made to understand the epidemic as a logical chain of events (details in Chapters 19, 52).

40.3.1 Working hypothesis

To examine the feed-borne hypothesis thoroughly [6; 7], the investigators focused on the question whether feeding practices of young calves during their first months of life and inclusion of MBM could be the main risk factor causing the epidemic in Great Britain. The design of a case-control study was chosen to test the following hypothesis: "There is an association between BSE and consumption of MBM during early calfhood". The conduct of this study, analyses, and interpretation of results will be described in the following subchapters.

40.3.2 Selection of case and control herds

According to the general procedure of a case-control study, two groups of study units (namely, herds) were compared: One group consisted of herds in which BSE cases were diagnosed (the so-called "case group" or "case herds"); the other group consisted of herds that were not affected by BSE (the so-called "control group" or "control herds") [8]. A comparison of these groups was supposed to show whether case and control herds differed with regard to the investigated risk factor (feeding of MBM to young calves). For case and control groups to be comparable, the same kind of information is needed for both groups, and data have to be gathered equally precisely and reliably for the two groups. A statistically significant difference of the two study groups would then indicate an association between the risk factor (MBM) and the disease (BSE).

In principle, case and control herds should be chosen that only might differ with regard to the factor underlying the feed-borne hypothesis (MBM: yes or no), in order to be comparable with respect to other factors not related to the investigated risk factor. Otherwise, it could happen that the results only provide a sham association with the investigated risk factor due to differences caused by other factors.

Herds from dairy and beef suckler farms from seven regions in Great Britain were chosen for the study groups, initially forming one case group and two control groups. The case group consisted of herds with confirmed BSE cases born between July 1, 1983, and June 30, 1984, and could be selected from the British BSE database.

The case group was compared with two control groups. It must be mentioned that in each analysis only one control group could be used; details can be found in the original literature [6].

40.3.3 Questions and data acquisition

The questionnaires used to interview the owners of the case and control herds included questions on the size of the herd, type of herd or its purpose (dairy cows, beef suckler cows), on the contact between cattle and sheep (yes or no), etc. The main emphasis, however, was on detailed questions on the diet of young calves to uncover the use of MBM in the feedstuffs. Details were collected on diets involving (i) mixed calf feedstuffs, (ii) feedstuffs used for the rearing of the animals, (iii) protein supplement added to farm feedstuffs, and (iv) mixed feed for adult animals. To clarify whether and how much MBM was used, a two-step investigation was carried out [6].

The first step involved getting details from the farmers on whether commercial mixed-feed products of the above-mentioned feed categories were used in 1983/1984 at all for feeding calves. Information from 626 herds with BSE cases and 374 control herds could be used for this first analysis. However, no information on whether these feedstuffs actually contained MBM was obtained; in other words, whether the animals were really exposed to the hypothesized risk factor. Feeding of commercial mixed-feed products to calves was therefore only a surrogate for their exposure to MBM.

The second step of data acquisition involved getting details from producers of animal feedstuffs. They were asked whether they delivered cattle feedstuffs containing MBM to the case and control herds under study in 1983/1984. If the answer was affirmative, they then had to indicate the amount and origin of the MBM used. Only with this information was it possible to assess the relevance the inclusion of MBM in animal feedstuffs had as a BSE risk factor. Since this was the more important part of the evaluation, the following subchapters will concentrate on the second step.

40.3.4 Analysis of the study data

Complete information on the inclusion of MBM in mixed feedstuffs used in 1983/1984 was available for 177 BSE case herds and 118 control herds. Odds ratios (as an approximation for risk ratios) were calculated to assess whether there was an association between the hypothetical risk factor "meat and bone meal" and the occurrence of BSE. Odds ratios were used to quantify the chances of becoming infected with BSE in herds that had been exposed to MBM compared with herds that had not been exposed to this factor. An odds ratio of approximately 1 would indicate that there were equal chances for both exposed and unexposed herds. An odds ratio greater than 1 would indicate an association between the risk factor studied and occurrence of BSE.

For the risk factor "inclusion of MBM in mixed feed consumed by calves", an odds ratio of 4.74 was calculated with a 95% confidence interval (CI) ranging from 2.25 to 10.20. Since in this study case and control herds could also differ with regard to region, type, and size of herd, as well as contact with sheep, adjusted odds ratios were estimated using logistic regression. An adjusted odds ratio of 5.89 with a 95% CI ranging from 2.53 to 13.1 was obtained for the alleged risk factor. The value 1 is not included in this interval; hence a statistically significant and 6-times higher[1] BSE risk was assumed in herds in which the calves consumed MBM included in feedstuffs during their first months of life [6].

40.3.5 Analyses using alternative approaches

The results of the study described above [6] were compared with an analysis of the study data carried out by a different set of authors [7; 9] based on an alternative approach. For this additional analysis, only herds consisting of dairy cows were investigated, which constitute the largest BSE subpopulation. The reason for concentrating on dairy cattle was that young calves that were brought up on dairy farms received a different diet from calves in beef suckler herds, with the amount of MBM included in feedstuffs being higher. In addition, it should be easier to compare case and control herds if they are derived from the same cow population, in particular with regard to the age structure of the herds. Furthermore, one of the seven[2] regions represented in the study was excluded from the additional analysis, as there were no longer any control herds when concentrating on dairy cattle.

Owing to these restrictions, the inclusion of MBM in mixed feedstuffs could be studied based on 168 BSE case herds and 96 control herds. Table 40.1 shows a two-by-two table giv-

[1] When interpreting odds ratios, it should be kept in mind that they denote an approximation to risk ratios. They indicate, e. g., how many times the probability of BSE occurrence increased by adding MBM to calf feedstuffs in comparison to herds in which the animals did not have any contact to such food additives. A significantly higher odds ratio due to a certain risk factor therefore does not mean that the BSE risk is fully explained by this factor or that there was no BSE risk for herds that were not exposed to the risk factor under study.

[2] region 1 = north; region 2 = mid and west; region 3 = east; region 4 = southeast; region 5 = southwest; region 6 = Wales; region 7 = Scotland.

Table 40.1: Case and control herds and exposure to meat and bone meal (MBM).

Exposure to MBM	BSE case herds	Control herds	Total
Yes	56	7	63
No	112	89	201
Total	168	96	264

ing the numbers of these case and control herds where calves had been exposed to MBM via mixed feedstuffs in 1983/1984 or not. A crude odds ratio of 6.36 and a 95 % confidence interval ranging from 2.9 to 13.60 were calculated using these data. The value 1 is not included in this interval. Hence, the risk of becoming affected by BSE was significantly higher for dairy herds in which calves were exposed to MBM in 1983/1984 compared with dairy herds in which calves were not exposed to this risk factor in 1983/1984.

Since the region, size of herd, and contact with sheep might also have an influence on the potential BSE risk, the relevance of MBM had to be assessed on the basis of "adjusted" odds ratios. The application of this procedure was based on two alternative models that described the association between the response (variable to be explained: case or control status of the herds) and the explaining variables (examined risk factor and other potential influences). The results and their interpretation were dependent on the chosen model. These alternative models were fit to the study data, again using logistic regression:

1. As in the original analyses carried out by Wilesmith and colleagues [6] (see Section 40.3.4), the first model assumed that the risk factor of interest ("inclusion of MBM") had a uniform effect in all subgroups (regions). As a result, this factor was regarded only as a main effect side by side with the other influencing factors "region", "size of herd", and "contact to sheep" in model 1.
2. Alternatively, the second approach supposed that the risk factor "inclusion of MBM" could have different effects in different regions (model 2). This approach assumed that MBM did not necessarily contain BSE pathogens. The "inclusion of MBM" was only a surrogate for the exposure of herds to the risk of taking up BSE pathogens or not. The appropriateness of this surrogate could depend on: (i) the technical production conditions of the rendering plants, (ii) the degree of scrapie or BSE contamination of the animal carcasses – which again depended on the scrapie or BSE prevalence in the region where the abattoir waste came from and which was not always identical with the region where the MBM was sold, and (iii) the dose (amount/kg) of MBM contained in the mixed feedstuffs. Therefore, in this alternative regression approach, the risk factor "MBM" was considered specifically for single regions, so that the degree to which MBM increased the BSE risk was assessed separately for each region.

Adjusted odds ratios and 95 % confidence intervals were calculated for the risk factor "inclusion of MBM in calf feedstuffs", and the two models were compared (Table 40.2).

The estimated odds ratio of Model 1 (first modeling approach of the additional analysis) is similar to the adjusted odds ratio calculated by Wilesmith, namely OR = 6.74 (95% CI 2.74–16.57) versus OR = 5.89 (95% CI 2.53–13.71) [6]. In general, this suggests that herds in which the calves had been fed with MBM-containing feedstuffs during their first months had a significantly higher (approximately 6-

Table 40.2: Adjusted odds ratios and 95% confidence intervals for "inclusion of MBM in calf feedstuffs".

Factor	Model 1	Model 2
MBM (all regions)	6.74 (2.74–16.57)	–
MBM in region 3 (east England)	–	1.43 (0.18–10.92)
MBM in region 5 (southwest England)	–	3.95 (1.18–13.17)

fold) BSE risk. Using the second model, however, for only the two most severely affected regions, east England (region 3) and southwest England (region 5), it was possible to calculate interpretable odds ratios.

For region 3, an odds ratio of 1.43 was derived. However, the 95% confidence interval overlaps value 1 (Table 40.2), and the result is thus statistically not significant. This means that it was not possible to deduce an increased risk of becoming infected with BSE for herds exposed to MBM, at least not on the basis of the small number of herds with which this region was represented in the study.

For region 5, which always had the highest region-specific incidence during the first 10 years of the BSE epidemic, an odds ratio of 3.95 was calculated. This indicates that the BSE risk of young calves from MBM-exposed herds was approximately 4-times higher than for herds that were not exposed to MBM-containing feedstuffs. The 95% confidence interval did not include the value 1 (Table 40.2). Hence, this result was statistically significant.

For the other regions, results were obtained that were difficult to interpret. However, this does not mean that MBM did not constitute a risk factor in these regions. The missing findings mainly resulted from difficulties in performing the analysis due to small numbers of herds.

Some of the case herds, or approximately 40% of the animals investigated here, seem not to have been exposed to the risk factor of interest during the first month after birth, presuming that the data available on the feeding habits were correct. The feeding history during their further life could not be studied in this investigation. These 40% might have become infected through the intake of MBM during the subsequent months of life [10; 11] or through other feedstuff categories (apart from calf-specific mixed feedstuffs). For example, the risk factor "feeding of certain cross-contaminated animal fats" – in particular in artificial milk replacers for calves – was not taken into account at the time of the investigation. In some of these cases, even maternal transmission might be considered as the cause of BSE infection (see Chapter 42.2.2).

The critical examination of this case-control study shows that several questions still remain unanswered. In particular, not all aspects involved in the feed-borne hypothesis were examined. Data were only collected for the first few months of the calves' lives; and the interpretation of the data was restricted to explicitly mixed feedstuffs for calves because an analysis of data provided for feedstuffs used for breeding, for protein supplements used to complement the feedstuffs used at the individual farms, and those provided for mixed feedstuffs for adult cows, was not possible.

According to the original evaluation undertaken by J. Wilesmith and the first approach of the supplementary analysis (Table 40.2), the exposure of young calves to MBM definitely bears a significant BSE risk. The second approach, which focused on the individual regions and involved fewer herds, could only determine an increased BSE risk for young calves brought up in southwest England. Nevertheless, given the fact that the feed bans have had a notable effect on the British BSE epidemic [12] (see Chapter 39.3.4), the case-control study clearly supports the hypothesis that infected MBM in mixed feedstuff poses a high BSE risk.

References

[1] Wells GAH, Scott AC, Johnson CT, et al. A novel progressive spongiform encephalopathy in cattle. Vet Rec 1987; 121:419–420.

[2] Wilesmith JW, Wells GAH, Cranwell MP, et al. Bovine spongiform encephalopathy: epidemiological studies. Vet Rec 1988; 123:638–644.

[3] Wilesmith JW, Ryan JB, Atkinson MJ. Bovine spongiform encephalopathy: epidemiological studies on the origin. Vet Rec 1991; 128:199–203.

[4] Taylor KC. The control of bovine spongiform encephalopathy in Great Britain. Vet Rec 1991; 129:522–526.

[5] Kimberlin RH. Bovine spongifrom encephalopathy. Animal Production and Health Paper, No. 109, FAO, Rome, 1993; pg. 8, ISSN 0254-6019.

[6] Wilesmith JW, Ryan JB, Hueston WD. Bovine spongiform encephalopathy: case-control studies of calf feeding practices and meat and bonemeal inclusion in proprietary concentrates. Res Vet Sci 1992; 52:325–331.

[7] Dahms S. Epidemiologische Studien zur Übertragung der Bovinen Spongiformen Enzephalopathie (BSE) – Anmerkungen aus biometrischer Sicht. Berl Münch Tierärztl Wochenschr 1997; 110(5):161–165.

[8] Schlesselman JJ. Case-control studies. Design, conduct, analysis. Oxford University Press, New York/Oxford, 1982.

[9] Dahms S, Baumann MPO, Wilesmith JW. Bovine spongiforme Enzephalopathie (BSE): Kritische Würdigung der epidemiologischen Untersuchungen zu Herkunft, Verbreitung und Übertragung. Bericht des 21. Kongresses der Deutschen Veterinärmedizinischen Gesellschaft e.V. Giessen. DVG, 1995:171–179.

[10] Wilesmith JW. BSE: epidemiological approaches, trials and tribulations. Prev Vet Med 1993; 18:33–42.

[11] Anderson RM, Donnelly CA, Ferguson NM, et al. Transmission dynamics and epidemiology of BSE in British cattle. Nature 1996; 382(6594): 779–788.

[12] Wilesmith JW, Hoinville LJ, Ryan JB, et al. Bovine spongiform encephalopathy: aspects of the clinical picture and analyses of possible changes 1986–1990. Vet Rec 1992; 130:197–201.

Topic VII: Transmissibility

41 The Experimental Transmissibility of Prions and Infectivity Distribution in the Body

Martin H. Groschup, Markus Geissen, and Anne Buschmann

41.1 Introduction

In a number of prion diseases, the causative agent is inefficiently transmitted under natural conditions, and transmissibility has only been proven in experimental studies. A detailed description of these experiments within a given species or across species barriers is presented in this chapter.

Up to the present time, no cell-culture system has been established that can be effectively infected with prions originating from the original host species (humans, cattle, sheep, goats, etc.), and currently only experimentally adapted prion strains can be propagated in cell culture [1].

Prions can only efficiently be replicated in animal studies. The agent may even be transmissible to "naturally resistant" species by choosing sufficiently high infectious doses and efficient routes of infection, e. g., direct injection into the brain. However, even in such experiments, there are certain species barriers that cannot be overcome. Thus, according to the experience gained so far, cattle-derived bovine spongiform encephalopathy (BSE) prions can be efficiently transmitted from cattle to mice, but not to hamsters.

Experimental transmission of various scrapie strains to mice and hamsters led to the development of animal models that have become indispensable in increasing the knowledge of prion diseases. The main objective of such studies is to gain insight into the pathogenesis of prion diseases and infectivity distribution in the body (WHO Tables$_{\text{Appendix}}$ IA–IC). Experimental transmission studies can also provide information on the natural transmissibility of the infectious agent within a certain species or to other species, as is described in detail in Chapter 42. A schematic graphical overview of the natural and experimental transmissibility of BSE, scrapie, and human prion diseases is provided in Figs. 42.1 to 42.3.

41.2 Brief historical overview

In 1936, the French scientists Cuillé and Chelle [2] first succeeded in transmitting scrapie by instillation of a homogenate from the spinal cord of a diseased sheep into the eye of another sheep. In contrast, the question of whether human prion diseases were transmissible remained unanswered for a long time: in the case of kuru until 1966 [3], Creutzfeldt–Jakob disease (CJD) until 1968 [4], and Gerstmann–Sträussler–Scheinker disease (GSS) until 1984 [5]. The transmissibility of fatal familial insomnia (FFI) was confirmed in 1995 [6; 7] and that of variant CJD (vCJD) in 1997 [8; 9].

In the meantime, other animal transmissible spongiform encephalopathies (TSEs) were also proven to be infectious by experimental transmission studies. Transmissible mink encephalopathy (TME) was experimentally transmitted for the first time in 1965 [10]. The infectious nature of chronic wasting disease (CWD), observed for the first time in North American cervids in 1967, was finally confirmed experimentally in the 1980s [11]. Shortly after the beginning of the BSE epidemic in the UK, the BSE agent was transmitted experimentally, and the results were described in 1988 [12].

41.3 Design of experimental transmission studies

In principle, transmission studies within a given species have to be distinguished from studies in which the infectious agent has to cross the species barrier. Experimental transmission studies can be designed as (i) contact experiments, (ii) oral transmission studies, or (iii) parenteral transmission studies with subcutaneous (s.c.), intraperitoneal (i.p.), intravenous (i.v.), or intracerebral (i.c.) inoculations. Contact experiments are carried out to examine the natural transmission potential of a TSE strain within one species or across species barriers.

41.3.1 Experimental transmission within one species

Experimental transmissions of mouse-adapted scrapie prions to wild-type mice (wt-mice) have demonstrated that different infection routes can vary in their efficiency. The i.c. or intraspinal injection of scrapie prions is 126,000 times more efficient than the oral inoculation [13]. The route of infection seems to be most effective when the agent can directly reach the target tissue, i. e., the central nervous system (CNS). Inefficient transmission routes, e. g., the oral route, may be overcome by using higher titers of infectivity in the inoculum (compare Fig. 1.6). The infectivity titer of a tissue can be determined by titration, e. g., in dilution steps of 1:10. Animals (usually mice) are inoculated through the i.c. route with the same quantity of inoculum for each dilution step. The dilution step that still causes an infection in 50% of the inoculated animals is called the infectious dose (ID_{50}) or lethal dose (LD_{50}). When indicating ID_{50} or LD_{50} values, one should always mention the route of inoculation and the animal species used for the titration. The infectivity titer can also be plotted against the incubation time for each dilution (with a low infectivity titer prolonging the incubation time), thus creating a titration curve. This can be useful in determining the infectivity titer of a tissue or suspension by comparing the incubation time obtained under standardized conditions with a previously determined titration curve [14].

Transmission of prions/strains with particularly long incubation times or transmission between different animal species that might also be associated with extremely long incubation times may cause false-negative transmission results. Thus, the agent may already have replicated in the brain of the recipient animal when the animal dies "naturally" due to its natural life-span (and not due to the infection), i. e., before the onset of clinical signs, as the infectivity titer in the brain was not sufficiently high (pseudo-resistance).

Comparison of scrapie incubation times between nontransgenic and transgenic mice, expressing a high amount of PrP^C, has demonstrated that the quantitative extent of the PrP expression in the brain has a profound influence on the incubation time (see Chapter 10). Two copies of the PrP gene (*Prnp*) are found in the genome of all individuals, one copy per chromosome set. In knock-out mice, both copies of *Prnp* have been destroyed genetically, so that these animals are unable to express PrP^C. Several mouse strains are available that express higher amounts of PrP than wt-mice. Additional copies of *Prnp* are inserted into the genome using genetic engineering techniques. For instance, transgenic mice designated Tg*a20* overexpress murine PrP^C by a factor of 10; Tg*a19* mice, by a factor of 4. The scrapie incubation times of these mice correlate inversely with the PrP^C expression rates. Knock-out mice ($Prnp^{o/o}$) are resistant to an infection with mouse-passaged scrapie, whereas wt-mice become ill after 131 days; Tg*a19* mice, after 87 days; and Tg*a20* mice, after 60 days. Crossbred animals from $Prnp^{o/o}$ and wt-mice become ill after 290 days [15]. These experiments confirm that the expression level of the PrP has a significant influence on the incubation time of prion diseases (see below).

41.3.2 Experimental transmission across species barriers

It is assumed that the relative efficiency of different routes of infection between species corresponds to that of transmissions within one species. In order to increase the transmission efficiency, different infection routes can be combined in the same animal, for instance the simultaneous i.c. (e. g., 20 μl) and i.p. (e. g., 100 μl) injection of a 10% brain homogenate.

A highly efficient barrier between different species leads to a low transmission efficiency in the primary passage. In practical terms, the effect of a highly efficient barrier prolongs the initial incubation time in comparison to an inefficient barrier. Sometimes the initial incubation time in a transmission across a species barrier is – theoretically – longer than the natural lifespan of the animal itself. The incubation time drops sharply in the subsequent secondary or tertiary passage in the new host species.

To obtain optimal transmission results across species barriers, the most efficient infection route (i. e., intracerebral) and highest reasonable challenge dose of infectivity should be chosen. As was shown in some cases, injection of infectious material into the tongue of the experimental animal is a relatively efficient peripheral transmission route [16].

The amino acid sequences of the prion protein of the donor and recipient animal species play a significant role in the species barrier (transmission barrier): in general, the higher the degree of similarity between the proteins' amino acid sequences, the lower the species barrier. Therefore, the phylogenetic relationship may have an influence on the transmissibility between different species (see Chapter 9). This effect was observed in in vitro conversion studies of PrP and in vivo transmission studies. Infection studies with chimeric mouse lines expressing hamster and mouse prion protein support the significance of the PrP amino acid sequence for the species barrier [17]. However, to date it is still unknown which part(s) of the amino acid sequence of the prion protein determines the level of the species barrier. Moreover, when transgenic mice expressing human PrP^C were created and challenged with human CJD infectivity, these mice were only susceptible when lacking endogenous mouse PrP^C expression. It was therefore postulated that a so-called "protein X" or "factor X" would function like a chaperone and would be required for the PrP conversion process itself. In transgenic mice co-expressing transgenic and wild-type murine PrP^C, protein X would bind preferentially to its natural binding partner and would not be available for the transgenic PrP conversion process. Hence, protein X could also have an impact on the species barrier.

41.4 Experimental transmissibility of human prion diseases

CJD and GSS were described for the first time at the beginning of the 20th century (see Chapter 1.6). However, major scientific interest in this new kind of disease did not arise until the end of the 1950s when the first reports of kuru in the Fore people were published (see Chapters 2, 13, 38); major public concern only arose with the emergence of vCJD in 1996.

First transmission studies of the kuru agent to chickens, mice, rats, guinea pigs, and rabbits were unsuccessful [18], and the experiments were ended 3 months post-inoculation, a time period which would have been adequately long to show clinical signs for most of the conventional infectious agents relevant at that time [19]. Gajdusek et al. did not consider the possibility of a scrapie-like infectious agent as the cause of kuru – for which a much longer incubation time would have been necessary – simply because the kuru researchers were unaware of the existence of the scrapie disease.

It was the veterinarian William J. Hadlow who pointed out in 1959 that there are considerable similarities in the neuropathology between kuru and scrapie in sheep [20]. He therefore suggested carrying out kuru transmission studies in nonhuman primates, which are phylogenetically closely related to humans (compare Chapter 9). Thus, Gajdusek, Gibbs, and Alpers used chimpanzees in their next study with brain samples of kuru and CJD victims and monitored the animals for several years. The apes developed the first clinical signs after 13 to 21 months, and the histopathological examination revealed typical spongiform changes in their brains. Reported in 1966, this was the first evidence of the transmissible nature of human spongiform encephalopathies [3; 4]. Since then 300 cases of sporadic, familial, and iatrogenic CJD, GSS, and kuru have been transmitted to nonhuman primates [21]. Besides chimpanzees, new world monkeys such as squirrel monkeys, capuchins, marmosets, and spider monkeys, as well as old world monkeys such as macaques, rhesus monkeys, crab-eating macaques, and blackish-green guenons were used.

In addition, CJD proved to be transmissible to cats [22], mice [23; 24], goats [25], and guinea pigs [26; 27], while other species such as sheep, dogs, minks, ferrets, rats, gerbils, rabbits, chickens, and ducks [22] were found not to be susceptible. GSS could be transmitted to nonhuman primates and to mice [5]. Kuru could be transmitted to cats, mice, goats, guinea pigs, and minks, but not to hamsters, pigs, rabbits, rats, and dogs [26]. About one decade after the first description of FFI [28], two working groups succeeded almost simultaneously (in 1995) in transmitting the disease to mice [6; 7]. Transmission of vCJD was successful to wt-mice [8], transgenic mice [9], and macaques [29; 30].

41.5 Experimental transmissibility of animal prion diseases

41.5.1 Experimental transmission of scrapie

Before the middle of the 20th century, only scrapie transmission studies in sheep and goats were published. The first successful experimental transmission of a scrapie isolate to mice was reported in the early 1960s [31; 32] (see Chapter 12). As transmission studies in mice are less expensive, require fewer personnel and less time than studies in sheep and goats, most transmission studies were thereafter carried out in rodents. In the 1970s, it was reported that the Syrian hamster was a favorable experimental animal to quantify scrapie infectivity titers. Other animal species, such as gerbils [22; 33], voles, and Chinese hamsters [33], can also be infected with scrapie. To date, it has not been possible to infect guinea pigs [33; 34], rats [34], and rabbits [34] with scrapie. Gajdusek also tried to transmit scrapie to nonhuman primates. In these studies, the chimpanzees, which were monitored for more than 12 years, proved to be resistant, while transmission to squirrel monkeys, spider monkeys, and crab-eating macaques was successful (26).

One of the approaches to simulate the origin of the BSE epidemic in an experimental transmission involved US scrapie isolates that were transmitted successfully to cattle. However, the clinical and histopathological picture (lesion profile) they produced was different from British BSE [35] (see Chapter 12).

41.5.2 Experimental transmissibility of BSE and FSE

The BSE agent is experimentally transmissible by parenteral inoculation (usually i.c. but sometimes combined with i.p. and/or i.v. inoculation) to a number of species: cattle [36], sheep and goats [37; 38], pigs [39] (further references see Chapter 24), mice [12], mink [40], marmosets [41], and macaques [30]. The i.c. transmission of BSE to domestic fowl [42] (further references see Chapter 25) was not possible.

BSE is orally transmissible to cattle [43; 44], sheep and goats [37; 38], and mice [45; 46], but not to pigs (see Chapter 24). In an attack-rate study, four groups of cattle were orally infected with different amounts of brain material derived from BSE-diseased cows (1 g, 10 g, 100 g, or 3×100 g given on 3 successive days). All animals in the 100-g and 3×100-g dosed groups, 7 out of 9 in the 10-g dosed group and 7 out of 10 in the 1-g dosed group developed BSE (Fig. 19.3). Interim results from a second experiment extending these findings to groups of calves orally dosed with 0.1, 0.01, and 0.001 g of BSE brain homogenate showed that even a dose of just 1 mg of crude brain-stem homogenate could lead to an infection in 1 out of 15 animals (Wells, personal communication; see Chapter 19.3.2 and 19.7).

Strain typing in mice has indicated that BSE prions may have been transmitted under field conditions to exotic ruminants (compare Table 19.1) and felines (compare Table 1.3) that were kept in zoological gardens, as well as to domestic cats: the experimental transmission of brain homogenates of the affected nyala, greater kudu, and domestic cats led to clinical signs after the same range of incubation times and produced an identical brain lesion profile in a defined set of mice as observed after experimental BSE infection [47; 48]. The study demonstrated that the feline spongiform encephalopathy (FSE) strain as well as the prion strain in the nyala and greater kudu was identical to the BSE strain, as described in detail in Chapter 50.3.

41.5.3 Experimental transmissibility of CWD

The CWD agent of deer is transmissible to minks, ferrets, new world monkeys, cervids, goats [11], and mice under experimental conditions. Hamsters, however, seem not to be susceptible [11].

41.5.4 Experimental transmissibility of TME

The TME agent is transmissible by i.c. inoculation to other minks, ferrets [49], Chinese and Syrian golden hamsters, cattle, sheep [50], goats [50; 51], skunks, raccoons, and nonhuman primates [52; 53]. Mice proved to be resistant to TME infection in several experiments [54; 55].

41.6 Infectivity distribution in peripheral organs

An excellent compilation of the available data for the infectivity distribution in human, bovine, and ovine prion diseases has been realized by a working group of the WHO[1].

41.6.1 CJD

The infectivity distribution in the body of sporadic CJD and kuru victims was determined in experiments with nonhuman primates. Infectivity was found in many tissues and body fluids including the brain, cerebrospinal fluid, cornea, lung, liver, kidney, spleen, and lymph nodes, but not in milk [56], saliva, urine, blood, peripheral nerves, bone marrow, heart adipose tissue, prostate, testes, placenta, amnion, and amniotic fluid [21; WHO Tables$_{\text{Appendix}}$ IA–IC].

41.6.2 vCJD

Remarkably higher PrP^{Sc} accumulation or infectivity titers were found in the case of vCJD, in particular in the sympathetic nervous system, but also in the tonsils [57–59] and other lymphatic organs/tissues [60] such as the spleen and gut-associated lymphoid tissue of the ileum and appendix (WHO Tables$_{\text{Appendix}}$ IA–IC). This is an indication that prion replication is possible outside of the nervous system. Several vCJD case reports have even indicated the potential risk of transfusion transmission of vCJD [61–64] (blood: see Chapter 51).

41.6.3 Scrapie

In scrapie-infected sheep [65] and goats [66; 67], infectivity was found in the CNS (brain, spinal cord), the placenta [68], the fetal membranes [69], blood [70], and lymphatic tissues (tonsils, lymph nodes, spleen; WHO Tables$_{\text{Appendix}}$ IA–IC).

41.6.4 BSE

The BSE infectivity distribution in cattle has been studied extensively by using either the conventional mouse or the "bovinized" transgenic mouse bioassay as well as by inoculation of cattle. Infectivity was found in brain, trigeminal ganglion (in the cranial cavity), spinal cord, spinal ganglia (in the intervertebral area), retina, distal ileum, etc. [38; 71] (compare OIE recommendations regarding specified risk material, SRM, in Chapter 53.3.1). Infectivity was also observed in the bone marrow of one cow suffering from BSE (Fig. 53.2) [72]. However, it is unclear whether this finding was caused by accidental contamination of the sample, e. g., with CNS tissue. Bioassays using transgenic "bovinized" mice have also revealed infectivity in the sciatic, facial, and optic nerves and in the Peyer's patches in the distal ileum of a BSE case from the field [73], whereas all other examined lymphatic tissues (including spleen) and reproductive organs were free of infectivity (WHO Tables$_{\text{Appendix}}$ IA–IC). Moreover, a minute amount of infectivity was also found in the Musculus semitendinosus, which is innervated by the Nervus tibialis originating in the sciatic nerve. A large range of other tissues and body fluids have led to negative results in conventional mouse bioassays, e. g., udder and milk

[1] WHO Tables$_{\text{Appendix}}$ IA–IC (see Appendix 1 of this book).

[56; 74], sperm, blood, etc. However, it is possible that some of the examined tissues contained minor amounts of infectivity, as the BSE agent had to cross the species barrier from cattle to mice. Although the extent of the respective species barrier is unknown, parallel infection experiments showed clearly that cattle are at least 1,000 times more susceptible to BSE of bovine origin than wt-mice [73; 75]. It is therefore necessary to re-assay all tissues from BSE-infected cattle that were believed to be free of infectivity by using cattle or transgenic "bovinized" mouse bioassays (compare Chapter 19.7.2). As cattle infection experiments are extremely laborious and time consuming, and as transgenic mice for the detection of BSE prions are now available (see Chapter 10), studies using bioassays with these mice are today the preferred models for detecting and measuring infectivity titer without a species barrier being involved [73; 76].

41.6.5 Major categories of infectivity

As mentioned above, the infectivity distribution in peripheral tissues and the CNS of patients affected with a human prion disease, of scrapie-infected sheep [77] and goats [78], and of BSE-infected cattle [46] are comprehensively summarized in the latest official data available: "Major Categories of Infectivity" from the World Health Organization, reprinted in the appendix of this book (WHO $\text{Tables}_{\text{Appendix}}$ IA–IC; ©).

Mouse inoculation experiments revealed clear differences between BSE-infected cattle, whose infectivity titer in the peripheral organs was always below 10^2 infectious units, and scrapie-infected sheep and goats, whose tissues of the peripheral nervous system and lymphatic organs such as tonsils may show infectivity titers of up to $10^{5.1}$ infectious units [79].

41.7 Pathogenesis studies

In affected vCJD patients, PrP^{Sc} is found not only in the brain but also in the lymphatic tissue, especially in the tonsils [60] and vessel walls [80]. Similar observations were made earlier in the spleen and tonsils of sheep suffering from scrapie [81; 82] and in the spleen of sheep infected experimentally with BSE [83]. In contrast, PrP^{Sc} was not detected in the lymphatic tissue of cattle suffering from BSE, and no infectivity was found in these tissues [38; 73]. The route via which the infectious agent reaches the CNS, and the importance of the lymphatic system in this process, are explained in Chapters 11 and 28. According to the observations mentioned above, the spread of the prions within the organism appears to differ substantially between species. In hamsters infected intraperitoneally with scrapie strain 263K, prions reach the spinal cord in the area of the thoracic spine and spread from there centripetally and centrifugally. In parallel, there seems to be a second pathway that does not involve the spinal cord, as PrP^{Sc} deposits can be found simultaneously in the spinal cord and in the brain [84]. This observation was reconfirmed by inoculation studies, in which hamsters were infected orally with scrapie. Finally, PrP^{Sc} was found in the dorsal motor nucleus of the vagus nerve, but could only be detected in the cervical spinal cord in a much later stage of the incubation time [84] (see Chapter 28). This observation clearly indicates that prions can also reach the brain via the vagus nerve.

The pathogenesis of scrapie in mice is slightly different, as the lymphoreticular system seems to play the crucial pathogenetic role in this species. In order to examine the role of the lymphatic system, immunodeficient mice, whose follicular dendritic cells are defective (SCID mice), were infected with scrapie by different routes. These animals developed the disease after intracerebral, but not after intraperitoneal infection. It was concluded that the follicular dendritic cells are essential for replication of the infectious agent in the periphery [85]. The significance of the different cell types of the immune system for the spread and replication of the infectious agent in the body can be examined by infection of transgenic mice deprived of defined immunocompetent cell subpopulations by molecular genetic intervention (knock-out mice). A detailed summary of these experiments and of their results is given in Chapters 10 and 11.

From transmission studies in transgenic mice it can be assumed that the BSE pathogenesis in cattle differs substantially from that in sheep [38; 73]. In order to further elucidate this phenomenon, large-scale oral BSE pathogenesis studies have been performed in the United Kingdom [86; 87] and in Germany [88] that confirm the hypothesis that the lymphatic system is not involved in the pathogenesis in cattle, but that the peripheral and vegetative nervous system play a major role in this process.

Experimental infection studies of the different forms or variants of prion diseases and prion strains or isolates sometimes cannot be performed in the authentic host species, e. g., for ethical considerations in humans or for reasons of cost and duration (incubation time) in the case of large farm animals. To circumvent these problems, mouse and hamster infection models have been developed: strain 263K in Syrian golden hamsters (incubation time approximately 60 days), strain RML (synonym: Chandler, 139A) in *s7s7* mice (C57BL, RIII, CD-1; incubation time approximately 150 days), strain ME7 in *s7s7* mice (incubation time approximately 170 days), and strain 301V in *p7p7* mice (VM mice incubation time approximately 100 days).

By means of genetic engineering, it was possible to create and breed knock-out mice [89–91] (*Prnp*$^{o/o}$; see Chapter 10). The resistance of these mice against infection with scrapie demonstrates the central role of PrP and *Prnp*, respectively, in the infection process and pathogenesis. Subsequently, the transgenic expression of homologous PrP or that of other species such as humans and cattle has opened a new avenue for the detection of infectivity in ovine, human, and bovine tissues and body liquids and for pathogenesis studies. In these transgenic mice, the incubation times of scrapie, vCJD/CJD, and BSE are significantly reduced.

References

[1] Klohn PC, Stoltze L, Flechsig E, et al. A quantitative, highly sensitive cell-based infectivity assay for mouse scrapie prions. Proc Natl Acad Sci USA 2003; 100:11666–11671.

[2] Cuillé J and Chelle PI. La maladie dite tremblante du mouton est-elle inoculable? C R Acad Sci Paris 1936; 203:1552–1554.

[3] Gajdusek DC, Gibbs CJ, Alpers M. Experimental transmission of a kuru-like syndrome to chimpanzees. Nature 1966; 209:794–796.

[4] Gibbs CJ Jr, Gajdusek DC, Asher DM, et al. Creutzfeldt–Jakob disease (spongiform encephalopathy): transmission to the chimpanzee. Science 1968; 161:388–389.

[5] Tateishi J, Sato Y, Nagara H, et al. Experimental transmission of human subacute spongiform encephalopathy to small rodents. IV. Positive transmission from a typical case of Gerstmann–Sträussler–Scheinker disease. Acta Neuropathologica 1984; 64:85–88.

[6] Collinge J, Palmer MS, Sidle KC, et al. Transmission of fatal familial insomnia to laboratory animals. Lancet 1995; 346:569–570.

[7] Tateishi J, Brown P, Kitamoto T, et al. First experimental transmission of fatal familial insomnia. Nature 1995; 376:434–435.

[8] Bruce ME, Will RG, Ironside JW, et al. Transmissions to mice indicate that new variant CJD is caused by the BSE agent. Nature 1997; 389:498–501.

[9] Hill AF, Desbruslais M, Joiner S, et al. The same prion strain causes vCJD and BSE. Nature 1997; 389:448–450.

[10] Burger D and Hartsough GR. Encephalopathy of mink. II. Experimental and natural transmission. J Infect Dis 1965; 115:393–399.

[11] Williams ES and Young S. Spongiform encephalopathies in cervidae. Rev Sci Tech Off Int Epiz 1992; 11:551–567.

[12] Fraser H, McConnell I, Wells GA. Transmission of bovine spongiform encephalopathy to mice. Vet Rec 1988; 123:472.

[13] Kimberlin RH and Walker CA. Pathogenesis of mouse scrapie: dynamics of agent replication in spleen, spinal cord and brain after infection by different routes. J Comp Pathol 1979; 89:551–562.

[14] Prusiner SB, Cochran SP, Groth DF, et al. Measurement of the scrapie agent using an incubation time interval assay. Ann Neurol 1982; 11:353–358.

[15] Fischer M, Rulicke T, Raeber A, et al. Prion protein (PrP) with amino-proximal deletions restoring susceptibility of PrP knockout mice to scrapie. EMBO J 1996; 15:1255–1264.

[16] Mulcahy ER, Bartz JC, Kincaid AE, et al. Prion infection of skeletal muscle cells and papillae in the tongue. J Virol 2004; 78:6792–6798.

[17] Raeber AJ, Race RE, Brandner S, et al. Astrocyte-specific expression of hamster prion protein (PrP) renders PrP knockout mice susceptible to hamster scrapie. EMBO J 1997; 16:6057–6065.
[18] Gajdusek DC and Zigas V. Untersuchungen über die Pathogenese von Kuru. Klin Wochenschr 1958; 36:445–459.
[19] Gajdusek DC and Gibbs CJ Jr. Attempts to demonstrate a transmissible agent in kuru, amyotropic lateral sclerosis, and other sub-acute and chronic nervous system degenerations of man. Nature 1964; 204:257–258.
[20] Hadlow WJ. Scrapie and kuru. Lancet 1959; 2:289–290.
[21] Brown P, Gibbs CJ Jr, Rodgers-Johnson P, et al. Human spongiform encephalopathy: the National Institutes of Health series of 300 cases of experimentally transmitted disease. Ann Neurol 1994; 35:513–529.
[22] Gibbs CJ Jr and Gajdusek DC. Experimental subacute spongiform virus encephalopathies in primates and other laboratory animals. Science 1973; 182:67–68.
[23] Brownell B, Campbell MJ, Grant DP, et al. The experimental transmission of Creutzfeldt–Jakob disease. Neuropathol Appl Neurobiol 1975; 1:398.
[24] Brownell B, Campbell MJ, Greenham LW, et al. Experimental transmission of Creutzfeldt–Jakob disease. Lancet 1975; 2:186–187.
[25] Hadlow WJ, Prusiner SB, Kennedy RC, et al. Brain tissue from persons dying of Creutzfeldt–Jakob disease causes scrapie-like encephalopathy in goats. Ann Neurol 1980; 8:628–632.
[26] Gibbs CJ Jr, Gajdusek DC, Amyx H. Strain variation in the viruses of Creutzfeldt–Jakob disease and kuru. Acad Press 1979; 2:87–110.
[27] Manuelidis EE, Kim J, Angelo JN, et al. Serial propagation of Creutzfeldt–Jakob disease in guinea pigs. Proc Natl Acad Sci USA 1976; 73:223–227.
[28] Lugaresi E, Montagna P, Baruzzi A, et al. Familial insomnia with a malignant course: a new thalamic disease. Rev Neurol (Paris) 1986; 142:791–792.
[29] Aguzzi A. Between cows and monkeys. Nature 1996; 381:734.
[30] Lasmezas CI, Deslys JP, Demaimay R, et al. BSE transmission to macaques. Nature 1996; 381: 743–744.
[31] Chandler RL. Encephalopathy in mice produced by inoculation with scrapie-brain material. Lancet 1961; 1:1378–1379.
[32] Zlotnik I and Stamp JT. Scrapie in a Dorset Down ram. A confirmation of the histological diagnosis by means of intracerebral inoculation of mice with formol fixed brain tissue. Vet Rec 1965; 77:1178–1179.
[33] Chandler RL and Turfrey BA. Inoculation of voles, Chinese hamsters, gerbils and guinea-pigs with scrapie brain material. Res Vet Sci 1972; 13:219–224.
[34] Barlow RM and Rennie JC. The fate of ME7 scrapie infection in rats, guinea-pigs and rabbits. Res Vet Sci 1976; 21:110–111.
[35] Clark WW, Hourrigan JL, Hadlow WJ. Encephalopathy in cattle experimentally infected with the scrapie agent. Am J Vet Res 1995; 56:606–612.
[36] Dawson M, Wells GAH, Parker BNJ. Preliminary evidence of the experimental transmissibility of bovine spongiform encephalopathy to cattle. Vet Rec 1990; 126:112–113.
[37] Foster JD, Hope J, Fraser H. Transmission of bovine spongiform encephalopathy to sheep and goats. Vet Rec 1993; 133:339–341.
[38] Fraser H and Foster JD. Transmission to mice, sheep and goats and bioassay of bovine tissues. In: Bradley R, Marchant B, editors. Transmissible spongiform encephalopathies. European Commission Agriculture, Brussels, 1993:145–160.
[39] Dawson M, Wells GA, Parker BN, et al. Primary parenteral transmission of bovine spongiform encephalopathy to the pig. Vet Rec 1990; 127:338.
[40] Robinson MM, Hadlow WJ, Huff TP, et al. Experimental infection of mink with bovine spongiform encephalopathy. J Gen Vir 1994; 75:2151–2155
[41] Baker HF, Ridley RM, Wells GA. Experimental transmission of BSE and scrapie to the common marmoset. Vet Rec 1993; 132:403–406.
[42] Dawson M, Wells GAH, Parker BNJ, et al. Transmission studies of BSE in cattle, pigs and domestic fowl. In: Bradley R, Marchant B, editors. Transmissible spongiform encephalopathies. European Commission Agriculture, Brussels, 1993:161–168.
[43] Wells GA, Dawson M, Hawkins SA, et al. Infectivity in the ileum of cattle challenged orally with bovine spongiform encephalopathy. Vet Rec 1994; 135:40–41.
[44] Wells GA, Hawkins SA, Green RB, et al. Preliminary observations on the pathogenesis of experimental bovine spongiform encephalopathy (BSE): an update. Vet Rec 1998; 142:103–106.
[45] Barlow RM and Middleton DJ. Dietary transmission of bovine spongiform encephalopathy to mice. Vet Rec 1990; 126:111–112.

[46] Fraser H, Bruce ME, Chree A, et al. Transmission of bovine spongiform encephalopathy and scrapie to mice. J Gen Virol 1992; 73:1891–1897.

[47] Jeffrey M, Scott JR, Williams A, et al. Ultrastructural features of spongiform encephalopathy transmitted to mice from three species of bovidae. Acta Neuropathol. 1992; 84:559–569.

[48] Fraser H, Pearson GR, McConnell I, et al. Transmission of feline spongiform encephalopathy to mice. Vet Rec 1994; 134:449.

[49] Bartz JC, McKenzie DI, Bessen RA, et al. Transmissible mink encephalopathy species barrier effect between ferret and mink: PrP gene and protein analysis. J Gen Virol 1994; 75:2947–2953.

[50] Hadlow WJ, Race RE, Kennedy RC. Experimental infection of sheep and goats with transmissible mink encephalopathy virus. Can J Vet Res 1987; 51:135–144.

[51] Zlotnik I and Barlow RM. The transmission of a specific encephalopathy of mink to the goat. Vet Rec 1967; 81:55–56.

[52] Marsh RF. Transmissible mink encephalopathy. In: Prusiner SB, et al., editors. Prion diseases of humans and animals. Ellis Horwood, New York, 1992:300–307.

[53] Marsh RF, Burger D, Eckroade R, et al. A preliminary report on the experimental host range of the transmissible mink encephalopathy agent. J Inf Dis 1969; 120:713–719.

[54] Hadlow WJ and Karstad L. Transmissible encephalopathy of mink in Ontario. Canad Vet J 1968; 9:193–196.

[55] Taylor DM, Dickinson AG, Fraser H, et al. Evidence that transmissible mink encephalopathy agent is biologically inactive in mice. Neuropathol Appl Neurobiol 1986; 12:207–215.

[56] EU. Report from the Scientific Veterinary Committee on the risk analysis for colostrum, milk and milk products. Leg Vet and Zoot VI/8197/96 Version J, 1996:1–26.

[57] Hill AF, Zeidler M, Ironside J, et al. Diagnosis of new variant Creutzfeldt–Jakob disease by tonsil biopsy. Lancet 1997; 349:99–100.

[58] Frosh A, Smith LC, Jackson CJ, et al. Analysis of 2000 consecutive UK tonsillectomy specimens for disease-related prion protein. Lancet 2004; 364:1260–1262.

[59] Yull HM, Ritchie DL, Langeveld JP, et al. Detection of type 1 prion protein in variant Creutzfeldt-Jakob disease. Am J Pathol 2006; 168:151–157.

[60] Glatzel M, Giger O, Seeger H, et al. Variant Creutzfeldt–Jakob disease: between lymphoid organs and brain. Trends Microbiol 2004; 12:51–53.

[61] Wilson K and Ricketts MN. The success of precaution? Managing the risk of transfusion transmission of variant Creutzfeldt–Jakob disease. J Neurol Sci 2004; 44:1475–1478.

[62] Bons N, Lehmann S, Mestre-Frances N, et al. Brain and buffy coat transmission of bovine spongiform encephalopathy to the primate Microcebus murinus. J Neurol Sci 2002; 42:513–516.

[63] Peden AH, Head MW, Ritchie DL, et al. Preclinical vCJD after blood transfusion in a PRNP codon 129 heterozygous patient. Lancet 2004; 364:527–529.

[64] Anonym. New case of transfusion-associated vCJD in the United Kingdom. Eur Surveill 2006; 11:E060209.2. Erratum in: Euro Surveill 2006; 11:E060209.2.

[65] Stamp JT, Brotherston JG, Zlotnik I, et al. Further studies on scrapie. J Comp Pathol 1959; 69:268–280.

[66] Pattison IH, Gordon WS, Millson GC. Experimental production of scrapie in goats. J Comp Pathol 1959; 69:300–312.

[67] Hadlow WJ, Eklund CM, Kennedy RC, et al. Course of experimental scrapie virus infection in the goat. J Infect Dis 1974; 129:559–567.

[68] Onodera T, Ikeda T, Muramatsu Y, et al. Isolation of scrapie agent from the placenta of sheep with natural scrapie in Japan. Microbiol Immunol 1993; 37:311–316.

[69] Pattison IH, Hoare MN, Jebbett JN, et al. Spread of scrapie to sheep and goats by oral dosing with foetal membranes from scrapie-affected sheep. Vet Rec 1972; 90:465–468.

[70] Hunter N, Foster J, Chong A, et al. Transmission of prion diseases by blood transfusion. J Gen Virol 2002; 83:2897–2905.

[71] MAFF. News release. Ministry of Agriculture, Fisheries and Food, London, December 3, 1997.

[72] Wells GA, Hawkins SA, Green RB, et al. Limited detection of sternal bone marrow infectivity in the clinical phase of experimental bovine spongiform encephalopathy (BSE). Vet Rec 1999; 144:292–294.

[73] Buschmann A and Groschup MH. Highly bovine spongiform encephalopathy-sensitive transgenic mice confirm the essential restriction of infectivity to the nervous system in clinically diseased cattle. J Infect Dis 2005; 192:934–942.

[74] Taylor DM, Ferguson CE, Bostock CJ, et al. Absence of disease in mice receiving milk from cows

with bovine spongiform encephalopathy. Vet Rec 1995; 136:592.

[75] Bradley R. Experimental transmission of bovine spongiform encephalopathy. In: Court L, et al., editors. Transmissible subacute spongiform encephalopathies. Elsevier, Amsterdam, 1996:51–56.

[76] Weissmann C and Flechsig E. PrP knock-out and PrP transgenic mice in prion research. Br Med Bull 2003; 66:43–60.

[77] Hadlow WJ, Kennedy RC, Race RE. Natural infection of Suffolk sheep with scrapie virus. J Inf Dis 1982; 146:657–664.

[78] Hadlow WJ, Kennedy RC, Race RE, et al. Virologic and neurohistologic findings in dairy goats affected with natural scrapie. Vet Path 1980; 17:187–199.

[79] Groschup MH, Weiland F, Straub OC, et al. Detection of scrapie agent in the peripheral nervous system of a diseased sheep. Neurobiol Dis 1996; 3:191–195.

[80] Koperek O, Kovacs GG, Ritchie D, et al. Disease-associated prion protein in vessel walls. Am J Pathol 2002; 161:1979–1984.

[81] Schreuder BE, van Keulen LJ, Vromans ME, et al. Preclinical test for prion diseases. Nature 1996; 381:563.

[82] van Keulen LJ, Schreuder BE, Meloen RH, et al. Immunohistochemical detection of prion protein in lymphoid tissues of sheep with natural scrapie. J Clin Microbiol 1996; 34:1228–12231.

[83] Foster JD, Bruce M, McConnell I, et al. Detection of BSE infectivity in brain and spleen of experimentally infected sheep. Vet Rec 1996; 138:546–548.

[84] Baldauf E, Beekes M, Diringer H. Evidence for an alternative direct route of access for the scrapie agent to the brain bypassing the spinal cord. J Gen Virol 1997; 78:1187–1197.

[85] Fraser H, Brown KL, Stewart K, et al. Replication of scrapie in spleens of SCID mice follows reconstitution with wild-type mouse bone marrow. J Gen Virol 1996; 77:1935–1940.

[86] Terry LA, Marsh S, Ryder SJ, et al. Detection of disease-specific PrP in the distal ileum of cattle exposed orally to the agent of bovine spongiform encephalopathy. Vet Rec 2003; 152:387–392.

[87] Wells GA, Spiropoulos J, Hawkins SA, et al. Pathogenesis of experimental bovine spongiform encephalopathy: preclinical infectivity in tonsil and observations of the distribution of lingual tonsil in slaughtered cattle. Vet Rec 2005; 156:401–407.

[88] Hoffmann C, Ziegler U, Buschmann A, et al. BSE prions spread via the autonomous nervous system from the gut to the CNS in incubating bovines. J Gen Virol 2006; in press

[89] Bueler H, Fischer M, Lang Y, et al. Normal development and behaviour of mice lacking the neuronal cell-surface PrP protein. Nature 1992; 356:577–582.

[90] Manson JC, Clarke AR, Hooper ML, et al. 129/Ola mice carrying a null mutation in PrP that abolishes mRNA production are developmentally normal. Mol Neurobiol 1994; 8:121–127.

[91] Sakaguchi S, Katamine S, Shigematsu K, et al. Accumulation of proteinase K-resistant prion protein (PrP) is restricted by the expression level of normal PrP in mice inoculated with a mouse-adapted strain of the Creutzfeldt–Jakob disease agent. J Virol 1995; 69:7586–7592.

42 Iatrogenic and "Natural" Transmissibility of Prion Diseases

Martin H. Groschup, Beat Hörnlimann, and Anne Buschmann

42.1 Introduction

This chapter summarizes selected examples of non-experimental transmission of prion diseases and discusses briefly the sources of infection. In contrast, Chapter 41 describes experimental infections in different animal species exposed to the causative agent of various prion diseases by particular transmission routes. Knowledge of the transmissibility of prion diseases provides a basis for developing strategies for their prevention and control.

42.1.1 Explanation of terms

Infectious agents fulfill, by definition, the criteria given in the Henle-Koch postulates: i) the agent must be found in body tissues or fluids as well as in the excretions of the affected organism; ii) it must be possible to subsequently isolate and cultivate the agent in pure culture; and iii) using the pure culture, it should be possible to reproduce the original disease. In the broader sense, pathogens causing prion diseases fulfill these postulates, inasmuch as the disease marker PrP^{Sc} can be isolated from various body tissues and fluids of diseased animals and humans (WHO $Tables_{Appendix}$ IA-IC). The isolated infectivity can then be used to reproduce clinical disease in experimental animals.

Natural transmission. In general, natural transmissions of infectious diseases can occur during birth, feeding, intensive contact between individuals, and under other circumstances of life. Thus, infections through person-to-person or animal-to-animal contact (direct) or the oral uptake of contaminated body tissues (see SRM in Chapter 53.3.1) or material such as meat-and-bone-meal (MBM; indirect), are regarded as "natural transmissions".

Horizontal transmission. This includes all non-germinal transmission of infectious agents to other individuals. An uptake of infectious tissues or body fluids and excretions can take place via two routes, oral or parenteral.

Oral transmission. This subgroup involves the ingestion of pathogens via the gastrointestinal tract, as for instance in kuru. It is assumed that primary (cow to human) variant CJD is acquired through oral transmission. In animals the best known example of a feed-borne prion disease is BSE. In the case of scrapie, oral transmission of the agent probably occurs through consumption of placenta or by licking and nibbling contaminated surfaces or objects.

Parenteral transmission. This subgroup includes intracerebral, intravenous, percutaneous, and perocular transmission, i. e., any means that is not alimentary. The only known natural parenteral transmission of prion disease in humans is kuru in Papua New Guinea (PNG). Here, percutaneous or perocular transmission may have played a role when people had intimate contact with infectious brain or spinal cord tissue during mourning rites in which infection may have occurred via skin lesions (percutaneous transmission) or via the mucous membranes or the eyes (see Chapters 13, 38).

Vertical transmission. Vertical transmission involves, in its proper sense, the germinal transmission from parents to their offspring. There is evidence for the vertical transmission of scrapie, i. e., from the ewe to her lambs (see Section 42.2.2).

Maternal transmission. This term embraces both horizontal (peri- and postnatal) and vertical (prenatal) transmission of the agent from mother to offspring. The following are possible:
1. Prior to birth (prenatal transmission), e. g., germinally or through oral uptake of infectivity by the fetus.
2. During or shortly after birth (perinatal transmission), e. g., by intimate contact between the fetus and fetal membranes during and after expulsion [1; 2].
3. After birth (postnatal transmission), e. g., by ingestion of placenta or by licking the contaminated holding or pasture.

As maternal transmission can occur either horizontally or vertically, the term is imprecise. However, since the two major transmission routes are potentially associated with reproduction and birth, they cannot be distinguished in terms of practical preventive measures under field conditions. Therefore this term is often used, but among prion diseases it is epidemiologically only relevant in scrapie.

Iatrogenic transmission. Cases of iatrogenic Creutzfeldt–Jakob disease (iCJD) or iatrogenic scrapie represent a type of horizontal transmission. They are caused by accidental transmissions in modern medicine, as summarized in Table 14.1 (see also Chapter 36.4.2). For example, iCJD has occurred after i) administration of contaminated human cadaver-derived pituitary hormones, ii) transplantation of an infectious cornea, or iii) implantation of contaminated dura mater. In all cases of iCJD and iatrogenic scrapie the agent was transmitted parenterally, mostly by intracerebral (i.c.), intraocular or percutaneous routes.

42.1.2 Models for transmission

Experiments to model transmission under field conditions must fulfill certain preconditions such as proof of transmissibility and proof of diagnosis by bioassays. The first experimental transmission of scrapie was performed in the 1930s [3], but only since the 1960s has it been possible to adequately model natural scrapie transmission routes in rodents [4]. The availability of bioassays has substantially simplified research on prion diseases [5; 6]. As a result, the

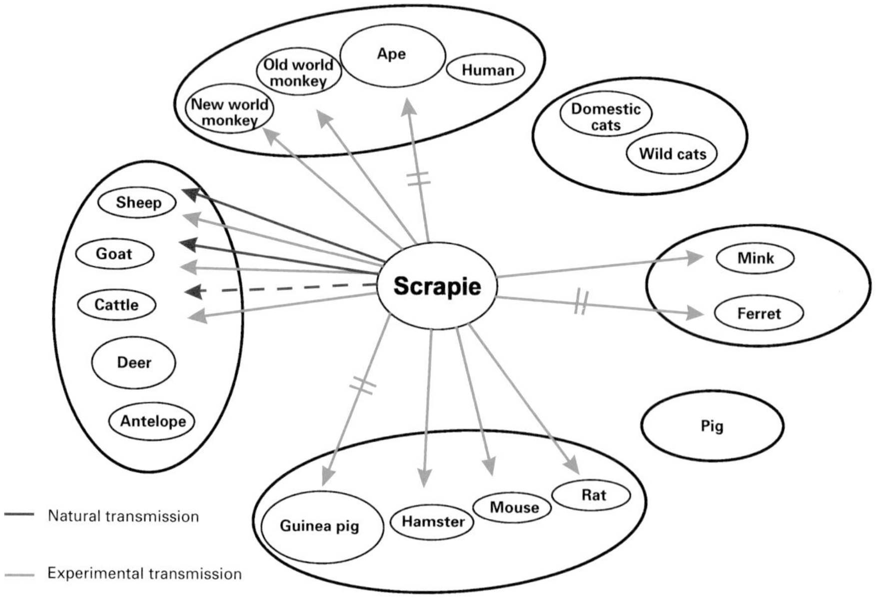

Fig. 42.1: Overview: transmission of scrapie from sheep to other animal species. Figure by M.H. Groschup.

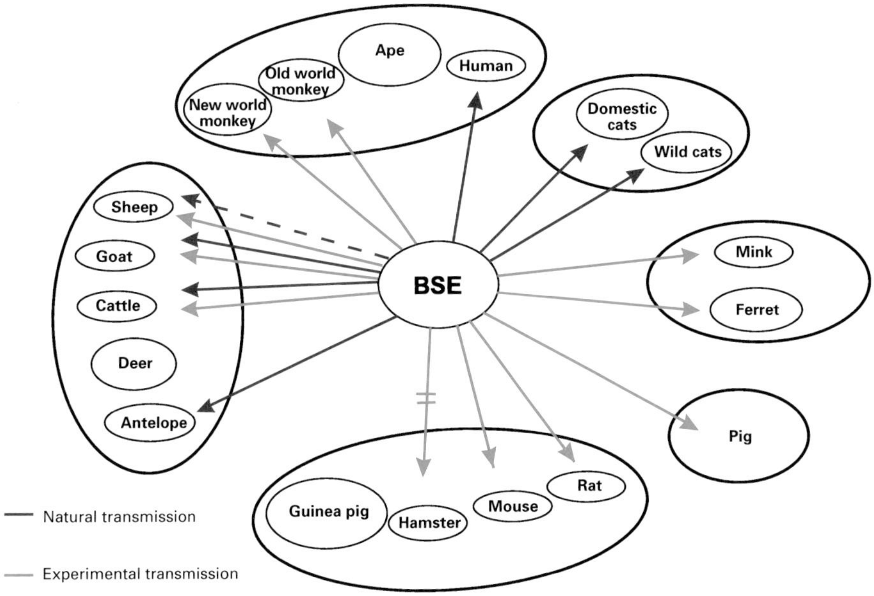

Fig. 42.2: Overview: transmission of BSE from cattle to other animal species. Figure by M.H. Groschup.

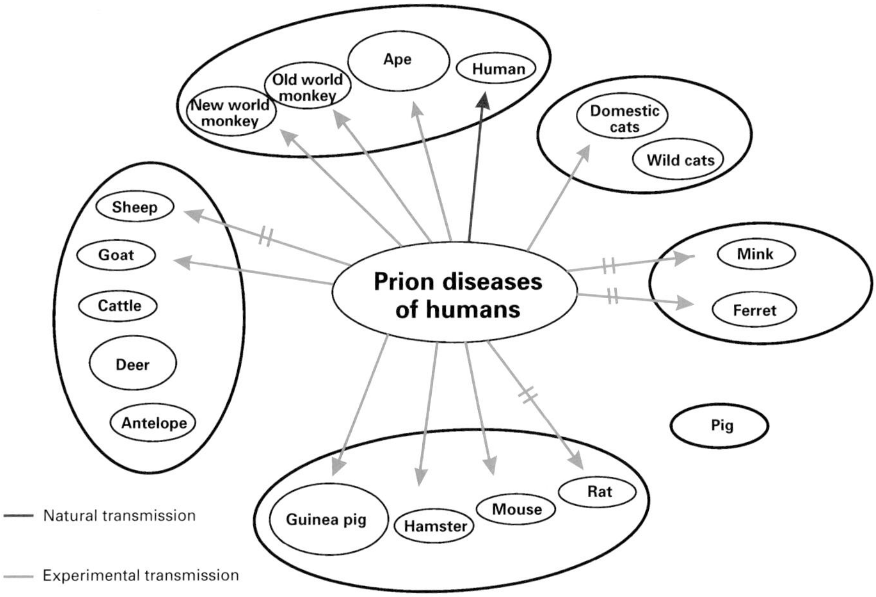

Fig. 42.3: Overview: transmission of human TSE to animals. Figure by M.H. Groschup.

quantitative determination of infectivity in different tissues and organs has become possible, enabling the categorization of organs according to the titer of infectivity (WHO Tables$_{\text{Appendix}}$ IA-IC), thus answering important pathogenetic questions (see Chapter 41).

Figures 42.1–42.3 summarize the natural and experimental transmissibility of bovine spongiform encephalopathy (BSE) and scrapie, as well as of human prion diseases.

41.1.3 Incubation time

Efficient iatrogenic or "natural" transmission of prions may lead to clinical disease, but only after a long incubation time, which in humans and animals lasts several months to years. However, in humans this is only relevant for the acquired forms iCJD, vCJD (Table 1.2) and kuru. iCJD has a mean incubation time of 15–16 months following direct intracerebral contact of infectious material by neurosurgery or stereotactic EEG electrodes; 6 years after dura mater implants, and 4.5–30 years after peripheral hormone injections [7–9] (Table 14.1; see also Chapter 51.3). The mean incubation time for vCJD appears to be about 15 years (see Chapter 1.9.3). The longest observed incubation time in humans is known from kuru, where it may exceed 45 years (see Chapters 13, 38). For the mean incubation times of animal prion diseases see Chapters 18 to 25.

42.2 Natural transmission within one species

42.2.1 Human prion diseases

CJD

Natural transmission has not been observed for CJD, not even between married couples or persons living closely together [10], and there is no risk for family members who care for CJD patients. There is, however, an occupational risk of CJD transmission for a few specific professional activities such as neuropathologists carrying out autopsies (brain / neuronal tissue) or researchers in laboratories handling prion-contaminated tissues. This has led to the implementation of a series of measures in these particular settings, in order to provide effective protection for the personnel in close contact with the infective tissue (see Chapters 47–49).

There is furthermore no epidemiological or experimental evidence for a maternal transmission of prion diseases [10; 11], including kuru [12; 13] (see Chapters 2, 13, 38). In transmission studies using nonhuman primates [14], guinea pigs [15] or mice [16–17], it has not been possible to detect infectivity in mammary tissue or in breast milk.

vCJD

To date, there has been no evidence of a natural mode of transmission of vCJD. However, human-to-human transmission of infection (and in two cases disease) has been reported by transfusion of red blood cells or blood from donors incubating vCJD (see iatrogenic transmission).

Kuru

Kuru prions were transmitted horizontally (mostly orally) [18; 19] and, in some cases, probably parenterally [12]. Both transmission routes have been confirmed experimentally [20]. Oral transmission occurred through ingestion of infectious tissue of persons who had died of kuru [19; 21] (see Chapters 2, 13, 38). Parenteral transmission may have played a role in women and children perhaps through scratching with contaminated hands, thus enabling the infected cerebral tissue to penetrate micro-lesions of the skin, the eye or the mucosa. Hands would have become contaminated through preparing the corpse of the dead relative for the mourning rites (endocannibalism).

Vertical and other modes of maternal transmission of kuru have never been reported from infected mothers to their children. The disease has never been diagnosed in children born after the cannibalistic rites were abandoned (around 1959), even if they were nursed by mothers who developed kuru during pregnancy or lactation [12; M. Alpers, personal communication]. Additionally, based on findings of animal experiments with nonhuman primates, a maternal transmission is considered unlikely [14; 16].

42.2.2 Prion diseases of animals

Scrapie in sheep and goats

Horizontal transmission. Classical scrapie in sheep occurs endemically or sporadically, whereas atypical scrapie cases in sheep and goats have so far – apart from few exceptions – been observed only in single animals [22]. For classical scrapie, the maternal or horizontal transmission is most important from an epidemiological point of view, i. e., pre-, peri-, and postnatally – from ewes to lambs, or from ewes to other sheep in the flock. High titers of infectivity have been detected in the placenta of preclinical or clinically affected sheep and goats [1; 2]. However, infectivity could not be detected in the placenta of some individual infected ewes [23; 24]. A study by Andreoletti et al. has shown that the genetic background of the sheep, e. g., whether or not an animal carries the so-called *Prnp*^{ARR} allele (see Section 42.6, Chapter 56), strongly affects the tissue distribution of infectivity and, thus, possibly also the transmissibility [25]. As in the case of a maternal transmission to lambs, the indirect horizontal spread between older sheep within the flock results from the expulsion of the infectious placenta. This occurrence has been well documented [1; 2; 23; 24; 26–30] (see Chapter 57).

Dickinson and colleagues examined the maternal and horizontal transmission of scrapie in sheep, and clearly demonstrated the significance of the transmission from adult sheep to flock mates and the peri- and postnatal transmission to lambs [31]. They reported that the incidence of scrapie increased in proportion to the time that lambs remained with their infected mothers. This correlation underlines the significance of the peri- and postnatal transmission [24].

Moreover, the scrapie agent may be transmitted indirectly via scarified skin or abrasions of the skin caused by scratching or rubbing [32]. Taylor demonstrated experimentally that percutaneous transmission of the scrapie agent is efficient [33]. Therefore, the risk of infection through scratching or rubbing against scrapie-contaminated objects or equipment is probable. Thus, pasture and environment such as grass, hay, floors, pens, etc. may be a lasting source of infection for indirect horizontal transmission. The scrapie agent is resistant enough to preserve its infectivity over several years thereby suggesting the possibility of an infectious scrapie reservoir in the environment. Brown and colleagues were able to recover a portion of scrapie infectivity which had been experimentally buried in garden soil for 3 years [34; 35].

Experiences in Iceland have indicated a spread of scrapie infectivity, possibly by automobile wheels, shoes, etc., even to distant pastures or barns [36]. Trading live sheep and goats from endemic scrapie areas represents a permanent risk for the introduction of scrapie into new flocks, regions, or importing countries [37; 38]. Some studies have suggested that hay mites may function as vectors [39, 40] and that *Sarcophagia spp.* larvae and pupae (carnivorous flies) may ingest and carry scrapie prions over the course of several weeks [41; 42]. However, the importance of these findings for the natural spread of scrapie remains unclear.

Vertical transmission. Contradictory results were found regarding a vertical transmission of scrapie by embryo transfer (ET). In a first attempt, embryos from experimentally infected sheep in the USA did transmit scrapie to the progeny [43], while in a second set of experiments performed by the same workgroup, scrapie did not transmit to either recipients or progeny [44], even if sheep of the highly susceptible PrP genotypes AA(136)QQ(171) were used [45] (see Chapter 42.6). In contrast in the UK, using *Sip*- or PrP-genotyped experimentally infected sheep and unwashed embryos, scrapie was seen in homozygous susceptible (sAsA) offspring within 979 days post inoculation [46; 47]. The question as to whether ET can be used to control natural scrapie in sheep is thus unresolved and further studies to investigate the effect of washing embryos would be needed [48].

BSE in cattle

Horizontal transmission. The ingestion of infectious meat-and-bone-meal (MBM) is regarded as the main – indirect horizontal – transmission route of BSE in the UK (see Chapter 19.9). Alternatively, BSE may have been transmitted via tallow (fat; see Chapter 46.4.2) in milk replacers [49]. The results of a recent statistical analysis in Bavaria suggests that, in comparison to control farms, a significantly higher proportion of BSE farms had fed proprietary concentrates and/or milk replacers containing bovine tallow to their calves [50].

Epidemiological studies in Great Britain have not revealed evidence of a direct horizontal BSE transmission (cattle-to-cattle) [51], although transmission rates $< 1\%$ would have been undetected [52; 53].

Maternal transmission. The role of maternal transmission of BSE seems to be virtually insignificant in the epidemic, i. e. the incidence of such transmission, even if it occurs, is insufficient to maintain the epidemic, at least in the UK. Unlike in scrapie, no infectivity has been detected in reproductive tissue (particularly placenta) or organs of BSE-infected cows or fetal membranes when these were tested by bioassay in either wild-type [54] or transgenic [55] (compare Chapter 19.7.2) mice.

Vertical transmission. In a study in the UK, 1,000 embryos were collected from confirmed BSE cases and some were implanted into 347 heifers imported from New Zealand (which is considered being free of BSE). These calves were subsequently held under quarantine conditions. None of the progeny from these embryo transfers developed BSE. Some embryos and uterine flushings were also inoculated into susceptible mice without resulting in disease [56].

Experimental transmission studies in mice failed to detect BSE infectivity in the semen or in the male sex glands of cattle [57].

BSE in exotic ruminants in zoos

In most cases of BSE or TSE in exotic ruminants (Table 1.3), it is probable that the infection was acquired via BSE-contaminated feed [58–59]. An exception might have been the first TSE-affected African ungulate observed, a nyala [60], which was detected about 5 months before the first BSE cases were confirmed, but this is purely speculative and based on the current knowledge there is more evidence for a feed-borne source (see Chapters 19.2.3, 20): at the time, this new disease in a nyala was not recognized as a form of BSE since this diagnosis did not yet exist; later, however, Bruce and colleagues reported that nyalas can indeed be infected with the BSE agent (see Chapter 50), as is the case for greater kudus [61] (see Chapter 42.4.6). Maternal transmission may have been responsible in at least one TSE case in a greater kudu [62; 63] (Tables 1.3 and 19.1). In summary, the evidence for an intraspecies maternal transmission in captive exotic ruminants is not strong, (compare Figs. 20.1 and 39.2), especially as the British feed ban is known to have been insufficiently applied until 1 August 1996.

It is of note that in the greater kudu BSE infectivity was distributed in lymphoreticular and neuronal tissues and was also detected in skin, conjunctiva, and salivary glands, as well as in tissues in which infectivity had not previously been detected in any naturally occurring animal prion disease [61] (WHO Tables$_{\text{Appendix}}$ IA-IC).

Feline spongiform encephalopathy in domestic and wild cats

Domestic [64] and captive wild [65–67] cats seem to be dead end hosts for prion diseases but, according to Kirkwood (see Chapter 20), there is no definitive answer to the question whether under natural conditions feline spongiform encephalopathy (FSE) is horizontally transmissible from cat to cat.

Chronic wasting disease in North American Cervidae

As with scrapie in sheep, chronic wasting disease (CWD) occurs endemically in the free-ranging deer population in some areas of North America [68–70]. According to US studies (see Chapter 21), CWD transmits horizontally between susceptible cervids, in free-ranging or captive populations. Early accumulation of

PrP^{Sc} in alimentary tract-associated lymphoid tissues during incubation suggests agent shedding in feces or saliva as possible transmission routes [71; 72]. Moreover, it is also possible that CWD prions are shed into the environment via the infectious placenta and fetal membranes.

Transmissible mink encephalopathy

The origin of transmissible mink encephalopathy (TME) is still unclear, but it is believed that outbreaks could have started by feeding TSE-affected ruminants to mink. TME does not seem to be transmitted maternally [73]. On the other hand, it is possible that particular horizontal transmission routes were responsible for the extremely high within-herd incidence observed in the few outbreaks in the USA (Table 22.1). Such routes may include feeding of dead mink as raw meat to their fellow animals in the same holding. Furthermore, mink may cannibalize fallen animals in their own cage [74]. Contaminated tissue may stick to the teeth and subsequently be transmitted parenterally to other animals through biting [75; 76].

42.3 Natural transmission of scrapie to other species

42.3.1 From sheep to cattle

There are two hypotheses for the origin of BSE: it was either caused by the accidental transmission of small ruminant scrapie to bovines [77] or by the intra-species recycling of a spontaneous TSE case in cattle [78]. The last hypothesis has become more plausible after the most recent discovery of so-called atypical BSE cases which may constitute spontaneous forms of prion diseases in cattle [79–81]. Scrapie challenge experiments with American isolates have revealed a significant difference between BSE and scrapie in experimentally infected cattle [82]. Whether scrapie was the original source of BSE, or whether it was BSE itself – the oral transmission mode was in any case indirectly horizontal, with MBM and/or contaminated tallow in milk replacers being the vehicles of transmission.

42.3.2 From sheep to goats / goats to sheep or to other ruminants

Contact experiments revealed that scrapie can be transmitted horizontally from sheep to goats [1; 83; 84]. In one study, when scrapie-free goats were introduced into a scrapie-infected sheep flock, 59% of the animals developed scrapie [85].

A horizontal transmission of scrapie from sheep to moufflons in zoological gardens or animal parks [86] and to cervid *spp.* in North America [70] has been discussed, but there is no well-founded literature proving this assumption.

42.3.3 From sheep or goats to mink

It is possible – but not experimentally proven – that oral uptake of scrapie-contaminated tissues may have caused TME in mink farms in the US [76; 87]. It has also been reported that TME outbreaks in the former Soviet Union may have been caused by feeding raw sheep carcasses to mink [88; 89] (compare Chapter 42.4.5).

42.4 Natural transmission of BSE to other species

42.4.1 From cattle to humans

Based on the epidemiological findings and on the experimental data on vCJD (biochemical examination and mouse bioassay) it can be assumed that BSE has crossed the species barrier between cattle and humans. The horizontal transmission route is supposed to be oral via BSE-contaminated cattle products (see Chapters 15.9; 36.4.1; 50).

42.4.2 From cattle to sheep and goats

At the present, there is no evidence that the BSE strain has been transmitted to sheep under field conditions, but more extensive surveillance and strain-typing studies need to be carried out to answer this question definitely [90; 91]. In contrast, a BSE case in a French goat which was diagnosed in 2005 has shown that goats are susceptible to BSE by the oral route under natural conditions [92].

42.4.3 From cattle to domestic or wild cats

BSE can be transmitted horizontally to cats, indirectly via the oral route. Based on the epidemiological findings and on the experimental strain-typing data on FSE (see Chapter 50), it is assumed that the BSE agent can cross the species barrier between cattle and cats. In view of the BSE-like features of the FSE agent in experimental studies, and the fact that FSE, like BSE, occurred in the UK and emerged a few years after the BSE epidemic (the first FSE case was reported in 1990), it is assumed that FSE is a feed-borne disease of domestic and wild cats [65–67] (Table 1.3). In particular, feeding of raw parts of the brain or spinal cord of British cattle in zoos or feeding of insufficiently heated components of commercial cat food, such as MBM or parts of SRM tissues, may have played a role as transmission vectors [93].

42.4.4 From deer to humans, cattle, or sheep

All molecular studies to date indicate the presence of a considerable species barrier for CWD between deer and other species [94]. Moreover, transmission studies of the CWD agent to transgenic humanized or cervidized mice showed low transmission efficiencies, if transmission occurred at all [95].

42.4.5 From cattle to mink

The original source of infection of TME is still unclear and may always remain so, but it is believed by some experts (see Section 42.3.3 and Chapter 22) that outbreaks could have started by feeding BSE-infective cattle tissue to farmed mink. Oral transmission of US scrapie to mink has never been successful in experiments.

42.4.6 From cattle to exotic captive ungulates

For details see Chapters 19.2.3, 20, 42.2.2.

42.5 Iatrogenic transmission in human and veterinary medicine

42.5.1 Humans

Iatrogenic transmissions of CJD to humans have mainly resulted from (for detailed lists see Chapters 14, 26, 36): (i) contaminated surgical instruments – from brain to brain, (ii) implantation or transplantation of infectious tissue (dura mater, cornea, etc.), and (iii) parenteral injection of cadaver-derived contaminated pituitary hormones. The iCJD cases have occurred since the 1970s at a time when TSEs, and the strong resistance of the agents to chemical disinfection and inactivation by heat, were incompletely understood. Brain or eye surgery [96] or the implantation / transplantation of infective dura mater or cornea etc. (see Chapter 36.4.2) led to transmission of the CJD agent through inadequately decontaminated surgical instruments [7; 97–102].

The largest number of iCJD transmissions resulted from human growth hormone or gonadotropin treatments. The hormones were extracted from pituitary glands or hypothalami collected from CJD-infected human cadavers [7; 103–107] (see Chapter 36.4.2). Contamination occurred when tissues were collected from several thousand cadavers and pooled for hormone extraction [108]. If one of the cadavers was infected with CJD, the entire extract could be contaminated. No adequate and selective inactivation measures exist for CJD prions for such hormone preparations. After administration of these hormone preparations by injection to growth-retarded children or to women with impaired fertility, some of the recipients developed iCJD after incubation periods ranging between 4.5 and 30.0 years [103] (Table 14.1). Today, such hormones can be produced by genetechnology and injected without any risk of TSE agent transmission.

Based on the data presently available, classical CJD is not transmitted through blood transfusions, whereas vCJD may be transmitted in this manner [109; 110]: the putative human-to-human transmission via blood transfusion and the known infectivity of some peripheral lymphoid tissue in vCJD (WHO Tables$_{\text{Appendix}}$ IA-

IC) raises the possibility of its iatrogenic transmissibility. This led to some additional prophylactic measures in regard to blood and blood products to provide effective protection. Because of its importance, this question is discussed in further detail in Chapter 51.

To date there has been no known case of human vCJD or animal BSE infection resulting from the use of commercially prepared vaccines or pharmaceuticals. In contrast, transmission of scrapie to sheep and goats has occurred as a result of the use of locally (non-commercially) prepared vaccines.

42.5.2 Sheep

In Scotland, the use of a scrapie-contaminated vaccine against the disease louping-ill resulted in a widespread iatrogenic scrapie infection in sheep. In an outbreak in the 1930s [111–114] approximately 1,800 out of about 18,000 vaccinated sheep developed scrapie. The vaccine had been produced from brain tissue from 8-month-old lambs and their dams. The latter belonged to a group used for a scrapie contact experiment. It may be that this incident has significantly contributed to the spread of scrapie in the UK. Another accidental intra- and interspecies transmission of scrapie agent occurred in Italy in 1997 and 1998, after the use of a scrapie-contaminated vaccine against *Mycoplasma agalactiae* [115; 116].

42.6 Genetically determined susceptibility

The susceptibility to prion diseases of individuals, particularly in humans (compare Chapters 14, 26, 36) and sheep (compare Chapter 56), is strongly influenced by host genetics. Furthermore strain-specific aspects (compare Chapter 12) and factors determining the species barrier also influence the susceptibility and, consequently, the transmissibility.

42.6.1 Humans

In human prion diseases, host-encoded genetic factors are the major risk factors for Gerstmann–Sträussler–Scheinker disease, familial fatal insomnia, and familial CJD; but even in iatrogenic CJD and vCJD [117] genetic factors may modulate the susceptibility and transmission – particularly due to methionine/methionine homozygosity of codon 129 of the PrP gene, which is associated with a shorter incubation period (compare Chapters 14, 26, 36). Genetic predisposing factors in humans are summarized in Tables 26.3 and 26.5. In the case of kuru, methionine/methionine homozygosity of codon 129 of the *PRNP* also appears to be associated with a shorter incubation period [18; 118] or, in other words, with an increased susceptibility to kuru.

42.6.2 Cattle

There is no evidence for genetic differences in susceptibility to BSE between different cattle breeds [119]. Therefore, all cattle must be considered susceptible to BSE. One recent publication postulates a breed-specific predisposition for one breed in Germany [120]; but this is based on statistical significance only and must be considered as a hypothesis until proven otherwise. In another comparative study on BSE diseased and control animals from Germany it was found that polymorphisms in the prion protein gene promoter modulate the prion protein expression level which may cause a higher BSE susceptibility of the carrier animal [121]. Some researchers suggested that a genetically induced higher susceptibility may increase the risk of offspring of BSE cases to "acquire" the disease from their mothers [122–124], but this is only a speculation.

42.6.3 Sheep and goats

The susceptibility of sheep to classical scrapie is genetically determined (compare Chapter 19.3 and Chapter 56). In goats, a genetic factor has been reported which influences the incubation period and thus the susceptibility to classical scrapie [125]. In the context of BSE in a French goat in 2005, it may be of note that the dimorphism in codon 142 of the caprine PrP gene appears to be associated with different incubation times in experimental infections with BSE or scrapie.

In contrast to classical scrapie, the genetic determination of the susceptibility of sheep to atypical scrapie is not well understood. Atypical scrapie also affects sheep carrying the PrP genotype conveying resistance to classical scrapie [126]. There are indications that polymorphisms at codons 141 and 154 in the ovine prion protein gene are associated with the susceptibility to atypical scrapie [127; 22].

References

[1] Pattison IH, Hoare MN, Jebbett JN, et al. Spread of scrapie to sheep and goats by oral dosing with foetal membranes from scrapie-affected sheep. Vet Rec 1972; 90:465–468.
[2] Pattison IH, Hoare MN, Jebbett JN, et al. Further observations on the production of scrapie in sheep by oral dosing with foetal membranes from scrapie-affected sheep. Br Vet J 1974; 130:65–67.
[3] Cuillé J and Chelle PI. La maladie dite tremblante du mouton est-elle inoculable? C R Acad Sci Paris 1936; 203:1552–1554.
[4] Chandler RL. Encephalopathy in mice produced by inoculation with scrapie brain material. Lancet 1961; 24:1378–1379.
[5] Renwick CC and Zlotnik I. The transmission of scrapie to mice by intracerebral inoculations of brain from an apparently normal lamb. Vet Rec 1965; 77:984–985.
[6] Zlotnik I and Stamp JT. Scrapie in a Dorset ram – a confirmation of the histological diagnosis by means of intracerebral inoculation of mice with formol fixed brain tissue. Vet Rec 1965; 77:1178–1179.
[7] Will RG. Acquired prion disease: iatrogenic CJD, variant CJD, kuru. Br Med Bull 2003; 66:255–265.
[8] Heckmann JG, Lang CJ, Petruch F, et al. Transmission of Creutzfeldt–Jakob disease via a corneal transplant. J Neurol Neurosurg Psychiat 1997; 63:388–390.
[9] Brown P, Cervenakova L, McShane L, et al. Ceutzfeld–Jakob disease in a husband and wife. Neurology 1998; 50:684–688.
[10] Brown P, Cathala F, Raubertas RF, et al. The epidemiology of Creutzfeldt–Jakob disease: conclusion of a 15-year investigation in France and review of the world literature. Neurology 1987; 37:895–904.
[11] Davanipour Z. Creutzfeldt–Jakob disease and other transmissible spongiform encephalopathies. In: Bastian FO, editor. Mosby Year Book, St. Louis, 1991:131–152.
[12] Gajdusek DC. Kuru in childhood: implications for the problem of whether bovine spongiform encephalopathy affects humans. In: Court L, Dodet B, editors. Transmissible subacute spongiform encephalopathies: prion diseases. Elsevier, Paris, 1996:15–26.
[13] Prusiner SB, Gajdusek DC, Alpers MP. Kuru with incubation period exceeding two decades. Ann Neurol 1982; 12:1–9.
[14] Amyx HL, Gibbs CJ Jr, Gajdusek DC, et al. Absence of vertical transmission of subacute spongiform viral encephalopathies in experimental primates. Proc Soc Exp Biol Med 1981; 166:469–471.
[15] Manuelidis EE and Manuelidis L. Experiments on maternal transmission of Creutzfeldt–Jakob disease in guinea pigs. Proc Soc Exp Biol Med 1979; 160:233–236.
[16] Brown P, Gibbs CJ Jr, Rodgers Johnson P, et al. Human spongiform encephalopathy: the National Institutes of Health series of 300 cases of experimentally transmitted disease. Ann Neurol 1994; 35:513–529.
[17] EU. Report from the scientific veterinary committee on the risk analysis for colostrum, milk and milk products. Leg Vet Zoot 1996;VI/8197/96 Version J:1–26.
[18] Goldfarb LG, Cervenakova L, Gajdusek DC. Genetic studies in relation to kuru: an overview. Curr Mol Med 2004; 4:375–384.
[19] Klitzman RL, Alpers MP, Gajdusek DC. The natural incubation period of kuru and the episodes of transmission in three clusters of patients. Neuroepidemiology 1984; 3:3–20.
[20] Gajdusek DC, Gibbs CJ Jr, Alpers M. Experimental transmission of a kuru-like syndrome to chimpanzees. Nature 1966; 209:794–796.
[21] Alpers MP. Kuru: implications of its transmissibility for the interpretation of its changing epidemiologic pattern. In: Bailey OT, Smith DE, editors. The central nervous system, some experimental models of neurological diseases. Int Acad Pathol Mono No 9. Proc 56th Annu Meet Int Acad Pathol, Washington, DC 12–15 Mar 1967. Williams & Wilkins, Baltimore, 1968:234–251.
[22] Lühken G, Buschmann A, Brandt H, et al. Epidemiological and genetical differences between classical and atypical scrapie cases. Vet Res 2006; in press

[23] Hadlow WJ, Kennedy RC, Race RE. Natural infection of Suffolk sheep with scrapie virus. J Infect Dis 1982; 146:657–664.
[24] Onodera T, Ikeda T, Muramatsu Y, et al. Isolation of scrapie agent from the placenta of sheep with natural scrapie in Japan. Microbiol Immunol 1993; 37:311–316.
[25] Andreoletti O, Lacroux C, Chabert A, et al. PrP(Sc) accumulation in placentas of ewes exposed to natural scrapie: influence of foetal PrP genotype and effect on ewe-to-lamb transmission. J Gen Virol 2002; 83:2607–2616.
[26] Hadlow WJ, Eklund CM, Kennedy RC, et al. Course of experimental scrapie virus infection in the goat. J Infect Dis 1974; 129:559–567.
[27] Detwiler LA. Scrapie. Rev Sci Tech Off int Epiz 1992; 11:491–537.
[28] Haralambiev H, Ivanov I, Vesselinova A, et al. An attempt to induce scrapie in local sheep in Bulgaria. Zbl Vet Med (B) 1973; 20:701–709.
[29] Ryder S, Dexter G, Bellworthy S, et al. Demonstration of lateral transmission of scrapie between sheep kept under natural conditions using lymphoid tissue biopsy. Res Vet Sci 2004; 76:211–217.
[30] Pattison IH and Millson GC. Scrapie produced experimentally in goats with special reference to the clinical syndrome. J Comp Pathol 1961; 71:101–108.
[31] Dickinson AG, Stamp JT, Renwick CC. Maternal and lateral transmission of scrapie in sheep. J Comp Pathol 1974; 84:19–25.
[32] Stamp JT, Brotherston JG, Zlotnik I, et al. Further studies on scrapie. J Comp Pathol 1959; 69:268–280.
[33] Taylor DM, McConnell I, Fraser H. Scrapie infection can be established readily through skin scarification in immunocompetent but not immunodeficient mice. J Gen Virol 1996; 77:1595–1599.
[34] Brown P and Gajdusek DC. Survival of scrapie virus after 3 years' internment. Lancet 1991; 337:269–270.
[35] Johnson CJ, Phillips KE, Schramm PT, et al. Prions adhere to soil minerals and remain infectious. PLoS Pathog 2006; 2:e32.
[36] Sigurdarson S. Epidemiology of scrapie in Iceland and experience with control measures. In: Bradley R, Savey M, Marchant B, editors. Subacute spongiform encephalopathies. Commission of European Communities. Kluwer Academic, Amsterdam, 1991:233–242.
[37] Brash AG. Scrapie in imported sheep in New Zealand. N Z Vet J 1952; 1:27–30.
[38] Parry HB. Scrapie disease in sheep. Oppenheimer DR, editor. Academic Press, London, 1983.
[39] Wisniewski HM, Sigurdarson S, Rubenstein R, et al. Mites as vectors for scrapie (letter). Lancet 1996; 347:1114.
[40] Rubinstein R, Kascsak RJ, Carp RI. Potential role of mites as a vector and/or reservoir for scrapie transmission. Alzheimer's Disease Review 1998; 3:52–56.
[41] Gruner L, Elsen JM, Vu Tien Khang J, et al. Nematode parasites and scrapie: experiments in sheep and mice. Parasitol Res 2004; 93:493–498.
[42] Post K, Riesner D, Walldorf V, et al. Fly larvae and pupae as vectors for scrapie. Lancet 1999; 354:1969–1970.
[43] Foote WC, Clark W, Maciulis A, et al. Prevention of scrapie transmission in sheep, using embryo transfer. Am J Vet Res 1993; 54: 1863–1868.
[44] Wang S, Foote WC, Sutton DL, et al. Preventing experimental vertical transmission of scrapie by embryo transfer. Theriogenology 2001; 56:315–327.
[45] Wang S, Cockett NE, Miller JM, et al. Polymorphic distribution of the ovine prion protein (PrP) gene in scrapie-infected sheep flocks in which embryo transfer was used to circumvent the transmissions of scrapie. Theriogenology 2002; 57:1865–1875.
[46] Foster JD, McKelvey WA, Mylne MJ, et al. Studies on maternal transmission of scrapie in sheep by embryo transfer. Vet Rec 1992; 130:341–343.
[47] Foster JD, Hunter N, Williams A, et al. Observations on the transmission of scrapie in experiments using embryo transfer. Vet Rec 1996; 138:559–562.
[48] Wrathall AE. Risks of transmitting scrapie and bovine spongiform encephalopathy by semen and embryos. Rev Sci Tech 1997; 16:240–264.
[49] Kamphues J. Alternatives to presently established forms of animal body removal: tolerated, intended and feared? Dtsch. Tierarztl. Wochenschr. 1997; 104:257–260.
[50] Clauss M, Sauter-Louis C, Chaher E, et al. Investigations of the potential risk factors associated with cases of bovine spongiform encephalopathy in Bavaria, Germany. Vet Rec 2006; 158:509–513.
[51] Anderson RM, Donnelly CA, Ferguson NM, et al. Transmission dynamics and epidemiology of BSE in British cattle. Nature 1996; 382:779–788.

[52] Lacey RW. Creutzfeldt–Jakob disease and bovine spongiform encephalopathy. Bovine spongiform encephalopathy is being maintained by vertical and horizontal transmission (letter). BMJ 1996; 312:180–181.
[53] Wilesmith JW. Creutzfeldt–Jakob disease and bovine spongiform encephalopathy: cohort study of cows is in progress (letter). BMJ 1996; 312:843.
[54] Bradley R. BSE transmission studies with particular reference to blood. Dev Biol Stand 1999; 99:35–40.
[55] Buschmann A and Groschup MH. Highly bovine spongiform encephalopathy-sensitive transgenic mice confirm the essential restriction of infectivity to the nervous system in clinically diseased cattle. J Infect Dis 2005; 192:934–942.
[56] Wrathall AE, Brown KF, Sayers AR, et al. Studies of embryo transfer from cattle clinically affected by bovine spongiform encephalopathy (BSE). Vet Rec 2002; 150:365–378.
[57] Curnow RN and Hau CM. The incidence of bovine spongiform encephalopathy in the progeny of affected sires and dams. Vet Rec 1996; 138:407–408.
[58] Kirkwood JK, Cunningham AA, Austin AR, et al. Spongiform encephalopathy in a greater kudu (Tragelaphus strepsiceros) introduced into an affected group. Vet Rec 1994; 134:167–168.
[59] Kirkwood JK, Cunningham AA, Wells GA. Spongiform encephalopathy in a herd of greater kudu (Tragelaphus strepsiceros): epidemiological observations. Vet Rec 1993; 133:360–364.
[60] Jeffrey M and Wells GAH. Spongiform encephalopathy in a nyala (Tragelaphus angasi). Vet Pathol 1988; 25:398–399.
[61] Cunningham AA, Kirkwood JK, Dawson M, et al. Bovine spongiform encephalopathy infectivity in greater kudu (Tragelaphus strepsiceros). Emerg Infect Dis 2004; 10:1044–1049.
[62] Kirkwood JK, Wells GA, Cunningham AA, et al. Scrapie-like encephalopathy in a greater kudu (Tragelaphus strepsiceros) which had not been fed ruminant-derived protein. Vet Rec 1992; 130:365–367.
[63] Aldhouse P. BSE. Maternal transmission in antelope. Nature 1990; 348:666.
[64] Wyatt JM, Pearson GR, Smerdon TN, et al. Spongiform encephalopathy in a cat. Vet Rec 1990; 126:513.
[65] Kirkwood JK and Cunningham AA. Epidemiological observations on spongiform encephalopathies in captive wild animals in the British Isles. Vet Rec 1994; 135:296–303.
[66] Young S and Slocombe RF. Prion-associated spongiform encephalopathy in an imported Asiatic golden cat (Catopuma temmincki). Aust Vet J 2003; 81:295–296.
[67] Willoughby K, Kelly DF, Lyon DG, et al. Spongiform encephalopathy in a captive puma (Felis concolor). Vet Rec 1992; 131:431–434.
[68] Williams ES and Young S. Chronic wasting disease of captive mule deer: a spongiform encephalopathy. J Wildl Dis 1980; 16:89–98.
[69] Williams ES and Young S. Spongiform encephalopathy of Rocky Mountain elk. J Wildl Dis 1982; 18:465–471.
[70] Williams ES and Young S. Spongiform encephalopathies in Cervidae. Rev Sci Tech Off Int Epiz 1992; 11:551–567.
[71] Miller MW and Williams ES. Chronic wasting disease of cervids. Curr Top Microbiol Immunol 2004; 284:193–214.
[72] Miller MW, Williams ES, Hobbs NT, et al. Environmental sources of prion transmission in mule deer. Emerg Infect Dis 2004; 10:1003–1006.
[73] Hanson RP and Marsh RF. Biology of transmissible mink encephalopathy and scrapie. In: Zeman W, Lennette EH, editors. Slow virus diseases. Williams & Wilkins, Baltimore, 1973:10–15.
[74] Burger D and Hartsough GR. Encephalopathy of mink. II. Experimental and natural transmission. J Infect Dis 1965; 115:393–399.
[75] Hadlow WJ, Race RE, Kennedy RC. Temporal distribution of transmissible mink encephalopathy virus in mink inoculated subcutaneously. J Virol 1987; 61:3235–3240.
[76] Marsh RF and Hanson RP. On the origin of transmissible mink encephalopathy. In: Prusiner SB, Hadlow WJ, editors. Slow transmissible diseases of the nervous system, Vol 1. Academic Press, New York, 1979:451–460.
[77] Wilesmith JW, Ryan JB, Atkinson MJ. Bovine spongiform encephalopathy: epidemiological studies on the origin. Vet Rec 1991; 128:199–203.
[78] Eddy RG. Origin of BSE. Vet Rec 1995: 137: 648.
[79] Buschmann A, Gretzschel A, Biacabe AG, et al. Atypical BSE cases in Germany. Proof of transmissibility and biochemical characterization. Vet Microbiol 2006; in press.
[80] Casalone C, Zanusso G, Acutis P, et al. Identification of a second bovine amyloidotic spongi-

form encephalopathy: molecular similarities with sporadic Creutzfeldt-Jakob disease. Proc Natl Acad Sci USA 2004; 101:3065–3070.
[81] Baron T and Biacabe AG. Origin of bovine spongiform encephalopathy. Lancet 2006; 367: 297–298.
[82] Clark WW, Hourrigan JL, Hadlow WJ. Encephalopathy in cattle experimentally infected with the scrapie agent. Am J Vet Res 1995; 56:606–612.
[83] Brotherston JG, Renwick CC, Stamp JT, et al. Spread of scrapie by contact to goats and sheep. J Comp Path 1968; 78:9–17.
[84] Hadlow WJ, Kennedy RC, Race RE, et al. Virologic and neurohistologic findings in dairy goats affected with natural scrapie. Vet Pathol 1980; 17:187–199.
[85] Kimberlin RH. Aetiology and genetic control of natural scrapie. Nature 1979; 278:303–304.
[86] Wood JL, Lund LJ, Done SH. The natural occurrence of scrapie in moufflon. Vet Rec 1992; 130:25–27.
[87] Marsh RF and Hadlow WJ. Transmissible mink encephalopathy. Rev Sci Tech Off Int Epiz 1992; 11:539–550.
[88] Gorham J. Viral and bacterial diseases of mink in Soviet Union. Fur Rancher 1991; 71:10–11.
[89] Gorham J. Viral and bacterial diseases of mink in Soviet Union. Fur Rancher 1991; 71:3.
[90] Bruce ME, Boyle A, Cousens S, et al. Strain characterization of natural sheep scrapie and comparison with BSE. J Gen Virol 2002; 83:695–704.
[91] Anonymous 2006: Regulation (EC) No 999/2001 of 22nd May 2001 laying down rules for the prevention, control and eradication of certain transmissible spongiform encephalopathies (latest amendment by Regulation (EC) 339/2006 of 24th February 2006).
[92] Eliot M, Adjou K, Coulpier M, et al. BSE agent signatures in a goat. Vet Rec 2005; 156:523–524. Erratum in: Vet Rec. 2005; 156:620.
[93] Hörnlimann B. Feline Spongiforme Enzephalopathie: Risikoeinschätzung für die Schweiz. IVI Report (unpublished), Mittelhäusern, 27 January 1998.
[94] Raymond GJ, Bossers A, Raymond LD, et al. Evidence of a molecular barrier limiting susceptibility of humans, cattle and sheep to chronic wasting disease. EMBO J 2000; 19: 4425–4430.
[95] Browning SR, Mason GL, Seward T, et al. Transmission of prions from mule deer and elk with chronic wasting disease to transgenic mice expressing cervid PrP. J Virol 2004; 78:13345–13350.
[96] Harbarth S, Alexiou A, Pittet D, et al. Creutzfeldt–Jakob Krankheit: Vorsichtsmassnahmen zur Prävention iatrogener Übertragungen im Spital. Swiss-NOSO 1996; 3:9–11.
[97] Duffy P, Wolf J, Collins G, et al. Possible person-to-person transmission of Creutzfeldt–Jakob disease (letter). N Engl J Med 1974; 290:692–693.
[98] Farrington M. Use of surgical instruments in Creutzfeldt–Jakob disease (letter). Lancet 1995; 345:194.
[99] Foncin JF, Gaches J, Cathala F, et al. Transmission iatrogène interhumaine possible de maladie de Creutzfeldt–Jakob avec atteinte des grains du cervelet. Rev Neurol Paris 1980; 136:280.
[100] Hogan RN and Cavanagh HD. Transplantation of corneal tissue from donors with diseases of the central nervous system. Cornea 1995; 14:547–553.
[101] Weber T, Tumani H, Holdorff B, et al. Transmission of Creutzfeldt–Jakob disease by handling of dura mater (letter). Lancet 1993; 341:123–124.
[102] Will RG and Matthews WB. Evidence for case-to-case transmission of Creutzfeldt–Jakob disease. J Neurol Neurosurg Psychiatry 1982; 45:235–238.
[103] Brown P. Human growth hormone therapy and Creutzfeldt–Jakob disease: a drama in three acts. Pediatrics 1988; 81:85–92.
[104] Cochius JI, Hyman N, Esiri MM. Creutzfeldt–Jakob disease in a recipient of human pituitary-derived gonadotropin: a second case. J Neurol Neurosurg Psychiatr 1992; 55:1094–1095.
[105] Gibbs CJ Jr, Joy A, Heffner R, et al. Clinical and pathological features and laboratory confirmation of Creutzfeldt–Jakob disease in a recipient of pituitary-derived human growth hormone. N Engl J Med 1985; 313:734–738.
[106] Koch TK, Berg BO, De Armond SJ, et al. Creutzfeldt–Jakob disease in a young adult with idiopathic hypopituitarism. Possible relation to the administration of cadaveric human growth hormone. N Engl J Med 1985; 313:731–733.
[107] Tintner R, Brown P, Hedley Whyte ET, et al. Neuropathologic verification of Creutzfeldt–Jakob disease in the exhumed American recipient of human pituitary growth hormone: epidemiologic and pathogenetic implications. Neurology 1986; 36:932–936.

[108] Cooke J. Cannibals, cows and the CJD catastrophe. Random House, Sidney, 1998.
[109] Peden AH, Head MW, Ritchie DL, et al. Preclinical vCJD after blood transfusion in a PRNP codon 129 heterozygous patient. Lancet 2004; 364:527–529.
[110] Anonym. New case of transfusion-associated vCJD in the United Kingdom. Eur Surveill 2006; 11:E060209.2. Erratum in: Euro Surveill. 2006; 11:E060209.2.
[111] Gordon WS, Brownlee A, Wilson DR. Louping-ill, tick-borne fever and scrapie. In: Dawson MH, editor. Third Int Congr Microbiol N Y, September 2–9, 1939. Int Assoc Microbiol N Y, 1940:262–263.
[112] Gordon WS. Advances in veterinary research. Louping-ill, tick born fever and scrapie. Vet Rec 1946; 58:516.
[113] Greig JR. Scrapie in sheep. J Comp Pathol 1950; 60:263–266.
[114] Taylor DM. Understanding the difficulty in inactivating TSE agents. A comprehensive update on the latest TSE research developments. London: IIR Conference Proceedings, March 9, 1998.
[115] Caramelli M, Ru G, Casalone C, et al. Evidence for the transmission of scrapie to sheep and goats from a vaccine against Mycoplasma agalactiae. Vet Rec 2001; 148:531–536.
[116] Zanusso G, Casalone C, Acutis P, et al. Molecular analysis of iatrogenic scrapie in Italy. J Gen Virol 2003; 84:1047–1052.
[117] Bishop MT, Hart P, Aitchison L, et al. Predicting susceptibility and incubation time of human-to-human transmission of vCJD. Lancet Neurol 2006; 5:393–398.
[118] Lee HS, Brown P, Cervenakova L, et al. Increased susceptibility to kuru of carriers of the PRNP 129 methionine/methionine genotype. J Infect Dis 2001; 183:192–196.
[119] Wijeratne WV and Curnow RN. A study of the inheritance of susceptibility to bovine spongiform encephalopathy. Vet Rec 1990; 126:5–8.
[120] Sauter-Louis C, Clauss M, Chaher E, et al. Breed predisposition for BSE: epidemiological evidence in Bavarian cattle. Schweiz Arch Tierheilk 2006; 148:245–250.
[121] Sander P, Hamann H, Grogemüller C, et al. Bovine prion protein gene (Prnp) promoter polymorphisms modulate Prnp expression and may be responsible for differences in bovine spongiform encephalopathy susceptibility. J Biol Chem 2005; 280:37408–37014.
[122] Donnelly CA, Ghani AC, Ferguson NM, et al. Analysis of the bovine spongiform encephalopathy maternal cohort study: evidence for direct maternal transmission. Appl Stat 1997; 46:321–344.
[123] Ferguson NM, Donnelly CA, Woolhouse MEJ, et al. Genetic interpretation of heightened risk of BSE in offspring of affected dams. Proc R Soc Lond Biol 1997; 264:1445–1455.
[124] Kimberlin RH. Bovine spongiform encephalopathy. Rev Sci Tech Off Int Epiz 1992; 11: 347–489.
[125] Goldmann W, Martin T, Foster J, et al. Novel polymorphisms in the caprine PrP gene: a codon 142 mutation associated with scrapie incubation period. (Published erratum in J Gen Virol 1997; 78:697) J Gen Virol 1996; 77:2885–2891.
[126] Buschmann A, Luhken G, Schultz J, et al. Neuronal accumulation of abnormal prion protein in sheep carrying a scrapie-resistant genotype (PrP ARR/ARR). J Gen Virol. 2004; 85:2727–2733.
[127] Moum T, Olsaker I, Hopp P, et al. Polymorphisms at codons 141 and 154 in the ovine prion protein gene are associated with scrapie Nor98 cases. J Gen Virol 2005; 86: 231–235.

Topic VIII: Agent Inactivation

43 Inactivation in Practice – Risk Assessment and Validation for Food Gelatin

Stuart C. MacDiarmid

43.1 Introduction

In 1996, human cases of a new transmissible spongiform encephalopathy (TSE) known as variant Creutzfeld–Jakob disease (vCJD) were reported and were soon shown to be caused by human infection with the bovine spongiform encephalopathy (BSE) agent. Because of fears that the BSE prion might be present in foods prepared from bovine tissues, including gelatin, precautionary measures to protect consumers were implemented around the world. However, in the intervening years much has been learned about BSE and the risk to human health, and in a number of countries some of the precautionary measures put in place after 1996 are being reviewed. This chapter, which complements Chapter 46.4.1, presents an assessment of the BSE risk posed by gelatin in food.

The current legislation summarized in Chapter 46/46.4.1 reflects the European attitude, which has been shaped by the intensity of public reaction and anxiety over BSE in recent years. However, food safety authorities in some countries are reviewing earlier precautionary policies adopted in the absence of good information on gelatin. The Terrestrial Animal Health Standards Commission of the World Organisation for Animal Health (OIE) has been asked to review the evidence with a view to moving gelatin into Article 2.3.13.1 (Reprint in Appendix 2 of this book) of the Terrestrial Animal Health Code [1], which lists those commodities that may be safely traded without specific BSE measures, regardless of the BSE status of the exporting country. Science and risk assessment support such a change, as outlined in this chapter.

43.2 Raw materials

The total world production of gelatin in 2003 was 278,300 tons [2]. Probably around 65% of this was produced from bovine materials, based on figures cited in [3]. Gelatin is made either from hides or bones, and although there are differences in the processes, both involve a series of chemical steps that inactivate the BSE agent [3; 4].

Gelatin is produced either from skin or bones of cattle and pigs (Fig. 46.1). The two raw materials are not mixed. Since BSE is not a disease of pigs (see Chapter 24), gelatin produced from porcine raw materials has never been of concern.

Hides are considered a safe source of raw material because BSE infectivity has not been found in skin, even in advanced clinical cases [5]. More gelatin is produced globally from skins than from bone [2]. BSE infectivity may be found in the bone marrow in some advanced clinical cases of the disease, based on figures cited in, and information from [6] (G. A. H. Wells, VLA Weybridge, UK, personal communication in April 2004), but has not been detected in bone marrow of infected cattle before they show clinical signs.

The overall global BSE epidemic is in decline, even though cases have recently been detected in North America. In Europe, the epidemic is declining in most countries. Occasional cases of BSE are still detected in countries with a history of feeding cattle on meat-and-bone-meal (MBM) containing BSE prions, and this may also be expected, in the future, in countries hitherto considered BSE-free.

The first step in processing the bones for gelatin manufacture is to grind them into pieces

(Fig. 46.1). Hides and skins are also chopped into small pieces. Hides may arrive at the gelatin plant in the form of "hide splits", a by-product of the tanning industry. Hide splits are the lower part of the cutis or corium.[1] The upper part of the cutis is used for leather production.

43.3 Dilution

As with all infections, with BSE there is a minimum level of infectivity necessary to transmit the infection. In the case of transmission to humans, there is also a species barrier to be surmounted.

Should a BSE-infected animal be tested negative by ante mortem inspection (false negative; see Chapter 35) and contribute raw materials to gelatin manufacture, its tissues would be diluted by those from a very large number of normal, uninfected animals. The average weight of raw material used from one animal, depending on age, is approximately 10–15 kg [3]. The normal batch size used in the industrial production of gelatin varies from around 20,000 to 100,000 kg.

This means that the dilution factor of raw material from one animal into an industrial batch is between $10^{3.5}$ and 10^5. So, regardless of any further treatment, in theory only in advanced clinical disease would there be detectable infectivity in the final product [3].

In the UK, the country that historically has experienced the highest BSE incidence rate with an epidemic peaking in 1992/1993, there were (only) 343 BSE cases in 2004 and 225 in 2005. That is equivalent to an annual BSE incidence rate of 68 (2004) and 46 (2005) affected cattle per million cattle over 24 months of age [7].

From the data on average weight of raw material per animal and average industrial batch size, it can be calculated that the average number of animals contributing to a single batch is 4,800 (standard deviation 1,880).[2] From these data and the annual UK BSE incidence rate (in 2004), it can be calculated using the software program @Risk (version 4.5.310; Palisade Corporation, Newfield, NY, USA) that the mean number of BSE cases that could contribute contamination to a batch of gelatin, assuming that they were to escape detection, would be 0.33 animals per batch (upper 95 percentile, 0.66 animals).

The New Zealand Food Safety Authority's Science Group developed [8] an @Risk simulation model to estimate the likelihood of BSE contamination of gelatin produced in the year 2004 in the UK from bones including the vertebral column.[3] BSE infectivity could be present in remnants of spinal cord and associated dorsal root ganglia (DRG). It has been estimated that the weight of DRG in a typical carcass is 30 g and the weight of spinal cord is 200 g [9]. The simulation model assumed that something between 0 and 100 % of the spinal cord might remain in vertebral columns used in gelatin production, but that the most likely quantity was 10 %. The model further assumed that all the DRG tissue would be included.

In an assessment of the risk to human health from inclusion of DRG in foods such as "T bone" steaks, Comer [10] assumed that the best estimate of the oral infectivity for humans of spinal cord and DRG derived from cases of BSE is 1 human oral ID_{50}/g with a confidence range of 0.0001 to 10 [10].

Using the data outlined above, the @Risk model estimated that the mean BSE contamination of raw material containing vertebral column of UK origin – an extremely unrealistic worst-case scenario – would be 9.8×10^{-7} human oral ID_{50}/kg (upper 95 % confidence interval of estimate, 3.03×10^{-6} human oral ID_{50}/kg).

It can be expected, on the basis of experimental studies described in Section 43.8, that this quantity of BSE infectivity is likely to be eliminated entirely by the processes used to manufacture gelatin.

---

[1] U. Seybold, DGF STOESS AG, Eberbach, Germany, personal communication on June 8, 2005.

[2] T. Ryan, Senior Adviser (Epidemiology & Public Health), New Zealand Food Safety Authority, personal communication on April 22, 2005.

[3] According to current European food legislation, vertebral column is defined as "specified risk material" (SRM) that has to be removed at the slaughterhouse depending on the age at slaughter and geographical origin of the cattle. Therefore, this model represents a worst-case scenario in terms of noncompliance with legislation.

43.3.1 Gelatin originating from skin: the removal of hair

After skin has been chopped into small pieces, hair is removed by tumbling in drums containing a mixture of sodium sulfide and lime.[4] This process not only removes the hair, but would also be expected to remove any surface contamination with tissues – such as brain – that might contain BSE prions. Hide splits, which are a by-product of the tanning industry, do not have hair.

43.3.2 Gelatin originating from bone: bone degreasing

Bone itself, i. e., without the bone marrow, is free from BSE infectivity. But infectivity has been detected in bone marrow in advanced clinical BSE (Fig. 53.2) [11]. Infectivity is, of course, present in high concentration in spinal cord and dorsal root ganglia, both of which can be expected to contaminate vertebral column used to produce the degreased chipped bone (DCB) used in the manufacture of gelatin.

If gelatin is produced from vertebral columns as raw material, BSE infectivity could potentially be present, depending on the BSE status of the country of origin. If the country of origin is any category other than GBR category 1 (BSE-free or BSE is highly unlikely; see Table 53.1), there is a chance that the raw material could be contaminated with BSE prions.

Before bone can be used to manufacture gelatin, bone tallow (fat) must be removed. This is done by crushing the bones to a particle size of less than 12 mm and then washing and degreasing the resulting chips with hot water to remove residues of fat, marrow, and other soft tissues such as spinal cord and dorsal root ganglia.[5]

Studies conducted on the ability of the degreasing process used to remove nervous tissue proteins from bones demonstrated that degreasing eliminated 98–99 % of such proteins [12]. It has been estimated that the degreasing process alone would reduce any BSE contamination of bone by a factor of approximately 10^2 [13].

43.4 Acid treatment

Before bone chips can be used to produce gelatin, the minerals calcium and phosphorus must be removed. This is achieved by immersion of the DCB in hydrochloric acid (approximately 4 %, pH < 1.5) for a period of at least 2 days. This intensive acid treatment changes the internal structure of the collagen protein, from which gelatin is extracted, and probably the structure of the BSE prion protein.

Acids with pH values below 3 lead to the neutralization of the amino acids glutamate and aspartate, which are negatively charged at pH 7. This also leads to the removal of salt bridges between different segments of the peptide chain and thus to the destabilization of the tertiary structure. It has been shown, however, that acid inactivation of prions is effective only at high concentrations of acid or at high temperatures [14].

On the basis of previous studies [15], this acid treatment would be expected to reduce the titer of any BSE infectivity that might be present. The demineralized bone is known as ossein.

43.5 Alkaline treatment

The next step in the production of gelatin (Fig. 46.1) is to soak the materials (pieces of skin or ossein) in a saturated lime solution, at pH > 12.5, for a period of between 20 and 50 days [3; 4].

Only certain strong alkali solutions, e. g., sodium hydroxide (NaOH; see Chapter 44.5.1); chaotrophic salts, e. g., GdnSCN (see Chapter 44.5.3); and oxidizing agents, e. g., sodium hypochlorite (NaOCl, javelle water, bleach; see Chapter 44.5.2) are able to inactivate the infectious prion protein completely.

A treatment sometimes used with a raw material known as hide splits (split form of skin) is to soak the material in 0.3 N sodium hydroxide (caustic soda) for around 14 days [compare 16].

---

[4] S. Ford, Purchasing Manager, Gelita New Zealand Ltd., personal communication on April 5, 2005.

[5] In some countries, where BSE is still present, vertebral columns from animals from a certain cut-off point (age limit) are classified as 'specified risk material' (see Chapter 53.3.1) and are not used in gelatin production.

As with the acid treatment referred to in Section 43.4, this alkaline treatment changes the internal structure of the collagen protein, and probably the structure of the BSE prion protein. On the basis of studies conducted into the destruction of TSE agents [17], it has long been expected that the length of time, temperature, and concentration of these alkaline treatments would significantly reduce the titer of any BSE infectivity present in the raw materials [3].

43.6 Further acid treatment

Some gelatin is also produced from ossein (demineralized bone) by an acid process, rather than by an alkaline one. In this acid process, the ossein is immersed for 12–24 hours in diluted acid at pH 2–3.5.[6]

43.7 Extraction of gelatin

After the skin or ossein has been subjected to alkaline or acid treatment, gelatin is extracted by a series of hot water steps. The gelatin extract is purified by filtration through diatomaceous earth and cellulose filter plates, and this process removes suspended particles [4].[7]

The purified gelatin solution is concentrated by evaporation in partial vacuum, and the concentrated solution is sterilized by UHT treatment of at least 138 °C for at least 4 seconds [4].

It is likely that the filtration and sterilization processes would also remove some BSE infectivity, in the unlikely event (see above) that any should still be present by this stage of production.

43.8 Experimental studies

The results of experimental studies have been published confirming the conclusions of earlier risk assessments [4; 18; 19]. Grobben and coworkers [4] developed an accurately scaled-down laboratory process to measure the effect of gelatin manufacturing processes on BSE infectivity. The experiment used crushed bones and intact calf vertebral columns. The crushed bone was smeared ("spiked") with mouse brain infected with the 301V strain of mouse-passaged BSE. The same brain was injected into the spinal cord of the vertebral columns. The 301V strain (see Chapter 12.3; Table 12.1) was selected because it has a high infectivity titer and is one of the most heat-resistant strains [4].

The BSE infectivity of the spiked starting material was $10^{8.4}$ mouse intracerebral ID_{50}/kg. Clearance factors of $10^{2.6}$ and $10^{3.7}$ ID_{50} were demonstrated for the first stage of the acid and alkaline processes (see Chapters 43.4 and 43.5), respectively. The complete acid and alkaline processes both reduced infectivity to undetectable levels, giving clearance factors of $\geq 10^{4.8}$ ID_{50} for the acid process and $\geq 10^{4.9}$ ID_{50} for the alkaline [4].

The contamination used in the experiment reflects worst-case conditions. The experiment did not take into account the very large effect of dilution of raw materials referred to in Section 43.3. Even if a BSE-infected animal contributed its vertebral column to an industrial batch of raw material, the concentration of BSE infectivity in the batch would not be as high as achieved by the "spiking" described in the experimental study [4].

43.9 Conclusions

In the years since the public health risk posed by BSE was first recognized, much has been learned about the disease. It is now clear that, fortunately, humans are relatively difficult to infect orally and that the BSE epidemic is largely under control on an international level. This means that any batch of raw material used to produce gelatin is highly unlikely to contain sufficient BSE to be able to infect humans consuming products made from it. Furthermore, recent experimental studies have confirmed what was long suspected; namely, the chemical processes used in the manufacture of gelatin are sufficient to inactivate any BSE infectivity that might have been present in the raw material from which the gelatin is made.

---

[6] On the basis of current information available, the use of a pH below 3 is suggested; compare [14] and see Chapter 44.

[7] S. Ford, Gelita New Zealand Ltd., personal communication on April 5, 2005.

In 2006, the European Food Safety Authority (EFSA) published a quantitative risk assessment on the human exposure to the BSE agent through gelatin [20]. That EFSA assessment also demonstrated a negligible risk, based on the clearance factors of the production processes and dilution factors.

Gelatin produced by modern industrial processes can thus be considered to pose no BSE risk to consumers, regardless of the source country from which it is derived.

References

[1] World Organisation for Animal Health (OIE). Report of the meeting of the OIE Terrestrial Animal Health Standards Commission, March 6–10, 2006. World Organisation for Animal Health, Paris, 2006:252pp.

[2] GEA Filtration. World production of gelatin. Webpage 2006: www.geafiltration.com/html/library/gelatin/gelatin_world_production.htm.

[3] Schrieber R and Seybold U. Gelatine production, the six steps to maximum safety – developments in biological standardization. In: Brown F, editor. Transmissible spongiform encephalopathies – impact on animal and human health. Developments in Biological Standardization 80, Karger, Basel, 1993.

[4] Grobben AH, Steele PJ, Somerville RA, et al. Inactivation of the bovine-spongiform-encephalopathy (BSE) agent by the acid and alkaline processes used in the manufacture of bone gelatine. Biotechnol Appl Biochem 2004; 39:329–338.

[5] EC. Update of the opinion on TSE infectivity distribution in ruminant tissues. Webpage 2006: http://europa.eu.int/comm/food/fs/sc/ssc/out296_en.pdf.

[6] EC. Food safety: from the farm to the fork. Outcome of discussions. Opinions. Webpage 2006: http://europa.eu.int./comm/food/fs/sc/ssc/outcome_en.html.

[7] World Organisation for Animal Health (OIE) 1989–2006. Bovine spongiform encephalopathy (BSE) – geographical distribution of countries that reported BSE confirmed cases since 1989. Webpage 2006: http://www.oie.int/eng/info/en_esb.htm.

[8] NZFSA. Officials' review of New Zealand's BSE country-categorisation measure. New Zealand Food Safety Authority, Wellington, 2005:35pp.

[9] EC. Food safety: from the farm to the fork. Opinion on BSE risk adopted by the Scientific Steering Committee at its plenary meeting of 26–27 March 1998, following a public consultation on the preliminary opinion adopted on 19–20 February 1998. Webpage 2006: http://europa.eu.int/comm/food/fs/sc/ssc/out13_en.html.

[10] Comer PJ. Assessment of risk from possible BSE infectivity in dorsal root ganglia. Risk assessment for the Ministry of Agriculture, Fisheries and Food and the Spongiform Encephalopathy Advisory Committee (report). Det Norske Veritas, Technical Consultancy Services, London, 1997:16pp.

[11] Wells GA, Spiropoulos J, Hawkins SA, et al. Pathogenesis of experimental bovine spongiform encephalopathy: preclinical infectivity in tonsil and observations on the distribution of lingual tonsil in slaughtered cattle. Vet Rec 2005; 156(13):401–407.

[12] Manzke U, Schlaf G, Poethke R, et al. On the removal of nervous proteins from materials used for gelatine manufacturing during processing. Pharmazeutische Industrie 1996; 58(9):837–841.

[13] Pharmaceutical Research & Manufacturers of America. BSE Committee. Assessment of the risk of bovine spongiform encephalopathy in pharmaceutical products. BioPharm 1998; 11(3):18–30.

[14] Appel TR, Lucassen R, Groschup MH, et al. Acid inactivation of prions – efficient at elevated temperature or high acid concentration. J Gen Virol 2006; 87(5):1385–1394.

[15] Brown P, Rohwer RG, Gajdusek DC. Newer data on the inactivation of scrapie virus or Creutzfeldt–Jakob disease virus in brain tissue. J Infect Dis 1986; 153:1145–1148.

[16] Taylor DM, Fernie K, McConnell I. Inactivation of the 22A strain of scrapie agent by autoclaving in sodium hydroxide. Vet Microbiol 1997; 58(2–4):87–91.

[17] Prusiner SB, McKinley MP, Groth DF, et al. Scrapie agent contains a hydrophobic protein. Proc Natl Acad Sci U S A 1981; 78:6675–6679.

[18] Grobben AH, Steele PJ, Somerville RA, et al. Inactivation of the BSE agent by the heat and pressure process for manufacturing gelatine. Vet Rec 2005; 157(10):277–289.

[19] Grobben AH, Steele PJ, Somerville RA, et al. Inactivation of transmissible spongiform encephalopathy agents during the manufacture of dicalcium phosphate from bone. Vet Rec 2006; 158(11):361–366.

[20] Opinion of the Scientific Panel on Biological Hazards of the European Food Safety Authority on the "Quantitative assessment of the human BSE risk posed by gelatine with respect to residual BSE risk". EFSA J 2006; 312:1–28.

44 Chemical Disinfection and Inactivation of Prions

Beat Hörnlimann, Walter J. Schulz-Schaeffer, Klaus Roth, Zheng-Xin Yan, Henrik Müller, Radulf C. Oberthür and Detlev Riesner

44.1 Introduction

This chapter summarizes the background and the application of chemical disinfectants that can be used effectively against prion pathogens. The details depend on the area of application and the nature of the work. For example, for the decontamination of surgical instruments in a hospital, work in a research laboratory, or the disinfection of bovine spongiform encephalopathy (BSE)-contaminated tools and rooms, different methods and means of chemical disinfection are required. For special applications, e. g., the decontamination of medical devices, the use of chemicals needs to be combined with thermal inactivation (heat, steam).

As early as 1953, without knowing the nature of the scrapie pathogen, D.R. Wilson (see Chapter 1.2) suggested that the pathogen exhibited very unusual features [1]: It was resistant to a 30-minute heat treatment at 100 °C and also against both formalin and chloroform [2]. In 1966, Tikvar Alper described for the first time the impossibility of inactivating the scrapie pathogen using radiation [3]. As a result, Alper argued that the pathogen cannot possess nucleic acids essential for infectivity (Fig. 4.2) [4], and it had to be expected that disinfectants that affect the viral nucleic acid do not have an effect on scrapie and other prions. In addition, unlike enveloped viruses, prions are not equipped with the usual viral surface spikes that serve as contact points and can be inactivated by chemical agents. Thus, transmission or intrusion of prions into the host cell cannot be prevented in this way.

In 1986 and 2000, researchers were able to show that the treatment of the infectious prion protein, the so-called PrP^{Sc}, with proteinase K only cleaved the first 59–74 N-terminal amino acids [5]. This led to protein fragments of 27–30 kDa (PrP27–30), which remained resistant to further treatment with the enzyme over several hours or even days [6; 7]. This result suggests that the resistance against proteinases and against disinfective agents is of a similar nature.

Prions are usually very stable against commonly used bactericide and virucide disinfectants and inactivating methods (Table 44.1). Because of a self-protecting structure, i. e., the formation of large aggregates, and a tailing phe-

Table 44.1: Chemical substances with no or insufficient inactivating effect on prions (or those that have proven unsuitable for inactivation).

Aldehydes[1)]: formaldehyde or formalin, glyoxal, succinic dialdehyde, glutaraldehyde
Acetone
Alcohols
β-Propiolactone
Chlorine dioxide
Diethylether
Ethylene oxide
Heptane
Hexane
Hydrogen peroxide
Iodine, iodide, iodophores
Potassium permanganate
Sodium-dichloro-isocyanurate (NaDCC)
Sodium periodate
Peracetic acid
Perchlorethylene
Petroleum ether, benzine
Phenoles[2)]
Urea

1) Note: All methods of steam sterilization (autoclaving) are less effective when the material is exposed to aldehydes or other preservative agents prior to sterilization.
2) Source: [40]

nomenon, i.e., the presence of particularly resistant prion subpopulations [8], the inactivation of prions cannot be achieved with mild decontamination technologies. Infectivity and pathogenicity of the prions can therefore neither be eliminated nor reduced with reagents that are regularly used in laboratories or hospitals for disinfection purposes.

Only certain strong alkali solutions (e.g., sodium hydroxide), chaotrophic salts (e.g., guanidinium-isothiocyanate), and oxidizing agents (e.g., sodium hypochlorite) are able to inactivate the infectious prion protein completely.

44.2 Basic knowledge regarding the chemical inactivation of prions

The tertiary structure of PrP^{Sc}, which is additionally stabilized by intermolecular interactions in more or less regular aggregates (scrapie-associated fibrils, SAF, Fig. 1.5b; prion rods), appears to be very important for the resistance against deactivation of prions. It is assumed, therefore, that the destruction or at least partial denaturation of the aggregate structure of prions (Fig. 1.5b) leads to a reduction of prion infectivity.

In principal, denaturation processes can either be reversible or irreversible. A reversion into the native state after inactivation would theoretically be possible for prions. However, PrP^{Sc} exhibits an extreme tendency to form complexes not only with other PrP^{Sc} molecules but also with other cell components like a polyglucose scaffold [9]. This contributes strongly to the stability of prions and makes the reversion practically impossible. At least a substantial and reproducible reactivation of inactivated prions has never been shown experimentally.

The following protein chemical characteristics must be taken into consideration when discussing the stability of prions. An intramolecular disulfide bridge in the PrP^{Sc} molecule holds two peptide segments closely together, thus fixing or at least constricting the tertiary structure of the prion protein molecule. Such S-S bridges can in principle also be formed intermolecularly and add to the stability of aggregates. Certain parts of the prion protein molecules are organized in α-helices and β-sheets, which are held together by hydrogen bonds. The tertiary structure of the PrP^{Sc} molecule and in particular the oligomeric and multimeric aggregates are supported by hydrophobic interactions between nonpolar amino acids. Charged side chains can also add to the internal stability of the molecule and to the intermolecular interactions in prions, particularly when they form salt bridges. One has to differentiate between chemicals that have an effect on covalent bonds within the protein structure and those that are able to destabilize non-covalent interactions and hydrogen bridges in the protein moiety [10].

Chaotrophic salts: guanidinium-hydrochloride (GdnHCl), guanidinium-isothiocyanate (GdnSCN), and urea lead to the destruction of hydrogen bonds within the polypeptide α-helices and β-sheets. Thus, the most important structural elements of the protein molecules are destabilized. In general, the destruction leads to an irregularly coiled polypeptide chain and unrestricted mobility of the amino acid elements within the chain [10]. Disulfide bridges (present in the PrP^{Sc} protein between amino acid residue 179 and 214) are not affected by either substance. GdnSCN is more effective than GdnHCl in terms of the destruction of hydrogen bonds.

Alkali substances at pH values above 12 lead to the neutralization of amine and guanidine groups (lysine and arginine), which are positively charged at pH 7. Thus, salt bridges and polar interactions to other segments of the peptide chain are no longer possible. This can lead to the destabilization of the tertiary structure. In addition, under the strong influence of alkali substances, the amide groups in glutamine and asparagine can be converted into the corresponding negatively charged carboxyl groups. The most effective chemical disinfection of prions is obtained by hydrolyzing the peptide bonds in PrP^{Sc}.

Acids with pH values below 3 lead to the neutralization of the amino acids glutamate and aspartate, which are negatively charged at pH 7. This also leads to the removal of salt bridges between different segments of the peptide chain and thus to the destabilization of the tertiary structure. It has been shown, however, that acid inactivation of prions is effective only at high concentrations of acid or high temperature [11].

Alcohols and dioxane add to the strength of the structure, i. e., they enhance the formation of α-helices in many proteins. In the case of PrP^{Sc}, even in precipitation activity prevails and thus adds to the stabilization of the molecule.

Detergents affect the tertiary structure of proteins, even at minute concentrations. They interact with hydrophilic and hydrophobic areas of the protein molecule and are largely responsible for loosening the hydrophobic center of the protein molecule. Thus, complex micelles are formed. Sodium dodecylsulfate (SDS) binds tightly to the peptide backbone and denatures proteins particularly when the interaction is enhanced by heat.

Oxidizing and reducing agents mainly exert an influence on the disulfide bridges [12]. Upon cleavage of intramolecular disulfide bridges, the stability of the protein conformation is lost and irreversible denaturation is facilitated.

Aldehydes, such as formaldehyde and glyoxal, can establish links between closely located segments of the peptide chains [12]. Although the structure is stabilized by these links, it is assumed that unspecific links block contact points to other PrP molecules that are essential for an infection.

44.3 Prerequisites for the efficiency of chemical disinfectants

When commercial products are used for disinfection, instructions, as provided by the manufacturer, must be clearly followed. However, the personnel trained to carry out disinfections must take special care – particularly when self-formulated disinfectants are used – that the effective disinfection depends on a number of factors that can mutually influence each other [13]. These include in the order of their importance: (i) concentration of the disinfectant, (ii) exposure time, (iii) pH optimum, (iv) exposure temperature, (v) stability or storage life of the disinfectant (vi) humidity and consistency of the surface area to be disinfected, and (vii) decrease in efficiency as a consequence of the addition of emulsifiers, proteins, etc.

In addition to the biological efficiency of the disinfection, the following points must also be taken into consideration: (i) potential health problems for the staff due to the method of disinfection, (ii) environmental damage and biological degradability of the disinfectant, (iii) compatibility of material and danger of corrosion through disinfectants, and (iv) financial aspects.

44.4 Testing for prion depletion and inactivation efficiency

44.4.1 Clinical background

Due to the occurrence of clinical cases of variant Creutzfeldt–Jakob disease (vCJD), new regulations have been set up in most European countries for the sterilization of surgical devices. Although national guidelines vary from country to country, most of them require that a combination of two procedures, which have proven an at least significant reduction in prion infectivity, has to be used for surgical instruments. In order to avoid cross-contamination from patient to patient, basic decontamination procedures need to inactivate also prions! The case history and the area of surgery have consequently to be taken into account to classify the risk group of each individual patient (see classification scheme in Chapter 37). When a CJD infection is known, utilized instruments need to be processed using two effective procedures for prion decontamination at a prolonged residence time or have to be destroyed.

44.4.2 Testing of the efficiency of processing procedures

Few disinfectants that are available commercially have proven effective against prions. In addition, the methods to test the efficiency of disinfectants are both time consuming and complex. The German Robert-Koch-Institute published a recommendation of a three-phase model [13a]. After all three phases have been successfully carried out, the efficiency of the disinfectant can be classified:

Phase 1: Prescreening for proteinase K (PK) resistance in Western blot

A brain homogenate of clinically ill animals is mixed and incubated with the disinfectant under study. Contact time, concentration, and temperature of the treatment are chosen as recommended by the manufacturer. After neutralization, the brain homogenate is separated from the disinfectant and a Western blot is performed to test for PK resistance. A sample without PK digestion serves as control. Standard procedures for sample preparation and Western blotting are applied. In the case of an efficient disinfectant, the PK-digested sample should not deliver a PrP-specific signal, whereas without PK digestion a signal for PrP^{Sc} can be obtained. This step can be achieved easily and rapidly without time-consuming *in vivo* tests in animals and consequently facilitates an effective prescreening of a huge number of formulations.

Phase 2: Quantitative suspension test

In order to evaluate a prion-inactivating effect, i. e., a reduction in the titer of prion infectivity, a quantitative comparison of prion infectivity before and after treatment with the disinfectant is performed. The current test method of choice is biotitration in animal models. It consists of five steps:

Step I: The test animals, mainly mice or hamsters, are intracerebrally (i.c.) injected with cerebral tissue of corresponding clinically ill animals.

Step II: After the incubation period, the test animals usually show typical transmissible spongiform encephalopathy (TSE) symptoms. Species-adapted infectious material is obtained from the brains of these animals with a sufficiently high reproducible infectivity titer.

Step III: To test the efficiency of the disinfectant, the standardized infectious TSE brain homogenate obtained in step II is exposed to the disinfectant as recommended by the manufacturer.

Step IV: The tested formulation is removed from the brain homogenate with washing solutions. Care has to be taken that infectious material is not removed in the cleaning step.

Step V: Samples from the brain homogenate are inoculated into a group of experimental animals. Concurrently, control animals are inoculated with untreated but contaminated brain homogenate that was also subjected to the washing procedure in step IV. Another control group is recommended in which noninfectious brain is treated identically in order to exclude pathological reactions of the disinfection that are not related to TSE. All groups are observed until the first TSE-specific clinical signs can be noted or for a sufficiently long period of time, which is several times longer than the incubation time without disinfection procedures. A reduction of infectivity of 4 $\log ID_{50}$ or more in animals inoculated with infectious brain homogenate subjected to the disinfectant compared with animals infected with the nontreated infectious brain homogenate is suggested to demonstrate prion inactivation. This can be achieved if animals are healthy after a doubled incubation time and no PrP^{Sc} is detectable in the brain with sufficient histochemical or biochemical methods.

Phase 3: Quantitative carrier test

The German Robert-Koch-Institute suggests testing not only the inactivation efficiency in suspensions but also the decontamination potential in a carrier test. Prions in tissue smeared and dried onto glass or metal surfaces are more resistant to inactivation by autoclaving than those in homogenates [14]. The infectious prion protein binds tightly to steel wires, rendering the wires infectious, and causes disease after the contaminated wires are implanted or even inserted for a short time in the brain of healthy animals [14a; 14b]. Hamsters with implanted wires contaminated by incubation in infectious brain homogenate become sick at the same time as hamsters infected intracerebrally with suspensions of the same brain homogenate. The contact time between infectious steel wire and the brain of the bioassay animal has a minor influence on the survival time [14b]. When the contaminated wires are treated with cleaning, disinfection, or sterilization procedures before implantation, the infectivity titer is reduced, which manifests directly as a prolonged surviving time of the test animals. Consequently, carrier tests can be utilized as a bioassay for validating reprocessing procedures for surgical instruments. The steel wire assay simulates problems of instrument decontamination more realistically and also allows the examination of substances that are toxic and cannot be adequately removed from tissue homogenates [15].

The carrier test is performed accordingly to steps I to V of *phase 2*. Briefly, stainless steel carriers (wires of 5 mm length and a diameter of 0.25 mm) are incubated in 10 % homogenate of clinically ill animals. After the incubation time, the wires are treated with the disinfectant or the process, respectively, to be tested and implanted into the brains of healthy hamsters. It is suggested that if animals survive implantation of treated steel wires two or more incubation times without clinical signs of disease or detectable PrP^{Sc} in the brain, and the experiment was done additionally to *phase 2*, the reprocessing process has a prion decontamination potential. It needs to be pointed out that a doubling of incubation time in *phase 2* or *phase 3* tests does not demonstrate complete inactivation or decontamination of prions because in ongoing experiments animals become ill during the third, fourth, and fifth incubation time [14b; 15a].

According to D. Taylor, it is important not to draw premature and definitive conclusions on the titer of the remaining infectivity when it is only based on the incubation time assay and extrapolated over orders of magnitude. Adequate tests for the efficiency of disinfectants must involve the final titration after different exposure times. Furthermore, in brain homogenates a small but very resistant subpopulation of the pathogen surviving the treatment has been observed [8; 16]. It is possible that some of the prion infectivity is "hidden" in the hydrophobic areas inside the PrP^{Sc} aggregates (SAF complexes, prion rods), which are difficult for the disinfectant to access [17].

44.5 Chemical disinfectants suitable for the inactivation of prions

The inactivation instructions described in the following paragraphs are based on empirical values. As a consequence of a relatively small set of data, it is recommended for efficient decontamination of potentially high levels of prion infectivity to use two accepted methods sequentially with an exposure time twice than that reported in efficient decontamination experiments.

44.5.1 Sodium hydroxide (NaOH)

A 1-hour disinfection with 2 N sodium hydroxide (8 %) is best suited for surfaces and materials that cannot be autoclaved but can resist alkaline treatment [18; 19]. The incubation period should usually last 2 × 30 minutes or overnight. NaOH at concentrations below 0.5 N is inclined to exert only detachment and destabilization effects, whereas at a concentration of 1 N a clear degradation effect of the prion protein can be observed [16]. NaOH attacks prions but has the disadvantage of corroding aluminum and zinc surfaces. Therefore, these surfaces must never come into contact with hydroxide ions. Instruments made of stainless steel can, however, be exposed to high-alkaline NaOH-containing solutions. In addition, it should not be forgotten that sodium hydroxide is highly caustic, and contact with skin, and in particular the eyes, must be avoided. Therefore, the use of safety goggles is mandatory (see Section 44.5.2, Point 3.).

44.5.2 Sodium hypochlorite (NaOCl; javelle water, bleach)

Sodium hypochlorite has proven highly effective for prion inactivation (20). Unfortunately, NaOCl corrodes not only aluminum and zinc but all oxidatively vulnerable metals and thus also stainless steel. It can, however, be used on

Table 44.2: Overview of effective chemical disinfection measures (at room temperature)[1].

Suitable method or substance	Concentration M (mol/l)	% (g/100 ml)	Duration	Suitable areas of application
Treatment with NaOH				
2 N (= 2 M)	2	8	1 h	Disinfection of (surface) areas and clothing[2]
1 N (= 1 M)	1	4	2 h	Surgical instruments
1 N (= 1 M)	1	4	5–10 min	Contaminated hands, rinse with H_2O afterwards
Treatment with sodium hypochlorite (NaOCl)[3),4)]				
5,900 ppm active chlorine	0.33	2.5	Short	Contaminated hands, rinse with H_2O afterwards
5,900 ppm active chlorine	0.33	2.5	1 h	Clothing[2]
11,800 pm active chlorine up	0.67	5	1 h	Surgical instruments
to 17,000 ppm[5] active chlorine	1	7.5	1 h	
Boiling in sodiumdodecylsulfate (SDS)				
Alone (at 100 °C)	0.10	3	10 min	Surgical instruments, also those containing aluminum
In combination with sterilization at 121 °C (1 h)	0.10	3	3 min	
Treatment with guanidinium-iso-thiocyanate (GdnSCN)[6]				
3 N (= 3 M)	3	35.4	24 h	Surgical instruments, also those containing aluminum
4 N (= 4 M)	4	46.9	1 h	In hospitals, dental practices, eye hospitals, for endoscopes
6 N (= 6 M)	6	70.8	15 min	

1) Only hypochlorite with at least 17,000 ppm active chlorine, autoclaving after or in 1 N NaOH, or boiling in 1 N NaOH appear to eliminate prion infectivity even under worst case conditions [14; 17; 19; 41].

2) Predisinfection of contaminated clothing, e. g., clothing contaminated with cerebrospinal fluid of patients suspected of suffering from TSE.

3) Shelf life: when the sodium hypochlorite solution is kept in opaque containers tightly closed, the available chlorine loss is directly proportional to the storage time, independent of temperature conditions. As a constant opening of containers appears to cause greater loss in chlorine concentration of diluted bleach solutions, a usage within the first month of the production date is recommended.

4) An irritant (on mucous membranes) and strongly corrosive (on eloxated material) substance; see explanations in Chapter 49.

5) 20,000 ppm are recommended.

6) Shelf life: > 1 year. The denaturing effect on prions is irreversible in concentrations over 3 N.

ceramics and mineral surfaces or for disinfecting large areas (e. g., walls and floors with contact to a large amount of potentially BSE-infected cattle or scrapie-infected sheep/goat material (see Chapter 57.4.2)).

NaOCl must always be freshly prepared (see below) and applied in a concentration of at least 5 % for a minimum of 1 hour [18; 19]. This corresponds to 1.18 % active chlorine or 11,800 ppm free chlorine (better 7.5 % NaOCl, 17,000 ppm free chlorine; Table 44.2). As it is not only the absolute amount of chlorine in a certain solution that is important but also the exposure time during which the chlorine is in contact with the medium, care must be taken that the disinfection of material contaminated with prions does not involve the use of other inorganic or organic chlorine compounds such as sodium-dichloro-isocyanurate (NaDCC) [21].

The inactivating principle of sodium hypochlorite is the production of hypochloric acid, which acts as a strong oxidizing agent. In solution, the hypochloric acid is unstable and rapidly disintegrates into nascent oxygen and hydrochloric acid according to the formula $2\,HOCl \rightarrow O_2 + 2\,HCl$. The effective part is thus neither chlorine nor the hypochloric acid but the nascent oxygen. In this most reactive form, the oxygen can trigger oxidative denaturing processes on the PrP^{Sc} protein. The exposure to NaOCl seems to lead to the irreversible inactivation through numerous covalent modifications in the PrP^{Sc} molecule [22]. When using NaOCl, special care must be taken in the following circumstances:

1. It is most important that the diluted solutions are freshly prepared. A 15 % aqueous sodium hypochlorite solution can be obtained as "javelle water" in pharmacies. This must be diluted at a ratio of 1:2 with cold water. It can be stored for a maximum of 4 weeks.
2. Before the solution is applied, it must be clarified whether pieces of equipment that cannot be removed will tolerate the treatment with NaOCl solution.
3. Apart from its corrosive effect (even on stainless steel), sodium hypochlorite has the disadvantage that it irritates the mucosa. Therefore, protective goggles must be worn at all times. If sodium hypochlorite is to be used in large quantities, e. g., when disinfecting scrapie contaminated sites or sites in which skulls of BSE-positive/BSE-suspected animals are handled, the use of a *gas mask* equipped with a safety filter that binds free chlorine is mandatory [23].
4. For environmental reasons, it is also mandatory to reduce the oxidizing effect of javelle water by adding the corresponding amount of sodium thiosulfate solution before it is allowed to enter the sewage system.

44.5.3 Guanidinium-isothiocyanate (GdnSCN)

GdnSCN has a partially denaturing effect in a 1.5 N solution (177 g/l) [24]. When it is used in concentrations over 3 N, the GdnSCN effect on prions is irreversible [25]. For decontamination purposes, a concentration of 4 N is recommended (469 g/l).

The inactivation of PrP^{Sc} in brain homogenates may be realized within 15 minutes using 4 N GdnSCN [26], but we recommend an exposure time of 1 hour. In addition to original experiments by Prusiner [22], which led in part to the prion hypothesis, decontamination experiments with GdnSCN were performed by Manuelidis [27], Tateishi [28], and Flechsig [29]. These experiments show that GdnSCN inactivates prions in a concentration of 3 N after immersion times much longer than 1 hour. In various experiments using a concentration of 4 N and at least 15 minutes residence time, no experimental animal was reported to have become ill after intracerebral inoculation. Lemmer et al. [15] showed with steel wire experiments that the effect of GdnSCN is based solely on the denaturing potential and not on degradation or detaching effects.

Again, the recommended exposure time is 1 hour [30] (Table 44.2). GdnSCN has an extraordinary long half-life of 1 year. Thus, it is *the solution of choice* as it has neither the disadvantages of sodium hydroxide nor the effect of sodium hypochlorite on sensitive instruments (see Section 44.5.2). The strong denaturing effect of GdnSCN can be inhibited by adding alcohol. The use of GdnSCN together with acids may lead to a release of toxic cyanide gas. After the decontamination process, crystallization of GdnSCN on surfaces of the instruments should be avoided by rinsing the instruments with water. Goggles should be worn and ingestion avoided.

44.5.4 Sodium dodecylsulfate (SDS)

Sodium dodecylsulfate is used by some laboratories as an alternative for the above-mentioned chemicals to prevent alkali burns on skin and mucosa and corrosion of faintly anodized aluminum or zinc surfaces. The emulsifying activity of this substance has proven particularly useful for doing so. Accordingly, the observed mechanism of decontamination appears to be a detachment and destabilization rather than a degradation of PrP^{Sc} [15].

The utensils are heated in an SDS solution (3 % or greater; ≈ 0.1 N) for at least 10 minutes

[19; 31]. Some residual infectivity remains after the treatment with SDS unless the material is also sterilized for 1 hour at 121 °C. Thus, in combination with autoclaving, a 3-minute exposure to SDS is sufficient.

Peretz et al. [32] suggest autoclaving in acidic SDS (2% SDS, 1% AcOH) at 121 °C or 134 °C for at least 15 minutes. Although they got good results for prion inactivation in homogenates, there was residual infectivity in the steel wire test (using 4% SDS, 1% AcOH at 134 °C for 30 minutes).

44.5.5 Formic acid (CH_2O_2)

In pathology, concentrated CH_2O_2 is used for the pretreatment of formalin-fixed tissue. Because of the corrosive effect and the safety requirements in handling the substance, this procedure is not useful for other applications. Concentrated formic acid was shown to reduce infectivity in formalin-fixed tissues after 1 hour exposure by at least 7.2 log ID_{50} and shows no remaining infectivity in the bioassay [33].

However, it only has a partially inactivating effect if the tissue blocks are too large in thickness (over 4 mm) or if the tissues are fixed in paraformaldehyde lysine periodate (PLP) [34; 35]. Some TSE strains are more resistant to formic acid after formol fixation than others [34]. Further details will be provided in Chapter 48.2.

Since it would not fit into the scope of this chapter, strongly denaturing substances, which are often used in research laboratories, and detergents, which are used for the purification of PrP27–30, are not dealt with here. Detailed information on this subject can be found in [36].

44.5.6 Hydrochloric acid (HCl)

Peptide bond hydrolysis with hydrochloric acid is only effective at concentrations higher than 5 N or at an elevated temperature of more than 60 °C [11]. A complete loss of infectivity was observed at 8 N HCl or 1 N HCl at a temperature of 80 °C.

44.5.7 Radiofrequency gas plasma treatment

Microwave excitation of a low-pressure gas mixture comprising oxygen, hydrogen, and argon produces ions and radicals affecting biomolecules by oxidizing them to gaseous products. Gas plasma sterilization is reported to be suitable for the sterilization of metals, silicone, and various polymers but appears to be inappropriate for liquids and biological tissues. In vivo testing showed that radiofrequency gas plasma treatment of metal spheres effectively removed all organic residues of scrapie-infected brain material from the surface of stainless steel, also eliminating the transmission of infectivity [37].

44.6 Chemical disinfectants unsuitable or less suitable for the inactivation of prions

44.6.1 Formaldehyde (CH_2O, formalin)

Aldehyde solutions cannot be recommended for the disinfection of prions (Table 44.1). Formaldehyde, which is a standard disinfectant used in medicine, is virtually ineffective in this case. The reuse of insufficiently disinfected CJD-contaminated surgical instruments led to the first documented iatrogenic CJD transmission; the instruments had been treated with formalin before their reuse [38]. Only a few cases are known worldwide that have been caused in this manner (Table 14.1), two of them in Switzerland. At that time, the problems of prion disinfection were not known. A number of different disinfection measures must be followed today to decontaminate surgical instruments that are used in neurosurgery, ophthalmology, etc. (see Chapter 47).

In 1982, Paul Brown and colleagues documented [39] that the widespread application of aldehydes against all kinds of pathogens could not be transferred to prions. The authors showed the limited inactivating effect of a 10% formalin solution on scrapie-infected tissue. They used an initial infectivity titer of 8.8 log ID_{50} and compared the effect to that of autoclaving and the combined effect of both processes. The following results were obtained:

1. When the tissue was only treated with formaldehyde for a period of 48 hours, the titer was reduced by 1.5 log ID_{50} units.

2. An identical treatment with formaldehyde followed by a 3-minute steam autoclaving step at 134 °C led to a reduction of the infectivity titer of 1.8 log ID_{50} units.
3. Autoclaving alone (30-minute steam sterilization at 134 °C) led to a reduction of the infectivity titer of 5.3 log ID_{50} units.
4. Steam autoclaving with subsequent formaldehyde treatment led to a reduction of 6.8 log ID_{50} units.

The sequence of the combined application of formaldehyde treatment and autoclaving has a considerable effect on the final titer (see Chapter 49). Treating the scrapie pathogen with formaldehyde reduces the sterilizing effect of subsequent autoclaving. When formalin-preserved protein samples are contaminated, for example in research laboratories, a decontamination *prior* to thermal treatment is necessary (see Chapters 48, 49).

A possible explanation of this phenomenon might be that brain cells containing prions or cell fragments present in the suspension are fixed through their exposure to formaldehyde and thus protect the scrapie infectivity. Another possibility might be that the pathogenic conformation of the PrP^{Sc} aggregates is fixed through exposure to formaldehyde. When the sample is autoclaved for 18 minutes at 134 °C before the exposure to formalin, a thermal denaturation of the PrP^{Sc} is achieved. This is even fixed by formaldehyde.

44.6.2 Peracetic acid

In the search for an alternative decontamination procedure for thermolabile instruments, peracetic acid came to notice in France because it was the highest rated of the listed chemical decontamination procedures. At this time, little was known about the antiprion efficiency of peracetic acid, but it was known for its reliably good antibacterial, antiviral, and antifungal effects. In the meantime, several animal experiments with operative peracetic acid concentrations up to 0.35 % showed an insufficient prion inactivation [15]. In the light of those results and since peracetic acid is not easy to handle in washers/disinfectors and also bears some incompatibilities with commercially used cleaners, especially the favored alkaline cleaners, it should no longer be propagated as a decontamination procedure for prions.

44.6.3 Alcohols

Alcohols such as ethanol and 1- and 2-propanol do not have a decontaminating effect on prions. This might be due to the stable conformation of the PrP^{Sc} and the general effect of alcohols in stabilizing the structure of prions (Table 44.1). One should keep in mind that PrP^{Sc} can be precipitated quantitatively with ethanol while preserving the infectivity. Therefore, the alcohol-based disinfectants that are generally used in pathology or research laboratory are not suitable for disinfecting hands that are contaminated with prions. Effective methods for disinfecting hands are described in the Chapters 48 and 49.

References

[1] Wilson DR, Anderson RD, Smith W. Studies in scrapie. J Comp Path 1950; 60:267–282.
[2] Pattison IH. A sideways look at the scrapie saga: 1732–1991. In: Prusiner SB, Collinge J, Powell J, et al., editors. Prion diseases of humans and animals. Ellis Horwood, New York, 1992:15–22.
[3] Alper T, Haig DA, Clarke MC. The exceptionally small size of the scrapie agent. Biochem Biophys Res Commun 1966; 22:278–284.
[4] Alper T, Cramp WA, Haig DA, et al. Does the agent of scrapie replicate without nucleic acid? Nature 1967; 214:764–766.
[5] Parchi P, Zou W, Wang W, et al. Genetic influence on the structural variations of the abnormal prion protein. Proc Natl Acad Sci U S A 2005; 97:10168–10172.
[6] Barry RA, Kent SB, McKinley MP, et al. Scrapie and cellular prion proteins share polypeptide epitopes. J Infect Dis 1986; 153:848–854.
[7] Prusiner SB and Hsiao KK. Prions causing transmissible neurodegenerative diseases. In: Schlossberg D, editor. Infections of the nervous system. Springer, New York, 1990.
[8] Brown P, Liberski PP, Wolff A, et al. Resistance of scrapie infectivity to steam autoclaving after formaldehyde fixation and limited survival after ashing at 360 degrees Celsius: practical and theo-

retical implications. J Infect Dis 1990; 161:467–472.

[9] Dumpitak C, Beekes M, Weinmann N, et al. The polysaccharide scaffold of PrP 27–30 is a common compound of natural prions and consists of alpha-linked polyglucose. Biol Chem 2005; 386(11):1149–1155.

[10] Tanford C. Protein denaturation. In: Anfinsen CB Jr, Anson ML, Edsall JT, et al, editors. Advances in protein chemistry. Academic Press, New York, 1968:121–128.

[11] Appel TR, Lucassen R, Groschup MH, et al. Acid inactivation of prions – efficient at elevated temperature or high acid concentration. J Gen Virol 2006; 87(5):1385–1394.

[12] Roberts JD and Caserio MC. Basic principles of organic chemistry. W.A. Benjamin, New York, 1965.

[13] Anonymous. Sichere Biotechnologie – Eingruppierung biologischer Agenzien: Stoffmerkblätter der BG Chemie. Merkblatt „Viren“ B004 9/96 ZH 1/344: Anhang 3, Abschnitt 3, 1996.

[13a] Bertram J, Mielke M, Beekes M, et al. Inactivation and removal of prions in producing medical products. A contribution to evaluation and declaration of possible methods [Inaktivierung und Entfernung von Prionen bei der Aufbereitung von Medizinprodukten]. Bundesgesundheitsblatt Gesundheitsforschung Gesundheitsschutz. 2004; 47:36–40.

[14] Taylor DM. Inactivation of transmissible degenerative encephalopathy agents: a review. Vet J 2000; 159(1):10–17.

[14a] Zobeley E, Flechsig F, Cozzio A, et al. Infectivity of scrapie prions bound to a stainless steel surface. Mol Med. 1999; 5:240–243.

[14b] Yan ZX, Stitz L, Heeg P, et al. Infectivity of prion protein bound to stainless steel wires: a model for testing decontamination procedures for transmissible spongiform encephalopathies. Infect Control Hosp Epidemiol 2004; 25:280–283.

[15] Lemmer K, Mielke M, Pauli G, et al. Decontamination of surgical instruments from prion proteins: in vitro studies on the detachment, destabilization and degradation of PrP^{Sc} bound to steel surfaces. J Gen Virol 2004; 85(12):3805–3816.

[15a] Fichet G, Comoy E, Duval C, et al. Novel methods for disinfection of prion-contaminated medical devices. Lancet 2004; 364:521–526.

[16] Brown P. BSE: the final resting place. Lancet 1998; 351:1146–1147.

[17] Taylor DM. Resistance of transmissible spongiform encephalopathy agents to decontamination. Contrib Microbiol 2004; 11:136–145.

[18] Brown P, Rohwer RG, Gajdusek DC. Newer data on the inactivation of scrapie virus or Creutzfeldt–Jakob disease virus in brain tissue. J Infect Dis 1986; 153:1145–1148.

[19] Taylor DM, Fraser H, McConnell I, et al. Decontamination studies with the agents of bovine spongiform encephalopathy and scrapie. Arch Virol 1994; 139:313–326.

[20] Brown P, Gibbs CJ Jr, Amyx HL, et al. Chemical disinfection of Creutzfeldt–Jakob disease virus. N Engl J Med 1982; 306:1279–1282.

[21] Taylor KC. The control of bovine spongiform encephalopathy in Great Britain. Vet Rec 1991; 129:522–526.

[22] Prusiner SB, Groth D, Serban A, et al. Attempts to restore scrapie prion infectivity after exposure to protein denaturants. Proc Natl Acad Sci U S A 1993; 90:2793–2797.

[23] Seddon HR. Scrapie. In: Seddon HR, editor. Diseases of domestic animals in Australia – part 4. A.J. Arthur, Commonwealth Government Printer, Canberra, 1958:176.

[24] Lapanje S. Physicochemical aspects of protein denaturation. John Wiley & Sons, New York, 1978.

[25] Bessen RA, Kocisko DA, Raymond GJ, et al. Non-genetic propagation of strain-specific properties of scrapie prion protein. Nature 1995; 375:698–700.

[26] Oesch B, Jensen M, Nilsson P, et al. Properties of the scrapie prion protein: quantitative analysis of protease resistance. Biochemistry 1994; 33:5926–5931.

[27] Manuelidis L. Decontamination of Creutzfeldt–Jakob disease and other transmissible agents. J Neurovirol 1997; 3(1):62–65.

[28] Tateishi J, Tashima T, Kitamoto T. Practical methods for chemical inactivation of Creutzfeldt–Jakob disease pathogen. Microbiol Immunol 1991; 35:163–166.

[29] Flechsig E, Hegyi I, Enari M, et al. Transmission of scrapie by steel-surface-bound prions. Mol Med 2001; 7(10):679–684.

[30] Anonymous. Desinfektion und Sterilisation von chirurgischen Instrumenten bei Verdacht auf Creutzfeldt–Jakob-Erkrankungen. Bundesgesundheitsblatt 1996; 39(8/96):282–283.

[31] Anonymous. Creutzfeldt–Jakob-Krankheit: Desinfektion und Sterilisation von chirurgischen Instrumenten (Empfehlungen einer Expertenberatung). Epidemiol Bulletin 1996; 27:182–184.

[32] Peretz D, Supattapone S, Giles K, et al. Inactivation of prions by acidic sodium dodecyl sulfate. J Virol 2006; 80(1):322–331.

[33] Brown P, Wolff A, Gajdusek DC. A simple and effective method for inactivating virus infectivity in formalin-fixed tissue samples from patients with Creutzfeldt–Jakob disease. Neurology 1990; 40:887–890.

[34] Taylor DM, Brown JM, Fernie K, et al. The effect of formic acid on BSE and scrapie infectivity in fixed and unfixed brain-tissue. Vet Microbiol 1997; 58(2–4):167–174.

[35] Gonzalez L, Jeffrey M, Sisó S, et al. Diagnosis of preclinical scrapie in samples of rectal mucosa. Vet Rec 2005; 156(26):846–847.

[36] Baker HF and Ridley RM. Prion diseases. Humana Press, Totowa, NJ, 1996.

[37] Baxter HC, Campbell GA, Whittaker AG, et al. Elimination of transmissible spongiform encephalopathy infectivity and decontamination of surgical instruments by using radio-frequency gas-plasma treatment. J Gen Virol 2005; 86(8): 2393–2399.

[38] Bernoulli C, Siegfried J, Baumgartner G, et al. Danger of accidental person-to-person transmission of Creutzfeldt–Jakob disease by surgery [letter]. Lancet 1977; 1(8009):478–479.

[39] Brown P, Rohwer RG, Green EM, et al. Effect of chemicals, heat, and histopathologic processing on high-infectivity hamster-adapted scrapie virus. J Infect Dis 1982; 145:683–687.

[40] ACDP. Transmissible spongiform encephalopathy agents: safe working and the prevention of infection. Advisory Committee on Dangerous Pathogens & Spongiform Encephalopathy Advisory Committee. Statutory Office, London 1998.

[41] Taylor DM, Fernie K, McConnell I. Inactivation of the 22A strain of scrapie agent by autoclaving in sodium hydroxide. Vet Microbiol 1997; 58(2–4):87–91.

45 Thermal Inactivation of Prions

Radulf C. Oberthür, Henrik Müller, and Detlev Riesner

45.1 Introduction

The thermostability of the scrapie agent has been recognized since the 1960s [1]. Indeed, it was one of the properties that led to the suggestion that it was an unconventional agent. Wallhäuser [2] classifies the unusual high resistance of prions to thermal treatment as "heat resistance stage VI", which is the highest degree in the spectrum. As a rule of thumb, autoclaving of prion-contaminated material in saturated steam at 132 °C for a period of 60 minutes is sufficient to inactivate prions.

It was the epidemic of BSE considered that stimulated research into the inactivation of prions. In 1988, it was suggested for the first time that BSE may have resulted from scrapie agent, which had not been adequately inactivated, being transmitted to cattle through feeding of animal proteins (see Chapters 19.9 and 52). In the early 1970s, the solvent extraction process had been abandoned as an adjunct of rendering. Instead, mild thermal conditions (see Chapter 19.9) were applied to render cattle tissues without it being recognized that prions in the resulting meat-and-bone-meal, which was used in animal feed as protein supplement, were not being completely inactivated [3]. Renderers were concerned mainly with ensuring the materials were free from *Salmonella* and *Enterobacteria*.

In contrast, since 1939 the German *Tierkörperbeseitigungsgesetz* – an animal by-product disposal law – had required that tissues and bones of dead animals as well as slaughterhouse by-products deemed unfit for human consumption be sterilized at a temperature of 132 °C at saturated steam pressure for at least 30 minutes [4] before being used as animal feed. This law was not enacted because of concerns over scrapie, but rather to inactivate the heat-resistant spores of anthrax and, later, other heat-resistant pathogens such as *Clostridium perfringens*. In 1975 these conditions were changed to 133 °C at saturated steam pressure of 3 bar for at least 20 minutes.

Despite the purely empirical origin of sterilization rules, this chapter deals with the biophysical chemistry of treating prions with heat and gives a rough assessment of potential prion inactivation measures.

45.2 Physical chemistry of heat inactivation of complex biological structures

Heat accelerates molecular movement, even to the point of destruction of molecular structures [5], such as the three-dimensional structure of proteins or even whole cells. High temperature promotes water-mediated hydrolysis as well as condensation reactions like the Maillard reaction linking carbohydrates and proteins. These reactions may lead to the destruction of the original function of a particular structure or system, such as enzyme activity, prion infectivity, proteins or whole cells.

On the other hand, several factors contribute to the stability of the original structure. These can include the adsorption to surfaces [6], immersion in fat [7; 8], or formation of larger aggregates (SAF, Fig. 1.5b, or prion rods) due to hydrophobic interactions (see Chapter 44.2). Structures that are chemically interlinked through sulfide bridges, such as is the case in keratin of bristles and feathers, are very resistant to denaturation by heat [9].

The destruction of biological structures is irreversible in most cases. Examples are enzymatic degradation or cooking, frying, and baking to make foodstuff more digestible. Thermal denaturation can, in principle, be reversible, as shown for ribonuclease [5] and other enzymes under specific conditions. Reversibility, however, becomes more difficult the more heterogeneous or complex a system is, i. e., the larger the number of potential binding partners.

45.3 Kinetics of thermal denaturation and inactivation

Most of the denaturation and inactivation reactions of complex systems or particles in liquids are first-order reactions. Such systems can be a bacterium, a virus, a protein molecule, a prion or another complex structure. The energy that is required to destroy the original structure is transmitted through the fluid in which the particles are dissolved. The concentration c of the original intact particles decreases exponentially over time:

$$-dc/dt = k \cdot c \tag{1}$$

$$c = c_0 \cdot \exp[-kt] \tag{2}$$

with c_0 denoting the initial concentration at time $t = 0$ and k the rate constant of the reaction.

In first-order reactions, the rate constant k is often replaced by the half-life $t_{1/2}$, e. g., in the case of radioactive decay, or decimation time D, e. g., in the sterilization kinetics of bacteria [2]. Half-life denotes the time required for half of the original concentration of particles to be inactivated; decimation time denotes the time during which 90 % of the original number of particles has reacted and only 10 % of the initial concentration remains.

Equation (2) leads to

$$t_{1/2} = \ln 2/k = 0.693/k, \tag{3}$$

$$D = \ln 10/k = 2.303/k. \tag{4}$$

In first-order reactions, the decimation time D is suitable for calculating the number of orders of magnitude n by which an original number of particles or a concentration, respectively, has decreased after the time t. According to Eqs. (2) and (4) this leads to

$$\log(c_0/c) = n = t/D. \tag{5}$$

In chemical reactions, rate constants are dependent on temperature or other conditions such as pH, ionic strength, or the presence of stabilizing or destabilizing substances. In general, rate constants increase with temperature. According to Arrhenius [5] the temperature dependence can be expressed by the following equation:

$$k = k_0 \cdot \exp[-E_A/RT] \tag{6}$$

with

E_A the activation energy of the reaction,
k_0 the rate constant without activation energy ($E_A = 0$),
R the gas constant ($R = 8.31$ J/[K·mol]), and
T the absolute temperature in Kelvin [K]

with T [K] = T [°C] + 273.15.

Typical chemical reactions involving only a small number of bonds are characterized by activation energies of 50–200 kJ/mol [10]. Enzymatic catalysis reduces the activation energy, resulting in typical E_A values of 10–60 kJ/mol [11]. In contrast, the thermal inactivation of enzymes [11] or the denaturation of proteins [12] in aqueous solutions has values of 200–750 kJ/mol.

In food technology, the calculation of the temperature dependence of bacterial inactivation kinetics generally relies on the temperature interval z, in which the decimation time D changes by a factor of 10 [11]. D as a function of temperature can thus be calculated from

$$D(T) = D(T_B) \cdot \exp[\ln 10 \cdot (T_B - T)/z], \tag{7}$$

with $D(T_B)$ the decimation time at the reference temperature T_B.

Equations (6) and (4) lead to

$$\begin{aligned} D(T) &= D(T_B) \cdot \exp[(E_A/R) \cdot (1/T - 1/T_B)], \\ &= D(T_B) \cdot \exp[(E_A/RTT_B) \cdot (T_B - T)]. \end{aligned} \tag{8}$$

The comparison of Eqs. (7) and (8) shows that the temperature interval z depends on E_A as well as on the absolute temperature T. This is demonstrated by the following equation:

$$z = (RT^2 \cdot \ln 10)/E_A \tag{9}$$

High z values, e. g., 40 K, are expected for low activation energies, as is typical for chemical reactions. Low z values, e. g., 7 K, are expected for high activation energies, as is typical for inactivation reactions of enzymes in aqueous solutions.

In order to characterize the inactivation of prions, the decimation time D at reference temperature T_B and the temperature difference z required to reduce the decimation time by a factor of 10, respectively, are used. Alternatively to $D(T_B)$, the temperature T_{mD} at which the number of prions is decimated by a factor of 10 in 1 minute [13–15] (there referred to as T_{SR}) can be used. T_{mD} is given by

$$T_{mD} = T_B + z \cdot \log(D(T_B)/\text{min}) \tag{10}$$

or

$$D(T_B)/\text{min} = \exp[\ln 10 \cdot (T_{mD} - T_B)/z]. \tag{11}$$

45.4 Experimental setup for the thermal inactivation of a specimen

In principle, there are two ways to inactivate pathogenic material. Either the specimen is placed into a nonagitated container and the heat transmission to the specimen is achieved by heat conduction only, or the specimen is constantly agitated by stirring or pumping and the heat is transmitted to the specimen just by mixing. In the first case, the temperature is measured in the center of the specimen, i. e., at its coldest point. In the second case, the temperature is measured in the streaming material. The time course of temperature is different in each case.

In the first example [16] (red curve in Fig. 45.1), 20 g of mixed raw material obtained from a rendering plant was put into a closed 30 ml glass flask and autoclaved at 133 °C for 20 minutes. The heat was transmitted from the steam via the glass wall of the bottle into the specimen. The steam temperature varied between 133–140 °C. In the second example, 70 g of a pig brain suspension, diluted to 50 % with physiological saline solution, was heated to the same temperature in a 100 ml steel autoclave equipped with stirrer and heating jacket for saturated steam heating [17]. In spite of a larger mass needing to be heated in the second example, the heating was accomplished faster because of stirring. Assuming that in both cases T_{mD} was 138 °C and z was 10 K, the actual degrees of inactivation can be obtained from integration over the two temperature courses according to Eqs. (5), (7), and (11). They are depicted as dashed curves. In theory, autoclaving of the material for a period of 20 minutes at 133 °C ought to lead to a sterilization of n = 6.3 orders of magnitude. The pre- and post-run sterilization temperatures, however, actually lead to n = 9.9 in the first example and n = 7.5 in the second example, respectively. In particular, the first example demonstrates that prerun and post-run phases contribute considerably to the entire sterilization process. If the time course of the temperature gradient within the material inside the flask were be taken into consideration, n would even be higher. With shorter sterilization times, more drastic differences become evident. Values obtained from such experiments can be used only for comparison with theoretical models (see Section 45.6) if the amount and the geometry of the specimen, as well as the exact temperature course at the cold spot of the specimen, are known. Experiments involving a stirred autoclave, in which

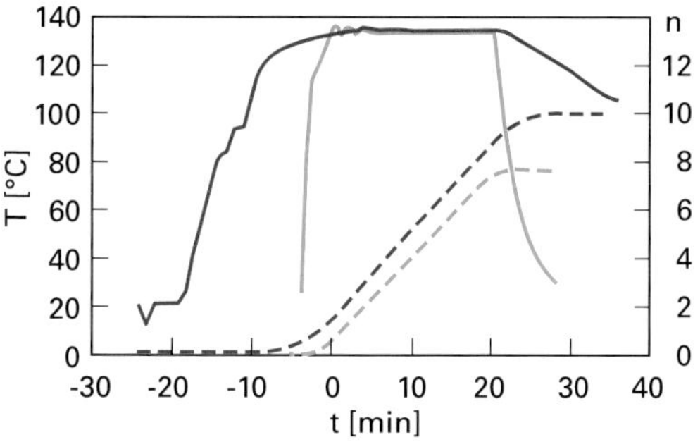

Fig. 45.1: Typical temperature course and extent of inactivation over time. Red: a 30 ml glass flask containing 20 g material was heated in a steam autoclave [16]. Blue: 70 g material was heated in a stirred autoclave equipped with a steam heating jacket [17]. The solid lines depict the temperature course. The dashed lines depict the inactivation in orders of magnitude calculated from the temperature course with T_{mD} = 138 °C and z = 10 K using Eqs. (5), (7), and (11), right scale. Final inactivation values are n = 7.5 (blue) and n = 9.9 (red). Figure by R. Oberthür ©.

Table 45.1: Analysis of the inactivation of a mouse-adapted scrapie strain [21] by dry heat used to determine the order of magnitude of inactivation n as function of temperature and time as well as the decimation time $D(T)$ as a function of temperature. A dose of 20 µl of a 100-fold diluted brain suspension was used ($m_D = 0.0002$ g).

Temperature (°C)	Time (min)	Mouse strain	Number of infected mice	$\log_{10}$ (ID_{50}/g)	Inactivation n	$D(T)$ (min)
20	n.a.	BRVR	(Titration)	−8.3	0	n.a.
160	60	BRVR	(Titration)	−7.0	1.3	46
160	240	A2G	5/8	−3.9	4.4	55
160	1440	BVR	7/8	−4.2	4.1	351
160	1440	A2G	2/5	−3.6	4.7	306
180	60	A2G	3/5	−3.8	4.5	13.3
200	20	A2G	4/4	−4.0	4.3	4.7

n.a. = not applicable

the temperature is distributed more evenly and heating and cooling phases are performed more rapidly, provide a more suitable basis for interpretation [18–20].

Taylor and colleagues [21] carried out an experiment on the inactivation of a mouse-adapted scrapie pathogen by dry heat that will be used as an example to evaluate inactivation experiments. They deposited 7 mg of macerated brain material on microscope slides and subjected them to dry heat using temperatures of between 120 °C and 200 °C for periods of between 20 minutes and 24 hours. The prion concentration in the brain macerate was determined by end point titration in indicator mice following intracerebral (i.c.) inoculation of 20 µl each of 10-times serial dilutions with physiological saline. In inactivation experiments, it is difficult to determine the concentration of prions from the incubation time assay as the dose–incubation time curve might be altered due to heat treatment [21].

The analysis of the experiment is summarized in Table 45.1. The two top rows show values for log(ID_{50}/g), which were determined from the end point titration obtained in [21] by the method of Spearman and Kärber. In contrast, in experiments in which the mice were injected only with diluted material and in which not all animals became ill, the log(ID_{50}/g) values were estimated from the fraction H of infected mice using Eq. (3) in Chapter 52:

$$\log(ID_{50}/g) = \log(m_D/g) + \log(-\ln 2/\ln(1-H)) \qquad (12)$$

with m_D denoting the administered dose of original brain material.

The experimental data shown in Table 45.1 for treatment at 160 °C can be used to validate the inactivation kinetics. According to Figure 45.2, deviations from first-order reactions are observed in case of long reaction times. Such a tailing phenomenon is often observed in inactivation experiments not only with prions but also with bacteria and viruses. A possible explanation is that some of the pathogens are located in a physicochemical environment that protects

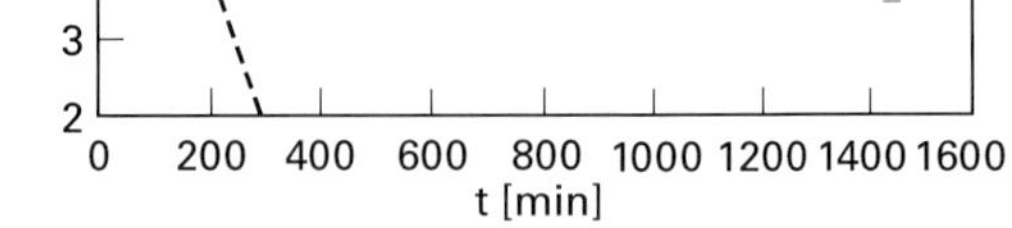

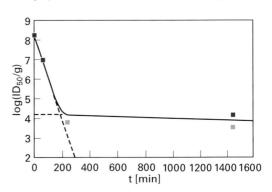

Fig. 45.2: Inactivation curve for a mouse-adapted scrapie strain under dry conditions at a temperature of 160 °C [21] (Table 45.1). The logarithm of the number of ID_{50} per g is plotted as function of time (in minutes). The pink squares refer to BRVR mice, the blue squares to A2G mice. The brown curve is adapted to experimental findings using two inactivation kinetics, namely a rapid kinetic with $D = 46$ min and a slow kinetic with $D = 3840$ min. The dashed lines depict the inactivation curves for the rapid and slow destruction, respectively. Figure by R. Oberthür ©.

them from thermal denaturation by the formation of aggregates or adsorption to surfaces. Another explanation might be that a subpopulation of the pathogen is more thermostable than the majority.

In the case of a heterogeneous physicochemical environment, an intense mixing of the sample during the heating process will lead to all pathogens being evenly exposed to different environments. In contrast, static heating may permit a portion of the pathogens to survive, as they receive a constant exposure to less inactivating conditions.

Prions might experience protection by adsorption to surfaces, as has been shown for the ME6 strain of scrapie in the presence of activated carbon under moist conditions [6]. It might be assumed that a small proportion of the prions is stabilized at the interfaces between hydrophobic and hydrophilic environments, such as the interface between the aqueous phase and the activated carbon phase, or the interface between the hydrophilic microscope slide and the dried, fatty brain material on the slide [21] (Fig. 45.2).

45.5 Results of inactivation studies

Prion strains vary significantly in their sensitivity to heat. Upon limited heat treatment, prions exhibit an initial rapid reduction in infectivity followed by a prolonged plateau during which little further inactivation takes place. The survival of a small subpopulation of infectivity is

Table 45.2: Data published on the prion inactivation by moist heat. These data are used for calculating the decimation time *D* as a function of temperature.

Prion strain	Medium	Temperature (°C)	Time (min)	Inactivation *n*	*D* (min)	Reference
263K	1 N NaOH	22	60	> 5.5	< 10.9	[29]
263K	1 N NaOH	22	120	> 5.1	< 23.5	[29]
22A	Saline solution	60	360	0.8	450	[30]
301V	Saline solution	60	360	1	360	[30]
22A	Saline solution	70	240	0.4	600	[30]
301V	Saline solution	70	240	1.6	150	[30]
22A	Saline solution	80	10	0.4	25	[30]
301V	Saline solution	80	10	0.9	11.1	[30]
22A	Saturated steam	100	30	0.8	37.5	[30]
301V	Saturated steam	100	30	0.8	37.5	[30]
263A	Saturated steam	100	120	2	60	[28]
Scrapie	Saturated steam	118	10	4	2.5	[1]
Scrapie	Saturated steam	121	60	7.5	8	[39]
263K	Saturated steam	121	5	7	0.7	[28]
CJD	Saturated steam	121	60	> 5	< 12	[40]
CJD	Saturated steam	121	60	> 8.3	< 7.2	[40]
Scrapie	Saturated steam	124	20	4.6	4.3	[41]
BSE	Meat pulp	125	15	1.3	11.5	[16]
BSE	Meat pulp	125	15	1.1	13.6	[16]
Scrapie	Meat pulp	125	15	2.7	5.6	[16]
Scrapie	Meat pulp	125	15	2.2	6.8	[16]
22A	Saturated steam	126	120	3	40	[27]
BSE	Brain tissue pulp	133	20	> 3.5	< 5.7	[16]
Scrapie	Brain tissue pulp	133	20	> 3.7	< 5.4	[16]
263K	Brain tissue pulp	134	18	> 7.2	< 2.5	[29]
263K	Brain tissue pulp	134	30	> 7.3	< 4.1	[29]
22A	Brain tissue pulp	136	4	> 6	< 0.7	[27]

probably induced by the heat-mediated conversion of prions into a more resistant form. In contrast to these large variations, there are only small differences between specimens of the same prion strain from brains of mice of different PrP genotypes [22; 23]. A systematic overview of the inactivation studies that have been described in the literature is given in Tables 45.1 and 45.2. The extent of inactivation is given in orders of magnitude n and decimation time D, as defined above.

Figure 45.3 depicts a semilogarithmic plot of $D(T)$ given in Tables 45.1 and 45.2 as a function of the temperature. One should keep in mind that according to Eq. (8) the thermodynamically exact linear interpolation is a function of $1/T$, with the absolute temperature T given in Kelvin. For practical reasons, and within the limits of error, the logarithm of $D(T)$ can also be plotted as function of temperature given

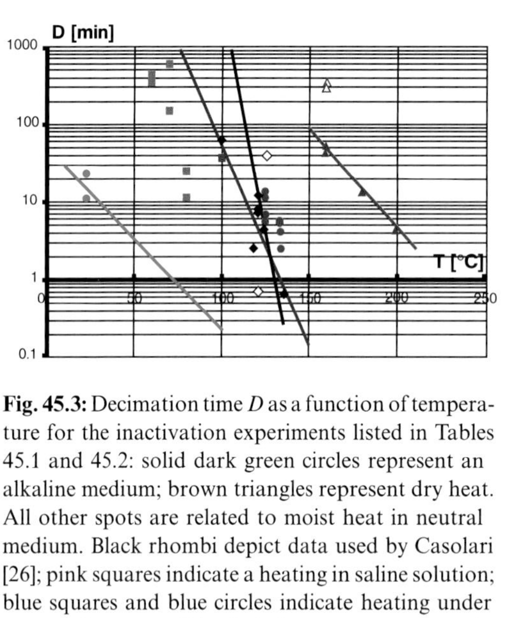

Fig. 45.3: Decimation time D as a function of temperature for the inactivation experiments listed in Tables 45.1 and 45.2: solid dark green circles represent an alkaline medium; brown triangles represent dry heat. All other spots are related to moist heat in neutral medium. Black rhombi depict data used by Casolari [26]; pink squares indicate a heating in saline solution; blue squares and blue circles indicate heating under saturated steam; solid red circles represent meat pulp heated in a closed flask. The green line corresponds to $z = 43$ K and $D(20\,°C) = 17$ min ($T_{mD} = 73\,°C$); the brown line corresponds to $z = 19.5$ K and $D(121\,°C) = 4.5$ min ($T_{mD} = 134\,°C$); and the red line to $z = 39$ K and $D(200\,°C) = 4.7$ min ($T_{mD} = 226\,°C$). The black line corresponds to Eq. (8) with $E_A = 347$ kJ/mol and $D(121\,°C) = 10$ min ($z = 8.6$ K and $T_{mD} = 226\,°C$). Figure by R. Oberthür ©.

in °C. Three distinct areas can be observed that can be interpreted corresponding to the theoretical considerations in Section 45.3.

The inactivation data obtained from dry heating (brown triangles in Fig. 45.3; data from Table 45.1) correspond well with $z = 39$ K. Open triangles denote values that are obtained from long-term inactivation experiments at a temperature of 160 °C and calculated from Eq. (5).

In the low temperature range, it can be predicted from Figure 45.3 that complete inactivation of prions ($n > 20$) will be achieved by boiling for 5 minutes in 1–2 N NaOH. Although this extrapolation has not been confirmed experimentally, the 22A strain of scrapie was completely inactivated after autoclaving for 30 minutes at 121 °C in 1–2 N NaOH [24]. In contrast, after the 139A strain of scrapie was subjected to autoclaving for 30 minutes at 126 °C in 1 N NaOH, an infectivity titer of $\log(ID_{50}/g) = 2.6$ remained [22].

The inactivation of prions under acidic conditions is effective only at temperatures higher than 60 °C. This finding has been interpreted as thermally induced disaggregation. A complete loss of infectivity due to a peptide bond hydrolysis with 1 N hydrochloric acid, for instance, was observed at a temperature of 80 °C [25].

The majority of experimental data are available for the central temperature segment of Figure 45.3. This reflects the results obtained from experiments using moist heat. Casolari [26] tried to systematize this knowledge and adjusted Eq. (7) to the data points shown in Figure 45.3 as solid black rhombi. He obtained the brown line with $z = 19.5$ K and $D(121\,°C) = 4.5$ minutes. The adjustment, however, does not include the data points presented as empty rhombi. This may be explained by a protective niche or a heat-resistant prion subpopulation that might have caused a similar tailing for the 120-minute measurement interval described by Kimberlin [27] as has been observed for the 1,440-minute measurement interval involving dry heat. The 5-minute measurement by Rohwer [28] might have led to an inactivation rate that was unrealistically high and did not account appropriately for the slow heating and cooling of the sample (Fig. 45.1). The distribution of a small proportion of the prions into a protective environment (aggregates) might also explain the higher values obtained by Schreuder and his colleagues (solid red circles) [16] and Taylor and colleagues (solid blue circles) [29]. In contrast, the value obtained by Taylor and colleagues [30] after a 30-minute heating period with saturated steam at 100 °C (solid blue square) seems to confirm the value obtained by Rohwer. The values obtained by Taylor and colleagues [30] at 100 °C are naturally associated with a large error factor due to the low inactivation rate. It is conspicuous, however, that all values are below the line determined by Casolari. Casolari's second suggestion (black line) for interpreting the prion inactivation data corresponds to Eq. (8). The activation energy of 347 kJ/mol corresponds to a typical inactivation reaction for enzymes. As suggested by the data in Figure 45.3, however, the heat-induced inactivation of prions might not only have been caused by protein denaturation and disaggregation but also by chemical reactions (see Section 45.6).

The data that are currently available on the inactivation of prions by moist heat are highly variable. Figure 45.3, however, provides a rough idea of the conditions leading to a complete inactivation of these pathogens.

In recent years, inactivation experiments have also been carried out in which the industrial processing of animal proteins as feed ingredients was simulated [18; 19; 31]. The first experiments carried out with BSE prions suggested that some of the rendering processes were safe. Only the analysis of subsequent scrapie experiments showed that sterilization at a temperature of 133 °C for a period of 20 minutes was necessary for an effective inactivation. In 1996, the European Union passed a directive requiring that the rendering of mammalian waste products must involve sterilization employing the parameters specified above [32]. Other heat processes resulted in nonuniform inactivation efficiencies. When material was dried in a fat bath under vacuum, an inactivation efficiency of $n = 0.1$ was obtained [18], whereas $n = 3.5$ was achieved when the material was separated from fat and dried under vacuum conditions [19; 31]. Only by close inspection of the rendering processes can the differences be explained roughly [31]. The following important findings were obtained:

Dry heating of the material to temperatures higher than 130 °C in a fat bath for a period longer than 60 minutes does not lead to any additional prion inactivation. This can easily be understood from the data depicted in Figure 45.3. In contrast, it has been observed that the amount of scrapie infectivity inactivated after 4 hours under vacuum at 72 °C is greater than that achieved over the same time at atmospheric pressure at an end temperature of 120 °C [19]. Although the data provided by Taylor and colleagues [30] suggest that a partial inactivation can be achieved at a temperature of 70 °C, such an intense inactivation might be explained by the overheating of the internal surface of a dryer equipped with an indirect heating. In order to gradually heat the particles in the center of such a dryer, nearly every particle would have to come into contact with the wall of the dryer, allowing an energy transmission from the high-temperature wall towards the low-temperature center. In 1996, Hahn and his colleagues [33; 34] tried to assess the effect on inactivation of surface overheating during the drying process, and this was developed further by Havekost [31]. In order to achieve a comprehensive interpretation of the data, however, further experiments and calculations on the temperature course and the drying process will have to be performed [18; 19].

Considering inactivation by autoclaving, a 20-minute sterilization period under saturated steam pressure at 133 °C seems to be the safest method for decontamination of prions (see Chapter 52). However, only autoclaving after treatment with, or in, 1 N NaOH appears to eliminate prion infectivity under worst-case conditions (see Chapter 44).

The inactivation by the major rendering processes used on animal carcasses and slaughter by-products in Great Britain during the 1980s, which involved a drying of the material under vacuum, resulted in an efficiency of approximately one order of magnitude [18; 19; 31; 33; 34]. In general, it can be assumed that any rendering process that relies only on drying and degreasing raw material will lead to an inactivation of prions of 1 to 2.5 orders of magnitude at best [31]. In contrast to previous assumptions [35], solvents used for degreasing at temperatures of 60 °C to 80 °C without treatment by dry heat and steam exert no effect on the inactivation of prions [30].

45.6 Inactivation of prions under oleochemical conditions

Peptide backbone degradation and infectivity inactivation of prions were studied systematically under the aspect of product safety in the oleochemical industry. In an early, more theoretical study, it was shown that the presence of lipids protects PrP27–30 against peptide hydrolysis [20]. It was confirmed that the presence of fats or fatty acids under realistic conditions of oleochemical processes has a profound capacity to protect the peptide backbone integrity of PrP27–30 against heat-enhanced degradation [36]. Glycerol is even more efficient in increasing the heat stability of PrP27–30. Furthermore, whereas the protective effect of fats is diminished with higher final temperatures, the protective effect of glycerol persists irrespective of temperature.

The influence of fats, fatty acids, glycerol, and aqueous mixtures thereof on the heat inactivation of prions has also been studied in bioassays. Measurements of peptide degradation by methods other than bioassays indicated minimum values for prion incactivation. However, it is of importance to assess prion inactivation, i. e. reduction in infectivity, with bioassays. Indeed, in all of these studies, the destruction of prion infectivity was achieved up to several orders of magnitude more efficiently than the integrity of the PrP peptide backbone was affected. This finding is depicted with an example in Figure 45.4 for the industrial conditions of the basic oleochemical process of hydrolytic fat splitting [37]. The results suggest a multistep mechanism of prion inactivation involving a transformation of infectious PrP27–30 into a noninfectious form, followed by a slower reaction step during which the noninfectious PrP is degraded (Müller et al., submitted).

Furthermore, in contrast to peptide hydrolysis, fats and fatty acids exert very little protective effect on prion infectivity compared to pure water conditions (Müller et al., submitted) and

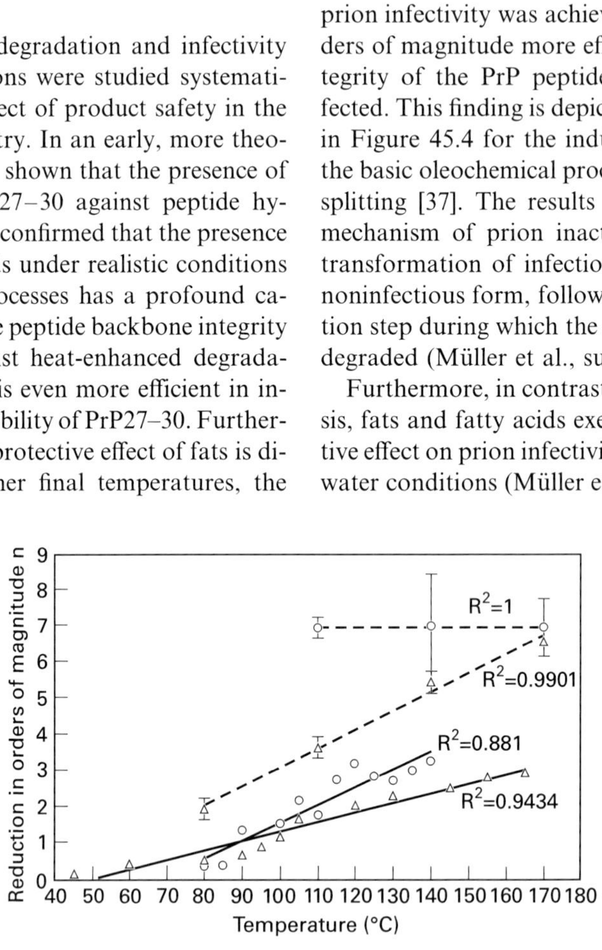

Fig. 45.4: Degradation of PrP27–30 and inactivation of prion infectivity obtained after heating prion rods (Fig. 1.5b) for 20 minutes under oleochemical process conditions. The reduction in peptide bone intactness and prion infectivity is given by orders of magnitude *n*. The conditions represent hydrolytic fat-splitting columns at the temperatures indicated in the upper part of the column (circles; 90 % bovine edible tallow, 10 % water) and the lower part of the column (triangles; 87.75 % water, 9.75 % glycerol, 2.5 % bovine edible tallow). The degradation data (solid lines) as well as the inactivation data (dotted lines) are interpolated by linear regression. Standard mean errors are indicated for each inactivation data point. Source: [37].

might have a comparatively stronger denaturing effect on prion infectivity at lower temperatures. Aqueous solutions of glycerol, however, increase the heat stability of prion infectivity over the whole temperature range, i. e., from 200 °C down to room temperature. These results are of particular importance as the fat content of mammalian tissues, and particularly of brain, is relatively high, constituting a heterogeneous environment for the heat inactivation of prions. At the temperatures used in industrial processes, however, the inactivation is sufficient for all conditions to guarantee prion-free products in case of a prion contamination of the raw materials [37] (Fig. 45.4). Experiments in which the distribution of PrP27–30 between the lipophilic and the hydrophilic phase was analyzed showed that the PrP27–30 molecule tends to accumulate in the interphase rather than in the lipophilic or the aqueous phases. Only by increasing temperature or salt concentration, and particularly by addition of detergent such as sodium dodecylsulfate, can a significant fraction of PrP27–30 be found in the water phase.

45.7 Practical and theoretical implications

The studies described above demonstrated that the inactivation of prions is a complex process involving many parameters. Inactivation is less efficient under dry conditions than under moist conditions involving saturated steam. Strong alkaline or strong acid solutions accelerate the inactivation process and are thus more effective than neutral solutions.

In practice, prions are found in a rather complex environment consisting of an aqueous phase, a fat phase, and different solid substances including the surface of decontamination containers. The distribution of the prions into different phases and interphases leads to different inactivation rates. In addition, adsorption onto surfaces can have stabilizing effects at least for a small proportion of prion particles. Oxidizing or reducing conditions might also have an effect on stabilizing the sulfide bridge of the prion protein (see Chapter 44.2).

A synergistic effect of several of the parameters discussed might also explain the low but still detectable residual infectivity that was detected by Brown and his colleagues [38] after heating infected hamster brain and fibrils in evacuated glass tubes to a temperature of 360 °C for 60 minutes.

The currently available knowledge suggests that the best inactivation is obtained when the medium is stirred or agitated, allowing an active exchange of different phases and a continuous formation of new interfaces. In technical experiments involving drying, i. e., under constant medium temperature but continuous heat transfer from hot interfaces along with an active interface exchange of material, special attention has to be paid to the temperature course from the interface into the medium and to the interface exchange of material, in order to understand the processes involved in prion inactivation. As temperature differences of up to 40 K may be observed between the center of the material and the surface of the heating jacket, it is evident that an overheating of the medium at the surface contributes substantially to inactivation. Because of the many parameters that may affect the inactivation of prions, as many experimental details as possible must be documented, such as quantity and composition of the environment, mode of stirring, exchange of substances between individual phases, temperature time course, and dimensions of the reaction vessel.

References

[1] Hunter GD and Millson GC. Studies on the heat stability and chromatographic behaviour of the scrapie agent. J Gen Microbiol 1964; 37:251–258.

[2] Wallhäuser KH. Praxis der Sterilisation – Desinfektion – Konservierung – Keimidentifizierung – Betriebshygiene. Georg Thieme, Stuttgart, 1988.

[3] Ockerman HW and Hansen CL. Animal by-product processing. VCH Verlagsgesellschaft, Weinheim, 1988.

[4] Von Ostertag R, Moegle E, Braun S. Die Tierkörperverwertung. Paul Parey, Berlin, 1958.

[5] Chang R. Physical chemistry with applications to biological systems. Macmillan Publishing, New York, 1981.

[6] Taylor DM. Impaired thermal inactivation of ME7 scrapie agent in the presence of carbon [letter]. Vet Microbiol 1991; 27:403–405.

[7] Senhaji AF and Loncin M. The protective effect of fat on the heat resistance of bacteria (I). J Fd Technol 1977; 12:203–216.
[8] Senhaji AF. The protective effect of fat on the heat resistance of bacteria (II). J Fd Technol 1977; 12:216–230.
[9] El Boushy ARY and van der Poel AFB. Poultry feed from waste. Processing and use. Chapman and Hall, London, 1994.
[10] Moelwyn Hughes EA. Physical chemistry. Pergamon Press, Oxford 1961.
[11] Belitz HD and Grosch W. Lehrbuch der Lebensmittelchemie. Springer, Berlin 1987.
[12] Joly M. A physical-chemical approach to the denaturation of proteins. Academic Press, London 1965.
[13] Riedinger OL. Untersuchungen über die aus hygienischen Gründen erforderliche Hitzebehandlung bei der Herstellung von Tiermehl in Dampfdruckanlagen, unter besonderer Berücksichtigung der Proteinschädigung durch Dampfdruckerhitzung [Dissertation]. Universität Hohenheim, 1979.
[14] Riedinger A. Untersuchungen über die Sterilisationseffektivität von Tierkörperbeseitigungs-Verfahren ohne Druckanwendung in heißem Fett [Dissertation]. Universität Hohenheim, 1981.
[15] Hoppe A. Die Hitzeresistenz verschiedener Mikroorganismen und ihre Beeinflussung durch verschiedene Faktoren, unter besonderer Berücksichtigung von experimentell an Sporen von *Bacillus anthracis* erarbeiteten Daten [Diplomarbeit]. Universität Stuttgart-Hohenheim, 1978.
[16] Schreuder BE, Geertsma RE, van Keulen LJ, et al. Studies on the efficacy of hyperbaric rendering procedures in inactivating bovine spongiform encephalopathy (BSE) and scrapie agents. Vet Rec 1998; 142:474–480.
[17] Oberthür RC, Havekost U, Steele PJ, et al. Unpublished measurements. Institute of Animal Health, Edinburgh, Scotland, 1998 (unpublished work).
[18] Taylor DM, Woodgate SL, Atkinson MJ. Inactivation of the bovine spongiform encephalopathy agent by rendering procedures. Vet Rec 1995; 137:605–610.
[19] Taylor DM, Woodgate SL, Fleetwood AJ, et al. The effect of rendering procedures on scrapie agent. Vet Rec 1997; 141:643–649.
[20] Appel T, Wolff M, von Rheinbaben F, et al. Heat stability of prion rods and recombinant prion protein in water, lipid and lipid–water mixtures. J Gen Virol 2001; 82(2):465–473.
[21] Taylor DM, McConnell I, Fernie K. Effect of dry heat on the ME7 strain of mouse-passaged scrapie agent. J Gen Virol 1996; 77:3161–3164.
[22] Taylor DM, Fernie K, Steele PJ, et al. Thermostability of mouse-passaged BSE and scrapie is independent of host PrP genotype: implications for the nature of the causal agents. J Gen Virol 2002; 83(12):3199–3204.
[23] Somerville RA, Oberthür RC, Havekost U, et al. Characterization of thermodynamic diversity between transmissible spongiform encephalopathy agent strains and its theoretical implications. J Biol Chem 2002; 277(13):11084–11089.
[24] Taylor DM, Fernie K, McConnell I. Inactivation of the 22A strain of scrapie agent by autoclaving in sodium hydroxide. Vet Microbiol 1997; 58(2–4):87–91.
[25] Appel TR, Lucassen R, Groschup MH, et al. Acid inactivation of prions – efficient at elevated temperature or high acid concentration. J Gen Virol 2006; 87(5):1385–1394.
[26] Casolari A. Heat resistance of prions and food processing. Food Microbiol 1998; 15:59–63.
[27] Kimberlin RH, Walker C, Millson GC, et al. Disinfection studies with two strains of mouse-passaged scrapie agent. Guidelines for Creutzfeldt–Jakob and related agents. J Neurol Sci 1983; 59:355–369.
[28] Rohwer RG. Virus-like sensitivity of the scrapie agent to heat inactivation. Science 1984; 223:600–602.
[29] Taylor DM, Fraser H, McConnell I, et al. Decontamination studies with the agents of bovine spongiform encephalopathy and scrapie. Arch Virol 1994; 139:313–326.
[30] Taylor DM, Fernie K, McConnell I, et al. Solvent extraction as an adjunct to rendering: the effect on BSE and scrapie agents of hot solvents followed by dry heat and steam. Vet Rec 1998; 143:6–9.
[31] Havekost U. Scrapie-Inaktivierung in der Fleischmehlindustrie [Diplomarbeit]. [Scrapie inactivation in the rendering industry.] Universität Gesamthochschule Paderborn, 1999.
[32] European Commission. Pressure cooking system for processing mammalian waste into MBM (inactivation of TSE agents). Directive 96/449 EC of 18 July 1996.
[33] Hahn F, Hogendoorn JA, van Swaaij WPM. BSE deactivation in rendering processes. Evaluation of experimental data and modelling of industrial processes (I). CHERA (Chemical Engineering Research and Advising), Enschede, Holland, 1996.

[34] Hahn F, Hogendoorn JA, van Swaaij WPM. Predicting BSE deactivation in rendering processes with sterilization 1.0 (II). CHERA (Chemical Engineering Research and Advising), Enschede, Holland, 1996.

[35] Wilesmith JW, Ryan JB, Atkinson MJ. Bovine spongiform encephalopathy: epidemiological studies on the origin. Vet Rec 1991; 128:199–203.

[36] Müller H and Riesner D. Thermal degradation of prions in presence of fats: implication for oleochemical processes. Eur J Lipid Sci Technol 2005; 107:833–839.

[37] Müller H, Stitz L, Riesner D. Risk assessment for fat derivatives in case of contamination with BSE. Eur J Lipid Sci Technol 2006, submitted.

[38] Brown P, Liberski PP, Wolff A, et al. Resistance of scrapie infectivity to steam autoclaving after formaldehyde fixation and limited survival after ashing at 360 degrees Celsius: practical and theoretical implications. J Infect Dis 1990; 161:467–472.

[39] Brown P, Rohwer RG, Green EM, et al. Effect of chemicals, heat, and histopathologic processing on high-infectivity hamster-adapted scrapie virus. J Infect Dis 1982; 145:683–687.

[40] Brown P, Rohwer RG, Gajdusek DC. Newer data on the inactivation of scrapie virus or Creutzfeldt–Jakob disease virus in brain tissue. J Infect Dis 1986; 153:1145–1148.

[41] Mould DL and Dawson AM. The response in mice to heat treated scrapie agent. J Comp Pathol 1970; 80:595–600.

Topic IX: Prevention

46 Prevention of Prion Diseases in the Production of Medicinal Products, Medical Devices, and Cosmetics

Michael Ruffing, Helena Windemann, and Jan Schaefer

46.1 Introduction

Many medicinal products and cosmetics still contain substances that are produced from bovine materials, although manufacturers make every effort to eliminate them.

Gelatin, mainly manufactured from bovine bones and hides, is used for the production of capsules, tablets, and as a stabilizer in solutions. Many drugs and cosmetics include substances such as magnesium stearate, glycerol,[1] and fatty alcohols that are produced from tallow. Tallow consists mainly of fats gained by rendering slaughter by-products. Cattle-derived products such as fetal calf serum and bovine serum albumin are still used for the production of some vaccines, recombinant proteins, monoclonal antibodies, and antibiotics. In addition, some medical devices contain bovine-derived materials, e. g., grafts from bone or collagen, which is used in cosmetic surgery.

Since most of the principles insuring safe food also apply to the production of medicinal products, medical devices, and cosmetics, these are at least as safe as food. Often the production process contributes additionally to safety.

It has been clear from the beginning of the current BSE crisis that specific aspects of these products have to be taken into account and a more flexible approach adopted. Therefore, guidelines for the individual scientific evaluation of risks from prion diseases (TSE risks) have been developed. Many medicinal products and medical devices are administered not by the oral route, but by parenteral routes. Special considerations apply when assessing these routes of administration. Animal experiments have shown that parenteral routes of administration of prions result in transmission rates up to 100,000-fold greater than the oral route [1]. In the case of ruminant products administered to ruminants, there is no species barrier to transmission (cattle-to-cattle; sheep-to-sheep; goat-to-goat). On the other hand, scientific evaluation of the production process may give reassurance that potential contamination by prions (TSE agents) is likely to be reduced or eliminated. The therapeutic benefit of medicinal products and medical devices has to be taken into consideration, and under exceptional circumstances materials with a potentially high level of infectivity may be used. The first part of this chapter focuses on the evolution of regulations and rules to minimize the risk from medicinal products and cosmetics following the emergence of BSE and variant Creutzfeldt–Jakob disease (vCJD). The development of these rules will be elucidated in a historical context. In the second part, criteria for evaluating the safety of animal-derived materials used in medicinal products and cosmetics will be explained. The third part of the chapter will focus on preventive measures taken with regard to pharmaceutical gelatin and tallow-derived products. The last part gives a short overview covering medical devices.

46.2 Regulations to prevent the transmission of prions by medicinal products and cosmetics

There is a risk that medicinal products and cosmetics produced from substances of animal origin could be contaminated by pathogens (Table 46.1), and so manufacturers must take preven-

[1] Glycerole Ph. Eur. Monograph 496 or 1,2,3-propantriol (glycerin).

Table 46.1: Examples of substances of bovine origin used in medicinal products and cosmetics.

Substance	Bovine organ/tissue	Application
• Aprotinin	• Lung	• Improvement of blood coagulation
• Bovine serum	• Blood	• Fermentation medium to produce recombinant proteins
• Gelatin	• Bones and skin	• Flexible drug capsules; blood plasma substitute
• Magnesium stearate, stearic acid	• Fats, bones, entrails	• Flow agent used in the production of tablets
• Lactose	• Milk	• Flow agent used in the production of tablets
• Cetostearyl alcohol	• Fats, bones, entrails	• Basis for ointments, cosmetics
• Lanolin	• Wool fat of sheep	• Basis for ointments, cosmetics
• Surfactants	• Lungs	• Respiratory aid for premature babies
• Ursodeoxy-cholic acid / Chenodeoxy-cholic acid	• Bile	• Treatment of gall stones
• Hyaluronidase	• Testicles (also from sheep)	• Softener of tissues before local anesthetics, surgery

Insulin (pancreas), heparin (intestines, lungs), and corticosteroids (bile) are no longer produced by the use of bovine materials.

tive measures to guarantee safety. Some principles of European legislation aimed at the prevention of prion contamination of food and feedstuffs also apply to medicinal products, medical devices, and cosmetics. One of the most important principles is laid down in Regulation (EC) No. 1774/2002, which regulates the hygienic sourcing of animal by-products [2]. Only materials obtained from healthy animals (e. g., milk, eggs) or from slaughtered animals officially declared as fit for human consumption are permitted in the production of substances that can be absorbed by the human body. Regulation (EC) No. 999/2001, which specifies measures against animal TSEs [3], defines the tissues to be removed at slaughter (specified risk materials, SRMs) and details procedures for testing slaughtered animals for TSE. Although officially not applicable to medicinal products, medical devices, and cosmetics, in practice the provisions of this regulation are also fulfilled for these products.

In the registration process required before a medicinal product can enter the market, it is examined for safety, efficacy, and quality. In the EC, medicinal products are registered according to Directive 2001/83/EC (human use) and Directive 2001/82/EC (animal use) or the respective national legislation of the Member States by which these directives were implemented. Centrally authorized products are approved by the EMEA (European Medicines Evaluation Agency) according to Regulation (EC) No. 726/2004. This legal framework is called Review 2004 [4–6]. If animal-derived substances are used for the production of drugs, the applicant for a marketing authorization has to provide detailed information on the sourcing of the material and the manufacturing processes.

For evaluation of TSE risks, an applicant must demonstrate that the medicinal product complies with the requirements of the latest version of the "Note for Guidance on Minimising the Risk of Transmitting Animal Spongiform Encephalopathy Agents via Human and Veterinary Medicinal Products" [7]. Since its first publication in 1991, this has been the central document for the evaluation of TSE risks from medicinal products in the EU. An applicant is required to submit documentation of measures taken to minimize TSE risks from the product in accordance with this guideline. The competent national authorities or the EMEA evaluate the documentation. There is an ongoing obligation in the marketing phase to update this information if changes occur in the TSE status of the product. Such changes must be indicated in the scope of so-called variation procedures.

A centralized procedure was also instituted in addition to that outlined above. The manufacturer of a pharmaceutical product can apply for a TSE certificate of suitability from the EDQM (European Directorate for the Quality of Medicines) in Strasbourg. Such certificates can be submitted within the scope of TSE risk evaluation procedures for medicinal products. They are acknowledged by all countries of the Council of Europe and are often accepted by other countries. Documentation by means of TSE certificates has proven to be especially appropriate for substances such as gelatin which are frequently used in the manufacture of medicinal products. The background of the TSE Note for Guidance and TSE risk evaluation of medicinal products is detailed below.

Since cosmetics are not intended to have a therapeutic effect and are used externally, they only have to meet the requirements of the guidelines with regard to composition, labeling, and packaging and are not subjected to an approval procedure.

46.2.1 The first CPMP Note for Guidance on Minimising the Risk of Transmitting Animal Spongiform Encephalopathy via Medicinal Products (1991)

The first cases of BSE gave rise to concerns that the agent might be transmissible to humans [8]. The European authorities responsible for the safety of drugs were thus obliged to take action. In 1991, the EU scientific committee for the licensing of human drugs (Committee for Proprietary Medicinal Products, CPMP; now: Committee for Medicinal Products for Human Use, CHMP) published a draft "Note for Guidance" to minimize the risk of transmitting animal spongiform encephalopathies [9]. It applied to all substances used in the manufacturing of a medicinal product that were derived from animal species naturally susceptible to TSE.

According to this guideline, which became effective in 1992, three criteria contribute to TSE safety:

- The selection of the animals used from countries with low or no TSE risk.
- The age and the tissues of the animals from which materials are sourced. This is because TSE infectivity accumulates over several years and tissues differ in their level of infectivity.
- The capacity of the production process to remove or inactivate TSE agents that might be present in the starting material.

46.2.2 The European Commission's urgent measures for protection against BSE

The occurrence of vCJD in GB was announced in March 1996. Expert opinion was that the most likely cause was the infection of humans with the BSE agent [10] (see Chapters 1, 15, 50). In response, the European Commission (EC) issued so-called emergency measures [11] involving the total ban on the export of live cattle from GB and cattle-derived products destined for the production of pharmaceuticals or cosmetics. This ban was amended several times and resulted in Council Decision 98/256/EC [12], which largely maintained the ban but defined safety measures under which derogations could be granted. This ban was repealed recently and measures to combat animal TSEs harmonized with the other EU countries.

46.2.3 Subsequent decisions of the European Commission

Further vCJD cases (Fig. 15.1) and growing evidence of its relation to BSE [13; 14] (see Chapter 50) led to public demand for more restrictive safety measures in the winter of 1996/97. Furthermore, inspections within EU Member States revealed deficiencies in control and compliance with feed bans [15].

To ensure the safety of cosmetics, the use of material derived from skull, including brain and eyes, tonsils, and spinal cord of cattle, sheep and goats older than 12 months, and the spleen of sheep and goats of all ages, was prohibited by amendments to the Cosmetics Directive [16–18]. That Directive was further amended in 2003 [19]. A reference to TSE Regulation 999/2001 was introduced prohibiting the use of the same SRM as in foods [3].

The technical requirements for medicinal products for human use were amended by Directive 1999/82/EC [20]. The applicant for a marketing authorization has to demonstrate that the product is manufactured according to the requirements of the CPMP's Note for Guidance [21] and subsequent updates. From July 1, 2000, any new application, and from March 1, 2001, all authorized medicinal products, had to comply with this annex of the Directive. A similar procedure was implemented for veterinary products [22].

Tight deadlines and the large number of marketing authorizations affected by this directive led to delays in the assessment of documents submitted to the authorities. It was therefore decided to take advantage of the experience gained with the certification of the quality of chemical substances that had been established at the EDQM in Strasbourg. In 2000, an equivalent certification system to assess products of ruminant origin was established. This system was considered to be the most effective way of reducing the work and harmonizing the assessment of the dossiers to prove compliance with the Note for Guidance.

46.2.4 Certification of suitability to the monographs of the European Pharmacopoeia

In 1992, a certification scheme was established to evaluate the chemical purity and microbiological quality of substances according to the standards of the European Pharmacopoeia (Ph. Eur.). The procedure was extended to products bearing a risk of transmitting animal TSE [23]. A new monograph (No. 1483) and general chapter (5.2.8), reflecting the Note for Guidance, were introduced in the European Pharmacopoeia [24]. The applicant for a TSE certificate must now submit a dossier containing detailed information about the origin of the animal-derived raw material and the type of tissues used. The process used to manufacture the substance must be detailed and the quality assurance system described. An important aspect is the avoidance of cross-contamination by other ruminant materials which have a higher risk of contamination by TSE agents. Information on the system to ensure the traceability of the raw materials used is required. The dossier is completed by an expert statement on how the risk of TSE transmission has been minimized for the relevant substance.

The dossiers are assessed by experts nominated by national regulatory authorities.

So far, more than 500 product TSE certificates of suitability have been granted to applicants in countries around the world. Most of the products certified are used as excipients for pharmaceutical use, e. g., tallow derivatives or gelatin. Some, such as bovine sera, are used in the production of active substances.

The EDQM [25] has initiated on-site inspections to make sure that the substance for which a certificate has been granted is manufactured according to the dossier and the quality assurance system or the principles of good manufacturing practice (GMP) are implemented properly.

46.2.5 CPMP/CVMP Note for Guidance on Minimising the Risk of Transmitting Animal Spongiform Encephalopathy Agents via Human and Veterinary Medicinal Products (2004)

The first version of the CPMP Note for Guidance [9] has been revised several times. The changes reflect advances in scientific knowledge and legislation. Since 1999, the Note for Guidance has been effective for products for human as well as veterinary use [21].

The second revision of the CPMP/CVMP (Committee for Proprietary Medicinal Products / Committee for Veterinary Medicinal Products) Note for Guidance [7], which took effect in July 2004, was given the force of law by the EC [5; 6; 26]. Applicants for a marketing authorization for human or veterinary medicinal products must demonstrate that they comply with the requirements of the latest version of this Note for Guidance. This also applies to medicinal products for which authorization has already been granted.

The assessment of risk from animal-derived material used in the production of medicinal products has become an essential part of the regulatory compliance process. An applicant has to consider all factors contributing to the risk of TSE transmission by the product and must follow the principles of the Note for Guidance to minimize any such risk.

Some materials, including collagen, gelatin, bovine blood derivatives, tallow derivatives, animal charcoal, milk and milk derivatives, wool derivatives, and amino acids are considered in compliance with the Note for Guidance if they meet certain specified requirements (see Section 46.4 for gelatin and tallow derivatives).

46.3 Evaluation of the risk of medicinal products transmitting prions

The BSE epidemic and occurrence of vCJD have led to drug producers being required to refrain from using substances made from ruminant tissues if alternatives are available [21; 27].

If materials derived from animal species naturally susceptible to infection by prions are used at any stage in the production of a medicinal product, an assessment of the risk of transmitting prions must be carried out. The Note for Guidance also covers materials that come into contact with the medicinal product or manufacturing equipment, as these might contaminate the final product.

The following criteria are important for the evaluation of drug safety:

- Species of source animals and country of origin;
- Source material (organs, tissues, body fluids);
- Production process and its validation for reduction of transmissible agents;
- Risk assessment.

46.3.1 Source animals and their country of origin

The risk of transmitting TSEs through medicinal products can be substantially reduced by careful selection of the animals from which raw materials are derived. Raw materials should be sourced from animals declared fit for human consumption following official ante and post mortem inspection.

Sourcing of cattle

A process to classify countries according to their Geographical BSE Risk (GBR) was developed by the Scientific Steering Committee (SSC) of the EC [28]. On the basis of information provided by a particular country, the stability of its cattle system and the external challenge posed by the importation of potentially infected animals and MBM was assessed. The structure of the ruminant population, animal trade, animal feed production, the implementation of feed bans, monitoring and reporting systems, as well as the disposal of animal carcasses were examined. The process takes into account the possibility that countries may have exported BSE prior to the presence of the disease being recognised.

The GBR process assigned each country to one of four levels indicating the present probability of cattle being infected with BSE (compare Table 53.1):

1. In GBR I countries, the presence of BSE infected cattle is looked upon as highly unlikely.
2. In GBR II countries, the occurrence of BSE cases is unlikely but not excluded.
3. In GBR III countries, the presence of BSE is confirmed (some infected / affected animals), or it is likely that BSE is present but there are no confirmed cases.
4. In GBR IV countries, BSE is present at a high level ($\geq$ 100 cases/1 million adult cattle per year).

By August 2006, 65 countries had been classified according to the GBR system (Table 46.2). The GBR process has been maintained and updated by the European Food Safety Authority and its Scientific Expert Working Group on the Assessment of the GBR [29; 30].

In principle, materials used in the production of medicinal products should be obtained from animals from GBR I or GBR II countries. However, it is acknowledged that geography should not be the only, or even the primary criterion, for safety of raw materials.

If specific materials, such as gelatin and tallow derivatives, meet the CPMP/CVMP requirements (see Section 46.4), sourcing from

Table 46.2: GBR classification of countries at 1 August 2006. For explanation of status I, II, III and IV compare Table 53.1. To see in which countries BSE cases have been detected by surveillance of any kind, check Table 19.2.

Countries assessed	GBR status	Country assessed (ctnd.)	GBR status
Argentina	I	Croatia	III
Australia	I	Cyprus	III
Iceland	I	Czech Republic	III
Singapore	I	Denmark	III
New Caledonia	I	Estonia	III
New Zealand	I	Finland	III
Panama	I	France	III
Paraguay	I	Germany	III
Uruguay	I	Greece	III
Vanuatu	I	Hungary	III
Botswana	II	Ireland	III
Brazil	II	Israel	III
Colombia	II	Italy	III
Costa Rica	II	Latvia	III
El Salvador	II	Lithuania	III
India	II	Luxembourg	III
Kenya	II	Macedonia (former	III
Mauritius	II	Yugoslavian Republic of...)	
Namibia	II	Malta	III
Nicaragua	II	Mexico	III
Nigeria	II	Netherlands	III
Norway	II	Poland	III
Pakistan	II	Republic of South Africa	III
Swaziland	II	Romania	III
Sweden	II	San Marino	III
Albania	III	Slovak Republic	III
Andorra	III	Slovenia	III
Austria	III	Spain	III
Belarus	III	Switzerland	III
Belgium	III	Turkey	III
Bulgaria	III	USA	III
Canada	III	Portugal*	IV
Chile	III	United Kingdom*	IV

* EFSA lately published the statements that GB and Portugal could be considered as countries with moderate BSE risks (= "controlled BSE risks" according to international standards by OIE; defined in Table 53.1).

countries of a higher GBR level may be acceptable. Sourcing from countries that have reported BSE cases may be acceptable so long as the animals come from "negligible risk (closed) bovine herds" [31]. Animals in such herds have never been exposed to any source of infection, such as through feeding of MBM, and have no epidemiological link with any BSE case. It must be demonstrated by reliable records that no animal has been introduced into the herd for at least 8 years unless from a herd with an equivalent status or from GBR I countries. For newly established herds, all animals should come from such safe sources.

The OIE (World Organisation for Animal Health) has also established criteria to determine the status of a country with respect to BSE risk [32] (OIE A. H. Code 2.3.13 on $\text{BSE}_{\text{Appenidex 2}}$). In the OIE's process, the status is determined on the basis of a risk analysis identifying all factors that could contribute to the appearance of BSE, including the consumption of ruminant-derived MBM or similar slaughter by-products and the importation of potentially contaminated MBM or infected animals. The OIE process takes into account any education program for personnel involved in breeding, raising, and slaughtering cattle, the reporting and examination of cattle showing clinical signs of BSE, the surveillance and monitoring system including laboratories examining samples of brain.

The OIE process has three risk categories (compare Table 53.1):

Category 1: Negligible BSE risk,
Category 2: Controlled BSE risk (= moderate BSE risk),
Category 3: Undetermined BSE risk.

Efforts to harmonize the European GBR process and that of OIE are ongoing (Table 53.1). This project is described in the TSE Road Map [33] for the future TSE policy of Europe. Countries already classified according to the GBR scheme are to be classified according to the OIE criteria. The TSE Regulation 999/2001 contains a clause for the future adoption of the international standards. As long as there is no harmonization, the Note for Guidance relies on the GBR process.

The application of post mortem tests to detect PrP^{Sc} in brain stem or spinal cord of slaughtered cattle was evaluated by the EC (see Chapter 35). In EU countries of GBR III and GBR IV, post mortem testing is carried out on cattle over 30 months of age or on a representative sample size of the population slaughtered annually.

Sourcing of small ruminants (sheep and goats)

Because scrapie occurs in many countries and sheep and goats hypothetically infected with BSE could be mistakenly diagnosed for scrapie, materials of small ruminants used in the production of medicinal products should preferably be sourced from countries free from scrapie, such as New Zealand and Australia. Alternatively, the CPMP/CVMP recommend establishing scrapie-free flocks of sheep analogous to the "negligible risk (closed) bovine herds" [31] composed of genotype(s) shown to be resistant to BSE/scrapie infection (see Chapter 56). Although experiments have shown BSE to be transmissible to sheep, no case has so far been detected in sheep in the field. Active surveillance has identified a single French goat infected with BSE. Monitoring of sheep and goats was recently increased in the EU [34], and a three-step testing scheme (rapid testing, discriminatory testing, and mouse bioassay testing) was introduced in order to obtain more data on the possible prevalence of BSE in small ruminants (compare OIE A H Code 2.4.8 on $scrapie_{\text{Appenidex 3}}$) [32a].

Except from products obtained from wool wax of sheep (lanolin, cholesterol, and cholesterol derivatives such as cholecalciferol – Vitamin D3), materials of small ruminants are seldom used for the manufacture of medicinal products and cosmetics.

46.3.2 Source material (organs, tissues, body fluids)

BSE infectivity has been detected in brain, spinal cord, retina, and nictitating membrane of naturally infected cattle. In experimental BSE, the dorsal root ganglia (DRG), trigeminal ganglia, and distal ileum of cattle infected by the oral route have also been shown to be infected. The bone marrow (Fig. 53.2) of some animals showed infectivity during advanced clinical illness, but this might have resulted from cross-contamination during the study. Other bovine tissues have not shown infectivity in the mouse bioassay, suggesting a more limited distribution of the BSE pathogen in bovine tissues than of the scrapie and BSE agent in sheep: for example, infectivity has been detected also in the blood of sheep with natural scrapie and of sheep experimentally infected with BSE. Some tissues have also been tested in a cattle bioassay to bypass the species barrier. Such bioassay has detected infectivity in the palatine tonsil of infected cows.

The first CPMP Note for Guidance with regard to minimizing the transmission risk of prion diseases [9] classifies organs, tissues, and body fluids of animals into four infectivity categories. This classification was based on the results of studies in which the distribution of scrapie pathogens was examined in sheep and goats [35; 36]. The classification was also adopted for the assessment of the BSE risk.

The second revision of the CPMP/CVMP Note for Guidance [7] revised the classification of tissues in the light of the latest data available in 2003 on the distribution of infectivity in BSE-infected cattle. Three categories, from highly infectious (category A) to no infectivity detectable (category C), are based on the WHO guidelines for biological and pharmaceutical products [37]. These guidelines have been supplemented recently by additional WHO guidelines on tissue infectivity distribution [38] with the same principal categorization of potential infectivity but updated results for some animal materials (WHO Tables$_{\text{Appendix}}$ IA-IC).

Category A comprises tissues of the central nervous system and of anatomically associated tissues and should not be used in the manufacturing of a medicinal product, unless its use can be justified.

Category B tissues pose a lower risk but have been found infectious in at least one form of TSE.

Category C includes tissues that have been examined for infectivity and shown to be negative.

According to Regulation 999/2001, ruminant tissues that could pose a BSE/TSE risk are defined as specified risk materials depending on the categorization of the country of origin and the age of the animals [3]. The definition of SRM is subject to frequent changes (see Chapter 53.3.1) and was last amended in 2006 [39]. In addition, there are differences in the definition of SRM between some countries (compare for example EU, Switzerland, USA), which may pose problems in the context of compliance with the restrictions described in this chapter.

The definition of bovine SRM in the EU

The skull excluding the mandible but including the brain and eyes, and the spinal cord of bovine animals aged over 12 months, the vertebral column excluding the vertebrae of the tail, the spinous and transverse processes of the cervical, thoracic, and lumbar vertebrae and the median sacral crest and wings of the sacrum, but including the dorsal root ganglia of bovine animals aged over 24 months, and the tonsils, the intestines from the duodenum to rectum and the mesentery of bovine animals of all ages (compare Chapter 53.3.1).

The definition of small ruminant SRM in the EU

The skull, including the brain and eyes, the tonsils, and the spinal cord of ovine and caprine animals aged over 12 months and the spleen and ileum of ovine and caprine animals of all ages (compare OIE A H Code 2.4.8 on scrapie$_{\text{Appenidex 3}}$).

Until recently, additional bovine SRM had to be removed in GB and Portugal. However, these requirements were repealed by amendments to Regulation 999/2001 [39]. The TSE measures in the EU were further harmonized by these steps. According to a recent statement of the European Food Safety Authority [40], GB can now be considered to be a country with a moderate BSE risk, according to international standards, as there has been a continuous decrease in BSE cases and strict control measures have been enforced.

Although Regulation 999/2001 does not apply to medicinal products, the use of SRM would be accepted by regulatory authorities under exceptional circumstances only [7].

Considering the risk of cross-contamination

In addition to the categorization with respect to infectivity, the risk of cross-contamination of the source tissue by tissues with a higher TSE risk has to be considered. The categorization of source material to the infectivity categories B (low infectivity) and C (no detectable infectivity; WHO Tables$_{\text{Appendix}}$ IA-IC) would need to be questioned if cross-contamination by highly infectious tissue seemed likely.

Certain slaughtering techniques, such as penetrative brain stunning using a pneumatic stunner that injects air, increase the risk of cross-contamination of peripheral organs, particularly the lungs and the heart. These organs may be contaminated by tissue particles of the central nervous system [41].

Infectivity of organs and tissues also depends on the age of the animals [35; 42; 43]. In young cattle, infectivity has been detected only in the intestine (Fig. 53.2) and – in the case of sheep and goat – in the reticuloendothelial system, e. g., spleen and lymph nodes (see Chapter 28).

46.3.3 The production process and its validation for reduction of transmissible agents

If animal tissues are used for the production of a medicinal product, any pathogens potentially present should be removed and/or inactivated in the manufacturing process. The validation of the complete manufacturing process or some of its steps with regard to their efficiency in reducing infectious agents is conducted by scaling down critical process parameters to laboratory scale, because for GMP reasons it is not permissible to deliberately introduce infectious agents into a production facility.

Regulatory authorities demand the validation of the production process if the producer claims the removal and/or inactivation of TSE agents in the production process [7].

In the process of validation infectious agents are added to the starting material or intermediate substances of the production process ("spiking"). The processing of the spiked material is carried out according to the usual manufacturing steps. The reduction factor for each manufacturing step is obtained by comparing the infectivity spiked into the starting material and with that detected in the product resulting from this step.

In validating the complete manufacturing process, one has to consider that maximal reduction of infectivity might be limited by the quantity of pathogen added at the start. For this reason, it might not be apparent to what extent each processing step contributes to the overall reduction of infectivity.

For this reason, it might be advantageous to consider each step of the process that may have the potential to reduce TSE contamination individually. Methods that might remove prions include precipitation, chromatographic fractionation, and nanofiltration. Procedures that contribute to the inactivation of prions, such as treatment with sodium hypochlorite, sodium hydroxide, or wet heating at 133 °C, 3,000 hPa [3 bars] for 20 minutes, are not suitable for the production of most medicinal products as it is likely to destroy biological activity.

However, in the validation of individual manufacturing step(s) it may not be possible to detect alterations of the prions in a way that might lead to the development or retention of a "prion subpopulation" which is more resistant to inactivation (see Chapter 44.1). In such a case, the overall reduction factor obtained by adding the logarithmic reduction factors of individual steps would imply a greater reduction than actually occurs in the full production process.

There are a number of additional issues to be considered in the design of a TSE validation study. In most studies, the scrapie strain 263K prepared as a brain homogenate from experimentally infected Syrian hamsters is used as spiking agent. Mouse-adapted scrapie strains used in validation studies include ME7 and 22A (see Chapter 12). Some validation studies have been conducted using mouse-adapted BSE strain 301V, e. g., for the re-validation of the gelatin production process (see below).

The spike material used in the validation study is also of great importance, as it should represent as closely as possible the actual level of contamination to be expected in the manufacturing process. Crude brain homogenate has been used as spiking material in many studies and has led to some concern over whether this reflects the type of contamination encountered in the actual manufacturing process of the respective medicinal product.

In validation experiments, prion infectivity is routinely detected in a mouse or hamster bioassay. Western blotting might be a faster and cheaper alternative, but this technique has been shown to be two to three logs less sensitive than the bioassay. It is, nevertheless, currently used to identify manufacturing steps that are effective in prion reduction.

Validation studies are currently regarded as an additional option to ensure the safety of a medicinal product, whereas the sourcing of animals and the nature of the tissue used remain the main factors in reducing TSE risk.

46.3.4 Risk assessment

If any materials from TSE susceptible animal species have been used in the manufacture of a medicinal product, the applicant for a marketing authorization must demonstrate that all risk factors have been considered and minimized by applying the principles of the CPMP/CVMP Note for Guidance.

The risk assessment should also consider the following parameters [7]:

- The route of administration of the medicinal product;
- The quantity of animal material present in the medicinal product;
- The maximum therapeutic dosage (daily dose and duration of treatment);
- The intended use of the medicinal product and its clinical benefit.

The assessment may be based on TSE certificates issued by the EDQM. The risk of cross-contamination with tissue of potentially higher infectivity should be considered in conjunction with the GBR assessment of the animals' country of origin, the age of the cattle, and post mortem BSE testing.

46.4 Regulations for specific materials used in the production of medicinal products

46.4.1 Gelatin

Gelatin consists of approximately 85% polypeptides, 8–12% water, and 2–4% mineral salts and is often applied as an excipient in medicinal products. It is used as capsule material, binding agent in tablets, stabilizer of emulsions, in hemostatic sponges, and as chemically modified gelatin preparation in blood plasma substitute solutions. The raw material for gelatin production is bones and hides of cattle or pigs. Depending on the source material and production process, different types of gelatin are obtained.

Origin and kind of the source material

In the 1990s, 245,000 tons of gelatin were produced annually worldwide (compare Chapter 43). The amount of gelatin produced by European gelatin manufacturers comprised 110,000 tons per year. This production requires 500,000 tons of raw material obtained from several thousand animals daily [44]. The material used in a gelatin factory is often obtained from different suppliers and originates from different abattoirs.

The careful sourcing of raw materials, where needed, in combination with appropriate processing, has been considered the most important factor for the production of a safe gelatin [21; 45–49], although studies published recently indicate that the normal industrial processes used in gelatin production are likely to eliminate any BSE infectivity that might be contaminating the raw material (see below).

The members of the Gelatine Manufacturers of Europe (GME) have not used bovine raw material of British origin since July 1994. The exclusion of such raw material has been regarded as prerequisite for the safety of gelatin in a statement given by the CPMP in 1996 [50]. At that time, the CPMP adopted a more risk averse position for gelatin than the European Commission, which allowed exceptions in urgent cases [51].

The SSC suggested higher standards with respect to sourcing for pharmaceutical gelatin than for gelatin used in cosmetics because pharmaceutical gelatin may be applied by parenteral administration or as an implanted device [48]. The SSC recommended sourcing the starting materials from GBR I countries (Table 46.2) or from "negligible risk (closed) herds" [31] from other countries.

Gelatin for pharmaceutical purposes[2] is mainly produced from bovine bones or hides. The current CPMP/CVMP Note for Guidance [7] recommends that both raw materials should be obtained from animals that have passed veterinary inspections as fit for human consumption. Bovine hides, irrespective of the animals' country of origin and the manufacturing process, do not pose a TSE risk.

Gelatin produced from bovine bones has a higher risk of contamination by TSE-infected tissues such as brain, spinal cord, and dorsal root ganglia (Fig. 53.2). Thus, the removal of

[2] Gelatina, according to Ph. Eur. Monograph 330.

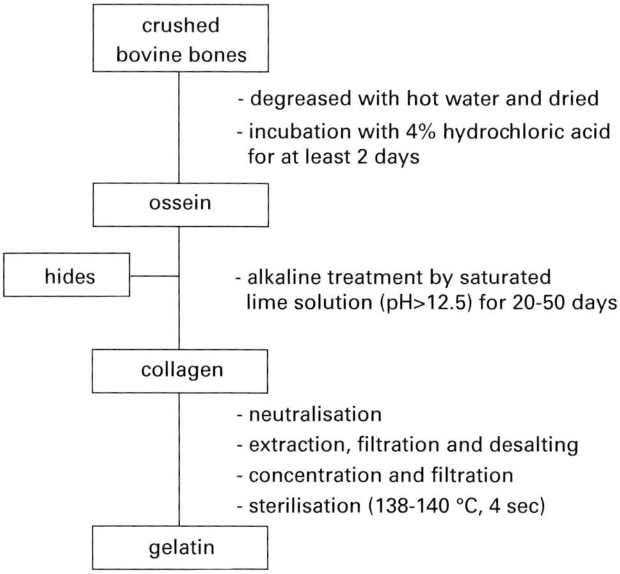

Fig. 46.1: Production of gelatin by a classic acid-and-lime process. Figure by M. Ruffing & H. Windemann ©.

skulls and spinal cords is considered the main factor contributing to the safety of bone gelatin. The removal of these parts during the slaughtering process is common practice in the EU, Switzerland and the USA. In addition, vertebrae should be removed from the bone material if sourced from category III countries. According to the draft revision of the CHMP/CVMP Note for Guidance, vertebrae should also be removed if cattle from GBR II countries are used [52].

The recent risk assessment by the European Food Safety Authority [53] questions the former SSC recommendation [48] to exclude certain bovine bones from the starting material. The human exposure due to gelatin produced from bovine bones including the skull and vertebral column would be very low, as current gelatin manufacturing processes reduce TSE infectivity considerably. However, this risk assessment is restricted to food, i. e., oral intake of gelatin. The current European food legislation [3] defines skulls, spinal cord, and vertebral column as SRM.

Manufacturing of gelatin from bovine bones

As mentioned, in addition to the careful selection of the bones, normal manufacturing processes contribute to the safety of gelatin [52] (see Chapter 43). Validation studies on gelatin manufacturing initiated by the GME in the 1990s were initially criticized because scrapie-infected brains instead of spiked bones were used [54]. Subsequently, further studies have been conducted and different gelatin manufacturing processes compared for their capacity to remove TSE agents. This re-validation has been performed under more realistic conditions. Vertebral columns spiked with BSE agents (mouse-adapted 301V strain) were processed in this extensive study. The results of the study have been published [55–57].

Acid gelatin is obtained by treating degreased bone chips with hydrochloric acid. The resulting ossein is soaked with acid at $pH < 4$. Further purification by filtration, ion exchange, and sterilization also contributes to reduction of infectivity, resulting in a clearance factor of at least 4.8 $\log_{10}$ for the complete acid process. Ossein obtained from bone chips may additionally be incubated with 0.3 M NaOH for 2 hours at pH 13, thereby further increasing the safety of acid gelatin.

In the classical acid-and-lime production process (Fig. 46.1), the acid treatment is followed by incubation with lime, filtration, ion exchange, and sterilization. For this process, a total clearance factor of at least 4.9 $\log_{10}$ has been reported.

The replacement of the liming step by treatment with 0.25–0.30 M NaOH for 5–7 days at $15 \pm 2\,^{\circ}C$ has been shown to inactivate TSE agents even more efficiently.

In the heat / pressure manufacturing process, degreased and dried bone chips were autoclaved with saturated steam at 133 °C, greater than 3,000 hPa [3 bars] for at least 20 minutes. After cooling, gelatin is extracted with hot water (80 °C). Autoclaving and extraction are repeated several times at lower temperatures and/or shorter times depending on the product intended to be obtained. Further processing comprises filtration, centrifugation, and sterilization. The clearance factor of heat / pressure gelatin manufacturing has been determined to be higher than 6.8 $\log_{10}$.

Experimental studies of gelatin production processes have shown a reduction capacity of more than 4.5 $\log_{10}$. In combination with the precautionary principles proposed [52] for the sourcing of bones, the safety of gelatin is deemed to be adequately ensured (see Chapter 43).

46.4.2 Tallow and tallow derivatives

The term "tallow" applies to fats produced by different processes depending on the starting materials and the intended use of the final product. Tallow is mainly produced by extraction from subcutaneous, abdominal, and intramuscular fats, entrails and bones of cattle. Food-standard tallow for human consumption is produced by gentle heat treatment of discrete adipose tissues. It is purified to remove proteinaceous components and may be subjected to a sterilization step. Since the purification of tallow is not 100% effective due to technical limitations, the residual proteins may contain infectivity (prions); because of these facts tallow is – as a precaution – considered as a potential risk in the context of this chapter. While unprocessed (food-standard) cattle tallow is seldom used, tallow derivatives are among the most important excipients in medicinal products and cosmetics. They are used as stabilizers, emulsifiers, and lubricants in tablets.

Tallow derivatives are mostly obtained from raw tallow produced by rendering plants that process animal by-products. This tallow is covered by Regulation 1774/2002, which specifies the requirements of procurement of the raw material and processing conditions for rendering [2]. If tallow or its derivatives is intended to be used in medicinal products, medical devices, or cosmetics, the fat must be sourced from animals that were subjected to ante- and post mortem inspection and declared fit for human consumption. Tallow exclusively for other technical purposes may be obtained from animal material not strictly fulfilling these conditions, but not from SRM. Many recommendations of the SSC [58] were included in the legislation for tallow production.

In validation studies, the ability of different industrial tallow production procedures to reduce TSE infectivity were tested on a pilot scale. No infectivity was detected by mouse bioassay in the final product [59–61].

Tallow derivatives, such as glycerol, fatty acids, and fatty acid esters, are produced by subjecting tallow to different transformation reactions. Some of these processes are considered to be so rigorous that the geographical origin and the nature of the cattle tissues can be neglected [7].

Cleavage of fats into glycerol and fatty acids, as well as the esterification and distillation of fatty acids at temperatures of at least 200 °C under pressure for at least 20 minutes, are rated as safe procedures. If saponification reactions are employed, 12 M sodium hydroxide must be used at 95 °C for at least 3 hours in batch processes or at least 140 °C under pressure for a period of not less than 8 minutes in continuous processes. Distillation at 200 °C is also included.

If these drastic manufacturing procedures are adhered to, tallow derivatives for medical purposes are considered to be safe. The same standards are required by the Cosmetics Directive as amended [16–19]. As a result of the BSE crisis, animal-derived tallow derivatives in medicinal products and cosmetics have tended to be replaced by those of plant origin.

Table 46.3: Examples of substances of bovine origin used in medical devices.

Substance/tissue	Bovine organ/tissue	Application
• Pericard	• Pericard	• Patches for substitution and supplement of connective tissue including dura mater substitution
• Vascular grafts	• Blood vessels (veins)	• Vascular surgery
• Bone matrix (spongiosa compacta)	• Bones	• Orthopedic and dental surgery
• Collagen	• Hides, tendons, (bones)	• Hemostatic sponges, substitution of soft tissue
• Gelatin	• Hides, bones	• Coating material for grafts

46.5 Regulations to prevent the transmission of prions by medical devices

The relevant medical devices are registered according to Directive 93/42/EEC [62] by Notified Bodies and must meet the essential requirements in a "conformity assessment procedure". Notified Bodies are facilities under private law that have a license by a Member State of the European Economic Area (EEA: EU plus Norway, Lichtenstein, Iceland) to conduct the conformity assessment procedure. Directive 93/42/EEC was amended by Directive 2003/32/EC [63] for laying down the TSE risk assessment procedure. As with the procedure for medicinal products, tight deadlines for a harmonized TSE assessment were given. Since April 1, 2004, new applications must be in conformity with this directive. For products already on the market, the TSE assessment had to be conducted by September 30, 2004. The following risk evaluation procedure is provided:

- The TSE documentation of the manufacturer is evaluated by the Notified Body after consideration of the clinical benefit and possible suitable alternatives.
- The Notified Body submits its evaluation to its competent national authority, which coordinates (with the competent authorities of the other Member States of the EEA) the so-called procedure of verification of compliance, i. e., all Member States participate in the assessment procedure on the opinion of the Notified Body.
- Finally the decision of issuing a CE certificate (European document of approval of technical products including medical devices) remains in the competence of the Notified Body.

In addition to the general advice (risk analysis and risk management) given in the annex of the directive, the guidelines provide further assistance. The document "Guidelines on Medical Devices" [64] is helpful for manufacturers and Notified Bodies in the interpretation of Directive 2003/32/EC. A reference is made to the European Standard EN 12442 "Animal tissues und their derivatives for the manufacture of medical devices" [65]. This standard and its draft were the central document for TSE risk evaluation before the directive took effect. For substances where a TSE certificate issued by EDQM exists (e. g., gelatin, collagen), these can be used as a basis for risk assessment. Further examples of substances of bovine origin used in medical devices are summariesed in Table 46.3.

Most of the relevant medical devices are implanted into the human body and persist there for a long time. Therefore, the safety measures applied must be essentially the same as required for parenterally administered medicinal products. In many cases the manufacturing process is unlikely to result in any substantial reduction of potential TSE infectivity. In such cases, the safety of the medical device relies on the safe origin of the raw materials used. According to an opinion of the Scientific Steering Committee [66], medical devices manufactured from bovine materials sourced from GBR I countries can be considered as safe, but this statement would not exclude other countries. As with medicinal products, bovine-derived materials used in the manufacture of medical devices are increasingly being replaced by alternatives.

References

[1] Kimberlin RH. European Commission Doc. IV/1838/94-Rev 1, Annex C. Brussels, 1994.

[2] European Community. Regulation (EC) No 1774/2002 of the European Parliament and of the Council of 3 October 2002 laying down health rules concerning animal by-products not intended for human consumption. Official Journal of the European Communities 2002;L 273:1–95.

[3] European Community. Regulation (EC) No 999/2001 of the European Parliament and of the Council of 22 May 2001 laying down rules for the prevention, control and eradication of certain transmissible spongiform encephalopathies. Official Journal of the European Communities 2001;L 147:1–40.

[4] European Community. Regulation (EC) No 726/2004 of the European Parliament and of the Council of 31 March 2004 laying down community procedures for the authorization and supervision of medicinal products for human and veterinary use and establishing a European Medicines Agency. Official Journal of the European Union 2004;L 136:1–33.

[5] European Community. Directive 2001/82/EC of the European Parliament and of the Council of 6 November 2001 on the community code relating to veterinary medicinal products. Official Journal of the European Communities 2001;L 311:1–66.

[6] European Community. Directive 2001/83/EC of the European Parliament and of the Council of 6 November 2001 on the community code relating to medicinal products for human use. Official Journal of the European Communities 2001;L 311:67–128.

[7] European Commission. Note for guidance on minimizing the risk of transmitting animal spongiform encephalopathy agents via human and veterinary medicinal products (EMEA/410/01 Rev. 2 – October 2003). Adopted by the Committee for Proprietary Medicinal Products (CPMP) and by the Committee for Veterinary Medicinal Products (CVMP). Official Journal of the European Union 2004; 2004/C 24/03:6–18.

[8] Johnson RT. Real and theoretical threats to human health posed by the epidemic of bovine spongiform encephalopathy. In: Gibbs CJ Jr, editor. Bovine spongiform encephalopathy – the BSE dilemma. Springer, New York, 1996:359–363.

[9] CPMP. Guidelines for minimizing the risk of transmitting agents causing spongiform encephalopathy via medicinal products (Guidelines). Committee for Proprietary Medicinal Products (CPMP). EEC regulatory document. Brussels, Belgium, 1991. Biologicals 1992; 20:155–158.

[10] Will RG, Ironside JW, Zeidler M, et al. A new variant of Creutzfeldt–Jakob disease in the UK. Lancet 1996; 347:921–925.

[11] European Commission. Commission Decision 96/239/EC of 27 March 1996 on emergency measures to protect against bovine spongiform encephalopathy. Official Journal of the European Communities 1996;L 78:47–48.

[12] European Community. Council Decision 98/256/EC of 16 March 1998 concerning emergency measures to protect against bovines spongiform encephalopathy, amending Decision 94/474/EC and repealing Decision 96/239/EC. Official Journal of the European Communities 1998;L 113:32–43.

[13] Lasmezas CI, Deslys JP, Demalmay R, et al. BSE transmission to macaques [Letter]. Nature 1996; 381:743–744.

[14] Collinge J, Sidle KC, Meads J, et al. Molecular analysis of prion strain variation and the aetiology of 'new variant' CJD. Nature 1996; 383:685–690.

[15] WHO. Zoonoses control – bovine spongiform encephalopathy. Wkly Epidemiol Rec 1996; 71(11):83–85.

[16] European Community. Council Directive 76/768/EEC of 27 July 1976 on the approximation of the laws of the Member States relating to cosmetic products. Official Journal of the European Communities 1976;L 262:1.

[17] European Commission. Twentieth Commission Directive 97/1/EC of 10 January 1997 adapting to technical progress Annexes II, III, IV and VII to Council Directive 76/768/EEC on the approximation of the laws of the Member States relating to cosmetic products. Official Journal of the European Communities 1997;L 16:85–86.

[18] European Commission. Twenty-second Commission Directive 98/16/EC of 5 March 1998 adapting to technical progress Annexes II, III, IV and VII to Council Directive 76/768/EEC on the approximation of the laws of the Member States relating to cosmetic products. Official Journal of the European Communities 1998;L 77:44–46.

[19] European Commission. Commission Directive 2003/1/EC of 6 January 2003 adapting to technical progress Annex II to Council Directive 76/768/EEC on the approximation of the laws of

the Member States relating to cosmetic products. Official Journal of the European Communities 2003;L 5:14–15.

[20] European Commission. Commission Directive 1999/82/EC of 8 September 1999 amending the Annex to Council Directive 75/318/EEC on the approximation of the laws of the Member States relating to analytical, pharmacotoxicological and clinical standards and protocols in respect of the testing of medicinal products. Official Journal of the European Communities 1999;L 243:7–8.

[21] CPMP. Note for guidance on minimising the risk of transmitting animal spongiform encephalopathy agents via medicinal products (guidelines). Committee for Proprietary Medicinal Products. Brussels: CPMP/BWP/1230/98, Revision October 1998.

[22] European Commission. Commission Directive 1999/104/EC of 22 December 1999 amending the Annex to Council Directive 81/852/EEC on the approximation of the laws of the Member States relating to analytical, pharmacotoxicological and clinical standards and protocols in respect of the testing of veterinary medicinal products. Official Journal of the European Communities 2000;L 3:18–20.

[23] Council of Europe PHC. Resolution AP CSP (99) 4 adopted by the Public Health Committee (Partial Agreement) (CD-P-SP) on 22 December 1999. Certification of suitability to the monographs of the European Pharmacopoeia (revised version). 1999.

[24] European Pharmacopoeia. 5th edition published by the European Directorate for the Quality of Medicines of the Council of Europe, Strasbourg, France, 2004.

[25] EDQM. European Directorate for the Quality of Medicines. Webpage 2006: www.pheur.org/site/page_628.php.

[26] European Commission. Commission Directive 2003/63/EC of 25 June 2003 amending Directive 2001/83/EC of the European Parliament and of the Council on the community code relating to medicinal products for human use. Official Journal of the European Union 2003;L 159:49–94.

[27] EMEA. European Agency for the Evaluation of Medicinal Products. Gelatin for use in pharmaceuticals: explanatory note (13 December 2000) on the manufacture of gelatin in relationship to the CPMP note for guidance on minimizing the risk of transmitting animal spongiform encephalopathy agents via medicinal products (CPMP/BWP/1230/98 Rev 1). 2000.

[28] European Commission. Update of the opinion of the Scientific Steering Committee on the geographical risk of bovine spongiform encephalopathy (GBR) [report]. Adopted on January 11, 2002.

[29] EFSA. European Food Safety Authority. Webpage 2006: www.efsa.eu.int/science/tse_assessments/gbr_assessments/catindex_en.html.

[30] European Commission. Opinions on the geographical risk of bovine spongiform encephalopathy (GBR) and official reports on the assessment of the geographical BSE-risks. Scientific Steering Commitee (SSC). Webpage 2006: http://europa eu int/comm/food/fs/sc/ssc/outcome_en html.

[31] European Commission. Scientific opinion on the conditions related to "BSE negligible risk (closed) bovine herds". Adopted by the Scientific Steering Committee at its meeting of July 22–23, 1999.

[32] Office International des Epizooties (OIE). BSE chapter. World Organisation for Animal Health. Terrestrial Animal Health Code 2005: Chapter 2.3.13. [Reprint in Appendix 2 of this book].

[32a] Office International des Epizooties (OIE). Scrapie chapter. World Organisation for Animal Health. Terrestrial Animal Health Code 2005: Chapter 2.4.8. [Reprint in Appendix 3 of this book].

[33] European Commission. The TSE Road Map, COM (2005) 322 final. Brussels, July 15, 2005.

[34] EFSA. European Food Safety Authority, Scientific Panel on Biological Hazards: Statement on the assessment of the safety with respect to the consumption of goat meat and goat meat products in relation to BSE/TSE. Brussels, January 28, 2005.

[35] Hadlow WJ, Kennedy RC, Race RE. Natural infection of Suffolk sheep with scrapie virus. J Infect Dis 1982; 146:657–664.

[36] Hadlow WJ, Kennedy RC, Race RE, et al. Virologic and neurohistologic findings in dairy goats affected with natural scrapie. Vet Pathol 1980; 17:187–199.

[37] WHO. Guidelines on transmissible spongiform encephalopathies in relation to biological and pharmaceutical products. WHO/BCT/QSD/03.01, February 2003.

[38] WHO Guidelines on Tissue Infectivity Distribution in Transmissible Spongiform Encephalopathies. ISBN 92-4-154701-4. Report of the WHO Consultation in Geneva, 14–16 Sept 2005. [Appendix 1: reprinted in this book]. See also Wepage 2006: www.who.int/bloodproducts/TSEREPORT-LoRes.pdf

[39] European Commission. Commission Regulation (EC) No 657/2006 of 10 April amending Regulation (EC) No 999/2001 of the European Parliament and of the Council as regards the United Kingdom and repealing Council Decision 98/256/EC and Decisions 98/351/EC and 1999/514/EC. Official Journal of the European Union 2006;L 116:9–12.

[40] EFSA. Statement of the BIOHAZ Panel RE: Technical advice on the United Kingdom application for application of moderate risk in terms of bovine spongiform encephalopathy. European Food Safety Authority, March 15, 2005.

[41] European Commission. Opinion on stunning methods and BSE risk. The risk of dissemination of brain particles into the blood and carcass when applying certain stunning methods [report]. Adopted by the Scientific Steering Committee at its meeting of January 10–11, 2002.

[42] Anderson RM, Donnelly CA, Ferguson NM, et al. Transmission dynamics and epidemiology of BSE in British cattle. Nature 1996; 382:779–788.

[43] Kimberlin RH and Walker C. Incubation periods in six models of intraperitoneally injected scrapie depend mainly on the dynamics of agent replication within the nervous system and not the lymphoreticular system. J Gen Virol 1988; 69: 2953–2960.

[44] Schrieber R and Seybold U. Gelatine production, the six steps to maximum safety. Dev Biol Stand 1993; 80:195–198.

[45] European Commission. Opinion on quantitative risk assessment on the use of the vertebral column for the production of gelatin and tallow [report]. Adopted by the Scientific Steering Committee at its meeting of April 13–14, 2000.

[46] European Commission. Opinion on the questions submitted by EC services following a request of 4 December 2000 by the EU Council of Agricultural Ministers regarding the safety with regard to BSE of certain bovine tissues and certain animal-derived products. Opinion of the Scientific Steering Committee adopted on January 12, 2001.

[47] European Commission. Updated opinion on the safety with regard to TSE risks of gelatin derived from ruminant bones or hides [report]. Adopted by the Scientific Steering Committee at its meeting of December 5–6, 2002.

[48] European Commission. Updated opinion on the safety with regard to TSE risks of gelatin derived from ruminant bones or hides. Adopted by the Scientific Steering Committee at its meeting of March 6–7, 2003.

[49] FDA. Guidance for industry: the sourcing and processing of gelatin to reduce the potential risk posed by bovine spongiform encephalopathy (BSE) in FDA-regulated products for human use. Food and Drug Administration, USA, 1997.

[50] EMEA. Opinion of the EMEA on the potential risk associated with medical products in relation to bovine spongiform encephalopathy. European Agency for the Evaluation of Medicinal Products, 1996.

[51] European Commission. Commission Decision 96/362/EC of 11 June 1996 amending Decision 96/239/EC on emergency measures to protect against bovine spongiform encephalopathy. Official Journal of the European Communities 1996;L 139:17–20.

[52] EMEA. Committee for medicinal products for human use (CHMP), committee for medicinal products for veterinary use (CVMP): Note for guidance on minimizing the risk of transmitting animal spongiform encephalopathy agents via human and veterinary medicinal products (amendments to sections 6.2 and 6.3). European Agency for the Evaluation of Medicinal Products, EMEA/410/01/rev.3 draft, 2004.

[53] Opinion of the Scientific Panel on Biological Hazards of the European Food Safety Authority on the "Quantitative assessment of the human BSE risk posed by gelatine with respect to residual BSE risk". EFSA Journal 2006; 312:1–28.

[54] European Commission. Opinion on the safety of gelatine – adopted at the Scientific Steering Committee. Plenary meeting. March 26–27, 1998.

[55] Grobben AH, Steele PJ, Somerville RA, et al. Inactivation of the bovine-spongiform-encephalopathy (BSE) agent by the acid and alkaline processes used in the manufacture of bone gelatine. Biotechnol Appl Biochem 2004; 39(3):329–338.

[56] Grobben AH, Steele PJ, Taylor DM, et al. Inactivation of the BSE agent by heat and pressure process for manufacturing gelatine. Vet Rec 2005; 157:277–289.

[57] Grobben AH, Steele PJ, Somerville RA, et al. Inactivation of transmissible spongiform encephalopathy agents during the manufacture of dicalcium phosphate from bone. Vet Rec 2006; 158:361–366.

[58] European Commission. Revised opinion and report on the safety of tallow obtained from ruminant slaughter by-products. Adopted by the Scientific Steering Committee at its meeting of June 28–29, 2001.

[59] Taylor DM, Woodgate SL, Atkinson MJ. Inactivation of the bovine spongiform encepha-

lopathy agent by rendering procedures. Vet Rec 1995; 137:605–610.

[60] Taylor DM, Woodgate SL, Fleetwood AJ, et al. The effect of rendering procedures on scrapie agent. Vet Rec 1997; 141:643–649.

[61] MAFF. Inactivation of the BSE and scrapie agents during the rendering process. Final report of the study contract No. 8001 [report]. Ministries of Agriculture and Fisheries, UK, 1997.

[62] European Community. Council Directive 93/42/EEC of 14 June 1993 concerning medical devices. Official Journal of the European Communities 1993;L 169:1–43.

[63] European Commission. Commission Directive 2003/32/EC of 23 April 2003 introducing detailed specifications as regards the requirements laid down in Council Directive 93/42/EEC with respect to medical devices manufactured utilising tissues of animal origin. Official Journal of the European Union 2003;L 105:18–23.

[64] European Commission. Guidelines on medical devices: application of Council Directive 93/42/EEC taking into account the Commission Directive 2003/32/EC for medical devices utilizing tissues or derivatives originating from animals for which a TSE risk is suspected. MEDDEV.2.11 Rev. April 1, 2005.

[65] CEN. European Standard EN 12442. Animal tissues and their derivatives utilized in the manufacture of medical devices Parts 1, 2 & 3. European Committee for Standardization, 2001.

[66] European Commission. Opinion on sourcing of ruminant materials from GBR I countries for medical devices – originally adopted by the Scientific Steering Committee at its meeting of 6–7 September 2001. Update adopted on November 29–30, 2001.

47 Prevention of the Transmission of Prion Diseases in Healthcare Settings

Beat Hörnlimann, Georg Pauli, Karin Lemmer, Michael Beekes, and Martin Mielke

47.1 Introduction

This chapter defines groups of patients, tissues, and medical procedures – all of which carry different risks of transmission of prion diseases [1]. Measures to prevent the accidental or iatrogenic transmission of transmissible spongiform encephalopathies (TSEs or CJD[1]) in healthcare settings as suggested by international (WHO) [2] as well as national recommendations from Great Britain [3–7], France [8–10], Switzerland [11–13], Germany [14; 15], Australia [16; 17], and the USA [18; 19] are summarized.

As somewhat different approaches – according to the specific epidemiological situation – can be found in various countries, this chapter will focus on basic principles. Therefore, it is of vital importance that the safety measures recommended for laboratory, hospital, and home care settings are continuously adapted to national conditions and newly available scientific knowledge.

With respect to the precautionary measures to be taken when caring for patients suffering from a prion disease or for those suspected or at risk of being infected, it should be noted that prion diseases are not transmitted via common social contacts. Iatrogenic transmission to humans so far has only been described with highly infective tissue (such as nerve tissues, CNS) and via highly efficient routes, such as intracerebral (i.c.) or intravenous (i.v.) application. While, by definition, one 50% i.c. infective dose ($ID_{50i.c.}$) is sufficient to infect animals intracerebrally with a probability of 50%, i.v. application of the same amount of a given TSE agent strain is 10 times less effective, i.e., requires approximately 10 ID_{50}, the intraperitoneal (i.p.) infection 100 ID_{50}, the subcutaneous infection 10^4 ID_{50}, and infection via the oral route approximately 10^5 ID_{50}. These figures have been obtained by experimental transmission within the same species. Transmission across species barriers generally requires higher doses.

The development and implementation of measures to cope with the potential risk of iatrogenic vCJD transmission has been a tremendous challenge for health care policy makers. Addressing this task, decision makers have to weigh the expected benefits of precautionary measures against possible negative effects in other domains of the public health system, e.g., among other factors, binding financial resources. Such weighing is often difficult given the still existing gaps in the knowledge of the epidemiology and pathogenesis of vCJD. This emphasizes the importance of further research into the complex subject of TSEs, although elaborate scientific activities have already provided a much better understanding during the past few years. However, there is still no practical diagnostic test available which would allow preclinical diagnosis of TSEs.

The potential risk of vCJD transmission within the human population, e.g., through blood, blood products or surgical instruments contaminated with neuronal or lymphatic tissue led to a series of prophylactic measures, which should provide effective protection against an inadvertent and unpredictable secondary spread of vCJD.

[1] Unless otherwise stated, the term "CJD" in this article encompasses all human prion diseases.

47.2 Patient care

Based on currently available data, hospital and home care of patients suffering from CJD or vCJD is not associated with a risk of transmission, provided that standard precautions of infection control are followed. This is also true for the handling of patient secretions or excreta. Exposure of intact skin to TSE agents can be considered as a "negligible" risk, but should nevertheless be avoided by barrier precautions. So far, no case has been reported in which a prion disease has been transmitted through contact with a TSE patient in a healthcare setting. Spills of blood should be treated according to standard measures for the prevention of common blood-borne infections.

Dishes and cutlery used by a TSE patient, as well as washing utensils, clothing and bed linen can be conventionally washed or handled according to standard hygienic measures.

In case of contamination with cerebrospinal fluid (CSF), disinfection with 2.5–5% sodium hypochlorite or 1–2 M sodium hydroxide solution for a period of 1 hour is recommended prior to normal cleaning (compare Chapter 44; Table 44.2).

Waste originating from noninvasive care of patients may be disposed of as conventional hospital waste.

47.3 Risk of accidental occupational transmission in nosocomial and other healthcare settings

No unequivocal cases of accidental occupational transmission in a nosocomial environment, including neuropathological laboratories, have been reported in the literature to date. There is no epidemiological evidence suggesting that the incidence of CJD is higher in medical professionals than in the general population; this holds true even for specific hospital environments and laboratories (including neurosurgical and neuropathological facilities), where exposure to TSE agents may be above average [20–25]. However, as Robert G. Will has emphasized, from an epidemiologist's perspective, only long-term internationally coordinated monitoring programmes will be able to shed further light on this question [25]. Although "classical" as well as BSE-related human TSEs have occurred in a variety of professionals potentially exposed to TSE agents (e. g., physicians, nurses, dentists, dental technicians, veterinarians, butchers, farmers, or cooks), retrospective investigations have largely failed to establish the workplace as a possible source of infection. Only three case reports of CJD, one in a neurosurgeon [26; 27] and two laboratory workers [28], imply that the disease may be attributed to occupational exposure to the infectious agent.

47.4 Iatrogenic transmission of human TSEs: retrospective findings and current risk assessment

So far, transmissions of TSEs between humans, with the exception of kuru, have only been observed in the context of iatrogenic cases. All cases of iatrogenic CJD could be attributed to implants or transplants of highly infectious tissue (dura mater, cornea), parenteral treatment with contaminated medications such as growth hormones or gonadotropin extracted from pituitary glands of cadavers, or neuroinvasive medical interventions using contaminated and inadequately reprocessed instruments. Until 2006, the following cases of iatrogenic CJD were reported [29]: 191 cases following growth hormone treatment, 4 following treatment with a sex hormone, 169 following implantation of dura mater, 4 following corneal transplantation, 5 following neurosurgery with insufficiently sterilized surgical instruments, and 2 in the context of insufficiently sterilized intracerebral EEG electrodes [25; 26; 30; 31] (see Table 14.1).

Thus, the majority of iatrogenic CJD cases that occurred in the past were caused by treatment with growth hormones, which were prepared from the pituitary gland of human cadavers. From 1963 until 1985, approximately 10,000 [32] recipients were given such medication. Although 8 M urea was used to process pituitary-derived growth hormone prior to its use in humans, this was found to be insufficient

Table 47.1: Classification of patients with different transmissible spongiform encephalopathy (TSE) transmission risks (prion diseases: for case definitions of definite, probable and possible TSEs please refer to Tables 29.1, 29.3, 31.1).

Group I: Patients with a high risk of contracting a TSE and/or developing a TSE / symptomatic patients:
- Patients with definite TSE
- Patients with clinical suspicion (probable or possible TSE)
- Carriers of pathogenic mutations in the gene encoding for the prion protein (*PRNP*)
- Members of a family with familial CJD (fCJD), GSS, FFI, even if the genotype has not been determined
- Patients suffering from symptomatic neurological disease of unknown etiology where the diagnosis of CJD is being actively considered

Group II: Patients with a heightened risk of having a TSE or developing a TSE:
- Patients with an unclarified, rapidly progressing disease of the CNS with or without dementia
- Members of families in which such kind of diseases have often occurred
- Recipients of human pituitary gland hormones (growth hormone and gonadotropines)
- Recipients of dura mater implants in the years 1972–1987

Group III: Patients with a low risk of suffering from a TSE or developing a TSE:
- All other people

Table 47.2: Tissue distribution of infectivity. (Compare WHO tables$_{\text{Appendix}}$ IA–IC).

Tissue	Presence of abnormal prion protein (PrP^{Sc}) and level of infectivity			
	CJD other than vCJD		vCJD	
	PrPres detected	Assumed level of infectivity	PrPres detected	Assumed level of infectivity
Brain	+	High	+	High
Spinal cord	+	High	+	High
Spinal ganglia	+	High	+	High
Dura mater	NT	High	NT	High
Cranial nerves	+	High	+	High
Cranial ganglia	+	High	+	High
Posterior eye	+	High	+	High
Anterior eye and cornea	–	Medium	–	Medium
Olfactory epithelium	+	Medium	NT	Medium
Tonsil	–	Low	+	Medium
Appendix	–	Low	+	Medium
Spleen	+*	??	+	Medium
Thymus	–	Low	+	Medium
Other lymphoid tissues	–	Low	NT	Medium
Peripheral nerve	–	Low	–	Low
Dental pulp	–	Low	NT	Low
Gingival tissue	NT	Low	NT	Low
Blood and bone marrow	NT	Low	NT	Low
CSF	–	Low	–	Low
Placenta	NT	Low	–	Low
Urine	NT	Low	NT	Low
Skeletal muscle	+*/**	??	+**	??

Notes: + = tested positive, – = tested negative, NT = not tested
* Reference [44]
** Reference [45]

for the inactivation of inadvertent contaminations with CJD agent [26].

As prophylactic measures specifically directed against the identified sources and routes of transmission underlying iatrogenic CJD, it was recommended that pituitary hormones of human origin should be replaced by other therapeutics or alternative treatments if available (since 1987 growth hormones and gonadotropins have been manufactured in vitro) [33]. Moreover, it was recommended that individuals with an elevated risk for CJD should be excluded as blood, tissue, or organ donors.

Iatrogenic cases of CJD show that risk has to be ascertained for a wide spectrum of possible medical scenarios. This can be done by assessing (i) patients (see Section 47.4.1; Table 47.1), (ii) tissues and body fluids (Table 47.2), and (iii) routes of exposure (see Section 47.4.2), for their infectious potential leading to a segregation of different risk categories. This enables those responsible for risk management to minimize the hazard of nosocomial transmission of TSEs [1].

47.4.1 Risk stratification of patients

A potentially elevated risk of human-to-human transmission is inherent in invasive medical interventions performed on:

1. symptomatic patients with a definite diagnosis of TSE or patients with clinical symptoms of a probable or possible TSE as defined according to published diagnostic criteria (group I) or
2. asymptomatic patients at risk of developing a TSE due to genetic mutations (familial forms of TSE) or previous iatrogenic exposure (group II; Table 47.1).

In Germany, for instance, the latter group (group II) consists of patients who received hormonal extracts from human pituitaries or who underwent dura mater implantation (prior to August 1992), as well as patients with a known risk for familial TSE (two or more blood relatives affected by CJD, another TSE, or a relative known to have a genetic mutation indicative of familial CJD or another genetic form of TSE such as GSS). The probability of infection in recipients of pituitary gland hormones is about 1 %, while recipients of dura mater implants have a probability of approximately 0.1 % [34].

Considering the unknown number of persons incubating vCJD the ACDP and SEAC extended group I by a further risk group in the UK, in 2003, i. e., patients with neurological disease of unknown etiology and who do not fit the criteria for possible CJD, but where a diagnosis of CJD is being actively considered. Furthermore, a risk group consisting of patients who have been exposed to instruments used on, or received blood, plasma derivatives, organs or tissues donated by a patient who later went on to develop CJD or vCJD, has been added to group II in the UK.

Compared to the group of patients presenting with symptoms or a suspect history, the collective of neurologically asymptomatic patients without any recognizable risk for CJD or vCJD (group III) is much higher. Patients belonging to this group may develop sCJD with an incidence of 1.0–2 cases per million inhabitants/year. The frequency of so far asymptomatic vCJD infections due to ingestion of BSE agent via the alimentary route cannot be precisely assessed [35–37].

47.4.2 Risk stratification of tissues and routes of exposure

Knowledge about the different infectivity of various tissues helps to identify possible sources of contamination and to assess the potential risk of transmission associated with specific medical interventions and treatments. Table 47.2 gives an overview on tissue distribution of infectivity (see also WHO Tables$_{\text{Appendix}}$IA–IC) [2]; the TSE agent has been detected in classical CJD, GSS, and FFI by bioassay most frequently and at the highest titers in the brain, spinal cord, and eye. Less frequently and in medium- or low-range titers, infectivity could also be detected in CSF, lungs, liver, kidney, spleen, placenta, lymph nodes, and olfactory epithelium [38–40]. In body fluids other than CSF or in excrement, TSE agent could not be detected. Controversial results exist concerning the presence of CJD agent in blood. Human-to-human transmission of classical CJD (or GSS and FFI) through blood or blood products has not been observed to date. Inoculation of blood or blood components from CJD patients into animals revealed no, or extremely low, infectious potential (compare Chapter 51). At present, the general consensus is that blood of patients with classical CJD may be considered as noninfectious material, at least in the context of infection control. However, the situation is different with respect to vCJD since it has been reported that blood from vCJD-infected humans has probably transmitted, in three cases so far, the vCJD agent from human to human [41–43].

In contrast to classical CJD, in which pathological prion protein as a biochemical marker for infectivity is hardly ever detectable outside the central nervous system (CNS) apart from spleen and muscle tissue [44; 45], it can be found in vCJD in a variety of peripheral tissues in addition to the CNS and eye (retina and posterior segment). Here, PrP^{Sc} has been detected consistently in tonsils (46–48) as well as in other components of the lymphatic system (spleen, lymph nodes, thymus, and appendix-associated lymphoid tissue) [46; 48], the optic nerve [46], in some parts of the peripheral nervous system [49] and in muscle tissue [45].

Adrenal glands and rectum were also positively tested for PrP^{Sc} [46]. Pathological prion protein was also found in two preserved specimens of the appendices from two vCJD patients who underwent appendectomy 8 and 24 months prior to the onset of the disease [50]. In addition, a more extended recent study found abnormal PrP accumulation in 3 out of 12,674 randomly tested appendix and tonsil samples [51].

The involvement of the lymphoreticular tissue in vCJD patients, together with experimental results pointing to a crucial – and meanwhile more comprehensively elucidated [52] – role of B cells in the neuroinvasion of TSE agents [53], and compelling evidence for the presence of infectivity in buffy coat or plasma of mice infected with a mouse-adapted strain of human TSE [54; 55], have raised fears that blood of vCJD patients could be infectious. Bioassay experiments have not been able to demonstrate infectious agent in the plasma or buffy coat of vCJD patients [56], nor was it possible to detect PrP^{Sc} in concentrated buffy coat by sensitive Western blotting [46]. Nonetheless, a risk of transmission by blood cannot be ruled out; recently three British patients developed vCJD after receiving a blood transfusion from a donor who developed vCJD after blood donation [41–43].

Thus, taken together, invasive interventions on the CNS (brain and spinal cord), the posterior segment of the eye, olfactory epithelium, and the organized lymphatic tissue bear the greatest risk for contamination of medical devices with human TSE agents (see Chapter 46). This is consistent with the observation that contact with these tissues of potentially high infectivity was common to all published iatrogenic transmissions of classic forms of human TSEs.

47.5 Preventive measures for handling CSF and tissue samples

Lumbar punctures, excisional biopsies, and punctures of inner organs and cavities should always be carried out by experienced personnel observing precautionary measures routinely used for the prevention of blood-borne infections like HBV, HCV, or HIV infection.

Special care should be taken to avoid incisions and stabs. Because of the danger of disseminating patient material, mouth (oral mucosa), and eyes (conjunctiva) should be protected from spills.

Disposable materials used for sampling must be disposed of with the medical waste, sharp and sharp-tipped items must be disposed of in closed suited containers (see Section 47.12). Blood samples should be handled according to routinely used measures able to prevent blood-borne infections.

47.6 Transmission through blood and blood products

Currently, blood and blood products are thought to pose no increased risk in the context of accidental or iatrogenic transmission of classical CJD. However, vCJD transmission by this route has been reported thus far (see Chapter 51). This has led to various precautionary recommendations and measures. In Germany for example, these include the removal or depletion of leukocytes from blood donations and the exclusion of blood or plasma donors who have spent more than a total of 6 months in the United Kingdom between 1980 and 1996 as well as donors with a risk of infection with human TSE. Details can be found for example in a recent report published by the German Federal Ministry of Health on the strategy of blood supply after the emergence of vCJD [57], prepared by a commission consisting of governmental representatives of the Paul Ehrlich Institute and Robert Koch Institute, as well as of external experts. This report also outlines new approaches aiming at further minimising the risk of vCJD transmission, such as more stringent criteria for the "optimal use" of blood and blood products.

47.7 Precautionary measures to minimize the risk of transmission via surgical interventions on patients with an evident or potential risk of CJD or vCJD

Conventional methods of disinfection and sterilization may not sufficiently inactivate TSE agents (see Chapter 44). Prions, in addition, exhibit a high affinity for steel [58]. Therefore, in accordance with current guidelines, invasive procedures on patients with an evident (recognizable) risk of TSE should be avoided. If an invasive procedure on those patients is unavoidable (Table 47.3), the use of disposable materials and instruments is recommended whenever possible. In case that this is impossible, all instruments and medical devices used in patients with an evident risk of CJD (groups I and II), and which may have been in contact with potentially infectious tissues (Table 47.2), must be processed after use, observing special precautions, which encompass (i) safe disposal by incineration, (ii) quarantine pending diagnosis and (iii) stringent decontamination procedures. Treatment with 1–2 M sodium hydroxide solution (for 24 h), 2.5–5% sodium hypochlorite solution (for 24 h) as well as 3, 4, or 6 M guanidiniumthiocyanate solution (for 24 h, 1 h or 15 min, respectively) followed by steam sterilization at 134 °C (for 18 min to 1 h) are considered as appropriate for decontamination but may pose special problems concerning material compatibility. Additional aspects that are specific with regard to effective methods for decontamination and inactivation are described in Chapter 44.

The broader distribution of the infectious agent in the body of vCJD patients than that in patients with classical CJD, as well as the potentially widespread alimentary exposure to the BSE agent in the UK, called for a reconsideration of the general practices of cleaning, disinfection, and sterilization of instruments and medical devices. An overview of British safety guidelines and recommendations on sterilization issues in vCJD is given in a report on behalf of the vCJD Consensus Group by Spencer et al. [59] and most recently in a guidance from the Advisory Committee on Dangerous Pathogens (ACDP) and the Spongiform Encephalopathy Advisory Committee (SEAC) [for updated information see 60]. In Germany, a task force initiated by the Robert Koch Institute together with the Scientific Advisory Board of the German Medical Association has developed recommendations for minimizing the risk of vCJD transmission in the context of invasive interventions [15].

Table 47.3: Measures required for unavoidable surgery on patients carrying a high risk of transmitting a TSE (definite, probable, or possible CJD or vCJD; patients at risk for familial or iatrogenic CJD; TSE; prion diseases).

Safety precautions
• The number of people assisting in a well-planned operation – typically at the end of a daily program – must be kept to a minimum
• Only well-trained personnel must undertake surgery
• Use of disposable surface covers (e. g., on the operation table, etc.)
• Use of disposable protective clothing (liquid repellent operating gown, apron) and disposable masks
• Use of two pairs of gloves (double gloving)
• Use of safety goggles or full-face visor
• Correct treatment of indispensable reusable materials (see Section 47.8; Table 44.2); use of single use items whenever possible
• Safe waste disposal (see Chapter 47.12)

In these recommendations the existing guidelines concerning instruments and medical devices used in patients with an evident risk of sporadic CJD have been tightened; there is now demand for the use of disposable instruments and devices, which may have come into contact with tissues with high or medium infectivity (CNS, eye, olfactory epithelium and organized lymphatic tissue). Such kinds of interventions are usually carried out by neurosurgeons, ophthalmologists, and ENT surgeons. This demand may cause technical or economical problems, such as in the case of flexible endoscopes. Therefore, in Germany a national pool of instruments which are available for particular interventions on TSE patients has been established [61].

47.8 Disinfection and sterilization of instruments and materials

According to expert recommendations, all surgical instruments contaminated with CNS, eye, olfactory epithelium, or lymphatic tissue from patients presenting a recognizable risk for CJD (groups I and II) carry the potential for transmission and should consequently be disposed of by incineration or stored in a rigid, sealed container until the diagnosis is confirmed (quarantine). All other instruments in contact with less infective tissues may be reprocessed according to the best currently available practice (see Section 47.9). Disposable instruments should be used whenever possible.

Based on epidemiological data, as well as on risk assessment and stratification, several European countries have developed risk management guidelines to minimize the hazard of iatrogenic transmission of CJD/vCJD through surgical instruments and medical devices. Due to the specific situation (e. g., degree of exposure to BSE-related agents) and existing safety regulations in various countries, these guidelines differ to some degree, but share the following considerations:

1. Consider the current guidelines on the control of infection by TSE agents in nosocomial settings jointly published by the ACDP and the SEAC: *TSE agents: Safe working and the prevention of Infection*. Among other health authorities, the British Department of Health has issued special guidance on the importance of cleaning to minimize the risk of transmission of TSE agents via nondisposable surgical instruments and medical devices.
2. Consult specialists for infection control whenever it is possible that a patient undergoing an operation or invasive medical intervention is "at risk" of having CJD or vCJD, or whenever a diagnosis of CJD is actively considered.
3. In these patients, use disposable surgical instruments and medical devices whenever practicable. However, this must not affect the results of the medical intervention or treatment. Discard surgical instruments and medical devices which are impossible to clean.
4. Quarantine all nondisposable equipment exposed to high or medium risk tissues applied on patients who are at risk of having CJD or vCJD until a clear diagnosis has been established.
5. Avoid any fixation of blood, mucus or tissues on the surface of nondisposable surgical instruments or medical devices by drying, exposure to fixating disinfectants (e. g., aldehydes, alcohols) or heat before cleaning. Fixation can be avoided by immediate incubation in a suitable cleaning solution (e. g., an alkaline cleaner containing detergents and possibly nonfixating disinfectants) [62; 63].
6. Reprocess nondisposable instruments and devices exposed to low risk tissues in a (central) facility suitable for the cleaning, disinfection, and sterilization of surgical instruments and medical devices which are (potentially) contaminated with TSE agents according to best practice (see Chapter 44).

47.9 Decontamination of instruments following surgery on patients without any specific signs or symptoms pointing to a risk of transmission

Due to the French epidemiological situation (Table 14.1, tablenote 3) it was recommended in France that all reusable thermostable surgical instruments be sterilized for 18 min at 134 °C in a porous load steam sterilizer irrespective of whether the patients present any particular signs or symptoms. Most other countries have not yet adopted this suggestion.

For patients without an evident risk of classical CJD (group III, Table 47.1), the likelihood of harboring or developing the disease, and therefore the hazard of transmitting the disease, is extremely low (worldwide incidence of sporadic CJD: 1–2 cases per 1,000,000 inhabitants/year). Concerning the prevention of iatrogenic transmission of classical CJD, it was deemed unnecessary to submit all instruments or medical devices to a decontamination and/or inactivation procedure aiming at TSE agents in addition to conventional infectious agents.

However, in light of the uncertain extent of alimentary exposure of the general population to the BSE agent, the potential transmission of vCJD from thus far asymptomatic carriers had to be reconsidered. For various reasons (e. g., insufficient epidemiological data) the hazardous potential of vCJD cannot yet be as clearly assessed as that of classical CJD. In particular, it is currently impossible, and will most probably remain so in the near future, to draw precise conclusions on the number of asymptomatic carriers or unrecognized cases of vCJD. The situation is even more complicated by the fact, that PrP^{Sc} – and thus vCJD infectivity – is already present in lymphatic tissues, such as the appendix and the tonsils, during preclinical phases [35; 48; 50; 64]. A recent study found PrP^{Sc} accumulation in 3 out of 12,674 randomly tested appendix and tonsil samples [51]. An extrapolation of these findings provides an estimate that approximately 3,800 individuals in the age group 10–30 years within the UK population may currently be incubating the disease.

In 2001, the UK Department of Health published a risk assessment for the transmission of vCJD by surgical instruments, which has been updated in 2005 [65; 37]. This assessment was founded on the premise that optimal routine decontamination procedures may lead to a limited reduction of infectivity. Accordingly, it was assumed that certain kinds of medical interventions are associated with an elevated risk for iatrogenic and accidental transmission of vCJD. In particular, these include interventions performed on tissues, which bear an inherent risk of contamination of surgical instruments or medical devices, with an initial load of 10^5 ID_{50} or more infectivity, i. e., operations on the CNS, the posterior segment of the eye, the olfactory epithelium, and the organized lymphatic tissue. According to the hazard analysis, the peak of potential iatrogenic transmissions associated with interventions on lymphatic tissue, first and foremost on tonsils, has been expected in the UK in 2005. However, owing to the pathogenesis of vCJD, the maximum risk for transmission resulting from brain or ophthalmologic operations should occur about 10 years later, i. e., around 2015. The total number of cases that will probably be attributed to iatrogenic transmission is estimated at approximately 5–10 % of the vCJD cases acquired by the alimentary route.

Considering the variety as well as the frequency of surgical interventions on organs and tissues potentially containing infectious agent, the German vCJD Task Force aimed to recommend generally applicable decontamination procedures that take into account the theoretical risk of vCJD transmission from unrecognized carriers on the one hand, and the risk of compromising the conventional processes for cleaning, disinfection and sterilization of surgical instruments and medical devices, on the other. In addition, the suggested procedure should take into account the potential cross-contamination of instruments during the cleaning process (e. g., in a washer/disinfector) as well as the necessity to be suitable for thermolabile medical devices. Thus, the maintenance of surgical instruments and other medical equipment according to best practice should comply with the recommendations of the German National Committee for Hospital Hygiene and Infection Prevention and the Federal Institute for Drugs and Medical Devices of November 2001 [66], and combine two or more methods suitable for the decontamination and/or inactivation of TSE agents.

In particular, such methods include cleaning under alkaline conditions ($pH > 10$)[2] and subsequent steam sterilization at 134° C according to European standards. (For details, please refer to the comprehensive report of the vCJD task force [15] (see also Chapters 44; 45). Medical products that should generally only be used once emcompass: scalpel blades; (biopsy) needles, or cannulas; CSF aspiration cannulas; equipment for spinal anesthesia and nerve blocks; knives and lancets (except diamond-made) for ophthalmic use; thermolabile endotracheal tubes; bone drills and screws that may have contact with the spinal cord or CSF; and implantables.

The following topics refer to the cleaning, disinfection, and sterilization of all medical equipment even when used on patients without any particular signs or symptoms (i. e., recognizable risk for CJD or vCJD) and, therefore, aim to minimize the "theoretical" risk of transmission from asymptomatic carriers. As outlined above, in the case of invasive interventions on patients with an evident risk for these diseases (including patients with an established diagnosis of human TSE), additional safety measures (see above) must be observed.

1. Avoid any fixation of blood, mucus or tissues on the surface of nondisposable surgical instruments or medical devices by drying, exposure to fixating disinfectants (e. g., aldehydes, alcohols) or heat before cleaning. Fixation can be avoided by immediate incubation in a suitable cleaning solution (e. g., an alkaline cleaner containing detergents and possibly nonfixating disinfectants) or rapid reprocessing.
2. Clean all instruments thoroughly, according to best practice, preferably using a validated or, at the very least, a standardized method (e. g., use of a certified washer/disinfector) prior to sterilization.
3. According to the recommendations of the German vCJD Task Force [15], any maintenance of surgical instruments and medical devices (even in the absence of a recognizable risk) should comply with the *Requirements to Hygiene in the Maintenance of Medical Devices* [66] and combine two or more methods suitable for an at least partial decontamination and/or inactivation of TSE agents. This could be achieved for example by use of:
   - a validated or, at least a standardized cleaning process, using a washer/disinfector and an alkaline solution (pH > 10) at an elevated but protein nonfixating temperature (usually about 55 °C);
   - followed by thorough rinsing and thermal disinfection.

   The stringency of the recommendation to use alkaline cleaners varies for different European countries. Other approaches may also result in an optimized cleaning process; alkaline residues may be harmful in ophthalmology.
4. Critical (according to the Spaulding classification) thermostable medical devices shall generally be sterilized by steam sterilization at 134 °C in a porous load steam sterilizer
   - for a period of 5 min (or European standards), or
   - for a period of 18 min in the case that alkaline cleaning is not feasible or currently used.

   Guidelines for the control of infection by TSE agents in nosocomial settings published by the ACDP and SEAC include a sterilization cycle of 18 min holding time at 134–137 °C in a properly functioning porous load steam sterilizer. This step is generally suggested for the reprocessing of thermostable devices in France in the Circulaire DGS/5C/E 2n° 2001–138; 14. 3. 2001 [9]. Consequently, many benchtop steam sterilizers currently on the market feature a so-called prion cycle (i. e., a cycle of 18 min holding time at 134–137 °C). Ambitious claims have been made concerning the effectiveness of this cycle for the inactivation of TSE agents. However, it must be noted that a sterilization cycle of

[2] NaOH or KOH mixed with tensides and used at an elevated but not protein-denaturating temperature are expected to be effective within 10 min. However, the destabilizing effect on PrP^{Sc} should be confirmed by appropriate tests. The corrosive potential of NaOH or KOH to instruments and medical devices may be decreased by specific additives. In ophthalmology, potential risks associated with alkaline cleaning need to be excluded. In any case, changes in cleaning procedures must not affect the results of operative and other medical interventions [15].

18 min holding time at 134–137 °C alone cannot guarantee complete inactivation of infectivity – because, among other reasons, thermostable subpopulations of prions are known to exist (see Chapter 44).
Specific considerations: In cases in which proper steam sterilization is not possible, cleaning and disinfection provide the only risk-reducing steps:

- Semi-critical (according to the Spaulding classification) devices, which have been used without contact to the CNS, ocular fundus, olfactory epithelium or lymphatic tissue (tonsils, adenoids, appendix, lymph node, spleen, thymus, adrenal gland, rectum) may be thoroughly cleaned by alkaline, enzymatic or other appropriate cleaners with a high cleaning efficacy and subsequently disinfected. As exemplified in the disinfection of flexible gastrointestinal endoscopes, even a "fixating" disinfectant such as glutaraldehyde or peracetic acid (see Chapter 44.6.2) [62; 67] may be used for this purpose because of its reliable activity against a variety of more widely distributed pathogens (such as HBV, HCV, or mycobacteria). In some countries (e. g., France), double cleaning and the use of peracetic acid as a disinfectant is favored. It has to be mentioned, however, that the prion-inactivating capacity of peracetic acid is insufficient (see Chapter 44.6.2).
- Critical (according to the Spaulding classification) thermolabile medical devices, which have been used without contact to the CNS, ocular fundus, olfactory epithelium or organized lymphatic tissue may be thoroughly cleaned by alkaline cleaners and subsequently disinfected by a non-fixating procedure. The efficacy of plasma sterilization to inactivate TSE agents to some extent needs to be further investigated.
  - Thermostable medical devices to be used in contact with the CNS, ocular fundus, olfactory epithelium, or organized lymphatic tissue (e. g., in ENT surgery), which cannot be reprocessed by mechanical cleaning under alkaline conditions in a manner that ensures reliable decontamination and thorough rinsing, may be reprocessed by other suitable standardized and well-established (e. g., enzymatic) cleaning procedures, provided that the final step is a steam sterilization at 134 °C for a period of at least 18 min. This is particularly relevant for devices to be applied in eye surgery, since here residues of alkali pose a considerable hazard. In the case that the outlined procedure is not applicable for these or other medical devices, a special procedure for reprocessing should be developed and validated.
  - Critical (according to the Spaulding classification) thermolabile medical devices, which generally come into contact with the CNS, ocular fundus, olfactory epithelium or organized lymphatic tissue and which are difficult to clean, represent a considerable problem in the context of infection control. Manufacturers of such devices should resolve this problem, for example, by improving the design of those devices with respect to better decontamination and precise instructions for efficient cleaning. In any case, the use of critical thermolabile medical devices requires a careful risk evaluation and well-balanced decision whether recycling is possible.

47.10 Prevention in specific areas

In this chapter clearly not all problems that are encountered in specific medical areas can be addressed. Expert associations of the EU, or of individual countries such as the USA, for instance, will, if necessary, provide detailed advice on this topic.

As delineated above, all prophylactic measures in nosocomial environments pursue a central purpose: to prevent exposure of patients and personnel to possible sources of risk such as contaminated tissues, surgical instruments, or other medical devices. In this endeavor, the efficient decontamination (i. e., removal and/or inactivation) of TSE agents by cleaning, chemical disinfection and, if applicable, sterilization of potentially contaminated materials that may act as a source of infection or vehicle for transmission is of utmost importance in addition to barrier precautions.

47.10.1 Internal medicine: use of flexible endoscopes in CJD patients

The use of reusable flexible endoscopes in patients who present an increased risk of TSE transmission (see Section 47.4.1) should be avoided in all fields of medicine.

If necessary, for example in cases when a percutaneous endogastric (PEG) tube is used, the following decontamination procedure may be applied for selected endoscopes; in some countries, national pools of instruments, which are available for particular interventions on TSE patients, have been established [61]:

After immediate rinsing with water and suitable cleaning solutions to avoid any fixation of debris, incubate the endoscope in a 4 M guanidinium thiocyanate (GdnSCN) solution for 1 h. After a 30-min immersion in the GdnSCN solution the channels should be cleaned mechanically with single use brushes and then immersed for further 30 min (see Chapter 44) using the same GdnSCN bath (suitable protection of the operator has to be observed). It is imperative that the GdnSCN solution reaches all internal and external surfaces of the endoscope. To achieve this, it is recommended that a syringe is used and that the GdnSCN solution be repeatedly squirted into the channel and drawn out again. This procedure must be repeated for each individual channel of the endoscope. Further cleaning steps are necessary, preferentially in a washer/disinfector equipped with a validated process controlling all relevant cleaning parameters. Treatment with aldehydes and alcohol must never be done before specific disinfection with GdnSCN (see Chapter 44). GdnSCN must be completely removed from the endoscope. It must also be clarified with the manufacturer whether the material can tolerate GdnSCN under the suggested operational conditions.

Future research may reveal new approaches, which employ alkaline or suitable enzymatic cleaners. Other instructions particular to operational safety are described in Chapter 44.

47.10.2 Neurology: use of myography needles

After myography, needles that have been used on patients who carry an increased risk of TSE transmission (see Section 47.4.1) are to be disposed of in the same way as other disposable materials.

47.10.3 Ophthalmology: use of tonometers

To measure the pressure of the eye in patients who carry an increased risk of TSE transmission (see Section 47.4.1), it is recommended that only tonometers, which do not need to be touched, are used (despite the fact that lachrymal fluid is not infectious). Further remarks on ophthalmological aspects are provided in the following references: [68–70].

47.10.4 Dentistry

The risks of transmission of infection via routine dental treatment are thought to be very low provided optimal conventional standards of infection control and decontamination are maintained [71–73].

47.10.5 Transplantation medicine: use of transplants

To prevent the transmission of TSE, it is recommended that the following persons not be accepted as organ or tissue donors (see also Chapter 51):

1. persons who carry an increased risk of transmitting TSE (group I and II; see Section 47.4.1),
2. persons who have died of unclear or undiagnosed diseases of the CNS, and
3. persons who have died in mental institutions.

47.10.6 Pediatrics

Please refer to reference [74].

47.11 Handling of corpses prior to interment

Corpses with an increased risk of carrying TSE agent (see Section 47.4.1) should not be used for teaching in anatomy and pathology courses. If TSE is suspected, it is recommended that the body, after autopsy, is sealed inside a plastic bag before transfer to an undertaker. In general, it is possible to lay out the corpse. For this purpose after autopsy, the body should be washed with 1–2 M sodium hydroxide.

Further details on autopsy and sample handling are provided in Chapters 29 and 48.

It is not necessary to take special precautions with deceased TSE patients that are uninjured. In the case of TSE-infected corpses that are injured, e. g., trauma to the skull, spinal cord, or eye, it is advised that manipulations of the body should be restricted to those that are absolutely necessary. Care should also be taken by people manipulating the body; they should follow standard precautions of infection control to prevent blood-borne infections, and to take care to avoid injuring themselves. Contaminated surfaces should be disinfected using 2 M sodium hydroxide or 5% sodium hypochlorite for a period of at least 1 hour.

It is also recommended not to embalm the body. Special instructions on how to transport and bury the body should be obtained from the respective authorities of the individual countries.

47.12 Waste disposal in hospitals and laboratories

Waste contaminated with secretions, excreta, blood, CSF, vomit, as well as swabs, cloths, syringes, and other disposable materials originating from the daily care and treatment of TSE patients may be disposed of as medical waste. Hypodermic needles and other pointed or sharp objects must be disposed of in shatterproof containers with a lid. Tissue derived from the brain, spinal cord, or the eye (except for lachrymal fluid), and all other materials that are likely to be contaminated with TSE agents are classified as infectious waste (EAC 180103) and must be incinerated [75]. Surfaces that may be potentially contaminated with tissue or CSF, e. g., after surgery of a TSE patient, should be cleaned using disposable gloves, single use towels, and an alkaline detergent, followed by a 1-h disinfection period employing 2 M sodium hydroxide or 5% sodium hypochlorite (compare Chapter 44).

References

[1] Beekes M, Mielke M, Pauli G, et al. Aspects of risk assessment and risk management of nosocomial transmission of classical and variant Creutzfeldt–Jakob disease with special attention to German regulations. Contrib Microbiol 2004; 11:117–135.

[2] WHO. Infection control guidelines for transmissible spongiform encephalopathies. WHO/CDS/CSR/APH/2000.3. 1999. Webpage 2006: www.who.int/csr/resources/publications/bse/whocdscsraph2003.pdf.

[3] Advisory Committee on Dangerous Pathogens & Spongiform Encephalopathy Advisory Committee. Transmissible spongiform encephalopathy agents: safe working and the prevention of infection. 1998. Webpage 2006: www.archive.official-documents.co.uk/document/doh/spongifm/report.htm.

[4] Department of Health (UK). Webpage 2006: www.dh.gov.uk/PublicationsAndStatistics/Publications/PublicationsPolicyAndGuidance/PublicationsPolicyAndGuidanceArticle/fs/en?CONTENT_ID= 4113541&chk=Ak1zhj.

[5] Medicines and Healthcare Products Regulatory Agency. Webpage 2006: www.mhra.gov.uk.

[6] British dental association (BDA). Webpage 2006: www.bda-dentistry.org.uk.

[7] Department of Health. UK committees. Webpage 2006: www.advisorybodies.doh.gov.uk.

[8] Fondation recherche médicale. Maladie de Creutzfeldt–Jakob et maladies à prions. Webpage 2006: www.frm.org/informez.

[9] Direction générale de la santé et Direction de l'hospitalisation et de l'organisation de soins. Circulaire DGS/5 C/DHOS/E 2 n° 2001–138 du 14 mars 2001 relative aux précautions à observer lors de soins en vue de réduire les risques de transmission d'agents transmissibles non conventionnels. 2001. Webpage 2006: www.sante.gouv.fr/adm/dagpb/bo/2001/01-11/a0110756.htm.

[10] Comité technique national des infections nosocomiales. Désinfection des dispositifs médicaux

en anesthésie et en réanimation. Webpage 2006: www.sfar.org/pdf/desinfdm.pdf.
[11] Swiss Public Health Office (BAG). Verordnung über die Prävention der Creutzfeldt–Jakob-Krankheit bei chirurgischen und medizinischen Eingriffen (CJK-Verordnung). Bull BAG 2006; Nr. 3: 28–30. Webpage 2006: www.bag.admin.ch/prionen/01875/index.html.
[12] Harbarth S, Alexiou A, Pittet D, et al. Creutzfeldt–Jakob Krankheit: Vorsichtsmassnahmen zur Prävention iatrogener Übertragungen im Spital. Swiss-NOSO 1996; 3(2):9–11. Webpage 2006: www.chuv.ch/swiss-noso/d94a1.htm.
[13] Iffenecker A and Ruef C. Übertragungsrisiko von Prionen: Stellungnahme zur Aufbereitung thermostabiler chirurgischer Instrumente vor der Sterilisation. Swiss-NOSO 2002; 9(4):25–28. Webpage 2006: www.chuv.ch/swiss-noso/d32a1.htm.
[14] Simon D and Pauli G. Krankenversorgung und Instrumentensterilisation bei CJK-Patienten und CJK-Verdachtsfällen. Bundesgesundhbl 1998; 41:279–285.
[15] Anonym. Die Variante der Creutzfeldt–Jakob-Krankheit (vCJK): Epidemiologie, Erkennung, Diagnostik und Prävention unter besonderer Berücksichtigung der Risikominimierung einer iatrogenen Übertragung durch Medizinprodukte, insbesondere chirurgische Instrumente. Abschlussbericht der Task Force vCJK zu diesem Thema. Bundesgesundheitsbl – Gesundheitsforsch – Gesundheitsschutz 2002; 45:376–394. Webpage 2006: www.rki.de/cln_006/nn_226786/DE/Content/Infekt/Krankenhaushygiene/Erreger__ausgewaehlt/CJK/CJK__pdf__02,templateId=raw,property=publicationFile.pdf/CJK_pdf_02.
[16] Anonymous. Creutzfeldt–Jakob disease and other human transmissible spongiform encephalopathies: guidelines on patient management and infection control. Working Party of the National Health Advisory Committee of the National Health and Medical Research Council. Canberra: Australian Government Publishing Service, 1996.
[17] Anonymous. Creutzfeldt–Jakob disease and other human transmissible spongiform encephalopathies – guidelines on patient management and infection control. 1995. Webpage 2006: www.nhmrc.gov.au/publications/_files/withdrawn/ic5.pdf.
[18] Steelman VM. Creutzfeld-Jakob disease: recommendations for infection control. Am J Infect Control 1994; 22:312–318.
[19] CDC. Prion diseases. Webpage 2006: www.cdc.gov/ncidod/dvrd/prions.
[20] Berger JR and David NJ. CJD in health care workers. Neurology 1993; 43:2421.
[21] Berger JR and David NJ. Creutzfeldt–Jakob disease in a physician: a review of the disorder in health care workers. Neurology 1993; 43:205–206.
[22] Gorman DG, Benson DF, Vogel DG, et al. Creutzfeldt–Jakob disease in a pathologist. Neurology 1992; 42:463.
[23] Mitrova E and Belay G. Creutzfeldt–Jakob disease in health professionals in Slovakia. Eur J Epidemiol 2000; 16(4):353–355.
[24] Sitwell L, Lach B, Attack E, et al. Creutzfeldt–Jakob disease in histopathology technicians. N Engl J Med 1988; 318:854.
[25] Will RG, Esmonde TFG, Matthews WB. Creutzfeldt–Jakob disease epidemiology. In: Prusiner SB, Collinge J, Powell J, et al., editors. Prion diseases of humans and animals. Ellis Horwood, New York, 1992:188–199.
[26] Cooke J. Cannibals, cows and the CJD catastrophe. Random House, Australia, 1998.
[27] Schoene WC, Masters CL, Gibbs CJ Jr, et al. Transmissible spongiform encephalopathy (Creutzfeldt–Jakob disease) – atypical clinical and pathological findings. Arch Neurol 1981; 38:473–477.
[28] Miller DC. Creutzfeldt–Jakob disease in histopathology technicians. N Engl J Med 1988; 318:853–854.
[29] Brown P. Transmissible spongiform encephalopathy – Chapter 43. In: Goetz C, editor. Textbook of Clinical Neurology. W. B. Saunders, Philadelphia, PA, 2006.
[30] Bernoulli C, Siegfried J, Baumgartner G, et al. Danger of accidental person-to-person transmission of Creutzfeldt–Jakob disease by surgery (letter). Lancet 1977;i:478–479.
[31] Croxson M, Brown P, Synek B, et al. A new case of Creutzfeldt–Jakob disease associated with human growth hormone therapy in New Zealand. Neurology 1988; 38:1128–1130.
[32] Rappaport EB. Iatrogenic Creutzfeldt–Jakob disease. Neurology 1987; 37:1520–1522.
[33] WHO. Consultation on medicinal and other products in relation to human and animal transmissible spongiform encephalopathies. With the participation of the Office International des Epizooties (OIE) (report). Document WHO/EMC/ZOO/97 3-WHO/BLG/97 2, World Health Organization, 1997.
[34] Sato T, Hoshi K, Yoshino H, et al. Creutzfeldt–Jakob disease associated with cadaveric dura ma-

ter grafts – Japan, January 1979 to May 1996. MMWR 1997; 46:1066–1069.

[35] Hilton DA, Ghani AC, Conyers L, et al. Accumulation of prion protein in tonsil and appendix: review of tissue samples. BMJ 2002; 325(7365): 633–634.

[36] Ghani AC, Ferguson NM, Donnelly CA, et al. Predicted vCJD mortality in Great Britain. Nature 2000; 406(6796):583–584.

[37] Department of Health. Assessing the risk of vCJD transmission via surgery: an interim review. 2005. Webpage 2006: www.dh.gov.uk.

[38] Zanusso G, Ferrari S, Cardone F, et al. Detection of pathologic prion protein in the olfactory epithelium in sporadic Creutzfeldt–Jakob disease. N Engl J Med 2003; 348(8):711–719.

[39] Reuber M, Al-Din AS, Baborie A, et al. New variant Creutzfeldt–Jakob disease presenting with loss of taste and smell. J Neurol Neurosurg Psychiatr 2001; 71(3):412–413.

[40] Pauli G. Tissue safety in view of CJD and variant CJD (review). Cell Tissue Bank 2005; 6:191–200.

[41] Llewelyn CA, Hewitt PE, Knight RS, et al. Possible transmission of variant Creutzfeldt–Jakob disease by blood transfusion. Lancet 2004; 363(9407):417–421.

[42] Peden AH, Head MW, Ritchie DL, et al. Preclinical vCJD after blood transfusion in a PRNP codon 129 heterozygous patient. Lancet 2004; 364(9433):527–529.

[43] Health Protection Agency (UK). New case of variant CJD associated with blood transfusion (press release). Webpage 2006: www.hpa.org.uk/hpa/news/articles/press_Releases/2006/060209_cjd.htm.

[44] Glatzel M, Abela E, Maissen M, et al. Extraneural pathologic prion protein in sporadic Creutzfeldt–Jakob disease. N Engl J Med 2003; 349(19):1812–1820.

[45] Peden AH, Ritchie DL, Head MW, et al. Detection and localization of PrPSc in the skeletal muscle of patients with variant, iatrogenic, and sporadic forms of Creutzfeldt–Jakob disease. Am J Pathol. 2006; 168(3):927–35.

[46] Wadsworth JD, Joiner S, Hill AF, et al. Tissue distribution of protease resistant prion protein in variant Creutzfeldt–Jakob disease using a highly sensitive immunoblotting assay. Lancet 2001; 358(9277):171–180.

[47] Hill AF, Zeidler M, Ironside J, et al. Diagnosis of new variant Creutzfeldt–Jakob disease by tonsil biopsy. Lancet 1997; 349:99–100.

[48] Hill AF, Butterworth RJ, Joiner S, et al. Investigation of variant Creutzfeldt–Jakob disease and other human prion diseases with tonsil biopsy samples. Lancet 1999; 353(9148): 183–189.

[49] Ironside JW. Pathology of variant Creutzfeldt–Jakob disease. Arch Virol Suppl 2000;(16):143–151.

[50] Hilton DA, Fathers E, Edwards P, et al. Prion immunoreactivity in appendix before clinical onset of variant Creutzfeldt–Jakob disease. Lancet 1998; 352(9129):703–704.

[51] Hilton DA, Ghani AC, Conyers L, et al. Prevalence of lymphoreticular prion protein accumulation in UK tissue samples. J Pathol 2004; Epub 21 May 2004. DOI 10.1002/path.1580.

[52] Mabbott NA and Bruce ME. The immunobiology of TSE diseases. J Gen Virol 2001; 82(Pt 10):2307–2318.

[53] Klein MA, Frigg R, Flechsig E, et al. A crucial role for B-cells in neuroinvasive scrapie. Nature 1997; 390:687–690.

[54] Brown P, Cervenakova L, McShane LM, et al. Further studies of blood infectivity in an experimental model of transmissible spongiform encephalopathy, with an explanation of why blood components do not transmit Creutzfeldt–Jakob disease in humans. Transfusion 1999; 39(11–12):1169–1178.

[55] Brown P, Rohwer RG, Dunstan BC, et al. The distribution of infectivity in blood components and plasma derivatives in experimental models of transmissible spongiform encephalopathy. Transfusion 1998; 38(9):810–816.

[56] Bruce ME, McConnell I, Will RG, et al. Detection of variant Creutzfeldt–Jakob disease infectivity in extraneural tissues. Lancet 2001; 358(9277):208–209.

[57] Anonym. Bericht der Arbeitsgruppe "Gesamtstrategie Blutversorgung angesichts vCJK". 2001. Webpage 2006: www.pei.de/SharedDocs/Downloads/blut/gesamtstrategie-bericht,templateId = raw,property = publicationFile.pdf/gesamtstrategie-bericht.pdf. [Update published on 13 April 2006; see Internet.]

[58] Flechsig E, Hegyi I, Enari M, et al. Transmission of scrapie by steel-surface-bound prions. Mol Med 2001; 7(10):679–684.

[59] Spencer RC and Ridgway GL. Sterilization issues in vCJD – towards a consensus: meeting between the Central Sterilizing Club and Hospital Infection Society. 12th September 2000. J Hosp Infect 2002; 51(3):168–174.

[60] Department of Health. CJD publications. Webpage 2006: www.dh.gov.uk/PolicyAndGuidance/HealthAndSocialCareTopics/CJD/fs/en.

[61] Anonymous. Creutzfeldt–Jakob lending devices program of the University of Göttingen. Endo-Praxis 2001; 2:38.
[62] Lemmer K, Mielke M, Pauli G, et al. Decontamination of surgical instruments from prion proteins: in vitro studies on the detachment, destabilization and degradation of PrPSc bound to steel surfaces. J Gen Virol 2004; 85(12):3805–3816.
[63] Baier M, Schwarz A, Mielke M. Activity of an alkaline 'cleaner' in the inactivation of the scrapie agent. J Hosp Infect 2004; 57(1):80–84.
[64] Head MW, Ritchie D, Smith N, et al. Peripheral tissue involvement in sporadic, iatrogenic, and variant Creutzfeldt–Jakob disease: an immunohistochemical, quantitative, and biochemical study. Am J Pathol 2004; 164(1):143–153.
[65] Department of Health, Economics and Operational Division. Risk assessment for transmission of vCJD via surgical instruments: a modelling approach and numerical scenarios. Government Operational Research Service, London, 2001.
[66] Anonym. Empfehlungen der Kommission für Krankenhaushygiene und Infektionsprävention – Anforderungen an die Hygiene bei der Aufbereitung von Medizinprodukten. Bundesgesundheitsbl – Gesundheitsforsch – Gesundheitsschutz 2001; 44:1115–1126.
[67] Kampf G, Bloss R, Martiny H. Surface fixation of dried blood by glutaraldehyde and peracetic acid. J Hosp Infect 2004; 57(2):139–143.
[68] Lim R, Dhillon B, Kurian KM, et al. Retention of corneal epithelial cells following Goldmann tonometry: implications for CJD risk. Br J Ophthalmol 2003; 87(5):583–586.
[69] Mehta JS, Osborne RJ, Bloom PA. Variant CJD and tonometry. Br J Ophthalmol 2004; 88(4): 597–598.
[70] Macalister GO and Buckley RJ. The risk of transmission of variant Creutzfeldt–Jakob disease via contact lenses and ophthalmic devices. Cont Lens Anterior Eye 2002; 25(3):104–136.
[71] Smith A and Bagg J. CJD and the dentist. Dent Update 2003; 30(4):180–186.
[72] Smith A. CJD and the dentist. Dent Update 2003; 30(8):462–463.
[73] Department of Health. Risk assessment for vCJD and dentistry 2003. Webpage 2006: www.dh.gov.uk/PublicationsAndStatistics/Publications/PublicationsPolicyAndGuidance/PublicationsPolicyAndGuidanceArticle/fs/en?CONTENT_ID = 4084662&chk = twiXsY.
[74] Whitley RJ, MacDonald N, Asher DM. American Academy of Pediatrics. Technical report: transmissible spongiform encephalopathies: a review for pediatricians. Committee on Infectious Diseases. Pediatrics 2000; 106(5):1160–1165.
[75] HSAC guidance on "Safe disposal on clinical waste" 1999; HSE Books ISBNO-7176-24492-7. Webpage 2006: www.advisorybodies.doh.gov.uk/acdp/tseguidance/tseguidancepart4.pdf.

48 Precautionary Measures for Autopsies Performed in Cases of Suspected Prion Disease

Walter J. Schulz-Schaeffer, Armin Giese, and Hans Kretzschmar

48.1 Introduction

Prion diseases are transmissible within as well as between species. However, with the possible exception of scrapie and chronic wasting disease (CWD), prion diseases demonstrate no natural intraspecies transmissibility. Scrapie and CWD can transmit in the same livestock (see Chapter 42.2.2). In humans, disease transmission through social contact with infected persons can evidently be ruled out (see Chapter 47.2). To date, only one case in which the spouse of a CJD[1] patient contracted Creutzfeldt–Jakob disease (CJD) has been reported [1]. Experimentally, prion diseases can be transmitted with varying efficiency via intracerebral, intravenous, intraperitoneal, intramuscular, subcutaneous, and oral application of the scrapie agent (see Chapter 41.5.1). Iatrogenic transmissions of prion diseases have been documented by pituitary hormone therapy, dura mater implantations, cornea transplantations, and by neurosurgical interventions (Table 14.1).

European statistics show that 1 in 6,000 deaths is due to prion diseases (i. e., approximately 150 cases of prion diseases in about 900,000 deaths per year in Germany; compare Chapter 37). Approximately 1 in 1,000 autopsies is CJD, reflecting a higher autopsy rate in prion disease-related deaths than in persons dying of other, more common, causes. In the approximately 60,000 autopsies carried out in Germany per year, about 80 cases of prion disease are confirmed in autoptic tissue. It can therefore be assumed that all pathologists working in an institution where autopsies are performed will, at some point during their professional career, have to deal with a suspected CJD case. Epidemiological data have thus far shown no evidence that pathologists or persons in medical occupations contract CJD significantly more frequently than the rest of the population [2]. Only two cases of CJD are known worldwide in two technical assistants who had regular contact with autoptic tissue while working in a histological laboratory [3; 4], and only one case of a pathologist with CJD has been reported [5].

Clinically, CJD can be diagnosed as "probable" or "possible" on the basis of Europe-wide standardized criteria (Tables 29.1; 31.1). At present, only a brain tissue biopsy or an autopsy with subsequent histopathological processing of the tissue can confirm the clinical diagnosis of a prion disorder [6]. In the future, it will be necessary to monitor the range of variation in prion diseases using autopsy examination so that new variants in humans, such as the variant CJD (vCJD), can be identified at an early stage. Due to the particular risk of transmissibility of prion diseases during autopsies (see Chapter 42.2.1), special precautionary measures for the protection of laboratory staff should be taken with patients with suspected prion diseases. A prion disease can be idiopathic (possibly spontaneous), acquired (i. e., iatrogenic or by oral intake; e. g., vCJD), or genetic. Precautionary measures should be taken in autopsies of patients in whom cerebellar symptoms or dementia have been documented in the course of the disease or where the risk of accidental exposure to PrP^{Sc} cannot be ruled out. Such cases include patients who have undergone dura mater implantation, pituitary hormone treatment, or

[1] Unless otherwise states, the term "CJD" in this Chapter encompasses all human prion diseases.

cornea transplantation [7; 8]. The risk of a prion disease is also present in patients with a dementing disease in whom a close relative suffering from a prion disease. Patients with a neuropsychiatric disorder who are known to have a mutation of the prion protein gene (*PRNP*) should be regarded as CJD cases [9] (see Chapter 26).

48.2 Performing the autopsy

It is recommended that precautionary measures be taken to protect those in contact with the corpse and the environment from contamination [10]. Previous recommendations that only the head should be examined at autopsy [11] are now obsolete. It is also unnecessary to use a special dedicated facility for the autopsy as the required precautionary measures can be carried out in any normal autopsy room. The only prerequisite is a room, which includes an autopsy table and sufficient space for at least two people to carry out the autopsy. Autopsies should be performed by a minimum of two persons, ideally three. Water must not be used for rinsing.

Precautionary measures for persons conducting the autopsy are the same as those used in any other potentially infectious autopsy case. Particular care should be taken to (i) wear eye and nose/mouth protection to avoid intake of droplets and (ii) prevent injuries by wearing Kevlar gloves (or chain gloves) under latex gloves. To avoid unnecessary contamination of the surroundings and instruments, it is recommended that:

1. The autopsy table be covered with a plastic sheet and the autopsy be performed on the body retained in an open body bag.
2. Bone transections should be carried out with a hand saw, not an oscillating saw, so that no airborne dust is produced (sawing with an oscillating saw under a plastic sheet [12] is also potentially dangerous and as such is strictly not recommended).
3. Fluids be absorbed immediately using absorbent material (i. e., cellulose wadding).
4. The brain removal should occur in the final step. After tissue samples have been taken for biochemical prion protein typing (and frozen until further investigation), the brain should be fixed for at least 2 weeks in 4 % buffered formalin. The resulting "hardened" tissue can then be processed more easily and is not significantly damaged, as in unfixed tissue, by decontamination with formic acid (see Chapter 44).
5. All necessary instruments should be kept ready at the autopsy table and must be decontaminated after the procedure (see below).
6. The table covering, cellulose material, and single-use material should be disposed of in specially labeled containers for the incineration of infectious materials.

In general, only the central nervous system (i. e., brain, spinal cord, and retina) must be considered highly infectious [13]. Dissection of the brain is performed as the final step of the autopsy in order to avoid contamination of less infectious tissues with potentially contaminated instruments. Infectivity in the spleen, in lymphatic tissues, and in muscle tissues cannot be ruled out with certainty. In patients with acquired prion disease, including vCJD, it is advisable to assume an increased risk of infectivity in the peripheral organs and blood (see Chapter 51).

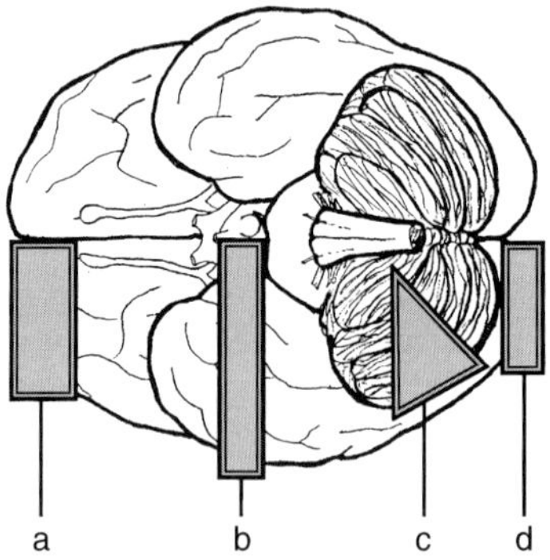

Fig. 48.1: Sampling scheme of frozen brain tissue in patients with suspected CJD: brain slices should be packed separately in suitable plastic bags and frozen on the flat surface of dry ice blocks. Snap freezing in liquid nitrogen is unnecessary. **(a)** Frontal pole of one hemisphere; **(b)** whole frontal slice of one hemisphere at the level of the mamillary body; **(c)** part of a cerebellar hemisphere; **(d)** occipital pole of one hemisphere.

It is recommended that the following tissues (Table 29.2) be removed for optimal examination:

1. The whole brain should be removed, then slices, approximately 1–2 cm thick should be taken from one of the cortical hemispheres. The slices should be made at the level of the mamillary bodies, at the frontal and at the occipital poles, and a parasagittal section of one cerebellar hemisphere (Fig. 48.1). The slices should then be frozen at −20 to −80 °C. The remaining brain tissue should be fixed in 4 % buffered formalin.
2. Specimens of the internal organs (placed in formalin or frozen).
3. Two tubes of blood (frozen at −20 °C) for the analysis of the prion protein gene.

Sectioning of the brain should be carried out after at least 2 weeks of fixation in formalin. It is important to note that formalin fixation does not result in the effective decontamination of the tissue [14]. Formalin fluid used for tissue fixation itself is infectious and must be disposed of in containers for the incineration of infectious materials. The brain sectioning should be performed on a table covered with two layers of plastic sheeting separated by a layer of cellulose; the topmost plastic sheet should be finely perforated so that the cellulose material in between can absorb the fluid. After slices have been obtained for histological examination, tissue specimens in the histology containers must be decontaminated in concentrated formic acid (=98 %) for 1 h and then fixed for 1–2 days in freshly prepared, 4 % buffered formalin and embedded in paraffin. Decontamination with formic acid reduces infectivity by a factor of at least 10^7 [15]. The morphology of fixed fatty tissues such as brain tissue is only slightly damaged by this treatment; connective tissue structures, however, suffer extensive damage. The instruments used for brain autopsy should be decontaminated in 2 N NaOH (for at least two 30-min periods exchanging fluids between periods, or alternatively overnight).

In European countries as well as in the USA and Canada, reference centers for spongiform encephalopathies (prion diseases) at departments of neuropathology provide consultative reports on tissue sections. Reference centers should be contacted regarding transport. The transport of prion-infected material must be carried out according to regulations for transporting dangerous goods; i. e., infectious human tissue.

48.3 Decontamination and resistance of the infectious agent

Brain tissue from CJD patients may contain up to 10^9 infectious units per gram of tissue based on the LD_{50} in animal experiments. The infectious agent is characterized by a high resistance to heat, detergents, and radiation and cannot be effectively inactivated by formalin, alcohol [16], etc. (for a complete list of substances with no or insufficient inactivating effect on prions: see Table 44.1). The following decontamination measures for instruments and surfaces allow for a virtually complete inactivation of the infectious agent:

1. Steam autoclaving of autoclavable material at 134 °C for 1.0 h [17], at 121 °C for 4.5 h [18], or at 134 °C for 1.0 h in two consecutive cycles of 30 min each.
2. Instruments that have been contaminated by formalin-fixed tissue not previously treated in formic acid cannot be effectively decontaminated by autoclaving [19]. They should be placed in 2 N NaOH for at least two 30 min periods exchanging fluids between periods, or alternatively overnight.
3. Surfaces and floors should be wiped with 2 N NaOH several times to insure a prolonged application of the solution. NaOH is effective on steel surfaces but cannot be used on Al or Zn. An alternative solution is NaOCl, which must contain at least 20,000 ppm of free Cl. Warning: this contains irritating gases; the solution must always be freshly prepared (Table 44.2).
4. Contaminated skin should be washed for 5–10 min in 1 N NaOH and then thoroughly rinsed with water [20] (see Chapter 49).
5. All material residues, single-use instruments and contaminated fluids need to be disposed of in specially labeled containers and carefully sealed for incineration (see Chapter 47).

Although CJD is a rare disease, affecting only 1–2 cases per million people per year, it should always be considered in the differential diagnosis of people with rapidly progressive dementia. Due to increasing awareness of prion diseases, the number of confirmed cases of CJD has doubled in the past few years in most European countries. Utilizing easily applied precautionary measures through which increased protection of people and the environment (see Chapters 44.3; 44.5) is possible, autopsies of suspected TSE cases can be performed without risk. A high autopsy rate in cases of clinically suspected prion disease is required for the early recognition of possible new variants, and for estimating the risk of transmitting animal TSEs to humans.

References

[1] Brown P, Cervenakova L, McShane L, et al. Creutzfeldt–Jakob disease in a husband and wife. Neurology 1998; 50:684–688.

[2] Duijn CM van, Delasnerie-Lauprêtre N, Masullo C, et al. Case-control study of risk factors of Creutzfeldt–Jakob disease in Europe during 1993–1995. Lancet 1998; 351(9109):1081–1085.

[3] Miller DC. Creutzfeldt–Jakob disease in histopathology technicians. N Engl J Med 1988; 318:853–854.

[4] Sitwell L, Lach B, Attack E, et al. Creutzfeldt–Jakob disease in histopathology technicians. N Engl J Med 1988; 318:854.

[5] Gorman DG, Benson DF, Vogel DG, et al. Creutzfeldt–Jakob disease in a pathologist. Neurology 1992; 42:463.

[6] Kretzschmar HA, Ironside JW, DeArmond SJ, et al. Diagnostic criteria for sporadic Creutzfeldt–Jakob disease. Arch Neurol 1996; 53:913–920.

[7] Brown P, Preece MA, Will RG. "Friendly fire" in medicine: hormones, homografts, and Creutzfeldt–Jakob disease. Lancet 1992; 340:24–27.

[8] Schulz-Schaeffer WJ, Giese A, Kretzschmar HA. Creutzfeldt–Jakob-Krankheit – neue Aspekte für die Rechtsmedizin. Rechtsmedizin 1998; 8:123–129.

[9] Budka H, Aguzzi A, Brown P, et al. Tissue handling in suspected Creutzfeldt–Jakob disease (CJD) and other human spongiform encephalopathies (prion diseases). Brain Pathol 1995; 5:319–322.

[10] Giese A, Schulz-Schaeffer WJ, Kretzschmar HA. Vorsichtsmassnahmen bei der Durchführung von Autopsien bei Verdacht auf Creutzfeldt–Jakob-Erkrankung. Verh Dtsch Ges Path 1995; 79:631.

[11] Pfeiffer J. Recommendations on the handling of tissue from patients with suspected Creutzfeldt–Jakob disease (last revised 1. 3. 1992). Pathologe 1993; 14:355.

[12] Bell JE and Ironside JW. How to tackle a possible Creutzfeldt–Jakob disease necropsy. J Clin Pathol 1993; 46:193–197.

[13] WHO. Infection control guidelines for transmissible spongiform encephalopathies. Report of a WHO consultation. Geneva, Switzerland 23–26 March. WHO/CDS/CSR/APH/2000 3 1999.

[14] Brown P, Gibbs CJ Jr, Gajdusek DC, et al. Transmission of Creutzfeldt–Jakob disease from formalin-fixed, paraffin-embedded human brain tissue (letter). N Engl J Med 1986; 315:1614–1615.

[15] Brown P, Wolff A, Gajdusek DC. A simple and effective method for inactivating virus infectivity in formalin-fixed tissue samples from patients with Creutzfeldt–Jakob disease. Neurology 1990; 40:887–890.

[16] Tateishi J, Koga M, Sato Y, et al. Properties of the transmissible agent derived from chronic spongiform encephalopathy. Ann Neurol. 1980; 7:390–391.

[17] Rosenberg RN, White CL, Brown P, et al. Precautions in handling tissues, fluids, and other contaminated materials from patients with documented or suspected Creutzfeldt–Jakob disease. Ann Neurol 1986; 19:75–77.

[18] Prusiner SB, Hsiao KK, Bredesen DE, et al. Human slow infections caused by prions. In: Gilden DH, Lipton HL, editors. Clinical and molecular aspects of neurotropic virus infection. Kluwer Academic Publishers, The Netherlands 1989: 423–467.

[19] Taylor DM and McConnell I. Autoclaving does not decontaminate formol-fixed scrapie tissues [letter]. Lancet 1988;i:1463–1464.

[20] Brown P, Rohwer RG, Gajdusek DC. Sodium hydroxide decontamination of Creutzfeldt–Jakob disease virus [letter]. N Engl J Med 1984; 310:727.

49 Prevention of Prion Diseases in Research Laboratories

Alex J. Raeber and Adriano Aguzzi

49.1 Introduction

Biosafety can be defined as the safe use of techniques and equipment that contain pathogens that are potentially harmful for humans. Common sense is the first and most important consideration. Good laboratory practice is the key to laboratory safety.

Biosafety is the responsibility of each person in the laboratory, and is mandatory for the work that is performed on prions. Proper and safe microbiological laboratory procedures are required to be implemented by all of the personnel. This chapter describes the risk classification for work with prions and prion proteins (PrP). The procedures presented in this chapter have been compiled from the guidelines *Transmissible Spongiform Encephalopathy Agents: Safe Working and the Prevention of Infection* [1], published by the Office of Public Health, UK. This chapter further describes the methods for the safe handling of prions in the laboratory, and presents the procedures for inactivation of prions in research laboratories. Finally, suggestions for a post-exposure prophylaxis after accidental inoculation with prions are discussed.

49.2 Risk categorization of prions

Biological agents are classified into four risk groups (and further subgroups if necessary) according to (i) the pathogenicity (disease-producing capability) of the organism to humans, animals, and plants and (ii) the likelihood that the pathogenic properties become relevant to laboratory workers as well as to the community, taking into account the availability of prophylaxis or effective treatment:

Risk group 1: Biological agents that are unlikely to cause human disease.

Risk group 2: Biological agents that can cause human disease and might be a hazard to workers; they are unlikely to spread to the community; there is usually effective prophylaxis or treatment available.

Risk group 3: Biological agents that can cause severe human disease and present a serious hazard to workers; they may present a risk of spreading to the community, but there is usually effective prophylaxis or treatment available.

Risk group 3:** Biological agents that are classified in risk group 3 but have a limited risk to be transmitted to workers by way of the aerosol route.

Risk group 4: Biological agents that cause severe human disease and present a serious hazard to workers; they may present a high risk of spreading to the community and there is usually no effective prophylaxis or treatment available.

According to the EU Directive 90/697/EEC amended by the Directive 97/65/EEC, prion diseases are listed under unconventional agents that are associated with transmissible spongiform encephalopathies (TSEs). Laboratory work with the Creutzfeldt–Jakob disease (CJD) agent, including new variant CJD (vCJD), bovine spongiform encephalopathy (BSE), and BSE-related agents is classified into risk group 3**. This classification does not apply to the scrapie agent, which is classified into risk group 2. The same classification applies to the defined isolates of scrapie passaged in mice and hamsters. In agreement with the EC regulations, the Swiss Expert Committee for Biosafety (SECB) of the Swiss Agency for the Environment, Forests, and Landscape (SAEFL) has adopted the same classification of TSE agents for Switzerland. Medical-diagnostic work (sampling of biopsy material and body fluids for biochemical and hematological analysis) has to be performed in accordance with general safety measures for work with human samples in the diagnostic laboratory (see Chapters 47 and 48).

49.3 Risk assessment for work with prions and prion proteins

The risk assessments for work with prions has to take into account the following considerations and scientific knowledge. In addition, the risk for work with natural and genetically modified prion proteins has to be assessed accordingly:

1. Experimental scrapie in mice and hamsters has been studied for several decades with no evidence presented that clearly shows a rodent-adapted scrapie agent that can spread to the community or infection that has occurred in workers through occupational exposure. Scrapie is not contagious and there is no rationale for handling rodents infected with scrapie any differently from the same animals infected with other natural rodent-specific pathogens (of risk group 2).
2. It has been demonstrated that the primary (amino-acid) structure of the prion protein is a major determinant of the species barrier for transmission of TSE agents [2]. On the amino-acid level, the bovine PrP gene shows a higher homology to human PrP than the mouse and hamster PrP [3] (see Chapter 9). Although cross-species transmission of the BSE agent to humans has occurred [4; 5], transmission of mouse and hamster prions to human beings is – based on lower sequence homology – considered extremely unlikely.
3. Considerable efforts worldwide to convert native or recombinant wild-type rodent prion proteins into an infectious isoform have been unsuccessful. Although several laboratories have shown that native and recombinant rodent prion proteins can be converted in vitro into a protease-resistant form reminiscent of PrP^{Sc}, neither recombinant PrP nor protease-resistant recombinant PrP have produced clinical disease when inoculated into wild-type mice. Chapter 8.5 discusses the possible reasons for the failure of generating prion infectivity by a wide variety of in vitro conversion assays. However, it has been shown recently that mutant recombinant PrP89–230 lacking the N-terminus could be refolded in vitro into amyloid and was subsequently able to elicit a transmissible prion disease in transgenic mice overexpressing PrP following intracerebral inoculation [6; 7]. Although it has not been possible so far to generate prion infectivity in vitro using a full-length recombinant PrP and wild-type mice, these experiments support the notion that spontaneous formation of prions can occur de novo.
4. Given the uncertainty surrounding the relationship between a TSE agent and PrP genes, certain genetic modification work of PrP genes should be approached with caution. Two areas of work are considered to pose a potentially increased risk to human health and safety of the environment: (i) cloning PrP genes into hosts capable of colonizing humans and (ii) expression of 'modified' PrP genes in prokaryotic and eukaryotic expression systems.

49.4 Risk classification for work with prions or prion proteins

On the basis of the risk assessment considerations outlined in Section 49.3, the following recommendations for work with prions and PrP genes are given.

Most of the work with TSE agents has to be performed under containment level 3**. Work with scrapie can be conducted under containment level 2. Containment level 3** is used in the EC Directive 90/697/EEC for organisms that have a low risk of infection for laboratory workers because infection by way of the airborne route is unlikely to occur according to the current state of scientific knowledge.

49.4.1 Animals

In general, experimental work with live animals infected with scrapie and rodent-adapted scrapie agents is classified into risk group 2. This work may include (i) handling of animals; (ii) invasive veterinary procedures such as inoculations with tissues containing prion infectivity; (iii) sampling of tissues and body fluids from animals infected with prions and; (iv) post mortem examination of infected animals.

Handling of mice and hamsters inoculated with BSE and BSE-related agents as well as human TSE agents is classified into risk group 3**. To reduce the risk of exposure to TSE agents from bites and scratches, the appropriate precautions such as the handling of animals with long forceps may be required. Invasive procedures, sampling of tissues, as well as post mortem examinations of animals infected with BSE, BSE-related, and human TSE agents are classified into risk group 3**.

49.4.2 Tissues

Any experimental work with lymphoid and neural tissues of animals infected with rodent-adapted scrapie prions is classified into risk group 2. Risk group 3** is required for experimental work with neural and lymphoid tissues derived from human TSE and animals infected with human or bovine prions.

49.4.3 Risk classification for work with genetically modified PrP genes

A risk assessment has to be performed for work with genetically modified PrP genes and the protein products derived from these genes. The risk assessment must include a review of all working procedures. In particular, it has to include the host in which the expression of the PrP gene is directed, the species from which the PrP gene is derived, and whether the PrP gene contains modifications or mutations that are linked to human TSEs.

49.5 Containment of laboratory work with prions

49.5.1 Basic precautions

In general, all laboratory work with TSE agents that are classified into risk group 3** has to be performed at containment level 3.

- Work that could lead to the release of aerosols has to be conducted in a biological safety cabinet class II.
- An autoclave that can reach a sterilization temperature of 134–137°C has to be in the building.
- The employer has to keep a list of employees exposed to prions for 40 years following the last known exposure risk.
- Washing and decontamination facilities have to be available in the dedicated working area.
- Liquid and dry waste has to be inactivated (see Chapter 47).
- Eye protection should be used for laboratory work.

49.5.2 Derogations

The following derogations from containment level 3 may be considered as part of any risk assessment process for work with TSE agents classified into risk group 3**:

- Negative air pressure in the laboratory;
- Filtration of the laboratory exhaust air by HEPA (high efficiency particulate arrestance) filtration if work can be conducted in a biological safety cabinet;
- Sealability of the laboratory for fumigation as a result of prions being largely unaffected by normal fumigants (see Table 44.1, ethylene oxide).

49.6 Inactivation of prions in research laboratories

The difficulties with the sterilization of prions (inactivating them so that no residual infectivity can be detected) are considerable. Most of the standard physical and chemical inactivation methods that are used for microorganisms such as bacteria and viruses are ineffective for the decontamination of prions (Table 44.1). The decontamination procedures for prions in the research laboratory include (i) chemical inactivation with 2 M NaOH for 1 hour [1; 8] (ii) autoclaving at 134–137°C for at least 18 minutes [1].

49.6.1 Inactivation of prions with sodium hydroxide

Decontamination of equipment that cannot be autoclaved, such as automatic pipetting devices, instruments, ultrasonic baths, etc., is performed by treatment with 2 M NaOH for 1 hour. This can be achieved by using paper towels soaked in 2 M NaOH and that are brought into contact with the surfaces to be cleaned for 1 hour [8; 9]. The decontamination should be followed by rinsing with water to remove any remaining NaOH (corrosive!). Similarly, surfaces (working bench, biosafety cabinets) are decontaminated daily, or after spilling occurs, by using a 2 M NaOH solution for 1 hour. Because of the corrosive nature of NaOH, preparation and handling of NaOH solutions always require the use of protective clothing, safety goggles, and gloves. It is critical that the final concentration of NaOH is at least 1 M during the decontamination. Therefore, it is recommended to use a 2 M NaOH stock solution. Decontamination of intact skin can be achieved with a 1 M NaOH solution [5–10] (see Section 49.7).

For the decontamination of delicate surfaces and surgical instruments, immersion of the equipment in a 4 M GdnSCN solution for 60 minutes is sufficient [10] (see Chapter 44).

49.6.2 Partial inactivation of prions with chemical disinfectants

Sodium dodecylsulfate (SDS)

Metal objects and devices with aluminum and zinc surfaces can easily corrode when coming into contact with NaOH. A partial inactivation of prions can be achieved by exposure to 3% SDS at 100 °C [8]. This is followed by rinsing with water to remove the SDS. Since SDS does not completely sterilize, equipment must still be considered potentially contaminated unless the SDS decontamination is followed by autoclaving at 121 °C for 1 hour. For this decontamination procedure, exposure to SDS for 3 minutes is sufficient.

Further disinfectants for prion decontamination

Further discussion of disinfectants for partial decontamination of prions, as well as of processes and chemicals that are ineffective against TSE agents, is found in Chapter 44.

49.6.3 Heat inactivation of prions by autoclaving

Disposable waste, glassware, surgical instruments as well as heat- and steam-resistant equipment can be decontaminated in a validated steam autoclave. Different times for autoclaving have been suggested. Processes in a porous load should consist of a single cycle at 134–137 °C and ≥ 3,000 hPa for a minimum holding time of 18 min (1 h is used commonly in practice) or six successive cycles at 134–137 °C and ≥ 3,000 hPa for 3 min each [1]. In some laboratories, autoclaving is carried out routinely at 134–137 °C for at least 1 hour.

Formalin-fixed material and tissue sections cannot be inactivated efficiently by autoclaving [11] (see Chapter 44). Chemical decontamination by way of the immersion of material in 2 M NaOH (2 × 30 min) is recommended and more efficient than heat inactivation.

49.7 Post-exposure prophylaxis following spills and accidents

If a spill occurs, surfaces have to be decontaminated immediately by covering the spill with paper towels soaked with 2 M NaOH as described in Section 49.6.1 (only 1 M NaOH is used on intact skin). All accidents resulting in injuries must be documented and reported to the Biosafety Officer (BSO).

49.7.1 Contamination of intact skin without penetrating wounds

Decontaminate skin with a fresh 2.5% sodium hypochlorite solution followed by extensive rinsing with water. Alternatively, affected skin can be decontaminated with 1 M NaOH for 5–10 minutes.

49.7.2 Contamination of penetrating wounds or cuts with ovine (scrapie) or bovine (BSE) prions

Wounds resulting from accidental stabs by syringe needles, contaminated sharp tips, or animal bites, etc., are forced to bleed profusely and then the wound is immersed in 2.5% sodium hypochlorite solution or 1 M NaOH for 5–10 minutes, followed by repeated washing with water.

49.7.3 Contamination of penetrating wounds or cuts with human prions

A surgical excision of the site of inoculation is recommended if the wound was contaminated with traces of highly infectious tissues (i.e., brain tissue or lymphatic tissues of vCJD). A surgical excision is *not* recommended in case the skin is penetrated with a needle only used for blood or CSF puncture in a CJD patient. A surgical excision should be accompanied by an immunosuppressive treatment with prednisone 60 mg/day for the first 7 days and 45 mg/day for an additional 7 days. This treatment should be accompanied by appropriate gastric protection with an H_2 antagonist and antibiotics [12].

49.8 Further useful information

The following guidance and annex related to the topic can be downloaded from [13]:

Guidance:
Part 1 – Introduction
Part 2 – Health and safety management of TSEs
Part 3 – Laboratory containment and control measures
Part 4 – Infection control of CJD and related disorders in the healthcare setting
Annex A.1 – Distribution of TSE infectivity in human tissues and body fluids
Annex A.2 – Distribution of infectivity in animal tissue and body fluids (update 2004)
Annex B – Diagnostic criteria
Annex C – Decontamination and waste disposal
Annex D – Transport of TSE infected material
Annex E – Quarantining of surgical instruments
Annex F – Decontamination of endoscopes (update 2004)
Annex H – After death

References

[1] ACDP. Transmissible spongiform encephalopathy agents: safe working and the prevention of infection. The Advisory Committee on Dangerous Pathogens (ACDP) and Spongiform Encephalopathy Advisory Committee (SEAC), UK, 1998.

[2] Prusiner SB, Scott M, Foster D, et al. Transgenetic studies implicate interactions between homologous PrP isoforms in scrapie prion replication. Cell 1990; 63(4):673–686.

[3] Gabriel JM, Oesch B, Kretzschmar HA, et al. Molecular cloning of a candidate chicken prion protein. Proc Natl Acad Sci USA 1992; 89:9097–9101.

[4] Bruce ME, Will RG, Ironside JW, et al. Transmissions to mice indicate that 'new variant' CJD is caused by the BSE agent [see comments]. Nature 1997; 389(6650):498–501.

[5] Hill AF, Desbruslais M, Joiner S, et al. The same prion strain causes vCJD and BSE. Nature 1997; 389(6650):448–450.

[6] Legname G, Baskakov IV, Nguyen HO, et al. Synthetic mammalian prions. Science 2004; 305(5684):673–676.

[7] Legname G, Nguyen HO, Baskakov IV, et al. Strain-specified characteristics of mouse synthetic prions. Proc Natl Acad Sci USA 2005; 102(6):2168–2173.

[8] Taylor DM, Fraser H, McConnell I, et al. Decontamination studies with the agents of bovine spongiform encephalopathy and scrapie. Arch Virol 1994; 139(3-4):313–326.

[9] Taylor DM. Inactivation of transmissible degenerative encephalopathy agents: a review. Vet J 2000; 159(1):10–17.

[10] Manuelidis L. Decontamination of Creutzfeldt–Jakob disease and other transmissible agents. J Neurovirol 1997; 3(1):62–65.

[11] Brown P, Liberski PP, Wolff A, et al. Resistance of scrapie infectivity to steam autoclaving after formaldehyde fixation and limited survival after ashing at 360 degrees C: practical and theoretical implications. J Infect Dis 1990; 161:467–472.

[12] Aguzzi A and Collinge J. Post-exposure prophylaxis after accidental prion inoculation. Lancet 1997; 350(9090):1519–1520.

[13] Department of Health, UK. Transmissible spongiform encephalopathy agents: safe working and the prevention of infection. Guidance from the Advisory Committee on Dangerous Pathogens and the Spongiform Encephalopathy Advisory Committee. December 15, 2003. Webpage 2006: www.dh.gov.uk/PolicyAndGuidance/HealthAndSocialCareTopics/CJD/CJDGeneralInformation/CJDGeneralArticle/fs/en?CONTENT_ID=4031067&chk=4gOe2r.

Topic X: Risk Assessment

50 Evidence for a Link between Variant Creutzfeldt–Jakob Disease and Bovine Spongiform Encephalopathy

Moira E. Bruce, Robert G. Will, James W. Ironside, and Hugh Fraser

50.1 Introduction

There are numerous phenotypically distinct strains of the agents that cause transmissible spongiform encephalopathies (TSEs) or prion diseases [1; 2]. This phenomenon of TSE strain variation has been studied most extensively in experimental mouse models. TSE strains differ in their incubation periods in panels of inbred mouse strains and in their targeting of pathological changes within the brains of these mice, features that have been developed into formal strain-typing protocols. There are also striking differences between TSE strains in the biochemical characteristics of the disease-associated form of the prion protein (PrP), for example, in the molecular size of the protease-resistant core, and in the relative prominence of the differently glycosylated forms of the protein [3–5]. Extensive studies of mouse-passaged laboratory TSE strains have shown that TSE agents carry some form of strain-specific information that is independent of the host [1], although the molecular basis of this informational component is still unclear.

Over many years, transmission and strain typing studies in mice have been used to investigate the relationships between TSEs occurring naturally in different animal species. These studies have been extended to include transmissions of Creutzfeldt–Jakob disease (CJD) from patients where there was a suspected dietary or occupational link with bovine spongiform encephalopathy or BSE [6]. This chapter documents the approaches used to demonstrate a link between variant CJD and BSE.

50.2 TSE strain discrimination in mice

In mice experimentally infected with TSE isolates, the incubation period of the disease, though long, is remarkably precise and repeatable (Fig. 50.1). This incubation period depends on a precise interaction between the strain of the TSE agent and genetic factors in the challenged mice [7]. The mouse *Sinc* (scrapie incubation) gene has long been known to exert a major influence on the incubation period [8]. The two alleles of *Sinc* (s7 and p7) can account for differences in incubation periods of hundreds of days when mice are infected with a single TSE strain. It is now established that the *Sinc* gene encodes the prion protein, PrP [9; 10]. The s7 and p7 alleles of *Sinc* correspond to the a and b alleles of the PrP gene, encoding proteins that differ in their sequence by two amino acids [11]. It is also known that other host genes influence the incubation period, particularly when TSEs are transmitted to mice from another species (see below).

TSE strains are routinely characterized by injection of a panel of inbred mouse strains, differing in their PrP (or *Sinc*) genotypes and also in respect to other genes. Each TSE strain has a characteristic and highly reproducible pattern of incubation periods in these mouse strains [2]. TSE strains also show dramatic and reproducible differences in the type, severity, and distribution of pathological changes they produce in the brains of infected mice. The most prominent change seen in routine histological sections is a vacuolation of the neuropil, which targets different parts of the brain, depending mainly on the strain of TSE, but also to some extent on PrP and other mouse genes. The distribution of vacuolar degeneration is the basis of a semi-

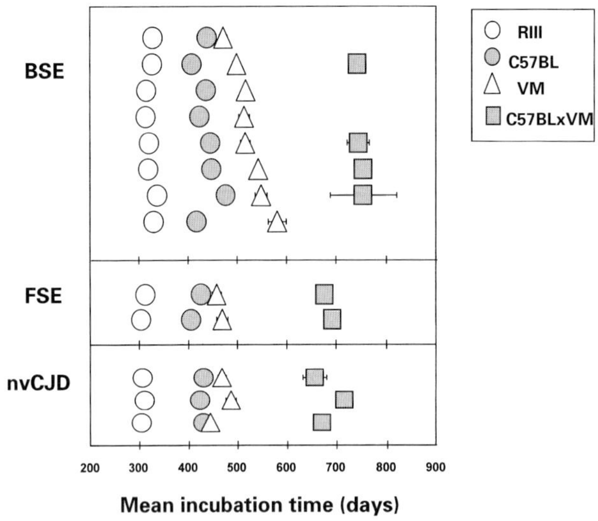

Fig. 50.1: Incubation periods in a panel of mouse strains challenged with eight sources of bovine spongiform encephalopathy, BSE, two sources of feline spongiform encephalopathy, FSE, and three sources of variant Creutzfeldt–Jakob disease, vCJD [6]. RIII and C57BL mice are of the PrP-a genotype, VM mice are of the PrP-b genotype, and C57BL × VM crossbred mice are PrP-ab heterozygotes. Figure by M. E. Bruce.

quantitative method of strain discrimination in which the severity of pathology is scored from coded sections in nine grey matter and three white matter brain areas to construct a lesion profile (Fig. 50.2) [12]. Each combination of TSE strain and mouse strain has a characteristic lesion profile [2].

These strain typing methods were developed originally for mouse-passaged laboratory strains of scrapie and have been used extensively in fundamental studies exploring the basis of agent strain variation. It is now clear that these methods can also be used at the primary transmission of a TSE from another species to mice, making it possible to strain type the TSE present in a naturally infected host.

50.3 Transmissions of animal TSEs to mice

BSE has been transmitted to mice from eight unrelated cattle sources, collected at different times during the epidemic and from widely separated geographical locations within the UK. All eight BSE sources produced similar incubation period patterns in a standard panel of mouse strains (Fig. 50.1) [13; 14]. There were large and consistent differences in incubation period between mouse strains of different PrP genotypes and also between mouse strains of the same PrP genotype. For example, the incubation period in RIII mice was consistently shorter than that in C57BL mice by about 100 days, even though these mouse strains are both of the PrP-a genotype. This shows that genes other than those encoding PrP can have a major influence on the progression of the disease. The eight BSE sources also produced a highly similar pattern of pathology in infected mice, as represented by the lesion profile (Fig. 50.2) [13; 14]. This uniformity of transmission results suggests that each of these cows was infected with the same major strain of agent. The consistency of the pathology reported in cattle with BSE [15; 16] also suggests that a single or a limited number of strains has been involved in the UK epidemic as a whole. However, to date, only a few BSE sources from a total of more than 180,000 cases have been strain typed, and the existence of other less common BSE strains remains possible. Indeed, a small number of cases of BSE with atypical features have now been identified, for example, in France [17], but these have not yet been characterized by mouse transmission.

Transmissions to mice have been achieved from other species with novel TSEs, which were suspected to be related to the BSE epidemic; the sources were three domestic cats, one greater kudu, and one nyala. The results of all five transmissions were strikingly similar to results

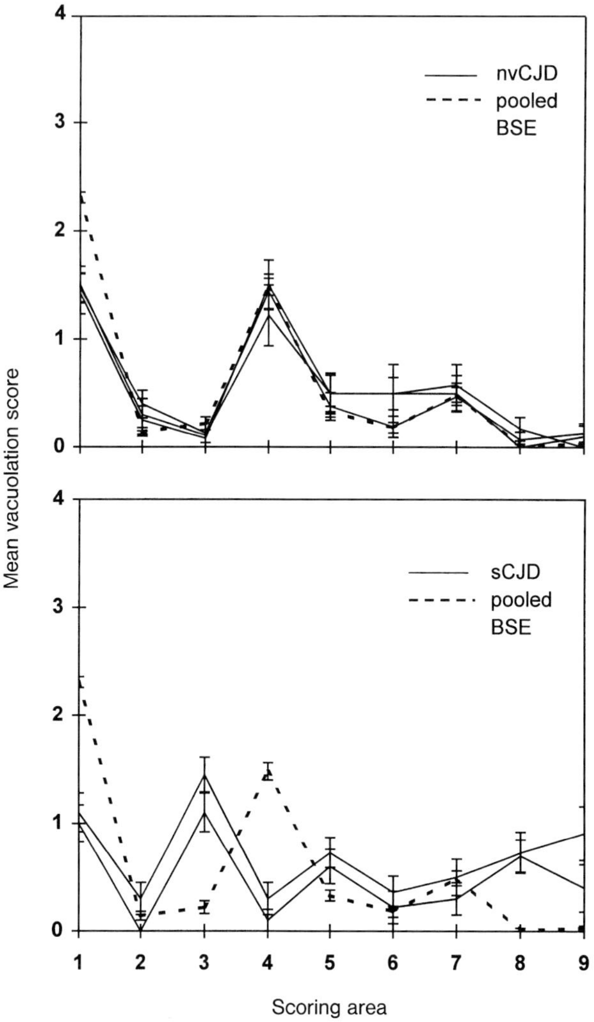

Fig. 50.2: Lesion profiles in RIII mice for transmissions of vCJD from three sources and sporadic CJD (sCJD) from two sources (a dairy farmer and a contemporary case), compared with a standard lesion profile for BSE based on pooled data from four transmissions [6]. Figure by M.E. Bruce.

from cattle sources (Fig. 50.1), indicating that these species were infected with the BSE strain and providing the first clear evidence for the natural spread of a TSE between species [13; 18]. BSE from cattle has also been transmitted experimentally to sheep (route: intracerebral, i.c., or oral), to goats and pigs (route: i.c.), and then from each of these species to mice. Again, the results of these mouse transmissions were closely similar to those in direct transmissions of BSE [13; 19]. These studies show, firstly, that the BSE agent is unchanged when passaged through a range of species and, secondly, that the donor species has little influence on the disease characteristics of BSE on transmission to mice.

In contrast to the above, sheep scrapie has transmitted poorly to mice, with most sources producing disease in only a proportion of the challenged mice [20]. In successful transmissions, the incubation periods and lesion profiles have been very variable, suggesting that there is agent strain variation in natural scrapie. None of the sheep scrapie sources tested so far has resembled BSE in terms of transmission characteristics.

50.4 Transmissions of vCJD and sCJD to mice

In the spring of 1996, approximately 10 years after the start of the BSE epidemic, the first 10 cases of a new variant of CJD (vCJD) were reported in the UK [21]. Up to May 2006 vCJD has affected 161 patients in the UK and 30 in France and other countries (see Chapter 1.9.3) [22; 23]. As vCJD is distinct from sporadic CJD (sCJD) in both its clinical and neuropathological characteristics and occurs in unusually young people, the epidemiology suggests that BSE may have spread to humans, possibly by dietary exposure. Circumstantial evidence of this came from the observation of florid amyloid plaques, identical to those seen in vCJD (Fig. 15.3a), in the brains of rhesus macaques experimentally infected with BSE [24]. Also, vCJD resembles BSE but differs from typical sporadic CJD (sCJD) in the molecular characteristics of the disease-specific isoform of PrP in brain extracts, i. e., in the relative prominence of the di-, mono-, and unglycosylated forms of the protein [3] and in the size of the protease-resistant core of the unglycosylated form [25].

To further investigate the possible link between vCJD and BSE, transmissions to mice were set up from the brains of three patients with vCJD, using an identical protocol to that used in the series of animal TSE transmissions described above [6]. For comparison, transmissions were also set up from six typical cases of sCJD. These included two dairy farmers who had had BSE in their herds and had therefore been potentially exposed to BSE-infected cattle or contaminated animal feed (26), two contemporary cases with no known occupational exposure to BSE, and two historical cases who had died before the onset of the BSE outbreak. All nine individuals were homozygous for methionine at codon 129 of the PrP gene and none had any of the PrP gene mutations associated with familial disease (fCJD; Tab. 26.5).

The vCJD transmissions from all three sources have given results in mice that are closely similar to those seen in transmissions of BSE from cattle and other species [6]. The incubation periods in all the mouse strains tested were consistent with those expected for very high doses of BSE (Fig. 50.1). The neuropathology in mice with vCJD was also closely similar to that seen in mice with BSE. For example, the lesion profiles in RIII mice for the three sources were almost superimposable and also closely similar to those in RIII mice with BSE (Fig. 50.2). As with BSE, a characteristic feature in all mouse strains with vCJD was a prominent involvement of the cochlear nucleus (an area that is not included in the lesion profile). In general, amyloid plaques were not a prominent feature of this pathology, except in periventricular areas in VM mice, as is also the case for BSE in this mouse strain [14]. These plaques were not of the florid type seen in patients with vCJD. Overall, the striking similarity between vCJD and BSE in mice, in terms of both incubation period and pathology, is very strong evidence that the same strain of agent is involved in the two species. Elsewhere, in a subsequent study, vCJD and BSE have produced closely similar disease characteristics when transmitted to transgenic mice overexpressing bovine PrP, adding to the evidence of a link [27].

In contrast to the results with the vCJD sources described above, no clinical neurological disease was seen in any mice challenged with sCJD, up to 800 days after injection. However, evidence for transmission from all six sCJD sources came from the presence of vacuolar degeneration typical of TSE infection in the brains of most mice, from about 400 days following challenge [6]. This diagnosis of sCJD was confirmed by the presence of the disease-associated form of PrP, demonstrated by Western blotting and immunocytochemistry. No such changes were seen in saline-injected control mice housed alongside the challenged groups, or in mice injected with human brain homogenates from patients with amyotrophic lateral sclerosis (ALS) or laryngeal carcinoma [28]. The neuropathological changes in mice with sCJD differed in severity between individual mice, but showed a consistent pattern between mouse strains and between sources. The lesion profile was strikingly different from that seen in BSE or vCJD transmissions to mice (Fig. 50.2) and also different from any seen in transmissions of sheep scrapie. These transmission results indicate that sCJD is associated with a different TSE strain from that causing vCJD or BSE. There is therefore no evidence from this study of a link between CJD in dairy farmers and BSE.

50.5 Conclusions

Strain typing based on transmission to mice has shown that:

1. vCJD is caused by the same strain of agent that has caused BSE and novel TSEs in cats and exotic ruminants,
2. vCJD is distinguishable from sCJD, and,
3. CJD in two dairy farmers is of the sCJD type and is not linked to the causative agent of BSE.

Supporting evidence that vCJD is caused by the BSE strain has come from Western blot analysis of the disease-associated isoform of PrP in each case. Epidemiological surveillance continues to indicate that vCJD is a novel condition occurring in comparatively young people, almost exclusively in the UK. The transmission studies described above, in combination with the surveillance data and other supporting evidence, provide compelling evidence of a link between BSE and vCJD.

References

[1] Bruce ME. Scrapie strain variation and mutation. Br Med Bull 1993; 49:822–838.
[2] Bruce ME, McConnell I, Fraser H, et al. The disease characteristics of different strains of scrapie in Sinc congenic mouse lines: implications for the nature of the agent and host control of pathogenesis. J Gen Virol 1991; 72:595–603.
[3] Collinge J, Sidle KC, Meads J, et al. Molecular analysis of prion strain variation and the aetiology of 'new variant' CJD. Nature 1996; 383:685–690.
[4] Somerville RA, Chong A, Mulqueen OU, et al. Biochemical typing of scrapie strains. Nature 1997; 386:564.
[5] Telling GC, Parchi P, DeArmond SJ, et al. Evidence for the conformation of the pathologic isoform of the prion protein enciphering and propagating prion diversity. Science 1996; 274:2079–2082.
[6] Bruce ME, Will RG, Ironside JW, et al. Transmissions to mice indicate that new variant CJD is caused by the BSE agent. Nature 1997; 389:498–501.
[7] Dickinson AG and Meikle VM. Host-genotype and agent effects in scrapie incubation: change in allelic interaction with different strains of agent. Mol Gen Genet 1971; 112:73–79.
[8] Dickinson AG, Meikle VM, Fraser H. Identification of a gene which controls the incubation period of some strains of scrapie in mice. J Comp Pathol 1968; 78:293–299.
[9] Hunter N, Dann JC, Bennett AD, et al. Somerville RA, McConnell I, Hope J. Are Sinc and the PrP gene congruent? Evidence from PrP gene analysis in Sinc congenic mice. J Gen Virol 1992; 73:2751–2755.
[10] Moore RC, Hope J, McBride PA, et al. Mice with gene targeted prion protein alterations show that Prnp, Sinc and Prni are congruent. Nat Genet 1998; 18:118–125.
[11] Westaway D, Goodman PA, Mirenda CA, et al. Distinct prion proteins in short and long scrapie incubation period mice. Cell 1987; 51:651–662.
[12] Fraser H and Dickinson AG. The sequential development of the brain lesions of scrapie in three strains of mice. J Comp Pathol 1968; 78:301–311.
[13] Bruce ME, Chree A, McConnell I, et al. Transmission of bovine spongiform encephalopathy and scrapie to mice: strain variation and species barrier. Philos Trans R Soc Lond Biol 1994; 343:405–411.
[14] Fraser H, Bruce ME, Chree A, et al. Transmission of bovine spongiform encephalopathy and scrapie to mice. J Gen Virol 1992; 73:1891–1897.
[15] Simmons MM, Harris P, Jeffrey M, et al. BSE in Great Britain: consistency of the neurohistopathological findings in two random annual samples of clinically suspect cases. Vet Rec 1996; 138:175–177.
[16] Wells GA, Hawkins SA, Hadlow WJ, et al. The discovery of bovine spongiform encephalopathy and observations on the vacuolar changes. In: Prusiner SB, Collinge J, Powell J, et al, editors. Prion diseases of humans and animals. Ellis Horwood, New York, 1992:256–274.
[17] Biacabe AG, Laplanche JL, Ryder S, et al. Distinct molecular phenotypes in bovine prion diseases. EMBO Rep 2004; 5:110–115.
[18] Fraser H, Pearson GR, McConnell I, et al. Transmission of feline spongiform encephalopathy to mice. Vet Rec 1994; 134:449.
[19] Foster JD, Bruce ME, McConnell I, et al. Detection of BSE infectivity in brain and spleen of experimentally infected sheep. Vet Rec 1996; 138:546–548.

[20] Bruce ME, Boyle A, Cousens S, et al. Strain characterization of natural sheep scrapie and comparison with BSE. J Gen Virol 2002; 83:695–704.

[21] Will RG, Ironside JW, Zeidler M, et al. A new variant of Creutzfeldt–Jakob disease in the UK. Lancet 1996; 347:921–925.

[22] Chazot G, Broussolle E, Lapras Cl, et al. New variant of Creutzfeldt–Jakob disease in a 26-year-old French man (letter). Lancet 1996; 347:1181.

[23] Ladogana A, Puopolo M, Croes EA, et al. Mortality from Creutzfeldt–Jakob disease and related disorders in Europe, Australia, and Canada. Neurology 2005; 64(9):1586–1591.

[24] Lasmezas CI, Deslys JP, Demalmay R, et al. BSE transmission to macaques (Letter). Nature 1996; 381:743–744.

[25] Hill AF, Desbruslais M, Joiner, S, et al. The same prion strain causes vCJD and BSE. Nature 1997; 389:448–450.

[26] Cousens SN, Zeidler M, Esmonde TF, et al. Sporadic Creutzfeldt–Jakob disease in the United Kingdom: analysis of epidemiological surveillance data for 1970–1996. Br Med J 1997; 315(7105):389–395.

[27] Scott MR, Will R, Ironside J, et al. Compelling transgenetic evidence for transmission of bovine spongiform encephalopathy prions to humans. Proc Natl Acad Sci USA 1999; 96(26):15137–15142.

[28] Fraser H, Behan W, Chree A, et al. Mouse inoculation studies reveal no transmissible agent in amyotrophic lateral sclerosis (see comments). Brain Pathol 1996; 6:89–99.

51 Risk Assessment of Transmitting Prion Diseases through Blood, Cornea, and Dura Mater

Johannes Löwer and Thomas R. Kreil

Preface

Laboratory evidence, animal experimentation, as well as clinical evidence have long suggested that the likelihood of transfusion-transmitted sporadic Creutzfeldt–Jakob disease (CJD) would be low or even hypothetical [1]. However, two cases of variant CJD (vCJD), believed to have been transmitted by blood transfusion, have been reported since the German edition of this book was published in 2001 [2; 3]. The potential significance of these recent developments within a very complex and still not fully understood subject is good reason for another systematic review of the pertinent evidence. The following section will include an update of the original discussion including information that has become available since 2001.

51.1 Introduction

It is widely accepted that human transmissible spongiform encephalopathies (TSEs or "prion diseases") can be transmitted through tissue or extracts that are derived directly from brain tissue, or which had been in intimate contact with such material (dura mater, pituitary gland hormones). Whether human prion diseases can also be transmitted through blood and/or blood products has been a matter of discussion among experts and also the general public. People are still very aware of the catastrophic events that resulted from the occurrence of the virus that causes AIDS (acquired immunodeficiency syndrome) in the blood supply. It is clear that everything possible must be done to prevent such a catastrophe from happening again with prion-related diseases.

In contrast to the well-documented transmission of the AIDS virus through blood and blood derivatives, the magnitude of risk of TSE diseases being transmitted through blood and/or blood products is less clear. Despite years of experimental work and rigorous clinical observation, there are currently no all-inclusive answers available.

There are two principal ways to determine whether a risk of transmission exists:

Epidemiological studies focus, for example, on the analysis of single case reports, the examination of specific at-risk populations, and case-control studies [4]. In addition, inferences regarding the situation in humans can be made following the observation of naturally infected animals.

Experimental studies can be performed to discover whether TSEs can be transmitted to laboratory animals by blood transfusion. Such studies will indicate the possible presence of infective agents in the blood of humans with TSEs.

51.2 Blood

51.2.1 Animal experiments

For more than 40 years now [5], animal experiments have focused on the question of whether TSE pathogens are present in blood. Generally, such examinations are carried out by administration of blood and tissue of naturally or experimentally infected animals, i. e., donor animals, to indicator animals, usually small laboratory animals such as mice, in order to determine the quantity of TSE agents required for infection (see Chapter 41). The interpretation of results from such experiments, summarized in

Table 51.1: Studies involving the crossing of the species barrier.

First author, year of publication	**Donor animals**			**Indicator animals**		Comment
	TSE strain, route of administration	Species	Tissue examined	Amount, route of administr.	Species	
Gibbs CJ Jr, 1965 [15]	Scrapie (natural infection)	Sheep	Blood serum		Mice	Infectivity in the serum is identical to that of the brain [cannot be reproduced (16)]
Hadlow WJ, 1980 [17]	Goat (kept together with scrapie-infected Suffolk sheep) with natural scrapie infection (in clinical stage)	3 goats (Nubian × Toggenburg breeds)	Blood serum and blood clots	30 µl i.c. (10-fold dilution, 10 mice per dilution)	Mice	Infectivity could not be detected in blood serum and blood clots
Hadlow WJ, 1982 [16]	Scrapie-exposed sheep (preclinical young sheep, and diseased sheep)	Suffolk sheep	Blood serum and blood clots	30 µl i.c. (10-fold dilution, 10 mice per dilution)	Mice	Infectivity could not be detected in blood serum and blood clots
Hadlow WJ, 1974 [20]	Scrapie Chandler strain i.c. $10^{7.3}$ mouse i.c. LD_{50} s.c. $10^{7.7}$ mouse i.c. LD_{50} i.c. $10^{6.6}$ mouse i.c. LD_{50}	Goats (Saanen breed)	20% blood clots	30 µl i.c.	Mice	Time course study, infectivity could not be detected in blood clots
Wells GAH, 1996 [106] Wells GAH, 1998 [6]	BSE (brain homogenate of 75 BSE cases) 100 g as single oral dose	Frisian/ Holstein calves	10% leukocyte layer ("buffy coat")	20 µl i.c. and 100 µl i.p.	RIII mice or C57BL -J6 mice	Time course study, infectivity could not be detected in the leukocyte layer up to 22 months after inoculation

Tables 51.1 and 51.2, requires cautious consideration of several aspects as described in the following paragraphs.

Host restrictions: species barrier, genotype

During an attempted transfer of TSE pathogens from one species to another, a rather pronounced species barrier is often encountered. This means that the transmission of TSE agents between animals of different species requires, if it is possible at all, orders of magnitude more infectious material than would be required for infection of an animal of the same species. For example, the titration of the BSE agent in calves is about 1,000 times more sensitive than titration of the same agent in certain mice [6]. Table 51.1 summarizes studies that were designed to transmit TSE agents across a species barrier. The shortening of the incubation period in the new host after serial transmission versus the original incubation time on first transmission defines the height of the species barrier. If this incubation period is longer than the natural lifespan of the species under observation, the species barrier will effectively prevent the development of clinical disease, i. e., the recipient species appears to be "resistant" against the TSE agent examined. In other cases, the TSE agent cannot be amplified in the new host species.

Studies of infection within the same species are inherently more sensitive and therefore also more relevant to our understanding of the transmissibility of prion diseases. However, in some species, notably in sheep, the susceptibility to infection by different strains of the TSE agent is profoundly influenced by the genotype of the PrP gene (*Prnp*). In sheep, polymorphisms at codon 136 [alanine (A) or valine (V)], codon 154 [arginine (R) or histidine (H)], and codon 171 [arginine (R) or glutamine (Q)] are of utmost importance. The genotype ARQ (in numerical order of the codons 136, 154, and 171) is highly susceptible to the scrapie agent, while the genotype AXQ (where X stands for either of the two possible amino acids encoded at 154) is most susceptible to the BSE agent [7; 8].

Investigations into the infectivity of blood within the same species are summarized in Table 51.2.

Routes of administration

Experimentally, samples from donor animals are examined for the presence of TSE agents by direct inoculation into the brain (intracerebral, i.c.) of indicator animals. The efficiency of this route of administration is generally greater than for parenteral routes (intravenous, i.v.; subcutaneous, s.c.) and/or oral infection; yet when a species barrier has to be overcome, this is not always the case.

Natural or experimental infection

The levels of infectivity within specific tissues, including the blood, are possibly different between naturally infected and experimentally infected animals. When animals are infected experimentally, usually a large amount of infectious material is injected intracerebrally or peripherally (e. g., into the abdominal cavity, intraperitoneally, i.p.). It is quite conceivable that under these circumstances infectivity occurs in the blood merely as a consequence of either damage to blood vessels or the uptake of infectious material by macrophages [9]. The observation of blood infectivity in several experiments [9–11] may thus simply represent recovery of the material inoculated into the indicator animals. This interpretation is also supported by recent data that indicate that a TSE inoculum can persist within an infected animal for months [12].

In Table 51.2, studies are arranged according to whether naturally or experimentally infected donor animals were used.

Limited inoculation dose

The significance of experimental studies is further limited by restrictions in the amount of material that can be inoculated, typically only a very small amount. Investigations in rodents, for example, usually employ sample volumes of no more than 50 μl or even less per animal. At infectivity levels of less than 20 infectious units (IU) per milliliter, there is a possibility that the inoculated volume could, by chance, be entirely free of TSE agents and hence yield a false-negative result. Studies published since the 1970s have thus tried to concentrate the infectious agent from blood. This was done either by enriching the white blood cells (buffy coat) [13; 14] or by preparing specific protein fractions [9; 10].

Cross-contamination

In early studies, the high resistance of TSE agents to inactivation had not been sufficiently appreciated and thus cross-contamination had not been adequately excluded. For analysis of a particular sample, incubation periods and percentages of infected animals that are entirely different from the values expected from the collective information established by other investigations, as well as lack of reproducibility, are indicators of such cross-contamination. A study that determined the blood infectivity titer of a naturally infected sheep as being equally high as otherwise expected for central nervous system (CNS) material may serve as an example [15]. The result could not be reconfirmed [16].

With these limitations in mind, the following section will evaluate studies that have been published in recent years, summarized in Tables 51.1 and 51.2.

Table 51.2: Studies in which the species barrier was not crossed.

First author, year of publication	**Donor animals**			**Indicator animals**		Comment
	TSE strain, route of administration	Species	Tissue examined	Amount, route of administr.	Species	
Pattison IH, 1962 [5]	Goat-adapted scrapie (3 × passaged in goats) (1 ml 10% brain homogenate) i.c.	Goat	Whole blood	1 ml i.c.	Goat	Time course study, infectivity was not detected in whole blood
Pattison IH, 1962 [5]	Scrapie-infected goats	Goat	1. Whole blood 2. Blood cells 3. Blood serum	1a. 1 ml i.c. 1b. 3 × 5 ml s.c. (weekly) 2. 1 ml i.c. 3. 1 ml i.c.	Goat	Infectivity was not detected
Marsh RF, 1969 [107]	TME 1. 1 ml 10% brain of naturally infected mink i.m. 2. 0.1 ml 10% brain (2. passage) i.c.	Mink	Blood serum	1 ml s.c. (dilution in steps of one order of magnitude)	Mink	Infectivity was not detected in the blood serum
Marsh RF, 1973 [37]	TME 1. Late clinical phase 10^5 LD_{50} i.c.	Mink	1. Whole blood, blood plasma, 10% erythrocytes, 10% thrombocytes, white blood cells (1.7×10^7/ml), PHA stimulated lymphocytes (1.5×10^7/ml) 2. Lymphocytes	0.1 ml i.c.	Mink	1. No infectivity found in the different preparations 2. Time course study, no infectivity in the lymphocytes, but in spleen and peripheral lymph nodes
Hadlow WJ, 1987 [23]	TME pathogens of the Idaho strain, second mink passage, 10^3 LD_{50} s.c.	Mink	Undiluted blood serum	100 µl i.c. (2 animals)	Mink	Time course study, serum (28 weeks after inoculation) found once to be infective in 1 of 2 animals
Clarke MC, 1967 [24]	Scrapie Chandler strain (mice) Scrapie Chandler and Fisher strain (rats)	Young mouse, Wistar rat	Serum from the blood pool of animals in the progressive stage of the disease 1. Blood from the chest cavity 2. Blood after heart punction	50 µl i.c. (mice and rats)	Young mouse, Wistar rat	More transmissions observed in experiment 1 than in experiment 2 (contamination through tissue?)
Eklund CM, 1967 [22]	Scrapie Chandler strain (4. mouse passage), $10^{5.7}$ LD_{50} s.c. (50 µl 10% mouse brain suspension)	Mouse	Blood clots Blood serum (pool of three mice)	30 µl i.c. (10-fold dilution, 6 mice per dilution)	Mouse	Time course study, no infectivity found in blood
Field EJ, 1968 [11]	Scrapie Chandler strain (50 µl 10% mouse brain suspension) 1. i.c. 2. i.p.	Mouse	Blood from the chest cavity 1. Blood serum 2. Whole blood	1. 50 µl i.c. 2. 50 µl i.c.	Mouse	Time course study for the first 18 hours p.i., infectivity was confirmed, in particular in whole blood
Dickinson AG, 1969 [21]	Scrapie ME7 strain 20 µl s.c. and 10 µl i.p. (1% brain suspension)	Different mouse strains	Blood (heart punction)	20 µl i.c. (10% suspension)	C57BL mouse (11 to 15 mice per sample)	Time course study, blood was occasionally found to be positive (at 3 of 12 time points, 1 or 2 of 11 to 15 mice)

Table 51.2 (continued)

First author, year of publication	Donor animals			Indicator animals		Comment
	TSE strain, route of administration	Species	Tissue examined	Amount, route of administr.	Species	
Diringer H, 1984 [10]	Scrapie 263K strain (100 µl 1% brain homogenate) i.p.	Hamster	Blood concentrate P_{215S}	50 µl (equivalent to 0.2 ml blood) i.c.	Hamster	Time course study, infectivity found in blood
Casaccia P, 1989 [9]	Scrapie 263K strain (50 µl 1% brain homogenate) i.p.	Hamster	Blood concentrate P_{215S}	50 µl (equivalent to 0.2 ml blood) i.c.	Hamster	Time course study, infectivity found in blood
Manuelidis EE, 1978 [14]	CJD (of 54-year-old patient) 5 × transferred into Guinea pigs (100 µl 1% brain homogenate) i.c.	Guinea pig (Hartely strain)	Leukocyte layer of 8 ml centrifuged blood	In total 0.4 ml (i.c. and s.c. and i.m. and i.p., 0.1 ml each) (corresponds to 6.4 ml blood)	Guinea pig	Irregular detection of infectivity
Tateishi J, 1980 [33]	CJD, serially transferred into mice (the same strain as used by Kuroda, 1983; 10 µml 15% brain homogenate) i.c.	Mouse	Whole blood	10 µl i.c.	Mouse	2/10 indicator mice were positive
Kuroda Y, 1983 [13]	"CJD": Fu strain (untypical case = GSS), 2 × serially transferred into mice (30 µl 10% brain homogenate) i.c.	BALB/c mouse	Leukocyte layer, blood serum, erythrocytes	30 µl leukocytes, i.c. 100 µl serum, i.p. 100 µl erythrocytes, i.p.	BALB/c mouse	Time course study, infectivity was found
Doi T, 1991 [108]	"CJD Fukuoka 1" strain (the same strain as that used by Kuroda, 1983), 5 × serially transferred into mice $10^{5.5}$ LD_{50} i.c. (20 µl)	ddY mouse	Whole blood (heart punction)	20 µl whole blood i.c.	ddY mouse	No infectivity was found in blood
Houston F, 2000 [19]	5 g BSE-affected cattle brain, orally	Cheviot sheep	Whole blood, taken preclinically	400 ml whole blood, i.v.	Cheviot sheep from MAFF scrapie-free flock	1 of 19 transfused animals developed BSE; see [8]
Hunter N, 2002 [8]	1) 5 g BSE-affected cattle brain, orally 2) Scrapie (natural infection)	Cheviot sheep	Whole blood or buffy coat, taken preclinically or during clinical stages	400 ml whole blood, or buffy coat, i.v.	Cheviot sheep from DEFRA scrapie-free flock	1) 2 of 17 (plus 2 more suspected) developed BSE from preclinical whole blood 2) 4 of 21 developed scrapie from preclinical or clinical blood
Bons N, 2002 [25]	Macaque-passaged and Microcebus-adapted BSE, i.c.	Prosimian microcebe (*Microcebus murinus*)	Buffy coat	Buffy coat from 1 ml of blood, i.c.	Prosimian microcebe (*Microcebus murinus*)	Incubation period similar to brain-inoculated animals
Cervenakova L, 2003 [35]	Mouse-adapted 1. vCJD, or 2. GSS (Fukuoka-1), i.c.	R III and Swiss mouse	Buffy coat, plasma, platelets, red cells	0.1–0.2 ml i.v., or 0.03 ml i.c.	R III and Swiss mouse	20–30 infectious doses/ml during pre- and clinical phases in buffy coat and plasma; similar between vCJD and GSS

51.2.2 TSE agents in the blood of naturally infected animals

When the German edition of this book appeared in 2001, only two studies had been reported that tried to find TSE infectivity in the blood of naturally infected animals, i. e., in scrapie-infected goats [17] and sheep [16] (Table 51.1). Although the animals were examined at different stages of the disease, TSE agents were not detected in the serum or in the blood clot. This is in contrast to Gibb's finding published in 1965 [15].

However, with the advent of more recent experimental evidence, these "... clouds of complacency ..." [18] have disappeared. After an initial finding on the transfusion-transmission of experimental BSE into one sheep [19], which will be discussed in the following paragraph, the same group subsequently reported the transmission of natural scrapie through blood transfusion in the same model [8]. This was the first demonstration of blood infectivity in a naturally TSE-infected donor, and, to date, the only other evidence in similar circumstances are the two presumed human vCJD transfusion-transmissions [2; 3]. From the sheep model [8], transfusions of 400-ml blood units derived from preclinical or scrapie-diseased donor animals were reported to transmit the disease to a total of 19 % of recipients. All indicator animals were of the genotype (VRQ/VRQ), which is highly susceptible to the scrapie agent (Section 51.2.1). In these experiments, the scrapie strain identity was confirmed by glycoform analysis.

51.2.3 TSE pathogens in the blood of experimentally infected animals

When goats experimentally infected with scrapie [20] and cattle perorally infected with BSE material were used to investigate the infectivity of individual tissue and body fluids of these animals during the incubation period and clinical stages of disease (Table 51.1), blood yielded negative results.

However, these experiments tried to detect infectivity by the use of indicator animals across a species barrier (i. e., mice), and by using blood samples of only 30 μl without prior concentration of the TSE agents. It can thus only be concluded that the blood did not contain infectivity concentrations high enough to reach the limit of detection in the approach used.

Experimental models enable investigations into the distribution of TSE agents in different tissues without the species barrier being an issue (Table 52.2). A number of these investigations have also dealt with the variation of infectivity during the incubation period.

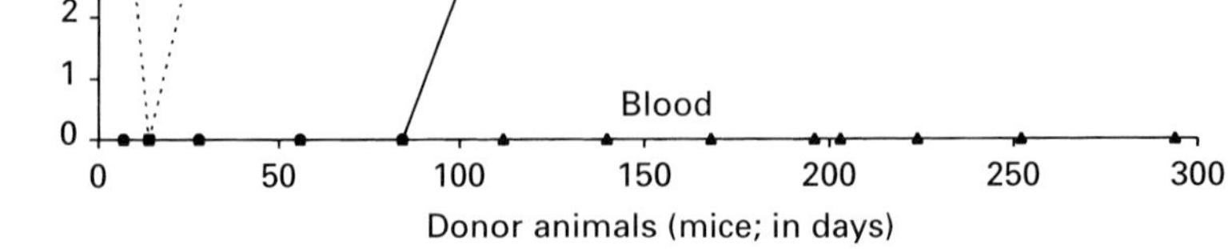

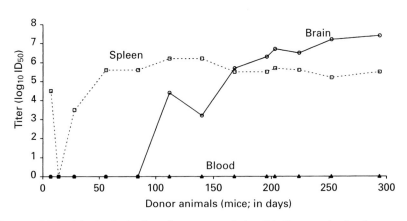

Fig. 51.1: Course of infectivity in the brain, spleen, and blood of mice that were infected with a mouse-adapted scrapie strain; the drawing was made following the data provided in publication [22]. Open symbols: all indicator animals of a group were infected; closed symbols: only a part or none of the indicator animals were infected. Figure by J. Löwer ©.

Scrapie-infected goats, mice, and hamsters, and TME-infected minks

Experiments with scrapie-infected goats [5] and mice [21; 22] (Fig. 51.1) and transmissible mink encephalopathy (TME)-infected minks [23] were carried out to investigate blood infectivity. The results showed that only isolated animals (Table 51.2) that had received blood developed disease [21; 23]. In a further series of experiments [24], the inoculation of blood samples from scrapie-infected and diseased mice resulted in disease in several of the recipient mice.

The situation is different in scrapie-infected hamsters [9; 10] where, at least after an enrichment of the TSE agent, blood infectivity has been detected. The continuous decline of blood infectivity titers [9] (Fig. 51.2) as determined by the length of the incubation period and the percentage of diseased animals per experimental group suggests recovery of the initial inoculum [12] rather than replication of the TSE agent in the recipients' blood. However, on the basis of available data, an exact comparison between the amount of infectivity within the inoculum and the total infectivity in the blood or within the organism infected is difficult. Evidently, though, the declining course of infectivity in blood is in marked contrast to the development of infectivity in the brain and spleen (Fig. 51.2). Other findings [11] can also be explained by the recovery of the inoculum.

BSE-infected sheep and prosimian microcebe

Two reports on experiments with BSE-infected sheep have shown that in this model transfusion transmission of the TSE agent is possible. Specifically, in the first report, a single sheep that had received a 400-ml blood transfusion from an asymptomatic BSE-infected sheep developed BSE [19]. The second report showed that another sheep had developed BSE after the same treatment, bringing the total to 2 per 19 transfused animals, or 17%, with 2 more animals suspected of developing BSE at the time of publication [8]. Again, the indicator animals were of the genotype AXQ/AXQ, which confers sensitivity to BSE (Section 51.2.1.). For both diseased animals, the identity of the agent in the recipient sheep was verified as BSE by glycoform analysis.

Consistent with these findings, in the primate *Microcebus murinus* model, a buffy coat donation obtained from a donor animal, clinically

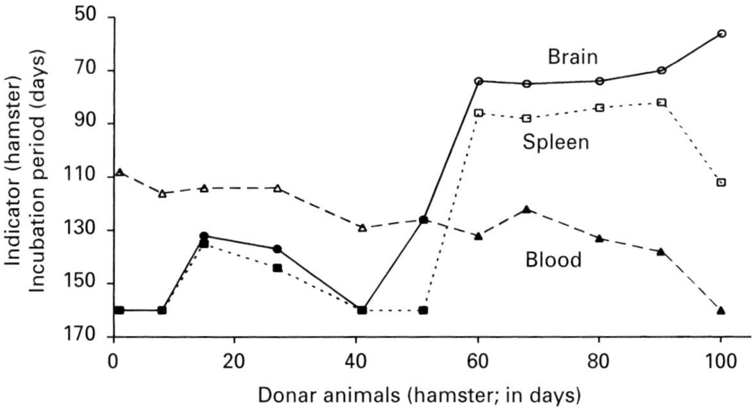

Fig. 51.2: Course of infectivity in the brain, spleen, and blood of hamsters that were infected with a hamster-adapted scrapie strain [9]. The assessment of infectivity based on the incubation periods in indicator animals led to the overestimation of low titer since the correlation between infection titer and incubation period has a biphasic course – steep in the case of high titer and flat in the case of low titer [21]. Open symbols indicate that all animals of a group were infected; closed symbols indicate that only part or none of the animals were infected. Drawing taken from [9], slightly modified, with kind permission of Springer.

infected with BSE, which had been passaged through macaques and then adapted to the prosimian microcebe, also induced the disease in a single recipient animal after i.c. inoculation [25].

CJD-/vCJD-infected guinea pigs, mice, and squirrel monkeys

While experiments studying human TSE agents in animal models are, in principle, interesting, a consistent picture has only arisen from more recent investigations. Initially, an investigation using a guinea pig model found CJD infectivity in blood, yet irregularly and biologically hardly plausible [14].

As further studies from the same laboratory [26–31] could not be reproduced, this observation remains doubtful. Data derived from a later study using a mouse model [13] (Fig. 51.3), however, prove consistent. The authors initially indicated that the human infectious agent was derived from an atypical CJD case, yet the clinical presentation of the case [32; 33] suggests that it was Gerstmann–Sträussler–Scheinker disease (GSS). As the disease progressed, increasing levels of infectivity were found in the blood.

Later, using a mouse-adapted strain of CJD in a mouse model, low levels of buffy coat infectivity were determined during preclinical stages of the disease (6–12 IU per ml), which increased toward the clinical stage (approximately 100 IU per ml) [1; 34]. When blood obtained during these late stages was fractionated, plasma was found to contain 20 IU per ml of plasma, cryoprecipitate (source material for plasma-derived factor VIII) 2 IU per ml, and less than 1 IU per ml was contained in fractions IV and V (source material for, amongst other plasma proteins, albumin).

Finally, in 2004, a study was published that compared mouse models of GSS and vCJD. Low levels of GSS and vCJD infectivity (20–30 IU per ml) could consistently be detected in the buffy coat and plasma of experimentally infected mice, during both preclinical and clinical stages of the disease [35]. Even lower levels of infectivity were observed in platelets (10 IU per ml), and the red blood cell samples tested were negative in this model. Incubation periods and brain infectivity titers were shorter and lower, respectively, in vCJD-infected mice compared to GSS-infected animals.

The development of CJD/vCJD was also investigated using a total of 80 squirrel monkeys.

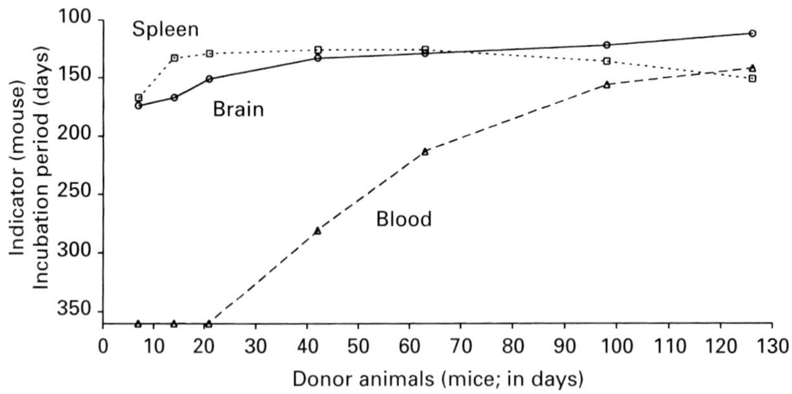

Fig. 51.3: Course of infectivity in the brain, spleen, and blood of mice that were infected with a mouse-adapted GSS strain; the drawing was made following the data provided in publication [13]. The assessment of infectivity based on incubation periods in indicator animals leads to an overestimation of lower titers, because the correlation between infection titer and incubation period has a biphasic course – steep in the case of high titers and flat in the case of low titers [21]. Open symbols indicate that all animals of a group were infected. Figure by J. Löwer ©.

In this study, animals infected i.c. with either CJD- or vCJD-derived human brain material were used as blood donors, with quarterly cumulative transfusions from donors to paired recipients. The recipients of the 1:10 diluted human brain specimens died approximately 21 and 24 months after infection from CJD and vCJD, respectively. However, the corresponding recipients of cumulatively 7–8 and 8–12 blood transfusions from these TSE-infected donor animals have remained asymptomatic for almost 6 and 5.5 years after receiving their first transfusion.[1]

Infectivity in lymphocytes

The suggestion that for virtually all animal models high levels of TSE infectivity can be found in the spleen [22; 36] also warrants discussion. It is argued that the spleen contains a large reservoir of lymphocytes that are supposedly exchanged with peripheral blood lymphocytes. However, in animal experiments, the occurrence of infectivity in the blood and the spleen do not correlate [9; 13; 22] (Figs. 51.1 and 51.2). This lack of a correlation between spleen and blood infectivity is further substantiated by studies that have determined that lymphocytes taken from the spleen of infected animals were infectious, whereas lymphocytes taken from the blood of the same animals did not carry any infectivity [37] (A. Aguzzi, personal communication). Although B lymphocytes play a certain role in the spread [38] of the TSE agents, blood B lymphocytes are not their likely carriers [39].

51.2.4 Overall assessment of animal experiments

An overall assessment of prion-related animal experiments shows that the results depend on four factors:[2] (i) The agent (scrapie, TME, BSE, CJD), (ii) the species and possibly also on the genotype of the host animal, (iii) the dose, and (iv) the route of administration [40].

It is therefore very difficult, and also associated with numerous uncertainties, to extrapolate from any one model to another, e. g., experimental animal data to the situation in humans. For example, PrP^{Sc} can be readily identified in the tonsils of infected sheep, whereas it cannot be found in the tonsils of BSE cattle.[3] An appropriate appraisal would be to say that little, if any, TSE infectivity can be found in the blood of TSE-infected model animals, which, under certain circumstances, may become measurable during the later stages of disease incubation [13; 21; 23; 24; 33]. This assessment is in agreement with more recent experimental information (R.G. Rohwer, personal communication), which was generated under conditions that exclude the recovery of the inoculum because the animals were only injected with a single infectious dose. From these results, a maximum TSE agent titer of 1–10 infectious units per ml of blood can be deduced. These titers are several orders of magnitude lower than the titers of viruses such as hepatitis B virus (HBV), hepatitis C virus (HCV), and human immunodeficiency virus (HIV; 10^5–10^{12} per ml), which are known to be transmitted via blood. It is impossible to infer from animal experiments whether such a low titer would be sufficient for the transmission of TSE via the transfusion of blood components.

51.2.5 Experimental examination of blood of sporadic and familial CJD patients

A few studies [41–44] have generated findings that should document the presence of CJD pathogens in the blood of CJD patients. It is difficult to evaluate these studies because either important experimental details are missing [41] or the results are not congruent [42–44]. In one case [44], the concentration of TSE infectivity found in urine, a body fluid which has in no other investigation been found to be infectious, was close to the titer usually found in the brain.

[1] Status May 2005; study conducted at the University of South Alabama, with principal investigators Dr. P. Brown, Dr. R.G. Will, and Dr. C. Abee; unpublished.
[2] These factors partially determine the species barrier.

[3] For preventive reasons, the tonsils of cattle are counted as specified risk material (SRM; see Chapter 53.3.1).

The same source [44] was initially unable to identify the CJD agent in the blood of two patients, yet it only took a three-fold concentration of the plasma to result in a titer that was similar to that of brain [43]. Other studies [42] are doubtful, as the same group also reported successfully transmitting the CJD agent from a patient with an undiagnosed neurological disease [30], an Alzheimer's patient [28], and even healthy persons [28; 29] to animals, yet these results could not be reproduced by others [26; 27; 31].

Opposing these case reports, a long-standing series of experiments has focused on examining the infectivity of human tissue and body fluids, including blood of kuru and CJD patients [40; 45]. These experiments generally used nonhuman primates (see Chapter 41) and, while still limited, a number of patients larger than that used in the reports described in the previous paragraphs. A transmission of TSE agents was not possible with blood, serum, and leukocytes, although it can be assumed that the species barrier between nonhuman primates (indicator animals) and humans (donors) would be relatively low (see Chapter 9.3). In addition, the human specimens tested were, at least in the majority of cases (although this is difficult to ascertain from the respective publications), injected intracerebrally, which, as mentioned previously, in many instances is the most effective way to achieve infection. One of these experiments is particularly remarkable in that more than 300 ml of blood taken from three different CJD patients was injected i.v. into three chimpanzees [46]. More than 25 years after receiving these transfusions, the recipient primates have still not developed a spongiform encephalopathy [45] (P. Brown, personal communication).

A critical appraisal of available publications does not provide convincing experimental evidence for the transmission of CJD through human blood or blood components. It must be remembered, though, that for obvious reasons a species barrier was involved in all these experiments. The use of transgenic mice in which the murine prion protein gene (*Prnp*) was rendered nonfunctional and which express the human *PRNP* homologue might improve the situation. But even using this approach, infectivity has not been found in the blood of two sporadic CJD cases and one familial CJD case (S. Prusiner, personal communication). On the basis of relatively few patients investigated to date, one has to conclude that the TSE agent is absent, or occurs at low levels, in the peripheral blood of clinically diseased patients in sporadic and familial CJD. Studies involving a much larger number of patients suffering from sporadic CJD, the variant CJD, or other human prion diseases (Table 1.1) are clearly desirable to substantiate this conclusion.

51.2.6 Epidemiological examinations of sporadic CJD

A number of epidemiological investigations have looked at whether there is evidence for the transmission of CJD through blood and blood products. Before going into any detail of the results obtained in these investigations, the major difficulties in conducting and interpreting such studies should be emphasized:

1. As mentioned, CJD is a rare disease with an incidence of about 1–2 per 1 million people per year (Table 36.2). For this reason, it is difficult to find a sufficiently high number of patients to include in a case-control study, and thus such a study would best be conducted through international, long-term cooperation (see Chapter 29.5).
2. The incubation periods of CJD can vary considerably. For example, patients who contracted CJD through cornea transplantations (see Section 51.3) developed the disease between 18 months and 30 years after transplantation [47; 48]. Significantly, varying incubation periods are also known from other iatrogenic CJD cases [49] (Table 14.1). The incubation period in turn is a reflection of the inoculum dose and the site of entrance. However, long and variable incubation periods may obscure the correlation between the appearance of the disease and prior infection (compare Chapter 2.14).

Within the most extensive epidemiological study available to date, 28 blood donors who were later diagnosed with CJD and 368 recipients of their respective blood products were identified [50]. As of October 2004, 342 of these recipients were either alive, or had died as a result of causes other than CJD, representing a total of almost 1,400 person years of follow-up. Within the limitations outlined above, this information seems still rather reassuring.

Sporadic CJD case reports

Published reports have, apart from an exception discussed later, not found an elevated risk of CJD in recipients of blood products obtained from donors who developed symptoms of CJD several years after their donation. Retrospective studies on people who were transfused with cellular blood components from donors who later developed CJD also failed to identify any CJD cases definitively linked to those transfusions [51; 52]. Of course, these studies involved relatively few patients; nevertheless, the time over which these patients were monitored was very long, in 9 cases, between 4 and 20 years had passed after transfusion [51].

An Australian study [53] that described four people who supposedly died from blood transfusion-associated CJD leaves many questions unanswered. Importantly, it is not known whether the blood donors themselves later suffered from CJD. A history of receiving transfusions of blood products would be expected also for a certain percentage of CJD patients. Whether or not CJD patients have received transfusions more frequently than a control group will be dealt with below.

In 1995, a patient developed CJD 2 years after she had undergone a liver transplant that involved the administration of albumin [54; 55]. Later it became known that the albumin had come from a plasma pool, to which a person who had subsequently died of CJD, had contributed. The finding was intensely debated in the scientific community [54–57], yet there are two major arguments against the suggestion that this particular batch of albumin was causally related to the woman's CJD:

1. The relatively short incubation period of only 2 years would suggest that a significant amount of infectivity had been received. This is at odds with current information that suggests that – even in a worst-case scenario – very low CJD infectivity titers might possibly occur in human albumin:
   - It is known that, although high titers of infectivity are observed in the brain of infected individuals, the levels of CJD infectivity in the plasma are several orders of magnitude lower, if detectable at all.
   - The implicated plasma sample would have been diluted with the plasma of several thousand other donations obtained from people who were not suffering from CJD.
   - The manufacturing processes of plasma derivatives have been shown experimentally to result in a very substantial reduction of different artificially added TSE agents [34; 58–62], and this reduction capacity is particularly high for albumin, i. e., between >7.7 and $>16.0 \log_{10}$ [63].
2. Development of CJD in other recipients of the same implicated albumin batch has not been observed. This is in contrast to the situation that was observed with recipients of CJD-contaminated pituitary gland hormones or dura mater implant batches (see Section 51.4). In the 1990s, it became obvious that blood donated by people who were subsequently diagnosed with CJD was relatively frequently processed into blood products, and consequently a certain number of further albumin batches contained CJD contamination. Nevertheless, there have been no further reports in which the use of albumin was suggested to be the cause of CJD [56].

To summarize, the publications do not give reason to assume that blood and blood products can transmit sporadic CJD. No single instance has been identified that would have supported the notion of regional or temporal clustering of CJD cases. If CJD pathogens could be transmitted by blood and blood derivatives, a higher incidence of CJD would have been observed in people dependent on the regular administration of blood or blood products (see Section 51.2.6). In addition, case-control studies designed specifically to determine the risk from transfusion would have detected blood and blood products as risk factors.

Examination of specific populations

People with hemophilia receive cumulatively large amounts of coagulation factor concentrates, on a regular basis, depending on the severity of their disease. The dreadful experience during the 1980s and 1990s showed that through treatment with blood plasma-derived factor concentrates, this group had been at a significant risk of contracting infections from blood-borne viruses. Consequently, a large number of people with hemophilia were infected with HIV, HBV, and HCV prior to the introduction of dedicated virus-inactivating procedures into the manufacturing processes of plasma derivatives. A study of people with hemophilia can also indicate any risk of blood-transmitted or -transmissible diseases. For example, only 1 year after AIDS was first recognized [64], clusters of this disease were also seen in people with hemophilia [65], and this despite an incubation period between infection and onset of disease of, on average, approximately 10 years.

If CJD cases were to occur at an increased frequency in people with hemophilia, this would be a potentially serious indication for the transmissibility of CJD through blood and blood derivatives. However, cases of CJD in people with hemophilia have not been reported to date [66]. Physicians unanimously agree that, to the best of their knowledge, no person with hemophilia has ever developed CNS diseases other than those associated with AIDS [67]. One report also asserts that CJD in people with hemophilia has not been misdiagnosed with the occurrence of AIDS-associated dementia [68]. A comprehensive study performed in the USA analyzed the causes of death as given in death certificates. The study did not identify any instance in which CJD had occurred in patients who had undergone long-term therapy with significant amounts of blood or blood derivatives, e. g., people affected with hemophilia A, hemophilia B, thalassemia, and sickle cell anemia [69].

It is sometimes argued that the long incubation period of CJD and the potentially restricted life expectancy of the patient groups mentioned above could suggest that the significance of these findings is limited. While it is true that it is likely that not all potential infections might have been identified, decades of treatment for people with hemophilia and other groups should have been sufficient for the identification of a few cases, if they had occurred.

Case-control studies

Case-control studies compare the frequency of certain risk factors in a specific group of patients with that in a control group of corresponding age and gender (compare Chapter 40.3.2). If a specific risk factor is significantly more frequent in the patient group compared to the control group, it must be considered whether this risk factor is causally associated with the development of the disease.

A number of studies [70–74] have focused on the question of whether CJD patients have received blood transfusions more frequently than individuals of control groups. The results clearly show that this is not the case. In contrast, studies involving sufficiently high numbers have found that blood transfusions are in fact less frequent in CJD patients than in the respective control groups. For the most comprehensive study, approximately 11 % of CJD patients versus 19% of the control group had received blood transfusions [74]. This discrepancy is probably due to the selection of the control group that consisted of hospital patients, i. e., persons who are generally in greater need of blood transfusions as compared to the normal population (selection bias). While it is recommended that future studies should also involve healthy people as controls, the results obtained so far suggest that blood transfusion has not been a significant factor in the pathogenesis of CJD [see also 4].

51.2.7 The risk associated with variant CJD

Assessing the risk associated with variant CJD is very different as far less experimental and epidemiological information is available. Also, the infectious agent is different from the CJD agent in terms of biochemical [75] and biological [76] properties and the clinical and pathological features of the disease it causes. A further discriminating feature seems to be a quantitatively different tissue distribution of vCJD, where PrP^{Sc}/infectivity occurs in lymphatic tissue such as tonsils [77] (see Chapter 47.4.2) or the appendix [78] more frequently, compared to the classical CJD forms where infectivity in these lymphatic tissues is rare [45].

It has been noted that the experimental studies mentioned in Section 51.2.3 show no correlation between the identification of TSE agents in the spleen versus the blood. Despite this, it had been prudently assumed that the risk of transmitting the vCJD agent through blood or blood derivatives would be higher than that observed in other human TSE forms. Since the German edition of this book appeared in 2001, the precautionary approach has certainly been justified with two reported presumed vCJD transmissions through labile blood products.

Late in December 2003, the first case was identified [2]. It involved a non-leukodepleted red cell concentrate, which had been collected approximately 3.5 years before disease onset in the donor. The recipient then developed symptoms of vCJD 6.5 years after receiving the transfusion.

Shortly thereafter, another case was reported – an elderly person had received a non-leukodepleted red cell concentrate from a donor who, 1.5 years after donation, developed vCJD. The recipient died 5 years later of an unrelated cause and without any evidence of neurological disease. However, as part of the UK vCJD surveillance program, an autopsy was performed and biochemical evidence for vCJD infection was found in the spleen and a cervical lymph node, though not in the brain [3]. It is worth noting that the person was heterozygous at the *PRNP* codon 129, marking the first time this genotype was reported in conjunction with vCJD. It remains to be seen whether this suggests that a second wave of vCJD in heterozygous individuals, which – based on the experience with kuru [79] – would be expected to occur later and in smaller numbers compared to the first wave. Alternatively, while the homozygous Met/Met 129 genotype may be required for infection after an oral exposure, heterozygous individuals may become infected via the intravenous route or may have a longer incubation time after oral exposure [18]. Met/Val 129 heterozygous patients might be carriers, with the infectious agent replicating in peripheral organs while the CNS would never be affected. An additional problem with Met/Val 129 heterozygous patients could be that the BSE agent might generate a disease that is clinically and pathologically similar to sporadic CJD and thus – in contrast to vCJD in Met/Met homozygotes – could not be identified as originating from BSE.

51.2.8 Risk assessment

An assessment of all available experimental and epidemiological data concludes that the transmission of CJD through blood or blood products is not a frequent occurrence, if it can be unequivocally documented at all. Epidemiological observations have not revealed any indication for the transmissibility of sporadic or familial CJD, although these reports would not have the power to completely dismiss the possibility of any risk. However, this residual risk would still be orders of magnitude lower than that posed by certain viruses such as HIV, HBV, and HCV, which had been transmitted through blood and blood products before the introduction of safety measures such as donor selection, donation testing, and dedicated virus reduction.

Experimental studies have, despite many of their shortcomings, indicated that at least in some species, particularly in mice, low titers of the CJD agent can be observed in the blood in the late phase of the disease. These results must not be extrapolated to other species, including humans, since TSE agents behave very differently in different species. Such extrapolation would also clearly be at odds with all the negative epidemiological findings so far. The low infectivity titers in these models, if they are present at all, would, according to current understanding, not be sufficient for an infection through the transfusion of blood.

The situation with vCJD is less certain, and particularly recent evidence has dismissed any false sense of security. Firstly, vCJD transmission through blood transfusion has been experimentally demonstrated in mice [35]. More importantly, though, in two instances transmission by blood transfusion seems very likely in humans [2; 3]. To date, the occurrence is limited to low numbers, and it has been suggested that it will remain so for several reasons [18]. vCJD is a rare disease, and so far only 18 blood donors were found among all later vCJD patients [2] through intensive screening programs. In addition, despite conclusions from epidemiological studies using very limited numbers [80], the statistically significant downward trend in vCJD incidence continues [81], although some uncertainty remains around heterozygous vCJD and its potential implications.

While a certain yet small risk remains for labile blood products for transfusion, stable blood products, which are produced by lengthy manufacturing processes, should be evaluated separately. The potential exposure of plasma pools for fractionation to a vCJD-implicated donation cannot be neglected, although the implemented deferral and recall requirements would minimize the potential consequences. This low level of infectivity would then be diluted into a pool of thousands of donations. Also, and clearly more significantly, experimental studies using artificially added prions and laboratory scale manufacturing processes have unanimously confirmed an intrinsic prion reduction capacity of these manufacturing processes that were initially designed to remove impurities from the final product. Whether or not the TSE spike materials used have been appropriate or not is a still continuing criticism of these studies. Arguably, though, regardless of the spike material used, the slightly different manufacturing processes investigated, the different test systems used, i. e., different sources and preparation of infectivity, and different in vitro/in vivo detection systems, and the different laboratories that have conducted the testing, all these studies have shown a certain prion reduction capacity of all manufacturing processes [63]. Given the very low levels of potential TSE agent intake, even lower $\log_{10}$ ID_{50} scale reduction levels following the manufacturing process, should provide some reassurance.

51.2.9 Worst-case scenario

As a matter of principle, it is virtually impossible to prove that sporadic CJD is not transmittable by blood or blood derivatives; the evidence for the transmissibility of vCJD to date remains relatively scarce. To establish unambiguous evidence would require studies involving large numbers of subjects and long observation periods. However, when dealing with a serious risk, preventive measures must be taken before the results of such studies are published. As a result, the least favorable circumstances need to be assumed, i. e., a worst-case scenario. After a review of present knowledge, for the current worst-case scenario, it would be assumed that while for the CJD agent there is no evidence to suggest occurrence in the blood, low amounts of the variant CJD agent seem to occur in human blood during the late phase of the disease.

51.2.10 Preventive measures concerning blood for transfusion

The deferral of potential donors who might be at an elevated risk of virus infection as identified by use of a questionnaire, or who have evidence of viral infection (not a TSE agent) based on laboratory tests, has significantly enhanced the safety of blood and blood derivatives.

A few analytical methods that can identify CJD by analyses of cerebrospinal fluid or serum samples have been developed during recent years, e. g., 14–3–3 [82] and S100 [83] (see Chapter 31). However, it is uncertain whether any of these tests would also detect CJD during the incubation period. For their use in blood donation schemes, it would also be necessary for tests to be performed on easily available examination material such as blood. Also, the potential ethical implications of results of such tests for an affected person need to be taken into consideration.

Until there are test methods that can reliably detect CJD infection during preclinical stages, any risk from CJD can only be minimized by interviewing donor applicants. Such donor questionnaires aim at identifying risk factors (as listed below) that may elevate the likelihood of such a donor developing CJD compared to the general population. Identification of such a risk factor would, as indicated earlier, lead to deferral of the person from blood donation. Any risk from sporadic CJD cannot be reduced by this approach, since predisposing factors are only known for iatrogenic and familial CJDs.

In agreement between the World Health Organization (WHO) [84], the European Council [85], the Council of the European Community [86], and national guidelines [87], persons should be deferred from donating blood who

- have been treated with human cadaveric pituitary gland extracts (growth hormones and FSH);
- have received a human dura mater graft;
- have been exposed to a iCJD risk factor;
- have a family history of a prion disease;
- have a family history of Creutzfeldt–Jakob disease, GSS, or fatal familial insomnia (FFI);
- suffer from a dementing disease or a TSE.

In addition, it might be worth considering excluding donors who suffer from any neurological disorder, and deferral of cornea recipients has been suggested as well [86; 87] (see Section 51.3). Also, the following measures are currently under discussion, or have been introduced in several countries.

Experiments transmitting the TSE agents from one species to another and also the spread of TSEs in different "naturally" infected animal species (see Chapter 42) have clearly shown that adaptation of the pathogen to a host increases the risk of transmission and facilitates further spread. Against this background, it was suggested that people who have received blood transfusions or tissue grafts should be excluded from giving blood. Initial studies in France had shown that this deferral criterion would only affect approximately 5% of the current blood supply, and meanwhile the measure has been implemented or is being implemented in several other countries.

There is still a debate as to whether TSE agents, if they occur in blood, are associated with leukocytes (see Section 51.2.3). Using specific nylon filters, leukocytes can, to a large degree, be removed from blood, a process called leukodepletion [88]. There are good reasons for the use of leukocyte-reduced blood components, i.e., prevention of alloimmunizations and reduction of the risk of transmitting intracellular pathogens such as cytomegalovirus, Epstein–Barr virus, and HTLV-I. CJD risk reduction might be another reason for introducing leukodepletion, and this reason was used to rationalize adopting the measure in France. Particularly the gaps in our current understanding about further developments with respect to vCJD induced decision-makers in Great Britain and several other countries to establish leukodepletion as a precautionary measure. The consulted expert panels have been well aware of the potential significance of the technique as well as the also possible lesser significance as a prion safety measure. More recent data obtained using a hamster model would now suggest that leukoreduction removes approximately half of the already low levels of TSE infectivity from the blood [61].

Leukoreduction filters with an additional prion reduction capacity are being developed at the time of writing, though the testing required to confirm the clinical efficacy and absence of side effects has not been completed.

These measures help keep, to as low a level as possible, the exposure of blood that is used for the production of blood components and blood derivatives to CJD pathogens. However, which steps should be taken if it becomes known at a later stage that a blood donor has developed CJD, or has violated the exclusion criteria (see above) at the time of blood donation?

There is agreement that in cases where individual blood components such as erythrocyte concentrates, blood platelet concentrates, frozen fresh plasma, or plasma for fractionation are still available, they must not be used.

Also, vCJD-implicated batches of plasma derivatives are currently recalled per regulatory guidance in both the US and the EU, although this topic may be revised for human albumin. Albumin has a generically particularly high prion reduction capacity [63] and at the same time disproportionately high impact on the supply of essential medicinal products such as vaccines.

Whether or not CJD-implicated batches of plasma products, which include coagulation factors, immunoglobulins, and albumin, should be recalled has been a matter of debate. Initially, several countries decided to require recall of plasma products if CJD exposure was later discovered. European countries, and since 1998 also the USA, have, following intense discussions, concluded that blood products do not present a risk for the transmission of sporadic CJD and hence that the recall of specific products is not necessary.

While scientific evaluations, including studies in the USA [89], have come to the conclusion that the risk of sporadic CJD transmission from blood derivatives is minimal, an inclination toward introducing the most stringent preventive measures seems understandable given that

- the public expects maximal safety with regard to blood derivatives;
- many people were infected with deadly viruses such as HIV, HBV, or HCV before the introduction of effective inactivation measures.

It should be noted though that even these preventive measures cannot reduce the theoretical risk to zero. Because of the long incubation period of prion diseases, it is likely that a recall would often come too late, with the majority of implicated products already used by the time the CJD risk of the donor became known. Data presented to a subcommittee of the American House of Representatives suggested that an average of only 5 % of any recalled lot of a plasma product could be retrieved from the market, whereas most of the remaining 95 % would already have been used.

51.3 Cornea

If an impairment of vision is caused by alterations of the cornea, eye surgery including replacement of the cornea with human cadaveric material is an accepted treatment option. Such cornea transplant operations are usually carried out in cooperation with large cornea banks, and each year more than 20,000 cadaveric corneas for transplantation are provided by the European Eye Bank Association [90].

Two cases have been reported in which the donor and the recipient of a cornea transplant both died from CJD (Table 14.1). In the first case, which was described in a publication in 1974 in the USA [47], the cornea had already been transplanted before the donor's disease could be histologically diagnosed. Eighteen months after transplantation, the recipient developed CJD. In another case, which was reported in Germany [48], the recipient of the cornea developed CJD 30 years after transplantation. When the cornea donor died from CJD in 1965, the transmissible nature of the disease was not properly understood [91].

A further publication from Japan [92] reported the case of a patient who contracted CJD 15 months after receiving a cornea transplant. While the disease status of the donor was not known, the transmission of CJD was suggested based on the incubation period, which was similar to the American case described above [47].

Animal experiments also indicate that TSE agents can be transmitted through cornea transplantation [44; 93; 94].

Although only a few cases of CJD transmission through cadaveric cornea material are known, it must still be assumed that due to the cornea's close vicinity to brain tissue, which includes the retina and the optic nerve, the cornea might be contaminated with the CJD agent and be infectious (Table 46.2).

Whether this contamination is present in the cornea when the donor is still alive, or whether it only appears there after removal of the material, which typically involves removing the entire eyeball, including dissection of the optic nerve, remains unknown. Given the very low number of cornea transplantation-associated CJD cases, it would appear exceedingly difficult to prove whether other surgical procedures, which would involve removal of only the cornea [95], would result in a lower risk for CJD transmission.

Neither the WHO, the European Council, nor other intergovernmental organizations have developed guidelines that govern the suitability of potential cornea donors. However, in January 1990, representatives of national cornea banks at the Third European Cornea Bank

Conference in Leiden unanimously agreed that cornea must not be taken from donors who died of CJD or another CNS disease of unknown etiology, such as multiple sclerosis, amyotrophic lateral sclerosis (ALS, MND) [96–99], or Alzheimer's disease. The German Cornea Work Group has also adopted this guideline. This policy, however, will not prevent rare instances in which a human cadaveric cornea might be obtained at a time when the donor is not known to suffer from CJD. Such an unfortunate case happened in Great Britain in 1997, when the body of a donor was only examined after the cornea had been used. As the accumulation of PrP^{Sc} in the retina has been shown for sporadic and variant CJD [100], testing the retina for the presence of PrP^{Sc} prior to using the cornea for transplantation might prevent such unfortunate situations, although it is not known when during the incubation PrP^{Sc} becomes detectable in the retina.

In summary, the medical benefits of cornea transplantations, which contribute to a significantly better quality of life for the recipient, clearly outweigh the extremely low residual risk of CJD transmission associated with cornea transplantations.

51.4 Dura mater

Owing to their mechanical features, dura mater grafts were frequently used in surgical interventions, in particular to close defects. Dura mater grafts are prepared from human cadaveric donors and are subjected to a multistep process including hydrogen peroxide, sodium hydroxide (see Chapter 44), gamma irradiation, and acetone for conservation and decontamination. After this treatment, dura mater grafts can be stored for up to 5 years.

In February 1987, a patient of around 30 years of age died of CJD only 21 months after an operation during which the patient had received a dura mater graft [101; 102]. By 2005, 169 [103] (Table 14.1) similar cases [104] had become known worldwide, with incubation times between surgery and the development of CJD between 1.5 and 23 years. Additional results were obtained from a survey [104] conducted in neurological, psychiatric, and neuropathological institutions in Japan, which identified a total of 43 people who had died of CJD between January 1979 and May 1996 following the application of a dura mater graft. With one exception, the same type of transplant [104] (LYODURA, B. Braun Melsungen AG, Germany) had been used, which had been "produced" before 1987. In contrast to the method employed by other manufacturers, the LYODURA transplants were made from dura mater material obtained from different donors. At that time, it was unusual to keep records about donors and the distribution of the material into different lots, so it was impossible to trace material back to the diseased donor. Nevertheless, it must be assumed that the CJD cases were caused through the dura mater grafts used. This is essentially based on the following considerations:

1. The majority of people affected were at an age at which CJD very rarely occurs sporadically. Therefore, the observed accumulation of cases cannot be attributed to a purely sporadic occurrence of the disease.
2. The vast majority of cases are associated with only one particular product.
3. Animal experiments using the hamster scrapie model have also shown that dura mater grafts might contain TSE agents [105].

These observations led to the fact that, from mid-1987 onwards, dura mater grafts were only produced from a single donor who had to meet certain exclusion criteria (e. g., no signs of CJD). In addition, and this is likely to be the pivotal element, dura mater grafts are treated with 1 N sodium hydroxide [104], which results in a very significant inactivation of TSE agents (see Chapter 44.5.1). The efficiency of this treatment was also confirmed in animal studies [105].

In spring 1997, an expert group on behalf of the WHO suggested [84] that dura mater grafts should no longer be used, in particular in neurosurgery, unless medically suitable alternatives were not available. This suggestion was based on the assumption that other tissue (e. g., muscle fascia) or synthetic products might be able to replace dura mater material. As this recommendation is quite controversial, most countries have not strictly prohibited the use of dura mater grafts. However, countries such as Germany strive to limit the use of dura mater transplants. This will lead to a reduction in demand, which should be easier to cover with material collected under strict requirements. For example, dura mater material can only be obtained from donors who died before the age of 50, and after their brains have been examined for the absence of any indication of CJD.

References

[1] Brown P, Cervenakova L, McShane LM, et al. Further studies of blood infectivity in an experimental model of transmissible spongiform encephalopathy, with an explanation of why blood components do not transmit Creutzfeldt–Jakob disease in humans. Transfusion 1999; 39(11–12):1169–1178.

[2] Llewelyn CA, Hewitt PE, Knight RS, et al. Possible transmission of variant Creutzfeldt–Jakob disease by blood transfusion. Lancet 2004; 363(9407):417–421.

[3] Peden AH, Head MW, Ritchie DL, et al. Preclinical vCJD after blood transfusion in a PRNP codon 129 heterozygous patient. Lancet 2004; 364(9433):527–529.

[4] Wilson K, Code C, Ricketts MN. Risk of acquiring Creutzfeldt–Jakob disease from blood transfusions: systematic review of case-control studies. BMJ 2000; 321(7252):17–19.

[5] Pattison IH and Millson GC. Distribution of the scrapie agent in the tissues of experimentally inoculated goats. J Comp Pathol 1962; 72:233–244.

[6] Wells GA, Hawkins SA, Green RB, et al. Preliminary observations on the pathogenesis of experimental bovine spongiform encephalopathy (BSE): an update. Vet Rec 1998; 142(5):103–106.

[7] Foster JD, Parnham D, Chong A, et al. Clinical signs, histopathology and genetics of experimental transmission of BSE and natural scrapie to sheep and goats. Vet Rec 2001; 148(6):165–171.

[8] Hunter N, Foster J, Chong A, et al. Transmission of prion diseases by blood transfusion. J Gen Virol 2002; 83(Pt 11):2897–2905.

[9] Casaccia P, Ladogana A, Xi YG, et al. Levels of infectivity in the blood throughout the incubation period of hamsters peripherally injected with scrapie. Arch Virol 1989; 108(1-2):145–149.

[10] Diringer H. Sustained viremia in experimental hamster scrapie. Brief report. Arch Virol 1984; 82(1-2):105–109.

[11] Field EJ, Caspary EA, Joyce G. Scrapie agent in blood. Vet Rec 1968; 83:109–110.

[12] Race R and Chesebro B. Scrapie infectivity found in resistant species. Nature 1998; 392 (6678):770.

[13] Kuroda Y, Gibbs CJ Jr, Amyx HL, et al. Creutzfeldt–Jakob disease in mice: persistent viremia and preferential replication of virus in low-density lymphocytes. Infect Immun 1983; 41(1): 154–161.

[14] Manuelidis EE, Gorgacs EJ, Manuelidis L. Viremia in experimental Creutzfeldt–Jakob disease. Science 1978; 200(4345):1069–1071.

[15] Gibbs CJ Jr, Gajdusek DC, Morris JA. Viral characteristics of the scrapie agent in mice. In: Gajdusek DC, editor. Slow, latent and temperate virus infections. US Department of Health, Education and Welfare, Bethesda, Maryland 1965:195–202.

[16] Hadlow WJ, Kennedy RC, Race RE. Natural infection of Suffolk sheep with scrapie virus. J Infect Dis 1982; 146(5):657–664.

[17] Hadlow WJ, Kennedy RC, Race RE, et al. Virologic and neurohistologic findings in dairy goats affected with natural scrapie. Vet Pathol 1980; 17(2):187–199.

[18] Brown P and Cervenakova L. The modern landscape of transfusion-related iatrogenic Creutzfeldt–Jakob disease and blood screening tests. Curr Opin Hematol 2004; 11(5):351–356.

[19] Houston F, Foster JD, Chong A, et al. Transmission of BSE by blood transfusion in sheep. Lancet 2000; 356(9234):999–1000.

[20] Hadlow WJ, Eklund CM, Kennedy RC, et al. Course of experimental scrapie virus infection in the goat. J Infect Dis 1974; 129(5):559–567.

[21] Dickinson AG, Meikle VM, Fraser H. Genetical control of the concentration of ME7 scrapie agent in the brain of mice. J Comp Pathol 1969; 79(1):15–22.

[22] Eklund CM, Kennedy RC, Hadlow WJ. Pathogenesis of scrapie virus infection in the mouse. J Infect Dis 1967; 117(1):15–22.

[23] Hadlow WJ, Race RE, Kennedy RC. Temporal distribution of transmissible mink encephalopathy virus in mink inoculated subcutaneously. J Virol 1987; 61(10):3235–3240.

[24] Clarke MC and Haig DA. Presence of the transmissible agent of scrapie in the serum of affected mice and rats. Vet Rec 1967; 80(16): 504.

[25] Bons N, Lehmann S, Mestre-Frances N, et al. Brain and buffy coat transmission of bovine spongiform encephalopathy to the primate *Microcebus murinus*. Transfusion 2002; 42(5): 513–516.

[26] Godec MS, Asher DM, Kozachuk WE, et al. Blood buffy coat from Alzheimer's disease patients and their relatives does not transmit spongiform encephalopathy to hamsters. Neurology 1994; 44(6):1111–1115.

[27] Godec MS, Asher DM, Masters CL, et al. Evidence against the transmissibility of Alzheimer's disease. Neurology 1991; 41(8):1320.

[28] Manuelidis EE, de Figueiredo JM, Kim JH, et al. Transmission studies from blood of Alzheimer disease patients and healthy relatives. Proc Natl Acad Sci USA 1988; 85(13):4898–4901.

[29] Manuelidis EE and Manuelidis L. A transmissible Creutzfeldt–Jakob disease-like agent is prevalent in the human population. Proc Natl Acad Sci USA 1993; 90(16):7724–7728.

[30] Manuelidis EE, Manuelidis L, Pincus JH, et al. Transmission, from man to hamster, of Creutzfeldt–Jakob disease with clinical recovery. Lancet 1978; 2(8079):40–42.

[31] Rohwer RG. Alzheimer's disease transmission: possible artifact due to intercurrent illness. Neurology 1992; 42(2):287–288.

[32] Tateishi J, Ohta M, Koga M, et al. Transmission of chronic spongiform encephalopathy with kuru plaques from humans to small rodents. Ann Neurol 1979; 5(6):581–584.

[33] Tateishi J, Sato Y, Koga M, et al. Experimental transmission of human subacute spongiform encephalopathy to small rodents. I. Clinical and histological observations. Acta Neuropathol (Berl) 1980; 51(2):127–134.

[34] Brown P, Rohwer RG, Dunstan BC, et al. The distribution of infectivity in blood components and plasma derivatives in experimental models of transmissible spongiform encephalopathy. Transfusion 1998; 38(9):810–816.

[35] Cervenakova L, Yakovleva O, McKenzie C, et al. Similar levels of infectivity in the blood of mice infected with human-derived vCJD and GSS strains of transmissible spongiform encephalopathy. Transfusion 2003; 43(12):1687–1694.

[36] Pattison IH and Millson GC. Further observations on the experimental production of scrapie in goats and sheep. J Comp Pathol 1960; 70:182–193.

[37] Marsh RF, Miller JM, Hanson RP. Transmissible mink encephalopathy: studies on the peripheral lymphocyte. Infect Immun 1973; 7(3):352–355.

[38] Klein MA, Frigg R, Flechsig E, et al. A crucial role for B cells in neuroinvasive scrapie. Nature 1997; 390(6661):687–690.

[39] Klein MA, Frigg R, Raeber AJ, et al. PrP expression in B lymphocytes is not required for prion neuroinvasion. Nat Med 1998; 4(12): 1429–1433.

[40] Asher DM, Gibbs CJ Jr, Gajdusek DC. Pathogenesis of subacute spongiform encephalopathies. Ann Clin Lab Sci 1976; 6(1):84–103.

[41] Deslys JP, Lasmezas C, Dormont D. Selection of specific strains in iatrogenic Creutzfeldt–Jakob disease. Lancet 1994; 343(8901):848–849.

[42] Manuelidis EE, Kim JH, Mericangas JR, et al. Transmission to animals of Creutzfeldt–Jakob disease from human blood. Lancet 1985; 2(8460):896–897.

[43] Tamai Y, Kojima H, Kitajima R, et al. Demonstration of the transmissible agent in tissue from a pregnant woman with Creutzfeldt–Jakob disease. N Engl J Med 1992; 327(9): 649.

[44] Tateishi J. Transmission of Creutzfeldt–Jakob disease from human blood and urine into mice. Lancet 1985; 2(8463):1074.

[45] Brown P, Gibbs CJ Jr, Rodgers-Johnson P, et al. Human spongiform encephalopathy: the National Institutes of Health series of 300 cases of experimentally transmitted disease. Ann Neurol 1994; 35(5):513–529.

[46] Gajdusek DC. Unconventional viruses and the origin and disappearance of kuru. Science 1977; 197(4307):943–960.

[47] Duffy P, Wolf J, Collins G, et al. Letter: possible person-to-person transmission of Creutzfeldt–Jakob disease. N Engl J Med 1974; 290(12): 692–693.

[48] Heckmann JG, Lang CJ, Petruch F, et al. Transmission of Creutzfeldt–Jakob disease via a corneal transplant. J Neurol Neurosurg Psychiatry 1997; 63(3):388–390.

[49] Brown P, Preece MA, Will RG. "Friendly fire" in medicine: hormones, homografts, and Creutzfeldt–Jakob disease. Lancet 1992; 340(8810):24–27.

[50] Dodd R. CJD Lookback Study (research study to assess the risk of blood borne transmission of CJD). Webpage 2006: www.fda.gov/ohrms/dockets/ac/04/slides/2004_4075OP1_06.ppt.

[51] Heye N, Hensen S, Muller N. Creutzfeldt–Jakob disease and blood transfusion. Lancet 1994; 343(8892):298–299.
[52] Sullivan MT. Creutzfeldt–Jakob Disease (CJD) Investigational Lookback Study. Transfusion 1997; 37:2S.
[53] Klein R and Dumble LJ. Transmission of Creutzfeldt–Jakob disease by blood transfusion. Lancet 1993; 341(8847):768.
[54] Creange A, Gray F, Cesaro P, et al. Creutzfeldt–Jakob disease after liver transplantation. Ann Neurol 1995; 38(2):269–272.
[55] Creange A, Gray F, Cesaro P, et al. Pooled plasma derivatives and Creutzfeldt–Jakob disease. Lancet 1996; 347(8999):482.
[56] Creange A, Gray F, Cesaro P, et al. Pooled plasma derivatives and Creutzfeldt–Jakob disease [author's reply]. Lancet 1996; 347(8999):967.
[57] De Silva R. Pooled plasma derivatives and Creutzfeldt–Jakob disease. Lancet 1996; 347 (9006):967.
[58] Lee DC, Stenland CJ, Miller JL, et al. A direct relationship between the partitioning of the pathogenic prion protein and transmissible spongiform encephalopathy infectivity during the purification of plasma proteins. Transfusion 2001; 41(4):449–455.
[59] Stenland CJ, Lee DC, Brown P, et al. Partitioning of human and sheep forms of the pathogenic prion protein during the purification of therapeutic proteins from human plasma. Transfusion 2002; 42(11):1497–1500.
[60] Vey M, Baron H, Weimer T, et al. Purity of spiking agent affects partitioning of prions in plasma protein purification. Biologicals 2002; 30(3):187–196.
[61] Gregori L, Maring JA, MacAuley C, et al. Partitioning of TSE infectivity during ethanol fractionation of human plasma. Biologicals 2004; 32(1):1–10.
[62] Foster PR, Welch AG, McLean C, et al. Studies on the removal of abnormal prion protein by processes used in the manufacture of human plasma products. Vox Sang 2000; 78(2):86–95.
[63] Kreil TR. Prion partitioning during plasma fractionation: a consolidated view. Webpage 2006: www.fda.gov/ohrms/dockets/ac/02/transcripts/3834t2_01.pdf.
[64] Anonymous. Pneumocystis pneumonia – Los Angeles. MMWR Morb Mortal Wkly Rep 1981; 30(21):250–252.
[65] Anonymous. *Pneumocystis carinii* pneumonia among persons with hemophilia A. MMWR Morb Mortal Wkly Rep 1982; 31(27):365–367.
[66] Evatt B, Austin H, Barnhart E, et al. Surveillance for Creutzfeldt–Jakob disease among persons with hemophilia. Transfusion 1998; 38(9):817–820.
[67] Lee CA, Ironside JW, Bell JE, et al. Retrospective neuropathological review of prion disease in UK haemophilic patients. Thromb Haemost 1998; 80(6):909–911.
[68] Operskalski EA and Mosley JW. Pooled plasma derivatives and Creutzfeldt–Jakob disease. Lancet 1995; 346(8984):1224.
[69] Holman RC, Khan AS, Belay ED, et al. Creutzfeldt–Jakob disease in the United States, 1979–1994: using national mortality data to assess the possible occurrence of variant cases. Emerg Infect Dis 1996; 2(4):333–337.
[70] Davanipour Z, Alter M, Sobel E, et al. Creutzfeldt–Jakob disease: possible medical risk factors. Neurology 1985; 35(10):1483–1486.
[71] Esmonde TF, Will RG, Slattery JM, et al. Creutzfeldt–Jakob disease and blood transfusion. Lancet 1993; 341(8839):205–207.
[72] Harries-Jones R, Knight R, Will RG, et al. Creutzfeldt–Jakob disease in England and Wales, 1980–1984: a case-control study of potential risk factors. J Neurol Neurosurg Psychiatry 1988; 51(9):1113–1119.
[73] Kondo K and Kuroiwa Y. A case control study of Creutzfeldt–Jakob disease: association with physical injuries. Ann Neurol 1982; 11(4):377–381.
[74] Van Duijn CM, Delasnerie-Laupretre N, Masullo C, et al. Case-control study of risk factors of Creutzfeldt–Jakob disease in Europe during 1993–95. European Union (EU) Collaborative Study Group of Creutzfeldt–Jakob disease (CJD). Lancet 1998; 351(9109):1081–1085.
[75] Collinge J, Sidle KC, Meads J, et al. Molecular analysis of prion strain variation and the aetiology of 'new variant' CJD. Nature 1996; 383(6602):685–690.
[76] Bruce ME, Will RG, Ironside JW, et al. Transmissions to mice indicate that 'new variant' CJD is caused by the BSE agent. Nature 1997; 389(6650):498–501.
[77] Hill AF, Zeidler M, Ironside J, et al. Diagnosis of new variant Creutzfeldt–Jakob disease by tonsil biopsy. Lancet 1997; 349(9045):99–100.
[78] Hilton DA, Fathers E, Edwards P, et al. Prion immunoreactivity in appendix before clinical onset of variant Creutzfeldt–Jakob disease. Lancet 1998; 352(9129):703–704.
[79] Lee HS, Brown P, Cervenakova L, et al. Increased susceptibility to kuru of carriers of the

PRNP 129 methionine/methionine genotype. J Infect Dis 2001; 183(2):192–196.
[80] Hilton DA, Ghani AC, Conyers L, et al. Prevalence of lymphoreticular prion protein accumulation in UK tissue samples. J Pathol 2004; 203(3):733–739.
[81] Andrews NJ. Incidence of variant Creutzfeldt–Jakob disease onsets and deaths in the UK. Webpage 2006: www.cjd.ed.ac.uk/vcjdq.htm.
[82] Hsich G, Kenney K, Gibbs CJ Jr, et al. The 14–3–3 brain protein in cerebrospinal fluid as a marker for transmissible spongiform encephalopathies. N Engl J Med 1996; 335(13): 924–930.
[83] Otto M, Wiltfang J, Schutz E, et al. Diagnosis of Creutzfeldt–Jakob disease by measurement of S100 protein in serum: prospective case-control study. BMJ 1998; 316(7131):577–582.
[84] World Health Organization. Consultation on medicinal and other products in relation to human and animal transmissible spongiform encephalopathies; with the participation of the Office International des Epizooties. WHO/EMC/ZOO/97, 1997.
[85] Council of Europe. Guide to the preparation, use and quality assurance of blood components, 11th ed. Council of Europe Publishing, 2005.
[86] Council of the European Commission. Recommendation of 29 Juni 1998 on the suitability of blood and plasma donors and the screening of donated blood in the European Community. 1998; L 203:98/463/EC.
[87] Wissenschaftlicher Beirat der Bundesärztekammer und des Paul-Ehrlich-Instituts. Richtlinien zur Blutgruppenbestimmung und Bluttransfusion (Hämotherapie). Deutscher Ärzteverlag, Köln, 1996.
[88] Williamson LM. Leucocyte depletion of the blood supply – how will patients benefit? Br J Haematol 2000; 110(2):256–272.
[89] Alter HJ. G-pers creepers, where'd you get those papers? A reassessment of the literature on the hepatitis G virus. Transfusion 1997; 37(6):569–572.
[90] Maas-Reijs J, Pels E, Tullo AB. Eye banking in Europe 1991–1995. Acta Ophthalmol Scand 1997; 75(5):541–543.
[91] Gibbs CJ Jr and Gajdusek DC. Infection as the etiology of spongiform encephalopathy (Creutzfeldt–Jakob disease). Science 1969; 165(897):1023–1025.
[92] Uchiyama K, Ishida C, Yago S, et al. An autopsy case of Creutzfeldt–Jakob disease associated with corneal transplantation. Dementia 1994; 8:466–473.
[93] Manuelidis EE, Angelo JN, Gorgacz EJ, et al. Experimental Creutzfeldt–Jakob disease transmitted via the eye with infected cornea. N Engl J Med 1977; 296(23):1334–1336.
[94] Marsh RF and Hanson RP. Transmissible mink encephalopathy: infectivity of corneal epithelium. Science 1975; 187(4177):656.
[95] Hudde T, Reinhard T, Moller M, et al. Corneoscleral transplant excision in the cadaver. Experiences of the North Rhine Westphalia Lions Cornea Bank 1995 and 1996. Ophthalmologe 1997; 94(11):780–784.
[96] Fraser H, Behan W, Chree A, et al. Mouse inoculation studies reveal no transmissible agent in amyotrophic lateral sclerosis. Brain Pathol 1996; 6(2):89–99.
[97] Roikhel VM, Fokina GI, Khokhlov AI, et al. Alterations of arginase activity in scrapie-infected mice and in amyotrophic lateral sclerosis. Acta Virol 1990; 34(6):545–553.
[98] Serratrice GT and Munsat TL. Pathogenesis and therapy of amyotrophic lateral sclerosis. Introduction. Adv Neurol 1995; 68:xix–xxx.
[99] Tavolato BF, Licandro AC, Saia A. Motor neurone disease: an immunological study. Eur Neurol 1975; 13(5):433–440.
[100] Head MW, Northcott V, Rennison K, et al. Prion protein accumulation in eyes of patients with sporadic and variant Creutzfeldt–Jakob disease. Invest Ophthalmol Vis Sci 2003; 44(1):342–346.
[101] Anonymous. Update: Creutzfeldt–Jakob disease in a patient receiving a cadaveric dura mater graft. MMWR Morb Mortal Wkly Rep 1987; 36(21):324–325.
[102] Anonymous. Rapidly progressive dementia in a patient who received a cadaveric dura mater graft. MMWR Morb Mortal Wkly Rep 1987; 36(4):49–50, 55.
[103] Brown P. Transmissible spongiform encephalopathy. In: Goetz C, editor. Textbook of clinical neurology. chapter 43, 3rd ed. W. B. Saunders, Philadelphia, 2006 in press.
[104] Anonymous. Creutzfeldt–Jakob disease associated with cadaveric dura mater grafts – Japan, January 1979–May 1996. MMWR Morb Mortal Wkly Rep 1997; 46(45):1066–1069.
[105] Diringer H and Braig HR. Infectivity of unconventional viruses in dura mater. Lancet 1989; 1(8635):439–440.
[106] Wells GA. Preliminary observations on the pathogenesis of experimental bovine spongi-

form encephalopathy. In: Gibbs CJ Jr, editor. Bovine spongiform encephalopathy – the BSE dilemma. Springer, New York, 1996.

[107] Marsh RF and Hanson RP. Transmissible mink encephalopathy: neuroglial response. Am J Vet Res 1969; 30(9):1637–1642.

[108] Doi T. Correlation between sequential organ distribution of infectivity in mice inoculated with Creutzfeldt–Jakob pathogenic agent and prion protein gene expression. Zasshi 1991; 66:104–114.

52 BSE Risk Assessment and Minimization

Radulf C. Oberthür, Aline A. de Koeijer, Bram E.C. Schreuder, and Stuart C. MacDiarmid

52.1 Introduction

Historically, prion diseases were not seen as important life-threatening diseases of animals and humans, mainly because they were relatively rare. For example, Creutzfeld–Jakob disease (CJD) generally occurs at a rate of only 1–2 person/s in a million. With the exception of scrapie, which has been endemic in several countries and has been the subject of sporadic scientific interest since the 18th century (see Chapters 1, 18), prion diseases have only attracted major attention in two instances: kuru and bovine spongiform encephalopathy (BSE). In the case of kuru, the disease was largely confined to members of the Fore people in Papua New Guinea (see Chapters 2, 13, 38). In the case of BSE, the disease started and propagated mainly within the British cattle herd (see Chapters 19, 39, 40). In both kuru and BSE, healthy individuals consumed infectious tissue of diseased or infected humans or animals, respectively. Depending on the dose of infectivity, the ingestion of infectious tissue caused the exposed animals or humans to produce further infective prions and, after a longer incubation period, these individuals also developed clinical disease and, after death, became a potential source of infection for further susceptible individuals.

The transmission of infectivity to several other individuals led to a general amplification of cases and increased presence of prions. This increase became particularly obvious in the explosive spread of BSE in Britain (Fig. 52.1). The enormous increase of infectious material within the British cattle herd in turn resulted in the spread of the BSE epidemic, though at a much lower rate, to other countries through trade in animal products containing small amounts of infectious material that were fed to bovines, and even to other susceptible species (see Chapter 50).

The three factors 1b, 2, and 3b, listed in Table 52.1, representing changes in *sourcing*, *processing*, and *use* of potentially infected material, seem to have provided the conditions for an exponential increase of BSE cases in Britain (Fig. 52.1). If every BSE case had been the result of the disease crossing from scrapie-infected sheep to cattle (scrapie → BSE; factors 1a and 3a), as the original hypothesis on the occurrence of BSE assumed (Fig. 52.2), then the incidence of BSE would have been much lower, and no exponential increase in incidence would have been observed.

Because of the high prevalence of scrapie in the large British sheep population (factors 0a and 0b), the "ignition" of the outbreak of BSE was, at first, attributed to a particular strain of sheep scrapie crossing the species barrier to cause a corresponding bovine disease. However, one or more cases of a hitherto-unrecognized bovine transmissible spongiform encephalopathy (TSE), analogous to spontaneously[1] occurring sporadic CJD (sCJD) in humans, cannot be excluded as the initial trigger for the BSE epidemic (factor 0c). Such a spontaneous occurrence of BSE would have led, assuming reproduction numbers (see below) increasing above 1 (Fig. 52.7) due to changes in the protein recycling system (factors 1, 2, and 3) and long-term, unnoticed "propagation" of prions in the cattle population (Fig. 52.2), to an epidemic course similar to that observed (Figs. 52.1 and 52.7). It should be noted that feeding sheep

[1] Definition of spontaneous cases (see Chapters 3 and 8).

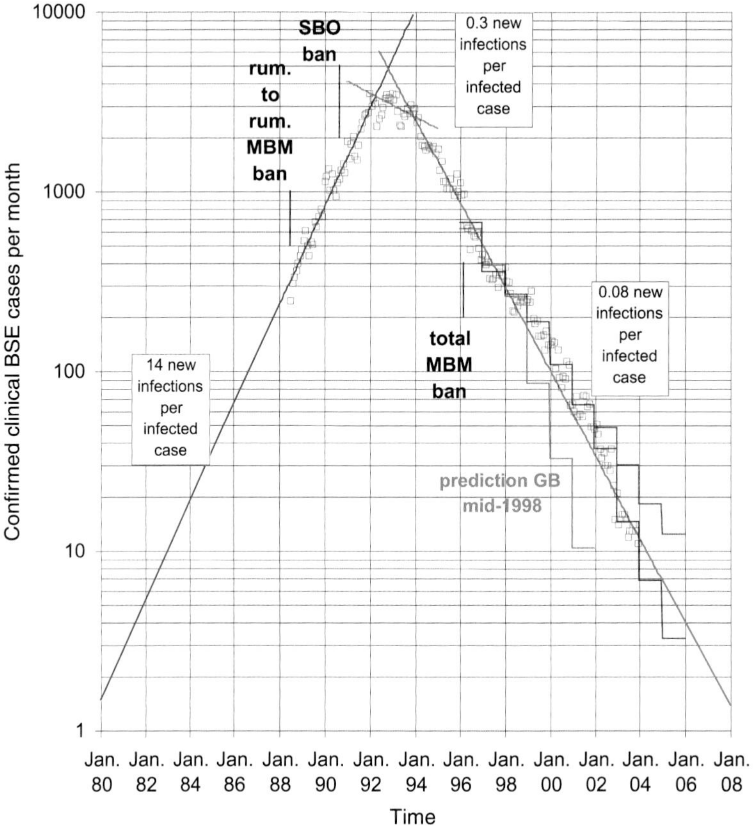

Fig. 52.1: Semilogarithmic presentation of the number of confirmed BSE cases per month from passive surveillance from 1988 to 2004 in Great Britain (blue open squares) and adaptation of the exponential growth with a growth rate of $r = 0.64$/year (from July 1988 until December 1991) and a negative growth rate of $r = -0.55$/year (from October 1993 to March 1998). As a result of the semilogarithmic presentation, the exponential functions become straight lines. Blue step function: Annual average of the monthly incidence of confirmed cases from 1996 to 2005 by passive surveillance. Brown step function: Annual average of the monthly incidence of BSE among cattle at risk (fallen stock and emergency slaughter) from active surveillance from 2002 to 2005 (data from [1]). Green step function: Prediction of the annual average of the monthly BSE incidence from passive surveillance using data from mid-1998 (data from [19]). Figure by R. Oberthür ©.

brains infected with various scrapie strains to calves in the UK has not, so far, resulted in a BSE-like condition.

Thus the epidemic chain reaction observed in Britain can only be explained by the recycling of cattle protein containing inefficiently inactivated BSE prions (factors 1b, 2, and 3b in Table 52.1) through the use as a bovine feed ingredient, resulting in more than one new infection, on average, from each infected animal. This spread can be described with classical mathematical methods [2]. The recycling of the BSE prions via the feed chain in the cattle population is equivalent to the transmission of the prions from cattle to cattle by means of a given vector. In the specific case of BSE, the vector is cattle feed. Transmission, however, was not restricted within a single herd, but rather included cattle on one farm being the producer of the vector, which then, after slaughter, ren-

Table 52.1: Overview of the factors that may have led to the BSE epidemic in Great Britain. The simultaneous occurrence of the factors 1b, 2, and 3b most probably has led to the observed exponential increase of the BSE incidence in Britain. Factor 0 (a + b or c) is responsible for the ignition of the explosive cycle. The factors associated with scrapie and printed in *italic*, which were initially assumed to be at the origin of BSE, are not relevant if a sporadic form of BSE occurs in cattle at a very low incidence (compare Chapter 19.9).

0. *a) High endemic scrapie prevalence in the British sheep population since the 18th century;*
   b) High sheep density in contrast to cattle density. The population density of sheep has always been very high in Britain (1980: approximately 175 sheep/km²);
   or
   c) A minimum undiscovered endemic presence of a kind of a sporadic form of BSE in cattle (e. g., <4 cases per year in Britain).
1. In the 1970s and 1980s, a constantly growing amount of potentially BSE-infectious material (specified risk material, see a and b below) together with other slaughter by-products such as intestines, bones, and adipose tissue **(sourcing)** was processed into animal protein concentrates (animal meal, meat-and-bone-meal, greaves) and used to feed farm animals:
   a) Skulls and backbones of sheep
   or
   b) Skulls and spinal cords of bovines.
2. Massive changes in rendering technology **(processing)** in Britain since approximately 1971: The transfer from *batch processing* to *continuous processing* involved the reduction of the temperature and inactivation time, as well as discontinuation of using solvents (e. g., hexane) for fat extraction (see Chapter 45).
3. Growing application of animal protein in concentrated cattle feed **(use)** to replace milk proteins for raising calves and to replace chemically modified soya protein for increasing the milk production (due to the high milk production of the high-bred cows; rumen-protected proteins) since the 1970s:
   a) Feeding of scrapie infected animal protein to cattle
   or
   b) Feeding of BSE-infected animal protein to cattle.

dering, and feed compounding at different locations, resulted eventually in the entire British cattle population being potentially acceptors of the vector. This whole system led to the exponential propagation of infectious load in British cattle (Figs. 52.1, 52.2, and 52.6a).

On the basis of British epidemiological BSE data (Fig. 52.1), three major phases can be distinguished and used for further calculations:

1. Exponential growth of BSE incidence from the beginning of 1988 to the end of 1991.
2. Stagnation of BSE incidence from the beginning of 1992 to the end of 1993.
3. Exponential decline of BSE incidence from the beginning of 1993 to the present (early 2006).

In the following sections, we show how the risk can be assessed quantitatively for the phases of the BSE epidemic listed above. With regard to interventions taken to reduce the BSE risk, we will elucidate how it might be achievable, at least theoretically, to keep the annual BSE incidence in a certain country below a value of 1 case per 1 million cattle (10^{-6}; compare to sCJD in humans).

52.2 Definition of the term "risk"

In this discussion, we define "risk" as the probability of the occurrence of an undesired event in the life of an individual (animal or human) or in the course of a certain period of time, e. g., 1 year, after the individual has been exposed to a certain hazard [5].

- For example, the probability of dying in a car accident within a period of 1 year is 1:10,000 for a person exposed to the hazard of driving 50,000 km per year [5].
- However, the risk of a man dying in the 1980s in Germany in his 56th year of life was 1:100, whereas the risk of a woman dying in her 56th year of life was "only" 1:300. It is important

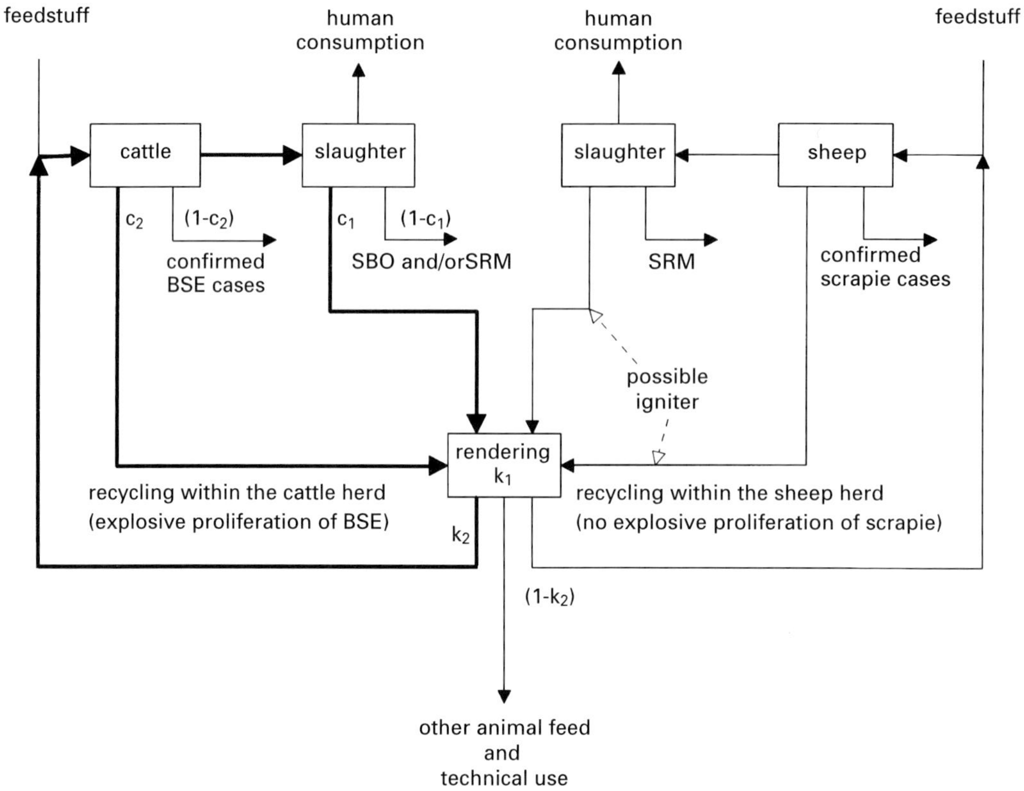

Fig. 52.2: Flow chart of the mass flow responsible for the origin of BSE. The trigger for the "explosive" exponential growth was thought to result from sheep scrapie, even though there was no similar "explosive" exponential growth of scrapie incidence in sheep, despite the same recycling processes (thin arrows). If "spontaneous" BSE cases occur in the cattle population, the exponential growth starts as soon as the reproduction number for the disease gets above 1 due to the recycling and propagation of prions in the cattle population (thick arrows). The mathematical model presented in this chapter basically refers to this recycling. The figure also presents residual amounts of prions in the mass flow within this cycle after the removal of SRM (c_1), and the segregation of BSE animals (c_2), after the thermal processing of bovine carcasses and slaughter by-products (k_1), and through the introduction of a ban prohibiting the feeding of ruminant protein to ruminants (k_2; explanations in the text). Figure by R. Oberthür; source: [4].

to note that the fact of being 55–56 years represented the hazard for the risk of dying at this particular age [6; 7].

- The risk for a cow in Great Britain being slaughtered before the age of 3, i. e., before reaching the age of becoming a dairy cow, is 1:7 [8], where the hazard is being born as a female calf in Great Britain.
- In contrast, in 1992 (the peak of the BSE epidemic), the risk for a cow or a bull in Great Britain reaching the clinical BSE stage was about 1:300 [8; 9], where the hazard was having been raised and living in this year as a cow or a bull in Great Britain, and having consumed the typical diet for a bovine in that country in the years before.

52.3 Dose–response relationship in BSE

An important aspect of quantitative risk assessment is the relationship between the dose, D, with which an individual was challenged and the probability of a particular response occurring as a direct consequence of this exposure [5]. In the case of BSE, the "successful"[2] exposure of an animal in a bioassay (see below) usually leads to the death of the animal, because the observation time in the bioassay is usually more than twice the average incubation period. Therefore, the response to being exposed to BSE prions in a bioassay is set equal to the probability of dying from BSE.

Dose refers to the applied quantity of pathogen-containing tissue. The unit of an infectious dose is defined as the dose that leads to the infection of an animal with a probability of 50% (ID_{50}). It has been shown in a hamster model that the infectious load of a certain tissue (number of ID_{50} per gram of tissue) increases proportionally to the concentration of infectious particles (PrP^{Sc}). A PrP^{Sc} molecule has a molecular mass of approximately 30 kDa. With this value and the amount of PrP^{Sc} in an ID_{50}, it was calculated that an intracerebral ID_{50} from hamster brain in the clinical stage of TSE contains about 1 million PrP^{Sc} molecules [10].

The following model reflects a dose–response relationship based on the hypothesis that each ingested infectious molecule can, although with a very small probability, P, result in infection and hence to a case of clinical BSE [11].[3] If P denotes the probability of an infection occurring from 1 infectious molecule, the probability of remaining free from infection is $(1 - P)$. Therefore, $(1 - P)^2$ describes the probability of not becoming infected by two infectious molecules, etc., up to $(1 - P)^N$. In the case of N infectious molecules, the probability of infection, H, is therefore

$$H(N) = 1 - (1 - P)^N. \quad (1)$$

For illustration, an example using dice may be useful: The probability of throwing at least one "6" with N throws is $H(N)$. The probability, P, of throwing a "6" in one throw is $P = 0.167$. This can be continued as follows:

$N = 2$ throws → $H(2) = 0.306$
$N = 3$ throws → $H(3) = 0.421$
$N = 4$ throws → $H(4) = 0.518$
$N = 6$ throws → $H(6) = 0.665$
$N = 12$ throws → $H(12) = 0.888$

For $N = 4$, a "6" can be thrown with a probability just greater than 50%. If the die is thrown 12 times, this will lead to a "6" with a probability of 89%.

In the case of small P and large N, which can be assumed in the case of a challenge of mice by prion-infected tissue, Eq. (1) can be expressed as exponential function:

$$H(N) = 1 - \exp(-N \cdot P) \quad (2)$$

The product of $N \cdot P$ can then be replaced by the quotient of D/ID_{50} with a constant factor ln2 because 50% = 1/2 probability of infection at ID_{50}:

$$\begin{aligned} H(D) &= 1 - \exp(-\ln 2 \cdot D/ID_{50}) \\ &= 1 - \exp(-\ln 2 \cdot N_{ID}) \end{aligned} \quad (3)$$

where $D/ID_{50} = N_{ID}$ denotes the number of ID_{50} that was administered to an animal (Fig. 52.3).

This *hypothetical* dose–response relationship has far-reaching consequences: there is no threshold value. That is, even a minimum dose can lead to an infection, although with extremely low probability. With regard to low doses, i. e., fractions of ID_{50}, the dose–response relationship is linear (Fig. 52.3a):

$$H(N_{ID}) = \ln 2 \cdot N_{ID} = 0.69 \cdot N_{ID} \quad (4)$$

If, for example, a group of 100 calves receives a total dose of 10 ID_{50} with their diet, each calf will, on average, receive a dose of $10/100 = 0.1$ ID_{50}. According to Eq. (4), $100 \cdot H = 100 \cdot 0.69 \cdot 0.1 \approx 7$ calves will become infected. If the same dose is administered to a group of 1,000 calves, each calf will only receive $10/1{,}000 = 0.01$ ID_{50}. However, according to Eq. (4), $1{,}000 \cdot H = 1{,}000 \cdot 0.69 \cdot 0.1 \approx 7$, it will once again result in seven infected calves. This shows

[2] "Successful" exposure is not restricted to an infection leading to clinical disease and death. For example, in a 14–18-month-old bovine used for meat production, which was infected with BSE as a calf, infection can be regarded as "successful", even though the incubation time has not been long enough for clinical manifestation or death (the animal was slaughtered). In the following epidemiological calculations, infectivity that was able to spread via subclinically infected animals is also taken into account.

[3] Depending on whether the animal is slaughtered before the end of the incubation time or not.

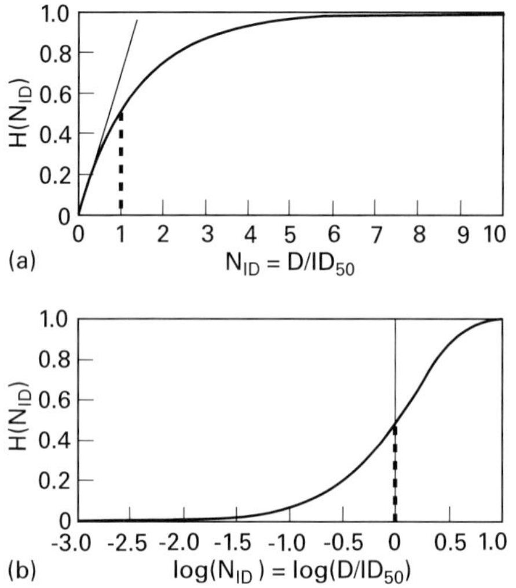

Fig. 52.3: (a) Probability of infection, H, as a function of N_{ID}, i. e., the number of administered infectious doses (ID_{50}). $H = 0.5$ is by definition at $N_{ID} = 1$. The thin line reflects the linear relationship of Eq. (4). **(b)** Probability of infection, H, as a function of N_{ID}, i. e., the number of administered infectious doses (ID_{50}) represented against the log of N_{ID}. $H = 0.5$ is by definition at $\log N_{ID} = 0$ (dotted vertical line). Figures by R. Oberthür ©.

that the dilution of infected material (because there is no threshold value) will not lead to a lower number in infections, if dilution means the spread of a given amount of infectious material over an increasing number of susceptible animals.

ID_{50} is usually determined by titration in a bioassay using mice or hamsters. A predefined number of animals are administered the specific dose perorally (p.o.), intracerebrally (i.c.), or intraperitoneally (i.p.), with the doses for the different groups of animals differing usually by a factor of 10. In order to present these dose ranges adequately, a logarithmic plot of the dose range of interest is chosen (Fig. 52.3b) instead of the linear plot shown in Fig. 52.3a.

Figure 52.4 shows how the dose–response curve can be applied to experimental data over a larger dose range. It can be seen that the prions exert different activities depending on the dose within a broad dose range. The figure also shows that it is necessary to distinguish, for any given doses, between peroral and intracerebral administration. If prions are administered directly into the brain, a small dose is sufficient to lead to an infection. In contrast, higher doses are necessary for infection following oral ad-

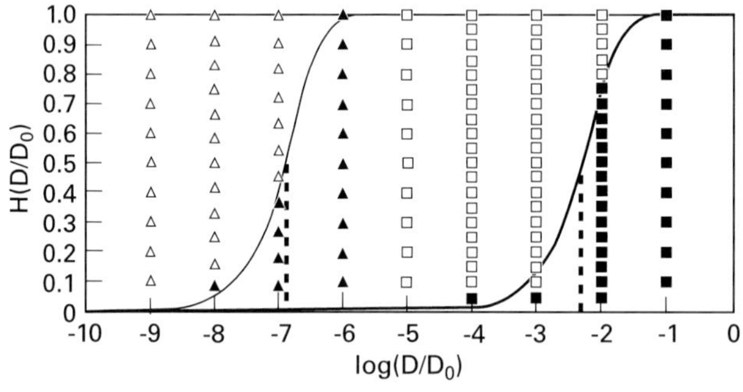

Fig. 52.4: Probability of infection, H, as a function of D/D_0, i. e., dose, D, taken from an original dose of infectious material D_0 and administered to hamsters in increasing dilutions. Full squares: infected hamsters; open squares: surviving hamsters after the oral administration of the respective dose. Full triangles: infected hamsters; open triangles: surviving hamsters after the intracerebral administration. Original dose: $D_0 = 100$ mg hamster brain in the terminal stage of TSE, 1st dilution: 100 mg of the supernatant (after 5 minutes centrifugation at 500 g) in a 10% suspension of the original dose diluted in buffer. The curves correspond to the dose–response curve according to Eq. (3) for oral administration (right curve) and for intracerebral administration (left curve). The dotted vertical line corresponds to an ID_{50} for oral (right line) and intracerebral (left line) administration. The difference between the two is 4.6 logarithmic units, i. e., a factor of 40,000. Figure by R. Oberthür ©.

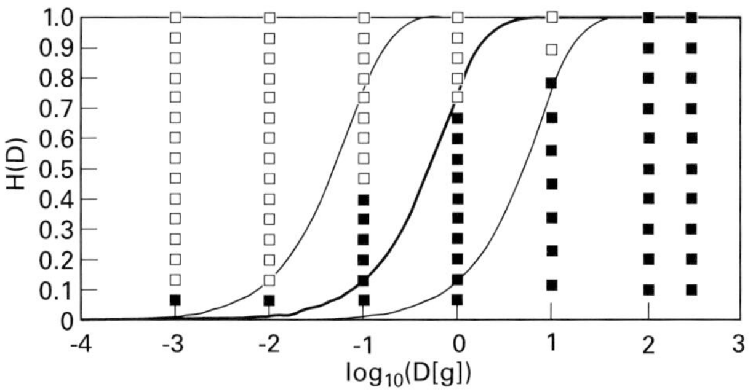

Fig. 52.5: Oral infectious dose for calves. Probability of infection, H, as a function of log (D/g), i.e., the common logarithm of dose, D, of brain of a terminally BSE-infected cow in gram administered orally to 4-month-old calves. Full squares: infected calves; open squares: surviving calves. The curves correspond to the dose–response curve according to Eq. (3). Thick curve corresponding to $ID_{50} = 500$ mg; thin curves corresponding to $ID_{50} = 50$ mg and $ID_{50} = 5$ g, respectively. Figure by R. Oberthür ©.

ministration, where they must pass the intestinal tract before reaching nervous tissue. Diringer and colleagues [12] have shown that the difference between oral and intracerebral administration in hamsters amounts to 4.7–5 logarithmic units (factor of 50,000–100,000) with a standard error of 0.3 logarithmic units (approximately a factor of 2, [13]).

The *oral ID_{50} for calves*, being of paramount interest for assessing the BSE risk in the cattle population, can be deduced from an experiment, similar to that shown in Fig. 52.4, in which four groups consisting of ten calves each received individual doses of 300 g, 100 g, 10 g, or 1 g brain tissue from a clinical BSE case in their feed [8] (Fig. 52.5). In a second experiment, five calves each received a dose of 1 g, while three groups of 15 calves each received individual doses of 0.1 g, 0.01 g, and 0.001 g brain tissue, respectively.

From the preliminary outcome of these experiments [14], as shown in Fig. 52.5, the oral calf ID_{50} of homogenized brain of BSE-infected cattle may be estimated as being about 500 mg [11],[4] but because of the atypical behavior of the data with respect to the proposed dose–response curve in Eq. (3), the experimental error might be still in the order of a factor of 10 (compare the two limiting curves in Fig. 52.5).

Assuming [15] that the brain of a cow weighs 470 g and there is a ratio of 2:1 regarding the total mass of brain and spinal cord, a total mass of approximately 700 g can be calculated for brain and spinal cord. Therefore, a BSE-infected cow would contain approximately 1,400 ID_{50}. The dilution of this material and distribution to many thousands of calves would lead to about 1,000 infected calves, e.g., $100{,}000\ H = 100{,}000 \cdot 0.69 \cdot 1{,}400/100{,}000 \approx 1{,}000$ according to Eq. (4). However, this would probably only be the case if the prions were ingested by 4-month-old calves. The susceptibility of older cattle to infection with BSE has been estimated to be 10-fold lower than that of 4-month-old calves [11].

52.4 Reproduction number

As demonstrated above, it is extremely important in a quantitative assessment of BSE risk to evaluate the mean number of animals that can be infected by one diseased animal under given conditions. This number is referred to as reproduction number, R_0, [2; 3]. In the case of BSE, in a country with a reproduction number $R_0 < 1$, the disease would be unable to spread, even should single cases occur spontaneously or through importation of infected live animals

[4] Applying the Kärber method to the data obtained with the assumption that at a dose of 0.0001 g no calf would be infected yields a geometrical mean value for $ID_{50} = 0.333$ g (G. Kärber. Arch. f. exp. Pathol. u. Pharm. 1931; 162:480.)

or infected cattle feed ingredients. An $R_0 > 1$, however, will lead to the explosive increase in infected animals, such as was observed in Great Britain between July 1988 and December 1991 (Fig. 52.1). Only appropriate measures to reduce R_0 to below 1 will lead to a decrease in the number of infected animals.

This is schematically shown for a hypothetical BSE epidemic in Fig. 52.6a,b. If a reproduction number of 3 is assumed, then the number of BSE cases will triple after each incubation period (about 4.5 years for BSE; at average). In reality, of course, the incubation period varies around the mean value of 4.5 years and has a minimum of approximately 2.5 years. Moreover, R_0 also represents an average value. Not all infected cattle reach the age of clinical manifestation of BSE. For example, a calf that becomes infected at the age of 4 months and is slaughtered at the age of 14–18 months, will never reach clinical manifestation. Nevertheless, such calves *do* contain infectious prions and can contribute to spreading the infection should parts of their remains be rendered into feed for other cattle. The individual R_0 is, however, much lower for such cases, as the concentration of prions in their tissues is less than in an adult with clinical BSE. On the other hand, a maximum value for R_0 has already been discussed in the previous paragraph.

The situation changes when the average R_0 is arbitrarily set to 0.5 as shown in Fig. 52.6b. Specific measures have led to a reduction in BSE incidence and the probability of any infected animal transmitting the disease to another is only 50 %. This means that it should be possible to eliminate the disease, even though it has occurred in epidemic proportions.

The events depicted in Fig. 52.6a,b lead to exponential growth or to exponential reduction of BSE cases. This was observed in Britain between 1988 and 1991 for the exponential growth (Fig. 52.1) and had probably occurred quite a long time before. Similarly, an exponential decay was observed between 1994 and 2005 (Fig. 52.1). Both phenomena can be described by the following equation:

$$N(t) = N_0 \cdot \exp(r \cdot t) \tag{5}$$

where N_0 denotes the number of cases at time $t = 0$ and r the positive ($r > 0$) or the negative ($r < 0$) growth rate.

For the simplified model shown in Fig. 52.6, R_0 is connected with r (Eq. (5)) through the following relationship:

$$R_0 \approx \exp(r \cdot \tau) = N(\tau)/N_0$$

or

$$r \approx \ln(R_0)/\tau \tag{6}$$

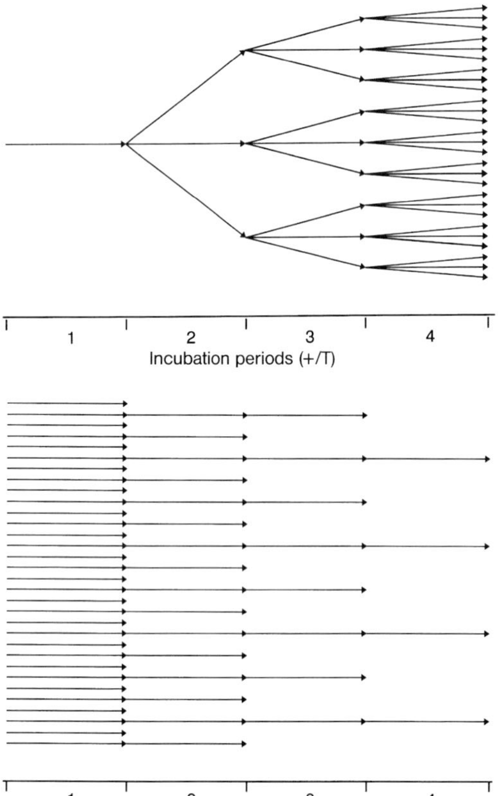

Fig. 52.6: (a) Schematic presentation of the explosive spread of BSE infections at a reproduction number of $R_0 = 3$ (avalanche effect or chain reaction). The horizontal axis represents the time course of the epidemic spread, one incubation period corresponding to approximately 4.5 years. In reality, a variable incubation period prevents the synchronous manifestation of BSE cases. **(b)** Schematic presentation of the gradual disappearance of BSE infections at a reproduction number of $R_0 = 0.5$. Figures by R. Oberthür ©.

where τ denotes the incubation period and $N(\tau)$ the number of newly infected animals calculated from the incubation period τ and the number of infected animals at time $t = 0$.

Therefore,

$$R_0 > 1 \Rightarrow r > 0$$

and

$$R_0 < 1 \Rightarrow r < 0$$

A plot of the logarithm of incidence as a function of time (semilogarithmic plot) has the advantage of yielding straight lines for the incidence with time if Eqs. (5) and (6) are obeyed. A positive slope corresponds to $R_0 > 1$ and a negative slope to $R_0 < 1$ (compare Figs. 52.1 and 52.7)

For the examples shown in Fig. 52.6a,b with $\tau = 4.5$ years, Eq. (6) leads to

$R_0 = 3 \qquad r = 0.244/\text{year}$
(exponential growth),

$R_0 = 0.5 \qquad r = \text{-}0.154/\text{year}$
(exponential decay).

From Eq. (5) and an initial value of $N_0 = 1$ after 13.5 years (three incubation periods), the following is obtained for the growth phase:

$$N = 1 \cdot \exp(0.244/\text{year} \cdot 13.5\ \text{years}) = 1 \cdot \exp(3.294) = 27$$

On the other hand, for an initial value of $N_0 = 32$ after 13.5 years (three incubation periods), the following is obtained for the decay phase:

$$N = 32 \cdot \exp(-0.154/\text{year} \cdot 13.5\ \text{years}) = 32 \cdot \exp(-2.079) = 4$$

In reality, the correlation between R_0 and r is much more complex than expressed in Eq. (6). For example, de Koeijer and colleagues [11; 16] took the following detailed parameters into account:

1. An age-dependent birth rate and death rate in cattle.
2. An age-dependent ID_{50} (age-dependent reduction in the probability that BSE develops following the oral intake of a specific dose from the ID_{50} of a 4-month-old calf down to 1/10 of that value for an adult cow).
3. The exponential "reproduction" of the infectious particles in the animal up to 1,400 ID_{50} per head of cattle that is in the final clinical stage with a doubling rate of 6 weeks until this clinical stage is reached.
4. A defined maternal transmission rate, which is proportional to the amount of infectious load in the dam (10% probability of transmitting the pathogen to the calf in the case of BSE-infected cows being in the last 6 months of the incubation period at the time of giving birth [17] (according to point 3, the cow carries an infectious load of 70–1,400 ID_{50}).
5. A defined "contact infection" rate, which includes all other potential paths of transmission.
6. A reduction in infectious load to the residual value k_1 following adequate heat treatment (*safer processing*) of animal proteins to be used in animal feed (Table 52.2; Fig. 52.2).
7. A reduction of the use of animal protein in cattle feed (*safer use*) to the residual fraction k_2 of cattle protein finding its way to cattle feed (Table 52.2; Fig. 52.2).
8. A reduction in the infectious load in offal from slaughtered cattle not recognized as being infected with BSE (subclinically infected animals) and used for the production of animal protein for use in animal feed (Table 52.2; Fig. 52.2) through SRM removal to the residual fraction c_1 (*safer sourcing*).
9. A reduction of the infectious load from BSE-infected fallen stock (i. e., cattle that died on the farm and, because of inadequate monitoring measures, were not identified as BSE cases and removed) being processed into animal protein to be used as animal feed (Table 52.2; Fig. 52.2) to the residual fraction c_2 (*safer sourcing*).
10. The probability of recognizing BSE relative to the length of time since infection.

These factors have all had an effect on the spread and decline of the BSE epidemic, although only crude assumptions can be made so long as no detailed experiments have been performed to determine at least some of the assumed dependencies.

Table 52.2: Reduction of prions in the material flow pathways of bovine tissue in Britain in different years in the course of the BSE epidemics as shown in Fig. 52.2. Measures taken to change the situation are also provided.

	1986	1990	1995	Measures
k_1	11%	11%	1%	Elevated processing temperatures in rendering
k_2	20%	2%	0.2%	Introduction of a ban prohibiting the feeding of ruminant protein to ruminants
c_1	70%	70%	7%	Removal of SRM*
c_2	85%	70%	7%	Segregation of BSE-diseased animals
Improved recognition skill for clinical BSE (further details in the text)	normal**	optimal	optimal	–

* SRM = specified risk material (see Chapter 53.3.1).
** Normal for inexperienced examiners and for the majority of examiners in countries with a low BSE incidence.

With regard to the last parameter, the probability of recognizing BSE relative to the length of time since infection, the identification of clinical cases assumes that the probability of detection increases by a factor of 6 when the veterinarians who are involved in BSE surveillance and monitoring have attained sufficient clinical experience (→ transition from *normal* to *optimal*; Table 52.2). The transition to *optimal* is most likely only applicable in countries with a very high BSE incidence (i. e., formerly in Great Britain). It must, however, be emphasized that in such countries the proportion of BSE is, relative to all neurological diseases observed in cattle, very high. This increases the probability of even inexperienced veterinarians being able to diagnose a clinical BSE case correctly.

The variables used in the model by de Koeijer and colleagues [11; 16] include the *age of the animals* and the *time since infection*. As the result of their model calculations, the *number of confirmed BSE cases* $N(t)$ is obtained as a function of the *time of detection*, and at the same time the reproduction number R_0 and the growth rate r in Eq. (5) allow plausible assumptions to be made on the BSE situation at any given time in a given country. This calculated growth rate is compared with the "experimentally" revealed growth rate obtained from a fit of Eq. (5) to the BSE incidence in a given country as a function of time.

In the next section, the BSE risk in Great Britain is assessed, based on the assumptions of de Koeijer and colleagues on how the disease could have spread in that country (see above), in agreement with the observed exponential increase and decrease of the British BSE epidemic (Fig. 52.1) fitted by a straight line according to Eq. (5) and yielding time-dependent values for r.

52.5 Reproduction number in Great Britain over time

The logarithmic graph of the BSE epidemic in Fig. 52.1 clearly depicts the linearity of the exponential increase in the first major phase and of the exponential decrease in the third major phase. A straight line with a slope $r = 0.64/\text{year}$ is obtained for the growth phase using a linear regression according to Eq. (5). From this value, a reproduction number $R_0 = 18$ new infections per infected animal can roughly be estimated using Eq. (5) and assuming $\tau = 4.5$ years. For the decay phase, $r = -0.55/\text{year}$ and $R_0 = 0.08$ is obtained.

The model calculations (11; 16) use plausible values for the values characterizing the transmission pathways introduced in Section 52.4 (compare Fig. 52.2 and Table 52.2). The changes in these values have a direct influence on the course of the epidemic as a consequence of the measures adopted to fight BSE.

Even the rough calculation with these characteristic values for BSE transmission will lead us to reproduction numbers, which can be compared with the reproduction numbers obtained from the slope of the growth and the decay phase of the BSE epidemic. Assuming for the situation before 1988 that

1. for a cow in the clinical stage of BSE a maximum reproduction number $R_0 = 1{,}000$ is adopted (compare Section 52.3) under the assumption that all the infectious material enters calf feed being fed to a multitude of calves without any infectivity reducing treatment, and without any removal of parts of the potentially infectious material to be used for other purposes (influence of *sourcing* of raw material for processed animal proteins);
2. during the rendering process, the infectivity was reduced to 11 % of the initial value (influence of *processing*);
3. about 20 % of the rendered bovine protein containing all the infectious material from that cow entered calf feed and was spread over a large number of calves (influence of *use*);

then a remaining reproduction number of $R_0 = 1{,}000 \cdot 0.11 \cdot 0.2 = 22$ is obtained, which is in the same order of magnitude as the value given for 1986 in Table 52.3 for the growth phase of BSE.

The law banning the feeding of ruminant protein to ruminants (*safer use*) became effective in Great Britain in July 1988. This ban may have led to a reduction of animal protein (with the exception of milk proteins) in ruminant feed to 1 % of the value before, i. e., from 20 % to a residue of 20 % $\cdot$ 0.01 = 0.2 % of the rendered bovine protein in ruminant feed. This residue may be due to cross-contamination. Within the above-mentioned model, this would have led to a reproduction number $R_0 = 22 \cdot 0.01 = 0.22$, a value again in the same order of magnitude as the value given for 1990 in Table 52.3.

In September 1990, it became obligatory to remove from food and feed chains all cattle suspected of suffering from BSE as well as those tissues considered capable of carrying the BSE prions (specified bovine offal, SBO) from cattle slaughtered at 6 months of age or older (*safer sourcing*). This should have led to a further reduction of the reproduction number by more than 90 %. A reproduction number of $R_0 = 0.22 \cdot 0.1 = 0.02$ should have been the result, which may be compared with the value for 1995 in Table 52.3.

Table 52.3: Reproduction number for BSE in Britain at different points in time during the course of the BSE epidemic.

	1986	1990	1995
R_0	14	0.3	0.08
Upper threshold of the 95 % confidence interval	26	0.5	0.1

In May 1994, following studies on the thermal inactivation of BSE prions [18] under various processing conditions in use at the time, the processing of animal proteins under conditions where prions were inactivated by less than about 99.5 % were prohibited in the European Union (Commission decision 94/384; effective January 1995). For the processes not prohibited, temperature/time conditions were defined under which BSE prions were inactivated by at least 99.5 % (*safer processing*). This should have led to an additional average reduction of the reproduction number to at least 5 % of the previous value, i. e., $R_0 = 0.02 \cdot 0.05 = 0.001$.

It becomes evident that this value is already much lower than the reproduction number $R_0 = 0.08$ calculated for the decay phase of the British BSE epidemic (Table 52.3; Figure 52.1). The prohibition of the use of any mammalian meat-and-bone-meal (MBM) for the purpose of feeding to farmed animals including horses and farmed fish in March 1996 should have led to a further decrease of the reproduction number (additional *safer use*). Furthermore, in April 1996, a scheme was introduced in Britain to ensure that all cattle over the age of 30 months at the time of slaughter did not enter the human food or animal feed chain (over thirty months scheme, OTMS), leading to an additional *safer sourcing*. All the prohibited tissues and their rendered products were required to be incinerated or used as a fuel for heat or power generation, leading to an additional *safer processing*.

On the basis of all the measures taken, Donelly and Ferguson (Table 11.1 in [19]) used data from the epidemic through mid-1998 to make calculations in a model without maternal or horizontal transmission. They predicted a decay from 1998 to the end of 2001 (green step function in Fig. 52.1) much steeper than was later actually observed. In contrast, the observed BSE incidence further has followed the same exponential decay with roughly the same slope since 1994 (green straight line in Fig. 52.1). This may be due to a remaining minor vector of transmission that so far has been overlooked and that might have become dominant in the decay phase of the British epidemic when all other vectors had been eliminated. However, it should be kept in mind that such a minor vector of transmission is detectable only because of the relatively large number of infected cattle still present.

On the other hand, as long as a large amount of infectivity was present in the British cattle herds, transmission to other countries by means of export of products of bovine origin used in cattle feed could have led to outbreaks of BSE, even in countries that never had a proliferation of BSE due to a reproduction number $R_0 < 1$.

The results of the detailed calculation, including all the described intercorrelations that could have had an effect on the epidemic (see above) and that had been taken into account by de Koeijer, Heesterbeek et al., is presented for three selected situations in Great Britain [11] in Table 52.3.

The curve in Fig. 52.7a is drawn on the basis of these three reproduction numbers (Table 52.3). It reflects the possible gradual change of R_0 in Great Britain over time and attempts to explain, on the basis of the reproduction numbers, the emergence of BSE in that country and its decline after the measures taken in 1988 and 1990. $R_0 = 1$ for 1970 was taken because it is the limiting number for a linear increase in incidence based on a constant incidence N_{sp} of spontaneous BSE cases

$$N(t) = N_{sp} \cdot t/\tau \text{ for } R_0 = 1 \qquad (7)$$

whereas for $R_0 < 1$, the incidence based on a constant incidence of spontaneous cases tends toward a limiting value of the incidence N_{lim} given by

$$N_{lim} = N_{sp} \;/\; (1 - R_0) \text{ for } R_0 < 1 \qquad (8)$$

The reproduction number before 1970 was probably less than 1. For illustration, a reproduction number of 0.9 would have led, according in Eq. (8), to a 10-fold incidence only with respect to a hypothetical spontaneous incidence and, on the other hand, a 10-fold increase in incidence at $R_0 = 1$ would have only been reached according to Eq. (7) after 45 years, 10 times the average incubation period.

In Fig. 52.7a, it is assumed that an average incidence of spontaneous cases potentially will be detected by a good passive surveillance system at a rate of <0.1 cases per month. This corresponds roughly to <10 cases per year, about 1 in 1 million of the national cattle herd of Great Britain, assuming that 2/3 of the cases are slaughtered before the possible manifestation of clinical BSE and only 1/3 is detected by passive, compared to active, surveillance. From such a low incidence, with an increasing reproduction number as shown in Fig. 52.7a by the blue curve, the incidence shown by the pink curve is obtained schematically according to Eqs. (5) and (6).

For the period 1971 to 1984, the transition from $R_0 = 1$ to $R_0 = 14$ was chosen with the background of information on the increase of continuous rendering in Britain provided in Fig. 19.2; the course of the transition from $R_0 = 14$ to $R_0 = 0.08$ was estimated and appears to be in close agreement with the estimates of Donelly and Ferguson (compare Fig. 5.5 in [19]).

The essentials of Fig. 52.7a for the situation in Great Britain are

- the increase of the reproduction number R_0 for BSE from below 1 to about 14;
- the maintenance at that level for several years until July 1988;
- the decrease to about 0.08 after the measures taken in 1988 and 1990;
- the maintenance of the reproduction number at a residual level since about 1994.

The consequences of such a course of a reproduction number are

- an incidence of infection of previously healthy animals proportional to the number of infected cases terminating their life and entering the recycling process and proportional to the reproduction number;
- an incidence of outbreak of the disease delayed by the incubation time;
- before the exponential growth, a possible rate of BSE cases below the rate of detection by passive surveillance;
- an exponential growth of BSE incidence observed by passive surveillance due to a reproduction number remarkably greater than 1;
- a slope of the exponential growth proportional to the reproduction number greater than 1;
- transition from growth to decline about 4.5 years (incubation time) after measures had been taken to reduce the reproduction number;

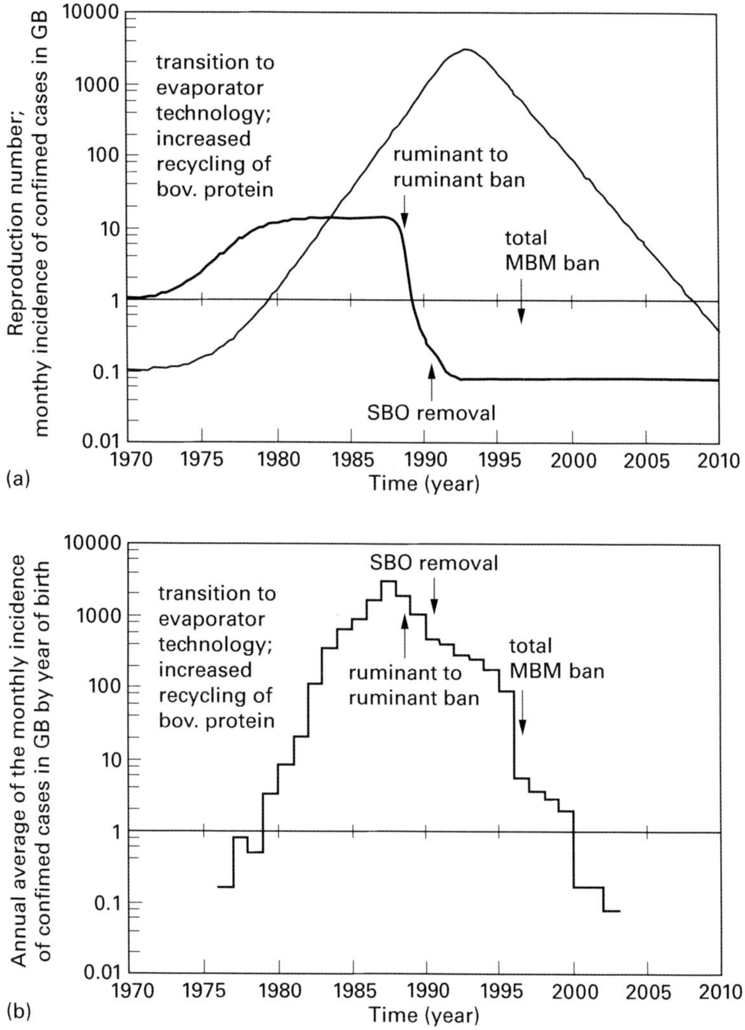

Fig. 52.7: (a) Reproduction number R_0 for BSE in Britain between 1970 and 2010 based on the epidemiological calculation made by de Koeijer et al. [11; 16] (blue curve) and the extrapolated monthly incidence of confirmed BSE cases by month of restriction based on the data of Fig. 52.1 and the hypothesis of an origin of BSE from spontaneous cases. **(b)** Monthly incidence of BSE cases in Great Britain by months of birth from passive surveillance. Figures by R. Oberthür; data taken from source [20].

- an exponential decline of BSE incidence observed by passive surveillance due to a residual reproduction number remarkably less than 1;
- a slope of the exponential decline proportional to the reproduction number less than 1;
- a tendency that a residual incidence of possibly spontaneous BSE cases only detectable by active surveillance of all animals at risk (bovine suspects, fallen stock, emergency slaughter over 30 months) will probably be reached by 2015.

Different information is provided by the incidence by time of birth of the BSE cases, again detected by passive surveillance, as shown in Figure 52.7b. From this curve the effects of the measures taken to reduce the incidence of infection are immediately apparent. The ruminant to ruminant feed ban in July 1988 especially led to the decrease in comparison to the previous year, and the change from a positive to a negative slope of the curve in the logarithmic plot. Unfortunately, the figures given in Fig. 52.7b are only available with a reasonable accuracy more than 6 years after the indicated year of birth due to the spread of incubation times and are still changing slightly even after this delay.

It is interesting that the measures taken to reduce the incidence of BSE all resulted in a remarkable decrease in the year the measure was taken; however, they never resulted in a complete disappearance of new infections (BAB, born after ban since July 1988, and BARB, born after reinforced ban since August 1996). It is especially interesting to observe that the number of cases drops immediately in the year a measure has been taken, but then continues to decrease with a similar slope in the logarithmic presentation shown in Fig. 52.7 as before the measure had been taken, only starting at a lower level of incidence. According to Eqs. (5) and (6), it seems as if part of the majority of the infected animals acting as a source of possible BSE proliferation, according to the given reproduction number for that time, had disappeared, and the propagation of the disease continues with the similar reproduction number but with a reduced number of infected animals as source for further infections. This phenomenon is especially pronounced for the year 1996, when the total MBM ban in farmed animal feed came into force and the slaughter of, and the processing and incineration of MBM from, animals older than 30 months (OTMS) started. The reduction of incidence in 2000 and the data for 2001 and 2002 are still too preliminary to be interpreted (as at the beginning of 2006).

52.6 BSE risk minimization within the cattle population

Of the two main factors responsible for the spread of BSE within the cattle population, the first is the actual number of infected cattle in a country and/or the amount of BSE-infectious load in bovine products produced and available for use in that country. The second is the reproduction number in a given country, yielding the number of new infections due to the specific system of sourcing, processing, and use of bovine products in that country. Measures for BSE risk minimization can target both factors.

As a guideline for an estimate of the number of infected cattle entering the transformation process per year in a given country, we may assume that 2/3 do not reach the age when BSE manifests and so cannot be detected with available tests. Of the other remaining infected animals, about one-half is slaughtered before clinical signs of BSE manifest. They are only found positive if the active surveillance extends over all slaughtered animals older than at least 30 months. The other half is found among animals at risk, i. e., BSE suspects, including selective cohort cull, as well as fallen stock and emergency slaughter. The total number of these animals at risk per year amounts to about 1.8 % of the total cattle population in any given European country. The limits of detection of a BSE incidence in a given country could then be assumed in the range of 1 detected BSE case per year within 100,000 cattle at risk. This corresponds roughly to 1 case of a BSE infected bovine ending its life per year in 1 million of a given cattle population (possible incidence of spontaneous BSE cases if BSE has the same incidence of spontaneous cases as CJD), counting both detectable and undetectable BSE infections.

For example, the incidence of BSE per 100,000 cattle at risk obtained from active surveillance in Great Britain in 2005 was 60, whereas in Germany in the same year it was 7. For comparison, the incidence of BSE in Great Britain in the years 1992/1993 at the peak of the epidemic would have been about 17,000 per 100,000 cattle at risk. Although this last figure is based on passive surveillance, it gives at least an impression that the range between maximum observed and just detectable annual BSE incidence covers more than four orders of magnitude (compare also Figs. 52.1 and 52.7).

The effect of measures to minimize the BSE risk in a given country depends on the BSE incidence and the reproduction number in that country.

In countries with a high incidence, and reproduction number $R_0 \gg 1$, i. e., in the case of an annually dramatically increasing BSE incidence, the first aim must be to reduce the reproduction number far below 1 by safer sourcing, safer processing, and safer use of products of bovine origin (compare Table 52.4). Measures to directly reduce the number of infected cases among the healthy bovines are much more difficult because there is as yet no method to detect infected live animals. However, the cull of animals over 30 months of age (see OTMS) to take them out of the recycling system certainly contributes to a reduction of infectivity in the cattle herd.

Table 52.4: Three measures related to animal feed that may lead to a significant reduction of the reproduction number R_0 for a system with mixed animal protein production and the production of animal feed for different species in the same production line.

- Safer sourcing: Removal of risk material (reduction factor 10 to 50).
- Safer processing: Processing at sufficiently high temperatures (possible reduction factor of 1,000 to 10,000)
- Safer use: Ban of feeding ruminant protein* to ruminants (reduction factor 50 to 500).

* The general ban of feeding mammalian protein to ruminants leads to greater safety. This measure can be supported by the fact that the analytical distinction between mammalian and other animal protein (e. g., from poultry or fish) is easier than the distinction between ruminant and other mammalian protein (e. g., from pig). Hence, the control of a ruminant-to-ruminant ban is more difficult than the control of a mammalian-to-ruminant ban.

In countries where BSE incidence is at the limit of detection, and reproduction number is low, never exceeding $R_0 = 0.05$, the main risk comes from the importation of contaminated bovine products from countries with a high BSE-incidence even if the reproduction number in that country has the same low value in the range of $R_0 = 0.05$. Although in the country with a high BSE incidence the annual incidence decreases, in the country of low incidence at the same time an increase in BSE cases would be observed. This incidence would, however, rapidly decrease if importation were to be stopped as a measure against BSE risk [21].

As a consequence of different measures to reduce the risk of BSE-transmission taken in various EU countries, from 2001 the European Union has temporarily banned the use of animal proteins in all feed of farmed animals. However, in 2002, the animal by-products regulation was decreed (Table 52.5), which provides a framework within which bovine protein might possibly be used again in the feed of farmed animals, so long as a system of good traceability capable of insuring safe sourcing and processing could be provided by the industry for the different products of animal protein.

The measures decreed by the EU provide reasonable safety against BSE transmission, mainly due to the complete removal of possibly BSE-contaminated products within Category I material, which is condemned to incineration, and to the ban on within-species feeding.

In this respect, it is interesting to remark no dramatic reduction of infectious load in cattle feed would have been necessary to obtain the reproduction number between 0.05 and 0.1 which so far continues in Great Britain. Rather, a reduction of infectious load in the feed for cattle by a factor of 200 to 300 would be sufficient to result in a drop of the reproduction number as observed in Great Britain after 1988. This also shows that countries other than Great Britain could easily have had a recycling system with a reproduction number only a factor 20 less than that of Great Britain before 1988 (com-

Table 52.5: General restrictions in the use of animal by-products for the production of processed animal protein (PAP) according to the European animal by-product regulation (EC 2002/1774) relevant for BSE proliferation.

A. Restricted **sourcing** through categorization of animal raw material		
	Category 1:	Potentially BSE-contaminated material
	Category 2:	Fallen stock and carcasses or by-products from slaughtered animals declared not fit for human consumption
	Category 3:	By-products of slaughtered animals after inspection declared fit for human consumption, but for economic reasons not used for human consumption
B. Restricted **processing** according to category		
	Category 1:	Incineration (without or with usable energy production)
	Category 2:	Pressure cooking at 133 °C for 20 minutes under saturated water vapour pressure (alternative processing with similar efficiency possible)
	Category 3:	Pressure cooking at 133 °C for 20 minutes under saturated water vapour pressure (alternative processing with similar efficiency possible)
C. Restricted **use** according to category		
	Category 1:	Fuel for energy production
	Category 2:	Fertilizer (not to be used on pasture land)
	Category 3:	Possible use in pet food, fur feed, or as feed ingredient for farmed animals destined for human consumption (if suitable control tools are available to verify the enforcement of the restriction in the finished feed), restricted by the *species-to-species* ban (interdiction of the feeding of animal protein within the same species with the exception of milk protein for mammals, and fish protein for fish)

pare Table 52.3), without observing a dramatic increase of BSE cases, because of a reproduction number less than 1, which kept the BSE incidence close to the limits of detection (compare Eq. (8)).

52.7 BSE risk minimization from cattle to humans

Concerns over the possible transmission of BSE from cattle to humans arose when a new variant of CJD (vCJD) was detected in Great Britain in 1996. Since then, the incidence of vCJD has gone through a maximum incidence in the year 2000, with 28 cases, and has since dropped to a value of 5 cases in 2005, i. e., fewer than 1 case per 10 million in Great Britain.

A study of the possible consumption of products contaminated with BSE infectivity has been carried out by Comer and Huntly [22]. They assumed a bovine oral infectious dose of $ID_{50} = 20$ mg brain tissue of a terminally BSE-infected cow, based on the data of Fig. 52.5, rather than the $ID_{50} = 500$ mg with an uncertainty of a factor of 10 used in this chapter.

With the ID_{50} value used in this chapter (dividing the original values of Comer and Huntly[5] by 25) based upon their assumptions, about 2 million bovine oral ID_{50} units would have been consumed by humans from 1980 until now with a peak of 440,000 ID_{50} units in 1993 and 0.1 ID_{50} units in 2001. From their inquiry, they assume that most of the infection from prime beef was due to brain tissue used in human food with a peak in 1989. The ban of the use of SBO for human consumption in November 1989 led to a rapid decrease of infectivity entering the human food chain from prime beef. The infectivity due to adult beef (older than 30 months) was found from the inquiry mainly due to mechanically recovered meat (MRM), which is derived by a process where bones with adherent meat were subjected to high pressure to separate the meat from the bone. This meat was used in a range of human food products, e. g., burgers. Since vertebral columns with adherent nervous tissue were part of the raw material for this process, remains of the spinal cord and dorsal root ganglia entered the food chain. This led to the assumed peak of total infectious load in the human food supply in 1993. The prohibition on the use of vertebral columns in the pro-

[5] See above; i. e., [22].

duction of MRM in December 1995 and the exclusion of bovines over 30 months from the human food chain led to the calculated dramatic reduction of BSE infectivity entering human food, which was mentioned in the beginning of this section.

If the peak of vCJD incidence in 2000, with a distribution from 1995 to 2005, (incidence < 5 cases per year) is compared with the peak of infectious load from prime beef with the peak in 1989 and a distribution from 1985 to 1992 (10% value of the peak infectious load), a mean incubation period of 11 years results. Comparison of the peak of vCJD incidence with the peak of total infectious load, including prime and adult beef, in 1993, with a distribution from 1986 to 1996 (10% value of the peak infectious load), yields a mean incubation period of 7 years (compare Chapter 1.9.3).

A cheaper method of reducing the human BSE risk than the British OTMS is the BSE testing of all animals older than 30 months being slaughtered for human consumption. This is done in Germany, where in 2005, among 1,839,573 tested cattle older than 24 months, 16 were found positive and removed from the human food chain. However, other countries with a similarly low incidence of BSE regard even this measure as disproportionate. They argue that all the measures already in place, especially the removal of specified risk material from the slaughtered carcass including the whole head, the spinal cord, and the vertebral column, together with a low incidence of BSE, contribute sufficiently to the protection of the population against transmission of BSE. In any case, even if dozens of bovine oral infectious doses would escape into human food in Germany, because of the species barrier, the oral ID_{50} for humans is probably several orders of magnitude higher than that for cattle. In this case, even without testing of every slaughtered animal older than 30 months, with the actual incidence of BSE, the risk to a human in Germany would be less than 1 in 1 million, which should be acceptable.

To date, almost all of the vCJD patients were homozygous (MM) at codon 129 of the PrP gene, whereas only about half of the normal population in Europe and the US are homozygous. Nothing is known about possible sensitivity to BSE transmission and incubation time of the other genotypes (MV or VV) in the human population. Since the exposure of the human population, mainly in Great Britain, to a large amount of BSE infectivity cannot be undone, it would be desirable that research be carried out into detection of infectivity in living individuals and the reversibility or the retardation of the development of infectious load. Compared with the cost of the measures to minimize the risk of the transmission of BSE to humans or animals, the amount of money spent on research is still rather small.

52.8 Control of the efficiency of BSE risk minimization

Most measures for BSE risk minimization involve high costs, which eventually are imposed on the whole community. Hence, it might be of interest to evaluate whether the money spent has really had the expected effects. Current measures to minimize BSE risk are limited by the fact that valuable products for consumption by humans and animals (e. g., cats and dogs) are still being produced by a national cattle herd that might contain up to thousands of undetected BSE infections, which cannot be detected in the live animal or in the dead animal younger than about 30 months. In the face of these limitations, it has been the aim of the measures to reduce the proliferation of BSE to an undetectable minimum rate.

From the course of the BSE epidemic in Great Britain (Fig. 52.1) and the schematic comparative presentation of risk of transmission and incidence of BSE in Figure 52.7a, it is obvious that the efficiency of a measure to fight BSE is only detectable after the average incubation period of about 4.5 years has elapsed. That is, the effect of the ruminant-to-ruminant feed ban from July 1988 became visible only in the course of 1992 (Fig. 52.1). The efficiency of the measures was then quantitatively detectible from the slope of the decline of the BSE incidence. Unfortunately, an examination of the difference in the slopes of increase and decrease of the epidemic reveals that proliferation was reduced only by a factor of roughly 200 (compare Table 52.3). Since it seems improbable that the accumulated effect of all the measures taken to remove possible infectious load in carcass tissues from bovine feed did not lead to a further decrease of the reproduction number, it appears probable that there is an as-yet-undetected vector of infection that does not involve carcass tissues of BSE-infected animals.

A better monitor of the efficiency of a measure taken to fight BSE is provided by the incidence of cases by time of birth, before and after the specific measure has been implemented. This is shown for Great Britain in Figure 52.7b. Again, due to the long incubation period of BSE, this method of monitoring can also only be used several years after the measure has been implemented. However, it shows immediately the effect of a given measure on the rate of new infections in the year the measure has been put into place and in the years that follow. If the measure had led to a reduction of transmission to an undetectable level, no new infection should have been found, but due to a remaining transmission the decrease of incidence by month of birth also shows a gradual decline similar to the decline in the incidence of confirmed BSE cases by time of restriction.

Compared to Fig. 52.1, the decrease of incidence in Fig. 52.7b is much more obvious. The most remarkable feature, of course, is the decrease of incidence in 1988 following immediately the ruminant-to-ruminant feed ban of July 1988 after a steady increase until 1987. However, there is only a gradual decrease in the born-after-ban BSE cases (BABs) and no reduction of transmission to an undetectable level. The next remarkable downward shift in 1990 may be attributed to the SBO removal coming into force in September 1990. Another shift in 1995 may have been caused by the stricter processing of animal proteins leading to a greater inactivation of prions since January 1995. A dramatic decrease of the incidence is finally observed in 1996 when probably all transmission of infectivity through animal carcass tissue was stopped, and when at the same time the number of infected animals over 30 months was reduced by the beginning slaughter of all animals over that age (OTMS) and their complete removal from human food and animal feed chain.

Nevertheless, after 1996, the transmission continues, albeit at a lower level, but with a similar slope. That is, there appears to be a remaining vector of transmission. However, even with this remaining vector, from the extrapolation of the slope of the course of the active surveillance data, it looks probable that the disease will reach the limits of detection in Great Britain between 2012 and 2015. Therefore, the active surveillance data on cattle at risk could still be used to judge the efficiency of measures in the next 5 years, especially, if they are combined with the incidence by year of birth.

References

[1] Defra. BSE – background to the disease and latest developments. Webpage 2006: www.defra.gov.uk/animalh/bse.

[2] Diekmann O and Heesterbeek JAP. Mathematical epidemiology of infectious diseases. Wiley, Chichester, 2000.

[3] Anderson RM and May RM. Population biology of infectious diseases. Report of the Dahlem Workshop on Population Biology of Infectious Disease Agents, Berlin, March 14–19, 1982. Springer, Berlin, 1982.

[4] World Health Organization. Consultation on public health and animal transmissible spongiform encephalopathies: epidemiology, risk and research requirements; with the participation of the Office International des Epizooties. Document WHO/CDS/CSR/APH/2000.2 [meeting report]. Geneva, December 1–3, 1999.

[5] Rodricks JV. Calculated risks – the toxicity and human health risks of chemicals in our environment. Cambridge University Press, Cambridge, 1994.

[6] Anonymous. Sterbetafel 1981/83 (Männer) und Sterbetafel 1981/83 (Frauen). In: Statistisches Jahrbuch der Bundesrepublik Deutschland, 1985.

[7] Hauser J. Bevölkerungslehre. UTB, Stuttgart, 1982.

[8] Anderson RM, Donnelly CA, Ferguson NM, et al. Transmission dynamics and epidemiology of BSE in British cattle. Nature 1996; 382(6594): 779–788.

[9] Kuipers P. Analyse van risicofactoren voor BSE in Nederland. Faculteit Diergeneeskunde, University of Utrecht, 1994.

[10] Beekes M, Baldauf E, Diringer H. Pathogenesis of scrapie in hamster after oral and intraperitoneal infection. In: Court L, Dodet B, editors. Transmissible subacute spongiform encephalopathies: prion diseases. Elsevier, Paris, 1996:143–149.

[11] de Koeijer AA, Heesterbeek JAP, Oberthür RC, et al. Calculation of the reproduction ratio for BSE infection among cattle. BSE risk assessment [scientific report]. ID-DLO Report. Institute for Animal Science and Health, Lelystad, The Netherlands, 1998:1–36.

[12] Diringer H, Roehmel J, Beekes M. Effect of repeated oral infection of hamsters with scrapie. J Gen Virol 1998; 79:609–612.

[13] Schreuder BE, Geertsma RE, van Keulen LJ, et al. Studies on the efficacy of hyperbaric rendering procedures in inactivating bovine spongiform encephalopathy (BSE) and scrapie agents. Vet Rec 1998; 142:474–480.

[14] Matthew D. Gaps and opportunities – what remains to be done. Proceedings of the Beef Improvement Federation 38th Annual Research Symposium and Annual Meeting. Choctaw, MS: Pearl River Resort, April 18–21, 2006.

[15] Nickel R, Schummer A, Seiferle E. Nervensystem, Sinnesorgane, Endokrine Drüsen. Lehrbuch der Anatomie der Haustiere – Band IV. Paul Parey, Berlin, 1992.

[16] de Koeijer AA, Heesterbeek H, Schreuder B, et al. Quantifying BSE control by calculating the basic reproduction ratio R0 for the infection among cattle. J Math Biol 2004; 48:1–22.

[17] Collee JG and Bradley R. BSE – a decade on. 2. [Review]. Lancet 1997; 349(9053):715–721.

[18] Taylor DM, Woodgate SL, Atkinson MJ. Inactivation of the bovine spongiform encephalopathy agent by rendering procedures. Vet Rec 1995; 137:605–610.

[19] Donnelly CA and Ferguson NM. Statistical aspects of BSE and vCJD – models for epidemics. Chapman & Hall/CRC, Boca Raton, 2000.

[20] Defra. BSE: Statistics. Confirmed cases of BSE in GB by year of birth where known. Webpage 2006: www.defra.gov.uk/animalh/bse/statistics/bse/yrbirth.html.

[21] Oberthür RC. Modellrechnung zur Erklärung der BSE-Inzidenz in Deutschland. [Model calculation to explain the BSE-incidence in Germany]. Berl Münch Tierärztl Wschr 2004; 117:230–242.

[22] Comer PJ and Huntly PJ. Exposure of the human population to BSE infectivity over the course of the BSE epidemic in Great Britain and the impact of changes to the Over Thirty Month Rule. J Risk Res 2004; 7:523–543.

53 BSE Control – Internationally Recommended Approaches

Stuart C. MacDiarmid, Paul Infanger, and Beat Hörnlimann

53.1 Introduction

The epidemic of bovine spongiform encephalopathy (BSE), which began in 1986 in the United Kingdom (UK), is now dwindling. However, it leaves in its wake an outbreak of human variant Creutzfeldt–Jakob disease (vCJD), most probably resulting from the consumption of beef products contaminated by infectious central nervous system tissue. Between its first clinical appearance in 1994 and 2006, vCJD cases in the United Kingdom averaged around 15 per year, but annual incidence has been declining since 2000, as described by Robert G. Will in Chapter 15.

Having affected an estimated three million or more cattle [1], of which around 190,000 cases have been reported worldwide (Table 19.2), it is – without a doubt – the most important animal epidemic of modern times. The occurrence of the epidemic and the challenges in combating it have influenced the implementation of food safety policies for animal-derived products aimed at increasing the level of consumer safety to an extent greater than any other epidemic in history.

This chapter considers the scope of policy options and measures to address the current BSE situation in the broader, worldwide context. Recommendations are made, taking into account the guidelines of the World Health Organization (WHO) and the World Organisation for Animal Health (OIE), so there is a summary of both public and animal health aspects. The chapter includes measures for the surveillance, prevention, and control of BSE [2]. The recommendations are based on the current state of scientific knowledge [compare 3], and it should be recognized that some of the unknown factors that generated public health anxiety during the BSE crisis in 1996 are now better understood. Fear of the unknown is no longer sufficient justification for the establishment of national and international policy. Resources should be directed toward the implementation of measures recommended here to protect human and animal health, and also the implementation of appropriate surveillance programs. All countries should collaborate in a common goal of worldwide BSE eradication.

53.2 Measures in response to the first case of BSE in a country

53.2.1 Epidemiological investigation

The epidemiological investigation into the origin of newly emerging BSE in a country gives rise to significant challenges, as the current situation is the result of events that may have occurred many years before. It is also highly resource intensive.

Infected animal(s) may have been imported from a BSE risk country at some time in the past. If such animals have not been detected, BSE-infectious tissue originating from them may have been rendered and fed to cattle, thus spreading and amplifying the BSE agent within the indigenous cattle population of the country into which they were imported.

The epidemiological investigation should trace all cattle which, during their first year of life, were reared with the animal found to have BSE during its first year of life and which may have consumed the same contaminated feed during that period (what is known as a "feed cohort"). Such animals represent a significant additional risk for further propagation of BSE within the country.

All the progeny of female BSE cases born within 2 years prior to or after clinical onset of the disease should also be traced. Risk materials from unidentified infected animals must be considered when developing policies for (i) the prevention of human infection through food, pharmaceuticals, or cosmetics (see Chapter 46) and (ii) the prevention of new infections of cattle through feed.

An effective cattle identification system is necessary in order to identify (i) the imported cattle, (ii) the "feed cohort", and (iii) the progeny of female cases. It may not be possible to trace every animal of these three groups in countries where an effective animal identification system is not in place. Hence, the quality and reliability of a cattle identification and traceability system in a specific country must be considered as a limiting factor for the effectiveness of BSE control and disease control in general.

Intensive efforts should be made to trace the potentially contaminated concentrated feedstuff to which the BSE case and the "feed cohort" were possibly exposed during their first year of life [4–6]. The feedstuffs should be traced back to the point of sale using the farmers' receipts, if available, as these are often kept for several years for tax purposes. From point of sale, the suspect feedstuff can be traced back to the feed mill where the meat-and-bone-meal (MBM) was added as an ingredient, and from there to the importer (if this is of relevance; see [7–8]) and/or even the rendering plant from where the involved MBM originates.

Examination of feed mill records should reveal the composition of the suspect feedstuff, allowing identification of the ingredient likely to have caused the BSE case. If the suspect ingredient, for example MBM, is still being produced in the same way as at the time when the BSE-affected animal was infected – i. e., most likely during its first year of life – all this material should be regarded as a BSE risk factor and removed entirely from the feed market (for all species including zoo animals and for all pet food). It should be incinerated or – in exceptions and only under strict control by the authorities – disposed of by deep burial.

53.2.2 Incineration of risk material and tracing of feed cohorts and by-products of animal groups at a higher risk

Incineration of carcasses of BSE cases

The carcasses of BSE cases and traced products such as MBM derived from BSE cases should be incinerated. The WHO recommends that tissues from cattle known to be infected with BSE should not enter the human food chain or any other commercial products (see Chapter 46). Neither should they be fed to zoo animals (see Chapter 20). Traced products include the BSE-related risk material, e. g., brain, spinal cord, and other tissues normally categorized as specified risk materials or SRMs (see Section 53.3.1).

Tracing of by-products arising from the SRM derived from BSE cases, involved "feed cohorts" and their progeny

The tracing of the rendered MBM that may have been contaminated with SRM from BSE cases, from their progeny (if the case is female) and from the "feed cohort" of BSE cases should be effective and appropriate. In some situations, extra precautions may be taken to assure that such contaminated material is destroyed and does not enter trade.

Tracing of products derived from cattle slaughtered on the same day and on the same premises as BSE cases

In the situation where a BSE case is identified in a slaughterhouse, the issue of cross-contamination should be considered. Products from cattle slaughtered on the same day may be (cross-)contaminated with BSE-related risk material. The sources of contamination are particularly the spinal cord, when the carcass is split into halves, or the brain, if aerosols may be produced (see Section 53.3.1) during certain slaughter processes (Fig. 53.1). For this reason, meat from such cattle should not enter the human food chain.

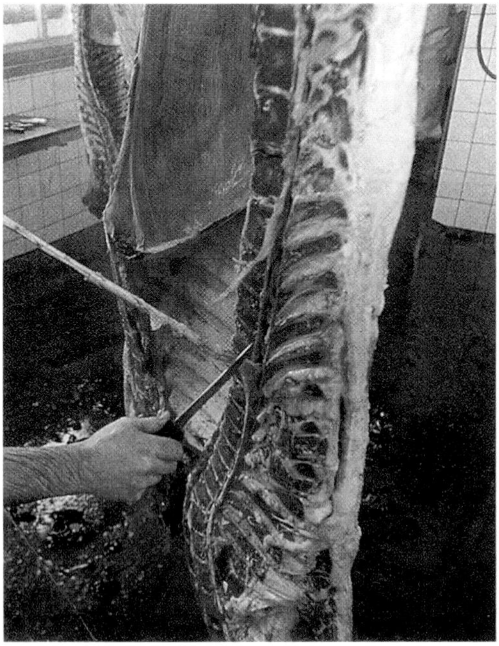

Fig. 53.1: Photograph showing a cattle carcass just after having been split in half; potential BSE cross-contamination by inappropriate knife hygiene (aerosol and smears from the saw splitting the carcass can also potentially put parts of the carcass at risk of cross-contamination). Figure © courtesy of M. Gyger; source: [2].

Culling of "feed cohort" animals and progeny of female BSE cases

Because BSE is not spread by direct contact among animals, culling the whole herd where a case occurs is not justified. The long incubation period of BSE and movements of livestock mean that the animal may well have been infected on a farm other than where the onset of clinical signs is observed. On the other hand, the OIE recommends [9] that animals from the same "feed cohort" should be traced and permanently identified and their movements controlled.

Culling strategies[1] are inevitably complex, especially because of the difficulty in identifying the "feed cohorts". This has been common in the majority of BSE cases identified worldwide. This means that the identification of a "feed cohort" may not always be possible, so the OIE recommends that in such circumstances, all cattle born in the same herd as, and within 12 months of, the birth of the BSE case be traced and permanently identified and their movements controlled. The OIE recommends that the progeny of the BSE cases born within the 2 years prior to, or after, the onset of clinical signs should be identified and their movements controlled.

With regard to the eventual fate of progeny of BSE cases and of "feed cohort" animals, the OIE does not recommend their immediate culling. Rather, when they die or are slaughtered they are completely destroyed and excluded from all food and feed chains. The reason for this, once again, is because BSE is not spread by direct contact among animals. When "feed cohort" animals or progeny are culled in numbers exceeding those recommended by OIE, the reasons behind the decisions should be made clear in public statements.

53.3 Measures for surveillance, prevention, and control of a BSE epidemic

The first case of BSE in a country should not be considered in isolation from international or regional trade and cattle production systems. Historical movements of cattle and their products, such as occurred in continental Europe or North America, mean such regions should be considered as a single entity as far as the BSE epidemic is concerned. The first BSE case in a country should not be dismissed as "an imported case" unless the evidence is compelling that the affected animal was imported within such a time frame that it is not possible for the disease to have been acquired within the importing country. Elucidating the most probable source of infection is one of the reasons why close collaboration between involved countries is essential, and for the proper risk management of the BSE problem to be tackled internationally.

[1] A study by R.M. Anderson allowed a critical appraisal of different culling policies for eradication of the disease [1].

Several policy actions should be adopted to achieve the following objectives:

1. Protection of public health: reduction of public health risk by eliminating SRM from foods.
2. Protection of animal (and public) health: reduction of risks to animals, and hence to humans, through control of rendering and implementation of appropriate feed bans (see Tables 39.2–3), thus limiting recycling and amplification of prions causing BSE.

3a. Import control: prevention of any inadvertent introduction of BSE from abroad.

3b. Export control: prevention of the spread of the epidemic worldwide (see Chapter 19.9).

4. Surveillance: establish efficient passive and active surveillance programs. Further effective measures are based on monitoring and analysis of all relevant BSE data from epidemiological investigations.

To achieve the objectives 1 to 4, a system of complementary barriers, and implementation and enforcement of all measures on the national level, is necessary. On the one hand, measures to meet the objectives 1, 2, and 4 are described below; on the other hand, measures to meet the objectives 3a and 3b are described in the sources indicated in the legend of Table 53.1 (geographic BSE risk assessment; GBRA).

The objectives cannot be achieved by governments alone; effective implementation of measures requires a shared commitment and action on the part of national and state governments, producers, industries, veterinary professionals, and consumers (see Chapter 32). Extensive national coordination and cooperation is necessary and should be extended to include the whole country, region or, in the frame of the global market, even the whole world; at least for certain aspects. A BSE task force, including governmental and nongovernmental stakeholders, should be established under the leadership of the competent national authorities responsible for disease control, thus insuring that policies are developed and implemented consistently, and are reviewed and reinforced as necessary, on a regular basis.

53.3.1 Specified risk materials – the single most important public health measure

The removal of SRMs

SRMs are those tissues that are considered to represent the greatest BSE exposure risk to humans and susceptible animals because they have been demonstrated to contain infectivity at some point during the incubation period in the WHO table set of 2005/2006 in the Annex of this book (WHO Tables$_{\text{Appendix}}$ IA-IC), as well as all relevant data and comprehensive tables from [3], form the scientific basis for the definitions and re-definitions of SRM. Figure data taken from [13] and updated from [11]; Crown copyright ©.

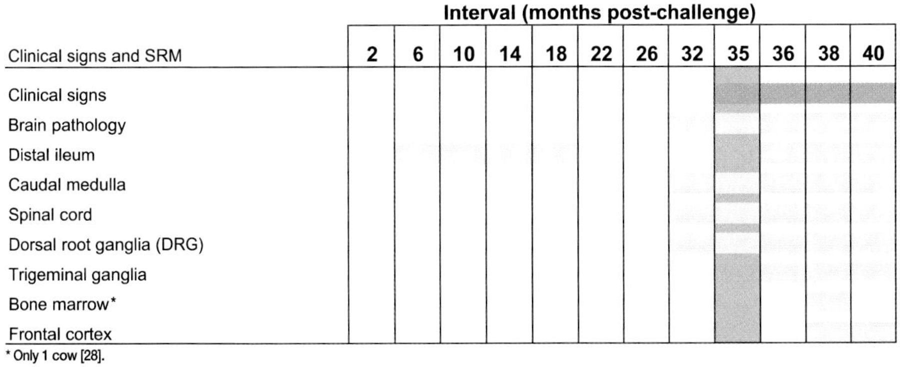

Fig. 53.2: Clinical signs, brain pathology, and tissue infectivity by interval from challenge during the pathogenesis of experimental BSE in cattle following oral exposure to infected brain. No animals were killed at 35 months post-challenge. Observations reported in [11; 13–15], the further studies referenced

(Fig. 53.2). Infectivity has been consistently shown to be present in the central nervous tissue (CNS) of naturally and experimentally infected cattle [3]. In experimentally infected cattle, ganglia associated with the CNS (trigeminal and dorsal root ganglia) have also been assayed and shown to be infectious. Distal ileum was also infectious at several time points during the incubation phase. In cattle experimentally inoculated by the oral route, BSE infectivity has been demonstrated in palatine tonsil 10 months after exposure [11]. Experimental results for tonsil and bone marrow remain to be fully interpreted [compare 3; 28].

Removal of all tissues with demonstrated potential for accumulation of BSE agent and strict attention to preventing cross-contamination of the carcass through stunning, slaughter, and processing practices are the internationally recognized standards for protection of public and animal health.

The ban on SRM removes the highest risk tissues (i. e., SRM from cattle over 30 months) from the human food supply. As understanding of the pathogenesis (see important details in Chapter 19.7) of BSE increases, the age limit might be subject to revision, therefore it should be in accordance with current OIE standards.

It may be that BSE infectivity is circulating in ruminant feed within a country, based on evidence from a geographical BSE risk assessment (GBRA) or from the tracing of concentrated feedstuff described in Section 53.2.1 [see also 7; 8], in response to the first (or one of the first) BSE cases in a country. If this is likely or almost certain, consideration should be given to excluding SRM from both the human food and animal feed supplies. This recommendation follows the current trend in international recommendations to stop amplification and limit exposure.

The OIE recommends [9] that the following SRM be excluded from human food, animal feed, fertilizers, and pharmaceuticals:

- From cattle of any age originating from a country of controlled, undetermined (or high) BSE risk: tonsils and distal ileum
- From cattle over 30 months of age at time of slaughter originating from a country of controlled BSE risk: brains, eyes, spinal cord, skull, and vertebral column (the latter are not inherently infected, but cannot be separated from dorsal root/trigeminal ganglia or from residual contamination with CNS tissue)
- From cattle over 12 months of age at time of slaughter originating from a country of undetermined (or high) BSE risk: brains, eyes, spinal cord, skull, and vertebral column

In the European Union, some ovine and caprine tissue have also been defined as SRM because of the theoretical risk that BSE could be circulating undetected in sheep and goats [12].

The OIE's recommended ages for exclusion of SRM are based on expert consultations facilitated by the OIE in 2004. Data sources used by experts are summarized in the legend of Figure 53.2. The detection of infectivity in the CNS of infected animals at approximately mid-incubation was extrapolated to 30 months for cattle, as this is half way through the mean incubation period of 60 months for BSE-affected cattle in the UK. In other words, CNS was assumed to be infectious in infected cattle aged 30 months or older. However, limited data from experimental oral exposure in cattle (Fig. 53.2) suggest that, in this species, BSE infectivity only becomes detectable after approximately 78 % of the incubation period [11; 13].

Nevertheless, a cut-off of 12 months recommended for countries of undetermined or high BSE risk takes into account the fact that some cattle under 30 months (see Chapter 39) of age may be slaughtered with infectivity present in the SRM. Examination of European Commission data on frequency of cases according to age sheds light on this issue. In the European Union, there were only two young cases (diagnosed outside of the UK), confirmed age 32 and 36 months, recorded in 2002 and 2003 European Commission data.

The question of infectivity entering the CNS at a time later than the mid-point of the incubation period is important in defining the appropriate age limit not only in the context of the SRM ban, but also the time at which the various diagnostic tests may be expected to be effective in detecting infected animals.

While the removal of SRM will significantly

reduce the BSE risk for humans, it must be re-emphasized that contamination of the carcass with SRM (specifically CNS) should also be avoided. Slaughter and carcass dressing procedures, including currently used stunning procedures and mechanical deboning processes that increase the risk of contaminating meat and meat products with CNS tissue (including associated ganglia), should be brought into line with international standards. For example, for countries with a controlled BSE risk, the OIE recommends that cattle not be subjected to a stunning process, prior to slaughter, with a device injecting compressed air or gas into the cranial cavity, or to a pithing process. For such countries, the OIE also recommends that meat should not be mechanically separated from the skull and vertebral column from cattle over 30 months of age. In countries of undetermined (or high) BSE risk, the OIE recommends ending the mechanical separation of meat from the skull and vertebral column from cattle over 12 months of age.

Exclusion and destruction of the high volume of raw material covered by SRM bans is a massive burden on all countries currently affected by BSE. Given the susceptibility of cattle to low-dose exposure (1 mg) [3] (G.A.H. Wells, personal communication) of infectious brain tissue (see Chapter 19), and the fact that none of the current processing systems guarantee destruction of infectivity in commercial processes, it is probable that restoration of traditional uses in feed may be impossible. Innovative solutions are required to enable the safe use of such materials in future [16]. This could include adding value through their use for purposes other than the manufacture of feed and fertilizers (e. g., as a fuel source).

Pathogenesis of BSE

An understanding of the pathogenesis of BSE and tissue distribution of infectivity at different ages is important in formulating measures to protect human and animal health. A summary of relevant scientific information in the context of SRM and age limit definitions is given in Chapter 19.7.

53.3.2 Feed restrictions

In countries where there is a BSE risk, consideration should be given to excluding SRM from all animal feed, including pet food.

Epidemiological investigations in the UK in particular highlighted the dangers to cattle of infection through the consumption of feed that had been (cross-)contaminated accidentally when manufactured in premises that legitimately used mammalian MBM in feed for monogastric species (pigs and poultry). This kind of contamination risk was, for example, also demonstrated in Switzerland [6] and other countries.

As mentioned, data from ongoing studies at the UK Veterinary Laboratories Agency show that cattle can be infected orally with as little as 1 mg [3] (G.A.H. Wells, personal communication) of infectious brain tissue (compare Chapter 19.3, 19.7). Experience in many countries has shown that cross-contamination with such small quantities is virtually impossible to prevent in situations where ruminant feedstuffs are produced in the same feed plants as pig/poultry feed or pet food containing MBM. There is a need for rigorous audit of compliance with feed controls, and it needs to be recognized that testing of feed and feed ingredients is unlikely to detect low levels of contamination due to the limitations of sampling techniques and test sensitivity [17].

While scientific evidence supports feed bans limited to the prohibition of ruminant-derived MBM in ruminant feed, practical difficulties of enforcement have led to more pragmatic and effective solutions. The prohibition of the use of all MBM (including avian) in ruminant feed has been justified in Europe partly on the basis of cross-contamination, as well as the current problems in differentiating mammalian and avian MBM. It also prevents the inclusion into animal feed of ruminant-derived protein, which is contained within the lumen of porcine or avian intestines at slaughter.

Cross-contamination must be prevented throughout the feed production process, from reception and transportation of feed ingredients, during the manufacturing process, through transportation and storage of finished feed, and on farms where mixing, blending, and feeding occur.

53.3.3 Nonambulatory (downer) cows

In those countries where there may be a BSE risk (i. e., categories 2–4 in Table 53.1), it has been shown that a ban prohibiting nonambulatory cattle (downers) from entering the food and feed chains is an important control measure. This is because nonambulatory cattle are more likely to be infected with BSE than are healthy cattle (see Chapter 41.7 of [6] on "The surveillance quality of countries is a determinant of reported country-specific BSE-incidences" (in German)). Such cattle, therefore, pose a greater risk to public and animal health. The goals for measures related to these cattle should be to (i) test them for surveillance purposes and (ii) prevent potentially infective tissues from entering the food and feed chains. However, if nonambulatory cattle are excluded from supervised slaughter at inspected slaughterhouses, this important subpopulation may no longer be available for the BSE surveillance program at these locations. Therefore, it is important that the competent national authority take additional steps to assure that facilitated pathways exist for the collection of samples and proper disposal of carcasses of dead and nonambulatory cattle.

In order to decrease the risk of these potentially infected cattle entering the normal slaughter process, supplementary measures to encourage compliance should be in place. Such measures could include financial incentives (or at least the removal of financial disincentives, such as extra disposal costs borne by the farmer) and the strengthening of ante mortem inspections on presentation at the slaughter plant to identify questionable animals. To further prevent these cattle from being brought into the normal slaughter process, consideration should be given to the random sampling of appropriate subpopulations of aged cattle that have passed ante mortem inspection on presentation at the slaughter plant.

Veterinary authorities should consider all possibilities, including education of the farmers and all professionals (see Chapter 32.2.2) in close contact with live cattle, sheep, and goats to insure that these people are able to recognize clinical signs that suggest the presence of prion diseases (see Chapters 33, 34). This is a part of their role as producers of safe food, in order to achieve maximum surveillance of this risk population.

Table 53.1: Current relationship between EC and OIE geographic BSE risk categories. According to the TSE Roadmap [16], EC and OIE classifications should be correlated by July 2007. Official opinions on the outcome of the European Union's Geographical BSE Risk Assessment (GBRA) have been published on the Internet [25], including for third countries [26]. In several countries, the SSC (formerly) or the EFSA (now), respectively, carried out the GBRA twice, at least. It is important to understand that the geographical BSE risk (GBR) for any country is dynamic and must be regularly reassessed. The final reports on the updated assessment of the GBR of many countries can be found on the Internet: see [27] and Table 46.2.

Categories	EC (GBR) categories	OIE categories [9]
1	GBR I: Highly unlikely	Negligible BSE risk
2	GBR II: Unlikely but not excluded	(–)*
3	GBR III: Likely but not confirmed, or confirmed at a lower level	Controlled BSE risk
4	GBR IV: Confirmed at a high level	Undetermined BSE risk

* In the former OIE definition, category 2 was described and subdivided as "provisionally free" or "no indigenous case(s)/ indigenous case(s)". This category does not exist any more.

Editorial note 1: OIE does NOT number its three categories; the authors have allocated the OIE categories to lines that they considered were closest to the GBR categories of the EC.

Editorial note 2: There are numerous other documents relating to an opinion on what should constitute a dossier for a GBRA (adopted February 19–20, 1998). Other documents that need to be consulted are those on the methods to be adopted (original March 22, 1998, and updated in January 2000).

53.3.4 Surveillance and laboratory diagnosis

A general introduction to surveillance for animal prion diseases is described in Chapter 32, whereas the following is pointing at a few specific key factors (see also Chapter 35).

According to the OIE's Terrestrial Animal Health Code [9], surveillance for BSE may have one or more goals, depending on the BSE risk category of a country [see OIE A.H. Code 2.3.13 on $BSE_{\text{Appendix 2 of book}}$]. Surveillance goals include detecting BSE, to a predetermined design prevalence; monitoring the evolution of a BSE epidemic; monitoring the effectiveness of a feed ban and/or other risk mitigation measures, in conjunction with auditing; supporting a claimed BSE status; and gaining or regaining a higher BSE status (see also Chapter 32.2.4). It has commonly been proposed that once it has been established that the BSE agent is circulating in the cattle population, the surveillance program should be significantly extended. However, in 2006, the OIE's Terrestrial Animal Health Standards Commission stated "...given the long incubation period of BSE, the number of cases which reflects the situation in the distant past, was not as important as the implementation of mitigation measures. Consequently, the expenditure of resources on testing more samples was considered to be less valuable than verifying that mitigation measures were currently being strictly enforced" [18].

Surveillance programs should be targeted at the animal population with highest risk of BSE. These have been shown to be those exhibiting signs compatible with BSE (passive surveillance), fallen stock (cattle that die on the farm or during transport), nonambulatory animals ("downers"; see Section 53.3.3), and cattle for emergency slaughter more than 30 months old. Surveillance systems targeting these subpopulations have been shown to be the most efficient at identifying BSE cases.

Active surveillance may be strengthened by requiring that cattle older than 30 months in the above risk populations be tested using a rapid post mortem BSE test. This will not only establish the prevalence of BSE but also build confidence both domestically and for trading partners. This could be achieved by means of a 1-year program, the outcome of which would then assist in designing future ongoing testing programs. The age limit "older than 30 months" could be subject to revision, in light of the new scientific knowledge on the pathogenesis of BSE or changes in the observed epidemiology of the disease, such as changes in age structure of affected animals.

An estimation of the age of cattle (to determine whether or not they are over 30 months of age) can be done most satisfactorily and cost-effectively by examination of dentition. This method is a well-established means of aging cattle and is completely adequate for the purpose of determining those cattle to be tested in the BSE surveillance program.

The testing of all cattle slaughtered for human consumption is usually considered unjustified in terms of protecting human and animal health. An exception might be in a country with a "high BSE incidence", according to the definition in Table 53.1, category 4. However, to support the overall surveillance system and encourage reporting at the farm level, testing of a random sample of healthy slaughter cattle over 30 months may be appropriate.

The comparison of findings from surveillance programs conducted by different countries will be facilitated if similar or equivalent tools, including diagnostic tests, are adopted. Today, rapid BSE tests are the primary screening tests for active surveillance (see Chapters 32, 35). Tests must be used in accordance with manufacturer's instructions.

Decentralization of testing facilities to appropriately trained staff in laboratories that can guarantee testing with a minimum of delay may be desirable for the surveillance program recommended. The delay between receipt of samples and testing clearly needs to be shorter when animals have been slaughtered for human consumption than when they are derived from risk animals. A number of laboratories throughout the country should be approved by the competent national authority to conduct screening tests as part of the national surveillance program. The BSE reference laboratory should remain under the control of the competent national authority, and should be responsible for confirmatory and proficiency testing (see Chapter 32.2.3).

53.3.5 Enforcement, education, and traceability

Control of the implementation of measures taken (enforcement), education, and traceability is important. Experiences of several countries have shown that control measures prescribed by law are not always implemented as intended. Regional officials are often responsible for implementation, and the quality and effectiveness of controls may vary greatly between regions. Therefore, quality assurance systems need to be implemented at all levels under the overall supervision of the competent national authorities.

The implementation of a national animal identification system that is appropriate to the farming practices of the specific country is important. Besides its value for the cost-effective and rapid tracing of animals for BSE control purposes (see above), it is also useful in programs designed to control other contagious diseases.

It is not possible to achieve compliance with legislative controls without an effective educational program. The dissemination of information on the variety of clinical signs of BSE is crucial, especially considering the subtle nature of many of the signs. Education should involve, for example, (i) the production and publication of comprehensible, practical videos [19–23], some of which may now be accessible on the Internet, and (ii) comprehensible detailed description of the clinical signs of BSE such as in Chapter 33 and – in the context of BSE in a goat [24] – Chapter 34.

The reasons for actions taken must be clearly explained so as not to create the wrong impression or appear to validate actions that are not supported by the scientific knowledge of the disease. This will ensure that the effectiveness of the surveillance program is improved by ensuring the capture of appropriate risk animals for testing. As access to current information and traceability increases, so does consumer confidence and effectiveness of the control and prevention measures. Thus the goal of protecting human and animal health worldwide can be achieved.

The important measures taken to prevent the spread of BSE to humans and/or animals are comprehensively summarized in Tables 39.2–3.

References

[1] Donnelly CA, Ferguson NM, Ghani AC, et al. Implications of BSE infection screening data for the scale of the British BSE epidemic and current European infection levels. Proc Biol Sci 2002; 269(1506):2179–2190.

[2] Hörnlimann B and Infanger P. The control of BSE – internationally recommended public and animal health measures [BSE-Bekämpfung zum Schutz der Verbraucher und der Tierpopulation: international zu empfehlende Massnahmen]. In: Hörnlimann B, Riesner D, Kretzschmar H, editors. Prionen und Prionkrankheiten. Walter de Gruyter, Berlin, 2001:471–493.

[3] EFSA. Quantitative assessment of the residual BSE risk in bovine-derived products. The EFSA Journal 2005; 307:1–135. See also Webpage 2006: http://www.efsa.eu.int/science/biohaz/biohaz_documents/1280/efsaqrareport2004_final20dec051.pdf.

[4] Wilesmith JW, Ryan JB, Atkinson MJ. Bovine spongiform encephalopathy: epidemiological studies on the origin. Vet Rec 1991; 128:199–203.

[5] Wilesmith JW, Ryan JB, Hueston WD. Bovine spongiform encephalopathy: case-control studies of calf feeding practices and meat and bonemeal inclusion in proprietary concentrates. Res Vet Sci 1992; 52:325–331.

[6] Hörnlimann B. BSE in non-UK countries: spread by live cattle and MBM imports [Bovine Spongiforme Enzephalopathie (BSE): BAB-Fälle in der Schweiz (summary of Master of Public Health thesis, ISPM, University of Zürich 1999)]. In: Hörnlimann B, Riesner D, Kretschmar H, editors. Prionen und Prionkrankheiten. Walter de Gruyter, Berlin 2001:337–358.

[7] Hörnlimann B, Guidon D, Griot C. Risikoeinschätzung für die Einschleppung von BSE [Risk assessment for the importation of BSE]. Dtsch tierärztl Wschr 1994; 101:295–298.

[8] Hörnlimann B and Guidon D. Import of meat and bone meal as main risk factor for BSE in Switzerland. Kenya Veterinarian 1994; 18:467–469.

[9] World Organisation for Animal Health (OIE). Terrestrial Animal Health Code 2005, Chapter 2.3.13, 2005:173–180. [See also OIE A.H. Code 2.3.13 on $BSE_{Appendix\ 2}$.]
[10] Brown P, Will RG, Bradley R, et al. Bovine spongiform encephalopathy and variant Creutzfeldt–Jakob disease: background, evolution, and current concerns. Webpage 2006: http://www.cdc.gov/ncidod/EID/vol7no1/brown.htm.
[11] Wells GA, Spiropoulos J, Hawkins SA, et al. Pathogenesis of experimental bovine spongiform encephalopathy: preclinical infectivity in tonsil and observations on the distribution of lingual tonsil in slaughtered cattle. Vet Rec 2005; 156(13):401–407.
[12] Food Standards Agency. Atypical scrapie in sheep and goats update. Webpage 2006: http://www.food.gov.uk/news/newsarchive/2006/feb/sheep.
[13] Wells GAH, Hawkins SA, Green RB, et al. Preliminary observations on the pathogenesis of experimental bovine spongiform encephalopathy (BSE): an update. Vet Rec 1998; 142:103–106.
[14] Hadlow WJ, Kennedy RC, Race RE. Natural infection of Suffolk sheep with scrapie virus. J Infect Dis 1982; 146:657–664.
[15] Hadlow WJ, Kennedy RC, Race RE, et al. Virologic and neurohistologic findings in dairy goats affected with natural scrapie. Vet Pathol 1980; 17:187–199.
[16] European Commission. The TSE Road Map. COM (2005) 322 final, Brussels 15 July 2005. Webpage 2006: http://europa.eu.int/comm/food/food/biosafety/bse/roadmap_en.pdf.
[17] Guidon D. Control of the compliance of the (Swiss) feed ban using microscopic tools [Die Kontrolle der Einhaltung des Tiermehl-Beimischungsverbotes durch Mikroskopie]. In: Hörnlimann B, Riesner D, Kretzschmar H, editors. Prionen und Prionkrankheiten. Walter de Gruyter, Berlin, 2001:494–496.
[18] World Organisation for Animal Health (OIE). Report of the Meeting of the OIE Terrestrial Animal Health Standards Commission, 6–10 March 2006. World Organisation for Animal Health, Paris, 2006:1–252.
[19] Braun U, Pusterla N, Schicker E. Clinical findings in cattle with bovine spongiform encephalopathy (BSE) [video available in different languages]. Webpage 2006: http://www.bse.unizh.ch/english/video/content.htm.
[20] Ulvund MJ. Skrapesjuke – kliniske symptom [Scrapie – clinical symptoms; video in Norwegian]. Statens Dyrehelsetilsyn, Oslo, 1987.
[21] Veterinary Laboratories Agency. United Kingdom – OIE and European Community reference laboratory – updated details of approved tests, plus additional guidance on clinical and postmortem diagnosis. Webpage 2006: http://www.vla.gov.uk.
[22] Veterinary Laboratories Agency. Clinical signs of scrapie in sheep. 2004. http://www.defra.gov.uk/corporate/vla/science/documents/science-scrapie-res.pdf.
[23] Veterinary Laboratories Agency. Clinical signs of bovine spongiform encephalopathy in cattle. 2004. Webpage 2006: www.defra.gov.uk/corporate/vla/science/documents/science-bse-res.pdf.
[24] Eloit M, Adjou K, Coulpier M, et al. BSE agent signatures in a goat. Vet Rec 2005; 156(16):523–524.
[25] European Commission. Opinions on the geographical risk of bovine spongiform encephalopathy (GBR) and official reports on the assessment of the geographical BSE-risks. Scientific Steering Committee (SSC) 2006. Webpage 2006: http://europa.eu.int/comm/food/fs/sc/ssc/outcome_en.html.
[26] European Commission. Opinions on GBR in third countries, 2003. Webpage 2006: http://europa.eu.int/comm/food/fs/bse/scientific_advice03_en.html.
[27] European Commission – Opinions. Outcome of Discussions. Webpage 2006: http://europa.eu.int./comm/food/fs/sc/ssc/outcome_en.html.
[28] Wells GA, Hawkins SA, Green RB, et al. Limited detection of sternal bone marrow infectivity in the clinical phase of experimental bovine spongiform encephalopathy (BSE). Vet Rec 1999; 144:292–294.

54 Atypical Scrapie–Nor98[1]

Sylvie L. Benestad and Bjørn Bratberg

54.1 Introduction

Up until 1998, scientists considered scrapie to be a single disease with one general phenotype showing minor variations caused by strain differences. In 1998, another scrapie phenotype, designated "Nor98" or "atypical scrapie".[1] was discovered in sheep in Norway [1; 2]. As a result, the type of scrapie previously described is now referred to as "classical" scrapie [3]. The identification of atypical cases of scrapie (Nor98) has considerably upset established understanding of the disease, because the newly recognized disease clearly differs from classical scrapie – and experimental bovine spongiform encephalopathy (BSE) in sheep – in many aspects.

On the basis of the current knowledge of atypical scrapie–Nor98, the Norwegian animal health authorities have decided to apply different sanitary policies to flocks where a case of classical scrapie has been diagnosed (see Chapter 57) compared to those applied where a case of atypical scrapie–Nor98 has been diagnosed.

54.2 TSE surveillance program launched for small ruminants

The first Norwegian sheep affected by atypical scrapie–Nor98 were showing clinical signs of the disease. The diagnosis was therefore based on these signs as well as on brain histopathology and PrP^{Sc} detection by immunohistochemistry (IHC) and Western blot investigation. Since 2001, when the transmissible spongiform encephalopathy (TSE) surveillance program was launched in Europe (see Chapter 55.1), an increasing number of atypical cases have been detected in most European countries [4–9] and, recently, even outside Europe, in the Falkland Islands [10]. These cases include a small number of homozygous ARR sheep in Germany, UK, France, Portugal, and Norway. In Norway, from 1998 to 2005 a total of 49 cases were found in sheep in the 56 scrapie flocks affected in this period (Fig. 57.1). In 2004, a total of 97 cases of atypical scrapie were officially reported to the European Commission (EC); but as mentioned earlier, this number must be treated with caution as there is no harmonized diagnostic protocol for atypical scrapie. However, some standardization should be achieved in the near future since the EC has recently published a classification of TSE in small ruminants that describes the diagnostic differences between the three groups, namely classical scrapie, atypical scrapie, and BSE in small ruminants [11].

54.3 Particularity of clinical signs of atypical scrapie

Atypical scrapie–Nor98 is generally found in older animals, the mean age in Norway being around 6 years. Clinical signs are *not always* present in the atypical scrapie–Nor98 animals, and when signs are present, they can be vague and can vary between behavioral changes, emaciation and in some cases, ataxia, and/or circling [1; 8]. Atypical scrapie–Nor98 sheep never show pruritus or wool loss, signs that could immediately be associated with classical scrapie (see Chapter 34). Disease course varies

[1] "Nor98" is currently a subgroup, and probably the largest subgroup, of atypical scrapie, but the common international designation outside Norway is "atypical scrapie". To serve the international and the national Norwegian audience, the editor decided to use the designation "atypical scrapie–Nor98".

between 6 weeks and 8 months and the differential diagnosis might vary from listeriosis, cerebral cortical necrosis, to simply aging.

54.4 Particularity of genetics

As discussed in Chapter 56, variations in the *Prnp* influence the susceptibility of sheep to classical scrapie. Atypical scrapie–Nor98, however, does *not* show the same relationship between PrP genotype and susceptibility, as atypical scrapie–Nor98 cases are found in animals carrying PrP genotypes usually associated with resistance to classical scrapie such as AHQ/xxx (compare Chapter 18.9). Furthermore, the ARR allele does not seem to confer any particular resistance to atypical scrapie–Nor98 and, lately, a new PrP polymorphism, at codon 141 from leucine to phenylalanine (F_{141}; Table 54.1), has been shown to be related to high susceptibility to atypical scrapie–Nor98 [2].

Atypical scrapie–Nor98 has also been reported in goats [5; 12], but any effects of PrP polymorphisms on the susceptibility to the disease have yet to be found in this species.

54.5 Particularity of the pathology

Another important difference between atypical scrapie–Nor98 and classical scrapie is their neuropathology, and this suggests distinct neuropathogenesis. In classical scrapie, it is well recognized that the agent enters the brain via the dorsal motor nucleus of the vagus nerve (DMNV), as this is the first site where PrP^{Sc} can be detected by IHC in early cases [14–18]. This is not so with the atypical scrapie–Nor98 cases in which the DMNV is unaffected. In atypical scrapie–Nor98 cases, brain histopathology, if present, reveals vacuolar lesions essentially in the neuropil. Vacuolation is prominent in the molecular layer of the cerebellum (Fig. 54.1), often in the cerebral cortex, and to a minor degree in the brain stem, especially in the midbrain, while generally no vacuolation is seen at the level of the obex. In addition, PrP^{Sc},

Table 54.1: Distribution of the PrP genotypes of the 49 Norwegian Nor98 cases (until January 2006) according to the classification from the National Scrapie Plan (NSP) in the UK (13) with five different categories of scrapie resistance. (From R1, the highest degree of resistance, to R5, the highest degree of susceptibility; see Chapter 56.4).

Category	Degree of susceptibility/resistance to classical scrapie	PrP genotype	Number of Nor98 cases with notification of F_{141} (atypical scrapie)
NSP1 (R1)	Sheep that are genetically most resistant to scrapie	ARR/ARR	1
NSP2 (R2)	Sheep that are genetically resistant to scrapie, but will need careful selection when used for further breeding	ARR/AHQ	6
		ARR/ARH	–
		ARR/ARQ	8 ARR/$AF_{141}RQ$
NSP3 (R3)	Sheep that genetically have little resistance to scrapie	AHQ/AHQ	11
		AHQ/ARH	1
		AHQ/ARQ	4 7 AHQ/$AF_{141}RQ$
		ARH/ARH	–
		ARH/ARQ	–
		ARQ/ARQ	2 ARQ/ARQ 6 $AF_{141}RQ/AF_{141}RQ$ 3 ARQ/$AF_{141}RQ$
NSP4 (R4)	Sheep that are genetically susceptible to scrapie and *should not be used for breeding*	ARR/VRQ	–
NSP5 (R5)	Sheep that are highly susceptible to scrapie and *should not be used for breeding*	AHQ/VRQ	–
		ARH/VRQ	–
		ARQ/VRQ	–
		VRQ/VRQ	–

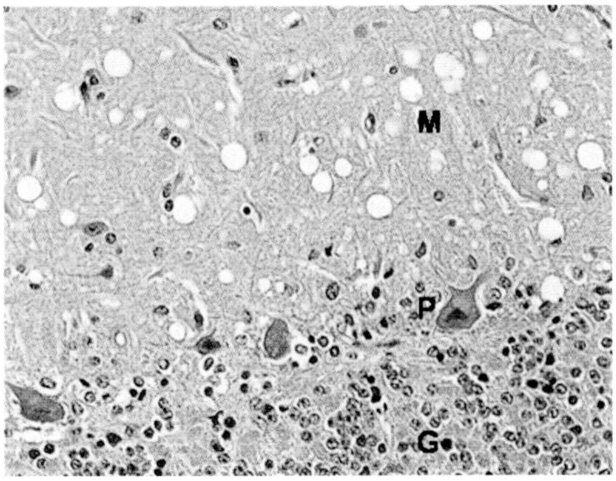

Fig. 54.1: Histopathological features of a Nor98 case ($AF_{141}RQ/AF_{141}RQ$). Note the severely vacuolated molecular layer (M) of the cerebellum in this case. However, the degree of cerebellar vacuolation varies considerably among the Nor98 cases, and spongiform change may even be absent. P, Purkinje cell layer; G, granular layer. H&E. Figure by Bjørn Bratberg ©.

if detectable at all by IHC at the level of the obex, is mostly confined to the nucleus of the spinal tract of the trigeminal nerve and/or in the white matter. In contrast, other anatomical brain regions like the cerebellum and the cerebral cortex generally display higher amounts of PrP^{Sc} staining. Nevertheless, it should be noted that, in some cases, the IHC staining can be minimal or absent both at the level of the obex and also in the cerebellum, suggesting that PrP^{Sc} IHC might not be optimal for the identification or confirmation of atypical scrapie–Nor98.

Furthermore, while in the majority of classical scrapie cases PrP^{Sc} is detected in lymphoid tissues at an even earlier stage of the disease than in the brain, until now no PrP^{Sc} has been detected outside the central nervous system for the animals affected by atypical scrapie–Nor98. This once again suggests that the pathogenesis of atypical scrapie–Nor98, even if scantly investigated up until now, could be different from that of classical scrapie.

54.6 Particularity of the diagnosis

Several research groups have shown that the atypical scrapie–Nor98-PrP^{Sc} is less resistant to protease digestion, such as Proteinase K (PK), than classical scrapie. This could partly explain the differences in sensitivity between diagnostic tests. In addition, the concentrations of PrP^{Sc} in the atypical scrapie–Nor98 cases as demonstrated by ELISA and Western blot are lower than in classical scrapie cases, especially in the medulla. This suggests that for efficient detection of atypical scrapie–Nor98 cases a sensitive test should be used, with effective purification and concentration of PrP^{Sc} and subsequent detection by antibodies with a high affinity to PrP^{Sc}. Appropriate sensitive diagnostic tests combined with sampling of the appropriate tissues (more than just the medulla) may overcome the diagnostic challenges presented by most of the atypical scrapie–Nor98 cases.

In addition to its unusual sensitivity to PK, the structure of PrP^{Sc} in atypical scrapie-Nor98 cases is also clearly distinct from the three-band pattern presented by the confirmatory Western blot method in classical scrapie and BSE strains. With a sensitive Western blot method, for example TeSeE Sheep/Goat Western blot (Bio-Rad Laboratories, Marnes-la-Coquette) that uses the SHa31 antibody, the atypical scrapie–Nor98 isolates are characterized by a lower molecular band at around 11–12 kDa as indicated by the arrow in Fig. 54.2 [see also 3].

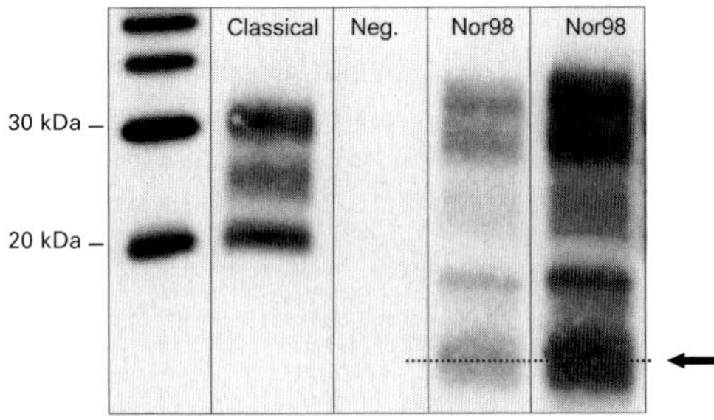

Fig. 54.2: Differences in Western blot profiles between a classical scrapie case (cerebral cortex, VRQ/ARQ), a scrapie-negative sheep, and Nor98 isolates (cerebral cortex, AHQ/AHQ) as shown by using the TeSeE Sheep/Goat Western blot (Bio-Rad Laboratories) with the SHa31 antibody. Figure by Sylvie L. Benestad ©.

54.7 The origin of atypical scrapie

The geographical distribution of atypical scrapie–Nor98 seems to be wider than that of classical scrapie, and seems to affect all types of sheep breeds throughout Europe. Furthermore, preliminary epidemiological studies have failed to detect connections between cases. For most of the atypical scrapie–Nor98 cases, just one *single animal* was tested PrP^{Sc} positive in the flock, but there have been a few exceptions in Germany, France, Ireland, and the UK, where one to three additional sheep were tested positive. If contagious at all, atypical scrapie–Nor98 seems to be less contagious than classical scrapie. Atypical scrapie–Nor98 is transmissible intracerebrally to transgenic ovine mice [19], and the discovery of several cases in the same flock could mean that atypical scrapie–Nor98 has been horizontally transmitted, or alternatively might be the result of the random statistical probability of finding more than one case in a large flock. The origin of atypical scrapie–Nor98 cases is still open to speculation, and one cannot exclude the possibility, based on the unusual neuropathology, structure of PrP^{Sc}, PrP genotypes, and epidemiology, that such cases could represent a spontaneous disorder of sheep, similar to sporadic Creutzfeldt–Jakob disease (sCJD) in humans.

References

[1] Benestad SL, Sarradin P, Thu B, et al. Cases of scrapie with unusual features in Norway and designation of a new type, Nor98. Vet Rec 2003; 153(7):202–208.

[2] Moum T, Olsaker I, Hopp P, et al. Polymorphisms at codons 141 and 154 in the ovine prion protein gene are associated with scrapie Nor98 cases. J Gen Virol 2005; 86(1):231–235.

[3] EFSA. Opinion on "Classification of atypical transmissible spongiform encephalopathy (TSE) cases in small ruminants". The EFSA Journal 2005; 276:1–30. See also Webpage 2006: www.efsa.europa.eu/etc/medialib/efsa/science/biohaz/biohaz_opinions/1216.Par.0001.File.dat/biohaz_op_ej_276_atypicalscrapiedefinition_en_vf1.pdf.

[4] Buschmann A, Luhken G, Schultz J, et al. Neuronal accumulation of abnormal prion protein in sheep carrying a scrapie-resistant genotype (PrPARR/ARR). J Gen Virol 2004; 85(9):2727–2733.

[5] Buschmann A, Biacabe AG, Ziegler U, et al. Atypical scrapie cases in Germany and France are identified by discrepant reaction patterns in BSE rapid tests. J Virol Methods 2004; 117(1):27–36.

[6] De Bosschere H, Roels S, Benestad SL, et al. Scrapie case similar to Nor98 diagnosed in Belgium via active surveillance. Vet Rec 2004; 155(22):707–708.

[7] Gavier-Widen D, Noremark M, Benestad S, et al. Recognition of the Nor98 variant of scrapie in the Swedish sheep population. J Vet Diagn Invest 2004; 16(6):562–567.

[8] Onnasch H, Gunn HM, Bradshaw BJ, et al. Two Irish cases of scrapie resembling Nor98. Vet Rec 2004; 155(20):636–637.

[9] Orge L, Galo A, Machado C, et al. Identification of putative atypical scrapie in sheep in Portugal. J Gen Virol 2004; 85(11):3487–3491.

[10] Epstein V, Pointing S, Halfacre S. Atypical scrapie in the Falkland Islands. Vet Rec 2005; 157(21):667–668.

[11] EFSA. Scientific report of the European Food Safety Authority on the evaluation of rapid post mortem TSE tests intended for small ruminants, N° EFSA-Q-2003-084. Adopted on 25 September 2005. Webpage 2006: www.efsa.europa.edu/etc/medialib/efsa/science/tse_assessments/bse_tse/1157.Par.0001.File.dat/biohaz_sr_ej49_smallruminanttsetests_en1.pdf.

[12] Food Standards Agency. Atypical scrapie in sheep and goats update. Webpage 2006: www.food.gov.uk/news/newsarchive/2006/feb/sheep.

[13] Defra and the State Veterinary Service (SVS). National scrapie plan for Great Britain. Compulsory scrapie flocks scheme [Booklet]. NSP 39, February 2006. Webpage 2006: www.defra.gov.uk/corporate/regulat/forms/ahealth/nsp/nsp39.pdf.

[14] Ligios C, Jeffrey M, Ryder SJ, et al. Distinction of scrapie phenotypes in sheep by lesion profiling. J Comp Pathol 2002; 127(1):45–57.

[15] Andreoletti O, Berthon P, Marc D, et al. Early accumulation of PrP(Sc) in gut-associated lymphoid and nervous tissues of susceptible sheep from a Romanov flock with natural scrapie. J Gen Virol 2000; 81(12):3115–3126.

[16] Ryder SJ, Spencer YI, Bellerby PJ, et al. Immunohistochemical detection of PrP in the medulla oblongata of sheep: the spectrum of staining in normal and scrapie-affected sheep. Vet Rec 2001; 148(1):7–13.

[17] van Keulen LJ, Schreuder BE, Meloen RH, et al. Immunohistochemical detection and localization of prion protein in brain tissue of sheep with natural scrapie. Vet Pathol 1995; 32:299–308.

[18] Wood JL, McGill IS, Done SH, et al. Neuropathology of scrapie: a study of the distribution patterns of brain lesions in 222 cases of natural scrapie in sheep, 1982–1991. Veterinary Record 1997; 140(7):167–174.

[19] Le Dur A., Beringue V, Andreoletti O, et al. A newly identified type of scrapie agent can naturally infect sheep with resistant PrP genotypes. Proc Natl Acad Sci USA 2005; 102(44):16031–16036.

55 Scrapie Control – Internationally Recommended Approaches

Marcus G. Doherr and Nora Hunter

55.1 Introduction

Scrapie, a transmissible spongiform encephalopathy (TSE) of sheep and goats (small ruminants), has been described as a clinical entity from some regions within Europe for more than 250 years. The disease has a worldwide distribution, and only Australia and New Zealand are currently considered as free of scrapie [1; 2]. Within Europe, until 2001, several countries claimed to be scrapie-free [3]. Despite its long history (see Chapter 1.3), many details of the geographic distribution and prevalence, the range of clinical symptoms, and the epidemiology of the disease are still unknown. In the context of the large British bovine spongiform encephalopathy (BSE) epidemic, and given the hypothesis of J. Wilesmith [4] that cattle BSE resulted from sheep scrapie, this lack of knowledge is of some concern. Experimental evidence that the cattle BSE agent can be transmitted experimentally to sheep has fueled these concerns. This agent has been shown to be infectious for cattle, sheep, and goats as well as other species including humans (Fig. 1.4). In 2004, a first field case of BSE was diagnosed in a goat in France [5]. It cannot be excluded that a certain number of sheep and goats were infected with the BSE agent, even though no further small ruminant BSE cases have been detected so far. The Wilesmith hypothesis on the origin of cattle BSE and the potential Veterinary Public Health (VPH) consequences of a BSE strain circulating in small ruminants have brought scrapie back to the attention of the scientific community.

Traditionally, most countries had a simple scrapie monitoring and control (surveillance) system that was based on the reporting of clinical suspects and control measures implemented whenever a case was indeed detected. Surveillance efforts were somewhat intensified after the occurrence of BSE outside of the UK, and even more after the crisis that arose around the first scientific reports of the new variant of Creutzfeld–Jakob disease (vCJD) in humans. Implementation levels, however, substantially differed between countries, thus making results very difficult to compare. On the basis of international and EU trade regulations, countries are not allowed to establish import restrictions (barriers) unless they can scientifically document that their animal disease status is better (lower prevalence, greater probability of freedom from disease) than that of the exporting country [6]. As a result, a need for a common Europe-wide scrapie monitoring approach arose. Since 1998, several EU-funded scientific exchange networks and research projects including the European Scrapie Network (NeuroPrion) [7], the Small Ruminant TSE Network [8], and the project Surveillance and Diagnosis of Ruminant TSE [9; 10] have been launched with the objective of establishing widely accepted and scientifically based protocols for the diagnosis, monitoring, and control of small ruminant TSE including scrapie.

The World Organisation for Animal Health (Office international des épizooties, OIE) has recently established internationally accepted guidelines for the monitoring and control of scrapie, and the conditions under which countries or regions can be classified as scrapie-free [11]. On the basis of the 2005 OIE Terrestrial Animal Health Code, the scrapie status of a country can be determined based on

(i) the outcome of a risk assessment that identifies all potential factors for scrapie occurrence and their historical perspective with particular emphasis on (a) the epidemiological situation concerning all animal TSEs in the country, (b) the importation or introduction of small ruminants potentially infected with scrapie, (c) the population structure and husbandry practices of sheep and goats, (d) feeding practices, including consumption of meat-and-bone-meal (MBM) or greaves derived from ruminants, (e) importation of MBM or greaves potentially contaminated with an animal TSE or feedstuffs containing either, (f) the origin and use of ruminant carcasses (including fallen stock), by-products, and slaughterhouse waste, the parameters of the rendering processes and the methods of animal feed manufacture, and (g) an ongoing awareness program for veterinarians, farmers, and workers to facilitate recognition and encourage reporting of all animals with clinical signs compatible with scrapie; and

(ii) a surveillance and monitoring system including (a) official veterinary surveillance, reporting, and regulatory control of scrapie cases, (b) compulsory notification and clinical investigation of all sheep and goats showing clinical signs compatible with scrapie, and (c) the examination of appropriate tissue from a defined number of sheep and goats older than 18 months displaying clinical signs compatible with scrapie.

On the basis of the results of that approach, countries may be classified as scrapie-free.

In 2002, in order to have a better insight into the distribution of small ruminant TSE (SRTSE, including scrapie) within Europe, the EU Member States implemented an extensive rapid-test monitoring program for the disease. On the basis of data from this intensive targeted surveillance program, SRTSE cases have been detected (see Chapter 1.10.2) in all old and some of the new EU Member States [12].

55.2 Criteria to assess the scrapie status of a country or region

Scientifically, the most important information needed to assess the scrapie status of a country or region is as follows:

I The quality of the scrapie monitoring program (its size and to what extent the sample of small ruminants examined for scrapie each year is representative, relative to the entire sheep population).

II The implementation of required control measures and thorough epidemiological investigations when detecting scrapie cases.

III The historical data on the disease in the country or region.

IV Past import of sheep and goats from scrapie-affected regions (import history).

V The current regulations and practices of importing sheep and goats.

VI Quality of the national animal identification and movement tracing system for sheep and goats.

VII Data on the distribution of scrapie-relevant resistance alleles within the major sheep breeds (see Chapter 56).

For countries or regions with endemic scrapie that intend to control (eradicate) the disease, items I, II, and V–VII are of paramount importance. For countries that want to be accepted as free of scrapie, the historical situation (disease detection, animal imports) needs to be considered in addition to the level of existing measures aimed at the rapid detection of cases and the prevention of importing them into the national flock. The longer a country with a good scrapie monitoring system does not detect any cases, the more likely it is to be considered free of the disease.

The items listed above will now be briefly addressed.

55.3 Disease monitoring

Item I: In the ideal situation, all sheep and goats with clinical signs indicative of scrapie (see Chapter 34) should be detected and reported to the respective veterinary authorities for closer examination. However, a broad range of diseases can lead to scrapie-like signs, and the disease can rarely be identified purely on a clinical basis. The post mortem laboratory diagnosis thus remains a very important confirmatory tool. Clinical suspect reporting thus is the broad screening instrument to identify animals in the field for further examination (testing). This mandatory reporting system needs to be supported by other measures such as (i) a guaranteed compensation of all control-related economic losses, (ii) routine flock level visits and examinations by state veterinarians and animal health services, and (iii) the routine examination of brain samples of emergency-slaughtered and fallen sheep and goats, using rapid and conventional diagnostic tests.

Because of the long incubation period (see Chapter 18.3) and pathogenesis (see Chapter 27.2) of scrapie and the use of diagnostic tests such as histopathology, immunohistochemistry (IHC), and the currently existing rapid screening assays that target detectable lesions and the accumulation of the infectious prion protein (PrP^{Sc}) in the central nervous system (CNS), mainly advanced stages of the incubation in older animals will be detected. Therefore, adult sheep and goats and, due to the complex epidemiology and transmission of the disease (see Chapter 42.2.2), offspring of ewes diagnosed with scrapie, and all contact between flocks should be traced in order to detect the disease. Reliable ante mortem (live animal) tests for scrapie are not yet available. Several research groups in the recent past have announced the development of such tests [13–16], but none of these tests has been shown to be "fit for purpose" in large-scale testing schemes. Once developed and validated, these tests should become important tools in scrapie monitoring and control programs.

In addition to screening, the national reference laboratories play an important role in the detection and subsequent confirmation of scrapie suspect cases. Currently, disease confirmation is based on the identification of spongiform lesions in histopathology, and the use of IHC and specific immunoblotting techniques (such as the OIE Western blot protocol) to detect the accumulation of PrP^{Sc} in specific CNS regions [1; 14]. All suspected cases should be confirmed in a national or international reference laboratory using accepted gold standard procedures. Those laboratories should participate in international collaboration and ring trials in order to maintain a high diagnostic standard. This will ensure a comparable quality in the detection and confirmation of scrapie cases across countries.

55.4 Measures to control scrapie in a country or region

Item II: The control of scrapie in endemic regions often is expensive, time consuming, and very frustrating. Despite all efforts, it seems impossible to eradicate the disease completely [1]. In case of a disease outbreak, all possible routes of agent (or disease) importation and spread need to be investigated and controlled. Because of the long incubation period of scrapie, however, outbreaks with very low numbers of infected animals as well as clinical cases per unit time can last for years or even decades. This results in difficulties in both the detection and the control of such outbreaks. Even the immediate culling of all animals in each case flock, as well as in all traced contact flocks, is not sufficient to eliminate the disease. Environmental contamination is suspected to be the main cause for that failure [see Chapter 42.2.2; compare for chronic wasting disease [17] and Chapter 21.9]. Thorough disinfection of affected premises, which includes the stables but also driveways and adjacent pastures, as well as an extended waiting time before restocking with sheep and goats from scrapie-free flocks, seems necessary to significantly reduce the risk of new outbreaks. In Iceland, there were 25 out of 500 cases in which sheep flocks experienced new scrapie cases after a short period of time despite whole flock culling, extensive cleaning and disinfection, and restocking with sheep from a scrapie-free region (S. Sigurdarson, personal communication). After a scrapie outbreak, newly established flocks should be closely monitored for several years to make sure that the disease has indeed been eliminated from the premises. More recently, genetic testing combined with selective culling, restocking, and breeding have been used in order to change the genetic profile of affected flocks toward scrapie resistance (Table 54.1, see Section 55.7). All these approaches, however, are very time consuming and require the dedication and collaboration between all parties involved. It is of paramount importance that those programs provide sufficient information and training for the farmers and veterinary practitioners, and that full compensation for all economic losses is provided.

55.5 Historical scrapie situation and potential import routes

Items III to V: Important criteria to assess the current scrapie status of a country or region are the historical data on scrapie within that area as well as the geographic origin of sheep and goats imported in the past. Current country-specific import regulations influence the risk of (re)introducing the disease into the national flock. Thus, a careful assessment of both risks and benefits is required in order to take decisions about trade. Imports limited to animals from closed certified "scrapie-free" flocks and from scrapie-free countries or regions as well as the long-term monitoring of such imported animals are ways to reduce the risk of disease importing and spread [18].

55.6 National animal identification and tracing system

Item VI: National animal identification and movement (tracing) systems (databases) have become very important tools in the context of disease outbreaks. This has recently been shown during the outbreaks of foot-and-mouth disease (FMD), classical swine fever (CSF), but also for BSE and scrapie. Therefore, individual sheep and goat identification and movement registration is considered to be an important part of a functional scrapie monitoring and control program. They allow (i) tracing back of affected (diseased) as well as contact animals (and their products) to their flocks of origin, (ii) the collection of reliable epidemiological data needed to investigate such outbreaks fully, and (iii) the successful control (eradication) of such diseases. In most European countries, individual animal identification and tracing systems with central (national) databases have now been established for cattle in the context of BSE. Their implementation is in progress (and more or less completed) for small ruminants.

55.7 Genetic influences

Item VII: The genotype of each individual sheep (and possibly goat?) is an important factor when considering the PrP gene (*Prnp*) and its influence on the susceptibility for scrapie, and therefore the susceptibility of whole breeds or flocks. Specific *Prnp* regions or codons are more important for scrapie susceptibility than others; Suffolk sheep that carry glutamine/glutamine on codon 171 (denoted QQ171) are very susceptible to classical scrapie (see Chapter 56). QQ171 carriers will develop clinical scrapie within a few years after exposure, while heterozygous animals (RQ171) and arginine-homozygous sheep (RR171) will rarely, or perhaps never, develop classical scrapie during their lifetime. Selective breeding with animals of resistant *Prnp* genotypes is a method through which, within a few years, scrapie-resistant flocks may be generated. The UK, the Netherlands, and subsequently the European Commission have adopted this approach of genotyping and selective breeding with resistance gene carriers in order to eradicate scrapie. Of all breeding-related approaches, this seems to be the most promising one to control scrapie (see Chapter 56.3–4). In previous years, the standard advice has been to cull the maternal line of affected animals. This approach is unlikely to result in disease eradication, but will lead to the unnecessary loss of blood lines. An alternative option is to kill every animal in an affected flock and to repopulate with sheep from elsewhere, but even this is no guarantee of freedom from disease, as the example of Iceland has demonstrated [19].

There are a number of potential disadvantages in changing the PrP genotype frequencies by the use, e. g., of AA136RR154RR171 rams. There might, by chance, be a loss of desirable breed characteristics. This problem is avoidable by means of selection on the basis both of PrP genetics and of breed-specific desired traits. A very relevant concern, however, is the fact that there are recently detected TSE strains that cause atypical scrapie in genotypes currently classified as resistant, even in RR171 sheep. Evidence that a considerable proportion of atypical scrapie cases (including Nor98, see Chapter 54) occurred in ARR/xxx sheep (Table 54.1) emphasizes this concern [20–23]. It is, however, not yet clear what role these atypical scrapie cases play in the epidemiology (transmission and maintenance) of scrapie at the flock or population level. In order to assess their impact on currently implemented scrapie control programs, more research is required.

References

[1] Detwiler LA. Scrapie. Rev Sci Tech Off Int Epiz 1992; 11:491–537.

[2] Hoinville LJ. A review of the epidemiology of scrapie in sheep. Rev Sci Tech Off Int Epiz 1996; 15:827–852.

[3] Schreuder BE. Epidemiological aspects of scrapie and BSE including a risk assessment study. Ph.D. dissertation. Department of Animal Health, University of Utrecht, The Netherlands, 1998.

[4] Wilesmith JW, Wells GA, Cranwell MP, et al. Bovine spongiform encephalopathy: epidemiological studies. Vet Rec 1988; 123:638–644.

[5] Eloit M, Adjou K, Coulpier M, et al. BSE agent signatures in a goat. Vet Rec 2005; 156(16):523–524.

[6] Hueston WD. Assessment of national systems for the surveillance and monitoring of animal health. Rev Sci Tech Off Int Epiz 1993; 1187–1196.

[7] NeuroPrion. European scrapie network "NeuroPrion". Webpage 2006: www.neuroprion.com/en/index.html.

[8] SRTSE Network. Small ruminant TSE network. Webpage 2006: www.srtse.net.

[9] European Commission. Food and feed safety – BSE/scrapie – introduction. Webpage 2006: http://ec.europa.eu/comm/food/food/biosafety/bse/index_en.htm.

[10] European Commission. Food and feed safety – BSE/scrapie – TSE in goats. (1) What is the new information about TSE in goats? Webpage 2006: ec.europa.eu/comm/food/food/biosafety/bse/goats_index_en.htm.

[11] World Organisation for Animal Health (OIE). OIE Terrestrial Animal Health Code 2005. Chapter 2.4.8. Scrapie. Webpage 2006: www.oie.int/eng/normes/mcode/en_chapitre_2.4.8.htm. [Reprint in appendix 3 of this book.]

[12] European Commission. Food and feed safety – BSE – monitoring results. Webpage 2006: www.europa.eu.int/comm/food/food/biosafety/bse/monitoring_en.htm.

[13] Schreuder BE, van Keulen LJ, Vromans ME, et al. Preclinical test for prion diseases [letter]. Nature 1996; 381:563.

[14] Schreuder BE, van Keulen LJ, Vromans ME, et al. Tonsillar biopsy and PrPSc detection in the preclinical diagnosis of scrapie. Vet Rec 1998; 142:564–568.

[15] O'Rourke KI, Baszler TV, Parish SM, et al. Preclinical detection of PrPSc in nictitating membrane lymphoid tissue of sheep. Vet Rec 1998; 142:489–491.

[16] Safar J, Wille H, Itri V, et al. Eight prion strains have PrPSc molecules with different conformations. Nat Med 1998; 4:1157–1165.

[17] Johnson CJ, Phillips KE, Schramm PT, et al. Prions adhere to soil minerals and remain infectious. PLoS Pathog 2006; 2(4):e32.

[18] MacDiarmid SC. Scrapie: the risk of its introduction and effects on trade. Aust Vet J 1996; 73:161–164.

[19] Sigurdarson S. Epidemiology of scrapie in Iceland and experience with control measures. In: Bradley R, Savey M, and Marchant B, eds. Subacute spongiform encephalopathies. Commission of European Communities. Kluwer Academic, Dordrecht, 1991:233–242.

[20] Buschmann A, Luhken G, Schultz J, et al. Neuronal accumulation of abnormal prion protein in sheep carrying a scrapie-resistant genotype (PrPARR/ARR). J Gen Virol 2004; 85:2727–2733.

[21] Buschmann A and Groschup MH. TSE eradication in small ruminants – quo vadis? Berl Münch Tierärztl Wochenschr 2005; 118:365–371.

[22] Gretzschel A, Buschmann A, Eiden M, et al. Strain typing of German transmissible spongiform encephalopathies field cases in small ruminants by biochemical methods. J Vet Med B Infect Dis Vet Public Health 2005; 52:55–63.

[23] Le Dur A, Beringue V, Andreoletti O, et al. A newly identified type of scrapie agent can naturally infect sheep with resistant PrP genotypes. Proc Natl Acad Sci USA 2005; 102:16031–16036.

56 The PrP Genotype as a Marker for Scrapie Susceptibility in Sheep

Nora Hunter and Alex Bossers

56.1 Introduction

Scrapie in sheep has been shown to be experimentally transmissible, but there is also a strong genetic component influencing the disease phenotype (such as incidence, incubation time, pathogenesis) – comparable to some forms of human transmissible spongiform encephalopathies (TSEs; see Chapter 26). There is overwhelming evidence that the major genetic component of the host determining this phenotype is the PrP gene (*Prnp*).

In this chapter, some basic aspects of PrP genetics in sheep and how they relate to the underlying molecular aspects of disease development are described. In addition, some comments are made on the high variability of the PrP coding region in relation to commonly used PrP genotyping methods in research or routine/high-throughput settings. The more genotyping and scrapie incidence information becomes available from various countries and different breeds, the higher the potential usefulness of these data to control the disease in sheep by selective breeding (see also Chapter 55).

56.2 Sheep PrP gene (*Prnp*) and its variations

In 1960, a flock of Cheviot sheep (NPU Cheviots) was founded and then selected into two lines based on susceptibility to disease following injection with scrapie (Scrapie Sheep Brain Pool number 1; SSBP/1). Response of infected sheep to SSBP/1 (a frequently used experimental source of scrapie) in NPU Cheviots was shown to be controlled by a single gene (*Sip* gene) with two alleles, sA (short incubation allele) and pA (prolonged incubation allele) [1]. More recent studies have provided evidence that the *Sip* gene (defined by disease phenotype) and *Prnp* (defined by molecular genetics) are one and the same [2]; see also literature on *Sinc* gene in mice).

Prnp lies on the *Prn*-locus that also contains *Prnd* (encoding the doppel of prion protein which has low amino acid sequence homology but has a highly similar structure) downstream of *Prnp* [3]. The organization and structure of *Prnp* are well conserved among mammalian species. In sheep its open reading frame (PrP-ORF) encodes a polypeptide of 256 amino acids. This ORF is completely localized within exon III of *Prnp* without being interrupted by intron sequences [4]. The primary translation product is usually processed into a mature 35–37 kDa membrane protein containing a disulphide bridge, N-linked glycosylation, and a C-terminal glycolipid anchor after cleavage of its N- and C-terminal signal sequences.

Prnp varies between individual sheep and between sheep breeds. Most striking is the high variability of the sheep PrP-ORF (mainly in the region corresponding to the protease resistant core of PrP^{Sc}). Only a minority of these polymorphisms could be significantly associated with different disease phenotypes. An overview of the currently described mutations (silent mutations and coding mutations) is provided in Figure 56.1. Sheep *Prnp* is very complex with more than 55 variations having been recorded in more than 39 different codons to date. One striking observation in this is that almost all mutations tested thus far (except some rare events) seem to be mutually exclusive (meaning, only one single point mutation is found on one single allele). Most routine PrP genotyping techniques

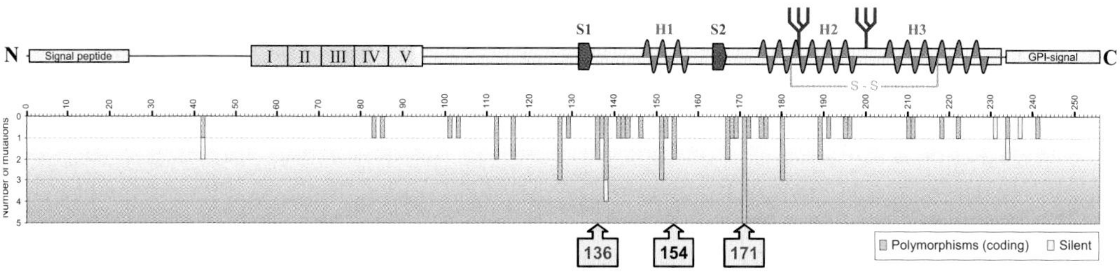

Mutation	Allele[1]	Remarks	References[2]
-	ARQ	Phylogenetic wildtype	[13, 21, 22, 35, 36], EMBL AJ000739
P42L	L$_{42}$ARQ	Bighorn sheep	Genbank DQ648476
P42P	P$_{42}$ARQ	Chinese breeds	[37]
G83G	G$_{83}$ARQ	Spanish breeds	[24]
G85R	R$_{85}$ARQ	Chinese breeds	[38]
Q101R	R$_{101}$ARQ	Spanish and Pakistani breeds	[24], Genbank DQ346682
N103N	N$_{103}$ARQ	Bighorn sheep	Genbank AY769957
M112I	I$_{112}$ARQ	Chinese breeds	[37]
M112T	T$_{112}$ARQ	Many breeds (Mainly Suffolk)	[14, 36, 39], EMBL AJ000735
A116P	P$_{116}$ARQ	Santa Inês, Mongolian, St Croix White	[40, 41]
G127A	A$_{127}$ARQ	US (MARC)	Genbank AY326330
G127S	S$_{127}$ARQ	Santa Inês, Chinese	[37, 40]
G127V	V$_{127}$ARQ	Mongolian, Chinese	[37, 41]
G129S	S$_{129}$ARQ	Chinese, tibetan sheep	Genbank AY723287
A136T	TRQ	Cypriotic/Greek	[29]
A136V	VRQ	Many	[4, 7, 8, 14, 22, 34, 36], EMBL AJ000738
M137T	AT$_{137}$RQ	Swifter, Icelandic	[22, 42], EMBL AJ000679
S138G	AG$_{138}$RQ	Bighorn sheep	Genbank DQ648472
S138N	AN$_{138}$RQ	Skandinavian	[42, 43]
S138R	AR$_{138}$RQ	Oklahoma	[44]
S138S	AS$_{138}$RQ	Chinese	[37]
L141F	AF$_{141}$RQ	Many	[9, 22, 26], EMBL AJ000680
I142T	AT$_{142}$RQ	Santa Inês, NZ sheep	[40, 45]
H143R	AR$_{143}$RQ	Oklahoma US, Santa Inês	[40, 44]
N146S	AS$_{146}$RQ	Chinese	[37]
R151C	AC$_{151}$RQ	Skandinavian, UK	[42, 43], Goldmann unpublished
R151G	AG$_{151}$RQ	Spanish	[24]
R151H	AH$_{151}$RQ	Spanish	[24]
Y152F	AF$_{152}$RQ	Chinese	[37]
R154H	AHQ	Many	[7, 9, 22, 36, 39, 43], EMBL AJ000737
R154L	ARL$_{154}$Q	Churra Spanish	[46]
R167S	ARS$_{167}$Q	Oklahoma US	[44]
R167T	ART$_{167}$Q	Santa Inês	[40]
P168L	ARL$_{168}$Q	Slovakian	Genbank AY822667
V169G	ARG$_{169}$Q	Slovakian	Genbank AY822667
Q171H	ARH	Many	[7, 22, 36, 39], EMBL AJ000734
Q171K	ARK	Chinese, US, Greek, Italian	[29, 44, 47, 48]
Q171R	ARR	Many	[7, 8, 13, 14, 22, 34, 35], Genbank M31313, EMBL AJ000736
Y172D	ARQD$_{172}$	Spanish, Santa Inês	[24, 40]
Q175E	ARQE$_{175}$	Ouessant	Bossers unpublished, Goldmann unpublished.
N176K	ARQK$_{176}$	Sarda	[23]
H180Y	ARQY$_{180}$	Oklahoma US	[44]
Q189L	ARQL$_{189}$	Mongolian, Santa Inês	[40, 41]
Q189R	ARQR$_{189}$	Mongolian	[41]
T191R	ARQR$_{191}$	Tibetan	Genbank AY723288
T195S	ARQS$_{195}$	Oklahoma US	[44]
T196S	ARQS$_{196}$	Oklahoma US	[44]
E210G	ARQG$_{210}$	Bighorn sheep	Genbank DQ648476
R211Q	ARQQ$_{211}$	Dutch crossbreed	[49], EMBL AJ000681
I218L	ARQL$_{218}$	Chinese	[37], Genbank AY304000
Q222K	ARQK$_{222}$	Chinese	[37], Genbank AY304008
R231R	ARQR$_{231}$	Chinese, US (MARC)	[37], Genbank AY326330
S234C	ARQC$_{234}$	Bighorn sheep	Genbank DQ648471
S234S	ARQS$_{234}$	Chinese, Tibetan	[37], Genbank AY723288
L237L	ARQL$_{237}$	Chinese, US (MARC)	[37], Genbank AY326330
P241S	ARQS$_{241}$	Chinese, US (MARC)	[37], Genbank AY304004

Fig. 56.1: Schematic representation of the sheep PrP-ORF and the described mutations [silent and coding (polymorphisms)]. **(top):** Structural elements generally recognized in sheep PrP are depicted at the top of the figure including the two putative N-linked glycosylation groups. N and C, N- and C-terminus of the protein; I–V, octarepeats; S1-2, Beta-sheet 1 and 2; H1-3 Alpha-helix 1-3; S-S, disulphide bridge. In yellow the fragment of PrP^{Sc} is shown that resists proteolysis; in grey the cleaved sequences in mature PrP^{C}. **(middle):** Frequency of mutations (silent and coding) at the codons of sheep PrP. Most important codons containing polymorphisms significantly associated with different disease phenotypes are indicated by the arrow box. **(bottom):** Table of mutations in sheep PrP, its corresponding allele name ([1]reconstructed based on the assumption that all mutations are mutually exclusive thus far), some remarks on breeds and how often it has been described, and finally the references describing or confirming the mutations ([2]papers and/or public database accession numbers). Figure by Alex Bossers ©.

use this assumption to reconstruct *Prnp*-allele and PrP genotypes from data obtained by rapid (high-throughput) tests at specific codons only (see paragraph 56.4). The precise combination of amino acid codons at numbered positions in the protein describes one allele of the gene and, two of these alleles together make up the genotype. The three major positions in the sheep PrP-ORF have been shown to be the codons 136, 154, and 171 (see next paragraph). Therefore these alleles are usually abbreviated using only the amino acids at these codons supplemented with another polymorphism if present (Fig. 56.1).

56.3 Sheep PrP genotypes and association with susceptibility to TSEs

Studies of natural scrapie and experimental TSE in sheep (SSBP/1, CH1641, BSE) have confirmed the importance of three codons in the sheep *Prnp* (**136**, **154** and **171**) regarding the control of susceptibility at the animal [5–9] but also at the molecular level of prion protein conversion [10, 11];

- codon 136: alanine (A) or valine (V)
- codon 154: arginine (R) or histidine (H)
- codon 171: glutamine (Q) or arginine (R).

There are breed differences in *Prnp*-allele frequencies and in disease-associated alleles; however, some clear genetic rules have emerged which are independent of the sheep breed. Sheep of genotypes which encode QQ171 (glutamine-glutamine homozygosity) are most susceptible to scrapie. For example in Cheviot sheep, the most susceptible genotype is $VV_{136}RR_{154}QQ_{171}$ (usually written in the format of VRQ/VRQ), whereas in Suffolk sheep, which do not usually have the VRQ allele, the ARQ/ARQ-animals are most susceptible to scrapie. There is much less genetic variation in the *Prnp* of Suffolk sheep than in some other breeds. The sheep PrP genotype most resistant to classical scrapie is ARR/ARR [12, 13]. Until very recently, out of many hundreds of scrapie-affected sheep worldwide, only one animal of this genotype (a Japanese Suffolk sheep) had been reported with clinical scrapie [14]. Sheep of ARR/ARR-genotype are also highly resistant to experimental challenge with both scrapie and BSE [6], although ARR/ARR-animals are capable of developing clinical BSE, this is only after intercerebral (i.c.) inoculation with high doses of BSE (3 out of 19), and only after a very long incubation time of about 1008–1127 days [15]. Recently prion protein deposits were also detected in one ARR/ARR preclinical experimental sheep BSE case (10 months incubation) out of six sheep that were orally dosed with two high doses of BSE [16]. In addition, at the molecular level the ARR protein could also be converted by ARQ scrapie and BSE but at very low efficiencies [10, 11]. Although the ARR/ARR sheep are susceptible to experimental TSEs (which is underlined by molecular conversion studies) they will probably not develop natural classical scrapie at high efficiencies.

Throughout Europe nowadays thousands of sheep are both, PrP-genotyped and tested for the presence of PrP^{Sc} protein in the brain. Samples of animals which develop clinical scrapie signs despite a resistant PrP genotype have been shown to have an unusual form of PrP^{Sc} (Fig. 54.2). Such a form of "atypical scrapie" commonly known as Nor98 was first detected in Norway in 1998 (as described in detail in Chapter 54). Atypical scrapie is characterized by a distinct pathology, genetics, and biochemistry [17–20] but is often confused with scrapie positivity in one type of rapid test but negativity in another. It is not known at present if atypical scrapie is a new form of scrapie or if this causative scrapie strain (compare Chapter 12) has always been present but remained undetected before the large-scale testing which is now mandatory in European countries to detect TSE in small ruminants (see Chapter 55). Thus far its transmissibility seems limited since the incidence is low within one flock (usually single *index cases*) and experimental transmission has proven to be difficult.

The difference in genotype association between several scrapie-affected breeds in various countries has given rise to a more general division of the sheep breeds / scrapie strains. Largely the breeds can be differentiated – on the basis of *Prnp*-alleles – as the so-called "valine breeds" such as Cheviots, Swaledales, Swifter, Romanov, and Shetlands that encode for valine at codon 136 on at least one allele (VRQ/xxx), and the "non-valine breeds" such as Suffolks and Sarda that do usually not encode for valine [8, 9, 13, 14, 21–23]. Sheep with the genotype VRQ/VRQ are usually extremely susceptible to scrapie in the valine-breeds; in contrast to ARQ/ARQ-, and vice versa. In an affected flock of the valine breeds, up to 100% of animals in the flock may be affected by scrapie, whereas the within-flock incidence is generally much lower in ARQ/ARQ sheep of non-valine breeds (incomplete penetrance). However, valine or non-valine breeds might respond differently to the present strain as has been demonstrated by experimental transmission studies in Cheviots using the SSBP/1 (VRQ-linked) and the CH1641 (non-VRQ linked) isolates of scrapie. Cheviots carrying VRQ are usually highly susceptible to inoculation with SSBP/1 (160 days) but resist inoculation with CH1641, and vice versa [6, 9].

The VRQ/VRQ itself seems to be a relatively rare genotype compared to the more common ARQ/ARQ, and VRQ/VRQ is more often found in sheep which are clinically affected by scrapie [12, 24, 25]. Therefore the question was also raised whether VRQ might induce a genetic instead of infectious form of the disease, but this was refuted by studies on sheep PrP genetics in a country free of scrapie, New Zealand, where the VRQ genotypes were found in comparable frequencies for the various breeds [26, 27]

The goat *Prnp* has also recently been shown to be complex [28–30] but clearly different from the sheep gene. Although a variation at codon 142 has been shown to be associated with experimental disease incubation time in goats [31] and the codon 222 might be associated with decreased susceptibility under natural conditions, further research is required to show whether breeding for scrapie resistance in goats is also possible [30]. Thus far none of the currently known sheep PrP alleles that are associated with differential scrapie susceptibility have been found in goats making the control of scrapie in goats by selective breeding impossible purely based on what is known of the sheep PrP genetics.

56.4 Methods of genotyping sheep

The UK, the Netherlands and subsequently the European Commission adopted the approach of genotyping and selective breeding to control scrapie. Of all breeding-related approaches this seems to be the most promising one. In previous years, the standard advice has been to cull out the maternal line of affected animals. This approach, however, is unlikely to result in disease eradication and will lead to the unnecessary loss of blood lines. An alternative option is to kill every animal in an affected flock and to repopulate with sheep from elsewhere, but even this is no guarantee of freedom from disease, as for example is demonstrated in the Icelandic situation [32].

A more practical scheme of genotype classification that is applicable across many breeds was put into operation in the UK [33]. It used a shortened system of nomenclature and gave a "risk assessment" according to breed and genotype, ranging from R1 (lowest risk) to R5 (highest risk) in individual sheep and their progeny (Table 54.1). The more genotyping information has become available, the greater the potential use for the sheep breeders (see Chapter 55.7). Many countries are now operating scrapie control policies along EU guidelines to breed for resistance or at least to counter select highly-susceptible PrP genotypes. In the following section, a few details on methods used in the PrP genotyping of sheep will be described.

Basically, all techniques can be divided into two categories; (i) dedicated techniques that only determine (several) known polymorphisms, and (ii) techniques to determine known but also unknown polymorphisms. Especially for experimental or field studies it is important to know which kind of technique is used since each technique has its own 'resolution' and an ARQ allele determined with technique A might still be different (less complete) from an ARQ determined by method B. Breeding programs are usually at large scale and these programs primarily focus on the three polymorphisms that are known to be associated with scrapie susceptibility (VRQ, AHQ, ARR and wildtype ARQ) to reduce costs and logistics in the labs.

Within these programs these polymorphisms are routinely determined in one or separate assays and the results are later combined (using the previous explained assumption that all mutations are still mutually exclusive) into the three-codon PrP genotype. All currently used routine techniques follow a few basic steps:

- chromosomal DNA isolation;
- PCR amplification of the PrP-ORF (or part thereof);
- detection of present polymorphisms in PCR product;
- reconstruct full PrP genotype.

56.4.1 DNA extraction

DNA can be extracted from any available tissue samples using methods described in standard literature. Usually it is more convenient to use blood because is does not need to be homogenized prior to extraction. Blood samples should be collected in an anticoagulant such as EDTA. Any other tissues (brain, liver, spleen, etc.) are also satisfactory if removed as biopsies or immediately after death and used as fresh tissue or quickly frozen tissue and stored for future extraction.

Various standard techniques are available for the isolation of chromosomal DNA needed for the PrP genotyping. Although general laborious (low scale) techniques are available in standard molecular biology books (i. e. Maniatis), most people use standardized kits of various manufacturers (i. e. Dynal and Promega). The advantage of such kits is the compatibility with more high-throughput systems and the outsourcing of quality control of chemicals.

56.4.2 Detection of variants at PrP codons 136, 154 and 171

Very small amounts of blood or tissue are required in order to analyze the genotypes of sheep, because usually the *Prnp* is amplified by the PCR-method, prior to the genetic analysis itself. The analysis may be carried out in a number of ways, for example, by allele-specific priming, digestion with restriction enzymes, allele-specific oligonucleotide hybridization, direct sequencing, cloning and sequencing, or analysis on denaturing gradient gels.

Every technique has its own resolution for analyzing mutations and every technique has its own capabilities/potential to be implemented in more high throughput settings. Therefore the choice of technique depends more or less on the specific needs of the laboratory involved, the number of samples to be analyzed, and the equipment that is or can be made available. Higher-throughput in most cases results in lower test resolution, meaning that with higher throughput generally fewer different mutations can be simultaneously analyzed. From the resolution point of view, the choice of technique ranges from Sanger sequencing and DGGE, which are able to analyse almost all mutations in the region of interest, to RFLP based or completely automated fluorescence based techniques like TaqMan or SNiP-it which have a resolution of only one specific mutation per test.

As easy start-up techniques, most labs consider the PCR-RFLP or Sanger sequencing. For more high-throughput classic techniques like allele specific amplification are also used, but they have largely been replaced by fluorescence real-time-PCR techniques. Denaturing gradient gels are one step further in the process as they allow screening for new and multiple mutations in addition to the known ones; however they require some training to get it going. The basic protocols for PCR-RFLP, ASA, and DGGE have been published [7, 22, 34] but high-throughput techniques are often shielded from publication by commercial interests. In the current genomics age, commercial Sanger sequencing facilities can provide easy access to high-throughput sequencing, generally at a low price. These services may be particularly useful for screening studies or conformational studies. However, this technique (which is considered the gold standard) also requires at least two reactions (one forward and one reverse directed) per sample for $>98\,\%$ confidence scoring of heterozygote samples. Furthermore, all techniques still have to work on the assumption of mutual exclusivity to reconstruct alleles and genotypes. Haplotyping by single allele cloning and sequencing has thus far revealed this assumption to be valid. However, if double mutations in the PrP alleles occur more frequently, it will restrict the reconstruction of alleles and genotypes.

Technical problems encountered when using these techniques for genotyping are largely those with which any molecular biologist is familiar. However, many potential problems can arise through attempts to simplify and cut the cost of the genotyping test. The best methods are undoubtedly those which provide the full DNA sequence giving a complete and accurate genotype. Many laboratories, however, do not use such intensive and expensive techniques routinely and instead rely on methods which are intrinsically less accurate with only some of the genotypes being recognized absolutely correctly. In addition, several variations can be misidentified if an additional (new) variation is present near an analyzed codon (like the 168, 172 and 175 mutation in sheep). Similar problems exist with other "short cut" methods but as long as these are kept in mind, and a random selection of samples are checked by full sequencing, useful information regarding scrapie risk can be produced.

References

[1] Hunter N, Foster JD, Dickinson AG, et al. Linkage of the gene for the scrapie-associated fibril protein (PrP) to the Sip gene in Cheviot sheep. Vet Rec 1989; 124(14):364–366.

[2] Moore RC, Hope J, McBride PA, et al. Mice with gene targetted prion protein alterations show that Prnp, Sinc and Prni are congruent. Nat Genet 1998; 18(2):118–125.

[3] Moore RC, Lee IY, Silverman GL, et al. Ataxia in prion protein (PrP)-deficient mice is associated with upregulation of the novel PrP-like protein doppel. J Mol Biol 1999; 292(4):797–817.

[4] Goldmann W. PrP gene and its association with spongiform encephalopathies. Br Med Bull 1993; 49(4):839–859.

[5] Goldmann W, Hunter N, Benson G, et al. Different scrapie-associated fibril proteins (PrP) are encoded by lines of sheep selected for different alleles of the Sip gene. J Gen Virol 1991; 72(10)2411–2417.

[6] Goldmann W, Hunter N, Smith G, et al. PrP genotype and agent effects in scrapie: change in allelic interaction with different isolates of agent in sheep, a natural host of scrapie. J Gen Virol 1994; 75(5)989–995.

[7] Belt PB, Muileman IH, Schreuder BE, et al. Identification of five allelic variants of the sheep PrP gene and their association with natural scrapie. J Gen Virol 1995; 76(3)509–517.

[8] Clouscard C, Beaudry P, Elsen JM, et al. Different allelic effects of the codons 136 and 171 of the prion protein gene in sheep with natural scrapie. J Gen Virol 1995; 76(8)2097–2101.

[9] Hunter N, Foster JD, Goldmann W, et al. Natural scrapie in a closed flock of Cheviot sheep occurs only in specific PrP genotypes. Arch Virol 1996; 141(5):809–824.

[10] Bossers A, Belt P, Raymond GJ, et al. Scrapie susceptibility-linked polymorphisms modulate the in vitro conversion of sheep prion protein to protease-resistant forms. Proc Natl Acad Sci USA 1997; 94(10):4931–4936.

[11] Bossers A, de Vries R, Smits MA. Susceptibility of sheep for scrapie as assessed by in vitro conversion of nine naturally occurring variants of PrP. J Virol 2000; 74(3):1407–1414.

[12] Hunter N, Moore L, Hosie BD, et al. Association between natural scrapie and PrP genotype in a flock of Suffolk sheep in Scotland. Vet Rec 1997; 140(3):59–63.

[13] Westaway D, Zuliani V, Cooper CM, et al. Homozygosity for prion protein alleles encoding glutamine-171 renders sheep susceptible to natural scrapie. Genes Dev 1994; 8(8):959–969.

[14] Ikeda T, Horiuchi M, Ishiguro N, et al. Amino acid polymorphisms of PrP with reference to onset of scrapie in Suffolk and Corriedale sheep in Japan. J Gen Virol 1995; 76(10)2577–2581.

[15] Houston F, Goldmann W, Chong A, et al. Prion diseases: BSE in sheep bred for resistance to infection. Nature 2003; 423(6939):498.

[16] Andreoletti O, Morel N, Lacroux C, et al. Bovine spongiform encephalopathy agent in spleen from an ARR/ARR orally exposed sheep. J Gen Virol 2006; 87(4):1043–1046.

[17] Benestad SL, Sarradin P, Thu B, et al. Cases of scrapie with unusual features in Norway and designation of a new type, Nor98. Vet Rec 2003; 153(7):202–208.

[18] Buschmann A, Biacabe AG, Ziegler U. Atypical scrapie cases in Germany and France are identified by discrepant reaction patterns in BSE rapid tests. J Virol Methods 2004; 117(1):27–36.

[19] Buschmann A, Luhken G, Schultz J, et al. Neuronal accumulation of abnormal prion protein in sheep carrying a scrapie-resistant genotype

(PrPARR/ARR). J Gen Virol 2004; 85(9):2727–2733.

[20] Everest SJ, Thorne L, Barnicle DA, et al. Atypical prion protein in sheep brain collected during the British scrapie-surveillance programme. J Gen Virol 2006; 87(2):471–477.

[21] Hunter N, Goldmann W, Smith G, et al. The association of a codon 136 PrP gene variant with the occurrence of natural scrapie. Arch Virol 1994; 137(1-2):171–177.

[22] Bossers A, Schreuder BE, Muileman IH, et al. PrP genotype contributes to determining survival times of sheep with natural scrapie. J Gen Virol 1996; 77(10)2669–2673.

[23] Vaccari G, Petraroli R, Agrimi U, et al. PrP genotype in Sarda breed sheep and its relevance to scrapie. Brief report. Arch Virol 2001; 146(10):2029–2037.

[24] Acin C, Martin-Burriel I, Goldmann W, et al. Prion protein gene polymorphisms in healthy and scrapie-affected Spanish sheep. J Gen Virol 2004; 85(7):2103–2110.

[25] Ridley RM, Baker HF. The myth of maternal transmission of spongiform encephalopathy. BMJ 1995; 311(7012):1071–1075; discussion 1075–1076.

[26] Bossers A, Harders FL, Smits MA. PrP genotype frequencies of the most dominant sheep breed in a country free from scrapie. Arch Virol 1999; 144(4):829–834.

[27] Hunter N, Cairns D, Foster JD, et al. Is scrapie solely a genetic disease? Nature 1997; 386(6621): 137.

[28] Acutis PL, Bossers A, Priem J, et al. Identification of prion protein gene polymorphisms in goats from Italian scrapie outbreaks. J Gen Virol 2006; 87(4):1029–1033.

[29] Billinis C, Psychas V, Leontides L, et al. Prion protein gene polymorphisms in healthy and scrapie-affected sheep in Greece. J Gen Virol 2004; 85(2):547–554.

[30] Goldmann W, Perucchini M, Smith A, et al. Genetic variability of the PrP gene in a goat herd in the UK. Vet Rec 2004; 155(6):177–178.

[31] Goldmann W, Martin T, Foster J, et al. Novel polymorphisms in the caprine PrP gene: a codon 142 mutation associated with scrapie incubation period. J Gen Virol 1996; 77(11)2885–2891.

[32] Sigurdarson S. Epidemiology of scrapie in Iceland and experience with control measures. In: Bradley R, Savey M, Marchant B, editors. Subacute spongiform encephalopathies. Commission of European Communities. Kluwer Academic, Netherlands, 1991:233–242.

[33] Dawson M, Hoinville LJ, Hosie BD, et al. Guidance on the use of PrP genotyping as an aid to the control of clinical scrapie. Scrapie Information Group. Vet Rec 1998; 142(23): 623–625.

[34] Hunter N, Goldmann W, Benson G, et al. Swaledale sheep affected by natural scrapie differ significantly in PrP genotype frequencies from healthy sheep and those selected for reduced incidence of scrapie. J Gen Virol 1993; 74(6)1025–1031.

[35] Goldmann W, Hunter N, Foster JD, et al. Two alleles of a neural protein gene linked to scrapie in sheep. Proc Natl Acad Sci USA 1990; 87(7): 2476–2480.

[36] Laplanche JL, Chatelain J, Westaway D, et al. PrP polymorphisms associated with natural scrapie discovered by denaturing gradient gel electrophoresis. Genomics 1993; 15(1):30–37.

[37] Zhang L, Li N, Fan B, et al. PRNP polymorphisms in Chinese ovine, caprine and bovine breeds. Anim Genet 2004; 35(6):457–461.

[38] Lan Z, Wang ZL, Liu Y, et al. Prion protein gene (PRNP) polymorphisms in Xinjiang local sheep breeds in China. Arch Virol 2006; 151(10):2095–2101.

[39] Goldmann W, Baylis M, Chihota C, et al. Frequencies of PrP gene haplotypes in British sheep flocks and the implications for breeding programmes. J Appl Microbiol 2005; 98(6):1294–1302.

[40] Lima ACB, Bossers A, Souza CEA, et al. PrP genotypes in a pedigree flock of Santa Inês sheep in Brazil. Vet Rec 2006; in press.

[41] Gombojav A, Ishiguro N, Horiuchi M, et al. Amino acid polymorphisms of PrP gene in Mongolian sheep. J Vet Med Sci 2003; 65(1):75–81.

[42] Thorgeirsdottir S, Sigurdarson S, Thorisson HM, et al. PrP gene polymorphism and natural scrapie in Icelandic sheep. J Gen Virol 1999; 80(9)2527–2534.

[43] Tranulis MA, Osland A, Bratberg B, et al. Prion protein gene polymorphisms in sheep with natural scrapie and healthy controls in Norway. J Gen Virol 1999; 80(4)1073–1077.

[44] DeSilva U, Guo X, Kupfer DM, et al. Allelic variants of ovine prion protein gene (PRNP) in Oklahoma sheep. Cytogenet Genome Res 2003; 102(1-4):89–94.

[45] Zhou H, Hickford JG, Fang Q. Technical note: determination of alleles of the ovine PRNP gene using PCR-single-strand conformational polymorphism analysis. J Anim Sci 2005; 83(4):745–749.

[46] Alvarez L, Arranz JJ, San Primitivo F. Identification of a new leucine haplotype (ALQ) at codon 154 in the ovine prion protein gene in Spanish sheep. J Anim Sci 2006; 84(2):259–265.
[47] Acutis PL, Sbaiz L, Verburg F, et al. Low frequency of the scrapie resistance-associated allele and presence of lysine-171 allele of the prion protein gene in Italian Biellese ovine breed. J Gen Virol 2004; 85(10):3165–3172.
[48] Guo X, Kupfer DM, Fitch GQ, et al. Identification of a novel lysine-171 allele in the ovine prion protein (PRNP) gene. Anim Genet 2003; 34(4):303–305.
[49] Belt PBGM, Bossers A, Schreuder BEC, et al. PrP allelic variants associated with natural scrapie. In: Gibbs CJ Jr, editor. Bovine spongiform encephalopathy; the BSE dilemma. Springer, New York, 1996:294–305.

57 Scrapie control at the National Level: The Norwegian Example

Kristin Ruud Alvseike, Ingrid Melkild, and Kristin Thorud

57.1 Introduction

This chapter focuses on specific measures through which it is possible to combat scrapie in sheep and goats on a national level, and effectively prevent animals from contracting the disease. Norway was chosen as an example, because the country successfully implemented a strict scrapie control program already in the 1990s.

57.2 Number of scrapie cases in Norway

Since the first case of scrapie in indigenous sheep in Norway was diagnosed in 1981, scrapie has been diagnosed in around 117 Norwegian sheep flocks (Fig. 57.1). In a 17-year period (1981–1998), scrapie was diagnosed in 55 different sheep flocks, 31 of these were detected in 1996 (Fig. 57.1). By that time, scrapie was found in five of the 19 counties in the country, however, 90 % of the affected farms were located in just two counties: Hordaland and Rogaland in South-West Norway.

In most flocks, the disease was suspected and diagnosed in one single sheep, although in several flocks, additional affected sheep were found. The average age of the affected animals was between three and four years (compare Chapter 34). The Rygja breed, the predominant sheep breed in Hordaland and Rogaland, was involved in about 80 % of the cases. Scrapie was never diagnosed in goats.

Based on clinical tests, histopathology, immunohistochemistry, and Western blot, a new type of scrapie, designated "Nor98" or "atypical scrapie"[1], was first identified in two Norwegian cases in 1998, as described in detail in Chapter 54. Internationally, the common designation for all the cases distinct from what is now designated as "classical scrapie" or "BSE in small ruminants" is "atypical scrapie" [4], and Nor98 is considered as the largest subgroup of atypical scrapie. All the atypical scrapie cases diagnosed in Norway so far were of the Nor98 type.

The number of atypical scrapie–Nor98 cases detected through the surveillance program increased, particularly from 2002 onwards after the implementation of a new diagnostic test (ELISA Platelia® Bio-Rad Laboratories), and the total number of all scrapie cases (classical and atypical) per annum peaked again (after the first peak in 1996).

From 1999 to 2005, a total of 47 atypical scrapie–Nor98 cases were found in the 54 scrapie flocks detected in this period, after the identification of the first two cases in two flocks in 1998.

The Western blot glycoprofile of the cases detected before 1998 (referred to as "unclassified" in Fig. 57.1) was not investigated because of the lack of appropriate fresh tissues, but the neuropathology and PrP genotypes of the affected cases indicate that the vast majority, if not all, were of the "classical type".

[1] "Nor98" is currently a sub-group, and probably the largest sub-group, of atypical scrapie, but the common international designation outside Norway is "atypical scrapie". To serve the international and the national Norwegian audience, the editors decided to use the designation "atypical scrapie–Nor98".

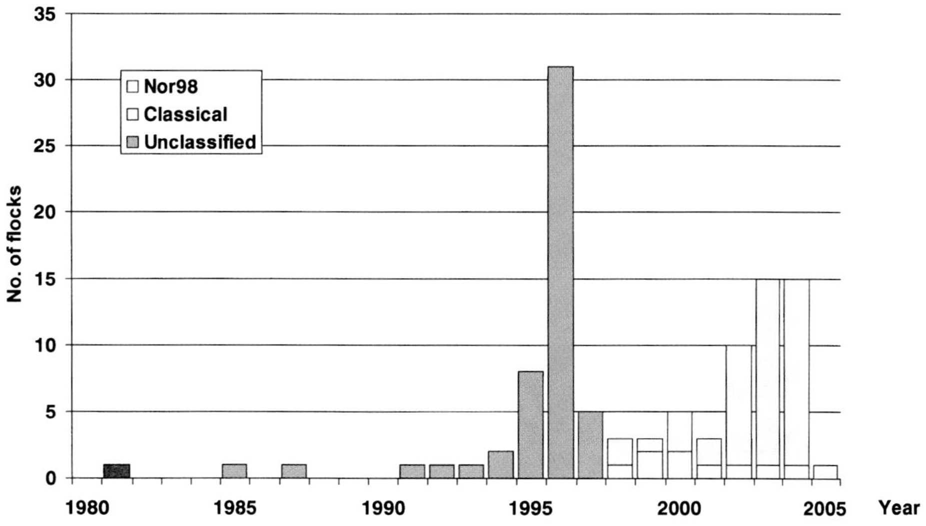

Fig. 57.1: Number of sheep flocks with scrapie in Norway by 31 August 2005. Figure by Petter Hopp ©.

57.3 Scrapie surveillance

57.3.1 System for identification of animals, animal movement, and animal health

The permanent identity marking of all sheep and goats, and recording of all flocks according to the provisions in the EU Council Directive 92/102/EEC, came into force on 1 January 1999, and still remain the basic elements of the program. Parts of the program were, however, already in force, due to voluntary recording of sheep, and voluntary and compulsory recordings of the flocks.

In addition to the stated requirement in the EU Directive, identity marking also includes the animal's year of birth. Recordings on the farms cover all individuals in the flock, all live births, the rams and ewes, all movements into and out of the flock, all deaths, including information on the possible cause of death for animals older than 7 months, all cases of disease treated by veterinarians, including information of signs consistent with scrapie for animals more than 7 months of age, and information on circumstances concerning grazing. Rams that are shared between flocks need individual identity cards. Records are to be kept for at least ten years.

57.3.2 Targeted surveillance of the sheep and goat population to detect cases

Scrapie is a notifiable disease. In the frame of the surveillance program all sheep and goats showing signs of nervous disorders or pruritus must be reported. All fallen stock older than 18 months shall also be reported.

Clinically suspect animals are examined by the local veterinary officers. Selected animals may be re-inspected after two weeks, or killed and submitted for testing. The animal may also be transferred to an isolated pen for further observation and later examination. Many of the cases investigated are not actual scrapie suspects or cases of scrapie, but are selected to rule out scrapie as a possible rather than probable cause of disease or death.

Thorough inspections of live animals to detect possible cases of scrapie are also being conducted by the farmers themselves, by the veterinarians on their regular visits to the farms, and by extended ante mortem examinations in abattoirs of all sheep older than 2 years. In flocks with official restrictions due to suspicion of scrapie, inspection has to be performed by the local veterinarians on the farm prior to transport to the abattoir.

Local veterinary officers inspect all sheep and goat flocks in Norway every third year. Farms with restrictions are visited every year. The visits include both inspections of the animals and the holding, as well as inspection of the identity marking of the animals and the records of the animals on the farms, in particular records concerning the health status, including veterinary treatments.

The veterinarians have been trained for these tasks in specific courses, and farmers have also been trained and motivated for the program through participation in specific courses held by the veterinary officers. The training courses include: details of the clinical scrapie history, risk factors for the spread of the disease in Norway [2], risks connected with imports, official measures to prevent scrapie, and measures taken when scrapie is suspected or has been diagnosed. A video showing different clinical manifestations of scrapie is shown to the farmers attending these courses [3].

57.3.3 Monitoring of the sheep population in abattoirs

When the scrapie control program was initiated, histopathological examination of brain stem material from approximately 3,000 slaughtered sheep per year was performed. The actual number of sheep examined was determined in order to detect an occurrence of scrapie lesions, if present, in at least 0.1% of the population of around 1.1 million winterfed sheep, at a confidence level of 95%. The number of samples from various regions was determined so that the relative number of samples was higher from the two counties where the majority of cases occurred. To further increase the probability of detecting animals with scrapie within the scope of this program, a lower age limit of 3 ½ years was chosen for the animals to be examined (Chapter 55).

Since 2001, the sampling of normal slaughtered animals has been carried out according to Regulation (EC) 999/2001 (TSE Regulation).

57.4 Control and eradication of scrapie

57.4.1 Classical scrapie; measures as a consequence of detecting suspect animals

Suspicion of the occurrence of classical scrapie may arise due to earlier contact with affected flocks, or clinically suspect animals found within a flock. Suspect animals are followed up by the local veterinary officers according to the procedure described in the surveillance program.

All farms or flocks from where sheep have been sold to an affected flock, or flocks containing sheep bought from affected flocks, or having kept sheep from affected farms temporarily on their farms within the last ten years prior to the diagnosis, are put under official restrictions for at least five years. The restrictions include a ban on transfer of live sheep to any other farm, or to exhibitions.

In farms where there have been ewes from affected flocks on their premises during the lambing period, the restriction includes a ban on taking the flock to pastures that are shared with other flocks. If these farmers agree to kill or slaughter all their small ruminants, i. e., sheep and goats, the authorities offer an economic compensation for each animal, provided that sanitary measures are carried out also in the indoor area after the culling.

Since 2003, all animals older than 18 months from flocks put under restrictions are tested for scrapie at the abattoir.

Sheep in some specific contact flocks are considered being of particular risk of developing classical scrapie. Three alternative criteria for being included in this group are:

1. if the flock has a close functional contact with an affected flock (compare Chapter 42.2.2), or
2. if the flock has recruited a large proportion (more than 30%) of its animals from the affected flock, or
3. if the sheep that developed classical scrapie in the affected flock originated from the contact flock.

Such flocks are treated in the same way as if classical scrapie was already confirmed in them.

57.4.2 Classical scrapie: measures as a consequence of scrapie confirmation

When classical scrapie has been confirmed in a flock, the entire flock of small ruminants is euthanized. The carcasses are treated as high-risk waste. The product thus obtained is burned or otherwise destroyed. Sheep in contact flocks that were born on the affected holding, and offspring of ewes that were born on the affected holding, are killed, destroyed, and compensation is paid by the authorities.

The prion protein genotype has since 2003 been determined for all animals that are killed, but even the sheep with the most resistant genotypes have been killed.

After an affected flock has been killed, extensive measures are taken to eliminate the infectious agent from the holding (Chapter 44). Depending on the possibilities for satisfactory cleaning and disinfection, buildings where sheep have been kept are either burned or subjected to strict sanitation measures.

The sanitation measures required for indoor areas include the removal of manure, removal and burning of all wooden materials and other material that has been in direct contact with the sheep (flooring, walls, drinking basins, etc.), cleaning and disinfection of remaining indoor areas, painting of at least the bottom 1.5 m of the walls of the building (including window panes), and fitting of new concrete floors, doors, walls, etc., according to the condition on the farm.

Sanitation measures for outdoor areas include changing of the upper layer on surrounding unpaved roads, painting of the outside wall of relevant buildings, ploughing and/or burning of grass on grazing areas, and fitting of new fences on areas that have been in contact with sheep.

Full compensation is paid to the owner to cover the value of the flock and the expenses related to the sanitation measures. Average compensation to a farmer with a classical scrapie flock exceeds 1 million Norwegian Crowns (NOK) (= 163,000 US Dollars). After completion of the sanitation measures, the farm must be left empty for at least two years, before restrictions are lifted and new sheep and goats are allowed to enter the farm. Since 2005 the period which the farm has to be left empty has been increased to three years according to the TSE Regulation of the EC.

57.4.3 Atypical scrapie–Nor98; new measures since 2005

Until 2004, confirmed cases of atypical scrapie–Nor98 were mainly handled in the same way as classical scrapie. However, as knowledge about atypical scrapie–Nor98 has increased (see also Chapter 54), the eradication measures have been adjusted. Atypical scrapie–Nor98 cases have appeared in single animals, with no additional cases found in the flocks. PrP^{Sc} is not found outside the central nervous system (CNS) in these sheep, and this may indicate lower infectivity and even sporadic appearance, comparable to sporadic Creutzfeldt–Jakob disease (sCJD; see Chapter 14). The current measures for atypical scrapie–Nor98 cases follow the provisions in the TSE Regulation of the EC. Instead of culling all animals in the affected flock, the genotype strategy in the TSE Regulation is applied. Measures are restricted to the *index flocks*, and only very rarely are flocks defined as contact flocks and put under restrictions. Until more knowledge is gained, the Norwegian Authorities have prohibited the movement of ARR/ARR animals from flocks put under restrictions due to atypical scrapie–Nor98, since several cases of atypical scrapie–Nor98 have been diagnosed in such animals, summarized in Table 54.1.

According to the TSE Regulation, certain measures may be applied if approved by the national authorities. The Norwegian Authorities have approved delayed destruction of animals where the frequency of ARR alleles within the breed or holding is low, or where it is deemed necessary to avoid inbreeding.

The Norwegian authorities believe the genotype strategy is not the optimal method for the control of atypical scrapie–Nor98. Because of the wide range of genotypes found in the atypical scrapie–Nor98 cases (Table 54.1), and the strong indications that atypical scrapie appears not to be (horizontally or laterally) transmissible under normal field conditions (as is the case for classical scrapie; compare Chapter 42.2.2), the Norwegian Food Safety Authority is suggesting that the EU changes the TSE Regulation regarding the eradication measures for atypical scrapie–Nor98.

57.5 Scrapie prevention

57.5.1 Feed-ban and measures to reduce the risk of spread of scrapie by animal trade and movement within Norway

Since June 1990, there has been a strictly enforced ban on the use of ruminant protein in ruminant feed. Animal fat, milk, and dairy products are not included in the feed for small ruminants, and due to lack of tradition, meat-and-bone-meal was not used in commercial Norwegian small ruminant feed even prior to the ban.

Since 1980, there has been a ban on the movement of live sheep and goats from one county to another. This ban is considered to have been an important factor in preventing the spread of scrapie from the main classical scrapie region in the south-western part of the country. The ban was primarily enforced as a measure to combat the maedi/visna disease in sheep. Since October 1995, specific regulations have been in force in the Hordaland and Rogaland counties, in order to further prevent the spread of scrapie within the Norwegian counties. These measures included a general ban on the transfer of live sheep between flocks, a general ban on exhibitions, markets and other forms of trade in sheep, and a ban on the transport of animals from different farms together in the same truck, restrictions on the grazing of animals from different farms on the same pasture, a general ban on the establishment of new local breeding cooperations that share rams between flocks, and a maximum limit of 4 flocks in each breeding group.

In August 1997, some of these regulations were followed up on a national basis and extended to cover the whole of Norway. These include a general ban on the sale of ewes, a ban on the expansion of existing or the establishment of new breeding groups. In addition to these national regulations, specific regulations covering certain parts of Norway are put into force when needed.

57.5.2 Measures at the border

There was for many years a general ban on the import of live sheep and goats. Live animals could only be imported if the Veterinary Authorities gave exemptions from the import ban. Diseases like scrapie, maedi/visna and paratuberculosis have long incubation times and are difficult to diagnose by laboratory methods. Because of this, only few exemptions were given, all including live animals from New Zealand, and these were quarantined for several years. From 1999 and onwards, trade with small ruminants was included in the EEA agreement. Hence, there is no longer a general ban on the import of live sheep and goats. However, following each import, the local veterinary officer evaluates whether further preventional health measures should be carried out. Additionally, a farmer's company (Koorimp) has specialized in advising farmers on all of the import matters.

57.5.3 Public health measures

All the carcasses from affected scrapie flocks, carcasses from contact flocks considered to be at particular risk for developing scrapie, and carcasses from sheep and offspring of ewes that originate from affected farms, are treated as high-risk waste. Therefore, the product obtained is incinerated or otherwise destroyed.

However, scrapie is not considered a risk to human health.

References

[1] Benestad SL, Sarradin P, Thu B, et al. Cases of scrapie with unusual features in Norway. Vet Rec; 2003:153:202–208.

[2] Hopp P, Ulvund MJ, Jarp J. A case-control study on scrapie in Norwegian sheep flocks. Prev Vet Med; 2001:51:183–198.

[3] Ulvund MJ. Scrapie – clinical symptoms. Thorud K, Statens Dyrehelsetilsyn, FIM VIDEO, Bergen 1997.

[4] EFSA. Opinion on "Classification of atypical transmissible spongiform encephalopathy (TSE) cases in small ruminants". EFSA J 2005; 276:1–30. See also Webpage 2006: www.efsa.eu.int/science/tse_assessments/bse_tse/983/biohaz_sr31_small-ruminanttsetests_en1.pdf.

Epilogue

The challenges presented by BSE and vCJD over the past 20 years have not yet been met, although the signs are good for the eventual control of BSE in Europe. Many critical scenarios lie ahead, particularly in the vCJD human health aspects of subclinical infection and human-to-human transmission through blood transfusions and surgical procedures. So far, the indications are that a major catastrophe has been averted, but it will be at least another 10 years before we have the confidence to assess the final risk.

In the meantime, we must not give up on the efforts to develop a sensitive blood and tissue assay system whilst pursuing rational therapeutic strategies at the same time. These research understandings will only come from a more complete analysis of the molecular basis of PrP infectivity and pathogenesis (neurodegeneration).

Much has been accomplished – witness the contributions in this book: but many surprises remain in store in our quest to remove this threat to both human and animal health.

Prof Colin L Masters, MD
Department of Pathology, The University of Melbourne, Parkville, 3010, Victoria, Australia.

Appendix 1
Major Categories of Infectivity: Tables IA, IB, IC

WHO Copyright

Editor's Note: The copyright for reprinting the following table set was kindly given by the World Health Organization (WHO), Department of Knowledge Management, Geneva. The major categories of infectivity – Tables IA-IC including Annex 1 (footnotes and references) – are a 1:1 reprint of the WHO Guidelines on Tissue Infectivity Distribution in Transmissible Spongiform Encephalopathies (report of the WHO Consultation in Geneva, 14–16 Sept. 2005). The original of the full document was published in both print and electronic formats on 22 June 2006 under www.who.int[1]. The table set is important to several chapters in this book[2], since it represents the most up-to-date officially acknowledged summary relating to TSE infectivity (vCJD, BSE, scrapie, and other prion diseases) at the time of publishing this book. The cross-reference to this appendix is (WHO Tables$_{\text{Appendix}}$IA-IC).

The information in these Tables is based exclusively upon observations of naturally occurring disease, or primary experimental infection by the oral route (in ruminants), and does not include data on models using strains of TSE adapted to experimental animals, because passaged strain phenotypes can differ significantly and unpredictably from those of naturally occurring disease. Also, because the detection of misfolded host prion protein (PrP^{TSE}) has proven to be a reliable indicator of infectivity, PrP^{TSE} testing results have been presented in parallel with bioassay data. Tissues are grouped into three major infectivity categories, irrespective of the stage of disease:

IA: **High-infectivity tissues:** Central Nervous System (CNS) tissues that attain a high titre of infectivity in the later stages of all TSEs, and certain tissues that are anatomically associated with the CNS.

IB: **Lower-infectivity tissues:** peripheral tissues that have tested positive for infectivity and/or PrP^{TSE} in at least one form of TSE.

IC: **Tissues with no detectable infectivity:** tissues that have been examined for infectivity and/or PrP^{TSE} with negative results.

Data entries are shown as follows:

- \+ Presence of infectivity or PrP^{TSE}
- – Absence of detectable infectivity or PrP^{TSE}
- NT Not tested
- NA Not applicable
- ? Controversial results
- ( ) Limited or preliminary data

The placement of a given tissue in one or another category can be disease-specific and subject to revision as new data accumulate from increasingly sensitive tests. In fact, it is conceivable that the detection of infectivity using transgenic mice that over-express genes encoding various prion proteins, or the detection of PrP^{TSE} using some newly developed amplification methods, might prove to be more sensitive than transmission studies in wild-type bioassay animals, and thus may not correlate with disease transmission in nature.

It is also important to understand that categories of infectivity are not the same as categories of risk, which require consideration not only of the level of infectivity in tissue, but also of the amount of tissue to which a person or

[1] See pages 17–25 of URL: http://www.who.int/bloodproducts/TSEREPORT_LoRes.pdf

[2] SRM-relevant data are shown in Figure 53.2.

animal is exposed, and the route by which infection is transmitted. For example, although the level of tissue infectivity (concentration of infectivity in tissue as reflected by titre) is the most important factor in estimating the risk of transmission by instrument cross-contamination during surgical procedures (e. g., neurosurgery versus general surgery), it is only one determinant of the risk of transmission by blood transfusions, in which a large amount of low-infectivity material is administered directly into the circulation, or the risk of transmission by food that, irrespective of high or low infectivity, involves the comparatively inefficient oral route of infection.

Table IA: High-infectivity tissues. CNS tissues that attain a high titer of infectivity in the later stages of TSE and certain tissues anatomically associated with the CNS.

Tissues	Human TSEs				Cattle		Sheep & goats	
	vCJD		Other TSEs		BSE		Scrapie	
	Infectivity[1]	PrP^{TSE}	Infectivity[1]	PrP^{TSE}	Infectivity[1]	PrP^{TSE}	Infectivity[1]	PrP^{TSE}
Brain	+	+	+	+	+	+	+	+
Spinal cord	+	+	+	+	+	+	+	+
Retina	NT	+	+	+	+	NT	NT	+
Optic nerve[2]	NT	+	NT	+	+	NT	NT	+
Spinal ganglia	+	+	NT	+	+	NT	NT	+
Trigeminal ganglia	+	+	NT	+	+	NT	NT	+
Pituitary gland[3]	NT	+	+	+	–	NT	+	NT
Dura mater[3]	NT	–	+	–	NT	NT	NT	NT

Table IB: Lower-infectivity tissues. Peripheral tissues that have tested positive for infectivity and/or PrP^{TSE} in at least one form of TSE.

Tissues	Human TSEs				Cattle		Sheep & goats	
	vCJD		Other TSEs		BSE		Scrapie	
	Infectivity	PrP^{TSE}	Infectivity	PrP^{TSE}	Infectivity	PrP^{TSE}	Infectivity	PrP^{TSE}
Peripheral Nervous system								
Peripheral nerves	+	+	(−)	+	+	+	+	+
Enteric plexuses[4]	NT	+	NT	(−)	NT	+	NT	+
Lymphoreticular tissues								
Spleen	+	+	+	+	−	−	+	+
Lymph nodes	+	+	+	−	−	−	+	+
Tonsil	+	+	NT	−	+	−	+	+
Nictitating membrane	NA	NA	NA	NA	+	−	NT	+
Thymus	NT	+	NT	−	−	NT	+	NT
Alimentary tract								
Esophagus	NT	−	NT	−	−	NT	NT	+
Fore-stomach[5] (ruminants only)	NA	NA	NA	NA	−	NT	NT	+
Stomach/ abomasum[6]	NT	−	NT	NT	−	NT	NT	+
Duodenum	NT	−	NT	NT	−	NT	NT	+
Jejunum[6]	NT	+	NT	−	−	NT	NT	+
Ileum[6,7]	NT	+	NT	−	+	+	+	+
Appendix	−	+	NT	−	NA	NA	NA	NA
Large intestine[6]	+	+	NT	−	−	NT	+	+
Reproductive tissues								
Placenta[8]	NT	−	(+)	−	−	NT	+	+
Other tissues								
Lung	NT	−	+	−	−	NT	−	−
Liver	NT	−	+	−	−	NT	+	NT
Kidney	NT	−	+	−	−	−	−	−
Adrenal	NT	+	−	−	NT	NT	+	NT
Pancreas	NT	−	NT	−	−	NT	+	NT
Bone marrow	−	−	(−)	−	(+)	NT	+	NT
Skeletal muscle[9]	NT	+	(−)	+	(+)	NT	−	+
Tongue[10]	NT	−	NT	−	−	NT	NT	+
Blood vessels	NT	+	NT	+	−	NT	NT	+
Nasal mucosa[11]	NT	NT	NT	+	−	NT	+	+
Salivary gland	NT	−	NT	NT	−	NT	+	NT
Cornea[12]	NT	−	+	−	NT	NT	NT	NT
Body fluids								
CSF	−	−	+	−	−	NT	+	NT
Blood[13]	+	?	−	?	−	?	+	?

Table IC: Tissues with no detected infectivity or PrP^{TSE}.

Tissues	Human TSEs				Cattle		Sheep & goats	
	vCJD		Other TSEs		BSE		Scrapie	
	Infectivity	PrP^{TSE}	Infectivity	PrP^{TSE}	Infectivity	PrP^{TSE}	Infectivity	PrP^{TSE}
Reproductive tissues								
Testis	NT	–	(–)	–	–	NT	–	NT
Prostate/ Epididymis/ Seminal vesicle	NT	–	(–)	–	–	NT	–	NT
Semen	NT	–	(–)	–	–	NT	NT	NT
Ovary	NT	–	NT	–	–	NT	–	NT
Uterus (Non-gravid)	NT	–	NT	–	–	NT	–	NT
Placenta fluids	NT	NT	(–)	NT	–	NT	NT	NT
Fetus[14]	NT	NT	NT	NT	–	NT	–	–
Embryos[14]	NT	NT	NT	NT	–	NT	?	NT
Musculo-skeletal tissues								
Bone	NT	NT	NT	NT	–	NT	NT	NT
Heart/ pericardium	NT	–	–	–	–	NT	–	NT
Tendon	NT	NT	NT	NT	–	NT	NT	NT
Other tissues								
Gingival tissue	NT	–	–	–	NT	NT	NT	NT
Dental pulp	NT	–	NT	–	NT	NT	NT	NT
Trachea	NT	–	NT	–	–	NT	NT	NT
Skin	NT	–	NT	–	–	NT	–	NT
Adipose tissue	NT	–	(–)	–	–	NT	NT	NT
Thyroid gland	NT	–	(–)	–	NT	NT	–	NT
Mammary gland/udder	NT	NT	NT	NT	–	NT	–	NT
Body fluids, secretions and excretions								
Milk[15]	NT	NT	(–)	NT	–	–	–	NT
Colostrum[16]	NT	NT	(–)	NT	(–)	–	–	NT
Cord blood[17]	NT	NT	(–)	NT	–	NT	NT	NT
Saliva	NT	–	–	NT	NT	NT	–	NT
Sweat	NT	NT	–	NT	NT	NT	NT	NT
Tears	NT	NT	–	NT	NT	NT	NT	NT
Nasal mucus	NT	–	–	NT	NT	NT	NT	NT
Bile	NT	NT	NT	NT	NT	NT	NT	NT
Urine[16,17]	NT	NT	–	–	–	NT	NT	NT
Feces	NT	NT	–	NT	–	NT	–	NT

Appendix 1 Footnotes

[1] Infectivity bioassays of human tissues have been conducted in either primates or mice (or both); bioassays of cattle tissues have been conducted in either cattle or mice (or both); and most bioassays of sheep and/or goat tissues have been conducted only in mice. In regard to sheep and goats, not all results are consistent for both species.

[2] In experimental models of TSE, the optic nerve has been shown to be a route of neuroinvasion and contains high titers of infectivity.

[3] No experimental data about infectivity in human pituitary gland or dura mater have been reported, but cadaveric dura mater allograft patches, and growth hormone derived from cadaveric pituitaries have transmitted disease to hundreds of people and therefore must be included in the category of high-risk tissues.

[4] In cattle, PrP^{TSE} was limited to enteric plexus in the distal ileum.

[5] Ruminant forestomachs (reticulum, rumen, and omasum) are widely consumed, as is the true stomach (abomasum). The abomasum of cattle (and sometimes sheep) is also a source of rennet.

[6] In vCJD, transmission to mice has so far been limited to rectal tissue [Dr. J. Wadsworth, unpublished data], and PrP^{TSE} was detected only in gut-associated lymphoid and nervous tissue (mucosa, muscle, and serosa were negative). In goats, PrP^{TSE} was also limited to gut-associated lymphoid and nervous tissue [Dr. O. Andreoletti, unpublished data]

[7] In cattle and sheep, only the distal ileum has been bioassayed for infectivity.

[8] A single report of transmission of CJD infectivity from human placenta has never been confirmed and is considered improbable.

[9] Muscle homogenates have not transmitted disease to primates from humans with sCJD, or to cattle from cattle with BSE. However, intracerebral inoculation of a semi-tendinosus muscle homogenate (including nervous and lymphatic elements) from a single cow with BSE has transmitted disease to over-expressing transgenic mice at a rate indicative of only trace levels of infectivity. Also, recent published and unpublished studies have reported the presence of PrP^{TSE} in skeletal muscle in experimental rodent models of scrapie and vCJD, in experimental and natural infections of sheep and goats, in sheep orally dosed with BSE [Dr. O. Andreoletti, unpublished data], and in humans with sCJD and vCJD [Prof. J. Ironside, unpublished data]. Bioassays to determine whether PrP^{TSE} is associated with transmissibility in these experimental or natural infections are in progress.

[10] In cattle, infectivity bioassay was negative, but the presence of PrP^{TSE} in palatine tonsil has raised concern about possible infectivity in lingual tonsillar tissue at the base of the tongue that may not be removed at slaughter.

[11] In sCJD, PrP^{TSE} is limited to olfactory mucosa. [Dr. G. Zanusso, unpublished data].

[12] Because only one or two cases of CJD have been plausibly attributed to corneal transplants among hundreds of thousands of recipients, cornea is categorised as a lower-risk tissue; other anterior chamber tissues (lens, aqueous humor, iris, conjunctiva) have been tested with a negative result both in vCJD and other human TSEs, and there is no epidemiological evidence that they have been associated with iatrogenic disease transmission.

[13] A wealth of data from studies of blood infectivity in experimental animal models of TSE has been extended by recent studies documenting infectivity in the blood of sheep with naturally occurring scrapie, and (from epidemiological observations) three blood-associated vCJD transmissions in humans. Blood has not been shown to transmit disease from patients with any form of 'classical' TSE, or from cattle with BSE (including fetal calf blood). However, several laboratories using new, highly sensitive methods to detect PrP^{TSE} claim success in studies of plasma and/or buffy coat in a variety of animal and human TSEs. Because the tests are all in a preliminary stage of development (and do not yet include results on blinded testing of specimens from naturally infected humans or animals), the Consultation felt that it was still too early to evaluate the validity of these tests with sufficient confidence to permit either a negative or positive conclusion.

[14] Embryos from BSE-affected cattle have not transmitted disease to mice, but no infectivity measurements have been made with fetal calf tissues other than blood (negative mouse bioassay). Calves born of dams that received embryos from BSE-affected cattle have survived for observations periods of up to seven years, and examination of the brains of both the unaffected dams and their offspring revealed no spongiform encephalopathy or PrP^{TSE}.

[15] Evidence that infectivity is not present in milk includes temporo-spatial epidemiologic observations failing to detect maternal transmission; clinical observations of over a hundred calves nursed by infected cows that have not developed BSE; and experimental observations that milk from infected cows has not transmitted disease when adminis-

tered intracerebrally or orally to mice. Also, PrP^{TSE} has not been detected in milk from cattle incubating BSE following experimental oral challenge.

[16] Early reports of transmission of CJD infectivity from human cord blood, colostrum, and urine have never been confirmed and are considered improbable. A recent bioassay in over-expressing transgenic mice of colostrum from a cow with BSE gave a negative result; and PrP^{TSE} has not been detected in colostrum from cattle incubating BSE following experimental oral challenge.

[17] IgG short chains mimicking the Western blot behavior of PrP^{TSE} have been identified in the urine of sporadic, variant, and familial CJD patients.

Table References

Tables IA, IB and IC are updated from tables published in an earlier consultation (WHO Guidelines on Transmissible Spongiform Encephalopathies in relation to Biological and Pharmaceutical Products. Report of a WHO Consultation, Geneva, Switzerland, 3–5 February 2003. WHO/BCT/QSD/03.01) by an *ad hoc* expert interim group composed of Dr. O. Andreoletti, Mr R. Bradley, Dr P. Brown, Prof. Dr H. Budka, Prof. Dr. M.H. Groschup, Dr. J. Ironside, Dr. M. Pocchiari, and Mr G.A.H. Wells. Dr P. Brown coordinated the group and consolidated the information for review by all Consultation participants.

Most of the observations that form the basis for the Tables have been published in original reports (or cited in reviews) that follow. Most studies published since the previous Consultation have been listed, but no attempt has been made to list the many earlier reports in which only one or two tissues were examined unless they concerned tissues of exceptional current interest. Also, a number of observations made by, or known to, members of the expert subcommittee that assembled the table, have not yet been published.

Human TSE

Brown P. Blood infectivity, processing, and screening tests in transmissible spongiform encephalopathy. *Vox Sang* 2005; **89**: 63–70

Brown P. Blood infectivity in the transmissible spongiform encephalopathies. In: Turner M, ed. *Creutzfeldt-Jakob disease: Managing the Risk of Transfusion Transmission*, AABB Press, Bethesda, Maryland, 2006, in press.

Brown P, Gibbs CJ Jr, Rodgers-Johnson P, Asher DM, Sulima MP, Bacote A, Goldfarb LG, Gajdusek DC. Human spongiform encephalopathy: the National Institutes of Health Series of 300 cases of experimentally transmitted disease. *Ann Neurol* 1994; **35**: 513–529

Bruce ME, McConnell I, Will RG, Ironside JW. Detection of variant Creutzfeldt-Jakob disease infectivity in extraneural tissues. *Lancet* 2001; **358**: 208–209

Glatzel M, Abela E, Maissen M and Aguzzi A. Extraneural Pathologic Prion Protein in Sporadic Creutzfeldt–Jakob Disease. *N Engl J Med* 2003; **349:** 1812–1820

Haïk S, Faucheux BA, Sazdovitch V, Privat N, Kemeny J-L, Perret-Liaudet A, Hauw J-J. The sympathetic nervous system is involved in variant Creutzfeldt-Jakob disease. *Nature Med* 2003; **9**: 1121–1123

Hainfellner JA, Budka H. Disease associated prion protein may deposit in the peripheral nervous system in human transmissible spongiform encephalopathies. *Acta Neuropathol* 1999; **98**: 458–460

Head MW, Kouverianou E, Taylor L, Green A, Knight R. Evaluation of urinary PrPSc as a diagnostic test for sporadic, variant, and familial CJD. *Neurology* 2005; **64**: 1794–1796

Head MW, Northcott V, Rennison K, Ritchie D, McCardle L, Bunn TJ, McLennan NF, Ironside JW, Tullo AB, Bonshek RE. Prion protein accumulation in eyes of patients with sporadic and variant Creutzfeldt-Jakob disease. *Invest Ophthalmol Vis Sci* 2003; **44**: 342–346

Head MW, Peden AH, Yull HM, Ritchie DL, Bonshek RE, Tullo AB, Ironside JW. Abnormal prion protein in the retina of the most commonly occurring subtype of sporadic Creutzfeldt-Jakob disease. *Br J Ophthalmol* 2005; **89**: 1131–1133

Head MW, Ritchie D, Smith N, McLoughlin V, Nailon W, Samad S, Masson S, Bishop M, McCardle L, Ironside JW. Peripheral tissue involvement in sporadic, iatrogenic and variant Creutzfeldt-Jakob disease: an immunohistochemical, quantitative and biochemical study. *Am J Pathol* 2004; **164**; 143–153

Ironside KW, Head MW, Bell JE, McCardle L, Will GR. Laboratory diagnosis of variant Creutzfeldt-Jakob disease. *Histopathology* 2000; **37**: 1-9

Kariv-Inbal Z, Halimi M, Dayan Y, Engelstein R, Gabizon R. Characterisation of light chain immunoglobulin in urine from animals and human infected with prion diseases. *J Neuroimmunol* 2005; **162**: 12–18

Kovacs GG, Lindeck-Pozza E, Chimelli L, Araújo AQC, Gabbai, AA, Ströbel T, Glatzel M, Aguzzi A, Budka H. Creutzfeldt-Jakob disease and inclusion body myositis: abundant disease-associated prion protein in muscle. *Ann Neurol* 2004; **55**: 121–125

Ironside JW, McCardle L, Horsburgh A, Lim Z, Head MW. Pathological diagnosis of variant Creutzfeldt-Jakob disease. *Acta Pathol Microbiol Immunol Scand (APMIS)* 2002; **110**: 79–87

Tabaton M, Monaco S, Cordone MP, Colucci M, Giaccone G, Tagliavini F, Zanusso G. *Prion deposition in olfactory biopsy of sporadic Creutzfeldt-Jakob disease Ann Neurol* 2004, **55**: 294–296

Thomzig A, Cardone F, Krüger D, Brown P, Pocchiari M, Beekes M. BSE- and vCJD-associated prion protein in muscles. *J Gen Virol*, 2006; **86**: in press.

Wadsworth JDF, Joiner S, Hill AF, Campbell TA, Desbruslais, M, Luthert PJ, Collinge J. Tissue distribution of protease resistant prion protein in var. Creutzfeldt-Jakob disease using a highly sensitive immunoblotting assay. *Lancet* 2001; **358**: 171–180

Wong BS, Green AJ, Li R, Xie Z, Pan T, Liu T, Chen SG, Gambetti P, Sy MS. Absence of protease resistant prion protein in the cerebrospinal fluid of Creutzfeldt-Jakob disease. *J Pathol* 2001; **194**: 9-14

Zanusso G, Ferrari S, Cardone F, Zampieri P, Gelati M, Fiorini M, Farinazzo A, Gardiman M, Cavallaro T, Bentivoglio M, Righetti PG, Pocchiari M, Rizzuto N and Monaco S. Detection of Pathologic Prion Protein in the Olfactory Epithelium in Sporadic Creutzfeldt–Jakob Disease. *N Engl J Med* 2003; **348**:711-9

Bovine Spongiform Encephalopathy

Buschmann A, Groschup MH. Highly BSE sensitive transgenic mice confirm essential restriction of infectivity to the nervous system in clinically diseased cattle. *J Infect Dis* 2005; **192**: 934–942

European commission (2002) SSC Of 8 November 2002. *Update Of The Opinion On TSE Infectivity Distribution In Ruminant Tissues.* Initially adopted by The Scientific Steering Committee at its Meeting of 10–11 January 2002 and amended at its Meeting of 7-8 November 2002, following the submission of (1) A Risk Assessment by the German Federal Ministry of Consumer Protection, Food and Agriculture and (2) New scientific evidence regarding BSE infectivity distribution in tonsils. Internet address: http://europa.eu.int/comm/food/fs/sc/ssc/outcome_en.html

Fraser H, Foster J. Transmission to mice, sheep and goats and bioassay of bovine tissues. In: Bradley R, Marchant B, eds. Transmissible Spongiform Encephalopathies. A Consultation on BSE with the Scientific Veterinary Committee of the Commission of the European Communities held in Brussels, September 14–15 1993. Document VI/4131/94-EN. Brussels, European Commission Agriculture, 1994:145–159.

Houston F, Foster J D, Chong A, Hunter N, Bostock CJ. Transmission of BSE by blood transfusion in sheep. *Lancet* 2000; **356**: 999–1000

Iwamaru Y, Okubo Y, Ikeda T, Hayashi H, Imamura M, Yokoyama T, Shinagawa M. PrP^{Sc} distribution of a natural case of bovine spongiform encephalopathy. In: Kitamoto T, ed. *Prions. Food and Drug Safety.* Springer Verlag, New York.

Middleton DJ, Barlow RM. Failure to transmit bovine spongiform encephalopathy to mice by feeding them with extraneural tissues of affected cattle. *Vet Rec* 1993; **132**:545-7.

SEAC. 88th Meeting of SEAC 30 June 2005. Discussion of the results of an experiment designed to determine if any PrP^{Sc} was detectable in milk and colostrum of cattle experimentally orally infected with BSE. Video recording of a presentation and the discussion on this subject can be found at: http://clients.westminster-digital.co.uk/seac/88thmeeting/

Taylor DM, Ferguson CE, Bostock CJ, Dawson M. Absence of disease in mice receiving milk from cows with bovine spongiform encephalopathy. *Vet Rec* 1995; **136**:592.

Terry LA, Marsh S, Ryder SJ, Hawkins SAC, Wells GAH, Spencer YI. Detection of disease-specific PrP in the distal ileum of cattle orally exposed to the BSE agent. *Vet Rec* 2003; **152**: 387–392.

Thomzig A, Cardone F, Krüger D, Brown P, Pocchiari M, Beekes M. BSE- and vCJD-associated prion protein in muscles. *J Gen Virol* 2006, in press

Wells GAH, Dawson M, Hawkins SAC, Austin AR, Green RB, Dexter I, Horigan MW, Simmons MM. Preliminary observations on the pathogenesis of experimental bovine spongiform encephalopathy. In: Gibbs CJ, ed. *Bovine Spongiform Encephalopathy: The BSE Dilemma.* Serono Symposia, Norwell, USA, Springer-Verlag, New York, 1996: 28–44.

Wells GAH, Hawkins SAC, Green RB, Austin AR, Dexter L, Spencer YI, Chaplin MJ, Stack MJ, Dawson M. Preliminary observations on the pathogenesis of experimental bovine spongiform encephalopathy (BSE): an update. *Vet Rec* 1998; **142**: 103–106

Wells GAH, Hawkins SAC, Green RB, Spencer YI, Dawson M. Limited detection of sternal bone marrow infectivity in the clinical phase of experimental bovine spongiform encephalopathy (BSE). *Vet Rec* 1999; **144**: 292–294.

Wells GAH, Spiropoulos J, Hawkins SAC, Ryder SJ. Pathogenesis of experimental bovine spongiform encephalopathy (BSE): pre-clinical infectivity in tonsil and observations on lingual tonsil in slaughtered cattle. *Vet Rec*, 2005; **156**: 401–407.

Wrathall AE, Brown KFD, Sayers AR, Wells GAH, Simmons MM, Farrelly SSJ, Bellerby P, Squirrell J, Spencer YI, Wells M, Stack MJ, Bastiman B, Pullar D, Scatcherd J, Heasman L, Parker J, Hannam DAR, Helliwell DW, Chree A, Fraser H. Studies of embryo transfer from cattle clinically affected by bovine spongiform encephalopathy (BSE). *Vet Rec* 2002, **150**:365–378.

Scrapie

Andreoletti O, Lacroux C, Chabert A, Monnereau L, Tabouret G, Lantier F, Berthon P, Eyenne F, Lafond-Benestad S, Elsen J-M, Schelcher F. PrP(Sc) accumulation in placentas of ewes exposed to natural scrapie: influence of foetal *PrP* genotype and effect on ewe-to-lamb transmission. *J Gen Virol* 2002; **83**: 2607–16.

Andreoletti O, Simon S, Lacroux C, Morel N, Tabouret G, Chabert A, Lugan S, Corbiere F, Ferre P, Foucras G, Laude H, Eychenne F, Grassi J, Schelcher F. PrP^{Sc} accumulation in myocytes from sheep incubating natural scrapie. *Nat Med* 2004; **10**: 591–93.

Andreoletti O, Berthon P, Marc D, Sarradin P, Grosclude J, van Keulen L, Schelcher F, Elsen J-M, Lantier F. Early accumulation of PrP(Sc) in gut-associated lymphoid and nervous tissues of susceptible sheep from a Romanov flock with natural scrapie. *J Gen Virol* 2000; **81**: 3115–26.

Casalone C, Corona C, Crescio MI, Martucci F, Mazza M, Ru G, Bozzetta E, Acutis PL, Caramelli M. Pathological prion protein in the tongues of sheep infected with naturally occurring scrapie. *J Virol* 2005, **79**: 5847–5849.

Groschup MH, Weiland F, Straub OC, Pfaff E. Detection of scrapie agent in the peripheral nervous system of a diseased sheep. *Neurobiol Disease* 1996; **3**: 191–195

Groschup MH, Beekes M, McBride PA, Hardt M, Hainfellner JA, Budka H. Deposition of disease-associated prion protein involves the peripheral nervous system in experimental scrapie. *Acta Neuropathol* 1999; **98**: 457–458

Hadlow WJ, Kennedy RC, Race RE. Natural infection of Suffolk sheep with scrapie virus. *J Infect Dis* 1982; **146**: 657–664

Hadlow WJ, Kennedy RC, Race RE, Eklund CM. Virologic and neurohistologic findings dairy goats affected with natural scrapie. *Vet Pathol* 1980; **17**: 187–199

Hardt M, Baron TMH, Groschup. A comparative study on the immunohistochemical detection of abnormal prion protein using mono and polyclonal antibodies. *J Comp Pathol* 2000; **122**: 43–53.

Hortells P, Monleón E, Luián L, Vargus A, Acín C, Monzón M, Bolea R, Badiola JJ. Study of Retina and Visual Pathways in Naturally Affected Scrapie Animals. *Proceedings of the 23rd Meeting of the European Society of Veterinary Pathology,* 2005, p16.

Hunter N, Foster J, Chong A, McCutcheon S, Parnham D, Eaton S, MacKenzie C, Houston F. Transmission of prion diseases by blood transfusion. *J Gen Virol* 2002; **83**: 2897–2905

Race R, Jenny A, Sutton D. Scrapie infectivity and proteinase K-resistant protein in sheep placenta, brain, spleen, and lymph node: implications for transmission and antemortem diagnosis. *J Infect Dis* 1998; **178**: 949–953

Somerville RA, Birkett CR, Farquhar CF, Hunter N, Goldmann W, Dornan J, Grover D, Hennion RM, Percy C, Foster J, Jeffrey M. Immunodetection of PrP Sc in spleens of some scrapie-infected sheep but not BSE-infected cows. *J Gen Virol* 1997; **78**: 2389–2396.

Thomzig A, Schulz-Schaeffer W, Kratzel C, Mai J, Beekes M. Preclinical deposition of pathological prion protein PrPSc in muscles of hamsters orally exposed to scrapie. *J Clin Invest* 2004; **113**: 1465–1472

Van Keulen LJM, Schreuder BEC, Vromans MEW, Langeveld JPM, Smits MA. Pathogenesis of natural scrapie in sheep. *Arch Virol* 2000; **16** (Suppl): 57–71

Appendix 2
Bovine Spongiform Encephalopathy

OIE Copyright

Editor's Note: The copyright for reprinting the following information on BSE was kindly given on 14 September 2006 by the World Organisation for Animal Health, OIE, Paris. Source: OIE. Terrestrial Animal Health Code 2005. http://www.oie.int/eng/normes/mcode/en_chapitre_2.3.13.htm.

Terrestrial Animal Health Code – CHAPTER 2.3.13

BOVINE SPONGIFORM ENCEPHALOPATHY

Article 2.3.13.1.

The recommendations in this Chapter are intended to manage the human and animal health risks associated with the presence of the bovine spongiform encephalopathy (BSE) agent in cattle (Bos taurus and B. indicus) only.

1. When authorising import or transit of the following commodities and any products made from these commodities and containing no other tissues from cattle, Veterinary Administrations should not require any BSE related conditions, regardless of the BSE risk status of the cattle population of the exporting country, zone or compartment:
   a. milk and milk products;
   b. semen and in vivo derived cattle embryos collected and handled in accordance with the recommendations of the International Embryo Transfer Society;
   c. hides and skins;
   d. gelatine and collagen prepared exclusively from hides and skins;
   e. protein-free tallow (maximum level of insoluble impurities of 0.15% in weight) and derivatives made from this tallow;
   f. dicalcium phosphate (with no trace of protein or fat);
   g. deboned skeletal muscle meat (excluding mechanically separated meat) from cattle 30 months of age or less, which were not subjected to a stunning process prior to slaughter, with a device injecting compressed air or gas into the cranial cavity or to a pithing process, and which passed ante-mortem and post-mortem inspections and which has been prepared in a manner to avoid contamination with tissues listed in Article 2.3.13.13.;
   h. blood and blood by-products, from cattle which were not subjected to a stunning process, prior to slaughter, with a device injecting compressed air or gas into the cranial cavity, or to a pithing process.
2. When authorising import or transit of other commodities listed in this Chapter, Veterinary Administrations should require the conditions prescribed in this Chapter relevant to the BSE risk status of the cattle population of the exporting country, zone or compartment.
   Standards for diagnostic tests are described in the Terrestrial Manual.

Article 2.3.13.2.

The BSE risk status of the cattle population of a country, zone or compartment should be determined on the basis of the following criteria:

1. the outcome of a risk assessment, based on Section 1.3., identifying all potential factors for BSE occurrence and their historic perspective. Countries should review the risk assessment annually to determine whether the situation has changed.
   a. Release assessment
      Release assessment consists of assessing, through consideration of the following, the likelihood that the BSE agent has either been introduced into the country, zone or compartment via commodities potentially contaminated with it, or is already present in the country, zone or compartment:
      i. the presence or absence of the BSE agent in the indigenous ruminant population of

the country, zone or compartment and, if present, evidence regarding its prevalence;
   ii. production of meat-and-bone meal or greaves from the indigenous ruminant population;
   iii. imported meat-and-bone meal or greaves;
   iv. imported cattle, sheep and goats;
   v. imported animal feed and feed ingredients;
   vi. imported products of ruminant origin for human consumption, which may have contained tissues listed in Article 2.3.13.13. and may have been fed to cattle;
   vii. imported products of ruminant origin intended for in vivo use in cattle.

   The results of any epidemiological investigation into the disposition of the commodities identified above should be taken into account in carrying out the assessment.
 b. Exposure assessment
   If the release assessment identifies a risk factor, an exposure assessment should be conducted, consisting of assessing the likelihood of cattle being exposed to the BSE agent, through a consideration of the following:
   i. recycling and amplification of the BSE agent through consumption by cattle of meat-and-bone meal or greaves of ruminant origin, or other feed or feed ingredients contaminated with these;
   ii. the use of ruminant carcasses (including from fallen stock), by-products and slaughterhouse waste, the parameters of the rendering processes and the methods of animal feed manufacture;
   iii. the feeding or not of ruminants with meat-and-bone meal and greaves derived from ruminants, including measures to prevent cross-contamination of animal feed;
   iv. the level of surveillance for BSE conducted on the cattle population up to that time and the results of that surveillance;
2. on-going awareness programme for veterinarians, farmers, and workers involved in transportation, marketing and slaughter of cattle to encourage reporting of all cases showing clinical signs consistent with BSE in target sub-populations as defined in Appendix 3.8.4.;
3. the compulsory notification and investigation of all cattle showing clinical signs consistent with BSE;
4. the examination in an approved laboratory of brain or other tissues collected within the framework of the aforementioned surveillance and monitoring system.

When the risk assessment demonstrates negligible risk, the country should conduct Type B surveillance in accordance with Appendix 3.8.4.

When the risk assessment fails to demonstrate negligible risk, the country should conduct Type A surveillance in accordance with Appendix 3.8.4.

Article 2.3.13.3.

Negligible BSE risk

Commodities from the cattle population of a country, zone or compartment pose a negligible risk of transmitting the BSE agent if the following conditions are met:

1. a risk assessment, as described in point 1 of Article 2.3.13.2., has been conducted in order to identify the historical and existing risk factors, and the country has demonstrated that appropriate specific measures have been taken for the relevant period of time defined below to manage each identified risk;
2. the country has demonstrated that Type B surveillance in accordance with Appendix 3.8.4. is in place and the relevant points target, in accordance with Table 1, has been met;
3. EITHER:
   a. there has been no case of BSE or, if there has been a case, every case of BSE has been demonstrated to have been imported and has been completely destroyed, and
      i. the criteria in points 2 to 4 of Article 2.3.13.2. have been complied with for at least 7 years; and
      ii. it has been demonstrated through an appropriate level of control and audit that for at least 8 years neither meat-and-bone meal nor greaves derived from ruminants has been fed to ruminants;

   OR
   b. if there has been an indigenous case, every indigenous case was born more than 11 years ago; and
      i. the criteria in points 2 to 4 of Article 2.3.13.2. have been complied with for at least 7 years; and
      ii. it has been demonstrated through an appropriate level of control and audit that for at least 8 years neither meat-and-bone meal nor greaves derived from ruminants has been fed to ruminants; and
      iii. all BSE cases, as well as:
         - all cattle which, during their first year of life, were reared with the BSE cases during their first year of life, and which in-

vestigation showed consumed the same potentially contaminated feed during that period, or
- if the results of the investigation are inconclusive, all cattle born in the same herd as, and within 12 months of the birth of, the BSE cases,

if alive in the country, zone or compartment, are permanently identified, and their movements controlled, and, when slaughtered or at death, are completely destroyed.

Article 2.3.13.4.

Controlled BSE risk

Commodities from the cattle population of a country, zone or compartment pose a controlled risk of transmitting the BSE agent if the following conditions are met:

1. a risk assessment, as described in point 1 of Article 2.3.13.2., has been conducted in order to identify the historical and existing risk factors, and the country has demonstrated that appropriate measures are being taken to manage all identified risks, but these measures have not been taken for the relevant period of time;
2. the country has demonstrated that Type A surveillance in accordance with Appendix 3.8.4. has been carried out and the relevant points target, in accordance with Table 1, has been met; Type B surveillance may replace Type A surveillance once the relevant points target is met;
3. EITHER:
   a. there has been no case of BSE or, if there has been a case, every case of BSE has been demonstrated to have been imported and has been completely destroyed, the criteria in points 2 to 4 of Article 2.3.13.2. are complied with, and it can be demonstrated through an appropriate level of control and audit that neither meat-and-bone meal nor greaves derived from ruminants has been fed to ruminants, but at least one of the following two conditions applies:
      i. the criteria in points 2 to 4 of Article 2.3.13. 2. have not been complied with for 7 years;
      ii. it cannot be demonstrated that controls over the feeding of meat-and-bone meal or greaves derived from ruminants to ruminants have been in place for 8 years;

   OR

   b. there has been an indigenous case of BSE, the criteria in points 2 to 4 of Article 2.3.13.2. are complied with, and it can be demonstrated through an appropriate level of control and audit that neither meat-and-bone meal nor greaves derived from ruminants has been fed to ruminants, but at least one of the following two conditions applies:
      i. the criteria in points 2 to 4 of Article 2.3.13. 2. have not been complied with for 7 years;
      ii. it cannot be demonstrated that controls over the feeding of meat-and-bone meal and greaves derived from ruminants to ruminants have been in place for 8 years;

   AND

   iii. all BSE cases, as well as:
      - all cattle which, during their first year of life, were reared with the BSE cases during their first year of life, and which investigation showed consumed the same potentially contaminated feed during that period, or
      - if the results of the investigation are inconclusive, all cattle born in the same herd as, and within 12 months of the birth of, the BSE cases,

   if alive in the country, zone or compartment, are permanently identified, and their movements controlled, and, when slaughtered or at death, are completely destroyed.

Article 2.3.13.5.

Undetermined BSE risk

The cattle population of a country, zone or compartment poses an undetermined BSE risk if it cannot be demonstrated that it meets the requirements of another category.

Article 2.3.13.6.

When importing from a country, zone or compartment posing a negligible BSE risk, Veterinary Administrations should require:

for all commodities from cattle not listed in point 1 of Article 2.3.13.1.

the presentation of an international veterinary certificate attesting that the country, zone or compartment complies with the conditions in Article 2.3.13.3.

Article 2.3.13.7.

When importing from a country, zone or compartment posing a controlled BSE risk, Veterinary Administrations should require:

for cattle

the presentation of an international veterinary certificate attesting that:

1. the country, zone or compartment complies with the conditions referred to in Article 2.3.13.4.;
2. cattle selected for export are identified by a permanent identification system enabling them to be traced back to the dam and herd of origin, and are not exposed cattle as described in point 3b)iii) of Article 2.3.13.4.;
3. in the case of a country, zone or compartment where there has been an indigenous case, cattle selected for export were born after the date from which the ban on the feeding of ruminants with meat-and-bone meal and greaves derived from ruminants was effectively enforced.

Article 2.3.13.8.

When importing from a country, zone or compartment with an undetermined BSE risk, Veterinary Administrations should require:
for cattle
the presentation of an international veterinary certificate attesting that:
1. the feeding of ruminants with meat-and-bone meal and greaves derived from ruminants has been banned and the ban has been effectively enforced;
2. all BSE cases, as well as:
   a. all cattle which, during their first year of life, were reared with the BSE cases during their first year of life, and, which investigation showed consumed the same potentially contaminated feed during that period, or
   b. if the results of the investigation are inconclusive, all cattle born in the same herd as, and within 12 months of the birth of, the BSE cases, if alive in the country, zone or compartment, are permanently identified, and their movements controlled, and, when slaughtered or at death, are completely destroyed;
3. cattle selected for export:
   a. are identified by a permanent identification system enabling them to be traced back to the dam and herd of origin and are not the progeny of BSE suspect or confirmed females;
   b. were born at least 2 years after the date from which the ban on the feeding of ruminants with meat-and-bone meal and greaves derived from ruminants was effectively enforced.

Article 2.3.13.9.

When importing from a country, zone or compartment posing a negligible BSE risk, Veterinary Administrations should require:
for fresh meat and meat products from cattle (other than those listed in point 1 of Article 2.3.13.1.)
the presentation of an international veterinary certificate attesting that:
1. the country, zone or compartment complies with the conditions in Article 2.3.13.3.;
2. the cattle from which the fresh meat and meat products were derived passed ante-mortem and post-mortem inspections.

Article 2.3.13.10.

When importing from a country, zone or compartment with an undetermined BSE risk, Veterinary Administrations should require:
for fresh meat and meat products from cattle (other than those listed in point 1 of Article 2.3.13.1.)
the presentation of an international veterinary certificate attesting that:
1. the country, zone or compartment complies with the conditions referred to in Article 2.3.13.4.;
2. the cattle from which the fresh meat and meat products were derived passed ante-mortem and post-mortem inspections;
3. cattle from which the fresh meat and meat products destined for export were derived were not subjected to a stunning process, prior to slaughter, with a device injecting compressed air or gas into the cranial cavity, or to a pithing process;
4. the fresh meat and meat products were produced and handled in a manner which ensures that such products do not contain and are not contaminated with:
   a. the tissues listed in points 1 and 2 of Article 2.3.13.13.,
   b. mechanically separated meat from the skull and vertebral column from cattle over 30 months of age.

Article 2.3.13.11.

When importing from a country, zone or compartment with an undetermined BSE risk, Veterinary Administrations should require:
for fresh meat and meat products from cattle (other than those listed in point 1 of Article 2.3.13.1.)
the presentation of an international veterinary certificate attesting that:
1. the cattle from which the fresh meat and meat products originate:
   a. have not been fed meat-and-bone meal or greaves derived from ruminants;
   b. passed ante-mortem and post-mortem inspections;

   c. were not subjected to a stunning process, prior to slaughter, with a device injecting compressed air or gas into the cranial cavity, or to a pithing process;
2. the fresh meat and meat products were produced and handled in a manner which ensures that such products do not contain and are not contaminated with:
   a. the tissues listed in points 1 and 3 of Article 2.3.13.13.,
   b. nervous and lymphatic tissues exposed during the deboning process,
   c. mechanically separated meat from the skull and vertebral column from cattle over 12 months of age.

Article 2.3.13.12.

Ruminant-derived meat-and-bone meal or greaves, or any commodities containing such products, which originate from a country, zone or compartment defined in Articles 2.3.13.4. and 2.3.13.5. should not be traded between countries.

Article 2.3.13.13.

1. From cattle of any age originating from a country, zone or compartment defined in Articles 2.3.13.4. and 2.3.13.5., the following commodities, and any commodity contaminated by them, should not be traded for the preparation of food, feed, fertilisers, cosmetics, pharmaceuticals including biologicals, or medical devices: tonsils and distal ileum. Protein products, food, feed, fertilisers, cosmetics, pharmaceuticals or medical devices prepared using these commodities (unless covered by other Articles in this Chapter) should also not be traded.
2. From cattle that were at the time of slaughter over 30 months of age originating from a country, zone or compartment defined in Article 2.3.13.4., the following commodities, and any commodity contaminated by them, should not be traded for the preparation of food, feed, fertilisers, cosmetics, pharmaceuticals including biologicals, or medical devices: brains, eyes, spinal cord, skull and vertebral column. Protein products, food, feed, fertilisers, cosmetics, pharmaceuticals or medical devices prepared using these commodities (unless covered by other Articles in this Chapter) should also not be traded.
3. From cattle that were at the time of slaughter over 12 months of age originating from a country, zone or compartment defined in Article 2.3.13.5., the following commodities, and any commodity contaminated by them, should not be traded for the preparation of food, feed, fertilisers, cosmetics, pharmaceuticals including biologicals, or medical devices: brains, eyes, spinal cord, skull and vertebral column. Protein products, food, feed, fertilisers, cosmetics, pharmaceuticals or medical devices prepared using these commodities (unless covered by other Articles in this Chapter) should also not be traded.

Article 2.3.13.14.

Veterinary Administrations of importing countries should require:

for gelatine and collagen prepared from bones and intended for food or feed, cosmetics, pharmaceuticals including biologicals, or medical devices

the presentation of an international veterinary certificate attesting that:

1. the commodities came from a country, zone or compartment posing a negligible BSE risk;

OR

2. they originate from a country, zone or compartment posing a controlled BSE risk and are derived from cattle which have passed ante-mortem and post-mortem inspections; and that
   a. skulls from cattle over 30 months of age at the time of slaughter have been excluded;
      the bones have been subjected to a process which includes all of the following steps:
      i. pressure washing (degreasing),
      ii. acid demineralisation,
      iii. acid or alkaline treatment,
      iv. filtration,
      v. sterilisation at $> 138\,^{\circ}C$ for a minimum of 4 seconds,
      or to an equivalent or better process in terms of infectivity reduction (such as high pressure heating);

OR

3. they originate from a country, zone or compartment posing an undetermined BSE risk and are derived from cattle which have passed ante-mortem and post-mortem inspections; and that
   a. skulls and vertebrae (except tail vertebrae) from cattle over 12 months of age at the time of slaughter have been excluded;
      the bones have been subjected to a process which includes all of the following steps:
      i. pressure washing (degreasing),
      ii. acid demineralisation,
      iii. acid or alkaline treatment,
      iv. filtration,
      v. sterilisation at $> 138\,^{\circ}C$ for a minimum of 4 seconds,

or to an equivalent or better process in terms of infectivity reduction (such as high pressure heating).

Article 2.3.13.15.

Veterinary Administrations of importing countries should require:
for tallow and dicalcium phosphate (other than as defined in Article 2.3.13.1.) intended for food, feed, fertilisers, cosmetics, pharmaceuticals including biologicals, or medical devices
the presentation of an international veterinary certificate attesting that:

1. the commodities came from a country, zone or compartment posing a negligible BSE risk; or
2. they originate from a country, zone or compartment posing a controlled BSE risk, are derived from cattle which have passed ante-mortem and post-mortem inspections, and have not been prepared using the tissues listed in points 1 and 2 of Article 2.3.13.13.

Article 2.3.13.16.

Veterinary Administrations of importing countries should require:
for tallow derivatives (other than those made from protein-free tallow as defined in Article 2.3.13.1.) intended for food, feed, fertilisers, cosmetics, pharmaceuticals including biologicals, or medical devices
the presentation of an international veterinary certificate attesting that:

1. they originate from a country, zone or compartment posing a negligible BSE risk; or
2. they are derived from tallow meeting the conditions referred to in Article 2.3.13.15.; or
3. they have been produced by hydrolysis, saponification or transesterification using high temperature and pressure.

Appendix 3
Scrapie

Terrestrial Animal Health Code – CHAPTER 2.4.8

SCRAPIE

Article 2.4.8.1.

Scrapie is a neurodegenerative disease of sheep and goats. The main mode of transmission is from mother to offspring immediately after birth and to other susceptible neonates exposed to the birth fluids and tissues of an infected animal. Transmission occurs at a much lower frequency to adults exposed to the birth fluids and tissues of an infected animal. A variation in genetic susceptibility of sheep has been recognised. The incubation period of the disease is variable, however it is usually measured in years. The duration in incubation period can be influenced by a number of factors including host genetics and strain of agent.

The recommendations in the present chapter are not intended, or sufficient, to manage the risks associated with the potential presence of the bovine spongiform encephalopathy agent in small ruminants.

Standards for diagnostic tests are described in the Terrestrial Manual.

Article 2.4.8.2.

The scrapie status of a country, a zone or an establishment can be determined on the basis of the following criteria:

1. the outcome of a risk assessment identifying all potential factors for scrapie occurrence and their historic perspective, in particular the:
   a. epidemiological situation concerning all animal transmissible spongiform encephalopathies (TSE) in the country, zone or establishment;
   b. importation or introduction of small ruminants or their embryos/oocytes potentially infected with scrapie;
   c. extent of knowledge of the population structure and husbandry practices of sheep and goats in the country or zone;
   d. feeding practices, including consumption of meat-and-bone meal or greaves derived from ruminants;
   e. importation of meat-and-bone meal or greaves potentially contaminated with an animal TSE or feedstuffs containing either;
   f. the origin and use of ruminant carcasses (including fallen stock), by-products and slaughterhouse waste, the parameters of the rendering processes and the methods of animal feed manufacture;
2. an on-going awareness programme for veterinarians, farmers, and workers involved in transportation, marketing and slaughter of sheep and goats to facilitate recognition and encourage reporting of all animals with clinical signs compatible with scrapie;
3. a surveillance and monitoring system including the following:
   a. official veterinary surveillance, reporting and regulatory control in accordance with the provisions of Appendix 3.8.1.;
   b. a Veterinary Administration with current knowledge of, and authority over, all establishments which contain sheep and goats in the whole country;
   c. compulsory notification and clinical investigation of all sheep and goats showing clinical signs compatible with scrapie;
   d. examination in an approved laboratory of appropriate material from sheep and goats older than 18 months displaying clinical signs compatible with scrapie taking into account the guidelines in Appendix X.X.X. (under study);
   e. maintenance of records including the number and results of all investigations for at least 7 years.

Article 2.4.8.3.

Scrapie free country or zone

Countries or zones may be considered free from scrapie if within the said territory:

1. a risk assessment, as described in point 1 of Article 2.4.8.2., has been conducted, and it has been demonstrated that appropriate measures have been taken for the relevant period of time to manage any risk identified;

AND EITHER

2. the country or the zone have demonstrated historical freedom taking into account the guidelines in Appendix 3.8.6.;

OR

3. for at least 7 years, a surveillance and monitoring system as referred to in Article 2.4.8.2. has been in place, and no case of scrapie has been reported during this period;

OR

4. for at least 7 years, a sufficient number of investigations has been carried out annually, to provide a 95% level of confidence of detecting scrapie if it is present at a prevalence rate exceeding 0.1% out of the total number of all chronic wasting conditions in the population of sheep and goats older than 18 months of age (under study) and no case of scrapie has been reported during this period; it is assumed that the occurrence rate of chronic wasting conditions within the population of sheep and goats older than 18 months of age is at least 1%;

OR

5. all establishments containing sheep or goats have been accredited free as described in Article 2.4.8.4.;

AND

6. the feeding to sheep and goats of meat-and-bone meal or greaves potentially contaminated with an animal TSE has been banned and effectively enforced in the whole country for at least 7 years;

AND

7. introductions of sheep and goats, semen and embryos/oocytes from countries or zones not free from scrapie are carried out in accordance with Articles 2.4.8.6., 2.4.8.7., 2.4.8.8. or 2.4.8.9., as relevant.

For maintenance of country or zone free status, the investigations referred to in point 4 above should be repeated every 7 years.

Article 2.4.8.4.

Scrapie free establishment

An establishment may be considered eligible for accreditation as a scrapie free establishment if:

1. in the country or zone where the establishment is situated, the following conditions are fulfilled:
   a. the disease is compulsorily notifiable;
   b. a surveillance and monitoring system as referred to in Article 2.4.8.2. is in place;
   c. affected sheep and goats are slaughtered and completely destroyed;
   d. the feeding to sheep and goats of meat-and-bone meal or greaves potentially contaminated with an animal TSE has been banned and effectively enforced in the whole country;
   e. an official accreditation scheme is in operation under the supervision of the Veterinary Administration, including the measures described in point 2 below;
2. in the establishment the following conditions have been complied with for at least 7 years:
   a. sheep and goats should be permanently identified and records maintained, to enable trace back to their establishment of birth;
   b. records of movements of sheep and goats in and out of the establishment are established and maintained;
   c. introductions of animals are allowed only from establishments of an equal or higher stage in the process of accreditation; however, rams and bucks complying with the provisions in point 2 of Article 2.4.8.8. may also be introduced;
   d. an Official Veterinarian inspects sheep and goats in the establishment and audits the records at least once a year;
   e. no case of scrapie has been reported;
   f. sheep and goats of the establishment should have no direct or indirect contact with sheep or goats from establishments of a lower status;
   g. all culled animals over 18 months of age are inspected by an Official Veterinarian, and a proportion of those exhibiting neurological or wasting signs are tested in a laboratory for scrapie. The selection of the animals to be tested should be made by the Official Veterinarian. Animals over 18 months of age that have died or have been killed for reasons other than routine slaughter should also be tested (including 'fallen' stock and emergency slaughter).

Article 2.4.8.5.

Regardless of the scrapie status of the exporting country, Veterinary Administrations should authorise without restriction the import or transit through their territory of meat (excluding materials as referred to in Article 2.4.8.11.), milk, milk products, wool and its derivatives, hides and skins, tallow, derivatives made from this tallow and dicalcium phosphate originating from sheep and goats.

Article 2.4.8.6.

When importing from countries not considered free from scrapie, Veterinary Administrations should require:
for sheep and goats for breeding or rearing
the presentation of an international veterinary certificate attesting that the animals come from a zone or an establishment free from scrapie as described in Article 2.4.8.3. and in Article 2.4.8.4.

Article 2.4.8.7.

When importing from countries or zones not considered free from scrapie, Veterinary Administrations should require:
for sheep and goats for slaughter
the presentation of an international veterinary certificate attesting that:

1. in the country or zone:
   a. the disease is compulsorily notifiable;
   b. a surveillance and monitoring system as referred to in Article 2.4.8.2. is in place;
   c. affected sheep and goats are slaughtered and completely destroyed;
2. the sheep and goats selected for export showed no clinical sign of scrapie on the day of shipment.

Article 2.4.8.8.

When importing from countries or zones not considered free from scrapie, Veterinary Administrations should require:
for semen of sheep and goats
the presentation of an international veterinary certificate attesting that:

1. in the country or zone:
   a. the disease is compulsorily notifiable;
   b. a surveillance and monitoring system as referred to in Article 2.4.8.2. is in place;
   c. affected sheep and goats are slaughtered and completely destroyed;
   d. the feeding of sheep and goats with meat-and-bone meal or greaves potentially contaminated with an animal TSE has been banned and effectively enforced in the whole country;
2. the donor animals:
   a. are permanently identified, to enable trace back to their establishment of origin;
   b. have been kept since birth in establishments in which no case of scrapie had been confirmed during their residency;
   c. showed no clinical sign of scrapie at the time of semen collection;
3. the semen was collected, processed and stored in conformity with the provisions of Appendix 3.2.1.

Article 2.4.8.9.

When importing from countries or zones not considered free from scrapie, Veterinary Administrations should require:
for embryos/oocytes of sheep and goats
the presentation of an international veterinary certificate attesting that:

1. in the country or zone:
   a. the disease is compulsorily notifiable;
   b. a surveillance and monitoring system as referred to in Article 2.4.8.2. is in place;
   c. affected sheep and goats are slaughtered and completely destroyed;
   d. the feeding to sheep and goats of meat-and-bone meal or greaves potentially contaminated with animal TSE has been banned and effectively enforced in the whole country;
2. the donor animals:
   a. are permanently identified, to enable trace back to their establishment of origin;
   b. have been kept since birth in establishments in which no case of scrapie had been confirmed during their residency;
   showed no clinical sign of scrapie at the time of embryo/oocyte collection;
3. the embryos/oocytes were collected, processed and stored in conformity with the provisions of Appendix 3.3.1.

Article 2.4.8.10.

Meat-and-bone meal containing any sheep or goat protein, or any feedstuffs containing that type of meat-and-bone meal, which originate from countries not considered free of scrapie should not be traded between countries for ruminant feeding.

Article 2.4.8.11.

When importing from countries or zones not considered free from scrapie, Veterinary Administrations should require:
for skulls including brains, ganglia and eyes, vertebral column including ganglia and spinal cord, tonsils, thymus, spleen, intestine, adrenal gland, pancreas, or liver, and protein products derived therefrom, from sheep and goats

the presentation of an international veterinary certificate attesting that:

1. in the country or zone:
   a. the disease is compulsorily notifiable;
   b. a surveillance and monitoring system as referred to in Article 2.4.8.2. is in place;
   c. affected sheep and goats are slaughtered and completely destroyed;
2. the materials come from sheep and goats that showed no clinical sign of scrapie on the day of slaughter.

Article 2.4.8.12.

Veterinary Administrations of importing countries should require:

for ovine and caprine materials destined for the preparation of biologicals

the presentation of an international veterinary certificate attesting that the products originate from sheep and goats born and raised in a scrapie free country, zone or establishment.

Authors Index

Aguzzi, Adriano
Institute of Neuropathology
University Hospital of Zurich
Schmelzbergstr. 12
CH–8091 Zurich, Switzerland
adriano.aguzzi@usz.ch
www.neuropathologie.usz.ch

Alpers, Michael P.
Centre for International Health
Division of Health Sciences
Shenton Park Campus
Curtin University of Technology
GPO Box U1987
Perth, WA 6845, Australia
m.alpers@curtin.edu.au
www.curtin.edu.au

The Kuru Surveillance Team
Papua New Guinea Institute of Medical Research
PO Box 60
Goroka, EHP 441, Papua New Guinea
www.pngimr.org.pg

MRC Prion Unit
University College London
Institute of Neurology
The National Hospital for Neurology and Neurosurgery
Queen Square
London WC1N 3BG, United Kingdom
www.ion.ucl.ac.uk

Alvseike, Kristin R.
Norwegian Food Safety Authority
Head office
PO Box 383
N–2381 Brumunddal, Norway
kristin.ruud.alvseike@ mattilsynet.no
www.mattilsynet.no

Bachmann, Jutta
Bachmann Consulting
Nokkefaret 12
N–1450 Nesoddtangen, Norway
info@jbachmann-consulting.com
www.jbachmann-consulting.com

Beekes, Michael
Robert-Koch-Institute
P24 – Transmissible Spongiform Encephalopathies
Nordufer 20
D–13353 Berlin, Germany
beekesM@rki.de
www.rki.de

Benestad, Sylvie L.
National Veterinary Institute
Department of Pathology
PO Box 8156 Dep.
N–0033 Oslo, Norway
sylvie.benestad@vetinst.no
www.vetinst.no

Bossers, Alex
Central Institute for Animal Disease Control (CIDC-Lelystad)
Wageningen University and Research Centre
PO Box 2004
NL–8203 AA Lelystad, The Netherlands
alex.bossers@wur.nl
www.cidc-lelystad.nl

Bradley, Ray
Private BSE Consultant
41 Marlyns Drive
Burpham
Guildford GU4 7LT, United Kingdom
raybradley@btinternet.com

Bratberg, Bjorn
National Veterinary Institute
Department of Pathology
PO Box 8156 Dep.
N–0033 Oslo, Norway
bjorn.bratberg@vetinst.no
www.vetinst.no

Braun, Ueli
Vetsuisse Faculty of the University of Zurich
Department of Farm Animals
Winterthurerstr. 260
CH–8057 Zurich, Switzerland
ubraun@vetclinics.unizh.ch
www.rind.unizh.ch

Bruce, Moira E.
Institute for Animal Health
Neuropathogenesis Unit
Ogston Building
West Mains Road
Edinburgh EH9 3JF, United Kingdom
moira.bruce@bbsrc.ac.uk
www.iah.bbsrc.ac.uk

Budka, Herbert
Austrian Reference Centre for Human Prion Diseases (ORPE)
Institute of Neurology
Medical University of Vienna
Waehringer Guertel 18–20
AKH 4J
A–1097 Vienna, Austria
herbert.budka@kin.at
www.meduniwien.ac.at

Buschmann, Anne
Institute for Novel and Emerging Infectious Diseases
Friedrich-Loeffler-Institut
Federal Research Institute for Animal Health
Boddenblick 5a
D–17493 Greifswald – Insel Riems, Germany
anne.buschmann@fli. bund.de
www.fli.bund.de

Cozzio, Antonio
Department of Dermatology
University Hospital of Zurich
Gloriastr. 31
CH–8091 Zurich, Switzerland
cozzio@usz.ch
www.dermatologie.usz.ch

Cunningham, Andrew A.
Reader & Head of Wildlife Epidemiology
Institute of Zoology
Zoological Society of London
Regent's Park
London NW1 4RY, United Kingdom
a.cunningham@ioz.ac.uk
www.zoo.cam.ac.uk

Dahms, Susanne
Clinical Statistics Europe
Global Medical Development
Schering AG
Muellerstr. 178
D–13353 Berlin, Germany
susanne.dahms@schering. de
www.schering.de

De Koeijer, Aline A.
Animal Sciences Group – WUR
Division of Infectious Diseases
PO Box 65
NL–8200 AB Lelystad, The Netherlands
aline.dekoeijer@wur.nl
www.asg.wur.nl

Doherr, Marcus G.
Department of Clinical Veterinary Medicine
Vetsuisse Faculty, University of Bern
Bremgartenstr. 109a
PO Box 8466
CH–3001 Bern, Switzerland
marcus.doherr@itn.unibe.ch
www.itn.vetsuisse.unibe.ch

Flechsig, Eckhard
Emmy Noether Research Group
Institute of Virology and Immunobiology
University Wuerzburg
Versbacherstr. 7
D–97078 Wuerzburg, Germany
flechsig@vim.uni-wuerzburg.de
www.uni-wuerzburg.de

Fraser, Hugh
Institute for Animal Health
Neuropathogenesis Unit
Ogston Building
West Mains Road
Edinburgh EH9 3JF, United Kingdom
hughfraser@jfraser60.fsnet.co.uk
www.iah.bbsrc.ac.uk

Geissen, Markus
Institute of Neuropathology
UKE Hamburg
Martinistr. 52
D–20246 Hamburg, Germany
m.geissen@uke.uni-hamburg.de
www.uke.uni-hamburg.de

Gelpi, Ellen
Austrian Reference Centre for Human Prion Diseases (ORPE)
Institute of Neurology
Medical University of Vienna
Waehringer Guertel 18–20
AKH 4J
A–1097 Vienna, Austria
ellen.gelpi@kin.at
www.kin.at

Giese, Armin
Center of Neuropathology and Prion Research
Ludwig-Maximilians-University Munich
Feodor-Lynen-Str. 23
D–81377 Munich, Germany
armin.giese@med.uni-muenchen.de
www.znp-muenchen.de

Glockshuber, Rudi
ETH Zurich – Swiss Federal Institute of Technology
Institute of Molecular Biology und Biophysics
ETH Hoenggerberg HPK E17
CH–8093 Zurich, Switzerland
rudi@mol.biol.ethz.ch
www.mol.biol.ethz.ch

Gretzschel, Anja
Institute for Novel and Emerging Infectious Diseases
Friedrich-Loeffler-Institut
Federal Research Institute for Animal Health
Boddenblick 5a
D–17493 Greifswald – Insel Riems, Germany
anja.gretzschel@fli.bund.de
www.fli.bund.de

Groschup, Martin H.
Institute for Novel and Emerging Infectious Diseases
Friedrich-Loeffler-Institut
Federal Research Institute for Animal Health
Boddenblick 5a
D–17493 Greifswald – Insel Riems, Germany
martin.groschup@fli.bund.de
www.fli.bund.de

Hadlow, William J.
Veterinary Pathologist
908 South Third Street
Hamilton, Montana
MT 59840–2924, USA

Hawkins, Stephen A.C.
Veterinary Laboratories Agency
New Haw
Addlestone
Surrey KT15 3NB, United Kingdom
s.a.c.hawkins@vla.defra.gsi.gov.uk
www.defra.gov.uk

Hegyi, Ivan
Department of Dermatology
University Hospital of Zurich
Gloriastr. 31
CH–8091 Zurich, Switzerland
ivan.hegyi@usz.ch
www.dermatologie.unispital.ch

Herms, Jochen W.
Center of Neuropathology and Prion Research
Ludwig-Maximilians-University Munich
Feodor-Lynen-Str. 23
D–81377 Munich, Germany
jochen.herms@med.uni-muenchen.de
www.znp-muenchen.de

Hewicker-Trautwein, Marion
Department of Pathology
University of Veterinary Medicine
Buenteweg 17
D–30559 Hannover, Germany
marion.hewicker-trautwein@tiho-hannover.de
www.tiho-hannover.de

Hörnlimann, Beat
BSE71–92/SVISS Consulting
on Animal & Public Health
PO Box 513
CH–6312 Steinhausen
Switzerland
b.h@sviss.net
www.prionone.ch

Hunter, Nora
Institute for Animal Health
Neuropathogenesis Unit
West Mains Road
Edinburgh EH9 3JF, United Kingdom
nora.hunter@bbsrc.ac.uk
www.iah.bbsrc.ac.uk

Infanger, Paul
CVO of Canton Lucerne (retired)
Adligenswilerstr. 113
CH–6006 Luzern, Switzerland
paul.infanger@bluewin.ch

Ironside, James W.
The National Creutzfeldt–Jakob Disease Surveillance Unit
Western General Hospital
Crewe Road
Edinburgh EH4 2XU, United Kingdom
james.ironside@ed.ac.uk
www.cjd.ed.ac.uk

Kirkwood, James K.
Universities Federation for Animal Welfare
The Old School
Brewhouse Hill
Wheathampstead
Hertfordshire AL4 8AN, United Kingdom
kirkwood@ufaw.org.uk
www.ufaw.org.uk

Konold, Timm
Veterinary Laboratories Agency
New Haw
Addlestone
Surrey KT15 3NB, United Kingdom
t.konold@vla.defra.gsi.gov.uk
www.defra.gov.uk

Kreil, Thomas R.
Global Pathogen Safety
Baxter BioScience
Benatzkygasse 2–6
A–1220 Vienna, Austria
thomas_kreil@baxter.com
www.baxter.com

Kretzschmar, Hans A.
Center of Neuropathology and Prion Research
Ludwig-Maximilians-University Munich
Feodor-Lynen-Str. 23
D–81377 Munich, Germany
hans.kretzschmar@med.uni-muenchen.de
www.znp-muenchen.de

Kuczius, Thorsten
Institute for Hygiene
University Hospital Muenster
Robert-Koch-Str. 41
D–48149 Muenster, Germany
tkuczius@uni-muenster.de
www.hygiene.uni-muenster.de

Lemmer, Karin
Robert-Koch-Institute
P24 – Transmissible Spongiform Encephalopathies
Nordufer 20
D–13353 Berlin, Germany
Lemmerk@rki.de
www.rki.de

Löwer, Johannes
Paul-Ehrlich-Institut
Federal Institute for Biological Medicinal Products
Paul-Ehrlich-Str. 51–59
D–63225 Langen, Germany
loejo@pei.de
www.pei.de

MacDiarmid, Stuart C.
Biosecurity New Zealand
PO Box 2526
Pastoral House, 25 The Terrace
Wellington, New Zealand
stuart.macDiarmid@maf.govt.nz
www.biosecurity.govt.nz

Masters, Colin L
Department of Pathology
The University of Melbourne, and
The Mental Health Research Institute of Victoria
Parkville, VIC 3010, Australia
c.masters@unimelb.edu.au
www.path.unimelb.edu.au
[*writer of epilogue]

Matthews, Danny
Veterinary Laboratories Agency
New Haw
Addlestone
Surrey KT15 3NB, United Kingdom
d.matthews@vla.defra.gsi.gov.uk
www.defra.gov.uk

Melkild, Ingrid
Norwegian Livestock Industry's Biosecurity Unit
PO Box 396 Oekern
N–0513 Oslo, Norway
ingrid.melkild@fagkjott.no
www.fagkjott.no

Mielke, Martin
Robert-Koch-Institute
Division for Applied Infection Control and Hospital Hygiene
Nordufer 20
D–13353 Berlin, Germany
mielkem@rki.de
www.rki.de

Miller, Michael W.
Wildlife Health Program
Colorado Division of Wildlife
Wildlife Research Center
317 West Prospect Road
Fort Collins
CO 80526–2097, USA
mike.miller@state.co.us
www.wildlife.state.co.us

Moser, Markus
Prionics AG
Wagistr. 27A
CH–8952 Schlieren, Switzerland
markus.moser@prionics.ch
www.prionics.com

Müller, Henrik
Heinrich-Heine-University Duesseldorf
Institute of Physical Biology
D–40225 Duesseldorf, Germany
mueller@biophys.uni-duesseldorf.de
www.biophys.uni-duesseldorf.de

Oberthür, Radulf C.
Labor Dr. Oberthür GmbH
Wahbruch 1
D–49844 Bawinkel, Germany
labor-oberthur@t-online.de
www.oberthur.de

Oesch, Bruno
Prionics AG
Wagistr. 27A
CH–8952 Schlieren, Switzerland
bruno.oesch@prionics.ch
www.prionics.com

Parchi, Piero
Laboratory of Neuropathology
Department of Neurological Sciences
University of Bologna
Via Foscolo 7
I–40123 Bologna, Italy
piero.parchi@unibo.it
www.neuro.unibo.it

Pauli, Georg
Robert-Koch-Institute
Center for Biological Safety (ZBS)
Highly Pathogenic Viral Agents (ZBS 1)
Nordufer 20
D–13353 Berlin, Germany
pauliG@rki.de
www.rki.de

Pohlenz, Joachim* †
National Animal Disease Center
2300 Dayton Avenue
Ames IA 50010, USA

Department of Pathology
University of Veterinary Medicine
Buenteweg 17
D–30559 Hannover, Germany

Poser, Sigrid †
Department of Neurology
Georg-August-University Goettingen
Robert-Koch-Str. 40
G–37075 Goettingen, Germany
www.neurologie.med.uni-goettingen.de

Prusiner, Stanley B.
University of California
Institute for Neurodegenerative Diseases HSE 774
513 Parnassus Avenue
San Francisco CA 94143–0518, USA
stanley@itsa.ucsf.edu
http://nobelprize.org/medicine/laureates/

Raeber, Alex J.
Prionics AG
Wagistr. 27A
CH–8952 Schlieren, Switzerland
alex.raeber@prionics.ch
www.prionics.ch

Riesner, Detlev
Heinrich-Heine-University Duesseldorf
Institute of Physical Biology
D–40225 Duesseldorf
Germany
riesner@biophys.uni-duesseldorf.de
www.biophys.uni-duesseldorf.de

Roth, Klaus
SMP GmbH Testen Validieren Forschen
Paul-Ehlich-Str. 40
D–72076 Tuebingen
Germany
kroth@smpgmbh.com
www.smpgmbh.com

Ruffing, Michael
Boehringer Ingelheim Pharma GmbH & Co. KG
ABP Quality & Compliance
Birkendorfer Str. 65
D–88397 Biberach an der Riss, Germany
michael.ruffing@bc.boehringer-ingelheim.com
www.boehringer-ingelheim.com

Ryan, Judith B.
Veterinary Laboratories Agency
New Haw
Addlestone
Surrey KT15 3NB, United Kingdom
j.ryan@vla.defra.gsi.gov.uk
www.defra.gov.uk

Ryder, Stephen J.
Animals (Scientific Procedures) Inspectorate
Mail point 1B, 1st Floor Seacole Building
2 Marsham Street
London SW1P 4DF, United Kingdom
stephen.ryder4@homeoffice.gsi.gov.uk

Schaefer, Jan
BfArM
Federal Institute for Drugs and Medical Devices
Kurt-Georg-Kiesinger-Allee 3
D–53175 Bonn, Germany
j.schaefer@bfarm.de
www.bfarm.de

Schatzl, Hermann M.
Institute of Virology
Technical University of Munich
Trogerstr. 30
D–81675 Munich, Germany
schaetzl@lrz.tum.de
www.virologie.med.tum.de

Schicker, Ernst
Am Arenenberg AG – Veterinary Practice
Arenenbergstr. 33
CH–8268 Salenstein, Switzerland
sc.er.fr@bluewin.ch
www.daktari-team.ch

Schreuder, Bram E.C.
Central Institute for Animal Disease Control
(CIDC-Lelystad)
PO Box 2004
NL–8203 AA Lelystad, The Netherlands
bram.schreuder@wur.nl
www.cidc-lelystad.wur.nl

Schulz-Schaeffer, Walter J.
Prion and Dementia Research Unit
Department of Neuropathology
Georg-August-University Goettingen
D–37075 Goettingen, Germany
wjschulz@med.uni-goettingen.de
www.prionforschung.de

Schwarzinger, Stephan
Department of Biopolymers
University of Bayreuth
Universitaetsstr. 30
D–95440 Bayreuth, Germany
stephan.schwarzinger@uni-bayreuth.de
www.uni-bt.de

Stöhr, Jan
Heinrich-Heine-University Duesseldorf
Institute of Physical Biology
D–40225 Duesseldorf, Germany
stoehr@biophys.uni-duesseldorf.de
www.biophys.uni-duesseldorf.de

Sturzenegger, Matthias
Department of Neurology
University Hospital, Inselspital
Freiburgstr. 30
CH–3010 Bern, Switzerland
matthias.sturzenegger@insel.ch
www.neuro-bern.ch

Thorud, Kristin
The Norwegian Ministry
of Fisheries and Coastal Affairs
PO Box 8118, Dep
N–0032 Oslo, Norway
kristin.thorud@fkd.dep.no
http://odin.dep.no

Ulvund, Martha J.
Department of Production Animal Clinical Sciences
Section for Small Ruminant Research
Norwegian School of Veterinary Science
Kyrkjevegen 332–334
N–4325 Sandnes, Norway
martha.ulvund@veths.no
www.veths.no

van Keulen, Lucien J.M.
Central Institute for Animal Disease Control
(CIDC-Lelystad)
Wageningen University and Research Centre
PO Box 2004
NL–8203 AA Lelystad, The Netherlands
lucien.vankeulen@wur.nl
www.cidc-lelystad.wur.nl

Weissmann, Charles
Scripps–Florida
Department of Infectology
5353 Parkside Drive, RF–2

Jupiter FL 33458, USA
charlesw@scripps.edu
www.scripps.edu/florida

Wells, Gerald A.H.
Veterinary Laboratories Agency
New Haw
Addlestone
Surrey KT15 3NB, United Kingdom
g.a.h.wells@vla.defra.gsi.gov.uk
www.defra.gov.uk

Consultant to:
Veterinary Laboratories Agency
New Haw
Addlestone
Surrey KT15 3NB, United Kingdom
g.a.h.wells@vla.defra.gsi.gov.uk
www.defra.gov.uk

Will, Robert G.
The National Creutzfeldt–Jakob Disease
Surveillance Unit
Western General Hospital
Crewe Road
Edinburgh EH4 2XU, United Kingdom
r.g.will@ed.ac.uk
www.cjd.ed.ac.uk

Willbold, Dieter
Heinrich-Heine-University Duesseldorf
Institute of Physical Biology
Research Centre Juelich
IBI–2 / NMR
D–52425 Juelich, Germany
d.willbold@fz-juelich.de
www.biophys.uni-duesseldorf.de

Williams, Elizabeth S. †
State University of Wyoming
Department of Veterinary Sciences
1174 Snowy Range Road
Laramie WY 82070, USA

Windemann, Helena
Swissmedic
Erlachstr. 8
CH–3000 Bern, Switzerland
helena.windemann@swissmedic.ch
www.swissmedic.ch

Windl, Otto
Veterinary Laboratories Agency
New Haw
Addlestone
Surrey KT15 3NB, United Kingdom
o.windl@vla.defra.gsi.gov.uk
www.defra.gov.uk

Yan, Zheng-Xin
SMP GmbH Testen Validieren Forschen
Paul-Ehlich-Str. 40
D–72076 Tuebingen, Germany
yan@smpgmbh.com
www.smpgmbh.com

Zerr, Inga
Department of Neurology
Georg-August-University Goettingen
Robert-Koch-Str. 40
D–37075 Goettingen, Germany
ingazerr@med.uni-goettingen.de
www.cjd-goettingen.de

Ziegler, Jan
Department of Biopolymers
University of Bayreuth
Universitaetsstr. 30
D–95440 Bayreuth, Germany
jan.ziegler@uni-bayreuth.de
www.uni-bt.de

The Editors

Dr. Beat Hörnlimann, MPH, born in 1958, studied veterinary medicine at the Universities of Basel, Bern and Zurich, Switzerland and later went on to obtain an additional university degree in public health (Master of Public Health). Having completed his DVM thesis, he joined the WHO Reference Laboratory for Rabies in Bern and afterwards the Institute of Animal Pathology at the University of Bern. Since 1990 he has dedicated himself to the study of prion diseases. At the Swiss Federal Veterinary Office in Bern he was mainly involved in the BSE eradication program and served as the scientific BSE/CJD-coordinator for Switzerland until early 1999. He carried out research on BSE epidemiology and investigations on FSE prevention at the Institute of Virology and Immunoprophylaxis, Mittelhäusern, with particular focus on the Swiss BSE cases born after the feed ban of December 1990. Parallel to these activities he contributed to the international harmonization of the Swiss CJD surveillance program and dealt with chemical disinfection – including conventional micro-organisms – for the Swiss Federal Public Health Office, Bern, until 1999. On several occasions between 1991 and 1999 he participated in WHO Consultations on prion diseases. From 1999 to 2003 he served as the Chief Veterinary Officer for Canton Zug. He founded SVISS Consulting in October 2000; a non-profit agency dedicated to networking with prion experts and publishing interdisciplinary information on prions in humans and animals, as exemplified in this book. SVISS Consulting also focuses on the history of research on prion diseases in humans and animals. Dr. Beat Hörnlimann's particular interests are the inactivation of prions using chemical disinfection and the containment of disease. He currently works in the food safety sector for the veterinary public health authorities.

Prof. Dr. Detlev Riesner, born in 1941, studied physics and biophysics at the University of Hannover, Germany. He obtained his Ph.D. on thermodynamics of ribonucleic acids with Prof. M. Eigen at the University of Braunschweig, Germany in 1970. He carried out postdoctoral research at the Society for Biotechnological Research in Braunschweig, Germany and at Princeton University, USA. In 1975, he obtained his 'Habilitation' in biophysical chemistry and molecular biology at the Hannover Medical School, Germany. Between 1977 and 1980 Detlev Riesner was professor of physical biochemistry at the Institute for Organic Chemistry and Biochemistry of the Institute of Technology Darmstadt, Germany. In 1980 he became a full professor and chairman of the Institute for Physical Biology at the Heinrich Heine University in Düsseldorf, Germany, where he also served as dean and vice-chancellor for research. His major interests focus on prion research and viroids. He is a cofounder of several successful biotech companies and works as an advisor to biotechnological start-up companies and international organizations. In 1985 began a close cooperation with Stanley B. Prusiner on prion diseases, resulting in both Detlev Riesner and Stanley B. Prusiner receiving the Max Planck Research Award for International Cooperation in 1992.

Prof. Dr. Hans Kretzschmar, FRCPath, born in 1953, studied human medicine at the University of Munich, Germany, where he was also trained in pathology. From 1983 to 1986 Hans Kretzschmar carried out research in the Department of Neuropathology at the University of California in San Francisco, USA, where he began his work on prion diseases in collaboration with Stanley B. Prusiner. After a one year research stay with Charles Weissmann's group at the Institute for Molecular Biology at the University of Zurich, Switzerland, he obtained his 'Habilitation' in neuropathology at the University of Munich. From 1992 until 2000 Hans Kretzschmar was director of the Department of Neuropathology at the University of Göttingen, Germany. He has been responsible for the German Reference Center for Prion Diseases since 1993. He has also served as advisor to the EC and WHO Committees concerned with prion diseases. In 2000 he became director of the Department of Neuropathology at the University of Munich, Germany, where he set up and directs the German Reference Center for Diseases of the CNS and the National Reference Center for Transmissible Spongiform Encephalopathies. In 1999 he received the Ernst Jung Award for his work on the molecular biology of the prion protein.

Index